Peter M.A. Sloot   C.J. Kenneth Tan
Jack J. Dongarra   Alfons G. Hoekstra (Eds.)

Computational Science – ICCS 2002

International Conference
Amsterdam, The Netherlands, April 21-24, 2002
Proceedings, Part III

Springer

Volume Editors

Peter M.A. Sloot
Alfons G. Hoekstra
University of Amsterdam, Faculty of Science, Section Computational Science
Kruislaan 403, 1098 SJ Amsterdam, The Netherlands
E-mail: {sloot,alfons}@science.uva.nl

C.J. Kenneth Tan
University of Western Ontario, Western Science Center, SHARCNET
London, Ontario, Canada N6A 5B7
E-mail: cjtan@acm.org

Jack J. Dongarra
University of Tennessee, Computer Science Department
Innovative Computing Laboratory
1122 Volunteer Blvd, Knoxville, TN 37996-3450, USA
E-mail: dongarra@cs.utk.edu

Cataloging-in-Publication Data applied for

Die Deutsche Bibliothek - CIP-Einheitsaufnahme

Computational science : international conference ; proceedings / ICCS 2002,
Amsterdam, The Netherlands, April 21 - 24, 2002. Peter M. A. Sloot (ed.). -
Berlin ; Heidelberg ; New York ; Barcelona ; Hong Kong ; London ; Milan ;
Paris ; Tokyo : Springer
Pt. 3 . - (2002)
(Lecture notes in computer science ; Vol. 2331)
ISBN 3-540-43594-8

CR Subject Classification (1998): D, F, G, H, I, J, C.2-3

ISSN 0302-9743

ISBN 3-540-43594-8 Springer-Verlag Berlin Heidelberg New York

Springer-Verlag Berlin Heidelberg New York
a member of BertelsmannSpringer Science+Business Media GmbH

http://www.springer.de

Printed in Germany

Typesetting: Camera-ready by author, data conversion by PTP-Berlin, Stefan Sossna e.K.
Printed on acid-free paper SPIN: 10869731 06/3142 5 4 3 2 1 0

Preface

Computational Science is the scientific discipline that aims at the development and understanding of new computational methods and techniques to model and simulate complex systems.

The area of application includes natural systems – such as biology, environmental and geo-sciences, physics, and chemistry – and synthetic systems such as electronics and financial and economic systems. The discipline is a bridge between 'classical' computer science – logic, complexity, architecture, algorithms – mathematics, and the use of computers in the aforementioned areas.

The relevance for society stems from the numerous challenges that exist in the various science and engineering disciplines, which can be tackled by advances made in this field. For instance new models and methods to study environmental issues like the quality of air, water, and soil, and weather and climate predictions through simulations, as well as the simulation-supported development of cars, airplanes, and medical and transport systems etc.

Paraphrasing R. Kenway (R.D. Kenway, Contemporary Physics. 1994): 'There is an important message to scientists, politicians, and industrialists: in the future science, the best industrial design and manufacture, the greatest medical progress, and the most accurate environmental monitoring and forecasting will be done by countries that most rapidly exploit the full potential of *computational science*'.

Nowadays we have access to high-end computer architectures and a large range of computing environments, mainly as a consequence of the enormous stimulus from the various international programs on advanced computing, e.g. HPCC (USA), HPCN (Europe), Real-World Computing (Japan), and ASCI (USA: Advanced Strategie Computing Initiative). The sequel to this, known as 'grid-systems' and 'grid-computing', will boost the computer, processing, and storage power even further. Today's supercomputing application may be tomorrow's desktop computing application.

The societal and industrial pulls have given a significant impulse to the rewriting of existing models and software. This has resulted among other things in a big 'clean-up' of often outdated software and new programming paradigms and verification techniques. With this make-up of arrears the road is paved for the study of real complex systems through computer simulations, and large scale problems that have long been intractable can now be tackled. However, the development of complexity reducing algorithms, numerical algorithms for large data sets, formal methods and associated modeling, as well as representation (i.e. visualization) techniques are still in their infancy. Deep understanding of the approaches required to model and simulate problems with increasing complexity and to efficiently exploit high performance computational techniques is still a big scientific challenge.

The International Conference on Computational Science (ICCS) series of conferences was started in May 2001 in San Francisco. The success of that meeting motivated the organization of the meeting held in Amsterdam from April 21–24, 2002.

These three volumes (Lecture Notes in Computer Science volumes 2329, 2330, and 2321) contain the proceedings of the ICCS 2002 meeting. The volumes consist of over 350 – peer reviewed – contributed and invited papers presented at the conference in the Science and Technology Center Watergraafsmeer (WTCW), in Amsterdam. The papers presented reflect the aims of the program committee to bring together major role players in the emerging field of computational science.

The conference was organized by The University of Amsterdam, Section Computational Science (http://www.science.uva.nl/research/scs/), SHARCNET, Canada (http://www.sharcnet.com), and the Innovative Computing Laboratory at The University of Tennessee.

The conference included 22 workshops, 7 keynote addresses, and over 350 contributed papers selected for oral presentation. Each paper was refereed by at least two referees.

We are deeply indebted to the members of the program committee, the workshop organizers, and all those in the community who helped us to organize a successful conference. Special thanks go to Alexander Bogdanov, Jerzy Wasniewski, and Marian Bubak for their help in the final phases of the review process. The invaluable administrative support of Manfred Stienstra, Alain Dankers, and Erik Hitipeuw is also acknowledged. Lodewijk Bos and his team were responsible for the local logistics and as always did a great job.

ICCS 2002 would not have been possible without the support of our sponsors: The University of Amsterdam, The Netherlands; Power Computing and Communication BV, The Netherlands; Elsevier Science Publishers, The Netherlands; Springer-Verlag, Germany; HPCN Foundation, The Netherlands; National Supercomputer Facilities (NCF), The Netherlands; Sun Microsystems, Inc., USA; SHARCNET, Canada; The Department of Computer Science, University of Calgary, Canada; and The School of Computer Science, The Queens University, Belfast, UK.

Amsterdam, April 2002

Peter M.A. Sloot,
Scientific Chair 2002,
on behalf of the co-editors:
C.J. Kenneth Tan
Jack J. Dongarra
Alfons G. Hoekstra

Organization

The 2002 International Conference on Computational Science was organized jointly by The University of Amsterdam, Section Computational Science, SHARCNET, Canada, and the University of Tennessee, Department of Computer Science.

Conference Chairs

Peter M.A. Sloot, Scientific and Overall Chair ICCS 2002 (University of Amsterdam, The Netherlands)
C.J. Kenneth Tan (SHARCNET, Canada)
Jack J. Dongarra (University of Tennessee, Knoxville, USA)

Workshops Organizing Chair

Alfons G. Hoekstra (University of Amsterdam, The Netherlands)

International Steering Committee

Vassil N. Alexandrov (University of Reading, UK)
J. A. Rod Blais (University of Calgary, Canada)
Alexander V. Bogdanov (Institute for High Performance Computing and Data Bases, Russia)
Marian Bubak (AGH, Poland)
Geoffrey Fox (Florida State University, USA)
Marina L. Gavrilova (University of Calgary, Canada)
Bob Hertzberger (University of Amsterdam, The Netherlands)
Anthony Hey (University of Southampton, UK)
Benjoe A. Juliano (California State University at Chico, USA)
James S. Pascoe (University of Reading, UK)
Rene S. Renner (California State University at Chico, USA)
Kokichi Sugihara (University of Tokyo, Japan)
Jerzy Wasniewski (Danish Computing Center for Research and Education, Denmark)
Albert Zomaya (University of Western Australia, Australia)

Local Organizing Committee

Alfons Hoekstra (University of Amsterdam, The Netherlands)
Alexander V. Bogdanov (Institute for High Performance Computing and Data Bases, Russia)
Marian Bubak (AGH, Poland)
Jerzy Wasniewski (Danish Computing Center for Research and Education, Denmark)

Local Advisory Committee

Patrick Aerts (National Computing Facilities (NCF), The Netherlands Organization for Scientific Research (NWO), The Netherlands
Jos Engelen (NIKHEF, The Netherlands)
Daan Frenkel (Amolf, The Netherlands)
Walter Hoogland (University of Amsterdam, The Netherlands)
Anwar Osseyran (SARA, The Netherlands)
Rik Maes (Faculty of Economics, University of Amsterdam, The Netherlands)
Gerard van Oortmerssen (CWI, The Netherlands)

Program Committee

Vassil N. Alexandrov (University of Reading, UK)
Hamid Arabnia (University of Georgia, USA)
J. A. Rod Blais (University of Calgary, Canada)
Alexander V. Bogdanov (Institute for High Performance Computing and Data Bases, Russia)
Marian Bubak (AGH, Poland)
Toni Cortes (University of Catalonia, Barcelona, Spain)
Brian J. d'Auriol (University of Texas at El Paso, USA)
Clint Dawson (University of Texas at Austin, USA)
Geoffrey Fox (Florida State University, USA)
Marina L. Gavrilova (University of Calgary, Canada)
James Glimm (SUNY Stony Brook, USA)
Paul Gray (University of Northern Iowa, USA)
Piet Hemker (CWI, The Netherlands)
Bob Hertzberger (University of Amsterdam, The Netherlands)
Chris Johnson (University of Utah, USA)
Dieter Kranzlmüller (Johannes Kepler University of Linz, Austria)
Antonio Lagana (University of Perugia, Italy)
Michael Mascagni (Florida State University, USA)
Jiri Nedoma (Academy of Sciences of the Czech Republic, Czech Republic)
Roman Neruda (Academy of Sciences of the Czech Republic, Czech Republic)
Jose M. Laginha M. Palma (University of Porto, Portugal)

Workshop Organizers

Automatic Differentiation and Applications
H. Martin Buecker (Aachen University of Technology, Germany)
Christian H. Bischof (Aachen University of Technology, Germany)
Computational Geometry and Applications
Marina Gavrilova (University of Calgary, Canada)
Computing in Medicine
Hans Reiber (Leiden University Medical Center, The Netherlands)
Rosemary Renaut (Arizona State University, USA)
High Performance Computing in Particle Accelerator Science and Technology
Andreas Adelmann (Paul Scherrer Institute, Switzerland)
Robert D. Ryne (Lawrence Berkeley National Laboratory, USA)
Geometric Numerical Algorithms: Theoretical Aspects and Applications
Nicoletta Del Buono (University of Bari, Italy)
Tiziano Politi (Politecnico-Bari, Italy)
Soft Computing: Systems and Applications
Renee Renner (California State University, USA)
PDE Software
Hans Petter Langtangen (University of Oslo, Norway)
Christoph Pflaum (University of Würzburg, Germany)
Ulrich Ruede (University of Erlangen-Nürnberg, Germany)
Stefan Turek (University of Dortmund, Germany)
Numerical Models in Geomechanics
R. Blaheta (Academy of Science, Czech Republic)
J. Nedoma (Academy of Science, Czech Republic)
Education in Computational Sciences
Rosie Renaut (Arizona State University, USA)
Computational Chemistry and Molecular Dynamics
Antonio Lagana (University of Perugia, Italy)
Geocomputation and Evolutionary Computation
Yong Xue (CAS, UK)
Narayana Jayaram (University of North London, UK)
Modeling and Simulation in Supercomputing and Telecommunications
Youngsong Mun (Korea)
Determinism, Randomness, Irreversibility, and Predictability
Guenri E. Norman (Russian Academy of Sciences, Russia)
Alexander V. Bogdanov (Institute of High Performance Computing and Information Systems, Russia)
Harald A. Pasch (University of Vienna, Austria)
Konstantin Korotenko (Shirshov Institute of Oceanology, Russia)

Sponsoring Organizations

The University of Amsterdam, The Netherlands
Power Computing and Communication BV, The Netherlands
Elsevier Science Publishers, The Netherlands
Springer-Verlag, Germany
HPCN Foundation, The Netherlands
National Supercomputer Facilities (NCF), The Netherlands
Sun Microsystems, Inc., USA
SHARCNET, Canada
Department of Computer Science, University of Calgary, Canada
School of Computer Science, The Queens University, Belfast, UK.

Local Organization and Logistics

Lodewijk Bos, MC-Consultancy
Jeanine Mulders, Registration Office, LGCE
Alain Dankers, University of Amsterdam
Manfred Stienstra, University of Amsterdam

Table of Contents, Part III

Workshop Papers II

Computational Geometry and Applications

Computing in Medicine

High Performance Computing in Particle Accelerator Science and Technology

Geometric Numerical Algorithms: Theoretical Aspects and Applications

Soft Computing: Systems and Applications

PDE Software

Numerical Models in Geomechanics

Education in Computational Sciences

Computational Chemistry and Molecular Dynamics

Geocomputation and Evolutionary Computation

Modeling and Simulation in Supercomputing and Telecommunications

Determinism, Randomness, Irreversibility, and Predictability

Table of Contents, Part I

Keynote Papers

Conference Papers

Computer Science – Information Retrieval

Complex Systems Applications 1

Computer Science – Computer Systems Models

Scientific Computing – Stochastic Algorithms

Complex Systems Applications 2

Computer Science – Networks

Scientific Computing – Domain Decomposition

Complex Systems Applications 3

Computer Science – Code Optimization

Methods for Complex Systems Simulation

Grid and Applications

Problem Solving Environment 1

Data Mining

Computer Science – Scheduling and Load Balancing

Problem Solving Environment 2

Computational Fluid Dynamics 1

Cellular Automata

Scientific Computing – Computational Methods 1

Problem Solving Environments 3

Computational Fluid Dynamics 2

Complex Systems Applications 4

Scientific Computing – Computational Methods 2

Scientific Computing – Computational Methods 3

Table of Contents, Part II

Workshop Papers I

Computer Graphics and Geometric Modeling

Modern Numerical Algorithms

Global and Collaborative Computing

Climate Systems Modelling

Parallel Computational Mechanics for Complex Systems

Tools for Program Development and Analysis

Automatic Differentiation and Applications

Automatic Differentiation and Applications

Workshop Papers II

Recent Developments in Motion Planning*

Mark H. Overmars

Institute of Information and Computing Sciences, Utrecht University, P.O. Box 80.089, 3508 TB Utrecht, the Netherlands. Email: markov@cs.uu.nl.

Abstract. Motion planning is becoming an important topic in many application areas, ranging from robotics to virtual environments and games. In this paper I review some recent results in motion planning, concentrating on the probabilistic roadmap approach that has proven to be very successful for many motion planning problems. After a brief description of the approach I indicate how the technique can be applied to various motion planning problems. Next I give a number of global techniques for improving the approach, and finally I describe some recent results on improving the quality of the resulting motions.

1 Introduction

Automated motion planning is rapidly gaining importance in various fields. Originally the problem was mainly studies in robotics. But in the past few years many new applications arise in fields such as animation, computer games, virtual environments, and maintenance planning and training in industrial CAD systems.

In its simplest form the motion planning problem can be formulated as follows: Given a moving object at a particular start position in an environment with a (large) number of obstacles and a required goal position, compute a collision free path for the object to the goal. Such a path should preferably be short, "nice", and feasible for the object.

The motion planning problem is normally formulated in the configuration space of the moving object. This is the space of all possible configurations of the object. For example, for a translating and rotating object in the plane the configuration space is a 3-dimensional space where the dimensions correspond to the x and y coordinate of the object and the rotation angle θ. For a robot arm consisting of n joints, the configuration space is n-dimensional space where each dimension corresponds to a joint position. A motion for the robot can be describes as a curve in the configuration space.

Over the years many different techniques for motion planning have been devised. See the book of Latombe[16] for an extensive overview of the situation up to 1991 and e.g. the proceedings of the yearly IEEE International Conference on Robotics and Automation for many more recent results.

Motion planning approaches can globally be subdivided in three classes: cell-decomposition methods, roadmap methods, and potential field (or local) methods. Cell decomposition methods try to divide the free part of the configuration

* This research has been supported by the ESPRIT LTR project MOLOG.

P.M.A. Sloot et al. (Eds.): ICCS 2002, LNCS 2331, pp. 3–13, 2002.

space (that is, those configurations that do not cause collisions) into a number of cells. Motion is than planned through these cells. Unfortunately, when the dimension of the configuration space gets higher or when the complexity of the scene is large, the number of cells required becomes too large to be practical. Roadmap methods try to construct a network of roads through the configuration space along which the object can move without collision. This roadmap can be seen as a graph, and the problem is reduced to graph searching. Unfortunately, computing an effective roadmap is very difficult. Potential field methods and other local methods steer the object by determining a direction of motion based on local properties of the scene around the moving object. The object tries to move in the direction of the goal while being pushed away by nearby obstacles. Because only local properties are used the object might move in the wrong direction, which can lead to dead-lock situations. Also there are some approaches based on neural networks (e.g. [26]) and genetic algorithms (e.g. [3]).

The *probabilistic roadmap planner (PRM)*, also called the probabilistic path planner (PPP), is a relatively new approach to motion planning, developed independently at different sites [2, 11, 13, 18, 22]. It is a roadmap technique but rather than constructing the roadmap in a deterministic way, a probabilistic technique is used. A big advantage of PRM is that its complexity tends to be dependent on the difficulty of the path, and much less on the global complexity of the scene or the dimension of the configuration space.

In the past few years the method has been successfully applied in many motion planning problems dealing with robot arms[14], car-like robots[23, 25], multiple robots[24], manipulation tasks[21] and even flexible objects[9, 15]. In all these cases the method is very efficient but, due to the probabilistic nature, it is difficult to analyze (see e.g. [12]).

In this paper I will give an overview of the probabilistic roadmap approach and indicate some of the recent achievements. After a brief description of the basic technique in Sect. 2 I will show how the approach can be used for solving various types of motion planning problems. Then, in Sect. 4, I will describe a number of interesting improvements that have been suggested. Finally, in Sect. 5, I will discuss a number of issues related to the quality of the resulting motions.

2 Probabilistic Roadmap Planner

The motion planning problem is normally formulated in terms of the *configuration space* $\mathcal{C}$, the space of all possible configurations of the robot. Each degree of freedom of the robot corresponds to a dimension of the configuration space. Each obstacle in the workspace, in which the robot moves, transforms into an obstacle in the configuration space. Together they form the forbidden part $\mathcal{C}_{\text{forb}}$ of the configuration space. A path of the robot corresponds to a curve in the configuration space connecting the start and the goal configuration. A path is collision-free if the corresponding curve does not intersect $\mathcal{C}_{\text{forb}}$, that is, it lies completely in the free part of the configuration space, denoted with $\mathcal{C}_{\text{free}}$.

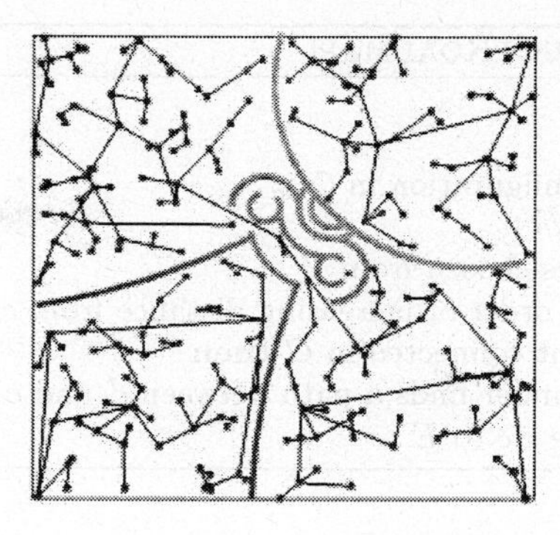

Fig. 1. A typical graph produced by PRM.

The probabilistic roadmap planner samples the configuration space for free configurations and tries to connect these configurations into a roadmap of feasible motions. There are a number of versions of PRM, but they all use the same underlying concepts. Here we base ourselves on the description in [22].

The global idea of PRM is to pick a collection of (random) configurations in the free space $\mathcal{C}_{\text{free}}$. These free configurations form the nodes of a graph $G = (V, E)$. A number of pairs of nodes are chosen and a simple local motion planner is used to try to connect these configurations by a path. When the local planner succeeds an edge is added to the graph. The local planner must be very fast, but is allowed to fail on difficult instances. (It must also be deterministic.) A typical choice is to use a simple interpolation between the two configurations, and then check whether the path is collision-free. See Fig. 1 for an example of a graph created with PRM in a simple 2-dimensional scene. (Because the configuration space is 3-dimensional, the graph should actually be drawn in this 3-dimensional space. In the figure and in all other figures in this paper we project the graph back into the workspace.)

Once the graph reflects the connectivity of $\mathcal{C}_{\text{free}}$ it can be used to answer motion planning queries. To find a motion between a start configuration and a goal configuration, both are added to the graph using the local planner. (Some authors use more complicated techniques to connect the start and goal to the graph, e.g. using bouncing motion.) Then a path in the graph is found which corresponds to a motion for the robot. In a post-processing step this path is then smoothed to improve its quality. The pseudo code for the algorithm for constructing the graph is shown in Algorithm ConstructRoadmap.

There are many details to fill in in this global scheme: which local planner to use, how to select promising pairs of nodes to connect, what distance measure to use, how to improve the resulting paths, etc. These typically depend on the type of motion planning problem we want to solve. See Sect. 3 for some information about this.

If we already know the start and goal configuration, we can first add them to the graph and continue the loop until a path between start and goal exists.

Algorithm 1 ConstructRoadmap

Let: $V \leftarrow \emptyset$; $E \leftarrow \emptyset$;
1: **loop**
2: $c \leftarrow$ a (random) configuration in $\mathcal{C}_{\text{free}}$
3: $V \leftarrow V \cup \{c\}$
4: $N_c \leftarrow$ a set of nodes chosen from V
5: **for all** $c' \in N_c$, in order of increasing distance from c **do**
6: **if** c' and c are not connected in G **then**
7: **if** the local planner finds a path between c' and c **then**
8: add the edge $c'c$ to E

Note that the test in line 6 guarantees that we never connect two nodes that are already connected in the graph. Although such connections are indeed not necessary to solve the problem, they can still be useful for creating shorter paths. See Sect. 5 for details.

The two time-consuming steps in this algorithm are line 2 where a free sample is generated, and line 7 where we test whether the local method can find a path between the new sample and a configuration in the graph. The geometric operations required for these steps dominate the work. So to improve the efficiency of PRM we need to implement these steps very efficiently and we need to avoid calls to them as much as possible. That is, we need to place samples at "useful" places and need to compute only "useful" edges. The problem is that it is not clear how to determine whether a node or edge is "useful". Many of the improvements described in Sect. 4 work this way.

Because of the probabilistic nature of the algorithm it is difficult to analyze it. The algorithm is not complete. It can never report that for certain no solution exists. But fortunately for most applications the algorithm is probabilistically complete, that is, when the running time goes to infinity, the chance that a solution is found goes to 1 (assuming a solution exists). Little is known about the speed of convergence[12]. In practice though solutions tend to be found fast in most cases.

3 Applications

The simplest application of PRM is an object that moves freely (translating and rotating) through a 2- or 3-dimensional workspace. In this case the configuration spaces is either 3-dimensional or 6-dimensional. As a local planner we can use a straight-line interpolation between the two configuration. (An interesting question here is how to represent the rotational degrees of freedom and how to interpolate between them but we won't go into detail here.) As distance between two configurations we must use a weighted sum of the translational distance and the amount of rotation. Typically, the rotation becomes more important when the moving object is large. With these details filled in the PRM approach can be applied without much difficulty. (See though the remarks in the next sections.) For other types of moving objects there is some more work to be done.

Car-like Robots A car-like robot has special so-called non-holonomic constraints than restrict its motion. For example, a car cannot move sideways. Still the configuration space is 3-dimensional because, given enough space, the car can get in each position in any orientation. Using a simple straight-line interpolation for the local planner no longer leads to valid paths for the robot. So we need to use a different local planner. One choice, used e.g. in [22, 23], is to let the local planner compute paths consisting of a circle arc of minimal turning radius, followed by a straight-line motion, followed by another circle arc. It was shown in [23] that such a local planner is powerful enough to solve the problem. The approach is probabilistically complete. Extensions have also been proposed towards other types of robots with non-holonomic constraints, like trucks with trailers[25].

Robot Arms A robot arm has a number of degrees of freedom depending on the number of joints. Typical robot arms have up to 6 joints, resulting in a 6-dimensional configuration space. Most of these are rotational degrees of freedom, often with limits on the angle. The PRM approach can be applied rather easily in this situation. As local method we can interpolate between the configurations (although there exist better methods, see [14]). When computing distances it is best to let the major axis of the robot play a larger role than the minor axis. Again the approach is probabilistically complete.

Multiple Robots When there are multiple moving robots or objects in the same environment we need to coordinate their motions. There are two basic approaches for this (see e.g. the book of Latombe[16]). When applying centralized planning the robots together are considered as one robotic system with many degrees of freedom. For example in the situation in Fig. 2 there are 6 robot arms with a total of 36 degrees of freedom. When applying decoupled planning we first compute the individual motions of the robots and then try to coordinate these over time. This is faster but can lead to deadlock. In [24] a solution based on PRM is proposed that lies between these two. Rather that coordinate the paths, the roadmaps themselves are coordinated, leading to a faster and probabilistically complete planner.

In a recent paper Sánchez and Latombe[19] show that with a number of improvements the PRM approach can be successfully applied to solve complicated motion planning with up to 6 robot arms, as shown in Fig. 2. When the number of robots is much larger the problem though remains unsolved.

Other Applications The PRM approach has been successfully applied in many other situations. Applications include motion planning for flexible objects[9, 15], motion planning with closed kinematic loops[7, 6], like two mobile robot arms that together hold an object, motion planning in the presence of dangerzones that preferably should be avoided[20], and manipulation tasks[21]. In all these cases one need to find the right representation of the degrees of freedom of the problem, construct an appropriate local planner and fill in the parameters of the PRM approach. It shows the versatility of the method.

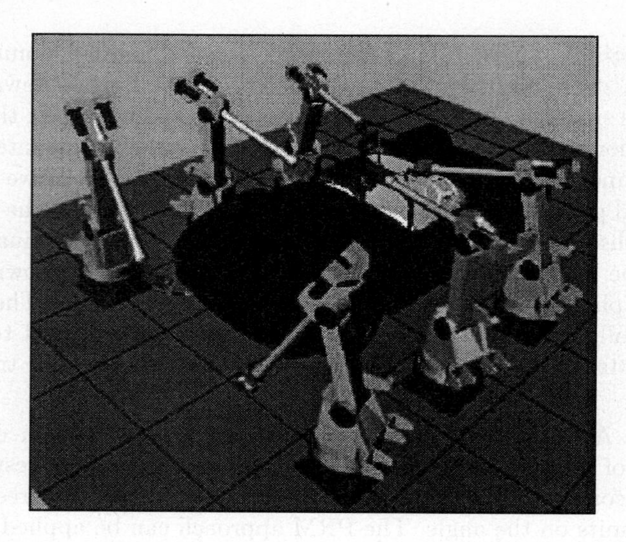

Fig. 2. An example where 6 robots must plan their motions together (taken from Sánchez and Latombe[19]).

4 Improving the Approach

Although the PRM approach can solve many different types of motion planning problems effectively there are a number of problematic issues. Here we discuss some improvements that have been proposed.

Sampling Strategy The default sampling approach samples the free space in a uniform way. This is fine when obstacle density is rather uniform over the scene but in practice this assumption is not correct. Some areas tend to be wide open while at other places there are narrow passages (in particular in configuration space). To obtain enough random samples in such narrow passages one would need way too many samples in total. So a number of authors have suggested ways to obtain more samples in difficult areas.

One of the early papers[14] suggested to maintain information on how often the local planner fails for certain nodes in the graph. When this number is large for a particular node this suggest that this node is located in a difficult area. The same is true when two nodes lie near to each other but no connection has been found between them. One can increase the number of samples in such areas.

Another approach is to place addition samples near to edges and vertices of obstacles[1, 23] or to allow for samples inside obstacles and pushing them to the outside[27, 10]. Such methods though require more complicated geometric operations on the obstacles.

An approach that avoids such geometric computations is the Gaussian sampling technique[5]. The approach works as follows. Rather than one sample we take two samples where the distance between the two samples is taken with respect to a Gaussian distribution. When both samples are forbidden we obviously

Fig. 3. A motion planning problem in a complex industrial environment with over 4000 obstacles. The left picture shows the nodes obtained with uniform sampling, and the right picture the nodes obtained with Gaussian sampling.

remove them. When both lie in the free space we also remove them because there is a high probability that they lie in an open area. When only one of the two samples is free we add this sample to the graph. It can be shown (see [5]) that this approach results in a sample distribution that corresponds to a Gaussian blur of the obstacles (in configuration space). The closer you are to an obstacle, the higher the change that a sample is placed there. See Fig. 3 for an example.

Roadmap Size As indicated above, computing paths using the local planner is the most time-consuming step in the PRM algorithm. We would like to avoid such computations as much as possible. One way to do this is to keep the roadmap as small as possible. The visibility based PRM[17] only adds a node to the roadmap if it either can be connected to two components of the graph or to no component at all. The reason is that a node that can be connected to just one component represents an area that can already be "seen" by the roadmap. It can be shown that the approach converges to a roadmap that covers the entire free space. The number of nodes tends to remain very small, unless the free space has a very complicated structure.

Another idea is not to test whether the paths are collision free unless they are really needed[4]. Such a lazy approach only checks whether the nodes are collision free and when nodes are close to each other they are connected with an edge. Only when an actual motion planning query must be solved we test whether the edges on the shortest path in the graph are collision-free. If not we try other edges, until a path is found. The rational behind this is that for most paths we only need to consider a small part of the graph before a solution is found. In [19] a similar idea is used. Here it is also argued and demonstrated that the chance that an edge is collision-free is large when the endpoints (the nodes) are collision-free and the length of the edge is short.

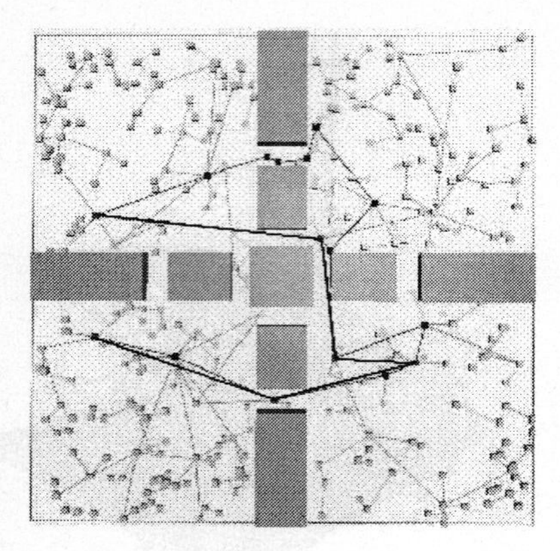

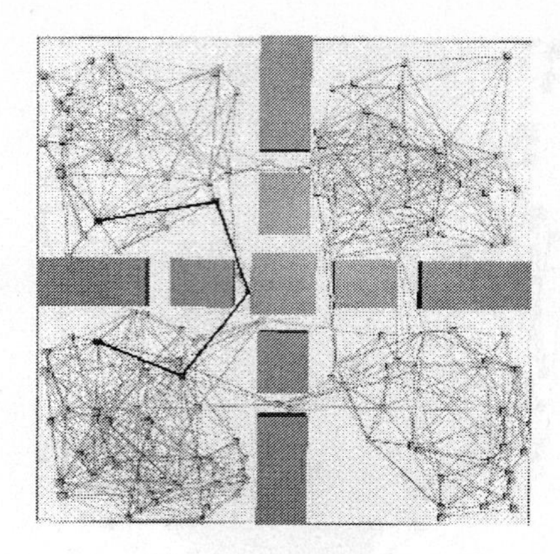

Fig. 4. The left picture shows the graph in the default algorithm. Here a long detour is made. In the right picture cycles are added and the length of the path is reduced considerably.

5 Path Quality

One of the problems of the PRM approach is that the resulting motions are ugly. This is due to the random nature of the samples. A resulting path can make long detours and contain many redundant motions. Also the path normally consists of straight-line motions (in the configuration space) leading to first-order discontinuities at the nodes of the graph. In most applications such ugly paths are unacceptable.

The standard method used to remedy these problems is to smooth the resulting path in a post-processing phase. This smoothing technique consists of taking random pairs (c_1, c_2) of configurations on the path (not necessarily nodes of the graph) and trying to replace the path between c_1 and c_2 by the path resulting from calling the local planner on (c_1, c_2), if this new path is collision free. Unfortunately, smoothing only partially solves the problem. It does reduce the length of the path in open areas but it often cannot correct long detours around obstacles. Also it does not make the path first-order continuous and the path can still include many redundant (rotational) motions, in particular in a 3-dimensional workspace.

In this section we will discuss some recent approaches to improving the path quality. More details will be given in an upcoming paper[8].

Length A prime reason why paths computed with PRM are too long is that a tree (or to be more precise, a forest) is used as roadmap. The advantage of this is that it will save computation time, because less calls to the local planner are required, while connectivity is maintained. So the obvious solution is to add additional edges, leading to cycles in the roadmap. This is easier said than done because we want to avoid calls to the local planner as much as possible (because this is the most time consuming operation in the algorithm). So we only want to create a cycle when it is "useful". We define useful as follows: Assume the algorithm is trying to add configuration c to the graph. Let c' be a node in the neighbor set N_c. We try to add an edge between c and c' when they are not yet connected in the graph or when the current distance d_G of the shortest path in the graph is larger than $k.d(c, c')$ for some given constant parameter k. So we only try to add the edge when it would improve the length of the shortest path with a factor at least k. The parameter k will determine how dense the graph will be. See Fig. 4 for an example.

There is one algorithmic problem left. To implement the approach we need to be able to compute a shortest path in the graph whenever we try to add an edge. This is rather expensive and would dominate the cost of the algorithm when the graph gets large (it will be called a quadratic number of times). The solution is based on the observation that we can stop searching the graph when shortest paths in the graph become longer than $k.d(c, c')$. We can than immediately decide to add the edge. This will prune the graph quite a bit. We can take this one step further by also taking the distance between the current node in the graph search and c' into account. This leads to some sort of A* algorithm that is a lot faster.

Smoothness Nodes in the graph introduce first-order discontinuities in the motion. We would like to avoid this. This can be achieved as follows. Let e and e' be two consecutive edges in the final path. Let p_m be the midpoint of e and p'_m be the midpoint of e'. We replace the part of the path between p_m and p'_m by a circle arc. This arc will have its center on the bisecting line of e and e', will touch e and e' and have either p_m or p'_m on its boundary. Doing this for each consecutive pair of edges results in a smooth path. The only problem is that the motion along the circle arc might collide with an obstacle. In this case we make the circle smaller, pushing it more towards the node between the edges. It is easy to verify that there always exists a circle arc between the edges that does not introduce collisions. Hence, the method is complete. See Fig. 5 for an example.

When the angle between two consecutive edges becomes small, the radius of the circle becomes small as well. We often like to avoid this. We are currently investigating how we can produce roadmaps that keep the angles as large as possible.

Redundant Motions Allowing cycles in graphs and smoothing the path improves the motion a lot. Still redundant motions can occur. For example, the object can continuously spin around its center. Such motion does not really increase the time it takes to execute the motion. Hence standard smoothing techniques tend not to work. One could add a penalty factor in the length of the path but this again does often not help.

There are a number of techniques that try to remedy this problem. One is to add many nodes with the same orientation. (Or stated in a more generic way, divide the degrees of freedom in major degrees of freedom and minor ones and generate many configurations with the same values for the minor degrees of

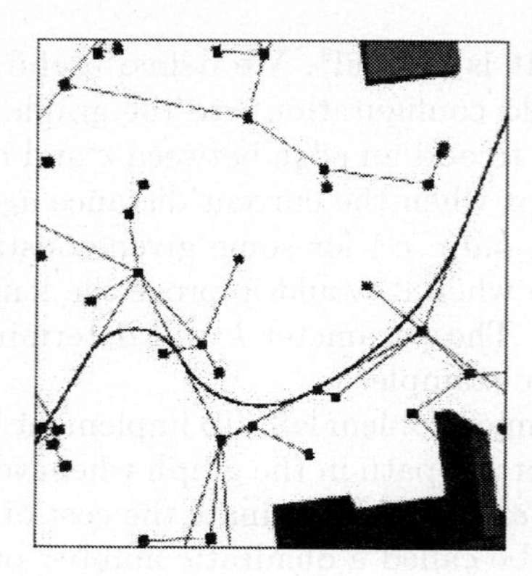

Fig. 5. An example of a part of a path with circular blends.

freedom.) A similar idea was used in a paper by Lamiraux and Kavraki on moving flexible objects[15]. A second approach is to do the smoothing in a different way. The standard smoothing technique replaces pieces of the path by calls to the local planner, that is, by a straight line in the configuration space. In this way all degrees of freedom are smoothed at the same moment. But some of them might be necessary while others are not. For example, the translational degrees of freedom might be necessary to get the object around an obstacle while the rotational degrees of freedom are not necessary. By smoothing the degrees of freedom one at a time we create better paths. Finally, we can try to find a better path by resampling the configuration space in a tube around the original path, similar to the technique in [25].

References

1. N. Amato, O. Bayazit, L. Dale, C. Jones, D. Vallejo, OBPRM: An obstacle-based PRM for 3D workspaces, in: P.K. Agarwal, L.E. Kavraki, M.T. Mason (eds.), *Robotics: The algorithmic perspective*, A.K. Peters, Natick, 1998, pp. 155–168.
2. N. Amato, Y. Wu, A randomized roadmap method for path and manipulation planning, *Proc. IEEE Int. Conf. on Robotics and Automation,* 1996, pp. 113–120.
3. P. Bessière. J.M. Ahuactzin, E.-G. Talbi, E. Mazer, The Ariadne's clew algorithm: Global planning with local methods, in: K. Goldberg et al. (eds.), *Algorithmic foundations of robotics*, A.K. Peters, 1995, pp. 39–47.
4. R. Bohlin, L.E. Kavraki, Path planning using lazy PRM, *Proc. IEEE Int. Conf. on Robotics and Automation,* 2000, pp. 521–528.
5. V. Boor, M.H. Overmars, A.F. van der Stappen, The Gaussian sampling strategy for probabilistic roadmap planners, *Proc. IEEE Int. Conf. on Robotics and Automation,* 1999, pp. 1018–1023.
6. J. Cortes, T. Simeon, J.P. Laumond, A random loop generator for planning the motions of closed kinematic chains using PRM methods, Rapport LAAS N01432, 2001.
7. L. Han, N. Amato, A kinematics-based probabilistic roadmap method for closed chain systems, *Proc. Workshop on Algorithmic Foundations of Robotics (WAFR'00)*, 2000, pp. 233–246.

8. O. Hofstra, D. Nieuwenhuisen, M.H. Overmars, Improving path quality for probabilistic roadmap planners, 2002, in preparation.
9. C. Holleman, L. Kavraki, J. Warren, Planning paths for a flexible surface patch, *Proc. IEEE Int. Conf. on Robotics and Automation,* 1998, pp. 21–26.
10. D. Hsu, L. Kavraki, J.C. Latombe, R. Motwani, S. Sorkin, On finding narrow passages with probabilistic roadmap planners, in: P.K. Agarwal, L.E. Kavraki, M.T. Mason (eds.), *Robotics: The algorithmic perspective,* A.K. Peters, Natick, 1998, pp. 141–154.
11. L. Kavraki, *Random networks in configuration space for fast path planning,* PhD thesis, Stanford University, 1995.
12. L. Kavraki, M. Kolountzakis, J.C. Latombe, Analysis of probabilistic roadmaps for path planning, *Proc. IEEE Int. Conf. on Robotics and Automation,* 1996, pp. 3020–3025.
13. L. Kavraki, J.C. Latombe, Randomized preprocessing of configuration space for fast path planning, *Proc. IEEE Int. Conf. on Robotics and Automation,* 1994, pp. 2138–2145.
14. L. Kavraki, P. Švestka, J-C. Latombe, M.H. Overmars, Probabilistic roadmaps for path planning in high-dimensional configuration spaces, *IEEE Trans. on Robotics and Automation* **12** (1996), pp. 566–580.
15. F. Lamiraux, L.E. Kavraki, Planning paths for elastic objects under manipulation constraints, *Int. Journal of Robotics Research* **20** (2001), pp. 188–208.
16. J-C. Latombe, *Robot motion planning,* Kluwer Academic Publishers, Boston, 1991.
17. C. Nissoux, T. Siméon, J.-P. Laumond, Visibility based probabilistic roadmaps, *Proc. IEEE Int. Conf. on Intelligent Robots and Systems,* 1999, pp. 1316–1321.
18. M.H. Overmars, *A random approach to motion planning,* Technical Report RUU-CS-92-32, Dept. Comput. Sci., Utrecht Univ., Utrecht, the Netherlands, October 1992.
19. G. Sánchez, J.-C. Latombe, A single-query bi-directional probabilistic roadmap planner with lazy collision checking, *Int. Journal of Robotics Research*, 2002, to appear.
20. D. Sent, M.H. Overmars, Motion planning in an environment with dangerzones, *Proc. IEEE Int. Conf. on Robotics and Automation,* 2001, pp. 1488–1493.
21. T. Simeon, J. Cortes, A. Sahbani, J.P. Laumond, A manipulation planner for pick and place operations under continuous grasps and placements, Rapport LAAS N01433, 2001.
22. P. Švestka, *Robot motion planning using probabilistic roadmaps,* PhD thesis, Utrecht Univ. 1997.
23. P. Švestka, M.H. Overmars, Motion planning for car-like robots, a probabilistic learning approach, *Int. Journal of Robotics Research* **16** (1997), pp. 119–143.
24. P. Švestka, M.H. Overmars, Coordinated path planning for multiple robots, *Robotics and Autonomous Systems* **23** (1998), pp. 125–152.
25. S. Sekhavat, P. Švestka, J.-P. Laumond, M.H. Overmars, Multilevel path planning for nonholonomic robots using semiholonomic subsystems, *Int. Journal of Robotics Research* **17** (1998), pp. 840–857.
26. J. Vleugels, J. Kok, M.H. Overmars, Motion planning with complete knowledge using a colored SOM, *Int. Journal of Neural Systems* **8** (1997), pp. 613–628.
27. S.A. Wilmarth, N.M. Amato, P.F. Stiller, MAPRM: A probabilistic roadmap planner with sampling on the medial axis of the free space, *Proc. IEEE Int. Conf. on Robotics and Automation,* 1999, pp. 1024–1031.

Extreme Distances in Multicolored Point Sets

Adrian Dumitrescu[1] and Sumanta Guha[1]

University of Wisconsin-Milwaukee,
Milwaukee, WI 53211, USA
{ad,guha}@cs.uwm.edu

Abstract. Given a set of n points in some d-dimensional Euclidean space, each point colored with one of $k(\geq 2)$ colors, a bichromatic closest (resp., farthest) pair is a closest (resp., farthest) pair of points of different colors. We present efficient algorithms to compute a bichromatic closest pair and a bichromatic farthest pair. We consider both static, and dynamic versions with respect to color flips. We also give some combinatorial bounds on the multiplicities of extreme distances in this setting.

1 Introduction

Given a collection of k pairwise disjoint sets with a total of n points in d-dimensional Euclidean space, we consider static and certain dynamic algorithms to compute the maximum (resp. minimum) distance between pairs of points in different sets. One may imagine each set colored by one of a palette of k colors – in which case we are considering distances between points of different colors (k is not fixed and may depend on n). In this paper, *distance* (or *length*) stands for Euclidean distance when not specified.

Given n (uncolored) points in d-dimensional Euclidean space, the problem of finding a *closest pair* is classical and, together with related problems, has been studied extensively. We refer the reader to recent surveys by Eppstein and Mitchell [10, 13]. In the following, we discuss the literature related to chromatic versions of the problem that is relevant to our paper.

The bichromatic case in two dimensions – a set of n points in the plane each colored red or blue – has been solved optimally in $O(n \log n)$ time, to find either the minimum distance between a bichromatic (red-blue) pair (i.e., the *bichromatic closest pair* or BCP problem [19, 6]), or the maximum distance between a bichromatic pair (i.e., the *bichromatic farthest pair* or BFP problem [21, 7]).

Extending to higher dimensions turns out to be more difficult if one seeks optimal algorithms. The approach of Bhattacharya-Toussaint [7] to the planar problem has been extended to higher dimensions by Robert [17], and reduces the BFP problem for n points in $\mathbb{R}^d$ to the problem of computing the diameters of c_d sets of points in $\mathbb{R}^d$, for some constant c_d depending exponentially on d.

The BCP problem is intimately related with that of computing an *Euclidean minimum spanning tree* (EMST). Similarly, the BFP problem is closely related

P.M.A. Sloot et al. (Eds.): ICCS 2002, LNCS 2331, pp. 14–25, 2002.

with that of computing an *Euclidean maximum spanning tree* (EXST). It is not difficult to verify that an EMST of a set of points each colored red or blue contains at least one edge joining a bichromatic closest pair, so after an EMST computation the BCP problem can be solved in further linear time. In the opposite direction Agarwal et al [1] show that if the BCP problem for a set of n red or blue points in $\mathbb{R}^d$ can be solved in time $T_d^{\min}(2, n)$, then an EMST of n points in $\mathbb{R}^d$ can be computed in time $O(T_d^{\min}(2, n) \log^d n)$. Their result is improved by Krznaric and Levcopoulos [12], where the authors show that the problem of computing an EMST and the BCP problem are, in fact, equivalent to within a constant factor.

Dynamic versions of the (uncolored) closest and farthest pairs problem, especially the former – the setting being an uncolored point set subject to insertion and deletion – have been of considerable interest as well and the literature is extensive. We refer the reader to a recent paper by Bespamyatnikh and the bibliography therein [5]. Dynamic versions of the bichromatic closest and farthest pairs problem have been studied as well [8, 9, 22], again from the point of view of inserting into and deleting from a point set. The best update times are polynomial in the current size of the point set.

In this paper we consider both static and dynamic bichromatic closest and farthest pairs problems, in the multicolor setting. In the dynamic case, the point set itself is fixed, but points change their color. To our knowledge ours is the first paper to consider this restricted dynamism and, not surprisingly, our update times are superior to and our algorithms less complicated than the best-known ones for the more general problem mentioned above where points themselves may be inserted and deleted.

Specifically, the input to our problem is a set of n points in $\mathbb{R}^d$, at *fixed* locations, colored using a palette of k colors, and the goal is to compute (resp., dynamically maintain after each color flip) a bichromatic closest pair and a bichromatic farthest pair (if exists).

The algorithms for the static version and the preprocessing involved in our dynamic algorithms are essentially EMST and EXST computations, so it is relevant to briefly discuss the current best-known times for these.

EMST and EXST Computations

Given a set S of n points in $\mathbb{R}^d$, an EMST (resp., EXST) is a spanning tree of S whose total edge length is minimum (resp., maximum) among all spanning trees of S, where the length of an edge is the Euclidean distance between its endpoints.

For two dimensions ($d = 2$), an optimal $O(n \log n)$ time algorithm to compute the EMST of n points is given by Shamos and Hoey [19]. Agarwal et al [1] show how to compute an EMST in d dimensions, for arbitrary d, in randomized expected time $O((n \log n)^{4/3})$ for $d = 3$, or deterministically in time $O(n^{2-\alpha_{\epsilon,d}})$ for $d \geq 4$ and any fixed $\epsilon > 0$ (here $\alpha_{\epsilon,d} = \frac{2}{\lceil d/2 \rceil + 1 + \epsilon}$). See also the two surveys mentioned above.

Monma et al [14] provide an optimal $O(n \log n)$ time algorithm for EXST computation in the plane. In higher dimensions, Agarwal et al [2] present subquadratic-time algorithms, based on efficient methods to solve the BFP problem: a randomized algorithm with expected time $O(n^{4/3} \log^{7/3} n)$ for $d = 3$, and $O(n^{2-\alpha_{\epsilon,d}})$ for $d \geq 4$ and any fixed $\epsilon > 0$ (here $\alpha_{\epsilon,d} = \frac{2}{\lceil d/2 \rceil + 1 + \epsilon}$). See also [10, 13].

Summary of Our Results. In this paper, we obtain several results on the theme of computing extreme distances in multicolored point sets, including:

(1) We relate the various time complexities of computing extreme distances in multicolored point sets in $\mathbb{R}^d$ with the time complexities for the bichromatic versions. We also discuss an extension of this problem for computing such extreme distances over an arbitrary set of color pairs.

(2) We show that the bichromatic closest (resp. farthest) pair of points in a multicolored point set in $\mathbb{R}^d$ can be maintained under dynamic color changes in logarithmic time and linear space after suitable preprocessing. These algorithms can, in fact, be extended to maintaining the bichromatic edge of minimum (resp., maximum) weight in an undirected weighted graph with multicolored vertices, when vertices dynamically change color.

(3) We present combinatorial bounds on the maximum number of extreme distances in multicolored planar point sets. Our bounds are tight up to multiplicative constant factors.

2 Algorithmic Implications on Computing Extreme Distances

We begin with a simple observation:

Observation 1. *Let S be a set of points in an Euclidean space, each colored with one of k colors. Then the Euclidean minimum spanning tree (EMST) of S contains at least one edge joining a bichromatic closest pair. Similarly, the Euclidean maximum spanning tree (EXST) of S contains at least one edge joining a bichromatic farthest pair.*

Proof. Assume that the minimum distance between a bichromatic pair from S, say between points p and q, is strictly smaller than that of each bichromatic edge (i.e., an edge joining points of different color) of the EMST T of S. Consider the unique path in T between p and q. Since p and q are of different colors there exists a bichromatic edge rs on this path. Exchanging edge rs for pq would reduce the cost of T, which is a contradiction.

The proof that the EXST of S contains at least one edge joining a bichromatic farthest pair is similar. □

In the static case, the only attempt (that we know of) to extend to the multicolor version, algorithms for the bichromatic version, appears in [3]. The authors present algorithms based on Voronoi diagrams computation, for the bichromatic closest pair (BCP) problem in the plane – in the multicolor setting

– that run in optimal $O(n \log n)$ time. In fact, within this time, they solve the more general *all bichromatic closest pairs problem* in the plane, where for each point, a closest point of different color is found. However the multicolor version of the BFP problem does not seem to have been investigated.

Let us first notice a different algorithm to solve the BCP problem within the same time bound, based on Observation 1. The algorithm first computes an EMST of the point set, and then performs a linear scan of its edges to extract a bichromatic closest pair. The same approach solves the BFP problem, and these algorithms generalize to higher dimensions. Their running times are dominated by EMST (resp., EXST) computations.

Next we consider the following generalization of this class of proximity problems. Instead of asking for the maximum (resp. minimum) distance between all pairs of points of different colors, we restrict the sets of pairs. To be precise, let G be an arbitrary graph on k vertices $\{S_1, \ldots, S_k\}$, where S_i, $(i = 1, \ldots, k)$ are the k sets, of different colors, comprising a total of n points. This extension of the BCP problem asks for a pair of points $p_i \in S_i$, $p_j \in S_j$ which realize a minimum distance over all pairs $S_i \sim_G S_j$. There is an analogous extension of the BFP problem.

Lemma 1. *The edge set of a graph on k vertices can expressed as a union of the sets of edges of less than k complete bipartite graphs on the same set of k vertices. This bound cannot be improved apart from a multiplicative constant. Moreover each such bipartition can be generated in linear time.*

Proof. Let $V = \{0, \ldots, k-1\}$ be the vertex set of G. For $i = 0, \ldots, k-2$, let $A_i = \{i\}$ and $B_i = \{j \in \{i+1, \ldots, k-1\} \mid i \sim_G j\}$ specify the bipartitions. Clearly each edge of G belongs to at least one complete bipartite graphs above. Also all edges of these bipartite graphs are present in G. One can easily see that certain sparse graphs (e.g C_k, the cycle on k vertices) require $\Omega(k)$ complete bipartite graphs in a decomposition. □

Lemma 1 offers algorithms for solving the extended versions of the BCP (resp., BFP) problem by making $O(k)$ calls to an algorithm which solves the corresponding bichromatic version.

Putting together the above facts we can relate the time complexities for computing extreme distances in bichromatic and multichromatic point sets. However, it is convenient first to define some notation.

Let $T_d^{EMST}(n)$ denote the best-known worst-case time to compute an EMST of n points lying in d-dimensional Euclidean space, and $T_d^{EXST}(n)$ denote the analogous time complexity to compute an EXST (see Section 1 for a discussion of these times). Let $T_d^{\min}(k, n)$ denote the best of the worst-case time complexities of known algorithms to solve the BCP problem for n points of at most k different colors lying in d-dimensional Euclidean space. Let $T_d^{\max}(k, n)$ denote the analogous time complexity for the BFP problem. Let $E_d^{\min}(k, n)$ (resp., $E_d^{\max}(k, n)$) be the analogous time complexities for the extended versions of the BCP (resp., BFP) problem.

Theorem 1. *The following relations hold between various time complexities:*

(i) $T_d^{\min}(k,n) = O(T_d^{\min}(2,n)) = O(T_d^{EMST}(n))$.
(ii) $T_d^{\max}(k,n) = O(T_d^{\max}(2,n)) = O(T_d^{\max}(n)) = O(T_d^{EXST}(n))$.
(iii) $E_d^{\min}(k,n) \leq O(k \cdot T_d^{\min}(2,n))$.
(iv) $E_d^{\max}(k,n) \leq O(k \cdot T_d^{\max}(2,n))$.

The algorithms implied by *(iii)* and *(iv)* represent improvements over the corresponding straightforward $O(dn^2)$ time algorithms (which look at all pairs of points), when k is not too large. For example, when $k = \sqrt{n}$ and $d = 4$, the algorithm for the extended version of the BCP would run in $o(n^2)$ time (making use of the algorithm in [1] for the bichromatic version).

For purpose of comparison, we present another approach based on graph decomposition. As usual, denote by K_n the complete graph on n vertices, and by $K_{m,n}$ the complete bipartite graph on m and n vertices.

Lemma 2. *The edge set of the complete graph on k vertices can expressed as a union of the sets of edges of $\lceil \log k \rceil$ complete bipartite graphs on the same set of k vertices. Moreover each such bipartition can be generated in linear time.*

Proof. Put $l = \lceil \log k \rceil$; l represents the number of bits necessary to represent in binary all integers in the range $\{0, \ldots, k-1\}$. For any such integer j, let j_i be its i-th bit. We assume that the vertex set of the complete graph is $\{0, \ldots, k-1\}$. For $i = 1, \ldots, l$, let $A_i = \{j \in \{0, \ldots, k-1\} \mid j_i = 0\}$ and $B_i = \{j \in \{0, \ldots, k-1\} \mid j_i = 1\}$ specify the bipartitions. It is easy to see that each edge of the complete graph belongs to at least one complete bipartite graphs above. Also all edges of these bipartite graphs are present in the complete graph, which concludes the proof. □

Lemma 2 offers us an algorithm for solving the BCP (resp. BFP) problem by making $O(\log k)$ calls to an algorithm which solves the corresponding bichromatic version. The algorithm in [3], as well as ours at the beginning of this section, have shown that $T_2^{\min}(k,n) = O(n \log n)$. Using Lemma 2, we get an algorithm for the BCP (resp., BFP) problem which runs in $O(T_d^{\min}(2,n) \log k)$ time (resp., $O(T_d^{\max}(2,n) \log k)$ time in d dimensions, e.g., in only $O(n \log n \log k) = O(n \log^2 n)$ time in the plane.

3 Dynamic color changes

The input to our problem is a set of n points in $\mathbb{R}^d$, at *fixed* locations, colored using a palette of k colors, and we maintain a bichromatic closest and a bichromatic farthest pair (if exists) as each change of a point color is performed. This, of course, maintains the distance between such pairs as well. When the point set becomes monochromatic that distance becomes ∞ and no pair is reported. Both our algorithms to maintain the bichromatic closest and farthest pairs run in logarithmic time and linear space after suitable preprocessing. These algorithms can, in fact, be extended to maintaining the bichromatic edge of minimum (resp., maximum) weight in an undirected weighted graph with multicolored vertices, when vertices dynamically change color. We first address the closest pair problem which is simpler.

3.1 Closest Pair

Our approach is based on the above observation. In the preprocessing step, compute T, an EMST of the point set, which takes $T_d^{EMST}(n)$ time. Insert all bichromatic edges in a minimum heap H with the Euclidean edge length (or its square) as a key. Maintain a list of pointers from each vertex (point) to elements of H that are bichromatic edges adjacent to it in T. In the planar case, the maximum degree of a vertex in T is at most 6. In d dimensions, it is bounded by c_d, a constant depending exponentially on d [18]. Thus the total space is $O(n)$.

To process a color change at point p, examine the (at most c_d) edges of T adjacent to p. For each such edge, update its bichromatic status, and consequently that edge may get deleted from or inserted into H (at this step we use the pointers to the bichromatic edges in H adjacent to p). The edge with a minimum key value is returned, which completes the update. When the set becomes monochromatic (i.e., the heap becomes empty), ∞ is returned as the minimum distance. Since there are at most c_d heap operations, the total update time $U(n) = O(\log n)$, for any fixed d. We have:

Theorem 2. *Given a multicolored set of n points in $\mathbb{R}^d$, a bichromatic closest pair can be maintained under dynamic color changes in $O(\log n)$ update time, after $O(T_d^{EMST}(n))$ time preprocessing, and using $O(n)$ space.*

3.2 Farthest Pair

We use a similar approach of computing an EXST of the point set and maintaining its subset of bichromatic edges for the purpose of reporting one of maximum length, based again on the observation made above. However, in this case matters are complicated by the fact that the maximum degree of T may be arbitrarily large and, therefore, we need new data structures and techniques.

In the preprocessing step, compute T, an EXST of the point set, which takes $T_d^{EXST}(n)$ time. View T as a rooted tree, such that for any non-root node v, $p(v)$ is its parent in T. Conceptually, we are identifying each edge $(v, p(v))$ of T with node v of T. Consider $[k] = \{1, 2, \ldots, k\}$ as the set of colors. The algorithm maintains the following data structures:

- For each node $v \in T$, a balanced binary search tree C_v, called the *color tree at v*, with node keys the set of colors of children of v in T. For example if node v has 10 children colored by $3, 3, 3, 3, 5, 8, 8, 9, 9, 9$, C_v has 4 nodes with keys $3, 5, 8, 9$.
- For each node $v \in T$ and for each color class c of the children of v, a max-heap $H_{v,c}$ containing edges (keyed by length) to those children of v colored c. In the above example, these heaps are $H_{v,3}, H_{v,5}, H_{v,8}, H_{v,9}$. The heaps $H_{v,c}$ for the color classes of children of v are accessible via pointers at nodes of C_v.

- A max-heap H containing a subset of bichromatic edges of T. In particular, for each node v and for each color class c, *distinct* from that of v, of the children of v, H contains one edge of maximum length from v to a child of color c. In other words, for each node v and each color c distinct from that of v, H contains one maximum length edge in $H_{v,c}$. For each node v (of color c), pointers to C_v, to the edge $(v, p(v))$ in H (if it exists there) and in $H_{p(v),c}$ are maintained.

The preprocessing step computes C_v and $H_{v,c}$, for each $v \in T$ and $c \in [k]$, as well as H, in $O(n \log n)$ total time. The preprocessing time-complexity is clearly dominated by the tree computation, thus it is $O(T_d^{EXST}(n))$.

Next we discuss how, after a color change at some point v, the data structures are updated in $O(\log n)$ time. Without loss of generality assume that v's color changes from 1 to 2. Let $u = p(v)$ and let j be the color of u. Assume first that v is not the root of T.

Step 1. Search for colors 1 and 2 in C_u and locate $H_{u,1}$ and $H_{u,2}$. Let e_1 (resp. e_2) be the maximum length edge in $H_{u,1}$ (resp. $H_{u,2}$). Recall that if any of these two edges is bichromatic, it also appears in H. Vertex v (edge (u, v)) is deleted from $H_{u,1}$ and inserted into $H_{u,2}$. The maximum is recomputed in $H_{u,1}$ and $H_{u,2}$. If $j = 1$, the maximum edge in $H_{u,2}$ updates the old one in H (i.e. e_2 is deleted from H and the maximum length edge in $H_{u,2}$ is inserted into H). If $j = 2$, the maximum edge in $H_{u,1}$ updates the old one in H (i.e. e_1 is deleted from H and the maximum length edge in $H_{u,1}$ is inserted into H). If $j > 2$, both maximum edges in $H_{u,1}$ and $H_{u,2}$ update the old ones in H.

Step 2. Search for colors 1 and 2 in C_v and locate $H_{v,1}$ and $H_{v,2}$. The maximum edge of $H_{v,1}$ is inserted into H, and the maximum edge of $H_{v,2}$ is deleted from H. Finally, the maximum bichromatic edge is recomputed in H and returned, which completes the update.

If v is the root of T, Step 1 in the above update sequence is simply omitted. One can see that the number of tree search and heap operations is bounded by a constant, thus the update time is $U(n) = O(\log n)$.

The total space used by the data structure is clearly $O(n)$ and we have:

Theorem 3. *Given a multicolored set of n points in $\mathbb{R}^d$, a bichromatic farthest pair can be maintained under dynamic color changes in $O(\log n)$ update time, after $O(T_d^{EXST}(n))$ time preprocessing, and using $O(n)$ space.*

Remark 1. As the approach for maintaining the farthest pair under dynamic color flips is more general, it applies to closest pair maintenance as well. Therefore, since the complexity of the first (simpler) approach to maintaining the closest pair increases exponentially with d, one may choose among these two depending on how large d is.

We further note that we have implicitly obtained an algorithm to maintain a bichromatic edge of minimum (resp., maximum) weight in general graphs.

Specifically, let $G = (V, E)$, $|V| = n$, $|E| = m$ be an undirected weighted graph whose vertices are k-colored, and $T^{MST}(n, m)$ be the time complexity of a minimum spanning tree computation on a graph with n vertices and m edges. Since for arbitrary graphs, the time complexity of a minimum spanning tree computation is the same as that of a maximum spanning tree computation, we have:

Theorem 4. *Given an undirected weighted graph on n multicolored vertices with m edges, a bichromatic edge of minimum (resp., maximum) weight can be maintained under dynamic color changes in $O(\log n)$ update time, after $O(T^{MST}(n, m))$ time preprocessing, and using $O(n)$ space.*

Open Problem. Given a multicolored set of n points in $\mathbb{R}^d$, a *bichromatic Euclidean spanning tree* is an Euclidean spanning tree where each edge joins points of different colors. Design an efficient algorithm to maintain a minimum bichromatic Euclidean spanning tree when colors change dynamically. Note that it may be the case that all its edges change after a small number of color flips.

4 Combinatorial Bounds in the Plane

In this section, we present some combinatorial bounds on the number of extreme distances in multicolored planar point sets. We refer the reader to [11] for such bounds in the bichromatic case in three dimensions.

Let $f_d^{\min}(k, n)$ be the maximum multiplicity of the minimum distance between two points of different colors, taken over all sets of n points in $\mathbb{R}^d$ colored by k colors. Similarly, let $f_d^{\max}(k, n)$ be the maximum multiplicity of the maximum distance between two points of different colors, taken over all sets of n points in $\mathbb{R}^d$ colored by k colors. For simplicity, in the monochromatic case, the argument which specifies the number of colors will be omitted.

A *geometric graph* $G = (V, E)$ [16] is a graph drawn in the plane so that the vertex set V consists of points in the plane and the edge set E consists of straight line segments between points of V.

4.1 Minimum Distance

It is well known that in the monochromatic case, $f_2^{\min}(n) = 3n - o(n)$ (see [16]). In the multicolored version, we have

Theorem 5. *The maximum multiplicity of a bichromatic minimum distance in multicolored point sets ($k \geq 2$) in the plane satisfies*

(i) $2n - o(n) \leq f_2^{\min}(2, n) \leq 2n - 4$.
(ii) For $k \neq 2$, $3n - o(n) \leq f_2^{\min}(k, n) \leq 3n - 6$.

Proof. Consider a set P of n points such that the minimum distance between two points of different colors is 1. Connect two points in P by a straight line segment, if they are of different colors and if their distance is exactly 1. We obtain a geometric graph G. It is easy to see that no two such segments can cross: if there were such a crossing, the resulting convex quadrilateral would have a pair of bichromatic opposite sides with total length strictly smaller than of the two diagonals which create the crossing; one of these sides would then have length strictly smaller than 1, which is a contradiction. Thus G is planar. This yields the upper bound in (ii). Since in (i), G is also bipartite, the upper bound in (i) is also implied. To show the lower bound in (i), place about $n/2$ red points in a $\sqrt{n/2}$ by $\sqrt{n/2}$ square grid, and place about $n/2$ blue points in the centers of the squares of the above red grid. To show the lower bound in (ii), it is enough to do so for $k = 3$ (for $k > 3$, recolor $k - 3$ of the points using a new color for each of them). Consider a hexagonal portion of the hexagonal grid, in which we color consecutive points in each row with red, blue and green, red, blue, green, etc., such that the (at most 6) neighbors of each point are colored by different colors. The degree of all but $o(n)$ of the points is 6 as desired. □

4.2 Maximum Distance

Two edges of a geometric graph are said to be *parallel*, if they are opposite sides of a convex quadrilateral. We will use the following result of Valtr to get a linear upper bound on $f_2^{\max}(n)$.

Theorem 6. (Valtr [23]) *Let $l \geq 2$ be a fixed positive integer. Then any geometric graph on n vertices with no l pairwise parallel edges has at most $O(n)$ edges.*

It is well known that in the monochromatic case (here $f_d^{\max}(n)$ is the maximum multiplicity of the diameter), $f_2^{\max}(n) = n$, (see [16]). In the multicolored version, we have

Theorem 7. *The maximum multiplicity of a bichromatic maximum distance in multicolored point sets ($k \geq 2$) in the plane satisfies*

$$f_2^{\max}(k, n) = \Theta(n).$$

Proof. For the lower bound, place $n - 1$ points at distance 1 from a point p in a small circular arc centered at p. Color p with color 1 and the rest of the points arbitrarily using up all the colors in $\{2, \ldots, k\}$. The maximum bichromatic distance occurs $n - 1$ times in this configuration.

Next we prove the upper bound. Consider a set P of n points such that the maximum distance between two points of different colors is 1. Connect two points in P by a straight line segment, if they are of different colors and if their distance is exactly 1. We obtain a geometric graph $G = (V, E)$. We claim that G has no 4 pairwise parallel edges. The result then follows by Theorem 6 above.

Denote by $c(v)$ the color of vertex v, $v \in V$. For any edge $e = \{u, v\}$, $u, v \in V$, let the *color set* of its endpoints be $A_e = \{c(u), c(v)\}$. Assume for contradiction that G has a subset of 4 pairwise parallel edges $E' = \{e_1, e_2, e_3, e_4\}$. Without loss of generality, we may assume e_1 is horizontal. Consider two parallel edges

$e_i, e_j \in E', (i \neq j)$. Let Δ_{ij} be the triangle obtained by extending e_i and e_j along their supporting lines until they meet. Let α_{ij} be the (interior) angle of Δ_{ij} corresponding to this intersection. (If the two edges are parallel in the strict standard terminology, Δ_{ij} is an infinite strip and $\alpha_{ij} = 0$.) The *circular sequence* (resp. *circular color sequence*) of e_i, e_j is the sequence of their four endpoints (resp. their colors), when the corresponding convex quadrilateral is traversed (in clockwise or counterclockwise order) starting at an arbitrary endpoint. Note that this sequence is not unique, but is invariant under circular shifts.

We make several observations:

(i) For all $i, j \in \{1, 2, 3, 4\}$, $i \neq j$, $\alpha_{ij} < 60°$. Refer to Figure 1: the supporting lines of the two edges $e_i = BD$ and $e_j = CE$ intersect in A, where $\angle A = \alpha_{ij}$. Assume for contradiction that $\alpha_{ij} \geq 60°$. Then one of the other two angles of $\Delta_{ij} = ABC$, say $\angle B$, is at most 60°. Put $x = |BC|$, $y = |AC|$. We have $x \geq y > 1$, thus $c(B) = c(C)$. Hence CD and BE are bichromatic and $|CD| + |BE| > |BD| + |CE| = 2$. So at least one of CD or BE is longer than 1, which is a contradiction. As a consequence, all edges in E' have slopes in the interval $(-\tan 60°, +\tan 60°) = (-\sqrt{3}, +\sqrt{3})$, in particular no two endpoints of an edge in E' have the same x-coordinate. For $e_i \in E'$, denote by l_i (resp. r_i), its left (resp. right) endpoint. We say that $e_i \in E'$ is *of type* $(c(l_i), c(r_i))$.

(ii) If the circular color sequence of e_i, e_j is $\langle c_1, c_2, c_3, c_4 \rangle$, then either $c_1 = c_3$ or $c_2 = c_4$. For, if neither of these is satisfied, the lengths of the two diagonals of the corresponding convex quadrilateral would sum to more than 2, so one of these diagonals would be a bichromatic edge longer than 1, giving a contradiction.

(iii) The circular sequence of e_i, e_j is $\langle l_i, r_i, r_j, l_j \rangle$. Assume for contradiction that the circular sequence of e_i, e_j is $\langle l_i, r_i, l_j, r_j \rangle$. But then by the slope condition (in observation (i)), the corresponding triangle Δ_{ij} would have $\alpha_{ij} \geq 60°$, contradicting the same observation.

As a consequence, if e_i and e_j have the same color set ($A_{e_i} = A_{e_j}$), they must be of opposite types, so there can be at most two of them. Assume for contradiction that e_i and e_j have (the same color set and) same type $\{1, 2\}$. Then the circular color sequence of e_i, e_j is $\langle 1, 2, 2, 1 \rangle$ by observation (iii), contradicting observation (ii).

We now prove our claim that that G has no 4 pairwise parallel edges (the set E'). We distinguish two cases:

Case 1: There exist two parallel edges with the same color set. Without loss of generality assume that e_1 is of type $(1, 2)$ and e_2 is of type $(2, 1)$. We claim that G has no 3 pairwise parallel edges. Without loss of generality, e_3 is of type $(2, 3)$, by observation (ii). By observation (iii), the circular color sequence of e_2, e_3 is $\langle 2, 1, 3, 2 \rangle$ which contradicts observation (ii).

Case 2: No two parallel edges have the same color set. We claim that G has no 4 pairwise parallel edges. Without loss of generality we may assume that e_1 is of type $(1, 2)$, and e_2 is of type $(3, 1)$ (that is 1 is the common color of e_1 and

e_2). To satisfy observations (ii) and (iii), e_3 is constrained to be of type $(2,3)$. Finally, one can check that there is no valid type choice for e_4, in a manner consistent with observations (ii) and (iii) and the assumption in this second case (e_4 would have an endpoint colored by a new color, say 4, and then it is easy to find two edges with disjoint color sets).

The claim follows completing the proof of the theorem. □

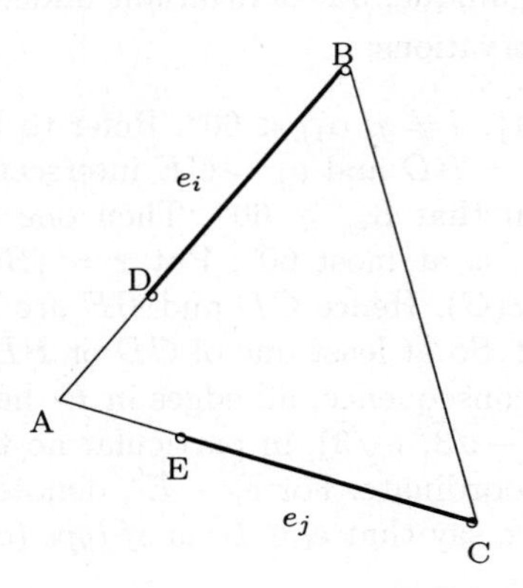

Fig. 1. Illustration for the proof of Theorem 7

Remark 2. It is not hard to show that $f_2^{\max}(2,n) = f_2^{\max}(n,n) = n$. A different approach than that taken in the proof of Theorem 7 leads to an upper bound of $2n$ [15, 20]. A lower bound of $\frac{3}{2}n - O(1)$ can be obtained for certain values of k. However determining an exact bound for the entire range $2 \le k \le n$ remains open.

References

1. P. K. Agarwal, H. Edelsbrunner, O. Schwarzkopf and Emo Welzl, Euclidean minimum spanning trees and bichromatic closest pairs, *Discrete & Computational Geometry*, **6** (1991), 407–422.
2. P. K. Agarwal, J. Matoušek and S. Suri, Farthest neighbors, maximum spanning trees and related problems in higher dimensions, *Computational Geometry: Theory and Applications*, **1** (1992), 189–201.
3. A. Aggarwal, H. Edelsbrunner, P. Raghavan and P. Tiwari, Optimal time bounds for some proximity problems in the plane, *Information Processing Letters*, **42**(1) (1992), 55–60.
4. J. L. Bentley and M. I. Shamos, Divide-and-conquer in multidimensional space, *Proceedings of the 8-th Annual Symposium on Theory of Computing*, 1976, 220–230.
5. S. N. Bespamyatnikh, An Optimal Algorithm for Closest-Pair Maintenance, *Discrete & Computational Geometry*, **19** (1998), 175–195.

6. B. K. Bhattacharya and G. T. Toussaint, Optimal algorithms for computing the minimum distance between two finite planar sets, *Pattern Recognition Letters*, **2** (1983), 79–82.
7. B. K. Bhattacharya and G. T. Toussaint, Efficient algorithms for computing the maximum distance between two finite planar sets, *Journal of Algorithms*, **4** (1983), 121–136.
8. D. Dobkin and S. Suri, Maintenance of geometric extrema, *Journal of the ACM*, **38** (1991), 275–298.
9. D. Eppstein, Dynamic Euclidean minimum spanning trees and extrema of binary functions, *Discrete & Computational Geometry*, **13** (1995), 111–122.
10. D. Eppstein, Spanning trees and spanners, in J.-R. Sack and J. Urrutia (Editors), *Handbook of Computational Geometry*, Elsevier, North-Holland, 2000, 425–461.
11. H. Edelsbrunner and M. Sharir, A hyperplane incidence problem with applications to counting distances, *Proceedings of the 1st Annual SIGAL International Symposium on Algorithms*, LNCS vol. 450, Springer Verlag, 1990, 419–428.
12. D. Krznaric, C. Levcopoulos, Minimum spanning trees in d dimensions, *Proceedings of the 5th European Symposium on Algorithms*, LNCS vol. 1248, Springer Verlag, 1997, 341–349.
13. J. S. B. Mitchell, Geometric shortest paths and geometric optimization, in J.-R. Sack and J. Urrutia (Editors), *Handbook of Computational Geometry*, Elsevier, North-Holland, 2000, 633–701.
14. C. Monma, M. Paterson, S. Suri and F. Yao, Computing Euclidean maximum spanning trees, *Algorithmica*, **5** (1990), 407–419.
15. J. Pach, Personal communication.
16. J. Pach and P.K. Agarwal, *Combinatorial Geometry*, John Wiley, New York, 1995.
17. J. M. Robert, Maximum distance between two sets of points in $\mathbb{R}^d$, *Pattern Recognition Letters*, **14** (1993), 733–735.
18. G. Robins and J. S. Salowe, On the Maximum Degree of Minimum Spanning Trees, *Proceedings of the 10-th Annual ACM Symposium on Computational Geometry*, 1994, 250–258.
19. M. I. Shamos and D. Hoey, Closest-point problems, *Proceedings of the 16-th Annual IEEE Symposium on Foundations of Computer Science*, 1975, 151–162.
20. G. Toth, Personal communication.
21. G. T. Toussaint and M. A. McAlear, A simple $O(n \log n)$ algorithm for finding the maximum distance between two finite planar sets, *Pattern Recognition Letters*, **1** (1982), 21–24.
22. P. M. Vaidya, Geometry helps in matching, *SIAM Journal on Computing*, **18** (1989), 1201–1225.
23. P. Valtr, On geometric graphs with no k pairwise parallel edges, *Discrete & Computational Geometry*, **19**(3) (1998), 461–469.

Balanced Partition of Minimum Spanning Trees

Mattias Andersson[1], Joachim Gudmundsson[2*], Christos Levcopoulos[1], and Giri Narasimhan[3]

[1] Department of Computer Science, Lund University, Box 118, 221 00 Lund, Sweden. christos@cs.lth.se, dat97mae@ludat.lth.se

[2] Department of Computer Science, Utrecht University, PO Box 80.089, 3508 TB Utrecht, the Netherlands. joachim@cs.uu.nl

[3] School of Computer Science, Florida International University, Miami, FL 33199, USA. giri@fiu.edu.

Abstract. To better handle situations where additional resources are available to carry out a task, many problems from the manufacturing industry involve "optimally" dividing a task into k smaller tasks. We consider the problem of partitioning a given set $\mathcal{S}$ of n points (in the plane) into k subsets, $\mathcal{S}_1, \ldots, \mathcal{S}_k$, such that $\max_{1 \leqslant i \leqslant k} |MST(\mathcal{S}_i)|$ is minimized. A variant of this problem arises in the shipbuilding industry [2].

1 Introduction

In one interesting application from the shipbuilding industry, the task is to use a robot to cut out a set of prespecified regions from a sheet of metal while minimizing the completion time. In another application, a salesperson needs to meet some potential buyers. Each buyer specifies a region (i.e., a *neighborhood*) within which the meeting needs to be held. A natural optimization problem is to find a salesperson tour of shortest length that visits all of the buyers' neighborhoods and finally returns to his initial departure point. Both these problems are related to the problem known in the literature as the *Traveling Salesperson problem with Neighborhoods* (TSPN) and which has been extensively studied [4, 5, 7–10]. The problem (TSPN) asks for the shortest tour that visits each of the neighborhoods. The problem was recently shown to be APX-hard [8].

Interesting generalizations of the TSPN problem arise when additional resources ($k > 1$ robots in the sheet cutting problem, or $k > 1$ salespersons in the second application above) are available. The k-TSPN problem is a generalization of the problem where we are given k salespersons and the aim is to minimize the completion time, i.e., minimize the distance traveled by the salespersons making the longest journey.

The need for partitioning the input set such that the optimal substructures are balanced gives rise to many interesting theoretical problems. In this paper we consider the problem of partitioning the input so that the sizes of the minimum

* Supported by the Swedish Foundation for International Cooperation in Research and Higher Education

P.M.A. Sloot et al. (Eds.): ICCS 2002, LNCS 2331, pp. 26–35, 2002.

spanning trees of the subsets are balanced. Also, we restrict our inputs to sets of points instead of regions. More formally, the *Balanced Partition Minimum Spanning Tree problem* (k-BPMST) is stated as follows:

Problem 1. Given a set of n points $\mathcal{S}$ in the plane, partition $\mathcal{S}$ into k sets $\mathcal{S}_1, \ldots, \mathcal{S}_k$ such that the weight of the largest minimum spanning tree,

$$W = \max_{1 \leqslant i \leqslant k} (|M(\mathcal{S}_i)|)$$

is minimized. Here $M(\mathcal{S}_i)$ is the minimum spanning tree of the subset $\mathcal{S}_i$ and $|M(\mathcal{S}_i)|$ is the weight of the minimum spanning tree of $\mathcal{S}_i$.

The paper is organized as follows. In section 2, we show that the problem is NP-hard. In section 3, we present an approximation algorithm with approximation factor $4/3 + \varepsilon$ for the case $k = 2$, and with an approximation factor $(2 + \varepsilon)$ for the case $k \geqslant 3$. The algorithm runs in time $O(n \log n)$.

2 NP hardness

In this section we show that the k-BPMST problem is NP-hard. In order to do this we need to state the recognition version of the k-BPMST problem:

Problem 2. Given a set of n points $\mathcal{S}$ in the plane, and a real number $\mathcal{L}$, does there exist a partition of $\mathcal{S}$ into k sets $\mathcal{S}_1, \ldots, \mathcal{S}_k$ such that the weight of the largest minimum spanning tree,

$$W = \max_{1 \leqslant i \leqslant k} (|M(\mathcal{S}_i)|) \leq \mathcal{L}?$$

In a computational model in which we can handle square roots in polynomial time, such as the real-RAM model (which will be used for simplicity), this formulation of the problem is sufficient in order to show that the k-BPMST problem is NP-hard. Note, however, that it may be inadequate in more realistic models, such as the Turing model, where efficient handling of square roots may not be possible. The computation of roots is necessary to determine the length of edges between points, which, in turn, is needed in order to calculate the weight of a minimum spanning tree. So in a realistic computational model the hardest part may not be to partition the points optimally, but instead to calculate precisely the length of the MST's. Thus, in these more realistic computational models we would like to restrict the problem to instances where the lengths of MST's are easy to compute. For example, this can be done by modifying the instances created in the reduction below, by adding some points so that the MST's considered only contain vertical and horizontal edges.

The proof is done (considering the real-RAM model) by a straight-forward polynomial reduction from the following recognition version of PARTITION.

Problem 3. Given integers $a = \{a_1 \leq \ldots \leq a_n\}$, the recognition version of the partition problem is: Does there exist a subset $P \subseteq I = \{1, 2, \ldots, n\}$ such that

$$\#P = \#I/P \quad \text{and} \sum_{j \in P} a_j = \sum_{j \in I/P} a_j$$

We will denote $\#P$ by h, $h = n/2$. This version of PARTITION is NP-hard [3].

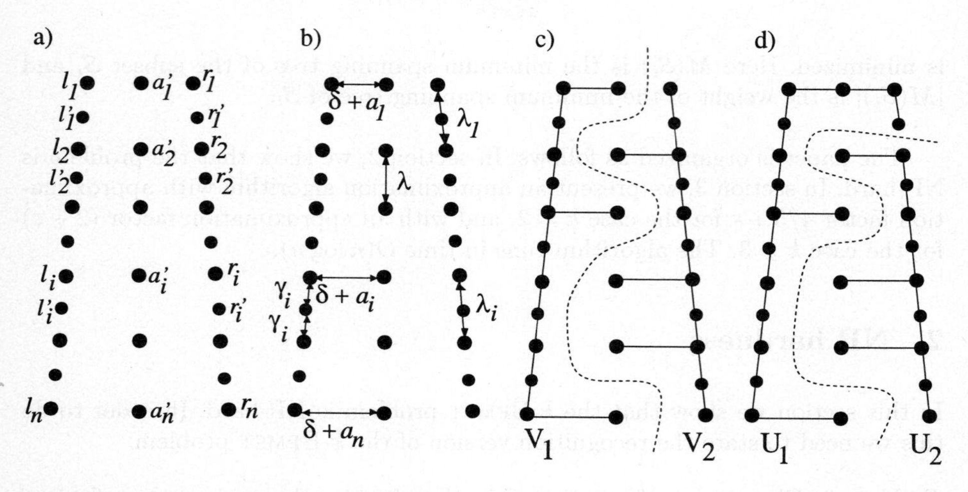

Fig. 1. The set of points $\mathcal{S}$ created for the reduction. In Figure (a) all notations for the points are given. Similarly, in Figure (b) the notations for the distances between points are given. Figure (c) illustrates a class 1 partition, and (d) illustrates a class 2 partition.

Lemma 1. *The k-BPMST problem is NP-hard.*

Proof. The reduction is done as follows. Given a PARTITION instance we create a 2-BPMST instance, in polynomial time, such that it is a yes-instance if, and only if, the PARTITION-instance is a yes-instance. Obviously PARTITION then polynomially reduces to 2-BPMST. Given that the PARTITION-instance contains n integers $a_1, \ldots, a_n$, we create the following 2-BPMST instance. A set of points $\mathcal{S}$, as shown in Figure 1a is created, with inter point distances as shown in Figure 1b. A closer description of these points and some additional definitions is given below:

- $a' = \{a'_1, \ldots, a'_n\}$, where $a'_i = (0, i\lambda)$,
- $l = \{l_1, \ldots, l_n\}$, where $l_i = (-\delta - a_i, i\lambda)$,
- $r = \{r_1, \ldots, r_n\}$, where $r_i = (\delta + a_i, i\lambda)$,
- $l' = \{l'_1, \ldots, l'_{n-1}\}$, where l'_i is the midpoint on the line between l_i and l_{i+1}, and

– $r' = \{r'_1, \ldots, r'_{n-1}\}$, where r'_i is the midpoint on the line between r_i and r_{i+1}

We also define the following set of points, $a^* = \{a'_{P[1]}, \ldots, a'_{P[h]}\}$. Further, let $\lambda = 11n(a_n + n)$ and let $\delta = 7n(a_n + n)$. Note that $\lambda_i^2 \leqslant \lambda^2 + a_n^2$ which implies that $\lambda_i \leqslant 12n(a_n + n)$, which means that $\gamma_i = \lambda_i/2 \leqslant 6n(a_n + n)$. Finally let (see definition 2)

$$\mathcal{L} = (\sum_{i \in I} a_i)/2 + n/2 \cdot \delta + \sum_{i=1}^{n-1} \lambda_i$$

Since the number of points in $\mathcal{S}$ is polynomial it is clear that this instance can be created in polynomial time. Next we consider the "if", and the "only if" parts separately.

If If P exists and we have a yes PARTITION-instance it is clear that the corresponding 2-BPMST instance is also a yes-instance. This follows when the partition $\mathcal{S}'_1 = a^* + l + l'$, $\mathcal{S}'_2 = \mathcal{S} - \mathcal{S}_1$ (a class 1 partition, as defined below) is considered. The general appearance of $M(\mathcal{S}'_1)$ and $M(\mathcal{S}'_2)$ (see Figure 1c) is determined as follows. The points $l + l'$ and the points $r + r'$ will be connected as illustrated in Figure 1c, which follows from the fact that $\gamma_i < \delta < \delta + a_1$. Next consider the remaining points a'. Any point a'_i will be connected to either l_i (in $M(\mathcal{S}'_1)$) or r_i (in $M(\mathcal{S}'_2)$), since r_i and l_i are the points located closest to a'_i (follows since $\lambda > \delta + a_n$). Thus,

$$|M(\mathcal{S}'_1)| = |M(\mathcal{S}'_2)| = (\sum_{i \in I} a_i)/2 + n/2 \cdot \delta + \sum_{i=1}^{n-1} \lambda_i$$

and we have that the created instance is a yes-instance.

Only if We have that P does not exist and we therefore want to show that the created 2-BPMST is a no-instance. For this two classes of partitions will be examined:

– All partitions $\mathcal{V}_1, \mathcal{V}_2$ such that $l + l' \subseteq \mathcal{V}_1$ and $r + r' \subseteq \mathcal{V}_2$
– All other partitions $\mathcal{U}_1, \mathcal{U}_2$ not belonging to class 1.

We start by examining the first class (illustrated by Figure 1c). Note that an optimal MST will contain the edges in $M(\mathcal{V}_1)$ and $M(\mathcal{V}_2)$ plus the edge between a'_1 and l_1 or r_1, hence $|M(\mathcal{S})| = |M(\mathcal{V}_1)| + |M(\mathcal{V}_2)| + \delta + a_1$. Note also that $|M(\mathcal{V}_1)| + |M(\mathcal{V}_2)| = 2 \cdot \mathcal{L}$. For all partitions $\mathcal{V}'_1 \subseteq \mathcal{V}_1, \mathcal{V}'_2 \subseteq \mathcal{V}_2$ such that each subset $\mathcal{V}'_1, \mathcal{V}'_2$ contains exactly $|a'|/2$ points from the set a' it is clear, since P does not exist, that $max\{|M(\mathcal{V}'_1)|, |M(\mathcal{V}'_2)|\} > \mathcal{L}$. This is true also for the partitions $\mathcal{V}^*_1 \subseteq \mathcal{V}_1, \mathcal{V}^*_2 \subseteq \mathcal{V}_2$ such that each subset does *not* contain exactly $|a'|/2$ points from the set a'. To see this consider any such partition and the corresponding subset $\mathcal{V}^*_i$ such that $|\mathcal{V}^*_i| = \max\{|\mathcal{V}^*_1|, |\mathcal{V}^*_2|\}$. We have that

$$|M(\mathcal{V}^*_i)| \geq \delta + n/2 \cdot \delta + \sum_{i=1}^{n-1} \lambda_i > (\sum_{i \in I} a_i) + n/2 \cdot \delta + \sum_{i=1}^{n-1} \lambda_i > \mathcal{L}$$

This implies that $\max\{|M(\mathcal{V}^*_1)|, |M(\mathcal{V}^*_2)|\} > \mathcal{L}$.

Next consider the class 2 partitions (illustrated by Figure 1d). There is always an edge of weight γ_i ($1 \leq i \leq n$) connecting the two point sets of any such partition. This means that there can not exist a class 2 partition $\mathcal{U}_1, \mathcal{U}_2$ such that $\max\{|M(\mathcal{U}_1)|, |M(\mathcal{U}_2)|\} \leq \mathcal{L}$, because we could then build a tree with weight at most $2 \cdot \mathcal{L} + \gamma_i < |M(\mathcal{V}_1)| + |M(\mathcal{V}_2)| + \delta + a_1 = |M(\mathcal{S})|$, which is a contradiction. Thus, $\max\{|M(\mathcal{U}_1)|, |M(\mathcal{U}_2)|\} > \mathcal{L}$, which concludes this lemma.

3 A $2 + \varepsilon$ approximation algorithm

In this section a $2 + \varepsilon$ approximation algorithm is presented. Note also that a straight-forward greedy algorithm, that partitions $M(\mathcal{S})$ into k sets by removing the $k - 1$ longest edges gives an approximation of k. The main idea of the $2 + \varepsilon$ approximation algorithm is to partition $\mathcal{S}$ into a constant number of small components, test all valid combinations of these components and give the best combination as output. As will be seen later, one will need an efficient partitioning algorithm, denoted ValidPartition or VP for short. A partition of a point set $\mathcal{S}$ into two subsets $\mathcal{S}_1$ and $\mathcal{S}_2$ is said to be valid if $\max(|M(\mathcal{S}_1)|, |M(\mathcal{S}_2)|) \leqslant 2/3 \cdot |M(\mathcal{S})|$. The following lemma is easily shown [1] using standard decomposition methods.

Lemma 2. *Given a set of points $\mathcal{S}$, VP divides $\mathcal{S}$ into two sets $\mathcal{S}_1$ and $\mathcal{S}_2$ such that (i) $\max\{|M(\mathcal{S}_1)|, |M(\mathcal{S}_2)|\} \leqslant \frac{2}{3}M(\mathcal{S})$, and (ii) $|M(\mathcal{S}_1)| + |M(\mathcal{S}_2)| \leqslant |M(\mathcal{S})|$. If VP is given a MST of $\mathcal{S}$ as input then it holds that the time needed for VP to compute a valid partition is $O(n)$.*

3.1 Repeated ValidPartition

ValidPartition will be used repeatedly in order to create the small components mentioned in the introduction of this section. Consider the following algorithm, given a MST of $\mathcal{S}$ and an integer m. First divide $M(\mathcal{S})$ into two components using VP. Next divide the largest of these two resulting components, once again using VP. Continue in this manner, always dividing the largest component created thus far, until m components have been created. Note that in each division the number of components increase by one. This algorithm will be denoted RepeatedValidPartition, or RVP for short. The following lemma expresses an important characteristic of RVP.

Lemma 3. *Given a minimum spanning tree of a set of points $\mathcal{S}$ and an integer m, RVP will partition $\mathcal{S}$ into m components $\mathcal{S}_1, \ldots, \mathcal{S}_m$ such that $\max(|M(\mathcal{S}_1)|, \ldots, |M(\mathcal{S}_m)|) \leqslant \frac{2}{m}|M(\mathcal{S})|$.*

Proof. Consider the following algorithm $\mathcal{A}$. Start with $M(\mathcal{S})$ and divide with VP until the weight of all components is less than or equal to $\frac{2}{m}|M(\mathcal{S})|$. The order in which the components are divided is arbitrary but when a component weighs less than or equal to $\frac{2}{m}|M(\mathcal{S})|$ it is not divided any further. If it now could be shown that the number of resulting components is at most m the lemma would follow.

This is seen when the dividing process of RVP is examined. Since RVP always divides the largest component created thus far, a component of weight at most $\frac{2}{m}|M(\mathcal{S})|$ would not be divided unless all other components also have weight at most $\frac{2}{m}|M(\mathcal{S})|$. Further, VP guarantees that the two components resulting from a division always have weights less than the divided component. Thus, when m components have been created by RVP these m components would also have weight less than or equal to $\frac{2}{m}|M(\mathcal{S})|$. Therefore, the aim is to show that algorithm $\mathcal{A}$, given $M(\mathcal{S})$, produces at most m components. The process can be represented as a tree. In this tree each node represents a component, with the root being $M(\mathcal{S})$. The children of a node represent the components created when that node is divided using VP. Note that the leaves of this tree represent the final components. Thus the aim is to show that the number of leaves do not exceed m. For this purpose we will divide the leaves into two categories. The first category is all leaves whose sibling is not a leaf. Assume that there are m_1 such leaves in the tree. The second category is all remaining leaves, that is, those who actually have a sibling leaf. Assume, correspondingly, that there are m_2 such leaves.

We start by examining the first category. Consider any leaf l_i of this category. Denote its corresponding sibling s_i and denote by p_i the parent of l_i and s_i. Further to each l_i we attach a weight $w(l_i)$ which is defined as $w(l_i) = |M(p_i)| - |M(s_i)|$. Since s_i is not a leaf it holds that $|M(s_i)| > \frac{2}{m}|M(\mathcal{S})|$, and since VP is used we know that $|M(s_i)| \leqslant \frac{2}{3}|M(p_i)|$. Thus, $|M(p_i)| > \frac{3}{m}|M(\mathcal{S})|$ which implies that $w(l_i) \geqslant \frac{1}{3}|M(p_i)| > \frac{1}{m}|M(\mathcal{S})|$ and $\sum_{i=1}^{m_1} w(l_i) > m_1 \cdot \frac{1}{m}|M(\mathcal{S})|$.

Next the second category of leaves is examined. Denote any such leaf l'_i and its corresponding parent p'_i. Since there are m_2 leaves of this category and each leaf has a leaf sibling these leaves have in total $m_2/2$ parent nodes. Further, for each such corresponding parent component $M(p'_i)$ we have that $|M(p'_i)| > \frac{2}{m}|M(\mathcal{S})|$ (they are not leaves). Thus, $\sum_{i=1}^{m_2} |M(p'_i)| > \frac{m_2}{2} \cdot \frac{2}{m}|M(\mathcal{S})| = m_2 \cdot \frac{1}{m}|M(\mathcal{S})|$.

Next consider the total weight of the components examined so far. We have that $m_1 \cdot \frac{1}{m}|M(\mathcal{S})| + m_2 \cdot \frac{1}{m}|M(\mathcal{S})| < \sum_{i=1}^{m_1} w(l_i) + \sum_{i=1}^{m_2} |M(p'_i)| \leqslant |M(\mathcal{S})|$, which implies that $m_1 + m_2 \leqslant m$. Thus, the number of leaves do not exceed m. □

3.2 The approximation algorithm

Now we are ready to state the algorithm CA. As input we are given a set $\mathcal{S}$ of n points, an integer k and a positive real constant ε. The algorithm differs in two separate cases, $k = 2$ and $k \geqslant 3$. First $k = 2$ is examined, in which the following steps are performed:

step 1: Divide $M(\mathcal{S})$ into $\frac{4}{\varepsilon'}$ components, using RVP, where $\varepsilon' = \frac{\varepsilon}{4/3+\varepsilon}$. The reason for the value of ε' will become clear below. Let W denote the heaviest component created and let w denote its weight.

step 2: Combine all components created in step 1, in all possible ways, into two groups.

step 3: For each combination tested in step 2, compute the MST for each of its two created groups.

step 4: Output the best tested combination

Theorem 1. *For $k = 2$ the approximation algorithm* CA *produces a partition which is within a factor $\frac{4}{3} + \varepsilon$ of the optimal in time $O(n \log n)$.*

Proof. Let V_1 and V_2 be the partition obtained from CA. Assume that $\mathcal{S}_1$ and $\mathcal{S}_2$ is the optimal partition, and let e be the shortest edge connecting $\mathcal{S}_1$ with $\mathcal{S}_2$. According to Lemma 3 it follows that $w \leqslant 2/(4/\varepsilon')|M(\mathcal{S})| = \frac{\varepsilon'}{2}|M(\mathcal{S})|$. We will have two cases, $|e| > w$, and $|e| \leqslant w$, which are illustrated in Figure 2 (a) and Figure 2 (b), respectively. In the first case every component is a subset of either $\mathcal{S}_1$ or $\mathcal{S}_2$. This follows since a component consisting of points from both $\mathcal{S}_1$ and $\mathcal{S}_2$ must include an edge with weight greater than w. Thus, no such component can exist among the components created in step 1. Further, this means that the partition $\mathcal{S}_1$ and $\mathcal{S}_2$ must have been tested in step 2 of CA and, hence, the optimal solution must have been found.

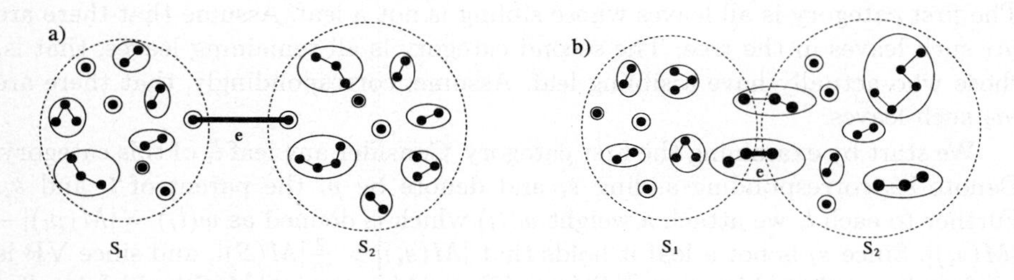

Fig. 2. The two cases for CA, $k = 2$. The edge e (marked) is the shortest edge connecting $\mathcal{S}_1$ with $\mathcal{S}_2$

In the second case, $|e| \leqslant w$, there may exist components consisting of points from both $\mathcal{S}_1$ and $\mathcal{S}_2$, see Fig. 2. To determine an upper bound of the approximation factor we start by examining an upper bound of CA. The dividing process in step 1 of CA starts with $M(\mathcal{S})$ being divided into 2 components $M(\mathcal{S}_1')$ and $M(\mathcal{S}_2')$, such that $\max(|M(\mathcal{S}_1')|, |M(\mathcal{S}_2')|) \leqslant \frac{2}{3}|M(\mathcal{S})|$. These two components are then divided into several smaller components. This immediately reveals an upper bound of $|CA| \leqslant \frac{2}{3}|M(S)|$. Next the lower bound is examined. We have:

$$|opt| \geqslant \frac{|M(\mathcal{S})| - |e|}{2} \geqslant \frac{|M(\mathcal{S})|}{2} - \frac{\varepsilon' \cdot M(\mathcal{S})}{2} \geqslant (1 - \varepsilon')\frac{M(\mathcal{S})}{2}.$$

Then, if the upper and lower bound are combined we get:

$$|CA|/|opt| \leqslant \frac{\frac{2}{3}|M(\mathcal{S})|}{(1 - \varepsilon')\frac{M(\mathcal{S})}{2}} \leqslant \frac{4/3}{1 - \varepsilon'} \leqslant 4/3 + \varepsilon.$$

In the third inequality we used the fact that $\varepsilon' \leqslant \frac{\varepsilon}{4/3+\varepsilon}$.

Next consider the complexity of CA. In step 1 $M(\mathcal{S})$ is divided into a constant number of components using VP. This takes $O(n)$ time. Then, in step 2, these components are combined in all possible ways. This takes $O(1)$ time since there are a constant number of components. For each tested combination there is a constant number of MST's to be computed in step 3. Further, since there are a constant number of combinations and $M(\mathcal{S})$ takes $O(n \log n)$ to compute, step 3 takes $O(n \log n)$ time. □

Next consider $k \geqslant 3$. In this case the following steps are performed:

step 1: Compute $M(\mathcal{S})$ and remove the $k-1$ heaviest edges $e_1, \ldots, e_{k-1}$ of $M(\mathcal{S})$, thus resulting in k separate trees $M(U'_1), \ldots, M(U'_k)$.

step 2: Divide each of the trees $M(U'_1), \ldots M(U'_k)$ into $\frac{k \cdot C}{\varepsilon'}$ components, using RVP. C is a positive constant and $\varepsilon' = \frac{\varepsilon}{2+\varepsilon}$. The reason for the value of ε' will become clear below. Denote the resulting components $M(U_1), \ldots, M(U_r)$, where $r = \frac{k \cdot C}{\varepsilon'} \cdot k$. Further set $w = \max\{|M(U_1)|, \ldots, |M(U_r)|\}$.

step 3: Combine $U_1, \ldots, U_r$ in all possible ways into $1, \ldots, k$ groups.

step 4: For each such combination do:

- Compute the MST for each of its corresponding groups.
- Divide each such MST in all possible ways, using RVP. That is, each MST is divided into $1, \ldots, i (i \leq k)$ components, such that the total number of components resulting from all the divided MST's equals k. Each such division defines a partition of $\mathcal{S}$ into k subsets.

step 5: Of all the tested partitions in step 4, output the best.

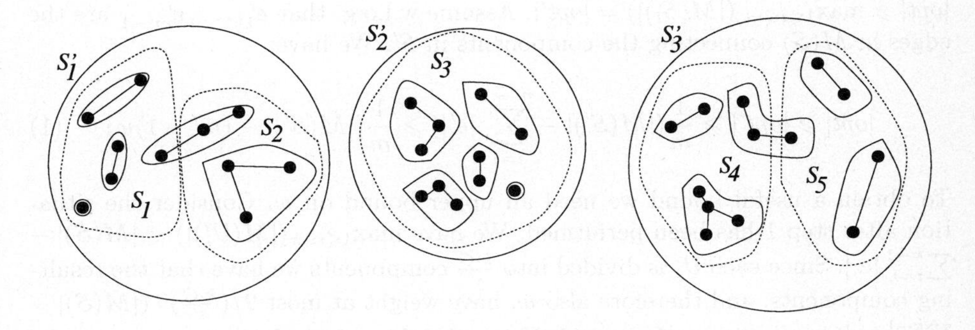

Fig. 3. $\mathcal{S}_1, \ldots, \mathcal{S}_k$ is an optimal partition of $\mathcal{S}$. All subsets that can be connected by edges of length at most w are merged, thus creating the new set $\mathcal{S}'_1, \ldots, \mathcal{S}'_{k'}$

Theorem 2. *For $k \geqslant 3$ the approximation algorithm CA produces a partition which is within a factor of $2 + \varepsilon$ of the optimal in time $O(n \log n)$*

Proof. The time complexity CA is the same as for the case $k = 2$. This follows as a constant number of components are created and a constant number of combinations and partitions are tested, hence the time complexity is $O(n \log n)$.

To prove the approximation factor we first give an upper bound on the weight of the solution produced by CA and then we provide a lower bound for an optimal solution. Combining the two results will conclude the theorem.

Consider an optimal partition of $\mathcal{S}$ into k subsets $\mathcal{S}_1, \ldots, \mathcal{S}_k$. Merge all subsets that can be connected by edges of length at most w. From this we obtain the sets $\mathcal{S}'_1, \ldots, \mathcal{S}'_{k'}$, where $k' \leqslant k$ (see Figure 3). Let m'_i denote the number of elements from $\mathcal{S}_1, \ldots, \mathcal{S}_k$ included in $\mathcal{S}'_i$. The purpose of studying these new sets is that every component created in step 2 of CA belongs to exactly one element in $\mathcal{S}'_1, \ldots, \mathcal{S}'_{k'}$. A direct consequence of this is that a combination into k' groups equal to $\mathcal{S}'_1, \ldots, \mathcal{S}'_{k'}$ must have been tested in step 3.

Step 4 guarantees that $M(\mathcal{S}'_1), \ldots, M(\mathcal{S}'_{k'})$ will be calculated, and that these MST's will be divided in all possible ways. Thus, a partition will be made such that each $M(\mathcal{S}'_i)$ will be divided into exactly m'_i components. This partitions $\mathcal{S}$ into k subsets $\mathcal{V}_1, \ldots, \mathcal{V}_k$. Let $\mathcal{V}$ be a set in $\mathcal{V}_1, \ldots, \mathcal{V}_k$ such that $|M(\mathcal{V})| = \max_{1 \leqslant i \leqslant k}(|M(\mathcal{V}_i)|)$. We wish to restrict our attention to exactly one element of the set $\mathcal{S}'_1, \ldots, \mathcal{S}'_{k'}$. Thus, we note that $\mathcal{V}$ is a subset of exactly one element $\mathcal{S}'$ in $\mathcal{S}'_1, \ldots, \mathcal{S}'_{k'}$. Assume that $M(\mathcal{V})$ was created in step 4 when $M(\mathcal{S}')$ was divided into m' components using RVP. Thus, $M(\mathcal{V}) \leqslant \frac{2}{m'}|M(\mathcal{S}')|$, according to Lemma 3. Since the partition $\mathcal{V}_1, \ldots, \mathcal{V}_k$ will always be tested we have that $|CA| \leqslant |M(\mathcal{V})| \leqslant \frac{2}{m'}|M(\mathcal{S}')|$.

Next a lower bound of an optimal solution is examined. Let $|opt'|$ be the value of an optimal solution for $\mathcal{S}'$ partitioned into m' subsets. Note that $\mathcal{S}'$ consists of m' elements from $\mathcal{S}_1, \ldots, \mathcal{S}_k$. Assume w.l.o.g that $\mathcal{S}' = \mathcal{S}_1 + \ldots + \mathcal{S}_{m'}$. This means that $\mathcal{S}_1, \ldots, \mathcal{S}_{m'}$ is a possible partition of $\mathcal{S}'$ into m' subsets. Thus, $|opt| \geqslant \max_{1 \leqslant i \leqslant m'}(|M(S_i)|) = |opt'|$. Assume w.l.o.g. that $e'_1, \ldots, e'_{m'-1}$ are the edges in $M(\mathcal{S})$ connecting the components in $\mathcal{S}'$. We have:

$$|opt| \geqslant |opt'| \geqslant \frac{1}{m}(|M(S')| - \sum_{i=1}^{m'-1} |e'_i|) \geqslant \frac{1}{m}(|M(S')| - (m'-1)w) \qquad (1)$$

To obtain a useful bound we need an upper bound on w. Consider the situation after step 1 has been performed. We have $\max_{1 \leqslant i \leqslant k}(|M(U'_i)|) \leqslant |M(\mathcal{S})| - \sum_{i=1}^{k-1} |e_i|$. Since each U'_i is divided into $\frac{k \cdot C}{\varepsilon'}$ components we have that the resulting components, and therefore also w, have weight at most $2/(\frac{k \cdot C}{\varepsilon'}) \cdot (|M(\mathcal{S})| - \sum_{i=1}^{k-1} |e_i|)$, according to Lemma 3. Using the above bound gives us:

$$\frac{w}{|opt|} \leqslant \frac{2/(\frac{k \cdot C}{\varepsilon'}) \cdot (|M(\mathcal{S})| - \sum_{i=1}^{k-1} |e_i|)}{\frac{1}{k}(|M(\mathcal{S})| - \sum_{i=1}^{k-1} |e_i|)} \leqslant \frac{2 \cdot \varepsilon'}{C} \Rightarrow w \leqslant \frac{2 \cdot \varepsilon'}{C}|opt| \qquad (2)$$

Setting $C \geqslant 2$ and combining 1 and 2 gives us:

$$|opt| \geqslant \frac{1}{m}\left(|M(S')| - (m'-1)\frac{2 \cdot \varepsilon'}{C}|opt|\right) \geqslant (1 - \varepsilon')\frac{|M(S')|}{m'}.$$

Combining the two bounds together with the fact that $\varepsilon' \leqslant \varepsilon/(2+\varepsilon)$ concludes the theorem.

$$|CA|/|opt| \leqslant \frac{\frac{2}{m'}|M(\mathcal{S}')|}{(1-\varepsilon')\frac{|M(\mathcal{S}')|}{m'}} \leqslant \frac{2}{1-\varepsilon'} \leqslant 2+\varepsilon.$$

□

4 Conclusion

In this paper it was first showed that the k-BPMST problem is NP-hard. After this had been determined the continued approach was to find an approximation algorithm for the problem. The algorithm is based on partitioning the point set into a constant number of smaller components and then trying all possible combinations of these small components. This approach revealed a $4/3+\varepsilon$ approximation in the case $k=2$, and an $2+\varepsilon$ approximation in the case $k \geqslant 3$. The time complexity of the algorithm is $O(n \log n)$.

References

1. M. Andersson. Balanced Partition of Minimum Spanning Trees, LUNDFD6/NFCS-5215/1–30/2001, Master thesis, Department of Computer Science, Lund University, 2001.
2. B. Shaleooi. Algoritmer för plåtskärning (*Eng. transl.* Algorithms for cutting sheets of metal), LUNDFD6/NFCS-5189/1–44/2001, Master thesis, Department of Computer Science, Lund University, 2001.
3. M. R. Garey and D. S. Johnson. Computers and Intractability: A guide to the theory of NP-completeness, W. H. Freeman and Company, San Francisco, 1979.
4. E. M. Arkin and R. Hassin. Approximation algorithms for the geometric covering salesman problem. Discrete Applied Mathematics, 55:197–218, 1994.
5. A. Dumitrescu and J. S. B. Mitchell. Approximation algorithms for TSP with neighborhoods in the plane. In *Proc. 12th Annual ACM-SIAM Symposium on Discrete Algorithms*, 2001.
6. M. R. Garey, R. L. Graham and D. S. Johnson. Some NP-complete geometric problems. In *Proc. 8th Annual ACM Symposium on Theory of Computing*, 1976.
7. J. Gudmundsson and C. Levcopoulos. A fast approximation algorithm for TSP with neighborhoods. Nordic Journal of Computing, 6:469-488, 1999.
8. J. Gudmundsson and C. Levcopoulos. Hardness Result for TSP with Neighborhoods, Technical report, LU-CS-TR:2000-216, Department of Computer Science, Lund University, Sweden, 2000.
9. C. Mata and J. S. B. Mitchell. Approximation algorithms for geometric tour and network design problems. In *Proc. 11th Annual ACM Symposium on Computational Geometry*, pages 360–369, 1995.
10. J. S. B. Mitchell. Guillotine Subdivisions Approximate Polygonal Subdivisions: A Simple Polynomial-Time Approximation Scheme for Geometric TSP, k-MST, and Related Problems. SIAM Journal on Computing, 28(4):1298–1309, 1999.

On the Quality of Partitions Based on Space-Filling Curves

Jan Hungershöfer and Jens-Michael Wierum

Paderborn Center for Parallel Computing, PC[2]
Fürstenallee 11, 33102 Paderborn, Germany
{hunger,jmwie}@upb.de
www.upb.de/pc2/

Abstract. This paper presents bounds on the quality of partitions induced by space-filling curves. We compare the surface that surrounds an arbitrary index range with the optimal partition in the grid, i. e. the square. It is shown that partitions induced by Lebesgue and Hilbert curves behave about 1.85 times worse with respect to the length of the surface. The Lebesgue indexing gives better results than the Hilbert indexing in worst case analysis. Furthermore, the surface of partitions based on the Lebesgue indexing are at most $\frac{5}{2\cdot\sqrt{3}}$ times larger than the optimal in average case.

1 Introduction

Data structures for maintaining sets of multidimensional points play an important role in many areas of computational geometry. While for example Voronoi diagrams have been established for efficient requests on neighborhood relationships, data structures based on space-filling curves are often used for requests on axis aligned bodies of arbitrary size. The aim of the requests is to find all points located in such multidimensional intervals. Those types of requests are needed in many applications like N-body simulations [12], image compression and browsing [10, 4], databases [2], and contact search in finite element analysis [5]. An overview on this and other techniques for range searching in computational geometry is given in [1]. Space-filling curves have other locality properties which are e. g. useful in parallel finite element simulations [3, 7].

Space-filling curves are geometric representations of bijective mappings $M : \{1, \ldots, N^m\} \to \{1, \ldots, N\}^m$. The curve M traverses all N^m cells in the m-dimensional grid of size N. An (historic) overview on space-filling curves is given in [11]. Experimental work and theoretical analysis have shown, that algorithms based on space-filling curves behave well on most inputs, while they are not well suited for some special inputs. Therefore, an analysis for the average case is often more important than for the worst case.

Due to the varying requirements on the *locality* properties, different metrics have been used to qualify, compare, and improve space-filling curves. A major metric for the analysis of the locality of space-filling curves is the ratio of index

P.M.A. Sloot et al. (Eds.): ICCS 2002, LNCS 2331, pp. 36–45, 2002.

interval to maximum distance within this index range. Results are published for different indexing schemes like Hilbert, Lebesgue, and H-indexing for Manhattan metric, Euclidean metric, and maximum metric [6, 9].

Other examinations concentrate on the number of index intervals which have to be determined for a given request region. Sharp results are given in [2] for squares in two-dimensional space. The costs for arbitrary shaped regions is discussed in [8].

Here we examine the surface of a partition which is induced by an interval of a space-filling curve. Practical results of this relationship for uniform grids and unstructured meshes can be found in [14]. We define a *quality coefficient* which represents a normed value for the quality of the induced partition in an uniform grid of size $N \times N$. We use the shape of an optimal partition, the square, as a reference:

Definition 1 (quality coefficient). *Let* curve *be an indexing scheme, p an index range, $S(p)$ the surface of a partition and $V(p) = |p|$ the size (volume) of it. $C^{curve}(p)$ defines the* quality coefficient *of the partition given by index range p:*

$$C^{curve}(p) = \frac{S^{curve}(p)}{4 \cdot \sqrt{V(p)}} \tag{1}$$

This formulation can be extended to a quality coefficient *of an indexing scheme:*

$$C^{curve}_{\max} = \max_p\{C^{curve}(p)\} \tag{2}$$

$$C^{curve}_{\text{avg}} = \text{avg}_p\{C^{curve}(p)\} \tag{3}$$

Definition 1 implies that $C(p) \geq 1$ for all indexing schemes.

2 Lebesgue Curves

Figure 1 illustrates the recursive definition of the Lebesgue indexing. The resulting curve is also known as *bit interleaving* or *Z-code*. In the following the edges of cells are assigned a level, depending on the step in which they were introduced during the recursive construction. The lines of the final step are of level 0. In the example shown dashed lines are of level 1, dotted lines of level 0. It is obvious that an arbitrary edge is of level l with an asymptotic probability of $2^{-(l+1)}$.

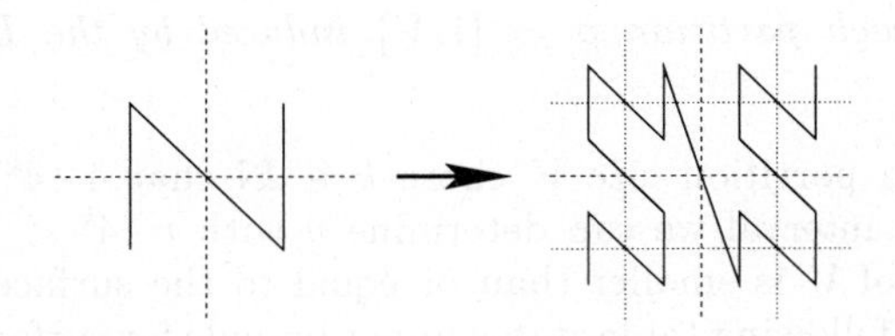

Fig. 1. Production rule for Lebesgue indexing.

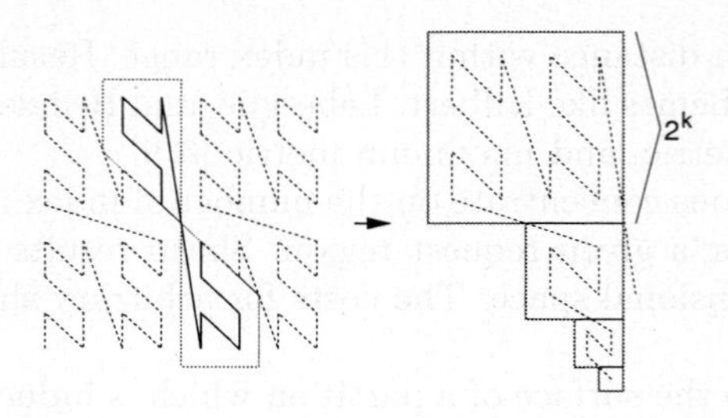

Fig. 2. Construction of a lower bound for Lebesgue indexing.

2.1 Lower Bound on Worst-Case Partitions

Theorem 1. *For the Lebesgue curve the quality coefficient is larger than or equal to* $3 \cdot \sqrt{\frac{3}{8}} - \varepsilon$ *with decreasing* ε *for increasing partition size.*

Proof. We construct a partition of size V and surface S which follows the Lebesgue curve and gives the stated bad quality: The symmetric partition is split by a high level border. Each half contains squares of size 4^k, 4^{k-1}, 4^{k-2}, ..., 4^1, 4^0. The first half of the partition is illustrated in Fig. 2. It follows:

$$V = 2 \cdot \frac{4^{k+1} - 1}{3} \text{ and } S = 2 \cdot 6 \cdot 2^k - 4 \ . \tag{4}$$

The quality coefficient of this partition is given by

$$\frac{S}{4 \cdot \sqrt{V}} > \frac{3}{2} \cdot \sqrt{\frac{3}{2}} \cdot \frac{2^k - 4}{2^k} = 3 \cdot \sqrt{\frac{3}{8}} - \varepsilon \approx 1.83 \ . \tag{5}$$

□

2.2 Upper Bound on Worst-Case Partitions

For the determination of an upper bound we examine partitions which start at the lower left corner of the grid. Due to the construction scheme the partition is always contiguous and its surface is equal to the surface of its bounding box.[1] We analyze the surface of those partitions with respect to a coarse granularity, to be able to examine a finite number of cases.

Lemma 1. *For each partition* $p = [1, V]$ *induced by the Lebesgue indexing* $C_{\max}^{\text{Lebesgue}} \leq \frac{12}{4 \cdot \sqrt{5}}$.

Proof. For a given partition size V chose $k \in \mathbb{N}$ that $4 \cdot 4^k < V \leq 16 \cdot 4^k$. For each V in the interval we can determine v with $v \cdot 4^k < V \leq (v+1) \cdot 4^k$. The surface $S(p)$ of V is smaller than or equal to the surface of the partition $[1, (v+1) \cdot 4^k]$. The following table states upper bounds for surfaces of partitions v in granularity 2^k called s with $s \cdot 2^k \geq S$ (values for $v < 4$ are used in Theorem 2):

[1] This fact does not apply to all indexing schemes, e. g. Hilbert indexing.

v	0	1	2	3	4	5	6	7	8	9	10	11	12	13	14	15
s	4	6	8	8	10	12	12	12	14	14	16	16	16	16	16	16

$$\frac{S}{4 \cdot \sqrt{V}} \leq \frac{s}{4 \cdot \sqrt{v}} \leq \frac{12}{4 \cdot \sqrt{5}} \approx 1.34 \tag{6}$$

It is obvious that the equation holds for all unexamined partitions smaller than $4 \cdot 4^0$, too. □

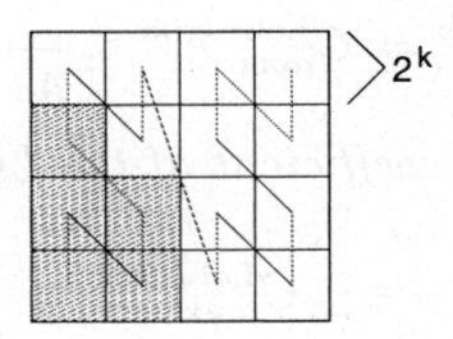

Fig. 3. Partition within a coarse structured Lebesgue indexing.

Example 1. Figure 3 shows the maximal partition induced by the Lebesgue indexing for $v = 4$. All partitions with size V, $4 \cdot 4^k < V \leq 5 \cdot 4^k$ are covered in this case. For all those partitions $S \leq 10 \cdot 2^k$ holds. It follows $s = 10$.

For the analysis of an arbitrary partition p within the Lebesgue indexing we use the fact that the curve is split at most into two sub-curves. The second part only has cells which lie in the same or a more right column *and* in the same or an upper row. It behaves like the partitions examined in Lemma 1. The same holds for the first part due to the symmetry of the curve. A partition p is examined as partitions p_1 and p_2 with $p = p_1 \circ p_2$ and $V = V_1 + V_2$. Again, the analysis is done on the coarse granularity used above.

Theorem 2. $C_{\max}^{\text{Lebesgue}} \leq \frac{7}{2 \cdot \sqrt{3}}$.

Proof. From $V = V_1 + V_2$ follows $3 \leq v_1 + v_2 \leq 15$ with $v_1, v_2 \in [0, 15]$. For the quality coefficient holds

$$C(p_1 \circ p_2) \leq \frac{s_1 + s_2}{4 \cdot \sqrt{v_1 + v_2}} . \tag{7}$$

The enumeration of all possible combinations for v_1 and v_2 shows that the maximum is achieved for $v_1 = 1$ and $v_2 = 2$. It follows $s_1 = 6$ and $s_2 = 8$ (compare table of Lemma 1) and

$$C(p_1 \circ p_2) \leq \frac{6 + 8}{4 \cdot \sqrt{1 + 2}} = \frac{7}{2 \cdot \sqrt{3}} \approx 2.02 . \tag{8}$$

□

The analysis of the upper bound is an enumeration of a finite number of cases with a maximum determination. We can shift the examined interval of partition size V by a refinement of the underlying granularity with the help of computational evaluation. For the listed refinement steps the result improves towards the following values:

$$4 \cdot 4^k < V \leq 16 \cdot 4^k \Rightarrow C_{\max}^{\text{Lebesgue}} \leq \frac{6+8}{4 \cdot \sqrt{1+2}} < 2.021 \tag{9}$$

$$16 \cdot 4^k < V \leq 64 \cdot 4^k \Rightarrow C_{\max}^{\text{Lebesgue}} \leq \frac{24+24}{4 \cdot \sqrt{21+21}} < 1.852 \tag{10}$$

$$256 \cdot 4^k < V \leq 1024 \cdot 4^k \Rightarrow C_{\max}^{\text{Lebesgue}} \leq \frac{192+192}{4 \cdot \sqrt{1365+1365}} < 1.838 \tag{11}$$

Corollary 1. *For the quality coefficient of the Lebesgue indexing holds:*

$$1.837 < 3 \cdot \sqrt{\frac{3}{8}} - \varepsilon \leq C_{\max}^{\text{Lebesgue}} \leq \frac{96}{\sqrt{2730}} < 1.838 \tag{12}$$

2.3 Upper Bound in Average Case

In this section we will focus on the average case. As stated in the introduction, most algorithms based on space-filling curves profit from a good behavior in the average case of all performed operations. Due to space limitations we present an asymptotical estimation. An exact but rather complex solution is presented in [13].

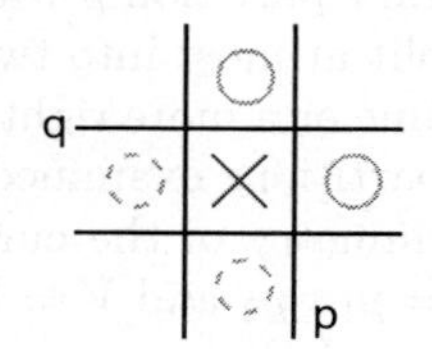

Fig. 4. Neighboring levels for an arbitrary cell within a grid structured by Lebesgue indexing.

For the evaluation of the surface the number of edges common to two cells is needed. It has to be subtracted twice from the number of all edges $4 \cdot V$. A cell has an inner edge on the right hand or upper side, if the index of the right or upper cell is small enough to be still a member of the partition. Given an arbitrary situation illustrated in Fig. 4 with levels p on the right hand and q on the upper side, the indices of the right and upper cell are $R_q = 4 \cdot \frac{4^q-1}{3} + 2$ and $U_p = 2 \cdot \frac{4^p-1}{3} + 1$.

Lemma 2. *For the surface of a partition induced by the Lebesgue curve holds in average case:*

$$S \leq \tfrac{3}{2^k} V + \tfrac{8}{3} \cdot 2^k - \tfrac{5}{3} \cdot \tfrac{1}{2^k} \quad \text{for } V \in \left[2\frac{4^k-1}{3} + 1, 4\frac{4^k-1}{3} + 2 \right[\quad \text{and}$$

$$S \leq \tfrac{2}{2^k} V + 4 \cdot 2^k - \tfrac{1}{2^k} \quad \text{for } V \in \left[4\frac{4^k-1}{3} + 2, 2\frac{4^{k+1}-1}{3} + 1 \right[\text{, with } k \in \mathbb{N}_0 \ .$$

Proof. The number of cells with a neighbor at level l is given by $\max\{V - R_l, 0\}$ and $\max\{V - U_l, 0\}$ for right and upper neighbor, resp. These terms can be used for a summation on all levels and its corresponding probabilities.

$$S \leq 4V - 2\sum_{i=0}^{\infty} \frac{1}{2^{i+1}} \max\{V - U_k, 0\} - 2\sum_{i=0}^{\infty} \frac{1}{2^{i+1}} \max\{V - R_k, 0\} \tag{13}$$

For further examinations of this formulation the evaluated space is split into two classes of intervals: $I_1 = [2\frac{4^k-1}{3}+1, 4\frac{4^k-1}{3}+2[$ and $I_2 = [4\frac{4^k-1}{3}+2, 2\frac{4^{k+1}-1}{3}+1[$. The size of the surface for all partitions p in intervals of class I_1 is given by:

$$\begin{aligned}
S &\leq 4V - \sum_{i=0}^{k} \frac{1}{2^i}\left(V - 2\frac{4^i-1}{3} - 1\right) - \sum_{i=0}^{k-1} \frac{1}{2^i}\left(V - 4\frac{4^i-1}{3} - 2\right) \\
&= 4V - \sum_{i=0}^{k}\left(\left(V - \frac{1}{3}\right)\frac{1}{2^i} - \frac{2}{3}\cdot 2^i\right) - \sum_{i=0}^{k-1}\left(\left(V - \frac{2}{3}\right)\frac{1}{2^i} - \frac{4}{3}\cdot 2^i\right) \\
&= 4V - \left(V - \frac{1}{3}\right)\left(2 - \frac{1}{2^k}\right) + \frac{2}{3}\left(2^{k+1} - 1\right) \\
&\quad - \left(V - \frac{2}{3}\right)\left(2 - \frac{1}{2^{k-1}}\right) + \frac{4}{3}\left(2^k - 1\right) \\
&= \frac{3}{2^k}V + \frac{8}{3}2^k - \frac{5}{3}\frac{1}{2^k}
\end{aligned} \tag{14}$$

Using the same arithmetic technique, for interval class I_2 holds:

$$\begin{aligned}
S &\leq 4V - \sum_{i=0}^{k} \frac{1}{2^i}\left(V - 2\frac{4^i-1}{3} - 1\right) - \sum_{i=0}^{k} \frac{1}{2^i}\left(V - 4\frac{4^i-1}{3} - 2\right) \\
&= \frac{2}{2^k}V + 4\cdot 2^k - \frac{1}{2^k}
\end{aligned} \tag{15}$$

□

It has to be kept in mind that the occurrence of the different edges of level l is not exactly $p(l) = 1/2^{l+1}$. For $l = 0$ the possibility is larger than $p(0)$, for all other levels it is smaller than $p(l)$. This results in an underestimation of inner edges and therefore in an overestimation for the size of the surface. For large grids the calculated values converge to the exact solutions (comp. [13]). However, the quality of the given estimation does *not* depend on the size of the partition.

Theorem 3. *The quality coefficient of the average case for the Lebesgue indexing scheme is less than or equal to* $\frac{5}{2\cdot\sqrt{3}}$.

Proof. For the determination of the upper bound for the average case the limits of the intervals of classes I_1 and I_2 has to be examined. For the upper limit of I_1 we get:

$$V = 4\frac{4^{k-1}-1}{3} + 2 \Leftrightarrow \sqrt{3V-2} = 2^k \tag{16}$$

Using the result of Lemma 2 gives:

$$S \le \frac{4}{2^k}V + 2\left(2^k - \frac{1}{2^k}\right) = \frac{4V-2}{\sqrt{3V-2}} + 2\cdot\sqrt{3V-2} = \frac{10V-6}{\sqrt{3V-2}}$$
$$\le \frac{10}{\sqrt{3}}\sqrt{V} \tag{17}$$

The corresponding lower limit of I_1 (eq. to upper limit of I_2) results in:

$$S \le \frac{7\sqrt{2}}{\sqrt{3}} \cdot \sqrt{V} \tag{18}$$

The surface size is obviously larger in the first case. The quality coefficient for the average case is:

$$C_{\text{avg}}^{\text{Lebesgue}} \le \frac{10}{\sqrt{3}} \cdot \frac{1}{4} = \frac{5}{2\cdot\sqrt{3}} \approx 1.44 \tag{19}$$

□

2.4 Summary

In Fig. 5 the analytical results are compared with computational results. Within a uniform 1024×1024 grid all possible partitions of size V (volume) are examined and the maximum, minimum, and average surface size is determined. The resulting values are plotted as solid lines while the analytical formulations for the worst case and average case are indicated by dashed lines. Two positions are tagged with an exclamation mark, where the computational results are very close to the analytical formulations.

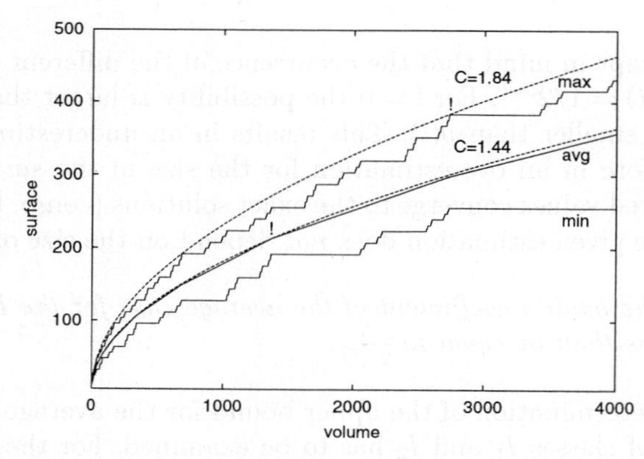

Fig. 5. Locality of partitions induced by the Lebesgue indexing.

3 Hilbert Curves

The Hilbert curve is presumably the most used and studied space-filling curve. It was introduced by Peano and Hilbert in the late 19th century [11]. It is known to be highly local in terms of several metrics mentioned in the introduction. The recursive definition of the curve is illustrated in Fig. 6. For the locality metric based on the quality coefficient this curve is much harder to analyze because the distance within the indexing for neighboring cells depends on the context during construction. An important result is that the lower bound on $C_{\max}^{\text{Hilbert}}$ is larger than the upper bound on $C_{\max}^{Lebesgue}$.

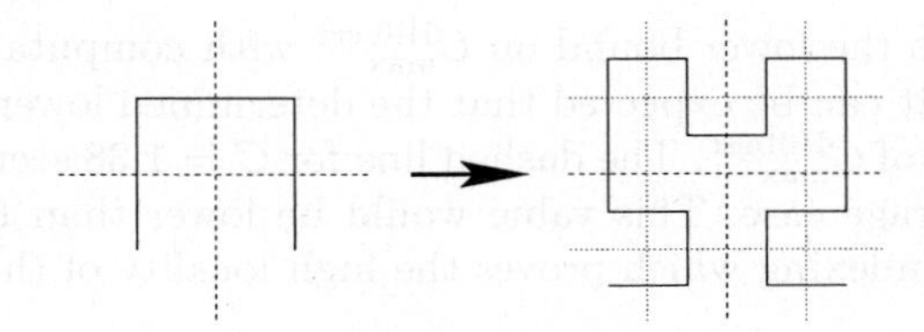

Fig. 6. Production rule for Hilbert indexing.

3.1 Lower Bound on Worst-Case Partitions

Theorem 4. *For the Hilbert curve the quality coefficient is larger than or equal to* $3\sqrt{\frac{5}{13}}$.

Proof. We construct a partition of size V and surface S which follows the Hilbert curve and gives the stated bad quality: Let k be an even number. The center of the partition is given by a square of size 4^{k+1}. On two sides of it 3 squares of sizes 4^k, 4^{k-2}, ..., 4^2, 4^0 are appended. Figure 7 shows the construction of the partition and its location within the Hilbert curve. It follows:

$$V = 4^{k+1} + 2\sum_{i=0}^{k/2} 3 \cdot 4^{2i} = \frac{52}{5}4^k - \frac{2}{5} \tag{20}$$

and

$$S = 8 \cdot 2^k + 12\sum_{i=0}^{k/2} 2^{2i} = 24 \cdot 2^k - 4 \ . \tag{21}$$

The quality coefficient of the partition is given by

$$\frac{S}{4 \cdot \sqrt{V}} = \frac{24 \cdot 2^k - 4}{4 \cdot \sqrt{\frac{52}{5} \cdot 4^k - \frac{2}{5}}} > \frac{24 \cdot 2^k}{4 \cdot \sqrt{\frac{52}{5}} \cdot 2^k} = 3 \cdot \sqrt{\frac{5}{13}} \approx 1.86 \ . \tag{22}$$

□

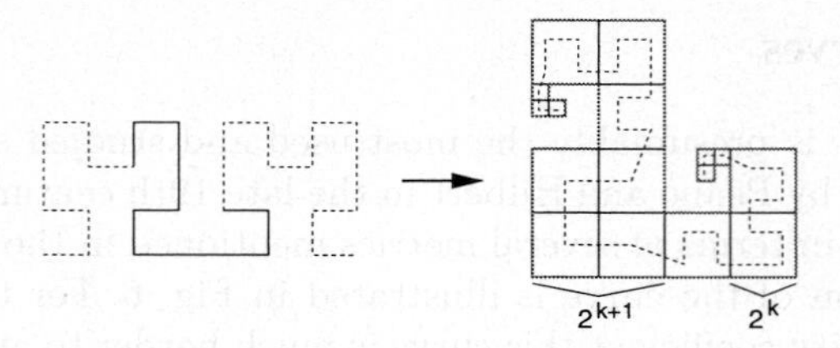

Fig. 7. Construction of lower bound for Hilbert indexing.

3.2 Summary

Figure 8 compares the lower bound on $C_{\max}^{\text{Hilbert}}$ with computational results in a 1024×1024 grid. It can be expected that the determined lower bound is close to the exact solution of $C_{\max}^{\text{Hilbert}}$. The dashed line for $C = 1.38$ seems to be an upper bound in the average case. This value would be lower than the corresponding for the Lebesgue indexing which proves the high locality of the Hilbert curve in another metric.

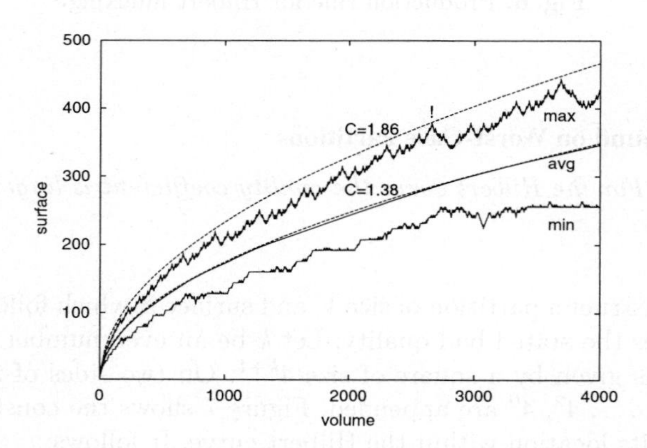

Fig. 8. Locality of partitions induced by the Hilbert indexing.

4 Concluding Remarks

The shown analytical results indicate that partitions based on the Lebesgue space-filling curve have good quality. We proved that they are slightly superior to Hilbert curves in the worst case. Computational results indicate that Hilbert curves behave better in average case. This is due to the fact that index intervals of the Hilbert curve are always connected.

It appears to be much harder to give sharp bounds for the Hilbert indexing than for Lebesgue indexing. We are near by an upper bound for $C_{\max}^{\text{Hilbert}} \leq \frac{26\frac{2}{3}}{8 \cdot \sqrt{\frac{41}{15}}} \approx 2.02$, which still means a weaker result than for the Lebesgue curve.

Obviously, an open question is the lower bound on the quality coefficient for an arbitrary indexing scheme. It is hard to argue whether this bound is closer to the coefficients of the Lebesgue and Hilbert indexings or to 1, the optimal value given by the square. It is easy to generate bad cases for very small partitions, e. g. $V = 3$. This partition has a surface of at least 8. It follows $C = \frac{8}{4 \cdot \sqrt{2}} = \sqrt{2}$. Excluding small volumes we can generate partitions with much lower quality coefficients. But it is an open question whether it is true for arbitrary partitions of an indexing scheme.

References

1. P. K. Agarwal and J. Erickson. Geometric range searching and its relatives. *Advances in Discrete and Computational Geometry*, 1998.
2. T. Asano, D. Ranjan, T. Roos, E. Welzl, and P. Widmayer. Space-filling curves and their use in the design of geometric data structures. *Theoretical Computer Science*, 181:3–15, 1997.
3. J. Behrens and J. Zimmermann. Parallelizing an unstructured grid generator with a space-filling curve approach. In A. Bode, T. Ludwig, W. Karl, and R. Wismüller, editors, *Euro-Par 2000*, LNCS 1900, pages 815–823. Springer, 2000.
4. S. Craver, B.-L. Yeo, and M. Yeung. Multilinearization data structure for image browsing. In *SPIE – The International Society for Optical Engineering*, pages 155–, 1998.
5. R. Diekmann, J. Hungershöfer, M. Lux, L. Taenzer, and J.-M. Wierum. Using space filling curves for efficient contact searching. In *Proc. IMACS*, 2000.
6. C. Gotsman and M. Lindenbaum. On the metric properties of discrete space-filling curves. *IEEE Transactions on Image Processing*, 5(5):794–797, May 1996.
7. M. Griebel and G. Zumbusch. Parallel multigrid in an adaptive PDE solver based on hashing and space-filling curves. *Parallel Computing*, 25:827–843, 1999.
8. B. Moon, H. V. Jagadish, C. Faloutsos, and J. H. Saltz. Analysis of the clustering properties of the Hilbert space-filling curve. *IEEE Transaction on Knowledge and Data Engineering*, 13(1), Jan/Feb 2001.
9. R. Niedermeier, K. Reinhardt, and P. Sanders. Towards optimal locality in mesh-indexings. *LNCS 1279*, 1997.
10. R. Pajarola and P. Widmayer. An image compression method for spatial search. *IEEE Transactions on Image Processing*, 9(3):357–365, 2000.
11. H. Sagan. *Space Filling Curves*. Springer, 1994.
12. S.-H. Teng. Provably good partitioning and load balancing algorithms for parallel adaptive N-body simulation. *SIAM Journal on Scientific Computing*, 19(2):635–656, 1998.
13. J.-M. Wierum. Average case quality of partitions induced by the Lebesgue indexing. Technical Report TR-002-01, Paderborn Center for Parallel Computing, www.upb.de/pc2/, 2001.
14. G. Zumbusch. On the quality of space-filling curve induced partitions. *Zeitschrift für Angewandte Mathematik und Mechanik*, 81, SUPP/1:25–28, 2001.

The Largest Empty Annulus Problem

J. M. Díaz-Báñez[1], F. Hurtado[2], H. Meijer[3], D. Rappaport[3] and T. Sellares[4]

[1] Universidad de Sevilla, Spain dbanez@cica.es
[2] Universitat Politècnica de Catalunya, Spain hurtado@ma2.upc.es
[3] Queen's University, Canada henk,daver@cs.queensu.ca
[4] Universitat de Girona, Spain sellares@ima.udg.es

Abstract. Given a set of n points S in the Euclidean plane, we address the problem of computing an annulus A, (open region between two concentric circles) of largest width such that no point $p \in S$ lies in the interior of A. This problem can be considered as a minimax facility location problem for n points such that the facility is a circumference. We give a characterization of the centres of annuli which are locally optimal and we show the the problem can be solved in $O(n^3 \log n)$ time and $O(n)$ space. We also consider the case in which the number of points in the inner circle is a fixed value k. When $k \in O(n)$ our algorithm runs in $O(n^3 \log n)$ time and $O(n)$ space. However if k is small, that is a fixed constant, we can solve the problem in $O(n \log n)$ time and O(n) space.

1 Introduction

Consider the placement of an undesirable circular route through a collection of facilities. We assume that the circumference of the circle contains at least one point in its interior. Applications of this problem occur in urban, industrial, military and robotic task planning. For example see [24], [14].

In recent years there has been an increasing interest in considering the location of obnoxious routes (transportation of toxic or obnoxious materials), most of the papers deal with models within an underlying discrete space. For example see [4, 6]. In the continuous case, in which the route can be located anywhere, there has been very little progress towards obtaining efficient algorithms. An iterative approach for finding the location of a polygonal route which maximizes the minimum distance to a set of points is proposed in [11], but efficient algorithms are not known.

Several problems on computing widest empty corridors have received attention within the area of computational geometry, for example considering empty strips, L-shapes, as well as many other possibilities, see [8, 9, 16–18, 7].

These notions can be cast in the setting where an annulus is used for separation. Given a set of points in the plane, we want to separate the set

P.M.A. Sloot et al. (Eds.): ICCS 2002, LNCS 2331, pp. 46–54, 2002.

into two subsets with the widest annulus. Related results are given in [5], [13], [21] [22].

A dual optimization problem is to find the location of a facility (circumference) which minimizes the maximum distance from all sites. In [10] this problem is termed the sphere-centre problem, and it corresponds to computing the smallest width annulus that contains a given set of points. Efficent algorithms for solving this problem are discussed in [12], [2].

We outline the rest of the paper. In section 2, we establish some notation and preliminary results. In section 3, we propose an algorithm to compute the centre of a largest empty annulus. In section 4, we address a particular case where we fix the number of points in the inner circle of the annulus. Finally, section 5 contains some concluding remarks and poses some open problems.

2 Characterization of candidate centres

We begin by introducing some notation. Let A denote an annulus, that is, the open region between two concentric circles. It will be convenient to access features of the annulus, thus we use $c(A)$ to denote the centre of the circles, $r(A)$ and $R(A)$ the radii of the circles, where it understood that $r(A) \leq R(A)$, and $o(A)$ and $O(A)$ the boundary of the circles, such that the radius of $o(A)$ is $r(A)$ and the radius of $O(A)$ is $R(A)$. Let $w(A) = R(A) - r(A)$ denote a quantity we call the *width* of A. We use $d(p,q)$ to denote the Euclidean distance between points p and q. Given a set, S, of n points in the Euclidean plane, we say that an annulus A is an *empty annulus* for S if the annulus induces a partition of S into two non-empty subsets $\mathrm{IN}(A,S) = \{s : d(s,c(A)) \leq r(A)\}$ and $\mathrm{OUT}(A,S) = \{s : d(s,c(A)) \geq R(A)\}$. Let $E(S)$ denote the set of all empty annuli for S, and let $\Gamma(S)$ denote the subset of $E(S)$ consisting of empty annuli of greatest width, defined precisely as:

$$\Gamma(S) = \{A \in E(S) : w(A) \geq w(B) \text{ for all } B \in E(S)\} \tag{1}$$

Observe that $\Gamma(S)$ is non-empty for any set of two or more points. We define $\omega(S)$ to be equal to $w(A)$ where A is any annulus in $\Gamma(S)$. We present an algorithm that determines the quantity $\omega(S)$. The algorithm will also be able to produce a witness annulus $A \in \Gamma(S)$. Although $\Gamma(S)$ may be infinitely big, our algorithm can also produce a concise description of $\Gamma(S)$. In this section we provide characterizations for largest empty annuli, that is, the annuli in $\Gamma(S)$.

To begin with we make the obvious observation that if $A \in \Gamma(S)$ then

$$|o(A) \cap S| \geq 1 \text{ and } |O(A) \cap S| \geq 1. \tag{2}$$

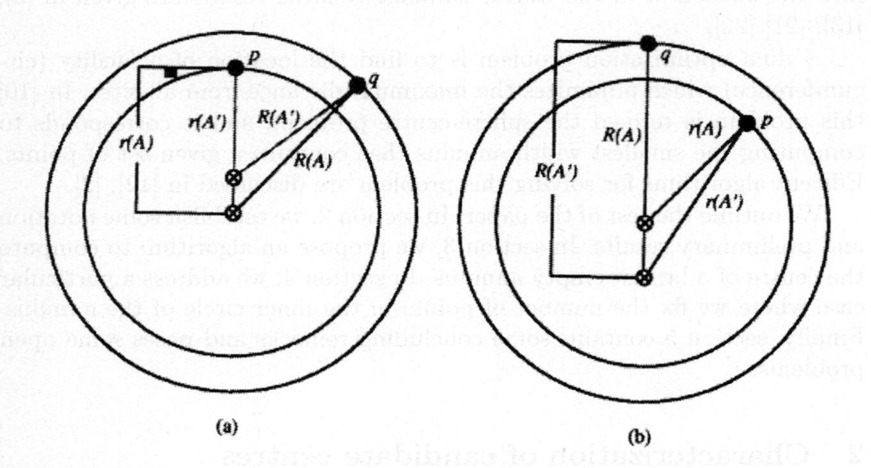

Fig. 1. If A is an optimal non-syzygy annulus then $|o(A) \cap S| \geq 2$ and $|O(A) \cap S| \geq 2$.

Consider the case where an annulus $A \in \Gamma(S)$ has the property that $|\mathrm{IN}(A, S)| = 1$. For a point $s \in S$ define $\alpha(s) = \min(d(s,t), t \in S - \{s\})$. In this case $\omega(S) = \max(\alpha(s), s \in S)$. That is $\omega(S)$ is realized by the farthest nearest neighbours. For example if q is the nearest neighbour of p then we can construct an annulus A such that $c(A) = p$, $r(A) = 0$, and $R(A) = d(p,q)$. The all nearest neighbour graph of a set of points is a well known structure that can be found in $O(n \log n)$ time [20, page 183]. Thus we can easily dispense with this special case.

From now on we restrict our attention to annuli where $|\mathrm{IN}(A, S)| > 1$.

A syzygy is an astronomical term meaning the points on the moon's orbit when the moon is in line with the earth and the sun. We borrow this term to define a *syzygy annulus* as an annulus $A \in \Gamma(S)$ such that there are points p, q, with $p \in S \cap o(A)$ and $q \in S \cap O(A)$, and p is contained in the open segment $(c(A), q)$.

Lemma 1. *Let* $A \in E(S)$, *where* A *is not a syzygy annulus. If* $|o(A) \cap S| = 1$ *or* $|O(A) \cap S| = 1$ *then* $A \notin \Gamma(S)$.

Proof: We begin by showing that if $A \in E(S)$ such that $|S \cap o(A)| = 1$ and $|S \cap O(A)| \geq 1$ then $A \notin \Gamma(S)$. Let $p = S \cap o(A)$. This implies that we can find an annulus $A' \in E(S)$ such that $c(A')$ is on the open segment $\{c(A), p\}$ and there is a $q \in S$ such that $q \in S \cap O(A)$ and $q \in S \cap O(A')$. See Figure 1 (a). Using the triangle inequality we have:

$$R(A') + (r(A) - r(A')) > R(A) \rightarrow R(A') - r(A') > R(A) - r(A). \quad (3)$$

Thus $w(A') > w(A)$ so $A \notin \Gamma(S)$.

Now suppose that $A \in E(S)$ and $|o(A) \cap S| \geq 2$ and $|O(A) \cap S| = 1$. Let $q = S \cap O(A)$. This time we construct an annulus $A' \in E(S)$ such that $c(A)$ is on the open segment $(c(A'), q)$ and there is a $p \in S$ such that $p \in S \cap o(A)$ and $p \in S \cap o(A')$. See Figure 1 (b). Again by the triangle inequality we have:

$$r(A) + (R(A') - R(A)) > r(A') \rightarrow R(A') - r(A') > R(A) - r(A). \quad (4)$$

Thus $w(A') > w(A)$ so $A \notin \Gamma$.

Thus we conclude that if A is not a syzygy annulus and $|o(A) \cap S| = 1$ or $|O(A) \cap S| = 1$, then $A \notin \Gamma$.

□

As a consequence of equation 2 together with lemma 1 we conclude that every optimal non-syzygy annulus A has $|o(A) \cap S| \geq 2$ and $|O(A) \cap S| \geq 2$.

We now deal with the syzygy annuli.

Lemma 2. *Suppose that A is a syzygy annulus such that $|\mathrm{IN}(A, S)| \geq 2$ and $A \in \Gamma(S)$. Then there exists an annulus $A' \in \Gamma(S)$ such that $|o(A') \cap S| \geq 2$.*

Proof: If $|o(A) \cap S| \geq 2$, then we set $A' = A$ and we are done. Otherwise, using the methods of lemma 1 equation 3 we can obtain an empty annulus A' such that $w(A) = w(A')$ and $|o(A') \cap S| \geq 2$. □

The preceding lemmas suggest the following theorem.

Theorem 1. *If there is an annulus $A \in \Gamma(S)$ such that $|IN(A, S)| \geq 2$ then there is an annulus $A' \in \Gamma(S)$ such that $|o(A') \cap S| \geq 2$.*

Proof: Follows immediately from Lemma 1 and Lemma 2. □

This theorem implies that the search space for the centre of a largest empty annulus can be limited to the right bisectors of pairs of points from S.

3 Finding a largest empty annulus

We describe an algorithm to determine the centre of a largest empty annulus that is constrained to lie on the right bisector of a pair of points p and q, $B(p, q)$. For convenience we adopt a Cartesian coordinate system such that $B(p, q)$ is the line $L : y = 0$. We denote the x and y coordinates of a point s as x_s and y_s. Now for every point $s \in S$ we determine the curve $L_s : y = \sqrt{(x_s - x)^2 + y_s^2}$.

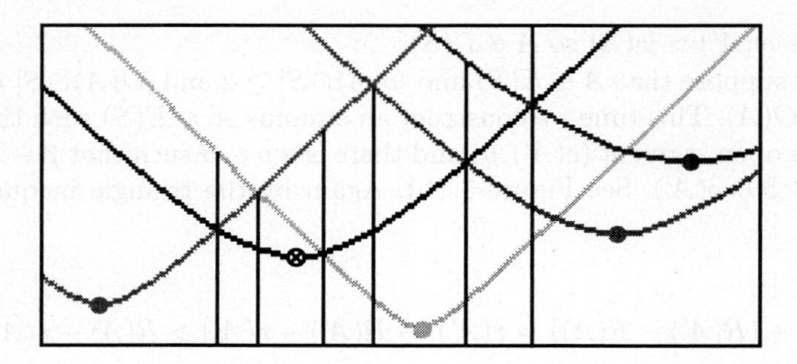

Fig. 2. An arrangement of curves, and a minimization partition of the bisector $B(p, q)$. The point marked with an x is representative of p and q.

Observe that for any two points s and t in S, the intersection of L_s and L_t satisfies

$$x = \frac{x_t^2 - x_s^2 + y_t^2 - y_s^2}{2(x_t - x_s)} \tag{5}$$

Thus

- L_s and L_t are coincident when $x_s = x_t$ and $y_s = -y_t$
- L_s and L_t have no common point when $x_s = x_t$ and $y_s \neq -y_t$
- L_s and L_t have one common point when $x_s \neq x_t$

For a fixed value of x let $L_s(x) = \sqrt{(x_s - x)^2 + y_s^2}$. Then

$$d_s(x) = \begin{cases} L_s(x) - L_p(x) & \text{if } L_s(x) > L_p(x) \\ \infty & \text{otherwise} \end{cases}$$

Let $S' = S - \{p, q\}$, and set

$$F(x) = \{s \in S' : d_s(x) \leq d_t(x) \text{ for all } t \in S'\}.$$

Observe that F induces a partition of L into intervals. Each equivalence class is a maximal interval of L such that for all points in the interval, $F(x)$ is the same subset of S. See Figure 2. We slightly modify the partition such that it only consists of single points and open intervals. Closed intervals of the form $[a, b]$ or a half-open interval $(a, b]$ or $[a, b)$ are replaced by $[a]$, (a, b), $[b]$, or by (a, b), $[b]$, or $[a]$, (a, b) respectively. The number of intervals in the partition is in $O(n)$, because the partition is a minimization partition of a pseudoline arrangement. See [15].

We can compute the intervals induced by F in $O(n \log n)$ time using a divide and conquer algorithm. The merge step simply scans two sorted lists of intervals. We then compute intersections of curves within overlapping intervals to merge in linear time.

We say that an interval is *good* if it contains a point $(x, 0)$ such that $d_s(x)$ is a positive finite value, for $s \in F(x)$. Then there is an empty annulus A such that $c(A) = (x, 0)$, $p, q \subseteq o(A) \cap S$ and $F(x) = O(A) \cap S$. If $d_s(x) = \infty$ for $s \in F(x)$ then a circle centred at $(x, 0)$ and passing through $\{p, q\}$ is a spanning circle of S. So this circle is not the inner circle $o(A)$ of an empty annulus A.

Once we have a determined the partition of L into intervals, we can determine for each good interval a locally maximal empty annulus. When an interval u is just a single point there is a unique annulus there. Let u be an interval that is not a single point and let z be a value in u. Let $s \in F(z)$. We need to find the x such that $d_s(x)$ is finite and maximal. It is easy to show that if this maximum exists, then it occurs at $x = (x_s y_p - x_p y_s)/(y_p - y_s)$. At this point x the distance between L_s and L_p is maximal, so either $d_s(x)$ is infinite or maximally finite. If $x \in u$ and $d_s(x)$ is finite, we have a syzygy annulus A centred at $(x, 0)$ with $\{p, q\} \subseteq o(A) \cap S$ and $F(x) = O(A) \cap S$. If $x \notin u$ then we know that any annulus A centred at a point within the interval u with $p, q \subseteq o(A) \cap S$ and $F(x) \subseteq O(A) \cap S$ is not optimal.

To summarize, by searching through bisectors for every pair of points in S, we can determine $\omega(S)$ in $O(n^3 \log n)$ time and $O(n)$ space. Furthermore we can characterize all annnuli that realize $\omega(S)$ by an ordered pair $(o(A) \cap S, O(A) \cap S)$.

4 Largest empty *k*-inner points annulus problem

An interesting related problem fixes the number of points in $\mathrm{IN}(A, S)$. In fact, in some situations we may be interested in separating exactly k points from the rest of the set S. Using a separation criteria of the widest circular corridor, leads to the problem of computing a largest empty annulus containing k points in its inner circle.

We adapt the notation of section 2 to handle the situation when the number of points in $\mathrm{IN}(A, S)$ is fixed. Let $E_k(S)$ denote the set of all empty annuli with k inner points, and let $\Gamma_k(S)$ denote the subset of $E_k(S)$ consisting of empty annuli with greatest width.

Theorem 2. *If there is an annulus $A \in \Gamma_k(S)$ with $k \geq 2$ then there is an annulus $A' \in \Gamma_k(S)$ such that $|o(A') \cap S| \geq 2$.*

Proof: The arguments of lemma 1 and lemma 2 do not modify the sets $\mathrm{IN}(A, S)$. Thus the results immediately hold for largest empty k-inner point annuli. □

We can apply the algorithm of section 3 with a simple modification. Recall, we constrain our search to empty annuli that lie on the right bisector, $B(p, q)$ of a pair of points from S, p and q . The simple modification is that we only consider empty annuli with k-inner points. Consider the arrangement of curves L_s for all $s \in S$. For every real value x we define

$$M(x) = |\{s \in S : L_x(s) \leq L_x(p)\}$$

Thus when $M(x) = k$ there is a circle centred at $(x, 0)$ passing through the points p and q and containing exactly k points in its interior and boundary. As before we have a set of intervals. Let us number the intervals from left to right as $I_0, I_1, \ldots, I_m$. If two points $(x_1, 0)$ and $(x_2, 0)$ are both in the same interval then $M(x_1) = M(x_2)$. Furthermore, it is easy to see that we can modify the algorithm that computes the intervals of L in such a way that it also computes $M(I_{j+1}) - M(I_j)$. Thus we can solve the empty annulus problem when k is fixed in the same time and space bound as before, that is, in $\mathrm{O}(n^3 \log n)$ time and $\mathrm{O}(n)$ space.

If k is a small fixed constant, then we can do considerably better by using a different approach. Given a set of points S, let $\mathrm{VD}_k(S)$ denote the order-k Voronoi diagram of S. Recall that $\mathrm{VD}_k(S)$ partitions the plane into cells, each of which is a locus of points closest to some k element subset of S. See [19]. Let C be a cell in $\mathrm{VD}_{k+1}(S)$ and let S_C denote the $k+1$ element subset of S associated with the cell C. Suppose A is a largest k-inner point empty annulus of S and $c(A)$, the centre of A, is in C. Then A is also a largest k-empty annulus of S_C with $|\mathrm{IN}(A, S_C)| = k$. Moreover, by Lemma 2, at least two points from S_C lie on $o(A)$, so $c(A)$ lies on a bisector of two points of S_C. This leads us to the following algorithm. We first find $\mathrm{VD}_{k+1}(S)$. In [19] D.T. Lee shows that $\mathrm{VD}_k(S)$ can be found in $\mathrm{O}(k^2 n \log n)$ time and $\mathrm{O}(k(n-k))$ space. For each cell C of $\mathrm{VD}_{k+1}(S)$ we find a point $c(A)$ in C that gives the largest empty k-inner point annulus. If k is a small fixed constant, then $\mathrm{VD}_{k+1}(S)$ can be computed in $O(n \log n)$ time and $\mathrm{O}(n)$ space. Finding the largest empty annulus in a cell C can be done in constant time by processing the bisectors of pairs of point from S_C. Therefore for k a fixed constant we can find the largest empty k-inner point annulus in $O(n \log n)$ time and $\mathrm{O}(n)$ space.

5 Conclusions and further research

In this paper we have dealt with the problem of computing an empty annulus of maximum width for a set of n points in the plane. We have characterized the centres of annuli which are locally optimal and we have proposed an algorithm to solve the problem in $O(n^3 \log n)$ time. We remark that the algorithm is easy to implement and produces a description of all the optimal annuli.

We have also presented an approach by using Voronoi diagrams to solve the case in which the number of interior points is a fixed constant k. When k is a fixed constant we obtain a substantially faster algorithm.

Finally, there are a set of open problems in this context where one may consider parabolic, or conic routes. Then we must find other geometric largest empty regions. Another issue to consider is attempting to find largest empty annuli in higher dimensions.

Acknowledgements

Ferran Hurtado is partially supported by Projects DGES-MEC PB98-0933, Gen.Cat. SGR1999-0356 and Gen.Cat. SGR2001-0224. Henk Meijer and David Rappaport acknowledge the support of NSERC of Canada research grants. Toni Sellares is partially supported by MEC-DGES-SEUID PB98-0933 and TIC2001-2392-C03-01 of the Spanish government.

References

1. P.K. AGARWAL AND M. SHARIR, Daventport-Schinzel sequences and their geometric applications. In Sacks and J.Urrutia, editors, "*Handbook of Computational geometry*", chapter 1, North Holland, 1999.
2. AGARWAL, P.K., SHARIR, M. AND TOLEDO S., Applications of parametric searching in geometric optimization. *Journal of Algorithms*, 17, 1994, 292–318.
3. AURENHAMMER, F. AND SCHWARZKOPF O., A simple randomized incremental algorithm for computing higher order Voronoi diagrams. In *Proc. 7th Annu. Sympos. Comput. Geom.*, 1995, 142–151.
4. BATTA, R. AND CHIU, S., Optimal obnoxious paths on a network: Transportation of hazardous materials. *Opns. Res.*, 36, 1988, 84–92.
5. BHATTACHARYA, B. K. Circular Separability of Planar Point Sets. *Computational Morphology, G.T. Toussaint ed. North Holland*, 1988.
6. BOFFEY, B. AND KARKAZIS, J., Optimal location of routes for vehicles: Transporting hazardous materials. *European J. Oper. Res.*, 1995, 201–215.
7. S. Chattopadhyay and P. Das. The k-dense corridor problems. *Pattern Recogn. Lett.*, 11:463–469, 1990.
8. S.-W. Cheng. Widest empty corridor with multiple links and right-angle turns. In *Proc. 6th Canad. Conf. Comput. Geom.*, pages 57–62, 1994.
9. S.-W. Cheng. Widest empty L-shaped corridor. *Inform. Process. Lett.*, 58:277–283, 1996.
10. DÍAZ-BÁÑEZ J.M., MESA J.A. AND SCHÖBEL A., Continuous Location of Dimensional Structures. *Manuscript.*
11. DREZNER, Z. AND WESOLOWSKY, G.O., Location of an obnoxious route. *Journal Operational Research Society*, 40, 1989, 1011–1018.
12. EBARA, H., NAKANO, H., NAKANISHI, Y. AND SANADA, T., A roundness algorithms using the Voronoi diagrams. *Transactions IEICE*, J70-A 1987, 620–624.
13. FISH S., Separating point sets by circles and the recognition of digital disks. *Pattern Analysis and Machine Intelligence* , 8 (4), 1986.

14. FOLLERT, F. Maxmin location of an anchored ray in 3-space and related problems. In *7th Canadian Conference on Computational Geometry, Quebec*, 1995.
15. HALPERIN, D. , *Handbook of Discrete and Computational Geometry.* Jacob E. Goodman and Joseph O'Rourke eds., CRC Press LLC, Boca Raton, FL, 389–412, 1997.
16. M. Houle and A. Maciel. Finding the widest empty corridor through a set of points. In *Snapshots of computational and discrete geometry*, pages 201–213. Dept. of Computer Science, McGill University, Montreal, Canada, 1988. Technical Report SOCS-88.11.
17. R. Janardan and F. Preparata. Widest-corridor problems. In *Proc. 5th Canad. Conf. Comput. Geom.*, pages 426–431, 1993.
18. R. Janardan and F. P. Preparata. Widest-corridor problems. *Nordic J. Comput.*, 1:231–245, 1994.
19. D. T. Lee. On k-nearest neighbor Voronoi diagrams in the plane. *IEEE Trans. Comput.*, vol. C-31, pp. 478487, 1982.
20. O'ROURKE J., *Computational Geometry in C.* Cambridge University Press, 2nd edition, 1998.
21. J. O'ROURKE, S.R. KOSARAJU AND N. MEGGIDO, Computing circular separability. *Discrete Computational Geometry*, 1 (1), 1986, 105–113.
22. T. J. RIVLIN, Approximation by circles. *Computing*, 21, 93–104 1979
23. M. SHARIR AND P.K. AGARWAL, "*Davenport-Schinzel sequences and their geometric applications*", Cambridge University Press 1995.
24. TOUSSAINT G. T., Computing largest empty circles with location constraints. *International Journal of Computer and Information Sciences*, 12, 1983, 347–358.

Mapping Graphs on the Sphere to the Finite Plane

Henk Bekker, Koen De Raedt

Institute for Mathematics and Computing Science, University of Groningen,
P.O.B. 800 9700 AV Groningen, The Netherlands,
bekker@cs.rug.nl, csg8042@wing.rug.nl

Abstract. A method is introduced to map a graph on the sphere to the finite plane. The method works by first mapping the graph on the sphere to a tetrahedron. Then the graph on the tetrahedron is mapped to the plane. Using this mapping, arc intersection on the sphere, overlaying subdivisions on the sphere and point location on the sphere may be done by using algorithms in the plane.

1 Introduction

In plane computational geometry three basic operations are line segment intersection, overlaying two subdivisions and point location. In a natural way these operations may also be defined on the sphere. For these problems, the correspondence between the plane and the sphere is so strong that it is tempting to try and solve the problems on the sphere by using algorithms working in the plane. That is however not possible because topologically the sphere differs from the plane. An obvious step to remedy this situation is to try and adapt the algorithms working in the plane, so that they work on a sphere. For naive and non-optimal implementations this might work. However, transforming sophisticated and optimised algorithms is not trivial. In fact, then every detail of the algorithm has to be reconsidered.

Instead of adapting algorithms we propose to adapt the problem, that is, we propose to map the graphs on the sphere to the plane, so that algorithms working in the plane may be used to solve the problems on the sphere. To make this scheme work the mapping has to fulfil three conditions.

1. The mapping has to be continuous and one-to-one.
2. The mapping has to be finite, that is, the image of the graphs on the sphere should not have points at infinity.
3. Each arc of a great circle on the sphere has to be mapped on one straight-line segments in the plane.

Let us comment on these three conditions.

1: The mapping has to be continuous because when the mapping is only piecewise continuous the image of the graph on the sphere would consist of patches in the plane. As a result, the operations in the plane would have to be done

P.M.A. Sloot et al. (Eds.): ICCS 2002, LNCS 2331, pp. 55–64, 2002.

on each of these patches. With a one-to-one mapping we mean a mapping that maps every point on one graph to a point on the other graph, so, we do not mean that the graph structure remains the same. The mapping has to be one-to-one because after the operations in the plane have been done the result has to be mapped back on the sphere in a unique way.
2:The mapping has to be finite because, in general, the algorithms working in the plane can not handle points at infinity.
3: Arcs of a great circle on the sphere have to be mapped on straight-line segments in the plane. This condition is not absolutely essential. We could use a mapping that maps an arc of a great circle on some curve segment in the plane, and then use an algorithm in the plane that works with curve segments instead of line segments. However, curve segment intersection, point location in curved graphs, and overlaying curved graphs is very inefficient compared with the corresponding straight-line segment algorithms.

In the literature no mapping is given that fulfils these three conditions. In this article we introduce a mapping that fulfils these three conditions almost. Only condition 3 is not met completely. In the mapping we propose some arcs are mapped on *two* or *three* connected line segments in the plane instead of *one* line segment. In the following section we describe the mapping in an operational way. After that we discuss some details and alternatives. As an example application we compute the overlay of two subdivisions on a sphere.

Problem motivation: During the past ages cartographers have proposed many mappings to map the sphere to the plane (e.g. Mercator-, cylindric-, and stereographic projection)[1]. None of these mappings maps an arc of a great circle on the sphere to a straight line segment in the plane. To our knowledge no such mapping exists. In this article we introduce a mapping that maps an arc of a great circle on the sphere most often to one line segment in the plane, and sometimes to two or three connected line segments in the plane. Our motivation for developing this mapping is in computational geometry. Some time ago we proposed [2] a linear time algorithm to compute the Minkowski sum of two convex polyhedra A and B. The crucial part of that algorithm consists of calculating the overlay of two subdivisions on the sphere, where the subdivisions are the slope diagrams of A and B [3]. By mapping this problem to the plane the 3D problem of calculating the Minkowski sum of two convex polyhedra is reduced to the problem of calculating an overlay in the plane. For this a linear time algorithm may be used [4], so, Minkowski addition of convex polyhedra in 3D may be done in linear time. In [5] algorithms are given to do overlay, point location and arc intersection on the sphere. However, these algorithms do not work in linear time, so, they can not be used for an efficient implementation of the Minkowski sum.

2 The mapping

We consider a sphere centered at the origin. On this sphere some structure is given consisting of points and arcs of great circles. To be a bit more concrete, and without loss of generality, in the sequel we assume that the structure is a subdivision on the sphere. We represent the subdivision by a graph SGA (Sphere Graph A). To keep things simple, we assume that the subdivision is connected, so SGA is also connected. To make SGA as similar as possible to the subdivision we embed SGA on the sphere so that nodes of SGA are embedded at positions of the corresponding points of the subdivision, and edges of SGA are embedded as arcs of great circles of the corresponding edges of the subdivision. In this way SGA and its generating subdivision are similar, so we can simply work with SGA. (See figure 2 upper.)

We want to map SGA to the plane, but as an intermediate step we first map SGA to a tetrahedron T with the following properties.

1. The base of T is parallel with the x,y plane.
2. T is almost regular.
3. T is centered at the origin.

Let us comment on these three properties.
1: T should have a unique top. This means that only one of the four vertices of T should have the maximum z coordinate in the positive z direction. By choosing the base of T in the x,y plane this condition is fulfilled.
2: We first construct a regular tetrahedron TTT with vertex coordinates $(0,0,0)(1,0,0)(0.5,\frac{\sqrt{3}}{2},0)(0.5,\frac{\sqrt{3}}{6},\frac{\sqrt{3}}{3})$. The first three vertices are in the x-y plane, the last vertex is the top. TTT has side length 1. Now we shift the first three vertices of TTT each with a random displacement in the x,y plane over a distance of say at most 0.1, and the top vertex with a random displacement in 3D over a distance of at most 0.1. This gives us a tetrahedron TT.
3:We vectorially add the vertex positions of TT, divide this vector by four, and shift TT over minus this vector. This gives us T.

Now we are going to map SGA to T. First we map all nodes of SGA to T, and then we add additional nodes. To this end, we copy the graph SGA in a new graph TGA (Tetrahedron Graph A) but we do not copy the node positions of SGA into TGA. Using central projection we map the nodes of SGA on T as follows. For every node n of SGA we construct a ray, starting at the origin and containing the position of node n. The intersection point p of this ray with one of the faces of T is assigned to the corresponding node position in TGA.

Every node of TGA is now on T. However, every edge of TGA represents a line segment, and not every of these line segments is on a face of T. That is because there are edges the endpoints of which are on different faces of T. Suppose that edge e has an endpoint on face f_i and an endpoint on face f_j, with $i \neq j$, and that f_i and f_j meet in edge e_T. We add a node to TGA, located on e. To calculate the position of the new node we construct a ray, starting at the origin, and intersecting e and e_T. The intersection point of this ray with e_T is the position of the new node in TGA. We mark these added nodes so that they can be deleted on the way back.

Now we are going to map TGA on the plane P containing the lower face of T. Crucial for this mapping is that TGA has no node located at the top vertex of

T. If there is a node at the top vertex a new T has to be generated, and SGA has to be mapped again on T. Because T is a random tetrahedron the probability that a node is mapped at the top vertex is virtually zero.

Having verified that no node of TGA is at the top vertex of T we determine a projection point *pp*. From *pp* the graph TGA will be projected onto P. *pp* has to fulfil two conditions.

1. *pp* has to be located inside T.
2. The z coordinate of *pp* should be greater than the z coordinate of the position of the highest node of TGA.

Let us comment on this.
1: That *pp* is located inside T implies that *pp* is not on the boundary of T. So, from *pp* the graph TGA is completely seen from the inside of T, and every node and edge of TGA is seen in a unique direction.
2: Choosing *pp* higher than the highest node position of TGA has the effect that by projecting from *pp*, every node and edge of TGA is projected on P.

We construct *pp* as follows. Call the z coordinate of the top vertex of T H_{tv} and the z coordinate of the highest node of TGA H_{hn}. We construct a plane *ppp* parallel with the x-y plane with $z = (H_{hn} + H_{tv})/2$, and a line *lpp* perpendicular to the x-y plane, containing the top vertex of T. The intersection point of *lpp* and *ppp* is *pp*. See figure 1.

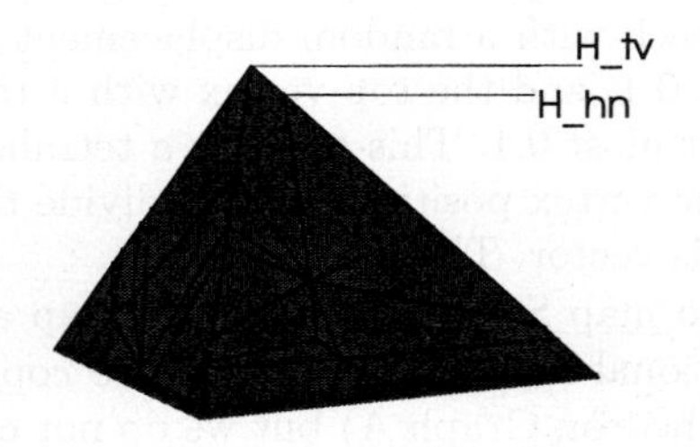

Fig. 1. The top of the tetrahedron T, the z coordinate H_{tv} of its top vertex and the z coordinate H_{hn} of the highest node of TGA. The projection point *pp* is located inside T, under the top vertex of T and at $z = (H_{hn} + H_{tv})/2$.

Using central projection from *pp* we project TGA in the plane P, resulting in graph PGA (Plane Graph A). PGA is constructed as follows. First we copy the graph TGA to PGA but not the node positions. For every node n of TGA we construct a line l containing *pp* and the node position n. The position of the intersection point of l with P is assigned to corresponding node of n in PGA.

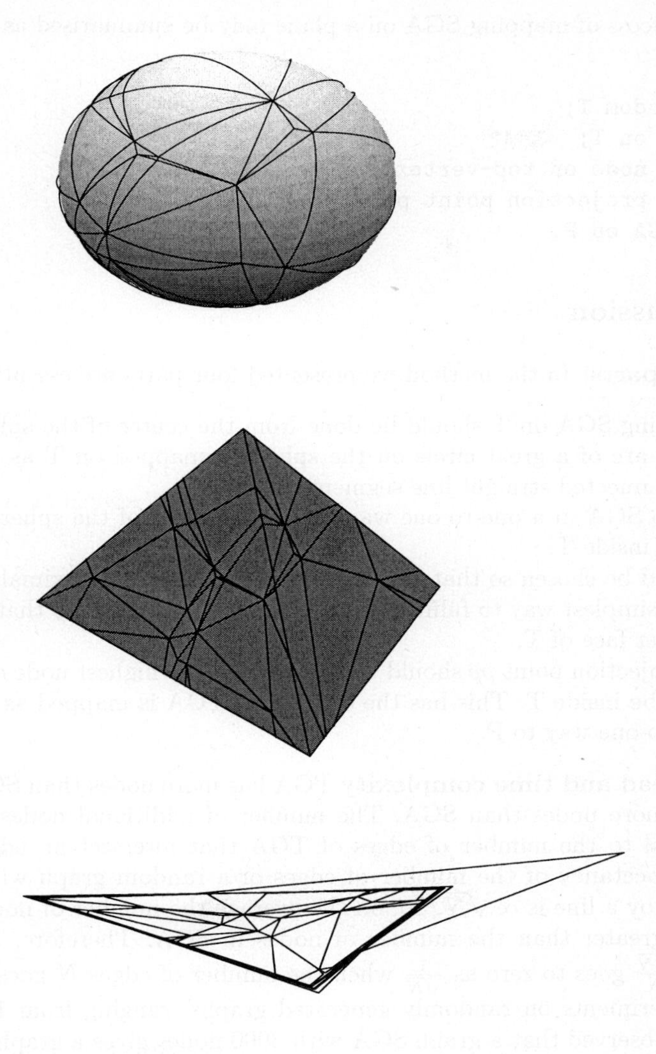

Fig. 2. The three stages of mapping a graph on a sphere to the plane. The view direction is the same for all three figures. Top: A sphere with a graph SGA. Middle: SGA mapped on a random tetrahedron T, giving the graph TGA. SGA can be recognised in TGA. Lower: The graph TGA mapped to a plane containing the lower face of T, giving the graph PGA. PGA is very inhomogeneous, and it is difficult to recognise TGA in PGA.

The process of mapping SGA on a plane may be summarised as follows.

```
repeat
  make random T;
  map SGA on T;
until (no node on top-vertex);
determine projection point pp;
project TGA on P;
```

3 Discussion

Essential parts In the method we presented four parts are essential.

1. Projecting SGA on T should be done from the center of the sphere. In this way an arc of a great circle on the sphere is mapped on T as one, two or three connected straight line segments.
2. To map SGA in a one-to-one way to T, the center of the sphere should be located inside T.
3. P should be chosen so that only one point of T has the maximal distance to P. The simplest way to fulfil this condition is to choose P so that it contains the lower face of T.
4. The projection point *pp* should be higher than the highest node of TGA and should be inside T. This has the effect that TGA is mapped as a whole, in a one-to-one way to P.

Overhead and time complexity TGA has more nodes than SGA, so, also PGA has more nodes than SGA. The number of additional nodes of TGA is proportional to the number of edges of TGA that intersect an edge of T. In 2D, the expectancy of the number of edges of a random graph with N edges intersected by a line is $\propto \sqrt{N}$. So, on the average the number of nodes in TGA is $\propto \sqrt{N}$ greater than the number of nodes in SGA. Therefore, the relative overhead $\frac{\sqrt{N}}{N}$ goes to zero as $\frac{1}{\sqrt{N}}$ when the number of edges N goes to infinity. In our experiments on randomly generated graphs, ranging from 10 to 10000 nodes, we observed that a graph SGA with 3000 nodes gives a graph TGA with $\approx 6\%$ more nodes. The time complexity of the method is linear in the number of nodes of SGA, so, also in the number of nodes of TGA and PGA. When TGA has a node at the top-vertex of T, T should be regenerated. In our experiments this never happened, so, we think the overhead related to this situation may be neglected.

Inhomogeneity of PGA It is difficult to recognise in PGA the original graph SGA. When SGA is more or less homogeneous TGA is also, but PGA is always very inhomogeneous. That is because that part of TGA that is on the lower face of T is mapped undistorted to P, while the parts of TGA that are on the other faces of T are projected on P in a direction that is almost parallel with these faces. To avoid numerical problems associated with the inhomogeneity of PGA, we implemented our algorithm in exact arithmetic, or more precise, we implemented our algorithm in C++ and LEDA [6]. When exact arithmetic is used for mapping the sphere on the plane, for the operations in the plane (for example overlaying, line intersection or point location), and for mapping back the result on the sphere, it is no problem that PGA is inhomogeneous.

4 Alternatives and improvements

Alternative position of T In section 2 we proposed to position T with its center at the origin. That is however not required. Any position of T will do as long as T contains the origin. This freedom may be used to improve the mapping somewhat. The alternative position of T is as follows. The x and y coordinates of T are the same as before, and the z position of T is chosen so that the origin is located slightly above the lower face. For example, the distance from the lower face to the origin could be chosen as $\frac{1}{100}H_T$, where H_T is the height of T in the z direction. Positioning T in this way has the effect that almost the whole lower half of SGA is mapped on the lower face of T. So, the density of edges and nodes of TGA on the other faces of T decreases. As a result, the number of edges TGA crossing edges of T decreases, so, the number of additional nodes decreases. Moreover, the density of nodes of TGA near the top vertex of T decreases, so, the probability that a node of TGA is located at the top vertex of T decreases. These effects can be seen in figure 3. The same SGA is used as in figure 2. It can be seen that the density of nodes and edges on the non-horizontal faces of T has decreased, and that the z coordinate of the highest node of TGA is lower.

The price we pay for this alternative is that in the middle of the lower face of T TGA has a high node and edge density. In fact, almost half the number of nodes and edges of SGA is mapped there. However, because we use exact arithmetic this is no problem.

Alternative mapping of TGA to P PGA is very inhomogeneous. The inhomogeneity can be strongly reduced by using another mapping of TGA to P. Unfortunately, due to limited space, we can not explain this mapping in detail. Compared with the previous mapping, those parts of PGA that are within the lower face of T remain unchanged, and the parts that were first mapped just outside the lower face of T are now stretched in a direction outward from the center of the lower face of T. The stretching is done over a distance of the order of the edge length of T. See figure 4. We implemented this mapping as a procedure that runs through all nodes of TGA and maps them one at a time on P. So, the time complexity of this mapping is proportional to the number of nodes of TGA, just like the mapping we discussed earlier. Also for this mapping, TGA should not have a node at the top vertex of T.

5 An example application

Until now we have been discussing how to map SGA on the plane. The main goal of our method is however not to map a single graph on the plane, but to

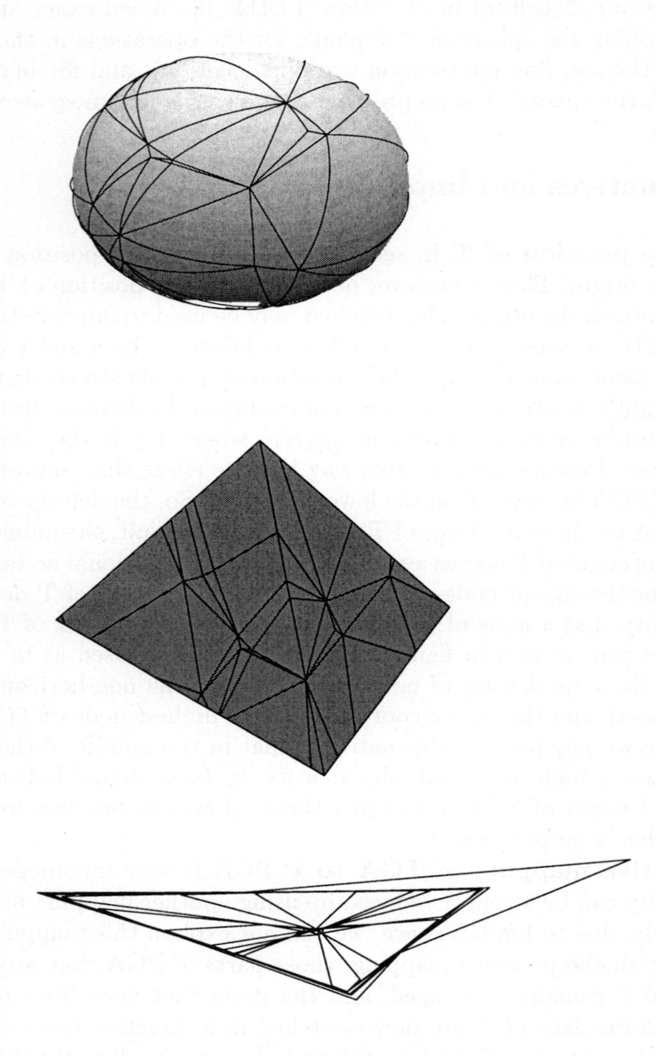

Fig. 3. The same initial graph SGA as in figure 2, mapped on the tetrahedron T in an alternative way. T is positioned so that the origin is slightly above the lower face. It can be seen that, compared with figure 2, the density of nodes and edges on the non-horizontal faces of T has decreased, and that the z coordinate of the highest node of TGA is lower. Also it can be seen that in PGA there is a high node and edge density in the middle.

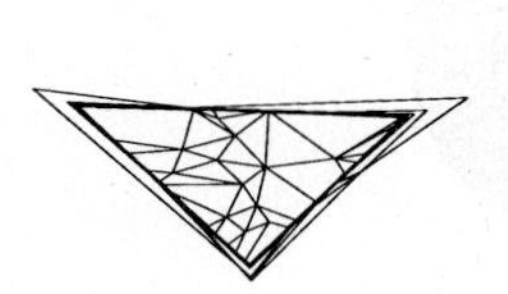
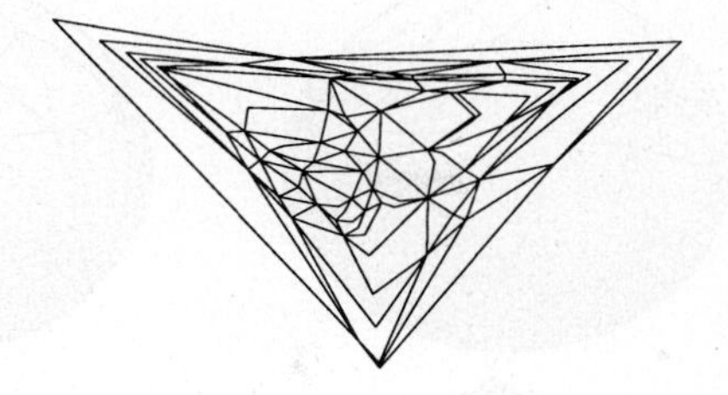

Fig. 4. Left: PGA resulting from the ordinary mapping. Right: PGA resulting from the alternative mapping, giving a more homogeneous graph. In the center of these two figures the graphs are the same.

map two graphs on the plane, to compute for example their overlay, and map the result back on the sphere. As an example we will do that. See figure 5.

We start with two graphs SGA and SGB on the sphere. These graphs are mapped on the tetrahedron, giving the graphs TGA and TGB. These graphs are mapped on the plane, giving PGA and PGB. The overlay of PGA and PGB is calculated with an algorithm implemented in LEDA, giving the graph PGAB. Then PGAB is mapped back on T giving TGAB, and TGAB is mapped back on the sphere giving SGAB. Finally, the marked nodes in SGAB that were created when mapping SGA and SGB to T are deleted when they do not coincide with other nodes. The whole process is shown in figure 4. When working with two graphs, both TGA and TGB have to be considered when it is checked whether a node is mapped on the top vertex of T, and when it is determined which node has the greatest z coordinate.

Literature

[1] D. H. Maling, Coordinate Systems and Map Projections. George Philop and Sons Limited, London, 1973.
[2] H. Bekker, J. B. T. M. Roerdink: An Efficient Algorithm to Calculate the Minkowski Sum of Convex 3D Polyhedra. Proc. of the Int. Conf. on Computational Science, San Francisco, CA, USA,2001
[3] A. V. Tuzikov, J. B. T. M. Roerdink, H. J. A. M. Heijmans: Similarity Measures for Convex Polyhedra Based on Minkowski Addition. Pattern Recognition 33 (2000) 979-995
[4] U. Finke, K. H. Hinrichs: Overlaying simply connected planar subdivisions in linear time. Proc. of the 11th Int. symposium on computational geometry, 1995.
[5] M. V. A. Andrade, J. Stolfi, Exact Algorithms for Circles on the Sphere. International Journal of Computational Geometry and Applications, Vol. 11, No. 3 (2001) 267-290.
[6] K. Melhorn, S. Näher: LEDA A Platform for Combinatorial and Geometric Computing. Cambridge University press,Cambridge. 1999

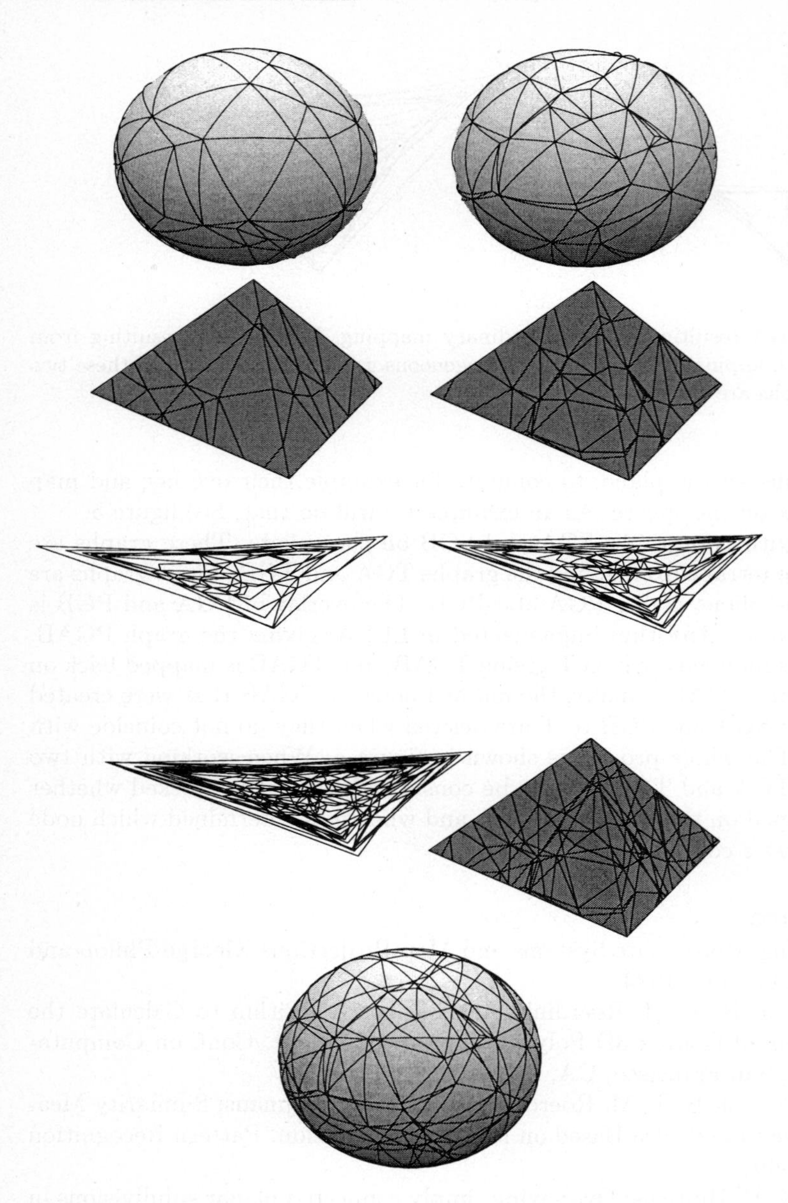

Fig. 5. The process of overlaying two graphs SGA and SGB on the sphere by mapping these graphs to the plane, calculating the overlay in the plane and mapping the result back to the sphere. Top row: SGA and SGB. Second row: TGA and TGB. Third row: PGA and PGB. The alternative mapping of TGA and TGB to P has been used, so PGA and PGB are not very inhomogeneous. Fourth row: the overlay of PGA and PGB, called PGAB, and PGAB mapped back to T, called TGAB. Fifth row: TGAB mapped to the sphere giving the overlay of SGA and SGB. In this figure the center of T coincides with the center of the sphere.

Improved Optimal Weighted Links Algorithms

Ovidiu Daescu

Department of Computer Science,
University of Texas at Dallas,
Richardson, TX 75083, USA
daescu@utdallas.edu

Abstract. In this paper, we present improved algorithms for computing an optimal link among weighted regions in the 2-dimensional (2-D) space. The weighted regions optimal link problem arises in several areas, such as geographic information systems (GIS), radiation therapy, geological exploration, environmental engineering and military applications. Our results are based on a (more general) theorem that characterizes a class of functions for which optimal solutions arise on the boundary of the feasible domain. A direct consequence of this theorem is that an optimal link goes through a vertex of the weighted subdivision. We also consider extensions and present results for the 3-D case. Our results imply significantly faster algorithms for solving the 3-D problem.

1 Introduction

We consider the 2-dimensional (2-D) weighted regions optimal link problem: Given a planar subdivision R, with m weighted regions R_i, $i = 1, 2, \ldots, m$, and a total of n vertices, find a *link* L such that: (1) L intersects two specified regions $R_s, R_t \in R$ and (2) the weighted sum $S(L) = \sum_{L \cap R_i \neq \phi} w_i * d_i(L)$ is minimized, where w_i is either the (positive) weight of R_i or zero and $d_i(L)$ is the length of L within region R_i. Depending on the application, the link L may be (a) unbounded (i.e., a line): the link L "passes through" the regions R_s and R_t; (b) bounded at one end (i.e., a ray): R_s is the source region of L and L passes through R_t and (c) bounded at both ends (i.e., a line segment): R_s is the source region of L and R_t is its destination region. Let R_L be the set of regions $\{R_{i_1}, \ldots, R_{i_k}\}$ intersected by a link L. Then, w_{i_1} and w_{i_k} are set to zero. This last condition ensures that the optimal solution is bounded when a source (and/or a destination) region is not specified (cases (a) and (b)) and allows the link to originate and end arbitrarily within the source and target regions (cases (b) and (c)).

The weighted regions optimal link problem arises in several areas such as GIS, radiation therapy, stereotactic brain surgery, geological exploration, environmental engineering and military applications. For example, in military applications the weight w_i may represent the probability to be seen by the enemy when moving through R_i, from a secured source region R_s to another secured target region R_t. In radiation therapy, it has been pointed out that finding the optimal

P.M.A. Sloot et al. (Eds.): ICCS 2002, LNCS 2331, pp. 65–74, 2002.

choice for the link (cases (a) and (b)) is one of the most difficult problems of medical treatment optimization [3]. In what follows, we will discuss case (c) of the problem that is, we want to compute a minimum weighted distance between R_s and R_t. The other two cases can be easily handled in a similar way.

Previous work. There are a few results in computational geometry that consider weighted region problems, for computing or approximating optimal shortest paths between pairs of points. We refer the reader to [1, 17, 18, 20] for results and further references. The optimal link problem however, has a different structure than the shortest path problem and its complexity is not known yet. In general, a k-link optimal path may have a much larger value than a minimum weighted length path joining the source and the target regions. The bicriteria path problem in a graph (does there exists a path from s to t with length less than L and weight less than W) is known to be NP-hard [14]. For the related problem of computing an Euclidean shortest path between two points within a polygonal domain (1 and ∞ weights only), constrained to have at most k-links, no exact solution is currently known. A key difficulty is that, in general, a minimum link path will not lie on any simple discrete graph. We refer the reader to [19] for more details and references on optimal path problems.

Note that case (c) of the link problem asks to find the minimum weighted distance d_{st} between R_s and R_t. From the definition of the link problem, the weights w_s and w_t of R_s and R_t are zero, and thus d_{st} is also the *optimal weighted bridge* between R_s and R_t. We mention that for the unweighted case, all the weights are one and optimal linear time algorithms are known for finding the minimum distance or bridge [4, 16].

Important steps toward solving cases (a) and (b) of the link problem have been first made in [7], where it has been proved that the 2-D problem can be reduced to a number of (at most $O(n^2)$) global optimization problems (GOPs), each of which asks to minimize a 2-variable function $f(x, y)$ over a convex domain D, where $f(x, y)$ is given as a sum of $O(n)$ fractional terms. Very recently, efficient parallel solutions for the optimal link problem have been reported in [10], where it has been shown that, if at most n processors are available, all GOPs can be generated using $O(n^2 \log n)$ work. Since the GOPs can be efficiently generated [7, 10], most of the time for computing an optimal link will be spent solving the GOPs. Using 2-D global optimization software to solve the GOPs may be too costly, however. In [8], it has been pointed out that, for simpler functions such as *sum of linear fractionals* (SOLF), while commercially available software performs well for the 1-D case, it has difficulties for the 2-D case. The experiments in [7] have shown that the optimal solution goes through one or even two vertices of the subdivision R (i.e., it is on the boundary or at a vertex of the feasible domain) or it is close to such vertex or vertices (which may be due to numerical errors). Following those experiments, it remained an open problem to prove or disprove that a global optimal solution can always be found on the boundary of the feasible domain.

Our results. In this paper, we affirmatively answer the question above. More specifically: (1) We prove that, given a global optimization problem instance as

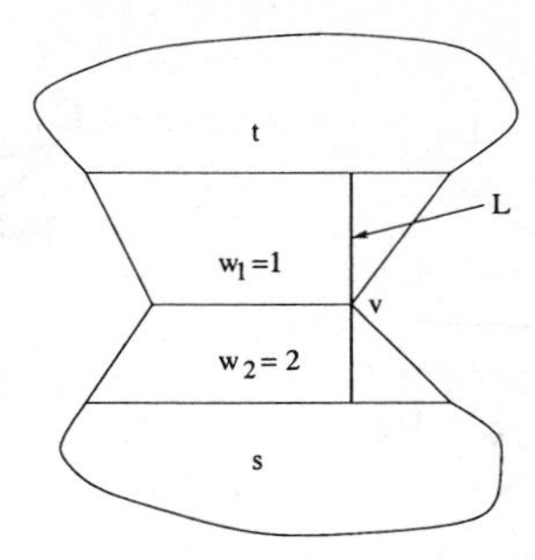

Fig. 1. An optimal link L goes through at most one vertex.

above, the 2-variable objective function attains its global optimum on the boundary of the feasible domain, thus reducing it to a 1-variable function. Accordingly, an optimal link goes through a vertex of the subdivision R. For this, we give a theorem that characterizes a class of functions for which optimal solutions can be found on the boundary of the feasible domain. Since some special instances of the 2-D SOLF problem fall in this class [8], significant speed-up over the general SOLF algorithms are possible for these instances. On the other hand, it is not hard to construct examples for which the optimal solution does not go through two vertices of the subdivision (i.e., it is in the interior of some boundary edge of the feasible domain). A simple example is given in Figure 1. (2) Our solution for case (c) of the link problem results in an efficient algorithm for finding the minimum weighted distance (and the minimum weighted bridge) d_{st} between R_s and R_t. (3) Using the same models of computation as in [10], very simple parallel algorithms for generating the GOPs can be developed. While matching the time/work bounds in [10], these new algorithms are expected to be faster in practice, since they do not require arrangement computation. Due to space limitation, we leave these results to the full paper. (4) We show that our results can be extended to the 3-D case. Consequently, 4-variable objective functions over 4-D domains are replaced with 2-variable functions over 2-D domains, and a relatively simple data structure can be used to generate the corresponding 2-dimensional optimization problems.

Some consequences of our results are: (1) Tremendous speed-up in solving the 2-D optimal link problem, when compared to the algorithms in [7] (and possible those in [8]): solve $O(n^2)$ 1-D global optimization problems as opposed to $O(n^2)$ 2-D global optimization problems in [7]. (2) Commercially available software such as Maple may be used even for large values of n to derive (local) optimal solutions. (3) Our results for the 3-D case (a) make an important step toward efficient solutions for the 3-D version of the optimal link problem and (b) can be used to obtain efficient solutions for the 3-D version in a semi-discrete setting. (4) Efficient approximation schemes can be developed, by discretizing the solution space. The inherent parallelism of such algorithms may be exploited to obtain fast, practical solutions.

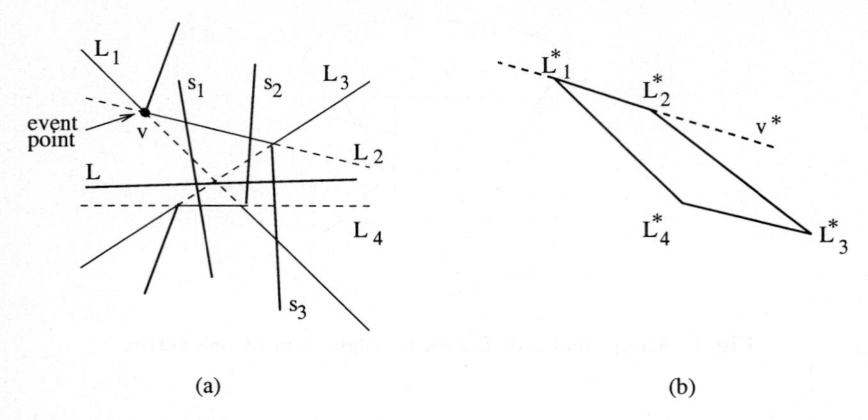

Fig. 2. (a) A link L intersecting s_1, s_2 and s_3 and (b) The corresponding dual cell.

2 Background and notations

In this section we introduce some notations and useful structures. Consider a planar weighted subdivision R, with a total of n vertices. Let L be a link intersecting the source and target regions, R_s and R_t. Let S be the set of line segments in the subdivision R and let $S_{st} = \{s_{i_1}, s_{i_2}, \ldots, s_{i_k}\}$ be the subset of line segments in S that are intersected by L. Consider rotating and translating L. We say that an event e_v occurs if L passes a vertex v of R. Such an event corresponds to some line segments (with an endpoint at v) entering or leaving S_{st} (see Fig. 2 (a)). As long as no event occurs, the formula describing the objective function $S(L)$ does not change and has the expression $S(L) = \sum_{i=i_1}^{i_{k-1}} w_i * d_i$. Here, d_i is the length of L inside region R_i. That is, d_i is the length of a line segment with endpoints on segments s_i and s_{i+1}, where s_i and s_{i+1} are on the boundary of R_i. We refer the reader to [7, 9] for more details.

Let $H = \{l_1, l_2, \ldots, l_n\}$ be a set of n straight lines in the plane. The lines in H partition the plane into a subdivision, called the *arrangement* $A(H)$ of H, that consists of a set of convex regions (cells), each bounded by some line segments on the lines in H. In general, $A(H)$ has $O(n^2)$ faces, edges and vertices and it can be optimally computed in $O(n^2)$ time and $O(n)$ space, by sweeping the plane with a pseudoline [13]. If one is interested only in the part of $A(H)$ inside a convex region, similar results are possible [2].

Assume the link L is a line, given by the equation $y = mx + p$ (case (a) of the optimal link problem). Then, by using the well known point-line duality transform that preserves the above/bellow relations (i.e., a point q above a line l dualizes to a line that is above the dual point of l), all lines intersecting the same subset of segments $S_{st} \in S$ correspond to a cell in the dual arrangement $A(R)$ of R (see Fig. 2 (b)). Here, $A(R)$ is defined by the line set $H_R = \{l_1, l_2, \ldots, l_n\}$, where $l_i \in H_R$ is the dual of vertex $v_i \in R$. The case of a semiline (case (b) of the link problem), and that of a line segment can be reduced to that of a line, by appropriately maintaining the set of line segments intersected by L and dropping those that arise before a segment in R_s or after a segment in R_t. This can be done sequentially in amortized constant time, by using the data structures in [7, 9]. Thus, one can produce the $O(n^2)$ global optimization problems (GOPs) by sweeping the arrangement $A(R)$: for each GOP, the feasible domain corresponds to some cell in $A(R)$, and the objective function is a 2-variable function defined as a sum of $O(n)$ fractional terms.

Simplifications. Rather than generating and sweeping the entire arrangement, observe that it is enough to compute the portion of $A(R)$ inside a (possibly unbounded) region D_{st}, bounded by two monotone chains: D_{st} is the dual of the set of lines intersecting both R_s and R_t. Thus, computing the cells of the arrangement defined by $A(R)$ that correspond to set of lines intersecting both R_s and R_t reduces to computing an arrangement of line segments in D_{st}. This can be done in $O(n \log n + k)$ time [6], resulting in an output sensitive, $O(n \log n + k)$ time algorithm for generating the GOPs, where k is the total complexity of the feasible domains of the GOPs and can be $O(n^2)$ in the worst case. For case (c) of the link problem, observe that all possible links between the source region R_s and the target region R_t can be found in the subdivision $R' = R \cap CH(R_s, R_t)$, where $CH(R_s, R_t)$ is the convex hull of $R_s \cup R_t$. $CH(R_s, R_t)$ and R' can be easily computed in $O(n \log n)$ time and we assume this "clean-up" computation done as a preprocessing step.

Lemma 1. *Given a planar weighted subdivision R with convex regions and a total of n vertices, all the GOPs associated with the optimal link problem can be generated in $O(n \log n + k)$ time, where k is the complexity of the feasible domains of the reported GOPs.*

In the lemma above, the convexity is required in order to apply the data structures in [7, 9] and obtain constant amortized time for generating a GOP. If the regions of R are not convex, R can be easily triangulated to satisfy the convexity condition.

3 GOP Analysis

As stated earlier, the optimal link problem can be reduced to solving a number of (at most $O(n^2)$) global optimization problems, where each GOP is generally defined as $\min_D \sum_{i=1}^{k-1} w_i * d_i$ or equivalently [7]

$$\min_{(x,y)\in D} f(x,y) = \sqrt{1+x^2} \sum_{i=1}^{k} \frac{a_i y + b_i}{x + c_i} \tag{1}$$

where D is a (convex) 2-D domain, a_i, b_i and c_i are constants and the variables x and y are the defining parameters for the link L (i.e., slope and intercept). Without loss of generality, we assume the denominators are positive over D. Observe that, with the L_1 and L_∞ metrics, the square root factor vanishes, resulting in a special case of the 2-D sum of linear fractionals (SOLF) problem [8]. Experimental results in [7] have shown the following fenomenon: the optimal solution for (1) is found on, or close to (but this may be due to numerical errors) the boundary of the feasible domain. It has remained an open problem to prove or disprove that a global optimal solution can always be found on the boundary of the feasible domain (i.e., it goes through a vertex of the subdivision R). In this section we consider a GOP as defined above and prove that indeed the 2-variable objective function attains its global optimum on the boundary of the feasible domain, thus reducing it to a 1-variable function. Accordingly, the optimal link goes through a vertex of the subdivision R.

Let D be a bounded, 2-dimensional domain and let $g(x, y)$ be a 2-variable function over D defined by the equation

$$g(x, y) = h(x) \sum_{i=1}^{k} \frac{q_i(y)}{r_i(x)} \tag{2}$$

and such that $Q(y) = \sum_{i=1}^{k} a_i q_i(y)$ is a monotone function over D, for any positive constants a_i, $i = 1, \ldots, k$. Without loss of generality, we assume that $r_i(x)$ is positive over D, for all $i = 1, 2, \ldots, k$. We then have the following theorem.

Theorem 1. *Given a function $g(x, y)$ defined as above, $\min_{(x,y)\in D} g(x, y)$ can be found on the boundary of the feasible domain D.*

Some of the cells associated with the link problem that appear on the boundary of D_{st} may not be convex. It is important to observe that Theorem 1 can be applied to such domains since it does not require the feasible domain to be convex (or even piecewise linear).

The lemma bellow settles the open problem from [7].

Lemma 2. *The optimal solution for a GOP instance associated with the 2-D optimal link problem can be found on the boundary of the feasible domain.*

Proof. It is enough to show that a GOP instance satisfies the framework of Theorem 1, from which the proof follows. Consider a GOP associated with the optimal link problem. We can set $h(x) = \sqrt{1 + x^2}$ and $q_i(y) = a_i y + b_i$. For a given value of x,

$$Q(y) = \sum_{i=1}^{k} q_i(y) \frac{1}{r_i(x)} = \sum_{i=1}^{k} (a_i y + b_i) \frac{1}{r_i(x)} = y \sum_{i=1}^{k} \frac{a_i}{r_i(x)} + \sum_{i=1}^{k} \frac{b_i}{r_i(x)} \tag{3}$$

is a linear function and thus monotone. Then, the framework of Theorem 1 holds. □

Corollary 1. *The 2-D optimal link problem has an optimal solution (i.e., link) that goes through a vertex of the subdivision R.*

Let e_j be a boundary edge of D and let $v_j(x_j, y_j)$ be the vertex of R that has the line supporting e_j as its dual. Edge e_j corresponds to links going through vertex v_j and, for e_j, the initial 2-dimensional GOP reduces to the following 1-dimensional GOP:

$$\min_{x \in D_x} f_x(x) = \sqrt{1+x^2}\left(d_0 + \sum_{i=1}^{k} \frac{d_i}{x+c_i}\right) \tag{4}$$

where d_0, c_i and d_i, $i = 1, 2, \ldots, k$, are constants, D_x is the interval on the X-axis bounded by the x-coordinates of the endpoints of e_j and k is $O(n)$ in the worst case. Summing over all boundary edges of D, a 2-D GOP is reduced to $O(r)$ 1-D GOPs, where r is the boundary complexity of D. Since we solve a 1-D GOP for each edge of $A(H)$, we solve $O(n^2)$ 1-D GOPs overall, rather than $O(n^2)$ 2-D GOPs in the previous solutions [7, 9].

There are a few things to note here. First, besides reducing the dimension of a GOP, the new objective function is as simple as the initial one (in fact, $O(k)$ fewer arithmetic operations are necessary to evaluate the new objective function). Having a simple objective function is important: for $n = 1000$, assuming an average of 100 iterations per GOP, $O(n^2)$ GOPs each with $O(n)$ terms in the objective function would easily amount for a few teraflops, even without counting the additional computation required by the global optimization software (e.g., computing the derivatives, etc.). Second, commercially available software such as Maple may be used even for large values of n to derive (local) optimal solutions. Further, when using similar optimization software as in [7, 8], we expect order of magnitude speed-up and increased accuracy in finding the optimal solutions. Third, observe that our results are more general and could be applied to other problems. For instance, we do not require convexity of the feasible domain D and the boundary of D may be described by some bounded degree functions given in explicit form, rather than a piecewise linear function.

If the L_1 or L_∞ metric is used instead of the L_2 metric, the objective function of a GOP has been shown in [8] to reduce to $f(x, y) = C + \sum_{i=1}^{k} \frac{a_i y + b_i}{x + c_i}$, where C, a_i, b_i and c_i, $i = 1, 2, \ldots, k$, are constants. In [8], a fast global optimization algorithm has been proposed for solving the 1 and 2-dimensional sum of linear fractionals (SOLF) problem:

$$\max_{(x_1,\ldots,x_d) \in D} f(x_1, \ldots, x_d) = \sum_{i=1}^{k} \frac{n_i(x_1, \ldots, x_d)}{d_i(x_1, \ldots, x_d)} \tag{5}$$

where $n_i(x_1, \ldots, x_d)$ and $d_i(x_1, \ldots, x_d)$ are linear d-dimensional functions. Their experimental results show that the proposed SOLF algorithm is orders of magnitude faster on 1-dimensional SOLFs than on 2-dimensional ones.

Lemma 3. *A 2-dimensional SOLF problem of the form* $\min_{(x,y) \in D} f(x, y) = \sum_{i=1}^{k} \frac{a_i y + b_i}{d_i x + c_i}$ *has an optimal solution on the boundary of the feasible domain.*

Consider a boundary edge e_j of the feasible domain D and let $y = ax + b$ be the equation of the line supporting that edge. Then, the initial 2-dimensional SOLF reduces to the following 1-dimensional SOLF:

$$\min_{x \in D_x} f_x(x) = d_0 + \sum_{i=1}^{k} \frac{d_i}{x + c_i} \tag{6}$$

where d_0, c_i and d_i, $i = 1, 2, \ldots, k$, are constants and D_x is the interval on the X-axis bounded by the x-coordinates of the endpoints of e_j. Note that convexity of D is not required and the boundary of D may be described by some bounded degree functions.

Then, for 2-dimensional SOLF problems as in Lemma 3, one can use the 1-dimensional version of the SOLF algorithm, and thus tremendous speed-up over the general 2-dimensional SOLF algorithm is possible for those instances. We refer the reader to the time charts in [8] for running time comparisons of the 1 and 2-dimensional SOLF algorithms.

4 Extensions

In this section, we discuss extensions of the 2-D results. Specifically, we show that the 3-D case of the problem can be reduced to solving a number of 2-dimensional GOPs, as opposed to 4-dimensional GOPs in [7], and thus expect significant time/space improvements in practice. The reduction to 2-dimensional problems allows to simplify the data structures and algorithms involved in generating the feasible domains and the objective functions for various GOPs. We also mention here that, using the same models of computation as in [10], very simple parallel algorithms for generating the GOPs can be developed for the link problem, matching the time/work bounds in [10].

Similar to the 2-D case, the 3-D version of the weighted regions optimal link problem is a generalization of the 3-D optimal weighted penetration problem [7]. Since a line in 3-D can be identified using four parameters (e.g., the two spherical coordinates (θ, ϕ) of the direction vector of L and the projection (u, v) onto the plane orthogonal to L and containing the origin [11]), it is expected that in this case the optimal solution can be found by solving some 4-variable GOPs. In [7], it has been proved that finding an optimal penetration (line or semiline) can be reduced to solving $O(n^4)$ GOPs, by constructing the 3-D visibility complex [11] of some transformed scene. The cells of the visibility complex can be constructed based on a direct enumeration of the vertices of the complex, followed by a sweep of these vertices. The total storage required by the visibility complex is $O(n^4)$ in the worst case, where n is the total number of edges of the polygons in R, and the complex is represented using a polytope structure: each face has pointers to its boundaries (faces of lower dimension) and to the faces of larger dimension adjacent to it (see [11] for details). After constructing the visibility complex, the GOPs can be obtained by a depth-first search traversal of (the 4-D faces of) the complex [7]. For a given GOP, the feasible domain corresponds to a cell of the visibility complex (a 4-D convex region) and the objective function, which is a 4-variable function defined as a sum of $k = O(n)$ fractional terms, has the form:

$$f(u, v, \theta, \phi) = \sqrt{1 + \theta^2 + \phi^2} \sum_{i=1}^{k} \frac{u + a_i v + b_i}{\phi + c_i \theta + e_i} \tag{7}$$

where a_i, b_i, c_i and e_i are constants and the variables u, v, θ and ϕ are the defining parameters for the link L. As with the 2-D version, the case of a line segment link (case (c) of the link problem) can be reduced to that of a line, by using the data structures in [7, 9]. We then have the following lemma.

Lemma 4. *Let $f(u, v, \theta, \phi)$ be a function defined as above, over some bounded feasible domain D. Then, $\min_{(u,v,\theta,\phi) \in D} f(u, v, \theta, \phi)$ can be found on the 2-D faces (2-faces) of D.*

As observed in [12], while valuable at a theoretical level and for some specific worst cases (e.g., grid-like scenes), the visibility complex is an intricate data structure. The algorithm that constructs the visibility complex requires to compute a 4-D subdivision and is rather complicated, and the traversal of the complex is difficult due to the multiple levels of adjacency.

From the preceding lemma, it results that it is enough to construct (and depth-first traverse) only the 2-D faces of the complex. These faces can be obtained using the *visibility skeleton* data structure and construction algorithm in [12]. The visibility skeleton is easy to build and its construction is based on standard geometric operations, such as line-plane intersections, which are already implemented and available as library functions (see CGAL [5], the computational geometry algorithms library, and the references therein). Once the skeleton is constructed, one can easily obtain the list of blocking objects between two specified regions (e.g., source and target) and thus the fractional terms in the corresponding objective function. Note also that, although the time and space bounds for constructing and storing the skeleton are comparable with those for the visibility complex, the experiments in [12] show its effective use for scenes with more than one thousand vertices.

References

1. L. Aleksandrov, M. Lanthier, A. Maheshwari, and J.-R. Sack, "An ϵ-approximation algorithm for weighted shortest paths on polyhedral surfaces," *Proc. of the 6th Scandinavian Workshop on Algorithm Theory*, pp. 11-22, 1998.
2. T. Asano, L.J. Guibas and T. Tokuyama, "Walking in an arrangement topologically," *Int. Journal of Computational Geometry and Applications*, Vol. 4, pp. 123-151, 1994.
3. A. Brahme, "Optimization of radiation therapy," *Int. Journal of Radiat. Oncol. Biol. Phys.*, Vol. 28, pp. 785-787, 1994.
4. L. Cai, Y. Xu and B. Zhu, "Computing the optimal bridge between two convex polygons," *Information processing letters*, Vol. 69, pp. 127-130, 1999.

5. The Computational Geometry Algorithms Library, web page at http://www.cgal.org.
6. B. Chazelle and H. Edelsbrunner, "An optimal algorithm for intersecting line segments in the plane," *Journal of ACM*, Vol. 39, pp. 1-54, 1992.
7. D.Z. Chen, O. Daescu, X. Hu, X. Wu and J. Xu, "Determining an optimal penetration among weighted regions in two and three dimensions," *Journal of Combinatorial Optimization*, Spec. Issue on Optimization Problems in Medical Applications, Vol. 5, No. 1, 2001, pp. 59-79.
8. D.Z. Chen, O. Daescu, Y. Dai, N. Katoh, X. Wu and J. Xu, "Optimizing the sum of linear fractional functions and applications," *Proceedings of the 11th ACM-SIAM Symposium on Discrete Algorithms*, pp. 707-716, 2000.
9. O. Daescu, "On geometric optimization problems", PhD Thesis, May 2000.
10. O. Daescu, "Parallel Optimal Weighted Links," *Proceedings of ICCS, Intl. Workshop on Comp. Geom. and Appl.*, pp. 649-657, 2001.
11. F. Durand, G. Drettakis, and C. Puech, "The 3D visibility complex, a new approach to the problems of accurate visibility," *Proc. of 7th Eurographic Workshop on Rendering*, pp. 245-257, 1996.
12. F. Durand, G. Drettakis, and C. Puech, "The visibility skeleton: a powerful and efficient multi-purpose global visibility tool," *Proc. ACM SIGGRAPH*, pp. 89-100, 1997.
13. H. Edelsbrunner, and L.J. Guibas, "Topologically sweeping an arrangement," *Journal of Computer and System Sciences*, Vol. 38, pp. 165-194, 1989.
14. M.R. Garey and D.S. Johnson, "Computers and Intractability: A Guide to the Theory of NP-Completeness," W.H. Freeman, New York, NY, 1979.
15. A. Gustafsson, B.K. Lind and A. Brahme, "A generalized pencil beam algorithm for optimization of radiation therapy," *Med. Phys.*, Vol. 21, pp. 343-356, 1994.
16. S.K. Kim and C.S. Shin, "Computing the optimal bridge between two polygons," Research Report HKUST-TCSC-1999-14, Hong-Kong University, 1999.
17. M. Lanthier, A. Maheshwari, and J.-R. Sack, "Approximating weighted shortest paths on polyhedral surfaces," *Proc. of the 13th ACM Symp. on Comp. Geom.*, pp. 274-283, 1997.
18. C. Mata, and J.S.B. Mitchell, "A new algorithm for computing shortest paths in weighted planar subdivisions," *Proc. of the 13th ACM Symp. on Comp. Geom.*, pp. 264-273, 1997.
19. J.S.B. Mitchell, "Geometric Shortest Paths and Network Optimization," *Handbook of Computational Geometry*, Elsevier Science (J.-R. Sack and J. Urrutia, eds.), 2000.
20. J.S.B. Mitchell and C.H. Papadimitriou, "The weighted region problem: Finding shortest paths through a weighted planar subdivision," *Journal of ACM*, Vol. 38, pp. 18-73, 1991.
21. A. Schweikard, J.R. Adler and J.C. Latombe, "Motion planning in stereotactic radiosurgery," *IEEE Transactions on Robotics and Automation*, Vol. 9, pp. 764-774, 1993.

A Linear Time Heuristics for Trapezoidation of GIS Polygons

Gian Paolo Lorenzetto and Amitava Datta

Department of Computer Science & Software Engineering
The University of Western Australia
Perth, W.A. 6009
Australia email : {gian,datta}@cs.uwa.edu.au

Abstract. The decomposition of planar polygons into triangles is a well studied area of Computer Graphics with particular relevance to GIS. Trapezoidation is often performed as a first step to triangulation. Though a linear time algorithm [2] for the decomposition of a simple polygon into triangles exists, it is extremely complicated and in practice $O(n \log n)$ algorithms are used. We present a very simple $O(n)$-time heuristics for the trapezoidation of simple polygons without holes. Such polygons commonly occur in Geographic Information Systems (GIS) databases.

1 Introduction

The decomposition of a simple polygon into triangles is a well studied field of computational geometry and computer graphics [1]. In computer graphics and GIS applications, it is much easier to render a simple shape like a triangle compared to a large polygon. Similarly in computational geometry, a large number of data structures related to polygons are based on triangulations of these polygons. For many triangulation algorithms, a *trapezoidal decomposition* of the polygon is computed as a first step for *triangulation.* A trapezoid is a four-sided polygon with a pair of opposite sides parallel to each other. The main difference between a trapezoidal decomposition and a triangulation is that the vertices of the trapezoids can be additional points on the boundary of the polygon, whereas, in a triangulation it is not allowed to introduce new vertices.

The best known triangulation algorithm is by Chazelle [2] and it runs in $O(n)$ time, where n is the number of vertices of the polygon. However, Chazelle's algorithm is extremely complicated and difficult to implement. Earlier, Garey *et al.* [5] designed an $O(n \log n)$ algorithm for triangulation by dividing the input polygon into monotone pieces. Even now, the algorithm by Garey *et al.* [5] is extensively used for its simplicity and programming ease. Later, Chazelle and Incerpi [3] improved upon this algorithm by introducing the notion of *sinuosity* which is the number of *spiraling* and *anti-spiraling* polygonal chains on the polygon boundary. The algorithm by Chazelle and Incerpi [3] runs in $O(n \log s)$ time, where s is the sinuosity of the polygon. The sinuosity of a polygon is usually a small constant in most practical cases and the algorithm in [3] is an

P.M.A. Sloot et al. (Eds.): ICCS 2002, LNCS 2331, pp. 75–84, 2002.

almost linear-time algorithm. However, the algorithm by Chazelle and Incerpi [3] is based on a divide-and-conquer strategy and still quite complex.

In this paper, we are interested in the trapezoidal decomposition of a simple polygon rather than a triangulation. In computational geometry, a triangulation in performed for building efficient data structures based on this triangulation. Such data structures are used for many applications e.g., for processing shortest path queries or ray shooting. On the other hand, in almost all applications in computer graphics and geographic information systems (GIS), the main requirement is the fast rendering of a polygon. A trapezoid is as simple a shape as a triangle and a trapezoid is easy to render like a triangle in raster graphics devices. In this paper, we present an $O(n)$-time simple heuristics for computing trapezoidal decomposition of a simple polygon without holes. Our heuristics can be used for fast rendering of large GIS data sets which quite often consist of polygons with thousands or hundreds of thousands of vertices.

The fastest known algorithm for trapezoidation of general GIS data sets is by Lorenzetto *et al.* [6]. Their algorithm improve upon a previously published algorithm by Žalik and Clapworthy [7] and can do trapezoidal decomposition of general GIS polygons with holes in $O(n \log n)$ time, where n is the total number of vertices for the polygon and all holes inside it. However, many GIS polygons do not contain holes inside them and it is interesting to investigate whether it is possible to design a simpler and faster algorithm for this problem when the polygons do not contain holes. Our algorithm is simple and terminates in $O(n)$ time for most polygons encountered in GIS applications, where n is the number of vertices in the polygon. We call the algorithm as a linear-time heuristics since we are unable to prove a tight theoretical upper bound for the running time of the algorithm. However for almost all polygons encountered in GIS data sets, the running time is $O(n)$ with the hidden constant in big-Oh very small. However, our algorithm always produces a correct trapezoidation and this is ensured through an in-built correctness test in the algorithm.

The rest of the paper is organized as follows. We discuss some preliminaries and an algorithm for trapezoidal decomposition of monotone polygons in Section 2. We present our $O(n)$ time heuristics in Section 3 and finally we discuss the complexity of the heuristics and our experimental results in Section 4.

2 Terminology

We consider only *simple polygons* without holes inside them and the edges of the polygons intersect only at end points. A *trapezoid* is a four-sided polygon with a pair of opposite sides parallel to each other. In this paper, all trapezoids have two sides parallel to the x-axis. We assume that no two vertices are on the same line parallel to the x or the y axis. In other words, no two vertices have the same y or x coordinates. This assumption simplifies our presentation considerably but we do not impose this restriction in our implementation. The algorithm processes vertices based on their type. We consider three types of vertices depending on the two end points of the two edges incident on the vertex. In the following, we use the notion of the *left* and *right* side of the plane for a directed edge $\overline{uv}$ of the polygon. If the edge is directed from u to v, we assume that an observer is standing at u, facing v and the notion of the left or right side is with respect to this observer.

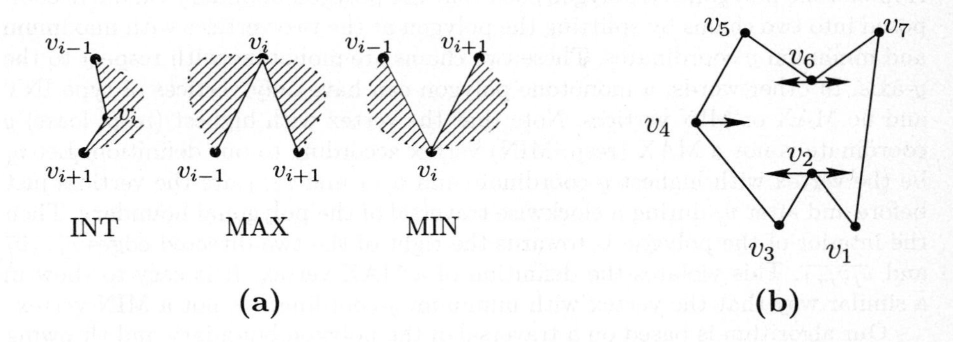

Fig. 1. **(a)** The three vertex types used in the algorithm. **(b)** The assignment of sweep lines. v_4 is of type INT, v_2 is of type MAX and v_6 is of type MIN. The other four vertices do not throw sweep lines.

Each vertex v_i is classified as one of the following types.

- **INT** if the neighboring vertices of v_i have a greater and a lower y-coordinate, respectively, than v_i. The interior of the polygon is to the left sides of the directed edges $\overline{v_{i-1}v_i}$ and $\overline{v_iv_{i+1}}$.
- **MIN** if both neighboring vertices have a greater y-coordinate than v_i. The interior of the polygon is to the right sides of the directed edges $\overline{v_{i-1}v_i}$ and $\overline{v_iv_{i+1}}$.
- **MAX** if both neighboring vertices have a lower y-coordinate than v_i. The interior of the polygon is to the left sides of the directed edges $\overline{v_{i-1}v_i}$ and $\overline{v_iv_{i+1}}$.

Note that all the vertices can be classified according to these three categories by traversing the polygon boundary once. In our algorithm, a vertex of type INT may become a corner vertex of a trapezoid. We indicate this by assigning a sweepline to an INT vertex. This sweepline starts at the INT vertex and goes either towards left or right depending on which side the interior of the polygon is. Similarly, a vertex of type MAX supports the bottom edge of a trapezoid and sweeplines may be thrown in both directions to determine the edge. A MIN vertex supports a top edge of a trapezoid and hence sweeplines are thrown in both directions. The different vertex types are shown in Figure 1(a). The shaded parts in Figure 1(a) denote the interior of the polygon. An example polygon with directions for sweeplines for its vertices is shown in Figure 1(b). Note that the directions of the sweeplines can be stored in the same pass of the polygon boundary while determining the vertex types. This computation takes $O(n)$ time when n is the number of vertices.

2.1 A trapezoidation algorithm for monotone polygons

A monotone polygon is a polygon such that the polygon boundary can be decomposed into two chains by splitting the polygon at the two vertices with maximum and minimum y coordinates. These two chains are monotone with respect to the y-axis. In other words, a monotone polygon can have only vertices of type INT and no MAX or MIN vertices. Note that the vertex with highest (resp. least) y coordinate is not a MAX (resp. MIN) vertex according to our definition. Let v_j be the vertex with highest y-coordinate and v_{j-1} and v_{j+1} are the vertices just before and after v_j during a clockwise traversal of the polygonal boundary. Then the interior of the polygon is towards the right of the two directed edges $\overline{v_{j-1}v_j}$ and $\overline{v_jv_{j+1}}$. This violates the definition of a MAX vertex. It is easy to show in a similar way that the vertex with minimum y-coordinate is not a MIN vertex.

Our algorithm is based on a traversal of the polygon boundary and throwing sweeplines from each vertex. We first determine the direction in which each vertex throws sweepline in a preprocessing step as explained before. Our main task is to determine where each of the sweeplines intersect the polygon boundary. To determine this, we traverse the polygon boundary twice starting from v_{min}, the vertex with the minimum y-coordinate. The traversals are first in the clockwise and then in the counter-clockwise direction. We use a stack for holding the vertices which we encounter. In the following, by v_{top} we mean the vertex at the top of the stack. The stack is empty initially. During the clockwise traversal of the polygon boundary, we push a vertex v_k on the stack if v_k throws a sweepline towards right. When we encounter vertex v_i, we execute the following steps.

- We check whether a horizontal sweepline towards right from v_{top} intersects the edge $\overline{v_iv_{i+1}}$. If there is an intersection, we mark the intersecting point as a vertex of a trapezoid. In this case, v_{top} is removed from the stack.
- If v_i throws sweepline to the right, we push v_i onto the stack.

We execute a similar counter-clockwise traversal for vertices which throw sweeplines towards left. The trapezoids are reported in the following way. We store the intersection of the sweeplines with the edges in a separate data structure. For example, for an edge e_i, all the intersections of sweeplines with e_i are stored as a linked list of intersection points. These intersection points can be kept ordered according to y-coordinate easily. We can report the trapezoids starting from v_{min} and checking these intersection points as we move upwards.

3 An $O(n)$-time algorithm for general polygons

The approach of the previous section does not work for more general polygons containing vertices of types MAX and MIN. The main problem is that the sweepline-edge intersections are not always correct.

In our algorithm, we use three stacks, one each for MAX, MIN, and INT vertex types. The only difference is that when pushing a vertex onto the stack, we check the vertex type and the vertex is pushed onto the correct stack. Similarly, previously an edge was tested against the top of the stack for an intersection, whereas now the edge must be tested against the top of all three stacks.

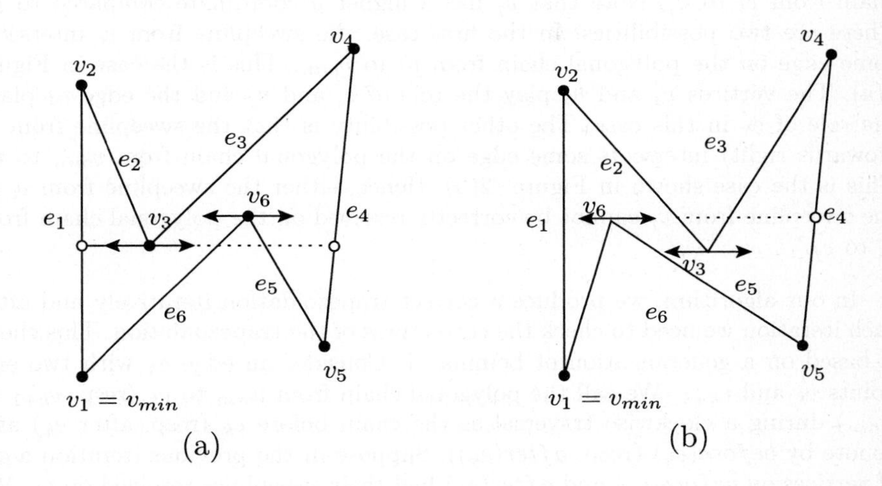

Fig. 2. Illustrations for incorrect trapezoidation. **(a)** After a clockwise traversal vertex v_3 has thrown a sweepline right and caused an incorrect intersection with edge e_4. **(b)** After a counter-clockwise traversal vertex v_6 has thrown a sweepline right and caused an incorrect intersection with e_4. Both traversals start at v_{min}.

Consider Figure 2(a) and in particular vertex v_3. During a counter-clockwise traversal v_3 correctly throws a sweepline to the left intersecting edge e_1. However it does not throw a sweepline to the right as v_3 is seen after edges e_6, e_5, and e_4 — the only edges a sweepline right from v_3 would intersect.

Note that in a clockwise traversal edge e_4 is seen before e_6. For correctness a sweepline right from v_3 must intersect edge e_6, but as e_4 is seen first, the right intersection from v_3 is calculated with e_4, and v_3 is then popped from the stack. In situations like this, we need to check whether the sweepline is correctly resolved, i,e., it intersects the correct edge. A similar situation is shown in Figure 2(b) when a counter-clockwise traversal causes a wrong intersection for vertex v_6. We denote the x and y coordinates of a point p_i by x_i and y_i. Consider a MIN vertex v_i and a MAX vertex v_j. Suppose the sweeplines towards right from v_i and v_j intersect an edge e_k at the points p_i and p_j respectively. Then the following lemma holds.

Lemma 1. *If $y_i < y_j$, then the sweepline towards right* **(i)** *either from v_i or* **(ii)** *from v_j, cannot be resolved correctly on the polygonal chain from v_i to v_j.*

Proof. First note that the sweepline towards right from the MAX vertex v_j has a greater y coordinate compared to the sweepline from the MIN vertex v_i. Also, v_j must have been encountered after v_i during a clockwise traversal starting from v_{min}. Otherwise, the sweepline from v_j would have intersected either the edge with v_i as an endpoint or some other edge before reaching v_i.

Suppose the sweepline from v_i is resolved on an edge e_k on the polygonal chain from v_i to v_j. Note that v_j has a higher y coordinate compared to v_i. There are two possibilities. In the first case, the sweepline from v_i intersects some edge on the polygonal chain from v_j to v_{min}. This is the case in Figure 2(a). The vertices v_3 and v_6 play the role of v_i and v_j and the edge e_4 plays the role of e_k in this case. The other possibility is that the sweepline from v_j (towards right) intersects some edge on the polygonal chain from v_{min} to v_i. This is the case shown in Figure 2(b). Hence, either the sweepline from v_i or the sweepline from v_j cannot be correctly resolved on the polygonal chain from v_i to v_j.

In our algorithm, we produce a correct trapezoidation iteratively and after each iteration we need to check the correctness of the trapezoidation. This check is based on a generalization of Lemma 1. Consider an edge e_k with two end points v_k and v_{k+1}. We call the polygonal chain from v_{min} to v_k (resp. v_{k+1} to v_{min}) during a clockwise traversal as the chain before e_k (resp. after e_k) and denote by $before(e_k)$ (resp. $after(e_k)$). Suppose in the previous iteration a set of vertices on $before(e_k)$ and $after(e_k)$ had their sweeplines resolved on e_k. We denote the minimum y-coordinate among all the vertices on $before(e_k)$ that are resolved on e_k by b_k and similarly the maximum y-coordinate among all the vertices on $after(e_k)$ that are resolved on e_k by a_k. Then the following lemma allows us to check the correctness of our trapezoidation.

Lemma 2. *A trapezoidation is correct if and only if for every edge* $e_k, 1 \leq k \leq n$, $b_k > a_k$.

Proof. First assume that there is at least one edge e_k such that $b_k < a_k$. From Lemma 1 it is clear that the vertex on $before(e_k)$ that is responsible for b_k has not resolved its sweepline correctly. Hence, the presence of at least one such edge e_k makes the trapezoidation incorrect.

To prove the other direction, assume that for every edge e_k, $b_k > a_k$. In that case, all the sweeplines are resolved correctly and hence the trapezoidation is correct.

In Figure 2, for clockwise traversal, the polygonal chain from v_1 to v_4 is $before(e_4)$ and the polygonal chain from v_5 to v_1 is $after(e_4)$. b_4 and a_4 are the y coordinates of the intersections on the edge e_4 by vertices on $before(e_4)$ and $after(e_4)$. In this case, b_4 is determined by v_3 and a_4 is determined by v_6. The trapezoidation is incorrect since $y_3 < y_6$.

It is clear that if we identify condition of Lemma 1, we cannot resolve one of the sweeplines, either from v_i or from v_j, on the polygonal chain from v_i to v_j. In the first case, we should try to resolve the sweepline from v_i only after crossing v_j during the clockwise traversal of the polygon. In the second case, we should try to resolve the sweepline from v_j only after crossing v_i during a counter-clockwise traversal of the polygon. A simple heuristics to overcome the problem described above is to extend the stack approach to *stack-of-stacks of vertices*, in effect restricting which edges see which sweeplines. In the above example from Figure 2 the problem is caused by the overlapping MAX and MIN vertices. This causes the order in which edges are seen to be incorrect.

3.1 Preprocessing

The preprocessing stage classifies each vertex based on type. Once a vertex has been classified it is assigned sweeplines. We assign an initial intersection point for each vertex in the preprocessing stage itself. In general some of these initial intersections are incorrect and they are refined in the next stage of the algorithm. Intersections are calculated for each vertex that throws a sweepline. The first vertex tested is v_{min}. From v_{min} each vertex is tested in order, in both clockwise and counter-clockwise directions. The polygon boundary is traversed twice for the following reason. For a sweepline-edge intersection to occur the vertex must be seen before the edge. This is necessary because the vertex must be on the stack when the edge is encountered. In some cases this will only occur during a clockwise traversal of the polygon boundary, and in others only during a counter-clockwise traversal. During the pre-processing stage only MAX and MIN vertices are considered, all other vertices are ignored. The following deals with MAX vertices, but the case for MIN is identical. As a sweeping MAX vertex is encountered the edge containing the intersection is updated to note that it has been intersected by a MAX vertex. Once both the clockwise and counter-clockwise passes are complete all edges intersected by MAX vertices are marked.

At the end of the preprocessing, each edge potentially stores two pieces of information. An array *min* of y-coordinates for intersections of all the sweeplines thrown by the vertices of type MIN and an array *max* of y-coordinates for intersections of all the sweeplines thrown by vertices of type MAX. These intersections occur according to increasing y coordinates for vertices of type MIN and according to decreasing y-coordinates for vertices of type MAX. Also, overall there are at most $2n$ intersections since each sweepline intersects only one edge. Hence, these two arrays are maintained in linear time during preprocessing.

By y_{min} we denote the minimum y-coordinate in the array *min* and by y_{max} the maximum y-coordinate in the array *max*. This information is used for checking the condition of Lemma 1. Note that not all intersections are correct at this stage. Consider again Figure 2. After pre-processing vertex v_3 still throws a sweepline to the right incorrectly, intersecting edge e_4.

3.2 Main iteration

We now discuss the main iterations of our algorithm. Each iteration has two parts, first a trapezoidation step and then a testing stage. We compute a trapezoidation first and then test whether the trapezoidation is correct by checking the condition of Lemma 2. We proceed to the next iteration if the trapezoidation is incorrect otherwise the algorithm terminates after reporting the trapezoids.

The trapezoidation in each iteration proceeds similar to the second part of the preprocessing stage, one clockwise and one counter-clockwise traversal of the polygon boundary. As in Section 2.1, we consider vertices of type INT which throw sweepline towards right during the clockwise traversal and vertices of type INT which throw sweeplines towards left during the counter-clockwise traversal. However, we consider vertices of types MAX and MIN during both the traversals. A vertex of type MIN or MAX is pushed onto the corresponding stack when encountered and popped from the stack when both the sweeplines from such a vertex is resolved. The MAX/MIN intersection information stored in the pre-processing stage is used to control which vertices are tested against each edge. A vertex can cause a sweepline-edge intersection left, or right, during both the clockwise and counter-clockwise traversals. As an edge e_i is encountered, we take the following actions with respect to v_{top}, the top vertices in each of the three stacks for MAX, MIN and INT vertices. We denote the y-coordinate of v_{top} by y_{top}. The following steps are taken when v_{top} is the top vertex in the MIN stack. The actions are similar when v_{top} is the top vertex in the other two stacks.

- We check whether a sweepline from v_{top} intersects e_i. If there is an intersection, we check whether the condition in Lemma 1 is violated. For example, if v_{top} is a MIN vertex, this condition will be violated if $y_{top} < y_{max}$. Suppose, a vertex v_j is responsible for the y_{max} value. We now initialize three new stacks for the three types of vertices. This action can be viewed as maintaining three data structures which are *stack-of-stacks*, one each for vertex types MIN, MAX and INT. We push three new empty stacks on top of each of the three stack-of-stacks. Note that the sweepline from the vertex v_{top} cannot be resolved until we encounter the vertex v_j, and hence we need to maintain these new stacks until we reach v_j.
- We delete y_{top} from the array *min* and if necessary, update y_{min}.
- If both (in case of MIN or MAX) or one (in case of INT) sweepline(s) of v_{top} is resolved, we pop v_{top} out of the stack-of-stacks. In other words, v_{top} is popped from the top-most stack of one of the three stack-of-stacks depending on its type.
- If the vertex v_i of edge e_i throws sweeplines, we push v_i into the correct stack-of-stacks depending on its type.

Once both the clockwise and counter-clockwise traversals are over, we test the condition of Lemma 2 for each edge of the polygon. This is done by comparing y_{min} and y_{max} for each of the edges. If at least one edge violates the condition, we proceed to the next iteration. Otherwise we report the trapezoids and the algorithm terminates.

Note that the motivation for using a stack-of-stacks data structure is the property discussed in Lemma 1. The current top stack in a stack-of-stacks takes care of the polygonal chain between a MIN vertex and a corresponding MAX vertex when these two vertices are interleaved, i.e., the y coordinate of the MAX vertex is higher than the y-coordinate of the MIN vertex as in Figure 2. An old stack is removed from the stack-of stacks when we reach a matching MAX vertex. For example, in Figure 2(a) when we are doing a clockwise traversal, we initialize a new stack when we are at edge e_4 and find the overlapping of MIN/MAX pair of v_3 and v_6. We remove this stack when we reach v_6 and the sweepline from v_3 is correctly resolved after that.

When the algorithm terminates, we report the trapezoids by traversing the arrays *min* and *max* for each of the edges in the polygon. These are the correct intersections of sweeplines and vertices and hence the supports of the trapezoids. We omit the details of the reporting of trapezoids in this version. and refer to Figure 3 for an example.

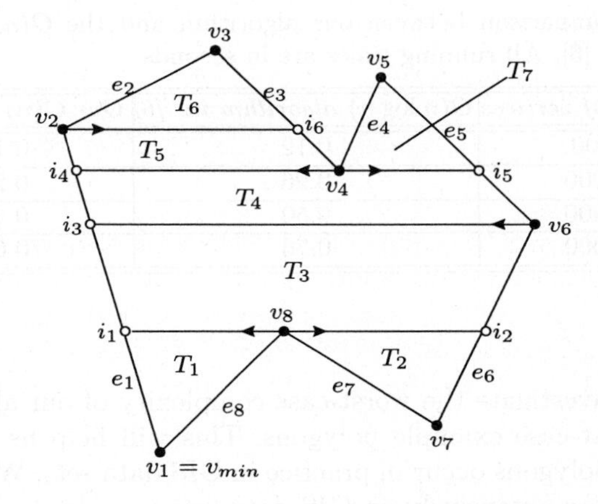

Fig. 3. An example polygon decomposed into trapezoids after the termination of our heuristics.

4 Experimental results and conclusion

The preprocessing stage as well as each iteration of the heuristics takes $O(n)$ time. The reporting of the trapezoids also takes $O(n)$ time. However, we are unable to prove a theoretical upper bound for the heuristics at this point. We strongly suspect that there is a close relationship between the number of iterations required for producing a correct trapezoidation and the sinuosity of the polygon as defined by Chazelle and Incerpi [3]. It is quite difficult to generate polygons with high sinuosity and we have tested our algorithm for polygons up to sinuosity 5 and the number of iterations required are always equal to the sinuosity of the polygon up to this point. Hence, we conjecture that the theoretical upper bound for the running time of our heuristics is $O(ns)$, where s is the sinuosity of the polygon.

However, most GIS polygons from the GIS Data Depot [4] are polygonal data for land subdivisions, property boundaries and roadways. It is not common to have a high sinuosity for such polygonal data. For all GIS polygons that we have used, the algorithm terminates in four or five iterations. We used a Celeron 400 PC with 328MB of RAM running Linux Redhat 6 as our experimental platform. We implemented the algorithm by Lorenzetto *et al.* [6] and the present algorithm, both in `C++` using the `g++` compiler. Some of the results are shown in Table 1. The present algorithm becomes more efficient for polygons with a large number of vertices and we expect that it will be useful in practice when it is necessary to do trapezoidal decomposition of a large number of polygons very fast.

Table 1. The comparison between our algorithm and the $O(n \log n)$ algorithm by Lorenzetto *et al.* [6]. All running times are in seconds.

Number of vertices	$O(n \log n)$ *algorithm in [6]*	*Our* $O(n)$ *heuristics*
5000	0.12	0.11
10000	0.26	0.21
20000	0.50	0.42
30000	0.76	0.64

We plan to investigate the worst-case complexity of our algorithm and construct some worst-case example polygons. This will help us to check whether such worst-case polygons occur in practice in GIS data sets. We also plan to test our algorithm more extensively on GIS data sets in order to optimize our code further.

References

1. M. Bern, *Handbook of Discrete and Computational Geometry*, **Chapter 22**, CRC Press, pp. 413-428, 1997.
2. B. Chazelle, "Triangulating a simple polygon in linear time", *Discrete and Computational Geometry*, **6**, pp. 485-524, 1991.
3. B. Chazelle and J. Incerpi, "Triangulation and shape complexity", *ACM Trans. Graphics*, **3**, pp. 153-174, 1984.
4. `http://www.gisdatadepot.com`
5. M. Garey, D. Johnson, F. Preparata and R. Tarjan, "Triangulating a simple polygon", *Inform. Proc. Lett.*, **7**, pp. 175-180, 1978.
6. G. Lorenzetto, A. Datta and R. Thomas, "A fast trapezoidation technique for planar polygons", *Computers & Graphics*, March 2002, to appear.
7. B. Žalik and G. J. Clapworthy, "A universal trapezoidation algorithm for planar polygons", *Computers & Graphics*, **23**, pp. 353-363, 1999.

The Morphology of Building Structures

Pieter Huybers [1]
[1] Assoc. Professor, Delft Univ. of Technology, Fac. CiTG,
2628CN, Stevinweg 1,Delft, The Netherlands.
p.huybers@hetnet.nl

Abstract.
The structural efficiency and the architectural appearance of building forms is becoming an increasingly important field of engineering, particularly because of the present wide-spread availability of computer facilities. The realisation of complex shapes comes into reach, that would not have been possible with the traditional means. In this contribution a technique is described, where structural forms are visualised starting from a geometry based on that of regular or semi-regular polyhedra, as they form the basis of most of the building structures that are utilised nowadays. The architectural use of these forms and their influence on our man-made environment is of general importance. They can either define the overall shape of the building structure or its internal configuration. In the first case the building has a centrally symmetric appearance, consisting of a faceted envelope as in geodesic sphere subdivisions, which may be adapted or deformed to match the required space or ground floor plan. Polyhedral shapes are also often combined so that they form conglomerates, such as in space frame systems.

1. Introduction

Polyhedra can be generated by the rotation of regular polygons around the centre of the coordinate system. Related figures, that are found by derivation from these polyhedra, can be formed in a similar way by rotating planar figures that differ from ordinary polygons. An inter-active program for personal computers is being developed - which at present is meant mainly for study purposes - with the help of which geometric data as well as visual presentations can be obtained of the regular and semi-regular polyhedra not only, but also of their derivatives. This method thus allows the rotation also of 3-dimensonal figures, that may be of arbitrary shape. The rotation procedure can eventually be used repeatedly, so that quite complex configurations can be described starting from one general concept. The outcome can be gained in graphical and in numerical form. The latter data can be used as the input for further elaboration, such as in external, currently available drawing or computation programs.

2. Definition of Polyhedra

A common definition of a polyhedron is [1]:
1) It consists of plane, regular polygons with 3, 4, 5, 6, 8 or 10 edges.
2) All vertices of a polyhedron lie on one circumscribed sphere.

P.M.A. Sloot et al. (Eds.): ICCS 2002, LNCS 2331, pp. 85–94, 2002.

3) All these vertices are identical. In a particular polyhedron the polygons are grouped around each vertex in the same number, kind and order of sequence.
4) The polygons meet in pairs at a common edge.
5) The dihedral angle at an edge is convex. In other words: the sum of the polygon face angles that meet at a vertex is always smaller than 360°.

3. Regular and Semi-Regular Polyhedra

Under these conditions a group of 5 regular and 13 semi-regular, principally different polyhedra is found. There are actually 15 semi-regular solids, as two of them exist in right- and left-handed versions. All uniform polyhedra consist of one or more - maximally 3 - sets of regular polygons.

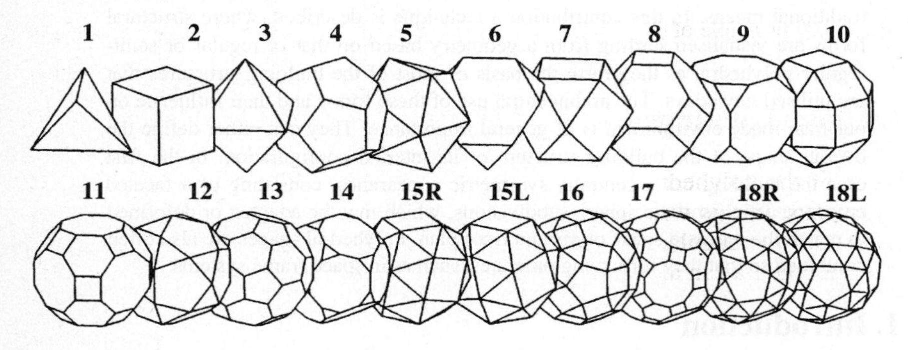

Fig. 1. Review of the regular (1 to 5) and semi-regular (6 to 18R) polyhedra

The five regular polyhedra have a direct mutual relationship: they are dual to each other in pairs. Fig. 3 shows this relation. All other polyhedra of Fig.2 also have dual or reciprocal partners [9]. This principle of duality is explained in Figs. 4 and 5.

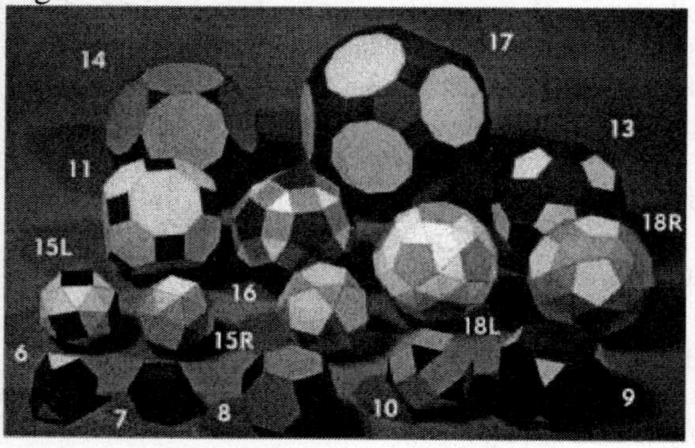

Fig. 2. Models of the 15 semi-regular polyhedra

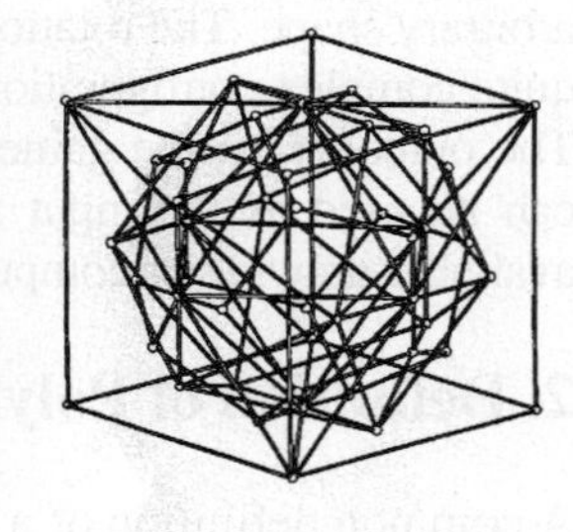

Fig. 3. The relations of the 5 regular solids

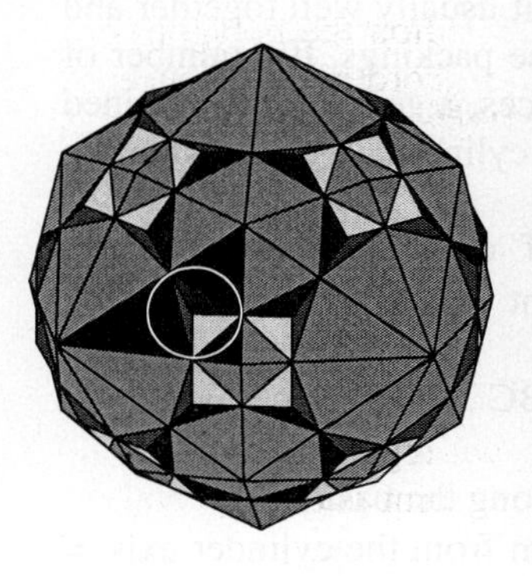

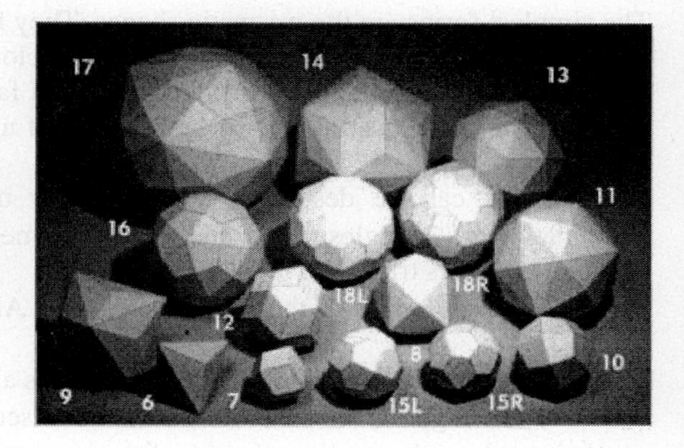

Fig. 4. The principle of duality **Fig. 5.** Models of the dual semi-regular polyhedra

4. Close-Packings

Some of the polyhedra lend themselves to being put together in tight packed formations. In this way quite complex forms can be realised. It is obvious hat cubes and rectangular prisms can be stacked most densely, but many of the other polyhedra can also be packed in certain combinations.

5. Prisms and Antiprisms

Other solids that also respond the previous definition of a polyhedron are the prisms and the antiprisms. Prisms have two parallel polygons like the lid and the bottom of a box and square side-faces; antiprisms are like the prisms but have one of the polygons slightly rotated so as to turn the side-faces into triangles.

Fig. 6 and 7. Models of prisms and antiprisms

The simplest forms are the prismatic shapes. They fit usually well together and they allow the formation of many variations of close-packings. If a number of antiprisms is put together against their polygonal faces, a geometry is obtained of which the outer mantle has the appearance of a cylindrical, concertina-like folded plane. [7]

These forms can be described with the help of only few parameters, a combination of 3 angles: α, β and γ. The element in Fig. 9A represents 2 adjacent isosceles triangles.

α = half the top angle of the isosceles triangle ABC with height a and base length $2b$.

γ = half the dihedral angle between the 2 triangles along the basis.

φ_n = half the angle under which this basis $2b$ is seen from the cylinder axis. = π/n, being n here the number of sides of the respective polygon.

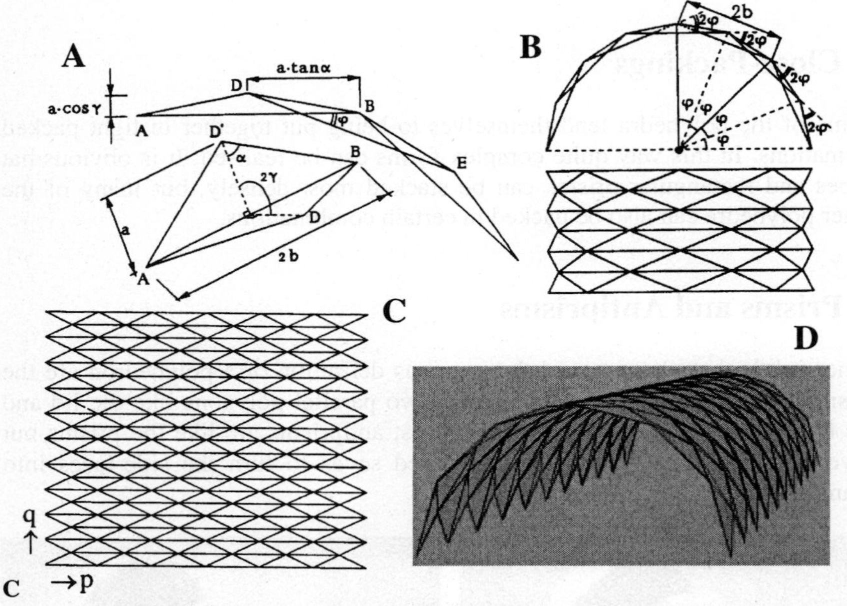

Fig. 8. Variables that define the shape of antiprismatic forms

The relation of these angles α, γ and φ_n [4]:

$$\tan \alpha = \cos \gamma \ \text{cotan} (\varphi_n/2) \qquad \{1\}$$

These three parameters define together with the base length $2b$ (or scale factor) the shape and the dimensions of a section in such a structure. This provides an interesting tool to describe any antiprismatic configuration. Two additional data must be given: the number of elements in transverse direction (p) and that in length direction (q).

6. Augmentation

Upon the regular faces of the polyhedra other figures can be placed that have the same basis as the respective polygon. In this way polyhedra can be 'pyramidized'. This means that shallow pyramids are put on top of the polyhedral faces, having their apexes on the circumscribed sphere of the whole figure. This can be considered as the first frequency subdivision of spheres. In 1582 Simon Stevin introduced the notion of 'augmentation' by adding pyramids, consisting of triangles and having a triangle, a square or a pentagon for its base, to the 5 regular polyhedra [2]. More recently, in 1990, D.G. Emmerich extended this idea to the semi-regular polyhedra (Fig.9). He suggested to use pyramids of 3-, 4-, 5-, 6-, 8- or 10-sided base, composed of regular polygons, and he found that 102 different combinations can be made. He called these: composite polyhedra [3].

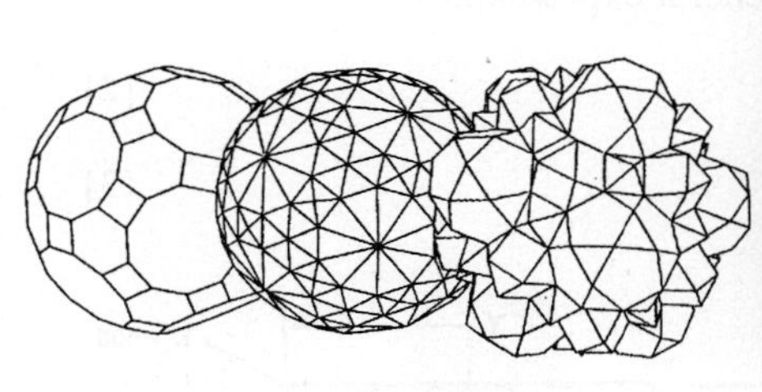

Fig. 9 A composite polyhedron (see also Fig. 21)

Fig. 10. Models of additions in the form of square or rhombic elements.

7. Sphere Subdivisions

For the further subdivision of spherical surfaces generally the Icosahedron – and in some cases the Tetrahedron or the Octahedron - are used as the starting point, because they consist of equilateral triangles that can be covered with a suitable pattern that is subsequently projected upon a sphere. This leads to economical kinds of subdivision up to high frequencies and with small numbers of different member lengths [8].

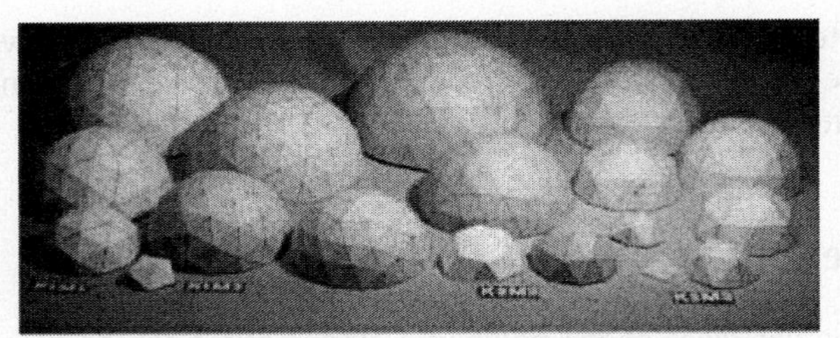

Fig. 11. Models of various dome subdivision methods

All other regular and semi-regular solids, and even their reciprocals as well as prisms and antiprisms can be used similarly [8]. The polygonal faces are first subdivided and then made spherical.

8. Sphere Deformation

The spherical co-ordinates can be written in a general form, so that the shape of the sphere may be modified by changing some of the variables. This leads to interesting new shapes, that all have a similar origin but that are governed by different parameters. According to H. Kenner [6] the equation of the sphere can be transformed into a set of two general expressions:

$$R_1 = E_1 / (E_1^{n1} \sin^{n1}\varphi + \cos^{n1}\varphi)^{1/n1} \qquad \{2\}$$

$$R_2 = R_1 E_2 / (E_2^{n2} \sin^{n2}\theta + R_1^{n2} \cos^{n2}\theta)^{1/n2} \qquad \{3\}$$

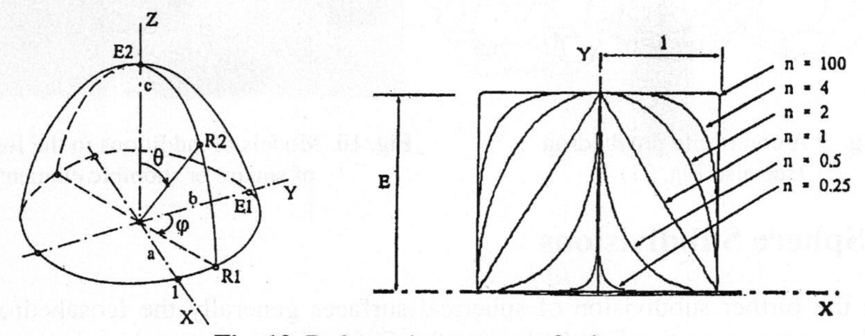

Fig. 12. Deformation process of spheres

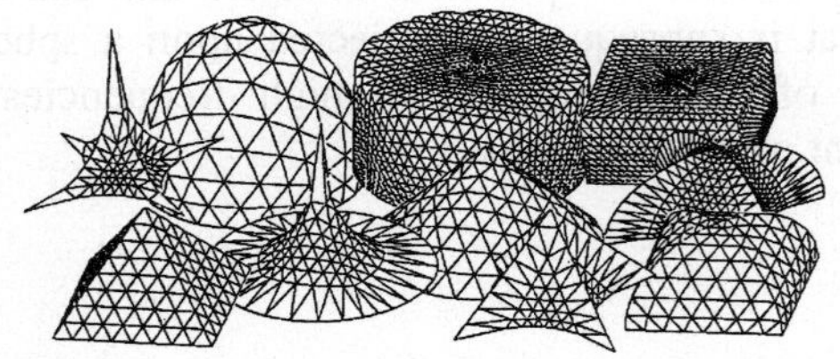

Fig. 13. Form variation of domes by the use of different variables

The variables n_1 and n_2 are the exponents of the horizontal and vertical ellipse and E_1 and E_2 are the ratios of their axes. The shape of the sphere can be altered in many ways, leading to a number of transformations. The curvature is a pure ellipse if $n = 2$, but if n is raised a form is found, which approximates the circumscribed rectangle. If n is decreased, the curvature flattens until $n = 1$ and the ellipse then has the form of a pure rhombus with straight sides, connecting the maxima on the co-ordinate axes. For $n < 1$ the curvature becomes concave and obtains a shape, reminiscing a hyperbola. For $n = 0$ the figure coincides completely with the X-, and Y-axes. By changing the value of both the horizontal and the vertical exponent the visual appearance of a hemispherical shape can be altered considerably [6, 8].

9. Stellation

Most polyhedra can be stellated. This means that their planes can be extended in space until they intersect again. This can sometimes be repeated one or more times. The Tetrahedron and the Cube are the only polyhedra that have no stellated forms, but all others do. The Octahedron has one stellated version, the Dodecahedron has three and the Icosahedron has as many as 59 stellations.

Fig. 14. Some stellated versions of the regular polyhedra

10. Polyhedra in Building

The role that polyhedra can play in the form-giving of buildings is very important, although this is not always fully acknowledged. Some possible or actual applications are referred to here briefly.

10.1. Cubic and Prismatic Shapes

Most of our present-day architectural forms are prismatic with the cube as the most generally adopted representant. Prisms are used in a vertical or in a horizontal position, in pure form or in distorted versions. This family of figures is therefore of utmost importance for building.

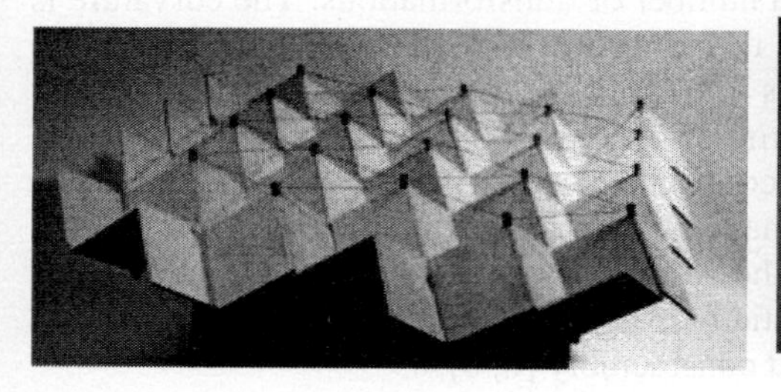

Fig. 15. Model of a space frame made of square plates in cubic arrangement

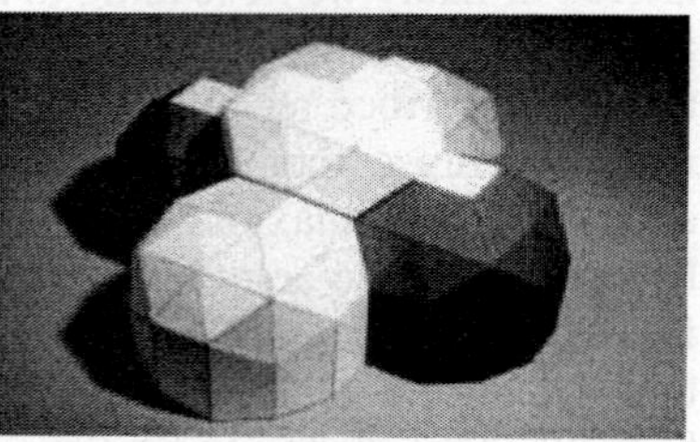

Fig. 16. Model of house, based on Rhombic Dodecahedron (honeycomb cell)

10.2. Solitary Polyhedra

Architecture can become more versatile and interesting with macro-forms, derived from one of the more complex polyhedra or from their reciprocal (dual) forms, although this has not often been done. Packings of augmented polyhedra form sometimes interesting alternatives for the traditional building shapes.

10.3. Combinations

Packings are also suitable as the basic configuration for space frames, because of their great uniformity. If these frames are based on tetrahedra or octahedra, all members are identical and meet at specific angles. Many of such structures have been built in the recent past and this has become a very important field of application. The members usually meet at joints having a spherical or a polyhedral shape.

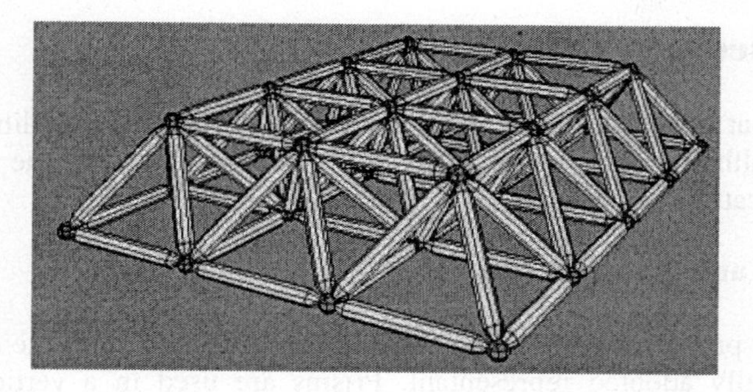

Fig. 17. Computer generated space frame made of identical struts

Fig. 18. Office building in Mali, based on three augmented rhombicuboctahedra and two tetrahedra, with a frame construction of palm wood

10.4. Domes

R.B. Fuller re-discovered the geodesic dome principle. This has proven to be of great importance for the developments in this field. Many domes have been built during the last decades, up to very large spans. A new group of materials with promising potential has been called after him, which has molecules that basically consist of 60 atoms, placed at the corners of a Truncated Icosahedron.

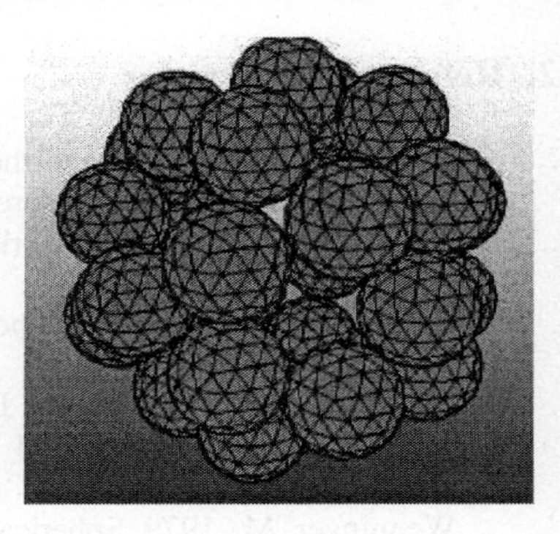

Fig. 19 and 20. Model and computer sketch of small 'Fullerene'

11. Stereographic Slide Presentation

The author will show during the conference a few applications with the help of a 3-D colour slide presentation. Two pictures with different orientation of the light wave directions are projected simultaneously on one screen. The screen must have a metal surface which maintains these two ways of polarisation, so that the two pictures can be observed with Polaroid spectacles that disentangle them again into a left and a right image. These pictures are made either analogously: with a normal photo camera on dia-positive film and with the two pictures token at a certain parallax or they are made by computer and subsequently printed and photographed or written directly onto positive film. This technique allows coloured pictures to be shown in a really three-dimensional way and gives thus a true impression of the spatial properties of the object [10]. This slide show will be done in cooperation with members of the Region West of the Dutch Association of Stereo Photography. Their assistance is gratefully acknowledged here.

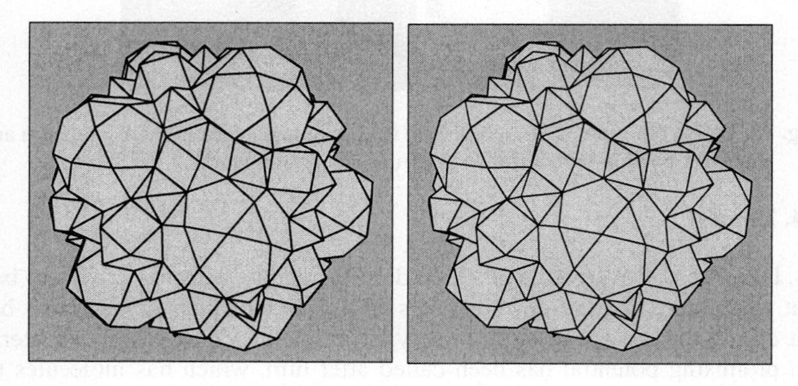

Fig. 21. Pair of stereoscopic pictures

12. References

1. Huybers, P., Polyhedra and their Reciprocals, Proc. IASS Conference on the Conceptual Design of Structures, 7-11 October, 1996, Stuttgart, 254-261.
2. Struik, D.J., The principle works of Simon Stevin, Vol. II, Swets & Seitlinger, Amsterdam, 1958.
3. Emmerich, D.G., Composite polyhedra. Int. Journal of Space Structures, 5, 1990, p. 281-296.
4. Huybers, P. and G. van der Ende: Prisms and Antiprisms, Proc. Int. IASS Conf. on Spatial, Lattice and Tension Structures, Atlanta, 24-28 april 1994, p.142-151.
5. Wenninger, M., 1979, Spherical Models, Cambridge University Press, USA
6. Kenner, H., Geodesic math and how to use it, University of California Press, London, 1976.
7. Huybers, P., Prismoidal Structures, The Mouchel Centenary Conf. on Innovation in Civil & Structural Engineering, Cambridge, p. 79-C88.
8. Huybers, P., and G. van der Ende, Polyhedral Sphere Subdivisions, Int. IASS Conf. on Spatial Structures, Milan 5-9 June, 1995, p. 189-198.
9. Huybers, P., The Polyhedral World, In: 'Beyond the Cube: The Architecture of Space frames and Polyhedra', J.F. Gabriel Ed., John Wiley and Sons, Inc., New York, 1997, p. 243-279.
10. Ferwerda, J.G., The World of 3-D, A practical Guide to Stereo Photography, 3-D Book Productions, Borger, 1987.

Voronoi and Radical Tessellations of Packings of Spheres

A. Gervois[1], L. Oger[2], P. Richard[2], and J.P. Troadec[2]

[1] Service de Physique Théorique, Direction des Sciences de la Matière,
CEA/Saclay, F91191 Gif-sur Yvette Cedex
gervois@spht.saclay.cea.fr

[2] Groupe Matière Condensée et Matériaux,
UMR CNRS 6626, Université de Rennes 1,
Campus de Beaulieu, Bâtiment 11A, F35042 Rennes Cedex
{Luc.Oger, Patrick.Richard, Jean-Paul.Troadec}@univ-rennes1.fr

Abstract. The Voronoi tessellation is used to study the geometrical arrangement of disordered packings of equal spheres. The statistics of the characteristics of the cells are compared to those of 3d natural foams. In the case of binary mixtures or polydisperse assemblies of spheres, the Voronoi tessellation is replaced by the radical tessellation. Important differences exist.

1 Introduction

An important element in the understanding of unconsolidated granular media is the description of the local arrangement of the grains and the possible correlations between them. The simplest model consists in monosize assemblies of spherical grains; the neighbourhood and steric environment of a grain can be described through the statistical properties of its Voronoi cell.

The Voronoi cells generate a partition of space (foam in three dimensions, mosaics in two dimensions) which may be considered without any reference to the underlying set of grains. Foams have been extensively studied in other contexts [1]. Some powerful experimental laws are known, at least for two-dimensional systems. We used the Voronoi construction to study the local arrangement of 2d disk assemblies [2] : results were qualitatively not too different from those observed on random classical mosaics but quantitative behaviour was much altered by steric constraints and the assembling procedure. We have performed the same study for equal spheres and verified that for our foams, the "universal" laws are again valid, again with quantitative steric constraints.

The next step consists in considering assemblies of unequal spheres. Several generalizations of the Voronoi construction were proposed, and their use depends on the problem under consideration. Though somewhat artificial, the radical tessellation [3, 4] keeps main topological features of the ordinary Voronoi tessellation and is well adapted to grain hindrance or segregation problems.

P.M.A. Sloot et al. (Eds.): ICCS 2002, LNCS 2331, pp. 95–104, 2002.

For a physicist, the main interest consists in analysing the randomness of the packing or the transition between a fluid-like and a solid-like packing. So, in the present paper, most attention is paid to **disordered** packings. It is intended as a small review on our contribution to the geometrical analysis of packings of spheres with emphasis on two points : the interpretation of the tessellation as a random froth and the case of binary mixtures. We first recall some classical algorithms for building sphere assemblies (Sec.2), then give geometrical properties of the packings of spheres, both in the monosize (Sec.3) and the binary cases (Sec.4).

2 Computer simulations

The main parameter is the packing fraction C

$$C = \frac{volume\ of\ spheres}{total\ volume} \tag{1}$$

For a 3-dimensional packing of monosize grains, it runs from $C = 0$ to $C = \pi/2\sqrt{3} = 0.7404..$ which is realized in the face centered cubic (FCC) and hexagonal compact (HCP) ordered packings. The lower limit is obtained by a generation of points inside a 3D space (Poisson distribution [5] or Meijering model [6]). For all packing fractions between these two limits, a lot of techniques exist in order to generate a packing. We recall here algorithms which we used to get a packing of spheres as disordered as possible. It is possible now to consider large packings (15000 spheres) and so to get good statistics.

These algorithms generally hold both for monosize or polydisperse assemblies and of course, in the bi-dimensional case (disk assemblies). A detailed presentation may be found in [7]. They can be classified in two well-defined groups : the static and the dynamic methods of construction.

In the **static case**, the grains are placed at a given time step and cannot move afterwards. The extension of the Poisson process is the Random Sequential Adsorption model (RSA) where the particle is no longer a point [8]. This procedure does not provide very dense packings, the jamming limits for monosize particles being close to 0.38 for a three-dimensional packing of spheres and to 0.55 for a two-dimensional packing of disks. If there is a binary or a polydisperse distribution of size, the jamming limit can vary drastically according to the way the rejection is made. We also use the Modified Random Sequential Adsorption algorithm (MRSA) [9] which provides a slightly higher packing fraction, but introduces some anisotropy.

In models which generate dense disordered packings of disks or spheres the grains are placed according to a "gravity" [10, 11] or a central field [12] and cannot move after this placement. These algorithms give a packing fraction close to 0.60, generally smaller than the maximum possible for disordered systems ($C_{RCP} \approx 0.64$). On the other hand, due to the building procedure, they clearly present some anisotropy.

The **dynamic approach** uses the initial position of the grains as an input parameter ; some short or long range interactions between grains generate displacements of the particles and reorganizations of the whole packing. The final positions of the particles and the possible values of the final packing fraction depend strongly on the process which can be collective or individual.

We specially used two algorithms :

- The Jodrey-Tory algorithm [13] and its modified version by Jullien et al [14], where any packing fraction value may be reached
- A Molecular Dynamics (event driven) code for granular media, specific of hard sphere systems [7] where all collisions are assumed to be instantaneous.

3 Monosize packings

3.1 Topological quantities

Distribution $p(f)$ For any packing fraction C, the probability $p(f)$ to have a f-faceted cell is a single-peaked function with a maximum close to $f = 15$ (Fig. 1), slightly asymmetric, with gaussian wings and it may be interpreted as the one which minimizes information at fixed averages $< f >$ and $< f^2 >$ (the brackets $< . >$ stand for an average on face statistics). When the C increases the distribution narrows rapidly and the dispersion μ_2 decreases. The average number of faces $< f >$ decreases too, from $2+48\ \pi^2/35 = 15.53...$ corresponding to the Random Poisson Voronoi (RVP) [5] or Meijering case [6]) to 14 (Fig. 2); notice that it is always larger than the maximum number (12) of contacts.

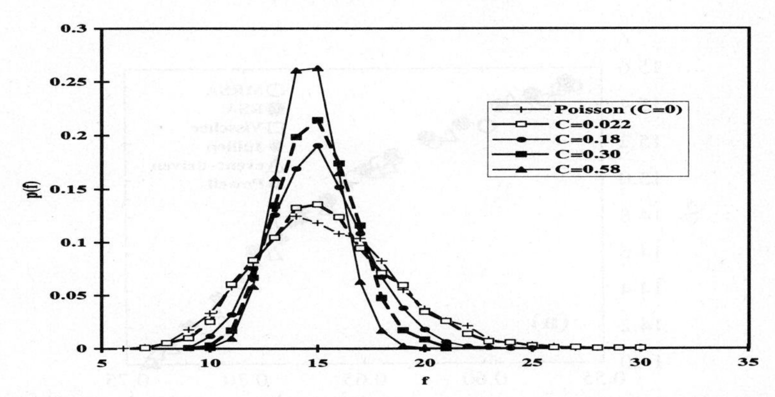

Fig. 1. Probability distribution $p(f)$ of f-faceted cell for different packing fractions C

These quantities depend not only on the packing fraction but also on the algorithm used, i.e. on the history of the packing: first, for different algorithms and the same packing fraction, $< f >$ is larger for algorithms with some anisotropy, such as MRSA [9] or Visscher-Bolsterli [11] and Powell [10] algorithms where the

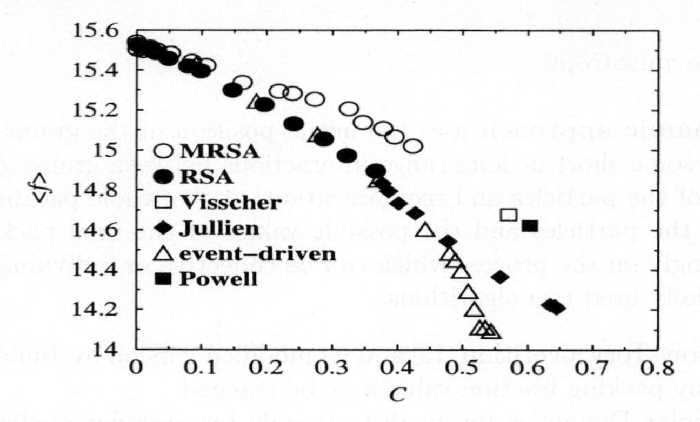

Fig. 2. Evolution of the mean number of faces versus the packing fraction, C, for different algorithms.

direction of gravity is favored. Secondly, for a given algorithm, it may be checked that cells become more isotropic when C increases. The dependence of f on anisotropy is in agreement with a theory developed by Rivier [15]. Let us point out the case of event-driven algorithm: for high packing fractions ($C > 0.545$) the system crystallizes [16]; the cells are then more isotropic than those of disordered packings at the same packing fraction and the limit value for $< f >$ is 14, i.e. the average number of neighbours (in the Voronoi sense) in slightly distorted FCC or HCP arrays [17].

So the packing fraction is clearly not a good quantity to describe the state of a foam and we look for a better parameter. On Fig. 3, we have plotted $< f >$ as a function of the sphericity coefficient K_{sph}

$$K_{sph} = 36\pi \left\langle V^2 \right\rangle / \left\langle A^3 \right\rangle . \quad (2)$$

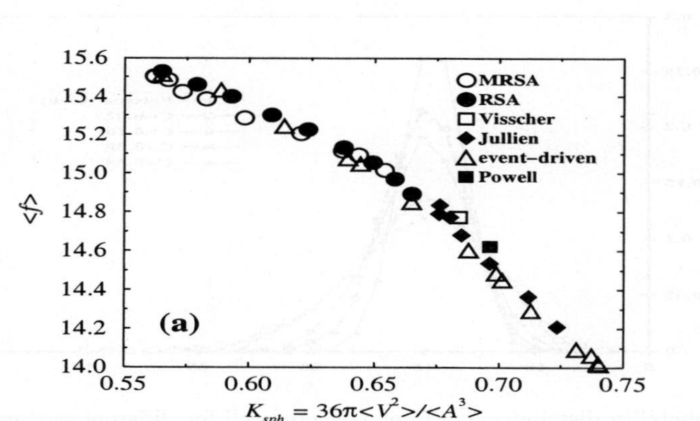

Fig. 3. Evolution of the mean number of faces versus the sphericity coefficient K_{sph} for all the algorithms used.

It turns out that points are positioned on a unique curve. Then, K_{sph} appears to be the relevant parameter [18].

Average number of edges For any f-faceted cell and any packing fraction, we have checked the relation $n(f) = 6 - 12/f$ and computed its average $< n(f) >$. It is actually nearly constant on the whole packing fraction range ($< n(f) > \approx 5.17$) because $< f >$ is a function slowly varying with C. As already noticed by Voloshin and Medvedev [19], most faces are 4, 5 and 6-sided (from 66% in the dilute case up to 80% in the most compact samples); however, 5-sided faces are not really preeminent (40% in the regular lattices).

The distribution of the number of sides per face is a very adequate tool for studying the transition from a disordered to an ordered packing in dense systems as the thermalization goes on. We have plotted on Fig.4 the fraction p_i of faces with $i = 4, 5$ and 6 edges in event driven systems as the packing fraction increases. In the C range where crystallization may occur, they behave quite differently in the stable branch (crystal) and the metastable one (supercooled liquid) [16].

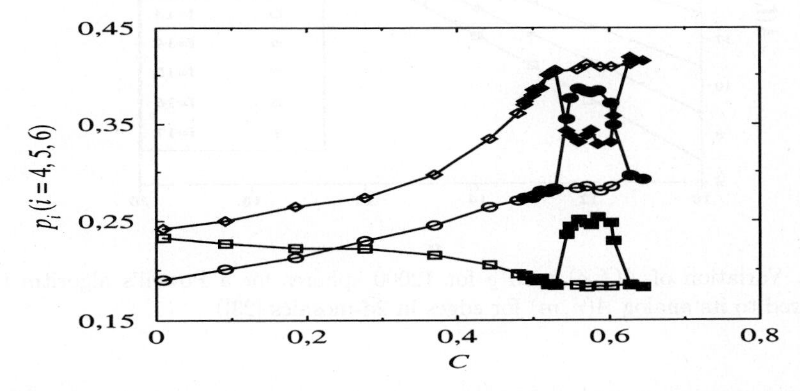

Fig. 4. Values of p_4 (□) p_5 (◇) and p_6 (○) for disordered packings (empty symbols) and for initially metastable packings (full symbol).

First neighbours The relations of a f-faceted cell with its neighbour cells is described by their average number of faces $m(f)$ and related quantities. The generalization in three dimensions of Weaire's identity [20] is easily checked.

$$< f\ m(f) > = < f^2 > \tag{3}$$

Moreover, at any packing fraction, the average number of faces $f\ m(f)$ of all neighbours is a linear function of f,

$$f\ m(f) = (< f > - a)f + a\ \ < f > + \mu_2 \tag{4}$$

which is precisely the Aboav-Weaire law [21] which has been shown to be exact in some particular 3d systems [22]. Like $< f >$ and μ_2, the parameter a depends on the packing fraction C.

For 2d foams, Peshkin et al. [23] suggested that the linear dependence on the number of neighbours of Aboav's law was a consequence of the MAXENT principle applied to the topological correlation numbers. The corresponding correlation quantities in 3d are the $A(f,g)$, related to the symmetric probability that two cells with f and g faces are neighbors. We measured them on several samples of 12000 objects both for Powell packings and RVP point systems. They are increasing functions of f and g (Figure 5), and although the precision is not very good, we can represent them by a linear dependence in f and g [24]

$$A(f,g) = \sigma(g - \langle f \rangle)(f - \langle f \rangle) + f + g - \langle f \rangle \,. \tag{5}$$

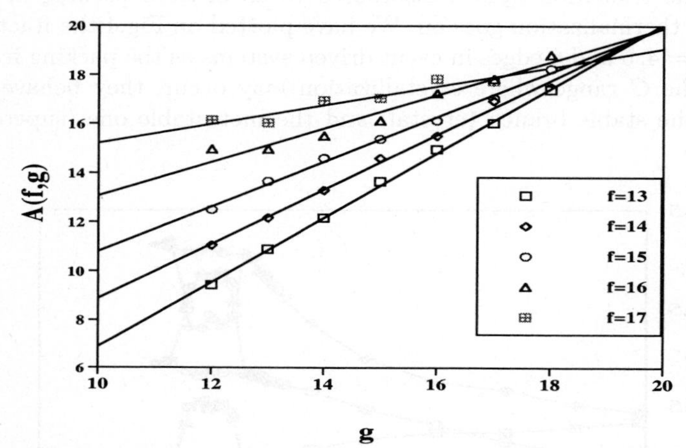

Fig. 5. Variation of $A(f,g)$ with g for 12000 spheres for a Powell's algoritm (to be compared to its analog $A(n,m)$ for edges in 2d-mosaics [23])

3.2 Metric properties

We focus mainly on the volume of the cells, first as a function of the number of faces, then globally. For any packing fraction, the cell volume $V(f)$, is a linear function of f

$$V(f) = V_0 \left[1 + (f - \langle f \rangle)/K_V\right] \tag{6}$$

where $V_0 = \sum_f V(f)\, p(f)$ is the average volume, and K_V is a parameter depending on the packing fraction C (see Fig. 6). Linear dependence on f exists also for the interfacial area and total length of edges. Thus, generalizations of Lewis [25] and Desch [26] laws are possible, even at high packing fractions, in opposition with the 2d case, where steric exclusion implied the existence of a minimum at medium and high packing fractions. This is probably due to the fact that the number of faces is always high ($f \geq 10$). The curves would probably not be linear any more if many cells had 6, 7 or 8 faces, since polyhedra with a small number of faces require a larger volume than higher order ones [27].

Notice that, at the same packing fraction, the slope of the normalized volume is larger than that of the area or of the length. Thus, everything else being equal, the linear volume law (Lewis) is the most relevant and discriminatory.

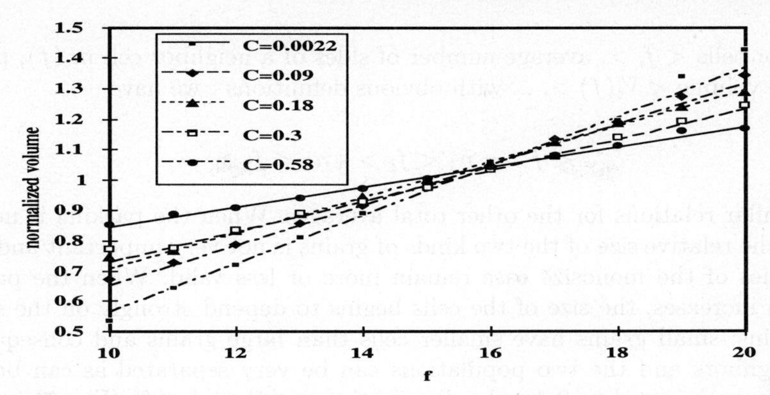

Fig. 6. Variations of the normalized volume of a f-faceted cell versus f for different packing fractions C (between 0.0022 and 0.58).

4 Binary mixtures

It is known that the arrangement of grains strongly depends on their geometry rather than on their physical properties, so that the relative size of a grain and their numerical proportion in the packing are the main relevant quantities. In binary mixtures, we have two species, with radii R_i, $i = 1, 2$ ($R_2 > R_1$), and numerical proportions n_i ($n_1 + n_2 = 1$) ; we choose $R_1/R_2 > \sqrt{3/2} - 1 = 0.224..$ so that small grains cannot be trapped in the tetrahedron made with four contacting large spheres without any increase in volume.

4.1 Radical tessellation

The first thing consists in partitioning space by cells containing one grain. The Voronoi tessellation is no longer adequate: in dense packings, the cell can "cut" the larger grain and touching spheres may not be neighbors in the Voronoi sense. Several generalizations have been proposed, which all reduce to the usual tessellation in the monosize case and are generalized without difficulty to the polydisperse case. The simplest consists in the radical tessellation [3, 4], where the bisecting plane is replaced by the radical plane (all the points in the radical plane have the same tangency length - or power- for the two spheres). The definition may be somewhat artificial, but many of the features of the Voronoi tessellation are maintained: the cells are convex polyhedrons with planar faces, each one containing one grain, two touching spheres have a common face (the tangent plane) and thus may be considered as neighbors. The incidence properties still hold : a face is generically common to two cells, an edge to three and a vertex to four cells. Moreover, big grains have in general larger cells than smaller ones, a way again for estimating their relative hindrance.

4.2 Topological and metric quantities

It may be interesting to consider each species separately. Then we define partial quantities such as the partial distributions $p_i(f)$, partial average number of neighbor cells $< f_i >$, average number of sides of a neighbor cell $m_i(f)$, partial average volume $< V_i(f) >$, ... with obvious definitions ; we have

$$< f >= n_1 < f_1 > + n_2 < f_2 > \tag{7}$$

and similar relations for the other total averages. When the packing is not too dense, the relative size of the two kinds of grains is not very important and most properties of the monosize case remain more or less valid. When the packing fraction increases, the size of the cells begins to depend strongly on the size of the grains: small grains have smaller cells than large grains and consequently less neighbors and the two populations can be very separated as can be seen for a size ratio equal to 2 for the distributions $p_i(f)$ and $p(f)$ (Fig. 7) and the volume distribution (Fig.8) respectively.

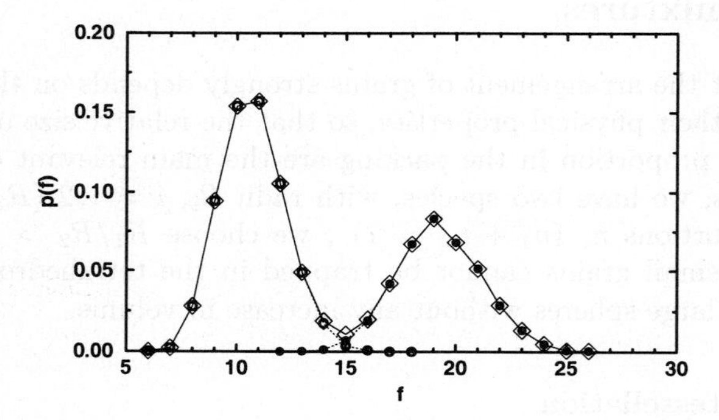

Fig. 7. Distribution $p(f)$ ($\diamond$) of the number of cell sides and weighted partial distributions $n_i p_i(f)$ ($\circ$: i=1, $\bullet$: i=2) for $k = 2$ and $n_1 = 0.6$ for $C = 0.61$

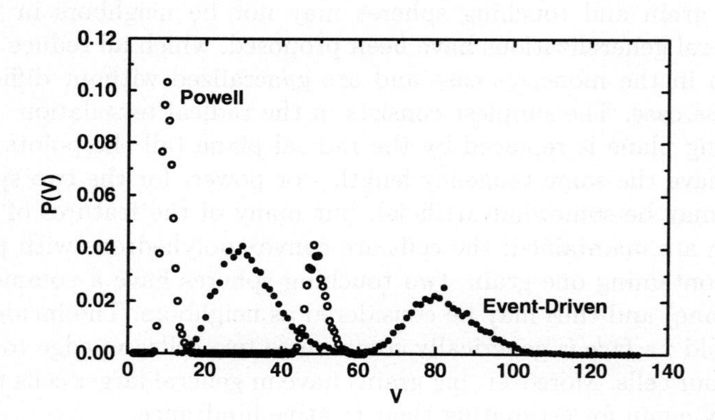

Fig. 8. Volume distribution P(V) at $C = 0.31$ (Event-Driven) and $C = 0.61$ (Powell) for n_1=0.6. The length unit is the value of the small sphere radius

We shall not insist very much on these results. Most of them may be found in [28]. Let us just list main features, assuming $R_2 > R_1$:

- distribution $p(f)$ is a one peaked function at low packing fractions, and separate into two distinct distributions as C increases. Then, $< f_1 > \leq < f_2 >$ and the total average coordination number may be smaller than 14.

- separately each species obeys Aboav-Weaire law, with different slopes. The whole packing **does not** obey Aboav law and the curve is S-shaped.

- partial volumes $V_i(f)$ and areas $A_i(f)$ **do not** verify Lewis nor Desch law and of course, the total averages do not either. The volume (resp. area) distribution is a two peaked-function.

The same behaviour is probably true for ternary mixtures, with possibly 3 peaks. When the number of species involved increases, a smoothing may appear, depending on the radii size distribution and polydisperse assemblies may provide an intermediate behaviour.

5 Conclusion

We have given here a description of the geometrical organization of disordered packings of spheres, analyzed through the statistical properties of their Voronoi (and radical) cells. In the case of equal spheres, the Voronoi tessellation behaves like an ordinary foam and follows classical universal laws (Aboav, Lewis,...) which are known from a long time in two dimensions and which we find to be true in 3d either ; numerical values depend on the packing fraction and the steric constraints due to the grain hindrance. Differences between disordered and ordered packings may be seen easily on very simple quantities like the distribution of the number of edges of the faces.

In the case of binary mixtures, the two species behave very differently at high density and the empirical laws for ordinary foams do not hold. On the other hand, radical tessellation is a good tool to test the possible segregation in the repartition of the two species [29]. For polydisperse assemblies, an intermediate behaviour is expected.

References

1. see for instance Weaire D., Rivier N., : Soap, Cells and Statistics - Random Patterns in two Dimensions, Contemp. Phys. **25** (1984) 59
2. Lemaitre J., Gervois A., Troadec J.-P., Rivier N., Ammi M., Oger L. and Bideau D.: Arrangement of cells in Voronoi tessellations of monosize packings of discs, Phil. Mag. B, **67** (1993) 347
3. Gellatly B.J. and Finney J.L.: Characterization of models of multicomponent amorphous metals : the radical alternative to the Voronoi polyhedron, J. Non Crystalline Solids **50** (1982) 313
4. Telley H., Liebling, T.M., and Mocellin A.: The Laguerre model of grain growth in two dimensions, Phil. Mag. B**73** (1996) 395-408
5. Okabe A., Boots B., Sugihara K. and Chiu S.N., in *Spatial Tessellations : concepts and applications of Voronoi diagrams*, Wiley (2000)
6. Meijering J.L.: Interface area, edge, length and number of vertices in crystal aggregates with random nucleation, Philips Res. Rep., **8** (1953) 270

7. Oger L., Troadec J.-P., Gervois A. and Medvedev N.N.: Computer simulations and tessellations of granular Materials, in *Foams and Emulsions* J.F. Sadoc and N. Rivier eds, Kluwer (1999), p527
8. Feder J.: Random Sequential Adsorption, J. Theor. Biol. **87** (1980) 237
9. Jullien R. and Meakin P.: Random sequential adsorption with restructuring in two dimensions, Journal of Physics A: Math. Gen., **25** (1982) L189-L194
10. Powell M.J.: Site percolation in randomly packed spheres, Phys. Rev. B **20** (1979) 4194
11. Visscher W.H. and Bolsterli H.: Random packing of equal and unequal spheres in two and three dimensions, Nature, **239** (1972) 504
12. Bennett C.H.: Serially deposited amorphous aggregates of hard spheres, J. Appl. Phys., **43** (1972) 2727
13. Jodrey W.S. and Tory E.M.: Computer simulation of random packings of equal spheres, Phys. Rev. A, **32** (1985) 2347
14. Jullien R., Jund P., Caprion D. and Quitman D.: Computer investigation of long range correlations and local order in random packings of spheres, Phys. Rev. E, **54** (1996) 6035
15. Rivier N.: Recent results on the ideal structure of glasses, Journal de Physique Colloque **C9** (1982) 91-95
16. Richard P., Oger L., Troadec J.-P. and Gervois A.: Geometrical characterization of hard sphere systems, Phys. Rev. E **60** (1999) 4551
17. Troadec J.-P., Gervois A. and Oger L.: Statistics of Voronoi cells of slightly perturbed FCC and HCP close packed lattices, Europhys. Lett. **42** (1998) 167
18. Richard P., Troadec J.-P., Oger L. and Gervois A.: Effect of the anisotropy of the cells on the topological properties of two- and three-dimensional froths, Phys. Rev. E **63** (2001) 062401
19. Voloshin V.P., Medvedev N.N. and Naberukin Yu. I.: Irreguler packing Voronoi Polyhedra I & II, Journal of Structural Chemistry, **26** (1985) 369 & 376
20. Weaire D.: Some remarks on the arrangement of grains in a polycrystal, Metallography, **7** (1974) 157
21. Aboav D.A.: The Arrangement of cells in a Net, Metallography, **13** (1980) 43
22. Fortes M.A.: Topological properties of cellular structures based on staggered packing of prisms, J. Phys. France, **50** (1989) 725,
23. Peshkin M.A., Strandburg K.J. and Rivier N.: Entropic prediction for cellular networks, Phys. Rev. Lett., **67** (1991) 1803
24. Oger L., Gervois A., Troadec J.-P. and Rivier N.: Voronoi tessellation of spheres : topological correlations and statistics, Phil. Mag. B **74** (1996) 177-197
25. Lewis F.T.: The correlation between cell division and the shapes and sizes of prismatic cells in the epidermis of cucumis, Anat. Record., **38** (1928) 341
26. Desch C.H.: The solidification of metals from the liquid state, J. Inst. Metals, **22** (1919) 24
27. Fortes M.A.: Applicability of the Lewis and Aboav-Weaire laws to 2D and 3D cellular structures based on Poisson partitions, J. Phys. A **28** (1995) 1055
28. Richard P., Oger L., Troadec J.-P. and Gervois A.: Tessellation of binary assemblies of spheres, Physica A **259** (1998) 205
29. Richard P., Oger L., Troadec J.P. and Gervois A.: A model of binary assemblies of spheres, EPJE (2001) in press

Collision Detection Optimization in a Multi-particle System

Marina L. Gavrilova and Jon Rokne

Dept of Comp. Science, University of Calgary, Calgary, AB, Canada, T2N1N4
marina@cpsc.ucalgary.ca, rokne@cpsc.ucalgary.ca

Abstract. Collision detection optimization algorithms in an event-driven simulation of a multi-particle system is one of crucial tasks, determining efficiency of simulation. We employ dynamic computational geometry data structures as a tool for collision detection optimization. The data structures under consideration are the dynamic generalized Voronoi diagram, the regular spatial subdivision, the regular spatial tree and the set of segment trees. Methods are studies in a framework of a granular-type materials system. Guidelines for selecting the most appropriate collision detection optimization technique summarize the paper.

1 Introduction

A particle system consists of physical objects (particles), whose movement and interaction are defined by physical laws. Studies conducted in the fields of robotics, computational geometry, molecular dynamics, computer graphics and computer simulation describe various approaches that can be applied to represent dynamics of particle systems [9,8]. Disks or spheres are commonly used as a simple and effective model to represent particles in such systems (ice, grain, atomic structures, biological systems). Granular-type material system is an example of a particle system. When a dynamic granular-material system is simulated, one of the most important and a time consuming tasks is predicting and scheduling collisions among particles.

Traditionally, the cell method was the most popular method employed for collision detection in molecular dynamics and granular mechanics [8]. Other advanced kinetic data structures, commonly employed in computational geometry for solving a variety of problems (such as point location, motion planning, nearest-neighbor searches [7,11,1]), were seldom considered in applied materials studies.

Among all hierarchical planar subdivisions, binary space partitions (BSP) are most often used in dynamic settings. Applications of binary space partitions for collision detection between two polygonal objects were considered in [1,4,6]. Range search trees, interval trees and OBB-trees were also proposed for CDO [6,3].

This paper presents an application of the weighted generalized dynamic Voronoi diagram method to solve the collision optimization problem. The idea of

P.M.A. Sloot et al. (Eds.): ICCS 2002, LNCS 2331, pp. 105–114, 2002.

employing the generalized dynamic Voronoi diagram for collision detection optimization was first proposed in [2]. The method is studied in general d-dimensional space and is compared against the regular spatial subdivision, the regular spatial tree and the set of segment trees methods. Results are summarized in a form of guidelenes on the selection of the most appropriate collision detection optimization method.

2 Dynamic event-driven simulation algorithm

As of today, most of the research on collision detection in particle systems is limited to consideration of a relatively simple simulation model. The idea is to discretize time into short intervals of fixed duration. At the end of each time interval, the new positions of the moving particles are computed. The state of the system is assumed to be invariant during the time interval between two consecutive time steps. The common problem with such methods is related to choosing the length of the interval. If the duration is too short, unnecessary computations take place while no topological changes occurred. If the duration is too large, some important for analysis of the model events can be omitted. A much more effective approach, the dynamic event-driven simulation of a particle system, relies on discrete events that can happen at any moment of time rather then during fixed time steps. This can be accommodated by introducing an event queue. We employ this scheme and suggest the following classification of events: *collision events*, *predict trajectory events* and *topological events*.

A set of n moving particles in R^d is given. The particles are approximated by spheres (disks in the plane). Collision event occurs when two particles come into contact with each other or with a boundary. A predict trajectory event occurs when the trajectory and the velocity of a particle is updated due to the recalculation of the system state. Between two consecutive predict trajectory events a particle travels along a trajectory defined by a function of time.

Collision detection algorithms optimize the task of detecting collisions by maintaining a set of neighbors for every particle in the set and only checking for collisions between neighboring particles. The algorithm is said to be correct if at the moment of collision the two colliding particles are neighbors of each other (i.e. the collision is not missed). The computational overhead associated with a CDO algorithm is related to the data structure maintenance.

The Event-Driven Simulation Algorithm

1. (Initialization) Construct the topological data structure; set the simulation clock to time $t_0 = 0$; schedule predict trajectory events for all particles and place them in the queue.
2. (Processing) While the event queue is not empty do:
   (a) Extract the next event e_i from the event queue, determine the type of the event (topological event, predict trajectory event or collision event);
   (b) Advance the simulation clock to the time of this event t_e

(c) Process event e_i and update the event queue:
   i. if the event is a topological event:
      - modify the topology of the data structure;
      - schedule new topological events;
      - check for collisions between new neighbors and schedule collision events (if any);
   ii. if the event is a collision event:
      - update the states of the particles participating in a collision;
      - delete all events involving these particles from the event queue;
      - schedule new predict trajectory events at current time for both particles;
   iii. if the event is a predict trajectory event:
      - compute the trajectory of the particle for the next time interval $(t_e, t_e + \Delta t]$;
      - schedule the next predict trajectory event for the particle at time $t_e + \Delta t$;
      - schedule new topological and collision events for the updated particle

Note, that new topological or collision events are never scheduled past the time of the next predict trajectory event of all particles involved.

The following criteria can be introduced to estimate the efficiency of a CDO algorithm: the total number of pairs of closest neighbors; the number of neighbors of a single particle; the number of topological events that can take place between two consecutive collision events (or predict trajectory events); computational cost of a topological event (scheduling and processing); computational cost of the data structure initialization; and space requirements.

3 Dynamic Computation Geometry Data Structures

Consider a problem of optimizing the collision detection for a set of moving particles in the context given above. In a straightforward approach each pair of particles is considered to be neighbors, i.e. there are $\frac{1}{2}n\,(n-1)$ neighbor pairs. For a large system the application of this method is very computationally expensive. Thus, our goal is to reduce the number of neighbors to be considered on each step.

3.1 The Dynamic Generalized Voronoi diagram for CDO

The dynamic generalized Voronoi diagram in Laguerre geometry is the first data structure applied for collision detection optimization.

Definition 1. A generalized Voronoi diagram (VD) for a set of spheres S in R^d is a set of Voronoi regions $GVor\,(P) = \{\mathbf{x}|\ d\,(\mathbf{x}, P) \leqslant d\,(\mathbf{x}, Q)\,, \forall Q \in S - \{P\}\}$, where $d\,(\mathbf{x}, P)$ is the distance function between point $\mathbf{x} \in R^d$ and particle $P \in S$.

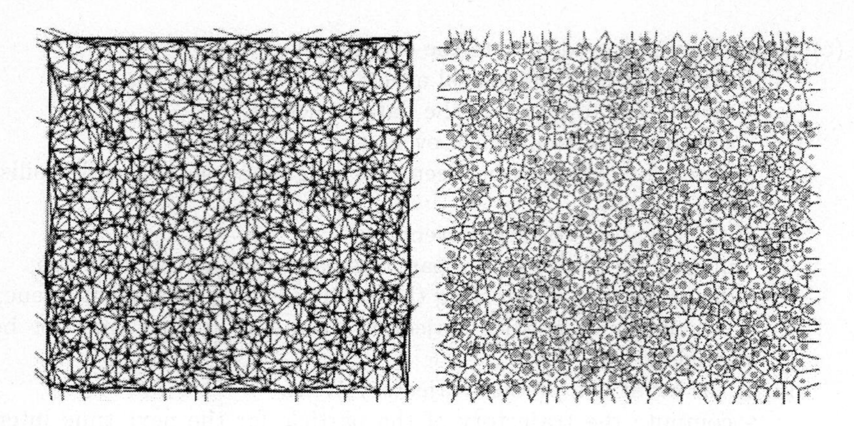

Fig. 1. The generalized Delaunay triangulation (left) and the Voronoi diagram (right) for 1000 particles in Laguerre geometry.

In Laguerre geometry, the distance between a point and a sphere is defined as $d(\mathbf{x}, P) = d(\mathbf{x}, \mathbf{p})^2 - r_P^2$, where $d(\mathbf{x}, p)$ is the Euclidean distance between $\mathbf{x}$ and $\mathbf{p}$ [10]. The generalized Voronoi diagram stores the topological information for a set of particles. Each Voronoi region represents the locus of points that are closer to the particle than to any other particle from set S. The dual to the Voronoi diagram, the Delaunay tessellation, contains the proximity information for the set of particles (see Fig. 1).

The following approach is implemented. The Delaunay tessellation (DT) is constructed in Laguerre geometry. The computation of topological events is incorporated in the Event-Driven Simulation Algorithm. To ensure the algorithms correctness, the following property should be satisfied: if two particles are in contact with each other, then there must be an edge in the Delaunay tessellation incident to these particles. Due to the fact that the nearest-neighbor property in DT in Laguerre geometry is satisfied, the dynamic generalized DT can be used for collision detection optimization.

According to [2], a topological event in the dynamic generalized VD occurs when the topological structure of the VD is modified and the proximity relationships between particles are altered. Handling of a topological event requires flipping a diagonal edge (or a facet) in a quadrilateral of the Delaunay tessellation and scheduling future topological events for the newly created quadrilaterals (a *swap* operation). A topological event occurs when the Delaunay tessellation sites comprising a quadrilateral become co-spherical.

Let $P_i = \{(x_i = x_i(t), y_i = y_i(t)), r_i\}$, $i = 1..d+2$ be a set of spheres with centers $(x_i(t), y_i(t))$ and radii r_i. The computation of the time of a topological event requires solving equation: $F(P_1(t), P_2(t), ..., P_{d+2}(t)) = 0$.

Lemma 1. *The time of the topological event in a Delaunay d-dimensional quadrilateral of $d+2$ spheres $P_i = \{(x_i = x_i(t), y_i = y_i(t)), r_i\}$, can be found in La-*

guerre geometry as the minimum real root t_0 of the equation

$$F(P_1, P_2, ..., P_{d+2}) = \begin{vmatrix} x_{11} & x_{12} & \cdots & x_{1d} & w_1 & 1 \\ x_{21} & x_{22} & \cdots & x_{2d} & w_2 & 1 \\ \cdots & \cdots & \cdots & \cdots & \cdots & \cdots \\ x_{d+1,1} & x_{d+1,2} & \cdots & x_{d+1,d} & w_{d+1} & 1 \\ x_{d+2,1} & x_{d+2,2} & \cdots & x_{d+2,d} & w_{d+2} & 1 \end{vmatrix} \quad (1)$$

where $w_i = x_{i,1}^2 + x_{i,2}^2 + ... + x_{i,d}^2 - r_i^2, i = 1..d+2$ *and the following condition is satisfied:*

$$CCW(P_1, P_2, ..., P_{d+1}) = \begin{vmatrix} x_{11} & x_{12} & \cdots & x_{1,d} & 1 \\ x_{21} & x_{22} & \cdots & x_{2,d} & 1 \\ \cdots & \cdots & \cdots & \cdots & \cdots \\ x_{d+1,1} & x_{d+1,2} & \cdots & x_{d+1,d} & 1 \end{vmatrix} > 0. \quad (2)$$

Performance Analysis

The detailed performance analysis for the CDO algorithm employing dynamic DT in Laguerre geometry follows. Some of the estimates apply to all CDO employing different data structures, while some are specific for the Delaunay triangulation based approach.

First, consider the planar Delaunay triangulation. During the preprocessing stage the DT is constructed in $O(n \log n)$ using the sweep-plane technique. The space required to store the data structure is $O(n)$. Placing the initial events into the event queue takes $O(n)$ (since they all occur at time $t = 0$).

The upper bound on the number of predict trajectory events at any moment of time in the queue is $O(n)$, since only one predict trajectory event is scheduled for each particle. The upper bound on the number of collision events at any moment of time in the queue is $O(n^2)$, since a possible collision can be scheduled for each pair of particles. It is independent of the CDO data structure. The upper bound on the number of topological events stored in the queue is $O(n)$ at any moment of time, since only one topological event is scheduled for every Delaunay triangulation edge.

The algorithm efficiency also depends on the number of *collision checks* that need to be performed in order to determine if a collision event needs to be scheduled. The number of collision checks that need to be performed after a predict trajectory event is the number of neighbors of the particle. This number is $O(n)$ in the worst case. However, for planar DT, the average number of neighbors of a particle is a constant. The maximum number of collision checks per topological event is equal to the number of new neighbors of a particle after the topological event occurs. For the dynamic DT, this number is exactly one (since one new edge appears in the DT due to a topological event).

Processing a topological event requires scheduling up to five new topological events (one for the new edge and one for each of the four neighboring quadrilaterals). It also requires performing one collision check and scheduling one possible collision event. Thus, the overall complexity of this step is $O(logn)$.

The most time consuming step in processing of a collision event in the worst case is deleting up to $O(n)$ previously scheduled events. If every particle contains a list of references to all events involving this particle, then they can be deleted in $O(n)$ time. Thus, the overall complexity of this step is $O(n)$.

Processing a predict trajectory event requires performing collision checks and scheduling new collision events (which is $O(n)$ in the worst-case). It also requires scheduling new topological events for the particle, which can result in scheduling new topological events for all edges of DT adjacent to this particle ($O(n)$ in total). Thus, the overall complexity of this step is $O(n \log n)$.

The total number of collisions between particles during the simulation cannot be bounded since the simulation time is unlimited. Hence, the overhead associated with the use of a particular CDO algorithm is usually estimated by the maximum number of topological events that can occur between two consecutive collisions. For a planar Delaunay triangulation, the number can be as large as $O\left(n^2 \lambda_s(n)\right)$, where $\lambda_s(n)$ is the maximum length of an (n, s)-Davenport-Schinzel sequence, or as low as $O(1)$ for densely packed systems. The above discussion is summarized in Table 1, Section 3.5.

In higher dimensions, only a few estimates differ. The worst-case number of topological events that can happen between two consecutive collisions is $O\left(n^d \lambda_s(n)\right)$, the initialization step takes $O\left(n^{\lceil d+1/2 \rceil}\right)$ (using an incremental construction technique), the space required is $O\left(n^{\lceil d/2 \rceil}\right)$.

3.2 The Regular Spatial Subdivision

The regular spatial subdivision is a data structure that is traditionally used for collision detection optimization. The performance of the method strongly depends on the ratio between the maximum size of the particle and the diameter of the cell. Our contribution is in establishing the condition under which the number of the particles in the cell is a constant, thus guaranteeing a successful application of this method for CDO.

The space is subdivided into axis-parallel hypercubes in R^d. These are generically called cells in the sequel. A particle is said to reside in a cell if its center belongs to the cell. Each cell contains a list of particles that currently reside in it. The set of neighbors of a particle comprises all particles residing in the same or any of the $3^d 1$ neighboring cells. To ensure correctness, the size of a cell must be greater or equal to the diameter of the largest particle. Then, if two particles are in contact, they are guaranteed to reside in the same or in the two neighboring cells. Each particle $P_i = (\mathbf{p}_i, r_i)$ is represented by a pair consisting of the coordinates of its center $\mathbf{p}_i = \mathbf{p}_i(t)$ and the radius r_i. Assume that the size of the simulation domain is such that there are k cells in each direction. Consider a d-dimensional box with a diameter l as a simulation domain. The size of a cell must exceed the diameter of the largest particle. Thus, k is defined as the diameter of the simulation domain divided by the diameter of a largest particle $M = \max\limits_{P_i \in S}(2r_i)$, i.e. $k = \lceil l/M \rceil$. The diameter of the smallest particle is denoted by $m = \min_{P_i \in S}(2r_i)$.

Assumption 1. The ratio $\gamma = M/m$ between the maximum and the minimum diameter is invariant of n.

Lemma 2. *Under Assumption 1, the maximum number n_c of particles within each cell is bounded by a constant.*

A topological event in the regular spatial subdivision occurs when the center of a particle moves from one cell to another. The time of a topological event can be determined exactly by computing the time when the center of the particle passes the boundary of a cell.

Performance analysis

The space required to store the data structure is $O\left(k^d + n\right)$. The regular spatial subdivision can be constructed by first allocating k^d cells and then placing each of the particles into the appropriate cell based on their coordinates at the moment $t = 0$. The cells are stored in a multi-dimensional array and are accessed directly in $O(1)$.

For each particle only one topological event can be scheduled at any particular moment of time. Therefore, the upper bound on a number of topological events stored in the queue is $O(n)$ at any moment of time. Upper bounds on collision and predict trajectory event are invariant of the data structure used.

Collision checks after topological event are performed between particles that reside in neighboring cells. Since the topological event occurs when a particle moves into one of these neighboring cells, collision checks with particles from some of these cells were computed previously. Thus, only particles in 3^{d-1} *new* neighboring cells must be checked for collisions. The total number of collision checks after topological event is the number of new neighbors of a particle and is $O(1)$ under Assumption 1. Therefore, the total number of collision checks per predict trajectory event is also a constant.

Processing a topological event requires scheduling one new topological event (move to a new cell). It also requires performing $O(1)$ collision checks with new neighbors and scheduling the detected collision events. Thus, the overall complexity of this step is $O(logn)$.

Processing a predict trajectory event requires performing collision checks and scheduling new collision events. Since each cell contains only a constant number of particles (according to Lemma 1), only a constant number of collision events will be scheduled. It also requires scheduling one new topological event. Thus, the overall complexity of this step is $O(logn)$. Following the same arguments as for the dynamic DT, processing of the collision event takes $O(n)$.

Finally, since a particle can cross maximum k cells before it collides with the boundary, the number of topological events between two collisions is $O(nk)$.

Note 2. The efficiency of this method strongly depends on the distribution of particle diameters. For the algorithm to perform well, the maximum number

n_c of particles that can fit into a cell must be smaller than the total number of particles n.

3.3 The Regular Spatial Tree

This approach is a modification of the regular spatial subdivision method that reduces the memory overhead at the expense of increased access time. We propose the following approach. The non-empty cells are stored in an AVL tree according to the lexicographical ordering on their coordinates. Each node of the tree is associated with a cell in the d-dimensional Euclidean space and implicitly with a non-empty set of particles $\{P_{i_1}, P_{i_2}, ..., P_{i_l}\}$, whose centers belong to the cell.

The method reduces the storage to $O(n)$, since the number of occupied cells cannot exceed the total number of particles in the system. On the other hand, each cell access now requires $O(\log n)$ time. All the complexity estimates obtained for the regular spatial subdivision method hold with the following exception. Any operation involving modifications of the data structure (such as moving a particle from one cell into another) will now require $O(\log n)$ time. Hence, each topological event requires $O(\log n)$ operations. The initial construction of the data structure takes $O(n \log n)$ time, independent of the dimension.

3.4 The Set of Segment Trees

This is an original method proposed for collision detection optimization. We maintain a set of trees of intersecting segments, obtained by projecting the bounding boxes of particles onto the coordinate axes. The particles are said to be neighbors if their bounding boxes intersect. The algorithm is correct since if two particles are in contact, then their bounding boxes intersect.

A segment tree, represented as an AVL tree, is maintained for each of the coordinate axis. The size of each tree is fixed, since the total number of particles does not change over time. For every particle, associated list of its neighbors is dynamically updated. A topological event occurs when two segment endpoints on one of the axes meet. This indicates that the bounding boxes of the two corresponding particles should be tested for intersection. Positive test identifies that the particles became neighbors, thus their neighbor lists are updated and collision check is performed.

As the particles move, it is necessary to maintain the sequence of segment endpoints in sorted order for each of the coordinate axes. When two neighboring endpoints collide, they are exchanged in the tree. Note that we do not rebalance the tree, but exchange references to segment endpoints.

Performance analysis

The segment tree is constructed initially by sorting the segment endpoints in $O(n \log n)$ time. The space required is $O(n)$.

The upper bound on the number of topological events stored in the event queue is $O(n)$ at any moment of time, since every segment endpoint can only collide with either of the neighboring endpoints of another segment, and there are $2d$ endpoints for every particle. Upper bounds on collision and predict trajectory event are the same as in Section 3.1.

At most one collision check is performed per topological event. Note that a half of the collisions between segment endpoints, when two segments start to intersect, might result in collision checks between particles. The other half correspond to topological events, when two segments stop intersecting (no collision checks are required for these events, though the neighbor lists are updated). The number of collision checks per predict trajectory event is estimated as follows.

Lemma 3. *Under Assumption 1, the total number of collision checks per predict trajectory event is* $O(1)$.

Processing a topological event requires scheduling up to two new topological event (one for each new neighboring segment endpoint). Thus, the overall complexity of this step is $O(logn)$. Processing a predict trajectory event requires scheduling up to $4d$ new topological events (two for every segment endpoint). Only a constant number of collision events will be scheduled. Thus, the complexity of this step is $O(logn)$. The overall complexity of processing a collision event is $O(n)$.

Lemma 4. *If the particles move along the straight-line trajectories, then the upper bound on the number of topological events that can take place between two consecutive collisions is* $O(n^2)$.

3.5 Summary of Performance Analysis

The complexities of the presented algorithms for the planar case are now summarized. The following notations are used:

A the upper bound on the number of neighbors of a particle
B maximum number of neighbors appearing due to a topological event
C time per topological event (excluding collision checks)
D time per predict trajectory event
E maximum number of topological events between two collisions
F initialization
G space

4 Conclusion

The analysis of algorithm performance is summarized as follows.

1. The worst-case number of topological events that can happen between two collisions is the largest for the Delaunay tessellation method. This method should only be used in particle systems with high collision rate.

Table 1. Algorithms performance in d-dimensions

Algorithm	A	B	C	D	E	F	G
Dynamic DT	$O(n)$	1	$O(\log n)$	$O(\log n)$	$O(n^2\lambda_s(n))$	$O\left(n^{\lceil d+1/2\rceil}\right)$	$O\left(n^{\lceil d/2\rceil}\right)$
Subdivision	$O(1)$	$O(1)$	$O(\log n)$	$O(\log n)$	$O(nk)$	$O(n+k^d)$	$O(n+k^d)$
Spatial tree	$O(1)$	$O(1)$	$O(\log n)$	$O(\log n)$	$O(nk)$	$O(n \log n)$	$O(n)$
Segment tree	$O(1)$	0 or 1	$O(\log n)$	$O(n)$	$O(n^2)$	$O(n \log n)$	$O(n)$

2. The regular spatial subdivision is the most space consuming method and should not be used if the memory resources are limited.
3. The regular spatial subdivision and the regular spatial tree methods perform worst for densely packed granular systems.
4. The Delaunay tessellation based method is the only method which performance is independent of the distribution of radii of the particles and their packing density.
5. In order to implement the regular subdivision method the size of the simulation space and the size of the largest particle in the system should be known in advance. This information is not required for other data structures considered.

References

1. Agarwal, P., Guibas, L., Murali, T. and Vitter, J. Cylindrical static and kinetic binary space partitions, in Proceedings of the 13th Annual Symposium on Computational Geometry (1997) 39–48.
2. Gavrilova,M., Rokne,J. and Gavrilov, D. Dynamic collision detection algorithms in computational geometry, 12th European Workshop on CG, (1996) 103–106.
3. Gottschalk,S., Lin,M. and Manocha,D. OBBtree: A hierarchical data structure for rapid interference detection, Computer Graphics Proc., (1996) 171–180.
4. Held,M., Klosowski,J. and Mitchell,J. Collision detection for fly-throughs in virtual environments, 12th Annual ACM Symp. on Comp. Geometry (1996) V13–V14
5. Hubbard, P. Approximating polyhedra with spheres for time-critical collision detection, ACM Transaction on Graphics, **15(3)** (1996) 179–210.
6. Kim, D-J., Guibas, L., Shin, S-Y. Fast collision detection among multiple moving spheres, IEEE Transactions on Visualization and Computer Graphics, **4(3)** (1998).
7. Lee, D.T., Yang, C.D., and Wong, C. K. Rectilinear Paths among Rectilinear Obstacles, ISAAC: 3rd Int. Symp. on Algorithms and Computation (1996)
8. Milenkovic,V. Position-based physics: Simulating a motion of many highly interactive spheres and polyhedra, Comp. Graph., Ann. Conf. Series (1996) 129–136.
9. Mirtich, B and Canny, J. Impulse-based simulation of rigid bodies, in Symposium on Interactive 3D Graphics, (1995) 181–188.
10. Okabe, A., Boots, B. and Sugihara, K. Spatial Tessellations: Concepts and Applications of Voronoi Diagrams, John Wiley Sons, England (1992).
11. van de Stappen, A.V., M.H. Overmars, M. de Berg, and J. Vleugels. Motion planning in environments with low obstacle density. Discrete Computational Geometry 20 (1998) 561–587.

Optimization Techniques in an Event-Driven Simulation of a Shaker Ball Mill

Marina Gavrilova[1], Jon Rokne[1], Dmitri Gavrilov[2], and Oleg Vinogradov[3]

[1] Dept of Comp. Science, Universit of Calgary, Calgary, AB, Canada, T2N1N4 marina@cpsc.ucalgary.ca, rokne@cpsc.ucalgary.ca

[2] Microsoft Corporation, Redmond, WA, USA dmitrig@microsoft.com

[3] Dept of Mech. and Manufacturing Eng., University of Calgary, Calgary, AB, Canada, T2N1N4 ovinogra@ucalgary.ca.

Abstract. The paper addresses issue of efficiency of an event-driven simulation of a granular materials system. Performance of a number of techniques for collision detection optimization is analyzed in the framework of a shaker ball mill model. Dynamic computational geometry data structures are employed for this purpose. The results of the study provide insights on how the parameters of the system, such as the number of particles, the distribution of their radii and the density of packing, influence simulation efficiency.

1 Introduction

Principles of mechanical alloying were first established in 1960's by J.S. Benjamin [3]. Since then a number of authors contributed to the theory of efficient ball mill design [12, 2]. Both a continuous and a discrete simulation models were considered in those studies [2, 4].

In any simulation model, one of the important aspects is scheduling collisions between particles. The collision detection optimization (CDO) in multi-particle systems has been proven to be a crucial task that can consume up to 80% of the simulation time and thus significantly influence the efficiency of the simulation [4]. Up until recently, the cell method was the most popular method for collision detection optimization in molecular dynamics, computational physics and granular materials simulation [10, 11, 13]. In this method, the simulated space is divided into rectangular subdomains and the collision detection performed only on the neighboring cells. The known downside of this method is its dependency on the distribution of particle sizes that can deem the approach completely inefficient. A variety of other methods based on hierarchical planar subdivisions were recently developed. Successful applications are found in motion planning, robotics and animation [7, 1, 9].

The paper presents the results of an experimental comparison of a number of collision detection algorithms as applied to the problem of efficient simulation of a mechanically alloyed system. The event-driven approach was employed to develop the shaker ball mill simulation model. The performance of algorithms

P.M.A. Sloot et al. (Eds.): ICCS 2002, LNCS 2331, pp. 115–124, 2002.

was studied for data sets containing monozized and polysized particles, with various particle distribution and configurations. To the best of our knowledge, this is the first systematic study focused on understanding the correlation between specific features of the developed algorithms and the parameters of the simulated system.

2 The shaker ball mill model

A *shaker ball mill* comprises a massive cylinder filled with steel balls oscillating in a vertical direction with a specified amplitude and frequency. Balls are moving along parabolic trajectories. Collisions between the balls and between the balls and the cylinder are assumed to be central and frictionless (the tangential velocity component of colliding particles is conserved). The cylinder is assumed to have an infinite mass, so that its velocity is not affected by the collisions. The inelastic nature of collisions can be represented by introducing restitution coefficients, which represent the ratio between particle velocity components before and after collision. The restitution coefficient ε is calculated as a function of particle velocity and size by using the Hertz contact theory to find the deformations of particles due to a collision [4]. It is assumed that the energy lost due to impact was spent on heating this volume. It is also assumed that the temperature returns to normal after each impact.
Dynamics of a mechanical alloyed system is described by a system of ordinary differential equations of time [14]. The solution of the system describes, in particular, the trajectories of bodies in the time interval $(t, t+\Delta t]$ and the velocities of the bodies in the next time step. In a shaker ball mill, balls move in a gravitational force field. The general equation of motion of a body in a force field is given according to the 2nd Newton's law. For the gravitational field, this equation can be solved analytically. The solution is

$$\mathbf{x}(t+\Delta t) = \mathbf{x}(t) + \mathbf{v}(t)\Delta t + \mathbf{g}\frac{\Delta t^2}{2}, \mathbf{v}(t+\Delta t) = \mathbf{v}(t) + \mathbf{g}\Delta t, \tag{1}$$

where $\mathbf{v}(t)$ is the velocity vector of the particle, $\mathbf{x}(t)$ is the position vector of the body, and $\mathbf{g}$ is the gravitational constant. The time of collision between two particles is found as the minimal positive root t_0 of the equation

$$\| \mathbf{x}_1(t) - \mathbf{x}_2(t) \| = r_1 + r_2, \tag{2}$$

where $\mathbf{x}_i(t), i = 1, 2$ are the positions of the centers of two particles, and $\mathbf{r}_i, i = 1, 2$ are the radii of the particles. For parabolic motion in the gravitational field, the root is found exactly by solving a second degree algebraic equation. If the root does not exist, the particles will not collide.
The normal velocities after collision has taken place are found as:

$$\begin{pmatrix} v_1^{after} \\ v_2^{after} \end{pmatrix} = \begin{bmatrix} A - \varepsilon & 1 - A + \varepsilon \\ A & 1 - A \end{bmatrix} \begin{pmatrix} v_1^{before} \\ v_2^{before} \end{pmatrix}, \; A = \frac{(1+\varepsilon)m_1}{m_1 + m_2}, \tag{3}$$

where m_1, m_2 are the masses of particles, and ε is the coefficient of restitution [4]. When the energy of the system is high, and there are no long-term interactions between particles, a common approach is to employ the event-driven simulation scheme, as described below.

3 Event-driven simulation scheme

As of today, most of the research on collision detection of particle systems is limited to consideration of a relatively simple simulation model. The idea is to discretize time into short intervals of fixed duration. At the end of each time interval, the new positions of the moving particles are computed. The common problem with such methods is related to choosing the length of the interval.

A much more precise and effective approach, the *dynamic event-driven simulation* of a particle system, relies on discrete events that can happen at any moment of time rather then on the fixed time steps [14]. This can be accommodated by introducing an event queue. We employ this scheme with the following events: collision events, predict trajectory events and topological events. The collision optimization problem is considered in the framework of the dynamic event-driven simulation.

In mechanically alloyed materials simulation, particles are usually approximated by balls (disks). A *collision event* occurs when two balls come into contact with each other or with a boundary. A *predict trajectory event* occurs when the trajectory and the velocity of a ball is updated due to re-calculation of the system state at the next time step. A particle travels along a trajectory defined by a function of time between two consecutive events. In most cases, the trajectories are piecewise linear.

The task of detecting collisions can be optimized by maintaining a set of neighboring particle for every particle in the simulated system. A computational overhead, related to collision detection algorithm, is related to *topological events.* New neighboring pairs appearing due to a topological event must be checked for collisions.

4 Dynamic Computation Geometry Data Structures

Consider a problem of optimizing the collision detection for a set of moving particles in the context given above. In a straightforward approach each pair of particles is considered to be neighbors, i.e. the neighbor graph contains $\frac{1}{2}n\,(n-1)$ edges. For a large scale computation the method performs poorly. The number of neighbors considered on each step can be reduced by employing a geometric data structure and by dynamically maintaining the list of neighbors.

The dynamic generalized Delaunay triangulation, the regular spatial subdivision, the regular spatial tree and the set of segment trees based methods were chosen as candidates for analysis. Dynamic generalized Delaunay triangulation in power metric was built using the sweep-plane technique. *INCIRCLE* tests allowed to dynamically maintain the the set of neighbors [6]. A topological event occurs when the proximity relationship in Delaunay triangulation changes. An important characteristic of this method is its versatility - its performance is independent of the distribution of particle sizes and their configuration.

In regular spatial subdivision, the space is divided onto cells. The topological event happens when a particle moves from one cell to another. The size of cells in the regular spatial subdivision is selected so that it guarantees that no more than a constant number of particles resides in each cell at any moment of time. This allows imposing limit on the number of neighbors for each particle, which ensures a good performance for monosized particle systems. A topological event happens when a particle moves from one cell to another. Introducing hierarchy for regular spatial subdivision reduces storage requirements. An AVL tree is used for this purpose. The method is called the regular spatial tree method.

The set of segment trees is the final data structure considered. A tree of intersecting segments, obtained as a projection of the bounding boxes of particles onto one of the coordinate axes, is dynamically maintained. The particles are said to be neighbors if their bounding boxes intersect. A topological event takes place when two segment endpoints on one of the axes meet. If the bounding boxes of the corresponding particles intersect, then the particles become neighbors. A detailed description and performance analysis of the above data structures can be found in [5].

5 Experimentations

The event-driven simulation environment for shaker ball mill was created at the Dept. of Mechanical and Manufacturing Engineering, University of Calgary. Algorithms were implemented in Object-Oriented Pascal in the Borland Delphi environment and run under Windows 2000 operating system. The size of the simulation space (shaker area) was set to 200 by 200 mm. Balls radii were in the range from 1 to 10 mm. The duration of the simulation run was set to 10 sec. Predict trajectory events were computed every 0.005 sec, which defines the duration of the timestep.

5.1 Monosized Data Sets

The first series of experiments considered data sets comprising 100 monosized balls (i.e. particles with the same radius). The density of the distribution is defined as the ratio between the combined volume of balls in the simulation space and the volume of the simulation space. The density was gradually increasing from 5% to 70% (see Table 1). The number of collisions and the number of predict trajectory events are independent of the collision optimization method. The number of collisions grows as a linear function on a packing density until it reaches 50

Table 1. Number of collision and predict trajectory events for monosized data sets

Packing density	5%	20%	33%	50%	70%
Ball radius (mm)	2.5	5	6.5	8	9.5
Collision events	174	387	661	1143	4170
Predict trajectory	199855	199459	198941	198009	192371

In Table 2, the elapsed time, the total number of topological events (TE) and the total number of collision checks (CC) performed during simulation run are recorded. From Table 2, observe that the number of topological events is the

Table 2. Experimental results for monosized data sets.

Density	5%	5%	5%	33%	33%	33%	70%	70%	70%
Algorithm	Time	TE	CC	Time	TE	CC	Time	TE	CC
Direct	1114.61	0	20645172	1218.85	0	20783008	1255.10	0	21899782
Regular sub.	96.50	1718	93137	117.71	582	625788	166.15	759	1443880
Spatial tree	103.04	1718	93137	128.36	582	625788	171.81	759	1443880
Segment trees	106.94	13038	11542	121.72	13362	116468	151.87	30094	337770
Dynamic DT	304.84	778	1926240	312.64	601	1925194	333.89	770	1979036

largest for the segment tree method. This can be justified by the fact that the topological event in a segment tree happens every time the projections of two particles collide on a coordinate axis. Thus, the number of topological events increases significantly for a densely packed system. The number of topological events is the smallest for the dynamic Delaunay triangulation method.

The number of collision checks for the straightforward method is approximately 10 times larger than the number of collision checks in dynamic Delaunay triangulation, 20 times larger than that of the regular subdivision and spatial tree methods and 100 times larger than that of the segment tree method. The smallest number of the collision checks for the segment tree method is justified by the fact that collision checks are only performed when bounding boxes of two balls intersect, which happens rarely. The number of the collision checks is the largest for the dynamic DT method, due to the number of neighbors. Note, that the DT method is the only method where this number does not depend on the density of the distribution.

The dependence of the time required to simulate the system of 100 particles on the density of packing is illustrated in Fig. 1. Note that the actual time required to process the simulation on a computer is significantly larger than the duration of the "theoretical" simulation run. Immediate conclusion that can be drawn is that the use of any collision optimization technique improves the performance of the simulation algorithm at least by an order of magnitude.

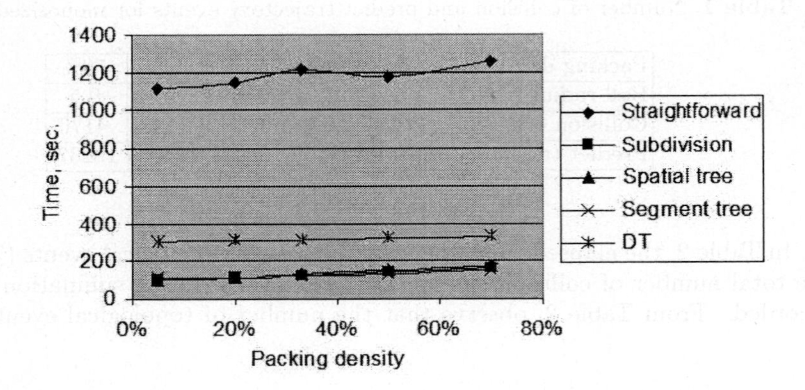

Fig. 1. Time vs. packing density for monosized data set.

The dynamic Delaunay triangulation method performs approximately 3 times slower than all the other collision optimization methods for distributions of low density. It is, however, almost twice as slow for high density distributions. Note that the difference in performance is partly due to the fact that every topological event for this method requires the solution of a 4th order polynomial equation. Also note that the number of collision checks is practically constant in the dynamic Delaunay triangulation, thus the performance remains unchanged as the density increases.

The spatial tree method is only 5% slower than the regular subdivision method, while it requires significantly less memory. The segment tree method is the most efficient among all methods considered for the monosized particle system.

5.2 Polysized Data Sets

The second series of experiments were conducted for polysized data sets of 100 balls. The radius of largest ball was approximately 10 times larger than the radius of the smallest ball. The packing density changed from 5% to 70%. The number of collision events and predict trajectory events were measured (see Table 3).

The comparison analysis on the number of topological events, collision checks and the elapsed time depending on the type of the CDO method and the density of the particle distribution is given in Table 4. As the packing density increases, the number of collision checks raises significantly for the regular subdivision and spatial tree methods. It is evident that these methods are inappropriatness for CDO for a system of polysized particles. Two of the best performing methods, the segment tree and the dynamic Delaunay triangulation, are now compared.

Table 3. Number of collision and predict trajectory events for monosized data sets.

Packing density	5%	20%	33%	50%	70%
Ball radius (mm)	0.4-4	1-10	1.8-18	3-27	7.5-50
Collision events	189	375	488	695	3127
Predict trajectory	199816	199416	199209	198912	220282

Table 4. Experimental results for for polysized data sets.

Density	5%	5%	5%	33%	33%	33%	70%	70%	70%
Algorithm	Time	TE	CC	Time	TE	CC	Time	TE	CC
Direct	1332.66	0	20657627	1284.71	0	20741959	1690.39	0	27485297
Regular sub.	103.26	1161	298667	242.66	257	3648103	1880.48	30	27358811
Spatial tree	111.99	1161	298667	247.72	257	3648103	1922.31	30	27358811
Segment tree	109.68	12432	14154	121.77	11050	91584	395.65	36450	575909
Dynamic PD	303.85	881	1934517	318.73	626	1934358	397.27	735	2221611

It was established that the number of collision checks for the dynamic DT is independent of the distribution density. Interestingly, this number approaches the number of collision checks for monosized data set. Number of collision checks for the segment tree method is approximately 14 times less than that of the dynamic DT method in case of low packing density. It is four times smaller than the number of collision checks in DT for distribution with high density of packing. The opposite trend can be noticed in regards to the number of topological events. This number twice smaller for DT method for the distribution with low packing density and almost 30 times smaller for the high density particle distribution.

The graph exploring elapsed time versus density of packing relationship is found in Fig. 2. It can be observed that the DT is a very steady method with the time behaving as a constant function. As the density increases over 45% , the regular subdivision and the spatial tree methods require more time than the dynamic DT or the segment tree method. The segment tree method outperforms the dynamic DT method for low densities. It matches the performance of the DT method at packing densities close to 70%.

5.3 Increasing the Number of Particles

In the final series of experiments, the number of particles gradually increases from 10 to 100 with their radii selected in the 5 to 10 mm range. Density of particle distribution varies from 4.5% to 35% (see Table 5). Graph in Fig. 3 illustrates the dependance of the elapsed time on the number of particles. It can be seen that the time required for the straightforward algorithm grows as a quadratic function while the time required by all the other methods shows just a linear growth. Once again the segment tree outperforms all other methods while

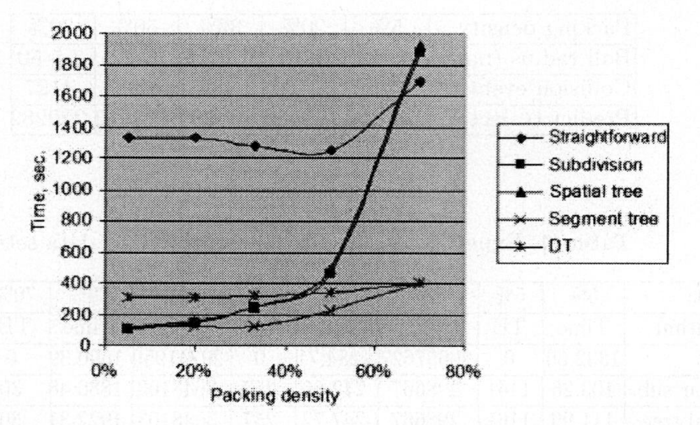

Fig. 2. Time vs. packing density for polysized data sets.

Table 5. Experiments for shaker ball mill data set.

Number of particles	10	20	30	40	50	60	70	80	90	100
Density	4.5%	8.5%	12.5%	15.0%	20.0%	23.0%	26.5%	31.0%	33.5%	35.0%
Collision events	3	5	8	24	31	52	47	79	121	128
Predict trajectory	4014	8021	12026	16006	20000	23985	27972	31902	35871	39854

the dynamic Delaunay triangulation method requires the most time among all collision detection optimization data structures.

6 Conclusions

Based on the results obtained, suggestions about the most appropriate data structure and algorithms for some other simulation problems can be made. They are applicable to any simulation model that can be described by specifying the number of particles, their size distribution, density of packing and functions defining their trajectories. The guidelines are:

1. The regular spatial subdivision, the regular spatial tree and the segment tree methods show similar performance for monosized particle systems with low packing density.
2. The segment tree method outperforms the other methods for monosized particle systems with high packing density (higher than 50%).
3. The dynamic Delaunay triangulation method performs worst for monosized distributions.

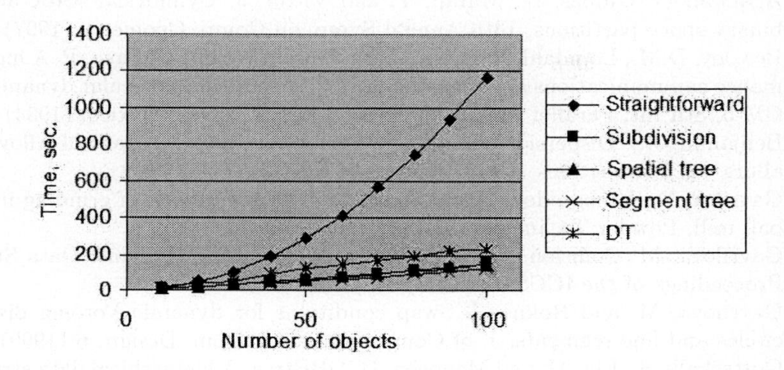

Fig. 3. Time vs. number of particles.

4. The dynamic Delaunay triangulation method is as efficient as the segment tree method for polysized high-density distributions (when the density of the distribution is 70% or higher). Both of them outperform other methods for this type of simulated system.
5. The segment tree method is the best method for polysized low-density distributions (for density lower than 70%).
6. The regular spatial subdivision and the regular spatial tree methods can perform almost as bad as the straightforward method for polysized high-density distributions (density higher than 50%) and thus shouldn't be used for collision detection optimization in such systems.
7. The regular spatial subdivision and the regular spatial tree show similar performance with regular subdivision method being 5% faster but requiring more space. The space required for DPT, spatial tree and segment tree method is practically the same.

Some general conclusions can be drawn from the above discussion. Based on the overall performance, the segment tree can be considered as the best candidate for almost all types of the systems considered. The dynamic generalized Delaunay triangulation method demonstrates consistent performance that practically does not depend on the distribution of radii of the particles or their packing density. It is shown that despite its complexity the method is a good candidate for collision detection optimization for polysized particle systems.

References

1. Agarwal, P., Guibas, L., Murali, T. and Vitter, J. Cylindrical static and kinetic binary space partitions, 13th Annual Symp. on Comp. Geometry (1997) 39–48.
2. Beazley, D.M., Lomdahl, P.S., Gronbech-Jensen, N. and Tamayo, P. A high performance communications and memory caching scheme for molecular dynamics on the CM-5, 8th Int. Parallel Processing Symposium, Cancun, Mexico, (1994) 800–809.
3. Benjamin, J.S. Dispersion Strengthened Superalloys by Mechanical Alloying, Metallurgical Transactions, **1** (1970) 2943–2951.
4. Gavrilov, D., Vinogradov, O and Shaw W. J. D. Simulation of grinding in a shaker ball mill, Powder Technology, **101(1)** (1999) 63–72.
5. Gavrilova, M. Collision Detection Optimization using Dynamic Data Structures, Proceedings of the ICCS'02 (2002), in print.
6. Gavrilova, M. and Rokne, J. Swap conditions for dynamic Voronoi diagram for circles and line segments, J. of Computer-Aided Geom. Design, **6** (1999) 89–106.
7. Gottschalk, S., Lin, M. and Manocha, D. OBBtree: A hierarchical data structure for rapid interference detection, Computer Graphics Proceedings, Annual Conference Series, (1996) 171–180.
8. Hubbard, P. Approximating polyhedra with spheres for time-critical collision detection, ACM Transaction on Graphics, **15(3)** (1996) 179–210.
9. Klosowski, J.T, Held, M., Mitchell, J.D.B, Sowizral, H. and Zikan, K. Efficient Collision Detection Using Bounding Volume Hierarchies of k-DOPs, IEEE Trans. Visualizat. Comput. Graph. **4(1)** (1998) 21–36.
10. Krantz, A. Analysis of an efficient algorithms for the hard-sphere problem, ACM Transaction on Modeling and Computer Simulation, **6(3)** (1996) 185–209.
11. Marin, M., Russo, D. and Cordero, P. Efficient algorithms for many-body hard particle molecular dynamics, Journal of Computer Physics, **109** (1993) 306–329.
12. McCormick, P.G. , Huang, H., Dallimore, M.P., Ding J. and Pan J. The Dynamics of Mechanical Alloying, in Proceedings of the 2nd International Conference on Structural Application of Mechanical Alloying, Vancouver, B.C. (1993) 45–50.
13. Medvedev, N.N. Voronoi-Delaunay method for non-crystalline structures, SB Russian Academy of Science, Novosibirsk (2000)
14. Vinogradov, O., Explicit equations of motion of interacting spherical particles, Recent Advances in Structural Mechanics, **248** (1992) 111–115.

Modified DAG Location for Delaunay Triangulation

Ivana Kolingerová [1]

Centre of Computer Graphics and Data Visualization, Department of Computer Science and Engineering,, University of West Bohemia, Plzeň
kolinger@kiv.zcu.cz, http://iason.zcu.cz/~kolinger

Abstract. The paper describes a modification of DAG-based location for incremental insertion Delaunay triangulation algorithm in E^2 and E^3. Instead of the whole simplices, only their separating faces are stored and tested. The proposed technique allows full reuse of results obtained on previous DAG levels. This enables to reduce time for point location and memory requirements. Asymptotic memory and time complexity is not changed.

1 Introduction

Incremental-insertion-based algorithms do not belong to the fastest but they are easy to understand, on-line, their implementation is usually quite simple and it is a bit easier to make them robust than other types of algorithms, e.g. divide-and-conquer-based. The complexity of incremental insertion algorithms is given as $n*C1$ where $C1$ is complexity needed to process one element and n is the number of elements. Therefore, they typically cannot achieve better complexity than $O(n^2)$ in the worst case and, with the help of auxiliary data structures, $O(n \log n)$ or even $O(n)$ in expected case for most type of input data distribution.

These general features fully pay also for incremental insertion algorithm for constructing Delaunay triangulation in E^2 and E^3. The algorithm is quite simple; to be effective, it is combined with such a data structure which enables efficient location of newly inserted points in the already finished part of triangulation; usually some kind of tree or uniform space subdivision. With such a data structure, this algorithm has $O(n^2)$ worst case and $O(n \log n)$ up to $O(n)$ expected case time complexity in E^2 and $O(n^3)$ worst case and $O(n \log n)$ up to $O(n)$ expected case time complexity in E^3. ($O(n^3)$ bound for E^3 can appear if the resulting triangulation has $O(n^2)$ simplices – it is possible but very rare, see [5, 6]).

One of the oldest and most often used data structure is a DAG, directed acyclic graph. One node of DAG contains one simplex. Using DAG and a randomization of the input data, optimal $O(\log n)$ location time per inserted point can be achieved in the expected case. Even better, its behaviour for various input data distribution is nearly identical. The most important disadvantage of DAG are big memory requirements – although in E^2 asymptotically linear, the memory for DAG reaches up to $9n$ according

[1] This work was supported by the Ministry of Education of The Czech Republic – project MSM 23 5200 005.

P.M.A. Sloot et al. (Eds.): ICCS 2002, LNCS 2331, pp. 125–134, 2002.

to our measurements (468*n* bytes in our implementation). For E^3, the memory is asymptotically quadratic because number of tetrahedra in a triangulation in E^3 can be quadratic; in practice, we measured 54*n* (3672*n* bytes in our implementation) as an upper limit for various input distributions and real data.

In this paper, we will show how time for point location and memory requirements of DAG can be reduced if, instead of the whole simplices, only separating faces are stored in DAG and tested, thus reducing redundancy in the hierarchical point-in-simplex tests. The proposed technique has some similarity to miniDAG location used for regular triangulation in E^3 in [7]; however, it was developed independently. In [12], we gave a slightly different version of this technique in E^2 as a practical algorithm for computer graphics and GIS applications. This paper shows further development of this technique and its E^3 generalization.

Section 2 gives all necessary definitions, briefly describes the fundamental algorithm and surveys previous work on fast location for incremental Delaunay insertion. Section 3 provides details about DAG for Delaunay triangulation. Section 4 explains the newly proposed modification and section 5 handling of singularities. Section 6 gives theoretical and experimental results. Section 7 concludes the paper.

2 Background

Given a point set *P* in E^d, a *k*-simplex, *k* *d*, is a convex combination of *k*+1 affinely independent points in *P*, called vertices of the simplex. An *s*-face of a simplex is the convex combination of a subset of *s*+1 vertices of the simplex. By 'face' we will mean *d*-1-face. [2]

A triangulation *T(P)* of a set of points *P* in E^d is the set of *d*-simplices such that:

1. a point *p* in E^d is a vertex of a simplex in *T(P)* if *p* *P*
2. the intersection of two simplices in *T(P)* is either empty or a common vertex or a common face
3. the set *T(P)* is maximal; it is not possible to add any simplex without violating the previous rules.

A triangulation *T(P)* in E^d is a Delanauy triangulation if the circumsphere of each *d*-simplex does not contain any point of the set *P*.

We suppose that the orientation of simplex faces is positive. To decide whether a point is inside a simplex, we use the orientation predicate, defined for E^d over *d*+1 points by the sign of *d*-dimensional determinant. For E^2 and E^3, these predicates can be as follows:

$$E^2\text{: Right}(a,b,c) = \text{Sgn} \begin{vmatrix} by & ay & cy & by \\ bx & ax & cx & bx \end{vmatrix}$$

$$E^3\text{: Right3D}(a,b,c,d) := \text{Sgn}\begin{vmatrix} dx & ax & dy & ay & dz & az \\ dx & bx & dy & by & dz & bz \\ dx & cx & dy & cy & dz & cz \end{vmatrix}$$

Let's point out that for positive orientation, these predicates return minus if c is to the left of ab in E^2 and d to the left of abc, it means, if it is inside the simplex whose face ab or abc is. The whole point-in-simplex test than calls the Right or Right3D function for the faces of the simplex. If the tested point is outside, we can stop the orientation test with the first positive sign; it saves a half of the tests in average. It means that 1.5 tests for a triangle and 2 tests for a tetrahedron are needed in average. If the point is inside, we have to test all faces. Singularities are indicated by at least one zero from orientation tests. Implementation of these predicates has to be robust enough to handle such cases, see e.g. [7].

Delaunay triangulation algorithm using DAG can be found for E^2 in [1, 9] and for E^3 in [10, 11, 7]. It achieves $O(\log n)$ per point location. We will not give a detailed explanation here because the algorithm is widely known. Brief description is as follows. At the beginning, all points of the input set P are emebeded into one big auxiliary simplex. Then points are inserted one at a time. For each inserted point, the simplex containing it has to be found. This simplex is then subdivided into several simplices; three in E^2 and four in E^3. Then Delaunay triangulation has to be restored. It is done by tests of new simplices: The outer face of the new simplex is tested and if the empty circumsphere property is not satisfied, simplices sharing this face are swapped, i.e., changed to the other possible configuration. This swap is easy in E^2 because instead of two old, two new triangles are created, and difficult in E^3 where three kinds of swaps are possible: either two tetrahedra are replaced by three, or vice versa; the last possibility is a special swap of such two or four tetrahedra which together form a pyramid or a double pyramid with four vertices in one plane. (More details about swappable and non-swappable configurations in E^3 can be found in [10, 11] and are not subject of this paper.) All simplices are stored as nodes of DAG. When all points of P have been inserted, all simplices containing the auxiliary points (which do not belong to P) are deleted and the final triangulation is extracted from DAG leaves.

Due to popularity of the incremental insertion algorithm, various point location techniques and data structures appeared, especially in E^2. Simple and popular is a strategy of walking, where the triangle containing the inserted point is found by visiting triangles between the inserted point and some starting point. This technique was for triangulations demonstrated in [8], achieving $O(n^{1/2})$ per point location in E^2. Further improvement uses the idea of 'better starting point' estimate by random sampling m middle centres of triangulation edges, $m = n^{1/3}$, and taking the point closest to the located point as a start of walking, see [14]. For randomly distributed points, $O(n^{1/3})$ in E^2 and $O(n^{1/4})$ in E^3 for one point location is achieved. Comparison of several walking strategies in Delaunay triangulation in E^2 and E^3 is given in [4].

[16] is a combination of walking, regular grid and presorting of points so that points neighbouring in the data structure are geometrically close. Tests show approximately $O(n^{1.1})$ time per point location in E^2.

[13] suggests a data structure with $O(\log n)$ levels, each level containing a random sample of triangles from the lower level of the data structure. At each level, Delaunay triangulation of the sample of points is computed and overlapping triangles at different levels are linked to ease a point-in-triangle location. One point location needs $O(\log^2 n)$ time. Simplification and improvement of this data structure can be found in [3] where $O(\log n)$ per point is achieved in E^2.

Another way to speed up point location is to use bucketing or quadtree data structures, see, e.g., [15, 18, 20]. Generally, this kind of data structure is more sensitive to input data distribution, although for more or less uniformly distributed points can be very fast.

The incremental insertion algorithm has also been implemented for non-euclidean metrics [19] and for constrained triangulation [18].

3 Details about DAG for Delaunay triangulation

DAG in the incremental insertion for Delaunay triangulation is a hierarchy in which a history of the triangulation is stored. For location, it is necessary to know links from parent nodes to sons to be able to traverse the DAG down to the leaves. Links in the opposite direction are not necessary. See example of a part of a triangulation history in a DAG in Fig.1.

In E^2, a DAG contains three types of inner nodes:

1. nodes with three sons obtained by point insertion – three new triangles are made instead of one old,
2. nodes with two sons obtained by an insertion a point incident to an existing triangulation edge – four new triangles are made instead of two old, and
3. nodes with two sons obtained by an edge swap between two triangles – two new triangles are made instead of two old.

In E^3, there are these six types of inner nodes:

1. nodes with four sons obtained by point insertion – four new tetrahedra are made instead of one old,
2. nodes with three sons obtained by an insertion a point lying in a tetrahedron face – six new tetrahedra are made instead of two old,
3. nodes with two sons obtained by an insertion of a point incident with an edge of the triangulation – each tetrahedron sharing this edge is divided into two tetrahedra; the total number of tetrahedra sharing an edge is usually a small constant but can be as high as $O(n)$,
4. nodes with three sons obtained by a face swap S23 – two tetrahedra are removed and replaced by three,
5. nodes with two sons obtained by a face swap S32 – three tetrahedra are removed and replaced by two,
6. nodes with two sons obtained by a face swap S44 or S22 – four or two new tetrahedra appear instead of the same number of old ones.

In any moment, leaves contain the current triangulation.

4 The proposed modification

We will suppose that the orientation test returns plus or minus, never zero; handling of singularities will be explained in section 5. We will call Test_*ij* the orientation test Right() (in E^2) or Righ3D() (in E^3) of the face separating a simplex *i* from simplex *j*, both simplices are sons of the same node.

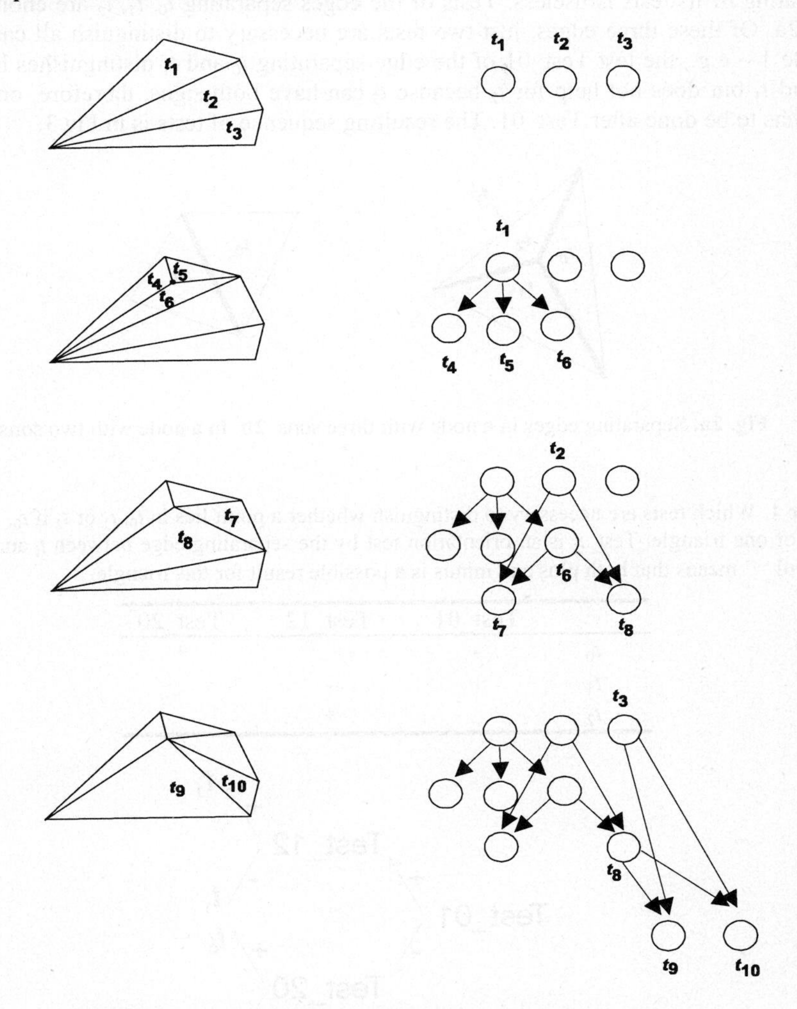

Fig. 1. Example of a part of a triangulation history in a DAG

For better readability, we will explain the proposed modification for E^2 case and then we will state differences for E^3.

Let's have such an inner node N_i of DAG which contains a triangle t_i that was found to contain the inserted point *q*. Let t_i have three sons t_0, t_1, t_2. All sons are tested by point-in-triangle tests, it means 9 orientation tests in the worst case and 4.5 in the average case. The first simple improvement that can be made is not to test the shared edges twice (once for each incident triangle), thus saving up to three tests for three sons. But the idea can be derived a bit further.

If we are testing the query point q against triangles t_0, t_1, t_2, we already know that q lies inside t_i because it has been tested in the previous step of the search traversal – the outer boundary of t_i is in this case identical with outer boundary of its sons and repeating of its tests is useless. Tests of the edges separating t_0, t_1, t_2 are enough, see Fig.2a. Of these three edges, just two tests are necessary to distinguish all cases, see Table 1 – e.g., the test Test_01 of the edge separating t_0 and t_1 distinguishes between t_0 and t_1 but does not help for t_2 because t_2 can have both signs, therefore, one more test has to be done after Test_01. The resulting sequence of tests is in Fig.3.

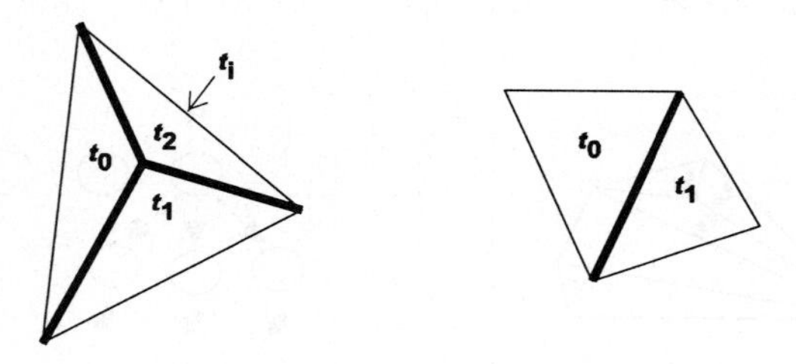

Fig. 2a. Separating edges in a node with three sons. **2b**. In a node with two sons

Table 1. Which tests are necessary to distinguish whether a point lies in t_0, t_1 or t_2 if t_0, t_1, t_2 are sons of one triangle. Test_*ij* is an orientation test by the separating edge between t_i and t_j. The symbol ' ' means that both plus and minus is a possible result for this triangle.

	Test_01	Test_12	Test_20
t_0	-		+
t_1	+	-	
t_2		+	-

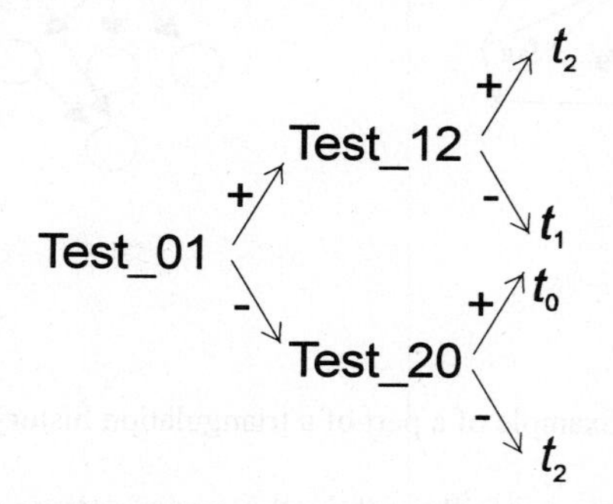

Fig. 3. A sequence of tests for 3 sons gives a structure of the corresponding algorithm

If N_i had two sons, the two triangles t_0 and t_1 would give together 6 and 3 orientation tests in the worst and average case, respectively. However, one test is enough to distinguish between t_0 and t_1, see Fig. 2b, even in case that they originated from a swap where new triangles are not covering the area of just one triangle but two, see t_2, t_7, t_8 or t_6, t_7, t_8 in Fig.1 for better understanding. The sequence of tests is trivial here, just Test_01.

In the average case we save 2.5 tests per triangle with 3 sons (2 instead of 4.5 tests) and 2 tests per triangle with 2 sons (1 instead of 3 tests). In the worst case we would save 7 tests for 3 sons and 5 tests for 2 sons but the average case is more probable (recall Section 2).

Now to the E^3 case. For 2 and 3 sons, the sequence of tests is the same as in E^2. For 4 sons, there exist 6 separating faces - we need to handle 6 mutual combinations of tetrahedra. Table 2 shows which test distinguishes between which pair of tetrahedra. Of these 6 tests, we need 3 to find unambiguously which tetrahedron contains the point, see Fig.4.

Table 2: Which tests are necessary to distinguish whether a point lies in t_0, t_1, t_2 or t_3 if t_0, t_1, t_2, t_3 are sons of one tetrahedron. Test_*ij* is the orientation test by the separating face between t_i and t_j. The symbol ' ' means that both plus and minus is a possible result for this tetrahedron.

	Test_01	Test_02	Test_03	Test_12	Test_13	Test_23
t_0	-	-	-			
t_1	+			-	-	
t_2		+		+		-
t_3			+		+	+

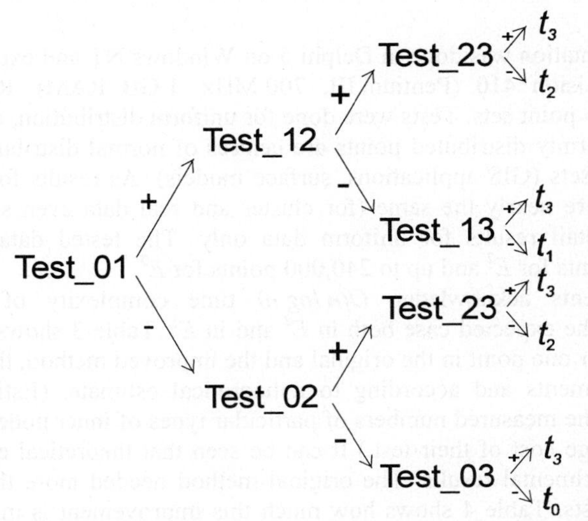

Fig. 4. A sequence of tests for 4 sons gives a structure of the corresponding algorithm

The proposed technique can be also used to reduce memory requirements of the DAG. It is not necessary to store all vertices of simplices if the simplices are not in the leaves; for DAG traversal, we need only part of vertices. Therefore, when a list becomes an inner node, we can deallocate part of its information. We described results of experiments with memory deallocation in [12] – in E^2, in our implementation about 28% memory savings were achieved, memory deallocation prolonged the computation 9% in average.

Before we will show estimated and measured savings, let us explain how to handle singularities – the inserted point lies on some face.

5 Singularities

It is necessary to recognize whether the tested point lies on some face to avoid creating simplices with empty volume. The orientation predicates return zero in such a case. However, if we are not in a leaf simplex, incidence of the tested point with some face is not substantial because the face may have been swapped and may not exist any more in the current triangulation. Therefore, it is useless to handle singularities earlier than at leaf level. For simplicity, we test singular cases by one extra call to point-in-simplex test, with full control of singular cases, at the end of DAG traversal, on leaf level. The amount of extra orientation tests caused by this suboptimal but simple solution is very small in comparison with the total number of tests (5% for n=1,000 down to 2.3% for n=1,000,000 in E^2 and 3.8% for n=1,000 down to 1.8% for n=240,000 in E^3).

6 Results

The implementation was done in Delphi 5 on Windows NT and experiments were run on Dell Precision 410 (Pentium III, 700 MHz, 1 GB RAM). Running times are averages for 5 point sets. Tests were done for uniform distribution, cluster distribution (several uniformly distributed points are centres of normal distribution, see [17] and for real data sets (GIS applications, surface models). As results for various types of input data were nearly the same (for cluster and real data even slightly better), we present in detail results for uniform data only. The tested data sets were up to 1,000,000 points for E^2 and up to 240,000 points for E^3.

Measurements acknowledge $O(n \log n)$ time complexity of the DAG-based algorithm in the expected case both in E^2 and in E^3. Table 3 shows number of tested edges in E^2 per one point in the original and the improved method, their rate according to the experiments and according to a theoretical estimate. (Estimates were done according to the measured numbers of particular types of inner nodes - see Section 3 - and the average cost of their test.) It can be seen that theoretical estimates are quite close to experimental results. The original method needed more than three times as many edge tests. Table 4 shows how much this improvement is mirrored in running times for location and for the whole incremental insertion algorithm. Running times for location are for higher n 2.56-2.85 times better than before; the whole running time is then 59-70% faster than before.

Table 3. Number of tested edges per one point in the original DAG tests, in the proposed modification (New) and their rate for uniformly distributed data in E^2

n	1,000	5,000	10,000	50,000	100,000	500,000	1,000,000
Original	86.23	113.95	130.008	156.20	163.88	194.25	204.68
New	29.40	37.23	41.57	48.97	51.64	59.80	62.93
Orig./New – exper.	2.93	3.06	3.13	3.19	3.17	3.25	3.25
Orig./New – theoret.	2.81	2.81	2.81	2.81	2.81	2.81	2.80

Table 4. Rate of running time for location and for the whole incremental insertion in E^2

n	1,000	5,000	10,000	50,000	100,000	500,000	1,000,000
Location; Orig./New	9.75	4.76	2.56	2.85	2.71	2.57	2.61
All; Orig./New	1.70	1.63	1.59	1.68	1.66	1.69	1.70

Tables 5 and 6 give results for E^3. Theoretical and practical results are in correspondence. Again, the original method needed more than three times as many face tests, however, location times are only 1.47-2.10 times better because some work needs to be spent to identify which face of the subdivided or swapped tetrahedron should be kept for further test and which not (in E^2 it is much easier to decide). The whole computation is improved only by 6-14% because in E^3, location is not so prevailing part of computation, taking up to 30-40% of the whole running time in comparison with 55-60% in E^2.

Table 5. Number of tested faces per one point in the original DAG tests, in the proposed modification (New) and their rate for uniformly distributed data in E^3

n	1,000	5,000	10,000	50,000	100,000	200,000	240,000
Original	155.82	220.64	242.23	303.00	334.60	357.88	361.27
New	52.43	71.13	77.37	94.70	103.51	109.68	110.91
Orig./New – exper.	2.97	3.10	3.13	3.20	3.23	3.26	3.26
Orig./New – theoret.	3.30	3.31	3.31	3.31	3.31	3.31	3.31

Table 6. Rate of running time for location and for the whole incremental insertion in E^3

n	1,000	5,000	10,000	50,000	100,000	200,000	240,000
Location; Orig./New	1.88	1.47	2.10	1.90	1.80	1.78	1.81
All; Orig./New	1.06	1.09	1.10	1.12	1.13	1.14	1.14

7 Conclusion

We proposed a simple modification of DAG-based location for Delaunay incremental insertion in E^2 and E^3. The proposed method reduces about two thirds of tests, thus bringing a speedup of location especially in E^2. As only part of geometrical information about former simplices is needed, memory savings can be also achieved.

References

1. de Berg M., van Kreveld M., Overmars M., Schwarzkopf O.: Computational Geometry. Algorithms and Applications, Springer-Verlag Berlin Heidelberg (1997)
2. Cignoni P., Montani C., Scopigno R.: DeWall: a Fast Divide and Conquer Delaunay Triangulation Algorithm in E^d, Computer-Aided Design, Vol. 30, No.5 (1998) 333-341
3. Devillers O.: Improved Incremental Randomized Delaunay Triangulation, ACM Symposium on Computational Geometry, Minneapolis, USA (1998) 106-115
4. Devillers O., Pion S., Teillaud M.: Walking in a Triangulation, SCG'01 proceedings, Medford, Massachusets, USA (2001) 106-114
5. Erickson J.: Dense Point Sets Have Sparse Delaunay Triangulations or "... But Not Too Nasty", To appear in Proceedings of the <u>*13th Annual ACM-SIAM Symposium on Discrete Algorithms*</u> (2002)
6. Erickson J.: Nice Point Sets Can Have Nasty Delaunay Triangulations, Proceedings of the <u>*17th Annual ACM Symposium on Computational Geometry*</u>, (2001) 96-105. Invited to the special issue of Discrete & Computational Geometry devoted to the symposium
7. Facello M.A.: Implementation of a Randomized Algorithm for Delaunay and Regular Triangulations in Three Dimensions, Computer Aided Geometric Design (1995) 349-370
8. Guibas L. J., Stolfi J.: Primitives for the Manipulation of General Subdivisions and the Computation of Voronoi Diagrams, ACM Transactions on Graphics, Vol. 4, No.2 (1985) 75-123
9. Guibas L. J., Knuth D. E., Sharir M.: Randomized Incremental Construction of Delaunay and Voronoi Diagrams, Algorithmica, Vol. 7 (1992) 381-413
10. Joe B.: Three-dimensional Triangulations from Local Transformations, SIAM J. Sci. Stat. Comput. 10 (1989) 718-741
11. Joe B.: Construction of Three Dimensional Triangulations Using Local Transformations, Computer Aided Geometric Design 8 (1989) 123-142
12. Kolingerová I., Žalik B.: Improvements to Randomized Incremental Delaunay Insertion, sumitted to Computers and Graphics (2001)
13. Mulmuley K.: Randomized Multidimensional Search Trees: Dynamic Sampling. In: Proc. 7th Annual ACM Sympos. Comput. Geom. (1991) 121-131
14. Mücke E. P., I. Saias I., Zhu B.: Fast Randomized Point Location Without Preprocessing in Two- and Three-dimensional Delaunay Triangulations, CompGeom'96, Philadelphia (1996) 274-283
15. Ohya T., Iri M., Murota K.: Improvements of the Incremental Method for the Voronoi Diagram with Computational Comparison of Various Algorithms, Journal of Operations Research Society of Japan, 27 (1984) 306-337
16. Sloan S. W.: A Fast Algorithm for Constructing Delaunay Triangulations in the Plane, Adv. Eng. Software, Vol. 9, No.1 (1987) 34-55
17. Su P., Drysdale R.L.S.: A Comparison of Sequential Delaunay Triangulation Algorithms, In: Proc. of the 11-th Annual Symposium on Computational Geometry (1995) 61-70, ACM
18. Vigo M.: An Improved Incremental Algorithm for Constructing Restricted Delaunay Triangulations, Computers & Graphics 21 (1997) 215-223
19. Vigo M., Pla N.: Computing Directional Constrained Delaunay Triangulations, Computers & Graphics 24 (2000) 181-190
20. Žalik B., Kolingerová I.: An Incremental Construction Algorithm for Delaunay Triangulation Based on Two-level Uniform Subdivision, submitted to International Journal of Geographical Information Science (2001)

TIN Meets CAD – Extending the TIN Concept in GIS

Rebecca, O.C. Tse and Christopher Gold

Department of Land Surveying and Geo-Informatics
Hong Kong Polytechnic University, Hung Hom, Kowloon, Hong Kong SAR
Tel: (852) 2766-5955; Fax: (852) 2330-2994
Rebecca.Tse@polyu.edu.hk, christophergold@voronoi.com

Abstract. Extending the current "2.5D" terrain model is a necessary development in GIS. Existing 2D surface information must be extended to 3D, while trying to preserve the topological integrity. Computer-Aided Design (CAD) on the other hand is often fully 3D, which may be excessive for our "extended 2D" mapping needs. We have used the "boundary representation" (b-rep) design from CAD systems, based on manifold models, to address these problems. Triangulated Irregular Network (TIN) models of the terrain surface are well known in Geographic Information System (GIS), but are unable to represent cliffs, caves or holes, which are required to represent complex buildings. The simple TIN b-rep structures may be extended using formal Euler Operators to add simple CAD functionality while guaranteeing the connectivity required for terrain modelling. We believe this provides a simple and reliable extension that is sufficient for many applications.

1. Introduction

Within the world of GIS the TIN has been well known for more than 20 years. It is one of the basic models for representing digital terrain. A piece of land on the earth's surface can be modelled as a terrain [3, 10]. Applications of TINs to problems of runoff, etc., have also been well studied in the computational geometry literature [13]. The difficulties with this "2.5D" model are also well known, in particular the fact that the structure itself is only two-dimensional, the third dimension being only an attribute. This limitation contrasts strongly with the situation in the CAD world, as well as in 3D graphics, where surface meshes of complex 3D objects (plausible and implausible) are routinely generated. What is the big problem?

It is fairly clear that GIS is still firmly two-dimensional in its implementation, despite various attempts to extend it, and despite frequent complaints about the inability to model non-planar networks, bridges, caves, etc. within the traditional TIN surface model. This paper attempts to alleviate this problem somewhat, by moving gently towards an extended view of the traditional TIN. The first step is to recognise that a terrain surface modelled by a TIN is not a two-dimensional entity – it is an air- (or water-) earth interface, the boundary between the "Polyhedral Earth" and the

P.M.A. Sloot et al. (Eds.): ICCS 2002, LNCS 2331, pp. 135–143, 2002.

exterior. Simply put: every TIN has an "underneath". This causes few difficulties in practice.

Once this polyhedral model is accepted, attention naturally turns to other disciplines' experience in polyhedral modelling. Computational geometry has developed a variety of tools for managing surface models [12]. In particular, we have been interested in the Quad-Edge structure [4, 5]. Computer graphics workers have developed many tools used extensively today. In particular, the field of CAD has been developing tools for solid model representation for many years, for example Baumgart [1] and Mantyla [7, 8].

In our current work we are using the CAD-type b-rep structure and Euler Operators to create a connected TIN model with holes (bridges or tunnels). Starting with the well-known TIN model, we have built a set of CAD-type Euler Operators, including the ability to form holes, which are easy to implement with the Quad-Edge structure [5]. Secondly we show that all basic TIN modification operations may be performed with Euler Operators. Thirdly, we show that additional Euler Operators may be used to modify the surface form in various ways, including the insertion and deletion of holes that give the basic forms of bridges and tunnels.

Thus our work consists of four stages:

- Definitions of three levels of operators to achieve our desired system;
- The use of Quad-Edge structures to implement CAD-type Euler Operators;
- The use of Euler Operators to implement basic triangulation functions;
- The extension of our triangulation models with additional Euler Operators.

2. TINs and CAD

The examination of CAD modelling techniques does not seem attractive at first sight. The intersection of cubes and tubes, as in Constructive Solid Geometry (CSG) modelling, hardly applies to terrain models. Modern non-manifold models of surfaces can have astoundingly complex data structures [6], and would be massive overkill for TIN enhancement. However, the simpler manifold-based "b-rep" CAD models appear to be feasible for our purposes. In particular, the careful specification of "Euler-Operators" that guarantee to preserve the topological validity of the bounding surface seems particularly appropriate. Euler Operators ensure the integrity of boundary models. As stated by Mantyla [8], "in a "b-rep" model, an object is represented indirectly by a description of its boundary. The boundary is divided into a finite set of faces, which in turn are represented by their bounding edges and vertices." According to his definition, b-reps are best suited for objects bounded by a compact (i.e. bounded and closed) manifold. However, if the main concern were building a non-manifold object, it would be better to use other models.

In 1988 Weiler [14] proposed non-manifold operators to manipulate topological data in non-manifold models. Moreover, they were not based on the basic Euler-Poincare formula, although some of the topological relationships still can be generalised by a new formula by Masuda [9], although it is more complicated. Nevertheless, it may be valuable to extend the b-rep with Euler Operators into a non-manifold GIS model in the future. We conclude that to extend the TIN, the b-rep model is the most appropriate at present.

3. Simplicity of Implementation – the Quad-Edge Structure

We have examined the basis for various sets of Euler operators mentioned in the CAD literature, and their data structures. This is usually the “Winged-Edge” data structure [1], which has a variety of supporting pointer structures, some of which, upon closer examination, are not necessary in order to maintain a TIN or triangulation-based b-rep surface model. In particular, if all faces are triangles, then no holes may be created in the interior of individual faces and thus some elements of both the Winged-Edge structure and the classical Euler Operators may be removed. We have used the Quad-Edge structure of Guibas and Stolfi [5]. More particularly, the Quad-Edge structure has been used to implement a set of Euler Operators that have been shown to suffice for the maintenance of surface triangulations. According to Weiler [13], if a topological representation contains enough information to recreate nine adjacency relationships without error or ambiguity, it can be considered a sufficient adjacency topology representation. These Euler Operators form the basis of the standard (two-dimensional) incremental triangulation algorithm. In addition, Euler Operators can serve to generate holes within our surfaces, thus permitting the modelling of bridges, overpasses etc. that are so conspicuously lacking in the traditional GIS TIN model. The individual Quad-Edge and Euler Operators take only a few lines of code each. “Make-Edge” and “Splice” (Figs. 1 & 2) are the two simple operations on the Quad-Edge structure, which is formed from four connected “Quad” objects, using the simple implementation of [4]. Every Quad has three pointers:

- N – link to next Quad (“Next”) anticlockwise around a face or vertex
- R – link to next ¼ Edge (“Rot”) anticlockwise around the four Quads
- V – link to vertex (or face)

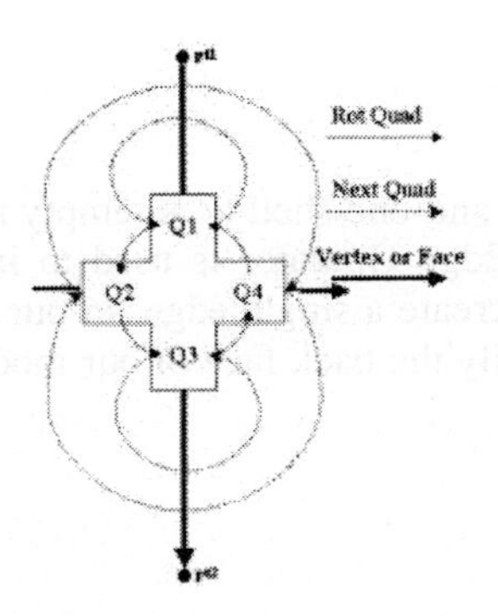

Fig. 1. Make-Edge

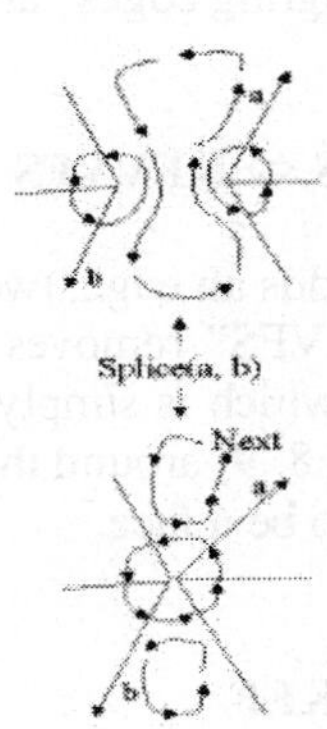

Fig. 2. Splice

The Make Edge operator creates a new independent edge. The Splice operation, which is its own inverse, either splits a face loop and merges two vertex loops (upper to lower parts of Fig 2) or else splits a vertex loop and merges two face loops (lower to upper). This operation suffices to maintain any connected graph on an orientable manifold, such as is the case for CAD b-rep models, and TINs.

4. Selection of a Set of Euler Operators

According to Braid [2], five spanning Euler Operators (plus the Euler-Poincare formula) suffice to specify the number of elements in any b-rep model. Conversely, the inverse formulation serves to identify the number of Euler Operators needed to build a specified model, and to validate its topological integrity. The six elements of the solid model are: vertices; edges; faces; loops (or rings); holes; and bodies (or shells), and are usually based on the Winged-Edge structure.

However, in TINs there are no loops (holes in individual faces), so these will not be considered, leaving five elements. Thus four spanning Euler Operators suffice for TINs with holes. In the traditional TIN models there are no holes, so three Euler Operators will serve for our initial model, to be extended to four in section 7.

5. Implementation of Euler Operators using Quad-Edge structures

The three initial Euler Operators are described below. "Make Edge Vertex Vertex Face Shell" (MEVVFS) creates the initial shell, and "Kill Edge Vertex Vertex Face Shell" (KEVVFS) removes it. "Make Edge Face" (MEF) and its inverse "Kill Edge Face" (KEF) create or kill an edge and a face. "Split Edge Make Vertex" (SEMV) splits an edge, and its inverse is "Join Edge Kill Vertex" (JEKV). While other operators are possible, this set of operators is appropriate for a triangulation model, where no "dangling edges" are permitted.

5.1. MEVVFS ⇔ KEVVFS

"MEVVFS" adds an edge, two vertices, one face and one shell to an empty model. Its inverse "KEVVFS" removes them. One Quad-Edge Operator is used to implement "MEVVFS" (which is simply "Make-Edge") to create a single edge. In our approach the "Loop" [7, 8, 9] around the single edge is really the back face of our model, so we consider this to be a face.

5.2. MEF ⇔ KEF

"MEF" and "KEF" are used to create an edge and a face, and to delete them. In "MEF" we need to give two quads as parameters to make a new face. In "KEF" we need to give an edge as a parameter for removing the edge, and one face will be destroyed as this edge is removed. Fig. 3 shows the operation of "MEF" and "KEF".

"MEF" inputs two quads "a" and "b", output is one new quad "e". "KEF" removes an edge "e" using two splices and one face is killed.

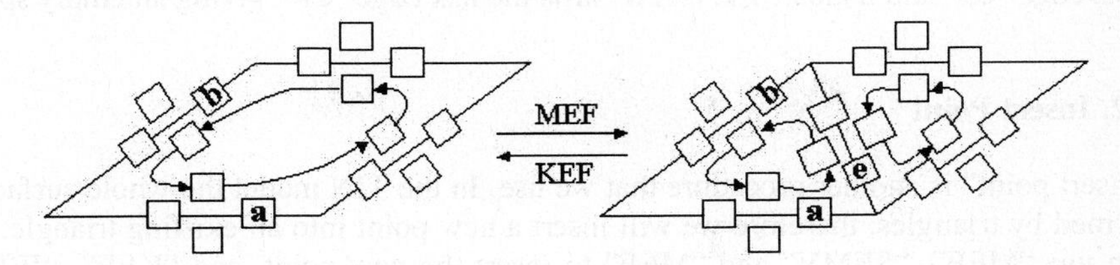

Fig. 3. "MEF" ⇔ "KEF"

5.3. SEMV ⇔ JEKV

We use "SEMV" and "JEKV" to split one edge into two pieces. This procedure adds or removes a point on an edge. It is useful for creating triangles without creating a dangling arc. Fig. 4 shows the operators of "SEMV" and "JEKV". "SEMV" splits edge "e" into two parts, and two parts are joined by "JEKV", using splice operations.

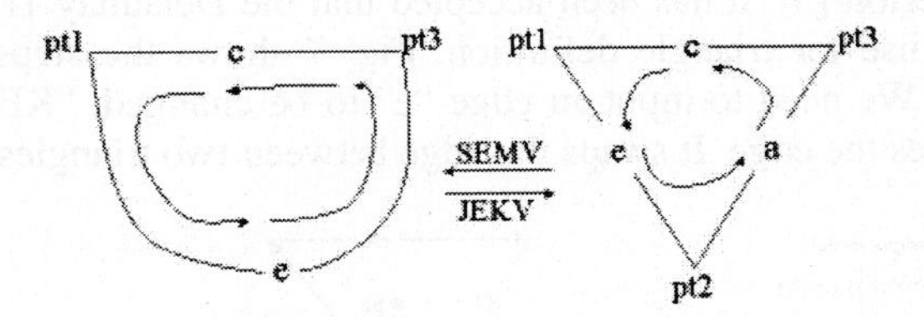

Fig. 4. "SEMV" ⇔ "JEMV"

6. Implementation of the TIN Model Using Euler Operators

In the TIN model we have three main functions, which are: create a first triangle; insert a point; and swap an edge. We use Euler Operators to implement triangulation functions. We use the functions "Big Triangle", "Insert Point" and "Swap". These also have inverses.

6.1. Big Triangle

In our model we limit ourselves to complete faces with no "dangling edges", so we begin from nothing, and create the first triangle using "Big Triangle". We kill the big triangle and go back to nothing by using the inverse functions as shown in Fig. 5. In creating the "Big Triangle", three points are needed as input. Three different Euler Operators are used: "MEVVFS", "MEF" and "SEMV". 3 points (pt1, pt2, and pt3) are input to create the Big Triangle and 3 edges (e1, e2, and e3) are the output.

"MEVVFS" creates the first edge "e1". "MEF" creates a new edge "e3". "SEMV" splits edge "e3". To kill the Big Triangle, "JEKV" joins edge "e2" and "e3". "KEF" kills edge "e3" and a face. "KEVVFS" kills the last edge "e1", giving an empty space.

6.2. Insert Point

"Insert point" is another procedure that we use. In the TIN model the whole surface is formed by triangles, therefore we will insert a new point into an existing triangle. We can use "MEF", "SEMV" and "MEF" to insert the new point, and "KEF", "JEKV" and "KEF" to delete the point. Fig. 6 shows the procedure of inserting a new point. Inserting a new point makes three edges and two faces. "MEF" creates an edge "N4". "SEMV" splits an edge "N4". Last step "MEF" creates a new edge "N6".

6.3. Swap

Swap is a procedure for swapping two edges inside the TIN model. For the Delaunay Triangulation, we use the "in-circle" test to test the triangle, and use the swap operator to change edges. The Delaunay Triangulation is based on the empty circumcircle criterion [5]. It has been accepted that the Delaunay Triangulation is the best criterion to use for triangle definition. Fig. 7 shows the steps for Swap using Euler Operators. We need to input an edge "e" to be changed. "KEF" kills the edge and "MEF" creates the edge. It swaps the edge between two triangles.

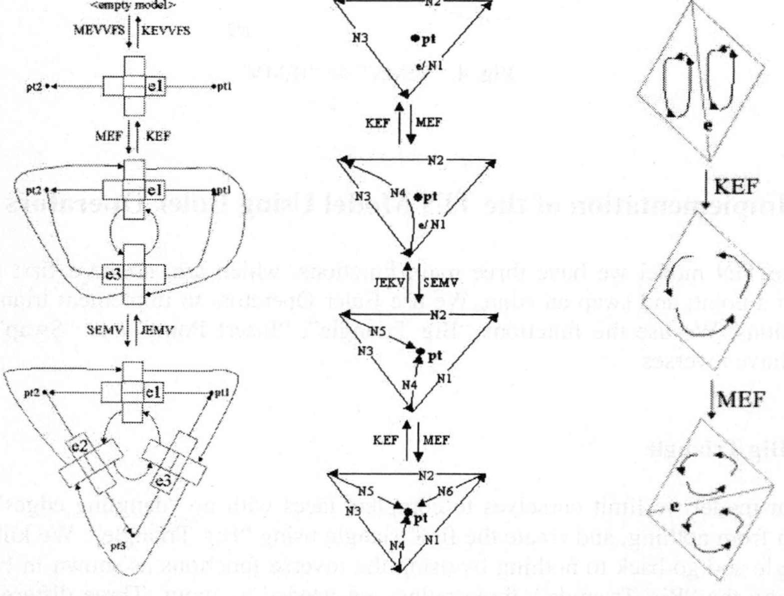

Fig. 5. "Big Triangle" **Fig. 6.** "Insert Point" **Fig. 7.** "Swap"

7. Extending the TIN Model Using Euler Operators

So far we have merely re-formulated the traditional incremental algorithm, but with this background we can use existing Euler Operators to modify the TIN surface in other ways, e.g. by the insertion and deletion of holes. The operator "Make Edge Hole Kill Face" (MEHKF) consists of exactly the same code as "MEF" [11]. However, instead of taking two Quad-Edges that are part of the same face loops as parameters, it takes Quad-Edges that are parts of separate face loops. The two triangles concerned are deleted, an edge is formed between the two triangles, and a new face is formed that loops through each of the deleted triangles and both sides of the new edge.

We select two triangles in an existing TIN model: their edges are ordered anti-clockwise from outside. Two original triangles from the TIN model are shown, looking from "inside" in Figs 10-12. The order of the three edges inside the two triangles is 1 => 2 => 3 => 1 and 4 => 5 => 6=> 4. We make a new edge between these two triangles and a hole is created. One face loop is created inside the hole and two triangle faces are killed.

Fig. 10 shows the result of the first step "MEHKF". The connection of the edges will be 1=>P =>5 =>6 =>4=> Q=> 2=> 3=> 1. Fig. 11 shows the result after performing "MEF". One new face is made and the connectivity of the edges will be 1 => P =>5 =>R =>1 (New face) and 3 => S => 6 => 4 => Q=> 2=> 3. Fig. 12 shows the result of the final "MEF". One new edge and face are created. There are now three faces inside the hole, but the connectivity is preserved and you can walk through the hole. We have therefore shown that the elementary Quad-Edge based Euler Operators are able to generate and modify the traditional TIN structure, permitting basic CAD operations. Three more "MEF"s are used to split the faces inside the hole into triangles. Fig. 13 shows the top view of a TIN model. Fig. 14 shows a hole and a bridge on the TIN.

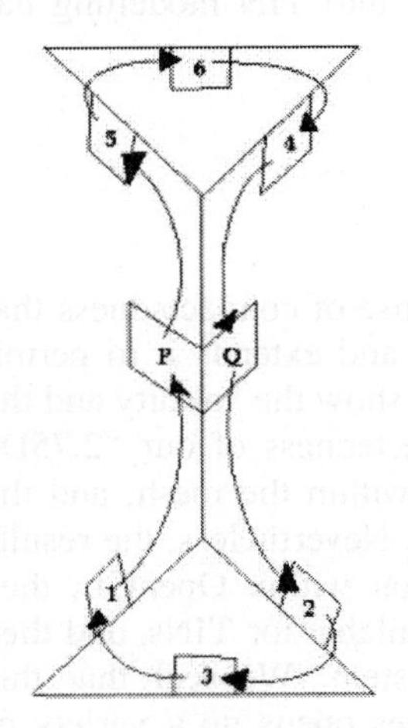

Fig. 10. "MEHKF"

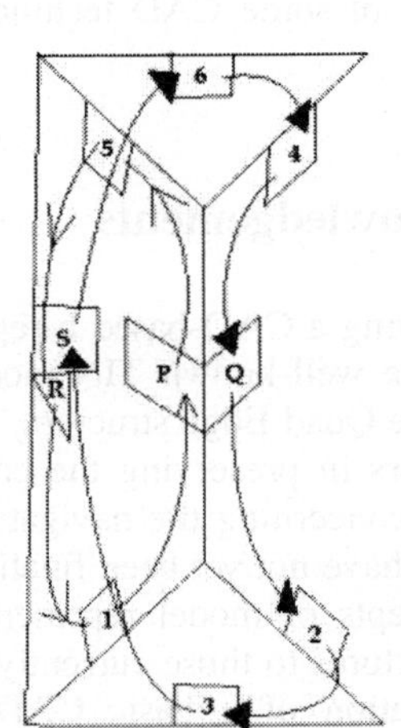

Fig. 11. "MEF"

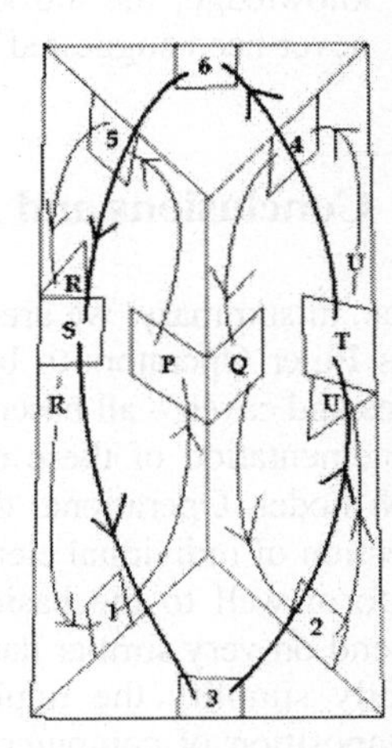

Fig. 12. Third "MEF"

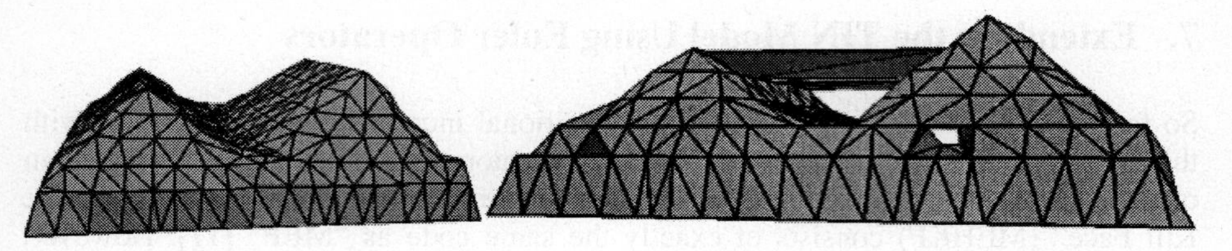

Fig. 13. Top view of TIN model **Fig. 14.** A Hole and a bridge on the TIN

8. Relation to Previous Work

The work we describe here has several steps, each of them new within a particular context:

1) Euler Operators, as used in CAD systems, are usually based on the Winged-Edge structure of [1]. The implementations of [7, 8] indicate the complexity of this approach. We show here that the Quad-Edge structure of [5] is a more natural, and much simpler, methodology. This is of potential interest to the CAD and Computer Graphics communities.
2) For the GIS community, we show the close relation between TIN modelling, b-rep structures from CAD, and Computer Graphics mesh generation. By eliminating one CAD primitive (holes in faces) and restricting ourselves to triangular elements, Euler Operators may be directly ported to GIS.
3) Once TIN operations are expressed as Euler Operators, additional topological operations may be performed directly, in particular the creation and deletion of holes permitting the interactive formation of bridges, tunnels, etc. To our knowledge, the introduction of some CAD techniques into TIN modelling has never been suggested before.

9. Conclusions and Acknowledgements

Thus, in summary, we are following a CAD-based b-rep sense of connectedness that uses Euler Operators to build the well-known TIN model, and extends it to permit holes and caves – all based on the Quad-Edge structure. We show the validity and the implementation of these operators in preserving the connectedness of our "2.75D" TIN model. Operational details concerning the navigation within the mesh, and the selection of individual elements, have not yet been finalized. Nevertheless, the results conform well to the basic concepts of model representation within OpenGL, they depend on very similar data structures to those currently available for TINs, and they greatly simplify the implementation of a basic CAD system. We feel that this juxtaposition of computer science, CAD and GIS techniques opens up a variety of interesting possibilities, not least the manipulation of terrain models to represent some of the more complicated forms of human landscape modification.

Having now based our TIN model on a small set of simply-implemented Euler Operators, we are free to follow the CAD approach and build additional high-level functions, to form specific features (e.g. buildings) in our triangulation structure.

The authors would like to acknowledge the partial support of this research by Research Grants Council, Hong Kong SAR, project B-Q405.

References

1. Baumgart, B.G. (1972): Winged edge polyhedron representation. Stanford University Computer Science Department, Stanford Artificial Intelligence Report No. CS-320.
2. Braid, I.C., R.C. Hillyard, and I.A. Stroud: Stepwise Construction of Polyhedra in Geometric Modelling. In: Mathematical Methods in Computer Graphics and Design, Ed K.W. Brodlie: Academic Press of Computer Laboratory, University of Leicester, Leicester, England, pp. 123-141.
3. Gold, C., (1979): Triangulation-Based Terrain Modelling-Where are we now? Proceedings: Auto-Carto 4, International Symposium on Cartography and Computing; Baltimore, MD, 1979, pp. 104-111.
4. Gold, C., (1998): The Quad-Arc Data Structure. Proceedings: 8th International Symposium on Spatial Data Handling; Vancouver, BC, pp. 713-724.
5. Guibas, L., and J. Stolfi (1985): Primitives for the Manipulation of General Subdivisions and the Computation of Voronoi Diagrams. ACM Transactions on Graphics, Vol. 4, No. 2, pp. 74-123.
6. Lee, K. (1999): Principles of CAD/ CAM/ CAE Systems. Seoul National University. Korea.
7. Mantyla, M. (1981): Methodological Background of the Geometric Workbench. Helsinki University of Technology, Finland.
8. Mantyla, M (1988): An Introduction to Solid Modeling. Helsinki University of Technology, Finland.
9. Masuda, H., Shimada, K., Numao, M., and Kawabe, S. (1990): A Mathematical Theory and Applications of Non-manifold Geometric Modelling. In: Advanced Geometric Modeling for Engineering Applications North-Holland, Amsterdam
10. Peucker, T. K., R. J., Fowler, J. J., Little, and D. M., Mark, 1978: The triangulated irregular network. In: Proceedings Digital Terrain Models Symposium (St. Louis: American Society for Photogrammetry), pp. 516-532.
11. Tse, O.C., and C. Gold (2001): Terrain, Dinosaurs and Cadastres - Options for Three-Dimensional Modelling. Proceedings, International Workshop on “3D Cadastres”, Delft, Netherlands, November, 2001, pp. 243-257.
12. van Kreveld. M. (1997): Digital elevation models and TIN algorithms. In: M. van Kreveld, J. Nievergelt, T. Roos, and P. Widmayer (eds.). Algorithmic Foundations of Geographic Information Systems, number 1340 in Lecture Notes in Computer Science (tutorials), pp. 37--78. Springer-Verlag, Berlin.
13. Weiler, K. J. (1986): Topological structures for geometric modeling. PhD thesis, Rensselaer Polytechnic Institute, University Microfilms International, New York, U.S.A.
14. Weiler, K. J. (1988): Boundary Graph Operators for Nonmanifold Geometric Modeling Topology Representations. In: Geometric Modeling for CAD Applications, Elsevier Science, Amsterdam.
15. Zeid, I., (1991): CAD/ CAM theory and practice. McGraw-Hill, New York, pp.368-388

Extracting Meaningful Slopes from Terrain Contours

Maciej Dakowicz and Christopher Gold

Department of Land Surveying and Geo-Informatics
Hong Kong Polytechnic University, Hung Hom, Kowloon, Hong Kong
Tel: (852) 2766-5955; Fax: (852) 2330-2994
maciej.dakowicz@dtm.prv.pl, christophergold@voronoi.com

Abstract. Good quality terrain models are becoming more and more important, as applications such as runoff modelling are being developed that demand better surface orientation information than is available from traditional interpolation techniques. A consequence is that poor-quality elevation grids must be massaged before they provide useable runoff models. Rather than using direct data acquisition, this project concentrated on using available contour data because, despite modern techniques, contour maps are still the most available form of elevation information. Recent work on the automatic reconstruction of curves from point samples, and the generation of medial axis transforms (skeletons) has greatly helped in expressing the spatial relationships between topographic sets of contours. With these techniques the insertion of skeleton points into a TIN model guarantees the elimination of all "flat triangles" where all three vertices have the same elevation. Additional assumptions about the local uniformity of slopes give us enough information to assign elevation values to these skeleton points. In addition, various interpolation techniques were compared using the enriched contour data. Examination of the quality and consistency of the resulting maps indicates the required properties of the interpolation method in order to produce terrain models with valid slopes. The result provides us with a surprisingly realistic model of the surface - that is, one that conforms well to our subjective interpretation of what a real landscape should look like.

1. Introduction

This paper concerns the generation of interpolated surfaces from contours. While this topic has been studied for many years [5], [6], [7], [11], [13], [15], the current project is interesting for a variety of reasons. Firstly, contour data remains the most readily available data source. Secondly, valid theorems for the sampling density along the contour lines have only recently been discovered [1]. Thirdly, the same publications provide simple methods for generating the medial axis transform, or skeleton, which definitively solves the "flat triangle" problem (which often occurs when triangulating contour data) by inserting additional points from this skeleton. Fourthly, the problem of assigning elevation values to these additional ridge or valley points can be resolved, using the geometric properties of this skeleton, in ways that may be associated with the geomorphological form of the landscape. In addition, comparisons of the methods used in a variety of weighted-average techniques throw light on the

P.M.A. Sloot et al. (Eds.): ICCS 2002, LNCS 2331, pp. 144–153, 2002.

key components of a good weighted-average interpolation method, using three-dimensional visualization tools to identify what should be "good" results – with particular emphasis being placed on reasonable slope values, and slope continuity. This last is often of more importance than the elevation itself, as many issues of runoff, slope stability and vegetation are dependent on slope and aspect – but unfortunately most interpolation methods can not claim satisfactory results for these properties. The techniques developed here are based on the simple point Voronoi diagram and the dual Delaunay triangulation.

2. Generation of Ridge and Valley Lines

Amenta *et al.* [1] examined the case where a set of points sampled from a curve, or polygon boundary, were triangulated, and then attempted to reconstruct the curve. They showed that this "crust" was formed from the triangle edges that did not cross the skeleton, and that if the sampling of the curve was less than 0.25 of the distance to the skeleton the crust was guaranteed to be correct. Gold [9] and Gold and Snoeyink [10] simplified their algorithm for extraction of the crust, showing that, in every Delaunay/Voronoi edge pair, either the Delaunay edge could be assigned to the crust or else the dual Voronoi edge could be assigned to the skeleton. The Delaunay edge belongs to crust when there exists a circle through its two vertices that does not contain either of its associated Voronoi vertices; if not then the corresponding Voronoi edge belongs to the skeleton. Skeleton points may be inserted into the original diagram, or not, as needed.

In our particular case the data is in the form of contour lines that we assume are sufficiently well sampled – perhaps derived from scanned maps. Despite modern satellite imaging, much of the world's data is still in this form. An additional property is not sufficiently appreciated – they are subjective, the result of human judgement at the time they were drawn. Thus they are clearly intended to convey information about the perceived form of the surface at a particular scale – and it would be desirable to preserve this, as derived ridges and valleys.

Fig. 1a shows our raw data set (which is completely imaginary), and Fig. 1b shows the resulting crust, which reconstructs the contour lines and the skeleton. Fig. 1c shows the crust and only those skeleton points that provide unique information – ridge and valley lines that separate points on the same contour, rather than merely those points that separate adjacent contours. Aumann *et al.* [3] produced somewhat similar results by raster processing.

Fig. 2a shows a close-up of the test data set, illustrating a key point of Amenta *et a.l*'s work: if crust edges (forming the contour boundary) may not cross the skeleton, then inserting the skeleton points will break up non-crust triangle edges. In particular, if the skeletons between different contours are ignored, then insertion of the remaining branch skeleton points will eliminate all "flat triangles" formed from points of the same elevation. Thus ridge and valley lines are readily generated automatically. The same is true in the case of closed summits (Fig. 2b). The challenge is to assign meaningful elevation values to skeleton points.

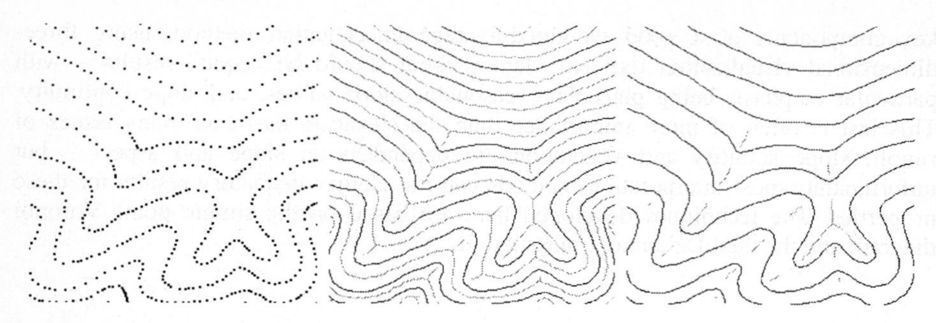

Fig. 1. Contours – a) data points; b) crust and skeleton; c) crust and skeleton branches

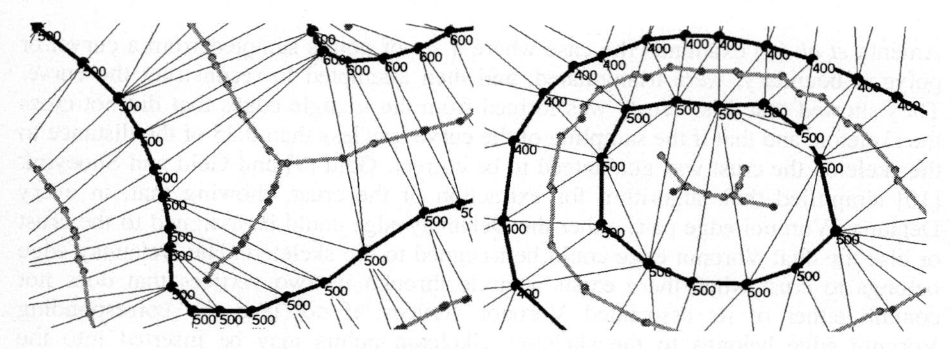

Fig. 2. Skeleton and "flat triangles" – a) ridge; b) summit

Two techniques have been developed for this, each with its own physical interpretation. The first, following Thibault and Gold [14], uses Blum's [4] concept of height as a function of distance from the curve or polygon boundary, with the highest elevations forming the crest at the skeleton line. This is illustrated in Figs. 3a and 3b, where points on a simple closed curve are used to generate the crust and skeleton.

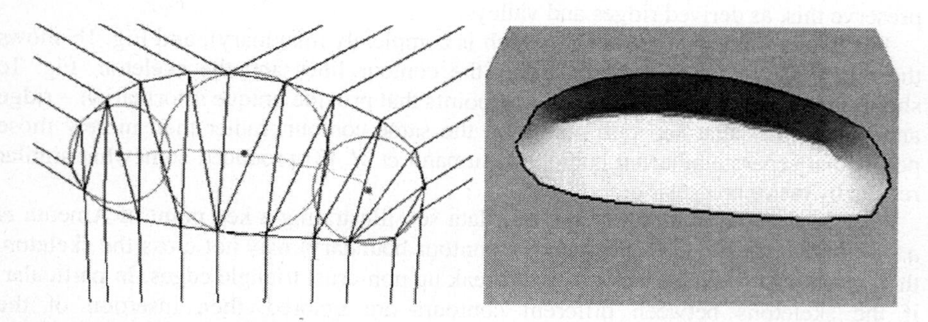

Fig. 3. Triangulation of a summit – a) skeleton and circumcentres; b) elevation model after adding skeleton vertices with assigned height values

In Fig. 3a, the circumcentres of the skeleton points are given a height above the previous contour level equal to the circumradius. The resulting interpolated model is shown in Fig. 3b. This model is based on the idea that all slopes are identical, and thus the radius is proportional to the height of the skeleton point. Of course, in the case of a real summit as in Fig. 2b, the slope would initially be unknown, and would be estimated using circumradii from the next contour level down – see [14].

In the case of a ridge or valley, the circumradius may also be used, as in Fig. 4a, to estimate skeleton heights based on the hypothesis of equal slopes. The larger circle, at the junction of the skeleton branches, has a known elevation – half way between the contours – and may be used to estimate the local slope. The elevation of the centre of the smaller circle is thus based on the ratio of the two radii. For more details see [14].

While this method is always available, it is not always the preferred solution where constant slope down the drainage valley, rather than constant valley-side slope, is more appropriate. In a second approach, illustrated in Fig. 4b, the line of the valley is determined by searching along the skeleton, and heights are assigned based on their relative distance along this line. This may be complicated where there are several valley branches – in which case the longest branch is used as the reference line. This involves careful programming of the search routines, although the concept is simple. In practice, an automated procedure has been developed, which uses the valley length approach where possible, and the side-slope method when no valley head can be detected, such as at summits and passes.

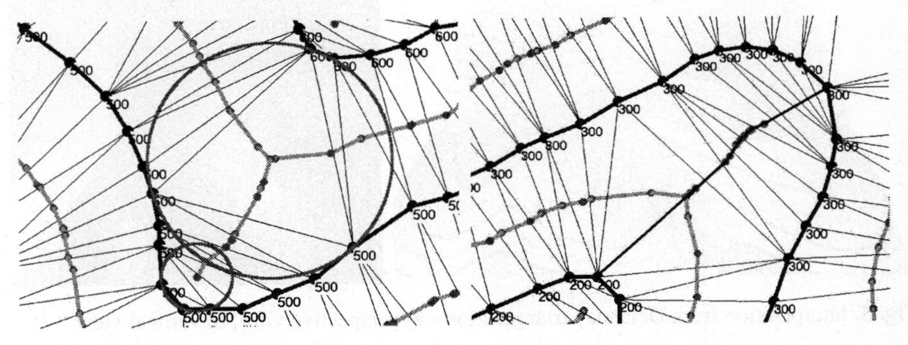

Fig. 4. Estimating skeleton heights – a) from circumradii; b) from valley length

3. Components of an Interpolation Model

On the basis of a sufficient set of data points, we now wanted to generate a terrain model with satisfactory elevations and slopes, as the basis of a valid rainfall runoff model. Our approach was to interpolate a height grid over the test area, and to view this with an appropriate terrain visualization tool. To obtain perspective views we used Genesis II, available from www.geomantics.com. Vertical views were generated using version 5 of the Manifold GIS, available from www.Manifold.net. We feel that 3D visualization has been under-utilized as a tool for testing terrain modelling algorithms, and the results are often more useful than a purely mathematical or statistical approach.

We have restricted ourselves to an evaluation of several weighted-average methods, as there are a variety of techniques in common that can be compared. All of the methods were programmed by ourselves – which left out the very popular Kriging approach, as too complicated, and not necessarily better. Nevertheless, many aspects of this study apply to this method as well, since it is a weighted-average method with the same problems of neighbour selection, etc., as the methods we attempted

In general, we may ask about three components of a weighted-average interpolation method. Firstly: what is the weighting process used? Secondly: what is the set of neighbours used to obtain the average? Thirdly: what is the elevation function being averaged? (Often it is the data point elevation alone, but sometimes it is a plane through the data point incorporating slope information as well.)

One simple weighted-average model is the triangulation, using the Delaunay triangulation. Fig. 5 shows the result, including the crust and skeleton draped over the terrain. The flat triangles are readily seen.. Fig. 6 shows the improved model when estimated skeleton points are added, and all flat triangles are removed.

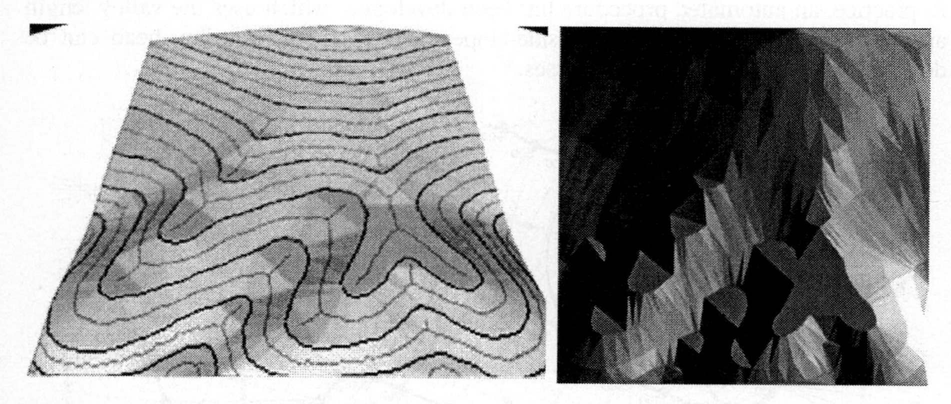

Fig. 5. Interpolation from Delaunay triangulation - a) perspective view; b) vertical view

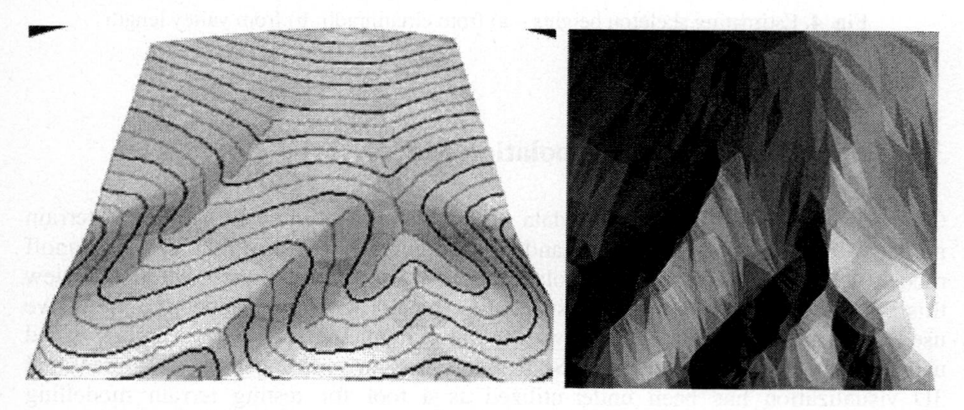

Fig. 6. Adding skeleton points to Fig. 5 - a) perspective view; b) vertical view

The other weighted average models that were tested were the traditional gravity model, and the more recent "area-stealing" or "natural neighbour" or perhaps more properly "Sibson" interpolation methods ([8], [12], [16]).

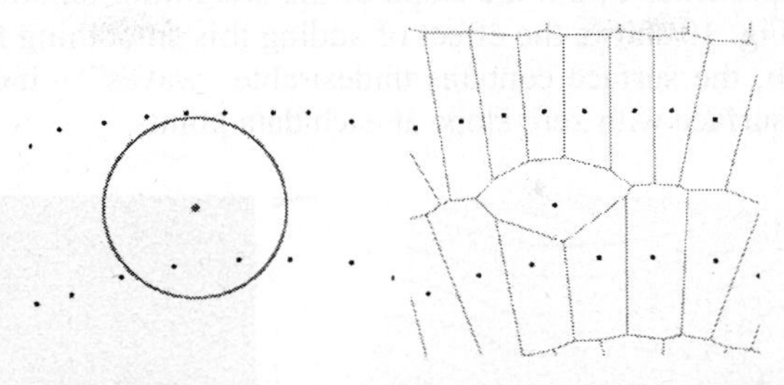

Fig. 7. Neighbour selection a) using a counting circle; b) using Voronoi neighbours

In the case of the gravity model the weighting of each data point used is inversely proportional to the square of the distance from the data point to the grid node being estimated, although other exponents have been used. There is no obvious set of data points to use, so one of a variety of forms of "counting circle" is used, as in Fig. 7a. When the data distribution is highly anisotropic there is considerable difficulty in finding a valid counting circle radius. Fig. 8 shows the resulting surface for a radius of about a quarter of the map. Data points form bumps or hollows. If the radius is reduced there may be holes in the surface where no data is found within the circle. If the radius is increased the surface becomes somewhat flattened, but the bumps remain. The result depends on the radius, and other selection properties, being used. Clearly, in addition, estimates of slope would be very poor, and very variable.

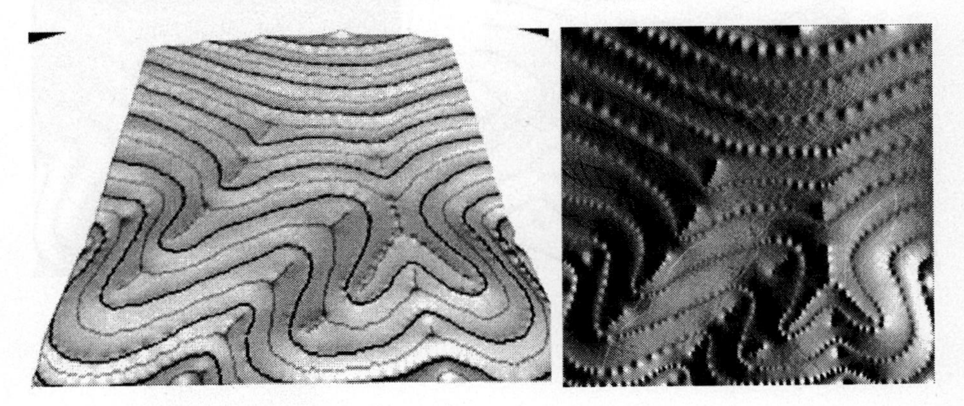

Fig. 8. Interpolation using the gravity model – a) perspective view; b) vertical view

The Sibson method is based on the idea of inserting each grid point temporarily into the Voronoi diagram of the data points, and measuring the area stolen from each of a well-defined set of neighbours. These stolen areas are the weights used for the weighted-average. The method is particularly appropriate for poor data distributions, as illustrated in Fig. 7b, as the number of neighbours used is well defined, but dependent on the data distribution. Fig. 9 shows the results of using Sibson interpolation. The surface behaves well, but is angular at ridges and valleys. Indeed, slopes are discontinuous at all data points (Sibson [12]). One solution is to re-weight the weights, so that the contribution of any one data point not only becomes zero as the grid point approaches it, but the slope of the weighting function approaches zero also (Gold [8]). Fig. 10 shows the effect of adding this smoothing function. While the surface is smooth, the surface contains undesirable "waves" – indeed, applying this function gives a surface with zero slope at each data point.

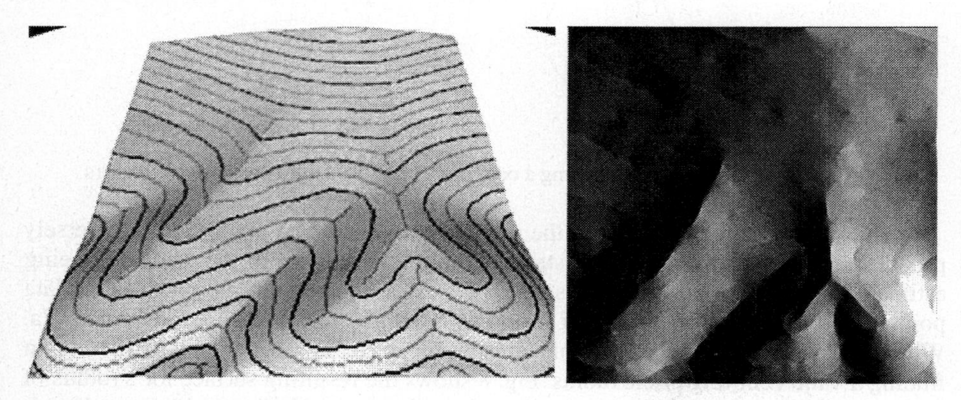

Fig. 9. Sibson interpolation - a) perspective view; b) vertical view

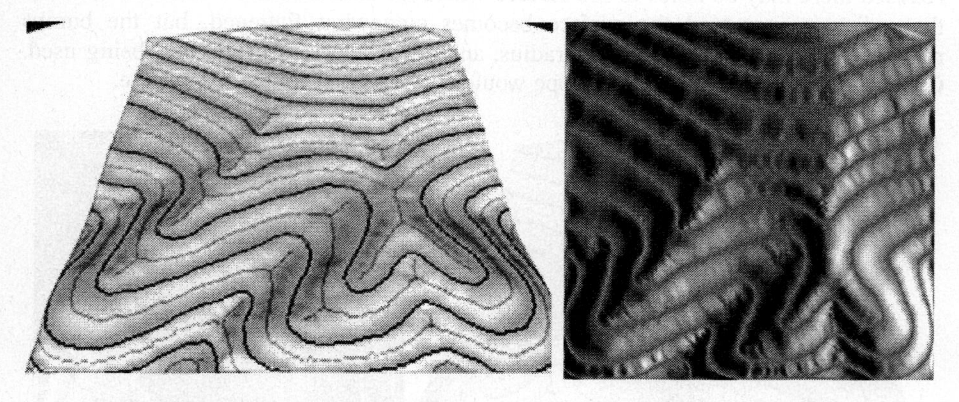

Fig. 10. Adding smoothing to Fig. 9 - a) perspective view; b) vertical view

4. Slopes – The Ignored Factor

This brings us to a subject often ignored in selecting a method for terrain modelling – the slope of the generated surface. In real applications, however, accuracy of slope is often more important than accuracy of elevation – for example in runoff modelling, erosion. Clearly an assumption of zero slope, as above, is inappropriate. However, in our weighted-average operation we can replace the height of a neighbouring data point by the value of a function defined at that data point – probably a planar function involving the data point height and local slopes. Thus at any grid node location we find the neighbouring points and evaluate their planar functions for the (x, y) of the grid node. These z estimates are then weighted and averaged as before.

Fig. 11 shows the result of using Sibson interpolation with data point slopes. The form is good, but slight breaks in slope can be seen at contour lines. When using smoothing and slope information together, the surface is smooth, but has unwanted oscillations, see Fig. 12. Clearly an improved smoothing function is desirable to eliminate these side-effects.

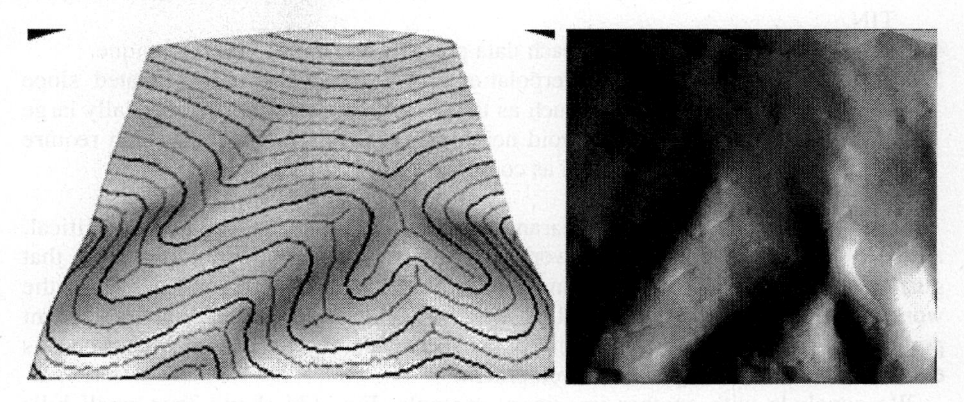

Fig. 11. Sibson interpolation using slopes at data points – a) perspective; b) vertical view

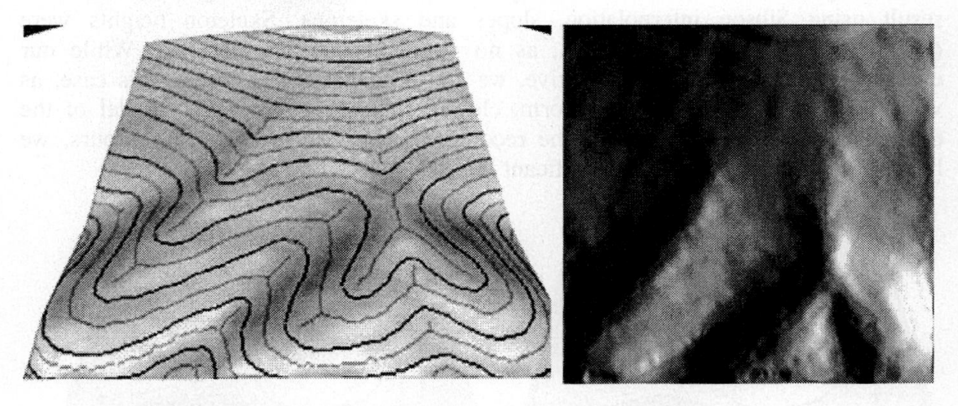

Fig. 12. Sibson interpolation using slopes and smoothing function at data points – a) perspective view; b) vertical view

Adding slopes to the simple TIN model (i.e. using the position in the triangle to provide the weights) produced results that were almost as good as the Sibson method when the sample points were closely spaced along the contours. However, the Sibson method is much superior for sparser data, or where the points do not form contour lines. The gravity model does not provide particularly good slope estimates, but even here including the data point slope function produces a significant improvement.

5. Proposed methodology and conclusions

For the common problem of deriving surfaces from contours, we propose a general approach:

1. Generate skeleton points along the ridges, valleys, pits, summits and passes by the method of Aumann *et al.* [3] or of Thibault and Gold [14].
2. Assign elevations to these skeleton points by the methods described here, or other suitable techniques.
3. Eliminate flat triangles by the insertion of these skeleton points into the original TIN.
4. Estimate slope information at each data point by any appropriate technique.
5. Perform weighted-average interpolation using the previously estimated slope information. Avoid methods, such as the gravity model, with exponentially large close-range weightings, and avoid neighbour selection techniques which require user-specified parameters, such as counting-circle radius.

Surprisingly, mathematically guaranteed slope continuity is not usually critical, although we are continuing to work on an improved smoothing function that guarantees both slope continuity and minimum curvature – probably based on the work of Anton *et al.* [2]. Nevertheless, the moral is clear: both for finding adjacent points and for skeleton extraction, a consistent definition of neighbourhood is essential for effective algorithm development.

We conclude with another imaginary example. Fig. 13a shows four small hills defined by their contours, modelled by a simple triangulation. Fig. 13b shows the result using Sibson interpolation, slopes and skeletons. Skeleton heights were obtained using circumcircle ratios, as no valley-heads were detected. While our evaluation was deliberately subjective, we consider that our results in this case, as with the previous imaginary landform, closely follow the perceptual model of the original interpretation. Thus, for the reconstruction of surfaces from contours, we believe that our methods are a significant improvement on previous work.

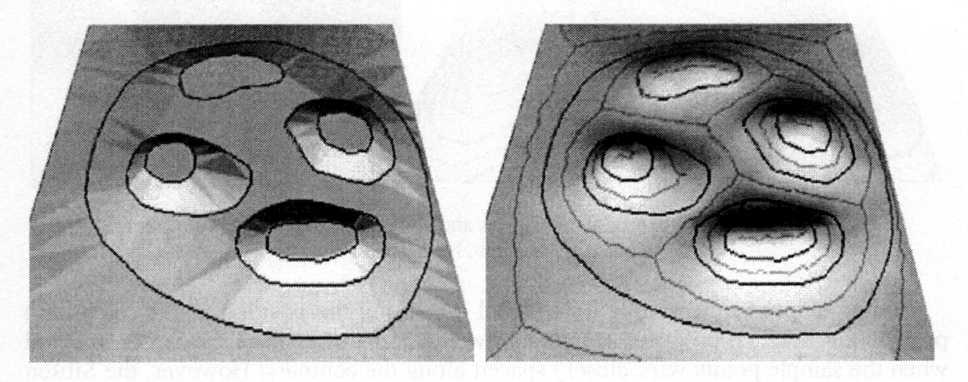

Fig. 13. Triangulation of several small hills – a) triangle based interpolation; b) Sibson interpolation with slopes

Acknowledgments

The work described in this paper was substantially supported by a grant from the Hong Kong Polytechnic University (Project No. PolyU A-PB79).

References

1. Amenta, N. Bern, M. and Eppstein, D. (1998) "The crust and the beta-skeleton: combinatorial curve reconstruction", Graphical Models and Image Processing, 60, 125-135.
2. Anton, F. Gold, C.M. and Mioc, D. (1998) "Local coordinates and interpolation in a Voronoi diagram for a set of points and line segments", Proceedings 2nd Voronoi Conference on Analytic Number Theory and Space Tillings, Kiev, Ukraine, 9-12.
3. Aumann, G. Ebner, H. and Tang, L. (1991) "Automatic derivation of skeleton lines from digitized contours", ISPRS Journal of Photogrammetry and Remote Sensing, 46, 259-268.
4. Blum, H. (1967) "A transformation for extracting new descriptors of shape", In: Whaten Dunn, W. (eds.), "Models for the Perception of Speech and Visual Form", 153-171, MIT Press.
5. Crain, I.K. (1970) "Computer interpolation and contouring of two-dimensional data: a review", Geoexploration, 8, 7x-86.
6. Davis, J. C. (1973) "Statistics and data analysis in geology", 313, New York, John Wiley and Sons.
7. Dayhoff, M.O. (1963) "A contour map program for X-ray crystallography", Communications of the Association for Computing Machinery, 6, 620-622.
8. Gold, C.M. (1989) "Chapter 3 - Surface interpolation, spatial adjacency and GIS", In: Raper, J. (eds.), "Three Dimensional Applications in Geographic Information Systems", 21-35, Taylor and Francis, Ltd., London.
9. Gold, C.M. (1999) "Crust and anti-crust: a one-step boundary and skeleton extraction algorithm", Proceedings of the ACM Conference on Computational Geometry, Miami, Florida, 189-196.
10. Gold, C. M. and Snoeyink, J. (2001) "A one-step crust and skeleton extraction algorithm", Algorithmica, 30, 144-163.
11. Peucker, T.K. (1978) "The triangulated irregular network", Proceedings, Digital Terrain Model Symposium, American Society of Photogrammetry, St. Louis.
12. Sibson, R. (1980) "A Vector Identity for the Dirichlet Tessellation", Math. Proc. Cambridge Philos. Soc., 87, 151-155.
13. Sibson, R. (1982) "A brief description of natural neighbour interpolation", In: Bamett, V. (eds), "Interpreting Multivariate Data", 21-36, John Wiley and Sons, London.
14. Thibault, D. and Gold, C.M. (2000) "Terrain Reconstruction from Contours by Skeleton Construction", GeoInformatica, 4, 349-373.
15. Walters, R.F. (1969) "Contouring by machine: a users' guide", American Association of Petroleum Geologists, Bulletin, 53, 2324-2340.
16. Watson, D.F. and Philip, G.M. (1987) "Neighborhood-based interpolation", Geobyte, 2, 12-160.

Duality in Disk Induced Flows*

Joachim Giesen and Matthias John

Institute for Theoretical Computer Science, ETH Zürich, CH-8092 Zürich
{giesen,john}@inf.ethz.ch

Abstract. We introduce a condition that establishes a duality known from Delaunay triangulations and Voronoi diagrams for diagrams associated with a class of dynamical systems defined via a set of disks in the plane. Under this condition the maximum geometric and worst case algorithmic complexities of the latter diagrams decrease. The condition is natural in the sense that it is automatically fulfilled by some important classes of sets of disks in the plane.

Keywords. Computational geometry, Voronoi diagram, Delaunay triangulation

1 Introduction

It is possible to associate a dynamical system with a set of disks in the plane which we want to call *disk induced flows.* In [4, 5] we studied the fixpoints of these dynamical systems, i.e. the minima, maxima and saddles of the flow. With a minimum m we associate all points that are connected to m by the disk induced flow, i.e. all points x that are connected to m by a curve such that all points in the interior of the curve flow into x under the disk induced flow. The set of all these points is a connected region in the plane that we call the Min region of m. Similarly Max regions are assigned to maxima. The collection of all Min regions defines a plane diagram that is closely related to the Delaunay triangulation of the same set of disks. The analogous diagram of the Max regions is closely related to the Voronoi diagram. The Delaunay diagram of a set of disks is the Voronoi diagram of a dual set of disks and vice versa. This is not the case for Min- and Max diagrams, i.e. the Min diagram of a set of disks need not be the Max diagram of the dual set of disks or vice versa. That is, in general a set of disks in the plane gives rise to four different flow diagrams, namely the Min- and Max diagrams of the original and the dual set of disks. There are further differences between Delaunay- and Min diagrams or Voronoi- and Max diagrams, respectively. The maximum geometric complexities of Min- and Max diagrams associated with n disks are $\Theta(n^2)$ vertices, $\Theta(n^2)$ edges and $\Theta(n)$ regions. This is in contrast to Voronoi- and Delaunay diagrams whose geometric complexities are: $\Theta(n)$ vertices, $\Theta(n)$ edges and $\Theta(n)$ regions. The worst case algorithmic

* Partly supported by the IST Programme of the EU as a Shared-cost RTD (FET Open) Project under Contract No IST-2000-26473 (ECG - Effective Computational Geometry for Curves and Surfaces).

P.M.A. Sloot et al. (Eds.): ICCS 2002, LNCS 2331, pp. 154–163, 2002.

complexity of Min- and Max diagrams is $\Theta(n^2)$ in contrast to the worst case algorithmic complexity of Voronoi- and Delaunay diagrams which is $\Theta(n \log n)$.

Here we introduce a natural condition under which one of the dualities between Voronoi- and Delaunay diagrams also holds for Min- and Max diagrams. This condition is for example always fulfilled for finite sets of points, i.e. disks of radius zero. We call the condition *strong duality condition* since it has some strong implications. In fact we will show that under the strong duality condition the following holds:

(1) The Max diagram of the original set of disks is the Min diagram of the dual set of disks.
(2) The Min diagram of the original set of disks is the Voronoi diagram of these disks.
(3) The maximum geometric complexities of the Min- and Max diagram of the original set of disk are $\Theta(n)$ vertices, $\Theta(n)$ edges and $\Theta(n)$ regions.
(4) The worst case algorithmic complexity the Min- and Max diagram of the original set of disk are $\Theta(n \log n)$.
(5) The Max diagram of the original set of disks coincides with a diagram associated with a discrete flow on the set of Delaunay triangles. This discrete flow was studied in [3] to identify and compute pockets in macromolecules.

The paper is organized as follows: In Section 2 we give basic definitions. In Section 3 we introduce disk induced flows and the associated notions of Max- and Min regions. Finally we introduce the strong duality condition and prove the results mentioned above.

2 Disks, diagrams and critical points

Disk. A disk is a pair $(z, r) \in \mathbb{R}^3$, where $z \in \mathbb{R}^2$ is the center of the disk and $\sqrt{r} \in \mathbb{C}$ its radius. Note that we also allow purely imaginary radii in which case the geometric intuition about disks is not helpful. We refer to r also as the power of the disk.

In the following we are going to consider finite sets of disks. Very often these disks are not in general position. That is, we do not assume general position unless stated differently.

Power distance. The power distance of a point $x \in \mathbb{R}^2$ from a disk (z, r) is $\pi(x) = \|x - z\|^2 - r$.

Voronoi diagram. Given a set B of disks. The *Voronoi cell* of a disk b_i under the power distance is given as

$$V_i = \{x \in \mathbb{R}^2 \, : \, \forall b_j \in B, \, \pi_i(x) \leq \pi_j(x)\}.$$

The sets V_i are convex polygons or empty since the set of points that have the same power distance from two disks forms a straight line. Closed line segments shared by two Voronoi cells are called *Voronoi edges* and the points shared by three or more Voronoi cells are called *Voronoi vertices*. The term *Voronoi object* can denote either a Voronoi cell, edge or vertex. The *Voronoi diagram* of B is the collection of Voronoi cells, edges and vertices. It defines a cell decomposition of $\mathbb{R}^2$.

Delaunay diagram. The *Delaunay diagram* of a set of disks B is dual to the Voronoi diagram of B. The convex hull of three or more center points defines a *Delaunay cell* if the intersection of the corresponding Voronoi cells is not empty and there exists no superset of center points with the same property. Analogously, the convex hull of two center points defines a *Delaunay edge* if the intersection of their corresponding Voronoi cells is not empty. Every center point is called *Delaunay vertex.* The term *Delaunay object* can denote either a Delaunay cell, edge or vertex. The Delaunay diagram defines a decomposition of the convex hull of all center points. This decomposition is a triangulation if the disks are in general position.

Sometimes it is convenient to introduce an additional disk with center at infinity and with power 0. This gives us a Voronoi vertex at infinity at the end of every unbounded Voronoi edge. One benefit of adding a disk at infinity is that it provides us with the following duality.

Duality. Let B be a set of disks. This set gives rise to another set of disks, namely for every Delaunay cell its orthodisk, see [2] for the details. The new disks are centered at Voronoi vertices. The set C exchanges Voronoi and Delaunay diagram.

Power height function. Let B be a set of disks in $\mathbb{R}^2$. The power height function is given as

$$h(x) = \min\{\pi_i(x) \ : \ b_i \in B\}. \tag{1}$$

Observe that the function h is continuous. It is smooth everywhere besides at points which have the same power distance from two or more disks, i.e. at points that lie on the boundary of Voronoi cells.

Regular- and critical points. Let B be a set of disks. The critical points of the power height function h are the intersection points of Voronoi objects V and their dual Delaunay object σ. The index of a critical point is the dimension of σ. All points which are not critical are called regular.

The power height function associated with a set of disks B and the power height function associated with the dual set of disks C have the same critical points. The index of a critical point in the dual is 2 minus its index in the primal, i.e. maxima and minima get exchanged.

3 Disk induced flow

Disk induced flow. Given a set B of disks, the induced flow ϕ is given as follows: Since the Voronoi diagram of B is a decomposition of the plane any point $x \in \mathbb{R}^2$ lies in some Voronoi object. Let V be the lowest dimensional Voronoi object that contains x. Assume that x is the intersection point of V and its dual Delaunay object. In this case we set:

$$\phi(t, x) = x \ , \ t \in [0, \infty)$$

Otherwise let σ be the dual Delaunay object of V and $y = \text{argmin}_{y' \in \sigma} \|x - y'\|$. Since σ is convex there is only one such y. If y also lies in a lower dimensional Delaunay object than σ then we replace V by the dual Voronoi object of the lowest dimensional Delaunay object. Let R be the ray originating at x and shooting in the direction $x - y$. Let z be the first point on R for which $\text{argmin}_{y' \in \tau} \|z - y'\|$ is different from y where τ denotes the dual Delaunay object of the lowest dimensional Voronoi object z lies in. Note that such a z need not exist in $\mathbb{R}^2$. In this case let z be the point at infinity. We set:

$$\phi(t, x) = x + t\frac{x - y}{\|x - y\|}\ , \ t \in [0, \|z - x\|]$$

For $t > \|z - x\|$ the flow is given by property (2) in the definition of flow, i.e.

$$\begin{aligned}\phi(t, x) &= \phi\left(t - \|z - x\| + \|z - x\|,\ x\right) \\ &= \phi\left(t - \|z - x\|,\ \phi\left(\|z - x\|,\ x\right)\right).\end{aligned}$$

It is not completely obvious but ϕ can be shown to be well defined on the whole of $[0, \infty) \times \mathbb{R}^2$. Furthermore, the following two properties of dynamical systems hold for disk induced flows,

(1) $\phi(0, x) = x$.
(2) $\phi(t + s, x) = \phi(t, \phi(s, x))$.

The following three observations are helpful to get a better understanding of disk induced flows and their their relationship to the power height function of the same set of disks.

(1) The fixpoints of ϕ are the critical points of the power height function.
(2) The orbits of ϕ are piecewise linear curves that are linear in Voronoi objects.
(3) The flow ϕ has no closed orbits.

Because of the first observation we want to refer to fixpoints of ϕ as minimum, saddle or maximum if the corresponding critical point of the power height function is a minimum, saddle or maximum, respectively.

Stable- and Unstable Manifolds. Given a disk induced flow ϕ in the plane. The stable manifold $S(x)$ of a fixpoint $x \in M$ contains all points that flow into x, i.e.

$$S(x) = \{y \in \mathbb{R}^2 \ :\ \lim_{t \to \infty} \phi_y(t) = x\}.$$

The unstable manifold $U(x)$ of a fixpoint x is a little bit more involved to define. Given a neighborhood U of x and let $V(U)$ be the set of all points which lie in an orbit that starts in U, i.e.

$$V(U) = \{y \in \mathbb{R}^2 \ :\ \exists z \in U,\, t \in [0, \infty) \text{ s.t. } \phi_z(t) = y \text{ or } \lim_{t \to \infty} \phi_z(t) = y\}.$$

Then $U(x)$ is given as the intersection of all such sets $V(U)$.

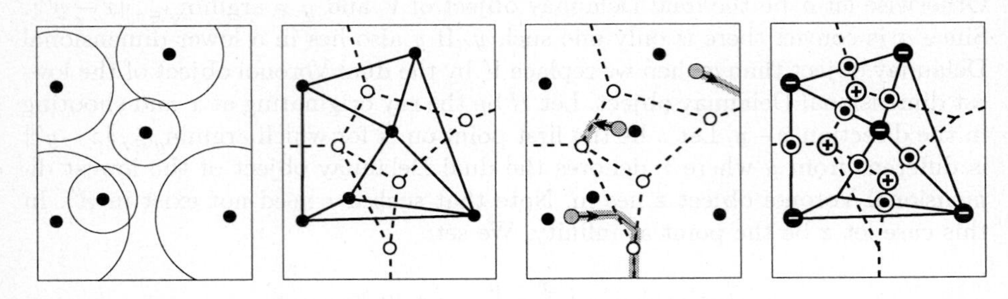

Fig. 1. A set of disk in the plane (on the left), its Voronoi- and Delaunay diagram (second from the left), some orbits of the flow induced by the disks (third form the left) and the minima ⊖, saddles ⊙ and maxima ⊕ of this flow (on the right).

Instead of working directly with unstable manifolds of minima and stable manifolds of maxima we introduce Max- and Min regions which have nicer properties.

Max- and Min regions. Given a disk induced flow system with finitely many fixpoints. If m is a maximum of this system we call the closure of the stable manifold $S(m)$ the Max region of m.

If m is a minimum of a disk induced flow we call the closure of the interior of the unstable manifold $U(m)$ the Min region of m.

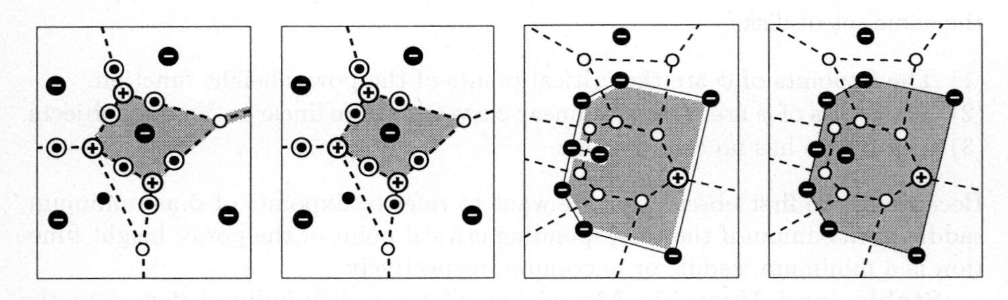

Fig. 2. An unstable manifold of a minimum ⊖ of the flow induced by the disks from Figure 1 (on the left) and its corresponding Min region (second from the left). The Min region is a true subset of the unstable manifold. A stable manifold of a maximum ⊕ of a different flow (second from the right) and its corresponding Max region (on the right). The Max region is a superset of the stable manifold.

Next we discuss stable and unstable manifolds of saddles which are a key tool to characterize and compute Max- and Min regions. Given a saddle s be a saddle of ϕ. We observe that

(1) The unstable manifold $U(s)$ of s is a piecewise linear curve.

(2) If the stable manifold $S(s)$ of s does not contain a Voronoi vertex then closure($S(s)$) is a piecewise linear curve.

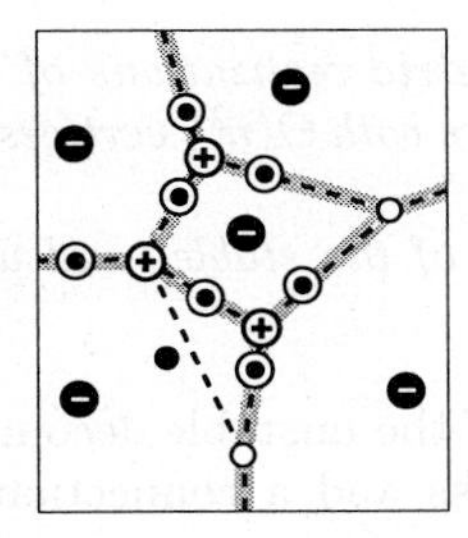

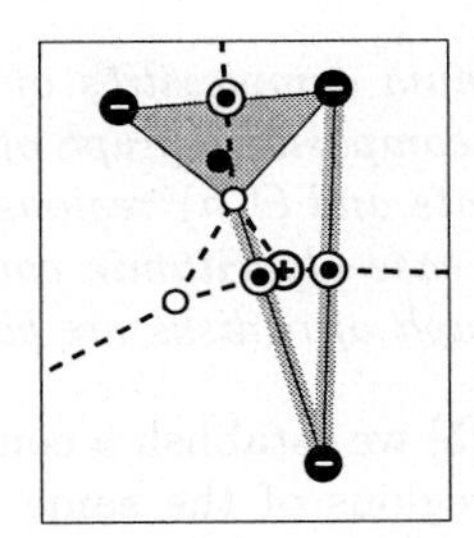

Fig. 3. The unstable manifolds of the saddles ⊙ of the flow induced by the disks from Figure 1 (on the left). An example that shows that the stable manifolds of saddles ⊙ need not be curves (on the right).

The second observation indicates that the stable manifolds of saddles can be more complicated to deal with than the unstable manifolds. In [5] we deal with this problems. Here we want to assume that the stable manifolds of all saddles do not contain any Voronoi vertex. This assumption is not really a restriction since it is automatically fulfilled under the *strong duality condition* which is the main topic of this paper.

The unstable manifolds of different saddles might join at some point or they might continue to infinity. The only points where two such unstable manifolds might join must lie on the boundaries of Voronoi cells, i.e. in the interior of Voronoi edges or at Voronoi vertices. If a point where two or more unstable manifolds of saddles join is not a maximum or saddle then we call it a *Steiner vertex.*

Unstable decomposition graph. Let G be the following graph: Its vertex set consists of all maxima, saddles and Steiner vertices. Two vertices are connected by an edge if there exists an unstable manifold $U(s)$ of some saddle s such that the two vertices are both contained in $U(s)$ and there is no other vertex in $U(s)$ between them. We add an additional vertex at infinity to the vertex set of G and add edges that connect this point to all vertices of G that flow under ϕ on a straight line to infinity. We refer to G as the unstable decomposition graph associated with a disk induced flow. Let S be the set of saddles of ϕ and $U(S) = \bigcup_{s\in S} U(s)$. We refer to $U(S)$ as the geometric realization of G.

Stable decomposition graph. Let G be the following graph: Its vertex set consists of all minima, saddles and joins. Two vertices are connected by an edge if there exists a stable manifold $S(s)$ of some saddle s such that the two vertices are both contained in closure($S(s)$) and there is no other vertex on $S(s)$ between them. We have to take the closure of the stable manifolds because otherwise the minima would not be contained in them. We refer to G as the stable decomposition graph associated with a disk induced flow. Let S be the set of saddles of ϕ and $\mathcal{S}(S) = \bigcup_{s \in S} \text{closure}(\mathcal{S}(s))$. We refer to $\mathcal{S}(S)$ as the geometric realization of G.

Theorem 1. *The following is true.*

(1) The maximum complexities of the geometric realizations of the stable- and unstable decomposition graph of n disks are both $\Theta(n^2)$ vertices, $\Theta(n^2)$ straight line segments and $\Theta(n)$ regions.
(2) The worst case algorithmic complexities of the stable- and unstable decomposition graph of n disks are both $\Theta(n^2)$. □

In [4] and [5] we establish a connection of the unstable decomposition graph and the Min regions of the same set of disks and a connection of the stable decomposition graph and the Max regions.

Theorem 2. *The following is true:*

(1) The regions of the unstable decomposition graph of a set of disks are exactly the Min regions of the flow associated with these disks.
(2) The regions of the stable decomposition graph of a set of disks are exactly the Max regions of the flow associated with these disks. □

4 Strong duality

Strong duality condition. A set B of disks in $\mathbb{R}^2$ obeys the strong duality condition if the center of each disk in B is contained in its dual Voronoi cell, i.e. every disk center is a minimum of the flow induced by B.

Observe that the strong duality condition is always obeyed by a set of points, i.e. a set of disks that all have radius zero.

The Gabriel graph of a set of disks is a well studied object that has applications in many areas. It turns out that the unstable decomposition graph is a Gabriel graph under the strong duality condition.

Gabriel graph. Given a set B of disks in the plane. The Gabriel graph of G is given as follows: Its vertices are the centers of the disks in B and its edges are given by Delaunay edges that intersect their dual Voronoi edge. The edges of the Gabriel graph are called Gabriel edges. From the Definition of critical points we know that there is a one to one correspondence between Gabriel edges and saddles of ϕ_B.

Theorem 3. *The following is true.*

(1) The maximum geometric complexity of a Gabriel graph of n disks is $\Theta(n)$ vertices, $\Theta(n)$ edges and $\Theta(n)$ regions.
(2) The worst case algorithmic complexity, i.e. the number of steps it takes to compute the Gabriel graph of n disks, is $\Theta(n \log n)$. □

Theorem 4. *Given a set B of disks that obeys the strong duality condition plus one additional disk at infinity. Let C be the set of dual disks and ϕ_B and ϕ_C be the flows induced by B and C, respectively. The following is true:*

(1) Every disk center of B is a minimum of ϕ_B and a maximum of ϕ_C.
(2) The geometric realizations of the unstable decomposition graph associated with ϕ_C and the stable decomposition graph associated with ϕ_B are both the Gabriel graph of B.

Proof. (1) This follows just from the definition of critical points. (2) Every saddle of ϕ_B is by definition also a saddle of ϕ_C and vice versa. The Delaunay edge (with respect to B) that contains some saddle s is a Gabriel edge by the definitions of saddles and Gabriel edges, respectively. This Gabriel edge is a Voronoi edge with respect to C and the endpoints of this edge are minima of ϕ_B and maxima of ϕ_C by the strong duality condition and the definition of duality. That is, s is connected by this Gabriel edge to two minima with respect to B and to two maxima with respect to C. Hence this Gabriel edge is both $U(s)$ (with respect to C) and $S(s)$ (with respect to B). This implies that the geometric realizations of the unstable decomposition graph associated with ϕ_C and the stable decomposition graph associated with ϕ_B are the Gabriel graph of B. □

Corollary 1. *The following is true:*

(1) The Max *regions of ϕ_B are exactly the bounded* Min *regions of ϕ_C.*
(2) The Max *regions of ϕ_B are a set of contiguous Delaunay cells of B separated by Gabriel edges.*
(3) The maximum complexities of the geometric realization of the stable decomposition graph associated with ϕ_B and the geometric realization of the unstable decomposition graph associated with ϕ_C are both $\Theta(n)$ vertices, $\Theta(n)$ edges and $\Theta(n)$ regions.
(4) The worst case algorithmic complexities of the geometric realization of the stable decomposition graph associated with ϕ_B and the geometric realization of the unstable decomposition graph associated with ϕ_C are both $\Theta(n \log n)$.
□

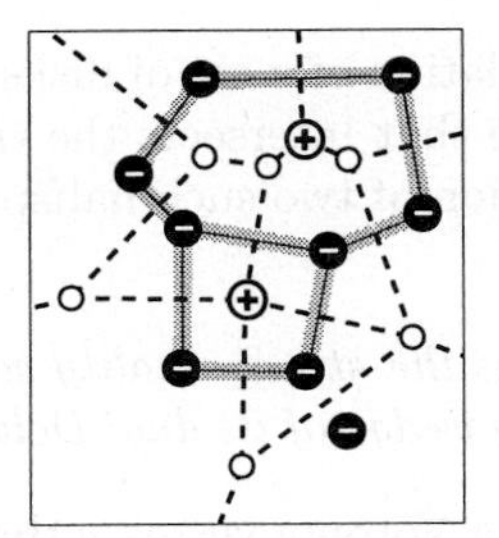

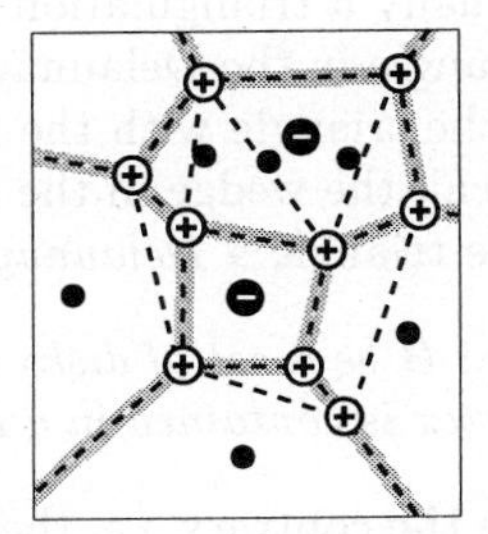

Fig. 4. Max regions of maxima ⊕ in the primal (on the left) are bounded Min regions of minima ⊖ in the dual (on the right) under the strong duality condition. Note that here the boundaries of these regions are emphasized.

The analogous result to the first observation for the bounded Min regions of ϕ_B and the Max regions of ϕ_c does not hold as can be seen easily from examples. Note that the the set C of dual disks need not fulfill the strong duality condition even if B does.

Under the strong duality condition we have a simpler characterization of the Min regions of ϕ_B.

Theorem 5. *Let B a set of disks that obeys the strong duality condition and ϕ the flow induced by B. The* Min *regions of ϕ are the Voronoi cells of B.*

Proof. Let z be the center of a disk in B and V its dual Voronoi region. By the strong duality condition z lies in the interior of V which implies that z is a minimum of ϕ. From the definition of ϕ we have that the interior of V is a subset of the Min region M of z. Furthermore, V is a subset of M since Min regions are always closed. Now assume that V is a true subset of M. The difference $M - V$ has to contain an open subset U of $\mathbb{R}^2$, because by definition Min regions are the closure of the interior of some subset of $\mathbb{R}^2$. The set U must have a non empty intersection with the interior of some Voronoi cell V' different from V. The interior of V' is part of the Min region of its dual Delaunay vertex m'. Let $x \in U \cap V'$. By the definition of ϕ the closure of the set $\{y \in \mathbb{R}^2 : \exists t \in [0, \infty) \text{ s.t. } \phi_y(t) = x\}$ is the line segment that connects m' with x. Thus x belongs neither to the unstable nor to the Min region of m. That is a contradiction to our assumption. Hence $M = V$. □

Corollary 2. *The following is true:*

(1) The maximum complexities of the geometric realization of the unstable decomposition graph associated with ϕ_B is $\Theta(n)$ vertices, $\Theta(n)$ edges and $\Theta(n)$ regions.
(2) The worst case algorithmic complexity of the geometric realization of the unstable decomposition graph associated with ϕ_B is $\Theta(n \log n)$. □

The strong duality condition allows to define a discrete flow, i.e. a discrete dynamical system on the set of triangles of the Delaunay triangulation [3]. In the following we assume that the disks are in general position, i.e. the Delaunay diagram is actually a triangulation.

Given a triangle in the Delaunay triangulation of a set of disks. We associate every edge of the triangle with the halfspace that intersects the triangle only in this edge. We call the wedge in the intersection of two such halfspaces that does not contain the triangle a *Delaunay wedge.*

Lemma 1. *Let B be a set of disks that obeys the strong duality condition. Then no Voronoi vertex is contained in a Delaunay wedge of its dual Delaunay triangle.*

Proof. Assume the contrary, i.e. there exists a Voronoi vertex v that is contained in a Delaunay wedge of its dual Delaunay triangle σ. Let x be the Delaunay vertex that is incident to the two edges of σ whose associated halfspaces define the wedge that contains x. By construction x cannot be contained in the wedge centered at v and bounded by the two Voronoi edges dual to the two Delaunay edges incident to x. That is, x is not contained in its dual Voronoi cell. That contradicts the strong duality condition. □

Triangle flow. Given a set B of disks that obeys the strong duality condition. Let $\mathcal{D}$ denote the set of all Delaunay triangles in the Delaunay triangulation of B together with an abstract triangle `inf` at infinity. The flow on $\mathcal{D}$ is a function $\Phi : \mathbb{N}_0 \times \mathcal{D} \to \mathcal{D}$ which satisfies

(1) $\Phi(0, \sigma) = \sigma$ for every $\sigma \in \mathcal{D}$.
(2) $\Phi(n + m, \sigma) = \Phi(n, \Phi(m, \sigma))$ for every $\sigma \in \mathcal{D}$.

That is, Φ is defined recursively. The recursion has to be anchored by defining $\Phi(1, \sigma)$. We set $\Phi(1, \sigma) = \sigma$ if $\sigma =$ `inf` or the dual Voronoi vertex v of σ is contained in σ. Otherwise v is contained in exactly one of the halfspaces associated with the edges of σ. If there exists a second Delaunay triangle σ' incident to the edge that corresponds to the halfspace that contains v we set $\Phi(1, \sigma) = \sigma'$. Otherwise we set $\Phi(1, \sigma) =$ `inf`.

We find directly from the definitions that the fixpoints besides `inf` of a triangle flow are exactly the dual triangles of the maxima of the continuous flow induced by the same set of disks.

Theorem 6. *Let B be a set of disks that obeys the strong duality condition and ϕ the flow induced by B. Let m be a maximum of ϕ and σ its dual Delaunay triangle. Then the* Max *region of m is the following set*

$$\{x \in \sigma' \in \mathcal{D} : \exists n \geq 0 \text{ such that } \Phi(n, \sigma') = \sigma\},$$

where Φ is the triangle flow on the triangle set $\mathcal{D}$ of the Delaunay triangulation of B.

Proof. First observe that the flow Φ has to be acyclic, because the power of the orthodisks of the Delaunay triangles increases in every time step of the flow Φ.

By construction the flow Φ cannot pass any Gabriel edge. Every region in the Gabriel graph of B contains exactly one maximum of ϕ, i.e. one fixpoint of Φ. Since Φ is acyclic all triangles in a region of the Gabriel graph have to flow into this unique fixpoint. From Corollary 1(2) we know that all points in this region flow into m under the flow ϕ. □

References

1. F. Aurenhammer and R. Klein. Voronoi Diagrams. In *Handbook of Computational Geometry,* J.-R. Sack and J. Urrutia (eds.), pp. 201–290, Elsevier (2000)
2. S.W. Cheng, H. Edelsbrunner, P. Fu and K.P. Lam. Design and Analysis of Planar Shape Deformation. Proc. 14th Symp. Comp. Geom. pp. 29–38, (1998)
3. H. Edelsbrunner, M. A. Facello and J. Liang. On the definition and the construction of pockets in macromolecules. *Discrete Apl. Math.* **88**, pp. 83–102, (1998)
4. J. Giesen and M. John. Dynamical Systems from Disks in the Plane. Manuscript (2001)
5. J. Giesen and M. John. A new Diagram from Disks in the Plane. Accepted for Proc. 19th Int. Symp. Theo. Aspects of Comp. Sc. (2002)

Improvement of Digital Terrain Model Interpolation Using SFS Techniques with Single Satellite Imagery

Mohammad A. Rajabi, J. A. Rod Blais

Dept. of Geomatics Eng., The University of Calgary, 2500, University Dr., NW, Calgary, Alberta, Canada, T2N 1N4

{marajabi, blais}@ucalgary.ca

Abstract. The technique of stereo measurements is mainly applied to extract Digital Terrain Model (DTM) height data from stereo images in photogrammetry and remote sensing. Tremendous amounts of local and global DTM data with different specifications are now available. However, there are numerous geoscience and engineering applications which need denser DTM grid data than available. Advanced space technology has provided much single (if not stereo) high-resolution satellite imageries almost worldwide. In cases where only monocular images are available, reconstruction of the object surfaces becomes more difficult. Shape from Shading (SFS) is one of the methods to derive the geometric information about the objects from the analysis of the monocular images. This paper discusses the use of SFS methods with single high resolution satellite imagery to densify regular grids of heights. Three different methodologies are explained and implemented with both simulated and real data. Very encouraging results are obtained and briefly discussed.

1. Introduction

Digital Terrain Models (DTMs) are simply regular grids of elevation measurements over the land surface. They are used for the analysis of topographical features in GISs and numerous engineering computations. Rajabi and Blais [1] briefly reviewed and referenced a number of sources for DTM and their applications in engineering as well as science.

Stereo measurements from a pair of aerial photographs or satellite images have mainly been used as the primary data in producing DTMs. Information from double or multiple images in overlap areas ensures reliable and stable models for geometric and radiometric processing. Especially recently, with the rapid improvement in remote sensing technology, automated analysis of stereo satellite data has been used to derive DTM data ([2], [3], and [4]).

Today, with the need for the better management of the limited natural resources, there are numerous geoscience and engineering applications which require denser DTM data than available. But due to some reasons such as cloud coverage, technical and/or political limitations, stereo satellite imagery is not available everywhere.

P.M.A. Sloot et al. (Eds.): ICCS 2002, LNCS 2331, pp. 164–173, 2002.

Obviously, collecting additional height data in the field is a solution, but if not impossible, is either expensive or time consuming or both. While interpolation techniques are fast and cheap, they have their own inherent difficulties and problems, especially in terms of accuracy of interpolation in rough terrain.

On the other hand, the availability of single satellite imagery for nearly all of the Earth is taken for granted nowadays. Unfortunately, reconstruction of objects from monocular images is very difficult, and in some cases, not possible at all. Inverse rendering or the procedure of recovering three-dimensional surfaces of unknown objects from two-dimensional images is an important task in computer vision research. Shape from Shading (SFS) [5] [6], [7] is one of the techniques used for inverse rendering which converts the reflectance characteristics in images to shape information.

This paper discusses the application of SFS techniques to improve the quality of the interpolated DTM grid data with single satellite imagery of better resolution than the DTM data. The idea is highly motivated by the wide availability of satellite remotely sensed imagery such as Landsat TM and SPOT HRV imagery. Section 2 briefly reviews the general SFS problem and the methods implemented in this paper. Section 3 discusses some implementation details of the methods explained in section 2 in more depth. Section 4 provides numerical examples to support the methodology. Last but not least, section 5 ends the paper with some remarks and conclusion.

2. Shape from Shading

SFS is one of the methods which transforms single or stereo 2D images to a 3D scene. Basically, it recovers the surface shape from gradual variations of shading in the image. The recovered shape can be expressed either in terrain height z(x,y) or surface normal $\vec{N}$ or surface gradient (p,q)= $(\partial z / \partial x, \partial z / \partial y)$.

Studying the image formation process is the key step to solve the SFS problem. A Lambertian model is the simplest one in which it is assumed that the gray level at each pixel depends only on light source direction and surface normal. Assuming that the surface is illuminated by a distant point source, we have the following equation for the image intensity:

$$R(x,y) = \rho\ \vec{N} \cdot \vec{L} = \rho \frac{pl_1 + ql_2 + l_3}{\sqrt{p^2 + q^2 + 1}} \tag{1}$$

where ρ is the surface albedo, $\vec{N}$ is the normal to the surface and $\vec{L} = (l_1, l_2, l_3)$ is the light source direction. Even with known ρ and $\vec{\mathrm{L}}$, the SFS problem will still be a challenging subject, as this is one nonlinear equation with two unknowns for each pixel in the image. Therefore, SFS is intrinsically an underdetermined problem and in order to get a unique solution, if there is any at all, we need to have some constraints.

Based on the conceptual differences in the algorithms, there are three different strategies to solve the SFS problem [1]: 1. Minimization (regularization) approaches 2. Propagation approaches, and 3. Local approaches. A more detailed survey of SFS methods can be found in [8]. The following subsections briefly review the minimization approach, which is widely used in to solve the SFS problem and the other variants of the minimization approach which are used here to enhance the solution.

2.1 Minimization Approach

Based on one of the earliest minimization methods, the SFS problem is formulated as a function of surface gradients, while brightness and smoothness constraints are added to ensure that a unique solution exists [10]. The brightness constraint ensures that the reconstructed shape produces the same brightness as the input image. The smoothness constraint in terms of second order surface gradients helps in reconstruction of a smooth surface.
Brooks and Horn [13] defined the error functional:

$$I = \iint \left\{ (E(x,y) - N \cdot L)^2 + \lambda \left(\|N_x\|^2 + \|N_y\|^2 \right) + \mu (\|N\|^2 - 1) \right\} dxdy \tag{2}$$

where E(x,y) is the gray level in the image, and the constants λ, N_x and N_y are the partial derivatives of the surface normal with respect to x and y directions respectively and μ are Lagrangian multipliers. As it can be seen the functional has three terms: 1) the brightness error which encourages data closeness of the measured images intensity and the reflectance function, 2) the regularizing term which imposes the smoothness on the recovered surface normals, and 3) the normalization constraint on the recovered normals.

The functional is minimized by applying variational calculus and solving the Euler equation: The resulting fixed-point iterative scheme for updating the estimated normal at the location of (i,j) and epoch k+1, using the previously available estimate from epoch k is:

$$N_{i,j}^{k+1} = \frac{1}{1 + \mu_{i,j}(\varepsilon^2 / 4\lambda)} \left(\tilde{N}_{i,j}^{k} + \frac{\varepsilon^2}{4\lambda} \left(E_{i,j} - N_{i,j}^{k} \cdot L \right) L \right) \tag{3}$$

where

$$\tilde{N}_{i,j}^{k} = \frac{1}{4} \left(N_{i+1,j}^{k} + N_{i-1,j}^{k} + N_{i,j+1}^{k} + N_{i,j-1}^{k} \right). \tag{4}$$

There are two comments about this update equation. First, it seems that one has to solve for the Lagrangian multiplier $\mu_{i,j}$ on a pixel-by-pixel basis. However, as it is seen $\mu_{i,j}$ enters the update equation as a multiplying factor which doesn't change the direction of the update normal, therefore we can replace that factor by a normalization step. The second comment is about the geometry of the update equation. As it is seen, the update equation is composed of two components. The first one comes from the smoothness constraint while the second one is a response to the physics of image irradiance equation.

The main disadvantage of the Brook and Horn method or any other similar minimization approach is the tendency of over smoothing the solution resulting in the loss of fine detail. Selecting a conservative value for the Lagrangian multiplier is a very challenging issue in this method. However, in an attempt to overcome this problem, Horn [11] starts the solution with a large value for the Lagrangian multiplier and reduces the influence of the smoothness constraint in each iteration as the final solution is approached.

2.2 Modified Minimization Approaches

As it was mentioned in the previous section, the update equation is composed of two components, the smoothness part and the data closeness part. As the first attempt to solve the over smoothing problem with the general minimization approach, an adaptive regularization parameter $\lambda(i,j)$ instead of a fixed λ is suggested to be used to adaptively control the smoothness over the image space [9]. In each iteration, the space varying regularization parameter at location (i,j) can be determined by the following function:

$$\lambda_{new} = (1 - e^{-\frac{c(i,j)}{V_T}})\lambda_{\min} + (e^{-\frac{c(i,j)}{V_T}})\lambda_{old}(i,j) \tag{5}$$

where c(i,j) is the control signal, V_T is a time-constant that regulates the rate of exponential decrease and $\lambda_{\min}$ is a preselected minimum value that $\lambda(i,j)$ may have. The control signal is defined as c(i,j)=abs{I(i,j)-R(i,j)}, where abs{-} denotes the absolute value and the function λ_{new} is an exponentially decreasing function with the following properties:

$$\lim_{c(i,j)\to 0} \lambda_{new} = \lambda_{old}(i,j) \quad and \quad \lim_{c(i,j)\to\infty} \lambda_{new} = \lambda_{\min} \tag{6}$$

so that the regularization parameter is only allowed to decrease with the iterations. Another method to solve the over smoothing problem is to use a robust error kernel in conjunction with curvature consistency instead of a quadratic smoothness. The robust regularizer constraint function can be defined as [14]:

$$\rho_\sigma(\|N_x\|) + \rho_\sigma(\|N_y\|) \tag{7}$$

where $\rho_\sigma(\eta)$ is a robust kernel defined on the residual η and with width parameter σ. Among different robust kernels, it is proved that the sigmodial-derivative M-estimator, a continuous version of Huber's estimator, has the best properties for handling surface discontinuities [14] and is defined by:

$$\rho_\sigma(\eta) = \frac{\sigma}{\pi} \log\cosh\left(\frac{\pi\eta}{\sigma}\right). \tag{8}$$

Applying calculus of variations to the constraint function using the above mentioned kernel results in the corresponding update equation. Here σ, the width parameter of the robust kernel, is computed based on the variance of the shape index. Based on Koenderink and Van Doorn [12] the shape index is another way of representing curvature information. It is a continuous measure which encodes the same curvature class information as the mean and Gauss curvature, but in an angular representation. In terms of surface normals, the shape index is defined as [12]:

$$\phi = \frac{2}{\pi} \arctan\left[\frac{(N_x)_1 + (N_y)_2}{\sqrt{((N_x)_1 - (N_y)_2)^2 + 4(N_x)_2 (N_y)_1}} \right] \tag{9}$$

where $(\ldots)_1$ and $(\ldots)_2$ denote the x and y components of the parenthesized vector, respectively. The variance dependence of the kernel is controlled using the exponential function:

$$\sigma = \sigma_0 \exp\left(-\left(\frac{1}{\mathrm{N}} \sum \frac{(\phi_1 - \phi_c)^2}{\Delta\phi_d^2} \right)^{1/2} \right) \tag{10}$$

where σ_0 is the reference kernel width which we set to one, ϕ_c is the shape index associated with the central normal of the neighborhood, $\mathrm{N}_{i,j}$, ϕ_1 is one of the neighboring shape index values and $\Delta\phi_d$ is the difference in the shape index between the center values of adjacent curvature classes which is equal to 1/8 [12].
The other modification that can be done on the minimization approach is on the data closeness part of the update equation. We know that the set of surface normals at a point which satisfy the image irradiance equation define a cone about the light source direction. In other words, the individual surface normals can only assume directions that fall on this cone. At each iteration the updated normal is free to move away from the cone under the action of the local smoothness. However, we can subsequently map it back onto the closest normal residing on the cone. This has not only numerical stability advantage but also all normal vectors in the intermediate states are all solutions of image irradiance equation. In other words, the update equation for the surface normals can be written as:

$$N_{i,j}^{k+1} = \Theta \tilde{N}_{i,j}^{k} \tag{91}$$

where $\tilde{N}^k_{i,j}$ is the surface normal that minimizes the smoothness constraint while Θ is the rotation matrix which maps the updated normal to the closest normal lying on the cone of ambiguity. The axis of rotation is found by taking the cross-product of the intermediate update with the light source direction:

$$\theta = -\cos^{-1}\left(\frac{\tilde{N}^k_{i,j} \cdot L}{\|\tilde{N}^k_{i,j}\| \|L\|}\right) + \cos^{-1} E \tag{102}$$

3. Implementation Details

The main goal of this investigation is to improve the accuracy of the interpolated DTM grid data by applying SFS techniques with the corresponding single satellite imagery, while the original DTM data are used as boundary constraints in the SFS problem.

The basic assumption here is that the satellite imagery has one dyadic order better resolution than the original DTM data. We also assume that 1) the surface is Lambertian (which is questionable in reality), 2) the surface albedo is known (by applying classification techniques to multispectral satellite imageries), 3) the surface is illuminated by a distant point source (sun), and finally 4) the position of the light source is known.

Our approach deals with a patch at a time (see Fig. 2) with forty nine points. Sixteen grid points have known heights (dark circles) and the other thirty three are points with the interpolated heights (unmarked grid points). Our main objective is to improve the accuracy of the interpolation for the five innermost unknown points. The idea of using a patch at a time for parallel processing techniques appears highly attractive for extensive applications

The method essentially consists of three stages: 1) preprocessing, 2) processing, and 3) postprocessing. The preprocessing stage itself has two steps. In the first step, using interpolation (bilinear) techniques, the heights of the unknown points in the patch are estimated. When dealing with the real height data, if there is a gap in height measurements due to any reason (such as existing of rivers or lakes), the whole patch is left untouched.

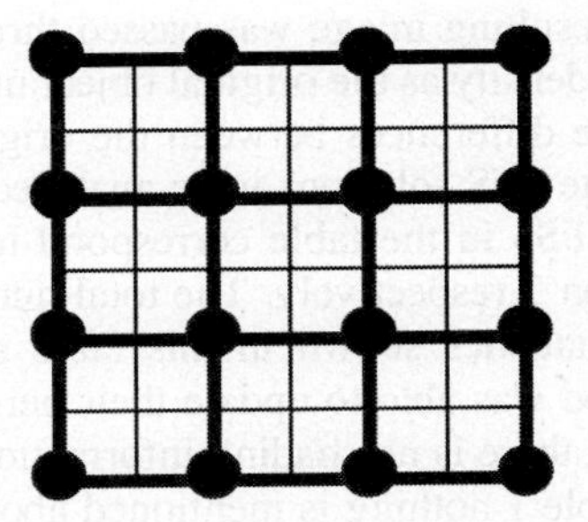

Fig. 1. A patch: Circles are the grid points with known heights and the unmarked ones are the points with the interpolated heights.

In the second step, using the known grid points, the relative orientation of the inner most square in the patch with respect to the light source is estimated. If this relative orientation implies that the patch is in the shadow, then there would be no useful shading information to improve the accuracy of the interpolated heights. Therefore, in this case the interpolated heights are considered the final height values.

Otherwise, the processing stage for each patch consists of three steps. In the first step, the smoothed surface normals (based on one of the three methods explained in Section (2) are computed. Then in the second step these surface normals are mapped onto the corresponding ambiguity cone. Finally in the third step, the surface normals are passed to an overdetermined (74 equations and 33 unknowns) linear adjustment process to solve for the heights. This is simply done by approximating p and q with finite differences in terms of heights. The control goes back to the first step of this stage unless the average difference between the calculated and original image gray values of all the pixels in the patch is less than a predetermined threshold.

The last stage, postprocessing, consists of taking arithmetic means of two solutions for the unknown heights located on the boundary of the innermost square in each patches coming from the neighboring patches, except for the outsides of the peripheral patches.

4. Numerical Examples

The methodologies described in Sections 2 and 3 have been tested using a number of numerical examples. One synthetic object with its synthetic imagery and one real DTM data and its corresponding satellite imagery were used in these experiments. With the synthetic object the orientation of the light source was considered as a variable to investigate the effects of relative positions of the object and the light source on the solution.

The synthetic object under investigation is a 1024 by 1024 pixel convex hemisphere with a radius of 250 units, sampled at each 0.5 unit. The corresponding DTM (one dyadic order less than the object, i.e., 512 by 512 pixels) was extracted out from the object. Meanwhile, the corresponding image of the object was created using a Lambertian reflectance model with a much (5 times) denser version of the object. The resulting image was passed through a smoothing filter to get an image with the same density as the original object under study.

The differences between the original object, the Interpolated Grid Solution (IGS) and the SFS solutions were analyzed. Table 1 summarizes these results. SFS1, SFS2, and SFS3 in the table correspond to the three different SFS solutions mentioned in Section 2 respectively. The total number of the patches in this experiment is 215296. The statistics shown in this table are computed with those pixels which our SFS method was able to update their height values. The other patches are either those for which there is no shading information or the SFS techniques failed to find a solution. In Table 1 nothing is mentioned about the azimuth of the light source. The symmetry of the synthetic object under investigation makes the process independent of the light source azimuth. As it can be seen from this table, SFS techniques have improved the standard deviation of the differences by an average rate of 43.1%.

Table 1. The convex hemisphere

Elev.	Object –IGS		Object –SFS1		Object-SFS2		Object –SFS3	
	Mean	Std	Mean	Std	Mean	Std	Mean	Std
30°	-0.04	0.32	-0.05	0.19	-0.04	0.18	-0.04	0.18
45°	-0.01	0.29	-0.03	0.17	-0.02	0.17	-0.03	0.16
60°	0.02	0.32	0.03	0.18	0.03	0.17	0.02	0.17
Patches not updated	-		32294		17223		10764	

The second test object is a real terrain data set from southern Alberta (Waterton), Canada with 25 metre spacing in UTM coordinate system. The original measured DTM data consists of 100 metre spacing grid in addition to the feature points. These measurements were used to interpolate the 25-metre grids. A 1024 by 1024 grid with more than 1300 metre height difference which was extracted out from the four quadrants of NTS 82H04 DTM data file.

The corresponding satellite imagery is a 3 channel SPOT data file with 20 m resolution which was originally georeferenced to an extended UTM coordinate system. By extracting the coordinates of the distinguished terrain features from the corresponding 1/20000 topographic map sheets, the SPOT imagery was georeferenced to the same coordinate system as the DTM data are. For this purpose 23 points with good distribution in the area under investigation were used. A second order polynomial was used for the purpose of georeferencing. The RMS of georeferencing in x and y directions were 2.96 m and 1.40 m respectively.

Using PCI software, a principal component transformation was applied and the first channel with 88.52% energy was selected for these experiments. Finally, the satellite imagery pixel size was changed from 20 m to 25 m using a bilinear interpolation method.

To test the efficiency of the SFS methods with the real data set, we tried to reconstruct the 25 m DTM from 50 m DTM and 25 m SPOT imagery. Similar to the simulated data the total number of the patches in this experiment is 215296. The statistics shown in this table are computed with those pixels which our SFS method was able to update their height values. The other patches are either those for which there is no shading information or the SFS techniques failed to find a solution. Moreover, there is no update in height values for those patches which contain natural features such as rivers or lakes where there is no measured height data. Table 2 summarizes the results of our experiment with this data set. As it can be seen from this table, the rate improvement in standard deviation can reach up to 48%.

Table 2. The real terrain data set

Object –IGS		Object –SFS1		Object-SFS2		Object –SFS3		
Mean (m)	Std (m)	Mean (m)	Std (m)	Mean (m)	Std (m)	Mean (m)	Std (m)	
0.03	14.95	-0.04	10.17	0.03	8.82	0.04	7.77	
-		32%		41%		48%		Improvement of Std
-		25835		2066		15716		Patches not updated

5. Remarks and Conclusions

SFS is intrinsically an underdetermined problem and in order to get a unique solution one has to implement some kind of constraint(s). Horn's method uses a single Lagrangian multiplier in the formulation of the constraint in the SFS problem. Selecting a conservative value for the Lagrangian multiplier is a very challenging issue in this method. Moreover, Horn's method and other similar approaches have the tendency of over smoothing the solution which results in the loss of fine detail. Obviously, selecting different Lagrangian multipliers for each pixel or cluster of pixels based on the roughness of the surface under study is the solution for this problem.

On the other hand, it seems that using the idea of mapping the computed smoothed normals in each iteration back onto the ambiguity cone as data closeness constraint works very well. This has not only numerical stability advantages but also all normal vectors in the intermediate states are all solutions of image irradiance equation.

Numerical examples show very encouraging results. The fact that the synthetic object is a much smoother surface than the real data set convinces us that the SFS methods should have given better results but this is not the case. The only apparent explanation is the way the synthetic image was constructed. It shows that our method of constructing the synthetic image does not resemble the real life case.

Satellite imagery used in this experiment has a very good quality without any with cloud or snow cover or any other major problem. However, to prove the efficiency of the above mentioned SFS methods, one should test them with different satellite imageries with different quality, resolution, sun position, spectral bands and last but not least different types of terrain.

Fine tuning different variable parameters in the SFS methods, applying classification techniques for having a more realistic value for the albedo, and using a more sophisticated reflectance model are obviously things that should be taken into consideration in future research and development.

6. References

1. Rajabi, M. A., Blais, J. A. R.: Densification of Digital Terrain Model Using Shape From Shading with Single Satellite Imagery. Lecture Notes in Computer Science, Vol. 2074, Springer (2001) 3-12
2. Gugan, D. J., Dowman, I. J.: Topographic Mapping from Spot Imagery. Photgrammetric Engineering and Remote Sensing 54(10) (1988):1409-1414
3. Simard, R., Rochon, G., Leclerc, A.: Mapping with SPOT Imagery and Integrated Data Sets. Invited paper presented at the 16th congress of the International Society for Photogrammetry and Remote Sensing held July 1988 in Kyoto, Japan
4. Tam, A. P.: Terrain Information Extraction from Digital SPOT Satellite Imagery. Dept. of Geomatics Eng., The University of Calgary (1990)
5. Horn, B. K. P.: Shape from Shading: A method for Obtaining the Shape of a Smooth Opaque from One View. Ph.D. Thesis, Massachusetts Ins. of Technology (1970)
6. Horn, B. K. P.: Height and gradient from shading. Int. J. Comput. Vision, 37-5 (1990)
7. Zhang, R., Tsai, P. S., Cryer, J. E., Shah, M.: Analysis of Shape from Shading Techniques. Proc. Computer Vision Pattern Recognition (1994): 377-384
8. Zhang, R., Tsai, P. S., Cryer, J. E., Shah, M.: Shape from Shading: A Survey. IEEE Transaction on Pattern Analysis and Machine Intelligence, Vol. 21, No. 8, August (1999): 690-706
9. Gultekin, A., Gokmen, M.: Adaptive Shape From Shading. ISCIS XI The Eleventh International Symposium on Computer and Information Sciences (1996): 83-92
10. Ikeuchi, K., Horn, B. K. P.: Numerical Shape from Shading and Occluding Boundaries. Artificial Intelligence, Vol. 17, Nos. 1-3 (1981): 141-184
11. Horn, B.K.P.: Height and Gradient from Shading. Int. J. Comput. Vision , Vol. 5, No. 1, (1990): 37-75
12. Koenderink, J. J., van Doorn, A.: Surface Shape and Curvature Scales. Image and Vision Computing, Vol. 10, No. 8, October (1992): 557-565
13. Brooks, M. J., Horn, B. K. P.: Shape and Source from Shading. International Joint Conference on Artificial Intelligence, (1985): 932-936
14. Worthington, P. L., Hancock, E. R.: Needle Map Recovery Using Robust Regularizers. Image and Vision Computing, Vol. 17, (1999): 545-557

Implementing an Augmented Scene Delivery System

James E. Mower

Department of Geography and Planning, ES 218
University at Albany, Albany, NY 12222 USA
jmower@albany.edu

Abstract. This paper addresses the core issues confronting the design and use of an augmented scene delivery system (ASDS). An augmented scene is a real-time, interactive, symbolized, perspective view of an environment that serves as a graphical index to an underlying spatial database. It allows a person in the field to interpret and navigate through the environment without reference to an external map. Augmented scenes will enable users with underdeveloped map use skills to effectively interpret and analyze their environment in professional, educational, and recreational contexts. This paper discusses an ASDS implementation that acquires imagery from a user-controlled webcam. It focuses on issues of data sampling and representation.

Introduction

Cartographic mapping can be understood as a series of transformations between views of geographic entities. Most of these transformations (modeling the earth as a solid, projecting its surface to a plane, etc.) are handled by the cartographer or geographic information system (GIS). One important transformation left to the user is map orientation. Faced with a map and a view of the terrain in the field, the user tries to understand the environment with the map as its model, gradually developing a cognitive transformation with elements of 3-dimensional rotation, perspective scale foreshortening, and hidden surface construction. Under the best viewing conditions, users differ sharply in their ability to interpret maps and orient them to their environment. As conditions degrade, even the most experienced user finds it difficult to determine her position or to find a path.

Most users find that they can reduce some of the cognitive burden of orientation by aligning the map with their direction of view, perhaps in association with a compass. A paper map can be rotated by hand; a computer-generated display with a simple coordinate transformation. The latter may be of further help if it can model the landform in perspective from an arbitrary position. Still, users are left with two independent views: that of the symbolic map and the raw visual scene.

Mower [8] discusses a theoretical and technical framework for removing the distinction between these views, unifying them into a new type of map, the augmented scene. An augmented scene places symbolic information directly over the user's current view of the environment, allowing her to query geographic features within the view through graphic selection. Employing a video camera as an imaging device, the user acquires an image of the landform. The augmented scene delivery

P.M.A. Sloot et al. (Eds.): ICCS 2002, LNCS 2331, pp. 174–183, 2002.

system (ASDS) applies the viewpoint coordinates and the user's viewing parameters to build a perspective model of the landform registered to the image.

This paper discusses a prototype implementation of an ASDS, focusing on landform elevation data structures and sampling issues. To reduce the complexity of the implementation, the author has substituted a user-controllable webcam for a hand-held video camera, eliminating the need for a GPS, digital compass, and digital inclinometer. All observations occur from a fixed location with full freedom of movement in horizontal and vertical viewing angles.

Previous Related Work

The introduction of affordable and sufficiently powerful computers to the cartographic community in the 1980s allowed for the dynamic overlay of acquired imagery on perspective maps [5]. The recent introduction of high-resolution position finding and geodetic angle acquisition devices has enabled cartographers to take this work a step further through the real-time overlay of map symbols onto acquired perspective imagery of the landform. Recent work in augmented reality has established a baseline for this research. At present, most augmented reality research emphasizes indoor, short-range applications [1]. Of these, Dorai and others [3] describe image and model registration techniques for industrial applications. Drasic and Milgram [4] discuss techniques for overlaying a user-controlled pointer on captured stereoscopic video imagery to locate objects in the 3D world for robotic navigation within lab settings. Schutz and Hugli [10] and Gleicher and Witkin [6] apply pattern recognition techniques and user intervention to improve model and image registration for short range applications. Recent papers by Neumann and others [9], Azuma and others [2], and Kim and others [7] discuss techniques for tracking objects for augmented reality applications in interior and exterior environments, generally over distances of up to several hundred meters. This project extends the usage of image acquisition, registration, and overlay operations to real-time perspective views of the general environment for geographic applications over long distances (to the user's horizon).

Implementation Design

The Camera, Mount, and Image Server

The camera, its mount, and control software were purchased from Perceptual Robotics, a supplier of user-controllable webcam systems. The camera is currently mounted on the roof of Mohawk Tower, a 24-story dormitory on the campus of the University at Albany. A Pentium II computer controls its operation from the floor below in the University's Climate Observatory. Camera and mount commands are sent through direct serial cable connections. A coaxial connection feeds video from the camera to the computer. The user controls camera, mount, and video functions through requests to the computer's web server that are formatted as URL strings.

ASDS Software Functions

The author has developed the current ASDS software implementation, *UrHere*, as a Microsoft Windows application written in C++. After loading a digital elevation model (DEM), the system creates a perspective viewing transformation that renders the DEM from the point of view of the camera. The area extending from the viewpoint to the horizon in all viewing directions will be referred to as the viewshed. A scene is defined as a portion of the viewshed rendered with respect to a fixed field of view, horizontal viewing angle (azimuth), and vertical viewing angle (altitude).

The user can select one or more geographic feature sets for labeling. Features are organized by theme (schools, businesses, landform features, etc.) and identified by their name, projected world coordinates (plus elevation), and other fields, including a URL specific to each feature. Any feature that occupies a point in the camera's current scene (modeled as a frustum) is a candidate for labeling. A user can change the orientation of the camera using one of three methods. First, she can elect to center the viewing plane on a given feature by clicking on its name in the feature database. Second, the user may enter viewing angles that move the camera horizontally (in azimuth) and vertically (in altitude). Azimuth angles are entered in degrees relative to true north; altitude angles are in degrees relative to 0 on the horizontal plane. The third method allows the user to center the camera on an arbitrary pixel in the current scene by picking it with a mouse click. The image coordinates of the pixel are employed to construct azimuth and altitude values that center the camera on its associated world coordinates.

Once the orientation of the camera has changed, features within the current feature sets having world coordinates intersecting those of the scene will be displayed as symbols at their projected locations, overlain on the returned camera image. The user can elect to render symbols as icons or pushbuttons. If the latter option is chosen, selection of a button launches a web browser that navigates to a feature-specific URL. All of the pushbutton URLs point to a single web server (independent of the camera hardware server) that returns a "home page" for each feature. Typically, the home page contains links to on-site and off-site sources of information. On-site sources can satisfy simple requests for text retrieval or more sophisticated text and graphic requests from a local internet map server such as ArcIMS or MapXtreme. Sample home pages are currently limited to text retrieval of attribute data. Figure 1 shows a cropped image of an augmented scene centered on the New York State Museum building, positioned approximately 7 km to the east of the camera. Its symbol is rendered as a pushbutton.

Fig 1. A feature symbol rendered as a pushbutton. The camera is viewing the New York State Museum in Albany, NY, USA, located approximately 7 km from the camera

To ensure consistent operation of the camera hardware, the hardware controller reboots each night, subsequently zeroing the azimuth and altitude drive motors. However, extended daytime use can cause the motors to "drift" slightly from their true bearings. *UrHere* provides a software registration procedure to fix drift. On the selection of this procedure, the system acquires an image at the current azimuth and altitude. It then overlays iconic symbols representing features in the scene that are members of a registration feature set. Features in this set are easily identifiable on the camera image in clear weather. To register the camera to ground coordinates, the user clicks a feature icon and then the icon's representative pixel on the image. The registration procedure determines the azimuth and altitude offset angles between the two locations in image space. Subsequent camera orientation requests add the offsets to compensate for drift in both the horizontal and vertical planes.

The registration procedure, as well as the procedures that aim the camera and place symbols on the acquired image, rely on the underlying surface model to provide a world frame of reference. The current model employs a triangulated, irregular network (TIN) with sampling density that decreases with distance from the camera viewpoint.

An earlier, variable density grid cell model was tried and rejected for this project. The structure of both models and their associated characteristics are addressed in the following section (Figure 2).

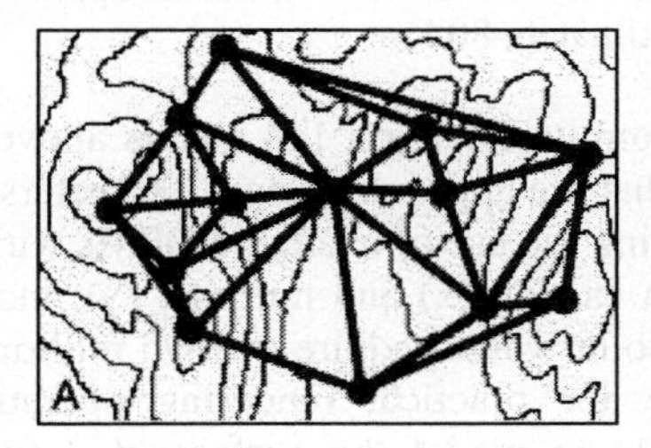

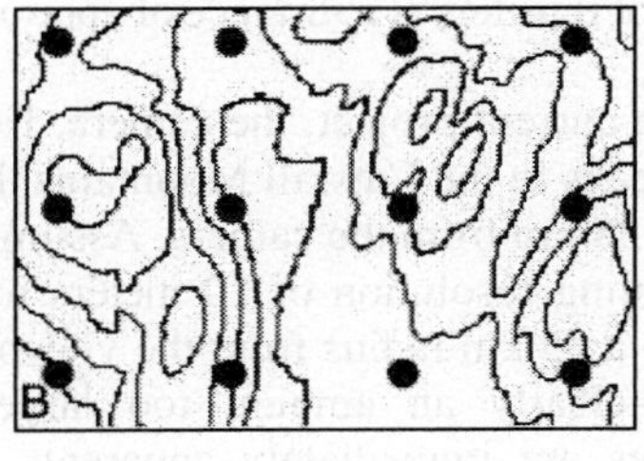

Fig 2. In A, the surface is sampled as a TIN at significant points. In B, the surface is sampled as a regular grid

Surface Data Structures

Rendering Perspective Scenes from Sampled Elevations

A mapping application that proposes to label features in a given scene must be able to model the world to its visible boundaries, determined by the elevations of the most distant features in the scene that intersect the plane of the viewing instrument (its "line of sight"). Here, feature elevation is measured relative to the surface of the sphere (the instrument horizon) with a radius equal to that of the earth plus the height of the viewing instrument. Since the instrument horizon is curved, the elevation of an intersecting feature must increase with distance from the viewpoint. Equation 1 provides the elevation of intersection (h) as a function of the radius of the instrument horizon (r) and angular distance from the viewpoint (θ). This calculation models the earth as a sphere and disregards atmospheric refraction. It returns the elevation of intersection in the same units provided for the radius.

$$h = (r / \cos(\theta)) - r \quad \textbf{(1)}$$

In Figure 3, a feature with an elevation of 500 meters above the instrument horizon will intersect its line of sight at a distance of approximately 80 km from the viewing instrument.

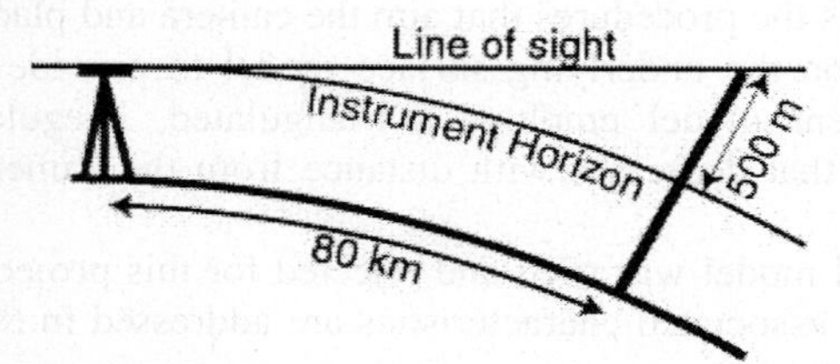

Fig 3. A feature rising 500 meters above the instrument horizon has an apparent relative elevation of 0 meters at a distance of approximately 80 km

For the current project, the camera, from its elevation 144 meters above sea level, can see peaks in the Catskill Mountains that are approximately 1200 meters above sea level and 55 km from the camera. Assuming the use of available DEMs with a regular grid sampling resolution of 10 meters in easting (x) and northing (y), modeling the surface to a 55 km radius from the viewpoint would require over 95 million elevation samples—clearly an amount too large for practical rendering operations. Two alternatives are immediately apparent: 1) to model the surface at a fixed, lower resolution, or 2) to vary the sampling resolution with distance from the viewpoint. Regardless of the applied sampling technique, the ground area covered by a rendered pixel in a perspective viewing model increases linearly with distance from the viewpoint. Given a half-angle θ for field of view and a distance d from the viewpoint, the width w of the viewing frustum at depth d is expressed in Equation 2.

$$w = 2dTan\theta \quad \textbf{(2)}$$

Assuming a camera viewing angle of 22° and a frame buffer 500 pixels in width, the width of a pixel representing a part of the scene at 1 km from the camera would represent approximately .8 meters on the ground. At 25 km its width would represent 19.4 meters. And at 55 km it would represent 42.8 meters.

Sampling Resolution

If sampling resolution is held constant, the number of samples per pixel will similarly increase with distance from the viewpoint. If the sampling resolution were fixed at 20 meters, for example, a pixel rendering the surface at 1 km from the viewpoint would cover .04 samples, at 25 km .97 samples, and at 55 km 2.13 samples. Pixels at the limit of the viewshed would over-sample a 1 sample per pixel ideal by a factor of 2, but pixels near the camera would be under-sampled by a factor of 25, enabling over-generalized renderings of foreground surfaces. To avoid this, a variable resolution sampling scheme would keep the per-pixel resolution constant with distance, neither under- nor over-representing any particular area.

A variable sampling technique adjusts the local resolution to account for perspective scale reduction with distance from the viewpoint. In the grid model, constant sampling resolution is maintained among bands surrounding the viewpoint. Resolution decreases at band borders as a stepwise linear function of distance from the viewpoint. Local irregularities cannot be accounted for without interrupting the integrity of the grid. In the TIN model, sampling resolution decreases continuously with distance from the viewpoint, modulated by local surface variability.

The variable resampling technique for grids partitions the viewshed into bands extending away from the viewpoint as square, non-overlapping bands (Figure 4). Elevations within a given band are resampled from a high resolution DEM (representing ground truth) as a linear function of the band's mean distance from the viewpoint. The effect of this technique is to create a viewing frustum that is partitioned into multiple regions, each characterized by successively lower sampling resolutions away from the viewpoint.

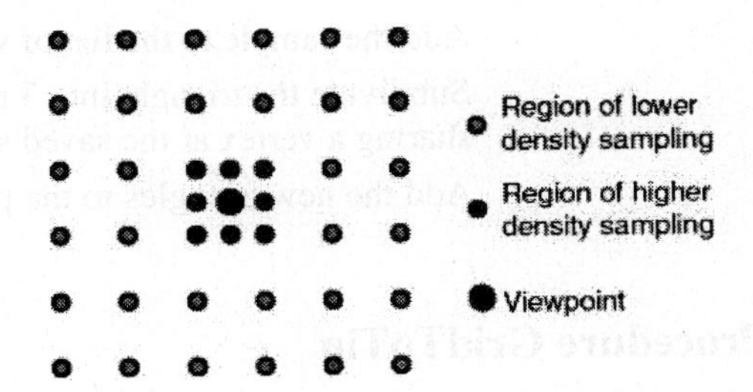

Fig 4. Variable resolution grid sampling

The grid model was tried but quickly abandoned for this project. Although resampled grids are relatively easy to construct, they have several disadvantages. This project works with a viewshed containing highly irregular features near the horizon—

the Catskill mountains to the south, the Taconics to the east, and the Adirondacks to the north. The resampling function was adjusted to ensure that such features were adequately rendered. The furthest visible feature, a peak in the Catskills approximately 55 km from the camera, needed to be rendered with a sampling of 43 meters per pixel to maintain a 1 sample per pixel ideal over a 500 pixel-wide frame buffer. Even if the entire data set were resampled at this resolution (thereby abandoning 1 sample per pixel representation in other parts of the viewshed) the entire viewshed would require over 5 million samples while still seriously underrepresenting terrain near the viewpoint. This analysis suggested that resampling techniques must account for surface variability as well as perspective scale reduction to enable efficient surface renderings.

Unlike the grid model, samples in a TIN elevation model generally represent topologically significant points on a surface. For this project, the procedure GridToTin was developed and implemented to extract such points from a grid cell DEM. The procedure iteratively subdivides a grid into triangles until no sample is further than a predetermined tolerance from the plane of its enclosing triangle.

Procedure GridToTin

Starting with a square grid of elevation samples and a base tolerance,

Form 2 triangles from the 4 samples at the grid corners. Add the samples to a list of significant samples and the triangles to the processing stack.

While there are triangles on the stack,

- Extract a triangle from the stack and find the equation of its plane.
- For each sample falling within the triangle's interior,
  - Compute the local tolerance as the base tolerance weighted by the sample's distance from the viewpoint.
  - Find the orthogonal distance of the sample to the plane.
  - If the distance exceeds the local tolerance,
    - Add the sample to the list of significant samples
    - Subdivide the triangle into 3 new triangles sharing a vertex at the saved sample
    - Add the new triangles to the processing stack

Discussion of Procedure GridToTin

The procedure becomes increasingly insensitive to deviations in sample elevations from their interpolated planar values as distance increases from the viewpoint. For rendering purposes, significant features such as peaks, ridgelines, and slope breaks can be retained out to the horizon with an appropriate base tolerance value and local tolerance function. A local tolerance value would be considered appropriate if 1) it does not create a triangle that covers an area less than that covered by a pixel and 2) if the further subdivision of a triangle would result in any of the child triangles sharing the shading value of its parent. The first criterion is identical to the 1 sample per pixel suggested for the grid model. The 2nd is more difficult to quantify, however. It ultimately depends on the color resolution of the rendering engine and the display controller. For this project, the base tolerance was set empirically by examining its effect on the selection of points near the horizon. The local tolerance was computed by multiplying the base tolerance by the distance of the sample to the viewpoint.

The expected running time of the procedure increases with the number of gridded elevation samples, the variance of the samples, and decreasing base selection tolerance values. Worst-case performance occurs when no sample within the interior of any triangle is within the local tolerance distance of its representative plane. This maximum grid partitioning is given in Equation 3 where T is the resulting number of triangles, r is the number of rows in the grid and c the number of columns.

$$T = 2(r-1)(c-1) \quad \textbf{(3)}$$

In practice, T will only equal the maximum partitioning if the local tolerance function returns zero for all distances from the viewpoint or if the elevation variance is very large.

At any given time during execution, the triangles on the processing stack represent mutually exclusive regions. Because such triangles can be processed independently of and simultaneously with one another, GridToTin can be adapted to a MIMD environment using a master/worker model. The master distributes triangles on the stack to available worker processors having read-only, shared access to the original gridded elevation data. If a worker finds a significant interior sample, it writes the sample to a local list, subdivides its triangle, and copies the resulting children to the master triangle-processing stack. After the stack is empty, each local sample list is written to a global list on the master.

Since most of the execution time will be spent processing triangles, the expected speedup for the MIMD version would be directly related to the number of available processors. Workload will remain balanced as long as the number of triangles on the stack exceeds the number of available workers. The only sequential bottlenecks occur when workers copy their child triangles to the master stack and when their significant sample lists are combined at the end of processing.

The expected running time of GridToTin is not very important to the fixed viewpoint prototype application since all of the viewsheds are modeled from a single TIN. In the mobile application, however, new TINs must be generated every time a shift in the user's field position extends the horizon or results in insufficient ground resolution near the viewpoint. To minimize the weight, complexity, and cost of a mobile ASDS, it would be reasonable to offload viewshed construction to a location-based service (LBS) provider. A GridToTin implementation would run on a server, returning a relatively small TIN to the user. A MIMD application would not be out of the question in such an environment.

Following the application of GridToTin, the selected samples are triangulated with *Triangle* [11] using the Delauney option.

The *UrHere* Implementation

The author wrote UrHere as a C++ program for the Microsoft Windows NT and 2000 environments using Visual C++ 6.0 and DirectX 8.0.

The elevation data for this project originated from grid cell DEMs produced by the New York State Department of Environmental Conservation (ENCON) at a horizontal ground resolution of 10 meters. The file data structure corresponds to that of the US Geological Survey 1:24,000 regular grid series. To simplify the TIN resampling procedure, the author first reformatted elevations across the ENCON data files covering the research area into 10 km by 10 km grids. As the TIN resampling procedure searches for significant samples, elevations are recomputed to account for curvature of the earth and the effects of atmospheric refraction with regard to the viewpoint.

To test the registration of surface world coordinates with image coordinates, the author used a GPS with post-processing to collect the coordinates of geographic features that occupy relatively small areas on the surface and that are clearly visible in camera imagery. With post-processing, the accuracy of the positions were estimated to be on the order of 3 meters or less from their true position. All surface and feature coordinates refer to Universal Transverse Mercator (UTM) zone 18 north, North American Datum (NAD) 1927.

Results

From the top of Mohawk tower on the campus of the University at Albany, the viewshed of the camera extends approximately 55 km to the south and somewhat less in other directions. To model the viewshed elevations, samples from ENCON DEMs at 10 meter horizontal resolution over a 90 km by 90 km region were converted to a TIN. Using a 2-meter vertical tolerance at the viewpoint that increases linearly to 133 meters at a distance of 55 km from the camera, the TIN renders the viewshed model with 39,329 triangles, far less than the maximum partitioning of approximately 162,000,000 triangles. The author found through inspection that the selected vertical tolerance range provided an adequate frame of reference for feature labeling. Running in Windows 2000 on a Pentium 4 at 1.3 GHz and rendering with an NVIDIA Quadro2 Pro accelerator rated at 30 million triangles per second, the model required 5 seconds to load (real time). Although profiling data were not available for this project, new scene renderings in the viewshed were displayed with no visible delay. The images in Figures 5 show a peak, High Point, at a distance of approximately 24 km from the camera, at magnification factors of 2X and 12X. Both images have been cropped from a larger scene. The original 24-bit color images have been converted to gray scale and reduced to 75% of their original dimensions.

Fig. 5a. High Point at 2X magnification

Fig. 5b. High Point at 12X magnification

Conclusion

This project has demonstrated the feasibility of using webcams from fixed locations for perspective mapping. Future work will extend the fixed model to a mobile platform, enabling a user to create viewshed maps on the fly. Such an application would benefit from recent work in wearable computing equipment, digital compasses and inclinometers, and location-based service technology.

References

1. Azuma, R. (1997). A Survey of Augmented Reality. In: Presence, Teleoperators and Virtual Environments. Vol. 6, No. 4, pp. 355-385.
2. Azuma, R., B. Hoff, H. Neely III, R. Sarfaty, M. Daily, G. Bishop, V. Chi, G. Welch, U. Neumann, S. You, R. Nichols, J. Cannon (1998). Making Augmented Reality Work Outdoors Requires Hybrid Tracking. Web address: http://www.cs.unc.edu/~azuma/ARpresence.pdf. Proceedings, 1st Int.Workshop on Augmented Reality, San Francisco.
3. Dorai, C. G. Wang, A. Jain, and C. Mercer (1998). Registration and Integration of Multiple Object Views for 3D Model Construction. In: IEEE Transactions on Pattern Analysis and Machine Intelligence. Vol. 20, pp. 83-89.
4. Drasic, D. and P. Milgram (1991). Positioning Accuracy of a Virtual Stereographic Pointer in a Real Stereoscopic Video World. In: SPIE Stereoscopic Displays and Applications II. Vol. 1457, pp. 302-312.
5. Faintich, M. (1986). Digital Cartographic Data Bases: Advanced Analysis and Display Technologies. In: Proc., 2nd Int. Symp. on Spatial Data Handling, Seattle, pp. 600-610.
6. Gleicher, M. and A. Witkin (1992). Through-the-Lens Camera Control. In: Computer Graphics. Vol. 26, No. 2, pp. 331-340.
7. Kim, J., H. Kim, B. Jang, J. Kim, D. Kim (1998). Augmented Reality Using GPS. In: Proceedings SPIE, Stereoscopic Displays and Virtual Systems V, Vol. 3295, pp. 421-428.
8. Mower, J. (1997). The Augmented Scene: Integrating the Map and the Environment. In: Proceedings, Auto-Carto 13, Seattle, pp. 42-51.
9. Neumann, U., S. You, Y. Cho, J. Lee, J. Park (1999). Augmented Reality Tracking in Natural Environments. In: Int. Symp. on Mixed Realities, Ch. 6, Ohmsha Ltd. & Springer-Verlag, Japan.
10. Schutz, C. and H. Hugli (1999). Augmented Reality Using Range Images. In: Proceedings SPIE, Stereoscopic Displays and Virtual Reality Systems IV, Vol. 3012, pp. 472-478.
11. Shewchuk, J. R (1996). Triangle: Engineering a 2D Quality Mesh Generator and Delaunay Triangulator. In: 1st Workshop on Applied Computational Geometry (ACM), Philadelphia, pp. 124-133.

Inspection Strategies for Complex Curved Surfaces Using CMM

R. Wirza[1], M.S. Bloor[2] and J. Fisher[2]

[1] Department of Computer Science, Faculty of Computer Science and Information Technology, University Putra Malaysia, Serdang, 43400, Selangor, Malaysia
rahmita@fsktm.upm.my

[2] School of Mechanical Engineering, Leeds University, U.K.
{m.s.bloor, j.fisher}@leeds.ac.uk

Abstract. The accurate measurement of complex surfaces is difficult. Accuracy demands precision in measuring technology, i.e., the measuring machine and also precise mathematical representation of complex geometries. This paper introduces a method of measuring a complex surface by using a Coordinate Measuring Machine and representing the measured surfaces mathematically. This enables comparison with other surfaces, e.g.. the as-designed surface or the original unworn surface. The measurement of the knee prosthesis was taken as a case study.

1 Background

The amount of wear particles released from the knee joint during operation within the human body is critical to the life expectancy of the joint. The knee joint contains a self-lubricating piece of Ultra High Molecular Weight Polyethylene (UHMWPE). By accurately measuring the UHMWPE before wear or approximating the original surface geometry and the measuring after wear and determining the difference, the volume of material lost can be calculated.

To measure accurately the complex surface, this paper proposes a method based on a surface fitting technique. From the many Polynomial based technique, the authors have decided to choose the Bézier technique [1]. In order to represent the complex surface with minimum measuring points, the author has chosen the technique of surface interpolation of scattered data and due to the complexity of the shape the use of triangular patches rather than rectangular patches is recommended. The work is an extension and combination of techniques suggested by Goodman and Said [1].

The need to measure accurately the amount of wear particles released from the knee joint implanted during operation within the human body is becoming important. This amount of wear is critical to the life expectancy of the joint. By accurately measuring the knee joint surface and subsequently determining the small difference between the measurements before and after wear testing, the volume of material can be calculated.

P.M.A. Sloot et al. (Eds.): ICCS 2002, LNCS 2331, pp. 184–193, 2002.

In this paper the author proposes a planning strategy for the measurement. The main objective of this proposed planning strategy is accurately to measure and represent a complex surface, e.g. the surface of the knee joint prosthesis, so that minimal changes of the surface can be detected. The proposed plan involves:

i. choosing several points on the knee, and naming them as initial points,
ii. with these initial points using the fitting technique to produce a set of interpolated points,
iii. producing a CMM program which uses the interpolated points which can measure one whole knee for each different design,
iv. approximating the original surface of the ex-plant, because in most cases, the surgeon does not have a blue-print or original design for the knee joint,
v. and finally, determining the small difference between the measurements after wear testing, and the original surface and calculating the lost volume of material.

There are many papers discussing the problems related to a planning strategy for the measurement. Cho and Kim [2], Lee, Kim and Kim [3], introduced their new inspection planning strategy for complex surfaces, which introducing a measuring point selection strategy, a probe path generation strategy, and movement of the probe approaching the surface. Since the measurement time and cost increase with the number of measuring points, and an equi-interval grid pattern measurement may result in insufficient sampling when there are sharp changes in surface curvature, or unnecessary sampling in a relatively flat region, they [2,3] decided to introduce a technique of choosing a suitable sample size as the first step. Their proposed technique was based on analysis of the mean curvature of the surface. The density of the measuring points will vary according to the magnitude of the mean curvature of the surface. They decided that by changing the density of the measuring points according to the curvature of the surface, the object can be measured more effectively. As in the paper reviewed by Legge, they also generated appropriate probing paths so that the surface can be inspected effectively with minimum measuring errors and within the required time period. They introduced a technique which can select the probing sequence of measuring points optimally so that the moving distance of the probe and the inspection time can be reduced.

Touch probe selection is one critical step for inspection to be carried out on a CMM. Moroni *et al.* [5] report a preliminary study and the prototypical realisation of an expert system to generate touch probe configurations. Ziemian and Medeiros [6] also concentrate more on probe selection and part set-up rather than introducing the whole planning system.

2 Measurement Planning Algorithm

The proposed measuring planning procedure involves two phases. Phase 1 is the identification of the geometry of the knee prosthesis by digitising a set of initial measurement points, reading the data file produced by CMM on-line software, interpolating these initial measurement points, and finally defining a set of surface points. Phase 2 involves calculating the partial derivative of the surface at the interpolation part, calculating the vector direction IJK and producing the CMM program for the whole knee prosthesis.

A CMM on-line program is produced in Phase 1 which is used to measure the knee joint prosthesis geometry. This early version of CMM on-line program instructs the user to move the probe manually and measure the geometry of each condyle, by measuring 20 points on the inside edges of the condyle, 15 points on the inside surface, one point near to the middle of the condyle, to identify the bottom surface of the knee condyle. For the purpose of the worn knee, at least 15 measurement points on the worn area were added, depending on the size of the worn area. The partial derivatives at each digitised point are calculated, using least square minimisation introduced by Renka and Cline [8]. The Delaunay triangulation diagrams are constructed and finally, the Bézier interpolation surface is produced, which consists of a set of interpolation points with 1 mm space between each of them in x and y axis directions.

2.1 Surface fitting

There are a variety of methods and techniques involved in surface fitting. The problem which will be discussed here is to construct a smooth surface which passes through scattered data, $\{(x_i, y_i, z_i)\}_{i=1}^{n}$. For the purpose of the proposed strategy, it was decided to apply the convex combination method, which was introduced by Goodman and Said [1] because it is a local interpolation method which can be implemented on scattered data with C^1 continuity and triangular patches which are the most suitable for the complex surfaces. This method, also known as the hybrid method, uses triangular patches, interpolating them to develop interpolation patches. A partial direction derivative was used on each edge of the triangular patches to make sure the surface fit satisfied the first order continuity, C^1 over each triangle boundary. To apply this method onto a set of scattered data points as $\{(x_i, y_i, z_i)\}_{i=1}^{n}$ where we wish to interpolate $z_i = F(x_i, y_i)$,i.e. z is a function of x and y, three pre-processing steps are needed:

1. Triangulate the domain of the data points, which is the initial points, by using triangulation techniques, such as Delaunay triangulation.
2. Estimate the first-order partial derivatives at the data points to satisfy C^1 smoothness [8]
3. Compute the interpolate points in each triangle by the triangular scheme.

For further explanation, the reader is referred to Goodman and Said [1].

2.2 The Probe Movement towards the Surface

During inspection, the probe approaches the surface and retracts after touching the surface. To minimise measuring errors, the probe should move along the normal direction. However, it is time consuming to vary the probe orientation to meet the normal direction of every point. Every time the probe orientation changes, calibration is necessary, which could add to the total error. As a compromise which reduces the measuring error without adding calibration error, the probe was moved towards the point in the normal direction, without changing the orientation of the probe. To achieve the normal movement, the author calculated the gradient vector for each interpolate point (Refer to figure 1).

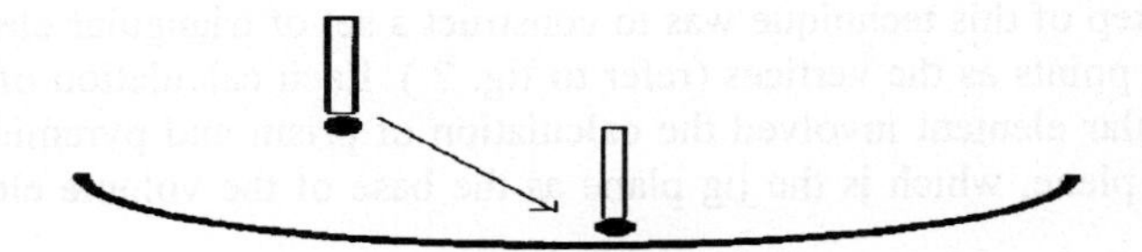

Fig. 1 -Vector movement of the probe towards the knee surface

The Algorithm

Step 1: [Measurement Procedure Phase 1]

Digitised 20 points on the inside edges of the condyle using CMM

Digitised 15 points randomly inside the condyle

Digitised 1 point near to the middle of the condyle(refer to fig. A.1 and fig. A.2).

Read the output data file and restructure them for the next stage.

Step 2: [Measurement Procedure Phase 2]

Triangulated the digitised points.

Interpolated the triangular patches.

Calculated the gradient vector for each interpolated points.

Produced the CMM on-line list of instruction.

Step 3: [Measurement Procedure Final Phase]

Run the on-line list of instruction.

Digitised a set of points on each condyle

Save the points coordinate in an output file

3 Analysis strategy for the measurement of knee joint surfaces

The knee implant design plays an important role in the rate of wear, which happens because of contact stress. A variety of research has been done regarding this problem, but sometimes the researchers did not have any or enough information on the original design or blueprint of the knee joint. It is hoped that the technique suggested in this paper can (1) give the approximate shape of the unworn surface of the knee prosthesis, and thus (2) give a good approximation for the volume of wear that has occurred. In Legge's review paper [4], in addition to the planning strategy for the measurement, an evaluation procedure is also involved. In the case of this study, the evaluation procedure included the worn volume approximation and the approximation of the original surface. The next step was introducing the technique for the calculation of the volume between the measuring points and the plane datum.

3.1 Volume under the measurement points

The first step of this technique was to construct a set of triangular elements with the measuring points as the vertices (refer to fig. 2). Each calculation of volume under the triangular element involved the calculation of prism and pyramid volume, with the datum plane, which is the jig plane as the base of the volume element (refer to fig. 3).

For the calculation of the prism volume the formula is:

$$V_{prism} = \text{the area of the domain triangular element} \times \text{height}_{prism}$$

$$= \frac{1}{2} \begin{vmatrix} x_1 & x_2 & x_3 \\ y_1 & y_2 & y_3 \\ 1 & 1 & 1 \end{vmatrix} \times \text{height}_{prism} \tag{1}$$

For the calculation of the pyramid the formula is;

$$V_{pyramid} = \tfrac{1}{3} \times \text{area of pyramid base} \times \text{height}_{pyramid}$$

$$= \frac{1}{3} \times \begin{vmatrix} x_2 & y_2 & z_2 \\ x_3 & y_3 & z_3 \\ x_2 & y_2 & z_{prism} \end{vmatrix} \times \left(\frac{|Ax_1 + By_1 + Cz_1 + D|}{\sqrt{A^2 + B^2 + C^2}} \right) \tag{2}$$

And finally for the whole element, (1) and (2) are added together for each element and all the elements under the chosen area totalled,

$$V_{total} = \sum_{i=1}^{N} V_i = \sum_{i}^{N} \left(V_{prism} + V_{pyramid} \right)_i .$$

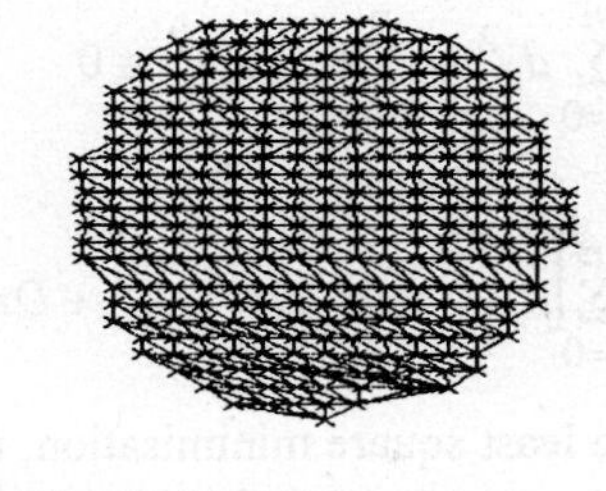

Fig. 2 - set of triangular elements on domain (x,y) of the digitised points of the knee prosthesis.

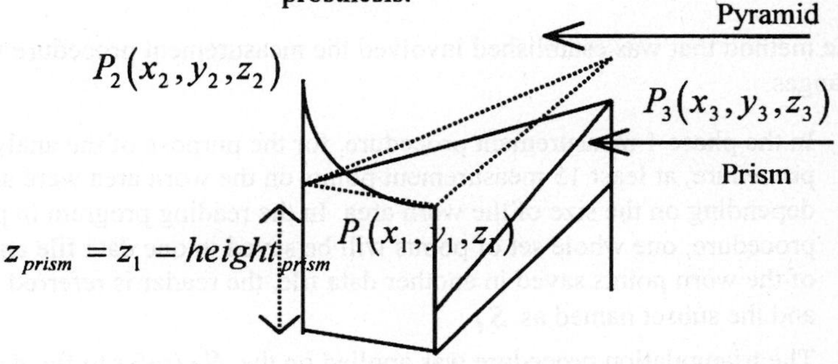

Fig. 3 - Volume under single triangular element

3.2 Approximation of the Original Unworn Surface Method

The geometry of the articular surfaces of a knee prosthesis is complex, varying from two discrete and different ellipsoids of revolution, to more complex anatomical shapes. There is a need to approximate the changes of shape due to creep and wear on the polymer. Before proceeding to the approximation method, the author needed to consider the algebraic surface that could best represent the surface of the knee condylar. In the case of this study, by looking i.e., visual inspection, the author assumed that each knee condylar shape is similar to an ellipsoid and proceeded with the proposed technique.

Assuming that the surface of the condyle is a semi-ellipsoid or an elliptic paraboloid, the surface equation will be given as

$$z = Ax^2 + Bxy + Cy^2 + Dx + Ey + F \tag{3}$$

with A, B, C, D, E, and F are unknown variables. If this assumption is acceptable then

$$\phi(A,B,C,D,E,F) = \sum_{i=0}^{n} d_i^{\,2} = \sum_{i=0}^{n} (z - z_i)^2 \cong 0 \tag{4}$$

$$\phi(A,B,C,D,E,F) = \sum_{i=0}^{n} \left[Ax_i^{\,2} + Bx_i y_i + Cy_i^{\,2} + Dx_i + Ey_i + F - z_i\right]^2 \cong 0 \tag{5}$$

By proceeding with the least square minimisation, finding the A,B,C,D,E and F, the algebraic surface equation to represent the unworn surface can be approximate.

3.3 Approximation of the Original Unworn Surface Procedure

The method that was established involved the measurement procedure with minor changes.

1. In the phase 1 measurement procedure, for the purpose of the analysing procedure, at least 15 measurement points on the worn area were added, depending on the size of the worn area. In the reading program in phase 1 procedure, one whole set of points will be saved in one data file and the subset of the worn points saved in another data file, the reader is referred to Wirza[14], and the subset named as S_1
2. The triangulation procedure was applied on the S_1 (refer to fig. 4), and the results saved in a data file. These triangulation data points were named set S_2
3. The phase 2 procedure was applied and the CMM on-line programming produced (refer to fig. 5). The on-line program was run, the results read and the points saved, under the name set S_3, in a data file as in phase 2 measurement. The volume procedure was applied.
4. By applying Cramer's rule on data set S_2 , points in S_3 were identified that digitised the unworn part and this set of points named as S_{temp}.
5. The least square approximation and Gaussian elimination technique were applied to S_{temp} to approximate the surface equation.
6. The surface equation was used to approximate the new z values for the worn area.

With the digitised z values for the unworn part and the approximate z values for the worn part, the approximation of the original surface was established. By applying the volume procedure, the volume for the worn part could be calculated.

4 Experimental Results

Before proposing and implementing the algorithm, it was necessary to do some experiments to check the accuracy and the capability of the proposed technique. This experiment started with an implant knee prosthesis which the author weighed before and after drilling the knee left condylar to ≈ 0.5 mm depth and radii of ≈ 10 mm. By assuming it was an explant, and that the original surface design and its weight was unknown the experiment proceeded.

The objective of this experiment was to check and show the ability of the proposed method to approximate the original surface and the lost volume of the worn knee prosthesis. With the information of the UHMW-PE density and weight value before and after the drilling process, a comparison could be made and the results evaluated. By weighing the knee before and after the drilling process, the results could be evaluated.

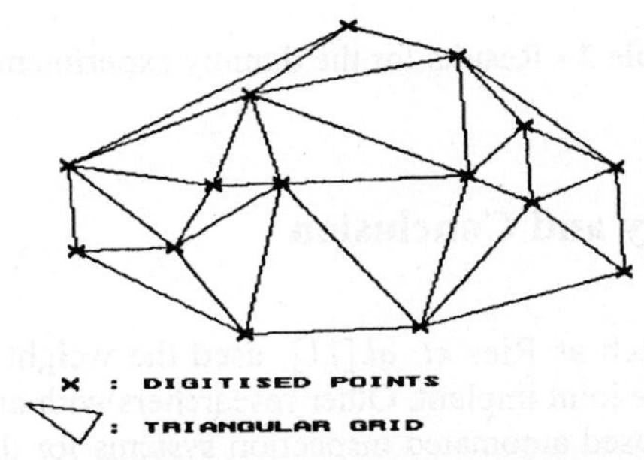

Fig. 4 - The triangular grid on the worn area

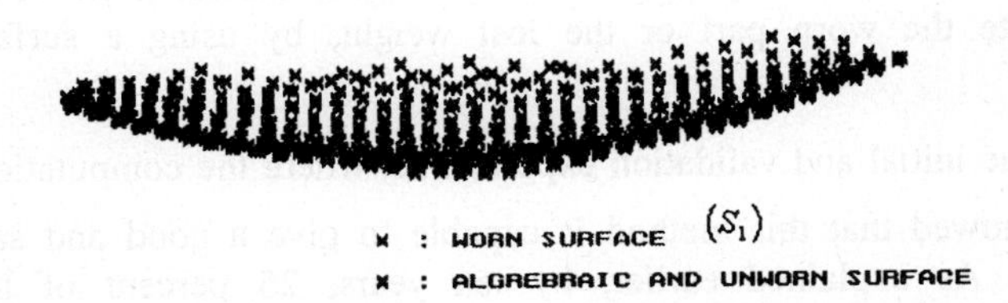

Fig. 5 - The digitised points from the side view

From the test certificate, production code : 701404, date 8/10/96 the density for the UHMW-PE is $0.934\ \mathrm{kg}/m^3$ and by taking the weight value for the knee before and after the drill process, the volume for the worn part can be calculated.

Weight (g)	1	2	3	4	5	Average
Before	25.84847	25.84846	25.84848	25.84844	25.84845	25.848
After	25.83025	25.83019	25.83032	25.83024	25.83014	25.830
drill out						0.018

Table 1 - a set of weight values before and after drilling

By approximating the volume lost using the proposed technique and weighting the lost volume the results for this experiment were:

Total volume for the digitised points	5542.242
Total volume for the approx. unworn points	5560.373
Differences, i.e. approx. volume lost	18.132
Volume lost calculate from the weight	19.486
Computation Error	1.354

Table 2 - Results for the dummy experiment (mm^3)

5 Summary and Conclusion

Most researchers such as Ries *et. al.*[11], used the weight method to measure the weight loss in a knee joint implant. Other researchers with an inspection background [2,3,4,6] have proposed automated inspection systems for the purpose of modelling the product and inspecting the milled complex surfaces to determine whether the standard tolerance is satisfied. This study, especially the experiment in this chapter, proposed a method that can: (1) approximate the original surface from the worn part, (2) approximate the worn part or the lost weight, by using a surface fitting technique.

Results from the initial and validation experiments, where the computation error is $1.354\,mm^3$, showed that this method is capable to give a good and satisfactory approximation. As explained earlier, by ten years, 25 percent of total knee replacements may look loose on x-ray, and about 10 percent will require re-operation [11,12,13]. Since recent work has shown that wear in a knee prosthesis is between 5 and 20 $mm^3/year$ [11], this means that by ten years, the wear in an explant of knee prosthesis might be between 50 and 200 mm^3. By assuming the depth of worn part is ≈ 0.5 mm, the proposed method will approximately contribute errors between 0.43 and 1.72 mm^3 (please refer to Wirza[14]). This is considered adequately small. With the digitised z value for the unworn part and the approximation z value for the worn part, the approximation of the original surface is established. This means the proposed method used in this experiment achieved its two objectives, as mentioned earlier.

References

1. Goodman T., and Said H. B.: A C^1 triangular interpolant suitable for scattered data interpolation. Communications in Applied Numerical Methods, Vol. 7. (1991) 479-485

2. Cho M. W., and Kim. K.: New Inspection Planning Strategy for Sculptured Surfaces using Coordinate Measuring Machine. Int. J. Prod. Res., Vol 33. no. 2. (1995) 427-444
3. Lee J. W., Kim M. K., and Kim K.: Optimal Probe Path Generation and New Guide Point selection Methods. Engng. Applic., Artif., Intell. Vol 7. No. 4 (1994) 439-445
4. Legge D. I.: Integration of design and Inspection System – a literature review. Int. J. Prod. Res., Vol 34. No. 5 (1996) 1221-1241
5. Moroni G., Polini W., and Semeraro Q.: Knowledge Based method for touch probe configuration in an automated inspection system. Journal of Materials Processing Technology, Vol 76. (1998) 153-160
6. Ziemian C. W., and Medeiros D. J.: Automated Feature Accessibility Algorithm for Inspection on a Coordinate Measuring Machine. Int. J. Prod. Res., Vol. 35. No. 10. (1997) 2839-2856
7. Brassel K. E., and Reif D.: A Procedure to Generate Thiessen Polygon. Geographical Analysis, Vol 11. No. 3. (1979) 289-303
8. Renka R. J., and Cline K.: A triangle-based C^1 surface interpolation. The Rocky Mountain Journal of Maths, Vol. 14. (1984) 223-237
9. Kim S., and Chang S.: The Development of the Off-Line Measurement Planning System for Inspection Automation. Computers Ind. Engng., Vol. 30. No. 3. (1996) 531-542
10. Derbyshire B., Hardaker C. S., Fisher J., and Brummitt K.: Assessment of the change in volume of acetabular cups using a Coordinate Measuring Machine. Proc Instn. Mech. Engrs., Vol. 208. (1994) 151-158
11. Ries M., Banks S., Sauer W., and Anthony M., (inprint): Abrasive Wear Simulation in Total Knee Arthroplasty. Consensus ®Knee. (1998)
12. http://www.hayesmed.com/products/consensus/conknee.htm.
13. http://www.hipandkneesurgery.com/tka.htm.
14. R Wirza: Inspection Strategies for Complex Curved Surfaces. PhD theses. The University of Leeds. April 2000

Appendix

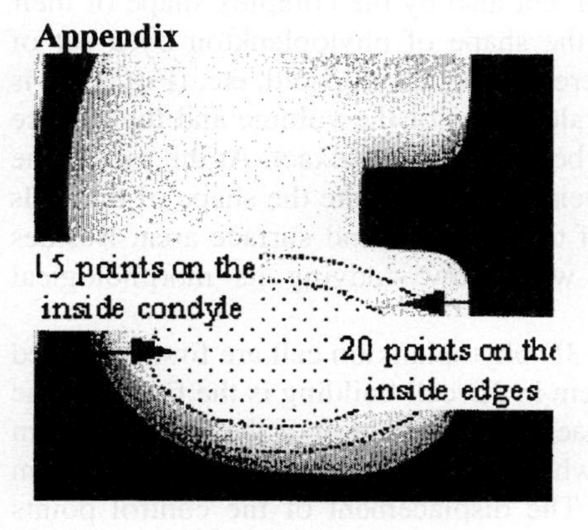

Fig A.1 – The measuring points planned on the knee condyle

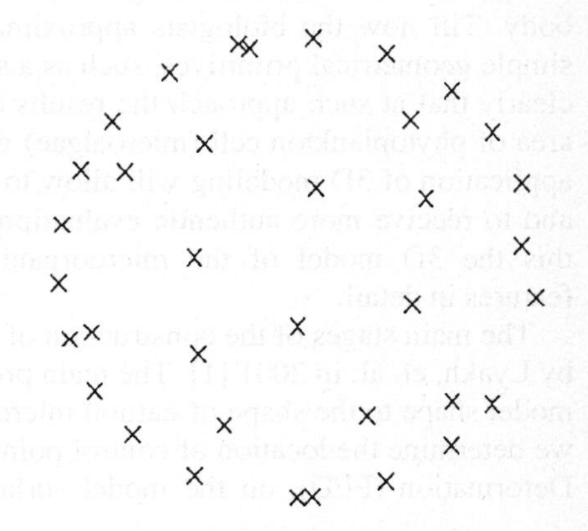

Fig. A.2 - The initial chosen measured points

The Free Form Deformation of Phytoplankton Models

Anton Lyakh

Dept. of Biophysical Ecology, Institute of Biology of the Southern Seas, 2, Nakhimov av., Sevastopol, 99011, Ukraine

anton@ibss.iuf.net, antonberlin@yahoo.com

Abstract. We construct the 3D models of phytoplankton. In presented article we solve the problem about deformation of the created models: How to deform the model so that its shape should conform to the shape of the natural microorganism? We use the Free Form Deformation (FFD) for the solving of this problem. In the first case, when the dimensions of the microorganism are given, the distances between control points of the FFD are equated to these dimensions. In the second case, when a scientist has the image of the microorganism, he should deform the model, displacing the control points, so that the image edge and the model's outer border has coincided. The control points location is used in the studying morphological changes of microorganisms and describing their shapes.

1 Introduction

The oceanic microorganisms (bacteria, phytoplankton) are the important part of the World Ocean food web, without which the life in the ocean will die out. Microorganisms' tiny dimensions (less than 1 mm) produce difficulties for their studying.

The volume and surface area of microorganisms need to be measured practically at all microbiological researches. Not only the tiny dimensions of microorganisms complicate the solution of the given problem, but also by the complex shape of their body. Till now the biologists approximate the shape of phytoplankton by a set of simple geometrical primitives, such as a sphere, a cone, an ellipsoid, etc. (Fig. 1). It is clearly that at such approach the results of calculation of the volume and the surface area of phytoplankton cell (microalgae) will be insufficiently exact. At the same time application of 3D modeling will allow to precisely approximate the shape of the cells and to receive more authentic evaluations of their volume and surface area. Besides this the 3D model of the microorganism will allow studying its morphological features in detail.

The main stages of the construction of the 3D phytoplankton cell are first described by Lyakh, et. al. in 2001 [1]. The main problem in 3D cell building is the fitting of the model shape to the shape of natural microalgae. For the solution of the given problem we determine the location of control points, which are the control points of Free Form Deformation (FFD), on the model surface. The displacement of the control points

P.M.A. Sloot et al. (Eds.): ICCS 2002, LNCS 2331, pp. 194–201, 2002.

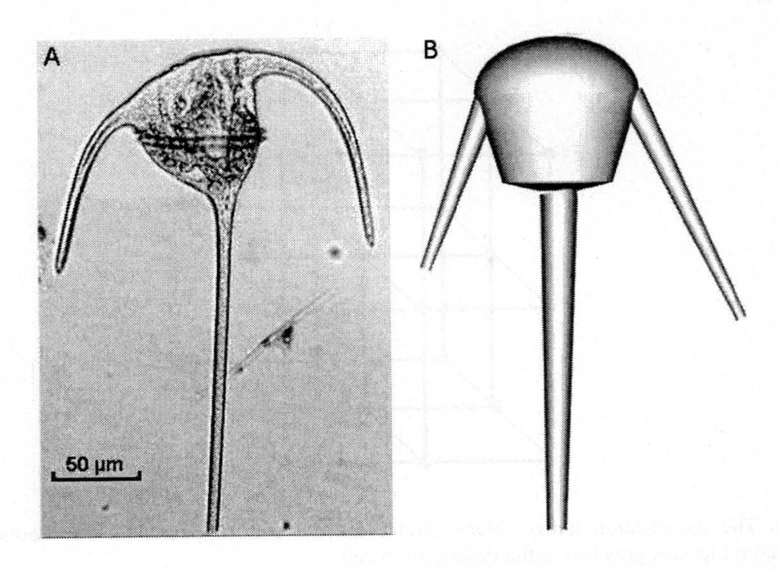

Fig. 1. The comparison of the phytoplankton cell's body shape with the geometric model, which is used for the cell volume and surface area calculation. A - the microphotograph of the phytoplankton cell *Ceratium tripods;* B - the 3D model created from different geometric primitives. In many cases the 3D model can be simpler

produces the deformation of the model that allows fitting the model shape to the shape of the natural cell.

2 Free Form Deformation

FFD is defined by parametric functions (3D splines) whose values are determined by the location of the control points (CP) of deformation lattice (Fig. 2). Once the control points are moved, the new location of object vertexes is determined by the weighted sum of the control points [2].

The FFD is a mapping operating from the world space of the model to the local space of the deformation lattice, and conversely to the world space:

$$\text{FFD}: R^3 \rightarrow R^3 \rightarrow R^3. \tag{1}$$

Let function F: $R^3 \rightarrow R^3$ maps the vertex X = (x, y, z) of the object in the world space to the vertex U = (u, v, w) of the object in the space of deformation lattice:

$$F(X) = U. \tag{2}$$

Let function F': $R^3 \rightarrow R^3$ maps the vertex U = (u, v, w) to the vertex X' = (x', y', z') of an object in the deformed world space:

$$F'(U) = X'. \tag{3}$$

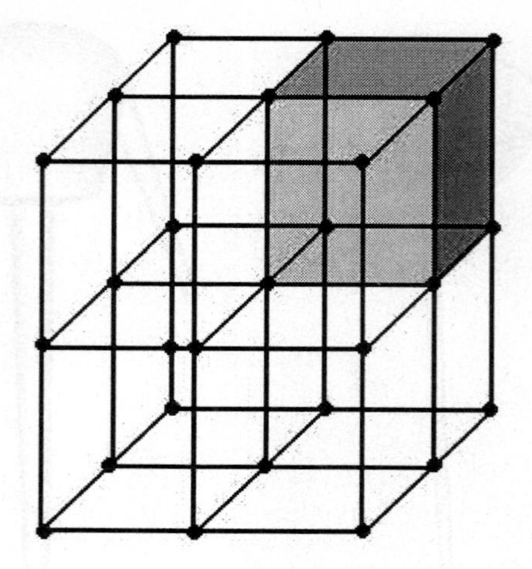

Fig. 2. The deformation lattice: black circles are the control points (or the knots) of the deformation lattice, gray box is the deformation cell

Then the composition of these functions defines FFD:

$$FFD\,(X) = F'\,(\,F\,(X)\,) = X'. \tag{4}$$

The form of the deformation lattice defines the function F, which assigns to each vertices of the object a set of local coordinates. We use the local parallelepiped-shape coordinate system of the deformation lattice. Therefore F is the transformation of the world coordinate system to the coordinate system of a parallelepiped, which sides are considered equal to unit length.

The function F' is a sum of control points $P_{i,,j,k}$ weighted by polynomial basis functions $B_i(\cdot)$:

$$F'(u,v,w) = \sum_{i=0}^{Ni}\sum_{j=0}^{Nj}\sum_{k=0}^{Nk} B_{i,Ni}(u)\cdot B_{j,Nj}(v)\cdot B_{k,Nk}(w)\cdot P_{i.j.k}\,. \tag{5}$$

The basis functions B are defined as follows:

$$B_{i,N}(u) = C_N^i \cdot (1-u)^{N-i} \cdot u^i \tag{6}$$

FFD proceeds in three steps:

1. Object vertices are assigned local coordinates *u,v,w* (F is applied). Since the local parallelepiped has sides with unit length, therefore the local coordinates *u,v,w* varies from 0 to 1.
2. Control points are displaced that cause the distortion of the local space. But the model's local coordinates don't change. They never change.
3. Equation 5 is applied to all of the object vertices to produce the deformation of the model (Fig. 3).

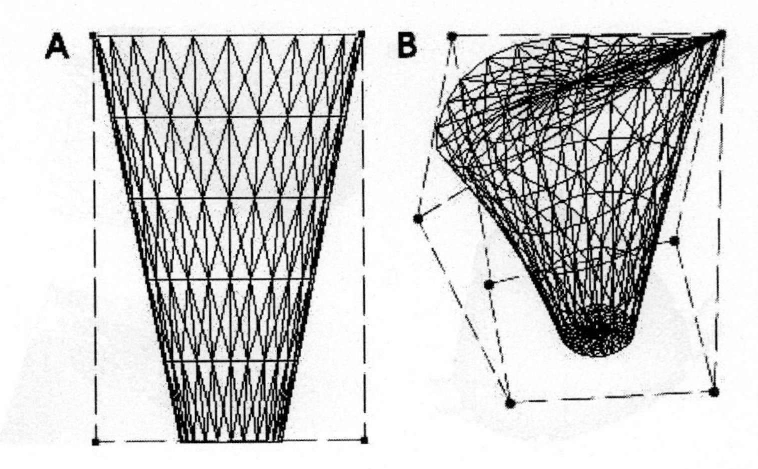

Fig. 3. The FFD of the cone: black circles are the FFD control points; dashed lines show the boundaries of parallelepiped-shape coordinate system. A – the source object; B – displacement of the control points causes the deformation of the cone

3 The Deformation of the Phytoplankton Model

The location of the control points is the important factor influenced on the model's deformation. Each control point influences only on adjacent deformation cells. Therefore, the dimension of deformation cells is the important factor too. The dimension of deformation cells determines a region, which will be deformed. Finally, the application of the one FFD operator to whole model will essentially hamper the process of model deformation. Therefore, the cell's model is divided into parts (elements), and a single FFD operator is applied to each model's part.

3.1 The location and classification of the control points

The initial location and amount of CP plays the defining role during the model deformation. With CP number increase the ability of the model to accept the necessary shape is increased. However, many CP will hamper the deformation.

The location of the control points. Not all control points of the model have an analog on the natural object. The location of the majority of CP is chosen so that the model of the microalgae would be convenient for deforming. Therefore control points are located on the following parts of the cell:

- One control point is located at concave and convex parts of the cell, at the top of the cell's growth, etc. (Fig. 4, A).

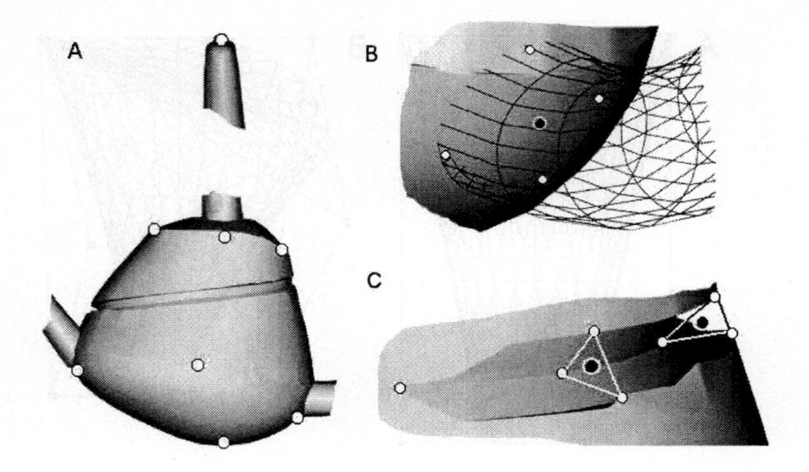

Fig. 4. The location of the control points (CP) on the different parts of phytoplankton cell. A – on the cell's growth and on the cell's body; B – in the zone of contact the cell's element with the cell's body; C – on the cell's groove: the groups of points, which are put along the groove's edges, are delineated by triangles; white circles are the ordinary CP; black circles are the main CP (see the text for detail)

- Five CP are located at the places where the cell's elements contact with the cell's body. One CP is located at the center of the contact zone; another four are located on the edges of the contact zone (Fig. 4, B).
- The group of the control points is uniformly distributed along the edges of the cell's grooves and keels. Two points are located at the groove/keel ends; two points at the groove/keel edges; one point in the hollow/crest; and the last one in the center of the formed triangle (Fig. 4, C).

The classification of the control points. The deformation of the cell element consists of the distortion of the element's surface and the deformation of element's medial axis. The distortion of the element's surface is caused by the displacement of *the ordinary* control points, which are situated on it (white circles on Fig. 4). *The changing of the ordinary CP position produces the immediate deformation of the model.* The deformation of the element's medial axis is caused by the displacement of the *main* control points, which are situated in the center of the contact zone, or along the axial line of the cell's groove/keel (black circles on Fig. 4). The *main* control point is connected with the *ordinary* control points. *The changing of the main CP position produces the shift of the ordinary CP that causes the deformation of a model.*

That is, the difference between the *ordinary* and the *main* CP is that the first one makes the direct deformation of the model, and the second one makes the indirect model deformation, by means of the first. Using the two types of the control points facilitates a 3D model deformation.

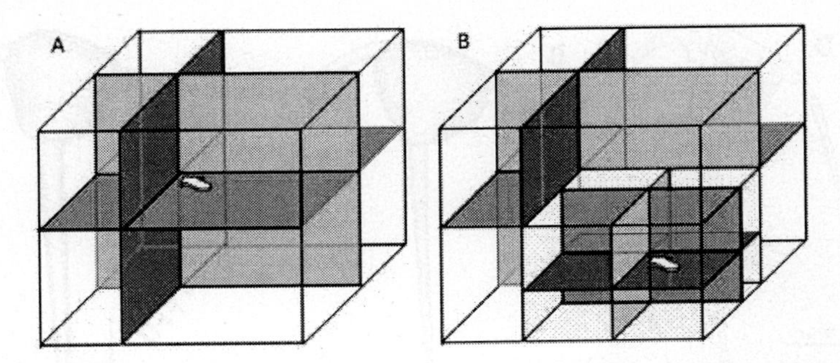

Fig. 5. The demonstration of the algorithm of the deformation lattice construction. White arrows show the active control point. The points of the intersection of the lines are the knots, which cannot be displaced by a scientist. A – the first step: three planes are passed through the CP till they reach the bounding parallelepiped; B – the following steps: three planes are passed through the CP till they reach another planes

3.2 The construction of the deformation lattice

It is important to determine dimensions of deformation cell so that the displacement of the control points not cause the effect unnecessary to a user. If the user expects the deformation of certain model's part, that this model's part must be deformed.

The traditional method of the deformation lattice construction not provides performance of this task. The traditional method assumes that the deformation cells have equal dimensions, and they are uniformly distributed inside the parallelepiped, which bounds the model (Fig. 2). Similar construction of the deformation lattice has the fault. In most cases the position of the deformation lattice knots not always coincides with the position of the model's control points. Therefore, for the movement of these control points it is necessary to displace the knots of the deformation lattice. But we must displace the control points not the knots. Therefore the position of the knots must coincide with the position of the control points. We use the following method of the deformation lattice construction, which provides the coincidence of the knots with the control points:

1. The bounded parallelepiped is circumscribed around the model, and an arbitrary control point is selected.
2. Three planes, parallel to the parallelepiped sides, are drawn through the first control point up to the bounds of the parallelepiped (Fig. 5, A).
3. The next control point is selected, and three planes, parallel to the parallelepiped sides, are drawn through this control point too. The planes are drawn up to intersection with other planes (Fig. 5, B).
4. Step three is repeated for each control point.

During the deformation lattice construction the new knots are created (the points of the intersection of the lines on Fig. 5, which are not marked by arrows). But these knots are not the control points. A user can't displace they. These knots only determine the bounds of deformation cells.

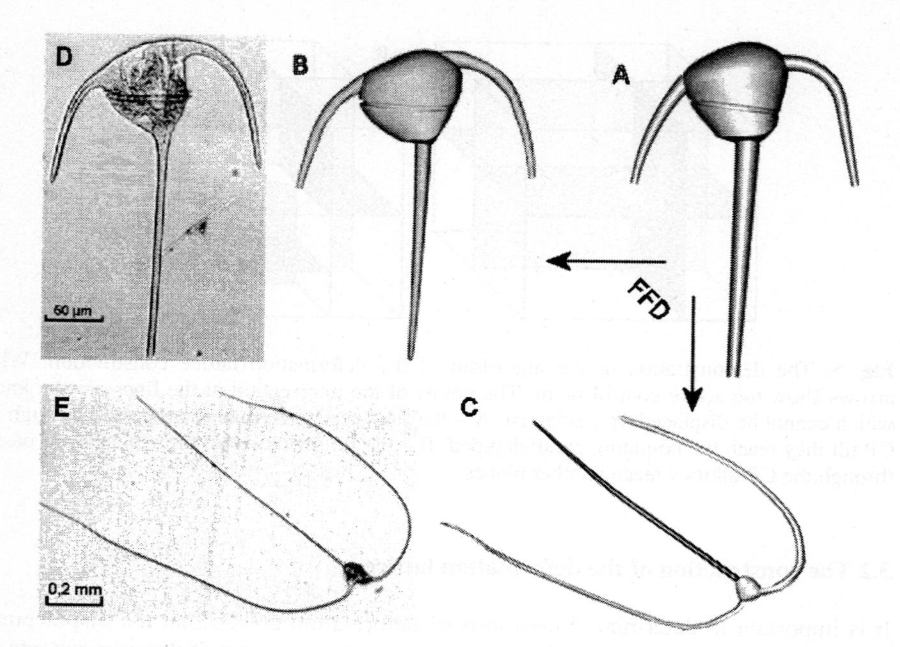

Fig. 6. Some examples of using the FFD operator to phytoplankton cells modeling. A – the base model, which is deformed by the user; B,C - the 3D models of phytoplankton cells created by the FFD procedure; D,E – the microphotographs of phytoplankton cell *Ceratium tripods* (D), and *Ceratium sp.* (E), which are used as patterns for the fitting of the base model

Besides the construction of the deformation lattice we apply several FFD operators to each element of a phytoplankton cell. The division of the cell's model into the single elements executes in accordance with the classification of microalgae parts used in biology. For example, the 3D model on Fig. 1 consists of four elements: the body, two lateral growths, and one apical growth. If one FFD operator is applied to whole model, the process of model deformation will be essentially hampered. On the other hand, if the single FFD operator is applied to the each model parts, the process of the model deformation will become easier.

3.3 Two methods of the phytoplankton model deformation

The FFD procedure is used in the construction of the phytoplankton models of those cells, which are observed through a microscope. The images of phytoplankton cells are used as patterns of their shapes. The process of the model deformation consists of the sequential movements of the control points (an application of the FFD operator).

The FFD is applied to the model till a correspondence between the model shape and the natural microorganism shape is achieved.

But how a scientist can achieves this correspondence? The initial data determines the methods of this problem solving:

1. In the first case, when the user cannot obtain the image of the phytoplankton cell, but can measure its dimensions, the measured sizes are used as the values of distances between the control points of the model. The setting of the distances causes the deformation of the model.
2. In the second case, when the user obtains the image of the phytoplankton cell, he should achieve coincidence of the model's borders with the outline of the microorganism on the image. The coincidence is achieved by displacing model's control points.

These two methods are the main methods of the application of the FFD operator to the phytoplankton cells modeling (Fig. 6). Initially a scientist has the *base model* of the phytoplankton cell, which represents an average shape of single phytoplankton species or genus. After that the user gets the initial data, deforms the base model, and obtains the model of specific phytoplankton species. Using this specific model he calculates the volume and the surface area of the phytoplankton cell and makes its morphological analysis. Undoubtedly, calculated surface and volume values will most exact.

4 Conclusion

The considered procedure of the phytoplankton models deformation is the main stage of the 3D phytoplankton cell building. The application of 3D simulation in biology allows receiving exact evaluations of the microorganisms' volumes and surface areas. The application of the FFD to the 3D phytoplankton model allows demonstrating microorganism morphological changes during its life cycle. The location of the phytoplankton model control points describes the shape of the cell, and it may be used at cell morphological analysis.

5 Acknowledgments

I express my thanks to Vladimir Mukhanov who has helped me during thinking over this paper.

References

1. Lyakh, A.M., Mukhanov, V.S., Kemp, R.B.: The Virtual Cell Project for the Investigation of Microalgal Morphology and Dispersity of Natural Phytoplankton. In: 17th European Workshop on Computational Geometry (CG 2001). Program & Abstracts (Freie Universität Berlin, March 2001). Berlin (2001) 57-58
2. Sederberg, T.W., Parry, S.R.: Free-Form Deformation of Solid Geometric Models. Computer Graphics 20 (1986) 151-160

Curvature Based Registration with Applications to MR-Mammography

Bernd Fischer and Jan Modersitzki

Institute of Mathematics
Medical University of Lübeck, 23560 Lübeck
Email: {fischer,modersitzki}@math.mu-luebeck.de

Abstract. We introduce a new non-linear registration model based on a curvature type regularizer. We show that affine linear transformations belong to the kernel of this regularizer. Consequently, an additional global registration is superfluous. Furthermore, we present an implementation of the new scheme based on the numerical solution of the underlying Euler-Lagrange equations. The real discrete cosine transform is the backbone of our implementation and leads to a stable and fast $\mathcal{O}(n \log n)$ algorithm, where n denotes the number of voxels. We demonstrate the advantages of the new technique for synthetic data sets. Moreover, first convincing results for the registration of MR-mammography images are presented.

1 Introduction

Registration of 2D or 3D medical images is necessary in order to study the evolution of a pathology of a patient, or to take full advantage of the complementary information coming from multimodal imagery. In our application, which is related to MR-mammography, the time evolution of an agent injection has to be studied subject to patient motion. Recent examples of the use of deformable models to perform a non-rigid, automatic registration include [5, 2, 1, 6, 3]. Most of these schemes may be viewed as a procedure which minimizes a suitable distance measure subject to a regularization term or some interpolation restrictions.

There are several problems with fully automatic registration approaches. One of which is the problem that the technique is sensitive to initial positioning of the images to be matched. If the initial rigid alignment is off by too much, the non-rigid matching procedure may perform poorly. Therefore it is desirable to incorporate the rigid alignment step, also known as global matching, into the non-rigid scheme.

In this note we propose a novel curvature based penalizing term which not only provides smooth solutions but also allows for automatic rigid alignment. Moreover, we devise a fast and stable implementation for a finite difference approximation of the underlying partial differential equation.

P.M.A. Sloot et al. (Eds.): ICCS 2002, LNCS 2331, pp. 202–206, 2002.

2 Approach

We refer to the template image as T and the reference as R. For a particular point $\mathbf{x} \in \Omega := [0,1]^d$, the value $T(\mathbf{x})$ is the intensity at $\mathbf{x}$. The registration algorithm described in this paper is applicable to images with any number of dimensions d.

The purpose of the registration is to determine a transformation of T onto R. Ideally, one wants to determine a displacement field $\mathbf{u} : \Omega \to \Omega$ such that $T(\mathbf{x} - \mathbf{u}(\mathbf{x})) = R(\mathbf{x})$. The question is how to find such a mapping $\mathbf{u} = (u_1, \ldots, u_d)$. A typical approach is the minimization of a measure $\mathcal{D}$, for example

$$\mathcal{D}[\mathbf{u}] = \frac{1}{2} \|R - T(\cdot - \mathbf{u})\|_{L_2}^2 = \frac{1}{2} \int_\Omega \left(T(\mathbf{x} - \mathbf{u}(\mathbf{x})) - R(\mathbf{x})\right)^2 d\mathbf{x}. \tag{1}$$

Of course other choices, like the mutual information based measure or the normalized cross correlation might be used as well, as long as a Gateaux derivative exists. A regularizing term $\mathcal{S}$ is introduced in order to rule out discontinuous and/or suboptimal solutions. The problem now reads, find a mapping $\mathbf{u}$ which minimizes the joint criterion $\mathcal{J}[\mathbf{u}] = \alpha\mathcal{S}[\mathbf{u}] + \mathcal{D}[\mathbf{u}]$. Since the choice of the regularization parameter α in not an issue here, we set $\alpha = 1$.

In this note, we investigate the novel smoothing term

$$\mathcal{S}^{\text{curv}}[\mathbf{u}] = \textstyle\sum_{\ell=1}^{d} \int_\Omega (\Delta u_\ell)^2 \, d\mathbf{x}. \tag{2}$$

The reason for this particular choice is twofold. The integral might be viewed as an approximation to the curvature of the ℓth component of the displacement field and therefore does penalize oscillations. Most interestingly, $\mathcal{S}^{\text{curv}}$ has a non-trivial kernel containing affine linear transformations, i.e.,

$$\mathcal{S}^{\text{curv}}[C\mathbf{x} + \mathbf{b}] = 0, \quad C \in \mathbb{R}^{d \times d}, \ \mathbf{b} \in \mathbb{R}^d.$$

Thus, in contrast to many other non-linear registration techniques, the new scheme does not require an additional affine linear pre-registration step to be successful.

To illustrate the difference between our new curvature based registration and the elastic registration approach [5] we consider an academic example. As the reference image a gray square on a white background positioned in the top left corner is used. In contrast, the considered template has the very same square in the bottom right corner. In other words, an appropriate affine linear transformation would produce a perfect registration result. It turns out, that both the curvature based, and the elastic registration, lead to a perfect registration, in the sense that the difference between the reference and deformed template vanishes. However, a tracking of the individual pixel reveals that the path towards the optimal registration is completely different. In Fig. 1 the templates as well as the interpolation grid, i.e., the points $\mathbf{x} - \mathbf{u}(\mathbf{x})$, are shown. As it is apparent from this figure, the curvature based registration finds the optimal registration result by computing an almost affine linear transformation, i.e. $\mathbf{u}(\mathbf{x}) \approx$ const. In

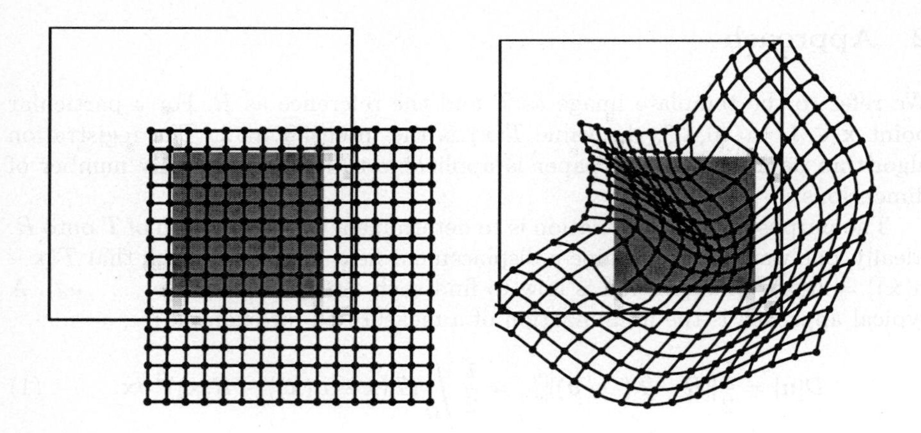

Fig. 1. Template image with interpolation grid; LEFT: after curvature based registration; RIGHT: after elastic registration.

contrast, the displacement computed by the elastic registration scheme is highly non-linear.

Also, this is a striking example for the fact that the similarity between the deformed template and the reference does not necessarily ensure a reliable registration.

We would like to point out, that there is a close relation between the new curvature based approach and the computation of thin-plate-splines, cf., e.g., [9]. These functions are defined as minimizer of the following smoothing term

$$\mathcal{S}^{\mathrm{TPS}}[\mathbf{u}] = \sum_{\ell=1}^{d} \int_{\mathbb{R}^d} \sum_{|\kappa|=q} c_\kappa \left(D^\kappa u_\ell\right)^2 d\mathbf{x},$$

subject to some interpolatory constraints. Here, the multi-index $|\kappa| = \kappa_1 + \cdots + \kappa_d$, and the set of coefficients $\{c_\kappa : |\kappa| = q\}$ are chosen such that $\mathcal{S}^{\mathrm{TPS}}$ is rotationally invariant. The smoothing terms $\mathcal{S}^{\mathrm{curv}}$ and $\mathcal{S}^{\mathrm{TPS}}$ are closely related. These similarities might be used to design a new algorithm which incorporates interpolation conditions (landmarks) into the registration scheme outlined in this note. This approach will be investigated in a forthcoming paper.

In accordance with the calculus of variations, a function $\mathbf{u}$ which minimizes the joint functional $\mathcal{J}$ for the particular choice $\mathcal{S} = \mathcal{S}^{\mathrm{curv}}$ has to satisfy the Euler-Lagrange equation

$$\mathbf{f}(\mathbf{x}, \mathbf{u}(\mathbf{x})) + \alpha \Delta^2 \mathbf{u}(\mathbf{x}) = 0 \text{ for } x \in \Omega, \tag{3}$$

subject to appropriate boundary conditions. The so-called force field $\mathbf{f}$ is the Gateaux derivative of the distance measure $\mathcal{D}$, i.e.,

$$\mathbf{f}(\mathbf{x}, \mathbf{u}(\mathbf{x})) = (T(\mathbf{x} - \mathbf{u}(\mathbf{x})) - S(\mathbf{x}))\ \nabla(T(\mathbf{x} - \mathbf{u}(\mathbf{x}))). \tag{4}$$

A popular approach to solve this PDE is to introduce an artificial time t and to compute the steady state solution of the time dependent PDE. Here, we employ the following semi-implicit iterative scheme,

$$\partial_t \mathbf{u}^{k+1}(\mathbf{x},t) - \Delta^2 \mathbf{u}^{k+1}(\mathbf{x},t) = \mathbf{f}(\mathbf{x},\mathbf{u}^k), \quad k = 1,2,\ldots, \tag{5}$$

where $\mathbf{u}^0$ is some initial deformation, typically $\mathbf{u}^0 = 0$.

The above fourth-order non-linear PDE is known as the bipotential or biharmonic equation and is well studied, see, e.g., [8]. To solve (5) numerically, we apply a finite difference discretization adapted to the particular simple geometry of the domain Ω. This approach results in a system of linear equations $\mathcal{A}\mathbf{u}^{k+1} = \mathbf{f}^{\,k}$. Consequently, the main work in the overall scheme is the repeated solution of this linear system. It can be shown that $\mathcal{A}$ is diagonalizable by cosine-transform matrices. Thus, a proper, real discrete cosine transform type technique leads to a fast and stable $\mathcal{O}(n \log n)$ implementation, where n denotes the number of voxels.

3 Experiments

To illustrate the performance of the new approach we present the registration of two clinical 2D magnet-resonance (MR) images of a female breast. We are indebted to Bruce L. Daniel (Department of Radiology, Stanford University) for providing the medical data. The task is to register low resolution MR-scans (256×256) of the wash-in and wash-out phase to a high resolution MR-scan (512×512), which are viewed as a gold-standard by the radiologist, taken just between the wash-in and wash-out phase. The overall goal is to study the dynamic behavior of the contrast agent in detail. Fig. 2 displays the arbitrarily chosen section 24 of the high resolution MR-scan and the difference to section 24 of a wash-in phase MR-scan before and after registration. Note that the difference has been reduced by about 30%.

References

1. Amit Y: A nonlinear variational problem for image matching. SIAM J. Sci. Comp., 15 (1994) 207–224
2. Bajcsy R, Kovačič S: Multiresolution elastic matching. Computer Vision, Graphics and Image processing, 46 (1989) 1–21
3. Bro-Nielsen M: Medical image registration and surgery simulation. PhD thesis, IMM, Technical University of Denmark, 1996
4. Bro-Nielsen M, Gramkow C: Fast fluid registration of medical images. Visualization in Biomedical Computing (VBC'96), Lecture Notes in Computer Science, Vol. 1131, Springer, (1996) 267–276
5. Broit C: Optimal registration of deformed images. PhD thesis, Computer and Information Science, University of Pennsylvania, 1981
6. Christensen GE: Deformable shape models for anatomy. PhD thesis, Sever Institute of Technology, Washington University, 1994
7. Fischer B, Modersitzki J: Fast inversion of matrices arising in image processing. Numerical Algorithms, 22 (1999) 1–11
8. Hackbusch W: Partial differential Equations. Teubner, Stuttgart, 1987
9. Rohr K: Landmark-based Image Analysis. Kluwer Academic Publisher, 2001

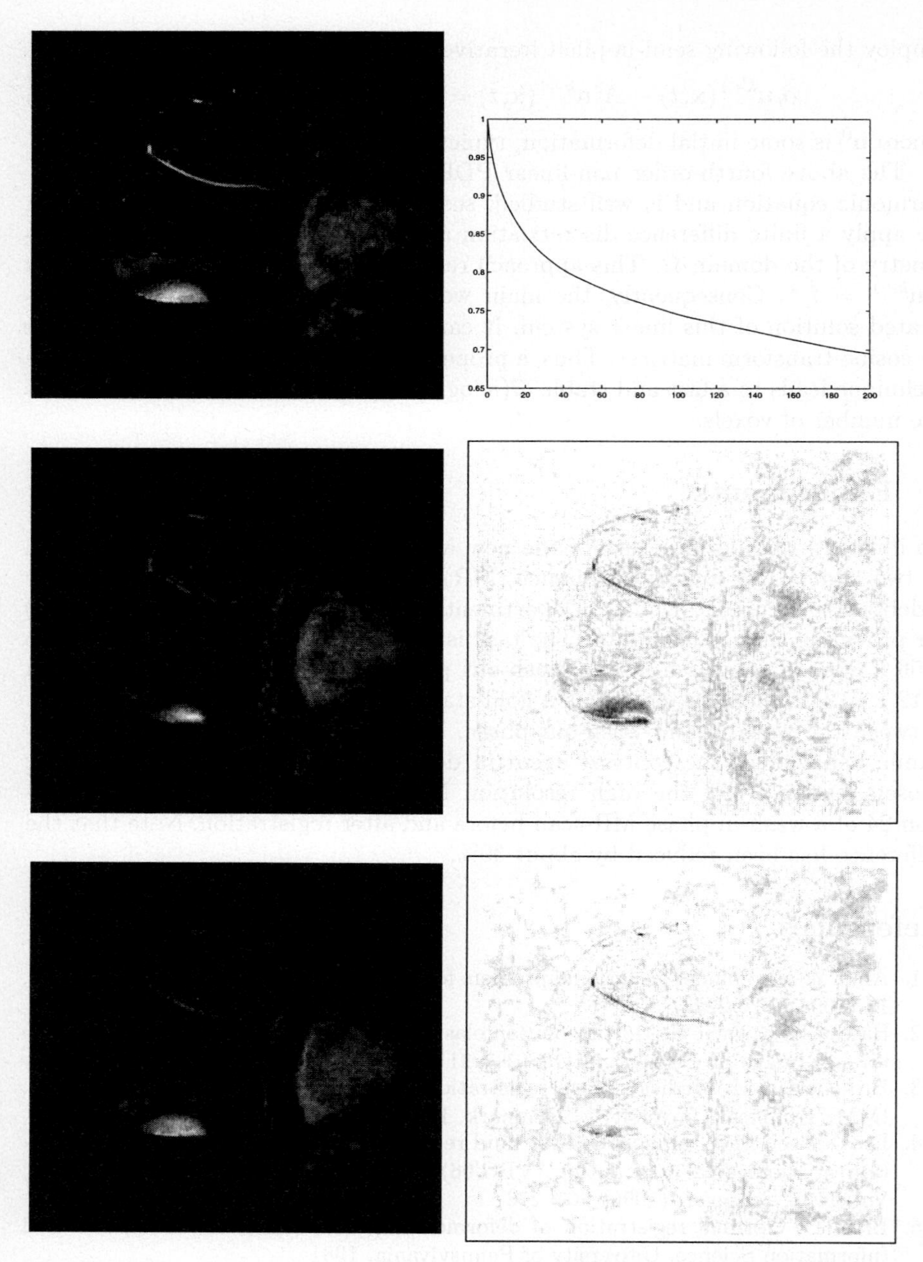

Fig. 2. TOP: Reference R (preprocessed section #24 from a high resolution image taken at optimal time-point); MIDDLE LEFT template T (preprocessed section #24 from the wash-in phase 20); MIDDLE RIGHT difference before registration, $|R - T| = 100\%$; BOTTOM LEFT template $\hat{T}$ after registration; BOTTOM RIGHT difference after registration, $|R - \hat{T}| = 69.5\%$; TOP RIGHT relative distance $|R - T_k|/|R - T|$ versus iteration.

Full Scale Nonlinear Electromagnetic Inversion for Biological Objects

Aria Abubakar and Peter M. van den Berg

Centre for Technical Geosciences, Delft University of Technology
Mekelweg 4, 2628 CD, Delft, The Netherlands
abubakar@its.tudelft.nl

Abstract. In this paper the contrast source inversion method using a multiplicative weighted L^2-norm total variation regularizer is applied to two- and three-dimensional image reconstructions from electromagnetic microwave tomography experiments. This iterative method avoids solving a full forward problem in each iteration which makes the method suitable to handle a large scale computational problem. The numerical results both from simulation and experimental data with high contrast biological phantom are presented and discussed.

1 Introduction

Image reconstruction is a complicated nonlinear problem in microwave tomography because both the material parameters and the field distribution in the investigation domain are unknowns. Serious efforts have been made to solve this problem in two-dimensional model where the scalar Helmholtz equation (TM electromagnetic polarization) can be applied. If the unknown object has small contrast of dielectric properties, a linearization procedure like the Born approximations [1] can be used to provide images with spatial resolution of a fraction of the wavelength. Computer codes based on these approximations demonstrate a high speed and can be used for almost real time imaging. Iterative procedures for the Born approximation have been used for intermediate contrast objects [4].

More complicated mathematical reconstruction algorithms have been developed for reconstruction objects with high contrast in dielectric properties [3, 5–8, 10]. These algorithms require large computer resources, but nevertheless theoretically provide a resolution with a sufficiently high signal-to-noise ratio. The Newton method [11] has been successfully applied to high contrast objects. The main bottlenecks of this Newton approach, especially discouraging in the three-dimensional case, are the multiple forward solutions needed to construct the Hessian matrix. Another type of method which avoids solving any forward problem in each iterative step is introduced in [9, 12]. This approach can solve the problem without dealing with a high-dimensional linear equation system but it requires a larger number of iterations.

Recently, this type of method has been armed with a total variation regularizer in order to handle high contrast objects [13]. Although the addition of

P.M.A. Sloot et al. (Eds.): ICCS 2002, LNCS 2331, pp. 207–216, 2002.

the total variation to the cost functional has a very positive effect on the quality of the reconstructions for both 'blocky' and smooth profiles, a drawback is the presence of an artificial weighting parameter in the cost functional, which can only be determined through considerable numerical experimentation and a priori information of the desired reconstruction. In [13], it was suggested to include the total variation as a multiplicative constraint, with the result that the original cost functional is the weighting parameter of the regularizer, so that this parameter is determined by the inversion procedure itself. This eliminates the choice of the artificial regularization parameters completely. The multiplicative type of regularization seems to handle noisy as well as limited data in a robust way without the usually necessary a priori information. In this paper the latter method using the new weighted L^2-norm total variation regularizer introduced in [14] is applied to handle more complicated biological objects embedded in a lossy medium. Numerical examples using simulation and experimental data demonstrate the ability of the presented method.

2 Problem Statement

We consider an object, B, of arbitrary bounded cross section with complex dielectric permittivity $\varepsilon(\boldsymbol{x})$ and arbitrary shape. The complex permittivity and the shape of this object B are unknown, but they are known to lie within a bounded simply connected object domain D. This object domain D is assumed to be embedded in the background medium, the immersion liquid, with permittivity ε_{b}. The position vector is denoted by $\boldsymbol{x}$. We assume a time harmonic dependence $\exp(\mathrm{j}\omega t)$, where $\mathrm{j}^2 = -1$, ω is angular frequency, and t is time. We also assume that the object is irradiated successively by a number of known incident electric fields $\boldsymbol{E}_s^{\mathrm{inc}}(\boldsymbol{x}) = \boldsymbol{E}^{\mathrm{inc}}(\boldsymbol{x}, \boldsymbol{x}_s^{\mathrm{S}})$, $s = 1, \cdots$, originating from source positions $\boldsymbol{x}_s^{\mathrm{S}}$.

For each incident field, the total electric field will be denoted by $\boldsymbol{E}_s$. Nowadays, it is well-known that the total field $\boldsymbol{E}_s$ and the scattered field $\boldsymbol{E}_s^{\mathrm{sct}}$ satisfy the following domain integral representations

$$\boldsymbol{E}_s(\boldsymbol{x}) = \boldsymbol{E}_s^{\mathrm{inc}}(\boldsymbol{x}) + (k_{\mathrm{b}}^2 + \boldsymbol{\nabla}\boldsymbol{\nabla}\cdot) \int_D g(\boldsymbol{x}-\boldsymbol{x}')\chi(\boldsymbol{x}')\boldsymbol{E}_s(\boldsymbol{x}')\,\mathrm{d}v(\boldsymbol{x}')\,, \quad \boldsymbol{x} \in D\,, \tag{1}$$

$$\boldsymbol{E}_s^{\mathrm{sct}}(\boldsymbol{x}) = (k_{\mathrm{b}}^2 + \boldsymbol{\nabla}\boldsymbol{\nabla}\cdot) \int_D g(\boldsymbol{x}-\boldsymbol{x}')\chi(\boldsymbol{x}')\boldsymbol{E}_s(\boldsymbol{x}')\,\mathrm{d}v(\boldsymbol{x}')\,, \quad \boldsymbol{x} \notin D\,, \tag{2}$$

where $\chi(\boldsymbol{x}) = \varepsilon(\boldsymbol{x})/\varepsilon_{\mathrm{b}} - 1$ is the contrast, $g(\boldsymbol{x}) = \exp(-\mathrm{j}\,k_{\mathrm{b}}|\boldsymbol{x}|)/4\pi|\boldsymbol{x}|$ is the Green's function, and $k_{\mathrm{b}} = \omega\,(\varepsilon_{\mathrm{b}}\mu_0)^{1/2}$ denotes the wavenumber in the embedding.

In the inverse scattering problem $\boldsymbol{F}_s = \boldsymbol{E}_s^{\mathrm{sct}}$ will be measured on some domain S outside D, so the integral representation in (2) for points exterior to D is written symbolically as the *data equation*,

$$\boldsymbol{F}_s = \mathcal{G}^S \chi \boldsymbol{E}_s\,, \quad \boldsymbol{x} \in S\,, \tag{3}$$

while the integral equation in (1) is written symbolically as the *object equation*,

$$\boldsymbol{E}_s^{\mathrm{inc}} = \boldsymbol{E}_s - \mathcal{G}^D \chi \boldsymbol{E}_s\,, \quad \boldsymbol{x} \in D\,, \tag{4}$$

where the operator $\mathcal{G}^S$ is an operator mapping from $L^2(D)$ into $L^2(S)$ and the operator $\mathcal{G}^D$ is an operator mapping $L^2(D)$ into itself.

The inverse scattering problem consists of determining χ from a knowledge of the incident fields, $\boldsymbol{E}_s^{\text{inc}}$, on D and the scattered fields, $\boldsymbol{F}_s$, on S. Because the total fields $\boldsymbol{E}_s$ on D are also unknown, this problem is non-linear. In practice it can only be solved iteratively due to a large number of unknowns.

3 Inversion Algorithm

In this method, one chooses to reconstruct the material contrast χ and the contrast sources $\boldsymbol{W}_s$ instead of the fields $\boldsymbol{E}_s$. The contrast sources are defined by

$$\boldsymbol{W}_s(\boldsymbol{x}) = \chi(\boldsymbol{x})\boldsymbol{E}_s(\boldsymbol{x}) . \tag{5}$$

Using (5) in (3), the data equation becomes

$$\boldsymbol{F}_s = \mathcal{G}^S \boldsymbol{W}_s \,, \quad \boldsymbol{x} \in S \,, \tag{6}$$

while the object equation becomes

$$\boldsymbol{E}_s = \boldsymbol{E}_s^{\text{inc}} + \mathcal{G}^D \boldsymbol{W}_s \,, \quad \boldsymbol{x} \in D \,. \tag{7}$$

Substituting (7) into (5), we obtain an object equation for $\boldsymbol{W}_s$

$$\chi \boldsymbol{E}_s^{\text{inc}} = \boldsymbol{W}_s - \chi \mathcal{G}^D \boldsymbol{W}_s \,, \quad \boldsymbol{x} \in D \,. \tag{8}$$

Then, in the Contrast Source Inversion (CSI) method [12] using the multiplicative weighted L^2-norm Total Variation (TV) regularization factor [14] (denoted by MR-CSI), the sequences of $\boldsymbol{W}_{s,n}$ and χ_n for $n = 1, 2, \cdots$, are iteratively found by minimizing a cost functional, viz.,

$$\begin{aligned} F_n(\boldsymbol{W}_s, \chi) &= \left[F^S(\boldsymbol{W}_s) + F_n^D(\boldsymbol{W}_s, \chi)\right] F_n^R(\chi) \,, \\ &= \left[\eta^S \textstyle\sum_s \|\boldsymbol{F}_s - \mathcal{G}^S \boldsymbol{W}_s\|_S^2 + \eta_n^D \textstyle\sum_s \|\chi \boldsymbol{E}_s^{\text{inc}} - \boldsymbol{W}_s + \chi \mathcal{G}^D \boldsymbol{W}_s\|_D^2\right] F_n^R(\chi) \,, \end{aligned} \tag{9}$$

where F^S and F_n^D are the errors in the data and object equations. The normalization factors in the cost functional are chosen as $\eta^S = \sum_s \|\boldsymbol{F}_s\|_S^2$ and $\eta_n^D = \sum_s \|\widetilde{\chi} \boldsymbol{E}_s^{\text{inc}}\|_D^2$, respectively. Further, $\|\cdot\|_{S,D}^2$ denotes the squared norm on S/D. The weighted L^2-norm TV factor is as introduced in [14],

$$F_n^R(\chi) = \frac{1}{\int_D \mathrm{d}v(\boldsymbol{x})} \int_D \frac{|\boldsymbol{\nabla}\chi|^2 + \delta_n^2}{|\boldsymbol{\nabla}\widetilde{\chi}|^2 + \delta_n^2} \, \mathrm{d}v(\boldsymbol{x}) = \|b_n \boldsymbol{\nabla}\chi\|_D^2 + \delta_n^2 \|b_n\|_D^2 \,, \tag{10}$$

where $b_n = \left[\int_D \mathrm{d}v(\boldsymbol{x}) \left(|\boldsymbol{\nabla}\widetilde{\chi}|^2 + \delta_n^2\right)\right]^{-1/2}$. In (9) and (10) $\widetilde{\chi}$ is some particular contrast value. Its choice will be determined later. Although the constant parameter δ_n^2 is introduced for restoring differentiability of the regularizer, it also controls the influence of the regularization. We therefore have chosen to increase the regularization as a function of the number of iterations by decreasing this parameter δ_n^2. Since the normalized object error term will decrease as a function of the number of iterations, we choose

$$\delta_n^2 = F_n^D(\boldsymbol{W}_{s,n}, \widetilde{\chi})\, \tilde{\Delta}^2 , \tag{11}$$

where $\tilde{\Delta}$ denotes the reciprocal mesh size of the discretized domain D. Its choice is inspired by the idea that in the first few iterations, we do not need the minimization of the regularizer and as the iterations proceed we want to increase the effect of the regularizer.

The structure of the cost functional in (9) is such that it will minimize the factor F_n^R with a large weighting parameter in the beginning of the optimization process, because the value of $F^S + F_n^D$ is still large, and that it will gradually minimize more and more the normalized errors in the data and object equations when the factor F_n^R remains a nearly constant value close to one. If noise is present in the data, the term F^S will remain at a large value during the whole optimization and therefore, the weight of the factor F_n^R will be more significant. Hence, the noise will, at all times, be suppressed in the reconstruction process.

This MR-CSI method starts with back propagation as the initial estimates for the contrast sources and the contrast [12]. Then, in each iteration we first update the contrast sources $\boldsymbol{W}_{s,n}$ using a conjugate gradient step

$$\boldsymbol{W}_s = \boldsymbol{W}_{s,n-1} + \alpha_n \boldsymbol{w}_{s,n} , \tag{12}$$

where the functions $\boldsymbol{w}_{s,n}$ are the Polak-Ribière conjugate gradient directions,

$$\boldsymbol{w}_{s,0} = \boldsymbol{0} , \quad \boldsymbol{w}_{s,n} = \partial\boldsymbol{w}_{s,n} + \frac{\sum_s \mathrm{Re}\langle \partial\boldsymbol{w}_{s,n}, \partial\boldsymbol{w}_{s,n} - \partial\boldsymbol{w}_{s,n-1}\rangle_D}{\sum_s \|\partial\boldsymbol{w}_{s,n-1}\|_D^2}\, \boldsymbol{w}_{s,n-1} . \tag{13}$$

The gradient $\partial\boldsymbol{w}_{s,n}$ of the cost functional $F_n(\boldsymbol{W}_s, \chi)$ in (9) with respect to $\boldsymbol{W}_s$ evaluated at $\boldsymbol{W}_{s,n-1}$ and χ_{n-1} is given by

$$\partial\boldsymbol{w}_{s,n} = -\eta^S \boldsymbol{\mathcal{G}}^{S*} \boldsymbol{\rho}_{s,n-1} - \eta_n^D \left(\boldsymbol{r}_{s,n-1} - \boldsymbol{\mathcal{G}}^{D*} \overline{\chi_{n-1}}\, \boldsymbol{r}_{s,n-1} \right) , \tag{14}$$

where $\boldsymbol{\mathcal{G}}^{S*}$ and $\boldsymbol{\mathcal{G}}^{D*}$ are the adjoint operators of $\boldsymbol{\mathcal{G}}^S$ and $\boldsymbol{\mathcal{G}}^D$, respectively, and the overbar denotes the complex conjugate. Further, the residuals are defined as

$$\boldsymbol{\rho}_{s,n} = \boldsymbol{F} - \boldsymbol{\mathcal{G}}^S \boldsymbol{W}_{s,n} \qquad \text{and} \qquad \boldsymbol{r}_{s,n} = \chi_n \boldsymbol{E}_s^{\mathrm{inc}} - \boldsymbol{W}_{s,n} + \chi_n \boldsymbol{\mathcal{G}}^D \boldsymbol{W}_{s,n} . \tag{15}$$

The real parameter α_n is found as minimizer of $F_n(\boldsymbol{W}_{s,n-1} + \alpha\, \boldsymbol{w}_{s,n}, \chi_{n-1})$,

$$\alpha_n = \frac{-\sum_s \mathrm{Re}\langle \partial\boldsymbol{w}_{s,n}, \boldsymbol{w}_{s,n}\rangle_D}{\eta^S \sum_s \|\boldsymbol{\mathcal{G}}^S \boldsymbol{w}_{s,n}\|_S^2 + \eta_n^D \sum_s \|\boldsymbol{w}_{s,n} - \chi_{n-1} \boldsymbol{\mathcal{G}}^D \boldsymbol{w}_{s,n}\|_D^2} . \tag{16}$$

Note that the cost functional in (9) is a quadratic function of α, and the minimizer is unique. In the updating scheme for the contrast sources $\widetilde{\chi}$ is chosen to be the contrast in the previous iteration, $\widetilde{\chi} = \chi_{n-1}$. Then $F_n^R(\chi_{n-1})$ is always equal to one during the updating of $\boldsymbol{W}_{s,n}$. Subsequently, we compute the field $\boldsymbol{E}_{s,n}$ by substituting (12) in (4).

After $\boldsymbol{W}_{s,n}$ have been obtained, in each iteration, we proceed with updating of χ_n. First, we observed that the closed-form expression of the contrast can be found if the regularization factor F_n^R is absent, viz.,

$$\chi_n = \arg\min_{\text{complex}\,\chi} \{F^S(\boldsymbol{W}_{s,n}) + F_n^D(\boldsymbol{W}_{s,n},\chi)\} = \frac{\sum_s \boldsymbol{W}_{s,n}\cdot\overline{\boldsymbol{E}_{s,n}}}{\sum_s |\boldsymbol{E}_{s,n}|^2}\,. \tag{17}$$

From this point we make an additional minimization step,

$$\chi_n^R = \chi_n + \beta_n d_n\,, \tag{18}$$

where χ_n is now given by (17) and d_n is the Polak-Ribière conjugate gradient

$$d_0 = 0\,, \qquad d_n = g_n^R + \frac{\mathrm{Re}\langle g_n^R, g_n^R - g_{n-1}^R\rangle_D}{\|g_{n-1}^R\|_D^2}\, d_{n-1}\,. \tag{19}$$

We remark that we prefer now a line minimization around the minimum of the cost functional $F^S + F_n^D$ (physical cost criterion). Then, during the updating of the contrast $\widetilde{\chi}$ is taken to be equal to χ_n in (17). In view of (17) we take g_n^R as

$$\begin{aligned} g_n^R &= \frac{\left[\frac{\partial F_n^D(\boldsymbol{W}_{s,n},\chi)}{\partial\chi} F_n^R(\chi) + \left[F^S(\boldsymbol{W}_{s,n}) + F_n^D(\boldsymbol{W}_{s,n},\chi)\right]\frac{\partial F_n^R(\chi)}{\partial\chi}\right]_{\chi=\chi_n}}{\sum_s |\boldsymbol{E}_{s,n}|^2} \\ &= \left[F^S(\boldsymbol{W}_{s,n}) + F_n^D(\boldsymbol{W}_{s,n},\chi_n)\right]\frac{\boldsymbol{\nabla}\cdot(b_n^2\boldsymbol{\nabla}\chi_n)}{\sum_s |\boldsymbol{E}_{s,n}|^2}\,, \end{aligned} \tag{20}$$

a preconditioned gradient of the cost functional $F_n(\boldsymbol{W}_{s,n},\chi)$ with respect to changes in the contrast around the point $\chi = \chi_n$. In view of the previous minimization step, the gradient of F_n^D with respect to changes in the contrast around the point $\chi = \chi_n$ vanishes. Hence, the gradient with respect to the contrast, in contrary to the previous approaches of the MR-CSI method [13, 14], contains only a contribution of the additionally imposed regularization.

The real parameter β_n is found from a line minimization as a minimizer of the cost functional in (9). The minimization of $F_n(\boldsymbol{W}_{s,n}, \chi_n + \beta d_n)$, which is a fourth-degree polynomial in β, can be performed analytically viz.,

$$F_n = (A + B\beta^2)(X + 2Y\beta + Z\beta^2)\,, \tag{21}$$

with

$$\begin{aligned} X &= \|b_n\boldsymbol{\nabla}\chi_n\|_D^2 + \delta_n^2\|b_n\|_D^2\,, & A &= F^S(\boldsymbol{W}_{s,n}) + F_n^D(\boldsymbol{W}_{s,n},\chi_n)\,, \\ Y &= \mathrm{Re}\langle b_n\boldsymbol{\nabla}\chi_n, b_n\boldsymbol{\nabla}d_n\rangle_D\,, & B &= \textstyle\sum_s\|d_n\boldsymbol{E}_{s,n}\|_D^2 / \sum_s\|\chi_{n-1}\boldsymbol{E}_s^{\mathrm{inc}}\|_D^2\,, \\ Z &= \|b_n\boldsymbol{\nabla}d_n\|_D^2\,. \end{aligned} \tag{22}$$

Then, differentiation with respect to β yields a cubic equation with one real root and two complex conjugate roots. The real root is the desired minimizer

β_n. The cost functional F_n in (21) is a convex function of real β and has one minimum for real β because its second derivative with respect to β is positive. The second derivative is obtained as $\partial^2\mathcal{F}_n/\partial\beta^2 = 12BZ\beta^2 + 12BY\beta + 2(BX+AZ)$. For $\beta = 0$ this function is positive, and remains positive when the right-hand side has no real zeros. This is the case if $Y^2 < 2/3(XZ + AZ^2/B)$. Since both A and B are functions of $\boldsymbol{w}_{s,n}$, we rather want a criterion in which only functions of contrast quantities occur. Further the term AZ^2/B is always non-negative. Hence a sufficient condition that the second derivative is a positive function of β is $Y^2 \le 2/3XZ$ or

$$\frac{[\mathrm{Re}\langle b_n \boldsymbol{\nabla}\chi_n, b_n \boldsymbol{\nabla} d_n\rangle_D]^2}{[\|b_n \boldsymbol{\nabla}\chi_n\|_D^2 + \delta_n^2 \|b_n\|_D^2]\ \|b_n \boldsymbol{\nabla} d_n\|_D^2} \le \frac{2}{3}\,. \tag{23}$$

But, by Cauchy-Schwarz, $[\mathrm{Re}\langle b_n \boldsymbol{\nabla}\chi_n, b_n \boldsymbol{\nabla} d_n\rangle_D]^2\ /(\|b_n \boldsymbol{\nabla}\chi_n\|_D^2 \|b_n \boldsymbol{\nabla} d_n\|_D^2) \le 1$. Thus

$$\delta_n^2 \ge \|b_n \boldsymbol{\nabla}\chi_n\|_D^2/(2\,\|b_n\|_D^2) \tag{24}$$

is a sufficient condition for the cost functional to be a convex function with one minimum. If the choice for the parameter δ_n^2 of (11) is less than the right-hand side of (24), we replace the value of δ_n^2 by the right-hand side of (24) in which we take $b_n = \left[\int_D \mathrm{d}v(\boldsymbol{x})\,(|\boldsymbol{\nabla}\chi_n|^2 + \delta_{n-1}^2)\right]^{-1/2}$. We have refrained from using this value for δ_n^2 in the whole iteration procedure because from our numerical observations it appears this value has a large variation at the beginning of the optimization procedure.

After we have obtained a new estimate χ_n^R for the contrast, we repeat again the updating of $\boldsymbol{W}_{s,n}$ (if the value of the cost functional is not small enough) starting with $\chi_{n-1} = \chi_{n-1}^R$ of the previous iteration.

4 Numerical Examples

4.1 Two-Dimensional Inversion From TM Electromagnetic Data

We first consider inversion from the 2D-TM polarization measurement. For this measurement there are experimental data available which have been measured using a circular microwave scanner operating at 2.33GHz. The scanner consists of a 12.5cm radius circular array of 64 water-immersed horn antennas [2]. Only the vertical componenent of the electric field parallel to the array axis (the x_3-axis) is measured. The measurement procedure records the total electric field values at the receiving antennas. If one antenna is transmitting, the fields are measured only with the 33 antennas located in front of the active source. The scattered fields are deduced from the total field by subtracting the incident field, measured in the absence of any targets. Further, the measured scattered fields have been calibrated so that a directed unit line source can be used as the model for the incident fields, viz., $E_{3,s}^{\mathrm{inc}}(\boldsymbol{x}) = -0.25\,\omega\mu_0 H_0^{(2)}(k_{\mathrm{b}}|\boldsymbol{x} - \boldsymbol{x}_s^{\mathrm{S}}|)$, where $\mu_0 = 4\pi\times10^{-7}$ is the permeability in vacuum.

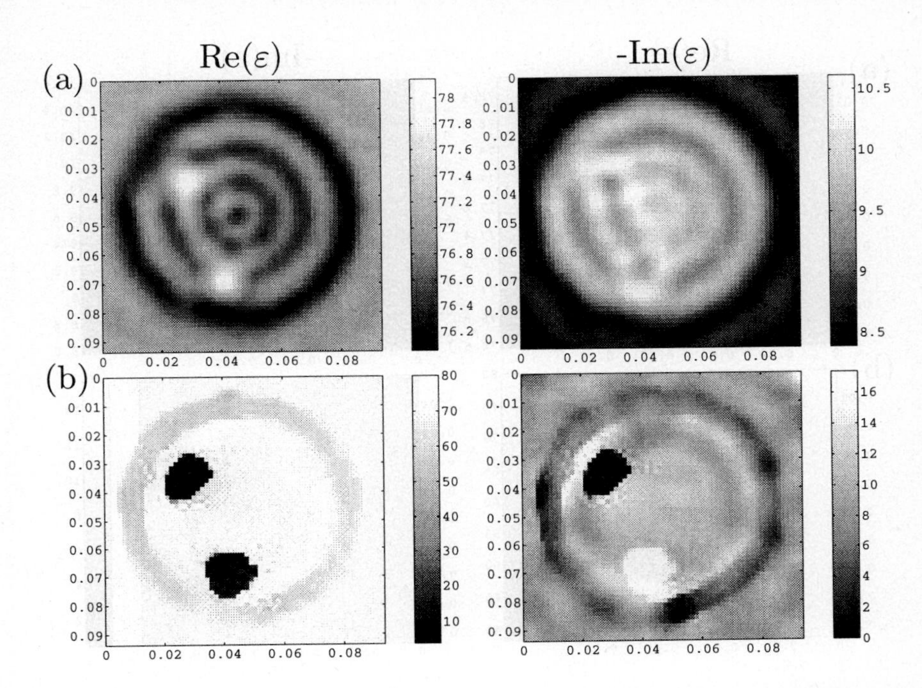

Fig. 1. Human arm phantom images: using back propagation (a) and MR-CSI (b).

In the inversion of experimental data we assumed that the unknown object is entirely located within a test domain D with dimension of 6.4λ by 6.4λ where λ is the wavelength in water. The permittivity of water is approximately $\epsilon_b = 77.3 - j\,8.66$ at frequency $f = 2.33$GHz. Hence, the wavelength $\lambda = 14.6$mm. The discrete form of the algorithm is obtained by dividing the test domain D into 64 by 64 subsquares. The discrete spatial convolutions are efficiently computed using FFT routines. The lower and upper bounds of the reconstructed complex permittivity in the inversion algorithm are enforced as $0 \leq \text{Re}[\varepsilon(\boldsymbol{x})]/\varepsilon_0 \leq 80$ and $0 \leq -\text{Im}[\varepsilon(\boldsymbol{x})]/\varepsilon_0 \leq 20$.

The first experimental data were obtained from a human arm phantom. The external layer (supposed to model the skin) and bones of the human arm phantom were made with PVC with complex permittivity $2.73 - j\,0.01$ and the muscle was $54.5 - j\,17.2$. We show first the results obtained from the initial estimates (back propagation). These results, approximately identical to those using the spectral diffraction tomography technique, are given in the top-plots of Fig. 1a. The results of the MR-CSI method after 1024 iterations are given in the bottom-plots of Fig. 1b. Although the total number of iterations is large, the total computation time is very limited. Note that we do not solve any forward problem at each iteration of the algorithm. One iteration of the MR-CSI method takes approximately 8 seconds on a personal computer with 600MHz Pentium III processor. After 1024 iterations the normalized data error F^S is already reduced to 6.33%, and adding more iterations does not change the result. From the results,

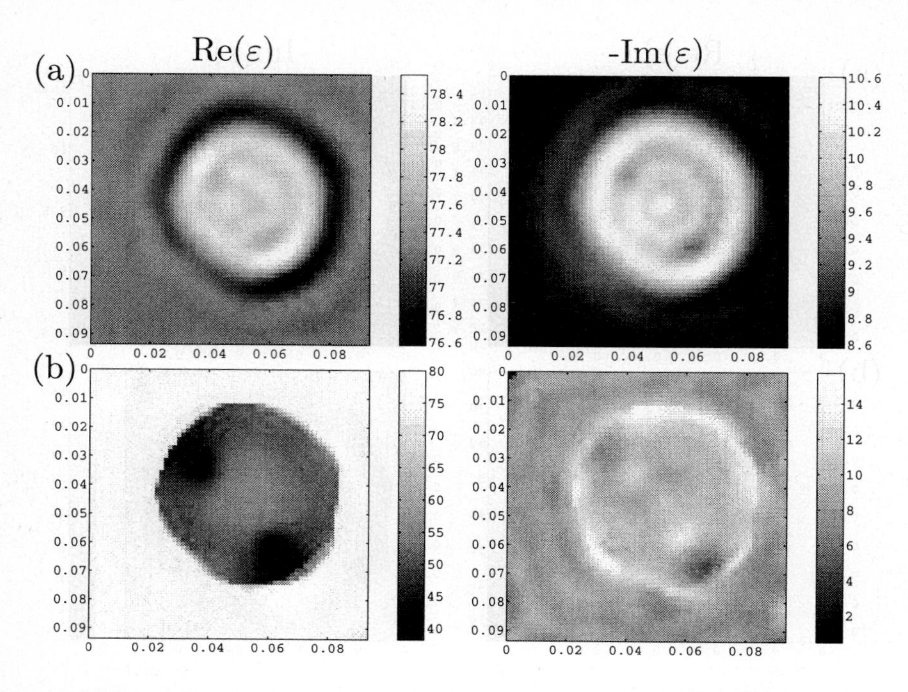

Fig. 2. Human forearm images: using back propagation (a) and MR-CSI method (b).

we observe that the bones are clear and sharp. The only drawback is that the reconstructed imaginary part of the complex permittivity for one of the bones is completely wrong. This can be caused by the presence of the noise in the experimental data.

For the second experiment we consider data that were taken from a human forearm. The back propagation results are given in the top-plots of Fig. 2a. The results of the MR-CSI method after 1024 iterations are given in Fig. 2b. After 1024 iterations the value of F^S is already reduced to 4.10%, and adding more iterations does not change the result. The reconstructed images show the positions of the two bones and the correct value of the muscle (approximately 54.5–i 17.2). Conversely, due to the water and tissue attenuation and the reduced dynamic range of the available data, the complex permittivity values of the bones are higher than the real ones (it should approximately be 5.5–i 0.59 at the present frequency of operation).

4.2 Three-Dimensional Inversion from Vectorial Data

As a test case for our full-vectorial 3D inversion algorithm we use the 3D neck model which is immersed in water with permittivity $\varepsilon_b = 78 - \mathrm{j}\,3.6$ at 1GHz. The original profile of this neck model is given in Fig. 3a. The neck model consists of fat tissue with permittivity $28 - \mathrm{j}\,13.5$, cartilage with $25 - \mathrm{j}\,10.78$, veins/arteries with $63 - \mathrm{j}\,20$, bone with $6.4 - \mathrm{j}\,2.16$, trachea 1, marrow $5.5 - \mathrm{j}\,0.59$,

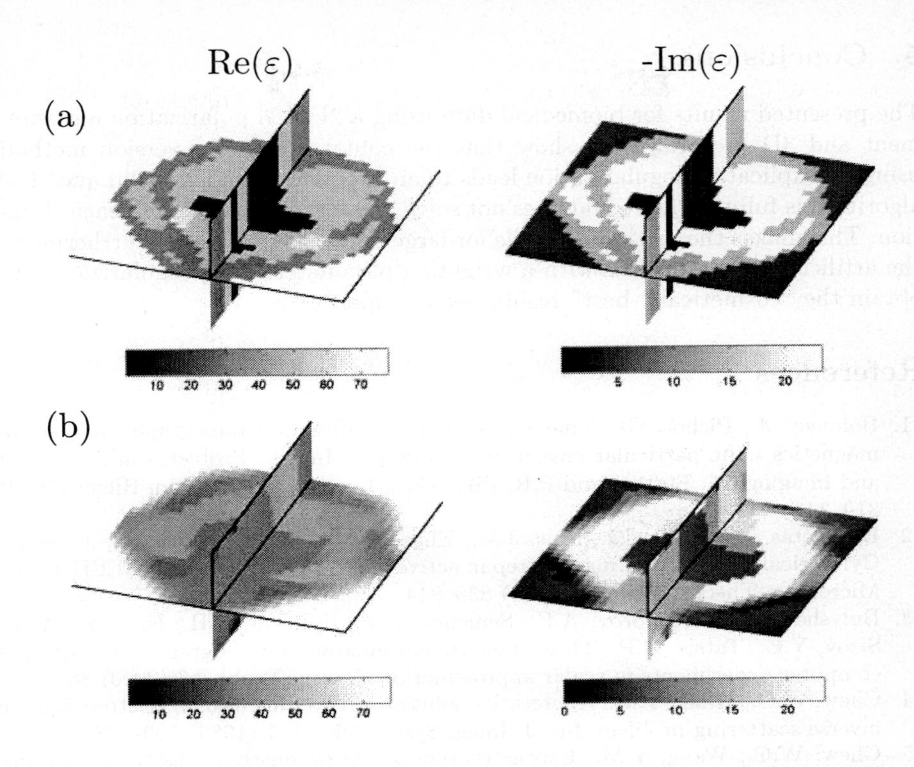

Fig. 3. Neck model: original 3D images (a) and reconstructed 3D images (b).

and muscle with 50−j 23.37. The size of the domain D in Fig. 3a is 3.1λ by 3.1λ by 1.55λ where $\lambda = 1/30$ is the wavelength in water. The measurement setup is similar to the one used by Bulyshev et al. [3]. We have three rings containing 30 transmitter antennas and six rings containing 90 receiver antennas. All receivers operate simultaneously while the transmitters operate one after another. Thus, in total we have 2700 data points. The receiver records all the components of the scattered electric field $\boldsymbol{E}_s^{\text{sct}}$. The transmitter is modeled using a point magnetic dipole directed in the x_3-direction, viz., $\boldsymbol{E}_s^{\text{inc}}(\boldsymbol{x}) = \mathrm{j}\omega\mu_0 \exp(-\mathrm{j}k_\mathrm{b}|\boldsymbol{x}-\boldsymbol{x}_s^\mathrm{S}|)/4\pi|\boldsymbol{x}-\boldsymbol{x}_s^\mathrm{S}|^3(-\mathrm{j}k_\mathrm{b}|\boldsymbol{x}-\boldsymbol{x}_s^\mathrm{S}| - 1)[(x_2-x_{2,s}^\mathrm{S})\boldsymbol{i}_1-(x_1-x_{1,s}^\mathrm{S})\boldsymbol{i}_2]$. The vertical positions of the transmitter rings are $x_3^\mathrm{S} = \pm 0.35\lambda$ and 0, and of the receiver rings are $x_3^\mathrm{S} = \pm\lambda$, $\pm 0.5\lambda$ and $\pm 0.25\lambda$.

The original profile is given in Fig. 3a. These plots are the volume slices at $x_1 = 0$ and $x_3 = 0$. After generation of synthetic data a 5% random additive white noise is added. In the inversion we discretize the test domain D into 31 by 31 by 15 cells, hence, the total number of unknowns is equal to 14415. The reconstruction results after 1024 iterations are given in Fig. 3b. Note that one iteration takes approximately 22 seconds on a personal computer with a 600MHz Pentium III processor. We observe that the results are quite satisfactory in spite of the use of very limited data.

5 Conclusions

The presented results for biomedical data using a 2D-TM polarization measurement and 3D vectorial data show that the contrast source inversion method using multiplicative regularization leads to an effective inversion technique. The algorithm is fully iterative and does not solve any forward problem in each iteration. This makes the method suitable for large scale computations. Furthermore, the artificial tuning process with a weighting parameter of the regularization to obtain the "cosmetically best" results seems superfluous.

References

1. Bolomey, J., Pichot, C.: Some applications of diffraction tomography to electromagnetics - the particular case of microwaves in: Inverse Problems in Scattering and Imaging, M. Bertero and E.R. Pike, Eds., London, U.K.: Adam Hilger (1982) 319–344
2. Broquetas, A., Romeu, J., Rius, J.M., Elias-Fuste, A.R., Cardama, A., Jofre, L.: Cylindrical geometry: A further step in active microwave tomography. IEEE Trans. Microwave Theory Tech. **39** (1991) 836–844
3. Bulyshev, A.E., Souvorov, A.E., Semenov, S.Y., Svenson, R.H., Nazarov, A.G., Sizov, Y.E., Tatsis, G.P.: Three-dimensional microwave tomography. Theory and computer experiments in scalar approximation. Inverse Probl. **16** (2000) 863-875
4. Chew, W.C., Wang, Y.M.: An iterative solution of two-dimensional electromagnetic inverse scattering problem. Int. J. Imag. Syst. Technol. **1** (1989) 100–108
5. Chew, W.C., Wang, Y.M.: Reconstruction of 2D permittivity distribution using the distorted Born iterative method. IEEE Trans. Med. Imag. **9** (1990) 218–225
6. Franchois A., Pichot, C.: Microwave imaging - complex permittivity reconstruction with a Levenberg-Marquadt method. IEEE Trans. Microwave Theory Tech. **46** (1997) 133–141
7. Harada, H., Wall, D., Takenaka, T., Tanaka, T.: Conjugate gradient method applied to inverse scattering problem. IEEE Trans. Antennas Propagat. **43** (1995) 784–792
8. Joachimowicz, N., Mallorqui, J.J., Bolomey, J.Ch., Broquetas, A.: Convergence and stability assessment of Newton-Kantorovich reconstruction algorithms for microwave tomography. IEEE Trans. Med. Imag. **17** (1998) 562–569
9. Kleinman, R.E., van den Berg, P.M.: An extended range modified gradient technique for profile inversion, Radio Sci. **28** (1993), 877–884
10. Meaney, P.M., Paulsen, K.D., Hartov, A., Crane, R.K.: Microwave imaging for tissue assessment: Initial evaluation in multitarget tissue equivalent phantoms. IEEE Trans. Biomed. Enq. **43** (1996) 878–890
11. Souvorov, A.E., Bulyshev, A.E., Semenov, S.Y., Svenson, R.H., Nazarov, A.G., Sizov, Y.E., Tatsis, G.P.: Microwave tomography: A two-dimensional Newton iterative scheme. IEEE Trans. Microwave Theory Tech. **46** (1998) 1654–1659
12. van den Berg, P.M., Kleinman, R.E.: A contrast source inversion method. Inverse Probl. **13** (1997) 1607–1620
13. van den Berg, P.M., van Broekhoven, A.L., Abubakar, A.: Extended contrast source inversion. Inverse Probl. **15** (1999) 1325–1344
14. van den Berg P.M., Abubakar, A.: Contrast source inversion method: State of art. Prog. in Electromag. Research **34** (2001) 189–218

Propagation of Excitation Waves and Their Mutual Interactions in the Surface Layer of the Ball with Fast Accessory Paths and the Pacemaker

Jiří Kroc

West Bohemia University, Research Center, Univezitní 8,
CZ 332 03, Plzeň, Czech Republic,
krocj@ntc.zcu.cz,
WWW home page: http://home.zcu.cz/~krocj/

Abstract. Propagation of excitation waves and their mutual interactions on the surface layer of a ball with fast accessory paths and pacemaker is studied by the cellular automaton (CA) computer simulation technique in three–dimensions. The existence of the conducting system created by fast accessory paths and the pacemaker in the excitable medium – i.e., in the medium having a morphology – is important for this study. The topology of the medium and the CA–model itself are kept quite simple in order to focus the main attention of this study into the spatio-temporal response of the excitable medium with respect to the position and size of excitatory disturbances and inclusions. Propagation of undistorted excitation waves in the homogeneous excitable medium and appearance of spirals in it – induced by initial excitatory disturbances – are studied as well.

1 Introduction

Propagation of excitatory process through the human heart [6] is a quite complex problem that deserves a great importance in non-invasive diagnosis of heart diseases. It should be emphasized that this complexity is manifold. The general aim of this article is to elucidate some parts of this complexity using a simple model and a simple topology.

The excitation of a myocardial cell leads to changes of the trans-membrane potential that triggers the action potential (AP) propagating parallel to the surface of the myocardial cell [7, 8]. This results in activation of neighbouring cells *via* electrotonic currents across the cell junctions. It is quite well known from experimental data that pathological changes of the heart, e.g. ischemia, dysplasia, etc., can change the shape of AP and/or coupling between cells. Macroscopically this leads to changes of the propagation of activation waves across the muscle tissue.

The model proposed in this paper contains several following simplifications. It works above the level of living cells. The simplified topology is used, i.e. a surface

P.M.A. Sloot et al. (Eds.): ICCS 2002, LNCS 2331, pp. 217–226, 2002.

layer of a ball with two fast accessory paths (the conducting system), and sino–atrial (SA) node that serves as the pacemaker. Muscle fibres – and anisotropy of excitation propagation resulting from it – are not taken into account.

It seems to be quite interesting to study development and interactions of propagating excitation fronts in three-dimensional (3D) excitable medium having a simple conducting system. Certain number of computational studies of spiral development and its mandering is known in the plane without the conducting system [4, 5] but no one with the simple conducting system to the best knowledge of the author. The conducting system is the element in excitable media that can brings new types of responses and effects. Therefore, it deserves a computational study.

CA–model discretize space into three–dimensional (3D) lattice of cubes. Cubes, i.e. elements of this lattice, are called cells. Every cell has defined a neighbourhood – usually a list of nearest neighbouring cells and the cell itself – that is uniform through the whole lattice. Every cell contains a list of variables, e.g. state, morphology, etc. Morphology is divided into three different classes: conducting system, muscle, and the SA–node. The evolution of the system is driven by a transition rule that computes new values of variables using values of cells laying in the neighbourhood of given cell from the previous CA–step. General information about the CA simulation technique with applications can be found in the book [1], in the articles of Vichniac [12], Toffoli [10] and in books [11, 13].

The topology of medium and the CA–model itself as mentioned before are kept quite simple in order to focus the main attention of this study into the spatio-temporal response of the excitable medium with respect to the position and size of excitatory disturbances and inclusions – having slower excitation. Propagation of undistorted excitation waves in the homogeneous excitable medium with the conducting system and appearance of spirals in it – induced by initial excitatory disturbances – are studied as well.

A new simulation cellular automata environment has been developed – as a part of this study that is important for future simulations – to simulate propagation of excitation in excitable media. This code is written in *C* and cooperates with the OpenGL graphical library [9] and runs in X-Windows window system under Linux. The OpenGL enables us, beside the other advantages, to display data using transparentness.

2 Cellular Automaton Model of Propagation of Excitation Waves

The proposed CA–model works with the uniform 3D–lattice of cube cells that are updated simultaneously according to a transition rule. As mentioned above, the evolution of states of one particular cell is controlled by the set of cells that form the neighbourhood or surrounding of this cell. In this model, the so called Moore neighbourhood, i.e. the twenty–six first and second nearest neighbours of an updated cell and this updated cell itself is used. The proposed probabilistic excitation wave front movement leads to approximately isotropic propagation of this front. The lattice size used is the cube composed of 200^3 cells – i.e., 8×10^6 cells in total – for large simulations, and 100^3 cells for small simulations. Smaller lattice size requires substantially smaller simulation time.

Generally, the transition rule can be split into several sequential steps that handle the evolution of different parts of the cellular automaton belonging into different morphological classes, i.e. the conducting system, muscle, and the SA–node. Due to computational reasons, one additional morphological class is defined with the empty value that should be understood in the following sense. A cell having the empty morphological state is not included into propagation of excitation events but is necessary because this CA–model works with the neighbourhood that strictly requires all neighbouring cells. Every surface cell has at least one cell with empty morphological state in its neighbourhood.

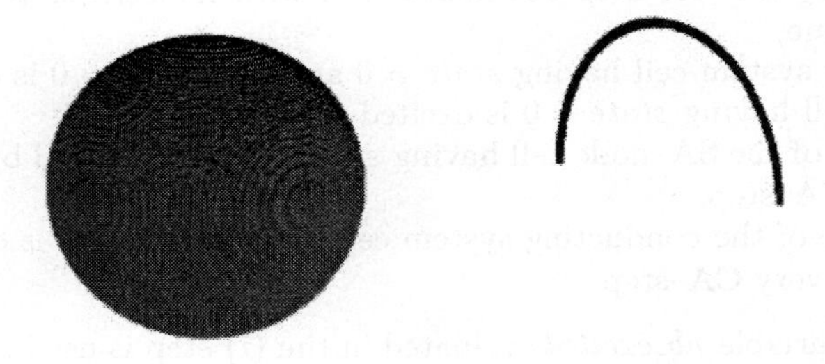

Fig. 1. The excitable medium is created by the ball surface layer, cf. the left side of the figure. Two fast accessory paths with the SA–node in between of them creates the conducting system, cf. the right side of the figure.

Each cell in the CA–model has two variables, namely, *state* and *morphology*. Variable *state* of the cell defines the level of excitation of the cell laying in the interval of $< 0, 21 >$. *State* equal to the value of 21 is the excited state. As the value of the state variable decreases, first absolutely refractory *states* and then relatively refractory *states* are reached. The resting *state*, i.e. the situation when the cell waits for an excitation event, has assigned the zero value. Every cell inside of the lattice belongs to one of the following morphology classes, i.e. empty, the conducting system, muscle, and the SA–node with values of 0, 1, 2, and 3, respectively. All cells with values between 1 and 3 defines excitable medium. Cells with value equal to zero, i.e. empty cells, does not influence the excitation process at all.

It is assumed in this model that the excitable medium is a ball surface layer[1] with two branches of conducting system laying in it, cf. Fig. 1. The SA–node

[1] The diameter of this ball is slightly below 200 cubes. Due to this reason, all figures presented in this work have a big granularity because sides of those cubes are visible.

serving as the pacemaker is located at the top of the ball and at the same time in the between of two branches of the conducting system. Excitation is generated by the SA–node, and then propagates through the conducting system and through the muscle tissue. Because the propagation speed in the conducting system is two or three times higher than in the muscle [2], the propagating excitation wave is not symmetric but is elongated in the direction of conducting system. Because the excitable medium has the shape of ball layer with empty inner part, the most of cells in the square lattice are in the empty state during the whole simulation.

The transition rule that handle the evolution of the whole cellular automaton is composed of the following logical steps:

(i) the total number of all excited neighbours are counted and saved into the temporary variable named *nb_excited*,
(ii) during every 3rd CA–step a muscle cell is excited with the probability of 50% if $state = 0$ and $nb\_exited > 0$,
(iii) during every 3rd CA–step the muscle cell with nonzero *state* value is decreased by one,
(iv) conducting system cell having $state = 0$ and $nb\_exited > 0$ is excited,
(v) SA–node cell having $state = 0$ is excited (self-excitation),
(vi) *state* value of the SA–node cell having $state > 0$ is decreased by one during every 10th CA–step,
(vii) *state* value of the conducting system cell having $state > 0$ is decreased by one during every CA–step.

Temporary variable *nb_excited* evaluated in the **(i)** step is used in the excitation steps **(ii)** and **(iv)**. The steps from **(ii)** to **(vii)** are evaluated independently. The steps **(ii)** and **(iii)** operates during every 3rd time step because it is assumed that propagation of the excitation front through the conducting system is three times higher than through the muscle tissue. The SA–node is independent of the rest of the excitable medium.

It has to be stressed out that a resting cell located next to the excitation wave front excite with probability of 50% (see step **(ii)**) if at least one of the neighbouring cells is excited. This leads to approximately isotropic propagation of the excitation wave front. If no probabilistic excitation of a cell is taken into account then anisotropic propagation of excitation waves occur that is caused by anisotropy of used lattice.

When inclusions – e.i. regions having slower excitability – are inserted then the morphology variable have to be extended by new state named inclusion. The morphology value of four is associated with an inclusion. The additional step should be added in the following form:

(viii) state value of a cell belonging to the inclusion having $state > 0$ is decreased by one during every 9th CA–step.

Propagation of excitation waves through the proposed topology, interactions of excitation waves with an excitatory disturbance, and interactions with an inclusion resulting in a response of such medium will be reported in the subsequent section.

3 Results and Discussion

As mentioned before, the topology of the excitable medium used in the present CA–model has the shape of the ball surface layer – several cells thick – with two simple arms of the conducting system and the pacemaker in between them, cf. Fig. 1. Such simplified topology serve as a toy–model that can helps to improve the understanding of the propagation, interactions and collisions of excitation waves, and interactions of those waves with excitatory disturbances and inclusions. Achieved results and experience can be used to construct models of the human heart.

3.1 Propagation of Undistorted Excitation Waves in the Anisotropic Excitable Medium

Excitation waves are generated – with given pacing rate – at the SA–node that is located at the top of the ball (cf. Fig. 1) and that operates as the pacemaker. The back propagation of excitation waves is protected due to non-excitability of cells having non-zero state variable, i.e. cells that are in any of the refractory states. Those excitation waves simple propagate through the excitable medium by excitation of cells in the resting state.

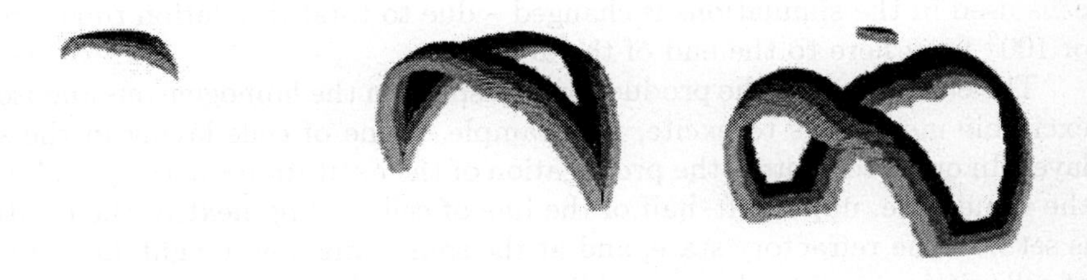

Fig. 2. A sequence of snapshots displaying the propagation of excitation waves through the anisotropic excitable medium created by the ball surface layer with the conducting system. The topology and orientation of the excitable medium are very same as in Fig. 1.

A sequence of excitation waves generated in the SA–node can be seen in Fig. 2 – generally sequences of figures start at the left-upper corner and continue in rows. From this figure, it is evident that propagation of the excitation wave front is anisotropic through given excitable medium. The excitation front propagates faster in the direction of arms of the conducting system as expected, compare to Fig. 1.

The excited state is associated with light grey colour. As the excitation of a given cell decreases, light grey colour becomes darker and then dark grey.

The final resting state of this cell is depicted by the invisible colour. Therefore, excited cells laying in the back side of the surface layer of the ball are visible through those cells having invisible colour laying in the front layer, see Fig. 2.

The excitation wave propagates from the top of the ball to the bottom through the surface layer. Finally, two parts of the excitation front collide and disappear due to lack of cells in the resting state. One excitation wave is generated after the other and they propagate through the excitable medium. In this case, no interactions between different excitation waves occurs. The question is what will happen if waves collide in some more topologically complicated cases of the excitable medium – with inclusions having slower excitability then the rest of media – or when an excitatory disturbance is applied?

3.2 Appearance of Spirals in the Excitable Medium with Initial Excitatory Disturbances

Production of spirals in the excitable media is of the great interest. Spirals in the human heart are associated with dangerous and often deadly arrhythmias. Generally, spirals can be created in an excitable medium using a special initial configuration of cells and their initial excitation states, or using an excitable medium with an inhomogeneous propagation speed of the excitatory process. The attention is focused to the influence of the initial pre–excitation of selected parts of media in this subsection. It should be mentioned that the number of cells used in the simulations is changed – due to total simulation time – to value of 100^3 from here to the end of the article.

The easiest way of the production of a spiral in the homogeneous and isotropic excitable medium is to excite, for example, a line of cells laying in the surface layer. In order to protect the propagation of the excitation waves up and down at the same time, upper left–half of the line of cells sitting next to the excited line is set into the refractory state, and at the same time, lower right–half of the line of cells sitting next to the excited line is set into the refractory state. The rest of the excitable medium – except the excited SA–node at the top – is set into the resting state. This leads to the propagation of one sole spiral having two arms through the excitable medium but this spiral spontaneously disappears after a certain period of time.

A another special initial pre-excitation of the excitable medium can be used to produce spirals, cf. Fig. 3. A line of cells laying in the surface layer is excited (dark grey) simultaneously with the SA–node. The line of cells sitting next to the excited line of cells on one side of the excited line is set into the refractory state (light grey). This produces two spirals in the excitable medium. Spirals overwhelms the activity of the SA–node and survive for ever as can be seen in Fig. 3. Please note that in this case, no special morphological disturbance or inhomogeneity of the excitable medium is necessary for initialization and surviving of the spiral.

A closer inspection of propagation of the excitation wave front in Fig. 3 gives the answer why spirals can survive for ever in this particular case. First, the excitation propagates from the SA–node and from the excited line as can be

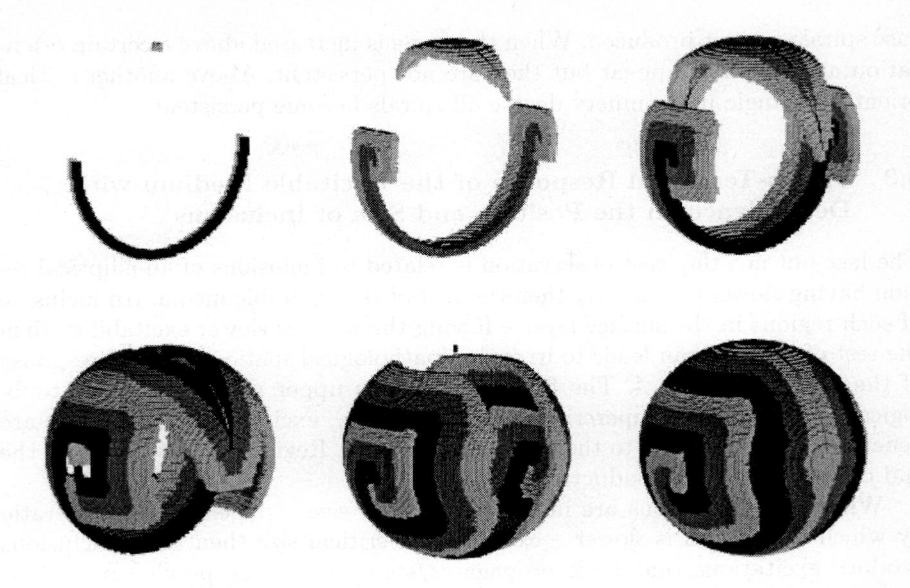

Fig. 3. A special initial pre–excitation of the medium is sufficient to produce spirals that can overwhelm the role of the SA–node. No special morphological disturbance of the excitable medium is necessary for initialization and surviving of the spiral.

seen in the first three figures. Two spirals having one arm are created at the ends of the initially pre–excited surface line. Consequently, on the first figure of the second row of figures, holes created by recovered cells and surrounded by cells in the refractory state – except a small gaps – appear around the ends of the originally excited line. Those holes are again invaded by few excited cells – from the above mentioned gaps – laying at the border of those holes. This activity overwhelms the pacemaking activity of the SA–node. From now, the centers of the spiral become the pacemakers.

From the above results, it is evident that spatio-temporal synchronization of excitation events and propagation of the excitatory process through the excitable medium with given topology plays the crucial role! Those observations cannot be neglected in any model of excitable media.

So far, we discussed the case where the SA-node is excited in synchrony with the pre–excited line of cells laying in the surface layer where the orientation of the conducting system – when we look from the top, i.e. from the position of the SA–node – is perpendicular to this pre–excited line as can be seen from the Fig. 3. Spirals are present in this case. A natural question arise what happen in the case where the conducting system and the initially pre–excited line are parallel or when their mutual angle is different from those two positions?

For simplicity, assume that in the 'parallel' case the angle between the conducting system and the pre–excitation is zero. It is observed that in the 'parallel' case spirals are not produced. When the angle is increased above a certain orientation angle, spirals appear but they are not persistent. Above another critical orientation angle up to ninety degree all spirals become persistent.

3.3 Spatio-Temporal Response of the Excitable Medium with Dependence on the Position and Size of Inclusions

The last but not the least observation is related to inclusions of an elliptical region having slower excitability then the rest of the excitable media. An inclusion of such regions in the surface layer – having three times slower excitability then the rest of media – can leads to irregular/pathological spatio–temporal response of the system, cf. Fig. 4. The first figure in the upper row displays the topological position of the supercritical inclusion – i.e., excitatory irregularities are generated by it – related to the conducting system. Region is located near of the end of one arm of the conducting system.

When such inclusions are inserted and their size – depending on the ratio by which this region is slower – exceed some critical size then those inclusions produce excitations that back–propagates/spreads into the previously excited excitable medium – now again recovered – irrespectively of the conducting system. It seem that excitation 'tunnels' the natural barrier created by recovering medium when the simulations is observed. The excitatory process operates at time and space where under normal conditions without the inclusion cannot be present. The overall activity again depends on the relative position of the inclusion to the conducting system.

One another aspect of 'tunnelling' process should be mentioned here. For larger sizes of inclusions, a distorted excitation followed by abnormal excitation of the conducting system without excitation from the pacemaker is observed. In details, one observe the following: (i) excitation of the conducting system triggered by the pacemaker, (ii) normal excitation of the media to the moment when the inclusion is reached, (iii) 'tunnelling' of excitation through the inclusion, (iv) abnormal back–propagation of excitation that activates the conducting system without excitation from the pacemaker, and (v) finally, all irregular excitation activity tends to disappear when a new excitation sequence starts at the pacemaker.

All the previously discussed aspects of the excitable media – or their modifications – and many others should be observed in the more complicated simulations where the topology of the human heart is used. All become more complicated when the anisotropy of the media related of the muscle fiber is used even for the simplified ball topology. For the human heart the excitatory maps can be compared to the experimentally measured ones [2].

4 Conclusion

A cellular automaton model describing propagation and mutual interactions of excitation waves on the surface layer of a ball with fast accessory paths and

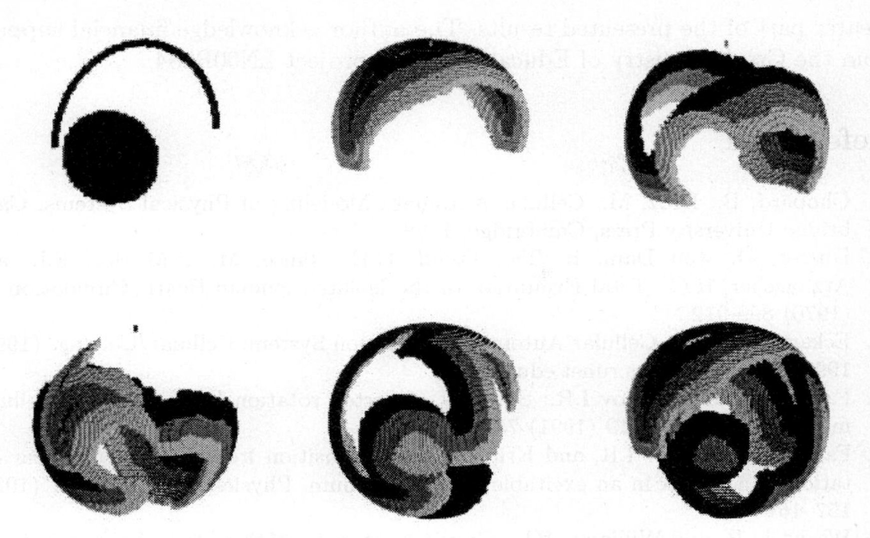

Fig. 4. Propagation of excitation in the surface layer with two different sizes of inclusion are showed in higher and lower rows. First figure displays the topology. The excitation invades the recovered medium in the third figure and distorts the subsequent evolution.

the pacemaker has been proposed. It is recognized that positions and sizes of excitatory disturbances and inclusions of the excitable medium are important and that they strongly influence the spatio-temporal response of this medium.

Propagation of undistorted excitation waves in the homogeneous excitable medium and appearance of spirals in it – induced by an applied initial excitatory disturbances, i.e. pre–excitations, are studied. Spirals do not appear for parallel orientation angle between the conducting system and the pre–excitation line. When the angle is increased above a certain orientation angle, spirals appear but they are not persistent. Above another critical orientation angle up to ninety degree all spirals become persistent.

The last but not the least observations are related to inclusions of elliptical regions having slower excitability then the rest of media. When such inclusions are inserted and their size – depending on the ratio by which this region is slower – exceeds some critical size then inclusions produce excitations that back–propagates into excitable medium irrespectively of the conducting system. The overall activity again depends on the relative position of the inclusion to the conducting system.

Acknowledgments

The author is greatly indebted to Prof. J.D. Eckart for his publicly available CA–environment called Cellular/Cellang [3] that have been used to compute greater part of the presented results. The author acknowledge financial support from the Czech Ministry of Education under project LN00B084.

References

1. Chopard, B., Droz, M.: Cellular Automata Modeling of Physical Systems. Cambridge University Press, Cambridge, 1998
2. Durrer, D., van Dam, R. Th., Freud, G.E., Janse, M.J., Meijler, F.L. and Arzbaecher, R.C.: Total Excitation of the Isolated Human Heart. Circulation **41** (1970) 899-912
3. Eckart, J.D.: The Cellular Automata Simulation System: Cellular/Cellang. (1991–1999) http://www.cs.runet.edu/~dana/
4. Fast, V.G. and Efimov I.R.: Stability of vortex rotation in an excitable cellular medium. Physica D **49** (1991) 75–81
5. Fast, V.G., Efimov I.R. and Krinsky, V.I.: Transition from circular to linear rotation of a vortex in an excitable cellular medium. Physics Letters A **151** (1990) 157–161
6. Warwick, R. and Williams, P.L.: Gray's Anatomy – 35th edition. Longman, Edinburgh, 1973
7. Guyton, A.C. and Hall, J.E.: Textbook of Medical Physiology. W.B. Saunders Company, Philadephia, London, Toronto, Montreal, Sydney, Tokyo, 1996
8. Hulín, I. et al.: Patofyziológia. Slovak Academic Press, Bratislava, 1998
9. Graphical library: OpenGL API. Silicon Graphics, Inc., California, USA, (1993–2002) http://www.sun.com/software/opengl/
10. Toffoli, T.: Cellular automata as an alternative to (rather than an approximation of) differential equations in modelling physics. Physica **10D** (1984) 117–127
11. Toffoli, T. and Margolus, N.: Cellular Automata Theory, The MIT Press, Cambridge, Massachusetts, London, England, 1987
12. Vichniac, G.Y.: Simulation physics with cellular automata, Physic **10D** (1984) 96–116
13. Weimar, J.R.: Simulation with Cellular Automata, Logos Verlag, Berlin, 1997
14. Wiener, N. and Rosenblueth, A.: The mathematical formulation of the problem of conduction of impulses in a network of connected excitable elements, specifically in cardiac muscle. Arch. Inst. Cardiol. Mexico **16** (1946) 205–265

Computing Optimal Trajectories for Medical Treatment Planning and Optimization

Ovidiu Daescu* and Ashish Bhatia

Department of Computer Science,
University of Texas at Dallas,
Richardson, TX 75083, USA,
{daescu,abhatia}@utdallas.edu

Abstract. In this paper we discuss approximation algorithms for the 2-dimensional weighted regions optimal penetration problem and propose a heuristic for speeding up the computation. The problem asks to find a ray (direction, trajectory) to access a target region in a weighted subdivision, such that some weighted distance function over the regions intersected by the ray is minimized.

1 Introduction

Recent developments in medical imaging have made possible to provide precise information on the anatomy of a 2-dimensional (2-D, planar) or 3-dimensional (3-D, spatial, volumetric) region of the human body and on the anatomical localization and extent of "target" subregions, such as tumors, contained within that region. For example, computed tomography and magnetic resonance imaging can be used to provide 2-D and 3-D data, and powerful image processing techniques can be employed to process the data for medical treatment planning (e.g., surgical planning, minimally invasive surgical methods). In this paper we assume that the planar or volumetric region has been preprocessed and partitioned into subregions, each of which has been associated a weight (based on some weighting criterion), and that one subregion has been identified as a target. The goal is to find a ray (direction, trajectory) to access the target such that some weighted distance function over the regions intersected by it is minimized.

More formally, we consider the weighted regions optimal penetration problem, introduced in [7]. The input is a subdivision R in the 2-D space, composed of r weighted regions R_i, $i = 1, 2, \ldots, r$, with a total of n vertices. The problem is to find a ray L that originates from outside R and intersects a specified *target region* $T \in \{R_1, R_2, \ldots, R_r\}$, such that the weighted sum $S(L) = \sum_{L \cap R_i \neq \phi} w_i * f_i(L)$ is minimized, where $f_i(L)$ is a function associated with the pair (R_i, L) and w_i is a positive integer weight associated with R_i. We will call L a penetrating ray with respect to the target, or penetration for short. The regions R_i, $i = 1, 2, \ldots, r$, are

* The research of this author is supported in part by the National Science Foundation under Grant EIA-0130847 and by founds from the Clark Foundation Research Initiation Grants Program.

P.M.A. Sloot et al. (Eds.): ICCS 2002, LNCS 2331, pp. 227–233, 2002.

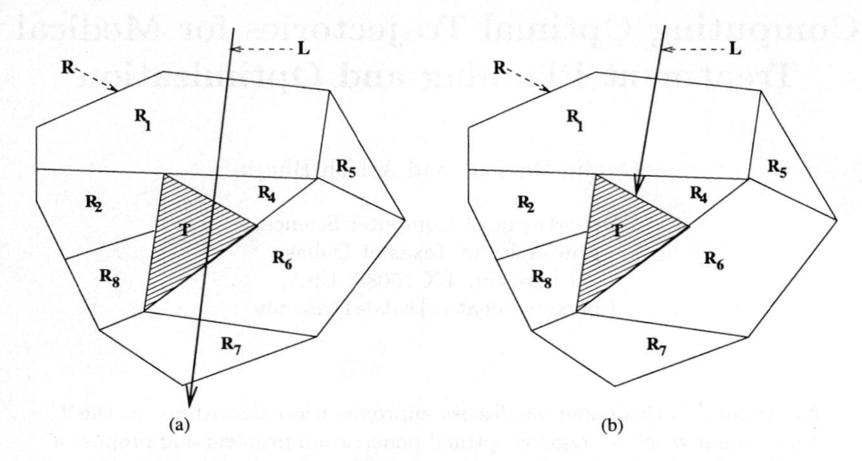

Fig. 1. The penetration L: (a) goes through the target; (b) stops at the target

all convex polygons in the 2-D space, and the weights of T and the complement $\overline{R}$ of R are zero ($\overline{R}$ is the free space outside R). Let R_L denote the set of all regions of R intersected by a penetration L and let d_i denote the Euclidean, L_2 length of L within $R_i \in R_L$. Two versions of this problem have been introduced in [7]. In the first version (P1), $f_i(L) = d_i$ for all $R_i \in R_L$. In the second version (P2), the penetration stops when it hits the target. Thus, $f_i(L)$ is either d_i or zero, depending on whether R_i is passed by L before or after L intersects the target region T (see Figure 1 for an example). Since the solutions to P1 and P2 are similar, we will only discuss the second version of the problem.

The penetration problem arises in several medical areas, such as radiation therapy, stereotactic brain surgery, minimally invasive computer assisted surgery and telerobotics applications in surgery [6, 3, 10, 13, 7, 14, 11, 2, 4, 5, 9, 12]. For example, in radiation therapy P1 is related to the use of a radiation ray (beam, pencil) directed at the target from an external source (*teletherapy*) while P2 can be used to model the process of delivering radiation by direct implantation of a radiation source into the target (*branchytherapy*). In practice, finding optimal directions and trajectories is one of the most difficult problems of medical treatment optimization. While there are quite a few 2-D and 3-D results on treatment optimization with respect to other parameters, significant progress on optimizing directions has been made only recently [7, 8].

In a companion paper, we have proved that the optimal penetration goes through a vertex of the subdivision in the 2-D case and that it is defined by a two dimensional set of lines in the 3-D case. In this paper, we make use of those results to design approximation algorithms for the 2-dimensional weighted regions optimal penetration problem. We also propose a heuristic for speeding up the computation of the approximation algorithms.

2 Previous 2-D Algorithms

A continuous ray space modeling for the 2-D optimal penetration problem have been presented in [7], where it has been proved that the problem can be reduced to solving $O(n^2)$ 2-D global optimization problems of the form

$$\min_{(x,y)\in D} f(x,y) = \sqrt{1+x^2}\sum_{i=1}^{k}\frac{a_i y + b_i}{x + c_i} \tag{1}$$

where D is a convex 2-D domain (a cell in an arrangement of lines), a_i, b_i and c_i are constants, the variables x and y are the defining parameters (slope and intercept) for the penetration L and k is $O(n)$ in the worst case.

Since the optimal penetration has been proved to go through a vertex of the subdivision R (i.e., the optimal solution lies on the boundary of some feasible domain D), the $O(n^2)$ 2-D global optimization problems can be reduced to solving $O(n^2)$ 1-D global optimization problems of the form

$$\min_{x\in D_x} f_x(x) = \sqrt{1+x^2}\Big(d_0 + \sum_{i=1}^{k}\frac{d_i}{x + c_i}\Big) \tag{2}$$

where d_0, c_i and d_i, $i = 1, 2, \ldots, k$, are constants, D_x is an interval on the X-axis defined by some bounding edge of D and k is $O(n)$ in the worst case.

When using general purpose global or local optimization software as in [7], the reduction above could result in orders of magnitude speed-up for solving the 2-D penetration problem. However, for computer based medical planning systems that require fast planning algorithms (e.g., on-line surgical planning) such speed-up may not be sufficient. In the next two sections, we use the knowledge that the optimal solution goes through a vertex of the subdivision and present simple and fast algorithms for finding an approximate solution.

3 Approximation Algorithms

In this section we present two approximation algorithms for the 2-D optimal penetration problem. Our algorithms are based on simple, yet efficient, edge and angle subdivision methods and can be refined to attend a user specified precision at an expense in computing time.

As mentioned in the previous section, a solution for the optimal penetration problem can be found by solving a number of global optimization problems. In [7, 8] it has been proved that the feasible domains for the 2-D global optimization problems can be generated using line (or line segment) arrangement traversal (construction) algorithms. Each 2-D feasible domain corresponds to a cell in the arrangement.

Lemma 1. *The 1-D global optimization problems can be generated in a total of $O(n\log n + k + km)$ time, where m is the maximum number of terms in any of the 1-D objective functions and k is the complexity of the arrangement.*

Proof. The boundary of each feasible domain is constructed and maintained by the arrangement traversal algorithm and thus the $O(k)$ 1-D feasible domains can be obtained in $O(k)$ time by traversing the corresponding arrangement structures (usually doubly-connected edge lists). For each boundary edge e, the objective function can be updated in linear time in the number of fractional terms in the body of the function. Thus, if m is the maximum number of terms in any of the 1-D functions, each update requires at most O(m) time. □

Alternatively, the following simple and practical algorithm can be used:

1. For each vertex $v \in R$, compute the two tangent lines to the target region T. Let $W(v,T)$ be the double wedge defined by v, T, and the tangents. This computation takes $O(n \log n)$ time.
2. For each $W(v,T)$, find the set of vertices $VW(v,T)$ of R that are in $W(v,T)$, which takes $O(n^2)$ time overall (a range searching data structure adapted from triangle range searching can be used for better theoretical bounds [1]).
3. For each vertex $v \in R$, sort the set $VW(v,T)$ around v and compute the intersection of the lines defined by v and $VW(v,T)$ with the boundary of T. This takes a total of $O(nl \log l)$ time, where l is the maximum size of any set $VW(v,T)$. Observe that this set of lines includes those defined by v and the vertices of T. Thus, we have obtained a set of $O(l)$ double wedges at v.
4. For each vertex $v \in R$, compute the objective function associated with the leftmost double wedge at v, then traverse the double wedges at v while updating the objective function. The overall computation in this step can be done in $O(nlm)$ time, where m is the maximum number of terms in any of the $O(nl)$ objective functions.

Lemma 2. *The algorithm above constructs the global optimization problems for finding an optimal penetration in $O(n^2 + nlm + nl \log l)$ time, using $O(nlm)$ space.*

Proof. Follows from the algorithm description. □

Some of the advantages of this algorithm are that it avoids using duality transforms and (line segment) arrangement computation for generating the optimization problems, which may lead to more robust implementations, and it can be easily adapted to generate only a subset of the optimization problems (e.g., only a subset of vertices of R may be considered; that subset can be specified by the user or it can be selected based on some vertex weighting criteria).

Both algorithms above have the following two features besides simplicity. First, they allow for easy, scalable, parallel computation of the optimization problems (e.g., they can be adapted for a coarse-grain parallel model of computation as in [8]). Second, as we will show bellow for the second algorithm, they can be easily modified to produce incrementally better approximations of the optimal solution.

Let $W(v,e)$ be one of the double wedges at v, where e is the line segment on the boundary of T defined by the double wedge (see Figure 2). Let ϵ be an

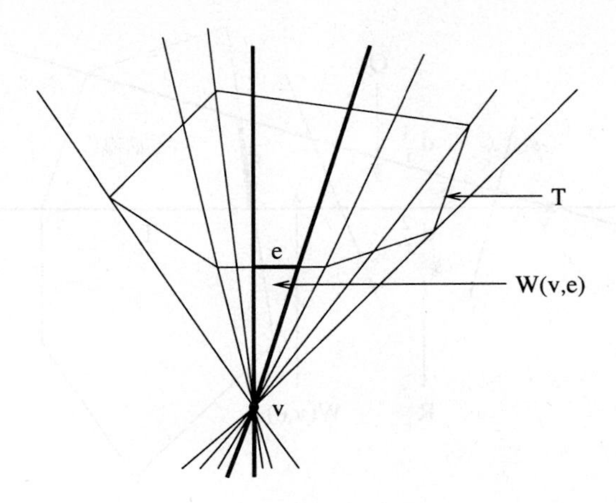

Fig. 2. The wedges at v and $W(v, e)$.

input parameter of the problem and let λ be the maximum length of any of the segments on T defined by double wedges at the vertices of R. Let $|e|$ be the Euclidean length of e. We propose two strategies for refining the quality of an optimal solution. The first one is a basic subdivision method for e: compute $\frac{|e|}{\epsilon} \leq \frac{\lambda}{\epsilon}$ evenly spaced points on e and the corresponding double wedges. Let $\delta = \frac{\lambda}{\epsilon}$. Thus, we obtain $O(\delta)$ double wedges and need to compute $O(\delta)$ values of the penetration at evenly spaced points (on the line segment e). Note that for the 1-D optimization problem associated with e we use the slope of a line through v as a parameter, and thus the corresponding feasible points are not evenly spaced, in general. The second strategy we propose is a discretization of the continuum of orientations at v that fall in the range of $W(v, e)$, resulting in $\delta = O(\frac{1}{\epsilon})$ evenly spaced directions at v, defining $O(\delta)$ double wedges. As before, this requires computing $O(\delta)$ values of the penetration at evenly spaced points (on the feasible domain).

Lemma 3. *Using the edge or angle subdivision methods, an approximate solution can be computed in* $O(nl(m + \delta)\log^2 m + nl\log l + n^2)$ *time.*

Proof. The last two terms in the time expression follow from steps 2 and 3 of the algorithm above. A double wedge computed by the algorithm is partitioned into $O(\delta)$ double wedges, each corresponding to *the same* objective function. Observe that an objective function is the product of two terms and the time to compute its value at a given point is dominated by the computation of the second one (see expression (2) in the previous section). That term is a sum of $O(m)$ 1-D linear fractional functions and can be expressed as the quotient of two polynomials of degree at most m. Thus, it can be evaluated at δ points in a total of $O((m+\delta)\log^2 m)$ time, using divide and conquer and Fast Fourier Transform (FFT). □

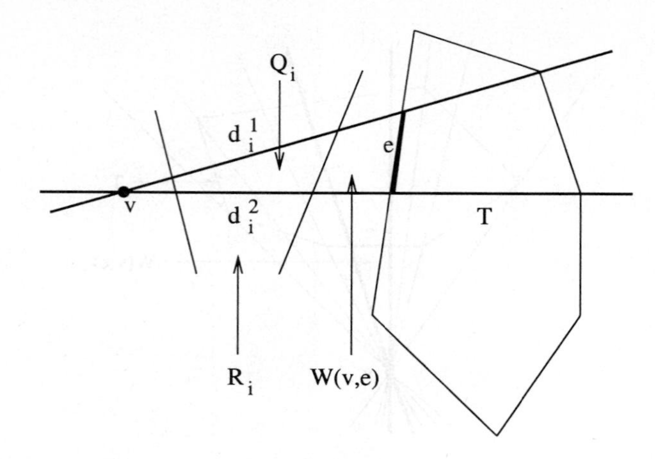

Fig. 3. A quadrilateral Q_i at v and its sides d_i^1 and d_i^2.

4 Making Good Choices

In this section we introduce a heuristic to help eliminate some of the optimization problems or reduce the number of approximate values computed by the approximation algorithms described earlier. The heuristic is used to *guide* the computation for approaching the optimal solution and can be easily incorporated into a recursive scheme for improving the quality of the approximation. More details are given bellow.

Before describing the method, we bound the quality of the approximation over a double wedge. Consider an optimization problem associated with some double wedge $W(v,e)$. Without loss of generality, let $R_L = \{R_1, R_2, \ldots, R_j\}$ be the set of regions of R intersected by any ray $L \in W(v,e)$. For a quadrilateral $Q_i = W(v,e) \cap R_i$, where $R_i \in \{R_1, R_2, \ldots, R_j\}$, let d_i^1 and d_i^2 be the two sides of Q_i that are on on the supporting lines of $W(v,e)$ (see Figure 3). Let d_i^{min} be the minimum length of any line segment s with endpoints on the other two sides of Q_i such that the supporting line of s goes through v. Similarly, let d_i^{max} be the maximum length. Then, we have the following Lemma.

Lemma 4. $\sum_{i=1}^{j} w_i d_i^{min} \leq L_{opt} \leq \sum_{i=1}^{j} w_i d_i^{max}$*, where* L_{opt} *is the value of an optimal penetration in double wedge* $W(v,e)$.

Let $L_v^{min} = \sum_{i=1}^{j} w_i d_i^{min}$ and $L_v^{max} = \sum_{i=1}^{j} w_i d_i^{max}$. We refer to L_v^{min} and L_v^{max} as the *lower bound* and *upper bound* at v, respectively. The heuristic we propose maintains the minimum L^{min} of the upper bounds over all optimization problems generated so far. A new optimization problem at some vertex v is accepted if $L_v^{min} \leq L^{min}$ and rejected otherwise.

Lemma 5. *The heuristic above rejects only optimization problems that cannot lead to the overall optimal solution.*

We say that an optimization problem is *active* if its lower bound is no larger than L^{min}. Optimization problems that have been accepted are maintained in a priority queue based on the value of the lower bound. Note that some of the accepted problems that are in the queue may become *inactive* as the computation progresses. Such problems can be eliminated from the queue by inspecting it each time the number of queued problems exceeds some predefined bound. Finally, the algorithm terminates when an optimization problem for some $W(v, e)$ is selected from the queue, such that $L^{min} - L_v^{min}$ is smaller than some user specified approximation error.

Currently, we are in the process of implementing our approximation algorithms and heuristic. Simulation results will be presented in the full paper.

References

1. P.K. Agarwal and J. Erickson, "Space-time tradeoffs for emptiness queries," Advances in Discrete and Comp. Geometry, (Bernard Chazelle, Jacob E. Goodman, and Richard Pollack, editors), *Contemporary Mathematics*, Vol. 223, pp. 1-56, 1999.
2. A.L. Boyer, T.R. Bortfeld, L. Kahler, and T.J. Waldron, "MLC modulation of x-ray beams in discrete steps," *Proc. of the 11th Conf. on the Use of Computers in Radiation Therapy*, pp. 178-179, 1994.
3. A. Brahme, "Optimization of radiation therapy," *Int. J. Radiat. Oncol. Biol. Phys.*, Vol. 28, pp. 785-787, 1994.
4. R.D. Bucholz, "Introduction to the Journal of Image Guided Surgery," *Journal of Image Guided Surgery*, Vol. 1, pp. 1-11, 1995.
5. C.W. Burckhardt, P. Flury and D. Glauser, "Stereotactic Brain Surgery," *IEEE Engineering in Medicine and Biology*, Vol. 14, pp. 314-317, 1995.
6. Y. Censor, M.D. Altschuler and W.D. Powlis, "A computational solution of the inverse problem in radiation-therapy treatment planning," *Applied Math. and Computation*, Vol. 25, pp. 57-87, 1988.
7. D.Z. Chen, O. Daescu, X. Hu, X. Wu and J. Xu, "Determining an optimal penetration among weighted regions in two and three dimensions," *Journal of Combinatorial Optimization*, Spec. Issue on Optimization Problems in Medical Applications, Vol. 5, No. 1, pp. 59-79, 2001.
8. O. Daescu, "Parallel Optimal Weighted Links," *Proceedings of ICCS, Intl. Workshop on Comp. Geom. and Appl.*, pp. 649-657, 2001.
9. D. Glauser, P. Flury, Ph. Durr, H. Funakubo, C.W. Burckhardt, J. Favre, P. Schnyder and H. Fankhauser, "Configuration of a Robot Dedicated to Stereotactic Surgery," *Stereotact. Func. Neurosurg.*, Vol. 54 and 55, pp. 468-470, 1990.
10. A. Gustafsson, B.K. Lind and A. Brahme, "A generalized pencil beam algorithm for optimization of radiation therapy," *Med. Phys.*, Vol. 21, pp. 343-356, 1994.
11. T. Holmes and T.R. Mackie, "A comparison of three inverse treatment planning algorithms," *Phys. Med. Biol.*, Vol. 39, pp. 91-106, 1994.
12. H. Koyama, T. Uchida, H. Funakubo, K. Takakura and H. Fankhauser, "Development of a New Microsurgical Robot for Stereotactic Neurosurgery," *Stereotact. Func. Neurosurg.*, Vol. 54 and 55, pp. 462-467, 1990.
13. A. Schweikard, J.R. Adler and J.C. Latombe, "Motion planning in stereotaxic radiosurgery," *IEEE Trans. on Robotics and Automation*, Vol. 9, pp. 764-774, 1993.
14. S. Webb, Optimizing the planning of intensity-modulated radiotherapy. *Phys. Med. Biol.*, Vol. 39, pp. 2229-2246, 1994.

CAD Recognition Using Three Mathematical Models

J. Martyniak, K.Stanisz-Wallis, and L.Walczycka

Deptartment of Biostatatistics & Medical Informatics,
Collegium Medicum Jagiellonian University, Cracow, Poland
mymartyn@cyf-kr.edu.pl, mywallis@kinga.cyf-kr.edu.pl

Abstract. In this paper we present the Bayesian diagnostic model and the logistic regression model with two different entrance strategies for determining factors predicting CAD risk. In the (R) methodological strategy the appropriate ratios lipids/apoproteins were considered as independent variables, in the (PC) the principal components analysis was used as a data reduction technique. The model based on Bayes' Theorem and the logistic regression model were good predictive models. It is the main objective of the project to try to design a tool for diagnosing.

1 The Aim

Coronary artery disease remains still the greatest cause of premature death for the Polish population. In the industrial countries it is responsible for 50% of all the death cases. The reasons of this are related to the adverse changes in the population's lifestyle resulting in the increase of the so called "risk factors" for development of arteriosclerosis, leading to coronary artery disease (CAD). The concept of "risk factor" comes from a long-term studies of Framingham population, in which the existence of several egzo- and endogenic phenomena has been shown, which are causally related to the CAD occurrence.

Taking into account the recent clinical studies, scientific achievements and more complete understanding of the peril and the action mechanism of the "risk factors" the International Commission for Prevention of Coronary Artery Disease, in cooperation with the International Society for Arteriosclerosis and Obesity, the European Society of Hypertension and the Diabetes Society, has prepared an accurate and comprehensive document concerning the primary and secondary prevention of coronary artery disease [2, 4].

The early detection and elimination of the risk factors of the CAD is particularly important, because in recent years the clinical form of the disease has been found in many young persons.

The earlier and greater is the awareness of the risk level for development of the disease in a given patient, the more appropriate and on-time decisions will be taken regarding the prevention of its development. The related changes in the patient's nourishment habits and lifestyle and the necessity of pharmacological treatment (choice of the appropriate medicine and its dose) depend on the global

P.M.A. Sloot et al. (Eds.): ICCS 2002, LNCS 2331, pp. 234–241, 2002.

risk assessment – the probability of coronary artery disease development in a given population, family and even particular patient.

Among more than 250 risk factors described up to date at least three have been recognized as independent i.e. individually responsible for development of the disease. These are: disorders in lipid level, uncured arterial hypertension and smoking. The lipid metabolism disorders have been found to be a primary risk factor for the CAD [8–10].

The aim of the study was to compare the anticipation accuracy of three different methods using a logistic regression analysis and the Bayesian diagnostic model.

2 The Data

The database used in the present study has been collected in the Dept. of Clinical Biochemistry, Chair of Clinical Biochemistry and Diagnosis in the Collegium Medicum of Jagiellonian University (CMUJ), Cracow, and it contained anonymous data of patients treated in the Clinic for Metabolic Diseases, CMUJ (including period 1985–1990), the Institute of Cardiology, CMUJ (1992–1998) and groups of patients from a selected population of healthy persons from the Tarnow Voiodship (the POL – MONICA project, realized in years 1985–1998).

Two hundred and twenty-eight subjects comprising 95 patients with arteriographically evidenced CAD (males, aged 47.99+11.12) and 133 healthy control (males, aged 47.11+7.29) without CAD were studied.

Blood samples were withdrawn after overnight fasting. Plasma levels of biochemical variables were assayed using automatic colorimetric methods, except insulin determined by radioimmunoassay.

Plasma lipid profile was determined as follows: total and free cholesterol and trigliceryde level in plasma (Ch, fCh, tg) and lipoproteins VLDL (VLDL-Ch, VLDL-fCh, VLDL-tg), LDL (LDL-Ch, LDL-fCh, LDL-tg) and HDL (HDL-Ch, HDL-fCh, HDL-tg).

Esterified cholesterol was calculated (eCh, VLDL-eCh, LDL-eCh, HLDL-eCh).

Apolipoprotein A1 level in plasma and HDL (apoA1, HDL-apoA1), alipoproteina B level in plasma and VLDL, LDL (apoB, VLDL-apoB, LDL-apoB); as well as protein B in VLDL, LDL and HDL (VB, LB, HB) were also determined.

Plasma glucose and insulin were determined in fasting state (Glu0, Ins0), and after 30, 60 and 120 min. during the oral glucose tolerance test (OGTT) (Glu30, Glu60, Glu120, Ins30, Ins60, Ins120 respectively).

Sum of glucose (SumGlu) and insulin (SumIns) were calculated as an area under glycemic and insulin kinetic curves estimated during OGTT respectively. Body mass index (BMI [kg/m^2]) was calculated.

3 Statistical Methods

A logistic regression model with two different entrance strategies for determining factors predicting CAD prevalence was used. In the first (R) methodological strategy the appropriate ratios lipids/apoproteins were considered as independent variables. Principal component analysis was used as data reduction techniques for the second strategy (PC). A model based on Bayes' theorem was used for calculating post-test probability.

The data set for the both methods was divided in two groups:

I – the learning set (121 patients – 72 persons in good health and 49 persons with CAD)

II – the test sets (107 patients – 57 persons in good health and 50 persons with CAD)

SAS and STATISTICA PL programmes were used for calculations [11, 12].

4 R Strategy

4.1 The Logistic Regression Model

For whole group of patients the relationship between five scores and CAD was explored by the logistic regression analysis. The probability of the appearance of the CAD (Y) conditional by the variable (x_p) can be calculated by means of the following form [3, 5].

$$P\,(Y = \text{the appearance of CAD}) = \frac{\exp\,(b_0 + b_1 x_1 + \ \ldots\ + b_p x_p)}{1 + \exp\,(b_0 + b_1 x_1 + \ \ldots\ + b_p x_p)} \qquad (1)$$

where: $x_1, x_2, \ldots, x_p$ – vector of independent variables – ratios, calculated from biochemical variables: SumGlu/SumIns, LDL-Ch/LDL-apoB, HDL-Ch/HB, HDL-tg/HDL-apoA1 and BMI

$b_0, b_1, \ldots, b_p$ – unknown coefficient of the regression

Y – dependent variable.

Table 1 shows the variables which significantly influence the appearance of the CAD.

Table 1. The significance of logistic regression coefficients assessed by the Wald chi-square statistic

Ratios of of variables	**Coefficient** b_i	**Standard error**	**p - Value (Wald's test)**
SumGlu/SumIns	–26,35	6,58	0,0001
HDL-Ch/HB	–6,42	3,23	0,049
HDL-tg/HDL-apoA1	14,51	6,16	0,02
b_0	4,50	2,39	0,062

5 PC Strategy

5.1 Principal Components Analysis

Principal components analysis is the multivariate data reduction technique [13].

This analysis selects the linear combination of the variables that best captures the variability of the data in a multidimensional space.

Given a data set with n numeric variables, m principal components can be computed. Each principal component is a linear combination of the original variables, with coefficients equal to the eigenvectors of the correlation or covariance matrix . The eigenvectors are customarily taken with unit length.

The principal components are sorted in descending order of the eigenvalues, which are equal to the variances of the components .

We have n variables, $X_1, X_2, \ldots, X_n$ all of which are standardised, and we desire to generate m composite variables of the form:

$$Y_i = \omega_{i1}X_1 + \omega_{i2}X_2 + \ \ldots \ + \omega_{in}X_n \quad \text{for} \quad i = 1, \ \ldots, \ m \tag{2}$$

where $m < n$ and ω_{ij} are selected to "explain" the maximum possible variance. That is Y_1 will have the largest possible variance (λ_1) subject to the restriction:

$$\sum_{j=1}^{n} \omega_{1j}^2 = 1$$

Y_2 will be uncorrelated with Y_1 and have the next largest possible variances (λ_2 with $\lambda_1 > \lambda_2$) etc., until we obtain m such uncorrelated composite variables all of which have weights normalized and variances $\lambda_1 > \lambda_2 > \ldots > \lambda_m$.

The percentage of the variance of the original n variables explained by the m composite variables is

$$100(\ \lambda_1 + \ldots + \lambda_m\ )/(\ \lambda_1 + \ldots + \lambda_n\ )$$

We call the m composite scores in this context component scores .

5.2 Data Reduction

The resulting scores are linear combination of the 34 original variables which are selected to be uncorrelated with each other.

We can retain only scores with eigenvalues greater than 1. In essence this is like saying that, unless a score extracts at least as much as the equivalent of one original variable, we drop it. Using this, we would retain 8 principal components.

The selected principal components were:
fraction V : VLDL-tg, VLDL-Ch, VB, VLDL-eCh, VLDL-fCh, tg
fraction L : fCh, LDL-eCh, LDL-Ch, eCh, Ch, apoB, LDL-apoB
fraction LH: LDL-fCh, LDL-apo B, LB, HDL-tg, HDL-fCh, height, BMI
fraction Totalgluc: Glu60, Glu120, SumGlu
fraction Ins: Ins30, Ins60, SumIns

fraction H : HDL-Ch, HDL-eCh, HB, apoA1, HDL-apoA1
fraction Weight: weight
fraction Glu0: Glu0, Glu30

The resulting components were included in a logistic regression that selected a model with eight scores .

5.3 Logistic Regression Model

For the whole group of patients the relationship between five scores and CAD was explored by the logistic regression analysis. The probability of the appearance of the CAD (Y) conditional by the variable (x_p) can be calculated by means of the previous form (1);
where:

$x_1, x_2, \ldots, x_p$ – vector of independent variables selected principal components:

- fraction V
- fraction L
- fraction LH
- fraction Totalgluc
- fraction Ins
- fraction H
- fraction Weight
- fraction Glu0.

The significance of logistic regression coefficients was assessed with the Wald chi-square statistic.

Table 2 shows the variables which significantly influence the appearance of the CAD.

Table 2. Significance levels of the Wald test

Principal Component	**Coefficient** b_i	**Standard error**	**p - Value (Wald's test)**
Fraction V	–1,45	0,45	0,0017
Fraction H	–1,80	0,45	0,0011
Fraction Ins	–3,24	0,68	0,000
Fraction Weight	0,99	0,4	0,016
Fraction Glu0	1,91	0,52	0,0004
b_0	0,13	0,38	0,743

6 The Bayes Theorem

Bayes' theorem is a quantitative method for calculating the posterior probability using the prior probability [6, 7].

Let $\{B_n\}$ be a countable measurable partition of Ω and $P(A) > 0$ and $B_1 \cup ... \cup B_n = \Omega$ and $B_i \cap B_j = \emptyset$ for $i \neq j$ and $P(B_i) > 0$ for $i = 1, \ldots, n$. Then

$$P(B_k|A) = \frac{P(A|B_k) \cdot P(B_k)}{\sum_{i=1}^{n} P(A|B_i) \cdot P(B_i)} \tag{3}$$

where $P(A|B)$is called the conditional probability of event A given event B.

We propose to use the theorem of Bayes as a tool to predict CAD for the individual patient. This model computes the posterior probability of an appearance CAD given a set of patients symptoms, syndromes or laboratory values.

Let $X = \{x_1, x_1, \ldots x_6\}$ will be a set of patient's syndromes, symptoms or laboratory values and $x_i \in \{0, 1\}$ and x_i represent the presence or the absence of those variables. The Bayes probability we can calculate as

$$P(CAD|X) = \frac{P(X|CAD) \cdot P(CAD)}{P(X|CAD) \cdot P(CAD) + P(X|No\,CAD) \cdot P(No\,CAD)} \tag{4}$$

where

$P(CAD)$ is the prior probability of CAD
$P(No\,CAD) = 1 - P(CAD)$
$P(X|CAD) = P(x_1|CAD) \cdot P(x_2|CAD) \cdot \ldots \cdot P(x_6|CAD)$ for $x_i = 1, \ldots, 6$.
$P(X|No\,CAD) = P(x_1|No\,CAD) \cdot P(x_2|No\,CAD) \cdot \ldots \cdot P(x_6|No\,CAD)$

The following independent (Pearson's chi-square test) syndromes, symptoms and laboratory values were selected: tg, Ch, LDL-Ch, HDL-Ch, BMI, tg/HDL-Ch.

7 Results

The results of examination of model quality are only for the test sets [1].

Table 3. The results of PC model and R model

PC model				**R model**			
	expected				**expected**		
		CAD	No CAD			CAD	NoCAD
observed	CAD	40	6	**observed**	CAD	32	14
	No CAD	10	51		No CAD	46	15

Sensitivity = 86,96% specificity = 83,61% Sensitivity = 69,58% specificity = 75,41%

Preliminary tests indicate that the PC model is of greater diagnostic relevance, which can undoubtedly be attributed to a greater number of variables introduced to the model than in the case of the R model.

8 Predictive Value

The predictive value of a test is simply the post - test probability that a disease is present based on the results of a test. The predictive value of a positive test depends on the test's sensitivity, specificity, prevalence and can be calculated of the following form [6]:

$$PV^{+} = \frac{(sensitivity)(prevalence)}{(sensitivity)(prevalence) + (1 - specifity)(1 - prevalence)} \quad (5)$$

The PV^{+} index is also one of the form of Bayes' theorem.

Table 4 shows the predictive value (PV^{+} index) of a positive test in three methods.

Table 4. Results of comparison of different predictive models

Model	Number of parameters	PV^{+} index
Bayesian analysis	6	0.841
PC model	34	0.807
R model	3	0.679

It is evident that, the PV^{+} index has the highest value for Bayes'model and the lowest for the R model.

9 Conclusion

The principal component analysis followed by stepwise logistic regression analysis showed the better predictive power of prevalence in the large and complex CAD data base than methodological strategy of ratios lipids/apoproteins considered as independent variables. The Bayes' theorem model also shows a relatively high predictive power compared with other two models.

References

1. Marshall G., Grover F., Henderson W., Hammermeister K.: Assessment of predictive models for binary outcomes: an empirical approach using operative death from cardiac surgery. Statist. Med. **13** (1994) 1501–1511

2. Buring J.E., O'Conner G.T., Goldhaber S.Z.: Risk factors for coronary artery disease: a study comparing hypercholesteronemia and hypertriglyceridemia. Eur.J.Clin.Invest. **19** (1989) 419–423
3. Hosmer D.W., Lemeshow S.: Applied Logistic Regression . 1st ed. New York, John Wiley & Sons (1989)
4. D'Agostino R.B., Belanger A.J., Markson E., Kelly-Hayes M. and Wolf P.A. : Development of health risk appraisal functions in the presence of multiple indicators: the Framingham Study nursing home institutionalization model. Statist. Med. **14** (1995) 1757–1770
5. Douglas G. Altman: Predictal statstics for Medical Research, CHAPMAN &HALL/CRC. rep.1999
6. Shortliffe E.H., Perreault L.E., Wiederhold G., Fagan L.M.: Medical Informatics. Springer-Verlag, New York (2001) 64–131
7. Grémy F., Salmon D. : Bases statistiques. Dunod, Paris (1969)
8. Wilhelmsen L., Wedel H., Tibblin G. : Multivariate Analysis of Risk Factors for Coronary Heart Disease. Circulation. **48** (1973) 950–958
9. Brand R.J., Rosenman R..H., Sholtz R.I., Friedman M. : Multivariate Prediction of Coronary Heart Disease in the Western Collaborative Study Compared to the Findings of the Framingham Study. Circulation **53** (1976) 348–355
10. Coronary Risk Handbook. Estimating Risk of Coronary Heart Disease in Daily Practice. American Heart Association (1973)
11. SAS Institute Inc., SAS Technical Report P-200, SAS/STAT Software: Logistic Procedure, Release 6.04, Cary. NC: SAS Institute Inc. (1990) 175–230
12. SAS Institute Inc. SAS User's Guide Statistics,Version 6.03 Edition, SAS Institute Inc., Cary North Carolina (1990)
13. Cureton E.E., D'Agostino R.B. : Factor Analysis, An Applied Approach. Erlbaum Publishers, New Jersey (1983)

3D Quantification Visualization of Vascular Structures in Magnetic Resonance Angiographic Images

J.A. Schaap MSc., P.J.H. de Koning MSc., J.P. Janssen MSc., J.J.M. Westenberg PhD., R.J. van der Geest MSc., and J.H.C. Reiber PhD.

Division for Image Processing, Dept. of Radiology, Leiden University Medical Center, Leiden, the Netherlands j.a.schaap@lumc.nl

Abstract. This paper describes a new method to segment vascular structures in 3D MRA data, based on the Wavefront Propagation algorithm. The center lumen line and the vessel boundary are detected automatically. Our 3D visualization and interaction platform will be prestended, which is used to aid the phycisian in the analysis of the MRA data. The results are compared to conventional X-ray DSA which is considered the current gold-standard. Provided that the diameter of the vessel is larger than 3 voxels, our method has similar result as X-ray DSA.

1 Introduction

Determination of vessel morphology along a segment of a vessel is important in grading vascular stenosis. Until recently most assessments of vascular stenosis were carried out using X-Ray Digital Subtraction Angiography (X-DSA). This technology has proven itself in the assessment of stenoses. However several problems exist when using this technology. Because X-DSA is a projection technique, over-projection of different vessels can occur even if the viewing angle is set optimally. Additionally, a contrast agent has to be injected, which requires an invasive procedure.

Magnetic Resonance Angiography (MRA) is a technique which acquires three dimensional (3D) images of vascular structures and the surrounding anatomy. Because of the 3D nature of the images, there is no problem of over-projection of different vessels. MRA images can be obtained without using a contrast agent, however by using a contrast agent the contrast-to-noise ratio increases. This contrast agent is normally applied intra-venously.

At the present time, evaluation of MRA images is commonly performed on two-dimensional (2D) Maximum Intensity Projections (MIPs), although it is known that this leads to under-estimation of the vessel diameter and a decreased contrast-to-noise ratio [2]. To improve upon the conventional analysis of MRA, it would be desirable to obtain quantitative morphological information directly from the 3D images and not from the MIPs. To accomplish this, accurate 3D

P.M.A. Sloot et al. (Eds.): ICCS 2002, LNCS 2331, pp. 242–254, 2002.

segmentation tools are required. Vessel segmentation of 3D images has been investigated by many researchers. However the majority of this research focussed on enhancing the 3D visualization of the vascular structures in the image and not on accurate quantification of these structures.
In this paper we describe a novel approach for quantitative vessel analysis of MRA images. Our approach uses knowledge about the image acquisition to accurately determine the vessel boundaries. The techniques we use operates on the full 3D images, and not the projections.
This paper is organized as follows. In Section 2 we present the algorithms which we use to segment a vessel, and the visualization platform is described. In Section 3 we describe the methods and materials we used for the validation of our approach. We conclude this paper with the discussion, future work and conclusions.

2 Methods

The focus of this section will be on the general outline of the used algorithms, and on the interaction between the user, the data and the algorithms. More information on the algorithms can be found in the references.

In short, the analysis trajectory is as follows: After loading and visually inspecting the data, the user places two points at the beginning and at the end of the vessel segment of interest. Using wavefront propagation and backtracking, a minimal cost path is found between these points through the vessel. This minimal cost path is then adjusted to form the center-lumen path. A tubular model constructed of a NURBS-surface is placed around the center-lumen path. The surface is then fitted to the data by balancing a set of internal and external forces that act on the surface. The result is presented visually, in graphs and can be exported to tables. The platform also provides means to generate snapshots and movies of any stage in the analysis. It should be noted that the only required user input is the placement of the two points in the vessel. The rest of the method is completely automated, thus minimizing inter- and intra user variability. We will now look into the different parts in more depth.

2.1 Wavefront Propagation

The 3D pathline detector is based on the Fast Marching Level Set Method (FMLSM) as described by Sethian et al.[7, 8, 5, 4]. The FMLSM calculates the propagation of a wave through a medium. Starting from one (or more) given location the propagation of the wave is calculated until a certain stop criterium is met. The stop criterium in most cases is the reaching of a given end point, however the propagation can continue until every point has been reached.

If we think of the image as an inhomogeneous medium, then the wave propagates through some parts of the image faster than through other parts. The idea is to make the wave propagate fast through the vessel, and slow through

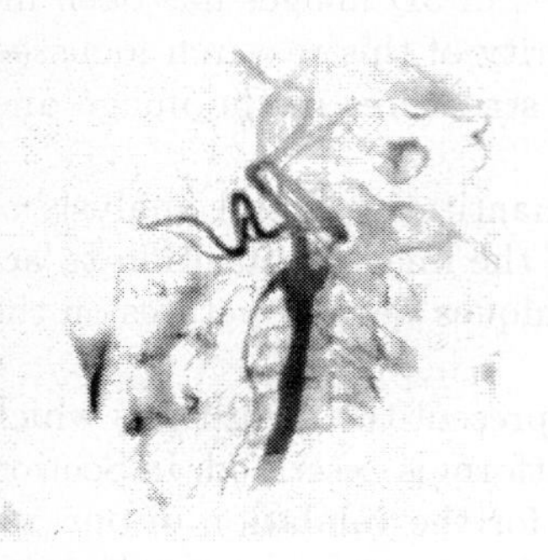

(a) *A filtered X-ray image of the coronary arteries.*

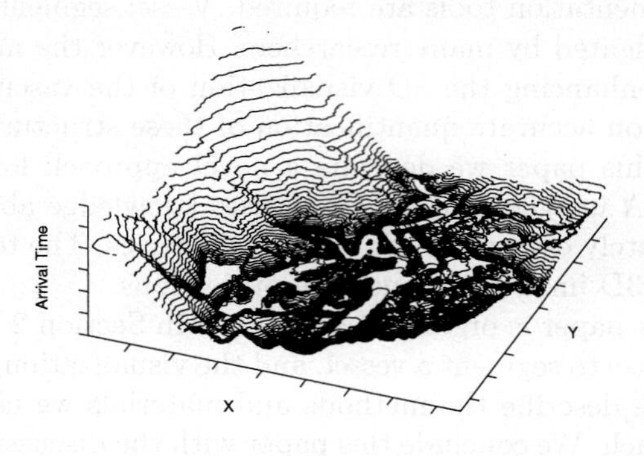

(b) *T surface after the wave has propagated. The starting point is located in the bottom right part of the image*

Fig. 1. *Example of the use of the WaveProp module.*

the rest of the image. This can be done by making the speed dependant on local image information like image intensity or image gradient.

Figure 1(a) and 1(b) shows an example: 1(a) shows the original image and 1(b) shows the associated T-surface. This surface represents the time at which the wave passed each point (x, y). An inverse linear speed function $speed(x, y) = Max - I(x, y)$ is used, where $I(x, y)$ is the image intensity at location (x, y), and where Max is the maximum intensity of the whole image. Therefore the wave will propagate faster through darker regions (i.e. the vessels) which translates itself in the valleys in the T-surface.

2.2 Back tracking

In order to find the path between two points p_0 and p_1, we use the arrival times that have been calculated by the FMLSM. Every point that has been reached during the propagation has an arrival time -the time at which the wavefront arrived at this point- associated with it. In figure 1(b) the arrival times of a 2D X-ray image are plotted as so-called T-surface. Using a steepest descent approach we move from $(p_1, T(p_1))$ to $(p_0, 0)$. The surface T has a convex like behavior in the sense that starting from any point $(p, T(p))$ on the surface and following the gradient descent direction, we will always converge to p_0. Given the point p_1, the path connecting p_0 (the point in T with the smallest value) and p_1, called the minimal path, is the curve $\widetilde{\mathcal{C}}(\sigma)$ starting at $\widetilde{\mathcal{C}}(0) = p_1$ and following the opposite gradient direction on T:

$$\frac{\partial \widetilde{\mathcal{C}}}{\partial \sigma} = -\nabla T \tag{1}$$

The back-tracking procedure is a simple steepest gradient descent approach. It is possible to make a simple implementation on a rectangular grid: given a point $q = (i, j)$, the next point in the chain connecting q to p is selected to be the grid neighbor (k, l) for which $T(k, l)$ is the minimal, and so forth.

2.3 Speed function

Since the speed function is essential to the algorithm, any manual adjustment in this speed function, may result in inter- and intra- user variability. Therefore, the speed function required by the FMLSM is based on the global image histogram. A sigmoid function is chosen for the speed function. The function is controlled by two parameters. L_{50} varies the position where the speed is 50% of the maximum speed (F_{MAX}) and slope modifies the slope of the speed function. F_{MAX} is set to 100.

2.4 Path correction

The minimal path obtained by the back tracking algorithm tends to cut the corners. We are interested however in the path through the center of the lumen. To correct the minimal path to form the center lumen line we use a modified Boundary Surface Shrinking (BSS) method [6]. This methods moves points along the gradient of a distance map [3]. The distance map contains the minimal distance to the object's boundary for every voxel, measured in some suitable metric. The gradient of the distance map at a point x will simply be denoted by $\nabla D(x)$ throughout this text.

The key idea of the original BSS algorithm is to iteratively translate the vertices of the initial surface along the gradient vectors at their current positions. Our modified version moves the points of our minimal path, instead of the vertices of the boundary surface.

The position of point i in iteration $n + 1$ is given by:

$$x_i^{(n+1)} := x_i^{(n)} + h\nabla D(x_i^{(n)},) \quad (2)$$

where h determines the size of a translation step, which is typically set at half the voxel diameter.

At the center lumen line, the magnitude of the gradient vectors becomes very small, however not exactly zero. To prevent the points of the center lumen line translating to the global maximum of the gradient distance map, we stop the translation when the magnitude of the gradient vector drops below a certain threshold.

2.5 Vessel model

Our vessel model is currently designed for vessel segments without changes in topology (such as bifurcations). We use Non-Uniform Rational B-Splines (NURBS) surfaces. A NURBS surface possesses many nice properties, such as local control of shape and the flexibility to describe both simple and complex objects. Simple surfaces can be described by less control points and the surface is smooth by construction. Complex surfaces require more control points and derivative smoothness constraints may be considered. The number of control points and the total area of the surface are not related, so the same model can work for all sizes of vessels.

2.6 Model Matching

The model is deformed to fit the underlying image data. To detect the lumen of the vessel we use the 3D extension of the Full Width 30% Maximum (FW30%M) criterion described in [9]. This method thresholds the image at 30% of the local center lumen line intensity. We adapted this criterion, by not only performing a simple threshold, but also placing restrictions on the way that the model deforms. This is done by making an energy function (3), which has to be minimized.
In order to make the threshold criterion less sensitive for noise, the array of image intesities at the center lumen line in smoothed. The first step is to initialize the surface as a tube centered around the previously detected center lumen line. The second step uses the conjugate gradient algorithm to minimize equation (3). To prevent the model from intersecting itself, we restrict the movement of the control points to a plane perpendicular to the center lumen line. This restriction is sufficient in practice, because the center lumen line has no sharp bends, compared to the diameter of the vessel.

$$\varepsilon = \varepsilon_{external} + \gamma_s \cdot \varepsilon_{stretching} + \gamma_b \cdot \varepsilon_{bending} \tag{3}$$

The external energy is based on the FW30%M criterion, and the stretching and bending energies are the internal energies that are needed to deform the NURBS-surface.

2.7 Visualization and interaction in 3D

We have developed a general visualization platform to visualize 3D images of any modality, and to intuitively interact with them, based on the VTK-software library [10]. It runs on a standard Windows PC, with minimal requirements of a PentiumII 500MHz, 128MB RAM, and an OpenGL video-card, such as a GeForce. The platform has an object oriented design that provides well known rendering techniques. We distinguish three types of data, voxel based (such as CT- and MRI-scans, as shown in Figure 2(a)), surface based (such as model-generated meshes, as shown in Figure 2(b)), and attribute data (such as scalars, streamlines, glyphs, vector fields, etc., as shown in Figure 2(c)). For each type of data we have developed an object that provides each of the three types of visualization: 3D rendering, projections and Multi Planar Reformatting. With these objects, a scene can be build, which can be visualized in any number of viewports, as shown in Figure 3.

(a) *Voxel data MRA data, thresholded and then volume rendered*

(b) *Surface data A mesh that describes the lumen*

(c) *Attribute data Each point represents the time that the wafefront passed*

Fig. 2. *Examples of the three different data types used in our visualization platform.*

Initially, the data is presented using the three orthogonal MIPs and a 3D view of the same MIPs as shown in the left part of Figure 3. As stated before, the segmentation algorithm needs two points to trace the vessel. It is not possible to position a point in 3D with a 2D mouse in just one action. However, when the MIPs are calculated, the depth of the voxel with maximum intensity is stored for each ray resulting in a depth image as shown in Figure 4(a). The user can depict a point in any of the three MIPs. The depth of the depicted 2D coordinate is read from the depth-image, resulting in a 3D position. This point is then shown in all three MIPs and in a 3D view, and can be moved interactively. This method provides an intuitive and fast way to place the two seedpoints in the vessel (see Figure 4(b)).

The result of the segmentation, i.e. the centerline and the vessel wall, are visualized using surface rendering, as shown in Figure 5. The user can interactively look at these structures from all directions, and combine the objects in one or more viewports. This provides the user a better understanding of the 3D shape of the vessel, and of the relationship with the surrounding anatomy. In order to inspect the segmentation result more closely, Multi Planar Reformatting (MPR) can be performed to cut through the objects, resulting in 2D contours, which can be compared with the original image data on the plane. Again, all these actions can be done at interactive speeds and from all directions. Finally, the detected centerline can be used as a path for the camera, thus generating a fly-through inside the vessel.

3 Validation

The centerline detection and lumen segmentation were validated using in-vitro and in-vivo data sets. The in-vivo data was used to test the centerline detection, while the in-vitro data was used to test the lumen segmentation. The lumen

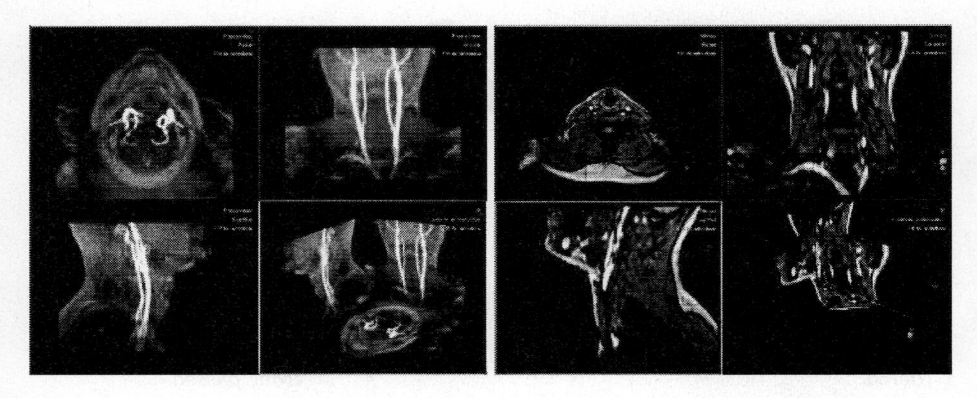

Fig. 3. *The two standard screen layouts. The left figure shows three orthogonal MIPs, both in 2D and 3D, while the right figure shows three orthogonal cross-sections, both in 2D and 3D. When new objects are added to the scene (such as the segmented vessel), these can be viewed together in any combination of 3D rendering, projections or slicing. It is also possible to have any number of viewports of any type to explore the data.*

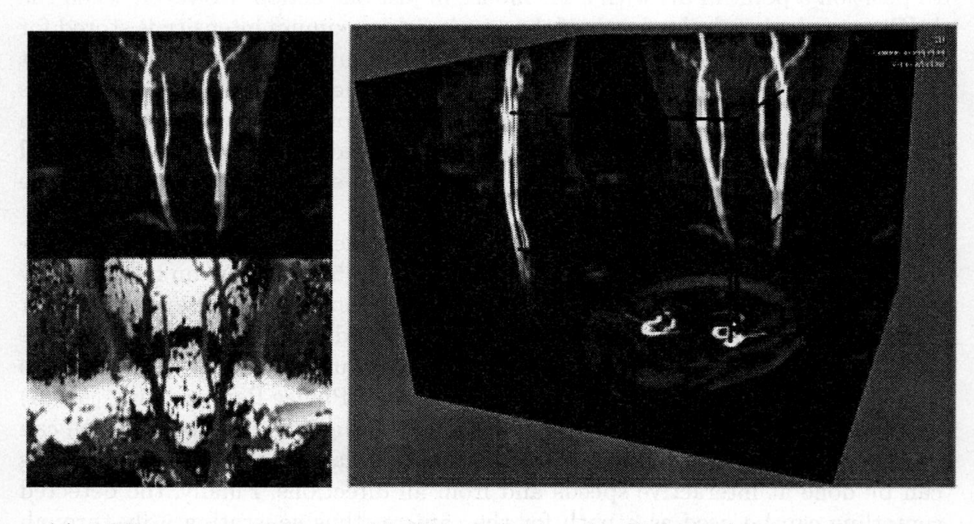

(a) *MIP and its depth-image*

(b) *The two points in 3D with the three orthogonal MIPs*

Fig. 4. *The only needed user interaction: the user needs to depict two 3D points at the beginning and the end of the vessel segment of interest. This can be done by placing a point in any of the 2D orthogonal MIPs; the third coordinate is then obtained from the associated depth-image.*

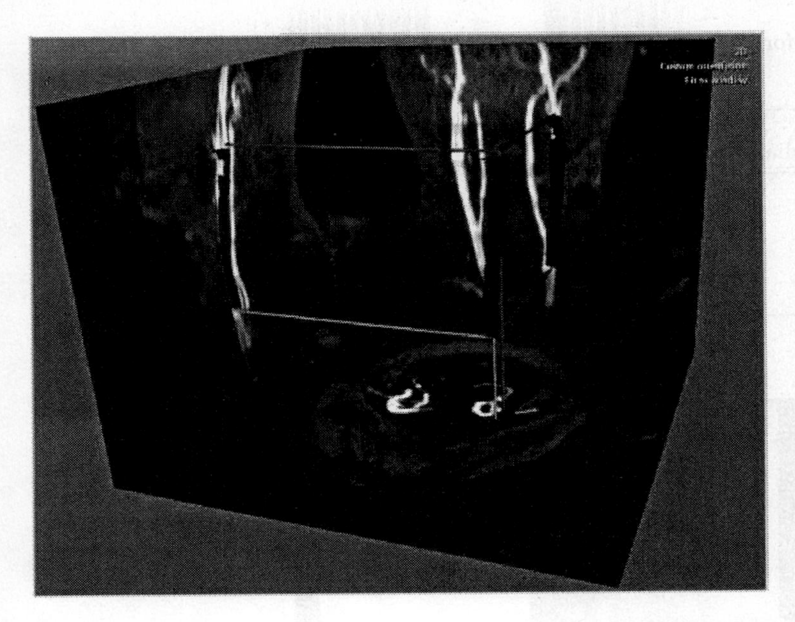

Fig. 5. *The segmentation result, rendered together with the MIPs of the MRA-data, the proximal and distal point of the vessel segment, and the centerline.*

segmentation was also compared to conventional X-ray angiography, which is considered the current gold-standard.

3.1 Materials

The in-vivo data included MRA CE studies of the lower abdomen and lower extremities and MRA TOF studies of the carotid arteries. The slice thickness varied from 2.37 to 4 mm, while the in-plane resolution varied from 0.67x0.67 to 1.76x1.76 mm.

The in-vitro data consisted of MRA Contrast Enhanced (CE) studies of several phantoms. These phantoms all had identical reference diameters and different obstruction diameters. The slice thickness was 2.37 mm with a slice-center-to-center-distance of 1.2 mm, while the in-plane resolution was 1.76x1.76 mm. The morphological parameters of the phantoms are listed in table 1. Additionally X-ray angiographic (XA) images of the phantoms were acquired. Further details about the phantom and the acquisition of the data can be found in [9].

3.2 Results

Our center lumen line detection was validated by detecting 43 centerlines in 22 studies by 2 observers. 40 centerlines (93%) were classified as correct (see table 2). The 3 failures were caused by brighter vessel running close and parallel to the vessel of interest and were identical for both observers.

Table 1. *Morphological parameters of the phantoms used in the in-vitro study.*

Phantom number	Reference diameter (mm)	Obstruction diameter (mm)	Percent diameter stenosis (%D)	Length of stenosis (mm)
1	6.80	5.58	18	7
2	6.80	4.69	31	7
3	6.80	3.47	49	7
4	6.80	2.92	57	7
5	6.80	1.97	71	7

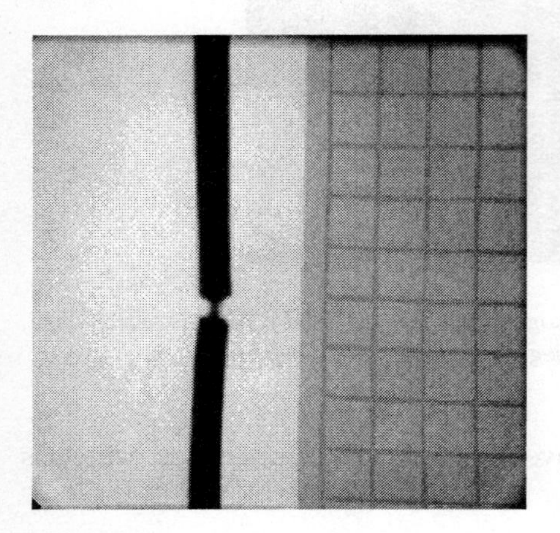

(a) *X-DSA image of the phantom*

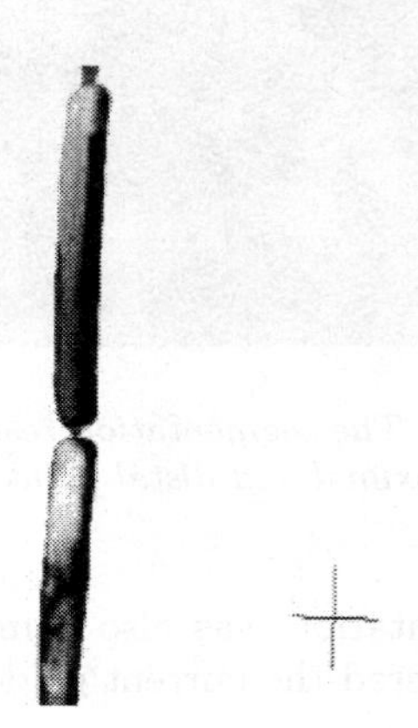

(b) *MRA image of the phantom*

Fig. 6. *Images of the phantom.*

The reference diameters of all phantoms were averaged for both MRA and XA. For the MRA data, several measurements were taken for a single phantom and averaged. See table 3 for the results.

The obstruction diameters of all phantoms were assessed and compared (see figure 8). From this figure, it becomes obvious that there is a lower bound on the diameter of the vessel that can be measured. If the diameter becomes smaller than 3 voxels, the lumen segmentation fails. If the diameter is larger then the error decreases to less than 1%.

4 Discussion

In this paper we discussed an approach for the automated quantification of contrast enhanced MRA studies. This approach involves the detection of a center lumen line by using the Fast Marching Level Set algorithm.

Table 2. *Results of the center lumen line detection validation.*

Analyst	Segments	Detected	Classified correctly	Success (%)
1	43	43	40	93.02
2	43	43	40	93.02

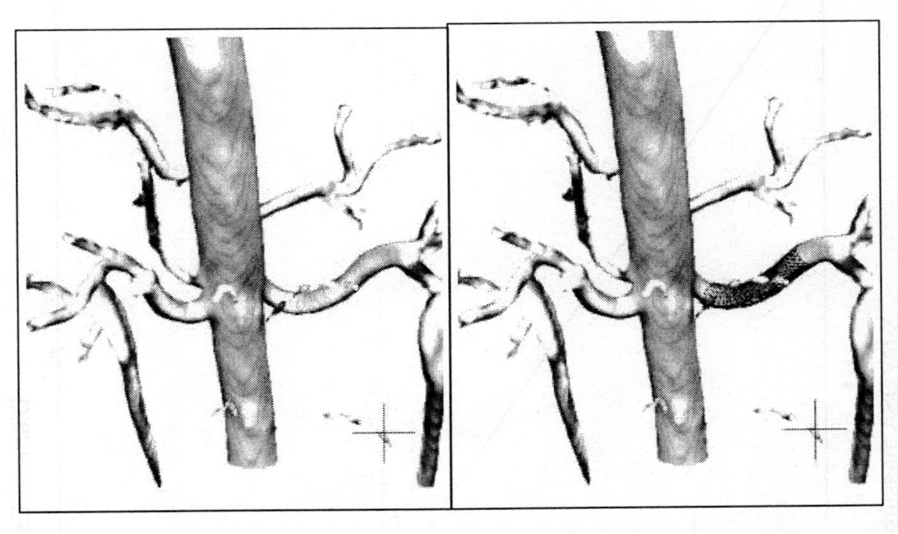

(a) *Original image* (b) *Segmented result and original image*

Fig. 7. *Images of a renal artery segmentation.*

The vessel boundery detection used a model based approach. A model of a vessel was created using NURBS surfaces and then modified to minimize an energy function using the conjugate gradient methods.
Our method performs quantitative analysis bases on the original 3D images. It is known from literature [xx] that assessment of stenoses based on MIPs tends to overestimate the degree of stenosis. De Marco et al. [xx] used multi-planar reformatting (MPR) images, which allow for better visualization of the vessel lumen in a plane perpendicular to vessel axis. De Marco et al. [xx] compared stenosis grading based on MIPs and MPR images of 3D TOF MRA studies, and used intra-arterial angiography (DSA) as a standard of reference. They reported a statistically significant difference between MIPs and DSA scores with an average absolute error of 9% (SD 14%). MPR images provided a better agreement and a negligible bias. Although this study suggests the potential benefit of MPR-based diagnosis, generation and inspection of MPRs is relatively time consuming.

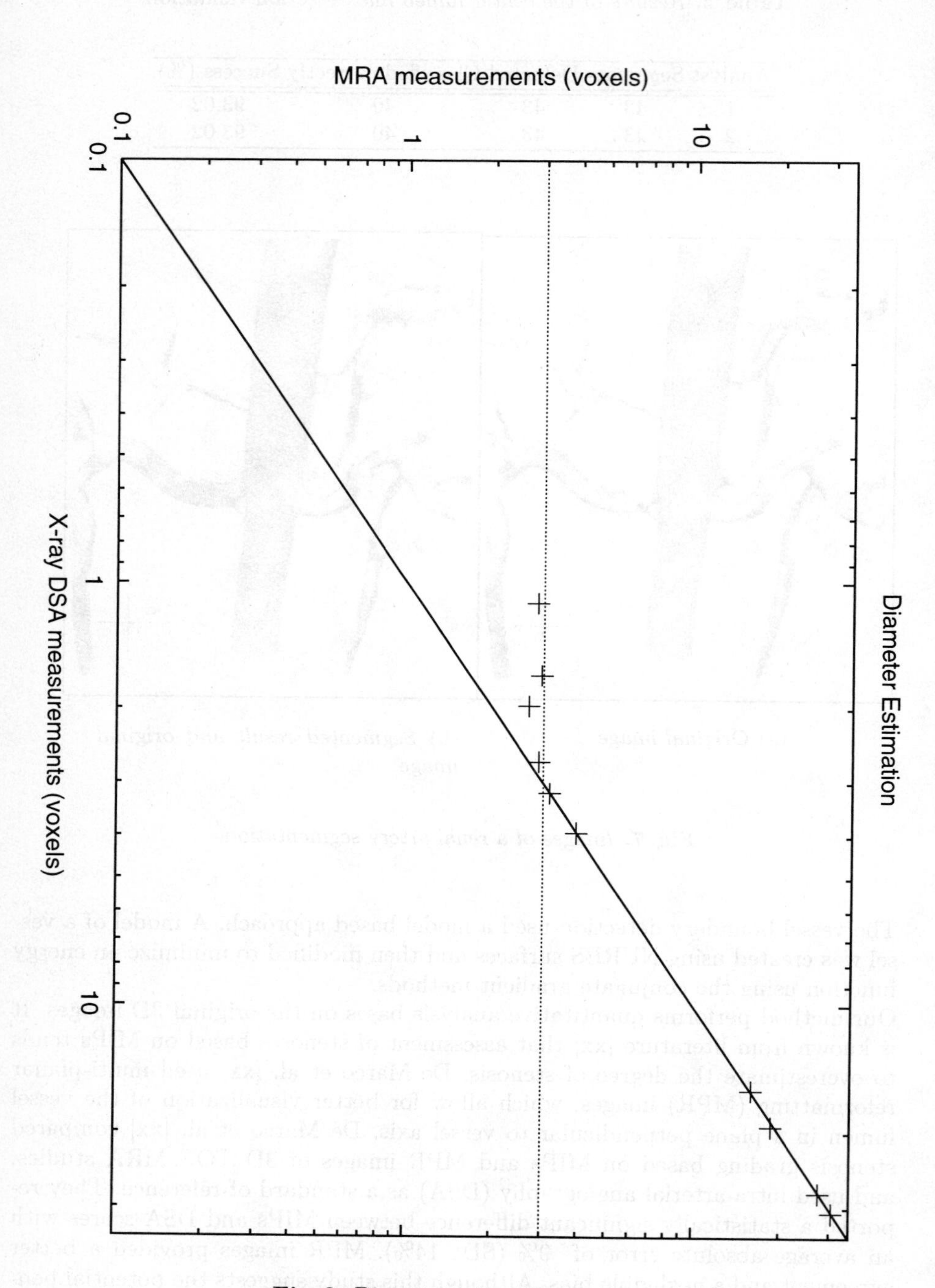

Fig. 8. *Diameter estimation errors*

Table 3. *Comparison X-ray DSA and MRA.*

True Diameter	X-ray DSA	MRA
6.8 mm	6.88 ± 0.19 mm	6.87 ± 0.25 mm

Our method shares the basic idea behind MPR based measurements. We apply an objective vessel diameter criterion in planes perpendicular to the vessel center lumen line, which is therefore similar to the radiologist's when analyzing MPR images. On the other hand, the method is objective (does not depend on window and level settings) and requires little interaction.
We used several in vivo studies to determine the accuracy and robustness of our center lumen line detection approach. In all cases a path was found, however in 3 cases the detected path was classified as incorrect by both analysts. Analysis of the cases showed a complex anatomy where several vessel were located close to each other, which resulted in a pathline which switched from one vessel to another and back again.
Using in vitro studies of a stenotic phantom, we investigated the accuracy of our vessel detection algorithm. We compared our method with X-ray DSA. The error in diameter estimation is less then 1%, provided the diameter is larger then 3 voxels. This limit is inherit to the approach used, in which a center intensity is needed in order to calculate the contours. In larger vessel such as the aorta, iliac and femoral arteries this criterium is fulfilled. In smaller vessels such as the carotid arteries this criterium is not always fulfilled, especially in stenotic regions.

5 Future Work

Our vessel segmentation method is currently limited to vessels without bifurcations and/or side-branches. We are currently designing a more general method which incorporates bifurcations, side-branches, and specific designs for stenoses and aneurysms.

Furthermore, we will continue our collaboration with the University of Amsterdam. One of their research topics is interactive simulation [1]. We have started a joint project in which we will connect our vessel analysis package to their computer grid. The vessel geometry generated by our package will serve as input for a real time fluid dynamics simulation program. The phycisian is then enabled to try different virtual interventions and see in real time the response of the blood flow.

6 Conclusion

We have demonstrated that our automated vessel detection in MRA data is able to detect the vessel wall and diameter with great accuracy, provided that the vessel diameter is larger than three voxels. The required user-interaction is limited to placing a proximal and a distal point. Furthermore we have presented a general visualization platform that can be used to visually inspect and interact with the data and the algorithms output.

References

1. R.G. Belleman and P.M.A. Sloot. Simulated vascular reconstruction in a virtual operating theater. *H.U. Lemke et al. editors, CARS Conference, Berlin*, pages 938–944, 2001.
2. D. Tsuruda L.G.Shapeero C.M. Anderson, J.S. Saloner and R.E. Lee. Artifacts in maximum-intesity0projection display of mr angiograms. *Amer. J. Roentgenol.*, 20(1):56–67, January 1990.
3. Olivier Cuisenaire. *Distance Transformations: Fast Algorithms and Applications to Medical Image Processing.* PhD thesis, Université catholique de Louvain, October 1999.
4. R. Malladi and J.A. Sethian. A real-time algorithm for medical shape recovery. In *Proceedings of International Conference on Computer Vision*, pages 304–310, 1998.
5. R. Malladi, J.A. Sethian, and B.C. Vemuri. Shape modeling with front propagation: A level set approach. *IEEE Transactions on Pattern Analysis and Machine Intelligence*, 17(2):158–175, February 1995.
6. Hartmut Schirmacher, Malte Zöckler, Detlev Stalling, and Hans-Christian Hege. Boundary surface shrinking - a continuous approach to 3D center line extraction. In B. Girod, H. Niemann, and H.-P. Seidel, editors, *Proc. IMDSP '98*, pages 25–28. Infix, ISBN 3-89601-011-5, 1998.
7. J. A. Sethian. A fast marching level set method for monotonically advancing fronts. *Proc. of the National Academy of Sciences of the USA*, 93(4):1591–1595, February 1996.
8. J. A. Sethian. *Level set methods: Evolving interfaces in geometry, fluid mechanics, computer vision, and materials science.* Number 3 in Cambridge monographs on applied and computational mathematics. Cambridge University Press, Cambridge, U.K., 1996. 218 pages.
9. J.J.M. Westenberg, R.J. van der Geest, M.N.J.M. Wasser, E.L. van der Linden, T. van Walsum, H.C. van Assen, A. de Roos, J. Vanderschoot, and J.H.C. Reiber. Vessel diameter measurements in gadolinium contrast enhanced three-dimensional mra of peripheral arteries. *Magnetic Resonance Imaging*, 18(1):13–22, January 2000.
10. Ken Martin Will Schroeder and Bill Lorensen. *The Visualization ToolKit.* Prentice Hall, 2nd edition, 1998, http://www.visualizationtoolkit.org.

Quantitative Methods for Comparisons between Velocity Encoded MR-Measurements and Finite Element Modeling in Phantom Models

Frieke M.A. Box[1], Marcel C.M. Rutten[3], Mark A. van Buchem[2], Joost Doornbos[2], Rob J. van der Geest[1], Patrick J.H. de Koning[1], Jorrit Schaap[1], Frans N. van de Vosse[3], and Johan H.C. Reiber[1]

[1] Division of Image Processing, Department of Radiology, Leiden University Medical Center, Leiden, the Netherlands

[2] Department of Radiology, Leiden University Medical Center, Leiden, the Netherlands

[3] Department of Biomedical Engineering, Eindhoven University of Technology, Eindhoven, the Netherlands

Abstract. Wall Shear Stress is a key factor in the development of atherosclerosis. To assess the WSS in-vivo, velocity encoded MRI is combined with geometry measurements by 3D MR-Angiography (MRA) and with blood flow calculations using the Finite Element Method (FEM). The 3D geometry extracted from the MRA data was converted to a mesh suitable for FEM calculations. Aiming at in-vivo studies the goal of this study was to quantify the differences between FEM calculations and MRI measurements. Two phantoms, a curved tube and a carotid bifurcation model were used. The geometry and the time-dependent flow-rate (measured by MRI) formed input for the FEM calculations. For good data quality, 2D velocity profiles were analyzed further by the Kolmogorov-Smirnov method. For the curved tube calculations and measurements matched well ($prob_{KS}$ approximately above 0.20). The carotid needs further investigation in segmentation and simulation to obtain similar results. It can be concluded that the error-analysis performs reliably.

1 Introduction

It is known that a correlation exists between the presence of atherosclerosis and the local Wall Shear Stresses (WSS) in arteries [2]. The WSS is defined as the mechanical frictional force exerted on the vessel wall by the flowing blood. The WSS τ_w is defined as wall shear rate $\dot{\gamma}$ multiplied by the dynamic viscosity η:

$$\tau_w = \eta \, \dot{\gamma}. \tag{1}$$

Near the wall $\dot{\gamma}$ may be expressed as the velocity gradient with respect to the outward normal $\boldsymbol{n}$ of the wall:

$$\tau_w = \eta \frac{\partial \boldsymbol{v}}{\partial \boldsymbol{n}}, \tag{2}$$

P.M.A. Sloot et al. (Eds.): ICCS 2002, LNCS 2331, pp. 255–264, 2002.

with $\boldsymbol{v}$ being the fluid velocity. To be able to assess the local WSS distribution from MRI-images of arteries a good approximation of the local velocity profiles is required. One of the major drawbacks of MRI-velocity-data is the relatively low resolution and the unknown error distribution (noise). To get around this problem, finite element(FEM) calculations may be used and compared with actual MRI-measurements. Crucial for precise and reliable measurements, however, is a thorough error analysis. Therefore, the goal of our study was to determine in a quantitative manner the correspondences and differences between actually measured time-dependent flow rates (flow(t)) and velocity profiles by MRI, with the corresponding flow(t) and velocity profiles derived from FEM calculations. The Kolmogorov-Smirnov method was applied to quantify the similarities between the 2D velocity profiles for FEM calculations and velocity encoded MRI measurements. Materials used for this study were two phantom models, one being a 90° curved tube and the other a carotid bifurcation. The curved tube was analyzed at three positions, i.e. at the inflow, in the middle section, and at the outflow. Data for the carotid phantom was assessed at two positions, i.e. at the entrance and just behind the bifurcation.

2 Materials and Methods

Two PMMA (polymethyl methacrylate) phantom models, one of a curved tube (diameter of 8 mm) and one of a carotid artery (inflow diameter of 8 mm, outflow diameters of 5.6 and 4.6 mm, for the internal carotid and external carotid arteries respectively), were connected to a MRI compatible pump (Shelley Medical Imaging Technologies), which can deliver an adjustable (pulsatile) flow profile flow(t). The blood emulating fluid is a Newtonian fluid with a viscosity of 2.5 mPa.s (Shelley) and MRI-compatible. The phantoms were connected to a straight and fixed tube with a length of 110 cm; as a result the inflow velocity profile is known and can be described analytically (Womersley-profile) [5, 13]. The phantoms were scanned and processed with a 1.5 T MR system (Philips Medical Systems) using a standard knee-coil in two ways:

1. The geometry of each phantom was obtained by means of a MR Angiographic acquisition protocol. The tubes were divided into 100 slices with a slice thickness of 1 mm, a TE/TR 6.8/21 ms, a field-of-view (FOV) of 256 mm, and a scan matrix of 512x512 pixels.

2. At different positions along the carotid- and curved tube model, the velocity and flow were assessed in a plane perpendicular to the major flow direction. Velocity encoded data were obtained by means of a gradient echo phase contrast imaging procedure. Triggering was applied during the acquisition and the simulated cardiac cycle was subdivided into 25 equidistant phases. The imaging parameters were: TE/TR 11.2/18.44 ms, flip angle 15 degrees, slice thickness 2 mm, FOV 150 mm, scan matrix 256x256 and velocity sensitivity 30 cm/s.

The finite element package that was used in this study is called SEPRAN [10]. Application of the package for the analysis of cardiovascular flow has been carried out in collaboration with the Department of Biomedical Engineering at the Eindhoven University of Technology, the Netherlands [8]. The mesh used for the FEM-calculations consisted of triquadratic bricks with 27 nodes [12]. The curved tube could be defined with a small addition, namely the description of the curvature, within the mesh generator of SEPRAN. The bifurcation needed more additions, and had to be parameterized, but was also generated within the package. This is called the standard mesh.

The geometry of the carotid bifurcation was segmented from the MR Angiographic data set (figure 1) using the analytical software package MRA-CMS® [9] [1]. Each segmented vessel is expressed as a stack of circles, while a bifurcation is expressed by two stacks of circles (figure 2). The circles of the MRA-CMS® geometry were transformed to rectangles with the same spatial distribution as the rectangles at the surface of the standard mesh. A solver for linear elasticity was applied to transform the standard mesh [11]. The Young's modulus E was set at a value of 100, while a small value was selected for the Poisson ratio ($\nu = 1.0e^{-4}$) to allow for independent motion of the mesh in each coordinate direction. The standard mesh was thus transformed until it matched the MRA-CMS® data (figure 3).

Fig. 1. Raw data from MRA

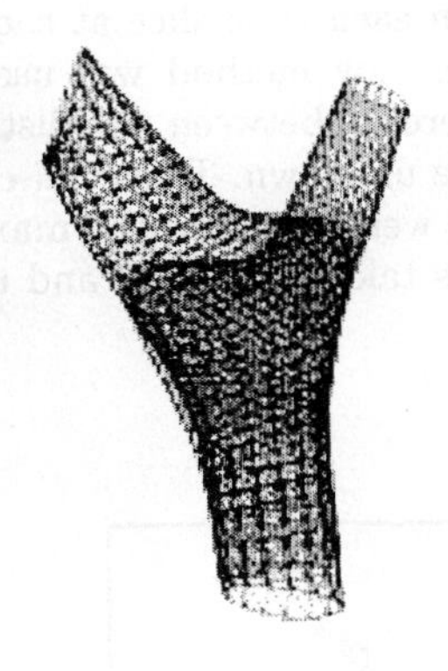

Fig. 2. Segmented carotid bifurcation phantom by MRA-CMS®

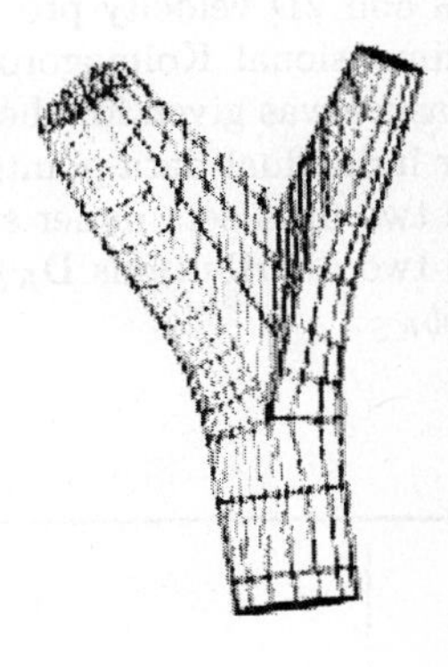

Fig. 3. Deformed mesh

The velocity encoded data were analyzed by means of the FLOW® analytical software package [3] [1]. FLOW® allows the quantification of the flow versus time and corresponding velocity profiles.

The flow values used as input for the calculations (flow_{MR}) were defined as follows: First, the velocity encoded data set was analyzed by means of the package FLOW®. The average velocity in the segmented area was calculated

[1] MEDIS medical imaging systems, Leiden, the Netherlands

and multiplied by the surface of this area, yielding the volume flow value. Next, the flow in the curved tube was measured at three positions: at the entrance (inflow$_{MR}$), in the middle (midflow$_{MR}$) and at the exit (outflow$_{MR}$). Data of the carotid phantom were assessed at two positions, at the entrance of the flow (inflow$_{MR}$) and just behind the bifurcation (intflow$_{MR}$ and extflow$_{MR}$).

Flow$_{MR}$(t) was Fourier transformed and the first 20 Fourier Coefficients were used as input in the FEM calculations. After solving the Navier-Stokes equations, the flow (in the primary flow-direction) was calculated at the position where the MR-measurement was performed (flow$_{calc}$). Flow$_{calc}$ has to be put in the same grid as the MR-measurement. Therefore, a gridding and interpolation procedure was carried out. Then the flow and velocity profiles can be presented in the same manner as the MR-velocity measurements (flow$_{image}$).

The calculation method was tested for the curved tube model under study at the Eindhoven University of Technology, the Netherlands [4]. The calculated flows and velocities were subsequently compared with the MRI-derived flow amounts and velocity profiles. For the purpose of comparing calculated with measured data [1], a special option was added to the standard FLOW® package. This option allows the assessment of the differences of calculated parameters and 2D velocity profiles in each time slice at each measured position. The 2-dimensional Kolmogorov Smirnov method was used for this purpose [7]. A measure was given for the difference between two distributions, when error bars (for individual data points) are unknown. Two semi-cumulative distributions of the two data sets under study were created. The maximum of the difference of the two distributions D_{KS} was taken (figure 4) and used for the calculation of $prob_{KS}$.

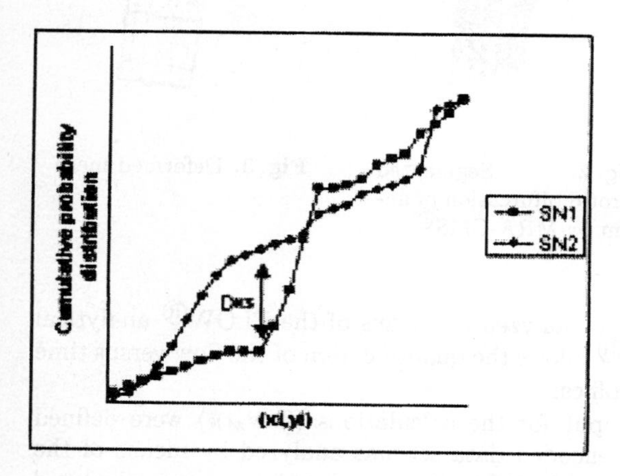

Fig. 4. D_{KS} gives the maximum difference between two arbitrary cumulative distributions SN1 and SN2 [6].

Prob_{KS} gives the significance level of the comparison. If prob_{KS} is larger than 0.20 the two datasets are not statistically significant different. If prob_{KS} equals 1, there is perfect correlation.

3 Results

The calculated flows and velocity profiles were compared with MR-measurements for the curved tube and the carotid phantom.

The results of the measurements and calculations of the curved tube are presented in figures 5 through 11. In figure 5 flow_{MR} is compared for the three measuring planes with flow_{calc}. In figure 6 flow_{image} is compared for different measuring planes to investigate the effect of gridding and interpolation. In figures 7 through 9 the velocity profiles at the three measuring planes are presented. The measurements and the simulations are plotted at the plane of symmetry for beginning, peak and end of systole, respectively (time slice nr 20, 24 and 5). The results of the KS-test D_{KS} and prob_{KS} are presented in figure 10 and 11, respectively.

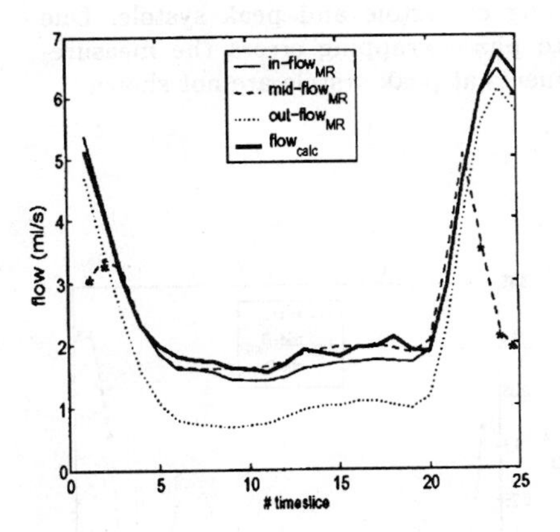

Fig. 5. Flow_{MR} and Flow_{calc} for inflow, central and outflow region of the curved tube. * Indicates a phase-wrapping for some pixels in the measurement.

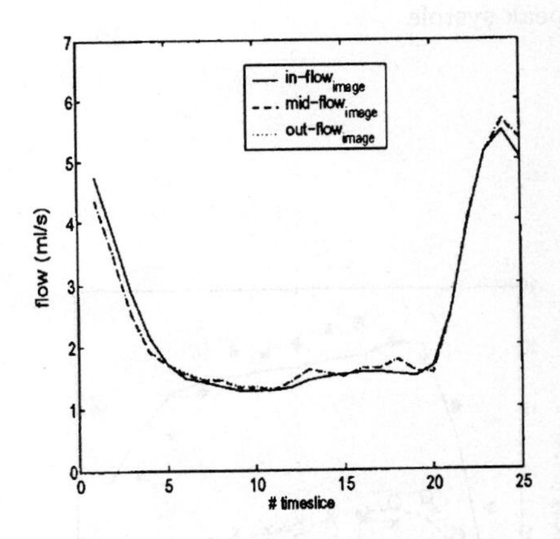

Fig. 6. Flow_{image} for inflow, central and outflow region of the curved tube.

For the carotid bifurcation the same procedure was used as for the curved tube. It has to be noted that some additional error sources are present here: 1) The outflow is stressfree, i.e. the probably present pressure difference between internal and external carotid is not taken into account; and 2) The MRA-CMS® package can segment single segments only. The segmented carotid bifurcation will therefore not exactly match the phantom (see figures 1 and 2) so the mesh

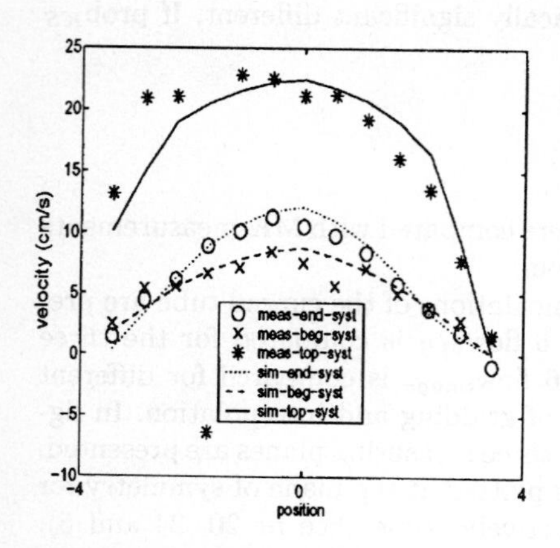

Fig. 7. Velocity profile in the plane of symmetry in the curved tube for inflow, at end systole, beginning of systole and peak systole

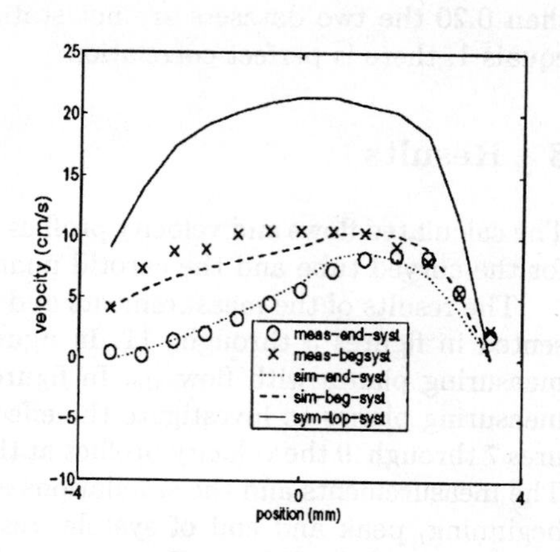

Fig. 8. Velocity profile in the plane of symmetry in the curved tube for the center of the bend, at end systole, beginning of systole and peak systole. Due to phase-wrapping errors the measurements at peak systole are not shown.

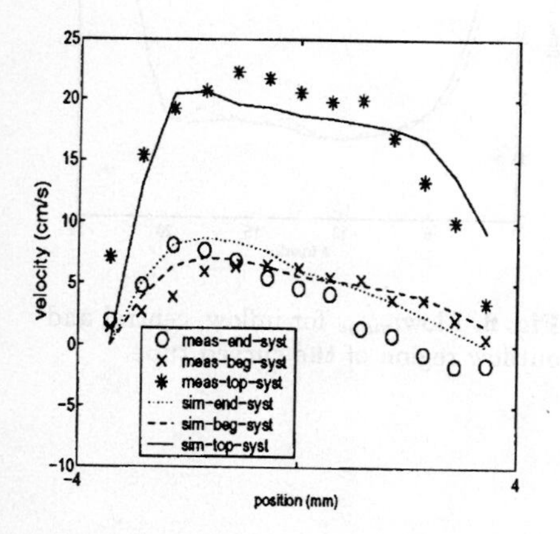

Fig. 9. Velocity profile in the plane of symmetry in the curved tube for outflow, at end systole, beginning of systole and peak systole.

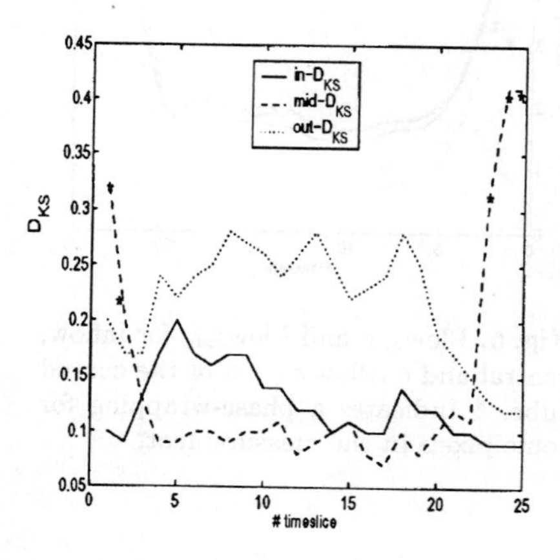

Fig. 10. D_{KS} values for inflow, central and outflow region of the curved tube. * Indicates a phase-wrapping for some pixels in the measurement.

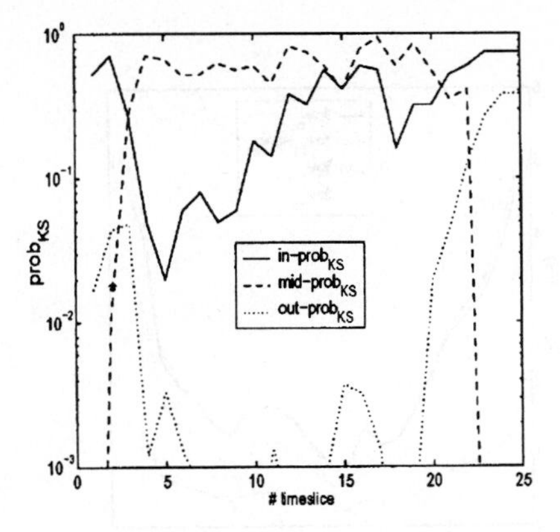

Fig. 11. prob$_{KS}$ values for inflow, central and outflow region of the curved tube. * Indicates a phase-wrapping for some pixels in the measurement.

cannot be optimal for the fluid dynamical calculations. For that reason, it is to be expected that the first results for the carotid bifurcation will not be as good as for the simple curved tube model. The carotid imaging planes were taken perpendicular to the inflow. The flow behind the bifurcation was split into two parts. One part for the external carotid and the other one for the internal carotid (see figures 12 and 13). Prob$_{KS}$ was very small (≤ 0.001) everywhere for the carotid. Even at the in-flow (which is assumed to be a Womersley profile) these small values were measured.

4 Discussion

The curved tube shows a good similarity between MR-measurements and calculations in the middle of the bend when the errors due to phase wrapping are excluded. For the inflow and the middle of the bend prob$_{KS}$ is almost everywhere above 0.20 indicating that the difference between measurements and calculations is not significant. The outflow does not match the other measurements and calculations. The in-flow and central-flow were taken in the Feet-Head direction and the outflow in the Left-Right direction. Perhaps this is causing the non-optimal out-flow-measurements, which seem to be of good quality at visual inspection. Out-FLOW$_{MR}$ gives lower values in all the time slices and also out-Prob$_{KS}$ indicates a poor similarity. In future research it will be inspected whether the direction of measurement has any effect on the results. Extra information for further analysis can probably be gathered from analysis of the secondary flows (vortices), which are also visible in the MR-data.

The carotid bifurcation demonstrates a difference between the calculations and the measurements as is illustrated in figure 13 and figure 12. This is under-

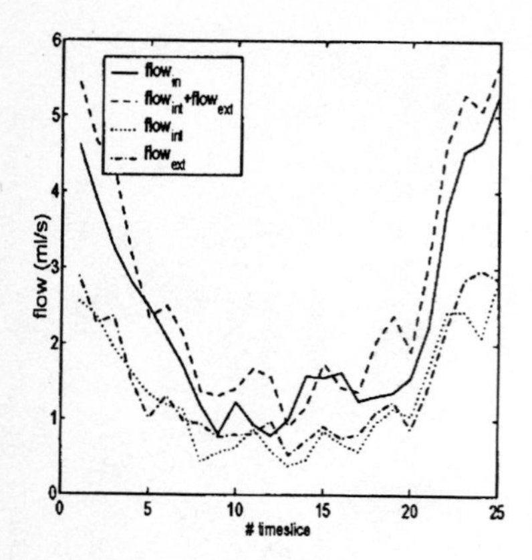

Fig. 12. Flow$_{MR}$. In the figure Flow$_{in}$ gives the inflow, and Flow$_{int}$ and Flow$_{ext}$ the flows in the internal and external carotid just behind the bifurcation respectively. Flow$_{int}$ + Flow$_{int}$ gives the total Flow$_{MR}$ measured behind the bifurcation.

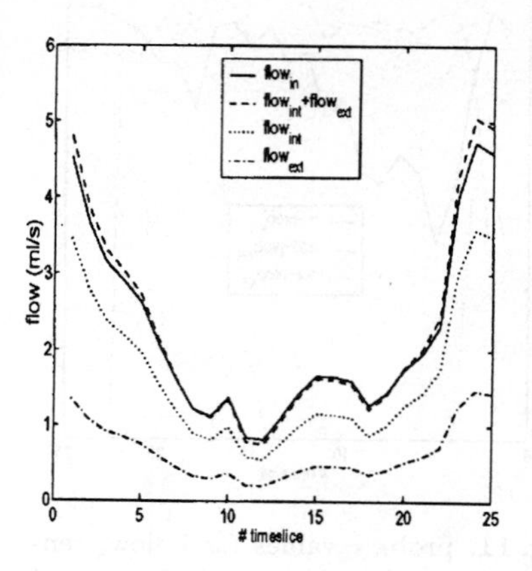

Fig. 13. Flow$_{calc}$. In the figure Flow$_{in}$ gives the inflow, and Flow$_{int}$ and Flow$_{ext}$ gives the flow in the internal and external carotid just behind the bifuraction respectively. Flow$_{int}$ + Flow$_{ext}$ gives the total Flow$_{calc}$ calculated behind the bifurcation.

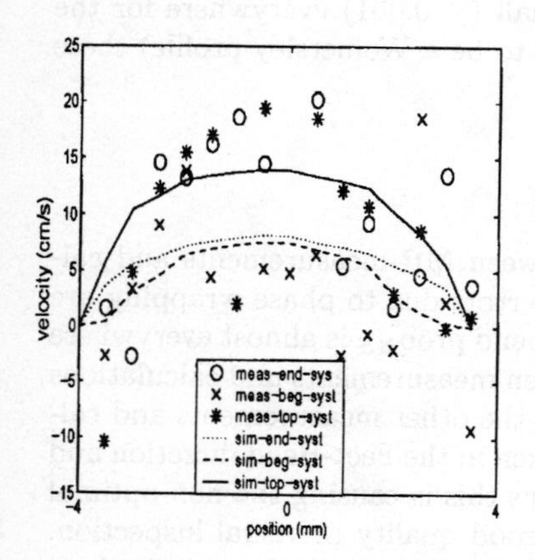

Fig. 14. The velocity of the inflow of the carotid is plotted for measurements and simulations for three time slices. For end of systole, beginning of systole and peak systole.

standable, because only the inflow conditions were pre-defined and no pressure or flow-profile for the outflow. The calculations for the carotid only indicate that the deformed mesh can be used for solving unsteady NS equations. For detailed error analysis additions have to be made in the near future. The MR-measurements suffered from much more noise than the measurements of the curved tube, which is shown in figure 14. Therefore even at the inflow it is seen that prob_{KS} is very small so that the KS data are not suitable. KS-analysis seems to be restricted to a certain noise level. With improved MRA bifurcation detection, measurements with less noise and prescribed outflow conditions, it is to be expected that KS-statistics in the carotid will become useable as well as for the curved tube.

5 Conclusion

The velocity profiles can be investigated in a quantitative manner with the KS-statistics and also be visualized individually. For the curved tube it was shown that the KS-statistics works well. Prob_{KS} is almost everywhere above 0.20 for the inflow and center regions, indicating that there is no statistically significant difference between measurements and calculations. The MRI flow measurements therefore are in good agreement with the calculated data. The computational method may be used to derive wall shear rates inside the chosen geometry [8, 4]. With the proper viscosity model for blood, the wall shear stresses may also be computed using (2) [4]. For noisy measurements KS-statistics are not suitable. The mesh deformation algorithm works fine and the deformed mesh can be used for fluid dynamical calculations of carotids. In summary it can be concluded that flow(t) and KS-results can indicate the amount of similarity between measurements and calculations. This approach opens the possibility for future *in-vivo* Wall Shear Stress measurements with MRI.

References

1. Box F.M.A., Spilt A., Van Buchem M.A., Reiber J.H.C., Van der Geest R.J.: Automatic model based contour detection and flow quantification of blood flow in small vessels with velocity encoded MRI. Proc. ISMRM 7, Philadelphia (1999), 571
2. Davies. P.F.: Flow-mediated endothelial mechanotransduction. Physiological Reviews, Vol 75, (1995) 519-560
3. Van der Geest R.J., Niezen R.A., Van der Wall E.E., de Roos A., Reiber J.H.C.: Automatic Measurements of Volume Flow in the Ascending Aorta Using MR Velocity Maps: Evaluation of Inter- and Intraobserver Variability in Healthy Volunteers. J. Comput. Assist. Tomogr. Vol. 22(6) (1998) 904-911
4. Gijsen, F.J.H., Allanic, E., Van de Vosse, F.N., Janssen, J.D.: The influence of non-Newtonian properties of blood on the flow in large arteries: unsteady flow in a 90 degrees curved tube. J. of Biomechanics. Vol.32(7), (1999) 705-713
5. Nichols W.W., O'Rourke M.F.: McDonald's Blood Flow in Arteries. Theoretical, experimental and clinical principles. Fourth edition. Oxford University Press, Inc. (1998) 36-40

6. Press W.H., Teukolsky S.A., Vetterling W.T., Flannery. B.R.: Numerical Recipes in C. Cambridge University Press (1988), 491
7. Press W.H., Teukolsky S.A., Vetterling W.T., Flannery. B.R.: Numerical Recipes in C. Cambridge University Press (1992), 645-64
8. Rutten M.C.M.: Fluid-solid interaction in large arteries. Thesis Technical University Eindhoven, the Netherlands (1998)
9. Schaap J.A., De Koning P.J.H., Van der Geest R.J., Reiber J.H.C.: 3D Quantification and visualization of MRA. Proc. 15th CARS (2000) 928-933
10. Segal G.: Ingenieursbureau SEPRA, Park Nabij 3, Leidschendam, the Netherlands
11. Johnson A., Tezduyar T.: Mesh update strategies in parallel finite element computations of flow problems with moving boundaries and interfaces. Computer methods in Applied Mechanics and Engineering, Vol 119, (1994) 73-94
12. Van de Vosse, F.N., Van Steenhoven, A.A., Segal, A., Janssen, J.D.: A finite element analysis of the steady laminar entrance flow in a 90 curve tube. Int. J. Num. Meth. In Fluids, Vol 9, (1989) 275-287
13. Womersley J.R.: An elastic tube theory of pulse transmission and oscillatory flow in mammalian arteries. Technical report, Wright Air Development Centre TR56-614, 1957.

High Performance Distributed Simulation for Interactive Simulated Vascular Reconstruction

Robert G. Belleman and Roman Shulakov

Section Computational Science, Faculty of Science, University of Amsterdam,
Kruislaan 403, 1098 SJ Amsterdam, the Netherlands.
(robbel|rshulako)@science.uva.nl

Abstract. Interactive distributed simulation environments consist of interconnected communicating components. The performance of such a system is determined by the execution time of the executing components and the amount of data that is exchanged between components. We describe an interactive distributed simulation system in the scope of a medical test case (simulated vascular reconstruction) and present a number of techniques to improve performance.

1 Introduction

Interactive simulation environments are dynamic systems that combine simulation, data presentation and interaction capabilities that together allow users to explore the results of computer simulation processes and influence the course of these simulations at run-time [4] (see also Fig. 1). The goal of these interactive environments is to shorten experimental cycles, decrease the cost of system resources and enhance the researcher's abilities for the exploration of data sets or problem spaces.

In a dynamic environment, the information presented to the user is regenerated periodically by the simulation process. The environment is expected to provide (1) a reliable and consistent representation of the results of the simulation at that moment and (2) mechanisms enabling the user to change parameters

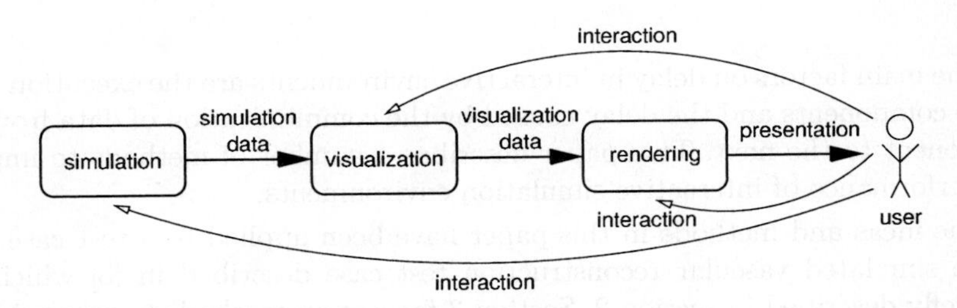

Fig. 1. Interactive simulation environments consist of a simulation, visualization and rendering component with which a user interacts to interactively explore data sets or problem spaces.

P.M.A. Sloot et al. (Eds.): ICCS 2002, LNCS 2331, pp. 265–274, 2002.

in the environment. An example of an application where this structure is used has previously been described in [5]. This test case environment allows vascular reconstruction procedures to be simulated using a fluid flow simulator that simulates the effect of a planned surgical procedure on a patient's blood circulation. An interactive immersive virtual reality environment allows a surgeon to study the effect of a reconstructive procedure and interactively explore alternatives for the best possible treatment for a specific patient.

1.1 Performance of interactive simulation environments

The most important factor in the performance of a dynamic simulation environment is *update time*; the delay between consecutive updates in the environment. Usability increases when update time is as short as possible, so for a highly responsive environment, delays should be minimized. Fig. 2 shows a schematic representation of the most typical delay imposing factors in a simple dynamic simulation environment.

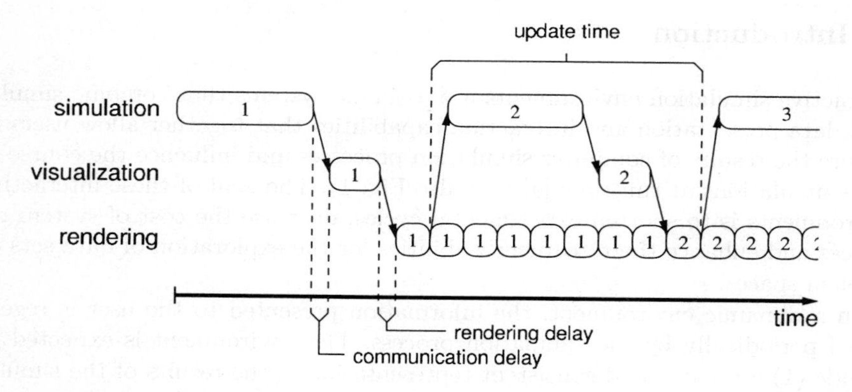

Fig. 2. Typical delays in a simple dynamic simulation environment.

The main factors on delay in interactive environments are the execution times of the components and the delay caused by the communication of data from one component to the next. This paper describes a number of methods to improve the performance of interactive simulation environments.

The ideas and methods in this paper have been applied to a test case, that of the simulated vascular reconstruction test case described in [5] which will be briefly described in section 2. Section 3 focuses on methods to maximize the performance of an interactive simulation environment. Section 4 presents some of the results obtained thus far. Finally, conclusions and future work are described in section 5.

2 Test case: interactive simulated vascular reconstruction in a virtual operating theater

The performance considerations described in the previous section are validated by analysis of a prototypical case study of *simulated vascular reconstruction*. This application combines simulation, visualization and interaction in an exemplary fashion. By a detailed analysis of the spatial and temporal characteristics of this test case we attempt to recognize generic elements for the design of an interactive simulation environment. Before we describe the methods that have been implemented to obtain a high performance interactive simulation environment, we begin with a description of the test case.

2.1 Introduction: interactive simulated vascular reconstruction

The purpose of vascular reconstruction is to redirect and augment blood flow or perhaps repair a weakened or aneurysmal vessel through a surgical procedure. The optimal procedure is often obvious but this is not always the case, for example, in a patient with complicated or multi-level disease. Pre-operative surgical planning will allow evaluation of different procedures *a priori*, under various physiological states such as rest and exercise, thereby increasing the chances of a positive outcome for the patient. The aim of this case study is to provide a surgeon with an environment in which he/she can explore the effect of a number of different vascular reconstruction procedures before it is put to practice. Our approach combines parallel flow simulation, interactive virtual reality and high performance computing and networking techniques into an interactive dynamic exploration environment that together allows human-in-the-loop types of experimentation [5].

2.2 Blood flow simulation: the lattice-Boltzmann method

The lattice-Boltzmann method (LBM) is a mesoscopic approach for simulating fluid flow based on the kinetic Boltzmann equation [7]. In this method fluid is modeled by particles moving on a regular lattice. At each time step, particles propagate to neighboring lattice points and re-distribute their velocities in a local collision phase. This inherent spatial and temporal locality of the update rules makes this method ideal for parallel computing [11]. During recent years, LBM has been successfully used for simulating many complex fluid-dynamical problems, such as suspension flows, multi-phase flows, and fluid flow in porous media [13]. All these problems are quite difficult to simulate by conventional methods [10, 12].

The data structures required by LBM (Cartesian grids) bare a great resemblance to the grids that come out of CT and MRI scanners. As a result, the amount of preprocessing can be kept to a minimum which reduces the risk of introducing errors due to data structure conversions. LBM has the benefit over other fluid flow simulation methods that flow around (or through) irregular geometries (like a vascular structure) can be simulated relatively easy. In addition, velocity fields, pressure and shear stress on the arteries can be calculated directly from the densities of the particle distributions [2]. This may be beneficial in cases where we want to interfere with the simulation while the velocity and the stress field are still developing, thus supporting fast data-updating given a proposed change in simulation parameters as a result of user interaction.

2.3 High performance interactive visualization

The simulated vascular reconstruction operating theater provides visualization and interaction methods to simulate a vascular reconstruction procedure and visualize the effect of that procedure on a patient's blood circulation in real time. The environment can be used in a CAVE [8] but also on low cost commodity hardware in conjunction with a projection display and tracking hardware [6]. Multi-modal interaction methods such as speech recognition, hand gestures, direct manipulation of virtual 3D objects and measurement tools allow researchers to explore simulation results [3].

3 High performance interactive simulation

The responsiveness of an interactive system is directly related to the rate at which updates are generated by each of the components in the system. To increase responsiveness, the delays between the consecutive components in the interactive system should be minimized. The accumulation of all delays is referred to as "update time". In an interactive system there will always be some delay from the moment interaction takes place until the moment that the environment has reacted to this interaction. This delay is referred to as "response time".

3.1 Update and response time

In a non-interactive environment, the update time T_U is the sum over the execution time for the different components (T_{sim} for simulation, T_{vis} for visualization and T_{ren} for rendering) and the communication delay between components ($T_{sim \rightarrow vis}$ and $T_{vis \rightarrow ren}$):

$$T_U = T_{sim} + T_{sim \rightarrow vis} + T_{vis} + T_{vis \rightarrow ren} + T_{ren}. \qquad (1)$$

Decreasing update time means that the delays imposed by the different components must be minimized. In the case of executing components this means that the time between the acceptance of input data and the production of results should be minimized. In the case of communication between components, the dominating factor on delay is the time that is required to transfer data from one component to the next.

In an interactive system, the response time T_R depends on which component the interaction is directed to since only this and subsequent components need to be updated. In case the user interacts with the simulation component, the response time will be $T_R = T_{i(sim)} + T_U$, where $T_{i(sim)}$ is the time required to "apply" the interaction to the simulation component.

3.2 Distributed pipelined execution

A dynamic system differs from a static system in that the simulation component is an iterative process that repeatedly produces (intermediate) results. Basically, the delay at which these intermediate results become available is given by equation (1). However, since the simulation component has no dependency on the execution of subsequent components, the update time of the whole environment can benefit from a pipelined execution model. In this model, a component resumes execution as soon as its output data has been accepted by the next (see Fig. 3). In this case simulation, visualization and rendering execute in parallel.

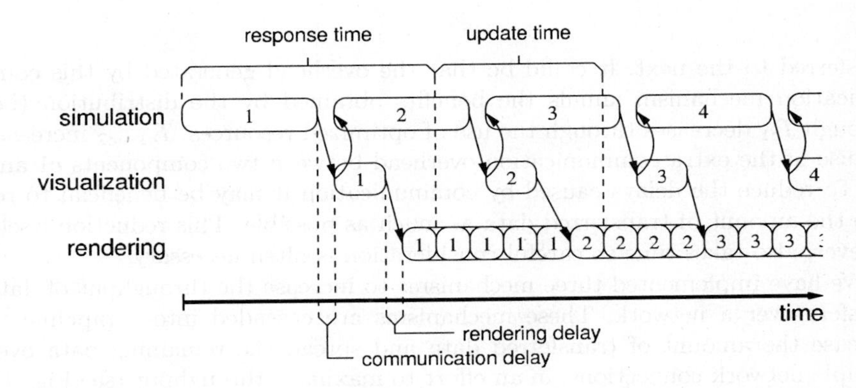

Fig. 3. Response time, update time and delays in a pipelined interactive simulation environment.

Once all components in the pipeline have executed at least once, and provided sufficient resources are available to execute all components simultaneously, the update time becomes

$$T_U = max(T_{sim} + T_{sim \to vis}, T_{vis} + T_{vis \to ren}, T_{ren}). \qquad (2)$$

Although the components in Fig. 1 show a close coupling between them, it is not necessary, or even beneficial, to execute all on the same computing system. It is often more efficient to execute the components on systems that have optimized resources available for the most time consuming type of operations. The flow simulation component described in section 2, for example, runs more efficient on a parallel system, while the visualization component performs better on a system with optimized visualization libraries and the rendering component performs better on a system with specialized graphics hardware on board.

3.3 High performance network communication

Distributing components over different systems means that some form of communication must be established to allow the output of one component to be

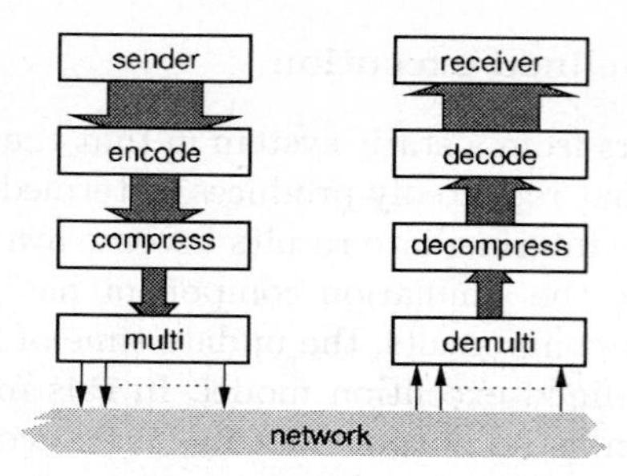

Fig. 4. The stages in the communication pipeline.

transferred to the next. It could be that the overhead generated by this communication mechanism annuls the benefits obtained by the distribution (i.e. although T_{c1} decreases through the use of optimized resources, $T_{c1 \to c2}$ increases because of the extra communication overhead between two components $c1$ and $c2$). To reduce the delays caused by communication it may be beneficial to reduce the amount of transferred data as much as possible. This reduction itself, however, also takes time so careful consideration is often necessary.

We have implemented three mechanisms to increase the throughput of data transfers over a network. These mechanisms are cascaded into a pipeline to decrease the amount of transferred data and spread the remaining data over multiple network connections, in an effort to maximize throughput (see Fig. 4).

Hiding latency by using multiple connections. This method increases communication throughput by using multiple network connections between peers at the same time [9]. There are two reasons why this increases throughput (see also Fig. 5). First; in the case of reliable network connections (such as connections using the TCP protocol), delays caused by waiting for acknowledgment packets can be "hidden" by serving other connections that are ready. Second; the technique can exploit "intelligent" network devices that spread communication over different routes. This technique therefore performs best when there are many such devices between peers (which is often the case when peers are geographically dispersed). In principle, data can travel along different routes from peer to peer, thereby circumventing congestion caused by other traffic on the network.

Note that the total volume of data that is transferred between peers is not decreased by this method. Instead, it increases throughput by exploiting as much available bandwidth as possible. The techniques described next aim to increase throughput by decreasing the volume of communicated data.

Data encoding. Data encoding is a data specific type of data reduction technique that aims to reduce the volume that is needed to represent the data to a minimum. This technique relies on the fact that the receiving side may not always be interested in the most accurate representation of the data that was

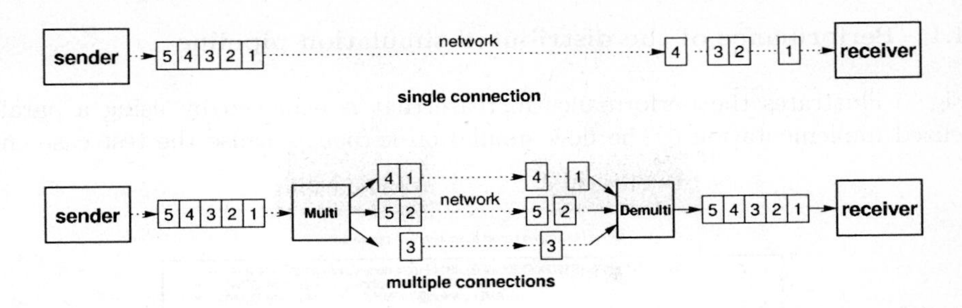

Fig. 5. Increasing communication throughput using multiple network connections.

calculated by the sender. Because this type of data reduction throws away unnecessary information, this method is frequently referred to as *lossy compression.*

Although a significant reduction in volume can be achieved using this technique, the receiver should be conscious of the fact that the data it has received is not of the same accuracy as was originally produced. Although this reduction may be acceptable under some circumstances, unexpected side effects may occur when the errors that are present in the data are accumulated due to an integrating method of analysis on the data. For example, an often used technique in vector field visualization is to represent the path of the flow using streamlines. These streamlines are created by integrating over individual vectors. Due to the accumulation of (small) errors in each individual vector, a streamline may follow a radically different path.

Data compression. Freely available compression libraries (such as *zlib* [1]) provide means of reducing data volumes very effectively. This data reduction does not make any assumption about the data contents and is therefore without any loss of information. Most compression libraries can be parameterized to indicate the level of compression that should be achieved at the expense of higher execution time. This type of data reduction is commonly referred to as *lossless compression* as the data is unchanged after decompression. The amount of compression depends on (1) the type of data and (2) the amount of data. In general, the compression ratio decreases when the amount of data decreases. Note that this is important in parallelized applications in which the original data volume is often decomposed into smaller subvolumes.

4 Results

The simulated vascular reconstruction environment described in section 2 has been implemented using the methods described in section 3. In this section we show some preliminary results on the performance of the environment.

4.1 Performance of the distributed simulation pipeline

Fig. 6 illustrates the performance increase that is achieved by using a parallelized implementation of the flow simulation kernel. Because the test case en-

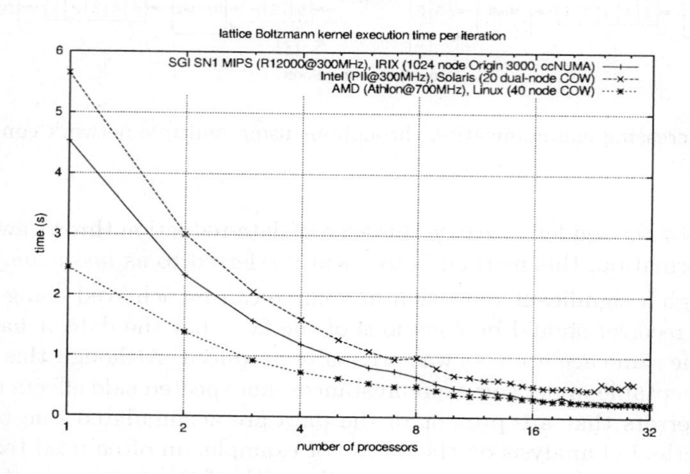

Fig. 6. Iteration time of the parallel lattice Boltzmann kernel on different multiprocessor systems.

vironment's architecture uses the distributed pipeline architecture described in section 3.2 and because the simulation component is, in general, the slowest executing component, the performance of the simulation environment in total is greatly increased (as shown by equation 2).

4.2 Performance of the network communication pipeline

Fig. 7 shows the mean throughput over 200 measurements of the multiple connection stage in the communication pipeline. As can be seen from this figure, the average throughput increases as more connections are used, but up to a maximum. Using more connections congests the network and no further increase in throughput can be obtained.

Fig. 8 shows the mean performance over 5 measurements of the complete network communication pipeline on a typical medical data set; using compression, multiple network connections, both with and without encoding. This figure illustrates that, on average, encoding doubles throughput. Although this figure shows the typical throughput that can be achieved, in some situations the total performance of the network communication pipeline resulted in a throughput of 62 Mbyte/s, which is over 5 times the bandwidth of the slowest network link (100 Mbit/s).

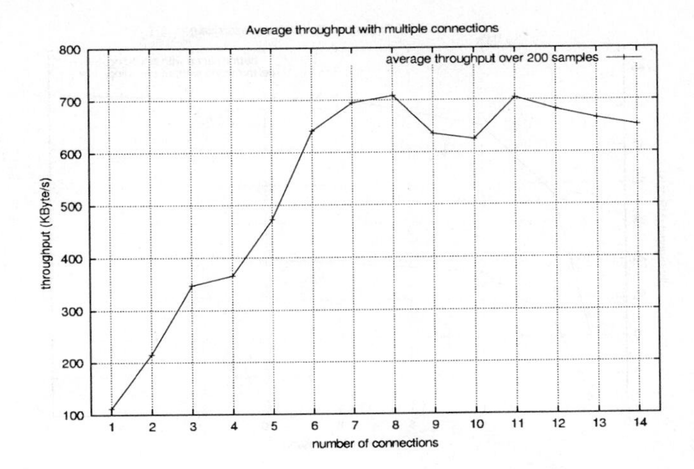

Fig. 7. Average network throughput when using multiple network connections on a 100 MBit/s link.

5 Conclusions and future work

Preliminary tests show a great performance increase over earlier versions of the interactive simulated vascular reconstruction environment. The results presented here were obtained using unoptimized algorithms; we expect future versions to increase performance even more.

The different stages in the communication pipeline require a certain amount of time to execute which adds delay to communication time. By tuning the parameters that influence this execution time, throughput can (in principle) be automatically optimized. For example; networks are generally shared by many institutes so that available bandwidth changes over time. The multiple connection technique can sense this change by measuring the effect of employing more or fewer connections during communication. An optimal number of connections can thus be determined dynamically (and transparently), while peers communicate. However, a different parameterization at one stage influences the execution time of subsequent stages which implies that parameter optimization is not a trivial task.

References

1. zlib homepage, 2001. On the web: `http://www.gzip.org/zlib/`.
2. A.M. Artoli, D. Kandhai, A.G. Hoekstra, and P.M.A. Sloot. Accuracy of shear stress calculations in the lattice Boltzmann method. Accepted for the 9th International Conference on Discrete Simulation of Fluid Dynamics.
3. R.G. Belleman, J.A. Kaandorp, D. Dijkman, and P.M.A. Sloot. GEOPROVE: Geometric probes for virtual environments. In P.M.A. Sloot et al., editors, *High Performance Computing and Networking (HPCN'99)*, pages 817–827, Amsterdam, 1999. Springer-Verlag.

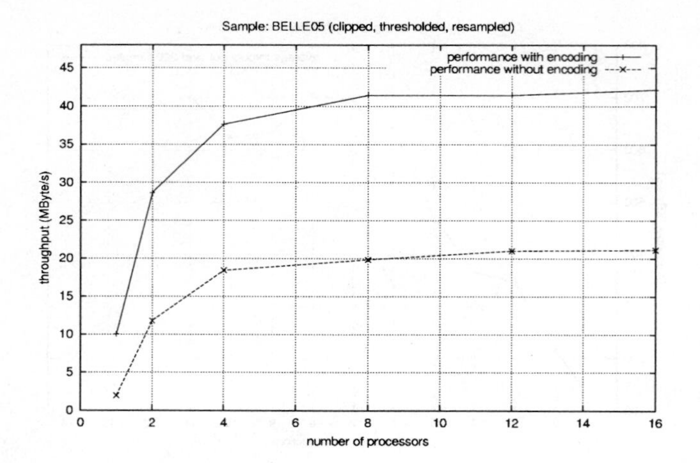

Fig. 8. Mean throughput of the network communication pipeline when used with the parallel LBM simulation kernel (with compression, shown with encoding and without encoding.

4. R.G. Belleman and P.M.A. Sloot. The design of dynamic exploration environments for computational steering simulations. In Marian Bubak et al., editors, *SGI Users' Conference*, pages 57–74, Kraków, 2000. CYFRONET AGH.
5. R.G. Belleman and P.M.A. Sloot. Simulated vascular reconstruction in a virtual operating theatre. In H.U. Lemke et al., editors, *CARS (Excerpta Medica, ICS-1230)*, pages 938–944, Berlin, 2001. Elsevier Science B.V.
6. R.G. Belleman, B. Stolk, and R. de Vries. Immersive virtual reality on commodity hardware. In R.L. Lagendijk et al., editors, *Seventh annual ASCI conference*, pages 297–304, Heijen, Netherlands, 2001. Advanced School for Computing and Imaging.
7. S. Chen and G.D. Doolen. Lattice Boltzmann method for fluid flows. *Annu. Rev. Fluid Mech.*, 30:329, 1998.
8. C. Cruz-Neira, D.J. Sandin, and T.A. DeFanti. Surround-screen projection-based virtual reality: The design and implementation of the CAVE. In *SIGGRAPH '93 Computer Graphics Conference*, pages 135–142. ACM SIGGRAPH, August 1993.
9. Open Channel Foundation. CAVERNsoft G2, A toolkit for high performance tele-immersive collaboration, 2001. On the web: `http://www.openchannelsoftware.org/projects/CAVERNsoft_G2/`.
10. D. Kandhai. *Large Scale Lattice-Boltzmann Simulations (Computational Methods and Applications)*. PhD thesis, Universiteit van Amsterdam, Amsterdam, the Netherlands, 1999.
11. D. Kandhai, A. Koponen, A.G. Hoekstra, M. Kataja, J. Timonen, and P.M.A. Sloot. Lattice Boltzmann hydrodynamics on parallel systems. *Computer Physics Communications*, 1998.
12. D. Kandhai, D. Vidal, A. Hoekstra, H. Hoefsloot, P. Iedema, and P.M.A. Sloot. Lattice-Boltzmann and finite element simulations of fluid flow in a SMRX mixer. *Int. J. Numer. Meth. Fluids*, 31:1019–1033, 1999.
13. A. Koponen, D. Kandhai, E. Hellen, M. Alava, A. Hoekstra, M. Kataja, K. Niskanen, P. Sloot, and J. Timonen. Permeability of three-dimensional random fiber webs. *Physical Review Letters*, 80(4):716–719, January 26 1998.

Fluid-Structure Interaction Modelling of Left Ventricular Filling

Pascal R. Verdonck, Jan A. Vierendeels

Institute Biomedical Technology
Ghent University
Sint-Pietersnieuwstraat 41
9000 Ghent, Belgium

Abstract. Non-invasive diagnosis of diastolic dysfunction remains difficult in clinical practice. Non-invasive assessment of the flow field within the left ventricle (LV) using color M-mode Doppler (CMD) echocardiography provides a potential technique that can differentiate between the normal and diseased heart.
A computer model is developed describing three-dimensional axi-symmetrical LV filling flow. The simulation results show that the hydrodynamical mechanism of LV flow wave propagation, as observed on 2D color and color M-mode Doppler echocardiograms, is the propagation of a vortex in the LV cavity.

1 Introduction

Recently color M-mode Doppler echocardiography (CMD) has been proposed as a useful method for the evaluation of LV flow wave propagation as an index of LV filling [1-3]. The objective of this study is to obtain insight in the hydrodynamics of the color M-mode Doppler flow propagation velocity v(p) based on a mathematical framework. For the development of fluid-structure interaction model of the left ventricular filling the unsteady Navier-Stokes flow equations are solved in a LV truncated ellipsoid geometry with moving LV walls including relaxation and compliance of the wall. The computed results confirm both intraventricular flow and pressure patterns during filling. Vortices are formed during the acceleration phases of the early and atrial filling waves. During the deceleration phases the vortices are amplified and convected into the ventricle. The vortices are recognized on the derived 2D color echocardiograms as in vivo. The propagation of this vortex determines the propagation of the maximum velocity observed in the color M-mode Doppler echocardiogram.

Peskin [4] was the first to model the blood-ventricular wall interaction using the immersed boundary method. Owen [5] developed a 1D model showing the importance of intraventricular pressure wave propagation during early filling. In our previous work, a 1D model was constructed in order to simulate intraventricular pressure gradients in the LV during filling [6,7]. The present model, based on a 3D axisymmetrical computational approach [8], evaluates the fluid dynamics of the LV

P.M.A. Sloot et al. (Eds.): ICCS 2002, LNCS 2331, pp. 275–284, 2002.

filling. The interaction between the LV wall and the blood is taken into account with a validated algorithm [8].

The calculated flow patterns in the LV are transformed into the format of 2D color and color M-mode Doppler echocardiograms as used in clinical practice. In this chapter we are studying in particular the relation between vortex center and position of maximum velocity along the centerline for both the early filling (E-wave) and atrial contraction wave (A-wave).

2 Computer model

2.1 Intraventricular flow model

The fluid domain of the LV is described by means of an axi-symmetrical model. The unsteady Navier-Stokes equations expressing conservation of momentum and the continuity equation expressing conservation of mass are solved in a LV truncated ellipsoid geometry with moving walls. The computation is started at the onset of LV relaxation. When the LV pressure drops below the atrial pressure, the mitral valve opens immediately. From this moment, a mitral velocity pattern is applied at the circular orifice (base) of the LV. It is assumed that forces of the mitral valve on the fluid are small and can be neglected once the mitral valve opens. After opening, pressure in the LV is determined by both the relaxation and compliance of the LV wall and the dynamics of the blood flow into the LV cavity. The computation ends before the LV contraction.
The mitral flow velocity pattern used as boundary condition and atrial pressure are derived from a separate calculation with Meisner's lumped parameter model [9].

2.2 Left ventricular wall model

The LV wall is described by a truncated ellipsoid in the zero stress state. At the zero stress state and with blood at rest, the transmural pressure is zero. The zero stress state is assumed to correspond with a cavity volume of 12 ml, diameter of mitral annulus of 1.5 cm and base to apex distance of 4 cm. These are physiological relevant parameters for a canine heart for which the model was validated [8].
Away from the zero stress state, the shape of the LV is computed from equilibrium equations for the LV wall. These equilibrium equations involve the circumferential and longitudinal cardiac stresses, the curvature of the heart wall and the transmural pressure difference. A non-linear extension of the thin shell equations is used [8]. The position of the mitral valve annulus is kept fixed.

2.3 Boundary conditions

We have used Meisner's model [9] at a heart rate of 80 beats per minute in order to obtain the necessary boundary conditions for our 3D model calculations: end systolic volume of 18 ml, end systolic pressure of 75 mmHg, time constant of relaxation τ_v=30 ms, left atrial pressure at opening of the mitral valve 6.75 mmHg and the computed mitral velocity pattern as shown in figure 1.

2.4 Coupling procedure

The coupling of the heart wall displacement and the LV filling dynamics is based on an iterative approach [8]. For each time step (1.5 ms), the procedure alternates between the following steps: 1) calculation of the movement of the LV wall, for a known intraventricular pressure distribution along the wall, 2) calculation of the flow field inside the moving ventricle, which results in an updated velocity pattern and a new approximation of the pressure field to be used in step 1) until convergence for the time step is achieved.

2.5 Computation

Given the cardiac wall properties together with the circulation characteristics, the cardiac contraction and relaxation are the driving forces for the computation. First, the mitral inflow pattern, ESV and ESP and the left atrial pressure at opening of the mitral valve are computed with Meisner's model. Then, with the present model and with the same characteristics for the LV, the wall motion and the intraventricular pressure and flow patterns are computed during filling respecting the conservation of mass and momentum of the blood flow and the equilibrium equations for the heart wall. All details of the model can be found in [8].

3 Results

3.1 Intraventricular pressure-time curves

Figure 1 shows calculated intraventricular pressures in function of time (bottom panel) for the imposed mitral velocity pattern (top panel). Pressures curves at base (full line) and at a position inside the LV 4 cm from the base (dotted line) are compared. The pressure rise during early filling first occurs at the apex and then at the base. At the onset of atrial contraction, there is first a pressure rise at the base and then at the apex.

3.2 Simulated 2D color Doppler echocardiograms

Figure 2 shows the corresponding flow velocity patterns and derived 2D color Doppler images at six different moments in time LV during filling. These moments are indicated with a black spot on the mitral velocity pattern: (1) before the immediate opening of the mitral valve, (2) at peak E velocity, (3) during deceleration of E-wave, (4) at diastasis, (5) at peak A velocity and (6) during deceleration of the A-wave. The flow pattern is shown and corresponds to the instantaneous velocity vectors in the LV cavity. The computed flow velocities are presented in a 2D color echo format for each of the six moments in the lower segment.

3.3 Vortex propagation

The filling is characterized by different vortex movements. A first vortex is formed before the maximum of the early filling wave occurs at the mitral annulus (figure 2, panel 2). The vortex is amplified during the deceleration phase of the early filling wave and moves into the LV towards the apex. Between E- and A-wave (diastasis) one vortex can be seen in the whole LV (panel 3). At the base a second vortex can be detected, with the direction of rotation opposite to the first vortex (panel 4). During the acceleration phase of the A-wave the second vortex grows, while a third vortex is formed, comparable to the first one (panel 5). During the deceleration phase of the A-wave the second vortex dissappears. The first vortex is now located at the apex and the third one fills the basal region (panel 6). The vortices can be recognized in the 2D color echocardiograms by the presence of the transition of red to blue color near the cardiac wall, indicating backflow during filling.
In figure 3 (left panel), the positions of the vortex center and the positions of the maximum flow velocity along the centerline as a function of time are compared for both the E- and the A-wave. The right panel of figure 3 compares the position of the maximum flow velocity and the position of the vortex for both the E- and the A-wave.

3.4 Simulated color M-mode Doppler Echocardiogram (CMD)

Figure 4 shows a CMD along the symmetry axis, calculated by the present model corresponding to the velocities in figure 3. Both the early and atrial filling wave can be seen. During the initial flow phase blood moves almost simultaneously in the whole LV, which can alse be seen in the top part of figure 3, panel 2. This phase is denoted as I in figure 4. The movement is caused by the passing of the pressure filling wave. The filling causes the generation of a vortex, as described earlier. The propagation of this vortex determines the propagation of the maximum velocity along the symmetry axis of the LV as can be seen in figures 3 and 4 (vortex propagation wave). The propagation of the maximum velocity or the propagation of the vortex is seen in the CMD and is denoted as phase II in figure 4.

4 Discussion

A computer model is developed describing three-dimensional axisymmetrical LV filling flow. The calculated pressure-time curves in figure 1 confirm the existence of intraventricular pressure gradients. These gradients are also observed in dog studies [11] and in a 1D computer simulation of LV filling [6,7]. The minimal pressure in the apex is achieved shortly after the arrival of the pressure filling wave at the apex (denoted as I on the CMD in figure 4). At that moment the pressure rises at the apex since the pressure wave is reflected (reflection at a closed end) as already described by Owen [5].
Vortex formation during filling is an experimentally observed phenomenon [12-14] and according to Bellhouse [12] the presence of a vortex ring provides a mechanism for early valve closure in diastole. More recently, Steen and Steen [14] showed in vitro that at the start of filling, blood moves simultaneously at all positions, behaving as an incompressible fluid column (phase I). Then they observed a flow wave propagating from the mitral orifice towards the apex (phase II). In this phase, a ring vortex was seen to travel from the orifice towards the apex. Both phases are clearly observed in our simulation (figure 4). Shortland et al. [15] studied vortex formation and travelling in a skeletal muscle ventricle. They suggested that vortex travelling is an important feature to decrease the residence time of blood cells in the apical region. It is clear from figure 3 that the propagation of the maximum blood velocity inside the LV, as it can be measured by CMD echocardiography, is associated to vortex propagation.
For the calculation shown in figure 4, the maximal blood velocity during the early filling wave is 70 cm/s and v(p) of the filling wave, which corresponds with the propagation velocity of the ring vortex, is 45 cm/s. The ratio is 1.56. This hemodynamic behaviour was also observed in vitro experiments [14] and in vivo measurements [1-3,16].

5 Conclusion

A computer model is developed describing three-dimensional axisymmetrical LV filling flow. The simulation results show that the hydrodynamical mechanism of LV flow wave propagation, as observed on 2D color and color M-mode Doppler echocardiograms, is the propagation of a vortex in the LV cavity.

Acknowledgements

The research reported here was granted by contract GOA-95003 of the concerted action programme of the Ghent university, supported by the Flemish government.

6 References

[1] Brun P, Tribouilloy C, Duval AM, Iseriu L, Meguira A, Pelle G, Dubois-Randé JL. Left ventricular flow propagation during early filling is related to wall relaxation: a color M-mode Doppler analysis. J. Am. Coll. Cardiol. 1992;20(2):420-32.

[2] Stugaard M, Risöe C, Halfdan I, Smiseth OA. Intracavitary filling pattern in the failing left ventricle assessed by color m-mode doppler echocardiography. J. Am. Coll. Cardiol. 1994;24(3):663-670.

[3] Takatsuji H, Mikami T, Urasawa K, Teranishi J-I, Onozuka H, Takagi C, Makita Y, Matsuo H, Kusuoka H, Kitabatake A. A new approach for evaluation of left ventricular diastolic function: Spatial and temporal analysis of ventricular filling flow propagation by color M-mode Doppler echocardiography. J. Am. Coll. Cardiol. 1996; 27(2):365-71.

[4] Peskin CS, McQueen DM. A three-dimensional computational method for blood flow in the heart I. immersed elastic fibers in a viscous incompressible fluid. J. Comp. Phys. 1989;81:372-405.

[5] Owen A. A numerical model of early diastolic filling: importance of intraventricular pressure wave propagation. Cardiovasc. Res. 1993;27:255-261.

[6] Vierendeels J, Verdonck P, Dick E. Intraventricular Pressure Gradients and the Role of Pressure Wave Propagation. J. Cardiovasc. Diagn. P. 1997;14(3):147-152.

[7] Verdonck PR, Vierendeels J, Riemslagh K, Dick E. Left Ventricular Pressure Gradients: a 1D Computer Model Simulation. Med. & Biol. Eng. & Comput. 1999;37(4):511-516.

[8] Vierendeels JA, Riemslagh K, Dick E, Verdonck PR. Computer Simulation of Intraventricular Flow and Pressure Gradients during Diastole, ASME J. Biomech. Eng. 2000;122(6):667-674.

[9] Meisner J. Left atrial role in left ventricular filling: dog and computer studies. Ph.D. thesis Albert Einstein College of Medicine. Yeshiva University, New York. 1986.

[10] Nishimura RA, Tajik AJ. Evaluation of diastolic filling of left ventricle in health and disease: Doppler Echocardiography is the clinician's Rosetta stone. J. Am. Coll. Cardiol. 1997;30:8-18.

[11] Courtois M, Kovács SJ Jr., Ludbrook PA. Transmitral pressure-flow velocity relation: importance of regional pressure gradients in the left ventricle during diastole. Circulation 1988;78:661-671.

[12] Bellhouse BJ. Fluid mechanics of a model mitral valve and left ventricle. Cardiovasc. Res. 1972;6:199-210.

[13] Lee CSF, Talbot L. A fluid-mechanical study of the closure of heart valves. J. Fluid. Mech. 1979;91(1):41-63.

[14] Steen T, Steen S. Filling of a model left ventricle studied by color M mode Doppler. Cardiovasc. Res. 1994;28:1821-1827.

[15] Shortland AP, Black RA, Jarvis JC, Henry FS, Iudicello F, Collins MW, Salmons S. Formation and travel of vortices in model ventricles: application to the design of skeletal muscle ventricles. J. Biomech. 1996;29:503-511.

[16] Greenberg NL, Vandervoort PM, Thomas JD. Instantaneous diastolic transmitral pressure difference from color Doppler M-mode echocardiography. Am. J. Physiol. 1996;271 (Heart Circ. Physiol. 40):H1267-H1276.

[17] Iudicello F, Henry FS, Collins MW, Salmons S, Sarti A, Lamberti C. Comparison of haemodynamic structures between a skeletal muscle ventricle and the human left ventricle, Internal Medicine, 1997;5:1-10.

[18] Nikolic S, Fenely M, Pajaro O, Rankin JS and Yellin E. Origin of regional pressure gradients in the left ventricle during early diastole. American Journal of Physiology, 1995;268:550-557.

Legends

Figure 1. Top: Velocity profile at mitral valve. Bottom: Computed pressures at mitral valve and at a position inside the LV 4 cm from the mitral valve.

Figure 2. Flow patterns and calculated 2D color Doppler echocardiograms at six different time steps during filling: before opening of mitral valve, peak E velocity, deceleration of E-wave, diastasis, peak A velocity and deceleration of the A-wave

Figure 3. Relation between vortex center and position of maximum velolicty along the centerline for both E- and A-wave.

Figure 4. Color M-mode Doppler echocardiogram derived from a 2D axisymmetrical flow simulation. The scanline corresponds with the axis of symmetry. Both the early and atrial filling wave can be seen, the initial flow phase with blood moving simultaneously in the LV is denoted as I (pressure filling wave), the propagation of the vortex wave is denoted as II.

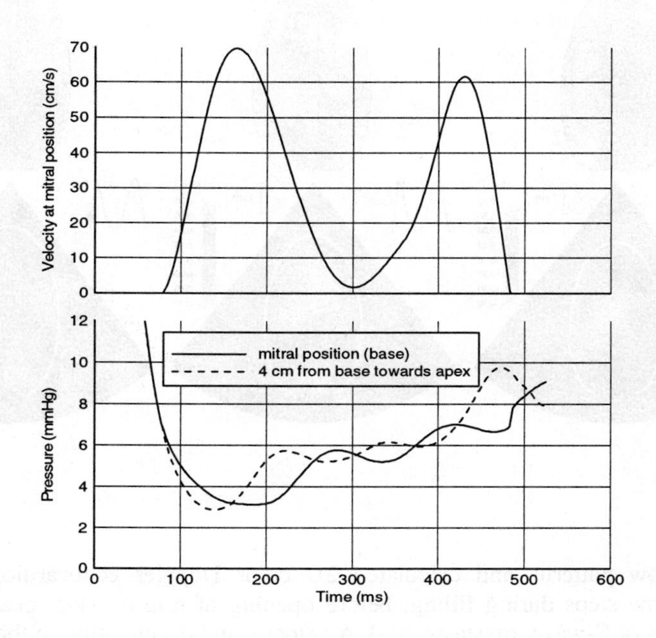

Fig. 1. Top: Velocity profile at mitral valve. Bottom: Computed pressures at mitral valve and at a position inside the LV 4 cm from the mitral valve.

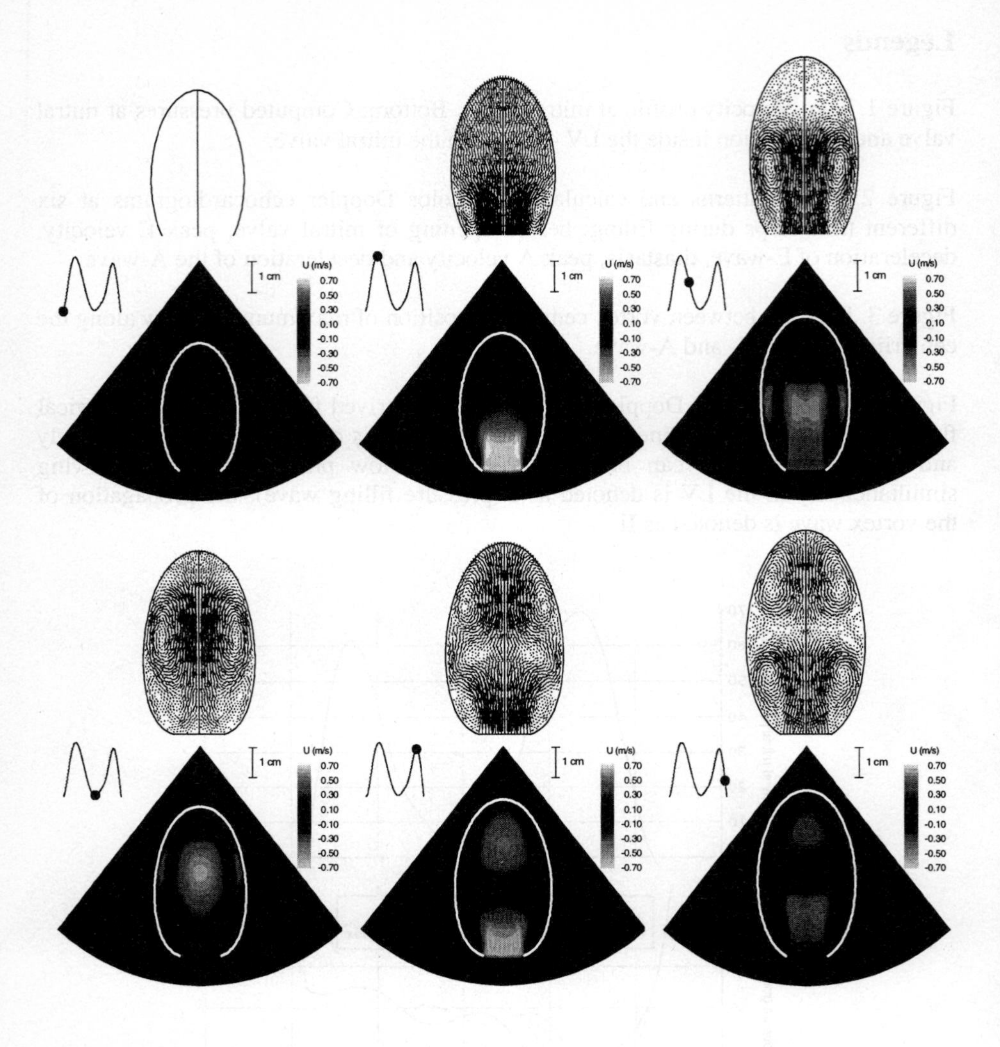

Fig. 2. Flow patterns and calculated 2D color Doppler echocardiograms at six different time steps during filling: before opening of mitral valve, peak E velocity, deceleration of E-wave, diastasis, peak A velocity and deceleration of the A-wave

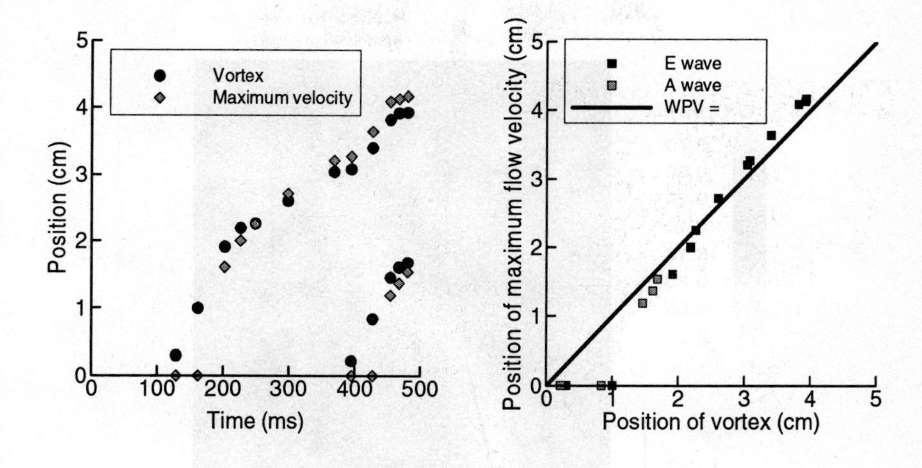

Fig. 3. Relation between vortex center and position of maximum velolicty along the centerline for both E- and A-wave.

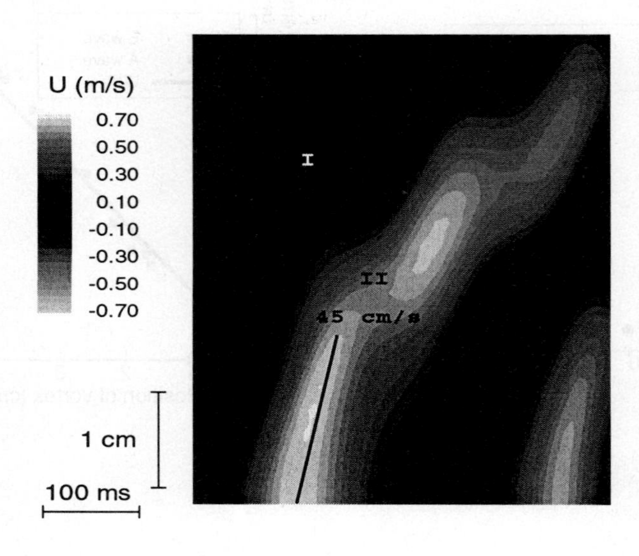

Fig 4. Color M-mode Doppler echocardiogram derived from a 2D axisymmetrical flow simulation. The scanline corresponds with the axis of symmetry. Both the early and atrial filling wave can be seen, the initial flow phase with blood moving simultaneously in the LV is denoted as I (pressure filling wave), the propagation of the vortex wave is denoted as II.

Motion Decoupling and Registration for 3D Magnetic Resonance Myocardial Perfusion Imaging

Nick Ablitt[1], Jianxin Gao[1], Peter Gatehouse[2], Guang-Zhong Yang[1,2]

[1]Royal Society/Wolfson Foundation MIC Laboratory, Imperial College, London, U.K.
[2]Royal Brompton Hospital, London, U.K.

{naa99,jxg,gzy@doc.ic.ac.uk}

Abstract. This paper presents a novel motion decoupling and registration method for 3D MR myocardial perfusion imaging. The technique uses tissue tagging for prospective through-plane motion correction of the left ventricle during 3D multi-slice image acquisition. The remaining in-plane distortion of the heart due to respiration was corrected by using a multi-resolution 2D free-form image registration method. Partial least square regression was adopted to recover the intrinsic relationship between respiration and cardiac deformation, both to speed up the registration process and to improve its internal consistency. Factor analysis is then applied to extract tracer characteristics of different regions of the myocardium. Both simulation data and *in vivo* images acquired from 8 normal subjects and 8 patients with coronary disease were used to validate the proposed techniques.

1. Introduction

Myocardial perfusion imaging is a valuable method in the evaluation of coronary artery disease. The technique is commonly performed in conjunction with coronary stress under coronary pharmacological vasodilatation when coronary blood flow is at its maximum. The technique provides several important haemodynamic parameters, including blood flow, volume, and mean transit time (the average time it takes for a tracer molecule to pass through the target tissue). The uptake of contrast agent within the myocardium is proportional to regional blood flow and the perfusion image series can be used to differentiate contrast agent uptake in regions of the myocardium supplied by normal or stenotic coronary arteries. The imaging method may be used to detect and distinguish between many different aspects of coronary heart disease. This includes ischaemia, infarction, reperfusion, myocardial viability, and detecting scar tissue in chronically infarcted myocardium. Traditionally, at the level of epicardial coronary arteries, angiographic anatomy is commonly used to identify hydraulically significant stenosis. At the level of coronary arteries, X-ray video-densitometry and electromagnetic and Doppler flow meters may be used to assess perfusion. Coronary venous thermodilution methods provide some insights into global or regional perfusion. At the arteriolar/capillary level, X-ray videodensitometer, contrast

P.M.A. Sloot et al. (Eds.): ICCS 2002, LNCS 2331, pp. 285–294, 2002.

echocardiography and microspheres may be employed. At the myocyte level, radionuclide and magnetic resonance imaging (MRI) methods can be used.

The recent development of MR myocardial perfusion imaging has extended the role of cardiovascular MRI in the evaluation of ischaemic heart disease beyond the situations where there have already been gross myocardial changes such as acute infarction or scarring. The ability to non-invasively evaluate cardiac perfusion abnormalities before pathologic effects occur, or as follow-up to therapy, is important to the management of patients with coronary artery disease. Early reperfusion of ischaemic myocardium has been shown to have a positive reversal effect on the ischaemic myocardium, which reduces mortality and morbidity. Differentiation of ischaemic but viable myocardium from infarcted regions requires detailed global quantitative assessment and modelling of myocardial perfusion characteristics. The prerequisite of such a study is the development of a spatially and temporally registered imaging strategy for a complete 3D coverage of the myocardium.

Hitherto, the assessment of myocardial perfusion using MRI has been based on first pass techniques using fast gradient echo (turboFLASH) [1-4] or echo-planar (EPI) sequences [5-7]. Quantitative results have been achieved in animal studies with intravascular agents (polylysine-Gd-DTPA)[8] as a macromolecular blood pool marker. At the same time, semi-quantitative results have also been established in human with conventional extracellular agents (Gd-DTPA) [9-11]. Either approach will have impact on detailed characterisation of the relationship between functional and perfusion abnormalities. Before comprehensive modelling can be achieved, however, it is necessary to address the imaging issues related to using MR for perfusion measurement. In 3D myocardial perfusion imaging, a complete volumetric data set has to be acquired for each cardiac cycle, and this can result in 150-300 such 3D data sets for studying the first pass of the contrast bolus. To ensure a comprehensive coverage of the myocardium and reasonably high resolution of the images, a typical data acquisition window of 100-200 *ms* per slice, and thus the overall acquisition time of more than 600 *ms* is required for each cardiac cycle to cover the entire volume of the left ventricle. When using multi-slice imaging, cardiac motion during this large acquisition window can cause the myocardium captured in different image planes to be mis-registered, *i.e.*, some parts of the myocardium may be imaged more than twice whereas other parts may be missed out completely. This type of mis-registration is difficult to correct by using post-processing techniques. With this study, we propose a novel motion decoupling technique based on motion tagging and real-time slice tracking. This resolves the movement of the myocardium along the long axis of the heart. A 2D free-form deformable model is then applied to correct for motion and deformation within image planes. By the use of motion decoupling, we turn the original four-dimensional problem (3D spatial + temporal) into multiple three-dimensional (2D spatial + temporal) subsets, and making this otherwise difficult task much more manageable.

2. Material and Methods

2.1 3D Motion Decoupling

Prior to perfusion imaging, a tagging sequence is used to highlight the short axis through plane motion of the heart. This consists of a vertical long axis (VLA) and horizontal long axis (HLA) tagged cine sequence within a single heartbeat, with the tag lines cutting through the short axis of the left ventricle (LV). The tags within these two cines are tracked throughout the cardiac cycle and a plane is fitted with least-mean-squares errors to the HLA and VLA short axis tag locations. The exact orientation of particular short axis planes within the myocardium is now known. Following from this, a standard multi-slice perfusion sequence is applied, which automatically adjusts the location and orientation of the imaging plane according to different trigger delays. The method assumes that the rate and extent of contraction of the LV is consistent between the motion tagging scan and subsequent perfusion imaging. The above procedures ensure that the through plane motion of the heart is captured during imaging. The decoupled in-plane motion can then be corrected for by using free-form 2D image registration techniques.

2.2 Free-Form Image Registration

The registration technique was adapted from our previous work on 2D electrophoresis image analysis [12]. It is a rapid multi-resolution free-form registration method based on a localized cross-correlation measure, as defined in Equation (1).

$$f(T)=\left|\frac{\sum_{\vec{x}\in\Omega}(I_1(\vec{x})-E\{I_1(\vec{x})\})(I_2(T(\vec{x}))-E\{I_2(T(\vec{x}))\})}{\sqrt{\sum_{\vec{x}\in\Omega}(I_1(\vec{x})-E\{I_1(\vec{x})\})^2}\sqrt{\sum_{\vec{x}\in\Omega}(I_2(T(\vec{x}))-E\{I_2(T(\vec{x}))\})^2}}\right| \quad (1)$$

where T is the applied deformation and Ω the overlapping pixels in the intersection region of the reference (I_1) and the transformed image (I_2). A linear tensor product B-spline as defined in Equation (2) was subsequently used as the transformation matrix as it enables adaptive automatic subdivision of the control grids when increasing the resolution, *i.e.*

$$T(\vec{p})=\sum_{i,j}\beta_{ij}(u,v)\vec{c}_{ij} \quad (2)$$

where $\beta_{ij}(u,v)$ is the coefficient applied to point $\vec{p}(u,v)$, and $\vec{c}_{ij}$ is the coordinate of control points surrounding $\vec{p}(u,v)$. The use of Equation (1) allows a close-form derivation of the derivatives of the similarity function such that quasi-Newton optimization techniques can be effectively used. The BFGS algorithm was used in this study to determine the optimum deformation vector of each grid control point with the following updating equations

$$x_{k+1} = x_k - H_k^{-1}\nabla f(x_k), \tag{3}$$
$$s_k = x_{k+1} - x_k, \quad y_k = \nabla f(x_{k+1}) - \nabla f(x_k),$$
$$H_{k+1} = H_k + \frac{y_k y_k^T}{y_k^T s_k} - \frac{H_k s_k s_k^T H_k}{s_k^T H_k s_k}$$

where H_k is the estimated Hessian matrix, and $\nabla f(x_k)$ is the vector of partial derivative of the function with respect to each parameter of T, dictated by the grid control points. In this study, the reference image was chosen to be at the end of the image series as all cardiac structures were clearly identifiable. Each image of the perfusion sequence was then registered to the reference image with a final grid resolution of 10×10.

2.3 Self-adaptive Learning and Prediction with Partial Least Squares Regression

After 3D motion decoupling with prospective slice tracking, the residual in-plane myocardial deformation is restricted to those caused by respiration. It is possible to exploit subject-specific near-linear relationships between respiration and cardiac deformation to increase the internal consistency of the registration process or to use the pattern of respiration as a prediction of myocardial deformation. To this end, the user is required to identify regions on one perfusion image, either through the chest or the diaphragm, where respiratory induced cyclic motion could be retrospectively recovered. The signals from these regions formed navigator traces and a partial least squares regression (PLSR) algorithm was used to recover the pattern of respiratory motion and the associated myocardial deformation extracted by the above free-form image registration process.

Let **X** and **Y** denote the respiratory motion and free-form deformation, respectively. The intrinsic relationship can be expressed as

$$\mathbf{Y} = \mathbf{XC} + \mathbf{E} \tag{4}$$

where **C** is the coefficient matrix, and **E** represents noise and higher order terms. In PLSR, both the input **X** and output **Y** are used for extracting the factors in forming the coefficient matrix **C**. To do so, **X** and **Y** are decomposed as

$$\mathbf{X} = \mathbf{TP} + \mathbf{E_1}$$
$$\mathbf{Y} = \mathbf{TQ} + \mathbf{E_2} \tag{5}$$

In Equation (5), **T** is the factor score matrix, **P** the factor loading matrix, **Q** the coefficient loading matrix, $\mathbf{E_1}$ and $\mathbf{E_2}$ represent, respectively, parts of **X** and **Y** that are unaccounted for. The latent vectors are the eigenvectors of the covariance matrix **(X'Y)'(X'Y),** through which a weighting matrix **W** can be computed iteratively such that **T=XW**. Once **P**, **Q** and **W** are found, the regression matrix in equation (4) can be determined as **C=WQ** [13]. In the current study, we used 30% of the image frames located at the end of the perfusion sequence for learning and extracting the regression matrix **C** using the first 4 principal components of the covariance matrix. Deformation fields for the remaining images were derived from the PLSR prediction.

2.4 Extraction of Regional Perfusion Abnormality with Factor Analysis

For the quantification of regional perfusion abnormality, both model-based and modeless approaches can be used. For the model-based approaches, tracer kinetics is commonly assumed to follow a Fermi function [14]. By using the intensity variation measured from the LV blood pool as the input signal, deconvolution is required to recover the impulse response of different regions of the myocardium. In practice, the signal from the LV blood pool may be significantly attenuated at peak amplitude due to the relatively high concentration of the Gd-DTPA used. In this case, the model based approach can introduce significant errors. Furthermore, the use of a single model for representing both normal and infracted regions requires further justification. In this study, factor analysis for dynamic image series was used to determine regional perfusion abnormalities.

Factor analysis is a valuable tool for extracting underlying characteristics of a region of interest (ROI) with different tissue types from dynamic image series without prior assumptions about tissue models. An image sequence S can be represented as the sum of K underlying images (spatial distribution) a_k, each weighted by their factors (temporal distribution) f_k, as follows [15-16],

$$S(p,t) = \sum_{k=1}^{K} a_k(p) \cdot f_k(t) + e(p,t) \tag{6}$$

where $e(p,t)$ is the residual error and K is the number of factors kept, usually is equal to the number of different tissue kinetics within the ROI. By neglecting the error term, the above expression can be turned into a matrix form

$$\mathbf{S} = \mathbf{FA} = \mathbf{U\tilde{\Lambda}V} \tag{7}$$

where **S** is a M×N matrix, **U** the M×K orthonormal column matrix composed of the first K principal kinetic curves, **V** the K×N orthonormal row matrix composed of K principal component images, $\tilde{\Lambda}$ the K×K diagonal matrix of the K roots of eigenvalue, M and N are the number of image frames and total number of pixels within the ROI, respectively. Since there is no unique solution to the above equation, additional constrains are necessary. Among them, the positive constraint is the most commonly adopted one. It is based on the fact that there are no negative values in the actual images and the corresponding time curves, *i.e.*, factors. This can be implemented by iterative oblique rotations of the factors and the corresponding factor images.

2.5 Validation and *in vivo* Data Acquisition

To assess the accuracy of the proposed techniques, both numerical simulation and *in vivo* data sets were used. For the synthetic dataset, the signal characteristics were generated based on an idealised myocardial first-pass perfusion model and a typical respiration pattern derived from normal subjects. The model includes two artificial defects in different parts of the myocardium, and this dataset is used to evaluate the proposed 2D free form registration, self-adaptive learning and prediction with PLSR, and perfusion characterisation based on factor analysis. *In vivo* results were acquired from both normal volunteers and patients with known coronary artery disease. These images were obtained using a 1.5T Siemens Sonata scanner (200 *T/m/s*; 40 *mT/m*) with a four-element phased-array receiver coil. For 3D tracking, the HLA and VLA tagged cine images were obtained in the same 16-cycle end-expiratory breath-hold, using segmented FLASH imaging. In each cardiac cycle, a multiplanar ("comb") saturation pulse was applied after the R-wave to tag the myocardium parallel to the short axis of the heart. The first-pass perfusion data was acquired from 8 patients with a FLASH sequence (Tr=3.7 *ms*) that consists of three saturation-recovery short-axis slices per cardiac cycle, for 50 cycles during the first-pass of Gd-DTPA (cubital vein, 0.1 *mmol/kg*, 3 *ml/s*) with a field of view of 400 *mm* (128 pixels) by 300 *mm* (64 pixels).

In addition to these, a special test pulse sequence was developed to examine the accuracy of the proposed prospective tracking method for 3D motion decoupling. The sequence used the same tagging pre-pulse as before, but the imaging part of the sequence was altered such that it imaged the same basal plane three times at 156 *ms* intervals. When slice tracking was enabled, the acquired images would always be in alignment with the tagging plane, and therefore the signal from the myocardium would be nullified. Otherwise, the acquired images would cover different parts of the myocardium, and therefore had an uneven intensity distribution. For the purpose of visualization, these images were subtracted from their counter parts without the use of the tagging pre-pulse. In this case, the intensity distribution of the myocardium would be uniformly bright if slice tracking had been accurate. Three normal subjects were studied for this purpose. A further study of five normal volunteers was carried out for 3D motion decoupling with a normal perfusion sequence but without the administration of Gd-DTPA bolus.

3. Results

Figure 1 illustrates one example of the slice tracking process for motion decoupling on one of the normal subjects studied. The first and second rows show the VLA and HLA images of the heart at three different phases of the cardiac cycle. The darker and brighter pilot grids illustrate the imaging planes in which the myocardium would be covered with and without motion tracking, respectively. The resultant short axis images acquired by the special test pulse sequence are shown at the two bottom rows. Without tracking, the acquired images covered different parts of the myocardium as indicated by both intra and inter-phase signal variations of the myocardium. By the use of effective slice tracking, the same part of the myocardium was always covered by the same image plane, resulting in a consistent intensity distribution within the myocardium. For the three subjects verified, the percentage changes in signal intensity varied from –59% to 74% (absolute mean±std = 45±9.8%) without tracking, and –9% to 22% (absolute mean±std = 9±3.9%) with tracking. For 3D motion decoupling based on the normal perfusion sequence, areas of the myocardium of the LV were measured for the five normal subjects studied. The three short axis slices were taken within the same cardiac cycle, with the most apical slice acquired first and the most basal slice last. The bar chart shown in Figure 2 indicates the absolute differences of measured regional errors in mm^2 without motion tracking.

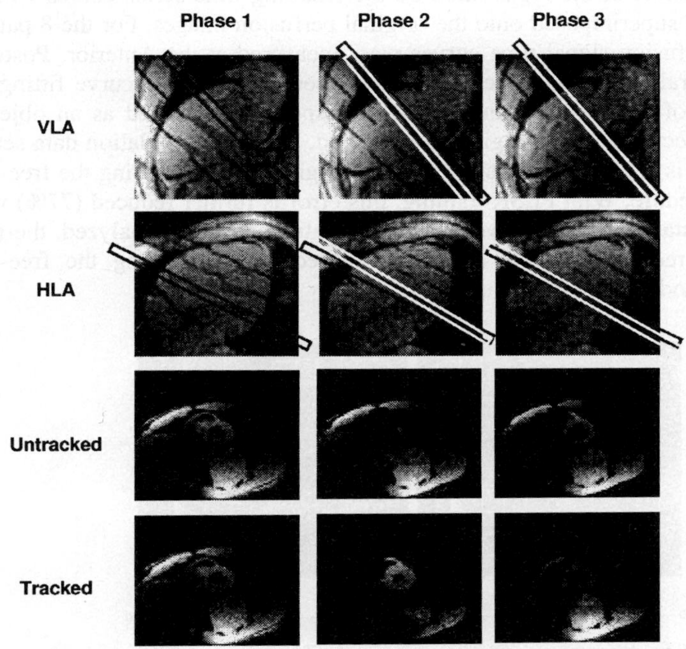

Fig 1. A schematic illustration of the basic procedures used for prospective tracking of through plane cardiac motion. Cine myocardial tagging is performed on both the VLA and HLA of the LV from which the positions of the tracking planes are derived. The two bottom rows demonstrate the effectiveness of slice tracking by using a special test pulse sequence. Signals form the myocardium should remain uniform if the imaging planes are accurately tracked over time.

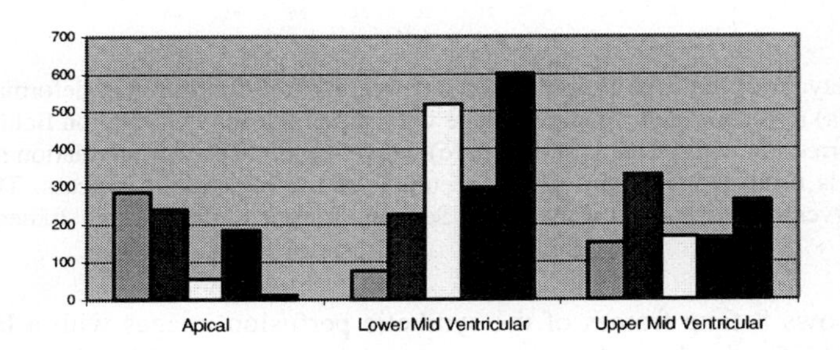

Fig 2. The amount of motion involved for the apical, lower and upper mid-ventricular slices measured from five normal subjects by using the short-axis area change as an indicator.

Figure 3 demonstrates the effectiveness of image registration by using the predicted myocardial deformation based on adaptive self-learning with PLSR. It shows an example of the effect of the registration on the resulting time series curves with the deformation grid superimposed onto the original perfusion images. For the 8 patients studied, four perfusion signal time curves were measured at the Anterior, Posterior, Septal, and Lateral Segments for each perfusion sequence. Spline curve fitting was applied to each of these curves and the mean dispersion was used as an objective measure of the accuracy of the registration method. With the simulation data set, the mean dispersion is reduced to about 85% of its original value by using the free-form registration procedure. With PLSR training, this error is further reduced (77%) while the actual computation time is halved. For the 8 patient data sets analyzed, the mean dispersion was reduced to 81% and 68%, respectively, by using the free-form registration method and the PLSR approach.

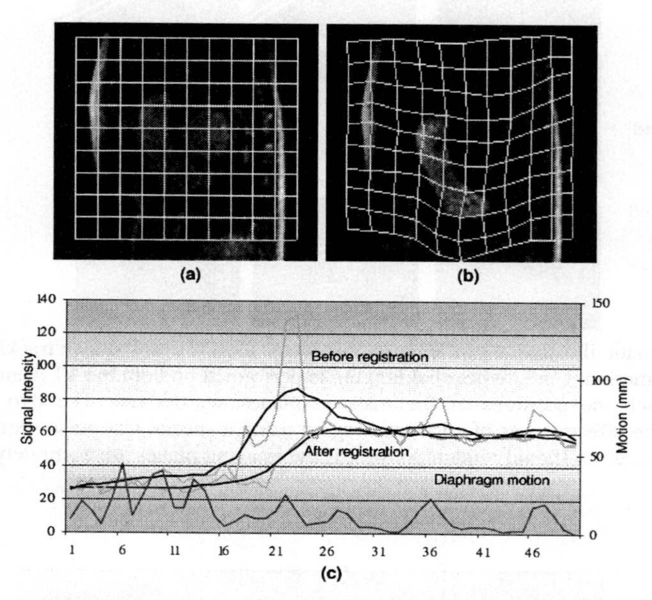

Fig 3. The layout of the control grid used for correcting free-form in-plan deformation due to respiration, (a) reference, and (b) target image with superimposed deformation field. The effect of motion correction with PLSR is shown in (c) where signal dispersion in relation to the spline fitted curve is used as a measure of the accuracy of the registration process. The example perfusion curves were measured at the Septal Segment for one of the patients studied.

Figure 4 shows 8 time frames of the synthetic perfusion images with a lateral sub-endocardial defect and another posterior transmural defect. The first three factor images of the perfusion sequence clearly indicate different tracer kinetic behaviours of the myocardium and the exact locations of the defects.

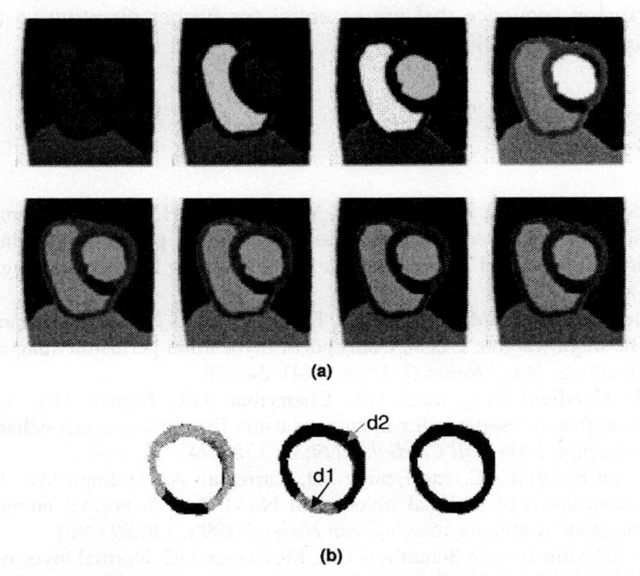

Fig 4. (a) Eight image frames from a synthetic perfusion sequence showing both sub-endocardial and transmural perfusion defects. (b) The derived factor images show different tracer characteristics of the myocardium and the locations of the defects (d1 and d2, pointed by the arrows).

4. Discussion and Conclusions

Magnetic resonance myocardial perfusion imaging is a promising method in the evaluation of coronary artery disease. Existing techniques for first-pass myocardial perfusion imaging analysis normally involve either single or multiple un-registered imaging planes based on manual delineation of ROIs at different regions of the myocardium. In-plane deformation can be restored to some extent by the labor-intensive manual correction process, albeit being error prone. The through plane cardiac motion due to extended imaging time involved in 3D perfusion imaging, however, is not recoverable by post-processing. The use of 3D motion decoupling effectively resolves this problem. It ensures the material captured by each imaging plane always covers the same part of the myocardium. The use of 2D free-form image registration removes any in-plane distortion and makes subsequent perfusion quantification fully automatic. Another major innovation of the current study is the use of PLSR for self-adaptive learning of subject-specific cardiac motion induced by respiratory motion. It has been shown that it not only substantially reduces the number of imaging frames to be registered but also makes the motion correction results much more consistent. Quantification is of significant importance in myocardial perfusion imaging as it allows the evaluation of further aspects of perfusion such as myocardial viability. Currently, there is very little evidence of the reproducibility of results. This paper presents a practical way of obtaining 3D registered perfusion sequence that are essential for further quantitative modelling of regional perfusion abnormalities.

References

1. Wilke N, Simm C, Zhang J, Ellermann J, Ya X, Merkle H, Path G, Ludemann H, Bache RJ, Ugurbil K. Contrast-enhanced first pass myocardial perfusion imaging: correlation between myocardial blood flow in dogs at rest and during hyperemia. *Magn Reson Med.* 1993;29(4): 485-97.
2. Wilke N, Jerosch-Herold M, Stillman AE, Kroll K, Tsekos N, Merkle H, Parrish T, Hu X, Wang Y, Bassingthwaighte J, et al. Concepts of myocardial perfusion imaging in magnetic resonance imaging. *Magn Reson Q.* 1994;10(4): 249-86.
3. Cullen JH, Horsfield MA, Reek CR, Cherryman GR, Barnett DB, Samani NJ. A myocardial perfusion reserve index in humans using first-pass contrast-enhanced magnetic resonance imaging. J *Am Coll Cardiol.* 1999;33: 1386-94.
4. Keijer JT, van Rossum AC, van Eenige MJ, Karreman AJ, Hofman MB, Valk J, Visser CA. Semiquantitation of regional myocardial blood flow in normal human subjects by first-pass magnetic resonance imaging. *Am Heart J.* 1995; 130:893-901.
5. Schwitter J, Debatin JF, von Schulthess GK, McKinnon GC. Normal myocardial perfusion assessed with multishot echo-planar imaging. *Magn Reson Med.* 1997;37(1): 140-7.
6. Beache GM, Kulke SF, Kantor HL, Niemi P, Campbell TA, Chesler DA, Gewirtz H, Rosen BR, Brady TJ, Weisskoff RM. Imaging perfusion deficits in ischemic heart disease with susceptibility-enhanced T2-weighted MRI: preliminary human studies. *Magn Reson Imaging.* 1998;16(1): 19-27.
7. Ding S, Wolff SD, Epstein FH. Improved coverage in dynamic contrast-enhanced cardiac MRI using interleaved gradient-echo EPI. *Magn Reson Med.* 1998;39(4): 514-9.
8. Wilke N, Kroll K, Merkle H, Wang Y, Ishibashi Y, Xu Y, Zhang J, Jerosch-Herold M, Muhler A, Stillman AE, et al. Regional myocardial blood volume and flow: first-pass MR imaging with polylysine-Gd-DTPA. *J Magn Reson Imaging.* 1995;5(2): 227-37.
9. Larsson HBW, Stubgaard M, Søndergaard L, Henriksen O. In vivo quantification of the unidirectional influx constant for Gd-DTPA diffusion across the myocardial capillaries with MR imaging. *J Magn Reson Imaging.* 1994; 4: 433-40.
10. Dendale P, Franken PR, Block P, Pratikakis Y, De Roos A. Contrast enhanced and functional magnetic resonance imaging for the detection of viable myocardium after infarction. *Am Heart J.* 1998; 135: 875-80.
11. Larsson HBW, Fritz-Hansen T, Rostrup Egill, Søndergaard L, Ring P, Henriksen O. Myocardial perfusion modeling using MRI. *Magn Reson Med.* 1996; 35: 716-26.
12. Veeser S, Dunn MJ, Yang GZ. Multiresolution image registration for two-dimensional gel electrophoresis. *Proteomics.* 2001; 1: 856-870.
13. Wold, H. Soft modelling with latent variables: the nonlinear iterative partial least squares approach. Perspectives in probability and Statistics: Papers in honour of M.S. Barlett, (J. Gani, ed). London: Academic Press. 1975: 114-142.
14. Jerosch-Herold, M, Wilke N, Stillman AE. Magnetic resonance quantification of the myocardial perfusion reserve with a Fermi function model for constrained deconvolution, *Medical Physics,* 1998; 25: 73-84.
15. Buvat I, Benali H, Paola RD. Statistical distribution of factors and factor images in factor analysis of medical image sequences. *Phys Med Biol.* 1998; 43: 1695–1711.
16. Martel AL, Moody AR, Allder SJ, Delay GS, Morgan PS. Extracting parametric images from dynamic contrast-enhanced MRI studies of the brain using factor analysis. *Med Img Analysis,* 2001; 5: 29-39.

A Comparison of Factorization-Free Eigensolvers with Application to Cavity Resonators

Peter Arbenz

Institute of Scientific Computing, ETH Zentrum, CH-8092 Zurich, Switzerland
arbenz@inf.ethz.ch

Abstract. We investigate eigensolvers for the generalized eigenvalue problem $A\mathbf{x} = \lambda M\mathbf{x}$ with symmetric A and symmetric positive definite M that do not require matrix factorizations. We compare various variants of Rayleigh quotient minimization and the Jacobi-Davidson algorithm by means large-scale finite element problems originating from the design of resonant cavities of particle accelerators.

1 Introduction

In this paper we consider the problem of computing a few of the smallest eigenvalues and corresponding eigenvectors of the generalized eigenvalue problem

$$A\mathbf{x} = \lambda M\mathbf{x} \tag{1}$$

without factorization of neither A nor M. Here, the real n-by-n matrices A and M are symmetric and symmetric positive definite, respectively.

It if is feasible the Lanczos algorithm combined with a spectral transformation [1] is the method of choice for solving (1). The spectral transformation requires the solution of a linear system $(A - \sigma M)\mathbf{x} = \mathbf{y}$, $\sigma \in \mathbb{R}$, which may be solved by a direct or an iterative system solver. The factorization of $A - \sigma M$ may be impossible due to memory constraints. The system of equations can be solved iteratively. However, the solution must be computed very accurately, in order that the Lanczos three-term recurrence remains correct.

In earlier studies [2, 3] we found that for large eigenvalue problems the Jacobi-Davidson algorithm [4] was superior to the Lanczos algorithm or the restarted Lanczos algorithm as implemented in ARPACK [5]. While the Jacobi-Davidson algorithm retains the high rate of convergence it only poses small accuracy requirements on the solution of the so-called correction equation, at least in the initial steps of iterations.

In this paper we continue our investigations on eigensolvers and their preconditioning. We include block Rayleigh quotient minimization algorithms [6] and the locally optimal block preconditioned conjugate gradient (LOBPCG) algorithm by Knyazev [7] in our comparison.

We conduct our experiments on an eigenvalue problem originating in the design of the new RF cavity of the 590 MeV ring cyclotron installed at the Paul Scherrer Institute (PSI) in Villigen, Switzerland.

P.M.A. Sloot et al. (Eds.): ICCS 2002, LNCS 2331, pp. 295–304, 2002.

2 The application: the cavity eigenvalue problem

After separation of time/space variables and after elimination of the magnetic field intensity the Maxwell equations become the eigenvalue problem

$$\mathbf{curl}\,\mathbf{curl}\,\mathbf{e}(\mathbf{x}) = \lambda \mathbf{e}(\mathbf{x}), \quad \operatorname{div} \mathbf{e} = 0, \quad \mathbf{x} \in \Omega, \qquad \mathbf{n} \times \mathbf{e} = \mathbf{0}, \quad \mathbf{x} \in \partial\Omega. \tag{2}$$

where $\mathbf{e}$ is the electric field intensity. We assume that Ω, the cavity, is a simply connected, bounded domain in $\mathbb{R}^3$ with a polyhedral boundary $\partial\Omega$. Its inside is all in vacuum and its metallic surfaces are perfectly conducting. Following Kikuchi [8] we discretize (2) as

$$\begin{array}{ll} \textit{Find } (\lambda_h, \mathbf{e}_h, p_h) \in \mathbb{R} \times N_h \times L_h \textit{ such that } \mathbf{e}_h \neq \mathbf{0} \textit{ and} & \\ \text{(a) } (\mathbf{curl}\,\mathbf{e}_h, \mathbf{curl}\,\mathbf{\Psi}_h) + (\mathbf{grad}\, p_h, \mathbf{\Psi}_h) = \lambda_h(\mathbf{e}_h, \mathbf{\Psi}_h), & \forall \mathbf{\Psi}_h \in N_h \\ \text{(b) } (\mathbf{e}_h, \mathbf{grad}\, q_h) = 0, & \forall q_h \in L_h \end{array} \tag{3}$$

where $N_h \subset H_0(\mathbf{curl};\Omega) = \{\mathbf{v} \in L^2(\Omega)^3 \mid \mathbf{curl}\,\mathbf{v} \in L^2(\Omega)^3, \mathbf{v} \times \mathbf{n} = \mathbf{0} \text{ on } \partial\Omega\}$ and $L_h \subset H_0^1(\Omega)$. The domain Ω is triangulated by tetrahedrons. In order to avoid spurious modes we choose the subspaces N_h and L_h, respectively, to be the Nédélec (or edge) elements $N_h^{(k)}$ of degree k [9–11] and the well-known Lagrange (or node-based) finite elements consisting of piecewise polynomials of degree $\leq k$. In this paper we exclusively deal with $k = 2$. In order to employ a multilevel preconditioner we use hierarchical bases and write [11]

$$N_h^{(2)} = N_h^{(1)} \oplus \bar{N}_h^{(2)}, \qquad L_h^{(2)} = L_h^{(1)} \oplus \bar{L}_h^{(2)}. \tag{4}$$

Let $\{\mathbf{\Phi}_i\}_{i=1}^n$ be a basis of $N_h^{(2)}$ and $\{\varphi_l\}_{l=1}^m$ be a basis of $L_h^{(2)}$. Then (3) defines

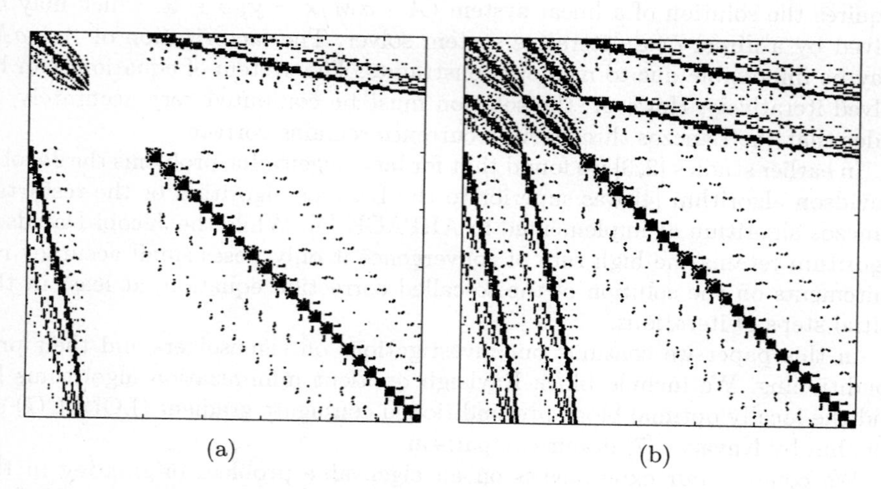

(a) (b)

Fig. 1. Example of matrices A (a) and M (b) for quadratic edge elements.

the matrix eigenvalue problem

$$\begin{bmatrix} A & C \\ C^T & O \end{bmatrix} \begin{pmatrix} \mathbf{x} \\ \mathbf{y} \end{pmatrix} = \lambda \begin{bmatrix} M & O \\ O & O \end{bmatrix} \begin{pmatrix} \mathbf{x} \\ \mathbf{y} \end{pmatrix}, \tag{5}$$

respectively, where A and M are n-by-n and C is n-by-m with elements

$$a_{i,j} = (\mathbf{curl}\,\mathbf{\Phi}_i, \mathbf{curl}\,\mathbf{\Phi}_j), \quad m_{i,j} = (\mathbf{\Phi}_i, \mathbf{\Phi}_j), \quad c_{i,l} = (\mathbf{\Phi}_i, \mathbf{grad}\,\varphi_l).$$

The non-zero structures of A and M are depicted in Fig. 1. The block of zero columns / columns in A corresponds to those curl-free basis functions of $\bar{N}_h^{(2)}$ that are the gradient of some $\varphi \in \bar{L}_h^{(2)}$. The non-zero block in the upper-left corner of A corresponding to the basis functions of $N_h^{(1)}$ is rank-deficient. The deficiency equals $\dim L_h^{(1)}$. The reason for approximating the electric field $\mathbf{e}$ by Nédélec elements and the Lagrange multipliers by Lagrange finite elements is that [9, §5.3]

$$\mathbf{grad}\, L_h^{(k)} = \left\{ \mathbf{v}_h \in N_h^{(k)} \mid \mathbf{curl}\,\mathbf{v}_h = \mathbf{0} \right\}. \tag{6}$$

By (3)(b), $\mathbf{e}_h$ is in the orthogonal complement of $\mathbf{grad}\, L_h^{(k)}$. On this subspace A in (5) is positive definite. Notice that $\mathbf{e}_h$ is divergence-free only in a discrete sense. Because of (6) we can write $\mathbf{grad}\,\varphi_l = \sum_{j=1}^{n} \eta_{jl} \mathbf{\Phi}_j$, whence

$$(\mathbf{\Phi}_i, \mathbf{grad}\,\varphi_l) = \sum_{j=1}^{n} (\mathbf{\Phi}_i, \mathbf{\Phi}_j)\eta_{jk} \qquad \text{or} \qquad C = MY, \tag{7}$$

where $Y = ((\eta_{jk})) \in \mathbb{R}^{n \times m}$. In a similar way one obtains

$$H := C^T Y = Y^T M Y, \qquad h_{kl} = (\mathbf{grad}\,\varphi_k, \mathbf{grad}\,\varphi_l). \tag{8}$$

H is the system matrix that is obtained when solving the Poisson equation with the Lagrange finite elements $L_h^{(k)}$. Notice that Y is very sparse. We have already mentioned that the gradient of a basis function $\varphi_k \in \bar{L}_h^{(2)}$ is an element of the first set of basis functions of $\bar{N}_h^{(2)}$. So, m_1 rows of Y have a single entry 1. The gradient of the piecewise linear basis function corresponding to vertex k, say, is a linear combination (with coefficients ± 1) of the basis functions of $N_h^{(1)}$ whose corresponding edge has vertex k as one of its endpoints.

Equation (7) means that $C^T \mathbf{x} = \mathbf{0}$ is equivalent to requiring $\mathbf{x}$ to be M-orthogonal to the eigenspace $\mathcal{N}(A) = \mathcal{R}(Y)$ corresponding to the eigenvalue 0. Thus, the solutions of (5) are precisely the eigenpairs of

$$A\mathbf{x} = \lambda M \mathbf{x} \tag{9}$$

corresponding to the *positive* eigenvalues. We could therefore compute the desired eigenpairs $(\lambda_j, \mathbf{x}_j)$ of (5) by means of (9) alone. The computed eigenvectors corresponding to positive eigenvalues would automatically satisfy the constraint $C^T \mathbf{x}_j = \mathbf{0}$. This is actually done if the linear systems of the form $(A - \sigma M)\mathbf{x} = \mathbf{y}$ that appear in the eigensolver can be solved by direct methods. Numerical experiments showed however that the high-dimensional zero eigenspace has a negative effect on the convergence rates if *preconditioned iterative* solvers have to be used.

3 Positive definite formulations of the eigenvalue problem

In this section we present two alternative formulations for the matrix eigenvalue problem (5) that should make its solution easier. The goal is to have generalized eigenvalue problems with both matrices positive definite.

3.1 The nullspace method

We can achieve this goal in a straightforward way by performing the computations in the space that is spanned by the eigenvectors corresponding to the positive eigenvalues. This space is $\mathcal{R}(Y)^{\perp_M} = \mathcal{R}(C)^{\perp} = \mathcal{N}(C^T)$, i.e. the nullspace of C^T whence its name. Formally, we solve

$$A|_{\mathcal{N}(C^T)}\mathbf{x} = \lambda M|_{\mathcal{N}(C^T)}\mathbf{x}. \tag{10}$$

In the iterative solvers, we apply the M-orthogonal projector $P_{\mathcal{N}(C^T)} = I - YH^{-1}C^T$ onto $\mathcal{N}(C^T)$ whenever a vector is not in this space, i.e. for generating starting vectors and after solving with the preconditioner. The eigenvalue problem (10) has dimension $n - m$. However, the vector $\mathbf{x} \in \mathcal{N}(C^T)$ has n components.

3.2 The approach of Arbenz-Drmač [12]

Let P be a permutation matrix and

$$\widehat{Y} = PY = \begin{bmatrix} \widehat{Y}_1 \\ \widehat{Y}_2 \end{bmatrix}, \qquad \widehat{Y}_2 \in \mathbb{R}^{m \times m} \text{ nonsingular.} \tag{11}$$

Let

$$\widehat{A} = P^T A P = \begin{bmatrix} \widehat{A}_{11} & \widehat{A}_{12} \\ \widehat{A}_{21} & \widehat{A}_{22} \end{bmatrix}, \quad \widehat{M} = P^T M P = \begin{bmatrix} \widehat{M}_{11} & \widehat{M}_{12} \\ \widehat{M}_{21} & \widehat{M}_{22} \end{bmatrix}, \quad \widehat{C} = P^T C = \begin{bmatrix} \widehat{C}_1 \\ \widehat{C}_2 \end{bmatrix}.$$

Then the $n - m$ eigenvalues of the eigenvalue problem

$$\widehat{A}_{11}\hat{\mathbf{x}} = \lambda(\widehat{M}_{11} - \widehat{C}_1 H^{-1} \widehat{C}_1^T)\hat{\mathbf{x}} \tag{12}$$

are precisely the $n - m$ positive eigenvalues of (5). The matrix on the right-hand side of (12) must not be formed explicitely as it is full.

3.3 Discussion

These approaches have in common that they require the solution of a system of equations involving the matrix H. In the present application this is the discretized Laplace operator in $W_h^{(2)}$. The order of H is about the same as the dimension of the 'coarse' space $V_h^{(1)}$. Therefore, we solve systems with H by a direct method. The method of Arbenz-Drmač has the advantage that the order of the eigenvalue problem is smaller by at least $1/8$. Thus, less memory space is required.

4 Solving the matrix eigenvalue problem

Factorization-free methods are the most promising for effectively solving very large eigenvalue problems

$$A\mathbf{x} = \lambda M\mathbf{x}. \qquad (13)$$

The factorization of A or M or a linear combination of them which is needed if a spectral transformation like shift-and-invert is applied requires way too much memory. If the shift-and-invert spectral transformation in a Lanczos type method is solved iteratively then high accuracy is required to establish the three-term recurrence. This is very time consuming even if the conjugate gradient method is applied with a good preconditioner. In most of our experiments the Jacobi-Davidson algorithm was much more effective than the implicitly restarted Lanczos algorithm as implemented in ARPACK [5] for solving the cavity and other eigenvalue problems [2, 3].

In this study we investigate three factorization-free algorithms for solving (13)

– **The symmetric Jacobi-Davidson algorithm (JDSYM)**
The Jacobi-Davidson algorithm has found considerable attention in recent years. It has been introduced by Sleijpen and van der Vorst [4]. There are variants for all types of eigenvalue problems [13]. Here, we use a modification of the algorithm JDQZ adapted to the generalized symmetric eigenvalue problem (13) as described in [2, 14].
In every step of JD, the search space is expanded by the solution of the so-called correction equation. This solution is closely related to the Newton correction of $\mathbf{grad}\,\rho(\mathbf{x}) = \mathbf{0}$ where $\rho(\mathbf{x}) = \mathbf{x}^T A\mathbf{x}/\mathbf{x}^T M\mathbf{x}$ is the Rayleigh quotient. The correction equation need not be solved to high accuracy. A few steps of MINRES is adequate except close to convergence (of the eigensolver) where high accuracy is required to get a decent convergence rate. By introducing a restart procedure the memory requirements can be bounded. Typically, besides the storage for the matrices A and M and the preconditioner, $(p + 3j_{\max})n$ memory locations are needed where $j_{\max}$ is the maximal dimension of the search space.
– **Block Rayleigh Quotient minimization (BRQMIN)**
BRQMIN is a proper subspace method. In the k-th iteration step the actual approximations of the eigenvectors corresponding to the p smallest eigenvalues of (13) are stored in X_k. X_{k+1} is obtained through the Rayleigh-Ritz procedure in the $2p$ dimensional space spanned by the columns $\mathbf{x}_j^{(k)}$ of X_k and P_k, the so-called search directions. The latter are obtained as $P_k = K^{-1}G_k + P_{k-1}B_k$. B_k is determined such that $P_{k-1}^T A P_k = O$, K is a preconditioner, and $G_k\mathbf{e}_j = \mathbf{grad}\,\rho(\mathbf{x}_j^{(k)})$. This is a cg-type algorithms for minimizing the Rayleigh quotient as proposed (but not implemented) 20 years ago by Longsine and McCormick [6]
– **The locally optimal block PCG method (LOBPCG)**
The locally optimal block preconditioned conjugate gradient algorithm has been introduced by Knyazev [7] as an improvement over BRQMIN. Local optimality is obtained by executing the Rayleigh-Ritz procedure in the $3p$ dimensional subspace $\mathcal{R}(X_k, K^{-1}G_k, P_{k-1}) \supset \mathcal{R}(X_k, P_k)$. For large eigenvalue problems, the

overhead introduced by the larger subproblem is negligible. The memory requirement is $10pn$ floating point numbers which is bearable as long as p is small.

5 The preconditioners

In the eigensolvers that we have discussed in the previous section a preconditioner, say K, is required to get a good convergence behavior. K should be a good approximation of A or $A - \theta M$ for some fixed or variable θ. For ease of presentation we set $\theta = 0$.

Let A in (5) be partitioned according to (4),

$$A = \begin{bmatrix} A_{11} & A_{12} \\ A_{21} & A_{22} \end{bmatrix}, \qquad A_{ij} \in \mathbb{R}^{n_i \times n_j}, \tag{14}$$

where the $(1,1)$-blocks correspond to the bilinear forms involving basis functions in $N_h^{(1)}$. Proceeding as Bank [15] we approximate the inverse of A in (14) by

$$K^{-1} = \begin{bmatrix} A_{11}^{-1} & O \\ O & \widetilde{A}_{22}^{-1} \end{bmatrix}, \tag{15}$$

where $\widetilde{A}_{22}^{-1}$ corresponds to a few (usually one) step of a stationary iteration method. K has similar properties as A. In particular, $KY = O$. This is important as it allows us to get a preconditioner $\widehat{K}_{11}$ for $\widehat{A}_{11}$ in (12) by choosing the (1,1)-block of $P^T KP$, where P is the permutation in (11).

In the nullspace method the inverse of the preconditioner has the form

$$K^{-1} = (I - YH^{-1}C^T)(A_1 - \theta M_1)^{-1} \tag{16}$$

where K is given in (15).

6 Numerical experiments

6.1 MATLAB experiments with a rectangular cavity

We first report on experiments that were executed in the MATLAB 6 (R12) environment on a Intel Pentium III (500 MHz, 512kB Cache, 512 MB RAM) running the Linux 2.2 operating system.

The test problem, a rectangular cavity, had the size $n = \dim(N_h^{(2)}) = 6292$. The number of constraints was $m = \dim(L_h^{(2)}) = 1155$. We used both the nullspace method and the Arbenz-Drmač approach. In the latter the problem size was $n - m = 5137$. The timings for computing $p = 8$ and $p = 16$ eigenpairs are listed in Table 1. The four columns give the execution times for the nullspace and Arbenz-Drmač (AD) approaches with the two-level preconditioner discussed in section 5 where $\widetilde{A}_{22}^{-1}$ in (15) was chosen to be one step of the Jacobi or symmetric Gauss-Seidel iteration (which is SSOR with relaxation parameter $\omega = 1$). Approximate eigenvectors $\mathbf{x}$ are considered converged if $\|A\mathbf{x} - \rho(\mathbf{x})M\mathbf{x}\| \le 10^{-8}$.

solver	nullspace Jacobi	SSOR(1)	Arbenz-Drmač Jacobi	SSOR(1)
$p = 8$				
LOBPCG	78.3	49.6	63.4	37.3
JDSYM (MINRES)	373.8	287.4	193.0	129.0
BRQMIN	78.5	51.7	63.4	40.0
$p = 16$				
LOBPCG	184.9	115.1	170.9	83.8
JDSYM (MINRES)	859.0	612.6	440.0	308.3
BRQMIN	172.2	119.3	144.2	100.2

Table 1. MATLAB timings for computing 8 and 16, respectively, eigenvalues of a problem of size $n = 6292$ with $m = 1155$.

The AD approach proved to be faster than the nullspace approach. LOBPCG is slightly faster than block Rayleigh quotient minimization in most cases. LOBPCG needs fewer iteration steps than BRQMIN until convergence, however a single step is more expensive as the dimension of the trial space has 1.5 times the dimension. Jacobi-Davidson does not perform so well. It is about three times slower than the other two solvers. JDSYM performs slightly better for $p = 16$. This conforms with the observation that it takes JDSYM a large number of iterations steps until it finds the first eigenpair. In these computations the dimension of trial spaces in JDSYM varied between p and $2.5p$. The correction equation was solved very inaccurately initially. Close to convergence the accuracy requirements are increased to obtain the high convergence rate [2]. The subspace dimension in LOBPCG and BRQMIN were p. There were no significant gains or losses if the subspace dimensions were chosen slightly bigger than p. It is however crucial for the performance that converged eigenvectors are locked.

6.2 The small model of the copper cavity

The next experiment concerned a coarse model of the future design of the copper cavity. Here, we computed $p = 10$ eigenpairs of a problem of size $n = 45'040$ with $m = 7824$ constraints. We used our C programs that make extensive use of the BLAS. The compute platform was a Sun Enterprise E3500 with 336 MHz UltraSPARC-II processors and 3 GB of main memory which was running the Solaris 2.6 operating system. We only used the nullspace method. The purpose of this experiment is to compare the two 2-level preconditioners used above with diagonal and no preconditioning. Timings are given in Table 2

For the three solvers t_{eig} gives the time spent for computing the 10 eigenpairs to the desired accuracy of $\|A\mathbf{x} - \rho(\mathbf{x})M\mathbf{x}\| \leq 10^{-8}$. t_{eig} does not include preparatory work like grid handling, matrix assembly etc. t_{11} gives the time that is spent for system solving by direct solvers. This concerns the $(1,1)$ block of the preconditioner (16) and the matrix H that appears in the projector in (16) in the nullspace method or in the matrix M in the AD approach, cf. (12). Notice that

Table 2. Timings for computing 10 eigenvalues of the problem of size $n = 45\,040$ with $m = 1155$ constraints. Times are in seconds.

Jacobi-Davidson				
preconditioner	it_{int}	it_{out}	t_{eig}	t_{11}
none	55	89	3680	
diagonal	19	86	1289	
hierarchical/Jacobi	8	82	900	237
hierarchical/SSOR(1)	4	76	593	137
LOBPCG				
preconditioner			t_{eig}	t_{11}
none			6135	
diagonal			1707	
hierarchical/Jacobi			644	163
hierarchical/SSOR(1)			413	95
BRQMIN				
preconditioner			t_{eig}	t_{11}
none			5862	
diagonal			1667	
hierarchical/Jacobi			502	129
hierarchical/SSOR(1)			384	89

the critical operation is not the factorization of these matrices but the forward and backward substitution in each iteration step. For JDSYM we also give the quantities it_{int} and it_{out}. The latter is the number of correction equations that had to be solved. The number it_{int} provides the average number of inner iterations, i.e., the steps until MINRES solver the correction equation to the desired accuracy. This number is quite low as the accuracy requirements is initially very loose and becomes more stringent as the outer iteration converges [2]. In this example we let the subspace dimensions in JDSYM vary between p and $2p$.

The Jacobi-Davidson performs best with the simple preconditioners while the simpler eigensolvers profit most from the more sophisticated preconditioners. With the two-level preconditioners the execution times of BRQMIN and LOBPCG are shorter than those of JDSYM by a factor of about 1.5.

With all three eigensolvers higher sophistication in the preconditioner decreases not only the iteration counts but also the execution times. This is in contrast to the experiments that we conducted with node elements in the box shaped cavity [16].

Numbers not given here show that with the hierarchical basis preconditioner it_{int} does not increase with the problem size as the analysis predicts. In fact, here it even decreased substantially.

Notice that t_{11} takes a considerable fraction of the solution of the eigenvalue problem. It is clear, that the present two-level preconditioner will not suffice for very problem sizes. A multi-level method will have to replace the direct solver that is used for solving systems involving the matrices A_{11} and H.

6.3 The big model of the copper cavity

Finally we discuss is a bigger model of the copper cavity with $n = 119'758$ and $m = 23'400$. We computed 10 eigenpairs with the nullspace method as well as with the Arbenz-Drmač approach. The convergence criterion was $\|A\mathbf{x} - \rho(\mathbf{x})M\mathbf{x}\| \leq 10^{-4}$. The best preconditioner of the previous examples was chosen. Timings are given in Table 3.

Table 3. Timings for computing 10 eigenvalues of the problem of size $n = 119'758$ with $m = 23'400$ constraints. Times are in seconds.

method, preconditioner	JDSYM	LOBPCG	BRQMIN
null space, 2-level/SSOR(1)	1803	849	1124
AD, 2-level/SSOR(1)	1024	668	990

LOBPCG turns out to be the fastest solver. In this example it is clearly superior to BRQMIN. The AD approach is much faster that the nullspace method. In the AD approach, JDSYM needs to solve (approximately) $it_{\text{out}} = 80$ correction equations with altogether $it_{\text{inner}} = 326$ iteration steps to extract the ten eigenpairs. This means that in the average four iteration steps are used per system of equations. LOBPCG in turn executes 33 block iteration steps which requires 251 calls of the preconditioner. This ratio explains to a large extent the ratio of the execution times. Notice that there are less than $33p = 330$ invocations of the preconditioner as converged eigenvectors are locked, i.e. kept fixed.

The main portions of the execution time of 1024 seconds of JDSYM are matrix-vector products (152" or 15%), applying the preconditioner (294" or 29%) and applying the projector (465" or 45%). The bulk of the latter is solving system involving H. The corresponding numbers for LOBPCG are 88" (13%) for matrix-vector products, 134" (20%) for the projector and 326" (49%) for matrix-vector product with $\widehat{M}_{11} - \widehat{C}_1 H^{-1} \widehat{C}_1^T$.

7 Conclusions

Jacobi-Davidson is in general a very powerful method that can be applied to every kind of eigenvalue problems. For the special class of symmetric matrices it can be outperformed by Rayleigh quotient minimization-type algorithms. Good preconditioners are needed to get fast eigensolvers. In the cavity eigenvalue problems diagonal preconditioning was not satisfactory. Hierarchical preconditioning makes iteration numbers insensitive to problem size. But our two-level approach turns out to be ineffective with large problem sizes as the 'coarse' systems are getting too big. We intend to replace the direct solvers by a conjugate gradient method with a algebraic multilevel preconditioner.

Acknowledgment

I thank Roman Geus for performing the experiments on the Sun.

References

1. Grimes, R., Lewis, J.G., Simon, H.: A shifted block Lanczos algorithm for solving sparse symmetric generalized eigenproblems. SIAM J. Matrix Anal. Appl. **15** (1994) 228–272
2. Arbenz, P., Geus, R.: A comparison of solvers for large eigenvalue problems originating from Maxwell's equations. Numer. Lin. Alg. Appl. **6** (1999) 3–16
3. Arbenz, P., Geus, R., Adam, S.: Solving Maxwell eigenvalue problems for accelerating cavities. Phys. Rev. ST Accel. Beams **4** (2001) 022001 (Electronic journal available from `http://prst-ab.aps.org/`).
4. Sleijpen, G.L.G., van der Vorst, H.A.: A Jacobi-Davidson iteration method for linear eigenvalue problems. SIAM J. Matrix Anal. Appl. **17** (1995) 401–425
5. Lehoucq, R.B., Sorensen, D.C., Yang, C.: ARPACK Users' Guide: Solution of Large-Scale Eigenvalue Problems by Implicitely Restarted Arnoldi Methods. SIAM, Philadelphia, PA (1998).
6. Longsine, D.E., McCormick, S.F.: Simultaneous Rayleigh–quotient minimization methods for $Ax = \lambda Bx$. Linear Algebra Appl. **34** (1980) 195–234
7. Knyazev, A.V.: Toward the optimal preconditioned eigensolver: Locally optimal block preconditioned conjugate gradient method. SIAM J. Sci. Comput. **23** (2001) 517–541
8. Kikuchi, F.: Mixed and penalty formulations for finite element analysis of an eigenvalue problem in electromagnetism. Comput. Methods Appl. Mech. Eng. **64** (1987) 509–521
9. Girault, V., Raviart, P.A.: Finite Element Methods for the Navier-Stokes Equations. Springer-Verlag, Berlin (1986) (Springer Series in Computational Mathematics, 5).
10. Nédélec, J.C.: Mixed finite elements in $\mathbb{R}^3$. Numer. Math. **35** (1980) 315–341
11. Silvester, P.P., Ferrari, R.L.: Finite Elements for Electrical Engineers. 3rd edn. Cambridge University Press, Cambridge (1996)
12. Arbenz, P., Drmač, Z.: On positive semidefinite matrices with known null space. Tech. Report 352, ETH Zürich, Computer Science Department (2000) (Available at URL `http://www.inf.ethz.ch/publications/`).
13. Fokkema, D.R., Sleijpen, G.L.G., van der Vorst, H.A.: Jacobi-Davidson style QR and QZ algorithms for the partial reduction of matrix pencils. SIAM J. Sci. Comput. **20** (1998) 94–125
14. Geus, R.: The Jacobi-Davidson algorithm for solving large sparse symmetric eigenvalue problems. PhD thesis, Computer Science Department, ETH Zurich (2002)
15. Bank, R.E.: Hierarchical bases and the finite element method. Acta Numerica **5** (1996) 1–43
16. Arbenz, P., Adam, S.: On solvinging Maxwellian eigenvalue problems for accelerating cavities (1998) Paper presented at the International Computational Accelerator Physics Conference (ICAP'98), Monterey CA, September 14–18, 1998. 5 pages. Available from `http://www.inf.ethz.ch/~arbenz/ICAP98.ps.gz`.

Direct Axisymmetric Vlasov Simulations of Space Charge Dominated Beams

F. Filbet[1], J.-L. Lemaire[2], E. Sonnendrücker[3]

[1] IRMA, Université Louis Pasteur, F-67084 Strasbourg Cedex,
filbet@math.u-strasbg.fr
[2] CEA, B.P.12, F-91680 Bruyères-le-Châtel, jlemaire@bruyeres.cea.fr
[3] IRMA, Université Louis Pasteur, F-67084 Strasbourg Cedex,
sonnen@math.u-strasbg.fr

Abstract. A numerical method for the direct simulation of the axisymmetric Vlasov equation is introduced. It is based on a modified formulation of the Vlasov equation using the invariance of the canonical angular momentum. This leads in particular to a straightforward and very efficient parallel algorithm. Then it is applied to simulations of a RMS-matched semi-Gaussian beam and a perturbed thermal equilibrium.

1 Introduction

Eulerian direct Vlasov simulation of space charge dominated beams has proven to be an efficient alternative to PIC methods as it is completely devoid of numerical noise. It enables in particular to get a better insight into phenomena happening at the edge of the beam where the distribution function is very small. These regions are generally described by too few particles in PIC simulations.

We shall describe a three-dimensional r, v_r, v_θ axisymmetric transverse solver involving fewer dimensions than a comparable cartesian solver, which would be four dimensional in phase space, enabling us to use a larger number of grid points.

The method is based on the use of the canonical angular momentum which is invariant, and thus only appears as a parameter in the equations. Thanks to the use of this invariant, a straightforward very efficient parallelization is achieved.

The code is then validated on two test-cases involving heavy ions, the evolution of a transverse space-charge wave in a RMS-matched semi-Gaussian beam and the formation of a halo in a beam where a perturbation from a Maxwell-Boltzmann thermal equilibrium is introduced.

The outline of the paper is as follows : We shall first recall the axisymmetric Vlasov equation and its properties. Then, we present the discretization of the axisymmetric Vlasov equation. And finally we present numerical results for the cases of a semi-Gaussian beam and a perturbed thermal equilibrium.

P.M.A. Sloot et al. (Eds.): ICCS 2002, LNCS 2331, pp. 305–314, 2002.

2 The axisymmetric Vlasov equation

We consider an axisymmetric beam uniform in the longitudinal direction. It can be represented by the axisymmetric Vlasov equation, which describes the evolution of a species of charged particles under applied and self-consistent fields, and reads

$$\frac{\partial f}{\partial t}+v_r\frac{\partial f}{\partial r}+\left(\frac{q}{m}(E_s+E_a)+\frac{q\,B_z}{m}v_\theta+\frac{v_\theta^2}{r}\right)\frac{\partial f}{\partial v_r}-\left(\frac{q\,B_z}{m}v_r+\frac{v_\theta\,v_r}{r}\right)\frac{\partial f}{\partial v_\theta}=0. \tag{1}$$

where the distribution function f is a function of radial position r, velocity (v_r, v_θ) and time t. We assume here that the applied magnetic field is longitudinal and uniform, i.e. $B=(0,0,B_z)$, where $B_z(t)$ only depends on time. The associated vector potential then has only a non vanishing A_θ component the value of which is $A_\theta=\frac{r}{2}B_z$. The self-consistent electric field $E_s(t,r)$, deriving from a scalar potential ϕ_s is given by the axisymmetric Poisson equation which reads

$$\frac{1}{r}\frac{\partial\, r E_s}{\partial r}=\rho(t,r)/\varepsilon_0,\quad \rho(t,r)=q\int_{\mathbb{R}^2}f(t,r,v_r,v_\theta)dv_r\,dv_\theta. \tag{2}$$

The characteristic curves of the axisymmetric Vlasov equation are the solutions of the following differential system

$$\begin{cases}\dot r=v_r\\ \dot v_r=\dfrac{v_\theta^2}{r}+\dfrac{q}{m}v_\theta B_z+\dfrac{q}{m}E_s\\ \dot v_\theta=-\dfrac{v_\theta v_r}{r}-\dfrac{q}{m}v_r B_z.\end{cases} \tag{3}$$

Classical invariants of the axisymmetric Vlasov equation (1) are the Hamiltonian

$$H(r,v_r,v_\theta)=\frac{1}{2}mv_r^2+\frac{1}{2m}(mv_\theta-q\frac{r}{2}B_z)^2+q\,\phi_s \tag{4}$$

and the canonical angular momentum

$$P(r,v_\theta)=mrv_\theta+\frac{r^2}{2}qB_z. \tag{5}$$

Let us make use of this last invariant, as suggested in [3], to simplify equation (1). Denoting by $I=\frac{P}{m}$ and making the change of variable $(r,v_r,v_\theta)\to(r,v_r,I)$ with

$$v_\theta=\frac{I}{r}-\frac{1}{2}\frac{q\,B_z}{m}r,$$

we get

$$\frac{\partial f}{\partial t}+v_r\frac{\partial f}{\partial r}+\left(\frac{q}{m}E_s(t,r)+\frac{I^2}{r^3}-\frac{1}{4}\left(\frac{q\,B_z}{m}\right)^2 r\right)\frac{\partial f}{\partial v_r}=0,\quad \forall I\in\mathbb{R}. \tag{6}$$

This new formulation of the axisymmetric Vlasov equation is particularly well adapted to parallelization as the variable I only plays the role of a parameter.

3 Discretization of the axisymmetric Vlasov equation

We use a grid in phase space (r, v_r, I). Yet it is necessary to take particular care of the I direction. Indeed, when the self consistent electric field is linear, as for the K-V distribution function, the characteristic curves associated to Eq. (6), along which the distribution function is constant, are of the form

$$\frac{\omega^2}{2} r^2 + v_r^2 + \frac{I^2}{r^2} = const.$$

Hence it is necessary to control the ratio I/r, therefore we discretize the I direction so that

$$I = \pm \omega\, r^2.$$

We then use the conservation of the distribution function along the characteristics to devise the numerical algorithm which will be based on the semi-Lagrangian methodology [5]. The new values are computed at the grid points in two steps: (i) compute the origin of the characteristic ending at the grid point one time step back, (ii) interpolate the value of the distribution function there, which is also the new value at the grid point, from the old values at the surrounding grid points. This method is not subject to a Courant condition on the time step which would be very restrictive near the axis $r = 0$.

In the axisymmetric Vlasov equation the I^2/r^3 factor acts like a repulsion potential with respect to the total electric field. This potential is largest near the axis, where the electric field is negligible. As usual for axisymmetric problems, the major difficulty in the discretization of the Vlasov equation in cylindrical coordinates lies in the handling of the equation near the axis $r = 0$. The most natural method would consist in separating the free transport part which can be solved explicitly from the self-consistent part. However, numerical errors would be generated near the axis and propagate inside the domain, and our goal here is to devise a very precise numerical method for which this is unacceptable. So, we shall go with a classical operator splitting method, and split between advection in r and advection in v_r.

On the time interval $[t^n, t^{n+1}]$ we proceed as follows: the distribution function at time t^n is given by $f^n(r, v_r, I)$, we first compute f^* such that

$$\begin{cases} \dfrac{\partial f^*}{\partial t} + v_r \dfrac{\partial f^*}{\partial r} = 0, \\ f^*(0, r, v_r, I) = f^n(r, v_r, I). \end{cases} \tag{7}$$

We then compute the self-consistent electric field E_s from the intermediate approximation $f^*(\Delta t, r, v_r, I)$. Then f^{**} such that,

$$\begin{cases} \dfrac{\partial f^{**}}{\partial t} + \left(\dfrac{q}{m} E_s(t, r) + \dfrac{I^2}{r^3} - \dfrac{1}{4} \left(\dfrac{q\, B_z}{m} \right)^2 r \right) \dfrac{\partial f^{**}}{\partial v_r} = 0, \\ f^{**}(0, r, v_r, I) = f^*(\Delta t, r, v_r, I). \end{cases} \tag{8}$$

Finally, $f^{n+1}(r, v_r, I) = f^{**}(\Delta t, r, v_r, I)$.

The discretization of equation (7) requires to apply artificial boundary conditions. Actually, for this equation, at $r = 0$ and for $v_r > 0$, the particle flux is incoming, whereas particles with velocity $v_r < 0$ leave the computational domain. Thus, we need to model how the particles cross the axis $r = 0$. This can be done by imposing specular reflection conditions:

$$f(0, v_r, I) = f(0, -v_r, I), \quad \forall\, v_r > 0.$$

The numerical resolution of transport equations (7) and (8) is then performed using a semi-Lagrangian method with a cubic Hermite interpolation, using the values of the function and its derivative at the end points of the interval. Let us describe it in details for equation (7). On an interval $[r_i, r_{i+1}]$, we approximate the derivative of the distribution function at each grid point by a fourth order finite difference scheme:

$$\partial_r f_i^n = \frac{1}{12\Delta r}\left[8\left[f_{i+1}^n - f_{i-1}^n\right] - \left[f_{i+2}^n - f_{i-2}^n\right]\right].$$

The polynomial reconstruction is then given on each interval $[r_i, r_{i+1}]$ by the cubic polynomial interpolating the distribution function and its derivatives on the grid

$$\begin{aligned} f^n(r) = f_i^n + (r - r_i)\,\partial_r f_i^n + (r - r_i)^2\left[3[f_{i+1}^n - f_i^n] - \Delta r\left[2\partial_x f_i^n + \partial_x f_{i+1}^n\right]\right] \\ + (r - r_i)^3\left[\Delta x\left[\partial_r f_{i+1}^n + \partial_x f_i^n\right] - 2\left[f_{i+1}^n - f_i^n\right]\right]. \end{aligned}$$

This formula allows to evaluate the distribution function anywhere on the grid. It only remains to use the characteristic curves which can be solved explicitly on each split step to compute the distribution function at the grid points at time t^{n+1}.

4 Numerical results

4.1 Semi-Gaussian beam

We want to study here the evolution of an axisymmetric semi-Gaussian beam. Therefore we solve the Vlasov-Poisson in cylindrical coordinates, with an applied uniform and constant longitudinal magnetic field B_z. Then the distribution satisfies the Vlasov equation 6. The initial distribution function describing a semi-Gaussian beam in is given by

$$f_0(x, y, v_x, v_y) = \frac{n_0}{(2\,\pi\, v_{th}^2)\,(\pi\, a^2)}\exp\left(-\frac{(v_x - \frac{q\,B_z}{2\,m}y)^2 + (v_y + \frac{q\,B_z}{2\,m}x)^2}{2\,v_{th}^2}\right),$$

for $x^2 + y^2 \leq a^2$ and $f_0(x, y, v_x, v_y) = 0$, if $x^2 + y^2 > a^2$. The magnetic field B_z and the thermal velocity v_{th} are computed from RMS quantities, so that the beam is equivalent to a matched K-V beam.

The beam particles are singly ionized potassium ($Z = 1$, $m = 39.1\,amu$). The density n_0 is computed from the current I=0.2 A and the beam velocity along the z-axis

$$v_z = c\sqrt{\frac{\gamma^2 - 1}{\gamma^2}}, \quad \gamma = 1 + \frac{q}{m\,c^2}K,$$

where K is the beam kinetic energy $K = 8.10^4$ eV. Finally the beam radius is a=0.02 m and the tune depression is $\omega/\omega_0 = 1/4$.

We observe a space charge wave starting from the edge of the beam, propagating inwards and finally being reflected on the axis $r = 0$. The initial self-consistent field is linear within the beam. The variations are relatively weak but sufficient to strongly perturb the density (Fig. 1). Solving the axisymmetric equation allows us to eliminate one direction and thus to use a finer mesh than in the cartesian case and describe the distribution function more precisely. Moreover our new formulation of the Vlasov equation conserves the invariant $I = r\,v_\theta + \frac{q\,B_z}{2\,m}r^2$.

This method gives very satisfying results for the present test case. The results are comparable to those obtained with a cartesian code, as presented in [6], for the same resolution and the code is much faster. Moreover it is possible to go to much finer resolutions and then diminish the numerical damping.

Table 1. *Computational time for a 2D× 2D cartesian and axisymmetric solvers.*

Number of processors	2D Cartesian solver PFC	Axisymmetric Solver
4 processors	178 min	59 min
8 processors	89 min	27 min

4.2 Perturbed thermal beam

We start now from a dimensionless Maxwell-Boltzmann distribution ($q = m = 1$)

$$f_0(r, v_r, v_\theta) = \frac{\alpha}{2\pi}\exp\left(-H\right), \tag{9}$$

where H is the dimensionless Hamiltonian obtained from (4), coupled with the Poisson equation

$$-\frac{1}{r}\frac{\partial}{\partial r}\left(r\frac{\partial \phi_s}{\partial r}\right) = \alpha\exp(-\phi_s - \frac{r^2}{4}). \tag{10}$$

When α is different from zero or one, there are no analytical solutions of Poisson's equation (10). Therefore, we approximate the potential ϕ_s using a finite

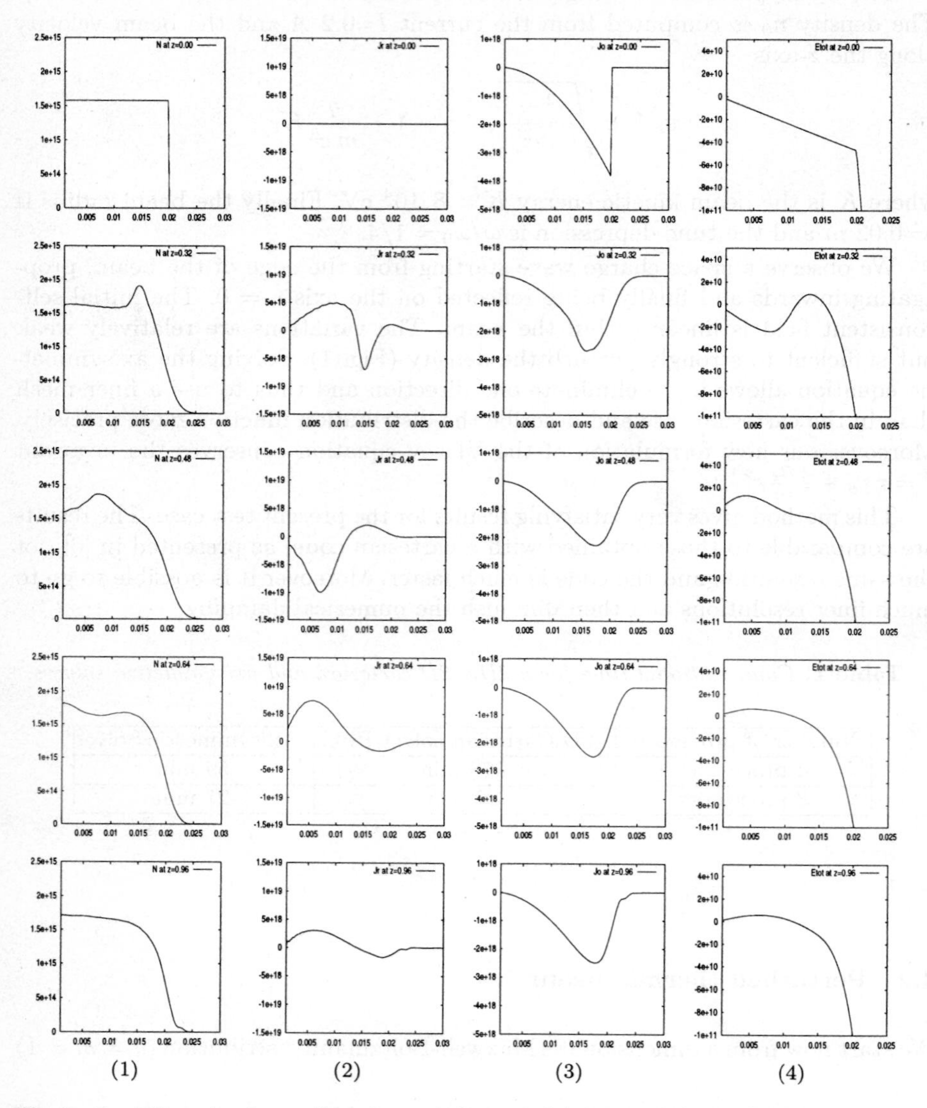

Fig. 1. *Semi-Gaussian beam. Snapshots of slices of: (1) density $N(t,r)$, (2) first moment in v_r, (3) first moment in v_θ, (4) total force field $E_s(t,r) - \left(\frac{q\,B_z}{2\,m}\right)^2 r$ inside the beam at times z=0, 0.32, 0.48, 0.64, 0.96 m.*

difference method : let $(r_i)_{i\geq 0}$ be a mesh of $]0, r_{max}]$

$$\phi_0 = 0,$$
$$\phi_1 = \alpha \Delta r^2/4,$$
$$\phi_i = \frac{r_{i-1}}{r_{i-1/2}}\left(2\,\phi_{i-1} - \phi_{i-2}\left(\frac{r_{i-3/2}}{r_{i-1}} - \alpha\Delta r^2 \exp(-\phi_{i-1} - r_{i-1}^2/4)\right)\right).$$

The parameter α is then determined by the tune-depression. Thus, we obtain a steady-state solution of the self-consistent Vlasov-Poisson system. The density is then increased by 50%. The beam parameters are the following : particles are singly ionized potassium ($Z = 1$, $m = 39.1\,amu$), current is $I = 0.2$ A, energy $K = 8.\ 10^4$ eV and radius $r_{max} = 0.01$ m. We first display snapshots of the RMS quantities

$$r_{rms} = (\overline{r^2})^{\frac{1}{2}}, \quad v_{r_{rms}} = (\overline{v_r^2})^{\frac{1}{2}}, \quad v_{\theta_{rms}} = (\overline{v_\theta^2})^{\frac{1}{2}},$$

and the RMS emittance ϵ_x, given by

$$\epsilon_x = (\overline{x^2 v_x^2} - \overline{x\,v_x}^2)^{\frac{1}{2}}.$$

As, $x = r\cos\theta$

$$\begin{aligned}\overline{x^2} &= \int_{\mathbb{R}^4} x^2\, f(t,x,y,v_x,v_y) dx\, dy\, dv_x\, dv_y \\ &= \int_{\mathbb{R}^3} r^2 f(r,v_r,I) dr\, dv_r\, dI \left(\int_0^{2\pi} (\cos\theta)^2 d\theta\right) \\ &= \pi\,\overline{r^2}.\end{aligned}$$

Then, as $v_x = v_r\cos\theta - v_\theta\sin\theta$

$$\begin{aligned}\overline{v_x^2} &= \int_{\mathbb{R}^4} v_x^2\, f(t,x,y,v_x,v_y) dx\, dy\, dv_x\, dv_y \\ &= \int_{\mathbb{R}^3} v_r^2 f(r,v_r,I) dr\, dv_r\, dI \left(\int_0^{2\pi} (\cos\theta)^2 d\theta\right) \\ &\quad + \int_{\mathbb{R}^3} v_\theta^2 f(r,v_r,I) dr\, dv_r\, dI \left(\int_0^{2\pi} (\sin\theta)^2 d\theta\right) \\ &= \pi\,(\overline{v_r^2} + \overline{v_\theta^2}).\end{aligned}$$

Finally, $x\,v_x = r\,v_r(\cos\theta)^2 - r\,v_\theta(\sin\theta)^2$,

$$\begin{aligned}\overline{x\,v_x} &= \int_{\mathbb{R}^4} x\,v_x\, f(t,x,y,v_x,v_y) dx\, dy\, dv_x\, dv_y \\ &= \pi\,\overline{r\,v_r}.\end{aligned}$$

which allows us to compute the emittance in cylindrical coordinates.

$$\epsilon_x = \pi\sqrt{\overline{r^2}\,(\overline{v_r^2} + \overline{v_\theta^2}) - \overline{r\,v_r}^2}.$$

Fig. 4 displays the evolution of the beam density, through slice plots on a logarithmic scale. It appears that, for a tune depression $\omega/\omega_0 = 1/2$, a plateau is formed at a density of around one thousandth of the core density. The snapshots are taken at times when the RMS value $v_{r_{rms}}$ is at an extremum as it is there where the halo can be best observed [4]. Our beam has a radius of 0.01 m and the dimension of the plateau is approximatively of 0.025 m, which corresponds to the maximal radius predicted by the empirical formula given in T.P. Wangler *et al.* [7] which is 0.0241 m.

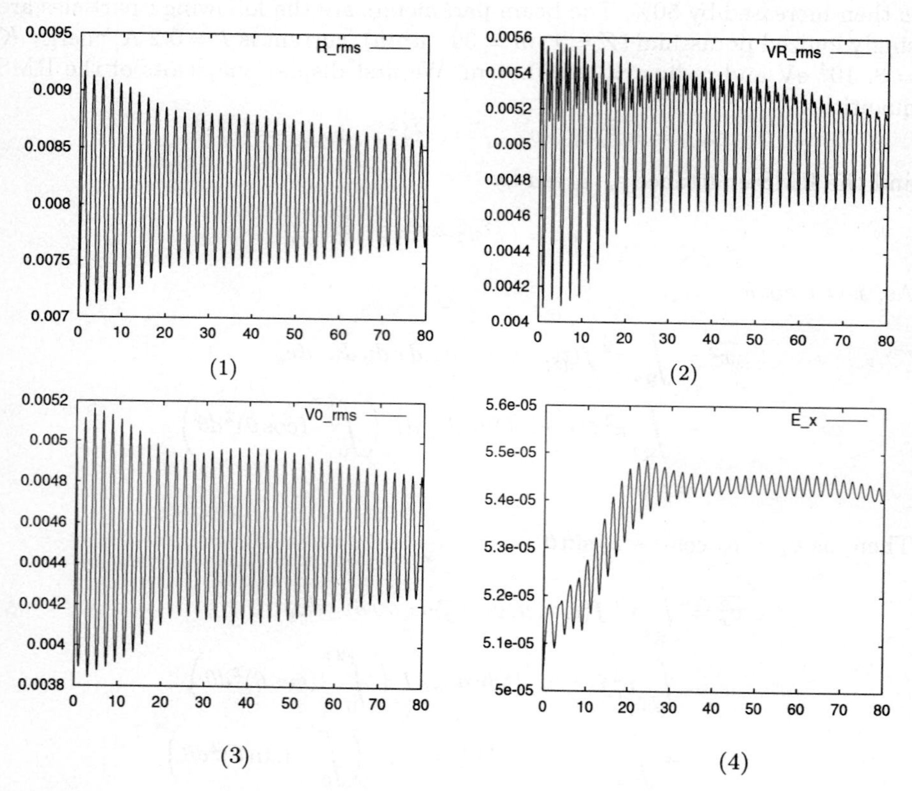

Fig. 2. *(1)* r_{rms}, *(2)* $v_{r_{rms}}$, *(3)* $v_{\theta_{rms}}$ *(4)* ϵ_x *for an axisymmetric Maxwell-Boltzmann beam.*

5 Conclusion

In this paper, we propose a new axisymmetric solver for the Vlasov equation. The formulation using invariants allows us to do staighforward and efficient parallel

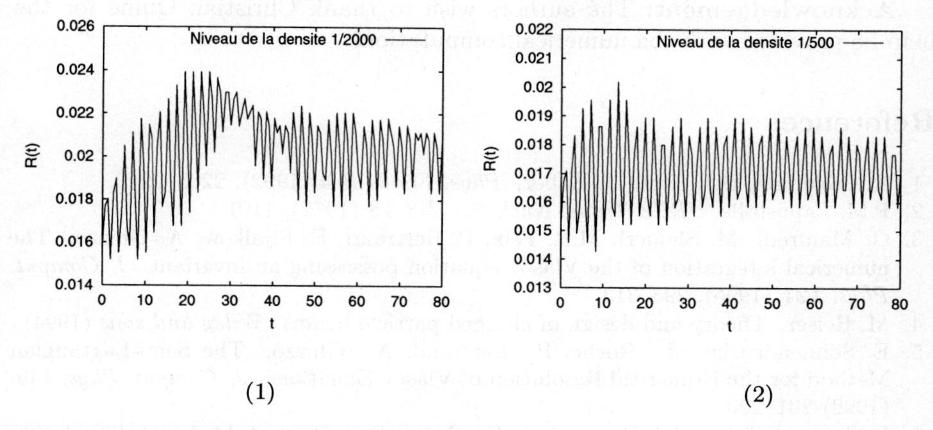

Fig. 3. *Axisymmetric Maxwell-Boltzmann beam. Snapshots of the isolines corresponding to: (1) one two thousandth of the total density, (2) one five hundredth of the total density.*

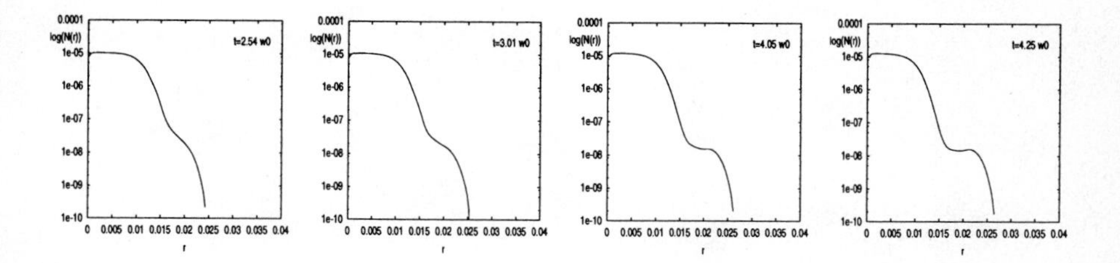

Fig. 4. *Axisymmetric Maxwell-Boltzmann beam. Snapshots of slice of density on a logarithmic scale.*

computations. The accuracy of the scheme is such that it is possible to observe halo formations and to get a good description of the distribution function in the phase space. These results are promising to treat more complex problems in accelerator physics.

Acknowledgement: The authors wish to thank Christian Quine for the help he provided with the numerical computations.

References

1. A. Friedman, D.P Grote, I. Haber, *Phys. Fluids B* **4** (1992), 2203–2210.
2. P.M. Lapostolle, *IEEE Trans. Nucl. Sci. NS* **18** (1971), 1101.
3. G. Manfredi, M. Shoucri, M.R. Feix, P. Bertrand, E. Fijalkow, A. Ghizzo, The numerical integration of the Vlasov equation possessing an invariant. *J. Comput. Phys.* **121** (1995), 298–313.
4. M. Reiser, Theory and design of charged particle beams. *Wiley and sons* (1994).
5. E. Sonnendrücker, J. Roche, P. Bertrand, A. Ghizzo, The Semi-Lagrangian Method for the Numerical Resolution of Vlasov Equations. *J. Comput. Phys.* 149: (1998) 201–220.
6. E. Sonnendrücker, J.J. Barnard, A. Friedman, D.P. Grote, S.M. Lund, Simulation of heavy ion beams with a semi-Lagrangian Vlasov solver, *Nuclear Instruments and Methods in Physics Research*, Section A, 464 (1-3), (2001), 653–661.
7. T.P. Wangler, R.W. Garnett, E.R. Gray, R.D. Ryne, T.S. Wang, Dynamics of beam halo in mismatched beams. *XVIII International Linac Conference, Genève, Suisse* (1996).

Fast Poisson Solver for Space Charge Dominated Beam Simulation Based on the Template Potential Technique

Leonid G. Vorobiev and Richard C. York

National Superconducting Cyclotron Laboratory, Michigan State University, East Lansing, MI 48824-1321, USA
{Vorobiev, York}@nscl.msu.edu
http://www.nscl.msu.edu

Abstract. A method for obtaining self-potentials and fields of charged particle beams called the template potential algorithm has been developed. The approach stems from the analytical Green's function formulation and is based on a discrete representation by auxiliary macro-elements or templates of the charge density distribution. Superposition of the potentials and fields including image forces for each template is used to reproduce the total potential and field of the original charge distribution within a conducting boundary. The technique is especially useful when the Poisson solver is being used repeatedly as with the simulation of charged particle beam dynamics in accelerators. Numerical results are presented and limitations of the method are discussed.

1 Introduction

The phrase fast Poisson Solver is most often used in relation to grid methods for solving the Poisson equation within a region $\mathcal{R}$,

$$\Delta u(\mathbf{x}) = -4\pi\rho(\mathbf{x}) \text{ for } \mathbf{x} = (x, y, z) \in \mathcal{R}$$

with specified conditions on the boundary $\partial\mathcal{R}$ using procedures based on the Fast Fourier Transform (FFT) or cyclic reduction for simple regions and multi-grid algorithms for more complicated boundary shapes. These techniques derive the grid potential $u(\mathbf{x})$ ($\mathbf{x} = (x, y, z)$) and field $E_{x,y,z}$ from a grid density $\rho(\mathbf{x})$ assumed known apriori by solving a set of finite-difference equations. The results usually have an accuracy of $\psi(h) = O(h^2 = h_x^2 + h_y^2 + h_z^2)$ (where $h_{x,y,z}$ are the mesh sizes). These solvers are in common use. See, e.g., [1] and references therein.

However, for charged particle beam simulation, at each step of the integration of the motion equations only the coordinates of N_p macro-particles are known. The spatial grid charge density $\rho(\mathbf{x})$ is derived from the position of the beam particles. Hence, reproduction of the grid density influences both the speed and the accuracy $\psi_\Sigma = \psi(h, N_p)$ of the calculated potential [2, 3]. Increasing the number of macro-particles, N_p, reduces the computational noise of the grid density at

P.M.A. Sloot et al. (Eds.): ICCS 2002, LNCS 2331, pp. 315–324, 2002.

the expense of computational speed. The use of fewer particles with numerical smoothing to reduce the noise, risks the masking of real physical phenomena.

This paper describes fast Poisson solvers for deriving self-potentials and fields from macro-particles coordinates for particle-in-cell (PIC) code simulation. We do not purpose to improve standard procedures such as density block or grid potential solvers. Instead, we introduce a new formulation based on the template potential concept [4, 5] for reconstruction of the total potential of charged particle beam in the presence of boundaries. This new approach allows a significant reduction in the number of macroparticles, N_p, and a sparser grid without concomitant loss of accuracy. The technique may be used for either envelope or PIC models in either two-dimensional (2D) or three-dimensional (3D) geometries. The template technique has been verified and shown to be appropriate for many practical space charge related applications.

2 Moment Method and Templates

A potential $u(\mathbf{x})$ generated by a charge density distribution $\rho(\mathbf{x})$ inside a volume $\mathcal{R}$ with specified boundary conditions on $\partial\mathcal{R}$ can be represented via the Green's function [6]. For free space, the potential $u(\mathbf{x})$ can be derived from a simple Green's function of the form $G_{free}(\mathbf{x}, \mathbf{x}') = 1/|\mathbf{x} - \mathbf{x}'|$. Difficulties arise in the presence of a conducting boundary. As a practical manner, the Green's function approach is limited to simple beam distributions and surface $\partial\mathcal{R}$ geometries.

Nevertheless the idea of the Green's function can be employed for numerical Poisson solvers, appropriate for rather general beam distributions and boundary shapes. In the moment method [7], the Green's function formulation is used as a part of computational technique, where the total potential is represented as $u_{total}(\mathbf{x}) = u_{free}(\mathbf{x}) + u_{image}(\mathbf{x}),\ \mathbf{x} \in \mathcal{R}$ satisfying on the boundary $u_{total}|_{\partial\mathcal{R}} = 0$:

$$u_{total}(\mathbf{x}) = \int_{\mathcal{R}} \rho(\mathbf{x}')G_{free}(\mathbf{x}, \mathbf{x}')d\mathbf{x}' + \int_{\partial\mathcal{R}} \sigma_{image}(\mathbf{x}')G_{free}(\mathbf{x}, \mathbf{x}')ds' \quad (1)$$

The potential u_{free} is produced by the charge density ρ in free space with $G_{free}(\mathbf{x}, \mathbf{x}')$. The corresponding image-potential, $u_{image}(\mathbf{x})$, defined by σ_{image} is found from a set of equations $\|A\|\, \sigma_{image} = -u_{free}$ that satisfy the constraint $u_{total}(\mathbf{x})|_{\mathbf{x}\in\partial\mathcal{R}} = 0$ on the conducting surface. Thus, the diffuculty in the construction of the proper Green's function for complex geometries is replaced by finding the image density.

Now the moment method with some modifications is known as the charge density method with numerous applications in, e.g., ion optics [8]. Both methods are slower than contemporary grid methods, but can be used in situations in which rapid grid algorithms may be difficult to employ.

In a previous paper [4], we introduced a numerical method, called the slice algorithm, based on the use of the template potential concept for space charge calculations of a 3D bunched beam. The beam bunch is represented by charged, infinitesimally thin disks, or slices, and the total beam potential is found by the superposition of the potentials from all slices. The space charge potential of an individual slice of radius R_{slice} within a conducting boundary is found by the moment method (1). Then, the matrix $||A||$ is calculated, σ_{image} is derived from the set of equations, and the total potential u_{total} obtained from:

$$u_{total}^{slice}(\mathbf{x}, S(z)) = u_{free}^{slice}(\mathbf{x}, S(z)) + u_{image}^{slice}(\mathbf{x}, S(z)) \quad (2)$$

where $S(z)$ is a shape function which defines the longitudinal z-profile of the beam for the case of round slices. For elliptical slices, there are two shape functions $S_{x,y}$, which determine the z-profiles. Fig. 1 illustrates the moment method.

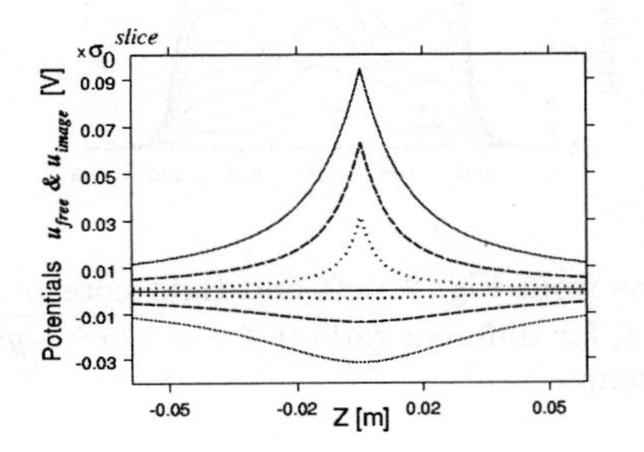

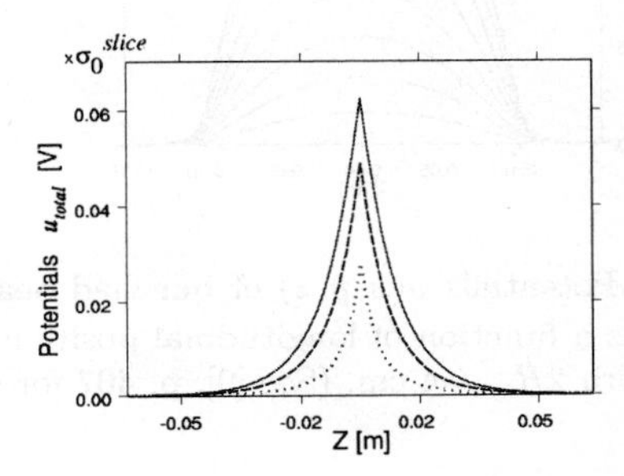

Fig. 1. Template potentials produced by three charged slices of different radii within a conducting beam pipe 4 cm in diameter respectively. Left: $u_{free}(\mathbf{x})$ (positive) and $u_{image}(\mathbf{x})$ (negative) with $\mathbf{x} = (0, 0, z)$, plotted with dotted, dashed and solid lines for slices of 1, 2 and 3 cm in diameter. Right: The corresponding total potentials $u_{total} = u_{free} + u_{image}$ for each slice respectively.

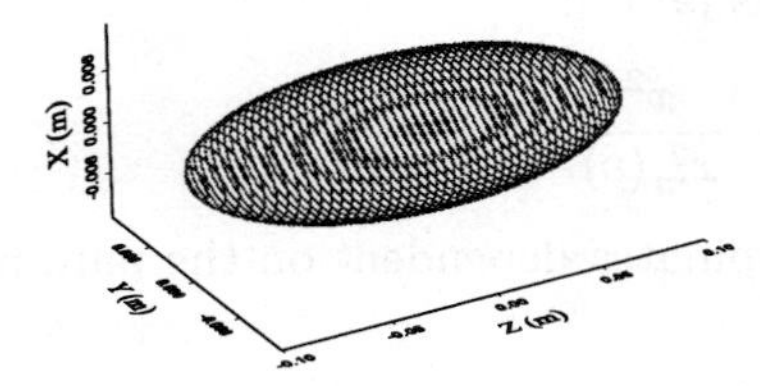

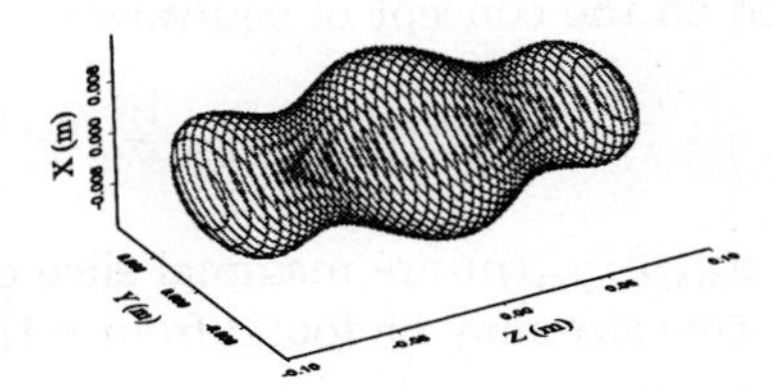

Fig. 2. Two possible 3D beams within a conducting beam pipe 4 cm in diameter represented by $N_s = 50$ charged slices. Left: Ellipsoid-like beam bunch with semiaxes $R_0 \times R_0 \times z_m$ =1 cm×1 cm×10 cm and the shape function $S(z) = R_0\sqrt{1 - (z/z_m)^2}$. Right: Beam bunch with more general longitudinal variation.

For this specific case all slices were assumed to have constant transverse density $\sigma^{slice}(r) \equiv \sigma_0^{slice}$. The potential for each template satisfies the zero potential boundary conditions and therefore, the superposition of all N_s slices representing the total beam bunch satisfies the Poisson equation. In Fig. 2 are shown two possible bunched beam geometries. In both cases, the beam was assumed to have a total charge of 10^{-11} C and be contained in a conducting chamber 4 cm in diameter. The potentials u_{total}^{bunch} shown in Fig. 3 were obtained by superposition of the potentials of N_s =50 slices, representing the bunch. These N_s slice potentials were found by appropriate scaling and interpolation of the tabulated template potentials.

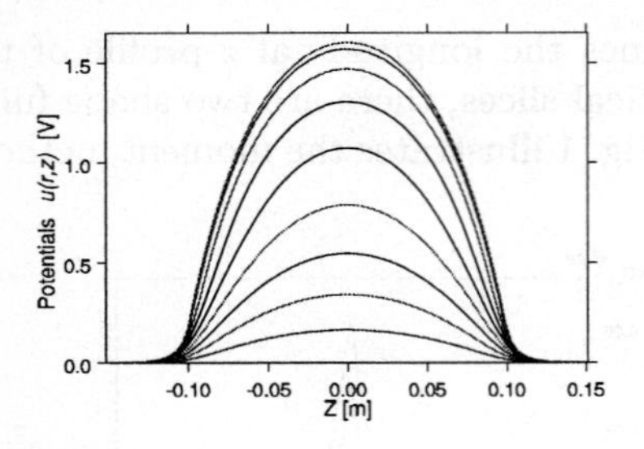

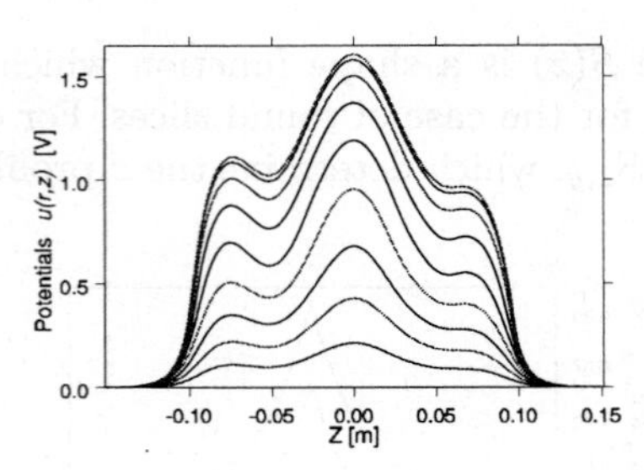

Fig. 3. Potentials $u(x, y, z)$ of bunched beams from Fig. 2 (left and right corespondingly) as a function of longitudinal position, z, for different radii $0 \leq r = \sqrt{x^2 + y^2} \leq R_{cyl}$, with $2R_{cyl}$=4 cm. (See [9], p. 407 for comparison).

3 Templates for Arbitrary Beams

The template approach can be used for variable transverse charge distributions and elliptical beam shapes common to quadrupole focusing channels. However, the beam is assumed centered within the conducting chamber.

A general two-dimensional model of the charge density, σ^{slice}, for templates is based on the concept of equivalent beams [9–11]:

$$\sigma^{slice}(x, y, p) = \sigma_m(p)\Big\{1 - \frac{x^2}{x_m^2(p)} - \frac{y^2}{y_m^2(p)}\Big\}^p \tag{3}$$

where $x_m(p), y_m(p)$ are maximal slice coordinates dependent on the parameter p. The rms size may be found from [11]:

$$< x^2 > (p) = \frac{\int_{x,y} x^2 \sigma^{slice}(x, y, p) dx dy}{\int_{x,y} \sigma(x, y, p) dx dy}$$

In Fig. 4, rms-matched 2D charge density distributions are plotted. For $p > 0$ the densities are maximum in the center going to zero near the edge. For $p = 0$, the charge density is constant, and for $p < 0$ hollow beams may be represented. For $p \to \infty$ the charge density is Gaussian [11]. Thus, this approach allows the representation of a broad range of possible charge densities $\sigma^{slice}(x, y, p)$ with rms-matched transverse dimensions.

The rms sizes for beams of elliptical symmetry are given by

$$< x^2 > (p) = \frac{\kappa^2 r_m^2(p)}{2(p+1)} \quad \text{and} \quad < y^2 > (p) = \frac{\kappa^{-2} r_m^2(p)}{2(p+1)} \tag{4}$$

where the semiaxes are $a_x = \kappa r_m(p), a_y = \kappa^{-1} r_m(p)$ with κ being the aspect ratio.

For the case of non-round ($\kappa \neq 1$) beams, it is necessary to calculate the off-axis potentials only for the first quadrant: $\phi \in [0, \pi/2]$. (See Fig. 4.) We found

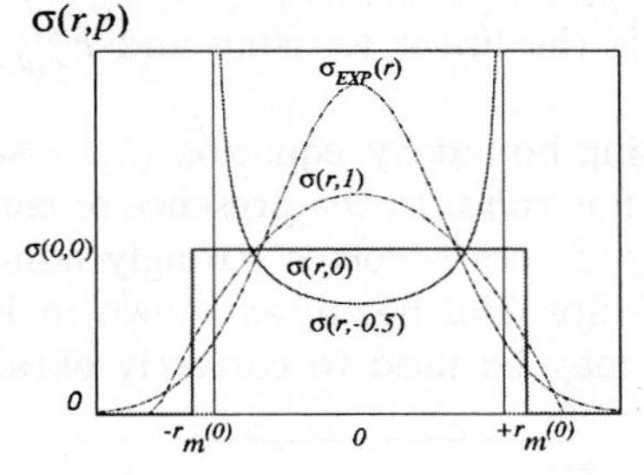

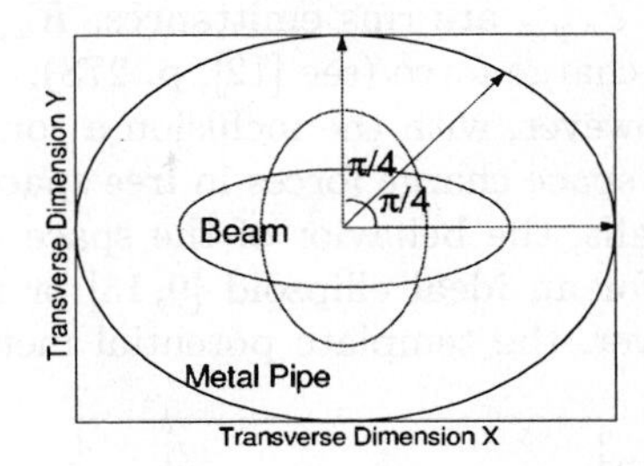

Fig. 4. Left: Rms-matched charge densities $\sigma(r,p)$ as functions of $r = \sqrt{x^2 + y^2}$ for different p. Right: Slice within elliptical conducting boundaries. For elliptical symmetry there will be horizontal and vertical shape functions $S_{x,y}(z)$. See section 4.

the template potentials along three rays ($\phi = 0, \pi/4, \pi/2$) to be sufficient to interpolate potentials for intermediate ϕ values. Further $10 - 15$ different aspect ratios (κ) were found to provide good accuracy for possible transverse beam configurations.

4 Beam Simulation, Using Templates

For the simulation of charged particle beams with significant space charge effects, either rms envelope equations or step-by-step PIC codes may be used. The envelope formalism though computationally fast, is appropriate only for beams with elliptical symmetry propagating through a linear focusing channel in the absence of image forces. The PIC methods are significantly slower, but accomodate arbitrary beam particle distributions, conducting chamber geometries, and focusing structures. The template technique may be applied with different degrees of generality to bridge the gap between rms envelopes and general PIC formulations:

4.1 Extension of 2D and 3D RMS Envelope Equations

In [11] it was shown how using the template technique the 2D rms envelope formalism [9, 12] can be extended to include the effects of a conducting elliptical chamber. In this context, charged cylinders, instead of disks, are used as templates. The difference between the more complete template approach and that presented by the free space KV formalism is most pronounced for elliptical bondaries with large aspect ratios. The rms envelope equations for $a_{x,y,z}$, may be also generalized for 3D ellipsoid-like beam bunch:

$$a''_{x,y,z} + K_{x,y,z}a_{x,y,z} - \frac{\varepsilon^2_{x,y,z}}{a^3_{x,y,z}} - F^{sc}_{x,y,z} = 0 \qquad (5)$$

where $\varepsilon_{x,y,z}$ are rms emittances, $K_{x,y,z}$ is the linear focusing and $F^{sc}_{x,y,z}$ the space charge force (see [12], p. 278).

However, with the inclusion a conducting boundary, equation (5), assuming linear space charge forces in free space, is not valid. In the presence of conducting walls, the behavior of the space charge forces becomes strongly non-linear even for an ideal ellipsoid [9, 13] or non-ellipsoidal beam, as shown in Fig. 5. However, the template potential method may be used to correctly obtain the

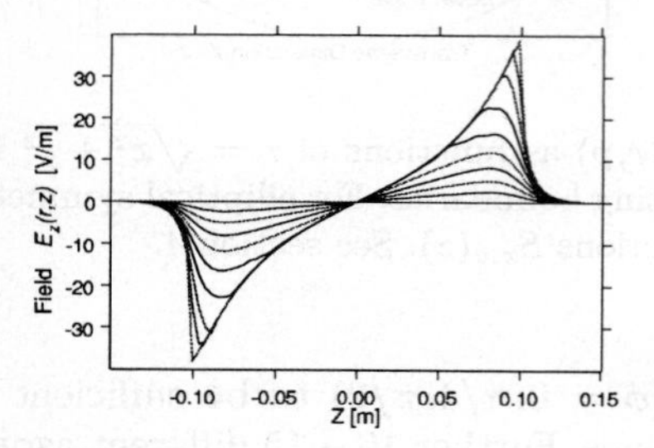

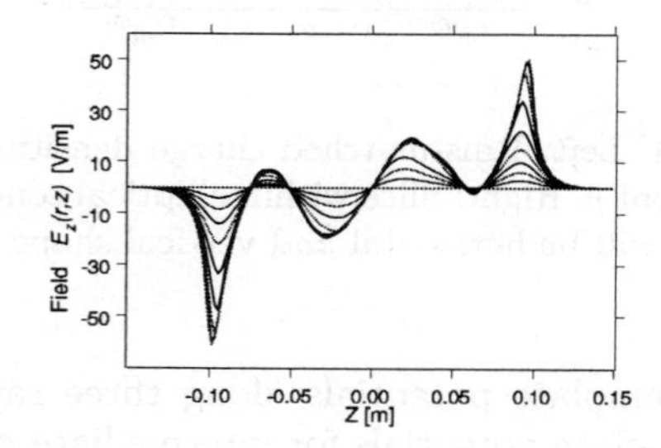

Fig. 5. Longitudinal space charge fields $E_z(x, y, z) = -\partial u/\partial z$ as a function of z, at different radii $0 \leq r = \sqrt{x^2 + y^2} \leq R_{cyl}$. Left: E_z for the ellipsoid-like beam. Right: E_z for the arbitrary beam from Fig. 2.

required fields, even for rather general beam distributions and boundary shapes. Both longitudinal F^{total}_z and transverse fields $F^{total}_{x,y}$ may be obtained from the potentials, by averaging and linearization of $F^{total}_{x,y,z}$. Substitution of this result in lieu of $F^{sc}_{x,y,z}$ in (5), provides a more self-consistent model.

4.2 3D Non-Ellipsoidal Beam

Arbitrary beams, like that of Fig. 2 (right), may not be appropriately accomodated by an envelope model. We need to include macro-particles $\{\mathbf{x}_i\}$, $i = 1, \ldots, N_p$ in the model. In this case, the template approach may again be applied to general beam distributions and conducting boundaries. The transverse rms beam dimensions $< x^2 >^{1/2} (z)$ and $< y^2 >^{1/2} (z)$, as functions of the longitudinal coordinate, z, are calculated. The shape function $S_{x,y}(z)$ is then found from (4) for a specific p. Previous analysis [11] has shown that for $0 < p < 3$ the result is not sensitive to the precise form of $S_{x,y}$. See Fig. 6. The space charge fields $E^{total}_{x,y,z}$ are derived from $u_{total}(x, y, z)$, $u_{total}(x \pm h_x, y \pm h_y, z \pm h_z)$ and define the space charge forces $\mathbf{F}^{total} = F^{total}_{x,y,z}$ on each particle. The integration of the motion equations:

$$\mathbf{x}''_i(s) + \mathbf{F}^{ext}(\mathbf{x}_i, s) - \mathbf{F}^{total}(\mathbf{x}_i, s) = 0 \qquad (6)$$

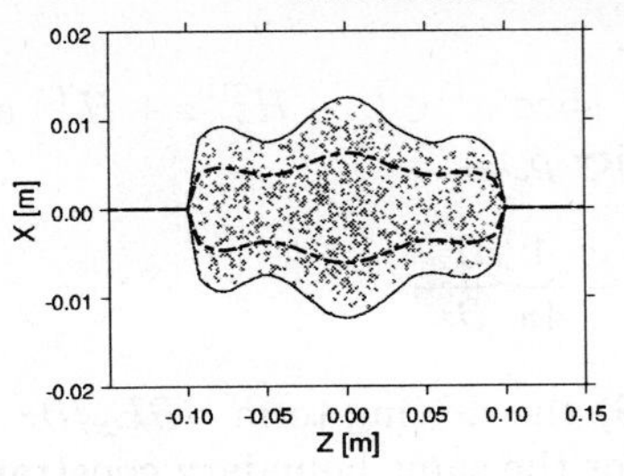

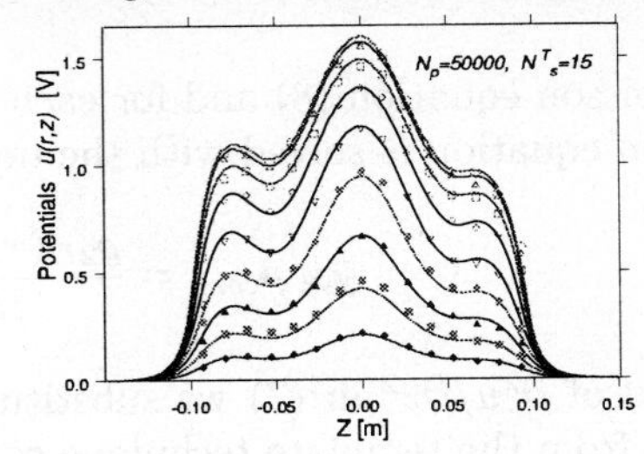

Fig. 6. Left: 3D bunched beam from Fig. 2 (right) within a conducting chamber of radius $R_{cyl} = 2$ cm, showing the macroparticles ensemble in ZX plane. The dashed line shows the rms beam size and the solid line corresponds the actual beam shape $S(z)$. Right: Potentials, found by the sub-3D Poisson Solver as a function of z, at different radii, $0 \leq r = \sqrt{x^2 + y^2} \leq R_{cyl}$. The solid lines are the solutions of the template technique and symbols are from the sub-3D Poisson Solver for N_z^T=15 "thick" slices.

can be implemented in a self-consistent manner even with arbitrary non-linear external forces $\mathbf{F}^{ext}$. After that, new coordinates $\{\mathbf{x}_i\}$ are found and new rms-profiles, shape functions, and total potentials obtained [14]. Note, that with this approach explicit calculation of charge density is not required. See Fig. 7.

The number of macroparticles required is relatively small, e.g., $N_p \propto 10^4$, since the macro-particles are used only to generate the shape function, $S_{x,y}(z)$, rather than the spatial density $\rho(\mathbf{x})$. In contrast, other PIC approaches such as those using cloud-in-cell (CIC) density algorithms, require $10^6 - 10^7$ macroparticles. As a result, the integration of equations (6) is very fast.

4.3 Sub-3D Poisson Solver for General Beam Simulation

The space charge potential in the region $\mathcal{R}$, satisfying zero boundary conditions at $\partial\mathcal{R}$ may be found from the 3D Poisson equation, which can be re-written as:

$$\frac{\partial^2 u}{\partial x^2} + \frac{\partial^2 u}{\partial y^2} = -4\pi\rho(x, y) - \frac{\partial^2 u}{\partial z^2}$$

Introducing the corrected charge density [5], we obtain:

$$\rho_{corr}(x, y, z) = \rho(x, y, z) + \frac{1}{4\pi}\frac{\partial^2 u}{\partial z^2} \tag{7}$$

Then the standard Poisson equation may be re-written as:

$$\frac{\partial^2 u}{\partial x^2} + \frac{\partial^2 u}{\partial y^2} = -4\pi\rho_{corr}(x, y) \tag{8}$$

Note, that the series of 2D solutions of (8) could be used to obtain the solution of the original Poisson equation if the term $\partial^2 u/\partial z^2$ would be known. Dividing the beam along the longitudinal, z, axis by N_z^T "thick" slices, we can rewrite the Poisson equation (8) and for each "thick" slice $z^T \in [z - H_z^T, z + H_z^T]$ a 2D Poisson equation is solved with the new density ρ_{corr}:

$$\rho_{corr} = \frac{\rho_{2D}(x, y, z^T)}{H_z^T} - \frac{1}{4\pi}\frac{\partial E_z}{\partial z} \tag{9}$$

Instead of $\partial^2 u/\partial z^2$ in (7) we substitute in (9) the driving term $-\partial E_z/\partial z$, obtained from the template technique solution for the same boundary constraints. Thus, the 2D Poisson solver for (8)-(9) for each "thick" slice finds the transverse self-potentials and electric fields $E_{x,y}$ with all possible generality, whereas the longitudinal fields E_z are supplied by the template technique. See Fig. 6. The corresponding space charge forces $F_{x,y,z}^{total}$ are substituted into the equation (6) for trajectories integration.

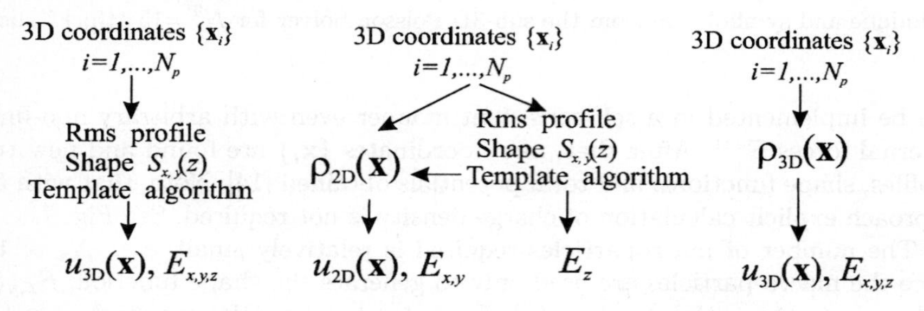

Fig. 7. Space charge calculations by different methods for one step of integration of the 3D beam motion equations. Left: Method from section 4.2. Middle: Sub-3D Poisson Solver from section 4.3. Right: Regular 3D Solver.

Note, that the generation of the transverse forces is separated from the longitudinal forces. The derivative $\partial E_z/\partial z$ participates in each of the 2D Solvers by correcting the 2D charge density. Without inclusion of the driving term $\partial^2 u/\partial z^2$ the calculation of $E_{x,y}$ would be in error [5]. Fig. 7 illustrates a step of integration using the sub-3D and a general 3D Poisson Solvers.

5 Templates as Special Functions

The tabulation of the template potentials prior to beam simulation follows the principle used for other special functions that do not have a simple analytical representation. During the actual space charge simulations, the template data is extracted from the table with appropriate scaling and interpolation. The storage requirements for the template table can be minimized. Since each template potential represents an even function, it is necessary to store only half of them, say $u(x, y, z)$ for $z > 0$. See Fig. 1. In addition, the template data could be parametrized and only the resulting coefficients stored. The number of templates required depends on the geometry of the problem. For the axially symmetric beam of Fig. 2, only 10 different radii were required. For each such template, 10 off-axis potentials should be calculated, resulting in a total of 100 templates. For cases with less symmetry, e.g., for beam and boundary with elliptical symmetry of Fig. 4 about of 3000 templates are required with the increase from necessary different azimuthal positions (3×) and aspect ratios (10×). Thus, the table of templates is of a moderate size. All intermediate quantities are obtained by interpolation and scaling of the tabulated data. For 2D cases, when there is no dependence on z, the memory demands are an order of magnitude less.

The accuracy of the sub-3D Poisson solver approaches is that of the general 3D PIC models. Nevertheless, the template formalism is not completely self-consistent because the pre-calculated data may not adequatedly reflect all possible evolutions of the particle distribution. Since the transverse 2D analysis can be used for arbitrary densities $\rho_{2D}(x, y,)$, the 2D problems are solved with all possible generality. The lack of self-consistency is in the replacement of $\partial^2 u/\partial z^2$ by $\partial E_z/\partial z$ in (9). Nonetheless, the quantitative analysis in [10,11] showed that for a large class of density distributions $\sigma(x, y, p)$ there is a relatively weak sensitivity of the longitudinal E_z fields to the details of transverse charge densities. As the result, the space charge fields: $E_{x,y}$ found from a series of 2D problems (8) and E_z, supplied by the slice algorithm, provide an accurate representation of space charge forces.

For beams, whose transverse densities may be described analytically (3), the template formalism is appropriate. However, for cases where the beam bunch has, e.g., isolated off-axis clusters, the template formalism is not appropriate and a general 3D PIC method should be employed.

6 Discussion and Conclusions

The template techique is oriented toward repeated calculations, e.g., for charged particle beam dynamics simulation. For those situations, where the beam pipe sizes are fixed or only slightly varying and the beam is "well-behaved" in the sense discussed above, the templates procedure significantly reduces simulation computational time. The verification of the method has shown a good agreement with general 3D grid solvers for a large class of charge density distributions. The proper inclusion of changing boundaries would require additional pre-calculated template data. This might be justified if only a few possible geometries are required. However, when the boundaries are complicated and/or changing significantly, and if the beam distribution is arbitrary (i.e. off-set from the axis, disintegrated into clusters, etc.) the template technique is likely inappropriate and conventional grid methods are recommended.

Preliminary numerical studies provide confidence that the template formalism is an efficient method for fast space charge calculations in the presence of conducting boundaries. It allows the extention of 2D and 3D beam rms-envelope equations including conducting boundaries. It provides a transition to self-consistent rather general sub-3D PIC, that is significantly faster than conventional 3D PIC formulations.

Acknowledgment

This work was supported by the U.S. Department of Energy under Contract No. DE-FG02-99ER41118.

References

1. Press, W.H., Teukolsky, S.A., Vetterling, W.,T., Flannery B.P.: Numerical Recipes. Second Edition. Cambridge (1992)
2. Hockney, R.W., Eastwood, J.W.: Computer Simulation Using Particles. McGraw-Hill, (1981)
3. Birdsall, C.K., Langdon, A.B.: Plasma Physics via Computer Similation. McGraw-Hill (1985)
4. Vorobiev, L.G., York, R.C.: Calculations of Longitudinal Fields of High-Current Beams within Conducting Chambers. In: Luccio, A., MacKay, W. (eds.): Proc. 1999 Particle Accelerator Conf., New York, IEEE, Piscataway, NJ (1999) 2781-2783
5. Vorobiev, L.G., York, R.C.: Space Charge Calculations for Sub-Three-Dimensional Particle-In-Cell Code. Phys. Rev. ST Accel. Beams **3**, 114201 (2000)
6. Jackson, J.D.: Classical Electrodynamics. Wiley, New York (1975)
7. Harrington, R.F.: Field Computation by Moment Methods. Macmillan Company, New York (1968)
8. Szilagyi, M.: Electron and Ion Optics. Plenum Press, New York (1988)
9. Reiser, M.: Theory and Design of Charged Particle Beams. Wiley, New York (1994)
10. Vorobiev, L.G., York, R.C.: Slice Algorithm – Advanced Version. Michigan State University Report MSUCL-1191, East Lansing (2001)
11. Vorobiev, L.G., York, R.C.: Method of Template Potentials to Find Space Charge Forces for High-Current Beam Dynamics Simulation. In: Lucas, P., Webber, S. (eds.): Proc. 2001 Particle Accelerators Conf., Chicago, IEEE, Piscataway, NJ (2001) 3075-3077
12. Wangler, T.P.: Principles of RF Accelerators. Wiley, New York (1998)
13. Vorobiev, L.G., York, R.C.: Numerical Technique to Determine Longitudinal Fields of Bunched Beams within Conducting Boundaries. Michigan State University Report MSUCL-1117, East Lansing (1998)
14. Vorobiev, L.G., York, R.C.: Template Potential Technique in Application to High-Current Beam Simulation. Michigan State University Report MSUCL-1225, East Lansing (2002)

Parallel Algorithms for Collective Processes in High Intensity Rings

Andrei Shishlo, Jeff Holmes, and Viatcheslav Danilov
Oak Ridge National Laboratory, SNS Project, 701 Scarboro Rd, MS-6473, Oak Ridge TN, USA 37830
{shishlo, vux, jzh}@sns.gov

Abstract. Computational three-dimensional space charge (3DSC) and wake field force algorithms were developed and implemented into the ORBIT computer code to simulate the dynamics of present and planned high intensity rings, such as PSR, Fermilab Booster, AGS Booster, Spallation Neutron Source (SNS), and proton driver. To provide affordable simulation times, the 3DSC algorithm developed for ORBIT has been parallelized and implemented as a separate module into the UAL 1.0 library, which supports a parallel environment based on MPI. The details of these algorithms and their parallel implementation are presented, and results demonstrating the scaling with problem size and number of processors are discussed.

1 Introduction

Collective beam dynamics will play a major role in determining losses in high intensity rings. The details of these processes are so complicated that a good understanding of the underlying physics will require careful computer modeling. In order to study the dynamics of high intensity rings, a task essential to the SNS project [1], we have developed direct space charge and impedance models in the macro-particle tracking computer code, ORBIT [2,3]. Initially, separate transverse space charge and longitudinal space charge/impedance models were developed, benchmarked, and applied to a number of problems [4,5]. We have now extended the impedance model to include the calculation of forces due to transverse impedances and, because such forces depend on the longitudinal variation of the beam dipole moments, the space charge model has been extended to three dimensions. In many cases, the resulting simulations including 3DSC calculations will require tracking tens of millions of interacting macro-particles for thousands of turns, which constitutes a legitimate high performance computing problem. There is little hope of carrying out such calculations in a single processor environment. In order to meet the need for credible simulations of collective processes in high intensity rings, we have developed and implemented the parallel algorithms for the calculation of these processes[1].

1.1 Parallel algorithms

The main goals of parallel computer simulations are to shorten the tracking time and to provide for the treatment of larger problems. There are two possible situations for

[1] Research on the Spallation Neutron Source is managed by UT-Battelle, LLC, under contract DE-AC05-00OR22725 for the U.S. Department of Energy.

P.M.A. Sloot et al. (Eds.): ICCS 2002, LNCS 2331, pp. 325–333, 2002.

tracking large numbers of particles with macro-particle tracking codes such as ORBIT. In the first case, particles are propagated through the accelerator structure independently without taking into account direct or indirect interactions among them, so there is no necessity for parallel programming. It is possible to run independent calculations using the same program with different macro-particles on different CPUs and to carry out the post-processing data analysis independently. In the opposite case there are collective processes, and we must provide communication between the CPUs where programs are running. Unfortunately, there is no universal efficient parallel algorithm that can provide communication for every type of collective processes. The best parallel flow logic will be defined by the mathematical approach describing the particular process and the ratio between computational and communication bandwidth. Therefore, our solutions for parallel algorithms cannot be optimal for every computational system.

Our implementation of parallel algorithms utilizes the Message-Passing Interface (MPI) library. The timing analysis has been carried out on the SNS Linux workstation cluster including six dual i586 CPUs with each having 512 kBytes L2 cache, 512 MB RAM and the 100 Mb/s Fast Ethernet switch for communication. The communication library MPICH version 1.2.1, a portable implementation of MPI, has been installed under the Red Hat 7.0 Linux operating system.

2 Transverse Impedance Model

The transverse impedance model in ORBIT [3] is based on an approach previously implemented in the longitudinal impedance model [5], which calculates the longitudinal kick by summing products of Fourier coefficients of the current with the corresponding impedance values, all taken at harmonics of the ring frequency [6]. A complication with the transverse impedance model arises due to the betatron motion, which has much higher frequency than synchrotron motion. Consequently, the harmonics of the dipole current must include the betatron sidebands of the revolution harmonics. Because of this and the fact that the number of transverse dimensions is two, the transverse impedance model requires four times as many arrays and calculations as does the longitudinal impedance. In the transverse impedance model, the kicks are taken to be delta-functions. This approximation is valid when the betatron phase advance over the physical extent of the impedance is small. If this is not the case, the impedance must be represented as a number of short elements, which is valid when the communication between elements is negligible. One exception to this rule is the resistive wall impedance. Because the resistive wake does not involve propagating waves, it can be treated as localized away from synchrobetatron resonances. When communication between elements is significant, then a more general Green's function approach, which is beyond the scope of this model, must be used.

2.1 Parallel Algorithm for Transverse Impedance Model

The parallel algorithm for ORBIT's transverse impedance model has been developed assuming that propagated macro-particles are arbitrarily distributed among CPUs. Typically, we must consider only a few transverse impedance elements in the accelerator lattice, and the resulting calculation time is small. Consequently, we derive this algorithm more for simplicity than for efficiency.

The parallel flow logic for the transverse impedance model is shown in Table 1. There are only two stages of calculation in which data is exchanged between CPUs. In the first stage the maximal and minimal longitudinal coordinates of all macro-particles must be determined. We used the MPI_Allreduce Collective Communication MPI function with the MPI_MAX parameter describing MPI-operation to find maximal and minimal values of the longitudinal coordinates for all CPUs. In the 5-th step we sum the array of transverse kick values for all CPUs and scatter results to all the processors by using the same MPI function with the MPI_SUM parameter.

Table 1. The parallel flow logic for the transverse impedance model. The "Communication" column indicates data exchange between CPUs

N step	Actions	Communication
1	Determine the extrema of longitudinal macro-particle coordinates and construct longitudinal grids for x and y dimensions	+
2	Distribute and accumulate the macro-particle transverse dipole moments for each direction onto the longitudinal grids	-
3	Calculate FFT values of total dipole moments in the mesh	-
4	Convolute the FFT coefficients with transverse impedance values to get transverse kick at each point in the longitudinal grids	-
5	Sum all transverse kicks across all CPUs	+
6	Apply resulting transverse kick to every macro-particle	-

A thorough timing of the transverse impedance parallel implementation was not made, because there are only a few impedance elements among several hundreds of elements in a typical case and their calculation consumes very little time. There is only one requirement, namely, the single processor version and the parallel version must give the same results. We have verified that this is the case to at least six significant figures in the coordinates of macro-particles for both codes.

3 Three-Dimensional Space Charge Model

The force in our three-dimensional space charge model is calculated as the derivative of a potential, both for longitudinal and transverse components. The potential is solved as a sequence of two-dimensional transverse problems, one for each fixed longitudinal coordinate. These separate solutions are tied together in the longitudinal direction by a conducting wall boundary condition $\Phi = 0$ on the beam pipe, thus resulting in a three-dimensional potential. This method depends for its legitimacy, especially in the calculation of the longitudinal force, on the assumptions that the bunch length is much greater than the transverse beam pipe size and that the beam pipe shields out the forces from longitudinally distant particles. Although our model is applicable to long bunches, and not to the spherical bunches of interest in many linac calculations, the three-dimensional space charge model adopted here is adequate to most calculations in rings.

The three-dimensional model implemented in ORBIT closely follows a method discussed by Hockney and Eastwood [7]. A three-dimensional rectangular grid, uniform in each direction, in the two transverse dimensions and in the longitudinal coordinate is used. The actual charge distribution is approximated on the grid by distributing the particles over the grid points according to a second order algorithm, called "triangular shaped cloud (TSC)" in [7]. Then, the potential is calculated independently on each transverse grid slice, corresponding to fixed longitudinal coordinate value, as a solution of a two-dimensional Poisson's equation. The charge distribution is taken from the distribution procedure and, for the two-dimensional equation, is treated as a line charge distribution. The two-dimensional Poisson equation for the potential is then solved using fast Fourier transforms and a Green's function formulation with periodic boundary conditions [8]. The periodic boundary conditions are used only to obtain an interim solution, and this solution is then adjusted to obey the desired conducting wall boundary conditions. These are imposed on a specified circular, elliptical, or rectangular beam pipe through a least squares minimization of the difference on the beam pipe between the periodic Poisson equation solution and a superposed homogeneous solution. The homogeneous solution is represented as a series constructed from a complete set of Laplace equation solutions with variable coefficients, as described in [9]. In addition to accounting for image forces from the beam pipe, these $\Phi = 0$ boundary conditions serve to tie together the independently solved potentials from the various longitudinal slices, resulting in a self-consistent three-dimensional potential.

Finally, with the potentials determined over the three-dimensional grid, the forces on each macro-particle are obtained by differentiating the potential at the location of the macro-particle using a second order interpolation scheme. The resulting forces include both the transverse and longitudinal components. The interpolating function for the potential is the same TSC function used to distribute the charge.

The detailed description of the three-dimensional space charge algorithm can be found in [10].

4 Parallel Algorithm for the 3D Space Charge Model

The approach to parallelization of the three-dimensional space charge algorithm is obvious. We distribute the two-dimensional space charge problems for solution to different CPUs. If the number of longitudinal slices is greater than the number of CPUs, then we must group the slices. To implement this scheme it is necessary to distribute the macro-particles among the CPUs before the solving two-dimensional problems. Then, after solving the two-dimensional problems, we must provide for the exchange of neighboring transverse grids (with potentials) between CPUs to carry out the second order interpolation scheme in the longitudinal coordinate necessary for calculating and applying the space charge force kick to the macro-particles. Therefore there should be a special module that distributes macro-particles between CPUs according their longitudinal positions. We call this module "The Bunch Distributor".

4.1 The Bunch Distributor Module

The "Bunch Distributor" module analyzes the longitudinal coordinates of macro-particles currently residing on the local CPU, determines which macro-particles don't belong to this particular CPU, and sends them to the right CPU. This means that the class describing the macro-particle bunch should be a resizable container including 6D coordinates of the macro-particle and an additional flag indicating macro-particles as "alive" or "dead". This additional flag provides the possibility to have spare space in the container and to avoid changing the size of container frequently.

The logic flow for the bunch distributor module is shown in Table 2. During the two first steps we define maximum and minimum longitudinal coordinates among all macro-particles in all CPUs. To eliminate the necessity of frequent changes in the longitudinal grid we add an additional 5% to each limit and save the result. During subsequent calls of the "Bunch Distributor" module we don't change the longitudinal limits unless necessary.

After defining the longitudinal grid, we sort macro-particles according the nearest grid point. Particles that no longer belong to the appropriate CPU are stored in an intermediate buffer together with additional information about where they belong. At the step 3 we define the exchange table $N_{ex}(i,j)$ where "i" is the index of the current CPU, "j" is the index of destination CPU, and the value is the number of macro-particles that should be sent from "i" to "j". After step 4 all CPUs know the number of macro-particles they will receive. The exchange table defines the sending and receiving procedures used in step 6; therefore we avoid a deadlock. Finally, all macro-particles are located in the correct CPUs, and we can start to solve the two-dimensional space charge problems on all CPUs.

Table 2. The flow logic for the "Bunch Distributor" module. The "Communication" column indicates data exchanging between CPUs

N stage	Actions	Communication
1	Determine the extrema of longitudinal macro-particle coordinates	-
2	Find the global longitudinal limits throughout all CPUs	+
3	Analyze macro-particle longitudinal coordinates to determine on which CPU they belong. Storing the 6D macro-particle coordinates to be exchanged in an intermediate buffer and mark these macro-particles as "dead". Define an exchange table $N_{ex}(i,j)$ (see text for the explanation)	-
4	Sum the exchange table throughout all CPUs by using the MPI_Allreduce MPI function with the MPI_SUM operation parameter	+
5	Check the spare place in the bunch container and resize it if necessary	-
6	Distribute the 6D macro-particle coordinates in the intermediate buffer to the correct CPUs according the exchange table. Store the received coordinates in the bunch container in the available places	+

4.2 Parallel 3D Space Charge Algorithm

In the parallel version of the three-dimensional space charge algorithm each CPU performs the same calculation of the potential on the transverse grids as the non-parallel version. There is no need for communication between CPUs, because the macro-particles have already been distributed between CPUs by the "Bunch Distributor" module and each CPU uses its own information to solve its own segment of the longitudinal grid. There is only one difference between parallel and non-parallel versions: In the parallel version there are two additional longitudinal slices beyond the ends of the CPU's own segment. Therefore the number of longitudinal slices for one CPU is $N_{slices}/N_{CPU}+2$ instead of N_{slices}/N_{CPU}, where N_{slices} is the total number of the transverse grid and N_{CPU} is the number of CPUs. The two additional slices are necessary because of the second order interpolation scheme. After the solution of the two-dimensional problems, the potential values from the two transverse grids on the ends of the segment should be sent to the CPU that is the neighbor according its index. In same fashion, the local CPU should obtain the potential values from its neighbors and add these potentials to its own. In this case the results of the parallel and non-parallel calculations will be the same.

5 Timing of Parallel Algorithm for the 3D Space Charge Model

Timings of the parallel algorithms were performed to elucidate the contributions of different stages in the total time of calculation and the parallel efficiency of their implementation. To avoid the effects of other jobs running on the same machine and other random factors, we did the timings on the Linux cluster with no other users and computed the average time for a number of iterations. We were able to use only five CPUs of our cluster because of the dual CPU effect.

5.1 Dual CPU Effect

Using two CPUs on the one node for parallel calculations drops the performance of our applications down by 20-30%. To clarify this situation we wrote a simple example that does not use communication between CPUs.

The executable code of the example

```
01: double xx[50000];
02: double r_arr [50000];
...
03: time_start = MPI_Wtime();
04: for( int j = 0 ; j < 275 ; j++){
05:  for (int i = 0; i < 50000 ; i++){
06:    x = x/(x+0.000001);
07:    r_arr[i] = x;
08:    xx[i] = x;
09:  }}
10:  time_stop = MPI_Wtime();
```

The execution time of the example is 1 sec for 1 CPU and 1.7 sec for 2 CPUs on the one dual CPU node. When we comment lines 07 and 08, the execution time does not depend on the number and sort of CPU's. This means that there is a competition between two CPUs with synchronized tasks on the one node for the access to the RAM if the 512 kBytes L2 cache of each CPU is not enough for data and code. To avoid this type of competition and the resulting performance drop, we use no more than 5 CPUs for each parallel run. This effect is significant for synchronized tasks only, so we can run two different parallel simulations at one time.

5.2 Timing of the Bunch Distributor Module

The timing of the bunch distributor module was carried out without including additional MPI functions in the code of the module. We measured the time needed to distribute macro-particles between CPUs according to their longitudinal positions when we have N_{part} previously distributed and N_{rand} undistributed macro-particles.

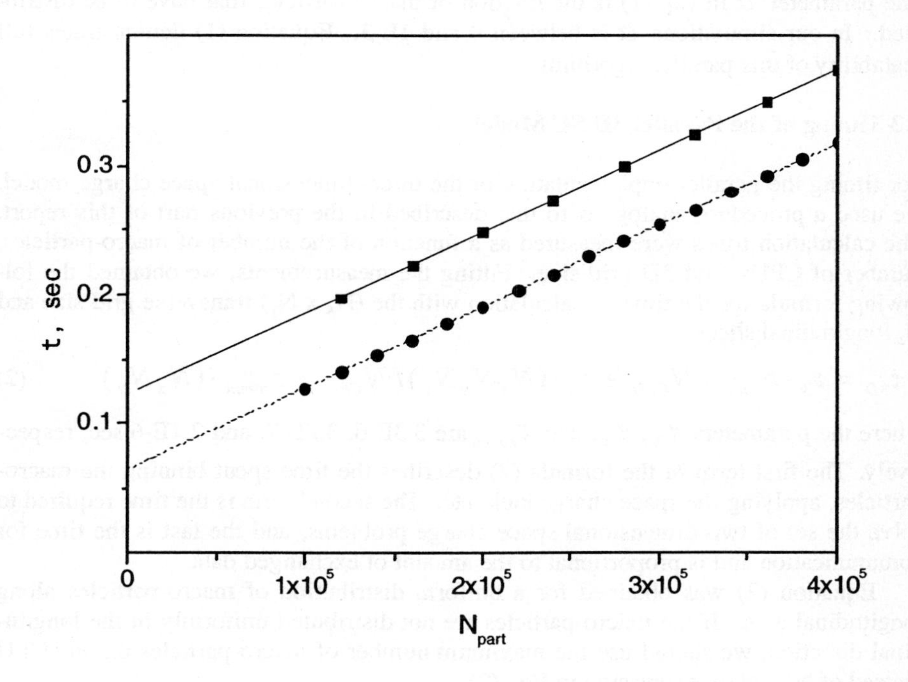

Fig. 1. The time required by the bunch distributor module to distribute N_{rand} between 2 CPUs in addition to N_{part} already distributed. The points are results of measurements, and the lines are linear approximations. The squares and circles denote N_{part} = 20000 and 10000 macro-particles accordingly.

Figure 1 shows the required time vs. N_{part} for 2 CPUs and N_{rand} = 20000 and 10000. As we expected, this time consists of two parts. The first part is proportional to the number of previously distributed particles. This is the time require for carrying out steps 1 and 3 in Table 2. The second part is proportional to the number of undistributed macro-particles that are distributed among CPUs during the step 6. Step 5 is normally carried out only once. The total execution time of steps 2 and 4 in Table 2 does not exceed 0.001 second for our case with $N_{CPU} < 6$. If the number of CPUs is large, for instance several tens, the execution time of step 4 could reduce the efficiency. In this case the parallel algorithm should be improved by using the fact that, for the long bunches found in rings, the macro-particles move very slowly along the longitudinal axis and the data exchange will be only between neighboring CPUs. This enables us to use the exchange table with 2 times N_{CPU} size instead of N_{CPU} times N_{CPU}. The analysis of graphs for several numbers of CPUs gives us the following approximation for the distribution time

$$t_{dist} = \tau_1 \cdot N_{part} / N_{CPU} + \tau_2 \cdot \alpha \cdot N_{part} \cdot (N_{CPU} - 1)/(N_{CPU} \cdot N_{CPU}) \qquad (1)$$

where the parameters τ_1 and τ_2 are equal to 1.35E-6 and 12.5E-6 sec accordingly. The parameter α in Eq. (1) is the fraction of macro-particles that have to be distributed. In our simulations α is between 0 and 1E-3. Equation (1) demonstrates full scalability of this parallel algorithm.

5.3 Timing of the Parallel 3D SC Model

For timing the parallel implementation of the three-dimensional space charge model, we used a procedure analogous to that described in the previous part of this report. The calculation times were measured as a function of the number of macro-particles, number of CPUs, and 3D grid size. Fitting the measurements, we obtained the following formula for the time of calculation with the (N_x x N_y) transverse grid size and N_z longitudinal slices

$$t_{3D} = \tau_3 \cdot N_{part} / N_{CPU} + \tau_4 \cdot (N_x N_y N_z)/ N_{CPU} + \tau_{comm} \cdot (N_x N_y) \qquad (2)$$

where the parameters τ_3, τ_4, and τ_{comm} are 3.3E-6, 3.8E-7, and 2.1E-6 sec, respectively. The first term in the formula (2) describes the time spent binning the macro-particles, applying the space charge kick, etc. The second term is the time required to solve the set of two-dimensional space charge problems, and the last is the time for communication and is proportional to the amount of exchanged data.

Equation (2) was obtained for a uniform distribution of macro-particles along longitudinal axis. If the macro-particles are not distributed uniformly in the longitudinal direction, we should use the maximum number of macro-particles on one CPU instead of N_{part}/N_{CPU} expression in Eq. (2).

5.4 Parallel Efficiency

Using Eqs. (1) and (2) we can define the parallel efficiency of the whole algorithm as follows:

$$\eta = 100\% \cdot (t_{dist}(N_{CPU} = 1) + t_{3D}(N_{CPU} = 1))/(N_{CPU} \cdot (t_{dist} + t_{3D})) \qquad (3)$$

For the cases of 64x64x64 grid, 200000 macro-particles, and 2,3,4, and 5 CPUs we obtained 98.6, 98, 97, and 96 %, respectively. These results are for a uniform distribution of the macro-particles along the longitudinal direction. If we suppose that one CPU contains 40% of all particles, instead of 20%, the parallel efficiency will be only 60%. To avoid this effect we should allocate the longitudinal slices between CPUs irregularly to provide a homogeneous load on all CPUs. This means that we must incorporate the timing results into the assignment of the longitudinal slices to the CPUs.

6 Conclusions

Parallel algorithms of the transverse impedance and the three-dimensional space charge models are developed. These algorithms provide close to 100% parallel efficiency for uniform longitudinal distributions of macro-particles. For uneven distributions of particles, the algorithms should be changed to achieve even loading and optimal performance.

7 Acknowledgments

The authors wish to thank Mike Blaskiewicz, John Galambos, Alexei Fedotov, Nikolay Malitsky, and Jie Wei for many useful discussions and suggestions during this investigation.

References

1. National Spallation Neutron Source Conceptual Design Report, Volumes 1 and 2, NSNS/CDR-2/V1, 2, (May, 1997)
2. J. Galambos, J. Holmes, D. Olsen, A. Luccio, and J. Beebe-Wang, ORBIT Users Manual, http://www.sns.gov//APGroup/Codes/Codes.htm
3. V.Danilov, J. Galambos, and J. Holmes, in Proceedings of the 2001 Particle Accelerator Conference, (Chicago, 2001)
4. J. A. Holmes, V. V. Danilov, J. D. Galambos, D. Jeon, and D. K. Olsen, Phys. Rev. Special Topics – AB 2, (1999) 114202
5. K. Woody, J. A. Holmes, V. Danilov, and J. D. Galambos, in Proceedings of the 2001 Particle Accelerator Conference, (Chicago, 2001)
6. J. A. MacLachlan, FNAL TechNote, FN-446, February (1987)
7. R. W. Hockney and J. W. Eastwood, "Computer Simulation Using Particles", Institute of Physics Publishing (Bristol: 1988)
8. J.A. Holmes, J. D. Galambos, D. Jeon, D. K. Olsen, J. W. Cobb, M. Blaskiewicz, A. U. Luccio, and J. Beebe-Wang, Proceedings of International Computational Accelerator Physics Conference, (Monterey, CA, September 1998)
9. F. W. Jones, in Proceedings of the 2000 European Particle Accelerator Conference, (Vienna, 2000) 1381
10. J. Holmes and V.Danilov, "Beam dynamics with transverse impedances: a comparison of analytic and simulated calculations", submitted to Phys. Rev. Special Topics – AB

VORPAL as a Tool for the Study of Laser Pulse Propagation in LWFA

Chet Nieter[1] and John R. Cary[1]

University of Colorado, Center for Integrated Plasma Studies,
Boulder, Colorado
{nieter, cary}@colorado.edu

Abstract. The dimension-free, parallel, plasma simulation code VORPAL, has been used to study laser pulse propagation in Laser Wakefield Acceleration (LWFA). VORPAL is a hybrid code that allows multiple particle models including a cold-fluid model. The fluid model implements a simplified flux corrected transport in the density update but solves the momentum advection directly to allow for simulations of zero density. An implementation of Zalesky's algorithm for FCT is in development. VORPAL simulations predict the rate of loss of energy by pulses propagation through plasma due to Raman scattering. A PIC model for the plasma particles is in development.

1 Introduction

The concept of Laser Wake Field Acceleration (LWFA) has generated considerable excitement in the physics community with its promise of generating extremely high accelerating gradients [1–3]. The basic idea is that a high intensity, short pulse laser is fired into a plasma. The laser pushes the electrons in the plasma out of the pulse by the pondermotive force. A restoring force is produced from the remaining positive charges which pulls the electrons back into region, setting up a plasma oscillation behind the laser pulse. The result is a wake traveling at relativistic speeds behind the laser pulse.

There are of course many technical obstacles that must be overcome before an actual Laser Wake Field Accelerator can be built. There has been on going experimental and theoretical work to explore the details of LWFA. Our plasma simulation code VORPAL is a powerful computational tool for work in this area. We present the results of low noise simulations of wake field generation in both two and three dimensions. Studies of pulse energy loss due to Raman scattering are shown as well. We also discuss additional features in development for VORPAL and their applications to LWFA.

2 VORPAL

VORPAL - Vlasov, Object-oriented, Relativistic, Plasma Analysis code with Lasers - was begun as a prototype code to explore the possibility of using modern

P.M.A. Sloot et al. (Eds.): ICCS 2002, LNCS 2331, pp. 334–341, 2002.

computing methods, in particular object-oriented design, to create a dimension free general plasma simulation code that would support multiple models.

Our code is dimension free, which means the dimension of the simulation is not hard coded but can be set at run time. This is done with the use of the templating feature of C++ and a method of coding we developed using recursion and template specialization. Three components are needed for dimension free coding. The first is a generalization of an iterator, which allows us to deal with the problem of indexing an array in an arbitrary dimension. The second is a class which holds a collection of iterators to perform a calculation on the grid. The last is a class which is responsible for updating the holder classes over a region of the grid.

A simple example of an iterator in one dimension is the bumping of an index for a one-dimensional array. One can bump the index either up or down. The index is then used to access the value stored at that location in the array. We generalize this by allowing our multi-dimensional iterator to be bumped in any direction. To implement a calculation on the grid, we use classes who contain collections of iterators and an update method that combines these iterators in a manner that produces the desired calculation. We refer to these classes as holders. They contain a iterator that points the result of the calculation, a set iterators that point to dependent fields, and an update method that implements the calculation.

A walker class, which is templated over dimension and direction, moves the holder class along the grid though recursion. The walker update method moves the iterator along the direction over which it was templated. While moving along that direction, the update method recursively calls the update method for the walker of the next lower direction. The walker for the lowest direction is specialized to perform the update of the holder class. Inlining these recursive calls provides the flexibility of setting the dimension at run time without loss to performance.

In addition to being dimension free, VORPAL is designed to run on most UNIX based platforms. It can be run on a single workstation as serial code or it can be run on parallel on both Linux Beowulf clusters and on high performance supercomputers. We have developed a general domain decomposition that allows static load balancing and gives the possibility of implementing dynamic load balancing with minimal re-engineering. This is done by using the idea of intersecting grid regions. Each domain has a physical region which it is responsible for updating and a extended region which is the physical region plus a layer of guard cells. To determine what needs to be passed from one processor to another, we just take the intersection of the receiving domain's extended region with the sending domain's physical region. This allows a decomposition into arbitrary box shaped regions.

Using the object oriented ideas of polymorphism and inheritance, VORPAL can incorporate multiple models for both the particles and fields in a plasma. At present we have cold fluid model implemented for the particles and a Yee mesh finite differencing for the fields. The fluid density update is done with a simplified flux corrected transport where the total flux leaving a cell in any direction is given an upper bound so it does allow the density in a cell to become negative. We are presently developing a full flux corrected transport for the density based on the algorithm developed by Zalesky. Rather than do a flux conservative update for the fluid momentum density as well, we directly advect the fluid momentum allowing us to simulate regions of zero density. A PIC model for the particles is also in development, which will allow us to run hybrid simulations for LWFA where the accelerated particles will be represented by PIC while the bulk plasma is modeled as a fluid to reduce noise.

3 Applications to LWFA

VORPAL was used to simulate the generation of a wake field by a laser pulse being launched into a ramped plasma in both two and three dimensions. In 2D the plasma is 40 μm in the direction of propagation and 100 μm in the transverse direction. Figure 1 shows the initial plasma density along along the direction of pulse propagation. The density is zero for the first 10 μm of the simulation, rises over 10 μm to 3.e25 m^{-3}, and is constant for the remaining 20 μm. There are 480 grid points in the direction of propagation and 100 grid in the direction transverse to propagation and the time step is 15 fs. An electromagnetic wave is injected from the boundary and propagates towards the ramp. The pulse is Gaussian in the transverse direction and is a half sine in the direction of propagation. The peak power of the laser pulse is 1.07 TW and the its total energy is 61.9 mJ. After 31 μm of propagation, a moving window is activated, so that the plasma appears to move to the left, while the laser pulse appears stationary. During injection, the laser causes peaking of the plasma by a factor of three or so, but no negative densities or numerical instabilities are observed.

In this simulation the pulse width was set to be half the wavelength of the plasma oscillations so a strong wake field is produced. In Fig. 2 we see a contour plot of the electric field parallel to the direction of propagation after the pulse has propagated 187 ps. The bowing of the wake field is due to nonlinear effects and has been seen in PIC simulations done with XOOPIC [4] and in other fluid simulations done with Brad Shadwick's fluid code [5]. In Fig. 3 we see a line plot of the electric field in the parallel direction of slice running down the middle of the simulation region. Since a fluid model is used for the plasma, the results are almost noiseless.

Due to the flexibility of VORPAL's dimension free coding, only minor changes are needed to the input file of the 2D run to generate a 3D wake field simulation. Using the same parameters as the 2D run with the two transverse directions having the same dimensions, we repeat the our wake field generation simulation in 3D. In Fig. 4 we see a contour plot of the plasma density along the plane transverse to the direction of propagation located at the edge of the plasma. In other words, we are seeing the plasma density along a slice that is half way between the point where the initial density starts to rise from zero and the point where it levels off. The concave region that appears in the density plot is where

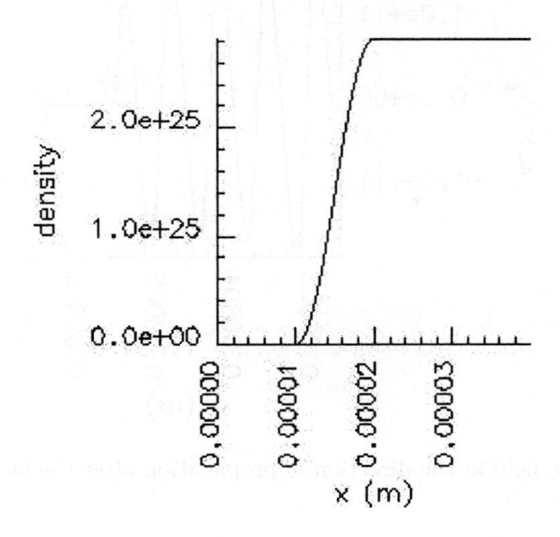

Fig. 1. The initial density of the plasma along the direction of propagation

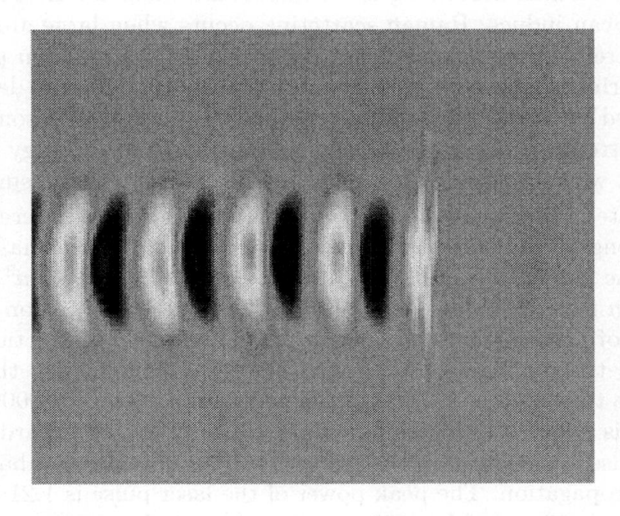

Fig. 2. The electric field in the direction of propagation after the laser pulse has propagated for 187.5 *ps*

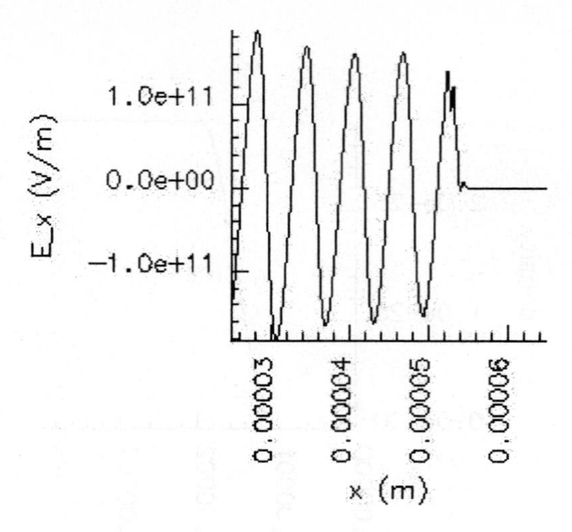

Fig. 3. The electric field in the direction of propagation after the laser pulse has propagated for 187.5 *ps*

the pondermotive force from the laser has pushed back the plasma, showing the region in the plasma where the electrons have been ejected. The electrons are pulled back into this region by the net positive charge from the plasma ions, which then sets up the plasma oscillations of the wake field.

Stimulated Raman scattering is an important effect for LWFA due to the energy loss it can induce. Raman scattering occurs when large amplitude light wave is scattered by the plasma into a light wave and a electron plasma wave. Raman scattering is triggered by small fluctuations in the plasma density such as those produced by noise. The electron plasma wave can then resonant with the scattered electromagnetic wave, generating an instability. Energy is transfered from the light wave to the electron plasma wave heating the plasma.

We simulated the propagation of a laser pulse in a plasma where pulse length is 2.5 times longer than the plasma wavelength. Again the plasma is ramped at one end of the simulation, rising to a peak density of 2.5e25 m^3 over 20 μm starting 20 μm from the edge of the plasma. The simulation region is 150 μm in the direction of propagation and 100 μm in the transverse direction with 2000 grid points in the propagation direction and 200 grid points in the transverse direction. The time step is 20 fs and the simulation is run for 9000 time steps. A laser pulse is injected from the boundary and propagates towards the plasma ramp. The pulse is Gaussian in the transverse direction and is a half sine in the direction of propagation. The peak power of the laser pulse is 1.21 TW and the its total energy is 70 mJ. After 127 μm a moving window is activated. In Fig. 5 we see the electric field in the direction of propagation after 4500 time steps. Behind the laser we see the electric field of the scattered electron plasma wave. Because we are using conducting boundary conditions, we see some reflections of the scattered wave from the boundaries. This does not affect the instability since

Fig. 4. The plasma density along a plane perpendicular to the direction of propagation shortly after the laser pulse has entered the plasma

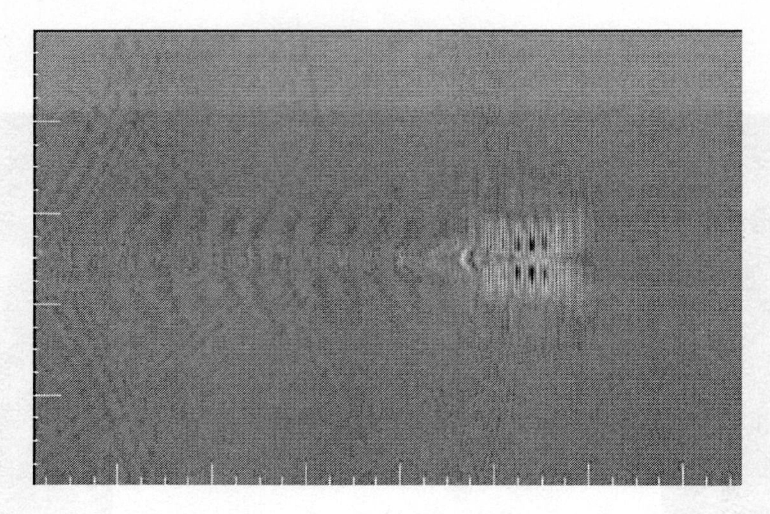

Fig. 5. The the electric field in the direction of propagation after 4500 time steps

the reflections occur well after the pulse. In Fig. 6 a line plot of the electric field along a plane running through the middle of the simulation shows the scattered wave a little clearer.

A method for controlling stimulated Raman scattering involving the use of chirped non-bandwidth limited laser pulse have been proposed [6] and recent experiments have been successful in reducing Raman scattering with this method. VORPAL's wave launching boundary has been modified to so chirped pulses can be simulated and we planning to apply our code to on going work in this area. Simulations of electron production in recent experimental work [7, 8] in chirped pulse propagation in plasma are planned once we have a PIC model for the plasma particles.

References

1. Tajima, T. and Dawson, J.M.: Laser electron accelerator. Phys. Rev. Lett. **43** (1979) 267–270
2. Sprangle, P., Esarey, E., Ting, A., Joyce, G.: Laser wakefield acceleration and relativistic optical guiding. Appl. Phys. Lett. **53** (1988) 2146–2148
3. Berezhinai, V.I. and Murusidze, I.G.: Relativistic wakefield generation by an intense laser pulse in a plasma. Phys. Lett. A **148** (1990) 338–340
4. Bruhwiler, D.L., Giacone, R.E., Cary, J.R., Verboncoeur,J.P., Mardahl,P. Esarey, E., Leemans, W.P., Shadwick, B.A.: Particle-in-cell simulations of plasma accelerators and electron-neutral collisions. Phys. Rev. ST-Accelerators and Beams, **4** (2001) 101302
5. Shadwick, B.A., Tarkenton, G.M., Esarey, E.H., and Leemans, W.P.: Fluid Modeling of Intense Laser-Plasma Interactions. In: Colestock, P. L., Kelly, S. (eds.): Advanced Accelerator Concepts, 9th workshop. AIP Conference Proceedings, Vol. 569. American Institute of Physics (2001)

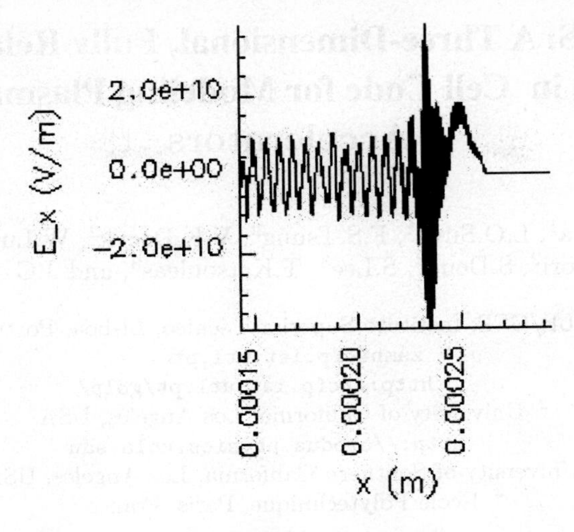

Fig. 6. The the electric field in the direction of propagation along a plane running through the middle of the simulation after 4500 time steps

6. Dodd, Evan S. and Umstadler, Donald: Coherent control of stimulated Raman scattering using chirped laser pulses. Phys. of Plasmas **8** (2001) 3531–3534
7. Leemans, W.P.: Experimental Studies of Self-Modulated and Standard Laser Wakefield Acceleators. Bull. Am. Phys. Soc. **45** No.7 (2000) 324
8. Marqués, J.-R.: Detailed Study of Electron Acceleration and Raman Instabilities in Self-Modulated Laser Wake Field excited by a 10 Hz - 500 mJ laser. Bull. Am. Phys. Soc. **45** No.7 (2000) 325

OSIRIS: A Three-Dimensional, Fully Relativistic Particle in Cell Code for Modeling Plasma Based Accelerators

R.A.Fonseca[1], L.O.Silva[1], F.S.Tsung[2], V.K.Decyk[2], W.Lu[2], C.Ren[2], W.B.Mori[2], S.Deng[3], S.Lee[3], T.Katsouleas[3], and J.C.Adam[4]

[1] GoLP/CFP, Instituto Superior Técnico, Lisboa, Portugal,
zamb@cfp.ist.utl.pt,
http://cfp.ist.utl.pt/golp/
[2] University of California, Los Angeles, USA
http://exodus.physics.ucla.edu
[3] University of Southern California, Los Angeles, USA
[4] École Polytechnique, Paris. France

Abstract. We describe OSIRIS, a three-dimensional, relativistic, massively parallel, object oriented particle-in-cell code for modeling plasma based accelerators. Developed in Fortran 90, the code runs on multiple platforms (Cray T3E, IBM SP, Mac clusters) and can be easily ported to new ones. Details on the code's capabilities are given. We discuss the object-oriented design of the code, the encapsulation of system dependent code and the parallelization of the algorithms involved. We also discuss the implementation of communications as a boundary condition problem and other key characteristics of the code, such as the moving window, open-space and thermal bath boundaries, arbitrary domain decomposition, 2D (cartesian and cylindric) and 3D simulation modes, electron sub-cycling, energy conservation and particle and field diagnostics. Finally results from three-dimensional simulations of particle and laser wakefield accelerators are presented, in connection with the data analysis and visualization infrastructure developed to post-process the scalar and vector results from PIC simulations.

1 Introduction

Based on the highly nonlinear and kinetic processes that occur during high-intensity particle and laser beam-plasma interactions, we use particle-in-cell (PIC) codes [1, 2], which are a subset of the particle-mesh techniques, for the modeling of these physical problems. In these codes the full set of Maxwell's equations are solved on a grid using currents and charge densities calculated by weighting discrete particles onto the grid. Each particle is pushed to a new position and momentum via self-consistently calculated fields. Therefore, to the extent that quantum mechanical effects can be neglected, these codes make no physics approximations and are ideally suited for studying complex systems with many degrees of freedom.

P.M.A. Sloot et al. (Eds.): ICCS 2002, LNCS 2331, pp. 342–351, 2002.

Achieving the goal of one to one, two and three dimensional modeling of laboratory experiments and astrophysical scenarios, requires state-of-the-art computing systems. The rapid increase in computing power and memory of these systems that has resulted from parallel computing has been at the expense of having to use more complicated computer architectures. In order to take full advantage of these developments it has become necessary to use more complex simulation codes. The added complexity arises for two reasons. One reason is that the realistic simulation of a problem requires a larger number of more complex algorithms interacting with each other than in a simulation of a rather simple model system. For example, initializing an arbitrary number of lasers or particle beams in 3D on a parallel computer is a much more difficult problem than initializing one beam in 1D or 2D on a single processor. The other reason is that the computer systems, e.g., memory management, threads, operating systems, are more complex and as a result the performance obtained from them can dramatically differ depending on the code strategy. Parallelized codes that handle the problems of parallel communications and parallel IO are examples of this. The best way to deal with this increased complexity is through an object-oriented programming style that divides the code and data structures into independent classes of objects. This programming style maximizes code reusability, reliability, and portability.

The goal of this code development project was to create a code that breaks up the large problem of a simulation into a set of essentially independent smaller problems that can be solved separately from each other. This allows individuals in a code development team to work independently. Object oriented programming achieves this by handling different aspects of the problem in different modules (classes) that communicate through well-defined interfaces.

This effort resulted in a new framework called OSIRIS, which is a fully parallelized, fully implicit, fully relativistic, and fully object-oriented PIC code, for modeling intense beam plasma interactions.

2 Development

The programming language chosen for this purpose was Fortran 90, mainly because it allows us to more easily integrate already available Fortran algorithms into this new framework that we call OSIRIS. We have also developed techniques where the Fortran 90 modules can interface to C and C++ libraries, allowing for the inclusion of other libraries that do not supply a Fortran interface. Although Fortran 90 is not an object-oriented language *per se*, object-oriented concepts can be easily implemented [3–5] by the use of polymorphic structures and function overloading.

In developing OSIRIS we followed a number of general principles in order to assure that we were building a framework that would achieve the goals stated above. In this sense all real physical quantities have a corresponding object in the code making the physics being modeled clear and therefore easier to maintain, modify and extend. Also, the code is written in a way such that it is largely independent from the dimensionality or the coordinate system used, with much of the code reused in all simulation modes.

Regarding the parallelization issues, the overall structure allows for an arbitrary domain decomposition in any of the spatial coordinates of the simulation, with an effective load balancing of the problems in study. The input file defines only the global physical problem to be simulated and the domain decomposition desired, so that the user can focus on the actual physical problem and does not need to worry about parallelization details. Furthermore, all classes and objects refer to a single node (with the obvious exception of the object responsible for maintaining the global parallel information), which can be realized by treating all communication between physical objects as a boundary value problem, as described below. This allows for new algorithms to be incorporated into the code, without a deep understanding of the underlying communication structure.

3 Design

3.1 Object-Oriented Hierarchy

Figure 1 shows the class hierarchy of OSIRIS. The main physical objects used are particle objects, electromagnetic field objects, and current field objects. The particle object is an aggregate of an arbitrary number of particle species objects. The most important support classes are the variable-dimensionality-field class, which is used by the electromagnetic and current field classes and encapsulates many aspects of the dimensionality of a simulation, and the domain-decomposition class, which handles all communication between nodes.

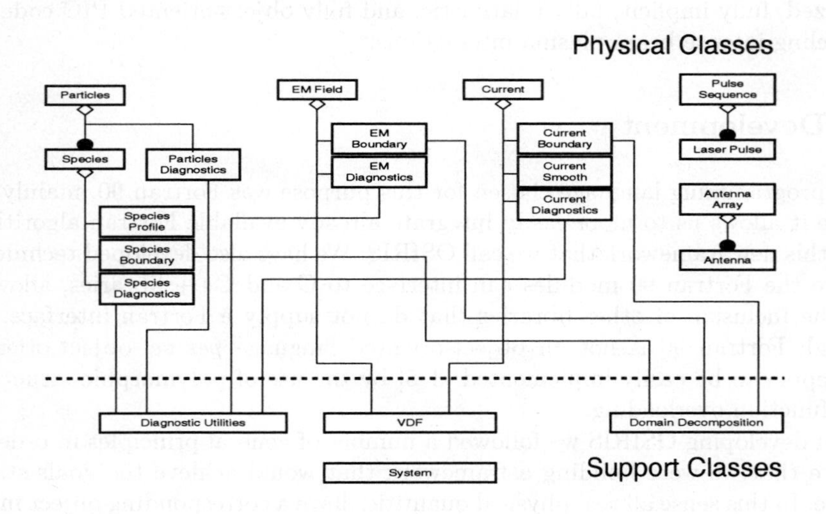

Fig. 1. Osiris main class hierarchy

Benchmarking of the code has indicated that the additional overhead from using an object oriented framework in Fortran 90 leads to only a 12% slowdown in speed.

3.2 Parallelization

The parallelization of the code is done for distributed memory systems, and it is based on the MPI message-passing interface [10]. We parallelize our algorithms by decomposing the simulation space evenly across the available computational nodes. This decomposition is done by dividing each spatial direction of the simulation into a fixed number of segments (N_1, N_2, N_3). The total number of nodes being used is therefore the product of these three quantities (or two quantities for the 2D simulations).

The communication pattern follows the usual procedure for a particle-mesh code [11]. The grid quantities are updated by exchanging (electric and magnetic fields) or adding (currents) the ghost cells between neighboring nodes. As for the particles, those crossing the node boundary are counted and copied to a temporary buffer. Two messages are then sent, the first with the number of particles, and the second with the actual particle data. This strategy allows for not setting an *a priori* limit on the number of particles being sent to another node, while maintaining a reduced number of messages. Because most of the message are small, we are generally limited by the latency of the network being used. To overcome this, and whenever possible, the messages being sent are packed into a single one, achieving in many cases twice the performance.

We also took great care in encapsulating all parallelization as boundary value routines. In this sense, the boundary conditions that each physical object has can either be some numerical implementation of the usual boundary conditions in these problems or simply a boundary to another node. The base classes that define grid and particle quantities already include the necessary routines to handle the later case, greatly simplifying the implementation of new quantities and algorithms.

3.3 Encapsulation of System Dependent Code

For ease in porting the code to different architectures, all code that is machine dependent is encapsulated in the system module. At present we have different versions of this module for running on the Cray T3E, the IBM SP, and for Macintosh clusters, running on both MacOS 9 (MacMPI [8]) and MacOS X (LAM/MPI [9]) clusters. The later is actually a fortran module that interfaces with a POSIX compliant C module and should therefore compile on most UNIX systems, allowing the code to run on PC-based (Beowulf) clusters. The MPI library has also been implemented on all these systems requiring no additional effort.

3.4 Code Flow

Figure 2 shows the flow of a single time step on a typical OSIRIS run. It closely follows the typical PIC cycle [2]. The loop begins by executing the diagnostic routines selected (diagnostics). It follows by pushing the particles using the updated values for the fields and depositing the current (advance deposit). After this step, the code updates the boundaries for particles and currents, communicating with neighboring nodes if necessary. A smoothing of the deposited currents, according to the specified input file, follows this step. Finally, the new values of the Electric and Magnetic field are calculated using the smoothed current values, and its boundaries are updated, again communicating with neighboring nodes, if necessary.

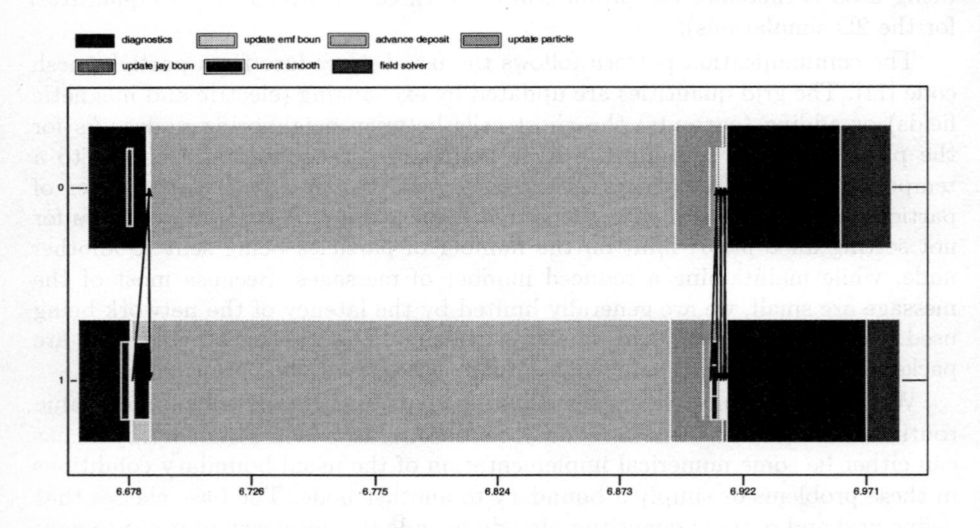

Fig. 2. A typical cycle, one time step, in an OSIRIS 2 node run. The arrows show the direction of communication between nodes.

If requested, at the end of each loop, the code will write restart information, allowing the simulation to be restarted later on at this time step.

4 OSIRIS Framework

The code is fully relativistic and it presently uses either the charge-conserving current deposition schemes from ISIS [6] or TRISTAN [7]. We have primarily adopted the charge-conserving current deposition algorithms because they allow the field solve to be done locally, i.e., there is no need for a Poisson solve. The code uses the Boris scheme to push the particles, and the field solve is done locally using a finite difference solver for the electric and magnetic fields in both space and time.

In its present state the code contains algorithms for 2D and 3D simulations in Cartesian coordinates and for 2D simulations in azimuthally symmetric cylindrical coordinates, all of which with 3 components in velocity (i.e. both 2D modes are indeed $2\frac{1}{2}$D or 2D3V algorithms). The loading of particles is done by distributing the particles evenly on the cell, and varying the individual charge of each particle according to the density profile stipulated. Below a given threshold no particles are loaded. The required profile can be specified by a set of multiplying piecewise linear functions and/or by specifying Gaussian profiles. The initial velocities of the particles are set according to the specified thermal distribution and fluid velocity. The code also allows for the definition of constant external electric and magnetic fields.

The boundary conditions we have implemented in OSIRIS are: conducting, and Lindmann open-space boundaries for the fields [17], and absorbing, reflective, and thermal bath boundaries for the particles (the later consists of reinjecting any particle leaving the box with a velocity taken from a thermal distribution). Furthermore, periodic boundary conditions for fields and particles are also implemented.

This code also has a moving window, which makes it ideal for modeling high-intensity beam plasma interactions where the beam is typically much shorter than the interaction length. In this situation simulation is done in the laboratory reference frame, and simulation data is shifted in the direction opposite to the window motion defined whenever an integer number of cells corresponds to this motion in the number of time steps elapsed. Since this window moves at the speed of light in vacuum no other operations are required. The shifting of data is done locally on each node, and boundaries are updated using the standard routines developed for handling boundaries, thus taking care of moving data between adjacent nodes. The particles leaving the box from the back are removed from the simulation and the new clean cells in the front of the box are initialized as described above.

OSIRIS also incorporates the ability to launch EM waves into the simulation, either by initializing the EM field of the simulation box accordingly, or by injecting them from the simulation boundaries (e.g. antennas). Moreover, a subcycling scheme [18] for heavier particles has been implemented, where the heavier species are only pushed after a number of time steps using the averaged fields over these time steps, thus significantly decreasing the total loop time.

A great deal of effort was also devoted to the development of diagnostics for this code that goes beyond the simple dumps of simulation quantities. For all the grid quantities envelope and boxcar averaged diagnostics are implemented; for the EM fields we implemented energy diagnostics, both spatially integrated and resolved; and for the particles phase space diagnostics, total energy and energy distribution function, and accelerated particle selection are available. The output data uses the HDF [12] file format. This is a standard, platform independent, self-contained file format, which gives us the possibility of adding extra information to the file, like data units and iteration number, greatly simplifying the data analysis process.

5 Visualization and Data-Analysis Infrastructure

It is not an exaggeration to say that visualization is a major part of a parallel computing lab. The data sets from current simulations are both large and complex. These sets can have up to five free parameters for field data: three spatial dimensions, time and the different components (i.e., E_x, E_y, and E_z). For particles, phase space has seven dimensions: three for space, three for momentum and one for time. Plots of y versus x are simply not enough. Sophisticated graphics are needed to present so much data in a manner that is easily accessible and understandable.

We developed a visualization and analysis infrastructure [13] based on IDL (Interactive Data Language). IDL is a 4GL language, with sophisticated graphics capabilities, and it is widely used in areas such as Atmospheric Sciences and Astronomy. It is also available on several platforms and supported in a number of systems, ranging from Solaris to the MacOS.

While developing this infrastructure we tried simplifying the visualization and data analysis as much as possible, making it user-friendly, automating as much of the process as possible, developing routines to batch process large sets of data and minimizing the effort of creating presentation quality graphics. We implemented a full set of visualization routines for one, two and three-dimensional scalar data and for two and three dimensional vector data. These include automatic scaling, dynamic zooming and axis scaling, integration of analysis tools, animation tools, and can be used either in batch mode or in interactive mode. We have also developed a comprehensive set of analysis routines that include scalar and vector algebra for single or multiple datasets, boxcar averaging, spectral analysis and spectral filtering, k-space distribution function, envelope analysis, mass centroid analysis and local peak tools.

6 Results

The code has been successfully used in the modeling of several problems in the field of plasma based accelerators, and has been run on a number of architectures. Table 1 shows the typical push times on two machines, one supercomputer and one computer cluster.

We have also established the energy conservation of the code to be better than 1 part in 10^5. This test was done in a simulation where we simply let a warm plasma evolve in time; in conditions where we inject high energy fluxes into the simulation (laser or beam plasma interaction runs) the results are better. Regarding the parallelization of the code, extensive testing was done on the EP^2 cluster [19] at the IST in Lisbon, Portugal. We get very high efficiency, (above 91% in any condition), proving that the parallelization strategy is appropriate.

Table 1. Typical push time for two machines, in two and three dimensions. Values are in μs/particle $\times$ node

Machine	2D push time	3D push time
Cray T3E-900	4.16	7.56
EP^2 Cluster	4.96	9.82

Also note that this is a computer cluster running a 100 Mbit/s network, and that the efficiency on machines such as the Cray T3E is even better

One example of a three-dimensional modeling of a plasma accelerator is presented on figure 3. This is a one-to-one modeling of the E-157 Experiment [14] done at the Stanford Linear Accelerator Center, where a 30 GeV beam is accelerated by 1 GeV. The figure shows the Lorentz forces acting on the laser beam e.g. $\mathbf{E} + \mathbf{z} \times \mathbf{B}$, where $\mathbf{z}$ is the beam propagation direction, and we can clearly identify the focusing /defocusing and accelerating/decelerating regions

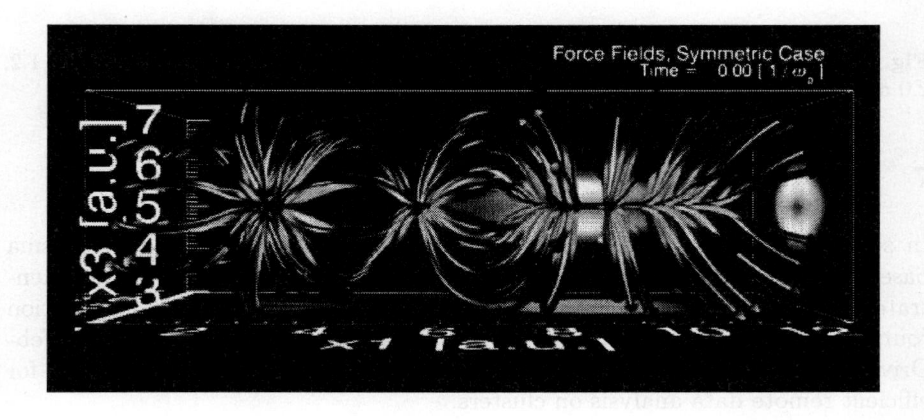

Fig. 3. Force field acting on the 30 GeV SLAC beam inside a plasma column.

Another example of the code capabilities is the modeling of the Laser Wakefield Accelerator (LWFA). In the LWFA a short ultrahigh intensity laser pulse drives a relativistic electron plasma wave. The wakefield driven most efficiently when the laser pulse length $L = c\tau$ is approximately the plasma wavelength $\lambda_p = 2\pi c/\omega_p$ - Tajima-Dawson mechanism [15]. Figure 4 shows the plasma wave produced by a 800 nm laser pulse with a normalized vector potential of 2.16, corresponding to an intensity of 10^{19}W/cm^2 on focus, and a duration of 30 fs, propagating in an underdense plasma.

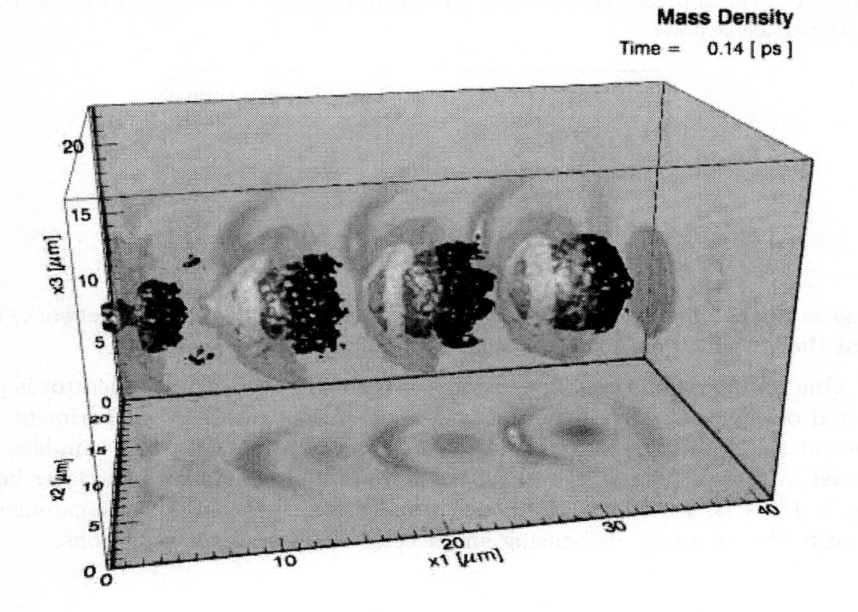

Fig. 4. Plasma Wave produced in the LWFA. Isosurfaces shown for values of 0.5, 1.2, 2.0 and 5.0 normalized to the plasma background density.

7 Future Work

In summary, we have presented the OSIRIS framework for modeling plasma based accelerators. This is an ongoing effort; future developments will concentrate on the implementation of true open-space boundaries [16] and ionization routines. Regarding the visualization and data analysis infrastructure, a Web-Driven visualization portal will be implemented on the near future, allowing for efficient remote data analysis on clusters.

8 Acknowledgements

This work was supported by DOE, NSF (USA), FLAD, GULBENKIAN, and by FCT (Portugal) under grants PESO/P/PRO/40144/2000, PESO/P/INF/40146 /2000, CERN/P/FIS/40132/2000, and POCTI/33605/FIS/2000.

References

1. Dawson, J.M.: Particle simulation of plasmas. Rev. Mod. Phys., vol.55, no. 2, April 1983, p. 403-47.

2. Birdsall, C.K., Langdon, A.B.: Plasma physics via computer simulation. Bristol, UK: Adam Hilger, 1991, xxvi+479 pp.
3. Decyk, V. K., Norton, C. D., Szymanski, B. K.: How to express C++ concepts in Fortran 90. Scientific Programming, Vol. 6, no. 4, 1998, p. 363.
4. Decyk, V. K., Norton, C. D., Szymanski, B. K.: How to support inheritance and run-time polymorphism in Fortran 90. Comp. Phys. Com., no. 115, 1998, pp. 9-17.
5. Gray, M. G., Roberts, R. M.: Object-Based Programming in Fortran 90. Computers in Physics, vol. 11, no. 4, 1997, pp. 355-361.
6. Morse, R.L., Nielson, C.W.: Numerical simulation of the Weibel instability in one and two dimensions. Phys. Fluids, vol.14, no.4, April 1971. pp.830-40.
7. Villasenor, J.; Buneman, O.: Rigorous charge conservation for local electromagnetic field solvers. Computer Physics Communications, vol.69, no. 2-3, March-April 1992.
8. Decyk, V.K., Dauger, D.E.: How to Build an AppleSeed: A Parallel Macintosh Cluster for Numerically Intensive Computing. Presented at the International School for Space Simulation ISSS-6, Garching, Germany September 2001; also at `http://exodus.physics.ucla.edu/appleseed/appleseed.html`
9. `http://www.lam-mpi.org/`
10. Message Passing Interface Forum.: MPI: A message-passing interface standard. International Journal of Supercomputer Applications, vol. 8, no. 3-4, 1994.
11. Gropp, W., Lusk, E., Skjellum, A.: Using MPI. MIT Press, 1999. xxii+371 pp.
12. `http://hdf.ncsa.uiuc.edu/`
13. Fonseca, R. *et al.*: Three-dimensional particle-in-cell simulations of the Weibel instability in electron-positron plasmas. IEEE transactions in plasma science Special Issues on Images in Plasma Science, 2002
14. Muggli, P. *et al.*: Nature, vol. 411, 3 May 2001
15. Tajima, T., Dawson, J. M.: Laser Electron Accelerator, Phys. Rev. Lett., vol. 43, 1979, pp. 267-270
16. Vay, J.L.: A new Absorbing Layer Boundary Condition for the Wave Equation. J. Comp. Phys., no. 165, 2000, pp. 511-521.
17. Lindmann, E. L.: Free-space boundary conditions for the time dependent wave equation. J. Comp. Phys., no. 18, 1975, pp. 66-78.
18. Adam, J. C.; A. Gourdin Serveniere, and A. B. Langdon: Electron sub-cycling in particle simulation of Plasmas J. Comp. Phys., no. 47, 1982, pp. 229-244.
19. `http://cfp.ist.utl.pt/golp/epp/`

Interactive Visualization of Particle Beams for Accelerator Design

Brett Wilson[1], Kwan-Liu Ma[1], Ji Qiang[2], and Robert Ryne[2]

[1] Department of Computer Science, University of California,
One Shields Avenue, Davis, CA 95616
[wilson, ma]@cs.ucdavis.edu
[2] Lawrence Berkeley National Laboratory,
One Cyclotron Road, Berkeley, CA 94720
[jQiang, RDRyne]@lbl.gov

Abstract. We describe a hybrid data-representation and rendering technique for visualizing large-scale particle data generated from numerical modeling of beam dynamics. The basis of the technique is mixing volume rendering and point rendering according to particle density distribution, visibility, and the user's instruction. A hierarchical representation of the data is created on a parallel computer, allowing real-time partitioning into high-density areas for volume rendering, and low-density areas for point rendering. This allows the beam to be interactively visualized while preserving the fine structure usually visible only with slow point-based rendering techniques.

1 Introduction

Particle accelerators are playing an increasingly important role in basic and applied sciences, such as high-energy physics, nuclear physics, materials science, biological science, and fusion energy. The design of next-generation accelerators requires high-resolution numerical modeling capabilities to reduce cost and technological risks, and to improve accelerator efficiency, performance, and reliability. While the use of massively-parallel supercomputers allows scientists to routinely perform simulations with hundreds of millions of particles [2], the resulting data typically requires terabytes of storage space, and overwhelms traditional data analysis and visualization tools.

The goal of beam dynamics simulations is to understand the beam's evolution inside the accelerator and, through that understanding, to design systems that meet certain performance requirements. These requirements may include, for example, minimizing beam loss, minimizing emittance growth, avoiding resonance phenomena that could lead to instabilities, etc. The most widely used method for modeling beam dynamics in accelerators involves numerical simulation using particles. In three-dimensional simulations each particle is represented by a six-vector in phase space, where each six-vector consists of three coordinates (X, Y, Z) and three momenta (P_x, P_y, P_z). The coordinates and momenta are updated as the particles are advanced through the components of the accelerator, each of which provides electromagnetic forces that guide and focus the particle beam. Furthermore, in high intensity accelerators, the beam's own self-fields are important. High intensity beams often exhibit a prounounced beam halo

P.M.A. Sloot et al. (Eds.): ICCS 2002, LNCS 2331, pp. 352–361, 2002.

that is evidenced by a low density region of charge far from the beam core. The halo is responsible for beam loss as stray particles strike the beam pipe, and may lead to radioactiviation of the accelerator components.

2 Particle Visualization

In the past, researchers visualized simulated particle data by either viewing the particles directly, or by converting the particles to volumetric data representing particle density [4]. Each of these techniques has disadvantages. Direct particle renderings takes too long for interactive exploration of large datasets. Benchmarks have shown that it takes approximately 50 seconds to render 300 million points on a Silicon Graphics Infinite-Reality engine, and PC workstations are unable to even hold this much data in their main memory.

Volume rendering can provide interactive framerates, even for PC-based workstations with commercial graphics cards. In this type of rendering, the range covered by the data is evenly divided into voxels, and each voxel value is assigned a density based on the number of points that fall inside of it. This data is then converted into an 8-bit paletted texture and rendered on the screen as a series of closely-spaced parallel texture-mapped planes. Taken together, these planes give the illusion of volume. A further advantage of volume-based rendering is that it allows realtime modification of a transfer function that maps density to color and opacity, since only the palette for the texture needs to be updated [5]. However, there are also limitations. In order to fit in a workstation's graphics memory, the resolution is typically limited to 256^3 (512^3 for large systems). This, as well as the low range of possible density values (256), can result in artifacts, and can hide fine structures, especially in the low-density halo region of the beam.

Ideally, a visualization tool would be able to interactively visualize the beam halo of a large simulation at very high resolutions. It would also provide realtime modification of the transfer function, and run on high-end PCs rather than a supercomputer. This tool would be used to quickly browse the data, or to locate regions of interest for further study. These regions could be rendered offline at even higher quality using a parallel supercomputer.

To address these needs, our system uses a combined particle- and volume-based rendering approach. The low-density beam halo is represented by directly rendering its constituent particles. This preserves all fine structures of the data, especially the lowest-density regions consisting of only one or two particles that would be invisible using a volumetric approach. The high-density beam core is represented by a low-resolution volumetric rendering. This area is of lesser importance, and is dense enough so that individual particles do not have a significant effect on the rendering. The volume-rendered area provides context for the particle rendering, and, with the right parameters, is not even perceived as a separate rendering style.

3 Data Representation

To prepare for rendering, a multi-resolution, hierarchical representation is generated from the original, unstructured, point data. The representation currently implemented is an octree, which is generated on a distributed-memory parallel computer, such as the PC cluster shown in Figure 1. This pre-processing step is performed once for each plot type desired (since there are six values per point, many different plots can be generated from each dataset). This data is later loaded by a viewing program for interactive visualization.

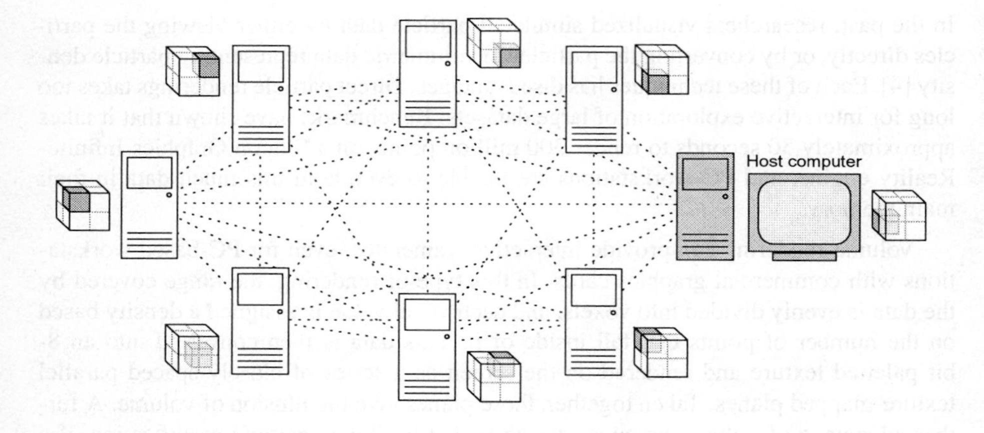

Fig. 1. The data is distributed to a parallel computer, such as a PC cluster, each processor of which is responsible for one octant of the data. After being read, the points are forwarded to the appropriate processor, which creates the octree for that section of data. Viewing is performed on one of the nodes with a graphics card.

The hierarchical data consists of two parts: the octree data, and the point data. At each octree node, we store the density of points in the node, and the minimum density of all sub-nodes. At the leaf octree nodes (the nodes at the finest level of subdivision), we store the index into the point data of the node's constituent points. The leaf nodes should be small enough so that the boundary between point-rendered nodes and volume-rendered nodes appears smooth on the screen. Simultaneously, the nodes need to be big enough to contain enough points to accurately calculate point density.

Since the size of the point data is several times the available memory on the workstation used for interaction, not all of the points can be loaded at once by the viewing program. Having to load points from disk to display each frame would result in a loss of interactivity. Instead, we take advantage of the fact that only low-density regions are rendered using the point-based method. High-density regions, consisting of the majority of points in the dataset, are only volume rendered, and the point data is never needed. Therefore, the points belonging to lower-density nodes are stored separately from the rest of the points in the volume. The preview program pre-loads these points from disk when it loads the data. It can then generate images entirely from in-core data as long as the display threshold for points does not exceed that chosen by the partitioning program. For this reason, the partitioning program generates approximately as much pre-loaded data as there is memory available on the viewing computer.

4 User Interaction

The preview program is used to view the partitioned data generated by the parallel computer. As shown in Figure 2, it displays the rendering of the selected volume in the main portion of the window, where it can be manipulated using the mouse. Controls for selecting the transfer functions for the point-based rendering and the volume-based rendering are located on the right panel.

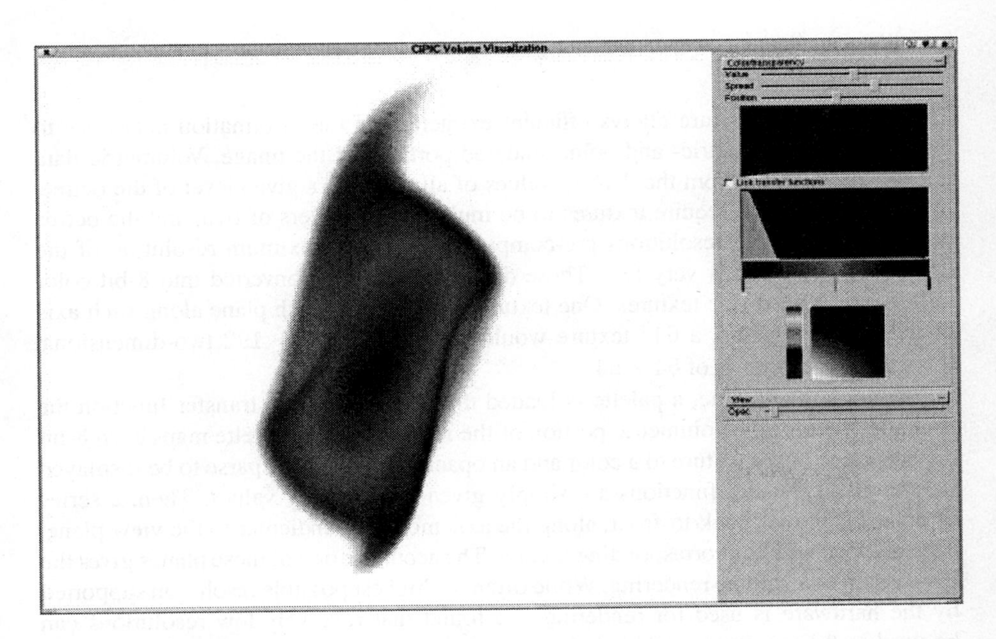

Fig. 2. The user interface, showing the volume transfer function (black box in the top right of the window) and the point transfer function (below it) with the phase plot (x, P_x, z) of frame 170 loaded. This image consists of 2.7 million points and displays in about 1 second on a GeForce 3.

The volume transfer function maps point density to color and opacity for the volume-rendered portion of the image. Typically, a step function is used to map low-density regions to 0 (fully transparent) and higher density regions to some low constant so that one can see inside the volume. The program also allows a ramp to transition between the high and low values, so the boundary of the volume-rendered region is less visible.

The point transfer function maps density to number of points rendered for the point-rendered portion of the image. Below a certain threshold density, the data is drawn rendered as points; above that threshold, no points are drawn. Intermediate values are mapped to the fraction of points drawn. When the transfer function's value is at 0.75 for some density, for example, it means that three out of every four points are drawn for areas of that density. This allows the user to see fewer points if too many points are obscuring important features, or to make rendering faster. It also allows a smooth transition between point-rendered portions of the image and non-point-rendered portions. Point opacity is given as a separate control, a feature that can be useful when many points are being drawn.

By default, the two transfer functions are inverses of each other. Changing one results in an equal and opposite change in the other. This way, there is always an even transition between volume- and point-rendered regions of the image. In many cases, this transition is not even visible. The user can unlink the functions, if desired, to provide more or less overlap between the regions.

5 Rendering

The octree data structure allows efficient extraction of the information necessary to draw both the volumetric- and point-rendered portions of the image. Volumetric data is extracted directly from the density values of all nodes at a given level of the octree. Most graphics cards require textures to be multiples of powers of two, and the octree contains all of these resolutions pre-computed up to the maximum resolution of the octree, so extraction is very fast. These density values are converted into 8-bit color indices and loaded into textures. One texture is created for each plane along each axis of the volume [1]. So, a 64^3 texture would require $64 \times 3 = 192$ two-dimensional textures at a resolution of 64×64.

To draw the volume, a palette is loaded that is based on the transfer function the user specified for the volumetric portion of the rendering. This palette maps each 8-bit density value of the texture to a color and an opacity; regions too sparse to be displayed for the given transfer functions are simply given zero opacity values. Then, a series of planes is drawn, back-to-front, along the axis most perpendicular to the view plane, each mapped with the corresponding texture. The accumulation of these planes gives the impression of a volume rendering. While often the highest possible resolution supported by the hardware is used for rendering, we found that relatively low resolutions can be used in this application. This is because the core of the beam is typically diffuse, rendered mostly transparent, and is obscured by points. All images in this paper were produced using a volume resolution of 64^3.

In contrast to the volume rendering, in which only the palette is changed in response to user input, point rendering requires that the appropriate points from the dataset be selected each time a frame is rendered. Therefore, we want to quickly eliminate regions that are too dense to require point rendering. When displaying a frame, we first calculate the maximum density a node must have to be visible in the point rendering, based on the transfer function given by the user. Since each octree node contains the minimum density of any of its sub-nodes, only octree paths leading to renderable leaf nodes must be traversed; octree nodes leading only to dense regions in the middle of the beam need never be expanded.

Once the program decides that a leaf node must be rendered, it uses the point transfer function to estimate the fraction of points to draw. Often, this value is one, but may be less than one depending on the transfer function specified by the user. It then processes the list of points, drawing every n-th one. The first point drawn is selected to be a random index between 0 and n. This eliminates possible visual artifacts resulting in the selection of a predictable subset of points from data that may have structure in the order it was originally written to disk. Figure 3 illustrates the two regions of the volume regarding to the image generation process.

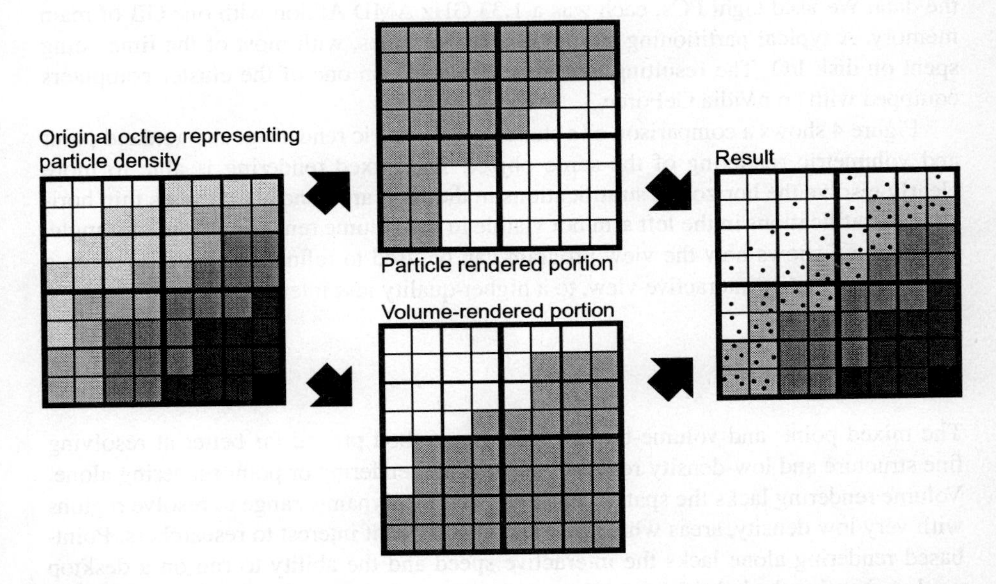

Fig. 3. The image is created by classifying each octree node as belonging to a volume-rendered region or a point-rendered region, depending on the transfer functions for each region (the regions can overlap, as in this example). The combination of the two regions defines the output image.

6 Results

The system was tested using the results from a self-consistent simulation of charged particle dynamics in an alternating-focused transport channel. The simulation, which was based on an actual experiment, was done using 100 million particles. Each particle was given the same charge-to-mass ratio as a real particle. The particles moving inside the channel were modeled, including the effects of external fields from magnetic quadrupoles and self-fields associated with the beam's space charge. The three-dimensional mean-field space-charge forces were calculated at each time step by solving the Poisson equation using the charge density from the particle distribution. The initial particle distribution was generated by sampling a 6D waterbag distribution (i.e.

a uniformly filled ellipsoid in 6D phase space). At the start of the simulation, the distribution was distorted to account for third-order nonlinear effects associated with the transport system upstream of the starting point of the simulation. In the simulation, as in the experiment, quadrupole settings at the start of the beamline were adjusted so as to generate a mismatched beam with a prounounced halo. The output of the simulation consisted of 360 frames of particle phase space data, where each frame contained phase space information at one time step.

Several frames of this data were moved onto a PC cluster for partitioning, although the data could have been partitioned on the large IBM SP that was used to generate the data. We used eight PCs, each was a 1.33 GHz AMD Athlon with one GB of main memory. A typical partitioning step took a few minutes, with most of the time being spent on disk I/O. The resulting data was visualized on one of the cluster computers equipped with an nVidia GeForce 3.

Figure 4 shows a comparison of a standard volumetric rendering, and a mixed point and volumetric rendering of the same object. The mixed rendering is able to more clearly resolve the horizontal stratifications in the right arm, and also reveals thin horizontal stratifications in the left arm not visible in the volume rendering from this angle.

Figure 5 shows how the view program can be used to refine the rendering from a low-quality, highly interactive view, to a higher-quality less interactive view.

7 Conclusions

The mixed point- and volume-based rendering method proved far better at resolving fine structure and low-density regions than volume rendering or point rendering alone. Volume rendering lacks the spatial resolution and the dynamic range to resolve regions with very low density, areas which may be of significant interest to researchers. Point-based rendering alone lacks the interactive speed and the ability to run on a desktop workstation that the hybrid approach provides.

Point-based rendering for low-density areas also provides more room for future enhancements. Because points are drawn dynamically, they could be drawn (in terms of color or opacity) based on some dynamically calculated property that the researcher is interested in, such as temperature or emittance. Volume-based rendering, because it is limited to pre-calculated data, can not allow dynamic changes like these.

8 Further Work

We plan to implement this hybrid particle data visualization method using the massively parallel computers and high-performance storage facility available at the Lawrence Berkeley National Laboratory. Through a desktop graphics PC and high-speed networks, accelerator scientists would be able to conduct interactive exploration of the highest resolution particle data stored.

As we begin to study the high resolution data (up to 1 billion points), the cost of volume rendering is not negligible any more. 3D texture volume rendering [6] will be thus used which offers better image quality with a much lower storage requirement.

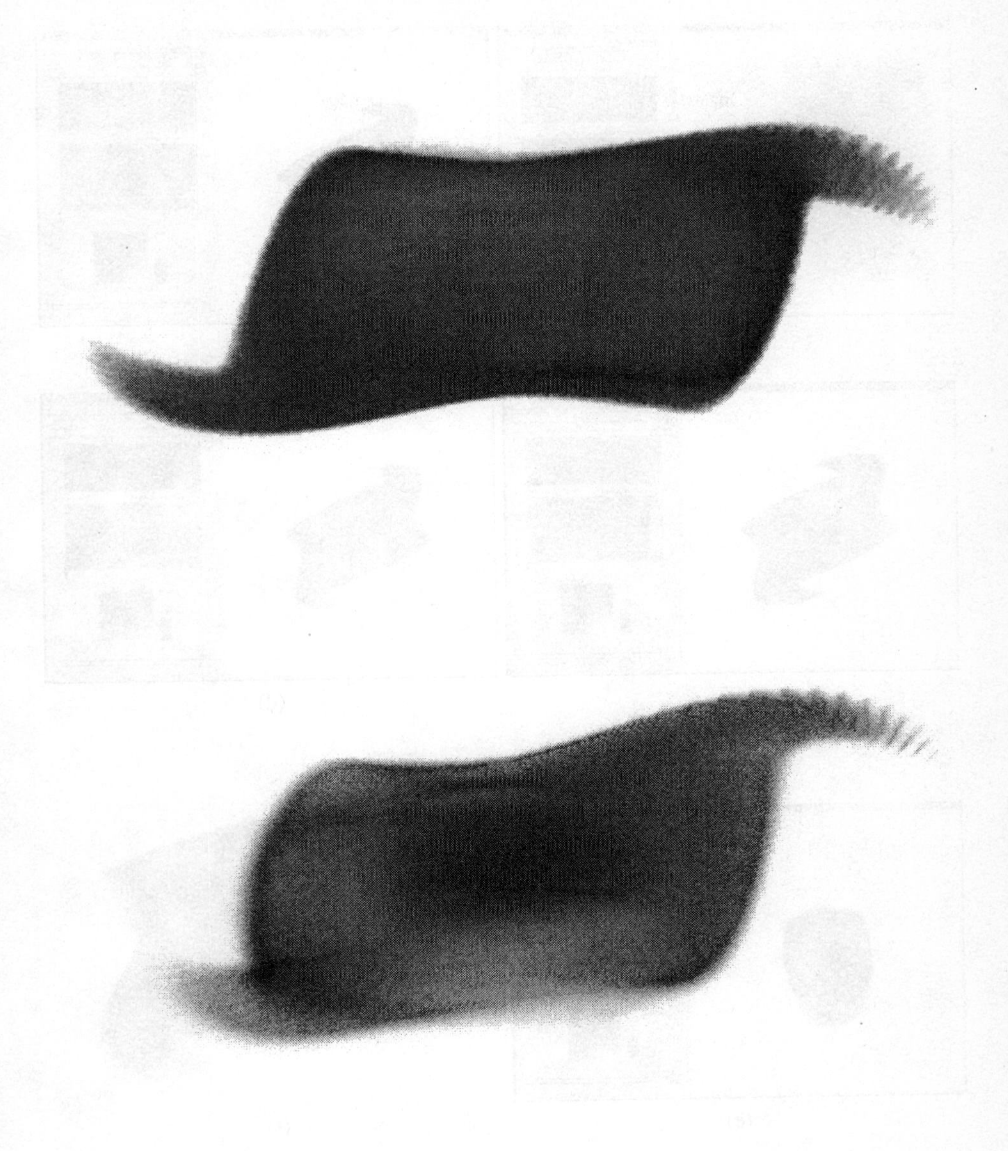

Fig. 4. Comparison of a volume rendering (top) and a mixed volume/point rendering (bottom) of the phase plot (x, P_x, y) of frame 170. The volume rendering has a resolution of 256^3. The mixed rendering has a volumetric resolution of 64^3, 2 million points, and displays at about 3 frames per second. The mixed rendering provides more detail than the volume rendering, especially in the lower-left arm.

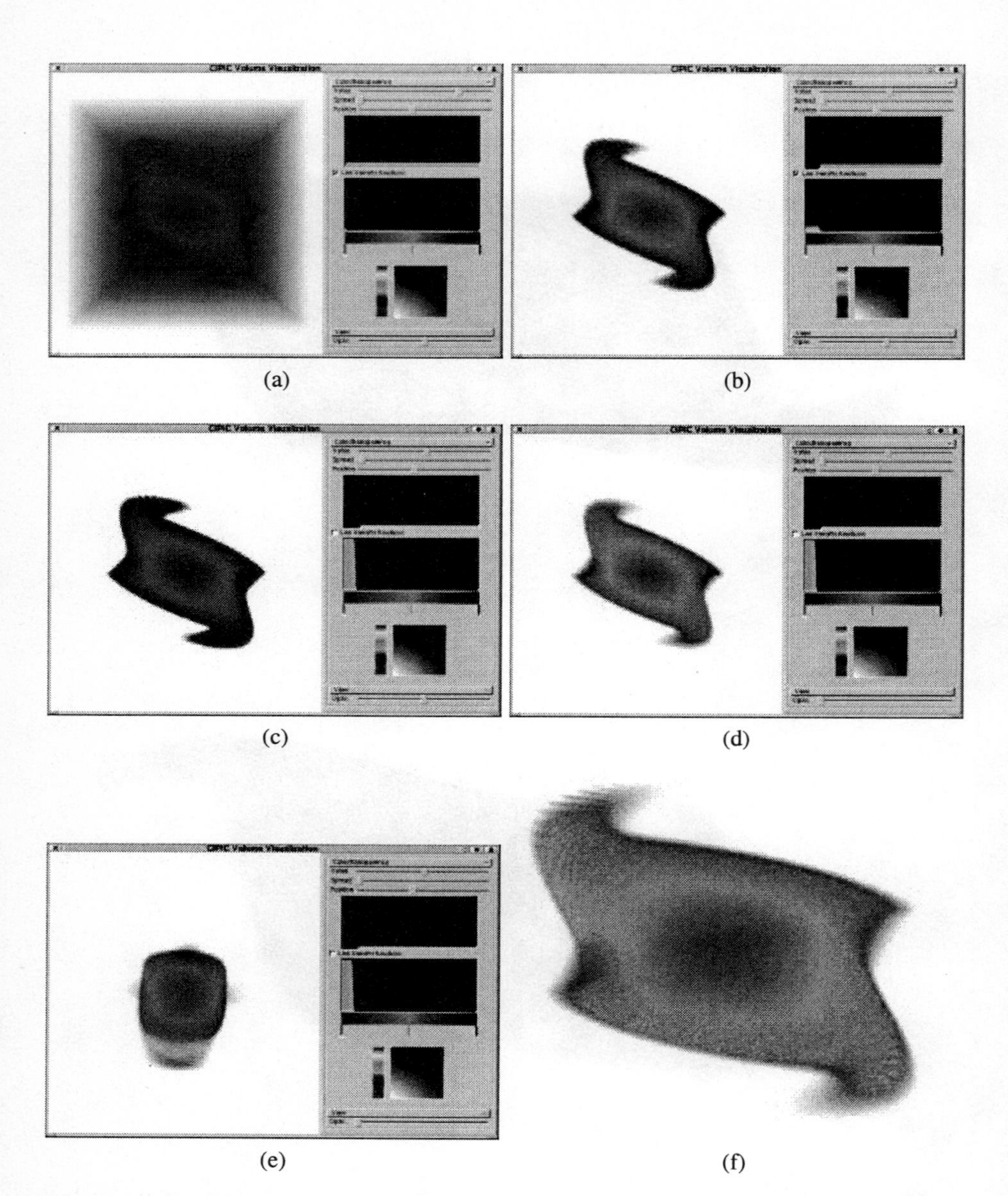

Fig. 5. A progression showing how exploration is performed. (a) Shows the initial screen, with a volume-only rendering. (b) The boundary between the high-density volume rendering and the low-density particle rendering has been moved to show more particles. (c) The transfer functions have been unlinked to show more particles while keeping the volume-rendered portion relatively transparent. (d) The point opacity has been lowered to reveal more structure. (e) The volume has been rotated to view it end-on. (f) A higher-resolution version similar to (d).

We will also investigate illumination methods to improve the quality of point-based rendering.

Acknowledgements

This work was performed under the auspices of the SciDAC project, “Advanced Computing for 21st Century Accelerator Science and Technology,” with support from the Office of Advanced Scientific Computing Research and the Office of High Energy and Nuclear Physics within the U.S. DOE Office of Science. The simulated data were generated using resources of the National Energy Research Scientific Computing Center, which is supported by the Office of Science under Contract No. DE-AC03-76SF00098. This work was also sponsored in part by the National Science Foundation under contracts ACI 9983641 (PECASE Award) and ACI 9982251 (LSSDSV).

References

1. B. Cabral, N. Cam, and J. Foran. Accelerated Volume Rendering and Tomographic Reconstruction using Texture Mapping Hardware. In 1994 Workshop on Volume Visualization, October 1994, pp. 91–98.
2. J. Qiang, R. Ryne, S. Habib, V. Decyk, ”An Object-Oriented Parallel Particle-In-Cell Code for Beam Dynamics Simulation in Linear Accelerators,” J. Comp. Phys. vol. 163, 434, (2000).
3. J. Qiang and R. Ryne, ”Beam Halo Studies Using a 3-Dimensional Particle-Core Model,” Physical Review Special Topics - Accelerators and Beams vol. 3, 064201 (2000).
4. P. S. McCormick, J. Qiang, and R. Ryne. Visualizing High-Resolution Accelerator Physics. Visualization Viewpoints (Editors: Lloyd Treinish and Theresa-Marie Rhyne), IEEE Computer Graphics and Applications, September/October 1999, pp. 11–13.
5. M. Meissner, U. Hoffmann, and W. Strasser. Enabling Classification and Shading for 3d Texture Mapping Based Volume Rending Using OpenGL and Extensions. In IEEE Visualization ’99 Conference Proceedings, October 1999, 207–214.
6. A. Van Gelder, and U. Hoffman. Direct Volume rendering with Shading via Three-Dimensional Textures. In ACM Symposium on Volume Visualization ’96 Conference Proceedings, October 1996, pp. 23–30.

Generic Large Scale 3D Visualization of Accelerators and Beam Lines

Andreas Adelmann and Derek Feichtinger

Paul Scherrer Institut (PSI),
CH-5323 Villigen, Switzerland
{Andreas.Adelmann, Derek.Feichtinger}@psi.ch
http://www.psi.ch

Abstract. We report on a generic 3D visualization system for accelerators and beam lines, in order to visualize and animate huge amount of multidimensional datasets.

The phase space data on together with survey information obtained from MAD9P runs, are post-processed and then translated into colored ray-traced POV-Ray movies. We use HPC for the beam dynamic calculation and for the trivially parallel task of ray-tracing a huge number of animation frames.

We show various movies of complicated beam lines and acceleration structure, and discuss the potential use of such tools in the design and operation process of future and present accelerators and beam transport systems.

1 Introduction

In the accelerator complex of the Paul Scherrer Institut the properties of the high intensity particle beams are strongly determined by space charge effects. The use of space charge effects to provide adequate beam matching in the PSI Injector II and to improve the beam quality in a cyclotron is unique in the world. MAD9P (**m**ethodical **a**ccelerator **d**esign version **9** - **p**arallel) is a general purpose parallel particle tracking program including 3D space charge calculation. A more detailed description of MAD9P and the presented calculations is given in [1]. MAD9P is used at PSI in the low energy 870 keV injection beam line and the separate sector 72 MeV isochronous cyclotron (Injector II), shown in Fig. 1, to investigate space charge dominated phenomena in particle beams.

2 The mad9p Particle Tracker

2.1 Governing Equations

In an accelerator/beam transport system, particles travel in vacuum, guided by electric or magnetic fields and accelerated by electric fields. In high-current accelerators and transport systems the repulsive coulomb forces due to the space

P.M.A. Sloot et al. (Eds.): ICCS 2002, LNCS 2331, pp. 362–371, 2002.

Fig. 1. PSI Injector II 72 MeV cyclotron with beam transfer lines

charge carried by the beam itself play an essential role in the design of the focusing system, especially at low energy. Starting with some definitions, we denote by $\Omega \in \mathcal{R}^3$ the spatial computational domain, which is cylindrical or rectilinear. $\Gamma = \Omega \times \mathcal{R}^3$ is the six dimensional phase space of position and momentum. The vectors $\mathbf{q}$ and $\mathbf{p}$ denote spatial and momentum coordinates. Due to the low particle density and the 'one pass' character of the cyclotron, we ignore any collisional effects and use the collisionless Vlasov Maxwell equation:

$$\partial_t f + \frac{\mathbf{p}}{m} \cdot \partial_{\mathbf{q}} f - (\partial_{\mathbf{q}} U + e\partial_{\mathbf{q}}\phi) \cdot \partial_{\mathbf{p}} f = 0. \tag{1}$$

Here the first term involving U represents the external forces due to electric and magnetic fields

$$U = \mathbf{E}(\mathbf{q};t) + \frac{\mathbf{p}}{m} \times \mathbf{B}(\mathbf{q};t) \tag{2}$$

and from Maxwell's equation we get:

$$\nabla \times \mathbf{E} + \frac{\partial \mathbf{B}}{\partial \mathbf{t}} = \mathbf{0}, \nabla \cdot \mathbf{B} = \mathbf{0}. \tag{3}$$

The external acting forces are given by a relativistic Hamiltonian $\mathcal{H}_{ext}$, where all canonical variables are small deviations from a reference value and the Hamiltonian can be expanded as a Taylor series. This is done automatically by the use of a *Truncated Power Series Algebra Package* [2], requiring no further analytical expansion.

The self-consistent Coulomb potential $\phi(\mathbf{q};t)$ can be expressed in terms of the charge density $\rho(\mathbf{q};t)$, which is proportional to the particle density

$$n(\mathbf{q};t) = \int d\mathbf{p} f(\mathbf{q},\mathbf{p};t) \text{ using } \rho(\mathbf{q},\mathbf{p};t) = en(\mathbf{q};t) \tag{4}$$

and we can write:

$$\phi(\mathbf{q};t) = \int_{\Omega} d\mathbf{q}' \frac{\rho(\mathbf{q}';\mathbf{t})}{|\mathbf{q}-\mathbf{q}'|}. \tag{5}$$

The self-fields due to space charge are coupled with Poisson's equation

$$\nabla \cdot \mathbf{E} = \int f(\mathbf{q},\mathbf{p};t) d\mathbf{p}. \tag{6}$$

2.2 Parallel Poisson Solver

The charges are assigned from the particle positions in continuum onto the grid using one of two available interpolation schemes: cloud in cell (CIC) or nearest grid point (NGP). The rectangular computation domain $\Omega := [-L_x, L_x] \times [-L_y, L_y] \times [-L_t, L_t]$, just big enough to include all particles, is segmented into a regular mesh of $M = M_x \times M_y \times M_t$ grid points. Let Ω^D be rectangular and spanned by $l \times n \times m$ with $l = 1 \dots M_x$, $n = 1 \dots M_x$ and $m = 1 \dots M_t$. The solution of the discretized Poisson equation with $\mathbf{k} = (l, n, m,)$ reads

$$\nabla^{2^D} \phi^D(\mathbf{k}) = -\frac{\rho^D(\mathbf{k})}{\epsilon_0}, \mathbf{k} \in \Omega^D. \tag{7}$$

The serial PM Solver Algorithm is summarized in the following algorithm:

PM Solver Algorithm

- ▷ Assign particle charges q_i to nearby mesh points to obtain ρ^D
- ▷ Use FFT on ρ^D and G^D (Green's function) to obtain $\widehat{\rho}^D$ and $\widehat{G}^D$
- ▷ Determine $\widehat{\phi}^D$ on the grid using $\phi^D = \rho^D * G^D$.
- ▷ Use inverse FFT on $\widehat{\phi}^D$ to obtain ϕ^D
- ▷ Compute $\mathbf{E}^D = -\nabla\phi^D$ by use of a second order finite difference method
- ▷ Interpolate $\mathbf{E}(\mathbf{q})$ at particle positions $\mathbf{q}$ from $\mathbf{E}^D$

The parallelization of the above outlined algorithm is done in two steps: first Ω^D is partitioned into subdomains Ω_k^D, $k = 1 \dots p$ where p denotes the number of processors. On each processor there are N/p particles using a spatial particle layout. The spatial layout will keep a particle on the same node as that which contains the section of the field in which the particle is located. If the particle moves to a new position, this layout will reassign it to a new node when necessary.

This will maintain locality between the particles and any field distributed using this field layout, and it will help keep particles which are spatially close to each other local to the same processor as well.
The second important part is the parallel Fourier Transformation, which allows us to speed up the above described serial PM solver algorithm. For more details on the implementation and performance see [3] [1].

To integrate the particle motion, we use a second order split-operator scheme [4]. This is based upon the assumption that one can split the total Hamiltonian in two solvable parts: $\mathcal{H}_{ext}$ and the field solver contribution $\mathcal{H}_{sc}$. For a step in the independent variable τ one can write:

$$\mathcal{H}(\tau) = \mathcal{H}_{ext}(\tau/2)\mathcal{H}_{sc}(\tau)\mathcal{H}_{ext}(\tau/2) + \mathcal{O}(\tau^3) \tag{8}$$

2.3 Design or Reference Orbit

In order to describe the motion of charged particles we use the local coordinate system seen in Fig. 2. The accelerator and/or beam line to be studied is described as a sequence of beam elements placed along a reference or design orbit. The

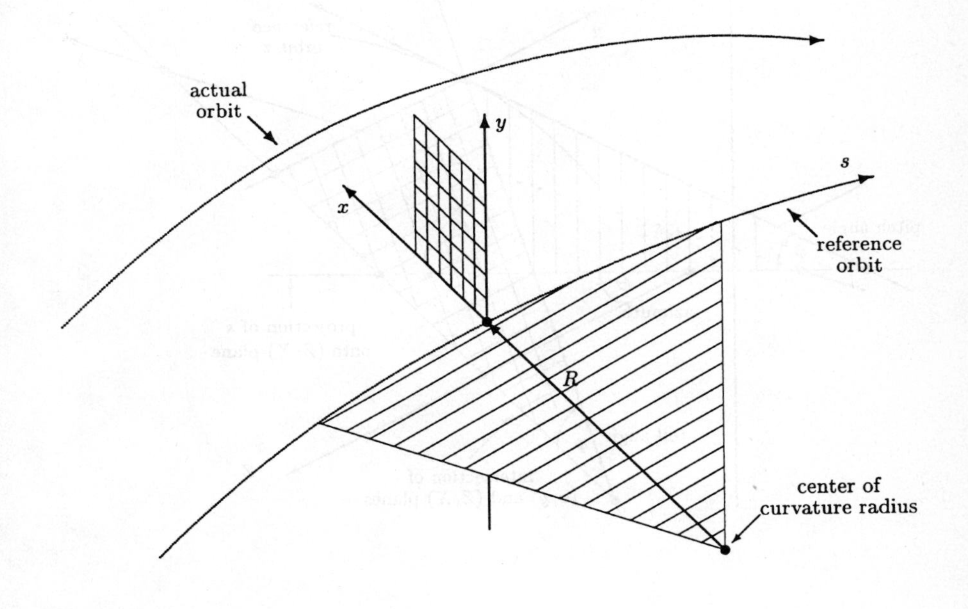

Fig. 2. Local Reference System

global reference orbit (see Fig. 3), also known as the design orbit, is the path of a charged particle having the central design momentum of the accelerator through idealized magnets with no fringe fields.

The reference orbit consists of a series of straight sections and circular arcs. It is defined under the assumption that all elements are perfectly aligned along the design orbit. The accompanying tripod (Dreibein) of the reference orbit spans a local curvilinear right handed system (x, y, s).

2.4 Global Reference System

The local reference system (x, y, s) may thus be referred to a global Cartesian coordinate system (X, Y, Z). The positions between beam elements are numbered $0, \ldots, i, \ldots n$. The local reference system (x_i, y_i, s_i) at position i, i.e. the displacement and direction of the reference orbit with respect to the system (X, Y, Z) are defined by three displacements (X_i, Y_i, Z_i) and three angles $(\Theta_i, \Phi_i, \Psi_i)$.

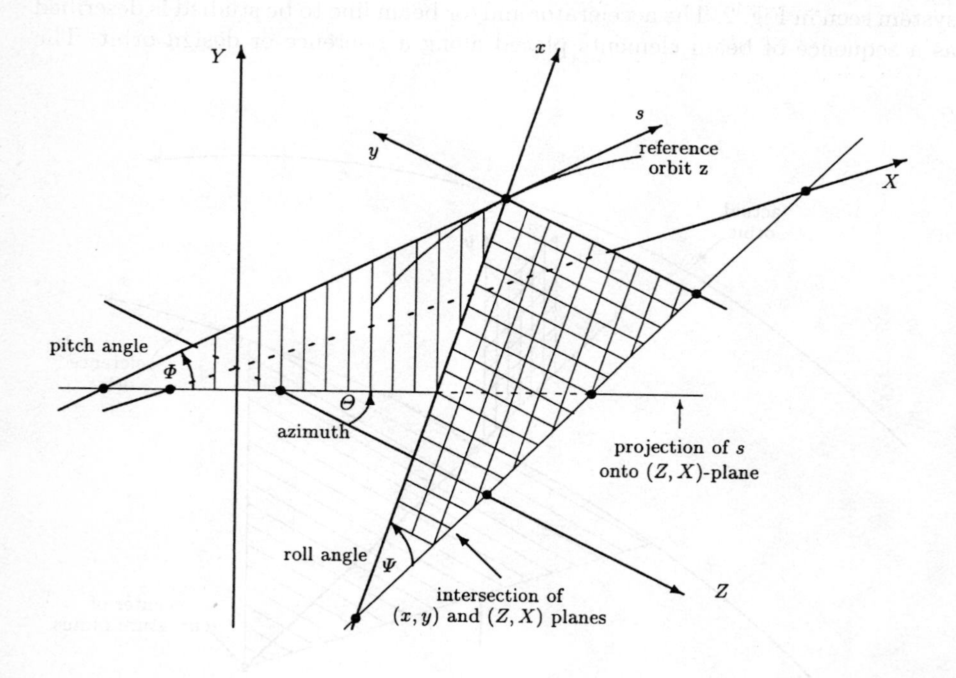

Fig. 3. Global Reference System

The above quantities X, Y and Z are displacements of the local origin in the respective direction. The angles (Θ, Φ, Ψ) are **not** the Euler angles. The reference orbit starts at the origin and points by default in the direction of the positive

Z-axis. The initial local axes (x, y, s) coincide with the global axes (X, Y, Z) in this order. The displacement is described by a vector $\mathbf{v}$ and the orientation by a unitary matrix $\mathcal{W}$. The column vectors of $\mathcal{W}$ are unit vectors spanning the local coordinate axes in the order (x, y, s). $\mathbf{v}$ and $\mathcal{W}$ have the values:

$$\mathbf{v} = \begin{pmatrix} X \\ Y \\ Z \end{pmatrix}, \qquad \mathcal{W} = \mathcal{S}\mathcal{T}\mathcal{U} \tag{9}$$

where

$$\mathcal{S} = \begin{pmatrix} \cos\Theta & 0 & -\sin\Theta \\ 0 & 1 & 0 \\ \sin\theta & 0 & \cos\Theta \end{pmatrix}, \quad \mathcal{T} = \begin{pmatrix} 1 & 0 & 0 \\ 0 & \cos\Phi & \sin\Phi \\ 0 & -\sin\Phi & \cos\Phi \end{pmatrix}, \tag{10}$$

$$\mathcal{U} = \begin{pmatrix} \cos\Psi & -\sin\Psi & 0 \\ \sin\Psi & \cos\Psi & 0 \\ 0 & 0 & 1 \end{pmatrix}. \tag{11}$$

Let the vector $\mathbf{r}_i$ be the displacement and the matrix $\mathcal{S}_i$ be the rotation of the local reference system at the exit of the element i with respect to the entrance of the same element.

When advancing through a beam element i, one can compute $\mathbf{v}_i$ and $\mathcal{W}_i$ by the recurrence relations

$$\mathbf{v}_i = \mathcal{W}_{i-1}\mathbf{r}_i + \mathbf{v}_{i-1}, \qquad \mathcal{W}_i = \mathcal{W}_{i-1}\mathcal{S}_i. \tag{12}$$

This relation (12) is used in the generation of ray-tracing movies.

3 Architecture of mad9p and accelVis

Today we use Linux Farms (also known as Beowulf clusters) with up to 500 Processors (Asgard ETHZ) as well as traditional symmetric multiprocessor (SMP's) machines like IBM SP-2 or SGI Origin 2000. Having such a wide variety of platforms available put some non negligible constraint on the software engineering part of a simulation code. MAD9P is based on two frameworks:[1] CLASSIC [6] and POOMA [3], shown schematically in Fig. 4. CLASSIC deals mainly with the accelerator physics including a polymorphic differential algebra (DA) package and the input language to specify general complicated accelerator systems. In order to ease the task of writing efficient parallel applications we rely on the POOMA framework which stands for Parallel Object-Oriented Methods and Applications. POOMA provides abstraction for mathematic/physical quantities

[1] We use the notion of framework in the following sense: a framework is a set of co-operating classes in a given problem frame. On this and other software engineering concepts see [5]

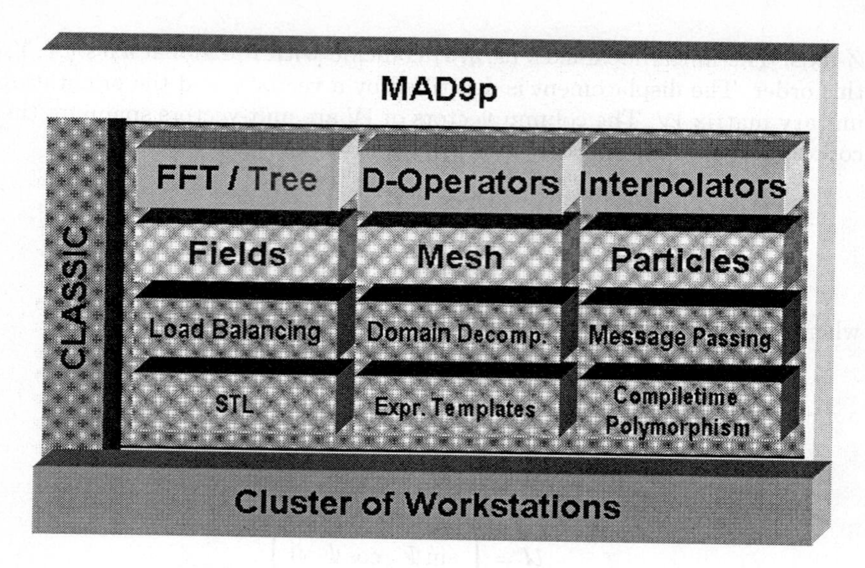

Fig. 4. Architectural overview on MAD9P

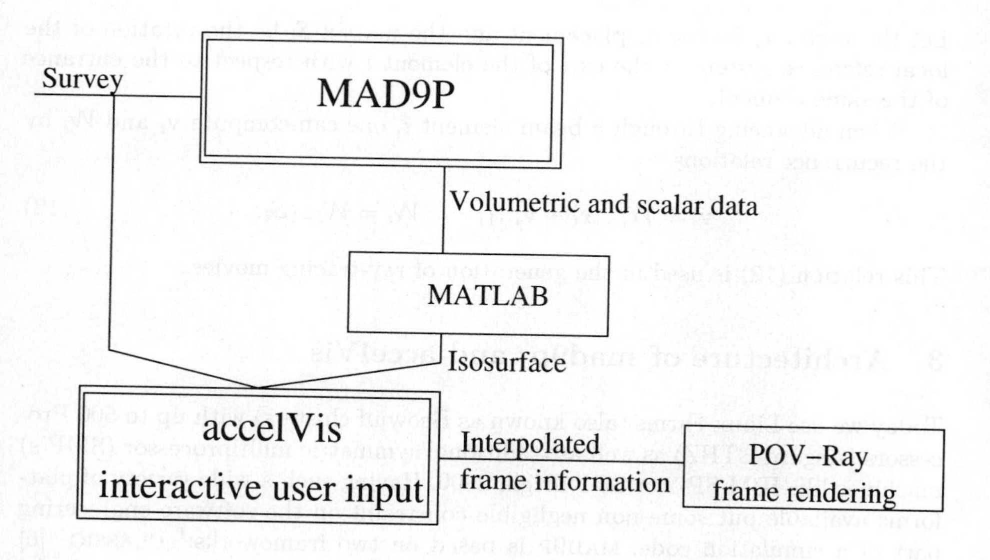

Fig. 5. Data flow between MAD9P

like particles, fields, meshes and differential operators. The object-oriented approach manages the complexity of explicit parallel programming; it encapsulates the data distribution and communication among real or virtual processors. POOMA and all the other components are implemented as a set of templated C++ classes. The computing nodes can be a single (real) cpu or a virtual node

(VNODE). Usually MAD9P uses the message passing interface MPI [7] in order to interconnect the individual nodes.

ACCELVIS is currently implemented using ANSI C. The program interfaces to the OpenGL graphics library and it's GLU and GLUT extensions to render the interactive 3D graphics. These libraries (or the compatible Mesa library) as well as POV-Ray [8] and MATLAB [9] are available on a wide range of platforms. Therefore, although the application was developed on a Red-Hat Linux System, only very minor modifications are necessary to transfer it to a variety of other architectures. The data-flow, involving MAD9P is shown schematically in Fig. 5.

3.1 Program capabilities

The ACCELVIS application enables the user to view a graphical interpretation of volumetric and scalar data provided by a MAD9P run. The reference trajectory and ISO-surfaces illustrating the particle density can be investigated interactively by gliding with a virtual camera through a representation of the accelerator (Fig. 6). By defining a trajectory for the camera the user is able to produce high quality animations for teaching and illustration purposes.

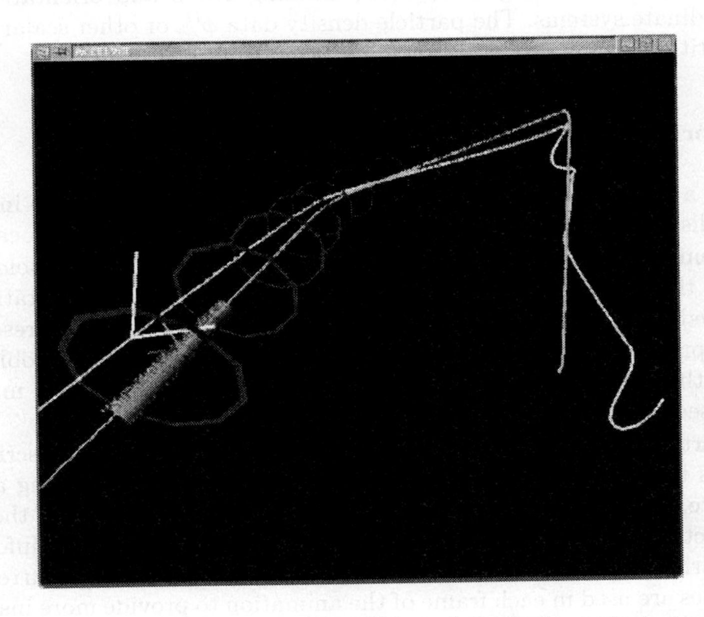

Fig. 6. ACCELVIS view of the particle cloud ISO-surface, the beam trajectory (red line), the camera trajectory (yellow line), and the camera viewing orientation (white lines)

Fig. 7. Animation frames generated from the ACCELVIS setup shown in Fig. 6. Two ISO-surfaces for cloud core (red) and halo (yellow) are used in visualizing the particle density

3.2 Program Input

The reference trajectory is read in as a sequence of displacement vectors v_i and the matching rotation angles Θ_i, Φ_i, Ψ_i defining origin and orientation of the local coordinate systems. The particle density data ϕ^D, or other scalar data like rms quantities (beam size, emittance), is taken from a MAD9P run.

3.3 Information Processing

To obtain a fluid animation of the particle clouds, it is necessary to interpolate between displacements as well as between rotations yielding the local coordinate systems along the reference trajectory. A simple spline interpolation was chosen for the displacement vectors of the reference particle. The rotations were interpolated through a spline interpolation of their quaternion representations since this provides smoother interpolation and avoids some of the problems that appear if the defining angles Θ_i, Φ_i, Ψ_i or elements of the rotation matrix are directly used [10].

The particle density is processed by interfacing to a MATLAB [9] script which transforms the data into a series of connecting triangles representing a density ISO-surface. To increase the smoothness of the generated graphics, the surface normal vectors at every triangle corner are also calculated (This information is customarily used by 3D visualization surface lighting models). Currently two ISO-surfaces are used in each frame of the animation to provide more insight into the density distribution. The surface gained from the higher iso value is termed the cloud core, the other the cloud halo. The halo is rendered translucent (Fig. 7).

The camera view is represented by yet another local coordinate system. For the production of the high quality animations a number of camera views are defined by interactively moving the camera to the desired position for the respective simulation frame. The camera views are then interpolated over the whole course of the simulation using the same procedure as described above for the interpolation of the reference trajectory orientations.

3.4 Generation of the Animations

The application creates input and command files for the free and commonly used POV-Ray [8] ray-tracing program. If desired a series of command files are produced where each one assigns a subset of the frames to be rendered to the nodes of a computing cluster. This trivially simple parallelization scheme enabled us to compile the 1600 frames ($320 * 240$ pixels each, 24 bit color depth) of this current animation in a rendering time of about 20 minutes on the 64 node Merlin Linux cluster at PSI. By using standard software the frames can be converted to one of the common movie formats (usually MPEG).

4 Application to the PSI Injector II Cyclotron

The use of high level visualization is one of the key aspects in interpretation of multi-dimensional datasets. In the presented approach, it was attempted to tightly couple large scale accelerator system simulations (using MAD9P) with advanced visualization techniques (ACCELVIS). Using principles of generality in the design of both components, one can easy adapt ACCELVIS to other accelerator system simulation frameworks.

First simulations of complicated structures, as shown in Fig. 1, were successful. The application area of such visualization ranges from education to the (re)design phase of existing or new machines. This might evolve into an indispensable tool for the use in the accelerator control-room.

References

1. Adelmann, A.: 3D Simulations of Space Charge Effects in Particle Beams. PhD thesis, ETH (2002)
2. Berz, M.: Modern Map Methods in Particle Beam Physics. Academic Press (1999)
3. Cummings, J., Humphrey, W.: Parallel particle simulations using the POOMA framework. In: 8th SIAM Conf. Parallel Processing for Scientific Computing. (1997)
4. J.M. Sanz-Serna, M.C.: Numerical Hamiltonian Problems. Chapman and Hall (1994)
5. E. Gamma, *et.al.*: Design Patterns. Addison Wesley (1995)
6. Iselin, F.: The classic project. Particle Accelerator **Vol. 54,55** (1996)
7. William Gropp, *et.al.*: Using MPI : portable parallel programming with the message-passing interface. Cambridge, Massachusetts : MIT Press (1999)
8. the POV-Team: Pov-ray 3.1 (1999) ray-tracing software.
9. MathWorks: MATLAB 6.1. URL: http://www.mathworks.com (2001)
10. E. B. Dam, M. Koch, M.L.: Quaternions, interpolation and animation. Technical Report DIKU-TR-98/5, Department of Computer Science, University of Copenhagen, Universitetsparken 1, DK-2100 Kbh, Denmark (1998)

Tracking Particles In Accelerator Optics With Crystal Elements

V. Biryukov[1], A. Drees[2], R.P. Fliller III[2], N. Malitsky[2], D. Trbojevic[2]

[1] IHEP Protvino, RU-142284, Russia
[2] BNL, Upton, 11973 NY, USA

Abstract. Bent channeling crystals as elements of accelerator optics with extreme, 1000-Tesla intracrystalline fields can find many applications in accelerator world from TeV down to MeV energies. Situated in accelerator ring, they serve for beam scraping or extraction, e.g. in RHIC and IHEP U70. Crystal itself is a miniature beamline with its own "strong focusing", beam loss mechanisms etc. We describe the algorithms implemented in the computer code CATCH used for simulation of particle channeling through crystal lattices and report the results of tracking with 100-GeV/u Au ions in RHIC and with 70-GeV and 1-GeV protons in U70. Recent success of IHEP where a tiny, 2-mm Si crystal has channeled a 10^{12} p/s beam of 70-GeV protons out of the ring with efficiency 85% followed the prediction of computer model.

1 Introduction

The idea to deflect proton beams using bent crystals, originally proposed by E.Tsyganov [1], was demonstrated in 1979 in Dubna on proton beams of a few GeV energy. The physics related to channneling mechanisms was studied in details, in the early 1980's, at St.Petersburg, in Dubna, at CERN, and at FNAL using proton beams of 1 to 800 GeV (see refs., e.g. in [2]). Recently, the range of bent crystal channeling was expanded down to MeV energy[3], now covering 6 decades of energy.

Crystal-assisted extraction from accelerator was demonstrated for the first time in 1984 in Dubna at 4-8 GeV and deeply tested at IHEP in Protvino starting from 1989 by exposing a silicon crystal bent by 85 mrad to the 70 GeV proton beam of U-70. The Protvino experiment eventually pioneered the first regular application of crystals for beam extraction: the Si crystal, originally installed in the vacuum chamber of U-70, served without replacement *over 10 years*, delivering beam for particle physicists all this time. However its channeling efficiency was never exceeding a fraction of %.

In the 1990's an important milestone was obtained at the CERN SPS. Protons diffusing from a 120 GeV beam were extracted at an angle of 8.5 mrad with a bent silicon crystal. Efficiencies of ~10%, orders of magnitude higher than the values achieved previously, were measured for the first time [4]. The extraction studies at SPS clarified several aspects of the technique. In addition, the extraction results were found in fair agreement with Monte Carlo predictions [2]. In

P.M.A. Sloot et al. (Eds.): ICCS 2002, LNCS 2331, pp. 372–380, 2002.

the late 1990's another success came from the Tevatron extraction experiment where a crystal was channeling a 900-GeV proton beam with an efficiency of ~30% [5]. During the FNAL test, the halo created by beam-beam interaction in the periphery of the circulating beam was extracted from the beam pipe without unduly affecting the backgrounds at the collider detectors.

Possible application of crystal channeling in modern hadron accelerators, like slow extraction and halo collimation, can be exploited in a broad range of energies, from sub-GeV cases (i.e. for medical accelerators) to multi-TeV machines (for high-energy research).

Crystal collimation is being experimentally studied at RHIC with gold ions and polarized protons of 100-250 GeV/u [6] and has been proposed and studied in simulations for the Tevatron (1000 GeV) [7], whilst crystal-assisted slow extraction is considered for AGS (25 GeV protons) [8]. In all cases, the critical issue is the channeling efficiency.

2 Crystal as a beamline

Let us understand how the crystal symmetry may be used for steering a particle beam. Any particle traversing an amorphous matter or a disaligned crystal experiences a number of uncorrelated collisions with single atoms. As these encounters may occur with any impact parameters, small or large ones, a variety of processes take place in the collision events. In disordered matter one may consider just a single collision, then simply make a correction for the matter density.

The first realization that the atomic order in crystals may be important for these processes dates back to 1912[9]. In early 1960s the channeling effect was discovered in computer simulations and experiments which observed abnormally large ranges of ions in crystals[10]. The orientational effects for charged particles traversing crystals were found for a number of processes requiring a small impact parameter in a particle–atom collision.

The theoretical explanation of the channeling effects has been given by Lindhard [11], who has shown that when a charged particle has a small incident angle with respect to the crystallographic axis (or plane) the successive collisions of the particle with the lattice atoms are correlated, and hence one has to consider the interaction of the charged particle with the atomic string (plane). In the low-angle approximation one may replace the potentials of the single atoms with an averaged *continuous potential.* If a particle is misaligned with respect to the atomic strings but moves at a small angle with respect to the crystallographic plane, one may take advantage of the continuous potential for the atomic plane, where averaging is made over the two planar coordinates:

$$U_{\mathrm{pl}}(x) = N d_{\mathrm{p}} \int_{-\infty}^{+\infty} \int_{-\infty}^{+\infty} V(x, y, z) \mathrm{d}y \mathrm{d}z \tag{1}$$

where $V(x,y,z)$ is the potential of a particle–atom interaction, N is the volume density of atoms, d_p is the interplanar spacing.

The atomic plane (string) gently steers a particle away from the atoms, thus suppressing the encounters with small impact parameters. The transverse motion of a particle incident at some small angle with respect to one of the crystal axes or planes is governed by the continuous potential of the crystal lattice. The fields of the atomic axes and planes form the potential wells, where the particle may be trapped. In this case one speaks of ***channeling*** of the particle: an ***axial channeling*** if the particle is bound with atomic strings, and a ***planar channeling*** if it is bound with atomic planes.

The interaction of the channeled particle with a medium is very different from a particle interaction with an amorphous solid. For instance, a channeled proton moves between two atomic planes (layers) and hence does not collide with nuclei; moreover, it moves in a medium of electrons with reduced density. In the channeling mode a particle may traverse many centimeters of crystal (in the ~100 GeV range of energy).

Leaving aside the details of channeling physics, it may be interesting to mention that accelerator physicist will find many familiar things there:

– Channeled particle oscillates in a transverse nonlinear field of a crystal channel, which is the same thing as the "***betatronic oscillations***" in accelerator, but on a much different scale (the wavelength is 0.1 mm at 1 TeV in silicon crystal). The analog of "beta function" is order of 10 μm in crystal. The number of oscillations per crystal length can be several thousand in practice. The concepts of beam emittance, or particle action have analogs in crystal channeling.
– The crystal nuclei arranged in crystallographic planes represent the "***vacuum chamber walls***". Any particle approached the nuclei is rapidly lost from channeling state. Notice a different scale again: the "vacuum chamber" size is ~2 Å.
– The well-channeled particles are confined far from nuclei (from "aperture"). They are lost then only due to scattering on electrons. This is analog to "***scattering on residual gas***". This may result in a gradual increase of the particle amplitude or just a catastrophic loss in a single scattering event.
– Like the real accelerator lattice may suffer from ***errors of alignment***, the lattice of real crystal may have dislocations too, causing an extra diffusion of particle amplitude or (more likely) a catastrophic loss.
– Accelerators tend to use low temperature, superconducting magnets. Interestingly, the crystals cooled to ***cryogenic temperatures*** are more efficient, too.

In simulations, the static-lattice potential is modified to take into account the thermal vibrations of the lattice atoms. Bending of the crystal has no effect on this potential. However, it causes a centrifugal force in the non-inertial frame related to the atomic planes. To solve the equation of motion in the potential $U(x)$ of the bent crystal, as a first approximation to the transport of a particle,

$$pv\frac{d^2x}{dz^2} = -\frac{dU(x)}{dx} - \frac{pv}{R(z)}, \tag{2}$$

(x being the transversal, z the longitudinal coordinate, pv the particle longitudinal momentum and velocity product, $R(z)$ the local radius of curvature), we use the fast form of the Verlet algorithm:

$$x_{i+1} - x_i = (\theta_i + 0.5f_i\delta z)\delta z\,, \tag{3}$$

$$\theta_{i+1} - \theta_i = 0.5(f_{i+1} + f_i)\delta z \tag{4}$$

with θ for dx/dz, f for the 'force', and δz for the step. It was chosen over the other second order algorithms for non-linear equations of motion, such as Euler-Cromer's and Beeman's, owing to the better conservation of the transverse energy shown in the potential motion.

Beam bending by a crystal is due to the trapping of some particles in the potential well $U(x)$, where they then follow the direction of the atomic planes. This simple picture is disturbed by scattering processes which could cause (as result of one or many acts) the trapped particle to come to a free state (feed out, or dechanneling process), and an initially free particle to be trapped in the channeled state (feed in, or volume capture).

Feed out is mostly due to scattering on electrons, because the channelled particles keep far from the nuclei. The fraction of the mean energy loss corresponding to single electronic collisions can be written as follows [12]:

$$-\frac{dE}{dz} = \frac{D}{2\beta^2}\rho_e(x)\ln\frac{T_{max}}{I}\,, \tag{5}$$

with $D = 4\pi N_A r_e^2 m_e c^2 z^2 \frac{Z}{A}\rho$, z for the charge of the incident particle (in units of e), ρ for the crystal density, Z and A for atomic number and weight, T_{max} the maximum energy transfer, and the other notation being standard [12]. It depends on the local density $\rho_e(x)$ (normalized on the amorphous one) of electrons. The angle of scattering in soft collisions can be computed as a random Gaussian with r.m.s. value $\theta^2_{rms} = \frac{m_e}{p^2}(\delta E)_{soft}$ where $(\delta E)_{soft}$ is the soft acts contribution. The probability of the hard collision (potentially causing immediate feed out) is computed at every step. The energy transfer T in such an act is generated according to the distribution function $P(T)$:

$$P(T) = \frac{D\rho_e(x)}{2\beta^2}\frac{1}{T^2}\,. \tag{6}$$

The transverse momentum transfer q is equal to $q = \sqrt{2m_eT + (T/c)^2}$. Its projections are used to modify the angles θ_x and θ_y of the particle.

The multiple Coulomb scattering on nuclei is computed by the approximation Kitagawa-Ohtsuki $\langle\theta^2_{sc}\rangle_{amorph} \cdot \rho_n(x)$, i.e. the mean angle of scattering squared is proportional to the local density of nuclei $\rho_n(x)$. The probability of nuclear collision, proportional to $\rho_n(x)$, is checked at every step.

3 Channeling of protons at IHEP Protvino

In crystal extraction, the circulating particles can cross several times the crystal without nuclear interactions. Unchanneled particles are deflected by multiple scattering and eventually have new chances of being channeled on later turns. The crystal size should be well matched to the beam energy to maximise the transmission efficiency. To clarify this mechanism an extraction experiment was started at IHEP Protvino at the end of 1997 [13].

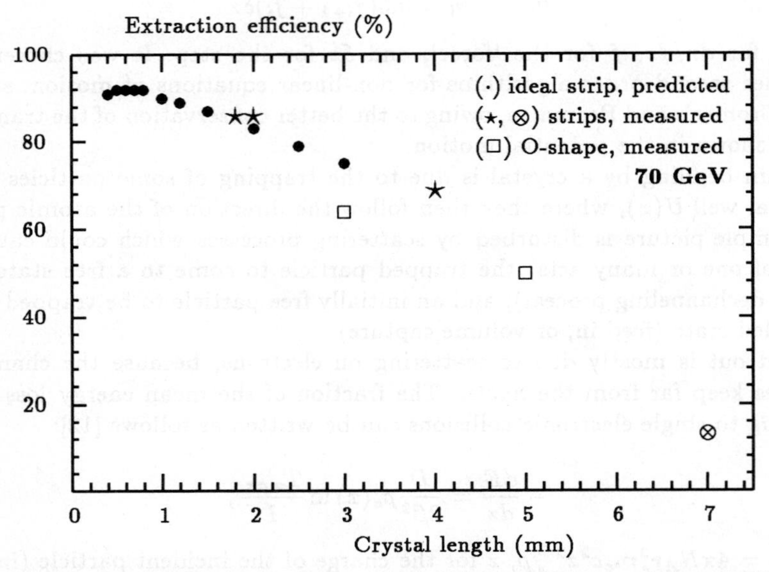

Fig. 1. Crystal extraction efficiency as measured for 70-GeV protons at IHEP (⋆, □, ⊗), and Monte Carlo prediction (o) for a perfect "strip" deflector.

As showed the simulation study of multi-turn crystal-asisted extraction taking into account the multiple encounters with crystal of the protons circulating in the ring, Fig.1, the crystal had to be quite miniature - just a few mm along the beam of 70 GeV protons - in order the extraction could benefit from crystal channeling.

Over the recent years, the experiment gradually approached the optimum found in the simulations. The recent extraction results, with 2 mm crystal of silicon, are rather close to the top of the theoretical curve. The experimentally demonstrated figure is excellent: 85% of all protons dumped onto the crystal were channeled and extracted out of the ring, in good accordance with prediction.

The channeling experiment was repeated with the same set-up at much different energy, 1.3 GeV. Here, no significant multiplicity of proton encounters with the crystal was expected due to strong scattering of particles in the same, 2-mm long crystal. The distribution of protons 20 m downstream of the crystal

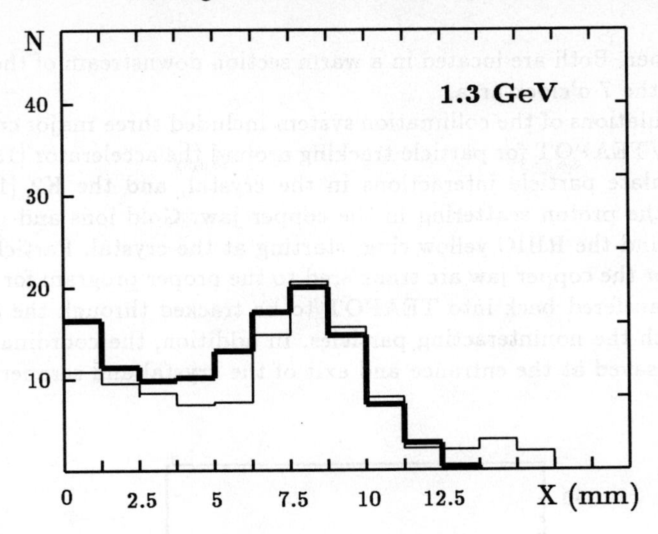

Fig. 2. The profile of 1.3 GeV protons on the collimator face as measured (thick line) and as predicted (thin) by simulations.

was observed on the face of a collimator, Fig.2. About half of the particles are found in the channeled peak. The distribution of 1.3 GeV protons is in good agreement with Monte Carlo predictions.

4 Channeling of gold ions at RHIC

In present day high energy colliders, the requirements of high luminosity and low backgrounds place strict requirements on the quality of the beams used. At facilities like RHIC, intra-beam scattering and other halo forming processes become a major concern[6]. Transverse beam growth not only leads to increased detector backgrounds, but also reduces dynamic aperture of the accelerator leading to particle losses at high beta locations. To minimize these effects, an efficient collimation system is needed.

The optics of two stage collimation systems have been reported numerous places[14]. The main disadvantage of the usual two stage system is that particles hitting the primary collimator with small impact parameters can scatter out of the material, causing a more diffuse halo. Using a bent crystal as the primary collimator in such a system, the channeled particles are placed into a well defined region of phase space. This allows the placement of a secondary collimator such that the impact parameters of the channeled particles are large enough to reduce the scattering probability, and most of the particles that hit the collimator are absorbed.

For the 2001 run, the yellow (counter-clockwise) ring had a two stage collimation system consisting of a 5 mm long crystal and a 450 mm long L-shaped copper scraper. Both are located in a warm section downstream of the IR triplet magnets in the 7 o'clock area.

The simulations of the collimation system included three major code components, UAL/TEAPOT for particle tracking around the accelerator [15], CATCH [16] to simulate particle interactions in the crystal, and the K2 [14] code to implement the proton scattering in the copper jaw. Gold ions and protons are tracked around the RHIC yellow ring, starting at the crystal. Particles that hit the crystal or the copper jaw are transfered to the proper program for simulation and then transfered back into TEAPOT to be tracked through the accelerator together with the noninteracting particles. In addition, the coordinates of each particle are saved at the entrance and exit of the crystal and scraper.

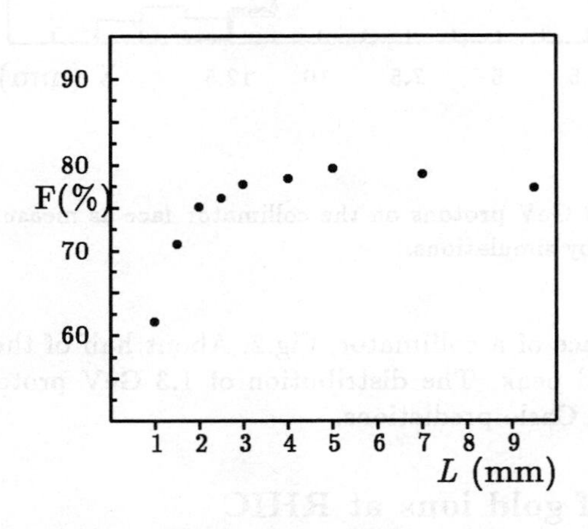

Fig. 3. Single-pass bending efficiency for 100-GeV/u Au ions vs crystal length, for 0.5 mrad bending.

The beam distribution at the entrance of the crystal was presented by the sample of fully stripped gold ions generated as described in [6]. We plot in Fig.3 how many particles were bent at least 0.1 mrad (this includes the particles channeled part of the crystal length as they are steered through the angles that might be sufficient for interception by the downstream collimator) when the incident particles are well within the acceptance of the crystal aligned to the beam envelope.

The gold ions tracked through the crystal and transported through the RHIC ring were eventually lost, at collimator and beyond. Fig.4 shows the losses around the RHIC rings from the particles scattered of the primary and secondary collimators and the losses from the particles deflected by the crystal. Two extreme cases are presented when the primary collimator downstream of the crystal is wide open and when it is set at 5 σ_x , the same horizontal distance as a front edge of the crystal.

The commissioning of the crystal collimator has occurred in the year 2001

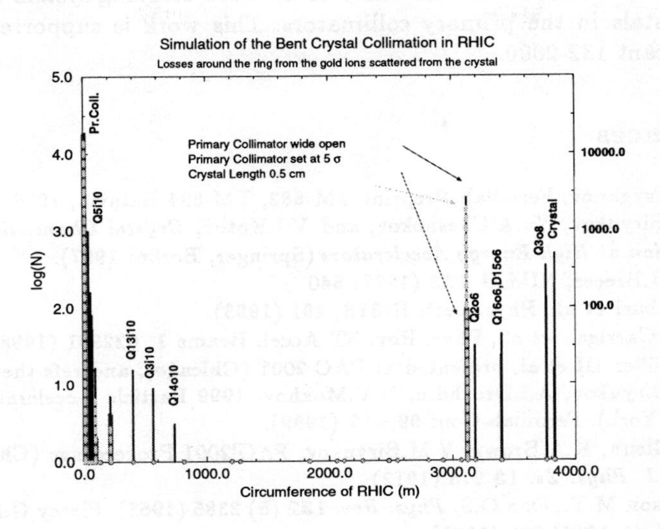

Fig. 4. Losses around the RHIC rings.

run. Once the efficiency of the crystal collimator has been determined, a second apparatus will be built for the blue ring. The collimator can be used in a variety of experiments to determine beam attributes such as size, angular profile, and diffusion rates. Experience gained at RHIC will be important for the plans to implement crystal extraction in the AGS [8] for a neutrino mass experiment.

5 Summary

The channeling crystal efficiency has reached unprecedented high values. The same 2 mm long crystal was used to channel 70 GeV protons with an efficiency of 85.3±2.8% and 1.3 GeV protons with an efficiency of 15-20%. The efficiency results well match the figures theoretically expected for ideal crystals. Theoretical analysis allows to plan for extraction and collimation with channeling efficiencies over 90-95%.

The high figures obtained in extraction and collimation provide a crucial support for the ideas to apply this technique in beam cleaning systems, for instance in RHIC and at the Tevatron. Earlier Tevatron scraping simulations [7] have shown that crystal scraper reduces accelerator-related background in CDF and D0 experiments by a factor of ∼10.

Besides the experience gained in crystal extraction and collimation at IHEP Protvino, first experimental data is coming from RHIC where crystal collimator [6] has been commissioned. This technique is potentially applicable also in LHC for instance to improve the efficiency of the LHC cleaning system by embedding bent crystals in the primary collimators. This work is supported by INTAS-CERN grant 132-2000.

References

1. E.N.Tsyganov, Fermilab Preprint TM-682, TM-684 Batavia, 1976
2. V.M.Biryukov, Yu.A.Chesnokov, and V.I.Kotov, *Crystal Channeling and its Application at High Energy Accelerators* (Springer, Berlin: 1997).
3. M.B.H.Breese, NIM B **132** (1997) 540
4. H. Akbari *et al.*, Phys. Lett. B **313**, 491 (1993).
5. R. A. Carrigan *et al.*, Phys. Rev. ST Accel. Beams 1, 022801 (1998).
6. R.P.Fliller III et al, presented at PAC 2001 (Chicago), and refs therein.
7. V.M.Biryukov, A.I.Drozhdin, N.V.Mokhov. 1999 Particle Accelerator Conference (New York). Fermilab-Conf-99/072 (1999).
8. J.W.Glenn, K.A.Brown, V.M.Biryukov, PAC'2001 Proceedings (Chicago).
9. Stark J. *Phys. Zs.* **13** 973 (1912)
10. Robinson M.T., Oen O.S. *Phys. Rev.* **132** (5) 2385 (1963). Piercy G.R., et al. *Phys. Rev. Lett.* **10**(4) 399 (1963)
11. J.Lindhard, Mat.Fys.Medd. Dan. Vid. Selsk., Vol. 34, 1 (1965).
12. Esbensen H. et al. *Phys. Rev. B* **18**, 1039 (1978)
13. A.G.Afonin, et al, Phys.Rev.Lett. 87, 094802 (2001)
14. T.Trenkler and J.B.Jeanneret. "K2. A software package for evaluating collimation systems in circular colliders." SL Note 94-105 (AP), December 1994.
15. N.Malitsky and R.Talman,"Unified Accelerator Libraries" CAP96; L.Schachinger and R.Talman, "A Thin Element Accelerator Program for Optics and Tracking", Particle Accelerators, 22, 1987.
16. V.Biryukov, "Crystal Channeling Simulation-CATCH 1.4 User's Guide", SL/Note 93-74(AP), CERN, 1993.

Precision Dynamic Aperture Tracking in Rings

F. Méot

CEA DSM DAPNIA SACM, F-91191 Saclay
{fmeot@cea.fr}

Abstract. The paper presents a variety of results concerning dynamic aperture tracking studies, including most recent ones, obtained using a highly symplectic numerical method based on truncated Taylors series. Comparisons with various codes, that had to be performed in numerous occasions, are also addressed.

1 Introduction

Precision is a strong concern in long term multiturn particle tracking in accelerators and storage rings. Considering the dramatic speed increase of informatics means, it became evident several years ago that there were no good reasons left for keeping using simplified field models and simplistic mapping methods, that both would lead, although fast, to erroneous results as to possible effects of field non-linearities on long-term particle motion.

This has motivated upgrading of the ray-tracing code Zgoubi [1], formerly developed by J. C. Faivre and D. Garreta at Saclay for the calculating trajectories in magnetic spectrometer field maps [2]. First multiturn ray-tracing trials concerned spin tracking for the purpose of studying the installation of a partial Siberian snake in the 3 GeV synchrotron Saturne ; the main task there was twofold, on the one hand assure symplectic transport throughout the about 10^4 turn lasting depolarizing resonance crossing, on the other hand satisfy $\sqrt{S_x^2+S_z^2+S_s^2}=1$ while tracking the all three spin components (S_x, S_z, S_s) in presence in particular of dipole and quadrupole fringe fields that have a major role in depolarization, which all proved to work quite well [3]. That led to cope with (much) larger size machines, at first reasonably close to first order behavior, and eventually including all sorts of more or less strong sources of non-linearities [4]-[8], without forgetting sophisticated non-linear motion in an electrostatic storage ring [9].

In the following, we first recall the principles of the integration method and field models used. Next we summarize some meaningful numerical results so obtained.

2 Numerical method

2.1 Integration

Zgoubi solves the Lorentz equation $d(m\boldsymbol{v})/dt = q(\boldsymbol{E}+\boldsymbol{v}\times\boldsymbol{B})$ by stepwise Taylor expansions of the vector position $\boldsymbol{R}$ and velocity $\boldsymbol{v}$, which writes

P.M.A. Sloot et al. (Eds.): ICCS 2002, LNCS 2331, pp. 381–390, 2002.

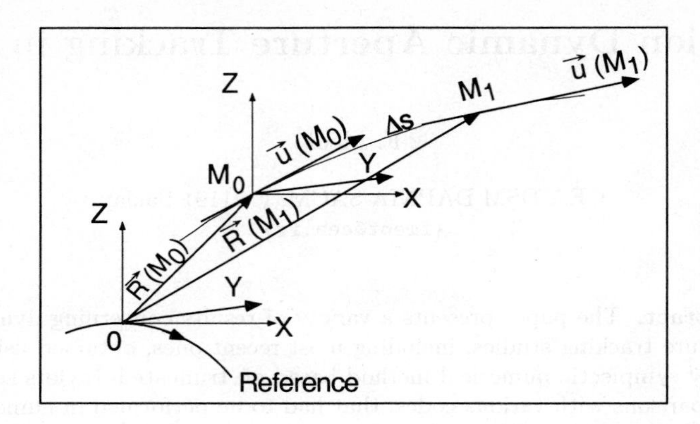

Fig. 1. Particle motion in the Zgoubi frame, and parameters used in the text.

$$\boldsymbol{R}(M_1) = \boldsymbol{R}(M_0) + \boldsymbol{u}(M_0)\Delta s + \boldsymbol{u}\,'(M_0)\ \Delta s^2/2! + \dots$$
$$\boldsymbol{u}(M_1) = \boldsymbol{u}(M_0) + \boldsymbol{u}\,'(M_0)\Delta s + \boldsymbol{u}\,''(M_0)\ \Delta s^2/2! + \dots$$

wherein $\boldsymbol{u} = \boldsymbol{v}/v$ with $v = |\boldsymbol{v}|$, $ds = v\ dt$, $\boldsymbol{u}\,' = d\boldsymbol{u}/ds$, and with $m\boldsymbol{v} = mv\boldsymbol{u} = q\ B\rho\ \boldsymbol{u}$, $B\rho$ = rigidity of the particle with mass m and charge q. The derivatives $\boldsymbol{u}^{(n)} = d^n\boldsymbol{u}/ds^n$ are obtained as functions of the field derivatives $d^n\boldsymbol{B}/ds^n$, $d^n\boldsymbol{E}/ds^n$ by recursive differentiation of the equation of motion, in the following way.

Magnetic fields : In purely magnetic optical elements the particle rigidity is constant and the recursive differentiation simply writes $B\rho\ \boldsymbol{u}' = \boldsymbol{u} \times \boldsymbol{B}$, $B\rho\ \boldsymbol{u}'' = \boldsymbol{u}' \times \boldsymbol{B} + \boldsymbol{u} \times \boldsymbol{B}'$, and so forth.

Electrostatic fields : In purely electric fields the rigidity varies and the recursive differentiation takes the less simple form $(B\rho)'\boldsymbol{u} + B\rho\ \boldsymbol{u}' = \boldsymbol{E}/v$, $(B\rho)''\boldsymbol{u} + 2(B\rho)'\boldsymbol{u}' + B\rho\ \boldsymbol{u}'' = (1/v)'\boldsymbol{E} + \boldsymbol{E}'/v$, etc., whereas the rigidity itself is also obtained by Taylor expansion

$$(B\rho)(M_1) = (B\rho)(M_0) + (B\rho)'(M_0)\Delta s + (B\rho)''(M_0)\ \Delta s^2/2! + \dots$$

The derivatives $(B\rho)^{(n)} = d^n(B\rho)/ds^n$ are in turn obtained by alternate recursive differentiation of, on the one hand $(B\rho)' = (\boldsymbol{e} \cdot \boldsymbol{u})/v$, and on the other hand $B\rho\ (1/v)' = (1/c^2)\ (\boldsymbol{e} \cdot \boldsymbol{u}) - (1/v)\ (B\rho)'$.

By principle these transformations are symplectic, in practice the Taylor series are truncated so that best precision is obtained when the higher order derivatives in the truncated series are zero (at least to machine accuracy) .

2.2 Field models

The major components in accelerators, at least relevant to DA studies, are multipoles or multipolar defects. Explicit analytical expressions of multipole fields and of their derivatives are drawn from the regular 3-D scalar potential (that holds for both magnetic and (skew-) electric multipoles)

$$V_n(s,x,z) = (n!)^2\{\sum_{q=0}^{\infty}(-)^q \frac{\alpha_{n,0}^{(2q)}(s)}{4^q q!(n+q)!}(x^2+z^2)^q\}\{\sum_{m=0}^{n}\frac{\sin(m\frac{\pi}{2})x^{n-m}z^m}{m!(n-m)!}\} \quad (1)$$

where s, x, z coordinates are respectively curvilinear, transverse horizontal and vertical, $\alpha_{n,0}(s)$ ($n = 1, 2, 3, etc.$) describe the longitudinal form of the field, including end fall-offs, and $\alpha_{n,0}^{(2q)} = d^{2q}\alpha_{n,0}/ds^{2q}$. Note that, within magnet body or as well when using hard edge field model, $d^{2q}\alpha_{n,0}/ds^{2q} \equiv 0$ ($\forall q \neq 0$) hence the field and derivatives derive from the simplified potentials

$$V_1(x,z) = G_1 z\ , \qquad V_2(x,z) = G_2 xz\ , \qquad V_3(x,z) = G_3(x^2 - z^2/3)z, \quad \text{etc.} \quad (2)$$

where $G_n/B\rho$ is the strength.

Field fall-off at magnet ends : As to the field fall-off on axis at magnet ends orthogonally to the effective field boundary (EFB), it is modeled by (after Ref. [10, page 240])

$$\alpha_{n,0}(d) = \frac{G_n}{1+\exp[P(d)]} \quad \text{with } P(d) = C_0 + C_1\frac{d}{\lambda_n} + C_2(\frac{d}{\lambda_n})^2 + \ldots + C_5(\frac{d}{\lambda_n})^5 \quad (3)$$

where d is the distance to the EFB and coefficients λ_n, $C_0 - C_5$ can be determined from prior matching with realistic numerical fringe field data.

More fields

Zgoubi is actually a genuine compendium of optical elements of all sorts, magnetic and/or electric, with fields derived from more or less sophisticated analytical models as above. This allows simulating with precision regular rings.

3 DA tracking

In the following various results drawn from (unpublished) reports are presented, with the aim of showing the accuracy and effectiveness of the ray-tracing method.

3.1 Effect of b_{10} in low-β quadrupoles in LHC [5]

The multipole defect b_{10} in LHC low-β quadrupoles derives from (Eq. 1, Fig. 2)

$$V_{10}(s,x,z) \approx \left(\alpha_{10,0} - \frac{\alpha''_{10,0}}{44}(x^2+z^2)\right)(10x^8 - 120x^6z^2 + 252x^4z^4 - 120x^2z^6 + 10z^8) \quad (4)$$

The goal in this problem was to assess the importance of the way b_{10} is distributed along the quadrupole. Three models are investigated (Fig. 2) : hard edge (a), a regular smooth fall-off at quadrupole ends (b), a lump model in

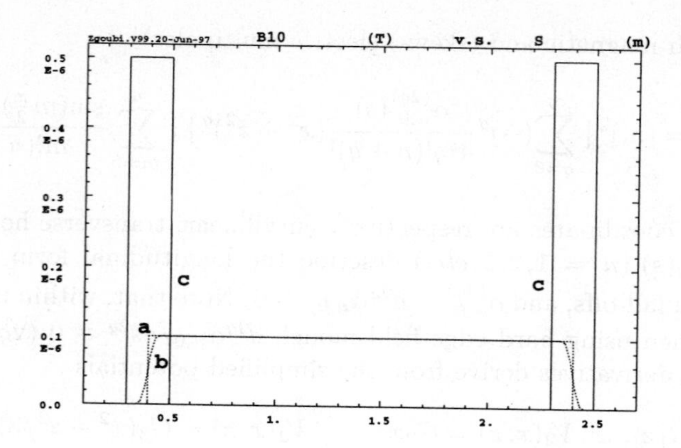

Fig. 2. Fringe field models used for assessing effect of b_{10} error on particle dynamics.

which b_{10} is zero in the body and the integral strength is shared between the two ends (c). In all three cases the overall field integral is the same.

Optical aberrations at IP5 : It can be seen from Fig. 3 that $b_{10} = -0.005\ 10^{-4}$ strongly distorts the aberration curves that would otherwise show a smooth, cubic shape. The aberration is of the form $x_{IP} \approx (\frac{x}{x'^3})x'^3_0 + (\frac{x}{x'^9})x'^9_0$ with x'_0 being the starting angle at point-to-point imaging location upstream of the interaction point (IP). The coefficient (x/x'^3) is mostly due to geometrical errors introduced

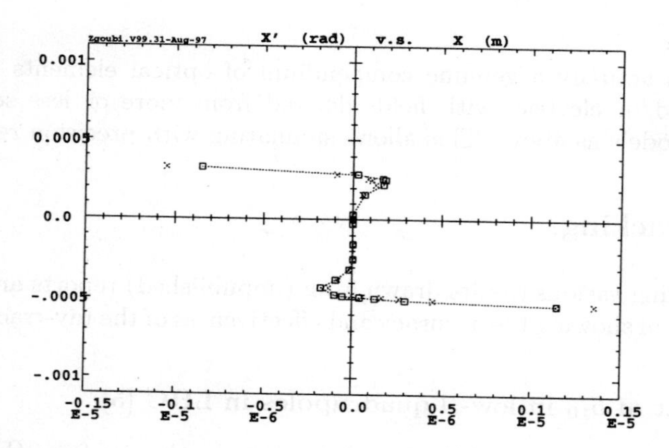

Fig. 3. Optical aberrations with inclined closed orbit at IP5 ($0.1\sqrt{2}$ mrad c.o. angle inclined 45°) ; fringe fields are set in separation dipoles D1/D2 and in the quadrupoles for the main component b_2. *Squares* : hard edge or fringe field model. *Crosses* : lump b_{10} model.

by the quadrupole and (x/x'^9) is due to b_{10} ; they have opposite signs and therefore act in opposite ways. The turn-round region between the two effects gets closer to the x-axis the stronger b_{10}. In particular with the present value of b_{10} a ± 1 μm extent at the image is reached with starting angle within -10 to $15\,\sigma_{x'_0}$, about twice smaller than without b_{10}.

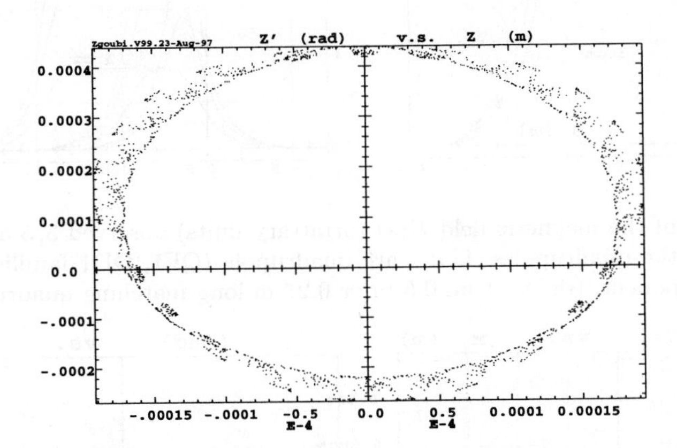

Fig. 4. Vertical phase space plot of a particle launched with x=x'=y=0 and $y' = 11.0\,\sigma$, in presence of inclined 0.28 mrad c.o. angles of identical signs at IP1 and IP5 simultaneously, with lumped b_{10} model (longitudinal distribution 'c' in Fig. 2).

DA tracking : Multiturn tracking of the dynamic aperture must stand the comparison. At first sight, considering the violent turn-round in the aberration curves (Fig. 3) and the fact that it occurs at $x'_0 \approx 9.5\,\sigma_{x'}$ whatever the longitudinal model for b_{10}, it can be expected that, on the one hand all three models provide similar DA, on the other hand the DA be about $9.5\,\sigma$ as well. This has been precisely confirmed by DA tracking, details can be found in Ref. [5]. As an illustration Fig. (4) provides a sample transverse phase space at 9.5σ DA.

3.2 Fringe field effects in the Fermilab 50 GeV muon storage ring [7]

The goal here was to examine possible effects of quadrupole fringe fields (Fig. 5) in the FERMILAB 50 GeV muon storage ring Feasibility I Study. An interesting outcome - amongst others - is a clear disagreement with similar studies subject to earlier publication.

Table 1 recalls the main machine parameters. Unprecedented apertures are required in the muon storage ring for a Neutrino Factory because of the exceptionally large emittances associated with the intense muon beams which must be accepted. The superconducting arc quadrupoles require a 14 cm bore, 3.6 T poletip field, which leads to strong, extended fringe fields.

Large acceptance motion in the absence of fringe fields is shown in Fig. 6. Particles are launched for a few hundred turns ray-tracing with initial coordinates either $x_0 = 1 - 4 \times 10^{-2}$ m $\approx 4\sigma_x$ and $\epsilon_z = 0$ (left column in the figure), or

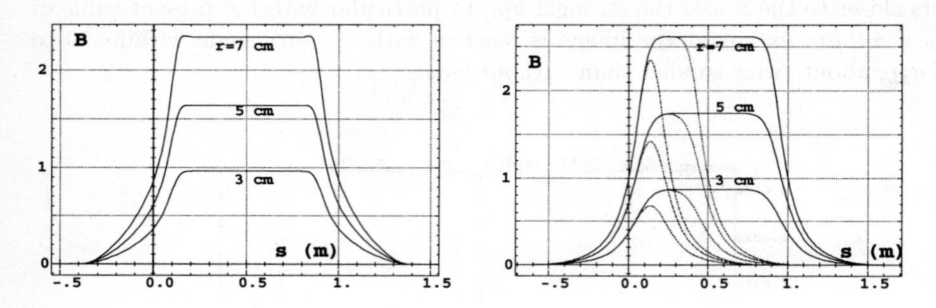

Fig. 5. Shape of the magnetic field $B(s)$ (arbitrary units) observed $3, 5$ or 7×10^{-2}m off-axis along the quadrupoles. Left : arc quadrupole (QF1, QD1 families) including sextupole component. Right : 1 m, 0.5 m or 0.27 m long matching quadrupoles.

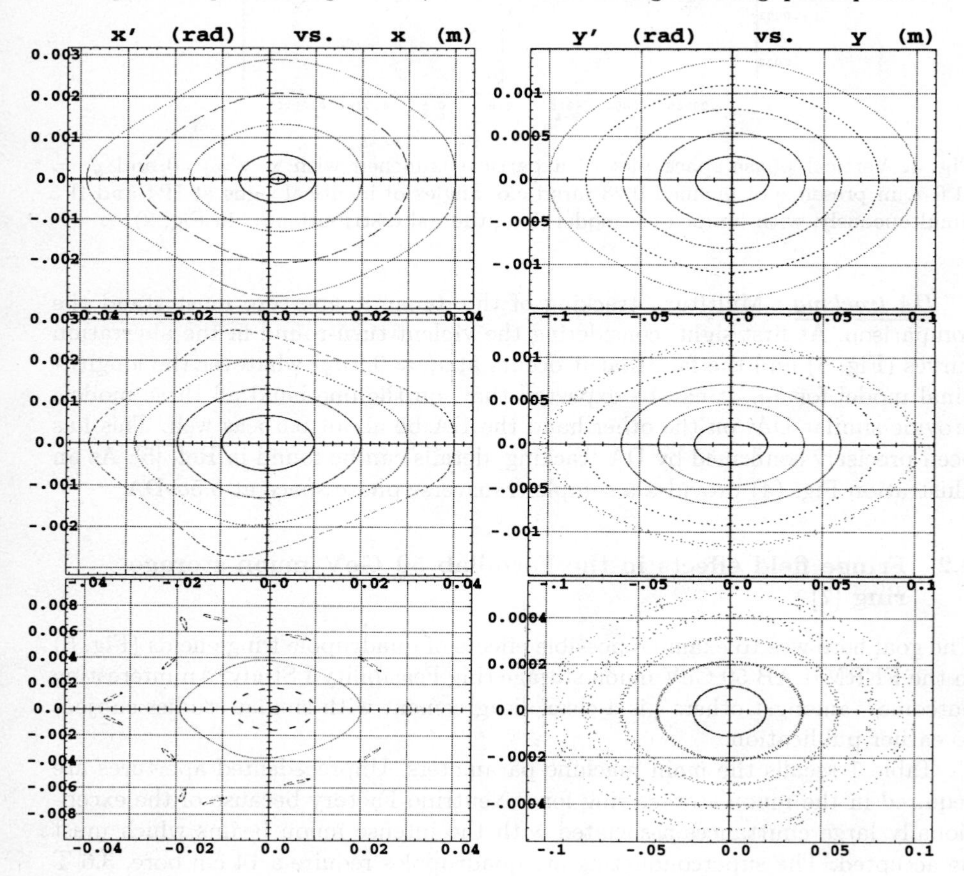

Fig. 6. Phase space plots up to DA region, no fringe fields, sextupoles are on. All particles survive except the largest amplitude one in the bottom left-hand plot (with initial conditions $x_0 = 0.04$ m $\approx 4\sigma_x$, $\delta p/p = -2\%$)

Table 1. Storage Ring Parameters at 50-GeV

Circumference	m	1752.8
matching and dispersion suppression	m	44.1
High-β FODO straight	m	688
$\beta_{xmax}/\beta_{zmax}$	m	435/484
ν_x/ν_z		13.63/13.31
natural chromaticity		-23.9/-23.9
$d\nu_x/d\frac{\epsilon_x}{\pi}/d\nu_z/d\frac{\epsilon_z}{\pi}$ (sextus on)		-3.8/-11

$z_0 = 1-10\times10^{-2}$ m $\approx 4\sigma_z$ and $\epsilon_x = 0$ (right col.), and with, top row : +2% off-momentum ; middle : on-momentum ; bottom row : -2% off-momentum. Setting the fringe fields leaves Fig. 6 practically unchanged [7], which is a good indication of the strong symplecticity of the method, considering the strong non-linearities so introduced (Fig. 5). It was therefore concluded to their quasi-innocuousness, contrary to Ref. [12] whose questionable results probably come from its using too low order mapping [7].

4 An electrostatic storage ring

This design study concerned a low energy electrostatic storage ring employed as a multiturn time of flight mass spectrometer (TOFMS), the principle being that multiturn storage is a convenient way to reach high resolution mass separation allied with small size apparatus. The ring is built from electrostatic parallel-plate mirrors which are highly non-linear optical elements, used for both focusing and bending.

Fig. 7 shows the geometry of the periodicity-2 ring of concern, built from the symmetric cell (LV, LH, MA, MB, LH, LV) wherein MA and MB are 3-electrode parallel-plate mirrors acting as 90 degrees deflector (i.e., bends), LV and LH are similar devices used as transmission vertical and horizontal lenses. The potential experienced by particles in these lenses and mirrors writes

$$V(X,Z) = \sum_{i=2}^{3} \frac{V_i - V_{i-1}}{\pi} \arctan \frac{\sinh(\pi(X - X_{i-1})/D)}{\cos(\pi Z/D)}$$

(Vi are the potentials at the 3 electrode pairs, Xi are the locations of the slits, X is the distance from the origin taken at the first slit, D is the inter-plate distance). In terms of non-linearities, a study of the second order coefficients of the cell reveals significant x-z, as well as z-$\delta p/p$ coupling. In addition particles are slowed down by the mirror effect so that their rigidity $B\rho$ undergoes dramatic variations (hence strong coefficients in its Taylor series).

A good indication of symplectic integration here is obtained by checking easily accessed constants, such as total energy (sum of the local kinetic and and potential (V(X,Y,Z)) energies), kinetic momentum, or else, calculated from

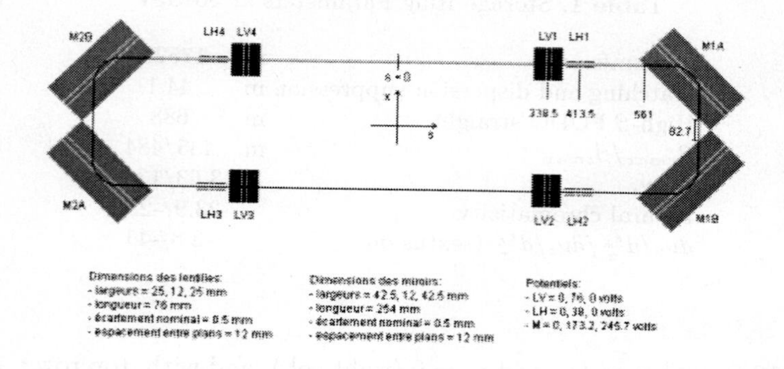

Fig. 7. The TOFMS ring. Quotations are in millimeters.

position $R(x, z, s)$ and speed $u(x, z, s)$. Such ckecks in particular led to the appropriate minimum value of the integration step Δs in the mirror bends.

DA tracking trials are shown in Fig. 9 for various potential values at LV. They were realized in order to assess the ring acceptance and prove to be well behaved, in particular free of any sign of spiral motion, in spite of the strongly non-linear fields. The horizontal acceptance comes out to be rather large, ± 8 mm as observed at $s = 0$, whereas the vertical one is drastically affected by the non-linearities and does not exceed ± 1 mm.

5 Conclusion

The Zgoubi ray-tracing method has been described, examples of its use have been given and show its very high accuracy that makes it an efficient tool for precision DA tracking.

References

1. The ray-tracing code Zgoubi, F. Méot, CERN SL/94-82 (AP) (1994), NIM A 427 (1999) 353-356.
   Zgoubi users' guide, F. Méot and S. Valero, Report CEA DSM/DAPNIA/SEA/97-13, and FERMILAB-TM-2010, Dec. 1997.
2. See for instance, Le spectromètre II, J. Thirion et P. Birien, Internal Report DPh-N/ME, CEA Saclay, 23 Déc. 1975.
3. A numerical method for combined spin tracking and ray-tracing of charged particles, F. Méot, NIM A 313 (1992) 492, and Proc. EPAC Conf. (1992) p.747.
4. On the effects of fringe fields in the LHC ring, F. Méot, Part. Acc., 1996, Vol. 55, pp.83-92.
5. Concerning effects of fringe fields and longitudinal distribution of b_{10} in low-β regions on dynamics in LHC, F. Méot and A. Parîs, Report FERMILAB-TM-2017, Aug. 1997.

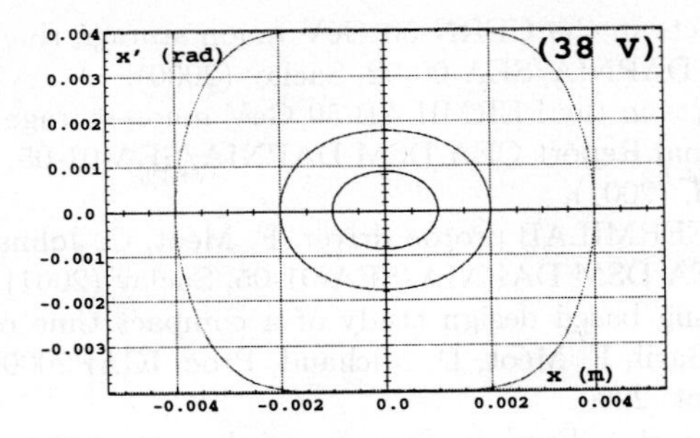

Fig. 8. 1000-turn horizontal acceptance as observed at $s = 0$, with $V_{LH} = 38$ Volts, $V_{LV} = 76$ Volts.

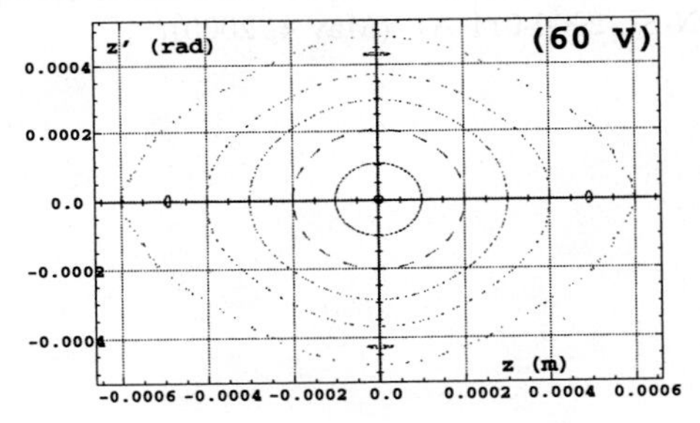

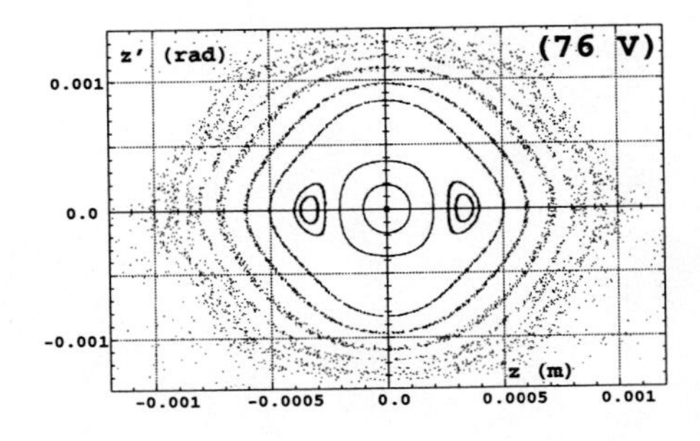

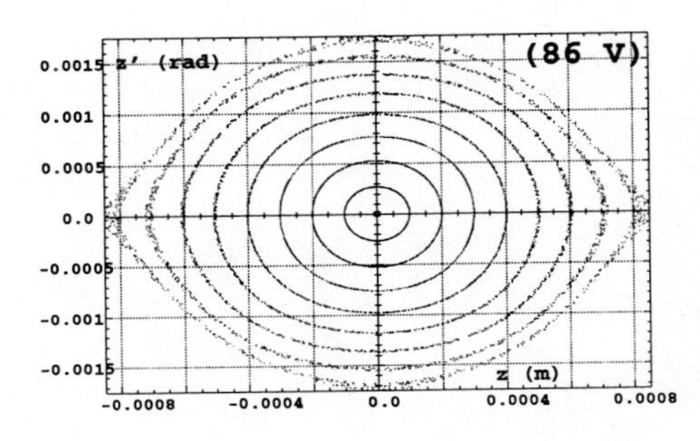

Fig. 9. 1000-turn vertical acceptance as observed at $s = 0$, for $V_{LV} = 60, 76$ and 86 Volts.

6. On fringe field effects in the CERN 50 GeV muon storage ring, F. Méot, Internal Report CEA DSM DAPNIA/SEA-00-02, Saclay (2000).
7. On fringe field effects in the FERMILAB 50 GeV muon storage ring, C. Johnstone and F. Méot, Internal Report CEA DSM DAPNIA/SEA-01-05, Saclay (2001) and Proc. PAC 01 Conf. (2001).
8. DA studies in the FERMILAB proton driver, F. Méot, C. Johnstone, A. Drozhdin, Internal Report CEA DSM DAPNIA/SEA-01-05, Saclay (2001).
9. Multiturn ray-tracing based design study of a compact time of flight mass spectrometer ring, M. Baril, F. Méot, D. Michaud, Proc. ICAP2000 Conf., Darmstadt, Germany, 11-14 Sept. 2000.
10. Deflecting magnets, H.A. Enge, in *Focusing of charged particles*, volume 2, A. Septier ed., Academic Press, New-York and London (1967).
11. A. Faus-Golfe and A. Verdier, Dynamic aperture limitations of the LHC in physics conditions due to low-β insertions, Proc. EPAC Conf. 1996.
12. Fringe fields and dynamic aperture in the FNAL muon storage ring, F. Zimmermann et als., CERN-SL-2000-011 AP (May 4, 2000).

Numerical Simulation of Hydro- and Magnetohydrodynamic Processes in the Muon Collider Target

Roman Samulyak

Center for Data Intensive Computing, Brookhaven National Laboratory,
Upton, NY 11973, USA,
rosamu@bnl.gov,
http://pubweb.bnl.gov/users/rosamu/www/r.samulyak.htm

Abstract. We have developed numerical methods and performed numerical simulations of the proposed Muon Collider target. The target will be designed as a pulsed jet of mercury interacting with strong proton beams in a 20 Tesla magnetic field. A numerical approach based on the method of front tracking for numerical simulation of magnetohydrodynamic flows in discontinuous media was implemented in FronTier, a hydrodynamics code with free interface support. The FronTier-MHD code was used to study the evolution of the mercury jet in the target magnet system. To model accurately the interaction of the mercury target with proton pulses, a realistic equation of state for mercury was created in a wide temperature - pressure domain. The mercury target - proton pulse interaction was simulated during 120 microseconds. Simulations predict that the mercury target will be broken into a set of droplets with velocities in the range 20 - 60 m/sec.

1 Introduction

In order to understand the fundamental structure of matter and energy, an advance in the energy frontier of particle accelerators is required. Advances in high energy particle physics are paced by advances in accelerator facilities. The majority of contemporary high-energy physics experiments utilize colliders. A study group was organized at Brookhaven National Laboratory to explore the feasibility of a high energy, high luminosity Muon-Muon Collider [11]. Such a collider offers the advantages of greatly increased particle energies over traditional electron-positron machines (linear colliders). However, several challenging technological problems remain to be solved. One of the most important is to create an effective target able to generate high-flux muon beams. The need to operate high atomic number material targets, that will be able to withstand intense thermal shocks, has led to the exploration of free liquid jets as potential target candidates for the proposed Muon Collider. The target will be designed as a pulsed jet of mercury (high Z-liquid) interacting in a strong magnetic field with a high energy proton beam [10, 11]. This paper presents results of numerical studies of hydro- and magnetohydrodynamic (MHD) processes in such a target.

P.M.A. Sloot et al. (Eds.): ICCS 2002, LNCS 2331, pp. 391–400, 2002.

We have developed an MHD code for multifluid systems based on FronTier, a hydrodynamics code with free interface support. FronTier is based on front tracking [2, 4, 5], a numerical method for solving systems of conservation laws in which the evolution of discontinuities is determined through solutions of the associated Riemann problems. We also have developed a realistic equation of state for mercury in a wide temperature - pressure domain.

The paper is organized as follows. In Section 2, we describe some details of the Muon Collider target design. In Section 3, we formulate the main system of MHD equations, boundary conditions and discuss some simplifying assumptions used in our numerical schemes. The numerical implementation of the MHD system in the FronTier code is presented in Section 4. Subsection 5 presents simulation results of the mercury target - proton pulse interaction. Subsection 5.2 contains results of the numerical simulation of conducting liquid jets in strong nonuniform magnetic fields and the evolution of the mercury jet in the target solenoid system. Finally, we conclude the paper with the discussion of our results and perspectives for the future work.

2 Muon Collider Target

The muon collider target [10] is shown schematically in Figure 1. It will contain a series of mercury jet pulses of about 0.5 cm in radius and 60 cm in length. Each pulse will be shot at a velocity of 30-35 m/sec into a 20 Tesla magnetic field at a small angle (0.1 rad) to the axis of the field. When the jet reaches the center of the magnet, it is hit with 3 ns proton pulses arriving with 20 ms time period; each proton pulse will deposit about 100 J/g of energy in the mercury.

The main issues of the target design addressed in our numerical studies are the distortion of the jet due to eddy currents as it propagates through the magnetic coil, the deformation of the jet surface due to strong pressure waves caused by the proton pulses and the probability of the jet breakup. Studying the state of the target during its interaction with proton pulses will help to achieve the maximal proton production rate and therefore an optimal target performance. The target behavior after the interaction with proton pulses during its motion outside the magnet determines the design of the chamber.

3 Magnetohydrodynamics of Liquid Metal Jets

The basic set of equations describing the interaction of a compressible conducting fluid flow and a magnetic field is contained in Maxwell's equations and in the equations of fluid dynamics suitably modified [1, 8]. Namely, the systems contains the mass, momentum and energy conservation equations for the fluid which have hyperbolic nature and a parabolic equation for the evolution of the magnetic field.

$$\frac{\partial \rho}{\partial t} = -\nabla \cdot (\rho \mathbf{u}), \tag{1}$$

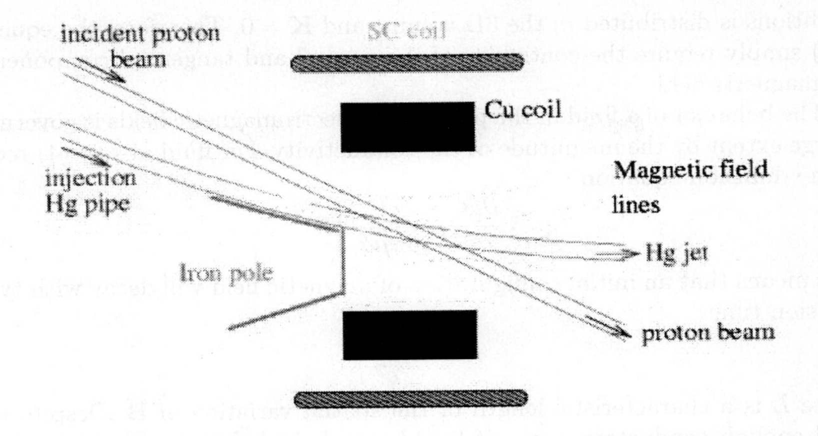

Fig. 1. Schematic of the muon collider target.

$$\rho\left(\frac{\partial}{\partial t}+\mathbf{u}\cdot\nabla\right)\mathbf{u}=-\nabla P+\rho\mathbf{X}+\frac{1}{c}(\mathbf{J}\times\mathbf{B}), \tag{2}$$

$$\rho\left(\frac{\partial}{\partial t}+\mathbf{u}\cdot\nabla\right)U=-P\nabla\cdot\mathbf{u}+\frac{1}{\sigma}\mathbf{J}^2-\frac{1}{c}\mathbf{u}\cdot(\mathbf{J}\times\mathbf{B}), \tag{3}$$

$$\frac{\partial\mathbf{B}}{\partial t}=\nabla\times(\mathbf{u}\times\mathbf{B})-\nabla\times(\frac{c^2}{4\pi\sigma}\nabla\times\mathbf{B}), \tag{4}$$

$$\nabla\cdot\mathbf{B}=0, \tag{5}$$

Here $\mathbf{u}$, ρ and U are the velocity, density and total energy of the fluid, respectively, P is the total stress tensor, $\mathbf{X}$ includes external forces of non-magnetic origin, $\mathbf{B}$ is the magnetic field induction, $\mathbf{J}$ is the current density distribution and σ is the fluid conductivity. The magnetic field $\mathbf{H}$ and magnetic induction $\mathbf{B}$ are related by the magnetic permeability coefficient μ: $\mathbf{B}=\mu\mathbf{H}$.

The system (1-4) must be closed with an equation of state. Equation of state models for the jet material and the ambient gas are discussed in Section 5.

The following boundary conditions must be satisfied at the jet surface:

i) the normal component of the velocity field is continuous across the material interface.

ii) the normal and tangential components of the magnetic field at the material interface are related as

$$\mathbf{n}\cdot(\mathbf{B}_2-\mathbf{B}_1)=0, \tag{6}$$

$$\mathbf{n}\times(\mathbf{H}_2-\mathbf{H}_1)=\frac{4\pi}{c}\mathbf{K}, \tag{7}$$

where $\mathbf{K}$ is the surface current density. The above jump conditions define the refraction of magnetic lines on the material interface. We can assume $\mu=1$ for most fluids. Notice that the surface current density $\mathbf{K}$ corresponds to a current localized in a thin fluid boundary layer (δ-functional current) which is non-zero only for superconducting materials. The current density in fluids at normal conditions is distributed in the 3D volume and $\mathbf{K} = 0$. Therefore, the equations (6,7) simply require the continuity of the normal and tangential components of the magnetic field.

The behavior of a fluid in the presence of electromagnetic fields is governed to a large extent by the magnitude of the conductivity. For fluid at rest (4) reduces to the diffusion equation

$$\frac{\partial \mathbf{B}}{\partial t} = \frac{c^2}{4\pi\mu\sigma}\Delta\mathbf{B} \tag{8}$$

This means that an initial configuration of magnetic field will decay with typical diffusion time

$$\tau = \frac{4\pi\mu\sigma L^2}{c^2},$$

where L is a characteristic length of the spatial variation of $\mathbf{B}$. Despite being good enough conductors, most of liquid metals including mercury are characterized by small diffusion times (33 microseconds for a mercury droplet of 1 cm radius) compared to some solid conductors (1 sec for a copper sphere of 1 cm radius). Therefore the magnetic field penetration in such liquid conductors can be considered as an instantaneous process.

Another crucial phenomena for MHD flows of compressible conducting fluids is the propagation of Alfven waves. For mercury at room temperature the Alfven velocity

$$\mathbf{v}_A = \frac{\mathbf{B}_0}{\sqrt{4\pi\rho_0}},$$

where $\mathbf{B}_0$ and ρ_0 are unperturbed (mean) values of the magnetic induction and density of the fluid, respectively, is $[B_0(\text{Gauss})/13.1]$ cm/sec. This is a small number compared with the speed of sound of 1.45×10^5 cm/sec even for the magnetic field of 20 T. In many cases, however, it is not desirable to compute Alfven waves explicitly in the system. If, in addition, both the magnetic field diffusion time and the eddy current induced magnetic field are small, an assumption of the constant in time magnetic field can be made. The current density distribution can be obtained in this case using Ohm's law

$$\mathbf{J} = \sigma\left(-\text{grad}\phi + \frac{1}{c}\mathbf{u} \times \mathbf{B}\right). \tag{9}$$

Here ϕ is the electric field potential. The potential ϕ satisfies the following Poisson equation

$$\Delta\phi = \frac{1}{c}\text{div}(\mathbf{u} \times \mathbf{B}), \tag{10}$$

and the Neumann boundary conditions

$$\left.\frac{\partial\phi}{\partial\mathbf{n}}\right|_{\Gamma} = \frac{1}{c}(\mathbf{u} \times \mathbf{B}) \cdot \mathbf{n},$$

where $\mathbf{n}$ is a normal vector at the fluid boundary Γ. This approach is applicable for the study of a liquid metal jet moving in a strong magnetic field.

We shall use also the following simplification for the modeling of a thin jet moving along the solenoid axis. Let us consider a ring of liquid metal of radius r that is inside a thin jet moving with velocity u_z along the axis of a solenoid magnet. The magnetic flux $\Phi = \pi r^2 B_z$ through the ring varies with time because the ring is moving through the spatially varying magnetic field, and because the radius of the ring is varying at rate $u_r = dr/dt$. Therefore, an azimuthal electric field is induced around the ring:

$$2\pi r E_\phi = -\frac{1}{c}\frac{d\Phi}{dt} = -\frac{\pi r^2}{c}\frac{dB_z}{dt} - \frac{2\pi r u_r B_z}{c}$$
$$= -\frac{\pi r^2 u_z}{c}\frac{\partial B_z}{\partial z} - \frac{2\pi r u_r B_z}{c}.$$

This electric field leads to an azimuthal current density

$$J_\phi = \sigma E_\phi = -\frac{\sigma r u_z}{2c}\frac{\partial B_z}{\partial z} - \frac{\sigma u_r B_z}{c}, \tag{11}$$

which defines the Lorentz force in the momentum equation (2) and leads to the distortion of the jet moving in a non-uniform magnetic field.

The linear stability analysis of thin conducting liquid jets moving along the axis of a uniform magnetic field [1] and the corresponding analysis for the Muon Collider target [6] show that an axial uniform field tends to stabilize the jet surface. The influence of a strong nonuniform field is studied below by means of the numerical simulation.

4 Numerical Implementation

In this section, we shall describe numerical ideas implemented in the FronTier MHD code. FronTier represents interfaces as lower dimensional meshes moving through a volume filling grid [4]. The traditional volume filling finite difference grid supports smooth solutions located in the region between interfaces. The location of the discontinuity and the jump in the solution variables are defined on the lower dimensional grid or interface. The dynamics of the interface comes from the mathematical theory of Riemann solutions, which is an idealized solution of a single jump discontinuity for a conservation law. Where surfaces intersect in lower dimensional objects (curves in three dimensions), the dynamics is defined by a theory of higher dimensional Riemann problems such as the theory of shock polars in gas dynamics. Nonlocal correlations to these idealized Riemann solutions provide the coupling between the values on these two grid systems.

The computation of a dynamically evolving interface requires the ability to detect and resolve changes in the topology of the moving front. A valid interface is one where each surface and curve is connected, surfaces only intersect along curves and curves only intersect at points. We say that such an interface is untangled. Two independent numerical algorithms, grid-based tracking and grid-free tracking, were developed [4, 5] to resolve the untangling problem for the moving interface. The advantages and deficiencies of the two methods are complementary and an improved algorithm combining them into a single hybrid method was implemented in the FronTier code and described in [5].

We solve the hyperbolic subsystem of the MHD equations, namely the equations (1-3), on a finite difference grid in both domains separated by the free surface using FronTier's interface tracking numerical techniques. Some features of the FronTier hyperbolic solvers include the use of high resolution methods such as MUSCL, Godunov and Lax-Wendroff with a large selection of Riemann solvers such as the exact Riemann solver, the Colella-Glaz approximate Riemann solver, the linear US/UP fit (Dukowich) Riemann solver, and the Gamma law fit. We use realistic models for the equation of state such as the polytropic and stiffened polytropic equation of state, the Gruneisen equation of state, and the SESAME tabular equation of state.

The evolution of the free fluid surface is obtained through the solution of the Riemann problem for compressible fluids [4, 9]. Notice that since we are primarily interested in the contact discontinuity propagation, we do not consider the Riemann problem for the MHD system and therefore neglect elementary waves typical for MHD Riemann solutions.

We have developed an approach for solving the equation (4) for the magnetic field evolution and the Poisson equation (10) using a mixed finite element method. The existence of a tracked surface, across which physical parameters and the fluid solution change discontinuously, has important implications for the solution of an elliptic or parabolic system by finite elements. Strong discontinuities require that the finite element mesh align with the tracked surface in order not to lose the resolution of the parameter/solution. A method for the generation of finite element mesh conforming to the interface and scalable elliptic/parabolic solvers will be presented in a forthcoming paper.

The aim of the present MHD studies is the numerical simulation of free thin liquid metal jets entering and leaving solenoid magnets using the above-mentioned simplifying assumptions. The eddy current induced magnetic field in such problem is negligible compared to the strong external magnetic field. We applied the approximation $d\mathbf{B}/dt = 0$ and implemented the MHD equations (1-3) in FronTier's hyperbolic solvers using the approximate analytical expression (11) for the current density distribution.

5 Results of Numerical Simulation

5.1 Interaction of Mercury Jet with Proton Pulses

Numerical simulations presented in this section aid in understanding the behavior of the target under the influence of the proton pulse, and in estimating the evolution of the pressure waves and surface instabilities. We have neglected the influence of the external magnetic field on the hydrodynamic processes driven by the proton energy deposition.

The influence of the proton pulse was modeled by adding the proton energy density distribution to the internal energy density of mercury at a single time step. The value and spatial distribution of the proton energy was calculated using the MARS code [10].

To model accurately the interaction of the mercury target with proton pulses, a tabulated equation of state for mercury was created in a wide temperature - pressure domain which includes the liquid-vapor phase transition and the critical point. The necessary data describing thermodynamic properties of mercury in such a domain were obtained courtesy of T. Trucano of Sandia National Laboratory. This equation of state was not directly used in the code for current simulations but it provided necessary input parameters for the stiffened polytropic eos model [9].

The evolution of the mercury target during 120 mks due to the interaction with a proton pulse is shown in Figure 2. Simulations predict that the mercury target will be broken into a set of droplets with velocities in the range 20 - 60 m/sec. The pressure field inside the jet developed small regions of negative pressure which indicates a possibility for the formation of cavities. Detailed studies of the cavitation phenomena in such a target will be done using the original tabulated equation of state for mercury.

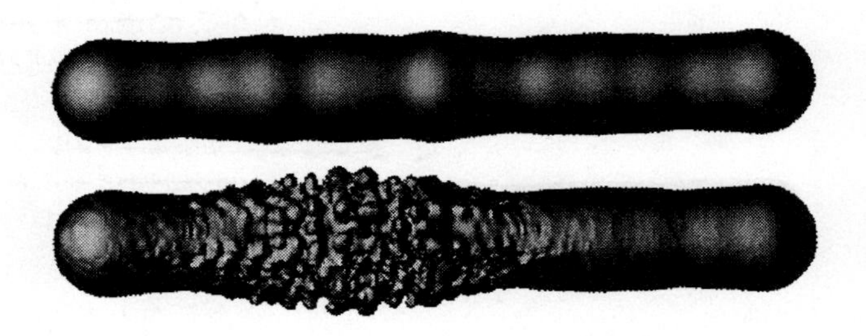

Fig. 2. Evolution of the target during 120 mks driven by the proton energy deposition.

5.2 Motion of Liquid Metal Jets in Magnetic Fields

In this section, we shall present numerical simulation results of thin jets of conducting fluid in highly nonuniform magnetic fields. In this numerical experiment, a 1 cm radius liquid jet is sent into a 20 T solenoid with the velocity 90 m/sec along the solenoid axis. The density of the liquid is $1\,g/cm^3$, the electric conductivity is 10^{16} in Gaussian units, and the initial pressure in the liquid is 1 atm. The electrically and magnetically neutral gas outside the jet has density 0.01 g/cm and the same initial pressure. The thermodynamic properties of the ambient gas were modeled using the polytropic equation of state [3] with the ratio of specific heat $\gamma = 1.4$ and the ideal gas constant $R = 1$. The properties of the liquid jet were modeled using the stiffened polytropic equation of state with the Gruneisen exponent = 5 and the stiffened gas constant $P_\infty = 3 \cdot 10^9 g/(cm \cdot sec^2)$.

The field of a magnetic coil of rectangular profile and $8 \times 8 \times 20$ cm size was calculated using exact analytical expressions.

A set of images describing the evolution of the liquid jet as it enters and leaves the solenoid is depicted in Figure 3. The strong nonuniform magnetic field near the solenoid entrance squeezes and distorts the jet. The magnetic field outside the solenoid stretches the jet which results in the jet breakup. Notice that the expression (11) is not valid for an asymmetrically deformed jet as well as for a system of liquid droplets. The simulation looses the quantitative accuracy when the jet is close to the breakup. The numerical algorithm based on finite elements conforming to the moving interface described briefly in Section 3 is designed for solving accurately the equation for the current density distribution and magnetic field evolution. The numerical simulation of the jet breakup phenomena using such approach will be presented in a forthcoming paper.

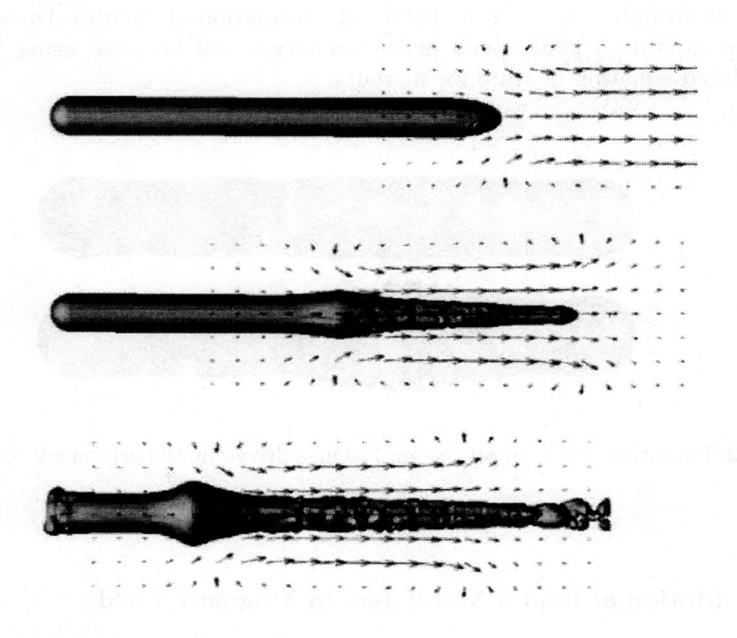

Fig. 3. Liquid metal jet in a 20 T solenoid.

The numerical simulation demonstrates the influence of strong magnetic field gradients on the jet distortion. To avoid such instabilities of the mercury target during its propagation in a 20 T magnetic field, a special magnet system was designed [10]. The nozzle of the mercury target was placed inside the main 20 T resistive magnetic coil. Therefore the jet will not meet strong magnetic field gradients before the interaction with the proton pulses. The magnetic field of superconducting and matching solenoids, needed for carrying muons to the linear accelerator (RF linac), essentially reduce the magnetic field gradient outside the main solenoid. Therefore the motion of the mercury target between the nozzle and the mercury dump is within a region of an almost uniform magnetic field.

We performed numerical simulations of the mercury jet motion in such magnetic system. Our results show that the magnetic field influence will not lead to significant jet deformations. However, the pressure waves, caused by the magnetic field forces, which propagate mainly along the jet axis speedup surface instabilities of the jet.

6 Conclusions

In this paper, we described a numerical approach based on the method of front tracking for the numerical simulation of magnetohydrodynamic free surface flows. The method was implemented in FronTier, a hydrodynamics code with free interface support. The FronTier MHD code was used for the numerical simulation of thin conducting liquid jets in strong nonuniform magnetic fields. Our results demonstrate a big influence of the Lorentz force on the stability of jets. The Loerntz force distorts a jet and stimulates the jet breakup in the region of nonuniform magnetic field behind a solenoid.

The method was also used for numerical simulation of the muon collider target. The magnetic field of the solenoid system designed for the muon collider experiments slowly decreases behind the main 20 T solenoid. This allows the mercury target to leave the interaction chamber without significant distortions. However, the magnetic forces speedup the natural instability and the pinchoff of the mercury jet.

We performed also numerical simulations of the interaction of the mercury jet with a proton pulse. Simulations predict that the mercury target will be broken into a set of droplets with velocities in the range 20 - 60 m/sec. An accurate analysis of the pressure field inside the jet indicates a possibility for the formation of cavities.

We have developed an algorithm for accurate numerical calculation of the magnetic field evolution and the current distribution in moving domains based on mixed finite elements dynamically conforming to evolving interfaces. Our future goals include studies of the jet breakup phenomena in nonuniform fields, the stabilizing effects of uniform magnetic field on liquid jets, and global numerical simulations of the mercury target including the target - proton beam interaction in the presence of a strong magnetic field.

Acknowledgments: The author is grateful to J. Glimm, K. McDonald, H. Kirk, R. Palmer and R. Weggel for fruitful discussions. Financial support has been provided by the USA Department of Energy, under contract number DE-AC02-98CH10886.

References

1. Chandrasekhar, S.: Hydrodynamics and Hydrodynamic stability. Clarendon Press, Oxford (1961)
2. Chern, I.R., Glimm, J., McBryan, O., Plohr, B., Yaniv, S.: Front tracking for Gas Dynamics. J. Comp. Phys. **62** (1986) 83-110
3. Courant, R., Friedrichs, K.: Supersonic Flows and Shock Waves. Interscience, New York (1948)
4. Glimm, J., Grove, J.W., Li, X.L., Shyue, K.M., Zhang, Q., Zeng, Y.: Three dimensional front tracking. SIAM J. Sci. Comp. **19** (1998) 703-727
5. Glimm, J., Grove, J.W., Li, X.L., Tan, D.: Robust computational algorithms for dynamic interface tracking in three dimensions. Los Alamos National Laboratory Report LA-UR-99-1780
6. Glimm, J., Kirk, H., Li, X.L., Pinezich, J., Samulyak, R., Simos, N.: Simulation of 3D fluid jets with application to the Muon Collider target design. Advances in Fluid Mechanics III (Editors: Rahman, M., Brebbia, C.), WIT Press, Southampton Boston (2000) 191 - 200
7. Kirk, H., et al.: Target studies with BNL E951 at the AGS. Particles and Accelerators 2001, June 18-22 (2001) Chicago IL
8. Landau, L.D., Lifshitz, E.M.: Electrodynamics of Continuous Media. Addison - Wesley Publishing Co., Reading Massachusetts (1960)
9. Menikoff, R., Plohr, B.: The Riemann problem for fluid flow of real materials. Rev. Mod. Phys. **61** (1989) 75-130
10. Ozaki, S., Palmer, R., Zisman, M., Gallardo, J. (editors) Feasibility Study-II of a Muon-Based Neutrino Source. BNL-52623 June 2001; available at http://www.cap.bnl.gov/mumu/studyii/FS2-report.html
11. Palmer, R.B.: Muon Collider design. Brookhaven National Laboratory report BNL-65242 CAP-200-MUON-98C (1998)

Superconducting RF Accelerating Cavity Developments

Evgeny Zaplatin

IKP, Forschungszentrum Juelich, D-52425 Juelich, Germany
e.zaplatine@fz-juelich.de

Abstract. For most practical accelerating rf cavities analytical solutions for the field distributions do not exist. Approximation by numerical methods provides a solution to the problem, reducing the time and expense of making and measuring prototype structures, and allowing rapid study and optimisation in the design stage. Many cavity design codes are available for sruding field distributions, interaction with charged particle beams, multipactor effects, stress analysis, etc. The types of calculations for different superconducting cavity design are discussed and illustrated by typical examples. The comparison of numerical simulations with some experimental results are shown.

1 Introduction

At present, many accelerators favour the use of superconducting (SC) cavities as accelerating RF structures[1]-[3]. For some of them, like long pulse Spallation Source or Transmutation Facility SC structures might be the only option. For the high energy parts of such accelerators the well-developed multi-cell elliptic cavities are the most optimal. For the low energy part the elliptic structures cannot be used because of their mechanic characteristics. The main working conditions of the SC cavities are as follow:

- Very high electromagnetic fields – maximum magnetic field on the inner cavity surface up to B_{pk}=100 mT, maximum electric field on the inner cavity surface up to E_{pk}=50 MV/m. These high fields together with small cavity wall thickness (2-4 mm) result in the strong Lorenz forces which cause the wall deformations;
- Low temperature down to 2K, that again causes wall displacements after cool down;
- The pulse regime of operation that results in the addition requirements on cavity rigidity;
- High vacuum conditions ($10^9 - 10^{10}$) and extra pressure on cavity walls from the helium tank also deform the cavity shape;
- High tolerances and quality surface requirements.

All deformations caused by these above mentioned reasons result in the working RF frequency shift in the range of hundreds Hz. Taking into account high

P.M.A. Sloot et al. (Eds.): ICCS 2002, LNCS 2331, pp. 401–410, 2002.

Q-factor of SC cavities (the resonance bandwidth in the range of Hz) such big frequency shift brings cavity out of operation. From the other hand, the use of any external tuning elements like plungers or trimmers are problematic as it results in the low down cavity acceleration efficiency. It means all these factors should be taken into account and complex electromagnetic simulations together with structural analysis should be provided during any real cavity design.

Here we present the scope of the possible RF accelerating structures, which can be used for different particle velocity β=v/c starting from β=0.09 and ending with β=1.0. The considered structures are quarter-wave and half-wave coaxial cavities, spoke cavity and based on spoke cavity geometry multi-cell H-cavities and 5-cell elliptic cavities, which have been developed for various projects.

Because of their low power dissipation, SC structures do not need to be designed to maximaze the shunt impedance, and new designs, which would be inefficient for a normal conducting cavity can be explored. For example, SC resonators can be designed with much larger apertures that wouldn't be practical for normal conducting resonators. SC accelerators can also be built from very short structures, resulting in a increase in flexibility and reliability. Finally, SC resonators have the demonstrated capability of operating continuously at high accelerating fields. That's why the criterium of cavity geometry optimisation is to reach the maximal accelerating electric field.

2 Elliptical Cavities

Elliptical cavities are the most simple shape SC resonators (Fig. 1). They can be in a single or multi-cell configuration. The length of the cell is defined by the particle speed β and cavity frequency f — $celllength = \beta c/2f$. This defines the limitation on this type cavity use for low energy (low β's) particles — at and below β=0.4-0.5 the cell geometry becomes narrow which doesn't fulfill the mechanic requirements. Another limitation comes from the frequency side — lower frequency bigger cavity. As the cavity is made out of rather expensive pure niobium, the lowest frequency which can be considered is about 350 MHz.

2.1 Middle Cell Geometry Optimisation

Usually, an elliptical cavity design is a compromise between various geometric parameters which should define a most optimal cavity shape in terms of an accelerator purpose (Fig. 1). Within a SC proton linac design there is a need of grouping of cavities with different $\beta = v/c$ values. It means that there should be several different cavities and the process of the accelerating structure design for SC linac becomes time consuming. The suggested in [4] procedure allows to get an optimal geometry optimisation leaving a freedom for the cavity parameters choice. As the cavity is strictly azimuthal symmetric these calculations can been done with a help of 2D cavity simulation code like SUPERFISH[5].

The main advantage of any SC cavity is a possibility of the high accelerating electric field (E_{acc}) creation. There are two characteristics which limit in principle an achievable value of E_{acc}. They are the peak surface electric field (E_{pk}),

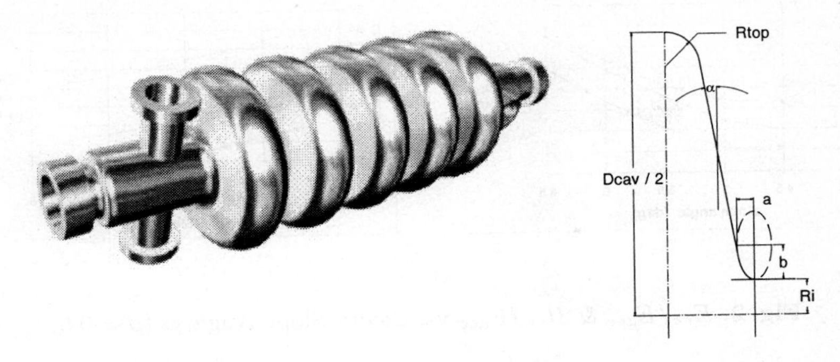

Fig. 1. Elliptical Cavity & Cell Geometry (1/4 part is shown)

which is allocated around cavity iris and the peak surface magnetic field (H_{pk}) in the cavity dome region. H_{pk} is important because a superconductor will quench above the critical magnetic field. Thus H_{pk} may not exceed this level. E_{pk} is important because of the danger of field emission in high electric field regions. All these mean that to maximise the accelerating field first of all it is therefore important during a cavity design to minimise the ratios of peak fields to the accelerating field. There are some more figures of merit to compare different designs such as power dissipation P_c, a quality factor Q and shunt impedance R_{sh}. But these parameters are not so crucial to the cavity accelerating efficiency and may be varied in some limits without any sufficient harm for a system in whole.

One of the basic figure which will influence on the further cell geometry is a cell-to-cell coupling coefficient in multicell cavity. This parameter defines field flatness along the cavity. This characteristic is obtained in conjunction with beam dynamic calculations and more or less is defined as a first. In practices the usual value for the coupling is above 1%.

Another cavity cell geometric limitation comes from the radius of the material curvature in the region of the cavity iris. The smallest radius estimated is to be 2-3 times bigger than a cavity wall thickness. An investigation of the plots presented on Figs. 2-3 with two limiting parameter characteristics crossing helps to make a proper choice of cell.

2.2 Multipacting

Making the cavity cell geometry optimisation one should take into account the possibility of a multipactor resonance discharge. At the moment several groups in the field had active programs or developments of programs that can be used to study potential multipacting effects in rf structures[6]-[8]. Three of the programs can be applied to 3D problems while the rest are applicable to 2D structures. Considering that the big fraction of SC cavities are rotationally symmetric, these 2D programs do still cover a wide range of interesting problems. Some of the groups are working on the extension of their codes to 3D capabilities.

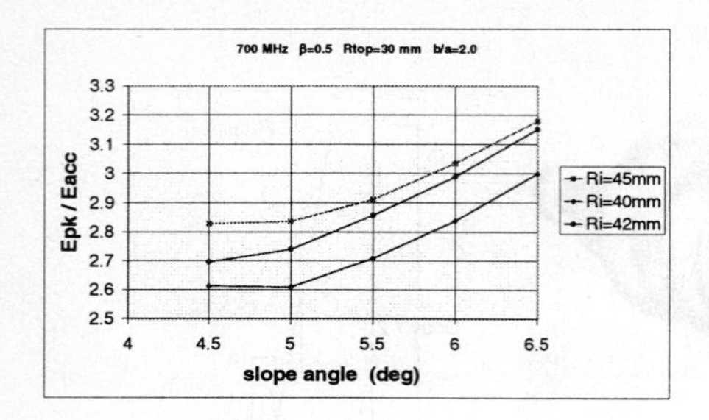

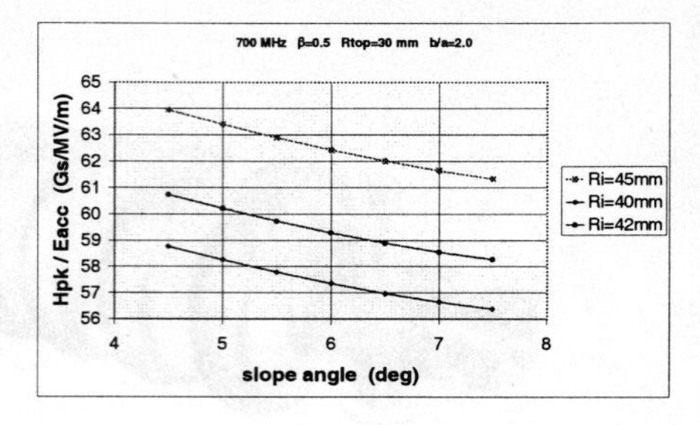

Fig. 2. E_{pk}/E_{acc} & H_{pk}/E_{acc} vs. Cavity Slope Angle α ($\beta = 0.5$)

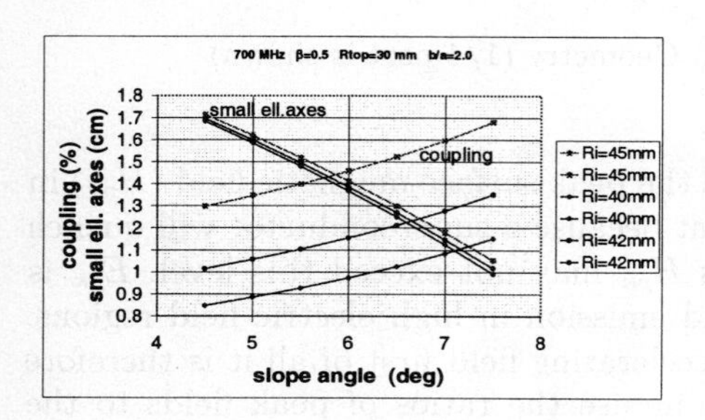

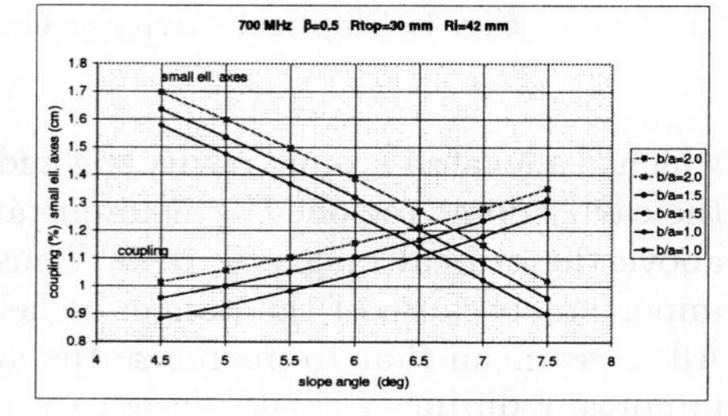

Fig. 3. Cell-to-Cell Coupling vs. Cavity Slope Angle α ($\beta = 0.5$)

Multipacting in rf structures is a resonant process in which a large number of electrons build up an multipacting, absorbing rf power so that it becomes impossible to increase the cavity fields by raising the incident power. The electrons collide with structure walls, leading to a large temperature rise and in the case of superconducting cavities to thermal breakdown. Basic tools of analysis of multipactor are counter functions. These functions are defined as follows. For a fixed field level a number of electrons are sent from different points on the cavity wall in different field phases and compute the electron trajectories. After an impact on the wall of cavity the trajectory calculation is continued if the phase of the field allows the secondary electrons to leave the wall. After a given number of impacts the number of free electrons (counter function) and total number of secondary electrons (enhanced counter function) are counted. Usually in elliptical cavity it happens around a dome equator region (Fig. 4). The proper cavity shape selection helps to avoid this resonance.

Here we present the results of such simulations[8] made by the Helsinki group for the same frequency (700 MHz) and different β=0.5-0.9. On Fig. 4 the possible dangerous regions of E_{acc} are shown. The tendency is that the lower β cavities because of their smaller dome radius are more affected by multipactor.

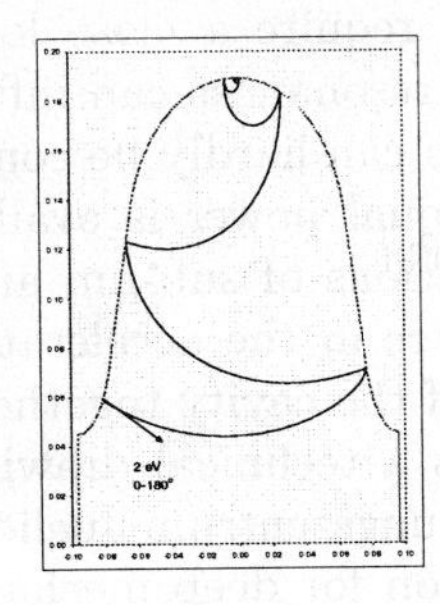

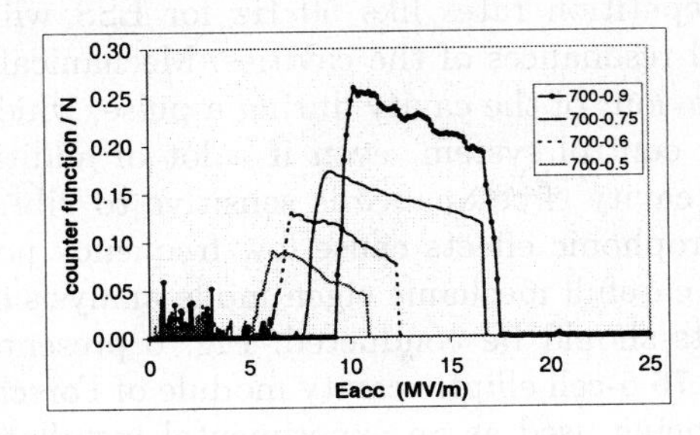

Fig. 4. Schematic drawing of multipactor resonance discharge & multipactor dangerous regions for different elliptic cavities (700 MHz, β=0.5–0.9).

2.3 Mechanics

All superconducting rf resonators are niobium cavities that are enclosed within helium vessels. These vessels are filled with liquid helium that floods the cavities and maintain the 2K operating temperature. Mechanical analysis consist of design calculations for all critical cavity assembly components, cavity tuning sensitivity analysis, active tuner and bench tuner load determination, cavity assembly cool-down analysis, natural frequency and random response analysis, inertia friction weld analysis and critical flaw size calculations.

As an example Fig. 5 presents results of cavity wall displacements caused by the extra pressure from the helium vessel. Similar deformations result from cool-down shape shrinking and the Lorenz forces originated by electromagnetic fields. To minimize the cavity deformation the stiffening rings are installed between cells still alowing enough cavity flexibility for structure tune.

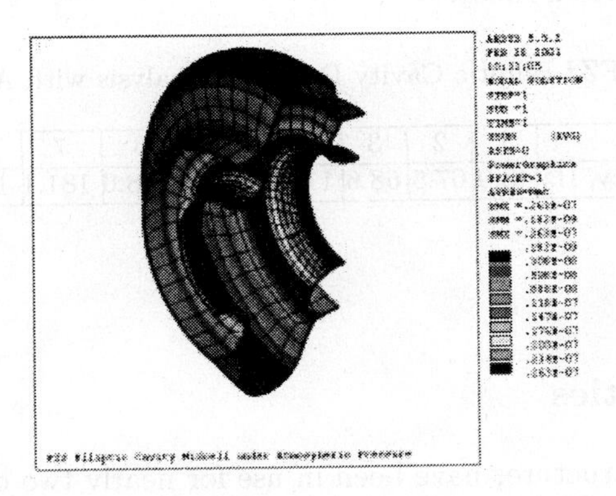

Fig. 5. Deformation of Elliptical Cavity Mid-Cell under Atmospheric Pressure

High repetition rates like 50 Hz for ESS will require a close look to the mechanical resonances of the cavities. Mechanical resonances can influence the phase behaviour of the cavity during a pulse, which can hardly be compensated by a good control system, even if a lot of additional power is available. Additionally, cavity rf resonance is sensitive to vibrations of sub-μm amplitudes. These microphonic effects cause low frequency noice in the accelerating fields. Therefore, carefull mechanic eigen mode analysis of the cavity together with its enviroments should be conducted. Fig. 6 presents a technical drawing of 500 MHz, β=0.75 5-cell elliptic cavity module of Forschungszentrum Juelich[3]. This facility is being used as an experimental installation for deep mechanic cavity property investigations. An experimental program comprises Lorenz force cavity detuning and its compensation and mechanical resonances and microphonic effect evaluations. On Fig. 6 a fast fourier transformation of the response to the time domain step function is shown. It reflects the first series of mechanical resonances of the structure. The dynamic analysis of the cavity with simplified cryoenviroments using ANSYS codes have been provided. The results are summarized in Table 1.

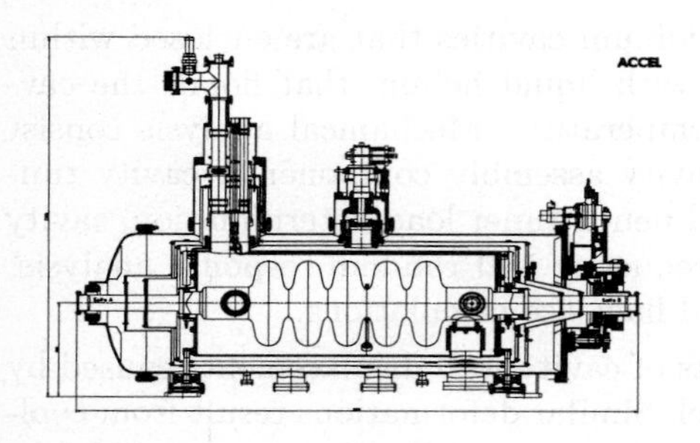

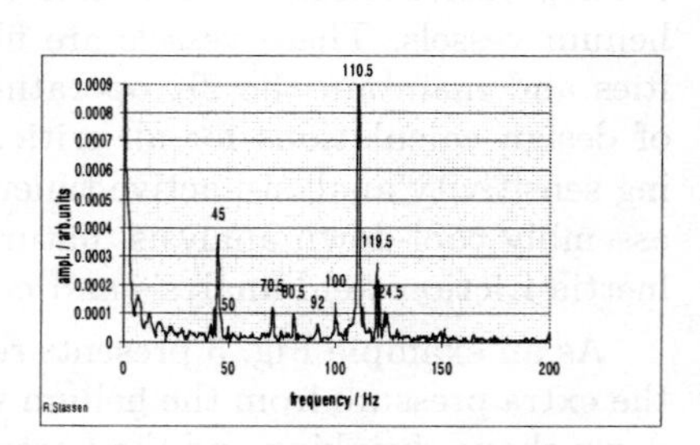

Fig. 6. FZJ Experimental Cryo Module with 5-cell Elliptic Cavity and Experimantal Results of Dynamic Modal Analysis

Table 1. FZJ Elliptic Cavity Dynamic Analysis with ANSYS

mode	1	2	3	4	5	6	7	8
frequency, Hz	48.9	67.5	68.6	110.8	131.9	158.0	181.1	191.0

3 Low-β Cavities

This type of SC rf structures have been in use for nearly two decades in heavy-ion booster linacs. Because of the requirement of short accelerating structures, low operating frequencies are used to make the structures as long as possible, so that the accelerating voltage per structure is maximized. The difference in use is defined by the type of the particle which has to be accelerated. The very first restriction on the structure choice for this range of particle energies (up to 20 MeV) for light particle (protons, deuterons) acceleration comes from the beam dynamics. Because of the phase shift between RF field and particle owing to acceleration it is impossible to use the cavities with a number of gaps bigger than two. A single-gap, so-called reentrant cavity, or elliptic cavities are not the best choice for such low energy range and especially for pulse mode operation because of its bad mechanic characteristics. This defines the use of coaxial quarter- (QWR) and half-wale length cavities for the β up to 0.2. The weak point of any QWR is its non-symmetry, which results in transversal components (especially magnetic) of rf field along the beam path, which is the serious problem for acceleration of light particles like protons. This can be eliminated by the half-wave cavity use. This type of structures comprises two well-known cavities, which are called coaxial half-wave resonator (HWR) and spoke cavity (Fig. 7). HWR is just a symmetric extension of quarter-wave cavity relative to the beam path. The simulation of this type of cavities require the use of real 3D codes, like for instance MAFIA[11].

3.1 Half-Wave Length Cavities

The range of half-wave structure application is for rather low $\beta \leq 0.2$ and fundamental frequency under 300 MHz. The resonant frequency is defined by the line length, inner-outer radius ratio and capacitance of an accelerating gaps. An accelerating field magnitude is limited by peak magnetic field that is defined mainly by the inner-outer electrode distance. This favours the use of the conical HWR (Fig. 7). The disadvantage of the cone cavity is its larger longitudinal extension. On the other hand this may be compensated by the cross-cavity installation.

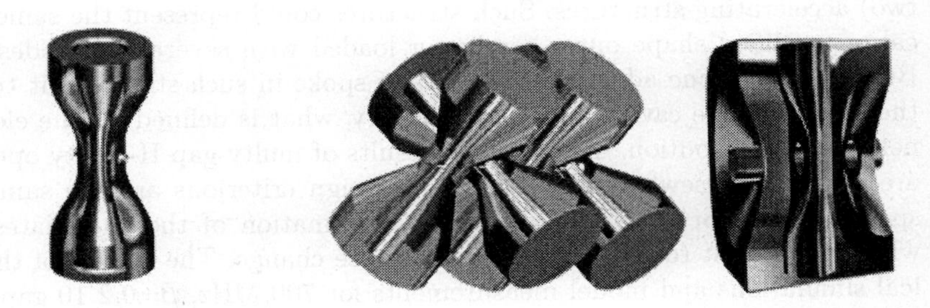

Fig. 7. Half-Wave Length Cavity, Cross-Cavity Installation & Spoke Cavity.

Another accelerating structure for this range of particle energy is a spoke resonator (Fig. 7). The spoke cavity by definition is a coaxial half-wave length cavity with an outer conductor turned on ninety degrees so that its axe is directed along the beam path. An equivalent capacitance of such cavity is defined by the distance between conductors in the center of the cavity along this axe. A distribution of an electromagnetic field in such cavity is the same like in coaxial cavity. The range of application of this cavity is from 100 to 800 MHz of fundamental frequency and β=0.1-0.6. The limitations of application are defined mainly by the resonance capacitance grow for low-β values which in its turn reduces cavity diameter. The spoke cavity acceleration efficiency is defined by the magnetic field magnitude on the inner electrode surface like in the coaxial resonators. The comparison of cross-cavity half-wave conical resonator installation with the set of spoke cavities in terms of maximal reachable accelerating field favours the use of the first option (Fig. 8)[12].

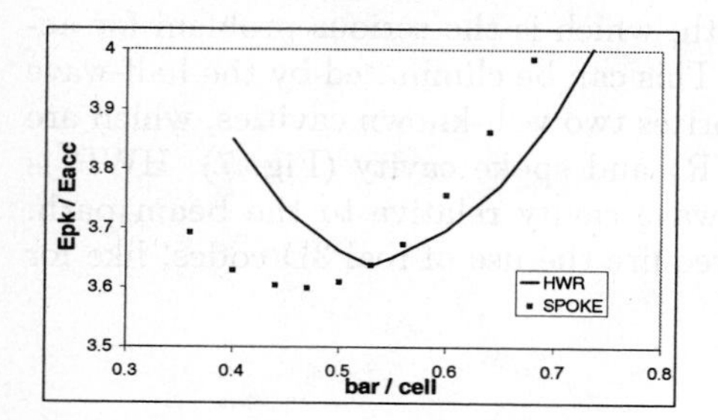

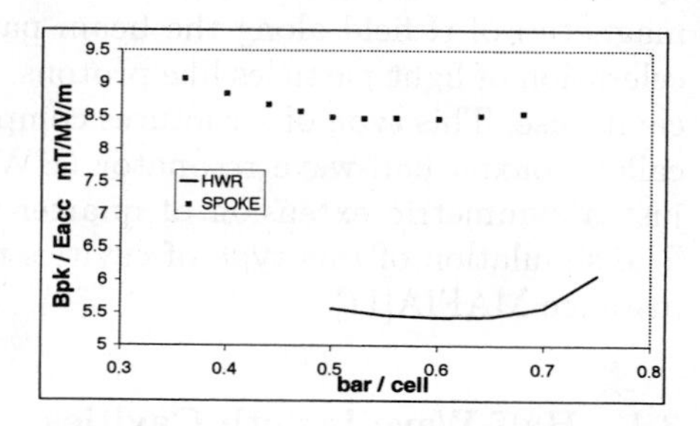

Fig. 8. Half-Wave Length Coaxial Cavity & Spoke Cavity Comparison (MAFIA Simulations).

3.2 Multi-Gap H-Cavity

Starting with the value β=0.2 there is a possibility to use multi-gap (more than two) accelerating structures. Such structures could represent the same cylindrical or modified shape outer conductor loaded with several electrodes (Fig. 9). But as soon as one adds at least another spoke in such structure it turns from the coaxial spoke cavity into H-type cavity, what is defined by the electromagnetic field distribution. The detailed results of multy-gap H-cavity optimisation are published elsewhere[13]. The main design criterions are the same like for spoke cavity. For the cavity tune the deformation of the end plates is used, which equivalent to the last gap capacitance change. The results of the numerical simulations and model measurements for 700 MHz, β=0.2 10-gap H-cavity are shown on Figs. 10-9. If to take into account that the whole cavity length is about 500 mm and end plate shift for tuning is within $\pm$1 mm, the high precision of simulations can be achieved.

3.3 Mechanics

The same structural analysis of low-β cavities has to be made to find the model predictions for peak stresses, deflections and flange reaction forces under vacuum

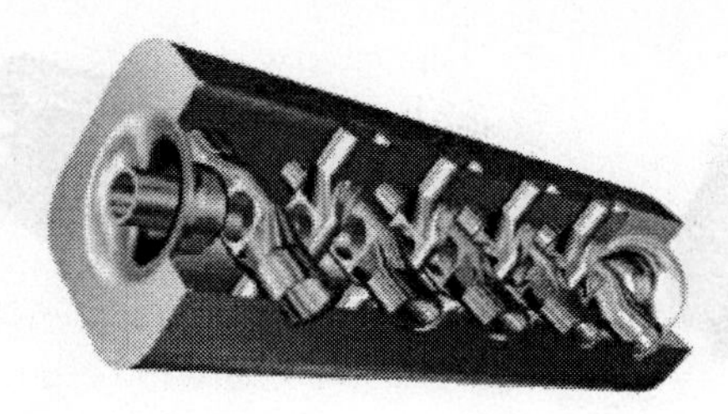

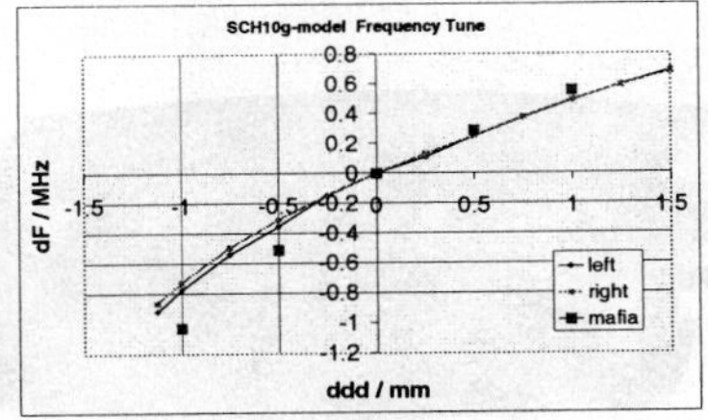

Fig. 9. 10-Gap H-Cavity and MAFIA Simulations & Experimental Data of Structure Tune.

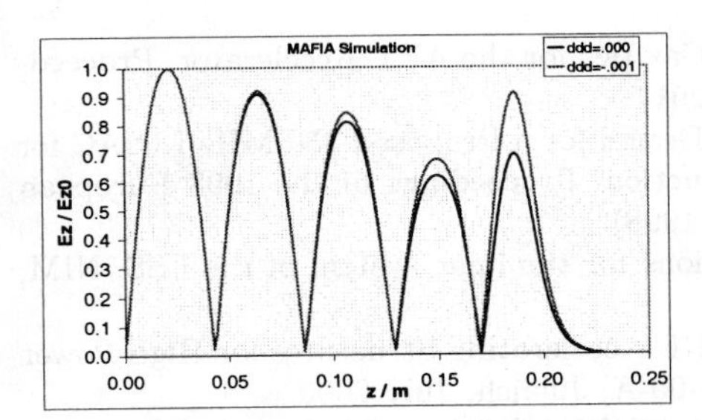

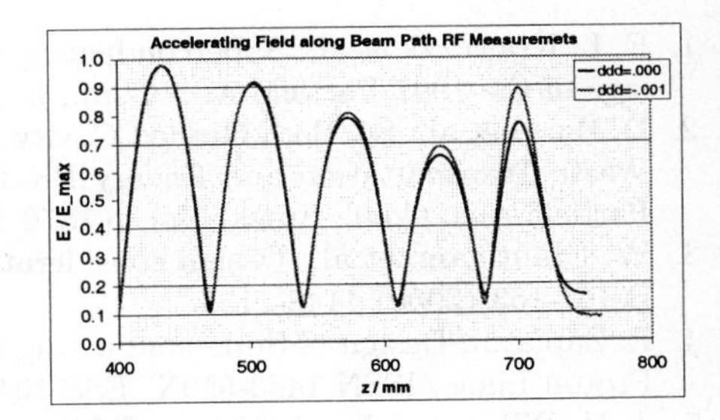

Fig. 10. MAFIA Simulation (left) and Experimental Results (right) of 10-Gap H-Cavity Electrical Field Distribution by Tuning

loads and room temperature, and also for forces required to produce a specify tuning deflection. The important part of simulations is devoted to the determination of resonant structural frequencies (Fig. 11). These structures differ from elliptic cavities by their more complicate 3D geometries that result in cavity higher rigidity and mechanical stability but complicate simulations. The main purpose of such calculations is as close as possible prediction of cavity frequency shift caused by structure deformations. This frequency shift should be later covered by cavity tuning range. The high structure frequency like 700 MHz defines smaller cavity dimensions, which result in smaller deformations and bigger tuning range. For spoke and 10-gap H-cavities (700 MHz, β=0.2) the tuning range is within 1 MHz/mm which easily can cover the possible wall deformation effects. For the frequency 320 MHz the tuning reduces down to 100-200 kHz/mm. This numbers are usually used as the minimal tuning range that should be reached. The coaxial quarter- and half-wave cavities with frequency 160 MHz can reach the same numbers. Making the proper structure design together with cryomodule the mechanical eigen resonances for all structures can be shifted well above 100 Hz.

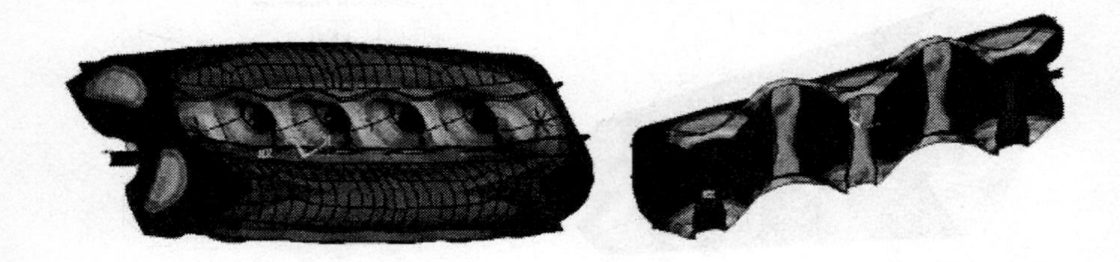

Fig. 11. 10-Gap H-Cavity under Extra Pressure and Modal Analysis.

References

1. F. L. Krawczyk et al.: Superconducting Cavities for the APT Accelerator. Proceedings of the 1997 Particle Acc. Conf., p. 2914.
2. D. Barni et al.: SC Beta Graded Cavity Design for a Proposed 350 MHz LINAC for Waste Transmutation and Energy Production. Proceedings of the 1998 European Particle Acc. Conf., Stockholm, p.1870 (1998).
3. W. Bräutigam, et al.: Design considerations for the linac system of the ESS. NIM, B 161-163 (2000) 1148-1153.
4. E. Zaplatin: Design of Superconducting RF Accelerating Structures for High Power Proton Linac. ISSN 1433-559X, ESS 104-00-A, Juelich, July 2000.
5. J. H. Billen and L. M. Young: POISSON/SUPERFISH on PC Compatibles. Proceedings of the 1993 Particle Acc. Conf., Vol. 2, p. 790.
6. R. Parodi: Preliminary analysis of Multipacting Barriers in the 500 MHz beta 0.75 ESS superconducting cavity. INFN, Genova, 2000.
7. S. Humphries Jr: Electron Multipactor Code for High-Power RF Window Development. Particle Accelerators, Vol. 62, pp. 139-163, 1999.
8. P. Ylä-Oijala: Multipacting Analysis for Five Elliptical Cavities. Rolf Nevanlinna Institute, University of Helsinki, Helsinki, September 22, 1999.
9. ANSYS is a trade mark of SAS IP, Inc., http://www.ansys.com
10. K. W. Shepard: Superconducting Heavy-Ion Accelerating Structures. Nuc. Instr. and Meth., A382 (1996) 125-131, North-Holland, Amsterdam.
11. M. Bartsch et al.: Solution of Maxwell's Equations. Computer Physics Comm., 72, 22-39 (1992).
12. E. Zaplatin et al.: Very Low-β Superconducting RF Cavities for Proton Linac. ISSN 1433-559X, ESS 104-00-A, Juelich, July 2000.
13. E. Zaplatin: Low-β SC RF Cavity Investigations. Workshop on RF Superconductivity SCRF2001, Tsukuba, Japan, 2001.

CEA Saclay Codes Review for High Intensities Linacs Computations

Romuald Duperrier, Nicolas Pichoff, Didier Uriot

CEA, 91191 Gif sur Yvette Cedex, France

rduperrier@cea.fr; npichoff@cea.fr; duriot@cea.fr

http://www.cea.fr

Abstract. More and more, computations are playing an important role in the theory and design of high intensities linacs. In this context, CEA Saclay is involved in several high power particles accelerators projects (SNS, ESS, IFMIF, CONCERT, EURISOL) and high intensities protons front end demonstrators (LEDA, IPHI). During these last years, several codes have been developed. This paper consists in a review of several of these codes especially: the linac generator able to design linac structures (computations coupled with SUPERFISH), the TOUTATIS code computing transport in RFQs in 3D grids (multigrid Poisson solver), the TraceWin code allowing to compute end to end simulation of a whole linac (automatic matching, up to 2 millions macroparticles run for halo studies with its PARTRAN module with 3D space charge routines, an errors studies module allowing large scale computations on several PCs using a multiparameters scheme based on a client/server architecture).

1 Introduction

A high power linacs can only work with very low structure activation. To design the accelerator, a very good estimation of losses location and emittances growth is necessary. In this goal, several transport codes have been developped at CEA Saclay. A lot of work has been performed in order to take into account several effects as space charge (3D routines), image forces, diffusion on the residual gas where it seems to be relevant. Applying basic matching rules, codes for linac generation has also been written and produce robust design. This paper is a review of these different tools.

2 Linac generation

2.1 Basic rules

For a transport with no emittance growth, a space charge driven beam has to be in equilibrium with the external focusing forces. When this external conditions are changed too abruptly (i.e. adiabatically), the beam reorganizes itself toward

P.M.A. Sloot et al. (Eds.): ICCS 2002, LNCS 2331, pp. 411–418, 2002.

a new equilibrium. This induces an amount of entropy (emittance growth). Moreover, when transition are smooth, the design is current independant (see figure 1). Linacs generator have been written taking into account such rules for each part of the linac. The following section gives an example with the DTL/SDTL generator.

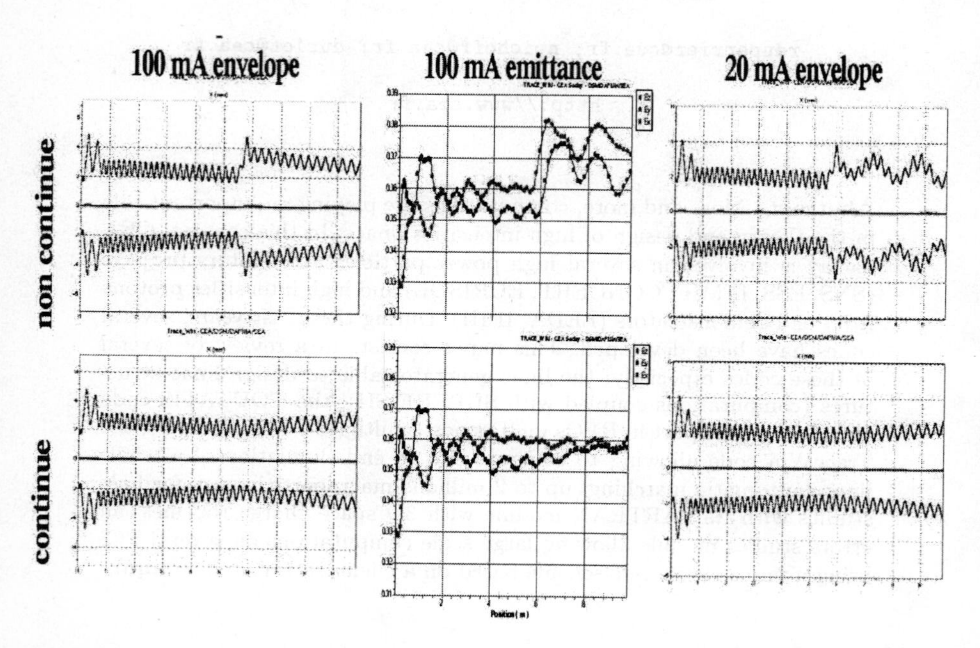

Fig. 1. Plot showing how emittance is affected by discontinuities of phase advance per meter and the current independance of a continuous channel.

2.2 Example of codes applying these rules

GenDTL is the DTL and SDTL generator [1]. This code generates a complete design. It may be compared to a pre and post processor for the code SUPERFISH [2]. This radio frequency solver is only used to mesh a particular design and to compute electrical field map. The transport of a synchronous particle through these field map accurately determine phase shift, synchronous phase laws and energy gain. No interpolation is necessary, each cell is computed. The figure 2 shows a snapshot of the generator front end. Applying the rules described above, DTL design without emittance blow up can be easily produced.

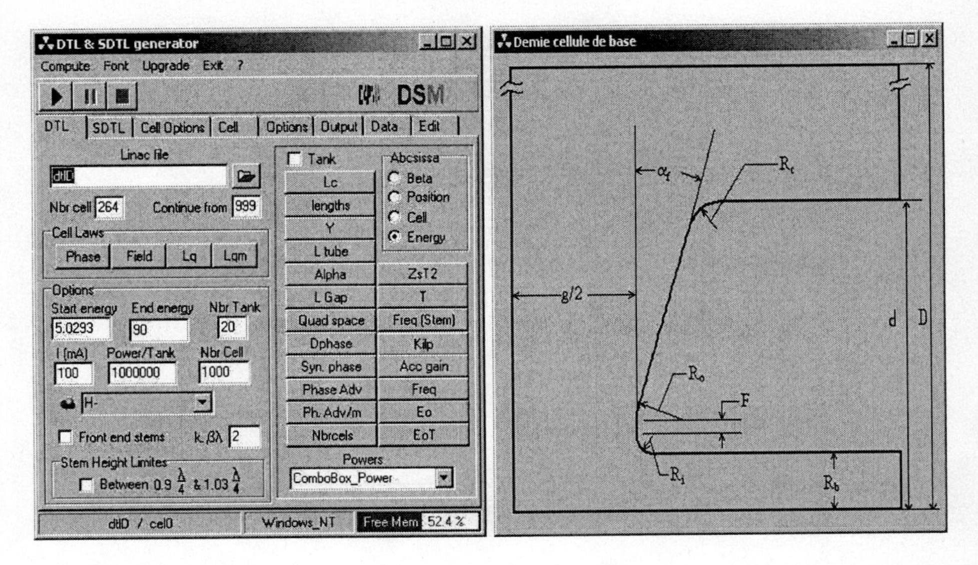

Fig. 2. Snapshot of the generator front end with a drift tube drawing.

3 TraceWin

This code is used for linear calculations with 2D and 3D beams [3]. It allows fast calculations. Its input file is written in a language describing linac element and beam characteristics (see figure 3). The current version is running only on win32 platform but have been successfully emulated on a linux machine. TraceWin is able to matched the beam in the whole linac exept the longitudinal plane for bunching RFQ. Either the beam is matched to a channel with lattices (figure 4), or gradient, field amplitude in cavity and element length are computed to obtain required beam parameters. Multiparticles codes can be run to validate the linear matching (TOUTATIS, PARTRAN, SCDYN, MONET). The criteria for matching is based on a smooth evolution of phase advance per meter.

The code can plot several results computed by multiparticle codes concerning the linac (lattice length, quads gradients, cavity fields, phase and power) and the beam (emittance, halo parameter, envelops, tune depression, energy and phase-space distributions). The figure 5 gives an example of such plot.

4 Multiparticles Codes

4.1 TOUTATIS

At low energy, the radio frequency quadrupole (RFQ) is an accelerator element that is very sensitive to losses (sparking). To simulate this structure, a high accuracy in field representation is required because the beam/aperture ratio is often very close to one. TOUTATIS aims to cross-check results and to obtain more

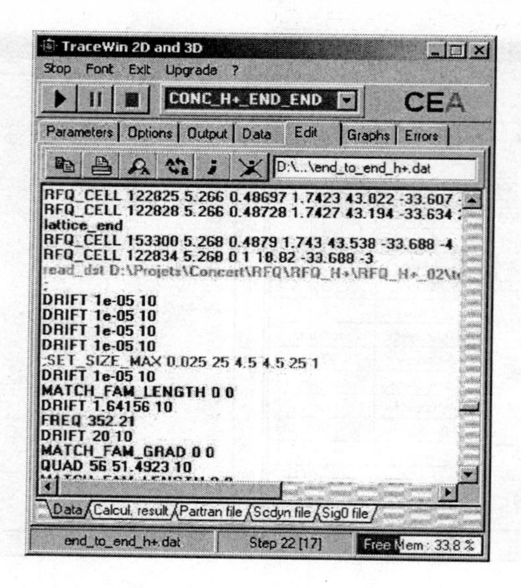

Fig. 3. Snapshot of the editor tabset of TraceWin showing an example for linac description.

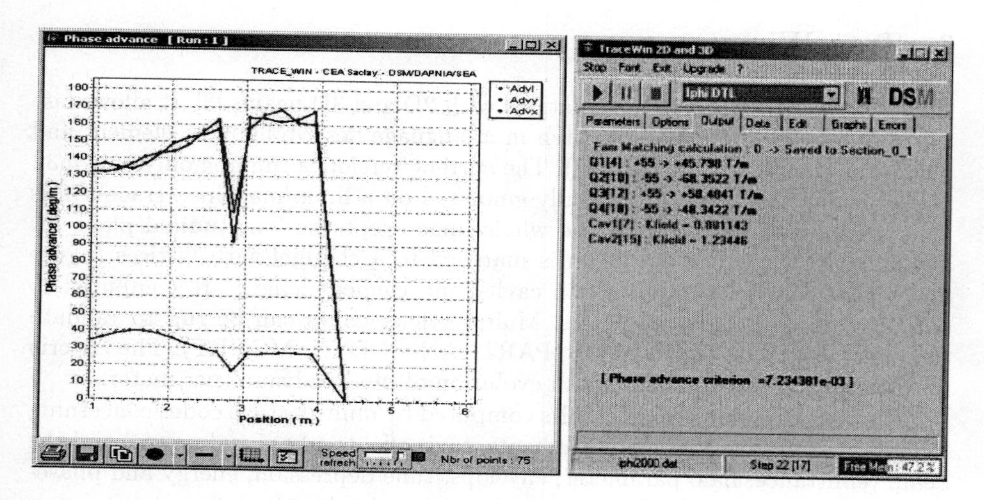

Fig. 4. The output tabset of TraceWin with phase advance per meter plot during a matching procedure.

reliable dynamics. Motion is solved with a symplectic integrator using time as independant parameter. The field is calculated through a Poisson solver and the vanes are fully described. The solver is accelerated by multigrids method (Fig. 6). An adaptive mesh for a fine description of the forces is included to compute accurately without being time consuming. This scheme allows to properly simulate the coupling gaps and the RFQs extremities. Theoretical and experimental

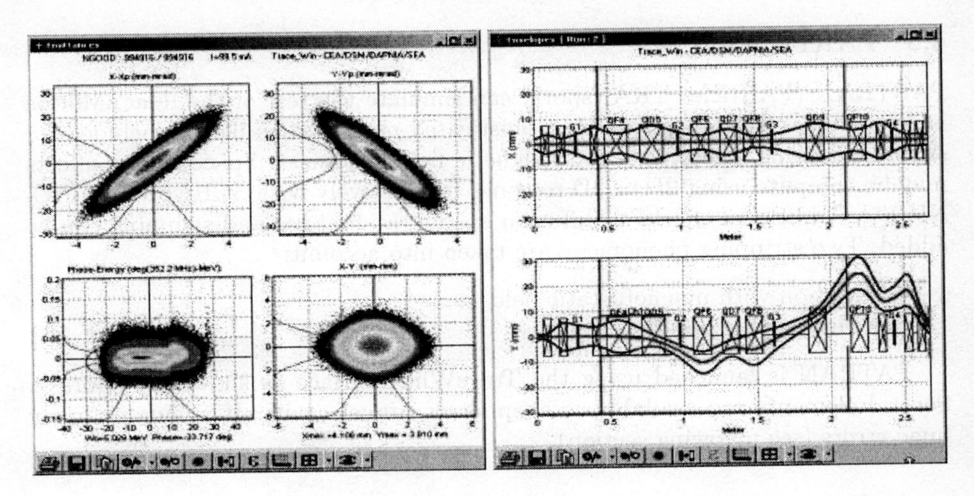

Fig. 5. An example of phase-space distribution plot and envelop evolution in a chopper line.

tests were carried out and showed a good agreement between simulations and reference cases [4].

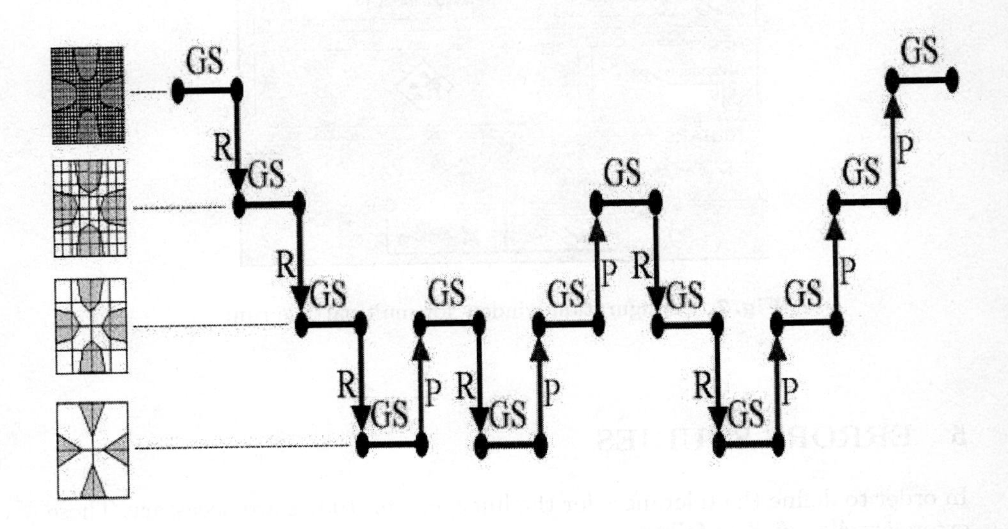

Fig. 6. Representation of the TOUTATIS cycle (GS = 3 Gauss-Seidel, R =Restriction, P = Prolongation).

4.2 PARTRAN

PARTRAN (PARticles TRANsport) can simulate the rest of the linac. Motion is calculated with a leap frog. A sophisticated model "sine-like" is included for superconducting cavity, external magnetic field are linear. Space charge effects may be compute using 2D or 3D routines (respectively with SCHEFF and PICNIC)[5]. Additional effects as diffusion on the residual gas and stripping can be added. Two stripping phenomena are taken into account:

- interaction with magnetostatic field
- interaction with residual gas

PATRAN is launched using the TraceWin interface as shown by figure 7. Several elements are available as chopper or funnel cavity. It simulates many linac errors (see following section).

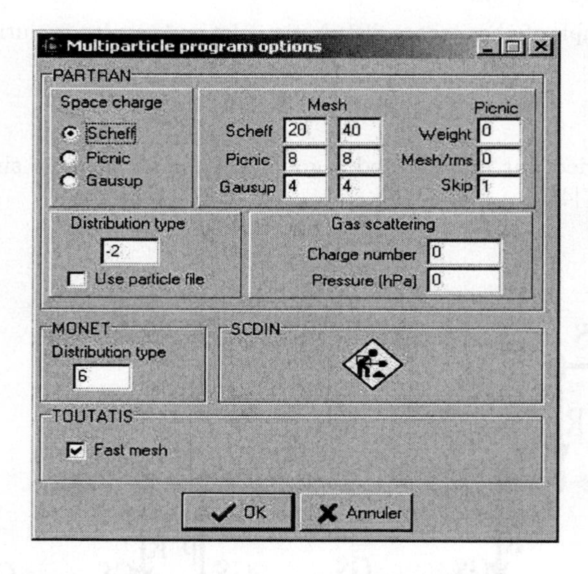

Fig. 7. Configuration window for multiparticle run.

5 ERRORS STUDIES

In order to define the tolerances for the linac, errors studies are necessary. These errors are classified as follow:

- Beam errors: displacements, emittance growth, mismatches, current.
- Statistics errors including:
  - Quadrupole errors: displacement and rotation in three directions, gradient, combination of errors.

- Cavity errors: displacement and rotation in two directions, field amplitude, field phase, combination of errors.
- Combination of quad and cavities errors.

Statistics errors may be static, dynamic or both. Dynamic means that correctors are included in the scheme.

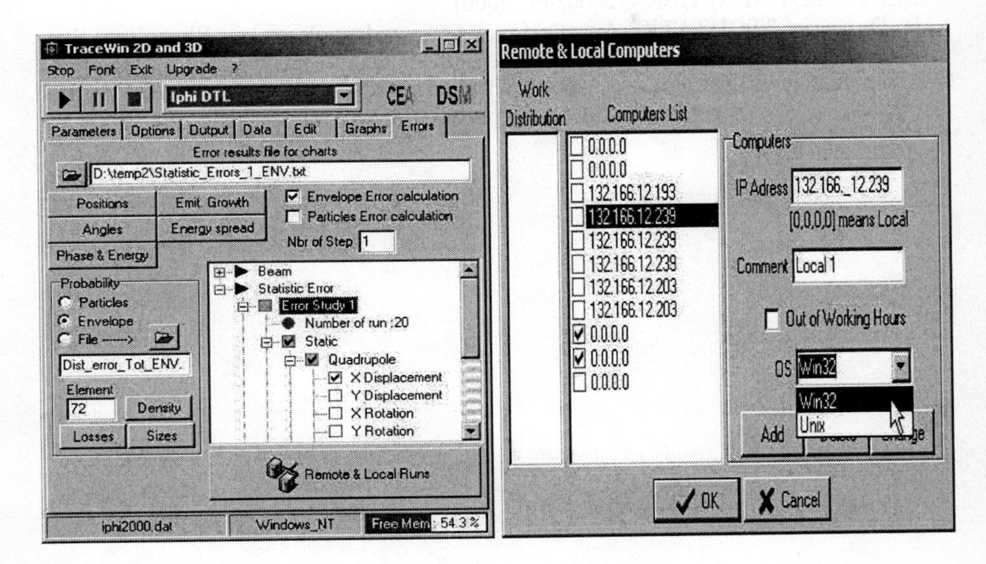

Fig. 8. Snapshot of the Errors tabset and remote runs window of TraceWin.

The TraceWin code is used as interface to manage this part (fig. 8). All these computations are distributed on a heterogen park of machines using a multiparameters scheme based on a client/server architecture. The client and the server are written in Tcl language. This allows to calculate on several platforms (win32, unix, mac) if executables to distribute are available. At present time, only TOUTATIS is multiplatorm but developments are in progress to port PARTRAN and TraceWin on unix platform.

6 Conclusion and prospects

Several tools have been developed at CEA Saclay for large scale computations on high power linac during this last years. Every parts of the linac can be simulated. Linac generator produces robust design with very low emittance growth and losses rate. Netherveless, it is still work to do. Space charge compensation can not be predicted at present time by these codes. The package is not completely multiplatform.

References

1. N. Pichoff, D. Uriot,"GenDTL", CEA internal report CEA/DSM/DAPNIA/SEA/2000/46 (2000).
2. J. Billen, L. Young, "Poisson Superfish", Los Alamos internal report LA-UR-96-1834 (1996).
3. N. Pichoff, D. Uriot,"TraceWin, documentation", internal report CEA/DSM/DAPNIA/SEA/2000/45 (2000).
4. R. Duperrier,"TOUTATIS: A radio frequency quadrupole code", Phys. Rev. Spec. Topics Acc. & Beams, Volume 3 (2000).
5. N. Pichoff et al.,"Simulation results with an alternate 3D space charge routine, PICNIC", MO4042, LINAC 1998 conference, Chicago.

Diagonalization of Time Varying Symmetric Matrices

Markus Baumann and Uwe Helmke

Mathematisches Institut, Universität Würzburg, Am Hubland, 97074 Würzburg
ms.b@gmx.de, helmke@mathematik.uni-wuerzburg.de

Abstract. We consider the task of diagonalizing symmetric time varying matrices $A(t)$. Based on the dynamic inversion technique developed by Getz and Marsden, a differential equation is proposed, whose solutions asymptotically track the diagonalizing transformation. In particular, one does not need to perfectly match the initial conditions, as the solutions converge exponentially towards the desired transformation. Thus, the proposed method is robust under perturbations.

1 Introduction

The aim of this paper is to establish an efficient algorithm for the diagonalization of time varying symmetric matrices. Thus we consider the task of determining orthogonal matrices $X(t)$ such that $X(t)'A(t)X(t)$ is diagonal. This is done by tracking a continuous time-dependent root of a suitable function $F_{A(t)}(X)$, whose roots $X_*(t)$ yield the desired solution. Therefore root finding methods, such as the dynamic inversion technique ([5]), can be applied to solve the problem. This approach leads to ordinary differential equations whose solutions converge exponentially fast to the diagonalizing transformations.

There exist already methods, which are able to track the diagonalizing transformation for a given symmetric matrix $A(t)$; see e.g. [1], [3], [4], [7].

These are given by the system of ODE's for D and X:

$$\dot{D} = X'\dot{A}X + DH - HD \tag{1}$$

$$\dot{X} = XH, \tag{2}$$

with H defined as: $H_{ij} = \begin{cases} 0, & i=j \\ \frac{(X'\dot{A}X)_{ij}}{d_j - d_i}, & i \neq j \end{cases}$

By solving this ODE, one obtains the diagonal matrix D and the corresponding orthogonal transformation matrix X. However, this algorithm only produces exact solutions, if it starts with perfect initial conditions. In particular, it does not perform any error-correction. In contrast, the dynamic inversion method achieves this purpose.

The paper is organized as follows. In the next section, a function $F(X,t)$ is introduced, which enables us to reformulate the diagonalization task as a root finding problem. Then, using the dynamic inversion technique, differential equations are proposed, that asymptotically compute the time-varying roots $X_*(t)$.

P.M.A. Sloot et al. (Eds.): ICCS 2002, LNCS 2331, pp. 419–428, 2002.

Examples are given in Section 3, which illustrate the convergence-properties of the new algorithms. Conclusions appear in Section 4.

2 Diagonalization of symmetric time-varying Matrices

Let $A(t) \in \mathbb{R}^{n\times n}$, $t \in \mathbb{R}$, be a continuously differentiable family of real symmetric matrices, with eigenvalues $\lambda_1(t), ..., \lambda_n(t)$. $A(t)$ is supposed to satisfy the conditions:

1. $\|A(t)\|$ and $\|\dot{A}(t)\|$ are uniformly bounded on $\mathbb{R}$.
2. There exists a constant $m > 0$ such that $|\lambda_i - \lambda_j| \geq m$ $(i \neq j)$ for all t.

Under these assumptions, there exists a continuously differentiable family of real orthogonal transformations $X_*(t)$, such that $X_*(t)'A(t)X_*(t)$ is diagonal for all $t \in \mathbb{R}$ ([6]). Our goal is to track such transformations by a suitable differential equation.

2.1 Reformulation of the problem

Consider the function

$$F : \mathbb{R}^{n\times n} \times \mathbb{R} \to \mathbb{R}^{n\times n} \tag{3}$$

defined by

$$F(X,t) = [N, X'A(t)X] + X'X - I, \tag{4}$$

where I is the identity matrix, $N = diag(1, ..., n)$ and $[,]$ is the Lie-Bracket product defined as $[A, B] := AB - BA$ for $A, B \in \mathbb{R}^{n\times n}$.

Lemma 1. *$F(X,t) = 0$ if and only if X is an orthogonal matrix such that $X'A(t)X$ is diagonal.*

Proof. Note that the first summand of F is skew symmetric while the second one is symmetric. Thus F vanishes if and only if the two summands vanish, i.e. if and only if X is orthogonal and

$$[N, X'AX] = 0.$$

Since N is diagonal with distinct eigenvalues, the result follows.

Using this function F, the task of finding an orthogonal transformation X such that $X'A(t)X$ is diagonal, is equivalent to that of finding a root of $F(X,t)$. Thus, the eigenvalue and eigenvector-problem is reformulated as a root-finding-problem.

To track those solutions $X_*(t)$, we apply the technique of dynamic inversion. In order to do so, certain technical assumptions made in [5] have to be checked. This is done in the next lemma.

Lemma 2. *Let $A(t)$ be as above. There exists a continuously differentiable isolated solution $X_*(t)$ to $F(X,t)=0$. $F(X,t)$ is C^∞ in X and C^1 in t. There exist constants $M, M_1 > 0$ such that*

(i) $\|D_1^2 F(X_*(t),t)\| \le M$,
(ii) $\|D_1 F(X_*(t),t)^{-1}\| \le M_1$,

holds for all $t \in \mathbb{R}$.

Proof. The claim concerning the differentiability properties of $F(X,t)$ is obvious.

The first and second partial derivatives of F w.r.to X are the linear and bilinear maps $D_1F(X,t)$ and $D_1^2F(X,t)$ respectively, given as

$$D_1F(X,t)\cdot H = [N, H'AX + X'AH] + H'X + X'H, \tag{5}$$

$$D_1^2F(X,t)\cdot(H,\hat{H}) = [N, H'A\hat{H} + \hat{H}'AH] + H'\hat{H} + \hat{H}'H, \tag{6}$$

and the partial derivative of $F(X,t)$ w.r.to t is

$$D_2F(X,t) = [N, X'\dot{A}(t)X]. \tag{7}$$

From this we deduce the operator norm estimates

$$\|D_1F(X,t)\| \le 2(1+2a\|N\|)\|X\|, \tag{8}$$

and $\|D_1^2F(X,t)\| \le 2(1+2a\|N\|)$, where a denotes the infinity-norm of A.

In particular, $D_1^2F(X,t)$ is uniformly bounded with respect to (X,t). This shows (i).

We next show, that the partial derivative operator $D_1F(X_*,t)$ is invertible for any solution (X_*,t) of $F(X,t)=0$. In particular, X_* is orthogonal and

$$D_1F(X_*,t)\cdot H = [N, H'AX_* + X_*'AH] + H'X_* + X_*'H. \tag{9}$$

Substituting $H = X_*\cdot\xi$, for $\xi \in \mathbb{R}^{n\times n}$ arbitrary, we obtain

$$\begin{aligned} D_1F(X_*,t)\cdot(X_*\xi) &= [N, \xi'X_*'AX_* + X_*'AX_*\xi] + \xi'X_*'X_* + X_*'X_*\xi. \\ &= [N, \xi'D + D\xi] + \xi' + \xi, \end{aligned} \tag{10}$$

where $D = X_*'AX_*$ is diagonal. Thus $X_*\xi$ is in the kernel of $D_1F(X_*,t)$ if and only if ξ is skew symmetric and $[N,[D,\xi]] = 0$. Hence $[D,\xi]$ must be diagonal and since D has distinct diagonal entries we conclude that $\xi = 0$. This shows that $D_1F(X_*,t)$ is invertible for any root of F. By the implicit function theorem it follows that for every orthogonal X_0 with $X_0'A(0)X_0$ diagonal, there exists a unique C^1-curve $X_*(t)$ of orthogonal matrices with $X_*(0) = X_0$. This shows the first claim.

To prove (ii), we derive a lower bound for the eigenvalues of $D_1F(X_*,t)$. Let ξ_{pq} denote the entry of ξ with the largest absolute value. Assuming that the norm of ξ is equal to one, the absolute value of ξ_{pq} is at least $\frac{1}{n^2}$. The smallest eigenvalue of $D_1F(X_*,t)$ is lower bounded by the sum of squares

$$(\xi_{pq}\lambda_p(p-q)+\xi_{qp}\lambda_q(p-q))^2+(\xi_{pq}+\xi_{qp})^2$$

of the pq-entries of $[N,\xi' D+D\xi]$ and $\xi'+\xi$. For $p=q$ this is lower bounded by $\frac{4}{n^4}$, while otherwise it is lower bounded by $(\xi_{pq}\lambda_p+\xi_{qp}\lambda_q)^2+(\xi_{pq}+\xi_{qp})^2$. The latter is a quadratic function in ξ_{qp} with minimum:

$$\frac{\xi_{pq}^2(\lambda_p-\lambda_q)^2}{1+\lambda_q^2}\geq\frac{(\lambda_q-\lambda_p)^2}{n^4(1+\lambda_q^2)}.$$

This is the desired lower bound for the eigenvalues of $D_1F(X_*,t)$. Thus (ii) follows with

$$M'=n^2\max(\frac{1}{2},\frac{1+a}{m}).$$

The next result follows from a straight forward perturbation argument; see e.g. [6].

Proposition 1. *Let $X_*(t)$ be a continuously differentiable orthogonal transformation, such that $X_*'(t)A(t)X_*(t)$ is diagonal for all t. Let $X(t)$ converge to $X_*(t)$ exponentially as t goes to infinity. Then $X'(t)A(t)X(t)\to X_*'(t)A(t)X_*(t)$ exponentially for $t\to\infty$.*

Main Results

A direct consequence of the Dynamic Inversion Theorem with vanishing error ([5], Theorem 2.3.5),which is applicable due to Lemma 2, can now be formulated as follows.

Theorem 1. *Let $A(t)$ be as above and X_0 be any orthogonal transformation that diagonalizes $A(0)$. Let $X_*(t)$ be the unique continuously differentiable solution to $F(X,t)=0$ with $X_*(0)=X_0$. For any $\mu>0$ sufficiently large, and any $X(0)$ sufficiently close to X_0, the solution $X(t)$ to*

$$D_1F(X,t)\dot{X}=-\mu([N,X'AX]+X'X-I)-[N,X'\dot{A}X]$$

converges exponentially to $X_*(t)$ as t goes to infinity. I.e. there exist some $k,b>0$ such, that $\|X(t)-X_*(t)\|\leq ke^{-bt}$. Moreover, any solution $X(t)$ is orthogonal for all t, provided $X(0)$ is orthogonal.

Remark 1. The constant μ in the previous theorem influences the rate of convergence of $X(t)$ to $X_*(t)$.

One difficulty with the above approach is that the differential equation is in an implicit form. In the following we show how to circumvent this problem by designing suitable explicit forms.

2.2 Dynamic Inversion using matrix representations of $D_1F(X,t)^{-1}$

Recall the function

$$F(X,t) = [N, X'AX] + X'X - I. \tag{11}$$

and its partial derivative given by

$$D_1F(X,t) \cdot H = [N, H'AX + X'AH] + H'X + X'H. \tag{12}$$

Let $B(X,t) \in \mathbb{R}^{n^2 \times n^2}$ denotes the matrix representation of $D_1F(X,t)$. Thus to compute the inverse of $D_1F(X,t)$, we have to invert a $n^2 \times n^2$ matrix. This difficulty can be avoided by augmenting a differential equation for the inverse of B.

In the following theorem let $\Gamma(t)$ denote an approximation for the inverse of $B(X,t)$, i.e. of the matrix representation for $D_1F(X,t)^{-1}$.

Theorem 2. *Let*

$$E(\Gamma, X, t) = -\Gamma vec([N, X'\dot{A}(t)X]),$$
$$F^{\Gamma}(\Gamma, X, t) := B(X,t)\Gamma - I,$$
$$E^{\Gamma}(\Gamma, X, t) := -\Gamma \frac{d}{dt} B(X,t)|_{\dot{X}=vec(E(\Gamma,X,t))}\Gamma.$$

Then the solution $(\Gamma(t), X(t))$ *of the ODE*

$$\begin{bmatrix} \dot{\Gamma} \\ vec(\dot{X}) \end{bmatrix} = -\mu \begin{bmatrix} \Gamma F^{\Gamma}(\Gamma, X, t) \\ \Gamma vec(F(X,t)) \end{bmatrix} + \begin{bmatrix} E^{\Gamma}(\Gamma, X, t) \\ E(\Gamma, X, t) \end{bmatrix}$$

converges exponentially to $(B(X_*(t),t), X_*(t))$, *assuming that* $(\Gamma(0), X(0))$ *is sufficient close to* $(B(X_*(0),0), X_*(0))$.

2.3 Dynamic Inversion by solving the Sylvester equation

We derive an explicit formula for the Sylvester equation associated with D_1F.

$$D_1F(X,t) \cdot H = [N, H'AX + X'AH] + X'H + H'X =: K + Y, \tag{13}$$

where K, Y are given skew-symmetric and symmetric matrices, respectively.

Equation (13) is equivalent to the following equations

$$X'H + H'X = Y, \tag{14}$$

$$[N, H'AX + X'AH] = K. \tag{15}$$

According to [2], a general solution to (14) is given by

$$H = P'(Z + \frac{1}{2}Q'YQ)Q^{-1}. \tag{16}$$

Here Z is skew-symmetric and arbitrary, and P, Q are arbitrary matrices satisfying $PXQ = I$.

For X sufficiently close to the orthogonal matrix X_*, X^{-1} exists, and we can choose $P = X^{-1}$ and $Q = I$. Hence (16) is equivalent to

$$H = (X^{-1})'(Z + \frac{1}{2}Y),$$

where $Z = -Z'$ is the only remaining variable. We plug this equation for H into (15) and obtain

$$[N, Z'X^{-1}AX + X'A(X^{-1})'Z] = K + \frac{1}{2}[Y'X^{-1}AX + X'A(X^{-1})'Y, N] \tag{17}$$

We replace X^{-1} by the approximation X' and $X'AX$ by $D := diag(X'AX) =: diag(d_1, ..., d_n)$. For $X = X_* + \epsilon$, these approximations are $O(\epsilon)$. With this, (17) is equivalent to

$$NZ'D + NDZ - Z'DN - DZN = K + [\frac{1}{2}(Y'D + DY), N],$$

which enables us to obtain Z_{ij}:

$$Z_{ij} = \frac{(K + [\frac{1}{2}(Y'D + DY), N])_{ij}}{n_i d_i - n_i d_j + d_j n_j - d_i n_j}.$$

Using simple algebraic manipulations we arrive at the following tracking algorithm. Note that the proposed algorithm differs from [3], [4] by an additional feedback term, which is necessary for the tracking property.

Theorem 3. *Let* $Y = \mu(X'X - I)$, $D := diag(X'AX)$, $d_i := D_{ii}$ *and* $\Omega := \tilde{\Omega} + \hat{\Omega}$, *where*

1. $\tilde{\Omega}_{ij} = \begin{cases} \frac{(X'\dot{A}X)_{ij}}{d_i - d_j}, & i \neq j, \\ 0, & i = j. \end{cases}$
2. $\hat{\Omega}_{ij} = \begin{cases} \frac{(\mu X'AX - YD)_{ij}}{d_i - d_j}, & i \neq j, \\ \frac{1}{2}Y_{ij}, & i = j. \end{cases}$

The solution to

$$\dot{X} = -X\Omega$$

converges exponentially to $X_*(t)$, *if* $X(0)$ *is sufficiently close to* $X_*(0)$ *and* $\mu > 0$ *is sufficiently large.*

Solving this ODE is considerably cheaper than solving the coupled system of theorem 2, as there is no need to compute matrices of size $n^2 \times n^2$. Of course, this also has a stabilizing impact on the numerical aspects of the algorithm.

3 Examples

We consider some numerical examples to demonstrate the new approach. In all subsequent figures the solid lines represent the theoretically exact solutions, while the dashed lines show the solutions of the ODE of Theorem 2 computed via a standard Runge-Kutta method.

Example 1. Let $A(t) = \begin{pmatrix} 10\sin(t) & 2 \\ 2 & 3\cos(t) \end{pmatrix}$

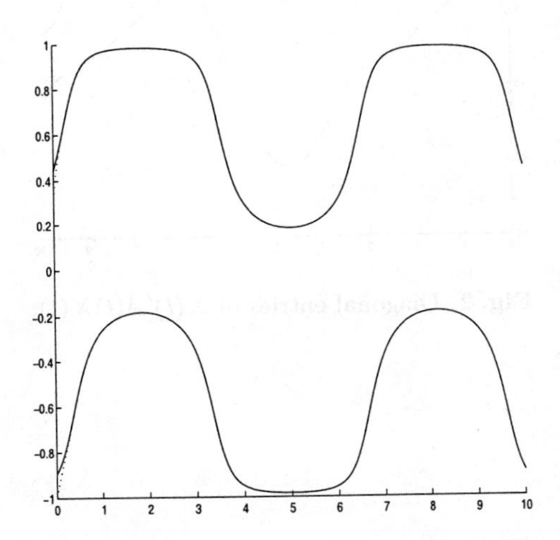

Fig. 1. Entries of the first column of $X(t)$ (dashed) and $X_*(t)$ (solid).

The figure shows a good approximation of the true eigenvectors/eigenvalues. This is to be expected, as the eigenvalues are well separated. In the next example, we will bring the eigenvalues very close together.

Example 2. Let $A(t) = \begin{pmatrix} t & 0.01 \\ 0.01 & 4-t \end{pmatrix}$ This example shows, that the algorithm reacts sensitively, if the eigenvalues get too close. Nevertheless, after a short period of time, the algorithm is stabilizing again.

For the same example, but using the last algorithm from Theorem 3 instead of the first one, the sensitivity problems are eliminated. Thus the last algorithm appears to be preferable.

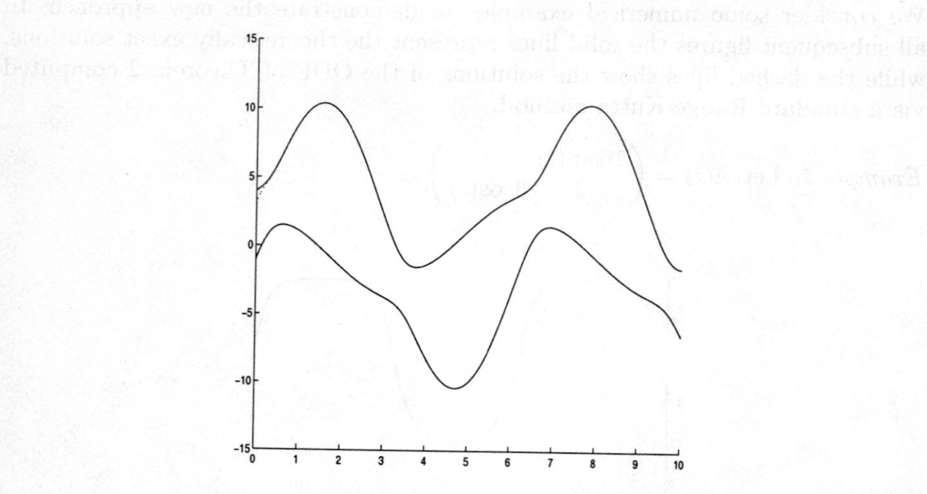

Fig. 2. Diagonal entries of $X(t)'A(t)X(t)$.

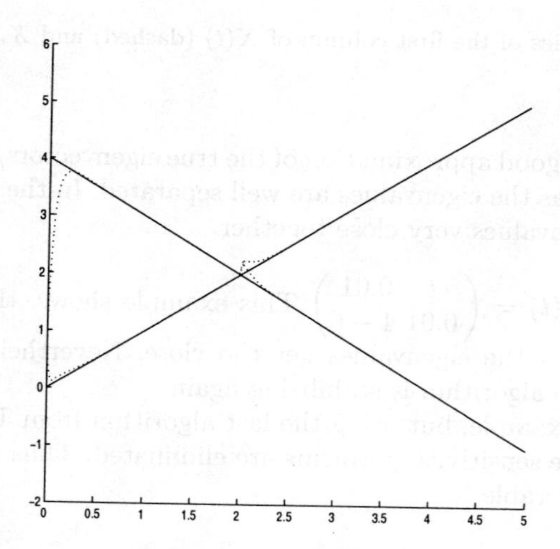

Fig. 3. Diagonal entries of $X(t)'A(t)X(t)$.

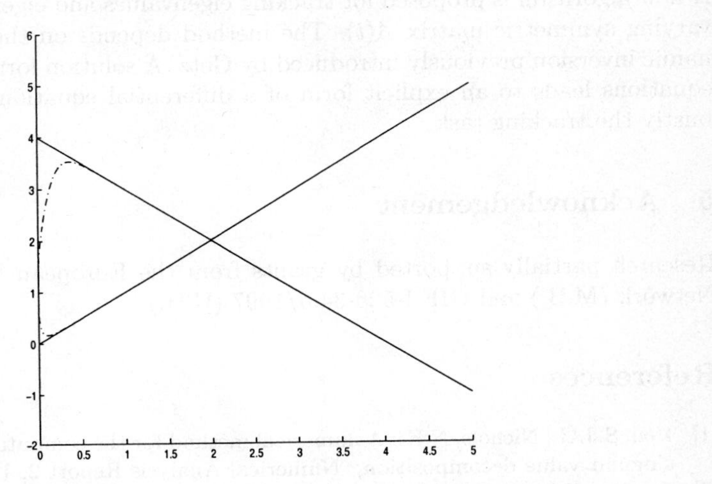

Fig. 4. Entries of the first column of $X(t)$ (dashed) and $X_*(t)$ (solid).

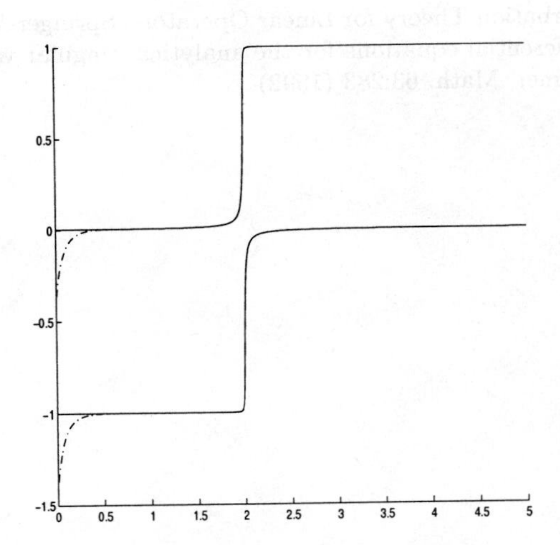

Fig. 5. Diagonal entries of $X(t)'A(t)X(t)$.

4 Conclusions

A new algorithm is proposed for tracking eigenvalues and eigenvectors of a time-varying symmetric matrix $A(t)$. The method depends on the technique of dynamic inversion previously introduced by Getz. A solution formula for Sylvester-equations leads to an explicit form of a differential equation that achieves robustly the tracking task.

5 Acknowledgement

Research partially supported by grants from the European Nonlinear Control Network (M.B.) and GIF I-526-34.6/1997 (U.H.).

References

[1] Bell, S.J.G., Nichols, N.K.: A numerical method for the computation of the analytic singular value decomposition, Numerical Analysis Report 2, 1994.
[2] Braden, H.W.: The equations $A'X \pm X'A = B$, SIAM Journal on Matrix Analysis and Applications, 20, (1998), 295–302.
[3] Dehaene, J.: Continuous-Time Matrix Algorithms, Systolic Algorithms and Adaptive Neural Networks, Ph.D. Thesis KU Leuven, 1995.
[4] Dieci, L., Eirola, T.: On smooth decompositions of matrices, SIAM Journal on Matrix Analysis and Applications, 20, (1999) 800-819.
[5] Getz, N.H.: Dynamic Inversion of Nonlinear Maps, Ph.D. Thesis, Univ. of California at Berkeley, 1995.
[6] Kato, T.: Perturbation Theory for Linear Operators, Springer-Verlag Berlin, 1966.
[7] Wright, K.: Differential equations for the analytical singular value decomposition of a matrix, Numer. Math. 63:283 (1992).

Conservation Properties of Symmetric BVMs Applied to Linear Hamiltonian Problems*

Pierluigi Amodio[1], Felice Iavernaro[1], and Donato Trigiante[2]

[1] Dipartimento di Matematica, Università di Bari,
Via E. Orabona 4, I-70125 Bari, Italy,
amodio@dm.uniba.it, felix@dm.uniba.it

[2] Dipartimento di Energetica, Università di Firenze,
Via C. Lombroso 6/17, I-50134 Firenze, Italy,
trigiant@cesit1.unifi.it

Abstract. We consider the application of symmetric Boundary Value Methods to linear autonomous Hamiltonian systems. The numerical approximation of the Hamiltonian function exhibits a superconvergence property, namely its order of convergence is $p+2$ for a p order symmetric method. We exploit this result to define a natural projection procedure that slightly modifies the numerical solution so that, without changing the convergence properties of the numerical method, it provides orbits lying on the same quadratic manifold as the continuous ones. A numerical test is also reported.

1 Introduction

Hamiltonian systems are not structurally stable against non-Hamiltonian perturbations, like those introduced by an ordinary numerical method during the discretization procedure. As a consequence, loss of some peculiar properties, such as the conservation of invariants or simplecticity, may be checked in the numerical solution, unless suitable classes of numerical integrators are used. This problem has led to the introduction of a number of methods and techniques to preserve the features of the Hamiltonian structure (see for example [5]; recent results on the subject may be found in [4] and references therein). We consider the linear Hamiltonian problem

$$\mathbf{y}' = L\mathbf{y}, \qquad t \in [t_0, t_0 + T] \tag{1}$$

where L is a Hamiltonian matrix of the form $L = JS$, S is a square real symmetric matrix of dimension $2m$ and

$$J = \begin{pmatrix} 0 & -I \\ I & 0 \end{pmatrix}$$

(I is the identity matrix of dimension m). We solve this problem numerically using a symmetric Boundary Value Method (BVM). The existence of a symplectic

* Work supported by GNCS.

P.M.A. Sloot et al. (Eds.): ICCS 2002, LNCS 2331, pp. 429–438, 2002.

generating matrix associated to a symmetric BVM has been shown in [3]. The main result of the present paper is a superconvergence property those schemes are proved to share. This will allow us to implement a trivial projection procedure that provides orbits lying on the same quadratic manifold as the continuous ones and preserving the order of the underlying method.

Introducing a uniform mesh $t_0, t_1, \ldots, t_N$ over the integration interval $[t_0, t_0 + T]$ with stepsize $h = T/N$, a symmetric BVM applied to (1) is defined by the linear multistep formula

$$\sum_{j=0}^{r} \alpha_j (\mathbf{y}_{n+j} - \mathbf{y}_{n-j-1}) - h \sum_{j=0}^{r} \beta_j L(\mathbf{y}_{n+j} + \mathbf{y}_{n-j-1}) = \mathbf{0}, \qquad (2)$$

that must be solved for $n = r+1, r+2, \ldots, N-r$. The additional conditions

$$\mathbf{y}_0, \ldots, \mathbf{y}_r \quad \text{and} \quad \mathbf{y}_{N-r+1}, \ldots, \mathbf{y}_N$$

are obtained adding the initial condition $\mathbf{y}(t_0) = \mathbf{y}_0$ and $2r$ extra multistep formulae called initial and final methods. The coefficients α_j and β_j are determined imposing that $\mathbf{y}_n$ is an approximation of order p of the continuous solution $\mathbf{y}(t_n)$; we will also assume $\sum_{j=-r_1}^{r_2} \beta_j = 1$ as a normalization condition to avoid the indetermination of the coefficients α_j and β_j. In matrix notation a symmetric BVM takes the form

$$(A \otimes I - hB \otimes L)Y = -\mathbf{a} \otimes \mathbf{y}_0 + h\mathbf{b} \otimes (L\mathbf{y}_0). \qquad (3)$$

where $Y = [\mathbf{y}_0^T, \mathbf{y}_1^T, \ldots, \mathbf{y}_N^T]^T$ is the solution vector, A and B are square matrices of dimension N, and the right hand side contains a known term that accounts for the initial condition. Apart from the initial and final blocks, each consisting of r rows containing the coefficients of the additional initial and final methods respectively, the matrices A and B have a Toeplitz structure defined by the $r+1$ coefficients of the main method; here is, for instance, how A looks like:

$$A = \begin{pmatrix} \boxed{\text{coefficients of the initial methods}} & & & & \\ -\alpha_{r-1} \ldots -\alpha_0 \; \alpha_0 & \ldots & \alpha_r & & \\ -\alpha_r \quad \ldots\ldots & -\alpha_0 \; \alpha_0 & \ldots \alpha_r & & \\ \ddots & \ddots & \ddots & \ddots & \\ & \ddots & \ddots \; \ddots & \ddots & \\ & & \ddots & \ddots \; \ddots & \ddots \\ & & -\alpha_r \; \ldots\ldots & & -\alpha_0 \; \alpha_0 \; \ldots \; \alpha_r \\ & & \boxed{\text{coefficients of the final methods}} & & \end{pmatrix},$$

and analogously for B. The vectors $\mathbf{a}$ and $\mathbf{b}$ contain the coefficients of the formulae in (3) that are combined with the initial condition $\mathbf{y}_0$; sometimes it is useful to insert such vectors as extra-columns in the matrices A and B. This is accomplished by defining the two extended matrices $\widetilde{A} = [\mathbf{a}, A]$, $\widetilde{B} = [\mathbf{b}, B]$, of size $N \times (N+1)$ and the vector $\widetilde{Y} = [\mathbf{y}_0^T, Y^T]^T$; in terms of these quantities, the system (3) becomes

$$\left(\widetilde{A} \otimes I - h\widetilde{B} \otimes L\right) \widetilde{Y} = \mathbf{0}. \tag{4}$$

The Extended Trapezoidal Rules of the first and second kind (ETRs and ETR$_2$s) and the Top Order Methods (TOMs) are three examples of classes of symmetric schemes (see [3] for their definition and properties). ETRs are defined as

$$\mathbf{y}_n - \mathbf{y}_{n-1} - h \sum_{j=0}^{r} \beta_j L(\mathbf{y}_{n+j} + \mathbf{y}_{n-j-1}) = \mathbf{0}, \tag{5}$$

and have order $p = 2r + 2$. The formula of order 4 is used in the last section in a numerical test.

2 Superconvergence of the numerical Hamiltonian function

We denote by $\sigma = (\mathbf{y}(t_n))^T S\mathbf{y}(t_n)$ the value (independent of n) of the Hamiltonian function over the continuous solution and by $\sigma_n = \mathbf{y}_n^T S\mathbf{y}_n$ the approximation of such value, generated by the numerical method.

Since $\mathbf{y}_n = \mathbf{y}(t_n) + O(h^p)$, it follows that

$$\sigma_n = \sigma + O(h^p). \tag{6}$$

The rate of convergence of the Hamiltonian function evaluated over the numerical solution, towards its limit value σ, is therefore (in general) inherited by the order of the underlying method. Hereafter, we prove that symmetric BVMs of even order p provide a two orders higher rate of convergence, that is $\sigma_n = \sigma + O(h^{p+2})$.

Given a vector $\mathbf{w}$, we will denote by $\mathbf{w}^k$ the vector whose nth component is $(w_n)^k$. Furthermore $\tilde{\mathbf{e}}$ and $\mathbf{e}$ will be two vectors with components equal to one and of length $N+1$ and N respectively, while $\mathbf{u} = [1, 2, \ldots, N]^T$ and $\tilde{\mathbf{u}} = [0, \mathbf{u}^T]^T$. The order conditions on a BVM may be recasted in block form:

$$\begin{array}{ll} \widetilde{A}\tilde{\mathbf{e}} = \mathbf{0}, \quad \widetilde{A}\tilde{\mathbf{u}} - \widetilde{B}\tilde{\mathbf{e}} = \mathbf{0} & \text{(consistency conditions)}, \\ \widetilde{A}\tilde{\mathbf{u}}^k - k\widetilde{B}\tilde{\mathbf{u}}^{k-1} = \mathbf{0}, & k = 2, \ldots, p. \end{array} \tag{7}$$

To begin with, we list some properties of the BVMs that will be exploited in the sequel. The proof of Lemma 1 is trivial and will be omitted.

Lemma 1. *For any BVM (3) the following relations hold true:*

(a) $\widetilde{B}\tilde{\mathbf{e}} = \mathbf{e}$; (b) $A^{-1}\mathbf{a} = -\mathbf{e}$;

(c) $A^{-1}\mathbf{b} = \mathbf{u} - A^{-1}B\mathbf{e}$; (d) $A^{-1}\mathbf{e} = \mathbf{u}$.

Lemma 2. *A BVM of order p satisfies the following $p-1$ conditions:*

$$(A^{-1}B)^{k-1}\mathbf{u} = \frac{1}{k!}\mathbf{u}^k, \qquad k = 2, \ldots, p. \tag{8}$$

Proof. For $k = 2, \ldots, p$ the order conditions (7) are simplified as follows

$$A\mathbf{u}^k - kB\mathbf{u}^{k-1} = \mathbf{0},$$

the first element of $\tilde{\mathbf{u}}$ being zero. Multiplying on the left by A^{-1} yields

$$\mathbf{u}^k = kA^{-1}B\mathbf{u}^{k-1} = k(k-1)(A^{-1}B)^2\mathbf{u}^{k-2} = \ldots = k!(A^{-1}B)^{k-1}\mathbf{u}$$

and (8) follows. □

Lemma 3. *Given a symmetric BVM with $2r+1$ steps and a vector $\boldsymbol{\xi}$ of length N and all null entries except (possibly) the first and the last r, one has:*

(a) $A^{-1}\boldsymbol{\xi} = g\mathbf{e} + \hat{\boldsymbol{\xi}}$, (b) $A^{-1}\hat{\boldsymbol{\xi}} = g_1\mathbf{e} + \hat{\boldsymbol{\xi}}_1$,

(c) $B\mathbf{u} = \mathbf{u} - \frac{1}{2}\mathbf{e} + \boldsymbol{\xi}$, (d) $A^{-1}\mathbf{u} = \frac{1}{2}(\mathbf{u}^2 + \mathbf{u}) - g\mathbf{e} - \hat{\boldsymbol{\xi}}$,

where g and g_1 are constant and $\hat{\boldsymbol{\xi}}$ and $\hat{\boldsymbol{\xi}}_1$ are vectors whose components decrease in modulus with an exponential rate when moving from the edge towards the inside of the vector (in the sequel the vectors denoted by $\boldsymbol{\xi}$ and $\hat{\boldsymbol{\xi}}$ are assumed to share the same kind of shape as described above).

Proof. The Toeplitz matrix associated to A has $r+1$ lower and r upper off-diagonals and is generated by the first characteristic polynomial of the basic method. This polynomial has r roots inside, r outside and one on the boundary of the unit circle (see [3]). It follows that, starting from the main diagonal, the entries in each column of the matrix A^{-1} tend to a constant value when we move downwards, and decrease exponentially in modulus when we move upwards [2]. From this consideration and the definitions of $\boldsymbol{\xi}$ and $\hat{\boldsymbol{\xi}}$, (a) and (b) immediately follow.

The n-th component of the vector $B\mathbf{u}$ is

$$(B\mathbf{u})_n = \sum_{j=0}^{r} \beta_j[(n+j) + (n-j-1)] = (2n-1)\sum_{j=0}^{r} \beta_j = n - \frac{1}{2},$$

which gives (c) (we notice that the non null elements of $\boldsymbol{\xi}$ depend on the initial and final methods). Since a consistent symmetric method has order at least two, we have that $A\mathbf{u}^2 - 2B\mathbf{u} = \mathbf{0}$ and exploiting in sequence (c), (d) of Lemma 1 and (a), we obtain (d). □

Since our goal is the computation of σ_n, we first derive an expression for $\mathbf{y}_n$ in terms of $\mathbf{y}_0$. From (3) we obtain (for small h):

$$\begin{aligned} Y &= (A \otimes I - hB \otimes L)^{-1}(-\mathbf{a} \otimes \mathbf{y}_0 + h\mathbf{b} \otimes (L\mathbf{y}_0)) \\ &= (I_N \otimes I - hA^{-1}B \otimes L)^{-1}(-A^{-1}\mathbf{a} \otimes \mathbf{y}_0 + hA^{-1}\mathbf{b} \otimes (L\mathbf{y}_0)) \\ &= \left(\sum_{j=0}^{\infty} h^j (A^{-1}B)^j \otimes L^j\right)(\mathbf{e} \otimes \mathbf{y}_0 + hA^{-1}\mathbf{b} \otimes (L\mathbf{y}_0)) \end{aligned}$$

$$= \sum_{j=0}^{\infty} h^j (A^{-1}B)^j \mathbf{e} \otimes L^j \mathbf{y}_0 + \sum_{j=0}^{\infty} h^{j+1} (A^{-1}B)^j A^{-1} \mathbf{b} \otimes L^{j+1} \mathbf{y}_0$$

$$= \mathbf{e} \otimes \mathbf{y}_0 + \sum_{j=1}^{\infty} h^j \left[(A^{-1}B)^j \mathbf{e} + (A^{-1}B)^{j-1} \mathbf{u} - (A^{-1}B)^j \mathbf{e} \right] \otimes L^j \mathbf{y}_0$$

$$= \mathbf{e} \otimes \mathbf{y}_0 + \sum_{j=1}^{\infty} h^j (A^{-1}B)^{j-1} \mathbf{u} \otimes L^j \mathbf{y}_0$$

(property (b) and (c) of Lemma 1 has been exploited to derive the third and fifth equalities). Denoting by $\mathbf{e}_n$ the n-th vector of the canonical basis on $\mathbb{R}^N$, we obtain

$$\mathbf{y}_n = (\mathbf{e}_n^T \otimes I) Y = \mathbf{y}_0 + \sum_{j=1}^{\infty} h^j (\mathbf{e}_n^T \otimes I) \left[(A^{-1}B)^{j-1} \mathbf{u} \otimes L^j \mathbf{y}_0 \right]$$

$$= \mathbf{y}_0 + \sum_{j=1}^{\infty} h^j \left(\mathbf{e}_n^T (A^{-1}B)^{j-1} \mathbf{u} \right) L^j \mathbf{y}_0.$$

For the computation of σ_n we will make use of the relation

$$\left(\sum_{j=1}^{\infty} \mathbf{v}_j \right)^T \left(\sum_{j=1}^{\infty} \mathbf{w}_j \right) = \sum_{j=1}^{\infty} \sum_{k=1}^{j} \mathbf{v}_k^T \mathbf{w}_{j-k+1}.$$

where $\{\mathbf{v}_j\}$ and $\{\mathbf{w}_j\}$ are two sequences of vectors whose related series are supposed to converge. We have

$$\sigma_n = \mathbf{y}_0^T S \mathbf{y}_0 + 2 \mathbf{y}_0^T S \left(\sum_{j=1}^{\infty} h^j \left(\mathbf{e}_n^T (A^{-1}B)^{j-1} \mathbf{u} \right) L^j \mathbf{y}_0 \right)$$

$$+ \sum_{j=1}^{\infty} \sum_{k=1}^{j} \left[h^k \left(\mathbf{e}_n^T (A^{-1}B)^{k-1} \mathbf{u} \right) \mathbf{y}_0^T (L^T)^k S h^{j-k+1} \left(\mathbf{e}_n^T (A^{-1}B)^{j-k} \mathbf{u} \right) L^{j-k+1} \mathbf{y}_0 \right]$$

$$= \sigma + 2 \sum_{j=1}^{\infty} h^j \left(\mathbf{e}_n^T (A^{-1}B)^{j-1} \mathbf{u} \right) \left(\mathbf{y}_0^T S (JS)^j \mathbf{y}_0 \right)$$

$$+ \sum_{j=1}^{\infty} h^{j+1} \sum_{k=1}^{j} \left[\left(\mathbf{e}_n^T (A^{-1}B)^{k-1} \mathbf{u} \right) \left(\mathbf{e}_n^T (A^{-1}B)^{j-k} \mathbf{u} \right) \left(\mathbf{y}_0^T (L^T)^k S L^{j-k+1} \mathbf{y}_0 \right) \right]$$

$$= \sigma + 2 \sum_{j=1}^{\infty} h^j \left(\mathbf{y}_0^T S (JS)^j \mathbf{y}_0 \right) \left(\mathbf{e}_n^T (A^{-1}B)^{j-1} \mathbf{u} \right)$$

$$+ \sum_{j=1}^{\infty} h^{j+1} \left(\mathbf{y}_0^T S (JS)^{j+1} \mathbf{y}_0 \right) \sum_{k=1}^{j} (-1)^k \left(\mathbf{e}_n^T (A^{-1}B)^{k-1} \mathbf{u} \right) \left(\mathbf{e}_n^T (A^{-1}B)^{j-k} \mathbf{u} \right)$$

$$= \sigma + 2h(\mathbf{y}_0^T SJS\mathbf{y}_0)(\mathbf{e}_n^T\mathbf{u}) + 2\sum_{j=2}^{\infty} h^j \left(\mathbf{y}_0^T S(JS)^j \mathbf{y}_0\right)\left(\mathbf{e}_n^T (A^{-1}B)^{j-1}\mathbf{u}\right)$$

$$+\sum_{j=2}^{\infty} h^j \left(\mathbf{y}_0^T S(JS)^j \mathbf{y}_0\right)\sum_{k=1}^{j-1}(-1)^k \left(\mathbf{e}_n^T (A^{-1}B)^{k-1}\mathbf{u}\right)\left(\mathbf{e}_n^T (A^{-1}B)^{j-k-1}\mathbf{u}\right).$$

For any vector $\mathbf{z} \in \mathbb{R}^{2m}$ one has $\mathbf{z}^T J\mathbf{z} = 0$, which gives $\mathbf{z}^T S(JS)^j\mathbf{z} = 0$ for any positive and odd integer j. A consequence is that the second term in the above expression of σ_n vanishes and both series will contain only terms with even powers in h:

$$\sigma_n = \sigma + \sum_{j=1}^{\infty} h^{2j} \left(\mathbf{y}_0^T S(JS)^{2j}\mathbf{y}_0\right)\left[2\left(\mathbf{e}_n^T (A^{-1}B)^{2j-1}\mathbf{u}\right)\right.$$

$$\left.+\sum_{k=1}^{2j-1}(-1)^k \left(\mathbf{e}_n^T (A^{-1}B)^{k-1}\mathbf{u}\right)\left(\mathbf{e}_n^T (A^{-1}B)^{2j-k-1}\mathbf{u}\right)\right].$$

From (6) we realize that the series cannot contain terms of order lower than h^p. Indeed we show that the first $p/2$ terms of the series vanish. For $j = 1, \dots, p/2$, considering the relations (8) (that can be extended to $p = 1$) we have

$$2\left(\mathbf{e}_n^T (A^{-1}B)^{2j-1}\mathbf{u}\right) + \sum_{k=1}^{2j-1}(-1)^k \left(\mathbf{e}_n^T (A^{-1}B)^{k-1}\mathbf{u}\right)\left(\mathbf{e}_n^T (A^{-1}B)^{2j-k-1}\mathbf{u}\right)$$

$$= \frac{2}{(2j)!}\mathbf{e}_n^T\mathbf{u}^{2j} + \sum_{k=1}^{2j-1}(-1)^k \frac{1}{k!(2j-k)!}\left(\mathbf{e}_n^T\mathbf{u}^k\right)\left(\mathbf{e}_n^T\mathbf{u}^{2j-k}\right)$$

$$= \frac{2}{(2j)!}n^{2j} + \sum_{k=1}^{2j-1}(-1)^k \frac{1}{k!(2j-k)!}n^k n^{2j-k}$$

$$= n^{2j}\sum_{k=0}^{2j}(-1)^k \frac{1}{k!(2j-k)!} = \frac{n^{2j}}{(2j)!}\sum_{k=0}^{2j}(-1)^k \binom{2j}{k} = 0.$$

The first non null term in the series is the one of order $p+2$:

$$h^{p+2}\left(\mathbf{y}_0^T S(JS)^{p+2}\mathbf{y}_0\right)\left[2\left(\mathbf{e}_n^T (A^{-1}B)^{p+1}\mathbf{u}\right)\right.$$
$$\left.+\sum_{k=1}^{p+1}(-1)^k \left(\mathbf{e}_n^T (A^{-1}B)^{k-1}\mathbf{u}\right)\left(\mathbf{e}_n^T (A^{-1}B)^{p-k+1}\mathbf{u}\right)\right]. \quad (9)$$

Since the dimension N of both matrices A and B is proportional to $1/h$, it is not possible to deduce so easily that such term is $O(h^{p+2})$. Such circumstance does hold true for symmetric methods (of even order). Theorem 1 will show that the term in square brackets in the above expression is indeed $O(1)$.

We begin with two lemmas.

Lemma 4. *A BVM of order p satisfies*

$$Au^{p+1} - (p+1)Bu^{p} = c_{p+1}\mathbf{e} + \boldsymbol{\xi}_{p+1}, \tag{10}$$

$$Au^{p+2} - (p+2)Bu^{p+1} = d_{p+2}\mathbf{u} + c_{p+2}\mathbf{e} + \boldsymbol{\xi}_{p+2}, \tag{11}$$

with $d_{p+2} = (p+2)c_{p+1}$ *and* $\boldsymbol{\xi}_{p+1}$ *and* $\boldsymbol{\xi}_{p+2}$ *with non null entries only in correspondence of the initial and final methods. If furthermore the basic method is symmetric, one has also* $d_{p+2} = -2c_{p+2}$.

Proof. We denote by $\boldsymbol{\alpha}$ and $\boldsymbol{\beta}$ the vectors of length k that contain the coefficients α_i and β_i of the main method. We consider the vector $\mathbf{w}(t) = [t-r+1, t-r, \ldots, t+r]^T$, with $t \in \mathbb{R}$. Since the method has order p, it follows that

$$\boldsymbol{\alpha}^T\mathbf{w}^p(t) - p\boldsymbol{\beta}^T\mathbf{w}^{p-1}(t) = 0.$$

Integrating with respect to t yields

$$\frac{1}{p+1}\boldsymbol{\alpha}^T\mathbf{w}^{p+1}(t) - \boldsymbol{\beta}^T\mathbf{w}^{p}(t) = \tilde{c}_{p+1}, \tag{12}$$

from which (10) follows with $c_{p+1} = (p+1)\tilde{c}_{p+1}$. Integrating (12) once again we obtain

$$\frac{1}{p+2}\boldsymbol{\alpha}^T\mathbf{w}^{p+2}(t) - \boldsymbol{\beta}^T\mathbf{w}^{p+1}(t) = c_{p+1}(t+\nu) + \tilde{c}_{p+2}, \tag{13}$$

with ν an arbitrary integer. Choosing suitably values of t and ν, the above expression is seen to be equivalent to the generic component of (11), with $d_{p+2} = (p+2)c_{p+1}$. In particular, if the main method is symmetric with $2r+1$ steps, one has $\boldsymbol{\alpha} = -P\boldsymbol{\alpha}$ and $\boldsymbol{\beta} = P\boldsymbol{\beta}$, with P the permutation matrix having unitary elements on the secondary main diagonal. In such a case, to obtain (11) we must choose $\nu = 0$. Let us set

$$\mathbf{w}_0 = \mathbf{w}(0) = [-(r+1), \ldots, -1, 0, \ldots, r]^T,$$

$$\mathbf{w}_1 = \mathbf{w}(1) = [-r, \ldots, 0, 1, \ldots, r+1]^T;$$

from (13) we have

$$d_{p+2} + c_{p+2} = \boldsymbol{\alpha}^T\mathbf{w}_1^{p+2} - (p+2)\boldsymbol{\beta}^T\mathbf{w}_1^{p+1},$$

$$c_{p+2} = \boldsymbol{\alpha}^T\mathbf{w}_0^{p+2} - (p+2)\boldsymbol{\beta}^T\mathbf{w}_0^{p+1}.$$

From the relation $\mathbf{w}_0^j = (-1)^j P\mathbf{w}_1^j$, that holds for any integer j we obtain

$$\begin{aligned} c_{p+2} &= \boldsymbol{\alpha}^T P\mathbf{w}_1^{p+2} + (p+2)\boldsymbol{\beta}^T P\mathbf{w}_1^{p+1} = -\boldsymbol{\alpha}^T\mathbf{w}_1^{p+2} + (p+2)\boldsymbol{\beta}^T\mathbf{w}_1^{p+1} \\ &= -(d_{p+2} + c_{p+2}), \end{aligned}$$

and hence the assertion. □

Lemma 5. *The extension of (8) to the indices $k = p+1$ and $k = p+2$ are respectively*

$$(A^{-1}B)^p\mathbf{u} = \frac{1}{(p+1)!}\left(\mathbf{u}^{p+1} + c_{12}\mathbf{u} + c_{11}\mathbf{e} + \hat{\boldsymbol{\xi}}_1\right), \tag{14}$$

and

$$(A^{-1}B)^{p+1}\mathbf{u} = \frac{1}{(p+2)!}\left(\mathbf{u}^{p+2} + c_{23}\mathbf{u}^2 + c_{22}\mathbf{u} + c_{21}\mathbf{e} + \hat{\boldsymbol{\xi}}_2\right), \tag{15}$$

where the constants c_{ij} satisfy the following relations:

$$c_{12} = -c_{p+1}, \qquad c_{23} = -(p+2)c_{p+1}, \qquad c_{22} = (p+2)c_{11}. \tag{16}$$

Proof. Multiplying (10) on the left by the inverse of A and using (d) of Lemma 1 and (a) of Lemma 3 we obtain

$$(A^{-1}B)\mathbf{u}^p = \frac{1}{p+1}(\mathbf{u}^{p+1} - c_{p+1}\mathbf{u} + c_{11}\mathbf{e} + \hat{\boldsymbol{\xi}}_1),$$

and considering Lemma 2 for $k = p$, we deduce

$$\begin{aligned}(A^{-1}B)^p\mathbf{u} = (A^{-1}B)(A^{-1}B)^{p-1}\mathbf{u} &= \frac{1}{p!}(A^{-1}B)\mathbf{u}^p \\ &= \frac{1}{(p+1)!}(\mathbf{u}^{p+1} - c_{p+1}\mathbf{u} + c_{11}\mathbf{e} + \hat{\boldsymbol{\xi}}_1),\end{aligned}$$

that coincides with (14), putting $c_{12} = -c_{p+1}$. With an analogous argument on formula (11) we derive

$$(A^{-1}B)\mathbf{u}^{p+1} = \frac{1}{p+2}(\mathbf{u}^{p+2} - d_{p+2}A^{-1}\mathbf{u} - c_{p+2}\mathbf{u} - g_1\mathbf{e} - \hat{\boldsymbol{\xi}}_{p+2}),$$

and using (14), Lemma 2 for $k = 1$ and (b)-(c) of Lemma 1,

$$\begin{aligned}(A^{-1}B)^{p+1}\mathbf{u} &= (A^{-1}B)(A^{-1}B)^p\mathbf{u} \\ &= \frac{1}{(p+1)!}(A^{-1}B)\left(\mathbf{u}^{p+1} + c_{12}\mathbf{u} + c_{11}\mathbf{e} + \hat{\boldsymbol{\xi}}_1\right) \\ &= \frac{1}{(p+2)!}\Big[\mathbf{u}^{p+2} - d_{p+2}A^{-1}\mathbf{u} - c_{p+2}\mathbf{u} - g_1\mathbf{e} - \hat{\boldsymbol{\xi}}_{p+2} \\ &\quad -(p+2)\frac{c_{p+1}}{2}u^2 + (p+2)c_{11}(\mathbf{u} - A^{-1}\mathbf{b}) + (p+2)A^{-1}B\hat{\boldsymbol{\xi}}_1\Big].\end{aligned}$$

Using (d), (a) and (c) of Lemma 3 for the terms $A^{-1}\mathbf{u}$, $A^{-1}\mathbf{b}$ and $A^{-1}B\hat{\boldsymbol{\xi}}_1$ respectively, we have

$$(A^{-1}B)^{p+1}\mathbf{u} = \frac{1}{(p+2)!}\Bigg[\mathbf{u}^{p+2} - \frac{1}{2}(d_{p+2} + (p+2)c_{p+1})\mathbf{u}^2 - \left(\frac{d_{p+2}}{2} + c_{p+2} - (p+2)c_{11}\right)\mathbf{u} + c_{21}\mathbf{e} + \hat{\boldsymbol{\xi}}_2\Bigg],$$

with c_{21} a suitable constant. Finally (15) follows exploiting the expressions for d_{p+2} obtained in Lemma 4. □

Now we proceed with the proof of the superconvergence property.

Theorem 1. *The solution of a symmetric BVM with* $2r+1$ *steps and even order* p *satisfies* $\sigma_n = \sigma + O(h^{p+2})$.

Proof. It is enough to prove that the term in square brackets in (9) is O(1). Considering once again the order conditions (8) and the extensions (14) and (15), this term is simplified as follows

$$\left[2\left(\mathbf{e}_n^T(A^{-1}B)^{p+1}\mathbf{u}\right) + \sum_{k=1}^{p+1}(-1)^k\left(\mathbf{e}_n^T(A^{-1}B)^{k-1}\mathbf{u}\right)\left(\mathbf{e}_n^T(A^{-1}B)^{p-k+1}\mathbf{u}\right)\right]$$

$$= \frac{2}{(p+2)!}\mathbf{e}_n^T(c_{23}\mathbf{u}^2 + c_{22}\mathbf{u} + c_{21}\mathbf{e} + \hat{\boldsymbol{\xi}}_2) - \frac{2}{(p+1)!}(\mathbf{e}_n^T\mathbf{u})\mathbf{e}_n^T(c_{12}\mathbf{u} + c_{11}\mathbf{e} + \hat{\boldsymbol{\xi}}_1)$$

$$= \frac{2}{(p+1)!}\left[\left(\frac{c_{23}}{p+2} - c_{12}\right)n^2 + \left(\frac{c_{22}}{p+2} - c_{11}\right)n + O(1)\right].$$

We arrive at the assertion considering that the coefficients of the terms in $\mathbf{u}^2$ and $\mathbf{u}$ vanish because of (16). □

The superconvergence property allows us to modify the numerical solution in order to obtain a new one preserving the value of the Hamiltonian function. Starting from the numerical solution $\mathbf{y}_n$, we define $\mathbf{z}_0 = \mathbf{y}_0$ and

$$\mathbf{z}_n = \left(\frac{\mathbf{y}_0^T S\mathbf{y}_0}{\mathbf{y}_n^T S\mathbf{y}_n}\right)^{1/2}\mathbf{y}_n, \qquad n = 2, \ldots, N. \tag{17}$$

Obviously now we have $\mathbf{z}_n^T S\mathbf{z}_n = \sigma$. The projection (17) together with formula (3) describes a new method sharing exactly the same convergence properties (order and error constants) of the original one. In fact, denoting by $\mathbf{g}_n(h) = \mathbf{y}_n - \mathbf{y}(t_n)$ the error function at step n ($\mathbf{g}_n(h) = O(h^p)$), from $\mathbf{y}_n^T S\mathbf{y}_n = \mathbf{y}_0^T S\mathbf{y}_0 + O(h^{p+2})$ it follows that

$$\mathbf{z}_n = \left(\frac{\sigma}{\sigma + O(h^{p+2})}\right)^{1/2}(\mathbf{y}(t_n) + \mathbf{g}_n(h))$$

$$= (1 + O(h^{p+2}))(\mathbf{y}(t_n) + \mathbf{g}_n(h)) = \mathbf{y}(t_n) + \tilde{\mathbf{g}}_n(h),$$

where $\tilde{\mathbf{g}}_n(h)$ and $\mathbf{g}_n(h)$ share the same $O(h^p)$ term.

3 A numerical test

We use the ETR of order 4 (see formula (5)) to solve the linear pendulum problem $\ddot{x} + \omega^2 x = 0$, $\omega = 10$ with initial condition $x(0) = 1$, $\dot{x}(0) = -2$, in the time interval $[0, 2\pi]$. At each run we halve the stepsize h, starting from $h = 2\pi/5$ (this will cause the doubling of the dimension N of the system). In the columns of Table 1 we report

- the maximum errors in the numerical solutions $\mathbf{y}_n$ and $\mathbf{z}_n$:

$$E(\mathbf{y}_n) = \max_{1 \le n \le N} ||\mathbf{y}(t_n) - \mathbf{y}_n||_\infty, \qquad E(\mathbf{z}_n) = \max_{1 \le n \le N} ||\mathbf{y}(t_n) - \mathbf{z}_n||_\infty,$$

  where $\mathbf{y}(t_n) = [x(t_n), \dot{x}(t_n)]^T$;
- the computed order of convergence of $\mathbf{z}_n$;
- the maximum error in the approximation of the Hamiltonian function obtained by $\mathbf{y}_n$:

$$H(\mathbf{y}_n) = \max_{1 \le n \le N} |\sigma - \sigma_n|;$$

- the computed order of convergence of σ_n towards σ.

As shown at the end of the last section due to the superconvergence (see last column in the table), $E(\mathbf{y}_n)$ and $E(\mathbf{z}_n)$ become eventually identical.

Table 1. Convergence properties of the solution of the linear pendulum problem obtained by the ETR of order 4

N	$E(\mathbf{y}_n)$	$E(\mathbf{z}_n)$	ord. $\mathbf{z}_n$	$H(\mathbf{y}_n)$	ord. $H(\mathbf{y}_n)$
5	$4.06103 \cdot 10^{0}$	$5.39728 \cdot 10^{0}$		$1.83 \cdot 10^{1}$	
10	$9.95836 \cdot 10^{-1}$	$1.01165 \cdot 10^{0}$	2.41	$1.52 \cdot 10^{0}$	3.59
20	$7.19667 \cdot 10^{-2}$	$7.19312 \cdot 10^{-2}$	3.81	$3.17 \cdot 10^{-2}$	5.58
40	$4.72723 \cdot 10^{-3}$	$4.72622 \cdot 10^{-3}$	3.93	$5.25 \cdot 10^{-4}$	5.92
80	$4.16611 \cdot 10^{-4}$	$4.16615 \cdot 10^{-4}$	3.50	$8.32 \cdot 10^{-6}$	5.98
160	$2.62860 \cdot 10^{-5}$	$2.62860 \cdot 10^{-5}$	3.99	$1.30 \cdot 10^{-7}$	6.00

References

1. Aceto, L., Trigiante, D.: Symmetric schemes, time reversal symmetry and conservative methods for Hamiltonian systems, J. Comput. Appl. Math. **107** (1999) 257–274
2. Amodio, P., Brugnano, L.: On the conditioning of Toeplitz band matrices, Mathematical and Computer Modelling **23 (10)** (1996) 29–42
3. Brugnano, L., Trigiante, D.: Solving ODEs by Linear Multistep Initial and Boundary Value Methods, Gordon & Breach, Amsterdam, 1998
4. Hairer, E., Lubich, Ch., Wanner, G.: Geometric Numerical Integration Structure-Preserving Algorithms for Ordinary Differential Equations. Springer Series in Computational Mathematics **31**, 2002
5. Sanz-Serna, J.M., Calvo, M.P.: Numerical Hamiltonian problems. Chapman and Hall, London, 1994

A Fixed Point Homotopy Method for Efficient Time-Domain Simulation of Power Electronic Circuits

E. Chiarantoni[1], G. Fornarelli[1], S. Vergura[1], T. Politi[2]

[1] Dipartimento di Elettrotecnica ed Elettronica, Politecnico di Bari, Via E. Orabona n° 4, 70125 Bari, Italy
chiarantoni@poliba.it, {fornarelli, vergura}@deemail.poliba.it
[2] Dipartimento di Matematica, Politecnico di Bari, Via E. Orabona n° 4, 70125 Bari, Italy
pptt@pascal.dm.uniba.it

Abstract. The time domain analysis of switching circuits is a time consuming process as the change of switch state produces sharp discontinuities in switch variables. In this paper a method for fast time-domain analysis of switching circuits is described. The proposed method is based on piecewise temporal transient analysis windows joined by DC analysis at the switching instants. The DC analysis is carried out by means of fixed point homotopy to join operating points between consecutive time windows. The proposed method guarantees accurate results reducing of the number of iterations needed to simulate the circuit.

1 Introduction

Modern power electronic is characterized by an extensive use of non linear devices as diodes, BJTs, MOSFETs and IGBTs, in the role of power switches. From a mathematical point of view, these elements can pose severe limitations in the time domain simulation. The switches can be modeled in detail or in a simplified form. With detailed models the circuits become normal analog circuits and can be handled by standard commercial simulators based on various versions of the Berkley SPICE [1]. In this case the simulation is very accurate, but takes very long time. On the other hand, in many cases it is advantageous to use simplified models for the switches (nonlinear resistors) for example in the first stage of circuit design and generally when, during the switching, the exact values of the currents and voltages across the switches are not required. In these cases the simulation users do not need a very accurate simulation, but they prefer a very quick one.
When simplified models are used for the switches, at the switching instants the switch conductances change instantaneously their values of some orders of magnitude. To compute each time point solution of the transient analysis, the SPICE-like simulators adopt time point solution obtained by the Newton-Raphson method (NR) and the solution at the previous time step. The sudden changes of switch conductances produce numerical difficulties to join, using NR, two temporal instants since the problem becomes stiff. When a stiff problem occurs NR fails to converge and it starts

P.M.A. Sloot et al. (Eds.): ICCS 2002, LNCS 2331, pp. 439–448, 2002.

again adopting a reduced time step to overcome the ill conditioned starting point. Hence very small time steps are used. For this reason, the iterations required by the simulation increase considerably and often even if a very little time step is adopted the simulation is stopped since the maximum number of iterations is exceeded.
To overcome this problem and to obtain fast simulations a big effort has been devoted and a new generation of Switch Mode Circuits analysis Methods (SMCM) (see [2], [3], [4], [5]) have been introduced.
To combine the advantages of simplified time analysis of SMCM and the efficiency of traditional simulators, in this paper we propose an hybrid approach to time domain analysis of switching circuits. The proposed method is based on piecewise temporal transient analysis windows joined by DC analysis at the switching instants. The switching instants are detected using a time-step analysis method while the DC analysis is carried out using fixed point homotopy. A great effort has been recently devoted to improve the performances of traditional simulators in the DC Analysis by means of homotopy methods (see [6], [7]) and global convergence results for Modified Nodal Equation (MNE) have been presented. The homotopy methods (also know as continuation methods) [10] are well know methods to solve convergence problems in NR equations when an appropriate starting point is unavailable.
These methods are based on a set of auxiliary equations to compute, with high efficiency, the solutions of nonlinear resistive circuits. In this paper it will be shown as the homotopy methods are useful also in transient analysis of switching circuits.

2 Time Domain Analysis of Electrical Circuits

In the time domain transient analysis of electrical circuits, the modern simulators use the Modified Nodal method to build a set of Non-linear Differential Algebraic Equations (NDAEs). In the following we assume that the circuit consists of b passive elements, M of which produce the auxiliary equations (current controlled elements and independent voltages generators) and $N+1$ nodes. Moreover we assume that the circuit consists only of independent sources and voltage controlled current sources and, for the sake of simplicity, there are no loops consisting only of independent voltage sources and no cut-set consisting only of independent current sources. Using these elements we obtain a set of NDAEs of the form:

$$\mathbf{f}(\dot{\mathbf{x}}(t), \mathbf{x}(t), t) = 0 \tag{1}$$

where

$$\mathbf{x} = \begin{bmatrix} \mathbf{v} \\ \mathbf{i} \end{bmatrix} \in \Re^{N+M} \tag{2}$$

is the vector of the solutions, and $\dot{\mathbf{x}}$ is its derivative respect to the time. In (2) $\mathbf{v} \in \Re^N$ denotes the node voltages to the datum node and $\mathbf{i} \in \Re^M$ denotes the branch currents of voltages sources (independent and dependent) and inductors. Moreover, we assume that voltage and current variables are adopted, even if charge and flux are usually

chosen in the numerical simulation for the numerical stability property (see [8], [9], [10]).
The equation (1) is solved using an integration method of variable step and variable order: Backward Euler, Trapezioidal or Gear formulae (see [9], [10], [11]). In the solution process, at each step of integration, after the discretization process, we obtain a set of Nonlinear Algebraic Equations (NAE) of the form:

$$\mathbf{f}(\mathbf{x}) = \mathbf{H}_1\mathbf{g}(\mathbf{H}_1^T\mathbf{x}) + \mathbf{H}_2\mathbf{x} + \mathbf{\acute{o}} \tag{3}$$

where $\mathbf{g}: \Re^K \rightarrow \Re^K$ is a continuos function representing the relation between the branch voltages $\mathbf{v}_b \in \Re^K$ and the branches currents $\mathbf{i}_b \in \Re^K$, excluding source branches, expressed as:

$$\mathbf{g}(\mathbf{v}_b, \mathbf{i}_b) = 0\,, \tag{4}$$

$\mathbf{H}_1$ is an $n \times K$ constant matrix represented as:

$$\mathbf{H}_1 = \begin{bmatrix} \mathbf{D}_g \\ 0 \end{bmatrix} \tag{5}$$

and $\mathbf{H}_2$ is an $n \times n$ matrix represented as:

$$\mathbf{H}_2 = \begin{bmatrix} 0 & \mathbf{D}_E \\ \mathbf{D}_E^T & 0 \end{bmatrix} \tag{6}$$

where $\mathbf{D}_g$ is an $N \times K$ incidence matrix for the $\mathbf{g}$ branches and $\mathbf{D}_E$ is an $N \times M$ reduced incidence matrix for the independent voltage sources branches. Moreover $\mathbf{\acute{o}} \in \Re^n$ is the source vector that is represented as:

$$\mathbf{\acute{o}} = \begin{bmatrix} \mathbf{J} \\ -\mathbf{E} \end{bmatrix} \tag{7}$$

where $\mathbf{J} \in \Re^N$ is the current vector of the independent current sources and $\mathbf{E} \in \Re^M$ is the voltage vector of the independent voltage sources. From (2), (5), (6) and (7), equation (3) can be written as:

$$\mathbf{f}_g(\mathbf{x}) = \mathbf{D}_g\mathbf{g}(\mathbf{D}_g^T\mathbf{v}) + \mathbf{D}_E\mathbf{i} + \mathbf{J} = 0 \tag{8a}$$

$$\mathbf{f}_E(\mathbf{v}) = \mathbf{D}_E^T\mathbf{v} - \mathbf{E} = 0\,. \tag{8b}$$

Equations (8) are usually solved adopting Newton-Raphson Method (NR), assuming as starting point the value of $\mathbf{x}$ at the previous integration step. When NR iterations fail to converge, a reduced time step is adopted to obtain new coefficients for (8a) and (8b) and a new series of NR iterations is started.

All the modern simulators adopt switch models based on resistive elements. The constrain equation for the switch j is normally written as:

$$g_{sj} = \begin{cases} G_{ON} & \text{if } v_j \geq V_{ON} \\ G_{OFF} & \text{if } v_j < V_{OFF} \end{cases} \tag{9}$$

where g_{sj} is the switch conductance, G_{ON} and G_{OFF} are respectively the conductance of the switch in the closed and open conditions respectively. The two values have usually different orders of magnitude and cannot be singular (0 or ∞). V_{ON} and V_{OFF} are threshold voltages for OFF to ON and vice-versa. This element is normally inserted into the matrix g of (8a). In the transient analysis this element produces a sharp discontinuity in the associated variables (current into and voltage across the switch) and, as consequence, numerical difficulties in NR. In fact the solution at previous time step cannot be used to produce a new temporal solution, even if the time step is considerably reduced. This behavior depends on the stiffness of the equations.
To overcome the above problem the new simulators (e.g. Pspice-Orcad Rel. 9) use a more complex switch model in which the conductance is a more smooth differentiable function, but this precaution is often not sufficient. Hence, when strong discontinuities are present in the circuit variables, the variable stepsize methods produce a sequence of consecutive step size reductions which produce a global decrement in the time step required to simulate the whole circuit, therefore, a number of iterations increasing with the number of discontinuities, i.e., with the number of switching.

3 The Piecewise Temporal Window Approach

In the proposed approach, given a circuit to analyze using transient analysis, a standard simulation is started using a standard routine (the simulation core of a 3F5 version of SPICE [11]). The standard simulation is carried out until the solution at time $t_2 = t_1 + \overline{\Delta t}$ requires 4 consecutive reductions of the time step (a time step control algorithm has been implemented); $\overline{\Delta t}$ is the last tried time step and t_1 the time of last solution found. If the above condition is met, the time step reduction routine is suspended, the solution at t_1 is saved and a check on the values of variables at the time $t_1 + \overline{\Delta t}$ is performed. We check if any switches result in inconsistent condition. A switch is in an inconsistent condition if the associated variables meet one of following conditions:

$$\begin{aligned} v_j \geq V_{ON} \quad &\text{and} \quad \left|V_{swj}\right| \geq \varepsilon \\ v_j \leq V_{OFF} \quad &\text{and} \quad \left|I_{swj}\right| \geq \varepsilon \end{aligned} \tag{10}$$

where ε is a small positive quantity and V_{swj}, I_{swj} are respectively the voltage across and the current trough the j-th switch.
If an inconsistent condition is found on some switches, we remove this inconsistency imposing to zero the voltage across the switches ON and the current into switches OFF. A new solution is then obtained, using the DC analysis with Fixed Point Homotopy Method (FPH) in a resistive circuit obtained by the original circuit substituting the capacitors C_i and inductors L_i by independent voltages and current generators whose values are the voltages across the capacitors, $V_{C_i}(t_1)$, and the current in the inductors, $I_{L_i}(t_1)$, at time t_1 and where all circuit parameters have the values of solution at time t_1, since the state of dynamic elements is unchanged between t_1 and $t_1 + \overline{\Delta t}$.
The fixed point homotopy method is based on an auxiliary equation of the form:

$$\mathbf{h}(\mathbf{x},\alpha) = \alpha\mathbf{f}(\mathbf{x}) + (1-\alpha)\mathbf{A}(\mathbf{x}-\mathbf{x}^0) \quad \mathbf{(11)}$$

where $\alpha \in [0,1]$ is the homotopy parameter and $\mathbf{A}$ is a non singular $n \times n$ matrix. The solution curve of the homotopy equation:

$$\mathbf{h}(\mathbf{x},\alpha) = \mathbf{0} \quad \mathbf{(12)}$$

is traced from the known solution $(\mathbf{x}^0,0)$ at $\alpha = 0$ to the final solution $(\mathbf{x}^*,1)$ at the $\alpha = 1$ hyperplane. In this case as starting point we consider the solution at time t_1, $\mathbf{x}^{t_1}$, if the FPH converges, the solution $\mathbf{x}^*$ is considered as bias point for the transient analysis at the time $t_2 = t_1 + \overline{\Delta t}$.
We assume that the resistive network associated to the circuit in the time t_1 is characterized by Lipschitz continuous functions and satisfies the following condition of uniform passivity:

Definition 1: A continuous function $\mathbf{g}: \Re^K \to \Re^K$ is said to be Uniformly Passive on $\mathbf{v}_b^0$ if there exist a $\gamma > 0$ such that $\left(\mathbf{v}_b - \mathbf{v}_b^0\right)^T \left(\mathbf{g}(\mathbf{v}_b) - \mathbf{g}(\mathbf{v}_b^0)\right) \geq \gamma \left\|\mathbf{v}_b - \mathbf{v}_b^0\right\|^2$ for all $\mathbf{v}_b \in \Re^K$.

These assumptions usually hold for a fairly class of resistive circuits containing MOS, BJT's, diodes and positive resistors [12].
With the previous hypotheses the convergence conditions are the following:

1. The initial point $\mathbf{x}^0$ is the unique solution of (3).
2. The solution curve starting from $(\mathbf{x}^0,0)$ does not return to the same point $(\mathbf{x}^0,0)$ for $\alpha \neq 0$.
3. There exists an $\varepsilon > 0$ such that for all $\mathbf{x} \in \left\{\mathbf{x} \in \Re^n \,\middle|\, \|\mathbf{x}\| \geq \varepsilon\right\}$ and $\alpha \in [0,1)$, $\mathbf{h}(\mathbf{x},\alpha) \neq 0$ holds.

For the fixed point homotopy the conditions 1 and 2 always hold. Condition 3 is related to the structure of the circuit equations and to the structure of matrix $\mathbf{A}$. In this case we have chosen the following matrix:

$$\mathbf{A} = \begin{bmatrix} \mathbf{1}_N & \mathbf{0} \\ \mathbf{0} & -\mathbf{1}_M \end{bmatrix} . \qquad \mathbf{(13)}$$

In [6] the global convergence results preserved even for MNE have been shown using (13). Even if the convergence of the proposed method for a class of power electronic circuit is preserved, a problem in the homotopy implementation is the computational effort required to step the homotopy parameter in its definition range. The procedure to vary the homotopy parameter could significantly reduce the effectiveness of the method if compared with the standard integration method.
In the proposed approach a linear method has been utilized to obtain the value of the parameter α_k for the k-th homotopy iteration:

$$\alpha_k = 0.1 \cdot k \quad k \in \Im; \quad k \in [0,10]; \ . \qquad \mathbf{(14)}$$

The simulations have shown as this parameter is not critical and other choices could be satisfactory utilized.
The new trial solution is then used as starting point to compute the solution at time t_2 in the suspended simulation. If the trial solution produce a solution without requiring a time step reduction in the integration routine, then the standard simulation is carried on. Otherwise we suspend the simulation and consider the circuit inconsistent.
In this way the DC analysis joints two switch cycles in succession, while during the switch cycle the transient analysis is not modified.
The proposed method avoids the iterations used by a standard simulation program to evaluate the switch variables during the switching. Finally, the time length of switch iterations is very short compared to switch cycle, therefore, the elimination of switching iterations does not yield any effect on the quality of final results.

4 Simulation and Comparison

To compare the aforesaid procedure with the standard simulators using the MNE, the method has been implemented modifying a kernel of the SPICE 3F5 source code [13], a new routine of time step reduction strategy has been added.
As example let us consider the transient analysis simulation of 2 ms on a standard buck converter with a power control element modeled by a switch internally driven. Figure 1 reports the values of the switch voltage V_{SW} (Fig. 1a) and switch current I_{SW} (Fig. 1b) as function of time, obtained by commercial simulation program, using the standard trapezoidal rule. The waveform are drawn between 0.945 ms and 0.963 ms of simulation.
We observe that the switch variables have sharp discontinuities. In particular when V_{SW} goes over 30 V there is the ON to OFF transition of the switch, while when V_{SW} goes above 30 V there is the OFF to ON transition. At the second transition we note a current pulse of I_{SW} (Fig. 1b). The switch current I_{SW} has been plotted in the range [0, 0.8] A.

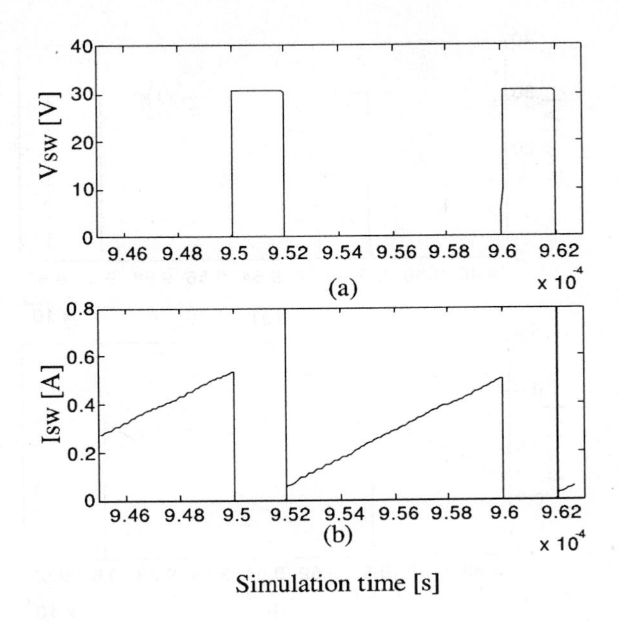

Fig. 1. Voltage (V_{SW} Fig. 1a) and current (I_{SW} Fig. 1b) into switch in a period of simulation for the standard method.

The same circuit has been simulated adopting the proposed method. The simulation parameters and the time window are the same of the first case. Figure 2 reports again the values of I_{SW} and V_{SW} obtained. Comparing Figures 1 and 2 we note that the obtained waveforms are almost identical. The information lost is only about critical parameters of switches (I_{SW} pulse), but, during the commutation, the extreme values of current or voltage pulse across the switches depend essentially by the switch model, therefore, are not realistic and they are not very useful in our analysis. The variable I_{SW} is the most different in two methods. Therefore the main information about all circuit variables have been preserved.
Figure 3 shows the values of the switch current I_{SW}, obtained by standard simulation program, with respect to iterates numerated from zero in previous time interval, while Figure 4 reports again I_{SW} with respect to iterates, obtained by proposed method.
If Figures 3 and 4 are compared it is easy to note the different number of iterates in switch cycle. In Figure 4 the calculation of a whole switch cycle need about 40% iterates less than the Figure 3. Moreover in these iterations the time-steps are very small.
In this simulation we have calculated the time steps distribution during the whole simulation time (2 ms). In the standard case a consistent number of iterations, about 65% is spent in very low time steps ($T_s < 0.5e\text{-}9$).

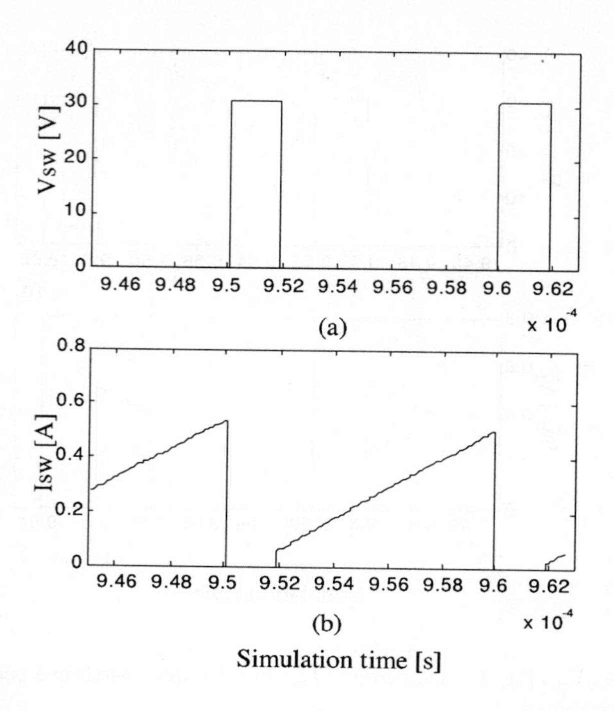

Fig. 2. Voltage (V_{SW} fig. 2a) and current (I_{SW} fig. 2b) into switch in a period of simulation for the proposed method.

The most of these iterations are produced by switching instants, during the switching regions a lot of simulation time is spent to handle the discontinuities introduced by the adopted switch model. The proposed approach avoids the calculation of switching regions and allows the time step to preserve an high value in the whole simulation with a notable reduction of simulation time, in the proposed approach the time steps with T_s < 0.5e-9 are about 30%. The simulation time reduction clearly depends on the circuit analyzed, the computer used, the comparing method. In this example we obtained a reduction about 25% of simulation time.

5 Conclusions

In transient analysis of switching circuits a lot of simulation time is spent to analyze the commutations of the switches, when the switches are modeled by means of nonlinear resistors. In fact, the time steps of analysis program during the switching are very small compared to the time interval of analysis. In this paper has been proposed a method based on piecewise temporal transient analysis windows joined by DC analysis carry out by means of FPH at the switching instants. The proposed approach allows the elimination of the calculation of switching iterations.

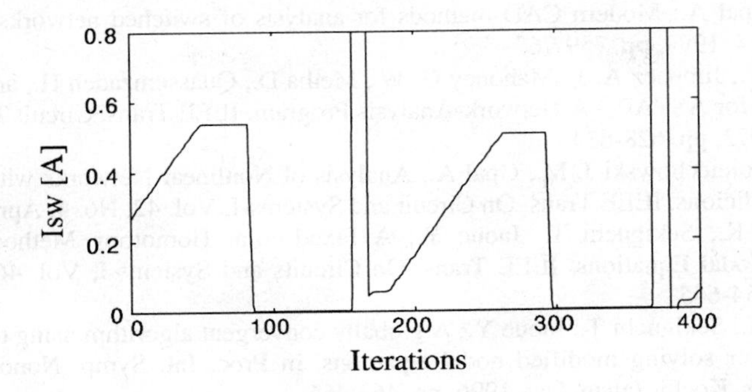

Fig. 3. Current I_{sw} into switch in a period of simulation with respect to iterate, for the standard method.

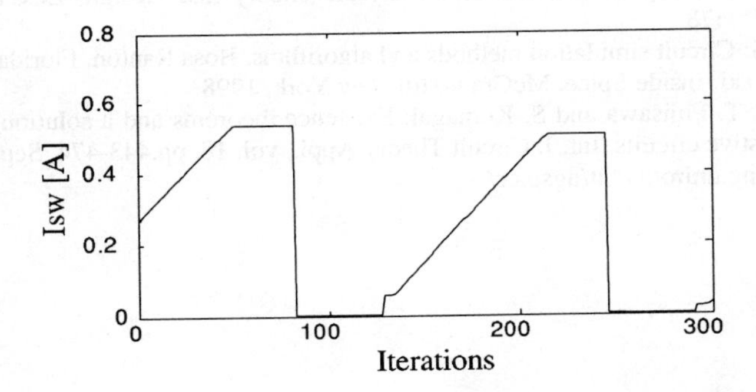

Fig. 4. Current I_{sw} into switch in a period of simulation with respect to iterate, for the proposed method.

Comparing the results obtained by proposed method and the results obtained by a standard simulation program the tests give the reduction of the number of iterations between 30% and 50%, depending on the circuit to analyze. Moreover the proposed method guarantees accurate results as well as standard methods, in fact, we have shown the elimination of switching iterations does not yield any effect on the quality of final results. Finally, the simulation of power circuits is carried out adopting the same accurate models of the PSPICE libraries.

References

1. Nagel L. W., and Pederson D., SPICE: Simulation Program with Integrated Circuits Emphasis. Berkely, Calif.: Univ. Of California, Electronic Research Laboratory, Mwmorandum ERL-M382, Apr. 12, 1973

2. Valsa J., and Vlach J.: SWANN-"A Programme for Analysis of switched Analogue Non-Linear Networks". International Journal of Circuits Theory and Applications, v. 23, 1995, pp. 369-379
3. Vlach J., Opal A.: Modern CAD methods for analysis of switched networks. IEEE Trans. CAS, vol. 44, 1997, pp. 759-762
4. Weeks W. T., Jimenez A. J., Mahoney G. W., Metha D., Quassemzadeh H., and Scott T. R.: Algorithms for ASTAP – A Network-Analysis Program. IEEE Trans. Circuit Theory, v. CT-20, Nov. 1973, pp. 628-634
5. Vlach J., Woiciechowski J. M., Opal A.: Analysis of Nonlinear Networks with Inconsistent Inizial Condicions. IEEE Trans. On Circuit and Systems-I, Vol. 42, No. 4, April 1995
6. Yamamura K., Sekiguchi T., Inoue Y.: A Fixed-Point Homotopy Method for Solving Modified Nodal Equations. IEEE Trans. On Circuits and Systems-I, Vol. 46, No. 6, June 1999, pp. 654-665
7 Yamamura K., Sekiguchi T., Inoue Y.: A globally convergent algorithm using the fixed-point homotopy for solving modified nodal equations. in Proc. Int. Symp. Nonolinear Theory Applications, Kochi, Japan, Oct. 1996, pp. 463-466
8. J. Ogrodzki: Transient Analysis of circuits with ideal switches using charge and flux substitute networks. Proc. Intl. Conf. Electr. Circ. Syst., Lisboa, Portugal, 1998
9. J. Ogrodzki, M. Skowron: The charge-flux method in simulation of first kind switched circuits. Proc. European Conference on Circuit Theory and Design, ECCTD'99, Stresa, Italy, pp. 475-478
10. J. Ogrodzki: Circuit simulation methods and algorithms. Bosa Ranton, Florida:CRC, 1994
11. R. Kielkowski. Inside Spice. McGraw-Hill New York, 1998
12. T. Ohtsuki, T. Fujisawa and S. Kumagai: Existence theorems and a solution algorithm for solving resistive circuits. Int. J. Circuit Theory Appl., vol. 18, pp.443-474, Sept. 1990
13. http://ieee.ing.uniroma1.it/ngspice/

A Fortran90 Routine for the Solution of Orthogonal Differential Problems

F. Diele[1], T. Politi[2], and I. Sgura[3]

[1] Istituto per Ricerche di Matematica Applicata-C.N.R., Via Amendola 122/I, I-70126 Bari (Italy). irmafd03@area.ba.cnr.it

[2] Dipartimento Interuniversitario di Matematica, Politecnico di Bari, Via Orabona 4, I-70125 Bari (Italy). pptt@pascal.dm.uniba.it

[3] Dipartimento di Matematica "E. De Giorgi", Università di Lecce, Via Prov. Lecce-Arnesano, 73100 Lecce (Italy). sgura@unile.it

Abstract. In this paper we describe a Fortran90 routine for the numerical integration of orthogonal differential systems based on the Cayley transform methods. Three different implementations of the methods are given: with restart, with restart at each step and in composed form. Numerical tests will show the performances of the solver for the solution of orthogonal test problems, of orthogonal rectangular problems and for the calculation of Lyapunov exponents in the linear and nonlinear cases, and finally for solving inverse eigenvalue problem for Toeplitz matrices. The results obtained using Cayley methods are compared with those given by Fortran90 version of Munthe-Kaas methods, which have been coded in a similar way.

1 Introduction

Recently there has been a growing interest in the numerical solution of matrix differential systems whose theoretical solutions preserve certain qualitative features of the initial condition, in particular the orthogonality (see [1], [4], [6], [9]). Let $\mathcal{O}_n(\mathbb{R}) = \{Y \in \mathbb{R}^{n\times n} | \, Y^T Y = I\}$ be the manifold of orthogonal matrices and $\mathbb{H} = \{A \in \mathbb{R}^{n\times n} | \, A^T + A = 0\}$ be the set of real skew-symmetric matrices. If the following matrix differential system is to be solved

$$Y'(t) = F(t, Y(t))Y(t), \qquad Y(0) = Y_0 \in \mathcal{O}_n(\mathbb{R}), \qquad t \in [t_0, t_f], \tag{1}$$

it is well-known that, if $F(t,Y) \in \mathbb{H}$ for $(t,Y) \in \mathbb{R}^+ \times \mathcal{O}_n(\mathbb{R})$, the theoretical solution of (1) lies in $\mathcal{O}_n(\mathbb{R})$ for all $t > 0$. Among the software tools developed to solve these problems we have to remember the Fortran77 code developed by Dieci and Van Vleck (see [5]) and the DiffMan toolbox developed by K. Engø, A. Marthinsen and H. Munthe-Kaas (see [8]), based on the Lie group methods (see [12]), that solves this kind of differential systems in MatLab environment. In this paper we describe a Fortran90 experimental software that gives the numerical solution of (1) solving the associated skew-symmetric ODE system obtained by the Cayley transform of $Y(t)$. For this aim it is rewritten as a vector system

P.M.A. Sloot et al. (Eds.): ICCS 2002, LNCS 2331, pp. 449–458, 2002.

and then solved using the DOPRI5 code (see [10]). Finally, a subroutine for the Cayley transform allows the computation of the orthogonal solution. The paper is organized as follows. In Section 2 we present Lie group orthogonal integrators such as Munthe-Kaas methods (see [13]) and the Cayley methods in three different versions: methods with restart, with restart at each step, and composed methods. Moreover the equivalence between the matrix and vector formulation of the skew-symmetric problems associated to (1) is also provided. In Section 3 we describe the general structure of the Fortran90 codes that implement the Cayley and Munthe Kaas methods. In particular, we explain how we have exploited and suitably modified the stepsize control technique used by DOPRI5 to compute, throughout the evolution, the values t_i where the restart is needed. In Section 4 the behaviour of all the methods applied to some test problems is shown.

2 Numerical Integration of Orthogonal Differential Systems

2.1 Lie group methods

These methods solve a problem more general than (1), in particular when $Y(t)$ evolves in a Lie group and $F(t, Y)$ belongs to the Lie algebra associated to the group. The manifold of orthogonal matrices is a particular Lie group and the set of skew-symmetric matrices is its Lie algebra. Many recent papers have been devoted to the formulation of numerical methods preserving the Lie group structure of the theoretical solution. The Munthe-Kaas methods exploit exponential map, defined by $\mathcal{R}_{\mathcal{M}}(z) = e^z$. This is a function mapping the Lie algebra into the Lie group. Given a partition of the integration interval, $t_0 < t_1 < t_2 < \ldots < t_N = t_f$ the solution $Y(t)$ of (1) may be given in the following form (see [12]):

$$Y(t) = \mathcal{R}_{\mathcal{M}}(\sigma(t))Y(t_i), \qquad t \in [t_i, t_{i+1}],\ i = 1, \ldots, N-1 \tag{2}$$

in some neighbourhood of $Y(t_i)$, where $\sigma(t)$ is the skew-symmetric solution of the initial value problem

$$\sigma'(t) = \sum_{m=0}^{\infty} f_m ad^m(F(t, Y(t)), \sigma) \qquad t \geq t_i, \qquad \sigma(t_i) = O. \tag{3}$$

The coefficients $\{f_k\}_{k=0}^{\infty}$ in (3) are

$$f_0 = 1, \qquad f_1 = B_1 + 1, \qquad f_m = \frac{B_m}{m!}, \qquad m \geq 2$$

with $\{B_m\}_0^{\infty}$ being the Bernoulli numbers. For a pair of matrices P and Q, the iterated commutator $\mathrm{ad}^m(\cdot, \cdot)$ is defined as

$$\mathrm{ad}^m(P, Q) = \begin{cases} P, & \text{if } m = 0 \\ [\mathrm{ad}^{m-1}(P, Q), Q], & \text{if } m \geq 1, \end{cases}$$

where $[P, Q] = PQ - QP$ is the commutator or Lie bracket product. Substituting (2) into (3) the following matrix differential equation is obtained:

$$\sigma'(t) = \sum_{m=0}^{\infty} f_m ad^m(F(t, \mathcal{R}_{\mathcal{M}}(\sigma(t))Y(t_i)), \sigma), \qquad t \in [t_i, t_{i+1}], \qquad \sigma(t_i) = O. \tag{4}$$

These methods find the approximation $\sigma_{i+1} \approx \sigma(t_{i+1})$ solving a $r+1$ term truncation of (4) , with $r = p - 2$, by means of a p order explicit Runge-Kutta method. Then the numerical approximation of $Y(t_{i+1})$, for $i = 0, 1, \ldots, N-1$, is obtained by (2).

2.2 Cayley methods

This class of methods is based on the Cayley map, the complex function defined by $\mathcal{R}_{\mathcal{C}}(z) = (1-z)^{-1}(1+z)$, for $z \in \mathbb{C} - \{1\}$. To guarantee that it can be applied, it is possible to find a partition $t_0 < t_1 < t_2 < \ldots < t_N = t_f$ of the integration interval, such that the solution $Y(t)$ of (1) may be given in the following form (see [6], [11]):

$$Y(t) = \mathcal{R}_{\mathcal{C}}(A(t))Y(t_i), \qquad t \in [t_i, t_{i+1}], \tag{5}$$

where $A(t)$ is the skew-symmetric solution of the differential system

$$A'(t) = \frac{1}{2}(I - A)F(t, Y(t))(I + A) \quad t \geq t_i, \quad A(t_i) = O. \tag{6}$$

If (5) is substituted in (6) the following differential matrix equation is obtained:

$$A'(t) = \frac{1}{2}(I - A)F(t, \mathcal{R}_{\mathcal{C}}(A(t))Y(t_i))(I + A) \quad t \geq t_i, \quad A(t_i) = O. \tag{7}$$

The skew-symmetric differential system (7) can be solved using classical numerical methods such as Runge-Kutta or linear multistep methods. In fact in [6] it has been proved that these methods preserve the skew-symmetry of the theoretical solution. The method proposed in [11] finds the approximation $A_{i+1} \approx A(t_{i+1})$ solving (7) thorough a p order explicit Runge-Kutta method. Hence the numerical approximation computed by these methods, $Y_{i+1} \approx Y(t_{i+1})$, for $i = 0, 1, \ldots, N-1$, is given by

$$Y_{i+1} = \mathcal{R}_{\mathcal{C}}(A_{i+1})Y_i$$

Accordingly to the choice of the interval partitioning, different methods can be deduced. If the subintervals $I_i = [t_i, t_{i+1}]$ are determined a priori by choosing a constant stepsize h, that is $t_{i+1} = t_i + h$, one-step Cayley methods named *Cayley methods with restart at each step* are obtained (see [11]). If a control technique on the boundedness of the solution $A(t)$, as studied in (7), is applied during the numerical evolution, the values $t_{i+1} = t_i + h_i$ are obtained and we have the *Cayley methods with restart.* The two previous approaches can be suitably combined considering a first coarse mesh of subintervals I_i and applying on each of them a Cayley method with restart. This technique is defined as *composed Cayley methods.*

2.3 Vector formulation of the problem

In order to apply standard and robust Fortran codes to solve the orthogonal problem (1), we have considered its vectorial formulation. If we rewrite the ODE problem (1) as a vector system, it results:

$$\dot{\boldsymbol{y}} = (I \otimes F(t, Y))\boldsymbol{y} \tag{8}$$

where $\boldsymbol{y} = vec(Y) = ({Y_1}^T, \ldots, {Y_n}^T)^T \in \mathbb{R}^{n^2}$, Y_i is the $i-$th column of matrix Y and $\otimes$ denotes the Kronecker product. The solution of the vector differential system starting from $\boldsymbol{y}_0 = vec(Y_0)$ coincides with the arrangement by columns of the matrix solution of (1). We show that the numerical approximation of the ODE in vectorial form (8) by means of a linear method coincides with the arrangement by columns of the matrix numerical approximation of (1). For example, let us consider an s stages explicit Runge-Kutta method applied to (1). Hence the following matrix algebraic equations hold:

$$Y_{k+1} = Y_k + h \sum_{i=1}^{s} b_i F(t_k + c_i h, K_i) K_i$$

$$K_i = Y_k + h \sum_{j=1}^{i-1} a_{ij} F(t_k + c_j h, K_j) K_j \qquad i = 1, \ldots, s.$$

Considering these matrices by columns, this is equivalent to solve

$$(Y_1^{k+1}, \ldots, Y_n^{k+1}) = (Y_1^k, \ldots, Y_n^k) + h \sum_{i=1}^{s} b_i F(t_k + c_i h, K_i)(K_1^i, \ldots, K_n^i)$$

$$(K_1^i, \ldots, K_n^i) = (Y_1^k, \ldots, Y_n^k) + h \sum_{j=1}^{i-1} a_{ij} F(t_k + c_j h, K_j)(K_1^j, \ldots, K_n^j)$$

for $i = 1, \ldots, s$. Then, for $l = 1, \ldots, n$, we have

$$Y_l^{k+1} = Y_l^k + h \sum_{i=1}^{s} b_i F(t_k + c_i h, K_i) K_l^i$$

$$K_l^i = Y_l^k + h \sum_{j=1}^{i-1} a_{ij} F(t_k + c_j h, K_j) K_l^j \qquad i = 1, \ldots, s.$$

Hence, setting $\boldsymbol{y}_k = vec(Y_k)$, and $\boldsymbol{k}_i = vec(K_i)$, the previous relations become

$$\boldsymbol{y}_{k+1} = \boldsymbol{y}_k + h \sum_{i=1}^{s} b_i (I \otimes F(t_k + c_i h, K_i))\boldsymbol{k}_i$$

$$\boldsymbol{k}_i = \boldsymbol{y}_k + h \sum_{j=1}^{i-1} a_{ij} (I \otimes F(t_k + c_j h, K_j))\boldsymbol{k}_j \qquad i = 1, \ldots, s.$$

It easy to see that these formulae correspond to apply the same Runge-Kutta method to the problem (8). With similar arguments it is possible to show that the application of an explicit Runge-Kutta method to the skew-symmetric system (7) obtained in the Cayley approach, is equivalent to apply the same Runge-Kutta method in vector formulation. Rewriting (7) as a vector systems, it is:

$$\dot{\boldsymbol{a}} = \frac{1}{2}(I \otimes ((I - A)F(t, \mathcal{R}(A)Y(t_i)))(\boldsymbol{i} + \boldsymbol{a}) \tag{9}$$

where $\boldsymbol{a} = vec(A) = ({A_1}^T, \ldots, {A_n}^T)^T \in \mathbb{R}^{n^2}$, and each A_l is the l–th column of A. Similarly $\boldsymbol{i} = vec(I) = (\boldsymbol{e}_1^T, \ldots, \boldsymbol{e}_n^T)^T \in \mathbb{R}^{n^2}$ where $\boldsymbol{e}_l$ is the l–th column of identity matrix. The solution of (9) starting from null initial condition gives the columns of the matrix solution of (7). In fact, applying an s stages explicit Runge Kutta method to (7) the following matrix algebraic equations hold:

$$A_{k+1} = A_k + \frac{h}{2}\sum_{m=1}^{s} b_m(I - K_m)F(t_k + c_m h, \mathcal{R}_C(K_m)Y(t_i))(I + K_m)$$
$$K_m = A_k + \frac{h}{2}\sum_{j=1}^{m-1} a_{m,j}(I - K_j)F(t_k + c_j h, \mathcal{R}_C(K_j)Y(t_i))(I + K_j)$$

for $m = 1, \ldots, s$. Again this is equivalent to solve, for $l = 1, \ldots, n$,

$$A_l^{k+1} = A_l^k + \frac{h}{2}\sum_{m=1}^{s} b_m(I - K_m)F(t_k + c_m h, \mathcal{R}_C(K_m)Y(t_i))(e_l + K_l^m)$$
$$K_l^m = A_l^k + \frac{h}{2}\sum_{j=1}^{m-1} a_{mj}(I - K_j)F(t_k + c_j h, \mathcal{R}_C(K_j)Y(t_i))(e_l + K_l^j)$$

for $m = 1, \ldots, s$. Hence putting $\boldsymbol{a}_k = vec(A_k)$ and $\boldsymbol{k}_m = vec(K_m)$ the previous relations become

$$\boldsymbol{a}_{k+1} = \boldsymbol{a}_k + \frac{h}{2}\sum_{m=1}^{s} b_m(I \otimes ((I - K_m)F(t_k + c_m h, \mathcal{R}_C(K_m)Y(t_i))(\boldsymbol{i} + \boldsymbol{k}_m)$$
$$\boldsymbol{k}_m = \boldsymbol{a}_k + \frac{h}{2}\sum_{j=1}^{m-1} a_{mj}(I \otimes ((I - K_j)F(t_k + c_j h, \mathcal{R}_C(K_j)Y(t_i))(\boldsymbol{i} + \boldsymbol{k}_j)$$

for $m = 1, \ldots, s$. These formulae correspond to apply the same Runge-Kutta method to (9). Moreover, it is possible to show that the same result holds for the vector form of (3).

3 Fortran90 Codes

Since results of previous section hold, we solve vector formulations (8) and (9) using a well known Fortran ODE routine. Hence, the philosophy underlying the code reflects the general structure of a classical Fortran ODE solver. The driver calls the subroutines defining the problem and its initial conditions, then the subroutine of the numerical method and finally the subroutine for the error estimate. Hence the subroutine defining the problem must provide only the matrix flow $F(t, Y(t)) \in \mathbb{R}^{n\times n}$, the integration interval $[t_0, t_f]$ and initial value $Y_0 \in \mathbb{R}^{n\times p}$ with $n \geq p$. In the section related to rectangular problems we recall some theoretical results of [11] and we illustrate how Cayley methods have been implemented for this class of problems. To calculate $\mathcal{R}_C(A) = (I - A)^{-1}(I + A)$ in (9) a subroutine named **cayt.f** has been written to furnish the solution of the linear systems $(I - A)Z = (I + A)$. The solutions of these systems are obtained calling once for step the routine **dec.f** that performs the LU factorization of the coefficient matrix and then n times for step the routine **sol.f** that solves the triangular systems. As explained in the previous section, we implement three different methods:

1. *Cayley methods with restart*: the Fortran77 code DOPRI5 based on the order five explicit Runge-Kutta formulae of Dormand and Prince with stepsize control has been used to solve, for all i, the problem (9). The values t_{i+1} of the interval partitioning are not given a priori, but they are dynamically determined by exploiting and suitably modifying the stepsize control technique of DOPRI5.
2. *Cayley methods with restart at each step*: one step of the explicit order five Runge-Kutta method used in DOPRI5, has been implemented to solve the problem (9) at $t_{i+1} = t_i + h$, $i = 1, \ldots, I$. The partitioning is determined by the costant stepsize h required as input value.
3. *composed Cayley method*: DOPRI5 code with variable stepsize, has been used to solve, for all i, the problem (9) on each subinterval $I_i = [t_i, t_{i+1}]$ of the partitioning determined by the constant stepsize h required as input value.

The Munthe-Kaas methods have been implemented in a similar way. To calculate the exponential matrix a subroutine, named **expt.f**, has been coded using the Padé approximation with the scaling and squaring technique.
In [6] it has been proved that if $||A(t)||$ becomes very large for some $t \in [t_i, t_f]$, then $\mathcal{R}_C(A(t))$ cannot be calculated. Hence a technique to control the behaviour of the solution $A(t)$ of (7) allows to find the values $t_{i+1} \in [t_i, t_f]$ and the approximation $A_{i+1} \simeq A(t_{i+1})$ where the Cayley map can be evaluated. For this aim we have used and modified the stepsize variation technique used by DOPRI5 to control the behaviour of the solution $A(t)$. We recall that when the control on stiffness and the control on stepsize are satisfied then the execution in DOPRI5 is stopped at the step t and the stepsize h that obey to:

$$0.1 \cdot |h| \leq |t| \cdot uround$$

where $uround = 2.3e - 16$. Hence the numerical approximation of $A(t)$, with $t \in [t_i, t_f]$, is stopped at the step t_{i+1} satisfing

$$0.1 \cdot |h| \leq |t_{i+1}| \cdot Tol_{\text{rest}} \tag{10}$$

where Tol_{rest} is a constant value. If this criterium is satisfied, we restart the numerical approximation of $A(t)$ with $t \in [t_{i+1}, t_f]$, where t_{i+1} verifies (10) as equality. The value assigned to Tol_{rest} permits to choose how far from an eventually critical step we want to stop the execution.

4 Numerical tests

We show the performance of the Fortran90 routines for Cayley and Munthe-Kaas methods for the problems in the following classes. For all experiments we define the following quantities:
$\mathbf{N_a}$: the number of the accepted steps for variable stepsize methods.
$\mathbf{N_r}$: the number of rejected steps for variable stepsize methods.
$\mathbf{N_f}$: the number of function evaluations.
$\mathbf{N_{rest}}$: the number of restarts performed with variable stepsize methods.
$\mathbf{N_i}$: the number of subintervals for composed methods.
$\mathbf{T_{CPU}}$: the CPU time for the calculation process.
$\mathbf{E_g}$: the global error, that is the maximum norm between the approximate and the exact solution. When the exact solution is not available, the approximate solution is compared with the solution provided by the same method applied with $h = 10^{-6}$.
$\mathbf{E_u}$: unitary error, that is $\|I - Y^TY\|_F$ where $\|\cdot\|_F$ is the Frobenius norm.
Tol_{rel} and Tol_{abs} are the tolerances required by DOPRI5 for relative and absolute error estimates respectively, while h is the step for constant stepsize methods or its initial value for the variable stepsize methods. In all tests, for the restart procedure we set $Tol_{\mathbf{rest}} = 10^{-6}$.

Example 1 (Rectangular problems). In [11] has been proved that, if the ODE matrix differential problem (1) is rectangular and F is a square skew-symmetric function for every $Y \in \mathcal{O}(n,p) = \{Y \in \mathbb{R}^{n\times p} | Y^TY = I_p\}$, $p < n$, then its solution $Y(t) \in \mathcal{O}(n,p)$, for $t > 0$. Similarly, if F is only weakly skew-symmetric, i.e.

$$Y^T[F^T(Y) + F(Y)]Y = O \qquad \forall Y \in \mathcal{O}(n,p) \tag{11}$$

then its solution $Y(t) \in \mathcal{O}(n,p)$ for $t > 0$. In this case we apply the Cayley method to skew-symmetric matrix A_{i+1} obtained solving the matrix differential equation

$$A'(t) = \frac{1}{2}(I - A)G(t, \mathcal{R}_C(A(t))Y(t_i))(I + A) \qquad t \geq t_i, \qquad A(t_i) = O$$

where $G(V) = -YY^TF^T(Y) + F(Y)YY^T$.
As example of rectangular problem we consider the orthonormal Procustes problem described in [3]. It leads to the following differential system

$$Y'(t) = \frac{Y}{2}(Y^TA^TB - B^TAY) - (I - YY^T)A^T(AY - B)$$

with $B = AY^* \in \mathbb{R}^{5\times 3}$, and

$$A = \begin{pmatrix} 0.2190 & 0.0470 & 0.6789 & 0.6793 & 0.9347 \\ 0.3835 & 0.5194 & 0.8310 & 0.0346 & 0.0535 \\ 0.5297 & 0.6711 & 0.0077 & 0.3834 & 0.0668 \\ 0.4175 & 0.6868 & 0.5890 & 0.9304 & 0.8462 \end{pmatrix}^T_{5\times 4}, \qquad Y^* = \begin{pmatrix} 0 & 1 & 0 \\ 1 & 0 & 0 \\ 0 & 0 & 1 \\ 0 & 0 & 0 \end{pmatrix}.$$

In this case the function $F(Y)$ satisfies the weak skew-symmetric condition (11). As initial condition we put $Y(0) = \begin{pmatrix} I_3 \\ 0 \end{pmatrix}$. The problem has exactly a global solution:

$$Y^{**} = \begin{pmatrix} 0.0094 & 0.3811 & -0.5768 \\ 0.9999 & 0.0094 & 0.0088 \\ 0.0088 & -0.5768 & 0.4625 \\ -0.0110 & 0.7225 & 0.6733 \end{pmatrix}^T .$$

The output values are examined at time interval $t_f = 100$ where the approximation of the global solution. Performances are shown in Table 1 for $t_f = 100$.

Example 2 (Lyapunov exponents). We have considered the continuous QR algorithm to compute the Lyapunov exponents of a linear dynamical system $y' = A(t)y$ or nonlinear dynamical system $y' = f(t, y)$ (see [4]). For this aim we have to solve (1) with

$$F_{i,j}(t, Y) = \begin{cases} -(YAY^T)_{i,j} & i > j \\ 0 & i = j \\ (YAY^T)_{j,i} & i < j. \end{cases}$$

In the nonlinear case $A(t)$ is the Jacobian of f calculated at a given solution trajectory $y(t)$. The Lyapunov exponents are defined as

$$\lambda_i = \lim_{t\to\infty} \frac{1}{t} \int_0^t \tilde{A}_{ii}(s)ds, \qquad \tilde{A}_{ii}(t) = (Y(t)A(t)Y(t)^T)_{i,i}$$

and calculated truncating the infinite limits to a given value t_f and approximating the integrals by the composite trapezoidal rule. Let us consider the van der Pol equation $u'' - k(1-u^2)u' + u = 0$ with $k > 0$. For this nonlinear problem one Lyapunov exponent is zero and the other is negative. Rewriting the equation as a first order differential system $w' = f(w)$, $w = (u, u')$, Lyapunov exponents are obtained by considering the matrix

$$A(t) = \frac{\delta f(w)}{\delta w} = \begin{pmatrix} 0 & 1 \\ -2kuv - 1 & k(1-u^2) \end{pmatrix}$$

All tests are performed for $k = 1$, initial condition $w_0 = (u(0), u'(0)) = (0, 2.1)$ and $t_f = 100$. All results are shown in Table 2.

Example 3 (Inverse Toeplitz Eigenvalue Problems). Another application of orthogonal flows is the solution of Inverse Toeplitz Eigenvalue Problems (ITEP) where a real symmetric Toeplitz matrix has to be built starting from a given set of eigenvalues. As showed in [2] and in [7], to numerically construct such a matrix the orthogonal flow (1) is considered where $Y_0 = I$ and $F(Y) = K(Y(t)L_0Y(t)^T)$, where $K(X)$ is the Chu's annihilator defined as:

$$k_{i,j} = \begin{cases} x_{i+1,j} - x_{i,j-1} & \text{if } 1 \le i < j \le n \\ 0 & \text{if } 1 \le i = j \le n \\ x_{i-1,j} - x_{i,j+1} & \text{if } 1 \le j < i \le n. \end{cases}$$

The equilibria of this system are also equilibria of the isospectral flow

$$L'(t) = [F(L), L], \qquad L_0 = L_0^T \tag{12}$$

whose solution $L(t)$ is given by $L(t) = Y(t)L_0Y(t)^T$, that numerically tends to a Toeplitz symmetric matrix with the same eigenvalues of L_0. Let us consider the problem in correspondence of $\lambda_1 = -3$, $\lambda_2 = \lambda_3 = 4 - \sqrt{11}$. Let us consider the isospectral flow (12) starting with L_0 equal to a diagonal matrix with eigenvalues λ_i, $i = 1, 2, 3$. The isospectral flow theoretically tends to the Toeplitz symmetric matrix given by

$$L^* = \begin{pmatrix} \frac{5-2\sqrt{11}}{3} & \frac{7-\sqrt{11}}{3} & \frac{\sqrt{11}-7}{3} \\ \frac{7-\sqrt{11}}{3} & \frac{5-2\sqrt{11}}{3} & \frac{7-\sqrt{11}}{3} \\ \frac{\sqrt{11}-7}{3} & \frac{7-\sqrt{11}}{3} & \frac{5-2\sqrt{11}}{3} \end{pmatrix}.$$

Performances of different methods are shown in Table 3 for $t_f = 100$.

Table 1. Example 1: $Tol_{rel} = Tol_{abs} = 10^{-6}$, $h = 0.1$

Method	N_a	N_r	N_f	N_{rest}	N_i	T_{CPU}	E_g	E_u
Cay with rest.	183	-	1100	-	-	$4.68 \cdot 10^{-2}$	$3.49 \cdot 10^{-7}$	$8.01 \cdot 10^{-16}$
Cay rest. at each step	1000	-	7000	1000	-	$3.5 \cdot 10^{-1}$	$2.88 \cdot 10^{-8}$	$1.36 \cdot 10^{-14}$
composed *Cay*	2010	-	14060	-	1000	$5.8 \cdot 10^{-1}$	$2.88 \cdot 10^{-8}$	$7.66 \cdot 10^{-15}$
composed *Cay*	317	-	2102	-	100	$8.88 \cdot 10^{-2}$	$5.01 \cdot 10^{-8}$	$2.26 \cdot 10^{-15}$
MK with rest.	838	53	5354	-	-	2.3	$8.5 \cdot 10^{-1}$	$2.5 \cdot 10^{-16}$
MK rest. at each step	1000	-	7000	1000	-	3.19	$2.88 \cdot 10^{-8}$	$1.73 \cdot 10^{-14}$
composed *MK*	2012	-	14072	-	1000	5.62	$2.88 \cdot 10^{-8}$	$1.05 \cdot 10^{-14}$
composed *MK*	325	-	2156	-	100	$8.82 \cdot 10^{-1}$	$3.94 \cdot 10^{-8}$	$3.45 \cdot 10^{-15}$

Table 2. Example 2: $Tol_{rel} = Tol_{abs} = 10^{-6}$, $h = 0.1$

Method	N_a	N_f	N_{rest}	T_{CPU}	E_u	λ_1	λ_2
Cay rest. at each step	1000	7000	1000	$1.2 \cdot 10^{-1}$	$7.56 \cdot 10^{-15}$	$7.8597 \cdot 10^{-3}$	-1.0636
composed *Cay*	2984	19904	-	$2.9 \cdot 10^{-1}$	$4.70 \cdot 10^{-15}$	$7.8476 \cdot 10^{-3}$	-1.0641
MK rest. at each step	1000	7000	1000	1.41	$8.68 \cdot 10^{-15}$	$7.8646 \cdot 10^{-3}$	-1.0636
composed *MK*	3062	20372	-	3.90	$6.72 \cdot 10^{-15}$	$7.8476 \cdot 10^{-3}$	-1.0641

5 Conclusions

Comparing Cayley and Munthe-Kaas methods we observe that in Cayley methods are usually less expensive than the others in term of CPU time, while com-

Table 3. Example 3: $Tol_{rel} = Tol_{abs} = 10^{-6}$, $h = 0.1$

Method	N_a	N_r	N_f	N_{rest}	N_i	T_{CPU}	E_g	E_u
Cay with rest.	238	2	1448	-	-	$3.31 \cdot 10^{-2}$	$1.56 \cdot 10^{-6}$	$3.89 \cdot 10^{-16}$
Cay rest. at each step	1000	-	7000	1000	-	0.18	$2.55 \cdot 10^{-15}$	$8.59 \cdot 10^{-15}$
composed *Cay*	2016	-	14096	-	1000	0.33	$1.77 \cdot 10^{-15}$	$4.25 \cdot 10^{-15}$
composed *Cay*	419	1	2920	-	200	$6.92 \cdot 10^{-2}$	$1.11 \cdot 10^{-15}$	$2.04 \cdot 10^{-15}$
MK with rest.	240	2	1460	-	-	0.56	$2.80 \cdot 10^{-6}$	$3.98 \cdot 10^{-16}$
MK rest. at each step	1000	-	7000	1000	-	2.61	$4.44 \cdot 10^{-16}$	$9.09 \cdot 10^{-16}$
composed *MK*	2018	-	14108	-	1000	5.05	$5.32 \cdot 10^{-15}$	$1.41 \cdot 10^{-14}$
composed *MK*	420	1	2926	-	200	1.05	$6.66 \cdot 10^{-16}$	$1.35 \cdot 10^{-15}$

paring different versions of Cayley methods we note that the method performs better when it is applied in composed form on few subintervals. Similar results hold for the different versions of Munthe-Kaas methods. We remark that, for controlled restart versions, when the restart is not required, the skew-symmetric problems (6) or (3) are solved on the whole interval of integration by the variable stepsize method. Finally the use of variable stepsize techniques improves the behaviour of the numerical methods.

References

1. Calvo, M.P., Iserles, A., Zanna, A.: Runge-Kutta methods for orthogonal and isospectral flows. Appl. Num. Math. **22** (1996) 153–164.
2. Chu, M.: Inverse eigenvalue problem. SIAM Review **40** (1998) 1–39.
3. Chu, M., Trendafilov, N.: The Penrose regression problem. Technical Report (1998)
4. Dieci, L., Russell, R.D., Van Vleck, E.: Unitary integration and applications to continuous orthonormalization techniques. SIAM J. Num. Anal. **31** (1994) 261–281.
5. Dieci, L., Van Vleck, E.: Computation of orthonormal factors for fundamental solution matrices. Numer. Math. **83** (1999) 599–620
6. Diele, F., Lopez, L., Peluso, R.: The Cayley transform in the numerical solution of unitary differential systems. Adv. in Comp. Math. **8** (1998) 317–334
7. Diele, F., Sgura, I.: An algorithm based on the Cayley flow for the inverse eigenvalue problem for Toeplitz matrices. Accepted for publication on BIT.
8. Engø, K., Marthinsen, A., Munthe-Kaas, H.: DiffMan, an object oriented MatLab toolbox for solving differential equations on manifolds. `http://www.math.ntnu.no/num/synode/`
9. Higham, D.: Runge-Kutta type methods for orthogonal integration. Appl. Num. Math. **22** (1996), 217–224.
10. Hairer, E., Wanner, G.: Solving Ordinary Differential Equation II. Springer Verlag, Berlin 1991
11. Lopez, L., Politi, T.: Applications of the Cayley approach in the numerical solution of matrix differential systems on quadratic groups. Appl. Num. Math. **36** (2001) 35–55
12. Munthe-Kaas, H.: Runge-Kutta methods on Lie groups. BIT **38** (1998) 92–111
13. Zanna, A.: Collocation and relaxed collocation for the Fer and the Magnus expansions. SIAM J. Numer. Anal. **36** (1999) 1145–1182

Two Step Runge-Kutta-Nyström Methods for $y'' = f(x, y)$ and P–Stability

Beatrice Paternoster

Dipartimento di Matematica e Informatica
Universitá di Salerno, Italy
beapat@unisa.it

Abstract. P-stability is an important requirement in the numerical integration of stiff oscillatory systems, but this desirable feature is not possessed by any class of numerical methods for $y'' = f(x, y)$. It is known, for example, that P–stable linear multistep methods have maximum order two and symmetric one step polynomial collocation methods can't be P-stable (Coleman 1992). In this note we show the existence of P-stable methods within a general class of two step Runge-Kutta-Nyström methods.

1 Introduction

Many physical problems arising from celestial mechanics, molecular dynamics, seismology and so on are modeled by Ordinary Differential Equations having periodic or oscillatory solutions of type

$$y''(t) = f(t, y(t)), \quad y(t_0) = y_0, \quad y'(t_0) = y_0', \qquad y(t), f(t, y) \in R^n, \tag{1}$$

Although it may be reduced into a first order system, the development of numerical methods for its direct integration seems more natural. Many linear multistep, hybrid and one step methods appeared in the literature: see for example [12, 4] for an extensive bibliography.

When the system is stiff some special stability properties are required, notably the P–stability. This concept was first introduced in [8], and it is of particular interest in the numerical treatment of periodic stiffness which is exhibited, for example, by Kramarz's system [7]. In this case two or more frequencies are involved, and the amplitude of the high frequency component is negligible or it is eliminated by the initial conditions. Then the choice of the step size is governed not only by accuracy demands, but also by stability requirements. P–stability ensures that the choice of the step size is independent of the values of frequencies, but it only depends on the desired accuracy [4, 10]. Moreover a necessary condition for a method to result P–stable is to be zero–dissipative. The property of nondissipativity is of primary interest in celestial mechanics for orbital computation, when it is desired that the computed orbits do not spiral inwards or outwards [12].

P.M.A. Sloot et al. (Eds.): ICCS 2002, LNCS 2331, pp. 459–466, 2002.

Only few numerical methods possess this desirable feature. It is worth mentioning that in the class of linear multistep methods for (1) P–stability can be reached only by methods of the second order and that the stability properties gradually deteriorate when the order increases. It is also known that symmetric one step polynomial collocation methods can't be P-stable [3], and no P–stable methods were found in the special class of two step collocation methods considered in [11].

We consider now a simple family of two step Runge–Kutta methods (TSRK) introduced in [6]:

$$\begin{aligned} y_{i+1} &= (1-\theta)y_i + \theta y_{i-1} + h\textstyle\sum_{j=1}^m (v_j f(x_{i-1}+c_j h, Y_{i-1}^j) + w_j f(x_i + c_j h, Y_i^j)), \\ Y_{i-1}^j &= y_{i-1} + h\textstyle\sum_{s=1}^m a_{js} f(x_{i-1}+c_s h, Y_{i-1}^s), \qquad j=1,\dots,m \\ Y_i^j &= y_i + h\textstyle\sum_{s=1}^m a_{js} f(x_i + c_s h, Y_i^s), \qquad j=1,\dots,m, \end{aligned} \tag{2}$$

for the not autonomous initial value problem

$$y'(t) = f(t, y(t)), \qquad y(t_0) = y_0, \tag{3}$$

where $f : \Re^q \to \Re^q$ is assumed to be sufficiently smooth. θ, v_j, w_j, a_{js}, b_{js}, $j, s, = 1, \dots, m$ are the coefficients of the methods.

It is known that the method is consistent if $\sum_{j=1}^m (v_j + w_j) = 1 + \theta$, and it is zero–stable if $-1 < \theta \le 1$ [6].

The method (2) belongs to the class of General Linear Methods introduced by Butcher [1], with the aim of giving an unifying description of numerical methods for ODEs. (2) can be represented by the following Butcher's array:

$$\begin{array}{c|c} \mathbf{c} & \mathbf{A} \\ \hline \theta & \mathbf{v}^T \\ & \mathbf{w}^T \end{array} \quad = \quad \begin{array}{c|cccc} c_1 & a_{11} & a_{12} & \cdots & a_{1m} \\ c_2 & a_{21} & a_{22} & \cdots & a_{2m} \\ \vdots & \vdots & \vdots & & \vdots \\ c_m & a_{m1} & a_{m2} & \cdots & a_{mm} \\ \hline & v_1 & v_2 & \cdots & v_m \\ \theta & & & & \\ & w_1 & w_2 & \cdots & w_m \end{array},$$

where $c_j = \sum_{j=1}^m a_{ij}$.

The reason of interest in this family lies in the fact that, advancing from x_i to x_{i+1} we only have to compute Y_i, because Y_{i-1} have already been evaluated in the previous step. Therefore the computational cost of the method depends on the matrix A, while the vector v adds extra degrees of freedom.

In this family A–stable methods exist [6], while no P–stable methods were found in [11] in the class of indirect collocation methods derived within family (2). We prove that P–stable methods, which are not collocation–based, can be constructed within the family (2) for the special second order ODEs (1).

2 Construction of the indirect method

To derive the method for the special second order system (1), from (2), following [5], we transform the system $y'' = f(x, y)$ into a first order differential equation of double dimension:

$$\begin{pmatrix} y \\ y' \end{pmatrix}' = \begin{pmatrix} y' \\ f(x,y) \end{pmatrix}, \qquad y(x_i) = y_i, \quad y'(x_i) = y'_i. \tag{3}$$

By making the interpretation

$$K_i^j = f(x_i + c_j h, Y_i^j),$$

which is usually done for Runge–Kutta methods, the following equivalent form of (2) is more convenient to our purpose:

$$\begin{aligned} y_{i+1} &= (1-\theta)y_i + \theta y_{i-1} + h\textstyle\sum_{j=1}^m (v_j K_{i-1}^j + w_j K_i^j), \\ K_{i-1}^j &= f(x_{i-1} + c_j h, y_{i-1} + h\textstyle\sum_{s=1}^m a_{js} K_{i-1}^s), \qquad j = 1, \ldots, m, \\ K_i^j &= f(x_i + c_j h, y_i + h\textstyle\sum_{s=1}^m a_{js} K_i^s), \qquad j = 1, \ldots, m, \end{aligned} \tag{4}$$

The application of the method (4) to the system (3) yields

$$\begin{aligned} K_{i-1}^j &= y'_{i-1} + h\textstyle\sum_{s=1}^m a_{js} K_{i-1}^{\prime s}, \\ K_i^j &= y'_i + h\textstyle\sum_{s=1}^m a_{js} K_i^{\prime s}, \\ K_{i-1}^{\prime j} &= f(x_{i-1} + c_j h, y_{i-1} + h\textstyle\sum_{s=1}^m a_{js} K_{i-1}^s), \\ K_i^{\prime j} &= f(x_i + c_j h, y_i + h\textstyle\sum_{s=1}^m a_{js} K_i^s), \\ y_{i+1} &= (1-\theta)y_i + \theta y_{i-1} + h\textstyle\sum_{j=1}^m (v_j K_{i-1}^j + w_j K_i^j), \\ y'_{i+1} &= (1-\theta)y'_i + \theta y'_{i-1} + h\textstyle\sum_{j=1}^m (v_j K_{i-1}^{\prime j} + w_j K_i^{\prime j}) \end{aligned} \tag{5}$$

If we insert the first two formulas of (5) into the others, we obtain

$$\begin{aligned} K_{i-1}^{\prime j} &= f(x_{i-1} + c_j h, y_{i-1} + h c_j y'_{i-1} + h^2 \textstyle\sum_{s=1}^m \bar{a}_{js} K_{i-1}^{\prime s}), \\ K_i^{\prime j} &= f(x_i + c_j h, y_i + h c_j y'_i + h^2 \textstyle\sum_{s=1}^m \bar{a}_{js} K_i^{\prime s}), \\ y_{i+1} &= (1-\theta)y_i + \theta y_{i-1} + h(\textstyle\sum_{j=1}^m v_j) y'_{i-1} + h(\textstyle\sum_{j=1}^m w_j) y'_i + \\ &\quad h^2 \left(\textstyle\sum_{s=1}^m (\bar{v}_s k_{i-1}^{\prime s} + \bar{w}_s K_i^{\prime s})\right), \\ y'_{i+1} &= (1-\theta)y'_i + \theta y'_{i-1} + h\textstyle\sum_{s=1}^m (v_s K_{i-1}^{\prime s} + w_s K_i^{\prime s}) \end{aligned} \tag{6}$$

where

$$\bar{a}_{js} = \sum_k a_{jk} a_{ks}, \qquad \bar{v}_s = \sum_k v_k a_{ks}, \qquad \bar{w}_s = \sum_k w_k a_{ks}. \tag{7}$$

From (7), setting $\bar{\mathbf{A}} = \mathbf{A}^2$, $\bar{\mathbf{v}} = \mathbf{v}^{\mathbf{T}}\mathbf{A}$, $\bar{\mathbf{w}} = \mathbf{w}^{\mathbf{T}}\mathbf{A}$, the direct method (6) for the second order system (1) takes the following form

$$
\begin{aligned}
Y_{i-1}^j &= y_{i-1} + hc_j y_{i-1}' + h^2 \sum_{s=1}^m \bar{a_{js}} f(x_{i-1} + c_s h, Y_{i-1}^s), \qquad j = 1, \dots, m,\\
Y_i^j &= y_i + hc_j y_i' + h^2 \sum_{s=1}^m \bar{a_{js}} f(x_i + c_s h, Y_i^s), \qquad j = 1, \dots, m,\\
y_{i+1} &= (1-\theta) y_i + \theta y_{i-1} + h \sum_{j=1}^m v_j y_{i-1}' + h \sum_{j=1}^m w_j y_i' +\\
&\quad h^2 \sum_{j=1}^m (\bar{v_j} f(x_{i-1} + c_j h, Y_{i-1}^j) + \bar{w_j} f(x_i + c_j h, Y_i^j)),\\
y_{i+1}' &= (1-\theta) y_i' + \theta y_{i-1}' + h \sum_{j=1}^m (v_j f(x_{i-1} + c_j h, Y_{i-1}^j) + w_j f(x_i + c_j h, Y_i^j)).
\end{aligned} \tag{8}
$$

and it is represented by the Butcher array

$$
\begin{array}{c|c}
\mathbf{c} & \mathbf{A^2} \\
\hline
 & \mathbf{v^T A} \\
 & \mathbf{w^T A} \\
\theta & \mathbf{v} \\
 & \mathbf{w}
\end{array} \; . \tag{9}
$$

Likewise to the one–step case, we call the method (8)–(9) two–step Runge–Kutta–Nyström (TSRKN) method.

3 Linear stability analysis

The homogeneous test equation for the linear stability analysis is

$$y'' = -\omega^2 y, \quad \omega \in \mathbf{R} \tag{10}$$

Following the analysis which has been performed in the one–step case (see [13]), the application of (8) to (10) yields the recursion

$$
\begin{aligned}
\mathbf{Y}_{i-1} &= \mathbf{N}^{-1}(y_{i-1}\mathbf{e} + hy_{i-1}'\mathbf{c})\\
\mathbf{Y}_i &= \mathbf{N}^{-1}(y_i\mathbf{e} + hy_i'\mathbf{c})\\
y_{i+1} &= (1-\theta) y_i + \theta y_{i-1} + h(\mathbf{v}^T\mathbf{e}\; y_{i-1}' + \mathbf{w}^T\mathbf{e}\; y_i') - z^2(\bar{\mathbf{v}}^T\mathbf{Y_{i-1}} + \bar{\mathbf{w}}^T\mathbf{Y_i})\\
hy_{i+1}' &= (1-\theta) hy_i' + \theta hy_{i-1}' - z^2(\mathbf{v}^T\mathbf{Y_{i-1}} + \mathbf{w}^T\mathbf{Y_i})
\end{aligned}
$$

where $z = \omega h, \quad \mathbf{e} = (1, \dots, 1)^T, \quad N = I + z^2 A.$

Elimination of the auxiliary vectors $\mathbf{Y_{i-1}}, \mathbf{Y_i}$ yields

$$\begin{aligned} y_{i+1} &= (1-\theta)y_i + \theta y_{i-1} + h(\mathbf{v}^T\mathbf{e}\, y'_{i-1} + \mathbf{w}^T\mathbf{e}\, y'_i) - \\ &\quad z^2(\bar{\mathbf{v}}^T\mathbf{N}^{-1}\mathbf{e}\, y_{i-1} + \bar{\mathbf{v}}^T\mathbf{N}^{-1}\mathbf{c}\, hy'_{i-1} + \bar{\mathbf{w}}^T\mathbf{N}^{-1}\mathbf{e}\, y_i + \bar{\mathbf{w}}^T\mathbf{N}^{-1}\mathbf{c}\, hy'_i) \\ hy'_{i+1} &= (1-\theta)hy'_i + \theta hy'_{i-1} - z^2(\mathbf{v}^T\mathbf{N}^{-1}\mathbf{e}\, y_{i-1} + \mathbf{v}^T\mathbf{N}^{-1}\mathbf{c}\, hy'_{i-1} + \\ &\quad \mathbf{w}^T\mathbf{N}^{-1}\mathbf{e}\, y_i + \mathbf{w}^T\mathbf{N}^{-1}\mathbf{c}\, hy'_i). \end{aligned}$$

The resulting recursion is

$$\begin{pmatrix} y_i \\ y_{i+1} \\ h\, y'_i \\ h\, y'_{i+1} \end{pmatrix} = M(z^2) \begin{pmatrix} y_{i-1} \\ y_i \\ h\, y'_{i-1} \\ h\, y'_i \end{pmatrix}$$

with

$$M(z^2) = \begin{pmatrix} 0 & 1 & 0 & 0 \\ \theta - z^2\bar{\mathbf{v}}^T\mathbf{N}^{-1}\mathbf{e} & \begin{matrix} 1-\theta- \\ z^2\bar{\mathbf{w}}^T\mathbf{N}^{-1}\mathbf{e} \end{matrix} & \mathbf{v}^T\mathbf{e} - z^2\bar{\mathbf{v}}^T\mathbf{N}^{-1}\mathbf{c} & \begin{matrix} \mathbf{w}^T\mathbf{e}- \\ z^2\bar{\mathbf{w}}^T\mathbf{N}^{-1}\mathbf{c} \end{matrix} \\ 0 & 0 & 0 & 1 \\ -z^2\mathbf{v}^T\mathbf{N}^{-1}\mathbf{e} & -z^2\mathbf{w}^T\mathbf{N}^{-1}\mathbf{e} & \begin{matrix} \theta- \\ z^2\mathbf{v}^T\mathbf{N}^{-1}\mathbf{c} \end{matrix} & \begin{matrix} 1-\theta- \\ z^2\mathbf{w}^T\mathbf{N}^{-1}\mathbf{c} \end{matrix} \end{pmatrix}. \tag{11}$$

$M(z^2)$ in (11) is the *stability* or *amplification* matrix for the two–step RKN methods (8). The stability properties of the method depend on the eigenvalues of the amplification matrix, whose elements are rational functions of the parameters of the method. Then the stability properties depend on the roots of the stability polynomial

$$\pi(\lambda) = det(M(z^2) - \lambda I). \tag{12}$$

For the sake of completeness, we recall now the following two definitions.

Definition 1. $(0, H_0^2)$ *is the interval of periodicity for the two step RKN method if,* $\forall z^2 \in (0, H_0^2)$, *the roots of the stability polynomial* $\pi(\lambda)$ *satisfy:*

$$r_1 = e^{i\phi(z)}, \quad r_2 = e^{-i\phi(z)}, \quad |r_{3,4}| \leq 1,$$

with $\phi(z)$ *real.*

Definition 2. *The two step RKN method is P–stable if its interval of periodicity is* $(0, +\infty)$.

For an A–stable method the eigenvalues of the amplification matrix are within the unit circle for all stepsizes and any choice of frequency in the test equations, and this ensures that the amplitude of the numerical solution of the test equation does not increase with time. If, what is more, there is no numerical dissipation, that is if the principal eigenvalues of the amplification matrix lie on the unit circle, then the method is P–stable [13].

4 One stage P-stable TSRKN method

Let us consider now the one stage TSRKN method (8) represented by the following Butcher array

$$\begin{array}{c|c} \mathrm{c} & a^2 \\ \hline & \mathrm{v\ a} \\ \theta & \mathrm{w\ a} \\ & v \\ & w \end{array} \tag{13}$$

The characteristic polynomial (12) of the method (13) is symmetric when, if λ is an eigenvalue of $M(z^2)$, then also $\frac{1}{\lambda}$ is an eigenvalue; in this case every stability interval is also an interval of periodicity. Therefore, to obtain a P–stable TSRKN method, it must be required that the characteristic polynomial (12) of the method is symmetric, and the periodicity interval is unlimited.

We can perform an analytical study of the inequalities representing the stability conditions for one stage two step RKN method (13). The TSRKN method (13) has order 1 if (see [6])

$$v + w = 1 + \theta, \quad -1 < \theta \leq 1.$$

The stability polynomial (12) of (13) can be written in the following way:

$$\pi(\lambda) = \lambda^4 + B(z^2)\lambda^3 + C(z^2)\lambda^2 + D(z^2)\lambda + E(z^2),$$

with

$$B(z^2) = \frac{(2a^2(\theta - 1) + w(a + c))z^2 + 2(\theta - 1)}{1 + az^2},$$

$$C(z^2) = \frac{1 - 4\theta + \theta^2 + z^2(a^2 + a^2\theta^2 + (v - w)(a + c) + w^2 + \theta(a(w - 4a) + cw))}{1 + az^2},$$

$$D(z^2) = \frac{2(1 - \theta)\theta + z^2(2a^2\theta(1 - \theta) + v(\theta - 1)(a + c) + w(2v - \theta(a + c)))}{1 + az^2},$$

$$E(z^2) = \frac{\theta^2 + z^2(a^2\theta^2 - (a + c)\theta v + v^2)}{1 + az^2}.$$

It is symmetric if and only if

$$E(z^2) \equiv 1, \qquad B(z^2) \equiv D(z^2).$$

The TSRKN method (13) has order 2 if the following conditions are satisfied [6]:

$$v + w = 1 + \theta, \qquad 2v(a-1) + 2wa = 1 - \theta. \tag{14}$$

If we ask now that $\pi(\lambda)$ is symmetric, then

$$E(z^2) \equiv 1 \quad \Longleftrightarrow \quad \theta^2 = 1,$$

from which $\theta = 1$ follows; indeed $\theta = -1$ violates the zero-stability of the method.

From (14), we must set

$$v = 2a, \qquad w = 2 - 2a.$$

Moreover $B(z^2) \equiv D(z^2)$ if and only if $c = a$ or $a = 1$.

We can choose a as free parameter and compute a value for it in such a way that the inequalities representing the stability conditions are satisfied (for example by using the Routh-Hurwutz criterion). In this way we conclude that the following method

$$\begin{array}{c|c} a & a^2 \\ \hline & 2a^2 \\ & 2a(1-a) \\ 1 & \\ & 2a \\ & 2(1-a) \end{array} \;,$$

with $a > \dfrac{1}{2}$ is an order 2, one stage, P–stable two step RKN method.

5 Concluding remark.

In [11] we did not find any P–stable method in the class of indirect collocation two step RKN methods. The result obtained in this note encourage us to proceed in our investigation on two–step Runge–Kutta–Nyström methods. Indeed it is certainly possible to derive P–stable methods within this family, as we have just shown, and we are hopeful that it is also possible to obtain high order P–stable methods within the class of two–step RKN methods, if we do not consider indirect collocation based methods. To derive P–stable methods with an increased number of stages the usage of symbolic computation will be very useful, as already done successfully in [2] in the one–step case.

References

1. Butcher, J.C.: The Numerical Analysis of Ordinary Differential Equations: Runge–Kutta and General Linear Methods, Wiley, New York (1987).

2. M. Cafaro and B. Paternoster, Analysis of stability of rational approximations through computer algebra, Computer Algebra in Scientific Computing CASC-99 (Munich 1999), V.G.Ganzha, E.W.Mayr, E.V.Vorozhtsov Eds., pp. 25-36, Springer Verlag, Berlin (1999).
3. Coleman, J.P.: Rational approximations for the cosine function; P–acceptability and order, Numerical Algorithms **3** (1992) 143–158.
4. Coleman, J.P., Ixaru, L.Gr.: P–stability and exponential–fitting methods for $y'' = f(x, y)$, IMA J. Numer. Anal.**16** (1996) 179–199.
5. Hairer, E., Norsett, S.P., Wanner, G.: Solving Ordinary Differential Equations I – Nonstiff Problems, Springer Series in Computational Mathematics **8** (1987) Springer–Verlag, Berlin.
6. Jackiewicz, Z., Renaut, R., Feldstein, A.: Two–step Runge–Kutta methods, SIAM J. Numer. Anal. **28**(4) (1991) 1165–1182.
7. Kramarz, L.: Stability of collocation methods for the numerical solution of $y'' = f(t, y)$, BIT **20** (1980) 215–222.
8. Lambert, J.D., Watson, I.A.: Symmetric multistep methods for periodic initial value problems, J. Inst. Math. Appl. **18** (1976) 189–202.
9. Paternoster, B.: Runge–Kutta(–Nyström) methods for ODEs with periodic solutions based on trigonometric polynomials, Appl. Numer. Math. **28**(2–4) (1998) 401–412.
10. Paternoster, B.: A phase–fitted collocation–based Runge–Kutta–Nyström method, Appl. Numer. Math. **35**(4) (2000) 239–355.
11. Paternoster, B.: General two–step Runge–Kutta methods based on algebraic and trigonometric polynomials, Int. J. Appl. Math. **6**(4) (2001) 347–362.
12. Petzold, L.R., Jay, L.O., Yen, J.: Numerical solution of highly oscillatory ordinary differential equations, Acta Numerica, Cambridge University Press,(1997) 437–483.
13. Van der Houwen, P.J.,Sommeijer, B.P.: Diagonally implicit Runge–Kutta–Nyström methods for oscillatory problems, SIAM J. Num. Anal. **26** (1989) 414–429.

Some Remarks on Numerical Methods for Second Order Differential Equations on the Orthogonal Matrix Group

Nicoletta Del Buono, Cinzia Elia

Dipartimento Interuniversitario di Matematica,
Università degli Studi di Bari
Via E.Orabona, 4 - I70125 Bari ITALY
[delbuono,elia]@dm.uniba.it

Abstract. In the last years several numerical methods have been developed to integrate matrix differential equations which preserve certain features of the theoretical solution such as orthogonality, eigenvalues, first integrals, etc. In this paper we approach the numerical solution of a second order matrix differential system whose solution evolves on the Lie- group of the orthogonal matrices $\mathcal{O}_n$. We study the orthogonality properties of classical Runge Kutta Nyström methods and non standard numerical procedures for second order ordinary differential equations.

1 Introduction

Recently, there has been an increasing interest in conservative numerical methods for solving ordinary differential equations which preserve certain features of the theoretical solution such as orthogonality, symplecticness, isospectrality, first integrals, etc. ([1], [2], [3], [10]). In this paper we shall concern with a system of special second order ordinary differential equations (ODEs) of dimension n, whose solutions remain for all t on the Lie group of the orthogonal matrices

$$\mathcal{O}_n = \{Y \in \mathbb{R}^{n\times n} | \ Y^T Y = I_n\},$$

where I_n is the unity matrix of dimension n.

Particularly, we are interested in solving differential equations of the following form:

$$\ddot{Y}(t) = C(t, Y(t))Y(t), \qquad t > 0, \qquad Y(0) = I_n, \qquad \dot{Y}(0) = B, \tag{1}$$

where B is a skew-symmetric matrix (i.e. $B^T = -B$) and the matrix function $C : \mathbb{R} \times \mathbb{R}^{n\times n} \to \mathbb{R}^{n\times n}$ is such that $Y(t) \in \mathcal{O}_n$, for all $t > 0$.

Second order ordinary differential equations evolving on $\mathcal{O}_n$ arise in several applications, for example in computation of embedded geodesics curve (see [5]). If we set $P(t) = \dot{Y}(t)$ the differential system (1) can be transformed into the equivalent first order system of twice the dimension

$$\begin{pmatrix} \dot{P}(t) \\ \dot{Y}(t) \end{pmatrix} = \begin{pmatrix} 0 & C(t, Y(t)) \\ I & 0 \end{pmatrix} \begin{pmatrix} P(t) \\ Y(t) \end{pmatrix}, \qquad \begin{pmatrix} P(0) \\ Y(0) \end{pmatrix} = \begin{pmatrix} B \\ I \end{pmatrix}, \tag{2}$$

P.M.A. Sloot et al. (Eds.): ICCS 2002, LNCS 2331, pp. 467–475, 2002.

thus usual methods for first order ODEs may be applied. However orthogonal preserving methods applied to (2) do not preserve the orthogonality of $Y(t)$. Moreover, it is well known that a direct solution of (2) may give computational advantages and standard numerical methods for second order ODEs are often considered to be more efficient for systems of the form (1) (see [6], [9]). In this paper after recalling some concepts on the geometrical structure of the Lie group $\mathcal{O}_n$, we particularize the second order differential equation we are interested in. In Section 3 we study when a general s-stage Runge Kutta Nyström method (hereafter abbreviated as RKN) is an orthogonal integrator for equation (1). In Section 4, we discuss the Cayley approach applied to second order orthogonal equations. Finally, we present some numerical tests to illustrate the behavior of the algorithms.

2 Background

In this section we review some concepts from differential geometry and provide the geometrical structure of the group of orthogonal matrices, which will be used throughout the rest of the paper. We also characterize a kind of second order ODEs on $\mathcal{O}_n$.

To begin with, assume $Y(t) \in \mathcal{O}_n$ for all t and denote by $\mathcal{T}_{Y(t)}\mathcal{O}_n$ the tangent space at $Y(t)$. The equation defining a tangent vector to $\mathcal{O}_n$ at the point Y is easily obtained by differentiating the constraint $YY^T = I$, i.e.

$$\dot{Y}Y^T + Y\dot{Y}^T = 0,$$

hence:

$$\mathcal{T}_Y\mathcal{O}_n = \{\Delta \in \mathbb{R}^{n\times n} | \Delta Y^T + Y\Delta^T = 0\},$$

Clearly, the Lie algebra o_n of $\mathcal{O}_n$ (i.e. the tangent space at the identity) is the set of all skew symmetric matrices:

$$o_n = \{B \in \mathbb{R}^{n\times n} | B^T + B = 0\}.$$

Furthermore, by differentiating twice the constraint, we get

$$\ddot{Y}Y^T + 2\dot{Y}\dot{Y}^T + Y\ddot{Y}^T = 0.$$

that is $\ddot{Y}$ belongs to the set:

$$\mathcal{N}_Y\mathcal{O}_n = \{\Omega \in \mathbb{R}^{n\times n} | \Omega Y^T + Y\Omega^T + 2\Delta\Delta^T = 0, \text{ with } \Delta \in \mathcal{T}_Y\mathcal{O}_n\}.$$

Theorem 1. *Let $Y(t)$ be the solution of (1). Then $Y(t)$ belongs to $\mathcal{O}_n$ for all $t > 0$, if and only if there exists $A : \mathbb{R} \times \mathbb{R}^{n\times n} \to o_n$ continuous and locally Lipschitz skew-symmetric matrix function such that*

$$C(t, Y) = \dot{A}(t, Y) + A_Y(t, \dot{Y}) + A(t, Y)A(t, Y), \tag{3}$$

with $A(0, Y(0)) = B$, where $\dot{A}$ denotes the derivative with respect to t and

$$A_Y(t, X) = \sum_{i,j=1}^{n} \frac{\partial A(t, Y)}{\partial Y_{i,j}} X_{i,j}.$$

Proof. Suppose that $C(t,Y)$ in (1) is given by (3), then being $A(t,V)$ a skew-symmetric matrix function, the first order differential system

$$\dot{V}(t) = A(t, V(t))V(t), \qquad V(0) = I_n, \tag{4}$$

has a solution $V(t)$ which is an orthogonal matrix for all $t > 0$. Moreover, by differentiating (4), we obtain that $V(t)$ satisfies the second order differential system:

$$\begin{aligned} \ddot{V}(t) &= [\dot{A}(t,V) + A_V(t,\dot{V}) + A(t,V)A(t,V)]V(t), \\ V(0) &= I_n, \qquad \dot{V} = B. \end{aligned} \tag{5}$$

From the uniqueness of the solution of (1) and (4), it follows that $V(t) = Y(t)$, for all $t > 0$. Now we assume that the solution of (1) is orthogonal. Then it satisfies a ordinary differential equation of the form

$$\dot{Y}(t) = A(t, Y(t))Y(t), \qquad Y(0) = I_n, \tag{6}$$

where $A : \mathbb{R} \times \mathbb{R}^{n \times n} \to o_n$ is continuous and locally Lipschitz skew-symmetric matrix function. Then the results follows by differentiating (6).

Now the following result may be easily proved.

Lemma 1. *Suppose that the unique solution $Y(t)$ of (1) is orthogonal and the matrix function $C(t)$ is independent on Y. Then $C(t) = \dot{A}(t) + A(t)A(t)$ for all $t > 0$, with $A(t)$ the skew-symmetric matrix function given by*

$$A(t) = B + \frac{1}{2}\int_0^t [C(s) - C^T(s)]\mathrm{d}s. \tag{7}$$

Proof. From the Theorem 1 it follows that there exists a skew-symmetric matrix function $A(t)$ such that:

$$C(t) = \dot{A}(t) + A(t)A(t), \quad t > 0. \tag{8}$$

Hence

$$-C^T(t) = -\dot{A}^T(t) - A^T(t)A^T(t). \tag{9}$$

Adding (8) to (9) and using the skew-symmetry of A and $\dot{A}$, then (7) follows.

Remark 1. From Theorem 1 it follows that the orthogonal solution of the second order differential equation (1) with $C(t)$ depending only on t, is equivalent to the solution of the first order differential equation of the same dimension. This leads to a computational advantage.

3 RKN methods and orthogonality

Let $h > 0$ be the step-size, $\{t_k\}$ the set of the step points and $Y_{k+1}, \dot{Y}_{k+1}$ denote the numerical approximation of $Y(t_{k+1})$ and $\dot{Y}(t_{k+1})$, respectively. A s-stage

RKN method for (1) is given by:

$$Y_{k+1} = Y_k + h\dot{Y}_k + h^2 \sum_{i=1}^{s} \bar{b}_i K'_i,$$
$$\dot{Y}_{k+1} = \dot{Y}_k + h \sum_{i=1}^{s} b_i K'_i, \tag{10}$$

with

$$K'_i = C(t_k + c_i h, Y_k + c_i h\dot{Y}_k + h^2 \sum_{j=1}^{s} \bar{a}_{ij} K'_j)(Y_k + c_i h\dot{Y}_k + h^2 \sum_{j=1}^{s} \bar{a}_{ij} K'_j),\ i = 1\dots, s$$

where a_{ij}, b_i, c_i, for $i, j = 1, \dots, s$ are real coefficients. Furthermore, introducing the $s \times s$ matrices $A = (a_{ij})$, $\bar{A} = (a_{ij})$, and the s-dimensional vectors $b^T = (b_1, \dots, b_s)$, $\bar{b}^T = (\bar{b}_1, \dots, \bar{b}_s)$ and $c = (c_1, \dots, c_s)^T$, the RKN scheme (10) can also be represented by the Butcher array

$$\begin{array}{c|c|c} c & \bar{A} & A \\ \hline & \bar{b}^T & b^T \end{array} \tag{11}$$

The method is said explicit if $a_{ij} = 0$ for $i \leq j$ and implicit otherwise.

If the coefficients A, b^T, c of the RKN method are equal to those of a Runge-Kutta method for first order ODE, then the Nyström method is said to be induced by this RK scheme and its coefficients satisfy

$$\bar{a}_{ij} = \sum_{k=1}^{s} a_{ik} a_{kj} \qquad \text{and} \qquad \bar{b}_i = \sum_{j=1}^{s} b_j a_{ji}. \tag{12}$$

Moreover, a RKN method is said to be a collocation scheme if it is obtained by applying collocation methods for first order differential equation (2).

We now investigate the properties a RKN scheme has to satisfy to be an orthogonal preserving scheme when applied to (1). We start with the following matrix differential system:

$$\ddot{Y}(t) = CY(t), \quad t > 0, \quad Y(0) = I_n, \quad \dot{Y}(0) = B, \tag{13}$$

where the matrix $C = B^2$ is symmetric seminegative definite and B is skew-symmetric.

Theorem 2. *The implicit RKN methods induced by the Gauss Legendre Runge Kutta schemes of order 2s are orthogonal integrators for differential systems (13).*

Proof. We give the proof for the Runge Kutta Nyström Gauss Legendre method with stage $s = 1$ and Butcher array

$$\begin{array}{c|c|c} 1/2 & 1/4 & 1/2 \\ \hline & 1/2 & 1 \end{array} \tag{14}$$

Applying the numerical scheme (14) to (13) we get:

$$\begin{aligned} K_1' &= C(Y_0 + \tfrac{h}{2}\dot{Y}_0 + \tfrac{h^2}{4}K_1'), \\ Y_1 &= Y_0 + h\dot{Y}_0 + \tfrac{1}{2}h^2 K_1', \\ \dot{Y}_1 &= \dot{Y}_0 + hK_1'. \end{aligned} \tag{15}$$

Hence

$$Y_1^T Y_1 = Y_0^T Y_0 + h(\dot{Y}^T Y_0 + Y_0^T \dot{Y}_0) + \tfrac{h^2}{2}(Y_0^T K_1' + K_1'^T Y_0 + 2\dot{Y}_0^T \dot{Y}_0) +$$

$$+ \tfrac{h^3}{2}(\dot{Y}_0^T K_1' + K_1'^T \dot{Y}_0) + \tfrac{h^4}{4} K_1'^T K_1'.$$

Substituting the expression of K_1' we get:

$$Y_1^T Y_1 = Y_0^T Y_0 + h(\dot{Y}^T Y_0 + Y_0^T \dot{Y}_0) + \tfrac{h^2}{2}(2Y_0^T C Y_0 + 2\dot{Y}_0^T \dot{Y}_0) +$$

$$+ \tfrac{h^3}{4}(Y_0^T C\dot{Y}_0 + \dot{Y}_0^T C Y_0) + \tfrac{h^3}{2}(\dot{Y}_0^T K_1' + K_1'^T \dot{Y}_0) +$$

$$+ \tfrac{h^4}{4}(Y_0^T C K_1' + K_1'^T C Y_0 + K_1'^T K_1').$$

By the initial conditions and observing that $C = B^2$, the terms in h^2 and $\frac{h^3}{4}(Y_0^T C\dot{Y}_0 + \dot{Y}_0^T C Y_0)$ vanish. Hence, substituting ricorsively the expression of K_1' all the powers of h vanish and so the result follows.

However, the positive result obtained by Theorem 2, are not still valid for more general second order nonlinear differential system.

Theorem 3. *The RKN Gauss Legendre schemes are not orthogonal integrators for the differential system (1).*

Proof. For the sake of simplicity we will give the proof in the linear nonautonomous case. Let us consider the differential system

$$\ddot{Y}(t) = C(t)Y(t), \quad t > 0, \quad Y(0) = I_n, \quad \dot{Y}(0) = B(0), \tag{16}$$

where $C(t) = \dot{B}(t) + B(t)B(t)$ and $B(t)$ is a skew-symmetric matrix. Applying the RKNGL scheme with $s = 1$ to (16), we get

$$\begin{aligned} K_1' &= C(\tfrac{h}{2})(Y_0 + \tfrac{1}{2}h\dot{Y}_0 + \tfrac{1}{4}h^2 K_1'), \\ Y_1 &= Y_0 + h\dot{Y}_0 + \tfrac{1}{2}h^2 K_1', \\ \dot{Y}_1 &= \dot{Y}_0 + hK_1'. \end{aligned} \tag{17}$$

Hence

$$Y_1^T Y_1 = Y_0^T Y_0 + h(\dot{Y}^T Y_0 + Y_0^T \dot{Y}_0) + \frac{h^2}{2}(Y_0^T C(\frac{h}{2})Y_0 + 2\dot{Y}_0^T \dot{Y}_0) + \dots.$$

Now, observe that the term in h^2, substituting the expression of $C(\frac{h}{2})$ and the values of Y_0 and of $\dot{Y}_0$, we get:

$$\frac{h^2}{2}\left(2\dot{B}(\frac{h}{2}) + 2B(\frac{h}{2})\dot{B}(\frac{h}{2}) + 2\dot{B}(0)^T \dot{B}(0)\right), \tag{18}$$

which, in general, does not vanish and so $Y_1^T Y_1 = Y_0^T Y_0 + \mathcal{O}(h^2)$.

4 Non standard RKN schemes

In this section we generalize to RKN methods some approaches uses to solve orthogonal first order differential equations.

4.1 Projected RKN methods

Following the ideas proposed in [3] for first order ODEs on $\mathcal{O}_n$, we describe here a projected integrators for equation (1).

A projected methods consists of a two steps procedure:

- firstly, an approximation $\tilde{Y}_{k+1}$ of the solution of (1), provided by any explicit s-stage RKN method, is computed using (10);
- then, the QR factorization of $\tilde{Y}_{k+1}$ by the modified Gram-Schmidt process is performed, that is

$$\tilde{Y}_{k+1} = Q_{k+1}R_{k+1},$$

and then the factor Q of the QR factorization is assumed as the approximation of the solution at t_{k+1}, i.e.,

$$Y_{k+1} = Q_{k+1}.$$

Proposition 1. *The projected RKN schemes preserve the order of accuracy of the RKN method they are based on.*

4.2 Cayley methods for second order ODEs

Another approach used for solving first order orthogonal systems is based on the transformation of the original system into a skew-symmetric one, obtained by continously applying the Cayley transform (see [4]).

Proposition 2. *If $Y(t)$ is an orthogonal matrix having for any t all eigenvalues different from -1, then there exists a unique smooth skew-symmetric matrix function $A(t)$ such that*

$$Y(t) = [I - A(t)]^{-1}[I + A(t)], \qquad t \geq 0. \tag{19}$$

The previous transformation is known as the Cayley transform of $Y(t)$.

An interesting remark is that for second order differential equations this approach does not lead to orthogonal schemes. In fact using (19), equation (1) can be transformed as in the following theorem.

Theorem 4. *Let $Y(t)$ be the solution of the differential systems (1), with all eigenvalue different from -1, for any t, then $A(t)$ given by (19) satisfies the second order differential system:*

$$\ddot{A} = H(A, \dot{A}), \qquad t > 0, \qquad A(0) = 0, \qquad \dot{A}(0) = \tfrac{1}{2}B, \tag{20}$$

where $H(A, \dot{A}) = \frac{1}{2}(I - A)C((I - A)^{-1}(I + A))(I - A) - 2\dot{A}(I - A)^{-1}\dot{A}$.

Proof. From (19) it follows that $(I-A)Y=(I+A)$; then differentiating twice

$$\dot{A}=\dot{Y}-\dot{A}Y-A\dot{Y}, \tag{21}$$

and

$$\ddot{A}=\ddot{Y}-\ddot{A}Y-2\dot{A}\dot{Y}-A\ddot{Y}. \tag{22}$$

Hence $(I-A)\ddot{Y}-2\dot{A}\dot{Y}=\ddot{A}(I+Y)$ and by (1)

$$\ddot{A}=(I-A)C(Y)(I+Y)^{-1}-2\dot{A}\dot{Y}(I+Y)^{-1}.$$

Moreover, from (19) we also obtain

$$(I+Y)^{-1}=\frac{1}{2}(I-A), \tag{23}$$

and from (21)

$$\dot{Y}=2(I-A)^{-1}\dot{A}(I-A)^{-1}. \tag{24}$$

Thus, substituting equality (23) and (24) into (22) and using the initial condition for equation (1) the statement follows.

Observe that a restarting procedure is required if there exists a τ such that $Y(\tau)$ has an eigenvalue equal to -1 (see [4]). (20).

Remark 2. We have to observe that the solution of (20) is not a skew-symmetric matrix function, because $H(\dot{A},A)$ is not a skew-symmetric. Indeed, this result was expected, in fact from (21), $\dot{A}$ is not a curve on $o(n)$ and therefore, when we consider its derivative with respect to time, i.e. $\ddot{A}$, this does not belong to $o(n)$. Hence, the Cayley approach does not lead to orthogonal schemes.

5 Numerical Tests

In this section we present some numerical tests in order to illustrate the properties of the geodesics based methods. All the numerical results have been obtained by Matlab codes implemented on a scalar computer Alpha 200 5/433 with 512 Mb RAM. We compare the considered methods in terms of accuracy, deviation of the numerical solution from orthogonal structure and CPU time. The deviation from the orthogonal manifold is measured by $\|I_k-Y_k^TY_k\|_F$, the accuracy by $\|Y(t_k)-Y_k\|_\infty$, where $\|\cdot\|_F$ and $\|\cdot\|_\infty$ denote respectively the Frobenius and the infinity norm on matrices and Y_k is the numerical approximation of the solution at the instant $t=t_k$.

Example 1. As first example we solve the constant linear second ordinary differential system

$$\ddot{Y}=CY,\quad Y(0)=I_2,\quad \dot{Y}(0)=B,$$

Table 1. Example 1 performance at $T = 1$.

h	Method	Global error	Orthogonal error	CPU time
0.01	RKNGL1	0.0035	3.3813e-16	0.06
	PRKN2	0.0017	2.6037e-16	0.27
	RKN2	0.0018	3.1754e-4	0.21
	RKNGL2	2.4443e-7	3.7974e-16	0.11
	PRKN4	6.1915e-7	1.0245e-16	0.38
	RKN4	3.1272e-6	1.4040e-7	0.32
0.005	RKNGL1	8.6470e-4	3.6873e-16	0.11
	PRKN2	4.4099e-4	2.4999e-16	0.55
	RKN2	4.3600e-4	3.9733e-5	0.21
	RKNGL2	1.5280e-8	5.4897e-15	0.22
	PRKN4	3.8504e-8	5.0354e-16	0.72
	RKN4	1.9520e-7	4.3891e-9	0.75

where $B \begin{pmatrix} 0 & 1 & -3 & -4 \\ -1 & 0 & 2 & 2 \\ 3 & -2 & 0 & -3 \\ 4 & -2 & 3 & 0 \end{pmatrix}$ and $C = B^2$. Table 1 summerizes the results obtained solving the problem with constant step, on the interval $[0, 1]$. All the error are estimated at the final point of the integration interval.

As shown in Table (1) both projective and Gauss Legendre RKN schemes preserve the orthogonality with a machine accuracy. Furthermore, the direct application of an explicit RKN to the system provides an orthogonal error of the same order of the scheme.

Example 2. As second example we solve the nonautonomous second ordinary differential system

$$\ddot{Y} = \begin{pmatrix} -\sin^2(t) & \cos(t) \\ -\cos(t) & -\sin^2(t) \end{pmatrix} Y, \quad Y(0) = I_2, \qquad \dot{Y}(0) = 0,$$

whose solution is $Y(t) = \begin{pmatrix} \cos(1-\cos(t)) & \sin(1-\cos(t)) \\ -\sin(1-\cos(t)) & \cos(1-\cos(t)) \end{pmatrix}$, ([3]).

As proved in Lemma 3, for generally nonautonomous orthogonal second order systems, the collocation RKNGL schemes do not preserve the orthogonality of the solution.

Conclusion

With the aim of solving second ordinary differential systems preserving the orthogonal structure, we have investigated the properties of Runge Kutta Nyström Gauss Legendre methods. These schemes are orthogonal preserving only for linear constant second order ODEs. To tackle the problem, we have also proposed

Table 2. Example 2 performance at $T = 5$.

h	Method	Global error	Orthogonal error	CPU time
0.01	RKNGL1	1.4456e-5	4.5856e-5	0.27
	PRKN2	0.0018	0	0.50
	RKN2	0.0019	0.0032	0.47
	RKNGL2	1.8464e-10	3.4853e-10	0.50
	PRKN4	3.3782e-11	6.2803e-16	0.94
	RKN4	9.1295e-11	4.1622e-10	0.96
0.005	RKNGL1	3.6140e-6	1.0213e-5	0.43
	PRKN2	4.7764e-4	0	0.98
	RK2	3.9814e-4	0.0013	0.91
	RKNGL2	1.1528e-11	2.1871e-11	0.98
	PRKN4	2.1917e-12	0	1.86
	RK4	6.1917e-12	3.2724e-10	1.95

a semi-explicit projection procedure based on the Gram-Smith factorization and we have pointed out that the good performance of the Cayley approach for first order differential orthogonal systems are not showed for second order one.

For further research, we intend to extend the study of orthogonal behavior to other RKN schemes, as for instance symplectic RKN, and investigate the exponential map approach.

References

1. Calvo, M.P., Iserles, A., Zanna, A.: Runge-Kutta methods for orthogonal and isospectral flows, Appl. Numer. Math. 22 (1996), 153–164.
2. Del Buono, N., Lopez, L.: Runge Kutta type methods based on geodesics for systems of ODEs on the Stiefel manifold, BIT 41, 5, (2001) 912–923.
3. Dieci, L., Russell, D., Van Vleck, E. S.: Unitary integration and applications to continuous orthonormalization techniques, SIAM J. Num. Anal. 31 (1994), 261–281.
4. Diele, F., Lopez, L., Peluso, R.: The Cayley transform in the numerical solution of unitary differential systems , Ad. Comp. Math. 8 (1998), 317–334.
5. Edelman, A., Arias, T. A., Smith, S. T.: The geometry of algorithms with orthogonality constraints, SIAM J. Matrix Anal. Appl. 20 (1998), 303–353.
6. Hairer, E., Nørsett, S.P., Wanner, G.: Solving ordinary differential equations, Vol. I, Springer, Berlin, 1987.
7. Iserles, A., Nørsett, S.P.: On the solution of linear differential equation on Lie groups, Cambridge University Tech. Rep. DAMTP 1997/NA3.
8. Iserles, A., Munthe-Kaas, H. Z., Nørsett, S.P., Zanna, A.: Lie-group methods, Acta Numerica 9, CUP, (2000), pp. 215–365.
9. Petzold, L.R., Lay, L.O., Yen, J.: Numerical solution of highly oscillatory ordinary differential equations, Acta Numerica (1997), 437-483, Cambridge University Press.
10. Sanz-Serna, J.M., Calvo, M.P., Numerical Hamiltonian Problems, Chapman Hall, London, 1994.

Numerical Comparison between Different Lie-Group Methods for Solving Linear Oscillatory ODEs

Fasma Diele[1] and Stefania Ragni[2]

[1] Istituto per Ricerche di Matematica Applicata, CNR, Bari, Italy
irmafd03@area.ba.cnr.it
[2] Facoltà di Economia, Università degli Studi di Bari, Bari, Italy
irmasr18@area.ba.cnr.it

Abstract. In this paper we deal with high oscillatory systems and numerical methods for the approximation of their solutions. Some classical schemes developed in the literature are recalled and a recent approach based on the expression of the oscillatory solution by means of the exponential map is considered. Moreover we introduce a new method based on the Cayley map and provide some numerical tests in order to compare the different approaches.

1 Introduction

Let us consider an initial-value ordinary differential system

$$\mathbf{y}' = \mathbf{f}(t, \mathbf{y}), \quad t \geq 0, \quad \mathbf{y}(0) = \mathbf{y}_0$$

whose solution oscillates with a timescale much shorter than the integration interval. We will refer to these kind of dynamical systems as *highly oscillatory* ones. Indeed this concept is very generic; a more precise definition can be found in a survey provided by L. R. Petzold et al. in 1997 (see [10]) where it was stated that such equations are characterized by a *fast* solution varying regularly about a *slow* solution.
High oscillatory systems arise in many applications such as vehicle simulations, molecular dynamics, circuit simulations, flexible body dynamics. In order to provide some examples of oscillatory differential problems, let us consider the solution of

$$y'' + ty = 0, \quad t \geq 0, \tag{1}$$

given by $y(t) = \pi[Ai(-t)Bi'(0) - Ai'(0)Bi(-t)]$, which satisfies the initial conditions $y(0) = 1$, $y'(0) = 0$, where $Ai(z)$ is the so-called *Airy function* and $Bi(z)$ represents the *Airy function of the second kind.* As another example, the *Bessel function* of index $\nu = 0$

$$J_0(t) = \sum_{k=0}^{\infty} \frac{(-1)^k}{(k!)^2} \left(\frac{t}{2}\right)^{2k} \tag{2}$$

P.M.A. Sloot et al. (Eds.): ICCS 2002, LNCS 2331, pp. 476–485, 2002.

is the solution of the following equation

$$t^2y'' + ty' + t^2y = 0, \quad t \geq 0. \tag{3}$$

In Figure 1 we plot the Airy function behaviour and the approximation of (2) obtained by means of the Besselj built-in Matlab code.

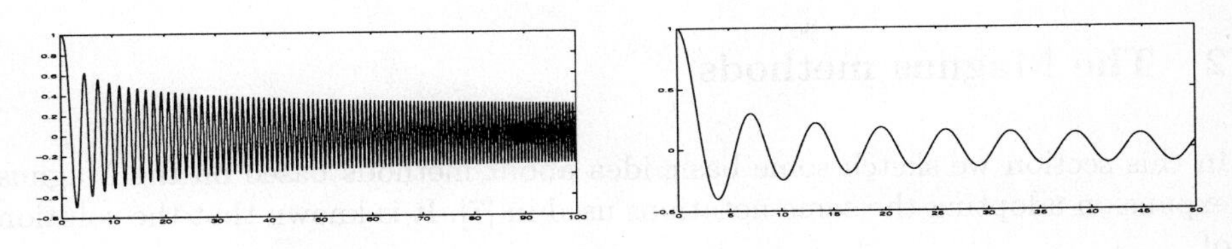

Fig. 1. Plot of the Airy (*left*) and the Bessel (*right*) functions.

Different approaches for the numerical solution of this kind of problems have been developed in the literature. With the aim to investigate the stability features of the classical numerical methods, such as multistep and Runge-Kutta ones, applied to oscillatory systems, the scalar *harmonic oscillator* equation

$$y'' = -\omega^2 y, \qquad \omega > 0$$

is usually chosen as standard test problem. In this respect, a suitable stability definition is due to Lambert and Watson who originally introduced the concept of P-stability (see [9]). We notice that, since the step length is constrained not only by stability but also by accuracy, then a stepsize of the same magnitude as the period of oscillation with highest frequency is required even for P-stable methods. The form and structure of the effective numerical methods is strongly application-dependent, so they vary according to specific classes of the considered applications. For instance, when the presence of forcing terms arises in some applications, it is interesting to account for the following inhomogeneous equation:

$$\mathbf{y}'' = A(t)\mathbf{y}(t) + \mathbf{b}(\mathbf{y}).$$

In the case when $A(t) = A$, the solution can be obtained by means of the *mollified impulse method* due to Garcia-Archilla, Sanz-Serna and Skeel [2]. Successively in [3] Hochbruch and Lubich introduced the *Gautschi-type methods* that reduce to solve exactly the equation with constant inhomogeneity.

Recently Iserles proposed a completely different approach based on the Magnus expansion (see [5]). More precisely the solution of the linear system $\mathbf{y}' = A(t)\mathbf{y}$, $\mathbf{y}(0) = \mathbf{y}_0$ is represented in terms of the exponential matrix $\mathbf{y}(t) = e^{\Omega(t)}\mathbf{y}_0$ where $\Omega(t)$ satisfies a suitable differential equation, as we recall in Section 2. By following this idea we propose to express the same solution in terms of the Cayley transform $\mathbf{y}(t) = (I - \frac{1}{2}\Omega(t))^{-1}(I + \frac{1}{2}\Omega(t))\mathbf{y}_0$ with $\Omega(t)$ satisfying a related equation, as shown in Section 3. We notice that these approaches arise in a completely different field such as the numerical approximation of conservative differential problems (see e.g. [7]). The idea to apply the Cayley transform in the framework of oscillatory systems is suggested by its competitiveness and cheapness with respect to the exponential map. In Section 4 we compare these two approaches verifying their effectiveness on the Airy and Bessel test problems. Finally, conclusion and suggestions for further research are discussed in Section 5.

2 The Magnus methods

In this section we sketch some basic idea about methods based on the Magnus expansion adopting the same notations used in [7]. It is known that the solution of

$$\mathbf{y}' = A(t)\mathbf{y}(t) \quad t \geq 0, \quad \mathbf{y}(0) = \mathbf{y}_0 \tag{4}$$

can be written in the form

$$\mathbf{y}(t) = e^{\Omega(t)}\mathbf{y}_0 \tag{5}$$

where Ω satisfies the following equation

$$\Omega' = \sum_{k=0}^{\infty} \frac{B_k}{k!} \mathrm{ad}_{\Omega}^k A, \quad t \geq t_0, \quad \Omega(t_0) = O$$

being $\{B_k\}_{k \in \mathbb{Z}^+}$ the Bernoulli numbers and

$$\begin{aligned} &\mathrm{ad}_{\Omega}^0 A = \Omega \\ &\mathrm{ad}_{\Omega}^k A = [A, \mathrm{ad}_{\Omega}^{k-1} A] = A\ \mathrm{ad}_{\Omega}^{k-1} A - \mathrm{ad}_{\Omega}^{k-1} A\ A, \qquad k > 0. \end{aligned}$$

The solution of the previous system is the so-called Magnus expansion of Ω given by

$$\begin{aligned} \Omega(t) = &\int_0^t A(\xi)d\xi - \frac{1}{2}\int_0^t \int_0^{\xi_1} [A(\xi_2), A(\xi_1)]d\xi_2 d\xi_1 \\ &+\frac{1}{12}\int_0^t \int_0^{\xi_1} \int_0^{\xi_1} [A(\xi_3), [A(\xi_2), A(\xi_1)]]d\xi_3 d\xi_2 d\xi_1 \\ &+\frac{1}{4}\int_0^t \int_0^{\xi_1} \int_0^{\xi_2} [[A(\xi_3), A(\xi_2)], A(\xi_1)]d\xi_3 d\xi_2 d\xi_1 + \dots \end{aligned} \tag{6}$$

In order to discretize the solution (5), it is necessary to truncate the infinite Magnus expansion (6) and to replace integrals by quadrature. Therefore the Magnus numerical scheme consists of advancing the Magnus expansion by step $h > 0$ and approximating $\mathbf{y}(t_{n+1}) = e^{\Omega_n(h)}\mathbf{y}(t_n)$, by

$$\mathbf{y}_{n+1} = e^{\tilde{\Omega}_n(h)}\mathbf{y}_n$$

with $\tilde{\Omega}_n(h)$ truncation of $\Omega_n(h)$, where the integrals are replaced by quadrature.

2.1 Modified Magnus methods

A modified version of the Magnus method can be designed explicitly for oscillatory systems as in [5]. The algorithm advances from t_n to $t_n + h$ by setting

$$\mathbf{y}(t) = \mathbf{e}^{(t-t_n)A(t_{n+1/2})}\mathbf{x}(t), \quad t \geq t_n$$

where $t_{n+1/2} = t_n + \frac{1}{2}h$ and the function $\mathbf{x}(t)$ satisfies

$$\mathbf{x}' = B(t)\mathbf{x}, \quad t \geq t_n, \quad \mathbf{x}(t_n) = \mathbf{y}(t_n) \tag{7}$$

with $B(t) = \mathbf{e}^{(t-t_n)A(t_{n+1/2})}[A(t) - A(t_{n+1/2})]\mathbf{e}^{(t-t_n)A(t_{n+1/2})}$. The latter equation is discretized by the standard Magnus method so that $\mathbf{x}(t) = \mathbf{e}^{\tilde{\Omega}_n(t)}\mathbf{y}_n$; therefore the global approximation is given by

$$\mathbf{y}_{n+1} = \mathbf{e}^{hA(t_{n+1/2})}\mathbf{e}^{\tilde{\Omega}_n(h)}\mathbf{y}_n, \quad n \in \mathbb{Z}_+.$$

Remark 1. The idea which is the basis of this algorithm, is that the oscillatory behaviour of (4) is locally well modelled by the linear equation with constant coefficients

$$\tilde{\mathbf{y}}' = A(t_{n+1/2})\tilde{\mathbf{y}}$$

whose solution is given by a matrix exponential.

3 The Cayley method

Following the same idea developed by Iserles, we apply a method based on the Cayley map in order to solve an oscillatory system. As a counter part of (5), the solution of (4) is given by

$$\mathbf{y}(t) = cay(\Omega(t))\mathbf{y}_0$$

where $cay(\Omega(t)) = [I - \frac{1}{2}\Omega(t)]^{-1}[I + \frac{1}{2}\Omega(t)]$ is the Cayley transform of the matrix $\Omega(t)$ that satisfies the following equation (see [1], [4])

$$\Omega' = A - \frac{1}{2}[\Omega, A] - \frac{1}{4}\Omega A \Omega, \quad t \geq t_0, \quad \Omega(t_0) = O. \tag{8}$$

Notice that evaluating the Cayley transform is cheaper than computing a matrix function exponential. It is possible to solve (8) by performing the Magnus expansion of Ω (see [4])

$$\begin{aligned}\Omega(t) = &\int_0^t A(\xi)d\xi - \frac{1}{2}\int_0^t \int_0^{\xi_1} [A(\xi_2), A(\xi_1)]d\xi_2 d\xi_1 \\ &-\frac{1}{4}\int_0^t \left[\int_0^{\xi_1} A(\xi_2)d\xi_2\right] A(\xi_1) \left[\int_0^{\xi_1} A(\xi_3)d\xi_3\right] d\xi_1 \\ &+\frac{1}{4}\int_0^t \int_0^{\xi_1} \int_0^{\xi_2} [[A(\xi_3), A(\xi_2)], A(\xi_1)]d\xi_3 d\xi_2 d\xi_1 + \ldots .\end{aligned} \tag{9}$$

Again, we can achieve the numerical solution by truncating the infinite Magnus expansion (9) and replacing integrals by quadrature. Therefore the Cayley numerical scheme is

$$\mathbf{y}_{n+1} = cay(\tilde{\Omega}_n(h))\mathbf{y}_n$$

with $h > 0$ and $\tilde{\Omega}_n(h)$ the truncation of $\Omega_n(h)$ obtained replacing the integrals by quadrature.

3.1 Modified Cayley method

As a counterpart of the modified Magnus method, we consider again the algorithm that advances from t_n to $t_n + h$ by defining

$$\mathbf{y}(t) = \mathbf{e}^{(t-t_n)A(t_{n+1/2})}\mathbf{x}(t), \quad t \geq t_n$$

where $\mathbf{x}(t)$ satisfies (7). In this case, we discretize equation (7) using the standard Cayley method; therefore we obtain the solution

$$\mathbf{y}_{n+1} = \mathbf{e}^{hA(t_{n+1/2})} cay(\tilde{\Omega}_n(h))\mathbf{y}_n, \quad n \in \mathbb{Z}_+.$$

4 Numerical schemes and results

We are going to list the methods described so far. More precisely, we consider the fourth order schemes for Magnus and Cayley methods already presented in [7]. In the sequel we will denote for each n

$$A_i = A(t_n + c_i h) \qquad i = 1, 2 \tag{10}$$

with $c_1 = \frac{1}{2} - \frac{\sqrt{3}}{6}$, $c_2 = \frac{1}{2} + \frac{\sqrt{3}}{6}$. Concerning the fourth order Magnus scheme (**Magnus4**), we compute $y_{n+1} = exp(\Omega_n)y_n$ where Ω_n is given by

$$\Omega_n = \frac{h}{2}(A_1 + A_2) - \frac{\sqrt{3}}{12}h^2[A_1, A_2].$$

As a counterpart, the fourth order method based on the Cayley expansion (**Cayley4**) provides the approximation $y_{n+1} = cay(\Omega_n)y_n$ by evaluating

$$\Omega_n = hB_0 + \frac{1}{12}h^2[B_1, B_0] - \frac{1}{12}h^3 B_0^3$$

where $B_0 = \frac{1}{2}(A_1 + A_2)$ and $B_1 = \sqrt{3}(A_2 - A_1)$. We implement the modified versions of the previous methods by following the same schemes given in [5].

Our aim is to validate the effectiveness of the Cayley approach in the field of oscillatory systems by applying it on the Airy and Bessel equations which we consider as test problems. Moreover, we are interested in the comparison between the different schemes based on the exponential and the Cayley maps. In each figure we plot, on different timescales, the errors in the solution of the test problems (1) and (3) by using Magnus and Cayley approaches with time

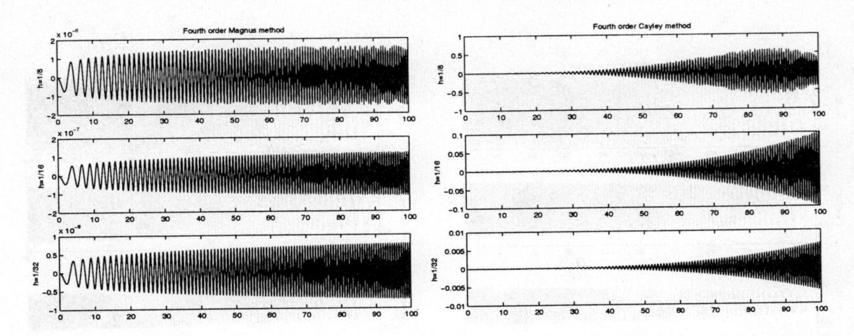

Fig. 2. The error in the solution of the Airy equation (1) by **Magnus4** (*left*) and **Cayley4** (*right*) in the time interval [0, 100].

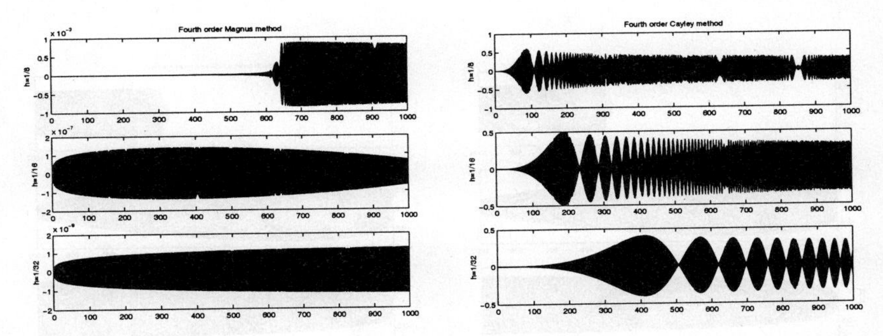

Fig. 3. The error in the solution of the Airy equation (1) by **Magnus4** (*left*) and **Cayley4** (*right*) in the time interval [0, 1000].

steps $h = \frac{1}{8}$, $h = \frac{1}{16}$ and $h = \frac{1}{32}$. In Figures 2 and 3, it is evident that the Magnus method performs better than the Cayley one on both timescales, when applied to the Airy equation; anyway, notice that it loses accuracy for the longer integration interval. Concerning the modified versions, as shown in Figures 4 and 5, both the approaches behave similarly. However, the Cayley approach takes a sharp improvement when implemented in modified version. Otherwise, this does not hold for the Magnus schemes. We notice that in [5] and in [6] a remarkable advantage is get out of the modified approach performing the Magnus expansion with exact integrals. We can make similar considerations regarding the performances of the considered schemes for the solution of the Bessel equation. Even if the results in Figures 6 and 7 obtained by applying **Cayley4** are acceptable, they are not competitive with respect to the **Magnus4** ones. Again, as shown in Figures 8 and 9, a clear improvement is achieved when the modified version of the Cayley method is used.

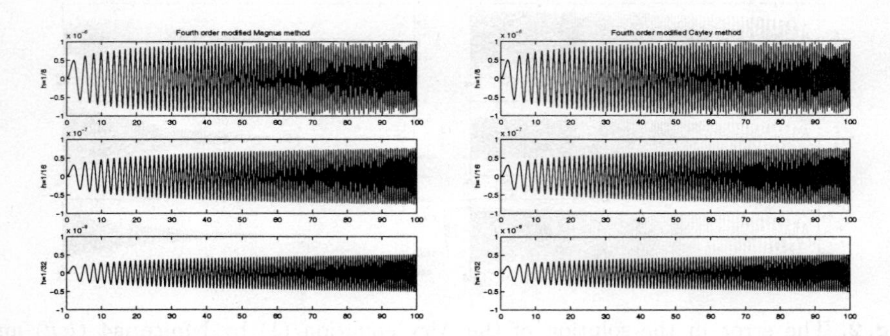

Fig. 4. The error in the solution of the Airy equation (1) by **modified Magnus4** (*left*) and **modified Cayley4** (*right*) in the time interval $[0, 100]$.

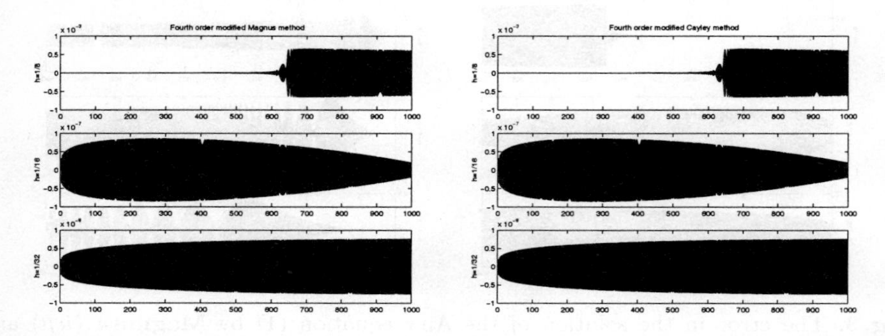

Fig. 5. The error in the solution of the Airy equation (1) by **modified Magnus4** (*left*) and **modified Cayley4** (*right*) in the time interval $[0, 1000]$.

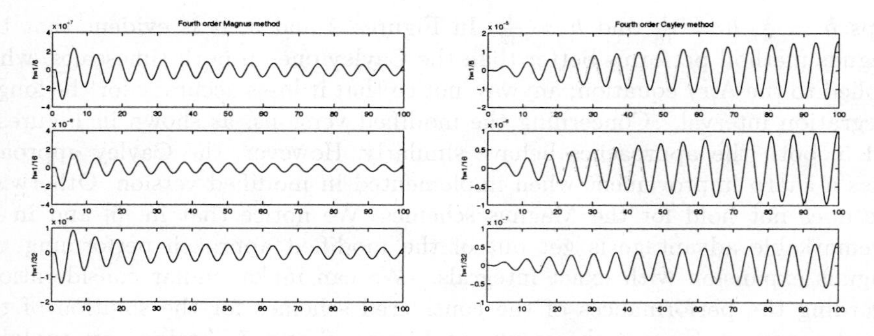

Fig. 6. The error in the solution of the Bessel equation (3) by **Magnus4** (*left*) and **Cayley4** (*right*) in the time interval $[1, 100]$.

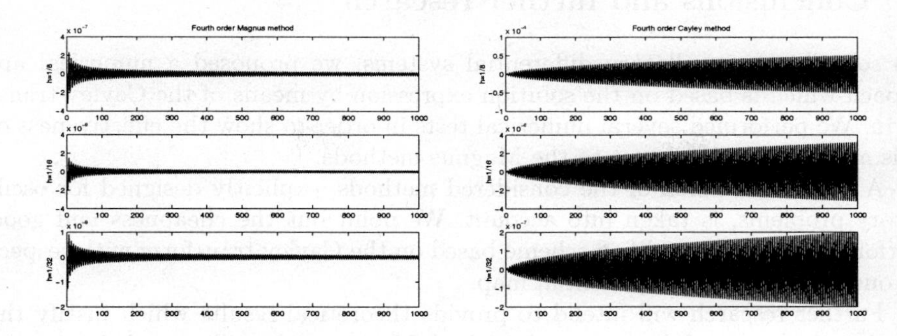

Fig. 7. The error in the solution of the Bessel equation (3) by **Magnus4** (*left*) and **Cayley4** (*right*) in the time interval [1,1000].

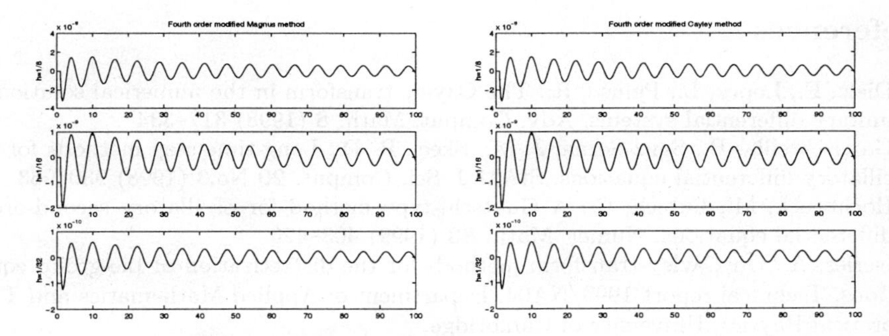

Fig. 8. The error in the solution of the Bessel equation (3) by **modified Magnus4** (*left*) and **modified Cayley4** (*right*) in the time interval [1, 100].

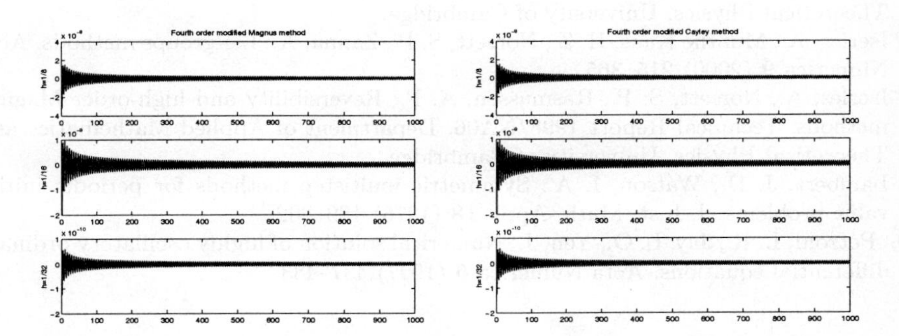

Fig. 9. The error in the solution of the Bessel equation (3) by **modified Magnus4** (*left*) and **modified Cayley4** (*right*) in the time interval [1, 1000].

5 Conclusions and further research

To solve highly oscillatory differential systems, we proposed a numerical approach which is based on the solution expression by means of the Cayley transform. We performed several numerical tests in order to show the effectiveness of this approach with respect to the Magnus methods.

A modified version of the considered methods, explicitly designed for oscillatory problems, is taken into account. We point out the cheapness and good performance of the modified scheme based on the Cayley transform with respect to ones based on the exponential map.

Further research will intend to provide theoretical results which justify the good performance of the proposed method. Moreover, we will extend the present approach to the non homogeneous case. It might be advantageous to use the Cayley method for the oscillatory part and following an approach like Gautschi-type to treat the non oscillatory term.

References

1. Diele, F., Lopez, L., Peluso, R.: The Cayley transform in the numerical solution of unitary differential systems. Adv. Comput. Math. **8** (1998) 317–334
2. Garca-Arcilla, B., Sanz-Serna, J. M., Skeel, R. D.: Long-time-step methods for oscillatory differential equations. SIAM J. Sci. Comput. **20** No.3 (1998) 930–963
3. Hochbruck, M., Lubich, C.: A Gautschi-type method for oscillatory second-order differential equations. Numer. Math. **83** (1999) 403–426
4. Iserles, A.: On Cayley-transform methods for the discretization of Lie-group equations. Technical report 1999/NA04. Department of Applied Mathematics and Theoretical Physics, University of Cambridge.
5. Iserles, A.: On the global error of discretization methods for highly-oscillatory ordinary differential equations. Technical report 2000/NA11. Department of Applied Mathematics and Theoretical Physics, University of Cambridge.
6. Iserles, A.: Think globally, act locally: solving highly-oscillatory ordinary differential equations. Technical report 2001/NA06. Department of Applied Mathematics and Theoretical Physics, University of Cambridge.
7. Iserles, A., Munthe Kaas, H. Z., Norsett, S. P., Zanna, A.: Lie-groups methods. Acta Numerica **9** (2000) 215–365
8. Iserles, A., Norsett, S. P., Rasmussen, A. F.: Reversibility and high-order Magnus methods. Technical Report 1998/NA06. Department of Applied Mathematics and Theoretical Physics, University of Cambridge.
9. Lambert, J. D., Watson, I. A.: Symmetric multistep methods for periodic initial value problems. J. Inst. Math. Appl. **18** (1976) 189–202
10. Petzold, L. R., Jay, L. O., Yen, J.: Numerical solution of highly oscillatory ordinary differential equations. Acta Numerica **6** (1997) 437–483

Appendix: MATLAB code

For the sake of completeness, herewith a MATLAB code for the fourth order modified Cayley method:

```
function [t,Y]=MCayley4(N, t0, tN, y0, h);

% Modified Cayley4
% Input:
%  N:   the spacing of solution points (typically N=1)
%  t0:  initial point
%  tN:  endpoint
%  y0:  initial value
%  h:   step size
% Output:
%  t = [t0:Nh:tN]
%  Y = [y(t0), y(t0+Nh), y(t0+2Nh), ..., y(tN)]
 c1 = (1/2 - sqrt(3)/6)*h;
 c2 = (1/2 + sqrt(3)/6)*h;
 h2 = h/2;
 t = t0;
 Y = y0;
 I = eye(length(y0));
 M = floor((tN - t0)/h + eps);
for n = 1:M
    A0 = Afun(t0 + h2);
    Th = expm(h*A0);
    T1 = expm(c1*A0);
    T2 = expm(c2*A0);
    A1 = Afun(t0 + c1);
    A2 = Afun(t0 + c2);
    U1 = T1\(A1 - A0)*T1;
    U2 = T2\(A2 - A0)*T2;
    B0 = 1/2*(U1 + U2);
    B1 = sqrt(3)*(U2 - U1);
    C = B1*B0 - B0*B1;
    Theta = h*B0 + h^2/12*C - 1/12*h^3*B0^3;
    I1 = (I - 1/2*Theta);
    I2 = (I + 1/2*Theta);
    U = I1\I2;
    y0 = Th*U*y0;
    t0 = t0 + h;
    if floor(n/N)*N == n
        Y = [Y,y0];
        t = [t,t0];
    end
end
```

where the function Afun.m provides the values of the matrix $A(t)$: for instance, for the Airy equation (1) it reads

```
function A = Afun(t) A = [0, 1; -t,0]
```

Multisymplectic Spectral Methods for the Gross-Pitaevskii Equation

A.L. Islas[1] and C.M. Schober[1]

Department of Mathematics and Statistics, Old Dominion University
cschober@lions.odu.edu

Abstract. Recently, Bridges and Reich introduced the concept of multi-symplectic spectral discretizations for Hamiltonian wave equations with periodic boundary conditions [5]. In this paper, we show that the 1D nonlinear Schrodinger equation and the 2D Gross-Pitaevskii equation are multi-symplectic and derive multi-symplectic spectral discretizations of these systems. The effectiveness of the discretizations is numerically tested using initial data for multi-phase solutions.

1 The Multisymplectic Approach

A Hamiltonian PDE with N space dimensions is said to be *multisymplectic* if it can be written as

$$\boldsymbol{M} z_t + \sum_{i=1}^{N} \boldsymbol{K}_i \, z_{x_i} = \boldsymbol{\nabla}_z S(z), \qquad z \in \mathbb{R}^d, \tag{1}$$

where $\boldsymbol{M}, \boldsymbol{K}_i \in \mathbb{R}^{d\times d}$ are skew–symmetric matrices and $S : \mathbb{R}^d \to \mathbb{R}$ is a smooth function [1]. Associated with (1) are the N+1 two–forms

$$\omega(U,V) = V^T \boldsymbol{M} U, \quad \kappa_i(U,V) = V^T \boldsymbol{K}_i U, \quad U, V \in \mathbb{R}^d \tag{2}$$

which define a symplectic structure associated with time and the space directions x_i, respectively. System (1) implies the existence of a multi–symplectic conservation law

$$\partial_t \omega + \sum_{i=1}^{N} \partial_{x_i} \kappa_i = 0, \tag{3}$$

when U, V are any two solutions of the variational equation associated with (1)

$$\boldsymbol{M} dz_t + \sum_{i=1}^{N} \boldsymbol{K}_i \, dz_{x_i} = D_{zz} S(z) dz.$$

One consequence of multi–symplecticity is that when the Hamiltonian $S(z)$ is independent of x_i and t, the PDE has an energy conservation law (ECL) [4]

$$\frac{\partial E}{\partial t} + \sum_{i=1}^{N} \frac{\partial F_i}{\partial x_i} = 0, \quad E = S(z) - \frac{1}{2} z^T \sum_{i=1}^{N} \boldsymbol{K}_i \, z_{x_i}, \quad F_i = \frac{1}{2} z^T \boldsymbol{K}_i \, z_t \tag{4}$$

P.M.A. Sloot et al. (Eds.): ICCS 2002, LNCS 2331, pp. 486–495, 2002.

as well as a momentum conservation law

$$\sum_{i=1}^{N}\frac{\partial I_i}{\partial t}+\sum_{i=1}^{N}\frac{\partial G}{\partial x_i}=0,\quad G=S(z)-\frac{1}{2}z^T\boldsymbol{M}\,z_t,\quad I_i=\frac{1}{2}z^T\boldsymbol{M}\,z_{x_i}. \tag{5}$$

When the local conservation laws are integrated over the domain in $\mathbb{R}^N$, using periodic boundary conditions, we obtain the global conservation of the total energy and total momentum.

Multi–symplectic integrators are approximations to (1) which conserve a discretization of the multi–symplectic conservation law (3). This newly emerging class of integrators has proven very promising as it includes simple and fast schemes with remarkable conservation properties for local as well as global invariants (cf. [1] - [5]).

The nonlinear Schrödinger (NLS) equation in two space dimensions with an external potential models the mean-field dynamics of a dilute-gas Bose Einstein condensate (BEC) [7]. In this case the equation is referred to as the Gross-Pitaevskii (GP) equation. Numerical experiments with the GP equation are used to provide insight into BEC stability properties. The NLS equation and the GP equation can be formulated as multi-symplectic systems. In order that the numerical discretization reflects the geometric properties of the PDE, we investigate the use of multi-symplectic spectral integrators for the NLS and GP equations.

1.1 The multi-symplectic form of the 1D nonlinear Schrödinger equation

The focusing one dimensional nonlinear Schrödinger (NLS) equation,

$$iu_t+u_{xx}+2|u|^2u=0, \tag{6}$$

can be written in multisymplectic form by letting $u=p+iq$ and introducing the new variables $v=p_x$, $w=q_x$ [2]. Separating (6) into real and imaginary parts, the multi-symplectic form (eq. (1) with N=1) for the NLS equation is obtained with

$$\boldsymbol{z}=\begin{pmatrix}p\\q\\v\\w\end{pmatrix},\qquad \boldsymbol{M}=\begin{pmatrix}0&1&0&0\\-1&0&0&0\\0&0&0&0\\0&0&0&0\end{pmatrix},\qquad \boldsymbol{K}=\begin{pmatrix}0&0&-1&0\\0&0&0&-1\\1&0&0&0\\0&1&0&0\end{pmatrix}$$

and Hamiltonian $S(z)=\frac{1}{2}\left[\left(p^2+q^2\right)^2+\left(v^2+w^2\right)\right]$. The multi-symplectic conservation law for the NLS equation is given by

$$\partial_t[\boldsymbol{d}p\wedge\boldsymbol{d}q]+\partial_x[\boldsymbol{d}p\wedge\boldsymbol{d}v+\boldsymbol{d}q\wedge\boldsymbol{d}w]=0. \tag{7}$$

Implementing relations (4)-(5) for the NLS equation we obtain the energy conservation law (ECL)

$$\frac{1}{2}\left[\left(p^2+q^2\right)^2-\left(v^2+w^2\right)\right]_t+\left[p_tv+q_tw\right]_x=0. \tag{8}$$

1.2 The multi-symplectic form of the 2D Gross-Pitaevskii equation

After rescaling the physical variables, the Gross-Pitaevskii (GP) equation is given by

$$iu_t = -\frac{1}{2}\left(u_{xx} + u_{yy}\right) + \alpha|u|^2 u + V(x,y)u \tag{9}$$

where $u(x,y,t)$ is the macroscopic wave function of the condensate and $V(\mathbf{x})$ is an experimentally generated macroscopic potential. The parameter α determines whether (9) is repulsive ($\alpha = 1$, defocusing nonlinearity), or attractive ($\alpha = -1$, focusing nonlinearity). Although in BEC applications both signs of α are relevant, here we will concentrate on (9) with repulsive nonlinearity. As in [7], we consider the family of periodic lattice potentials given by

$$V(x,y) = -\left(A_1 \mathrm{sn}_1^2 + B_1\right)\left(A_2 \mathrm{sn}_2^2 + B_2\right) + \left(m_1 k_1 \mathrm{sn}_1\right)^2 + \left(m_2 k_2 \mathrm{sn}_2\right)^2 \tag{10}$$

where $\mathrm{sn}_i = \mathrm{sn}(m_i x,\, k_i)$ denote the Jacobian elliptic sine functions, with elliptic moduli k_i. A nice feature of this potential is that it has some closed form solutions which can be used for comparative purposes.

The GP equation can be reformulated as a multisymplectic PDE by letting $u = p + iq$ and $v_1 = p_x$, $v_2 = p_y$, $w_1 = q_x$, $w_2 = q_y$. Then, with the state vector $z = (p, q, v_1, w_1, v_2, w_2)$, the skew-symmetric matrices

$$\boldsymbol{M} = \begin{pmatrix} 0 & 1 & 0 & 0 & 0 & 0 \\ -1 & 0 & 0 & 0 & 0 & 0 \\ 0 & 0 & 0 & 0 & 0 & 0 \\ 0 & 0 & 0 & 0 & 0 & 0 \\ 0 & 0 & 0 & 0 & 0 & 0 \\ 0 & 0 & 0 & 0 & 0 & 0 \end{pmatrix}, \quad \boldsymbol{K_1} = \begin{pmatrix} 0 & 0 & -\frac{1}{2} & 0 & 0 & 0 \\ 0 & 0 & 0 & -\frac{1}{2} & 0 & 0 \\ \frac{1}{2} & 0 & 0 & 0 & 0 & 0 \\ 0 & \frac{1}{2} & 0 & 0 & 0 & 0 \\ 0 & 0 & 0 & 0 & 0 & 0 \\ 0 & 0 & 0 & 0 & 0 & 0 \end{pmatrix}, \quad \boldsymbol{K_2} = \begin{pmatrix} 0 & 0 & 0 & 0 & -\frac{1}{2} & 0 \\ 0 & 0 & 0 & 0 & 0 & -\frac{1}{2} \\ 0 & 0 & 0 & 0 & 0 & 0 \\ 0 & 0 & 0 & 0 & 0 & 0 \\ \frac{1}{2} & 0 & 0 & 0 & 0 & 0 \\ 0 & \frac{1}{2} & 0 & 0 & 0 & 0 \end{pmatrix}$$

and Hamiltonian $S(z) = -\frac{1}{4}\left[\left(p^2+q^2\right)^2 + 2V\left(p^2+q^2\right) - \left(v_1^2 + w_1^2 + v_2^2 + w_2^2\right)\right]$, the system can be written in the canonical form (1).

The multi–symplectic conservation law

$$\partial_t \omega + \partial_x \kappa_1 + \partial_y \kappa_2 = 0 \tag{11}$$

where ω and κ_i are given by (2).

The energy conservation law is

$$\frac{\partial e}{\partial t} + \frac{\partial f_1}{\partial x} + \frac{\partial f}{\partial y} = 0 \tag{12}$$

with energy density

$$e = -\frac{1}{4}\left[\left(p^2+q^2\right)^2 + 2V\left(p^2+q^2\right) - \left(pv_{1x} + qw_{1x} + pv_{2y} + qw_{2y}\right)\right] \tag{13}$$

and the two energy fluxes

$$f_1 = -\frac{1}{4}\left(pv_{1t} + qw_{1t} - p_t v_1 - q_t w_1\right), \quad f_2 = -\frac{1}{4}\left(pv_{2t} + qw_{2t} - p_t v_2 - q_t w_2\right).$$

In this case the ECL for the GP equation becomes

$$\left[\left(p^2+q^2\right)^2+2V\left(p^2+q^2\right)+\left(v_1^2+w_1^2+v_2^2+w_2^2\right)\right]_t$$
$$-2\left[\left(p_t v_1+q_t w_1\right)_x+\left(p_t v_2+q_t w_2\right)_y\right]=0. \quad (14)$$

2 The Spectral Discretization

Bridges and Reich [5] have shown that using the Fourier transforms leaves the multi–symplectic nature of a PDE unchanged and that the discrete Fourier system recovers the standard spectral discretizations leading to a system of Hamiltonian ODEs which can be integrated by standard symplectic integrators. We briefly summarize these results.

Consider the space $L_2(I)$ of L-periodic, square integrable functions in $I=[-L/2,L/2]$ and let $U=\mathcal{F}u$ denote the discrete Fourier transform of $u\in L_2(I)$. Here $\mathcal{F}: L_2\to l_2$ denotes the Fourier operator which gives the complex–valued Fourier coefficients $U_k\in\mathbb{C}$, $k=-\infty,\dots,-1,0,1,\dots,\infty$, which we collect in the infinite–dimensional vector $U=(\dots,U_{-1},U_0,U_1,\dots)\in l_2$. Note that $U_{-k}=U_k^*$. We also introduce the L_2 inner product, which we denote by (u,v) and the l_2 inner product, which we donote by $\langle U,V\rangle$. The inverse Fourier operator $\mathcal{F}^{-1}: l^2\to L^2$ is defined by $\langle V,\mathcal{F}u\rangle=(\mathcal{F}^{-1}V,u)$. Furthermore, partial differentiation with respect to $x\in I$ simply reduces to $\partial_x u=\mathcal{F}^{-1}\Theta U$ where $\Theta: l_2\to l_2$ is the diagonal spectral operator with entries $\theta_k=i2\pi k/L$.

These definitions can be generalized to vector–valued functions $z\in L_2^d(I)$. Let $\hat{\mathcal{F}}: L_2^d(I)\to l_2^d$ be defined such that $Z=(Z^1,\dots,Z^d)=\hat{\mathcal{F}}z=(\mathcal{F}z^1,\dots,\mathcal{F}z^d)$. Thus with a slight abuse of notations and after dropping the hats, we have $Z=\mathcal{F}z$, $z=\mathcal{F}^{-1}Z$, and $\partial_x z=(\partial_x z^1,\dots,\partial_x z^d)=(\mathcal{F}^{-1}\Theta Z^1,\dots,\mathcal{F}^{-1}\Theta Z^d)=\mathcal{F}^{-1}\Theta Z$

Applying these operators to the multi–symplectic PDE (1), one obtains an infinite dimensional system of ODEs

$$\boldsymbol{M}\partial_t Z+\boldsymbol{K}\Theta Z=\boldsymbol{\nabla}_Z\bar{S}(Z),\quad \bar{S}(Z)=\int_{-L}^{L}S(\mathcal{F}^{-1}Z)\,dx. \quad (15)$$

This equation can appropriately be called a multi–symplectic spectral PDE with associated multi–symplectic and energy conservation laws

$$\partial_t\Omega_k+\theta_k\mathcal{K}_k=0,\qquad \Omega=\mathcal{F}\omega,\quad \mathcal{K}=\mathcal{F}\kappa, \quad (16)$$
$$\partial_t E_k+\theta_k F_k=0,\qquad E=\mathcal{F}e,\quad F=\mathcal{F}f. \quad (17)$$

Bridges and Reich show that the truncated Fourier series,

$$U_k=\frac{1}{\sqrt{N}}\sum_{l=1}^{N}u_l\,e^{-\theta_k(l-1)\Delta x},\ u_l=u(x_l),\ x_l=-\frac{L}{2}+(l-1)\Delta x,\ \Delta x=\frac{L}{N},$$

with

$$\theta_k = \begin{cases} i\frac{2\pi}{L}(k-1) & \text{for } k = 1,\dots,N/2, \\ 0 & \text{for } k = N/2+1 \\ -\theta_{N-k+2} & \text{for } k = N/2+2,\dots,N \end{cases}$$

gives a multi–symplectic spatial discretization [5]. Therefore to mantain the multi–symplecticity a discrete integrator that is symplectic in time should be used, such as the implicit midpoint method.

The 1D NLS equation The beauty of the spectral multi–symplectic scheme (15) is that in many cases, such as those considered in this paper, one can recover the standard spectral discretization of the original equation in complex form. That is, using a spectral discretization of the spatial derivatives one obtains a multi–symplectic Hamiltonian system of ODEs which can be integrated using standard symplectic methods such as the implicit midpoint rule.

Let $\mathcal{D}^n(u)$ be the spectral discretization of $\frac{\partial^n u}{\partial x^n}$, $\mathcal{D}^n(u) = \mathcal{F}^{-1}\{\Theta^n \mathcal{F} u\}$, and $\mathcal{C}(u) = 2|u|^2 u$. Then the multi–symplectic spectral method for the 1D NLS, using the implicit midpoint rule in time, is given by

$$i\frac{u^1 - u^0}{\Delta t} + \mathcal{D}^2(u^{1/2}) + \mathcal{C}(u^{1/2}) = 0, \tag{18}$$

where $u^{1/2} = (u^1 + u^0)/2$ and $u^j = u(x, j\Delta t)$. Scheme (18) is denoted as MS-S in the numerical experiments.

The numerically induced residual R_i of the ECL is given by

$$R_i = \frac{E_i^1 - E_i^0}{\Delta t} + D^1(F_i^{1/2}), \tag{19}$$

where $E^n = \frac{1}{2}\left(|u^n|^4 - |D^1(u^n)|^2\right)$ and $F^{1/2} = Re\left\{\left(\frac{u^1-u^0}{\Delta t}\right)^* D^1(u^{1/2})\right\}$. This residual of the ECL is due to *local* non–conservation of energy under numerical discretization, and it can be compared with the global energy error

$$\Delta E^j = |E^j - E^0|.$$

Note the relation

$$\frac{E^j - E^{j-1}}{\Delta t} = \sum_i R_i^j$$

.

The 2D GP equation As before, we let $\mathcal{D}_i^n(x)$ be the spectral discretization of $\frac{\partial^n u}{\partial x_i^n}$, $\mathcal{D}_i^n(u) = \mathcal{F}^{-1}\{\Theta_i^n \mathcal{F} u\}$ and $\mathcal{C}(u) = |u|^2 u - V(x,y)u$. Then the multi–symplectic spectral method for the 2D GP equation, using the second order implicit midpoint rule in time, is given by

$$i\frac{u^1 - u^0}{\Delta t} + \mathcal{D}_1^2(u^{1/2}) + \mathcal{D}_2^2(u^{1/2}) + \mathcal{C}(u^{1/2}) = 0, \tag{20}$$

where $u^{1/2} = (u^1 + u^0)/2$ and $u^j = u(x, y, j\Delta t)$. Scheme (20) is denoted as GP-MS in the numerical experiments.

The numerically induced residual R_i of the ECL is given by

$$R = \frac{E^1 - E^0}{\Delta t} - 2\left(D_1^1\left(F_1^{1/2}\right) + D_2^1\left(F_2^{1/2}\right)\right)$$

with

$$E^j = \left(|u^j|^4 + 2V|u|^2 + |D_1^1(u^j)|^2 + |D_2^1(u^j)|^2\right), \tag{21}$$

and

$$F_i^{1/2} = Re\left\{\left(\frac{u^1 - u^0}{\Delta t}\right)^* D_i^1(u^{1/2})\right\}. \tag{22}$$

We are interested in simulating multi-phase quasi-periodic (in time) solutions to the NLS and GP equations under periodic boundary conditions.

3 Numerical Experiments

The 1D NLS equation: We consider initial conditions of the form

$$u_0(x) = a(1 + \epsilon \cos \mu x) \tag{23}$$

where $a = 0.5$, $\epsilon = 0.1$, $\mu = 2\pi/L$ and L is either (a) $L = 2\sqrt{2}\pi$ or (b) $L = 4\sqrt{2}\pi$ [2]. Initial data (a) and (b) correspond to multi-phase solutions, near the plane wave, which are characterized by either one or two excited modes, respectively. We refer to these cases as the one mode and two mode case.

In [2], a multi-symplectic centered cell discretization (obtained by concatenating two implicit midpoint schemes) as well as an integrable-symplectic discretization (an integrable spatial discretization with symplectic integrator in time) were developed for the NLS equation. The multi-symplectic centered cell discretization is denoted as MS-CC in the subsequent discussion. The geometric integrators were shown to be more efficient than standard integrators in preservation of geometric features of the system such as local and global conserved quantities, quasiperiodic character of the motion and qualitative features of the waveform. However, among the geometric integrators, performance varied. The integrable-symplectic scheme reproduced more faithfully the qualitative features of the wave profile than the MS-CC scheme.

In this paper, we show that the MS spectral discretization captures the qualitative features of the waveform better than the MS centered cell discretization. We begin with initial data (a) for the one mode case. Figures 1 show the conservation of the residual energy $R(x,t)$ using the MS-S and the MS-CC discretizations, respectively, with $N = 64$, $\Delta t = 2.5 \times 10^{-3}$, $T = 450 - 500$. The error in the ECL obtained using the spectral scheme is several orders of magnitude smaller than the ECL obtained with the centered cell scheme (similarly for the corresponding error in the global energy, not shown). The surfaces of the one-mode case obtained with MS-S and MS-CC are essentially identical and the global momentum and norm are conserved exactly by both schemes, up to the error criterion of 10^{-14} in the iteration procedure in these implicit schemes (not shown).

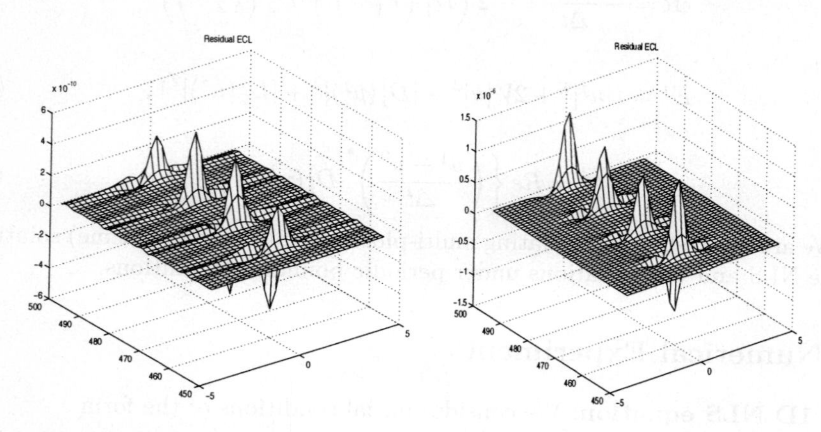

Fig. 1. The residual ECL, $R(x,t)$, of the NLS one mode case obtained using a) the MS-S scheme and b) the MS-CC sceme with $N = 64$, $\Delta t = 2.5 \times 10^{-3}$, $T = 450 - 500$.

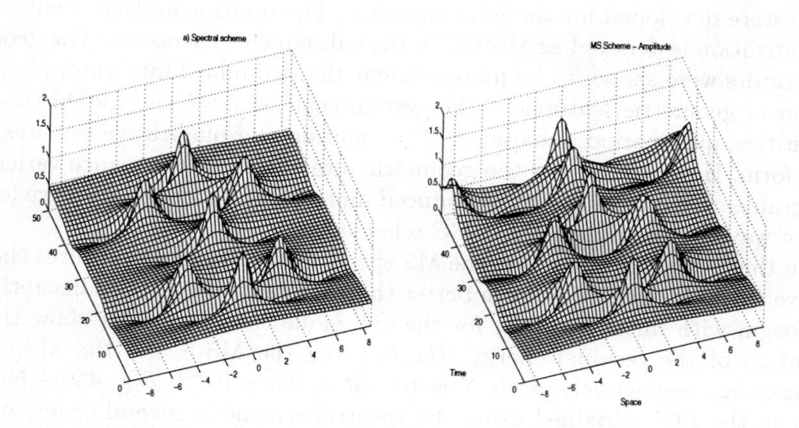

Fig. 2. The surface $|u(x,t)|$ of the NLS two mode case obtained using a) the MS-S scheme and b) the MS-CC sceme with $N = 64$, $\Delta t = 5 \times 10^{-3}$, $T = 0 - 50$.

Figures 2 show the surfaces $|u(x,t)|$ of the waveform obtained using MS-S and MS-CC, respectively for initial data (b) with discretization parameters $N = 64$, $\Delta t = 5 \times 10^{-3}$, $T = 0 - 50$. The MS-S scheme correctly captures the quasiperiodic motion and produces results which are comparable to those obtained using the integrable-symplectic scheme (see [2]). On the other hand, using the MS-CC integrator, the onset of numerically induced temporal chaos is observed at approximately $t = 25$. The temporal chaos is characterized by a random switching in time of the location of the spatial excitations in the waveform, see Figure 2(b). However, for the duration of the simulation $0 < t < 500$, switching in the spatial excitations does not occur using the MS-S scheme. As in the one mode case, the ECL is preserved better by the MS-S scheme. Since a significant improvement in the qualitative features of the solution is obtained with the MS-S scheme in 1D, MS spectral schemes should prove to be a valuable tool in integrating multi-dimensional PDEs.

The 2D GP equation: In the following experiments periodic boundary conditions in x and y are imposed and we use a fixed $N \times N$ spatial lattice with $N = 32$. The time step used throughout is $\Delta t = 2 \times 10^{-3}$. We begin by considering solutions of (9) with an elliptic function potential (10) that has the following choice of constants: $k_1 = k_2 = 1/2$, $m_1 = m_2 = 1$, $A_1 = A_2 = -1$, $B_1 = B_2 = -A_1/k_1^2$ and the initial condition

$$u_0(x,y) = \sqrt{B_1}\sqrt{B_2}\mathrm{dn}(m_1 x,\, k_1)\mathrm{dn}(m_2 x,\, k_2). \tag{24}$$

This is initial data for a linearly stable stationary solution of the GP equation.

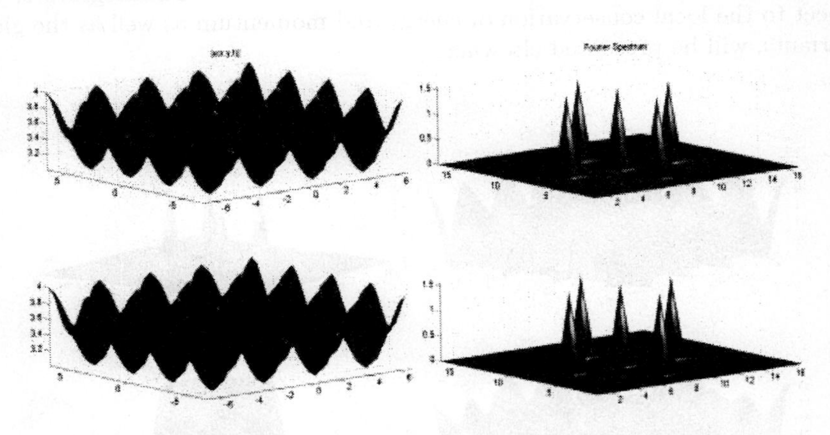

Fig. 3. Stable periodic potential with constants $k_1 = k_2 = 1/2$, $m_1 = m_2 = 1$, $A_1 = A_2 = -1$, $B_1 = B_2 = -A_1/k_1^2$ and initial condition (24): a) Surface; b) Fourier spectrum.

The evolution of the solution obtained using the GP-MS scheme (20) is shown in Figure 3. The surface $|u(x,y,t)|$ is given in the first column and the fourier spectrum is given in the second column. The plots are at $t = 0$ and $t = 60$, top and bottom figures respectively. As analytically determined in [7], this solution is clearly stable. In further numerical simulations (not shown) for $0 < t < 1000$, the solution obtained with GP-MS (20) remains stable with no growth in the Fourier modes. The ECL is preserved on the order of 10^{-3} and the error in the global energy oscillates in a bounded fashion as is typical of the behavior of a symplectic integrator.

Next, we examine the solution obtained using the elliptic function potential (10) with the values of the constants now specified to be $k_1 = k_2 = 1/2$, $m_1 = m_2 = 1$, $A_1 = A_2 = 1$, $B_1 = B_2 = -A_1$ and the initial condition

$$u_0(x, y) = \sqrt{B_1}\sqrt{B_2}\text{cn}(m_1 x,\, k_1)\text{cn}(m_2 x,\, k_2). \tag{25}$$

Figure 3 shows the surface of the waveform $|u(x, y, t)|$ and the fourier spectrum obtained using the GP-MS scheme (20) at $t = 0$ and $t = 60$, in the same order as before.

Obviously, this solution is unstable, as reported in [7]. The onset of the instability occurs between $t = 15$ and $t = 20$ and by $t = 60$ a significant number of additional Fourier modes have become excited. The ECL, as well as the global invaariants are well preserved by the GP-MS discretization. Even working with this coarse lattice, we are able to recover the main qualitative features of the solution. As for the 1D NLS equation, for two dimensional systems we already see the power of multi-symplectic spectral integrators. A more detailed study of the performance of the GP-MS scheme, relative to standard integrators, with respect to the local conservation of energy and momentum as well as the global invariants, will be presented elsewhere.

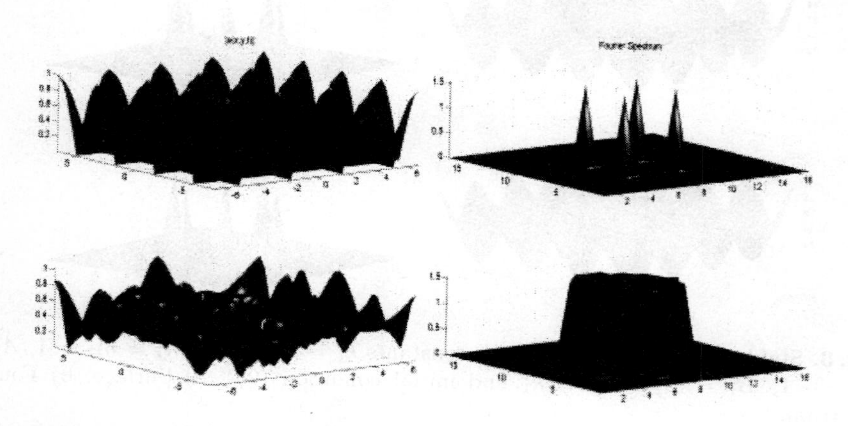

Fig. 4. Unstable periodic potential with constants $k_1 = k_2 = 1/2$, $m_1 = m_2 = 1$, $A_1 = A_2 = 1$, $B_1 = B_2 = -A1$ and initial condition (25): a) Surface; b) Fourier spectrum.

References

1. T.J. Bridges, Multisymplectic structures and wave propagation, Math. Proc. Cambridge Philos. Soc. **121**, 147 (1997).
2. A.L. Islas, D.A. Karpeev and C.M. Schober, Geometric Integrators for the Nonlinear Schrödinger Equation, J. of Comp. Phys. **173**, 116–148 (2001).
3. S. Reich, Multisymplectic Runge–Kutta Collocation Methods for Hamiltonian Wave Equations, J. of Comp. Phys. **157**, 473–499 (2000).
4. T.J. Bridges and S. Reich, Multisymplectic Integrators: numerical schemes for Hamiltonian PDEs that conserve symplecticity, University of Surrey, Technical Report (1999).
5. T.J. Bridges and S. Reich, Multisymplectic Spectral Discretizations for the Zakharov–Kuznetsov and shallow water equations, University of Surrey, Technical Report (2000).
6. P.J. Channell and C. Scovel, Symplectic integration of Hamiltonian systems, Nonlinearity **3**, 1–13 (1990).
7. B. Deconinck, B.A. Frigyik and J.N. Kutz, Stability of exact solutions of the defocusing nonlinear Schrödinger equation with periodic potential in two dimensions, Physics Letters A, submitted April 2001.
8. J.E. Marsden, G.P. Patrick and S. Shkoller, Multisymplectic geometry, variational integrators, and nonlinear PDEs, Comm. in Math. Phys. **199**, 351–395 (1999).
9. J.E. Marsden and S. Shkoller, Multisymplectic geometry, covariant Hamiltonians and water waves, Math. Proc. Camb. Phil. Soc. **125**, 553–575 (1999).

Solving Orthogonal Matrix Differential Systems in *Mathematica*

Mark Sofroniou[1] and Giulia Spaletta[2]

[1] Wolfram Research, Champaign, Illinois, USA. marks@wolfram.com
[2] Mathematics Department, Bologna University, Italy. giulia@cs.unibo.it

Abstract. A component of a new environment for the numerical solution of ordinary differential equations in *Mathematica* is outlined. We briefly describe how special purpose integration methods can be constructed to solve structured dynamical systems. In particular we focus on the solution of orthogonal matrix differential systems using projection. Examples are given to illustrate the advantages of a projection scheme over conventional integration methods.
Keywords. Geometric numerical integration; orthogonal projection; ordinary differential equations; computer algebra systems.
AMS. 65L05; 65L06; 68Q40.

1 Features of the framework

The *Mathematica* function **NDSolve** can be used to find numerical solutions of a wide range of ordinary as well as some partial differential equations. A drawback with the current function is that the design and implementation are in the form of a 'black-box' and there is only a single one-step numerical method available, an outdated explicit Runge-Kutta pair of Fehlberg. Since the function is not modular, it cannot be used to take advantage of new research developments.

One of the aims of an ongoing overhaul is to make a number of differential equations solvers available in a uniform, integrated and extensible environment. Many one-step integrators are being developed: explicit, implicit, linearly implicit Euler and Midpoint; embedded explicit Runge Kutta pairs of various orders; Gauss, Lobatto (IIIA, IIIB, IIIC), Radau (IA, IIA) implicit Runge Kutta methods; extrapolation methods.

In recent years there has been a considerable growth of interest in studying and numerically preserving a variety of dynamical properties, leading to so called geometric integrators (see for example [9], [10], [17], [18], [21]). The new **NDSolve** allows built-in methods to be exploited as building blocks for the efficient construction of special purpose (compound) integrators. The framework is also hierarchical, meaning that one method can call another at each step of an integration. These features facilitate the construction of geometric integrators and the implementation of one specific method is given here as demonstration.

This paper is organized as follows. Section 2 describes the class of problems of interest and various strategies for their numerical solution. Amongst the possible

P.M.A. Sloot et al. (Eds.): ICCS 2002, LNCS 2331, pp. 496–505, 2002.

choices, an iterative method is selected and an algorithm for the implementation is discussed together with appropriate stopping criteria. Section 3 describes the implementation of a projected integration method, **OrthogonalProjection**, written in top-level *Mathematica* code. Examples of improved qualitative behavior over conventional integrators are given by considering the solution of square and rectangular orthogonal matrix differential systems in Section 4 and Section 5. Some issues relating to potential extensions are given in Section 6 together with a motivating example.

2 Orthogonal projection

Consider the matrix differential equation:

$$y'(t) = f(t, y(t)), \quad t > 0, \tag{1}$$

where the initial value $y_0 = y(0) \in \mathbb{R}^{m \times p}$ is given and satisfies $y_0^T\, y_0 = I$, where I is the $p \times p$ identity matrix. Assume that the solution preserves orthonormality, $y^T\, y = I$, and that it has full rank $p < m$ for all $t \geq 0$.

From a numerical perspective, a key issue is how to integrate (1) in such a way that the approximate solution remains orthogonal. Several strategies are possible. One approach, presented in [4], is to use an implicit Runge-Kutta method such as the Gauss scheme. These methods, however, are computationally expensive and furthermore there are some problem classes for which no standard discretization scheme can preserve orthonormality [5]. Some alternative strategies are described in [3] and [6]. The approach that will be taken up here is to use any reasonable numerical integrator and then post-process using a projective procedure at the end of each integration step. It is also possible to project the solution at the end of the integration instead of at each step, although the observed end point global errors are often larger [13].

Given a matrix, a nearby orthogonal matrix can be found via a direct algorithm such as QR decomposition or singular value decomposition (see for example [4], [13]). The following definitions are useful for the direct construction of orthonormal matrices [8].

Definition 1 (Thin Singular Value Decomposition (SVD)). *Given a matrix $A \in \mathbb{R}^{m \times p}$ with $m \geq p$, there exist two matrices $U \in \mathbb{R}^{m \times p}$ and $V \in \mathbb{R}^{p \times p}$ such that $U^T A V$ is the diagonal matrix of singular values of A, $\Sigma = diag(\sigma_1, \ldots, \sigma_p) \in \mathbb{R}^{p \times p}$, where $\sigma_1 \geq \cdots \geq \sigma_p \geq 0$. U has orthonormal columns and V is orthogonal.*

Definition 2 (Polar Decomposition). *Given a matrix A and its singular value decomposition $U\, \Sigma\, V^T$, the polar decomposition of A is given by the product of two matrices Z and P where $Z = U\, V^T$ and $P = V\, \Sigma\, V^T$. Z has orthonormal columns and P is symmetric positive semidefinite.*

If A has full rank then its polar decomposition is unique. The *orthonormal polar factor* Z of A is the matrix that, for any unitary norm, solves the minimization problem [16]:

$$\|A - Z\| = \min_{Y \in \mathbb{R}^{m \times p}} \{\|A - Y\| \,:\, Y^T Y = I\}. \qquad (2)$$

QR decomposition is cheaper than SVD, roughly by a factor of two, but it does not provide the best orthonormal approximation.

Locally quadratically convergent iterative methods for computing the orthonormal polar factor also exist, such as Newton or Schulz iteration [13]. For a projected numerical integrator, the number of iterations required to accurately approximate (2) varies depending on the local error tolerances used in the integration. For many differential equations solved in IEEE double precision, however, one or two iterations are often sufficient to obtain convergence to the orthonormal polar factor. This means that Newton or Schulz methods can be competitive with QR or SVD [13]. Iterative methods also have an advantage in that they can produce smaller errors than direct methods (see Figure 2 for example).

The application of Newton's method to the matrix function $A^T A - I$ leads to the following iteration for computing the orthonormal polar factor of $A \in \mathbb{R}^{m \times m}$:

$$Y_{i+1} = (Y_i + Y_i^{-T})/2, \quad Y_0 = A.$$

For an $m \times p$ matrix with $m > p$ the process needs to be preceded by QR decomposition, which is expensive. A more attractive scheme, that works for any $m \times p$ matrix A with $m \geq p$, is the Schulz iteration [15]:

$$Y_{i+1} = Y_i + Y_i \,(I - Y_i^T \,Y_i)/2, \quad Y_0 = A. \qquad (3)$$

The Schulz iteration has an arithmetic operation count per iteration of $2\,m^2\,p + 2\,m\,p^2$ floating point operations, but is rich in matrix multiplication [13]. In a practical implementation, gemm level 3 BLAS of LAPACK [19] can be used in conjunction with architecture specific optimizations via the Automatically Tuned Linear Algebra Software (ATLAS) [22]. Such considerations mean that the arithmetic operation count of the Schulz iteration is not necessarily an accurate reflection of the observed computational cost.

A useful bound on the departure from orthonormality of A in (2) is [14]:

$$\|A^T \,A - I\|_F \,. \qquad (4)$$

By comparing (4) and the term in parentheses in (3), a simple stopping criterion for the Schulz iteration is $\|A^T \,A - I\|_F \leq \tau$ for some tolerance τ.

Assume that an initial value y_n for the current solution is given, together with a solution $y_{n+1} = y_n + \Delta y_n$ from a one-step numerical integration method. Assume that an absolute tolerance τ for controlling the Schulz iteration is also prescribed. The following algorithm can be used for implementation.

Algorithm 1 (Standard formulation)

1. *Set* $Y_0 = y_{n+1}$ *and* $i = 0$.
2. *Compute* $E = I - Y_i^T Y_i$
3. *Compute* $Y_{i+1} = Y_i + Y_i E/2$.
4. *If* $\|E\|_F \leq \tau$ *or* $i = imax$ *then return* Y_{i+1}.
5. *Set* $i = i + 1$ *and go to step 2.*

NDSolve uses compensated summation to reduce the effect of rounding errors made in repeatedly adding the contribution of small quantities Δy_n to y_n at each integration step [16]. Therefore the increment Δy_n is returned by the base integrator. An appropriate orthogonal correction ΔY_i for the projective iteration can be determined using the following algorithm.

Algorithm 2 (Increment formulation)

1. *Set* $\Delta Y_0 = 0$ *and* $i = 0$.
2. *Set* $Y_i = \Delta Y_i + y_{n+1}$
3. *Compute* $E = I - Y_i^T Y_i$
4. *Compute* $\Delta Y_{i+1} = \Delta Y_i + Y_i E/2$.
5. *If* $\|E\|_F \leq \tau$ *or* $i = imax$ *then return* $\Delta Y_{i+1} + \Delta y_n$.
6. *Set* $i = i + 1$ *and go to step 2.*

This modified algorithm is used in **OrthogonalProjection** and shows an advantage of using an iterative process over a direct process, since it is not obvious how an orthogonal correction can be derived for direct methods.

3 Implementation

The projected orthogonal integrator **OrthogonalProjection** has three basic components, each of which is a separate routine:

- initialize the basic numerical method to use in the integration;
- invoke the base integration method at each step;
- perform an orthogonal projection.

Initialization of the base integrator involves constructing its 'state'. Each method in the new **NDSolve** framework has its own data object which encapsulates information that is needed for the invocation of the method. This includes, but is not limited to, method coefficients, workspaces, step size control parameters, step size acceptance/rejection information, Jacobian matrices. The initialization phase is performed once, before any actual integration is carried out, and the resulting data object is validated for efficiency so that it does not need to be checked at each integration step.

Options can be used to modify the stopping criteria for the Schulz iteration. One option provided by our code is **IterationSafetyFactor** which allows control over the tolerance τ of the iteration. The factor is combined with a Unit in the

Last Place, determined according to the working precision used in the integration (ULP $\approx 2.22045\,10^{-16}$ for IEEE double precision). The Frobenius norm used for the stopping criterion can be efficiently computed via the LAPACK LANGE functions [19]. An option **MaxIterations** controls the maximum number of iterations $imax$ that should be carried out.

The integration and projection phase are performed sequentially at each time step. During the projection phase various checks are performed, such as confirming that the basic integration proceeded correctly (for example a step rejection did not occur). After each projection, control returns to a central time stepping routine which is a new component of **NDSolve**. The central routine advances the solution and reinvokes the integration method.

An important feature of our implementation is that the basic integrator can be any built-in numerical method, or even a user-defined procedure. An explicit Runge-Kutta pair is often used as the basic time stepping integrator but if higher local accuracy is required an extrapolation method could be selected by simply specifying an appropriate option.

All numerical experiments in the sequel have been carried out using the default options of **NDSolve**. The appropriate initial step size and method order are selected automatically by the code (see for example [1], [7] and [9]). The step size may vary throughout the integration interval in order to satisfy local absolute and relative error tolerances. Order and tolerances can also be specified using options. With the default settings the examples of Section 4 and Section 5 require exactly two Schulz iterations per integration step.

4 Square systems

Consider the orthogonal group $O_m(\mathbb{R}) = \{Y \in \mathbb{R}^{m\times m} \; : \; Y^T Y = I\}$. The following example involves the solution of a matrix differential system on $O_3(\mathbb{R})$ [23].

$$\begin{aligned} Y' &= \qquad F(Y)\,Y \\ &= \left(A + \left(I - Y\,Y^T\right)\right)Y \end{aligned} \quad \text{where} \quad A = \begin{pmatrix} 0 & -1 & 1 \\ 1 & 0 & 1 \\ -1 & -1 & 0 \end{pmatrix}. \tag{5}$$

The matrix A is skew-symmetric. Setting $Y(0) = I$, the solution evolves as $Y(t) = \exp[t\,A]$ and has eigenvalues:

$$\lambda_1 = 1,\ \lambda_2 = \exp\left(t\,i\,\sqrt{3}\right),\ \lambda_3 = \exp\left(-\,t\,i\,\sqrt{3}\right).$$

As t approaches $\pi/\sqrt{3}$ two of the eigenvalues of $Y(t)$ approach -1. The interval of integration is $[0, 2]$.

The solution is first computed using an explicit Runge-Kutta method. Figure 1 shows the orthogonal error (4) at grid points in the numerical integration. The error is of the order of the local accuracy of the numerical method.

The orthogonal error in the solution computed using **OrthogonalProjection**, with the same explicit Runge-Kutta method as the base integration scheme,

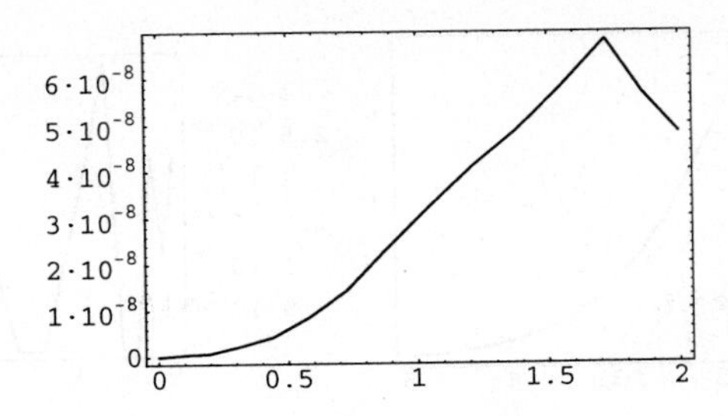

Fig. 1. Orthogonal error $\|Y^T Y - I\|_F$ vs time for an explicit Runge Kutta method applied to (5).

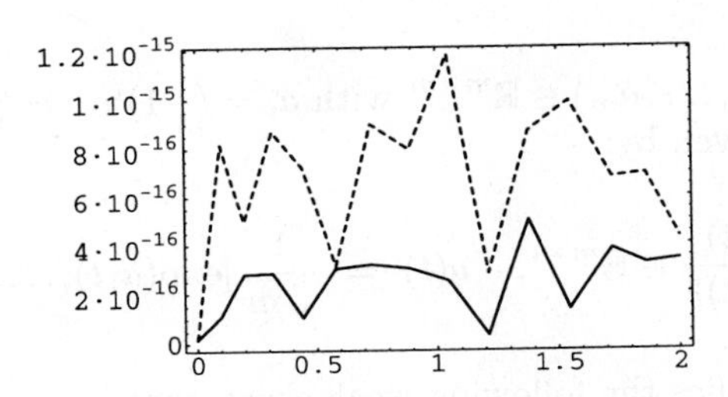

Fig. 2. Orthogonal error $\|Y^T Y - I\|_F$ vs time for projected orthogonal integrators applied to (5). The dashed line corresponds to forming the polar factor directly via SVD. The solid line corresponds to the Schulz iteration in **OrthogonalProjection**.

is illustrated in Figure 2. The errors in the orthonormal polar factor formed directly from the SVD is also given. The initial step size and method order are the same as in Figure 1, but the step size sequences in the integration are different. The orthogonal errors in the direct decomposition are larger than those of the iterative method, which are reduced to approximately the level of roundoff in IEEE double precision arithmetic.

5 Rectangular systems

OrthogonalProjection also works for rectangular matrix differential systems. Formally stated, we are interested in solving ordinary differential equations on the Stiefel manifold $V_{m,p}(\mathbb{R}) = \{Y \in \mathbb{R}^{m \times p} \ : \ Y^T Y = I\}$ of matrices of dimension $m \times p$, with $1 \leq p < m$. Solutions that evolve on the Stiefel manifold find numerous applications such as eigenvalue problems in numerical linear algebra, computation of Lyapunov exponents for dynamical systems and signal processing. Consider an example adapted from [3]:

$$q'(t) = A\,q(t), \quad t > 0, \quad q(0) = \frac{1}{\sqrt{m}}\,[1, \ldots, 1]^T, \tag{6}$$

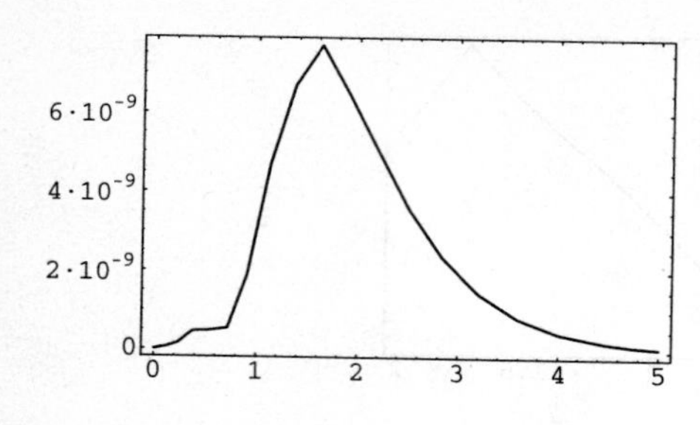

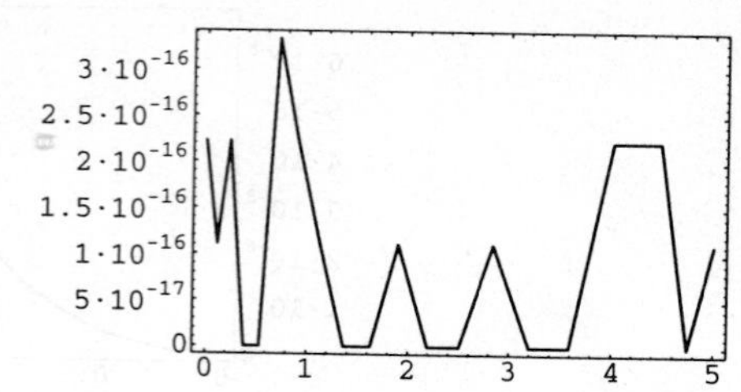

Fig. 3. Orthogonal error $\|Y^T Y - I\|_F$ vs time for (6) using **ExplicitRungeKutta** (left) and **OrthogonalProjection** (right).

where $A = diag(a_1, \ldots, a_m) \in \mathbb{R}^{m \times m}$ with $a_i = (-1)^i \alpha,\ \alpha > 0$. The normalized exact solution is given by:

$$Y(t) = \frac{q(t)}{\|q(t)\|} \in \mathbb{R}^{m \times 1}, \quad q(t) = \frac{1}{\sqrt{m}} \left[\exp(a_1 t), \ldots, \exp(a_m t)\right]^T .$$

$Y(t)$ therefore satisfies the following weak skew-symmetric system on $V_{m,1}(\mathbb{R})$:

$$\begin{aligned} Y' &= \quad F(Y)\, Y \\ &= (I - Y Y^T)\, A Y \end{aligned}$$

The system is solved on the interval $[0, 5]$ with $\alpha = 9/10$ and dimension $m = 2$.

The orthogonal error in the solution has been computed using an explicit Runge-Kutta pair and using **OrthogonalProjection** with the same explicit Runge-Kutta pair for the basic integrator. Figure 3 gives the orthogonal error at points sampled during the numerical integration. For **ExplicitRungeKutta** the error is of the order of the local accuracy. Using **OrthogonalProjection** the deviation from the Stiefel manifold is reduced to the level of roundoff.

Since the exact solution in known, it is possible to compute the component-wise absolute global error at the end of the integration interval. The results are displayed in Table 1.

Method	Errors
ExplicitRungeKutta	$(-2.38973\ 10^{-9},\ 4.14548\ 10^{-11})$
OrthogonalProjection	$(-2.38974\ 10^{-9},\ 2.94986\ 10^{-13})$

Table 1. Absolute global integration errors for (6).

6 Future work

OrthogonalProjection indicates how it is possible to extend the developmental **NDSolve** environment to add new numerical integrators. The method works by numerically solving a differential system and post-processing the solution at each step via an orthogonal projective procedure.

In some systems there may be constraints that are not equivalent to the conservation of orthogonality. An example is provided by Euler's equations for rigid body motion (see [12] and [20]):

$$\begin{pmatrix} \dot{y}_1 \\ \dot{y}_2 \\ \dot{y}_3 \end{pmatrix} = \begin{pmatrix} 0 & \frac{y_3}{I_3} & -\frac{y_2}{I_2} \\ -\frac{y_3}{I_3} & 0 & \frac{y_1}{I_1} \\ \frac{y_2}{I_2} & -\frac{y_1}{I_1} & 0 \end{pmatrix} \begin{pmatrix} y_1 \\ y_2 \\ y_3 \end{pmatrix}. \tag{7}$$

Two quadratic first integrals of the system are:

$$I(y) = y_1^2 + y_2^2 + y_3^2, \tag{8}$$

and

$$H(y) = \frac{1}{2}\left(\frac{y_1^2}{I_1} + \frac{y_2^2}{I_2} + \frac{y_3^2}{I_3}\right). \tag{9}$$

Constraint (8) is conserved by orthogonality and has the effect of confining the motion from $\mathbb{R}^3$ to a sphere. Constraint (9) represents the kinetic energy of the system and, in conjunction with (8), confines the motion to ellipsoids on the sphere. Certain numerical methods, such as the implicit midpoint rule or 1-stage Gauss implicit Runge-Kutta scheme, preserve quadratic invariants exactly (see for example [2]). Figure 4 shows three solutions of (7) computed on the interval $[0, 32]$ with constant step size $1/10$, using the initial data:

$$I_1 = 2,\ I_2 = 1,\ I_3 = \frac{2}{3},\ y_1(0) = \cos(\frac{11}{10}),\ y_2(0) = 0,\ y_3(0) = \sin(\frac{11}{10}).$$

For the explicit Euler method solutions do not lie on the unit sphere. **OrthogonalProjection**, with the explicit Euler method as the base integrator, preserves orthogonality but not the quadratic invariant (9), so that trajectories evolve on the sphere but are not closed. The implicit midpoint method conserves both (8) and (9) so that solutions evolve as ellipsoids on the sphere.

Runge-Kutta methods cannot conserve all polynomial invariants that are neither linear or quadratic [12, Theorem 3.3]. In such cases, however, the local solution from any one-step numerical scheme can be post-processed using a generalized projection based on Newton iteration (see for example [12, Section IV.4] and [10, Section VII.2]). In order to address these issues, a multiple constraint method, **Projection**, is currently under development. If the differential system is ρ-reversible in time then a *symmetric* projection process has also been shown to be beneficial [11].

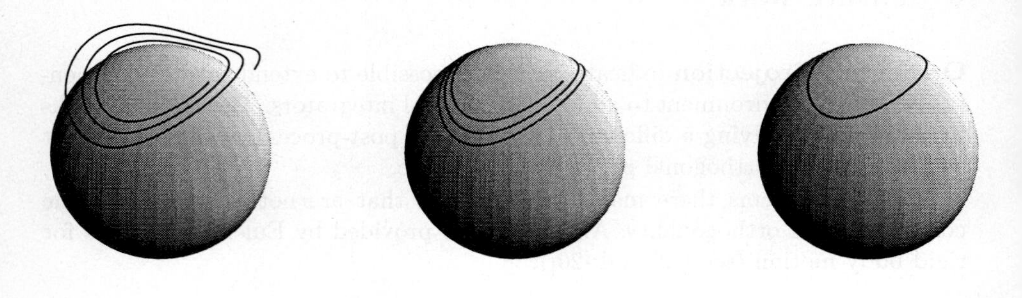

Fig. 4. Solutions of (7) using the explicit Euler method (left), **OrthogonalProjection** (center) and the implicit midpoint method (right).

Acknowledgements

The authors are grateful to Robert Knapp for his work on many aspects of **NDSolve** and to Ernst Hairer for his lectures on geometric integration and for pointing out the system (7).

References

1. Butcher, J. C.: Order, stepsize and stiffness switching. Computing. **44** (1990) 209–220.
2. Cooper, G. J.: Stability of Runge-Kutta methods for trajectory problems. IMA J. Numer. Anal. **7** (1987) 1–13.
3. Del Buono, N., Lopez, L.: Runge-Kutta type methods based on geodesics for systems of ODEs on the Stiefel manifold. BIT. **41** (5) (2001) 912–923.
4. Dieci, L., Russel, R. D., Van Vleck, E. S.: Unitary integrators and applications to continuous orthonormalization techniques. SIAM J. Num. Anal. **31** (1994) 261–281.
5. Dieci, L., Van Vleck, E. S.: Computation of a few Lyapunov exponents for continuous and discrete dynamical systems. Appl. Numer. Math. **17** (3) (1995) 275–291.
6. Dieci, L., Van Vleck, E. S.: Computation of orthonormal factors for fundamental solution matrices. Numer. Math. **83** (1999) 591–620.
7. Gladwell, I., Shampine, L. F., Brankin, R. W.: Automatic selection of the initial step size for an ODE solver. J. Comp. Appl. Math. **18** (1987) 175–192.
8. Golub, G. H., Van Loan, C. F.: Matrix computations. 3rd edn. Johns Hopkins University Press, Baltimore (1996).
9. Hairer, E., Nørsett, S. P., Wanner, G.: Solving ordinary differential equations I: nonstiff problems. 2nd edn. Springer-Verlag, New York (1993).
10. Hairer, E., Wanner, G.: Solving ordinary differential equations II: stiff and differential algebraic problems. 2nd edn. Springer-Verlag, New York (1996).
11. Hairer, E.: Symmetric projection methods for differential equations on manifolds. BIT. **40** (4) (2000) 726–734.

12. Hairer, E., Lubich, C., Wanner, G.: Geometric numerical integration: structure preserving algorithms for ordinary differential equations. Springer-Verlag, New York, draft version June 25 (2001).
13. Higham, D.: Time-stepping and preserving orthonormality. BIT. **37** (1) (1997) 241–36.
14. Higham, N. J.: Matrix nearness problems and applications. In: Gover, M. J. C., Barnett, S. (eds.): Applications of Matrix Theory. Oxford University Press, Oxford (1989) 1–27.
15. Higham, N. J., Schreiber, R. S.: Fast polar decomposition of an arbitrary matrix. SIAM J. Sci. Stat. Comput. **11** (4) (1990) 648–655.
16. Higham, N. J.: Accuracy and stability of numerical algorithms. SIAM, Philadelphia (1996).
17. Iserles, A., Munthe-Kaas, H. Z., Nørsett, S. P., Zanna, A.: Lie-group methods. Acta Numerica. **9** (2000) 215–365.
18. McLachlan, R. I., Quispel, G. R. W.: Six lectures on the geometric integration of ODEs. In: DeVore, R. A., Iserles, A., Süli, E. (eds.): Foundations of Computational Mathematics. Cambridge University Press. Cambridge. (2001) 155–210.
19. Anderson, E., Bai, Z., Bischof, C., Blackford, S., Demmel, J., Dongarra, J., Du Croz, J., Greenbaum, A., Hammarling, S., McKenney, A., Sorenson, D.: LAPACK Users' Guide. 3rd edn. SIAM, Philadelphia (1999).
20. Marsden, J. E., Ratiu, T.: Introduction to mechanics and symmetry. Texts in Applied Mathematics, Vol. 17. 2nd edn. Springer-Verlag, New York (1999).
21. Sanz-Serna, J. M., Calvo, M. P.: Numerical Hamiltonian problems. Chapman and Hall, London (1994).
22. Whaley, R. C., Petitet, A., Dongarra, J. J.: Automatated empirical optimization of software and the ATLAS project. available electronically from http://math-atlas.sourceforge.net/
23. Zanna, A.: On the numerical solution of isospectral flows. Ph. D. Thesis, DAMTP, Cambridge University (1998).

Symplectic Methods for Separable Hamiltonian Systems

Mark Sofroniou[1] and Giulia Spaletta[2]

[1] Wolfram Research, Champaign, Illinois, USA. marks@wolfram.com
[2] Mathematics Department, Bologna University, Italy. giulia@cs.unibo.it

Abstract. This paper focuses on the solution of separable Hamiltonian systems using explicit symplectic integration methods. Strategies for reducing the effect of cumulative rounding errors are outlined and advantages over a standard formulation are demonstrated. Procedures for automatically choosing appropriate methods are also described.
Keywords. Geometric numerical integration; separable Hamiltonian differential equations; symplectic methods; computer algebra systems.
AMS. 65L05; 68Q40.

1 Introduction

The phase space of a Hamiltonian system is a symplectic manifold on which there exists a natural symplectic structure in the canonically conjugate coordinates. The time evolution of the system is such that the Poincaré integral invariants associated with the symplectic structure are preserved. A symplectic integrator is advantageous because it computes exactly, assuming infinite precision arithmetic, the evolution of a nearby Hamiltonian, whose phase space structure is close to that of the original Hamiltonian system [11].

Symplectic integration methods for general Hamiltonians are implicit, but for separable Hamiltonians explicit methods exist and are much more efficient [23]. The aim of this work is to describe a uniform framework that provides a variety of numerical solvers for separable Hamiltonian differential equations in a modular, extensible way. Furthermore, the effect of rounding errors has not received a great deal of attention, so the framework is used to explore this issue in more detail.

This paper is organized as follows. In Section 2 separable Hamiltonian systems are defined together with a standard algorithm for implementing an efficient class of symplectic integrators. Practical algorithms for reducing the effect of cumulative rounding errors are presented in Section 3. Section 4 contains a description of the methods that have been implemented, along with algorithms for automatically selecting between different orders and a procedure for adaptively refining coefficients for high precision computation. Section 5 contains some numerical experiments that summarize the behavior of the various algorithms presented. Section 6 concludes with some potential enhancements and suggestions for future work.

P.M.A. Sloot et al. (Eds.): ICCS 2002, LNCS 2331, pp. 506–515, 2002.

2 Definitions

Let Ω be a nonempty, open, connected subset in the oriented Euclidean space $\mathbb{R}^{2d}$ of the points $(\mathbf{p}, \mathbf{q}) = (p_1, \ldots, p_d, q_1, \ldots, q_d)$. Denote by I an open interval of the real line. If $H = H(\mathbf{p}, \mathbf{q}, t)$ is a sufficiently smooth real function defined on the product $\Omega \times I$, then the *Hamiltonian system* of differential equations with Hamiltonian H is:

$$\frac{dp_i}{dt} = -\frac{\partial H}{\partial q_i}, \quad \frac{dq_i}{dt} = \frac{\partial H}{\partial p_i}, \qquad i = 1, \ldots, d.$$

The integer dimension d is referred to as the *number of degrees of freedom* and Ω as the *phase space.*

Many practical problems can be modeled by a *separable Hamiltonian* where: $H(\mathbf{p}, \mathbf{q}, t) = T(\mathbf{p}) + V(\mathbf{q}, t)$. The Hamiltonian system can then be expressed in partitioned form by means of two functions $\mathbf{f}$ and $\mathbf{g}$:

$$\frac{dp_i}{dt} = \mathbf{f}(q_i, t) = -\frac{\partial V(\mathbf{q}, t)}{\partial q_i}, \quad \frac{dq_i}{dt} = \mathbf{g}(p_i) = \frac{\partial T(\mathbf{p})}{\partial p_i}, \qquad i = 1, \ldots, d. \tag{1}$$

A Partitioned Runge Kutta method (PRK) can be used to numerically integrate (1). In most practical situations the cost of evaluating $\mathbf{f}$ dominates the cost of evaluating $\mathbf{g}$.

Symplecticity is a characterization of Hamiltonian systems in terms of their solutions and it is advantageous if a numerical integration scheme applied to (1) is also symplectic. A Symplectic Partitioned Runge Kutta (SPRK) method involves constraints on the coefficients of a PRK method which results in a reduction in the number of order conditions that need to be satisfied [20]. Symplecticity also gives rise to a particularly simple implementation [21]. Following [23], denote the coefficients of an s stage SPRK method as $[b_1, b_2, \ldots, b_s](B_1, B_2, \ldots, B_s)$. Algorithm 1 yields an explicit integration procedure starting from initial conditions $\mathbf{p}_n$, $\mathbf{q}_n$.

Algorithm 1 (Standard formulation)
$\mathbf{P}_0 = \mathbf{p}_n$
$\mathbf{Q}_1 = \mathbf{q}_n$
for $i = 1, \ldots, s$
  $\mathbf{P}_i = \mathbf{P}_{i-1} + h_{n+1}\, b_i\, \mathbf{f}(\mathbf{Q}_i, t_n + C_i\, h_{n+1})$
  $\mathbf{Q}_{i+1} = \mathbf{Q}_i + h_{n+1}\, B_i\, \mathbf{g}(\mathbf{P}_i)$

The algorithm returns $\mathbf{p}_{n+1} = \mathbf{P}_s$ and $\mathbf{q}_{n+1} = \mathbf{Q}_{s+1}$. The time weights are given by:

$$C_j = \sum_{i=1}^{j-1} B_i, \quad j = 1, \ldots, s.$$

Two d dimensional vectors can be used to implement Algorithm 1 [21], although practically three vectors may be necessary if the function call is implemented as a subroutine and cannot safely overwrite the argument data. If $B_s = 0$ then Algorithm 1 effectively reduces to an $s-1$ stage scheme since it has a First Same As Last (FSAL) property.

3 Rounding error accumulation

Errors are asymptotically damped when numerically integrating dissipative systems. Hamiltonian systems, on the contrary, are conservative and the Hamiltonian is a constant, or invariant, of the motion. Consequently, an issue when numerically integrating such systems is that errors committed at each integration step can accumulate. Furthermore, solutions of Hamiltonian systems often require very long time integrations so that the cumulative roundoff error can become important. Finally, high order symplectic integration methods also involve many basic sub steps and the form of Algorithm 1 means that rounding errors are compounded during each integration step. For these reasons it is useful to look for alternatives. In certain cases, Lattice symplectic methods exist and can avoid step by step roundoff accumulation, but such an approach is not always possible [6].

A technique for reducing the effect of cumulative error growth in an additive process is *compensated summation* (see [14] for a summary). In IEEE double precision, compensated summation uses two variables to represent a sum and an error, which has the effect of doubling the working precision. As illustration consider n steps of a numerical integration using Euler's method, with a fixed step size h, applied to an autonomous system. The updates $\mathbf{y}+h\,\mathbf{f}(\mathbf{y})$ are replaced by results of the following algorithm.

Algorithm 2 (Compensated summation)
$\mathbf{yerr} = \mathbf{0}$
for $i = 1$ *to* n
  $\mathbf{\Delta y} = h\,\mathbf{f}(\mathbf{y}) + \mathbf{yerr}$
  $\mathbf{ynew} = \mathbf{y} + \mathbf{\Delta y}$
  $\mathbf{yerr} = (\mathbf{y} - \mathbf{ynew}) + \mathbf{\Delta y}$
  $\mathbf{y} = \mathbf{ynew}$

Traditionally compensated summation has been considered for dissipative systems when the step size is small [4], [14]. However the technique can be particularly useful for conservative systems, where errors are not damped asymptotically, even when the step size is relatively large [15]. For Hamiltonian systems, Algorithm 2 requires an extra two d dimensional vectors over Algorithm 1 in order to store the rounding errors in updating $\mathbf{p}$ and $\mathbf{q}$. In a practical implementation in C, the *volatile* declaration avoids computations being carried out in extended precision registers.

Compensated summation can be used directly at each of the internal stages in Algorithm 1. However, other possibilities exist. Our preference is to use Algorithm 3 below with compensated summation applied to the final results. This approach yields some small arithmetic savings and is more modular: it also applicable if the basic integrator is a symplectic implicit Runge-Kutta method.

As an example, consider evaluating $\sum_{k=1}^{n} x$ in IEEE double precision, with $x = 0.1$ and $n = 10^6$. Since 0.1 is not exactly representable in binary, a representation error is made at the outset. More importantly, the scale of the cumulative sum increases and this causes successive bits from each term added in

Table 1. Default SPRK methods and a summary of their properties.

Order	f Evals	Method	Symmetric	FSAL
1	1	Symplectic Euler	No	No
2	1	Symplectic pseudo Leapfrog	Yes	Yes
3	3	McLachlan-Atela [16]	No	No
4	3	Forest-Ruth [7], Yoshida [25], Candy-Rozmus [5]	Yes	Yes
6	7	Yoshida A [25]	Yes	Yes
8	15	Yoshida D [25]	Yes	Yes

to be neglected. Without compensated summation the result that we obtain is 100000.0000013329, while with compensated summation it is 100000.

A reduction in compound numerical errors can be accomplished for SPRK methods by introducing the *increments* $\mathbf{\Delta P}_i = \mathbf{P}_i - \mathbf{p}_n$ and $\mathbf{\Delta Q}_i = \mathbf{Q}_i - \mathbf{q}_n$ and using the following modified algorithm.

Algorithm 3 (Increment formulation)
$\mathbf{\Delta P}_0 = \mathbf{0}$
$\mathbf{\Delta Q}_1 = \mathbf{0}$
for $i = 1, \ldots, s$

$$\mathbf{\Delta P}_i = \mathbf{\Delta P}_{i-1} + h_{n+1}\, b_i\, \mathbf{f}(\mathbf{q}_n + \mathbf{\Delta Q}_i, t_n + C_i\, h_{n+1})$$
$$\mathbf{\Delta Q}_{i+1} = \mathbf{\Delta Q}_i + h_{n+1}\, B_i\, \mathbf{g}(\mathbf{p}_n + \mathbf{\Delta P}_i)$$

Algorithm 3 can be implemented using three extra d dimensional storage vectors over Algorithm 1, as well as some additional elementary arithmetic operations. Two vectors are sufficient if the function can safely overwrite the argument data. An advantage over Algorithm 1 is that each of the quantities added in the main loop are now of magnitude $O(h)$. Furthermore instead of returning $\mathbf{p}_n + \mathbf{\Delta P}_s$ and $\mathbf{q}_n + \mathbf{\Delta Q}_{s+1}$, Algorithm 3 returns the increments of the new solutions and these can be added to the initial values using compensated summation. The additions in the main loop in Algorithm 3 could also be carried out employing compensated summation, but our experiments have shown that this adds a non-negligible overhead for a relatively small gain in accuracy.

4 Methods and order selection

The default SPRK methods at each order in our current implementation are given in Table 1. The modularity also makes it possible for a user to select an alternative method by simply entering the coefficients and indicating the order. Only methods of order four or less in Table 1 possess a closed form solution. Higher order methods are given as machine precision coefficients. Since our numerical integration solver also works for arbitrary precision, we need a process for obtaining the coefficients to the same precision as that to be used in the solver. When the closed form of the coefficients is not available, the order equations for the symmetric composition coefficients can be refined in arbitrary precision using the Secant method, starting from the known machine precision solution.

In our framework, users can select the order of an SPRK method by specifying an option, but an automatic selection is also provided. To accomplish this, two cases need to be considered, leading to Algorithm 4 and Algorithm 5 below. The first ingredient for both algorithms is a procedure for estimating the starting step size h_k to be taken for a method of order k (see [8], [10]). The initial step size is chosen to satisfy user specified absolute and relative local tolerances. The next ingredient is a measure of work for a method.

Definition 1 (Work per unit step). *Given a step size h_k and a work estimate W_k for one integration step with a method of order k, the work per unit step is given by W_k/h_k.*

For an SPRK method of order k the work per step is the number of function evaluations, effectively given by the number of stages s ($s-1$ if the FSAL device is used). The last ingredient is a specification of the methods that are available. Let Π be the non empty, ordered set of method orders that are to be selected amongst. A comparison of work for the methods in Table 1 gives the choice $\Pi = \{2,4,6,8\}$. Denote Π_k as the k-th element of Π and $|\Pi|$ as the cardinality.

Of the two cases to be considered, probably the most common is when the step size can be freely chosen. The task is to compute a starting step and balance it with an estimate of the cost of the method. By bootstrapping from low order, Algorithm 4 finds the order that locally minimizes the work per unit step.

Algorithm 4 (h free)

set $W = \infty$
for $k = 1$ *to* $|\Pi|$
  compute h_k
  if $W > W_k/h_k$ *set* $W = W_k/h_k$
  else if $k = |\Pi|$ *return* Π_k
  else return Π_{k-1}

The second case to be considered is when the starting step size estimate h is given. Algorithm 5 provides the order of the method that minimizes the computational cost while satisfying the given tolerances.

Algorithm 5 (h specified)

for $k = 1$ *to* $|\Pi|$
  compute h_k
  if $h_k \geq h$ *or* $k = |\Pi|$ *return* Π_k

The computation of h_1 usually involves estimation of derivatives obtained from a low order integration (see [10, Section II.4]). The derivative estimates are the same for all $k > 1$ and can therefore be computed only once and stored. Computing h_k, for $k > 1$, then involves just a few basic arithmetic operations which are independent of the cost of function evaluation of the differential system.

Algorithms 4 and 5 are heuristic since the optimal step size and order may change through the integration, but symplectic integration commonly involves

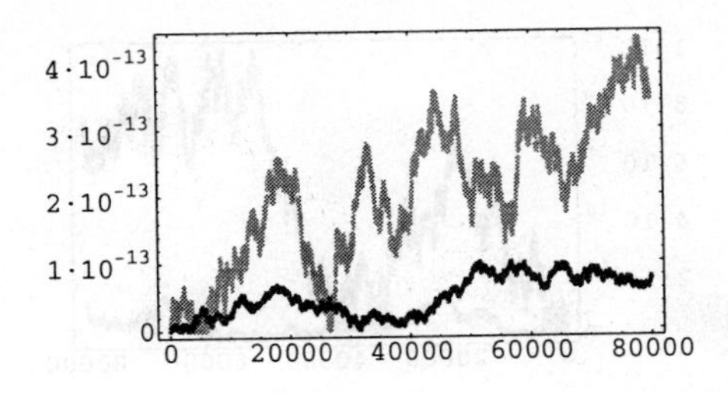

Fig. 1. Harmonic oscillator. Relative error growth vs time using Algorithm 1 (above) and Algorithm 3 (below).

fixed choices. In spite of this, both algorithms incorporate salient integration information, such as local error tolerances, system dimension and initial conditions, to avoid a poor default choice.

5 Numerical examples

The coefficients given in [25] are only accurate to 14 decimal digits, while those used in this section are accurate to full IEEE double precision. All computations are carried out using a developmental version of *Mathematica*.

5.1 The Harmonic oscillator

The Harmonic oscillator is a simple Hamiltonian problem that models a material point attached to a spring. For unitary mass and spring constant, the Hamiltonian is $H(p,q) = (p^2 + q^2)/2$, for which the differential system is:

$$q'(t) = p(t), \quad p'(t) = -q(t), \qquad q(0) = 1, \quad p(0) = 0.$$

The constant step size taken is $h = 1/25$ and the integration is performed over the interval $[0, 80000]$ with the 8th order integrator in Table 1. The error in the Hamiltonian is sampled every 200 integration steps. The exact solution evolves on the unit circle, but a dissipative numerical integrator produces a trajectory that spirals to the fixed point at the origin and exhibits a linear error growth in the Hamiltonian.

Figures 1, 2 and 3 show the successive improvement in the propagated error growth using Algorithm 1, Algorithm 3 without and with compensated summation and Algorithm 3 with compensated summation using arbitrary precision software arithmetic with 32 decimal digits.

In order to explain the observed behavior, consider a one dimensional random walk with equal probability of a deviation [22], [9]. In the numerical integration process considered here, the deviation corresponds to a rounding or truncation error of one half of a unit in the last place, which is approximately $\epsilon = 1.11 \times$

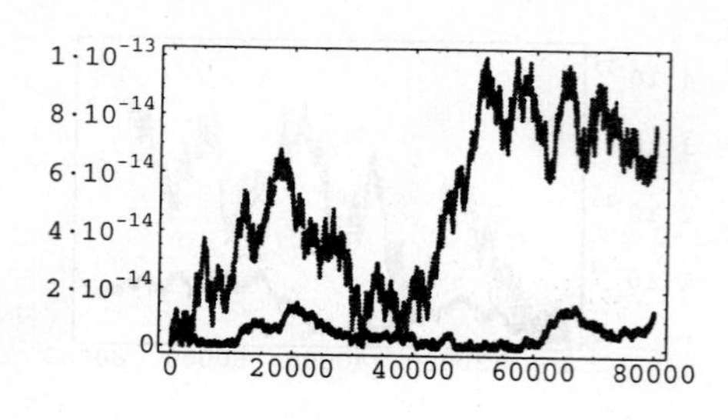

Fig. 2. Harmonic oscillator. Relative error growth vs time using Algorithm 3 without (above) and with (below) compensated summation.

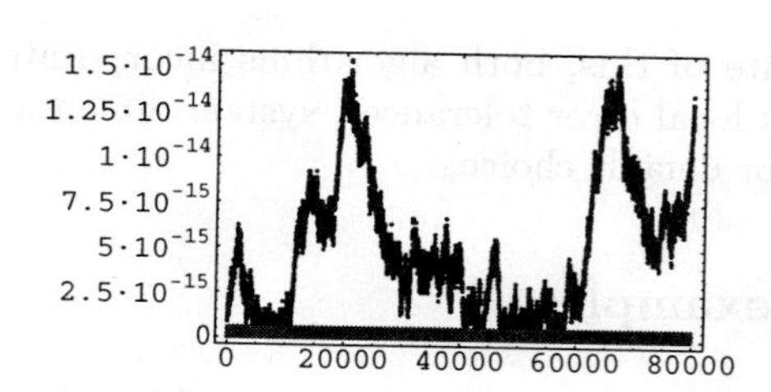

Fig. 3. Harmonic oscillator. Relative error growth vs time using Algorithm 3 with compensated summation using IEEE double precision (above) and using 32 digit software arithmetic (below).

10^{-16} in IEEE double precision arithmetic. The expected absolute distance of a random walk after N steps is given by $\sqrt{2\,N/\pi}$.

The integration for 2×10^6 steps carried out with the 8th order 15 stage method in Table 1, implemented using Algorithm 1, corresponds to $N = 3 \times 10^7$; therefore the expected absolute distance is 4.85×10^{-13} which is in good agreement with the value 4.4×10^{-13} that can be observed in Figure 1. In the incremental Algorithm 3 the internal stages are all of the order of the step size and the only significant rounding error occurs at the end of each integration step; thus $N = 2 \times 10^6$, leading to an expected absolute distance of 1.25×10^{-13} which again agrees with the value $1. \times 10^{-13}$ that can be observed in Figure 2. This shows that for Algorithm 3, with sufficiently small step sizes, the rounding error growth is independent of the number of stages of the method, which is particularly advantageous for high order. Using compensated summation with Algorithm 3 the error growth appears to satisfy a random walk with deviation $h\,\epsilon$.

Similar results have been observed taking the same number of integration steps using both the 6th order method in Table 1, with step size 1/160, and the 10th order method in [15], with the base integrator of order 2 in Table 1 and step size 1/10.

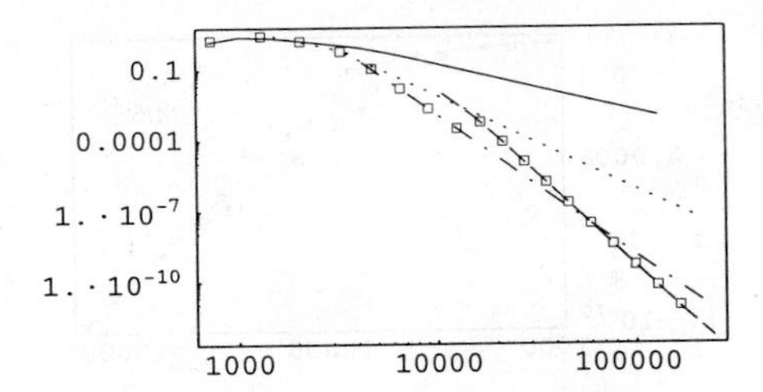

Fig. 4. Kepler problem. A log-log plot of the maximum absolute phase error vs number of evaluations of **f**, using SPRK methods of order 2 (solid line), order 4 (dotted line), order 6 (dashed-dot line) and order 8 (dashed line). The methods selected automatically at various tolerances using Algorithm 4 are displayed with the symbol □.

5.2 The Kepler problem

Kepler's problem describes the motion in the configuration plane of a material point that is attracted towards the origin with a force inversely proportional to the square of the distance. In non-dimensional form the Hamiltonian is:

$$H(\mathbf{p},\mathbf{q}) = \frac{1}{2}(p_1^2 + p_2^2) - \frac{1}{\sqrt{q_1^2 + q_2^2}}$$

The initial conditions are chosen as $p_1(0) = 0$, $p_2(0) = \sqrt{(1+e)/(1-e)}$, $q_1(0) = 1 - e$, $q_2(0) = 0$, where the eccentricity is $e = 3/5$. The orbit has period $2\,\pi$ and the integration is carried out on the interval $[0, 20\,\pi]$.

Figure 4 shows some SPRK methods together with the methods chosen automatically at various tolerances according to Algorithm 4, which clearly finds a close to optimal step and order combination. The automatic selection switches to order 8 slightly earlier than necessary, which can be explained by the fact that the starting step size is based on low order derivative estimation and this may not be ideal for selecting high order methods.

Figure 5 shows the methods chosen automatically at various fixed step sizes according to Algorithm 5. With the local tolerance and step size fixed the code can only choose the order of the method. For large step sizes a high order method is selected, whereas for small step sizes a low order method is selected. In each case the method chosen minimizes the work to achieve the given tolerance.

6 Summary and future work

We have illustrated a few techniques that can be used to reduce the effect of cumulative rounding error in symplectic integration. If one is willing to accept the additional memory requirements, then the improved accuracy even in the low dimensional examples of Section 5 can be obtained at an increased execution time of at most a few percent. In comparison, our implementation in arbitrary

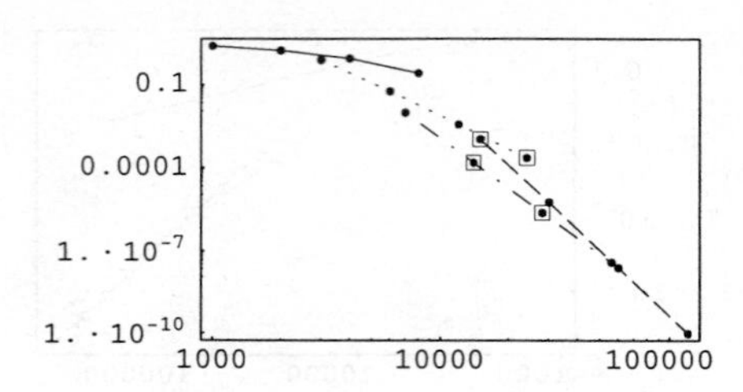

Fig. 5. Kepler problem. A log-log plot of the maximum absolute phase error vs number of evaluations of $\mathbf{f}$, using SPRK methods of order 2 (solid line), order 4 (dotted line), order 6 (dashed-dot line) and order 8 (dashed line) with step sizes 1/16, 1/32, 1/64, 1/128. The methods selected automatically by Algorithm 5 using an absolute local error tolerance of 10^{-9} are displayed with the symbol $\Box$.

precision using 32 decimal digit arithmetic was around an order of magnitude slower than IEEE double precision arithmetic.

Compensated summation is a general tool that can be used to improve the rounding error properties of many numerical integration methods. Furthermore, an increment formulation such as that outlined in Algorithm 3 can be beneficial if the numerical method involves a large number of basic steps. We have had similar success formulating increment algorithms for extrapolation methods. Runge-Kutta methods based on Chebyshev nodes (see [12] and [1] for a summary) also appear to be amenable to an increment formulation. We are currently investigating the application of techniques for rounding error reduction to some integration problems in celestial mechanics (see [22], [24] for example).

Algorithms for automatically selecting between a family of methods have been presented and have been shown to work well in practice. In order to test our implementation, established method coefficients have been chosen. It remains to carry out a more extensive selection amongst coefficients for better methods from recent work outlined in [2], [3], [15] [17], [18], [19].

Symplectic Runge Kutta Nyström methods are more efficient for the common class of separable Hamiltonian systems having $T(\mathbf{p}) = \mathbf{p}^T M^{-1}\mathbf{p}/2$, where M denotes a constant symmetric matrix of masses. Moreover processing or composition techniques can be used to improve the efficiency of Runge-Kutta methods (see [4] and [2] and the references therein). Variable step size in symplectic integration has not been discussed, see [13] for a way of modifying the Hamiltonian system to accomplish this.

Acknowledgements

Thanks to Ernst Hairer for suggesting investigation of error growth using the random walk model, to Robert McLachlan for a copy of [19] and to Robert Skeel for pointing out [6].

References

1. Abdulle, A.: Chebyshev methods based on orthogonal polynomials. Ph. D. Thesis, Section de Mathématiques, Université de Genève (2001).
2. Blanes, S., Casas, F., Ros, J.: Symplectic integration with processing: a general study. SIAM J. Sci. Comput. **21** (1999) 711–727.
3. Blanes, S., Moan, P. C.: Practical symplectic partitioned Runge Kutta and Runge Kutta Nyström methods. DAMTP report NA13, Cambridge University (2000).
4. Butcher, J. C.: The numerical analysis of ordinary differential equations: Runge Kutta and general linear methods. John Wiley, Chichester (1987).
5. Candy, J., Rozmus, R.: A symplectic integration algorithm for separable Hamiltonian functions. J. Comput. Phys. **92** (1991) 230–256.
6. Earn, D. J. D., Tremaine, S.: Exact numerical studies of Hamiltonian maps: iterating without roundoff error. Physica D. **56** (1992) 1–22.
7. Forest, E., Ruth, R. D.: Fourth order symplectic integration. Physica D. **43** (1990) 105–117.
8. Gladwell, I., Shampine, L. F., Brankin, R. W.: Automatic selection of the initial step size for an ODE solver. J. Comp. Appl. Math. **18** (1987) 175–192.
9. Gladman, B., Duncan, M., Candy, J.: Symplectic integrators for long-term integrations in celestial mechanics. Celest. Mech. **52** (1991) 221–240.
10. Hairer, E., Nørsett, S. P., Wanner, G.: Solving ordinary differential equations I: nonstiff problems. 2nd edn. Springer-Verlag, New York (1993).
11. Hairer, E.: Backward analysis of numerical integrators and symplectic methods. Annals of Numerical Mathematics, **1** (1984) 107–132.
12. Hairer, E., Wanner, G.: Solving ordinary differential equations II: stiff and differential algebraic problems. 2nd edn. Springer-Verlag, New York (1996).
13. Hairer, E.: Variable time step integration with symplectic methods. Appl. Numer. Math. **25** (1997) 219–227.
14. Higham, N. J.: Accuracy and stability of numerical algorithms. SIAM, Phil. (1996).
15. Kahan, W. H., Li, R. C.: Composition constants for raising the order of unconventional schemes for ordinary differential equations. Math. Comp. **66** (1997) 1089–1099.
16. McLachlan, R. I., Atela, P.: The accuracy of symplectic integrators. Nonlinearity. **5** (1992) 541–562.
17. McLachlan, R. I.: On the numerical integration of ordinary differential equations by symmetric composition methods. SIAM J. Sci. Comp. **16** (1995) 151–168.
18. McLachlan, R. I.: Composition methods in the presence of small parameters. BIT. **35** (1995) 258–268.
19. McLachlan, R. I.: Families of high-order composition methods (preprint).
20. Murua, A.: On order conditions for partitioned symplectic methods. SIAM J. Numer. Anal. **34** (6) (1997) 2204–2211.
21. Okunbor, D. I., Skeel, R. D.: Explicit canonical methods for Hamiltonian systems. Math. Comp. **59** (1992) 439–455.
22. Quinn, T., Tremaine, S.: Roundoff error in long-term planetary orbit integrations. Astron. J. **99** (3) (1990) 1016–1023.
23. Sanz-Serna, J. M., Calvo, M. P.: Numerical Hamiltonian Problems. Chapman and Hall, London (1994).
24. Wisdom, J., Holman, M.: Symplectic maps for the N-body problem. Astron. J. **102** (1991) 1528–1538.
25. Yoshida, H.: Construction of high order symplectic integrators. Phys. Lett. A. **150** (1990) 262–268.

Numerical Treatment of the Rotation Number for the Forced Pendulum

Raffaella Pavani

Dipartimento di Matematica, Politecnico di Milano
20133 Milano - ITALY
rafpav@mate.polimi.it

Abstract. The aim of this paper is to study (from a numerical point of view) the behavior of orbits of the forced pendulum. We resort mainly to the numerical method by the author, which allows to distinguish efficiently the numerical approximation of a rational rotation number from that one of an irrational rotation number, so we can characterize respectively the periodic and non-periodic behavior of the orbits. Moreover we study numerically the appearance of chaotic behavior depending on the forcing parameter and we show how the numerical results provide useful and reliable information about such an event.

1 Introduction to the problem

We consider the motion of a forced pendulum that moves on a vertical plane; in absence of friction, the displacement angle ϑ from the vertical rest position of the pendulum satisfies the following second order differential equation:

$$\frac{d^2\vartheta}{dt^2} + \frac{g}{l}\,\alpha(t)\sin\vartheta = 0 \tag{1}$$

where g is the gravitational constant and l is the length of the pendulum; without loss of generality now we assume $g = l$; moreover we assume $\alpha(t) = 1 + \varepsilon\cos t$, where ε is the forcing parameter and is supposed to be very small (i.e. $\varepsilon << 1$). Introducing the variables

$$x_1(t) = \vartheta(t), \qquad x_2(t) = \frac{d\vartheta}{dt} \tag{2}$$

equation (1) becomes

$$\frac{dx_1}{dt} = x_2 \tag{3}$$

$$\frac{dx_2}{dt} = -(1 + \varepsilon\cos t)\sin x_1 \tag{4}$$

At first we assume $\varepsilon = 0$.

We are interested in the orbits of the pendulum away from the equilibrium point $(n\pi, 0)$,where n is any integer point. As x_1 is $2\pi-$periodic, we will consider

P.M.A. Sloot et al. (Eds.): ICCS 2002, LNCS 2331, pp. 516–525, 2002.

the vertical strip of the plane with $0 \leq x_1 \leq 2/pi$. Moreover for the same reason, it suffices to take initial data on the x_2-axis, say x_2^0. Therefore we need to determine the shapes of the curves of the form

$$\frac{1}{2}(x_2)^2 + (1 - \cos x_1) = \frac{1}{2}(x_2^0)^2 \tag{5}$$

where x_2^0 is the initial data on the x_2-axis.

For any x_2^0, the curve defined by equation (5) is symmetric with respect to x_1-axis; so we consider only the curves

$$x_2 = \sqrt{(x_2^0)^2 - 2(1 - \cos x_1)}. \tag{6}$$

If the following inequality

$$(x_2^0)^2 > 2(1 - \cos x_1), \tag{7}$$

is satisfied, that is if the values of x_2^0 are such that $(x_2^0)^2 > 4$, then the range of x_1 is unrestricted and the curve defined by equation (6) is 2π-periodic graph over the x_1-axis. There are no equilibria on these curves and they correspond to the orbits of the motions of the pendulum with initial velocity so large that the pendulum revolves around and around without end.

As the behavior of this kind of orbits is well known for $\varepsilon = 0$, here our purpose is to study the behavior of these orbits when the pendulum is forced, that is in the case $\varepsilon \neq 0$, by means of numerical methods.

2 The rotation number for the forced pendulum

For our purpose we resort to the rotation number ρ, which is well-known to have strong implications for dynamical systems (e.g. [5]). Its relevance for the study of the considered orbits of the forced pendulum can be summed up in the following way:

- If a unique ρ exists and is rational, then the related orbit is periodic.
- If a unique ρ exists and is irrational, then exists an invariant set of rotation number ρ.
- If a unique ρ does not exist, then the forced pendulum exhibits topological chaos, i.e. positive topological entropy [1]; for more details see [6].

We remark that the last statement holds at least for circle maps [5]; however the considered orbit of the forced pendulum is topologically equivalent to a circle map, as it will be enlightened in the next Section.

Definition 1. *(Poincaré-1885) For a monotone map $g : \mathbb{R} \to \mathbb{R}$ with $g(\theta+1) = g(\theta) + 1$,*

$$\rho = \lim_{n\to\infty} (g^n(\theta) - \theta)/n \tag{8}$$

exists for all $\theta \in \mathbb{R}$ and is independent of θ and is called ***rotation number*** *of g.*

At first we consider $\varepsilon = 0$ in (3),(4), that is we deal with a simple pendulum.

As pointed out in the previous Section, x_1 is the displacement angle of the pendulum which, in absence of friction, is a monotone (increasing) function of the time. When we numerically integrate (3), (4) with step size $h = 1$, with any initial conditions, say $x_1(0)$, $x_2(0)$, we obtain the solutions $x_1(n), x_2(n)$, $n = 1, 2, 3, 4, ...$ In the following we will assume $x_1(0) = 0$ without loss of generality. For physical reasons, two solutions with the same x_2 that differ in x_1 by 2π, have to be considered the same (actually equations (3),(4) remain the same under the change of variables $(x_1, x_2) \to (x_1 + 2\pi, x_2)$); therefore if we normalize the period to 1 considering $f^n(0) = x_1(n) \,/\, 2\pi$, $n = 1, 2, 3....$, it can be viewed as the $n - th$ iteration of a monotone map with $\theta = x_1(0) = 0$ satisfying Definition 1. Therefore when the pendulum is simple the rotation number always exists and is unique.

What happens when $\varepsilon \neq 0$? It is known that for the forced pendulum the limit providing the rotation number does not always exist; therefore two main questions arise:

1. when does the rotation number exist for the forced pendulum with a given parameter ε?
2. if the rotation number exists for the forced pendulum with given initial conditions and a given parameter ε, is it the same as for the simple pendulum with the same initial conditions? does it remains rational or irrational?

Here we answer at once the first question, whereas we will deal with the second question in Section 4, where numerical examples will provide numerical evidence about the answer.

If we choose as initial conditions $x_1(0) = 0, x_2(0) = x_2^0 > 2$ and $0 < \varepsilon << 1$, as already pointed out, the numerical integration of (3), (4) allows to compute the rotation number ρ if and only if the following finite limit exists

$$\lim_{n\to\infty} \frac{x_1(n)}{2\pi n} = \rho \tag{9}$$

Obviously the used integration stepsize is much less than 1.

We remark that we always have $\rho \neq 0$ and positive

Then we can derive a **Numerical Convergence Condition**.

Condition. *If numerically integrating equations (3),(4) with given initial conditions and given parameter ε, we always obtain non-decreasing values for the displacement angle $x_1(n)/2\pi$ for increasing n until n is "large enough", then a unique rotation number ρ does exist.*

Proof. At first, we remark that the sequence $\frac{x_1(n)}{2\pi}\frac{1}{n}$ has a finite limit ρ, when $n \to \infty$, if and only if the non-decreasing sequence $\frac{x_1(n)}{2\pi}$ has a linear increasing asymptotic behavior, that is if $\frac{x_1(n)}{2\pi} \to \rho n$ when $n \to \infty$. Actually, if the sequence $\frac{x_1(n)}{2\pi}$ is bounded, then the sequence $\frac{x_1(n)}{2\pi}\frac{1}{n}$ converges to 0, and this means that the rotation number does not exist. Therefore, if $\frac{x_1(n+1)}{2\pi} > \frac{x_1(n)}{2\pi}$, for any considered n, the convergence of $\frac{x_1(n)}{2\pi} \frac{1}{n}$ to a finite (nonnull) limit can be achieved.

Moreover, we remark that from the numerical point of view, we have to know the accuracy of the results obtained by means of the used numerical integration algorithm, say η; therefore if $\frac{x_1(n+1)}{2\pi} - \frac{x_1(n)}{2\pi} > \eta$ for any n, then the rotation number does exists.

In other words we can state that:
If numerically integrating equations (3),(4) with given initial conditions and given parameter ε, we always obtain non-decreasing values for the displacement angle $x_1(n)/2\pi$ for increasing n until n is "large enough", then topological chaos does **not** appear in the orbits of the considered forced pendulum.

Obviously we cannot carry out the numerical integration until n is as large as we want, but only until n is "large enough" and we remind that the determination of when n is "large enough" is a questionable point (see Section 4).

3 Numerical computation of the rotation number

From a numerical point of view, it is clear that the main problems are:

– numerical integration of equations (3),(4) over a long time interval,
– numerical estimation of the limit (9), under the condition that it exists,

Let us treat the problems separately.

From the previous Section, it is clear that the numerical method used to integrate (3),(4) is definitively relevant. As here we deal with numerical integration of a perturbed Hamiltonian system over a long time interval (i.e. $n >> 1$), we choose a symplectic method such as the implicit midpoint method, which is second order, symmetric, and A-stable. This choice allows the numerical discretization to retain the property of symplectic map provided by the considered Hamiltonian system. Indeed the failure of well-known methods in mimicking Hamiltonian dynamics is due to the fact that they do not preserve symplecticity. In practice the midpoint solution is guaranteed to lie on the same orbit as the exact solution, whereas for example classical fourth order Runge-Kutta method does not; this fact is absolutely important for our kind of numerical problem, as we will see in the following. However here we will not give more details about symplectic methods (see e.g. [11], [4]).

In order to provide a numerical estimation of the limit (9), that is to numerically compute the rotation number, we resorted to some different methods, each of whom uses a different approach to the numerical problem.

1 - Method Ml

Under the condition that the limit (9) exists, the ratio $\frac{x_1(n)}{2\pi n}$ tends to be a constant, when n is large enough. Therefore **Method Ml** computes the numerical sequence $\frac{x_1(n)}{2\pi n}$, which provides a numerical approximation of the rotation number when $\frac{x_1(n)}{2\pi n}$ tends to remain constant for increasing n. As already pointed out, we cannot know *a priori* when n is large enough; in practice we make sure that, if for example $\eta = 0.5 \cdot 10^{-3}$, for small values of i we have $\left|\frac{x_1(n+i)}{2\pi(n+i)} - \frac{x_1(n)}{2\pi n}\right| < \eta$; this means that all the subsequent values coincide with $\frac{x_1(n)}{2\pi n}$ at least in three decimal digits which are consequently considered three correct digits in the exact value of the limit ρ

2 - Method Mg

When n is large enough (with the same caution as above), we have $\frac{x_1(n)}{2\pi} \simeq \rho n$. Therefore for a fixed large N, **Method Mg** computes the linear least square approximation of $x_1(n)\,/\,2\pi$, $0 \le n \le N$; from the angular coefficient of the best fitting line we obtain an approximation of the rotation number.

3 - Method Mp ([7], [8])

It exploits the order by which the iterates of a diffeomorphism of the circle are generated and is based on two theorems which can be summed up in the following

Theorem 1. *Let f be a circle map with rotation number ρ; for each integer $n > 0$, we call $X_1 = \{f(X_0)\}, \dots, X_n = \{f(X_{n-1})\} = \{f^n(X_0)\}$ the fractional parts of the first n iterates of X_0. Then for each $N > 0$, the geometric order of the set of $X_i = \{f^i(X_0)\}, 1 \le i \le N$, allows to define four integers A, a, B, b which are consecutive terms of the Farey sequence (therefore $|Ab - aB| = 1$) and provide the value of the rotation number ρ by means of one of the following expressions*

a) if ρ is irrational then $A/a < \rho < B/b$;

b) if ρ is rational (unreducible), say $\rho = p/q$ and $N < q$, then $A/a < \rho < B/b$;

c) if ρ is rational (unreducible), say $\rho = p/q$, and $N \ge q$, then either $A/a = \rho < B/b$ or $A/a < \rho = B/b$.

In all the cases $\Delta = |A/a - B/b|$ is such that $1/N^2 < \Delta < 1/N$.

In order to use this theorem, we need to clarify two points.

Remark 1. Equations (3) and (4) remain the same under the change of variable $x_1(t) = x_1(t + 2\pi)$; in such a way the orbit takes place on the circle $\mathbb{R}/\,2\pi\mathbb{Z}$, where the diffeomorphism of the circle $x_1(t)$ is defined.

Remark 2. The rotation number depends on the geometric order only; therefore it does not depend on the circle $\mathbb{R},/\,k\mathbb{Z}$, where k is any real number. Actually the points of the orbits along circles with different k, can be obtained the ones from the others by means of a simple homotety.

Because of these two remarks, the previous Theorem which applies to maps of the circle of length 1, can be applied to maps of the circle of any length. For more details [8], [9], [10].

Actually the numerical approximation ρ_c of an irrational rotation number ρ is computed by the following expression:

$$\rho_c = \frac{A + B}{a + b} \tag{10}$$

Provided that methods **Ml** and **Mp** work only if the rotation number ρ exists, it is easy to state the following **Numerical Comparison Criterion.**

Criterion. *A unique rotation number exists if and only if all the three mentioned methods provide as a result the same numerical value within the used precision.*

It is clear that all the presented methods are affected by the same integration errors due to (3), (4). Actually even though the used implicit midpoint method provides solutions lying on the exact orbit, it is not very accurate for N large because of the accumulation of truncation errors. Therefore a major problem is the numerical error estimation Δ of the difference between the exact value of the rotation number ρ and its numerical approximation ρ_c. Method **Mg** can be considered the most reliable as it goes through a least square linear approximation of the behavior of $\frac{x_1(n)}{2\pi}$ using linear polynomial $p(n) = s \cdot n + q$, where $q = 0$ (as the polynomial goes through the origin); the computed value of parameter s provides ρ_c. The advantage of this method is that it can be considered as a filter of the numerical integration errors. Moreover, as well-known, we can use the quantity $E = \sqrt{\frac{1}{n}\sum_{i=1}^{n}[\frac{x_1(n)}{2\pi} - p(n)]^2}$ as an estimation of the accuracy of the method; indeed when $E = 0$, it means that all th $\frac{x_1(n)}{2\pi}$ lie on the line $p(n)$, for all the considered n.

Method **Ml** provides an error estimation only in the sense seen before. In practice if we assume for example $\eta = 0.5 \cdot 10^{-3}$, and we have that there exists a $\overline{N}$ such that for all $N \geq \overline{N}$, we have $\left|\frac{x_1(N+1)}{2\pi(N+1)} - \frac{{}_1(N)}{2\pi N}\right| < 0.5 \cdot 10^{-3}$, then all the subsequent values coincide with $\frac{x_1(\overline{N})}{2\pi \overline{N}}$ at least in the first three decimal digits. This allows us to state that these digits are even the first three correct digits in ρ.

A major drawback of methods **Mg** and **Ml** is that they do not allow to detect whether the rotation number is rational or irrational, whereas another drawback is that their rate of convergence cannot be known *a priori*, even though from a theoretical point of view, it is known to be less than linear.

On the contrary method **Mp** can distinguish between rational and irrational rotation number and in the case of rational rotation number its error is equal to 0.

Method **Mp** provides an accurate error estimation. The method is based on a normalized continued fraction expansion of the rotation number. In literature other methods based on continued fraction expansions were presented (see [2] and [5]); however the rate of convergence can vary significantly when different continued fraction approximations are used. For example when the rotation number is a Liouville number, the convergence of method in [5] is so slow that it becomes questionable from a numerical point of view to achieve 7 significant digits, whereas method **Mp** converges very fast. From a theoretical point of view we can guarantee that method **Mp** approximates any irrational rotation number with an accuracy between $1/N$ and $1/N^2$, but experimentally we found that the rate of convergence is usually very close to be quadratic.

We remark that the only drawback of method **Mp** is that when the four integers A, a, B, b fail to be consecutive terms of the Farey sequence, due to machine accuracy, then the method is not reliable any longer; however as this event can be detected very easily (it happens when $Ab - aB \neq \pm 1$), the method can be controlled, improving for example the numerical integration accuracy.

4 Numerical examples

Here we present some numerical examples, which are significant but far to be exhaustive.

In the following Tables we report the values of ρ_c (rounded to 3 digits) for $x_1(0) = 0$, $N = 200$; the used integration step for the implicit midpoint rule was $h = 0.005$; Δ_{Ml} indicates the value of $\left|\frac{x_1(k+1)}{2\pi k} - \frac{x_1(k)}{2\pi k}\right|$ when $k = 195$ and we experimentally checked that for all k,$196 \leq k \leq 199$, such differences were $\leq \Delta_{Ml}$; therefore Δ_{Ml} can be considered an error estimation for method **Ml**; Δ_{Mp} refers to the actual error estimation by method **Mp**, therefore $\frac{1}{2}(\frac{B}{b} - \frac{A}{a})$, Δ_{Mg} reports the computed value of E, given above.

We emphasize that when the rotation number exists and is irrational, all the methods provide the same rounded value, as it is expected from Numerical Comparison Criterion. However only method **Mp** provides the rational rotation number (see Table 1), even though in this case we need to use $N > 209$.

Crosses indicate when the rotation number does not exist.

Now we can empirically answer the question 2 in Section 2: if the rotation number exists for the forced pendulum with given initial conditions and given ε, is it the same as for the simple pendulum with the same initial conditions?

From Table 3, we see that for given initial conditions, when ε increases significantly, the rotation number changes slightly, but clearly, therefore for $\varepsilon > 10^{-3}$ the numerical evidence provides a negative answer to the above question.

Table 1. $\varepsilon = 0$

$x_2(0)$	Ml	Δ_{Ml}	Mg	Δ_{Mg}	Mp	Δ_{Mp}
2.005	.124	4E-4	.124	1E-1	.124	3E-5
2.2	.237	2E-4	.237	5E-2	.237	8E-5
2.4	.290	4E-5	.290	3E-2	.290	8E-5
2.6	.335	3E-4	.335	2E-2	70/209	0
2.8	.376	2E-4	.376	2E-2	.376	2E-5
3.0	.414	2E-4	.414	2E-2	.414	3E-5
3.2	.452	7E-5	.452	1E-2	.452	8E-5

Table 2. $\varepsilon = 0.1$

$x_2(0)$	Ml	Δ_{Ml}	Mg	Δ_{Mg}	Mp	Δ_{Mp}
2.005	–	–	-2E-4	3E-1	–	–
2.2	–	–	-1E-2	3.1	–	–
2.4	.272	5E-4	.272	4E-2	.272	3E-4
2.6	.320	4E-4	.320	1E-1	.320	7E-4
2.8	.366	3E-4	.366	2E-2	.366	3E-4
3.0	.406	3E-4	.406	2E-2	.406	n.c.
3.2	.445	2E-4	.445	1E-4	.445	3E-4

Table 3. $\varepsilon = 0.1$

$x_2(0) \setminus \varepsilon$	0	0.001	0.005	0.01	0.05	0.1	0.5
2.005	.124	.124	×	×	×	×	×
2.2	.237	.237	.235	.234	.217	×	×
2.4	.290	.290	.289	.289	.282	.272	×
2.6	70/209	.335	.334	.334	.329	.320	×
2.8	.376	.376	.375	.375	.371	.366	.318
3.0	.414	.414	.414	.414	.410	.406	.365
3.2	.452	.452	.452	.451	.448	.445	.412

From Table 2 it is clear when we say that the rotation number does not exist; here "n.c." means that the error is not computable because method **Mp** fails to find four integers belonging to Farey sequence.

There is a numerical evidence that for each value of $x_2(0)$ there exists a value of ε^* such that for $\varepsilon > \varepsilon^*$ the rotation number does not exist. The value of ε^* does not increase linearly with $x_2(0)$. Moreover we can see that for any fixed value of $x_2(0)$ and for increasing values of ε from 0, the rotation number is always not increasing.

We emphasize that the existence of the rotation number is not affected by the "graphical" behavior of the orbits, in the sense that the rotation number exists even when orbits clearly split and do not overlap any longer. For example Fig. 1 and 2 show the orbits in the phase space for $x_2(0) = 2.4$ and $\varepsilon = 0.005$, 0.1, respectively (see Table 3); in spite of what appears in Fig. 2, in both cases the rotation number exists and chaos does not happen; indeed as the rotation number depends on the geometric order of the iterates of the map, it means that a kind of geometric order is preserved, even though it is not immediately detectable from the graphical representation, as in Fig.2. On the other hand, Fig. 3 shows the orbits for $x_2(0) = 2.4$ and $\varepsilon = 0.5$, that is the case when the rotation number does not exist at all and chaos happens, as no geometric order is preserved any longer. In Table 3 bold characters indicate rotation numbers referring to these cases of splitted orbits, whose rotation number exists anyway.

We point out that this new fact can be interesting to KAM theory and can be detected here just because we have approximated numerically the rotation numbers.

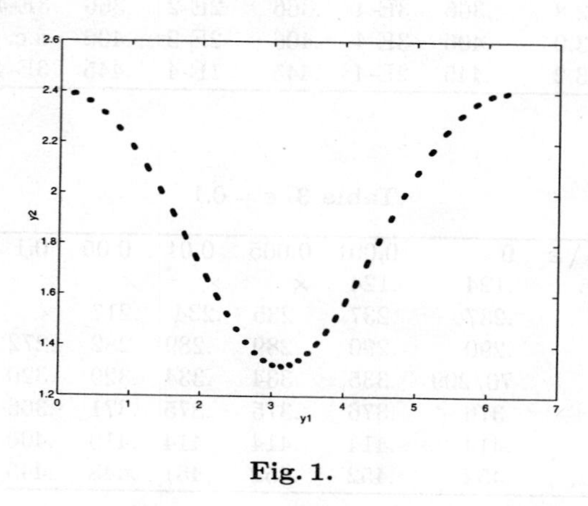

Fig. 1.

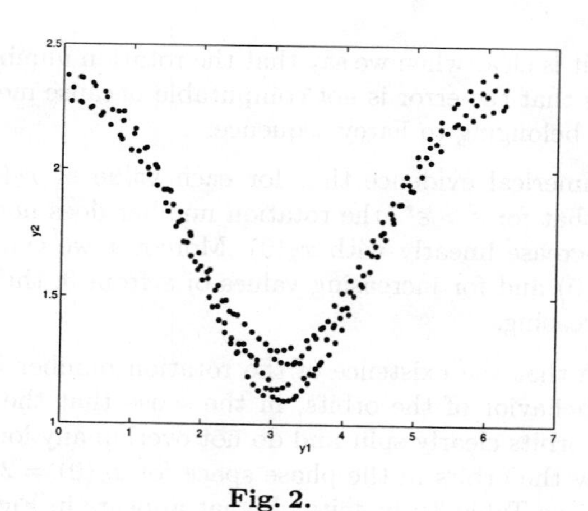

Fig. 2.

5 Conclusion

In [3] the rotation number was already presented as a quantitative measure of chaos, however the method given there to compute the rotation number did not provide any numerical result. Here referring to the forced pendulum, we numerically support their statement. Indeed we present a comparison among

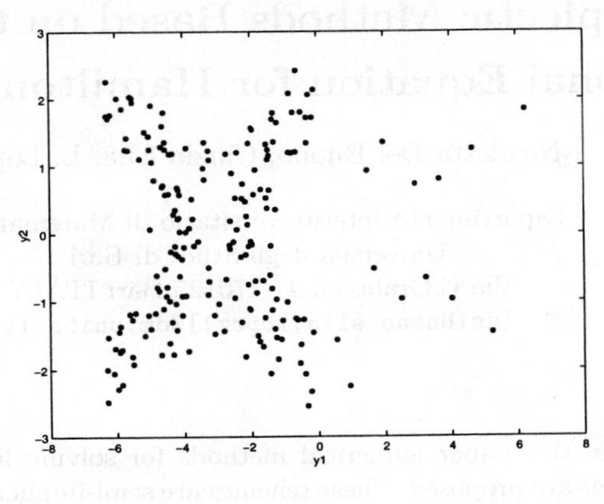

Fig. 3.

some different numerical methods and show how they can provide a reliable estimation of the rotation number, when it exists. Moreover we emphasize that the existence of the rotation number does not depend on the "graphical" behavior of the orbits, but on the geometric order of the iterates only. When the rotation number does not exist, topological chaos appears; therefore the rotation number can be used as a reliable quantitative index of chaos itself.

References

1. Block, L., Guckenheimer, J., Misiurewicz, M., Young, L. S.: Periodic points and topological entropy of one-dimensional maps. Springer Lecture Notes in Math. **819** (1980).
2. Bruin, H.: Numerical determination of the continued fraction expansion of the rotation number. Physica D **59** (1992) 158–168.
3. Gambaudo, J.M., Glendinning, P., Tresser, C.: The rotation interval as a computable measure of chaos.Physics Letters **105A** 3, (1984) 97–100.
4. Hairer, E., Lubich, C., Wanner, G.: Geometric Numerical Integration. Springer-Verlag (to appear).
5. MacKay, R.S.: Rotation interval from a time series. J. Phys. A, **20** (1987), 587–592.
6. MacKay, R.S., Tresser, C.: Transition to topological chaos for circle maps. Physica D **19**, (1986), 206–237.
7. Pavani, R., Talamo, R.: Conjugating the Poincaré-map to a rotation. Ann. Mat. Pura Appl. IV **166**, (1994), 381–394.
8. Pavani, R.: A numerical approximation of the rotation number. Appl. Math. Comp. **73**, (1995), 191–201.
9. Pavani, R.: The numerical approximation of the rotation number of planar maps. Computers Math. Appl. **33**, (1997), 103–110.
10. Pavani, R.: Numerical study of the rotation number for the forced pendulum. Quad. Dip. Mat. Politecnico di Milano n.449/P (2001).
11. Sanz-Serna, J.M., Calvo, M.P.: Numerical Hamiltonian Problems. Chapman & Hall, London New York (1994).

Symplectic Methods Based on the Matrix Variational Equation for Hamiltonian Systems

Nicoletta Del Buono, Cinzia Elia, L. Lopez

Dipartimento Interuniversitario di Matematica,
Università degli Studi di Bari
Via E.Orabona, 4 - I70125 Bari ITALY
[delbuono,elia,lopezl]@dm.uniba.it

Abstract. In this paper numerical methods for solving linear Hamiltonian systems are proposed. These schemes are semi-implicit procedures where no iteration is required and which are based on the numerical solution of the variational equation by means of symplectic-preserving procedures. Applications of these methods to certain nonlinear Hamiltonian systems and numerical tests are reported.

1 Introduction

In recent years several numerical methods have been developed to integrate systems of ODEs whose solutions preserve certain qualitative features such as orthogonality, symplecticity, isospectrality, etc (see [2], [3], [4], [6] [7]). In particular there has been a big growth of the literature on symplectic integration of Hamiltonian problems, that is models of physical systems which preserve energy (see [6],[7]). Two important features characterize Hamiltonian systems: the symplecticness of the evolution operator and the preservation of the Hamiltonian along the trajectories, hence numerical methods for such problems should inherit these properties. However in general, it is not possible to construct schemes which preserve these properties simultaneously with the same order of accuracy. Some results on symplectic methods (see [6], [7], [8]) seem to suggest that there is a better preservation of the Hamiltonian in symplectic computations over long time intervals than one expects from nonsymplectic schemes. Since, in the linear case, the solution of the associated variational equation evolves on the matrix symplectic group, here we consider semi-implicit procedures obtained by solving this variational equation by symplectic preserving methods. These procedures will be derived by projection techniques on the symplectic group or by transformation methods on the Lie algebra of the symplectic group. We also consider the nonlinear case in which such methods could be applied.

2 Numerical methods for the non-autonomous linear case

In this section we introduce our numerical methods in the linear case. Let us consider the Hamiltonian system

$$u'(t) = JA(t)u(t), \qquad t > t_0, \qquad u(t_0) = u_0, \tag{1}$$

P.M.A. Sloot et al. (Eds.): ICCS 2002, LNCS 2331, pp. 526–535, 2002.

with $u(t) \in \mathbb{R}^{2n}$ and $A(t) \in \mathbb{R}^{2n} \times \mathbb{R}^{2n}$ symmetric matrix for each t and J is the canonical symplectic matrix. Let $D(t) = \frac{\partial S(t)u_0}{\partial u_0}$, that is the derivative of the evolution operator $S(t)$ with respect to the initial data u_0. It is known that $D(t)$ evolves on the symplectic group $Sp(2n, \mathbb{R}) = \{Y \in \mathbb{R}^{2n\times 2n} \mid Y \text{ non singular and } Y^T JY = J\}$ and satisfies the variational equation:

$$D'(t) = JA(t)D(t), \qquad t > t_0, \qquad D(t_0) = I, \tag{2}$$

(see [7], [8]). Moreover, in the linear case we have $D(t) = S(t)$, thus the solution may be given in term of the solution of the variational equation, that is

$$u(t) = D(t)u_0, \qquad t \geq t_0. \tag{3}$$

Then numerical methods may be derived from a computational form of the previous equation. In particular given an approximation u_k of $u(t_k)$, we compute the numerical approximation D_{k+1} of $D(t_{k+1})$ by solving the variational equation

$$D'(t) = JA(t)D(t), \qquad t \in [t_k, t_k + h], \qquad D(t_k) = I, \tag{4}$$

and then we set

$$u_{k+1} = D_{k+1}u_k. \tag{5}$$

When D_{k+1} is computed by a symplectic-preserving method, that is a method for which $D_{k+1}^T JD_{k+1} = J$, then (5) leads to a symplectic method for (1). In this section we consider some numerical methods which preserve during the integration the symplecticness of D.

2.1 Automatic symplectic integrators

Let us consider the v-stage RK method of order p, defined by the coefficients $\mathbf{c} = (c_1, \ldots, c_v)^T$, $\mathbf{b} = (b_1, \ldots, b_v)$, $\mathbf{A} = (a_{ij})$. It is well known that a Runge Kutta method applied to matrix differential system (4) is a symplectic-preserving scheme if and only if the coefficient matrix $M = (b_i a_{ij} + b_j a_{ji} - b_i b_j)$ is the zero matrix. In particular, Gauss Legendre Runge Kutta methods (GLRKv) of order $2v$ are symplectic-preserving integrators (see [5], [6]). Thus applying a GLRKv method to equation (4), we get

$$\begin{aligned} D_{k+1} &= I + h\textstyle\sum_{i=1}^{v} b_i JA(t_k + c_i h)D_{ki}, \\ D_{ki} &= I + h\textstyle\sum_{j=1}^{v} a_{ij} JA(t_k + c_j h)D_{kj}, \qquad i = 1, \ldots, v. \end{aligned} \tag{6}$$

Hence, to find the matrix D_{k+1} a linear system of size nv must be solved at each step k in order to obtain the internal stages $(D_{k1}, \ldots, D_{kv})$. To reduce the computational costs of GLRK methods one can consider symplectic-preserving schemes constructed by explicit integrators.

2.2 Projected methods

In the same spirit of the projection methods on the orthogonal manifold proposed in [2], we may consider similar procedures on the symplectic manifold. A projected method consists of a two step procedure: in the first step we solve (4) by a v-stage explicit RK method of order p, that is

$$\begin{aligned} D_{k+1} &= I + h \textstyle\sum_{i=1}^{v} b_i JA(t_k + c_i h) D_{ki}, \\ D_{ki} &= I + h \textstyle\sum_{j=1}^{i-1} a_{ij} JA(t_k + c_j h) D_{kj}, \qquad i = 1, \ldots, v, \end{aligned} \tag{7}$$

and then, since D_{k+1} does not belong to $Sp(2n, \mathbb{R})$, it must be replaced by an approximation which is a symplectic matrix, for instance by a symplectic factorization of D_{k+1}.

Lemma 1. *(see [1]) Let D be a real $2n \times 2n$ matrix. Then D has a factorization $D = SR$, where S is unitary and symplectic, and*

$$R = \begin{bmatrix} R_{11} & R_{12} \\ R_{21} & R_{22} \end{bmatrix},$$

with R_{11}, R_{21} upper triangular matrices, and $R_{12}, R_{22} \in \mathbb{R}^{n \times n}$. Here R_{11} has nonnegative diagonal entries and R_{21} has a zero diagonal. Moreover, if D is symplectic then $R_{21} = 0$ and $R_{22} = R_{11}^{-T}$.

Given D_{k+1}, by using the results of the previous Lemma we compute the ortho-symplectic factorization $D_{k+1} = S_{k+1} R_{k+1}$ and we set

$$u_{k+1} = S_{k+1} u_k, \qquad k \geq 0. \tag{8}$$

It is easy to show that: the projected method (8) is of the same order of the basic explicit RK scheme.

2.3 Cayley methods

A different way of devising numerical methods for symplectic flows is to take in account the matrix group structure of the set of symplectic matrices. Matrix group methods associate to the equation (4), evolving on $Sp(2n, \mathbb{R})$, another ODE on the Lie algebra of the Hamiltonian matrices, that is the set

$$\mathrm{sp}(2n, \mathbb{R}) = \{A \in \mathbb{R}^{n \times n} | A^T J + JA = 0\},$$

and solve the ODE on $\mathrm{sp}(2n, \mathbb{R})$ by any explicit method. To return to $Sp(2n, \mathbb{R})$ a suitable map is adopted. Among these methods we consider the ones based on the Cayley map for details) where the theoretical solution $D(t)$ is given in the following form:

$$D(t) = (I - U(t))^{-1}(I + U(t)), \tag{9}$$

with $U(t)$ satisfying the following ODE on the set of Hamiltonian matrices:

$$U'(t) = \frac{1}{2}[I - U(t)]JA(t)[I + U(t)], \qquad t \in [t_k, t_{k+1}), \qquad U(t_k) = 0. \tag{10}$$

If the previous differential equation is solved by an explicit RK method, we get

$$\begin{aligned} U_{k+1} &= \tfrac{h}{2}\textstyle\sum_{i=1}^{v} b_i(I-U_{ki})JA(t_k+c_ih)(I+U_{ki}), \\ U_{ki} &= \tfrac{h}{2}\textstyle\sum_{j=1}^{i-1} a_{ij}(I-U_{kj})JA(t_k+c_jh)(I+U_{kj}), \qquad i=1,\ldots,v, \end{aligned} \tag{11}$$

and D_{k+1} may be computed by (9), that is

$$D_{k+1} = (I-U_{k+1})^{-1}(I+U_{k+1}). \tag{12}$$

Finally, the numerical solution of (1) may be obtained by

$$u_{k+1} = D_{k+1}u_k. \tag{13}$$

We will denote the group method (11)-(12) by ERKCAYp and regarding its order the following result may be derived.

Theorem 1. *Let $D(t)$ be the solution of (4). Suppose that $\|D(t)\|$ is bounded by a constant c_1, where $\|\cdot\|$ is the 2−norm on matrices. Then the method (13) based on the v-stage explicit Runge Kutta method (11) of order p, provides for $h \to 0$ the following estimate for the local truncation error*

$$\|u(t_{k+1}) - u_{k+1}\| \le ch^{p+1}$$

where c is a constant independent on h.

Proof. Let $D(h)$ and D_1 be respectively the theoretical and approximated solution of (4) at h, and let $U(h)$ and U_1 be the theoretical and approximated solution of the associated system (10) at h. Since the numerical method for (10) has order p we get:

$$U(h) - U_1 = O(h^{p+1})$$

with

$$(I-U(h))D(h) = (I+U(h)), \qquad (I-U_1)D_1 = (I+U_1),$$

from which it follows that

$$\begin{aligned} (I-U(h))D(h) - (I-U_1)D_1 &= U(h) - U_1, \\ D(h) - D_1 - U(h)D(h) + U_1D_1 + U_1D(h) - U_1D(h) &= [U(h) - U_1], \\ (I-U_1)(D(h)-D_1) &= (U(h)-U_1)(I+D(h)). \end{aligned} \tag{14}$$

Hence, it follows that

$$\begin{aligned} \|D(h) - D_1\| &\le \|(I-U_1)^{-1}\| \; \|U(h) - U_1\| \|I + D(h)\| \\ &\le (1+c_1)\|(I-U_1)^{-1}\| \|U(h) - U_1\| \end{aligned}$$

that is the local error on the solution $D(h)$ is of the same order of that of $U(h)$ only if $\|(I-U_1)^{-1}\|$ is bounded by a constant independent by h, for $h \to 0$. In the case of the symplectic group, the eigenvalues of the matrix U_1 may have real part different from zero, then the above norm can not be uniformly bounded for any value of h, as in the case of the orthogonal group. Nevertheless, $\|(I-U_1)^{-1}\|$ is bounded by a constant independent of h for $h \to 0$.

3 Hamiltonian nonlinear systems

The results of the previous sections may be applied to Hamiltonian nonlinear differential systems by using linearizing techniques. Let us consider the Hamiltonian nonlinear differential system

$$u'(t) = J\nabla H(u(t)), \qquad t > t_0, \qquad u(t_0) = u_0, \tag{15}$$

where $H \in C^2(\mathbb{R}^{2n} \to \mathbb{R})$ is the Hamiltonian function. The solution $u(t)$ may again be given as $u(t) = S(t)u_0$, that is in terms of evolution operator $S(t)$, but $S(t)$ now different from the Jacobian matrix function $D(t) = \frac{\partial u(t)}{\partial u_0}$, which satisfies the following variational equation on $\mathcal{S}p(2n, \mathbb{R})$:

$$D'(t) = J\nabla^2 H(u(t))D(t), \qquad t > t_0, D(t_0) = I, \tag{16}$$

where $\nabla^2 H(u(t))$ denotes the Hessian matrix of $H(u)$ evaluated at the exact solution $u(t)$.

3.1 Linearization around a stable equilibrium point

Now, let $\tilde{u}$ a point of $\mathbb{R}^n$ such that $\nabla H(\tilde{u}) = 0$, that is $\tilde{u}$ let a stable equilibrium point for the system. Then we can consider the linearized problem:

$$v'(t) = J\nabla^2 H(\tilde{u})v(t), \qquad t \in [t_k, t_{k+1}], \qquad v(t_k) = u_k, \tag{17}$$

where $v(t) = u(t) - \tilde{u}$ and where $D(t)$ satisfies its variational equation

$$D'(t) = J\nabla^2 H(\tilde{u})D(t), \qquad t \in [t_k, t_{k+1}], \qquad D(t_k) = I. \tag{18}$$

Hence we can set

$$u_{k+1} = D_{k+1}(u_k - \tilde{u}) + \tilde{u}, \tag{19}$$

where D_{k+1} is an approximation of (18) computed by a symplectic method, for instance ERKCAY.

3.2 Transformation methods

We suppose that our nonlinear Hamiltonian system (15) may be written in the following form

$$u' = A(u)u, \qquad t > t_0, \qquad u(t_0) = u_0, \tag{20}$$

with $A(u)$ Hamiltonian matrix for all $u \in \mathbb{R}^{2n}$. Then, if $\tilde{u}(t)$ is an approximation of $u(t)$ on $[t_k, t_{k+1}]$ of order p, we may consider the linearized system

$$u' = A(\tilde{u})u, \qquad t \in [t_k, t_{k+1}], \qquad u(t_k) = u_k, \tag{21}$$

the solution operator $D(t)$ of which evolves on the symplectic group. Thus we can apply to (21) the group method ERKCAY, that is, we first compute

$$\begin{aligned} U_{k+1} &= \tfrac{h}{2} \textstyle\sum_{i=1}^{v} b_i (I - U_{ki}) A(u_{ki})(I + U_{ki}), \\ U_{ki} &= \tfrac{h}{2} \textstyle\sum_{j=1}^{i-1} a_{ij} (I - U_{kj}) A(u_{kj})(I + U_{kj}), \qquad i = 1, \ldots, v, \end{aligned} \tag{22}$$

where u_{ki} may be given by

$$u_{ki} = (I - U_{ki})^{-1}(I + U_{ki})u_k, \qquad i = 1, \ldots, v, \tag{23}$$

or, to avoid the inverse matrices of (23), the numerical solution u_{ki} may be computed using the same explicit RK method of ERKCAY applied to (20), that is

$$u_{ki} = u_k + h \sum_{j=1}^{i-1} a_{ij} A(u_{kj}) u_{kj}, \qquad i = 1, \ldots, v. \tag{24}$$

Finally, D_{k+1} and u_{k+1} will be computed as in (12) and (13) respectively.

We observe that the Hamiltonian matrix $A(u)$ may be written in the following form

$$A(u) = \begin{pmatrix} A_{11}(u) & A_{12}(u) \\ A_{21}(u) & -A_{11}^T(u) \end{pmatrix},$$

where A_{12} and A_{21} must be symmetric matrices. Thus, since $u = (p, q)^T$, the condition $A(u)u = J\nabla H(u)$ is equivalent to the system

$$A_{11}(u)p + A_{12}(u)q = -\frac{\partial H}{\partial q}, \qquad A_{21}(u)p + A_{22}(u)q = \frac{\partial H}{\partial p}. \tag{25}$$

Thus Hamiltonian systems for which the previous system has a solution may be transformed in the form (20). Of course, this is possible when $H(u)$ is quadratic with respect to u, that is $H(u) = \frac{1}{2}u^T Su$, where S is a symmetric matrix. A different case is when $H(p, q)$ is separable, that is $H(p, q) = T(p) + V(q)$, with

$$\frac{\partial V}{\partial q} = -A_{12}(u)q, \qquad \frac{\partial T}{\partial p} = A_{21}(u)p. \tag{26}$$

In this case if we assume $A_{11} = 0$, and A_{12}, A_{21} given by (26) we obtain a solution of (25).

4 Numerical tests

In this section we test the proposed methods on linear and nonlinear Hamiltonian problems. All the numerical results have been obtained by Matlab codes implemented on a scalar computer Alpha 200 5/433 with 512 Mb RAM. Comparisons with methods of the same order are made in terms of symplectic error, energy error and CPU time. The symplectic error is measured by $S_k = \|D_k^T J D_k - J\|_F$ and the energy error by $E_k = |H(p_k, q_k) - H(p_0, q_0)|$, where $\|\cdot\|_F$ is the Frobenius norm on matrices, p_k, q_k are the numerical solutions at the instant t_k and p_0, q_0 are the vectors of the initial conditions. Note that we have denoted by PRKv the projected method of order v based on the SR factorization.

Example 1. We consider the harmonic oscillator with the following separable Hamiltonian energy function $H(p, q) = T(p) + V(q)$, with $T(p) = \frac{p^2}{2m}$, and

$V(q) = \frac{k(t)q^2}{2}$, with $k(t) = 1 + 10^{-4}\cos(t)$, $m = 2$ and taking as initial condition $p_0 = q_0 = -1$ at $t = 0$. In Table 1 and 2 we show the CPU-time required, the symplectic error S_k at the end of the integration, the maximum and the minimum value of the energy error E_k on the integration interval $[0, 100]$ for both second and fourth order methods applied with step-size $h = 1$. We note that the

Table 1.

Method	CPU-time	S_k at 100	$\max_k E_k$	$\min_k E_k$
ERKCAY2	0.0667	2.2204e-16	9.7728e-05	1.0351e-06
GLRK1	0.0833	2.2204e-16	1.0225e-04	8.6402e-09
PRK2	0.1667	0	0.2640	0.0019

behavior of the methods is very similar, but the ERKCAY2 and GLRK1 show a better conservation of the Hamiltonian than the projected methods. Moreover, ERKCAY2 requires less CPU-time.

Table 2.

Method	CPU-time	S_k at 100	$\max_k E_k$	$\min_k E_k$
ERKCAY4	0.1167	0	1.0010e-04	3.2177e-07
GLRK2	0.1333	2.2204e-16	9.9861e-05	3.0505e-07
PRK4	0.1867	0	0.2520	7.3342e-05

Example 2.

In this example we consider the equation of the motion of a pendulum in the absence of friction:

$$p'(t) = -\sin q(t), \qquad q'(t) = p(t), \tag{27}$$

with initial conditions $q(0) = q_0$ and $p(0) = p_0$ and Hamiltonian function

$$H(p, q) = -\frac{1}{2}p^2 + \cos(q) - 1.$$

The equilibrium points of (27) are $(p, q) = (0, n\pi)$ for any integer n. Let us consider the linearization around the stable equilibrium point $(p^*, q^*) = (0, 0)$. The linearized system may be written as

$$v' = J\begin{pmatrix} -1 & 0 \\ 0 & \cos(q^*) \end{pmatrix} v. \tag{28}$$

Table 3.

(p_0, q_0)	$\max_k E_k$	$\min_k E_k$
$(1, 1)$	0.1156	2.2971e-04
$(\frac{1}{2}, \frac{1}{2})$	0.0077	1.4905e-05
$(0.2, 0.1)$	6.6578e-05	1.9315e-07
$(0, -0.4)$	0.0011	4.4668e-09

Table 3 shows the maximum and the minimum of the energy error E_k obtained by ERKCAY2 solving the linearized systems (28) on the interval $[0, 100]$ with $h = 0.1$ for different values of starting points. We have to observe that only if the initial points are close to the equilibrium point $(0, 0)$ the solution preserves the Hamiltonian.

Fig. 1 (a) and Fig. 2 (a) plot the numerical solution of the nonlinear equation (27) with initial values $(p_0, q_0) = (0.2, 0.1)$ and $(p_0, q_0) = (1, 1.5)$ respectively, obtained by GLRK1 on the interval $[0, 100]$ with $h = 0.1$. Fig. 1 (b) and Fig. 2 (b) plot the numerical solution of the linearized system (28) obtained by ERKCAY2 starting from the same initial points with $h = 0.1$. Note that only starting from initial values in a neighborhood of the equilibrium point $(0, 0)$ the solution of the linearized system (28) coincides with the solution of the nonlinear problem (27).

Table 4.

Method	CPU-time	S_k at 100	$\max_k E_k$	$\min_k E_k$
ERKCAY2	0.0833	0	1.0830	0.007
PRK2	0.1833	2.2204e-16	1.0402	0.0089
ERKCAY4	0.1533	0	1.0121	0.0063
PRK4	0.4000	2.2204e-16	1.0095	0.0034

Example 3. We now consider the nonlinear pendulum with Hamiltonian function $H(p, q) = \frac{1}{2}p^2 + q^2(1 - \cos q)$. Using (26) we may write the differential system in the following form

$$\begin{bmatrix} p \\ q \end{bmatrix}' = A(p, q) \begin{bmatrix} p \\ q \end{bmatrix},$$

where

$$A = \begin{pmatrix} 0 & -2(1 - \cos q) - q \sin q \\ 1 & 0 \end{pmatrix} \in \mathrm{sp}(2, \mathbb{R}), \qquad p, q \in \mathbb{R}.$$

Table 4 reports the performance of second and fourth order methods on the interval $[0, 10]$ with step-size $h = 0.1$ and initial values $p_0 = q_0 = -1$ at $t = 0$.

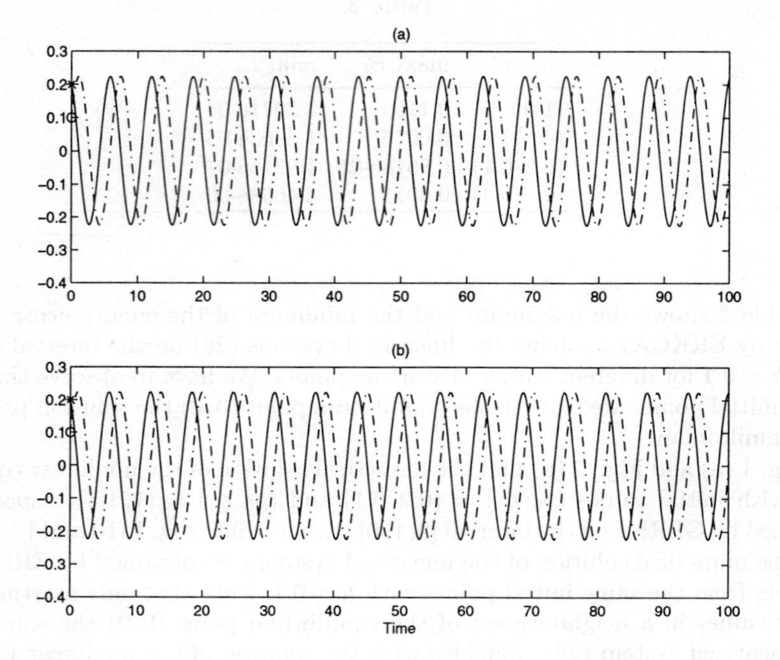

Fig. 1.

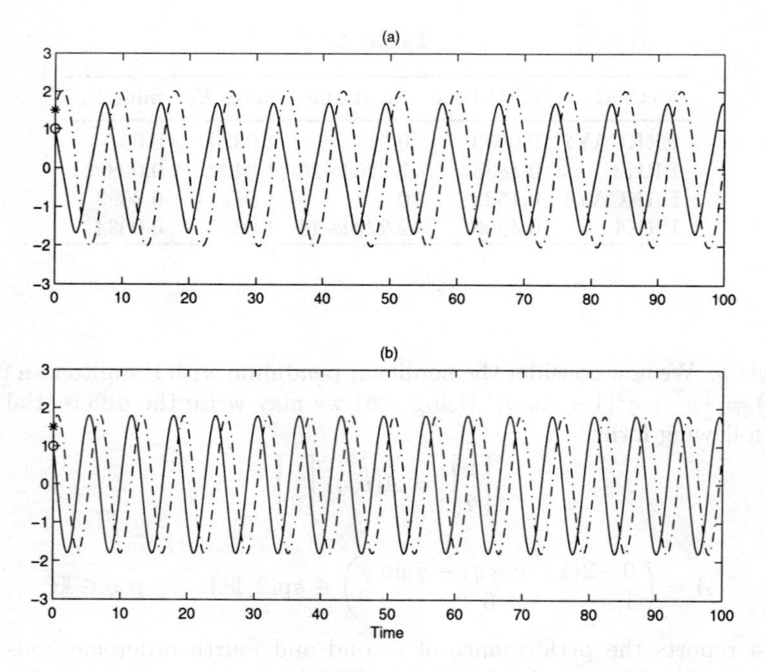

Fig. 2.

All methods preserve the symplecticity of the matrix D, while the preservation of the Hamiltonian function is not satisfactory as in previous examples.

References

1. Bunse-Gerstner, A.: Matrix factorization for symplectic QR-Like methods, Linear Alg. Appl. 83 (1986) 49-77.
2. Dieci, L., Russell, D., Van Vleck, E.: Unitary integration and applications to continuous orthonormalization techniques, SIAM J. Num. Anal. 31 (1994), 261–281.
3. Diele, F., Lopez, L., Peluso, R.: The Cayley transform in the numerical solution of unitary differential systems, Advances in Comp. Math. 8 (1998). 317–334.
4. Hairer, E., Wanner, G.: Solving ordinary differential equations, Vol. II, Springer, Berlin (1991).
5. Lasagni, F.M.: Canonical Runge-Kutta methods, ZAMP 39 (1988), 952-953.
6. Sanz-Serna, J.M.: Runge Kutta schemes for Hamiltonian systems, BIT 28 (1988), 877-883.
7. Sanz-Serna, J.M., Calvo, M.P.: Numerical Hamiltonian Problems, Chapman Hall, London, (1994).
8. Stuart, A.M., Humphries, A.R.: Dynamical Systems and Numerical Analysis, Cambridge University Press (1996).

Variants of Learning Algorithm Based on Kolmogorov Theorem

Roman Neruda, Arnošt Štědrý, and Jitka Drkošová *

Institute of Computer Science, Academy of Sciences of the Czech Republic, P.O. Box 5, 18207 Prague, Czech Republic
roman@cs.cas.cz

Abstract. A thorough analysis of theoretical and computational properties of kolmogorov learning algorithm for feedforward neural networks lead us to proposal of efficient sequential and parallel implementation. A novel approach to parallelization which combines our previous results in order to achieve higher parallel speedup is presented.

1 Introduction

In 1957 Kolmogorov [4] has proven a theorem stating that any continuous function of n variables can be exactly represented by superpositions and sums of continuous functions of only one variable. The first, who came with the idea to make use of this result in the neural networks area was Hecht-Nielsen [1]. Kůrková [2] has shown that it is possible to modify the original construction for the case of approximation of functions. Thus, one can use a perceptron network with two hidden layers containing a larger number of units with standard sigmoids to approximate any continuous function with arbitrary precision.

In [5] we have proposed a serial implementation of the Kolmogorov algorithm. Both the theoretical results and practical experiments have shown that the time complexity is relatively big. On the other hand, the algorithm has been suitable for parallelization. In [6] our first parallel implementation has been proposed. This quite efficient parallel algorithm with very low communication is suitable to be implemented on the cluster of workstation architecture. Its drawback is a limited number of processors which can be employed.

In this contribution we propose a different parallel implementation that is based on theoretical properties of the algorithm. This implementation can achieve higher degree of parallelism while still maintaining good efficiency. Communicational demands of the processors are very small in comparison to computations. Moreover, this implementation can make use of the efficient serial implementation we have developed earlier. The time complexity analysis of this algorithm is provided based on the BSP (Bulk Synchronous Parallel) model of a realistic distributed memory parallel machine. Our first experiments have been done in the PVM (Parallel Virtual Machine) envirinment.

* This research has been partially supported by Grant Agency of the Academy of Sciences of the Czech Republic under grants B1030006 and A2030801.

P.M.A. Sloot et al. (Eds.): ICCS 2002, LNCS 2331, pp. 536–543, 2002.

2 Preliminaries

By $\mathcal{R}$ we denote the set of real numbers, $\mathcal{N}$ means the set of positive integers, $\mathcal{I} = [0, 1]$ and thus $\mathcal{I}^n$ is the n-dimensional unit cube.

Definition 1. *The sequence* $\{\lambda_k\}$ *is* integrally independent *if* $\sum_p t_p \lambda_p \neq 0$ *for any finite selection of integers* t_p *for which* $\sum_p |t_p| \neq 0$

Definition 2. *By* sigmoidal function *we mean any function* $\sigma : \mathcal{R} \to \mathcal{I}$ *with the following limits:* $\lim_{t \to -\infty} \sigma(t) = 0 \quad \lim_{t \to \infty} \sigma(t) = 1$

Definition 3. *A function* $\omega_f : (0, \infty) \to \mathcal{R}$ *is called a* modulus of continuity *of a function* $f : \mathcal{I}^n \to \mathcal{R}$ *if* $\omega_f(\delta) = sup\{|f(x_1, \ldots, x_n) - f(y_1, \ldots, y_n)|; (x_1, \ldots, x_n), (y_1, \ldots, y_n) \in \mathcal{I}^n$ *with* $|x_p - y_p| < \delta$ *for every* $p = 1, \ldots, n\}$.

The original Kolmogorov result shows that every continuous function defined on n-dimensional unit cube can be represented by superpositions and sums of one-dimensional continuous functions.

Theorem 1 (Kolmogorov). *For each integer* $n \geq 2$ *there are* $n(2n + 1)$ *continuous monotonically increasing functions* ψ_{pq} *with the following property: For every real-valued continuous function* $f : \mathcal{I}^n \to \mathcal{R}$ *there are continuous functions* ϕ_q *such that*

$$f(x_1, \ldots, x_n) = \sum_{q=0}^{2n} \phi_q \left[\sum_{p=1}^{n} \psi_{pq}(x_p) \right] . \tag{1}$$

Further improvements by Sprecher provide a form that is more suitable for computational algorithm. Namely, the set of functions ψ_{pq} is replaced by shifts of a fixed function ψ which is moreover independent on a dimension. The overall quite complicated structure is further simplified by suitable parameterizations and making use of constants such as λ_p, β, etc.

Theorem 2 (Sprecher). *Let* $\{\lambda_k\}$ *be a sequence of positive integrally independent numbers. There exists a continuous monotonically increasing function* $\psi : \mathcal{R}^+ \to \mathcal{R}^+$ *having the following property: For every real-valued continuous function* $f : \mathcal{I}^n \to \mathcal{R}$ *with* $n \geq 2$ *there are continuous functions* ϕ *and a constant* β *such that:*

$$\xi(\mathbf{x}_q) = \sum_{p=1}^{n} \lambda_p \psi(x_p + q\beta) \tag{2}$$

$$f(\mathbf{x}) = \sum_{q=0}^{2n} \phi_q \circ \xi(\mathbf{x}_q) \tag{3}$$

3 Algorithm Proposal

Based on Theorem 2 Sprecher in [7] proposed a numerical algorithm whose core consists of four steps:

Let n be the input dimension. Let f is a known continuous function, $e_0 \equiv f$. For each iteration r the number $k = k(r)$ is determined according to previous approximation f_{r-1}. Note that the number $k(r)$ determines the dissection of the unit cube $\mathcal{I}^n$ and plays an important role for convergence of the algorithm.

1. Construct the mesh $\mathcal{Q}^n$ of rational points $\mathbf{d}_k$ dissecting the unit cube $\mathcal{I}^n$.

$$\mathcal{Q}^n = \{\mathbf{d}_k = (d_{k1}, \ldots, d_{kn}); d_{kp} \in \mathcal{Q}, p = 1, \ldots, n\}, \tag{4}$$

   where $\mathcal{Q} = \left\{ d_{kp} = \sum_{s=1}^{k} i_s \gamma^{-s}; i_s \in \{0, 1, \ldots, \gamma - 1\}, k \in \mathcal{N} \right\}$ and integers m, γ are such that $m \geq 2n$ and $\gamma \geq m + 2$.
2. Construct monotonically increasing functions $\psi : \mathcal{Q} \to \mathcal{I}$ whose linear combination defines the functions ξ in the points $\mathbf{d}_k^q$. The functions $\xi(\mathbf{d}_k^q)$are expressed by equation

$$\xi(\mathbf{d}_k^q) = \sum_{p=1}^{n} \lambda_p \psi(d_{kp}^q) \tag{5}$$

   where $\psi(d_{kp}^q)$ are defined by means of representation of 6 and real coefficients λ_p are the results of infinite sums dependent on γ (for details cf. [5]).
   For $q = 0, 1, \ldots, m$ numbers $\mathbf{d}_k^q \in \mathcal{Q}^n$ are constructed as:

$$d_{kp}^q = d_{kp} + q \sum_{s=1}^{k} \gamma^{-s}. \tag{6}$$

   Obviously $d_{kp}^q \in \mathcal{Q}$ for $p = 1, 2, \ldots, n$.
3. Create sigmoidal steps θ by means of a pair of shifted sigmoidal functions σ

$$\begin{aligned} \theta(\mathbf{d}_k^q; y_q) = &\ \sigma(\gamma^{\beta(k+1)}(y_q - \xi(\mathbf{d}_k^q)) + 1) \\ &- \sigma(\gamma^{\beta(k+1)}(y_q - \xi(\mathbf{d}_k^q) - (\gamma - 2)b_k), \end{aligned} \tag{7}$$

   where real numbers b_k are results of infinite sums dependent on γ (for details cf. [5]).
   Functions θ and previous iteration error function e_{r-1} are then used to compile the outer functions ϕ_q^r for $q = 0, \ldots, m$

$$\phi_q^r(y_q) = \frac{1}{m+1} \sum_{\mathbf{d}_k^q} e_{r-1}(\mathbf{d}_k)\ \theta(\mathbf{d}_k^q; y_q), \tag{8}$$

   Note that due to the definition of sigmoidal function σ (2) $\theta(\mathbf{d}_k; y_q) \in [0, 1]$, $\theta(\mathbf{d}_k; y_q) = 1$ for $y_q \in [\xi(\mathbf{d}_k^q); \xi(\mathbf{d}_k^q) + (\gamma - 2)b_k]$.
   The functions $\phi_q(y_q)$ from theorem 2 are then expressed as: $\phi_q(y_q) = \lim_{r\to\infty} \sum_{j=1}^{r} \phi_q^j(y_q)$.

4. Construct the r-th approximation f_r of original function f, and the r-th approximation error e_r.

$$e_r(\mathbf{x}) = e_{r-1}(\mathbf{x}) - \sum_{q=0}^{m} \phi_q^r \circ \xi(\mathbf{x}_q), \tag{9}$$

where $\mathbf{x} \in \mathcal{I}^n$, $\mathbf{x}_q = (x_1 + q\beta, \ldots, x_n + q\beta)$, and $\beta = \gamma(\gamma - 1)^{-1}$.
The r-th approximation f_r to f is then given by:

$$f_r(\mathbf{x}) = f_{r-1} + \sum_{q=0}^{m} \phi_q^j \circ \xi(\mathbf{x}_q). \tag{10}$$

We conclude this section by several notes and observations mostly taken from [7].

The convergence of the above alorithm is assured if $k(r)$ is set such that the dissection of the mesh $\mathcal{Q}^n$ in each step r corresponds to the modulus of continuity $\omega_{f_{r-1}}$.

Another assumption used in the proof is that the uniform norm f as well as the values of f on an everywhere dense set in $\mathcal{I}^n$ are known. Then we can restrict ourselves to the mesh $\mathcal{Q}^n$ and no other values than $f(\mathbf{d}_k^q)$ are required.

In practice, function f is typically given only by a prescribed set of arguments and values $\{\mathbf{x}_t, f(\mathbf{x}_t)\}$, the so-called *training set*.

For each k and q denote the family of closed intervals

$$E_k^q(d_k^q) = \left[\frac{-q}{\gamma - 1}\gamma^{-k} + d_k, \frac{\gamma - 2 - q}{\gamma - 1}\gamma^{-k} + d_k\right]. \tag{11}$$

Note that for fixed q and k these intervals are separated by gaps. Cartesian product of these intervals

$$S_k^q(\mathbf{d}_k^q) = E_k^q(d_{k,1}^q) \times E_k^q(d_{k,2}^q) \times \ldots \times E_k^q(d_{k,n}^q) \quad \text{for } q = 0, 1, \ldots, m \tag{12}$$

is a closed cube.

If the training set is appropriate and the set Q^n is fine enough, i.e. for known value $f(\mathbf{x}_k^q)$ there exists surrounding $S_k^q(\mathbf{d}_k)$ of $\mathbf{d}_k$ such that $\mathbf{x}_k^q \in S_k^q(\mathbf{d}_k)$) then the algorithm converges.

In some cases, the set of known values can be too irregular. Then, some special techniques that involve, among others, the use of interpolating functions, can be applied; see [3]. The disadvantage of this approach is its high computational cost.

4 Serial Implementation

In the following, we consider our practical implementation of the above described algorithm, which introduces several additional computational improvements and simplifications. First, the infinite sums are replaced by sums up to the sufficiently big constant K. Since the sums vanish quickly, usually, $K = 8$ suffices to reach machine precision limit. For the sake of simplicity we also do not deal with modulus of continuity of the function f and keep the number $k(r)$ fixed to k. Again, small values of $k = 3, 4, 5$ show to be satisfiable.

Note that all computations of f are performed on numbers $\mathbf{d}_k^q$ taken from the dense support set $\mathcal{Q}^n$. Due to the nature of functions ψ and ξ, it shows advantageous to represent numbers $\mathbf{d}_k^q$ by means of ordinal values of their coordinates $d_k \in \mathcal{Q}$ instead of only their floating point decimal value. It allows quite efficient computation of $\xi(\mathbf{d}_k^q)$ and moreover, it enables caching.

Caching of already computed values is quite a straightforward technique that can save reasonable amount of computations compared to formulae presented above in our sketch of algorithm. In our serial implementation caching takes place in the computations of values of $\psi(d_k^q)$, and $\xi(\mathbf{d}_k^q)$.

Taking all these computational issues into account, we have analyzed the time complexity of this serial implementation. The following theorem 3 summarizes our estimation of the total amount of computational time for one iteration (cf. [5]).

Theorem 3. *The computational requirements in one iteration of the sequential algorithm is estimated as follows.*

$$T_S = \mathcal{O}\left(m\, n\, \gamma^{nk}(n^k + k^3)\right). \tag{13}$$

5 Parallel Implementation I

One possible way how to parallelize the above described algorithm is to take the outermost sum from 10 and distribute the computations of mutually independent terms $\Phi_0^r, \dots, \Phi_m^r$ on separate processors. Additionally, one master processor is used to compute e_r and f_r, respectively, and to distribute the computed values back to the nodes.

We have performed our analysis and experiments for parallel architecture of cluster of workstations. No special network topology is considered, and we assume that (within a reasonable range of data amounts) the cost of sending one floating point number over the network is constant. Let us denote t_c such a time needed to communicate one float, expressed w.r.t. the cost of one arithmetic operation. Note that, considering a common Pentium processors connected via fast Ethernet, $t_c \approx 10^1 - 10^2$.

An analysis of the parallel time complexity is summarized in the following result (cf. [6]).

Theorem 4. *The total time for one iteration of the parallel algorithm is estimated as follows.*

$$T_{P_1} = \mathcal{O}\left(n\, \gamma^{nk}(n^k + k^3) + m\gamma^{nk}t_c\right). \tag{14}$$

Comparison of theorems 3 and 4 shows that this method of parallelization is very efficient, yet limited. The efficiency means that, employing m additional processors, one achieves speedup in the order of m. Considering reasonable values of t_c, this holds even for quite small values of parameters n, m and k. In our practical experiments, the communication time was always only a tiny fraction of the overall computation.

The drawback of this approach is the limited degree of parallelism. One cannot utilize more than $m \approx 2n$ processors. This can be overcome by a different way of parallelization which is described in the following section.

6 Parallel Implementation - II

Another natural approach to parallelization of the algorithm is to distribute subsets of $\mathcal{Q}^n$ among processors. Functions ξ map the closed cubes $S_k^q(\mathbf{d}_k^q) = E_k^q(d_{k,1}^1) \times \ldots \times E_k^q(d_{k,n}^1)$ into disjunct closed intervals $T_k^q(\mathbf{d}_k^q)$. Therefore, we can provide computations on $S_k^q(\mathbf{d}_k^q)$ in a mutually independent way. One master process dissects $\mathcal{I}^n$ into $S_k^q(\mathbf{d}_k^q)$ and distributes these subdomains to slave processors. They compute their corresponding contribution to approximation and deliver the result back to master.

This algorithm allows for much higher degree of parallelization — it can utilize up to γ^{nk} processors. The communication takes place only at the beginning and the end of the algorithm. The iterations can be done independently.

Theorem 5. *The total time for one iteration of the parallel algorithm is estimated as follows.*

$$T_{P_2} = \mathcal{O}\left(m\, n\, (n^k + k^3)\right). \tag{15}$$

It is not obvious that the computations on each $S_k^q(\mathbf{d}_k^q)$ can be done independently. Since the definitions of sigmoidal steps θ are global, it might seem that the opposite is true. But, a closer observation of equations 7 and 8 reveals that the global approximation is composed only of local steps. Although the sum in 8 is defined over all $\mathbf{d}_k^q$, the function $\theta(\mathbf{d}_k^q; y_q)$ gives non-zero values only on $S_k^q(\mathbf{d}_k^q)$.

Slave nodes can make use of a caching technique from the serial algorithm for their local computations. Each computation of one processor is repeated for $q = 0, \ldots, m$. Between each inner cycle iteration nodes do not clear their cache. Because the cubes S_k^q intersect for any two $q \in \{0, \ldots, m\}$ this precaution leads to computational saving.

In the following we present sketch of this parallel algorithm more thoroughly.

Parallel algorithm II

Let n be the input dimension. Let f is a known continuous function, $e_0 \equiv f$, k is fixed.

1. On the master node, construct the mesh $\mathcal{Q}^n$ of rational points $\mathbf{d}_k$ dissecting the unit cube $\mathcal{I}^n$ according to 4. Create the family of closed cubes $C(\mathbf{d}_k) = \cup_q S_k^q(\mathbf{d}_k^q)$ for all $\mathbf{d}_k^q) \in \mathcal{Q})$, where $S_k^q(\mathbf{d}_k^q)$ is defined by 12.
2. Distribute cubes $C(\mathbf{d}_k)$ among processors. Note, that $S_k^q(\mathbf{d}_k^q) \in C(\mathbf{d}_k)$ for any $q \in \{0, 1, \ldots m\}$.
3. for $q = 1, \ldots, m$
   (a) On each slave node, perform sequential computation (cf. section 4) of 10 and 9 on $S_k^q(\mathbf{d}_k^q)$. Store the results, and compute local partial sum over $S_k^q(\mathbf{d}_k^q)$.
   (b) Send the result back to the master node.
4. Master node computes the function f as a sum over all cubes $C(\mathbf{d}_k)$.

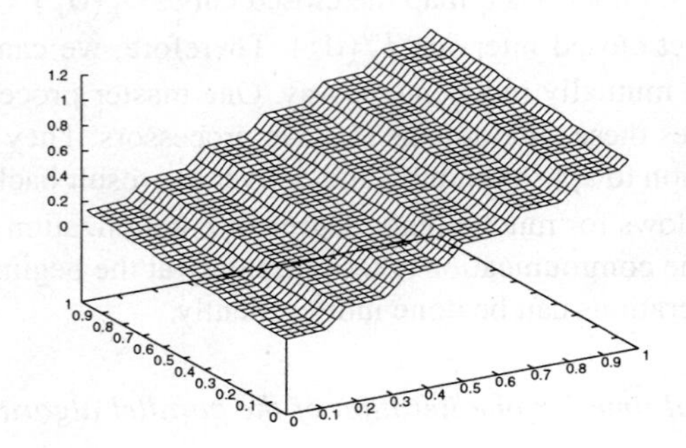

Fig. 1. Function ξ maps the closed cubes $S_k^q(\mathbf{d}_k^q) = E_k^q(d_{k,1}^1) \times \ldots \times E_k^q(d_{k,n}^1)$ into disjunct closed intervals on $\mathcal{I}$.

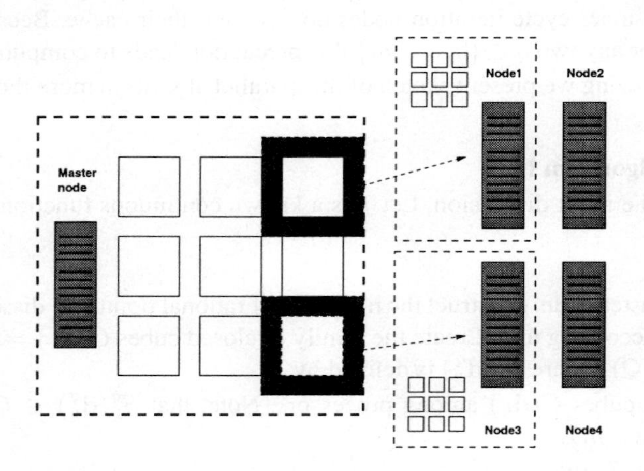

Fig. 2. Schema of paralelization II. After the decomposition master node sends the domain to the slave nodes which provide the local computation over $S_k^q(\mathbf{d}_k^q)$.

7 Conclusion

Comparing the two above parallel implementations, one can see the pros and cons of both approaches. The first parallel algorithm is quite simple, and very efficient with low communication demands. The degree of parallelism is limited by $2 * n$, though. On the other hand, the second parellel algorithm achieves much better degree of parallelism. Moreover, for the local case, it can make use of the efficient serial implementation (with features like local cache). The dissadvantage is in quite a complicated nature of the algorithm.

In our future work we would like to focus on these problems: What is the best way of expressing the algorithm for a general training set. So far, we still consider the data to be from the $\mathcal{Q}^n$. It might be better to use some interpolation scheme and avoid long lasting iterations. We have also considered a fixed value of parameter k over one iteration. The second parallel algorithm allows for a different value of k even among processors, as long as they communicate the value they have used for computations.

In case of relatively large number of processors, one can consider even more dynamic version of parallel algorithm, where the processors are assigned their domains during the second or third iteration. A similar effect can be achieved by increasing the parameter γ. All this can be done independently for each processor, since the individual computations are truly local.

References

1. Robert Hecht-Nielsen. Kolmogorov's mapping neural network existence theorem. In *Proceedings of the International Conference on Neural Networks*, pages 11–14, New York, 1987. IEEE Press.
2. Věra Kůrková. Kolmogorov's theorem is relevant. *Neural Computation*, 3, 1991.
3. Hidefumi Katsuura and David A. Sprecher. Computational aspects of kolmogorov's superposition theorem. *neural Netowrks*, 7:455–461, 1994.
4. A. N. Kolmogorov. On the representation of continuous function of many variables by superpositions of continuous functions of one variable and addition. *Doklady Akademii Nauk USSR*, 114(5):953–956, 1957.
5. R. Neruda, A. Štědrý, and J. Drkošová. Implementation of kolmogorov learning algorithm for feedforward neural networks. In *Procceedings of the International Conference on Computational Science*, pages 986–995. Springer, LNCS 2074, 2001.
6. R. Neruda, A. Štědrý, and J. Drkošová. Kolmogorov learning for feedforward networks. In *Procceedings of the International Joint Conference on Neural Networks*, pages 986–995, Washington,DC, 2001. IEEE.
7. David A. Sprecher. A numerical implementation of Kolmogorov's superpositions II. *Neural Networks*, 10(3):447–457, 1997.

Genetic Neighborhood Search

Juan José Domínguez[1], Sebastián Lozano[2] , Marcos Calle[2]

[1] E. S. Ingeniería, Dpto. Lenguajes y Sistemas Informáticos, University of Cadiz
C/ Chile, nº1, 11003 Cádiz, Spain
juanjose.dominguez@uca.es

[2] E. S. Ingenieros, Dpto. Organización Industrial y Gestión de Empresas, University of Sevilla
Camino de los Descubrimientos, s/n, 41092 Sevilla, Spain
slozano@us.es mcalle@esi.us.es

Abstract. The Genetic Neighborhood Search (GNS) is a hybrid method for combinatorial optimization problems. A main feature of the approach is the iterative use of local search on extended neighborhoods, where the better solution will be the center of a new extended neighborhood. We propose using a genetic algorithm to explore the extended neighborhoods. Computational experiences show that this approach is robust with respect to the starting point and that high quality solutions are obtained in reasonable time.

1 Introduction

Combinatorial optimization deals with the problem of minimizing (or maximizing) a function where the solution set is often represented by a set of decision variables, whose values can have certain ranges [1]. A solution is represented by a value assigned to these variables. To find a globally optimal solution consists in finding a solution between the set of feasible solutions. Mathematically, a problem of combinatorial optimization is defined by the pair *(F, f)* [2]:

$$min\ f(T), \quad T \in F \qquad \textbf{(1)}$$

Where F is the set of feasible solutions for the problem, and the function $f(T)$ measure the quality of the solution, such that:

$$f : T \to \Re \qquad \textbf{(2)}$$

Many heuristic methods for combinatorial optimization problems are based in local search. The use of a local search algorithm implies the definition of a neighborhood. A neighborhood function N is a mapping,

$$N : F \to 2^F \qquad \textbf{(3)}$$

Which defines for each solution $i \in F$, a set $N(i) \subseteq F$ of solutions, called the neighborhood of F, that are in some sense close to i. The meaning of "close to i" is that they can be reached from i by a single *move*. This single-move concept can be considered as a first-order neighborhood while performing several consecutive moves

P.M.A. Sloot et al. (Eds.): ICCS 2002, LNCS 2331, pp. 544–553, 2002.

allows the definition of extended neighborhood, the order of the extended neighborhood corresponding to the number of consecutive moves. A move is an operator, which transforms one solution onto another via small modifications [3]. A solution i is locally optimal with respect to the neighborhood N, if:

$$f(i) \leq f(x), \forall x \in N(i) \tag{4}$$

A local search process can be viewed as the procedure of minimizing the cost function f in a number of successive steps in each of which the current solution i is being replaced by a solution j such that:

$$f(j) < f(i),\ j \in N(i) \tag{5}$$

There are many different ways to conduct local search in the neighborhoods [1]: *Best improvement local search* replaces the current solution with the solution that improves most in cost after searching the whole neighborhood, and *first improvement local search* accepts a better solution when it is found.

The strategy presented with the name *Genetic Neighborhood Search* (GNS) is a new heuristic based in local search but of a special mode, where a Genetic Algorithm (GA) is used for explore the neighborhoods ([3] and [4]).

The rest of this paper is structured as follows. In section 2 the basic GNS algorithm is depicted. In section 3 we give an account of the application of the method to the Symmetric Traveling Salesman Problem. Finally, in section 4, some conclusions are drawn.

2 Genetic Neighborhood Search

The Genetic Neighborhood Search algorithm we propose has the following schema: given a initial solution, x^0, and an extended neighborhood N of order L around x^0, the GA proposed will perform an exploration of this neighborhood. The best solution obtained by the GA will become the center of a new extended neighborhood which will, in turn, be explored using the GA. The GNS is an iterative improvement local search, which pseudocode description follows:

```
Optimization: x^0 × L × A × • → x*
k ← 0
optimum ← LOCAL
Neighborhood ← L
do
    x^(k+1) ← GNS (Neighborhood, x^k)
    if ( Convergence (x^k, x^(k+1), •)) then Neighborhood ← L
    else if (Can_Increase (A))
    then Neighborhood ← Neighborhood + L
    else optimum ← GLOBAL
    k ← k + 1
while  (optimum = LOCAL)
return x^k
```

Note that when a GNS doesn't improve the center of the neighborhood, the order of the neighborhood is increased and this can be done up to A times. The objective is to avoid the local optimum. The maximum order of the extended neighborhood will be:

$$L_{max} = L \times A \quad \textbf{(6)}$$

The order of the neighborhood L is an input parameter of the algorithm. Since it determines the size of the search region, other parameters of the algorithm depend of the value of L. So, the size of the population and the number of generation of the GA are proportional to L:

$$Size\_Pob = L/\sigma \quad \textbf{(7)}$$

$$N\_Gen = Size\_Pob * \theta$$

2.1 Representation

The individuals are coded as a succession of L movements from the center point of the neighborhood. The movements are dependent of the specific combinatorial problem. In the section 3 we show how the individuals can be coded for the Traveling Salesman Problem.

2.2 Fitness

The evaluation of the individual consist in calculate the value of the objective function for each move in case of a maximization problem and minus the objective function for a minimization problem. The better solution in the trajectory of the individual is the value of the fitness. Such that, if $Fitness_{i,j}$ is the value of the objective function for the individual i after the movement j, the fitness of the individual is:

$$Fitness_i = max\{Fitness_{i,j},\ j = 1,\ldots,L\} \quad \textbf{(8)}$$

Note that the fitness of the individual is not the value in the last movement. Moreover, since the individual represents L possible solutions, the GA population implicitly contains $Size\_Pob \times L$ solutions.

2.3 Operators

The GA employs three operators for mutation and two for crossover. A steady state GA has been implemented. In each iteration either mutation or crossover is applied, so the probability of mutation (P_M) and the probability of crossover (P_C) are $P_M + P_C = 1$.

Each operator for mutation has a probability, such that the sum of these probabilities is P_M. The three operators for mutation are:

- *Exchange Mutation* consists in to exchange the position of some pairs of movements. The number of exchange is $L \times P_{EM}$, where P_{EM} is the probability of this operator.
- *Modify Mutation* consists in alter the value of some movements. The number of modify is $L \times P_{MM}$, where P_{MM} is the probability of this operator.
- *Mutation Directed by Fitness* consists in applying the Modify Mutation but only to the movements following the move that defines the fitness of the individual. The number of mutation is $L \times P_{MDF}$, where P_{MDF} is the probability of this operator.

Each operator for crossover has a probability, such that the sum of these probabilities is P_C. The operators for crossover are:

- *Uniform Crossover* consists in to generate a new individual with the movements, select randomly, of the two parents.
- *Crossover Directed by Fitness* consists in exchanging only the movements following the move that defines the fitness of the individual. Only the better child is introduced in the new population. The figure 1 shows this operator, where the gray zones are the moves prior to the one that defines the fitness.

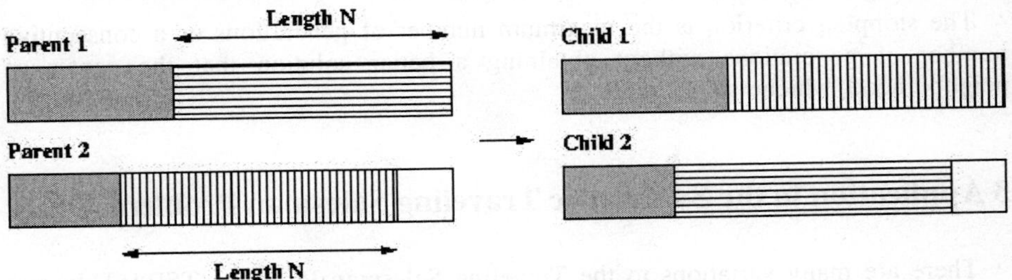

Fig. 1. The Crossover Directed by Fitness.

2.4 Selection

The operators for mutation and crossover require the selection of the parents, and the individual to be deleted from the population. In both operators (crossover or mutation), the selection of individual for replacement is fitness-based, where the individuals with worse fitness have a bigger probability of selection than the individuals with better fitness.

In the case of mutation, the individual necessary for the operator is randomly selected, i.e. the selection is not fitness-based. In the crossover, the two progenitors are selected proportional to their fitness, namely using the roulette wheel procedure [4], where the individuals with worse fitness have a smaller probability of selection than the individuals with better fitness.

2.5 The Genetic Algorithm

The pseudocode of GA employ for the exploration of a neighborhood is the following:

```
GNS: S_POP × P_C × P_CU × P_M × P_MI × P_MA × G × L × A × x^0 → x*
t ← 0
P_t ← Initialize_Population (S_POB, L, x^0 )
while (t < G)
do
    t← t + 1
    op ← Select_Operator (P_C, P_CU, P_M, P_MI, P_MA)
    [ind1, ind2 ] ← Select_Individual (P_t, op)
    new_ind ← Apply_Operator (P_t, op, ind1, ind2)
    old_ind ← Select_Individual_Deletion (P_t)
    Insert_Individual (P_t, old_ind, new_ind)
done
return Best_Individual (P_t)
```

The stopping criterion is the maximum number of generations or a consecutive number of generations without obtaining a better solution that the center of neighborhood, whichever occurs first.

3 Application to the Symmetric Traveling Salesman Problem

There are many variations to the Traveling Salesman Problem (TSP)[6]. In this report, we consider the classic Symmetric TSP. The problem is defined by *N* cities and a symmetric matrix *D*, of dimensions $N \times N$, where $d_{ij}=d_{ji}$ gives the distance between any two cities *i* and *j*, $i,j=1,...,N$. The goal in TSP is to find a tour of minimal length which visits each city exactly once [1]. That is, the objective is to find a cyclic permutation μ on the *N* cities, such that minimize the cost of the tour:

$$F(\mu)=\sum_{i=1}^{N-1} d(c_{\mu(i)},c_{\mu(i+1)})+d(c_{\mu(n)},c_{\mu(1)}) \tag{9}$$

3.1 The 2-exchange move

In section 2.1 we explain that the individuals on GNS are a succession of *L* movements. In the TSP, the move employed is the *2-exchange*. It consists in randomly select two cities, *P* and *Q*, in such a way as the arcs *(P,Π(P))* and *(Q,Π(Q))* are deleted, and the new arcs *(P,Q)* and *(Π(P),Π(Q))* are introducing. The figure 2 shows this movement.

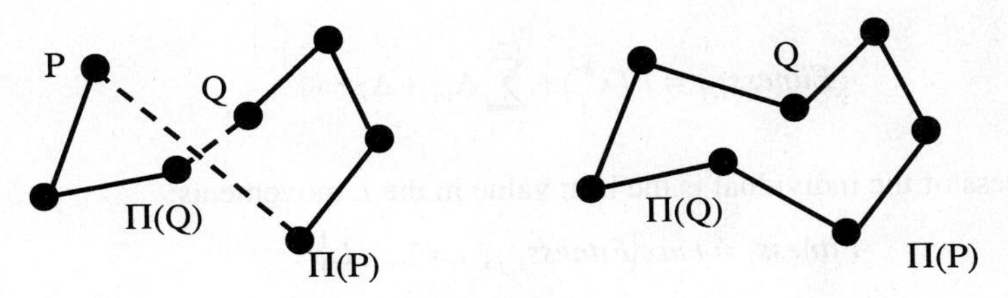

Fig. 2. The 2-exchange move.

The coding of this type of movement is a vector of two components, which represent the two arcs to delete. Each arc is coded with a vector of two elements: the origin city and the direction of the arc. Although the Symmetric TSP the tour has not a direction, for the codification of individuals the tour has a direction. So, the code for the direction of an arc has a value of 1 if it's the same direction, or 0 in other case. The figure 3 shows a tour, where the direction is clockwise. In this example, the arc between the cities 3 and 7 is coded *[3,1]*, while that the arc between the cities 3 and 2 is *[3,0]*.

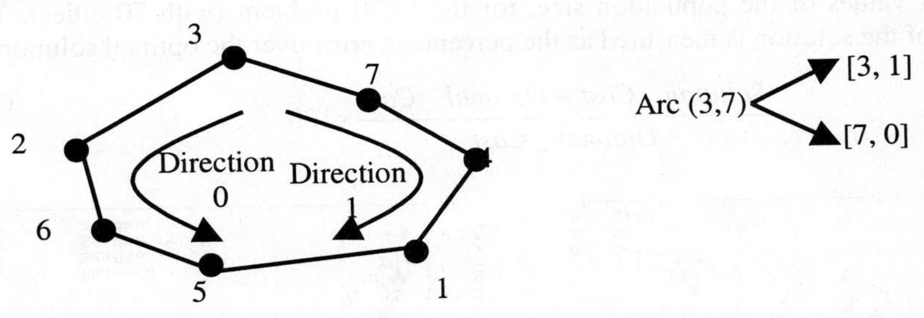

Fig. 3. The coded of an arc in a movement

The problem of this representation is that an arc can have two codes. The solution consists in using the code involving the lower-labeled city. In the example of the figure 3, the arc between the cities 3 and 7, with coded *[3,1]* and *[7,0]*, is only represented by *[3,1]*.

3.2 Fitness of the individuals

Each movement of an individual has a cost due to the two arcs deleted and the two new arcs. So, the individual i has an increment in the objective function due to the movement j:

$$\Delta_{ij} = d_{p,q} + d_{\Pi(p),\Pi(q)} - d_{p,\Pi(p)} - d_{q,\Pi(q)} \tag{10}$$

The value of the individual i in the k GNS associated to the movement j is, where the center individual of the neighborhood is x^k:

$$Fitness_{i,j} = F(x^k) + \sum_{h=1}^{j-1} \Delta_{ih} + \Delta_{ij} \tag{11}$$

The fitness of the individual is the best value in the L movements:

$$Fitness_i = max\{Fitness_{i,j},\ j = 1,\dots,L\} \tag{12}$$

3.3 Computational experiences

To test the performance of the GNS, the TSPLIB [7] was used. All the tests have been run on a 450 MHz PC Pentium III. What we report here are the results obtained with the proposed approach for some of the problems tested. A thorough study to benchmark GNS will be the subject of another paper.

Figure 4 shows the results obtained by the GNS in 100 runs performed with uniform probability for each operator and an order of the neighborhood L=15, with different values of the population size, for the ST70 problem (with 70 cities). The quality of the solution is measured as the percentage error over the optimal solution:

$$\frac{Solution\_Cost - Optimal\_Cost}{Optimal\_Cost} \times 100 \tag{13}$$

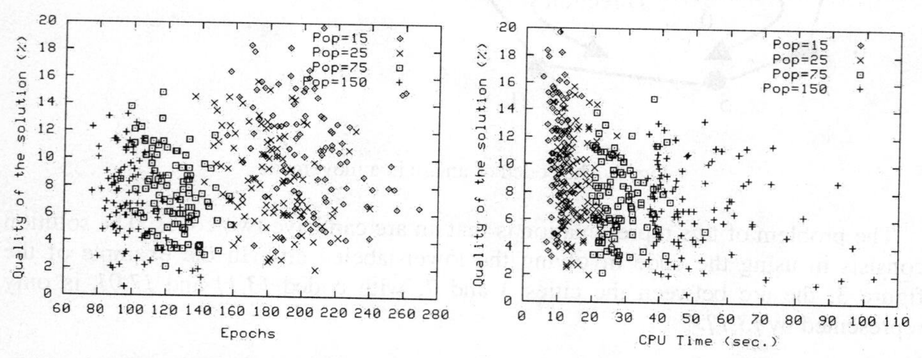

Fig. 4. Epochs and CPU time vs quality of solution for different size of population, ST70 problem.

Note that a small population size requires more epochs than a larger size. Obviously, a large population requires more computing time because it needs to evaluate more points and more generations (equation 7). However, a rise in the computing time doesn't always translate into better solutions unless an effective trade-off between intensification and diversification is maintained. The computational experiences show that the optimum population size is around 15-20% of number of cities.

The order L of the extended neighborhood is the main parameter of the algorithm. Figure 5 shows, for problems EIL51 and ST70 (51 and 70 cities, respectively), the results obtained by GNS for fifty runs carried out, using uniform probability for each operator, a population size of 20% of number of cities and different values of the order of the neighborhood.

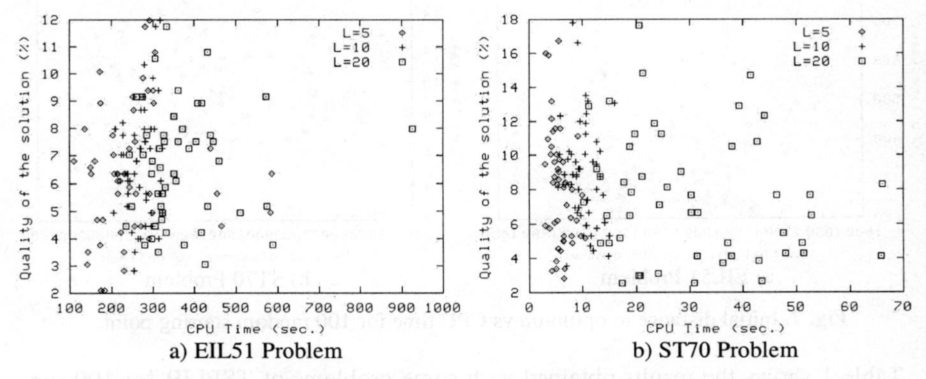

a) EIL51 Problem b) ST70 Problem

Fig. 5. CPU time vs quality solution by different order of neighborhood.

Note that a large neighborhood order seems to lead to better solutions than a smaller neighborhood order. This may be due to GNS with a large neighborhood order neighborhood more effectively avoiding getting trapped in local optima. However, a small order of neighborhood has a lower computing time. It is thus necessary to reach a trade-off between computing time and quality of the solution. The computational experiences show that the optimal neighborhood order is between 10 and 15, allowing for at most 6 or 10 rises.

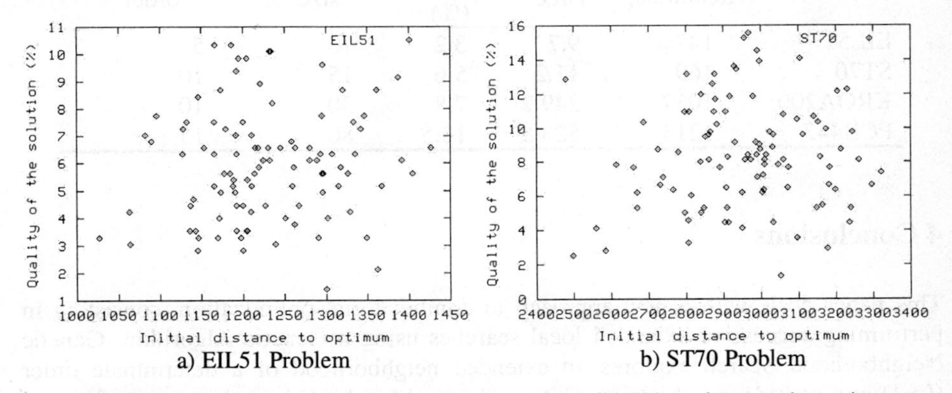

a) EIL51 Problem b) ST70 Problem

Fig. 6. Initial distance to optimum vs quality solution for 100 random starting point.

The proposed algorithm depends on the initial solution or starting tour. Figure 6 show the quality of solution obtained with GNS for EIL51 and ST70 starting from 100 random tours and the figure 7 shows the CPU time versus the initial distance to the optimum. The population size used is 20% of number of cities (i.e. 10 and 15 respectively). The neighborhood order is 10 and the probabilities of the different operators are uniform. Note that, independently of the starting point, GNS reaches a good approximate solution to the optimum. Similarly, the CPU time required, does not depend too much on the starting point.

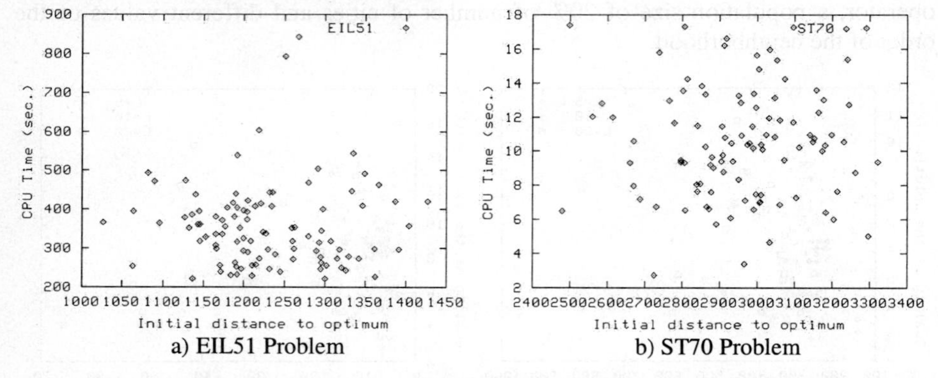

a) EIL51 Problem b) ST70 Problem

Fig. 7. Initial distance to optimum vs CPU time for 100 random starting point.

Table 1 shows the results obtained with some problems of TSPLIB for 100 runs. The probabilities employed are P_C=0.8 (P_{CDF}=0.64 y P_{UC}=0.16) y P_M=0.2 (P_{EM}=0.04, P_{MM}=0.04 y P_{MDF}=0.12). The size of the population is fixed to 20% of the number of cities, except in KROA200 that it is the 15%.

Table 1. Problem TSPLIB computational results.

Problem	No. Iterations	CPU Time	Average error (%)	Population size	Neighborhood order
EIL51	147	9.7	3.2	10	5
ST70	169	11.2	5.6	15	10
KROA200	1057	249.2	7.8	30	10
PCB442	1213	523.5	13.5	80	15

4 Conclusions

This paper deals with a new approach to combinatorial optimization, consisting in performing successive extended local searches using a Genetic Algorithm. Genetic Neighborhood Search explores an extended neighborhood of a determinate order (L>1) and considering the best solution obtained by the GA as the center of a new extended neighborhood. The originality of the method is the exploration of higher-order neighborhoods using a GA and the evaluation of the fitness of the individuals as the best solution found along the L-move trajectory. Both the codification of the moves as well as the mutation and crossover operators depend on the problem being solved. The experiments carried out for the symmetric TSP show that GNS is robust with respect the starting point and generally obtain solutions of good quality in a reasonable CPU time.

References

1. Aarts, E., Lenstra, J.K.: *Local Search in Combinatorial Optimization.* John Wiley & Sons, Ltd. (1997)
2. Michalewicz, Z., Fogel, D.B.: *How to solve it: Modern Heuristics.* Springer-Verlag, Berlin Heidelberg New York. (1998)
3. Voudouris, C., Tsang, E.: *Guided Local Search.* Technical Report CSM-247, Dep. of Computer Science, University of Essex. (1995)
4. Goldberg, D.E.: *Genetic Algorithms in Search, Optimization and Machine Learning.* Addison-Wesley, Reading. (1989)
5. Michalewicz, Z.: *Genetic Algorithms + Data Structures = Evolution Programs.* 3rd edn. Springer-Verlag, Berlin Heidelberg New York. (1996)
6. Lawler, E.L., Lenstra, L.K., Rinnooy Kan, A.H.G., Shmoys, D.B. (Eds.): *The Travelling Salesman Problem: A guided tour in combinatorial optimization.* John Wiley & Sons. (1985)
7. Reinelt, G.:TSPLIB 95. http://www.iwr.uni-heidelberg.de/groups/comopt/software/TSPLIB95 (1995)

Application of Neural Networks Optimized by Genetic Algorithms to Higgs Boson Search

František Hakl*, Marek Hlaváček**, and Roman Kalous**

*Institute of Computer Science AS CR, Pod Vodárenskou věží 2, 182 07 Prague 8, Czech Republic
**Mathematics department, Faculty of Nuclear Science and Physical Engineering, Czech Technical University, Prague, Czech Republic

Abstract. This paper describe an application of a neural network approach to SM (standard model) and MSSM (minimal supersymetry standard model) Higgs search in the associated production $t\bar{t}H$ with $H \to b\bar{b}$. This decay channel is considered as a discovery channel for Higgs scenarios for Higgs boson masses in the range 80 - 130 GeV. Neural network model with a special type of data flow is used to separate $t\bar{t}jj$ background from $H \to b\bar{b}$ events. Used neural network combine together a classical neural network approach and linear decision tree separation process. Parameters of these neural networks are randomly generated and population of predefined size of those networks is learned to get initial generation for the following genetic algorithm optimization process. A genetic algorithm principles are used to tune parameters of further neural network individuals derived from previous neural networks by GA operations of crossover and mutation. The goal of this GA process is optimization of the final neural network performance.
Our results show that NN approach is applicable to the problem of Higgs boson detection. Neural network filters can be used to emphasize difference of M_{bb} distribution for events accepted by filter (with better $\frac{signal}{background}$ rate) and M_{bb} distribution for original events (with original $\frac{signal}{background}$ rate) under condition that there is no loss of significance. This improvement of the shape of M_{bb} distribution can be used as a criterion of existence of Higgs boson decay in considered discovery channel.

1 Introduction

This work is devoted to application of neural network to high energy physic. There is a broad consensus in physics community that newly building Large Hadron Collider (LHC) detector at CERN, Geneve, should be able to produce showers of particles, whose will have capability to confirm presence of Higgs boson. Using theoretical background of high energy physic there is possible to postulate theoretical properties of some distributions of selected values whose

* This work is supported by grant of Ministry of Trade and Industry of the Czech Republic, Project No. RP-4210/69/97 and by grant of Ministry of Education of the Czech Republic, Project No. LN00B096.

P.M.A. Sloot et al. (Eds.): ICCS 2002, LNCS 2331, pp. 554–563, 2002.

describe properties of Higgs boson decay in the LHC. Using simulated output from LHC based on simulation package PYTHIA, we have available two sets of shower parameters, one set corresponding to case with Higgs boson decay (we denote this as signal) and second one corresponding to case without Higgs boson decay (we denote this as background). So we can see the problem of Higgs boson search as a classical problem of pattern recognition with this exception that the quality of separation is measured as difference between two distribution curves corresponding to separated sets.

Neural nets are broadly used in pattern recognition problems and functions approximation tools. There are many types of artificial neural networks whose differ in architecture, in the type of implemented transfer functions and strategy of learning. Regard to universal approximation property we declined to use a special kind of neural nets, namely neural network with switching units [1], [2], [3] to solve pattern recognition problem postulated above. To reach better performance of these neural networks we tune topology and parameters of such networks via genetic algorithm optimization.

2 Description of neural network with switching units

Neural network with switching is a combination of classical neural network architecture and decision tree. This network is actually an oriented acyclic graph (see Fig. 1) which node are structures called as building blocks. This acyclic graph will be referenced as outer graph.

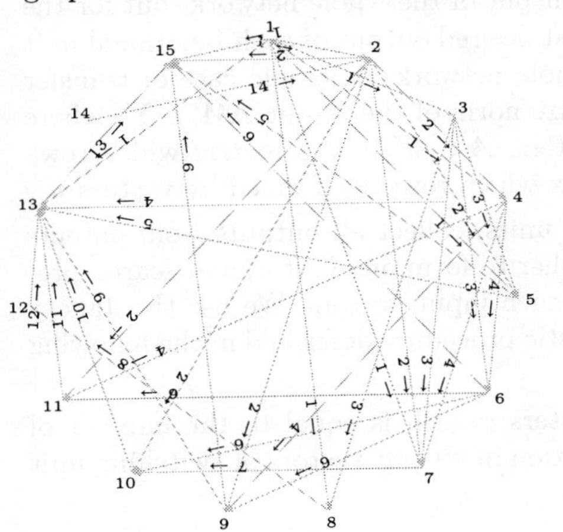

Fig. 1. Schema of connection between building blocks.

Each building block is a neural network consisting from two types of nodes. These nodes are connected together in such a way that they formed an acyclic graph again but with the restriction that outputs dimension of building block is the same for all building blocks in outer graph. First type of node, we refer this node as functional units, makes predefined mapping from input space corresponding to this node to output space of this node. Hence such node can be described by tuple of integers, input vector dimension and output vector dimension, and by transfer function. Definition of this transfer function include parameters of this functional unit (weight vectors, threshold etc. in current neural networks terminology).

Functional units map corresponding inputs to outputs by internal transfer function. This transfer function can differ for each functional unit. For example,

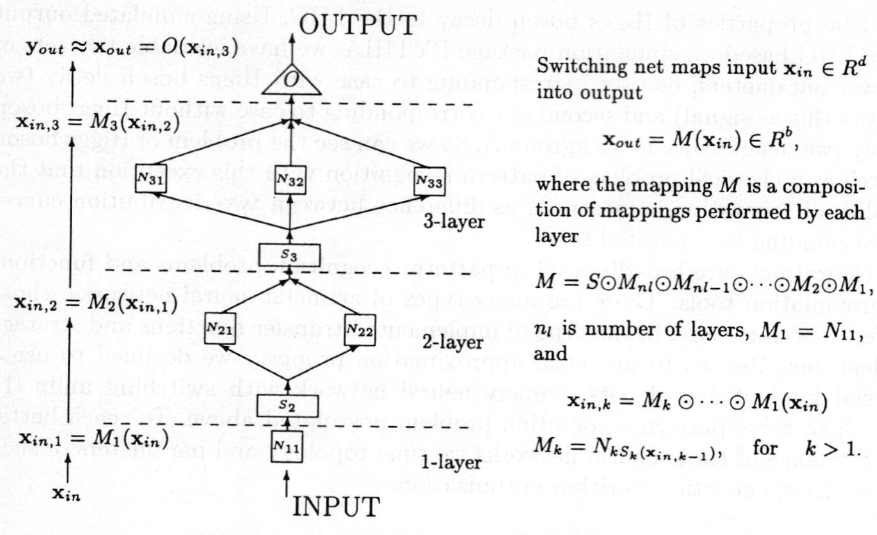

Fig. 2. A topology of simplified building block.

let now we describe transfer function currently implemented. Let $\boldsymbol{x}_i \in \Re^n$, $i \in \{1, \ldots, p\}$ are input patterns into functional unit, corresponding desired outputs let be denoted as $\boldsymbol{y}_i \in \Re^m$, $i \in \{1, \ldots, p\}$. Desired output of functional unit can be generally different from desired output of the whole network, but for the sake of explanation clarity we assume that desired output of each functional unit is the same as desired output of the whole network. A simple case of transfer function is linear mapping which minimize norm of the vector $\boldsymbol{AW} - \boldsymbol{Y}$, where $\boldsymbol{W} \in \Re^{m \times n}$ is matrix of weight parameters, $\boldsymbol{A} \in \Re^{n \times p}$ is a matrix which rows are vectors $\boldsymbol{x}_i$ and $\boldsymbol{Y} \in \Re^{m \times p}$ is a matrix which rows are formed by vectors $\boldsymbol{y}_i$.

The second type of nodes, switching units, collect all outputs from parents functional units, concatenate them together to form one vector, and search predefined number of clusters in the set of such input vectors. We use the Jancey cluster algorithm which is non-deterministic procedure described in the following schema:

(let d be the number of desired clusters (which is equal to the number of switching unit children), $\boldsymbol{z}$ is concatenation of output vectors of switching unit parents.

1. for randomly chosen sequence $1 \leq j_1 < j_2 < \cdots < j_d \leq p$ set

$$\boldsymbol{c}_q^{new} = \boldsymbol{c}_q^{old} = \boldsymbol{z}_{j_q} \quad \text{and} \quad \bar{S}_q^{new} = \bar{S}_q^{old} = \{\boldsymbol{z}_{j_q}\}, q \in \{1, \ldots, d\}$$

2. let $r_1, \cdots, r_p$ is random permutation of the $1, \cdots, p$,

3. FOR ALL $k = r_1, \cdots, r_p$
   DO
   let q be such index that $\boldsymbol{z}_k \in \bar{S}_q^{old}$,
   $i = \min\left\{v \,\middle|\, \left\|\boldsymbol{c}_v^{old} - \boldsymbol{z}_k\right\|_E = \min_{q\in\{1,\dots,h\}}\left\{\left\|\boldsymbol{c}_h^{old} - \boldsymbol{z}_k\right\|_E\right\}\right\}$,
   $c_q^{old} = \boldsymbol{c}_q^{old} - \frac{\boldsymbol{z}_k - \boldsymbol{c}_q^{old}}{|S_q^{old}|}$, $c_i^{old} = \boldsymbol{c}_i^{old} + \frac{\boldsymbol{z}_k - \boldsymbol{c}_i^{old}}{|S_i^{old}|}$
   $\bar{S}_q^{old} = \bar{S}_q^{old} \setminus \{\boldsymbol{z}_k\}$, $\bar{S}_i^{old} = \bar{S}_i^{old} \cup \{\boldsymbol{z}_k\}$,
   END
4. IF $(\exists q)(\bar{S}_q^{new} \neq \bar{S}_q^{old})$
   THEN for all such q let $\boldsymbol{c}_q^{new} = \boldsymbol{c}_q^{old}$, $\bar{S}_q^{new} = \bar{S}_q^{old}$ and GOTO 2
5. STOP

After clustering each cluster is jointed with a corresponding child functional unit and consequently parameters of this functional unit are adjusted with regard patterns in the corresponding cluster only. In fact, division of input patterns into two or more disjoint sets, and consecutive learning over these subsets of patterns, put a separation hypersurface into the input space. The type of these hypersurfaces is defined by the type of transfer functions of switching unit parents.

So each building block is learned, output from each building block is propagated to all children and output of the top building block is considered as final output from the neural network.

3 Tuning of neural nets via GA procedure

Computational power of neural network depends in general on two aspects

1. structure and connection state (topology of the net)
2. learning method.

The second one is more or less question of amount of learning data and 'quality' of learning method. Main goal of each learning method is optimization of some fitness function which is defined on the set of all possible neural net parameters. We can insight this problem as global optimization of fitness function. There are many gradient algorithm based methods whose provide a possibility to find a local solution (minimum or maximum). But these methods does not allow to find a global extreme point of fitness function due to non-continuous substance of some parameters (types of transfer function, number of inputs, graph connections, etc.). We need to use some nongradient global optimization method to do this. But many deterministic techniques of global optimization, like divide and conquer, or analytic extreme, search aren't the most efficient on problems like acyclic graph construction and operations on it. It could be convenient to use some more natural ways to do so. Therefore we decided use genetic algorithms [7], [6] to set up parameters and topology of neural net. Theoretical analysis of GA follows that no best solution is reached but an average fitness increases at all in new generations. This is done due a special property of GA which is known as 'implicit parallelism' (see [5]). Shortly implicit parallelism theorem imply that an exponentially large sets of parameter space are searched for domains with above-average fitness in polynomial time.

About the implementation, a queue (FIFO) of randomly generated networks is constructed (this queue has user predefined length). Then every network is learned, tested and evaluated with fitness, at this moment implemented in following ways (denote S_a accepted signal, e.g. all signals correctly classified, B_a accepted background, S_r rejected signal, e.g. all misclassified signal, B_r rejected background,):

1. maximal enrichment factor, e.g. max $\frac{S_a}{S_r}$ under condition that the amount of accepted signal will be statistically significant
2. minimal loss of signal under condition on amount of rejected background, e.g. max S_a and B_r =predefined value
3. maximal rejection of signal under condition on amount of accepted signal, e.g. max B_r and S_a =predefined value
4. maximization of quality factor, e.g.
$$\max \frac{S_a}{\sqrt{S_a + B_a}}.$$
5. negative value of total mean square error over all events

The values of fitness of the network in queue builds actually the base for probability, according to which the parents of newborn network are chosen. After parents are given, some analysis comes to figure out, how those can be crossovered. Crossover itself is considered as an interchange of corresponding blocks (note that blocks have the same number of inputs and the same number of outputs, so they are replaceable), whose represent here the 'separated gen'. Since mutation should be of tiny use and its effect weak at all, we have some operation (edge removing, edge adding, activation function change, etc.) for disposal. The new structures is grown, then learned, tested and evaluated. And so on, until some criteria aren't reached. As those conditions may serve given generation count, acceptable level of specified fitness, etc.

3.1 Paralleling of neural net

The merit of our work lies in the application of GA procedures of crossover and mutation to find better topology and parameters of neural networks. A reasonable applicability of this approach assume to take a great population of individuals (learned and tested neural networks) and many generation of such populations. It is well known that learning of a neural net on large set of learning data should be very time consuming, hence GA procedures over such individuals should be time consuming naturally too. This leads us to such implementation which should be able to run effectively on more processors. Primarily we intend run our experiments on cluster of Linux PCs which is available in ISC AC CR. Todays, we have at our disposal 20-node cluster of PCs (from 600MHz one processor machines to 1GHz two processors machines) running Linux operating system. This allows us to run our experiments on parallel mode under

Portable Batch System environment, which is able to distribute separated tasks to processors with regard to their load.

To improve overall performance we implemented parallel version of neural net learning process also. First question about paralleling of our learning process solves the problem of level of paralleling. In our case, because of incomparable complicated optimization of transfer function with regard to other operations in neuron learning process, we decided to use the paralleling at this point. Program uses Parallel Virtual Machines library (PVM), so it could be executed, only if PVM installed. The count of slaves (whose are machines on PC-cluster or processors on multiprocessors system) is configured in PVM console. Significant advantage of paralleling based on PVM is that this allow us to run our networks effectively on wide range of architectures.

4 Application of neural network with switching units and genetic algorithms to Higgs boson search

As we already mentioned objective of our work is a search of $H \to b\bar{b}$ decay. Those data are produced during proton–proton collision with energy 14 TeV (in centroid mass system). There is a certain probability that Higgs bosons are produced in this collision (see Fig. 3, a)). In the case that mass of this Higgs boson is $m \leq 200 GeV/c^2$ a decay $H \to b\bar{b}$ is dominating. The main background of process above is process without production of Higgs boson but with the same final state (see Fig. 3, b)). Instead of Higgs boson in this case a gluon is radiated and produce a $b\bar{b}$ pair.

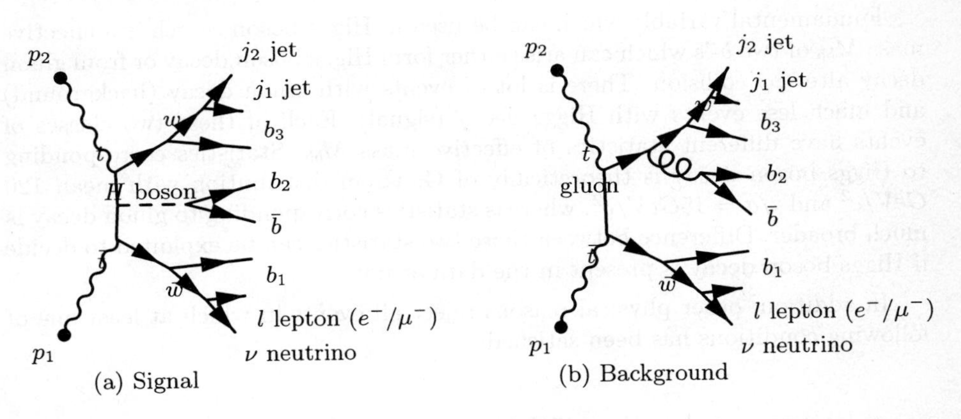

Fig. 3. Feynman diagram of decay trees.

Thus product of this type of collision are 2 jets from one w decay, 4 b-jets and one lepton (electron or muon) plus missing energy from unobserved neutrino. Each visible particle (2 jets, 4 b-jets and lepton) is described by three values P_T – transversal momentum $[GeV/c^2]$, η – pseudorapidity and ϕ – direction angles $[rad]$, corresponding to particle vectors with negligible mass. Neutrino is described by two missing energy values E_x^{miss}, E_y^{miss}. Pseudorapidity is an angular variable defined by

$$\eta = -\widetilde{ln}\left(\widetilde{\tan}\left(\frac{\theta}{2}\right)\right)$$

whose inverse function is

$$\theta = 2\widetilde{arctan}\,(e^{-\eta}).$$

Cartesian coordinates of each such particle can be expressed as $p_x = P_T \cos\phi$, $p_y = P_T \sin\phi$, $p_z = \frac{P_T}{\tan\theta}$. Using these values, we can evaluate values of energy E_i for each jet

$$E_i = \sqrt{(p_x)_i^2 + (p_y)_i^2 + (p_z)_i^2}.$$

Those values allow evaluate effective masses $M_{i,j}$

$$M_{i,j} = \sqrt{(E_i + E_j)^2 - \left((p_x)_i + (p_x)_j\right)^2 - \left((p_y)_i + (p_y)_j\right)^2 - \left((p_z)_i + (p_z)_j\right)^2}$$

for all sensible tuples of particles.

Fundamental variable which can be used in Higgs boson search is a effective mass M_{bb} of two b-'s which can arise either form Higgs boson decay or from gluon decay after $p\bar{p}$ collision. There is lot of events with gluon decay (background) and much less events with Higgs decay (signal). Each of these two classes of events have different statistics of effective mass M_{bb}. Statistics corresponding to Higgs boson decay is theoretically of Gaussian distribution with mean 120 GeV/c^2 and $\sqrt{\sigma} = 15GeV/c^2$, whereas statistics corresponding to gluon decay is much broader. Difference between those two statistics can be exploited to decide if Higgs boson decay is present in the data or not.

In addition, other physical reasons reject all events in which at least one of following conditions has been satisfied:

- at least one jets has $P_T < 15GeV$,
- at least one jet has pseudorapidity out of the range $(-2.5, 2.5)$,
- lepton is electron and $P_T^{lep} < 20GeV$
- lepton is muon and $P_T^{lep} < 6GeV$

All events passed those restrictions form histogram on Fig. 4.

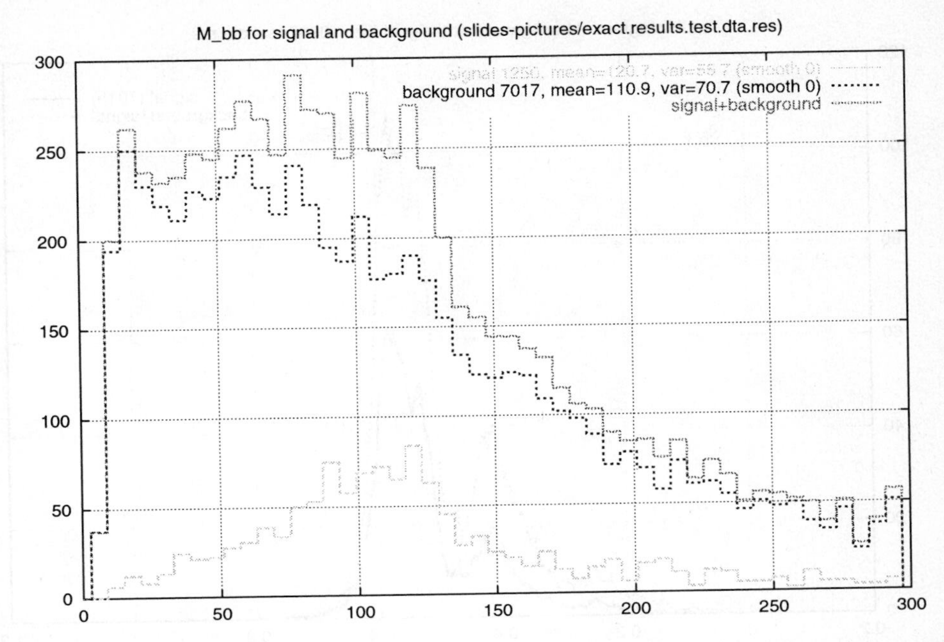

Fig. 4. Histogram of $M_{b\bar{b}}$ for signal and background (first number in the legend is number of events accepted by rejecting algorithm, means is average value of M_{bb}, var is mean square error and smooth is a smoothing factor used to plot the histogram).

Data really measured do not provide information about presence of Higgs decay in the event, hence for real data we have available total distribution of M_{bb} (see Fig. 4, upper curve) only. For data simulated, we can plot two histograms of M_{bb}, one for background only and the second one for plain signal (see Fig. 4, two bellow curves).

Application of neural networks covers the case when we know distribution of separated signal and separated background (e.g. below curves in the Fig. 4) because neural networks should provide information if a given event is signal or background (up to some misclassification, of course). So the main idea how to exploit neural network to confirm Higgs decay presence is based on filtering of events in such a way that percentage of signal will be increased after filtering and at the same time significance $\frac{S_a}{\sqrt{S_a+B_a}}$ will stay on the same level.

5 Results and Conclusions

Some experiments were performed with signal and background data described in section 4. We use raw data as they were produced by package PYTHIA. Our first results are demonstrated on the plot 5.

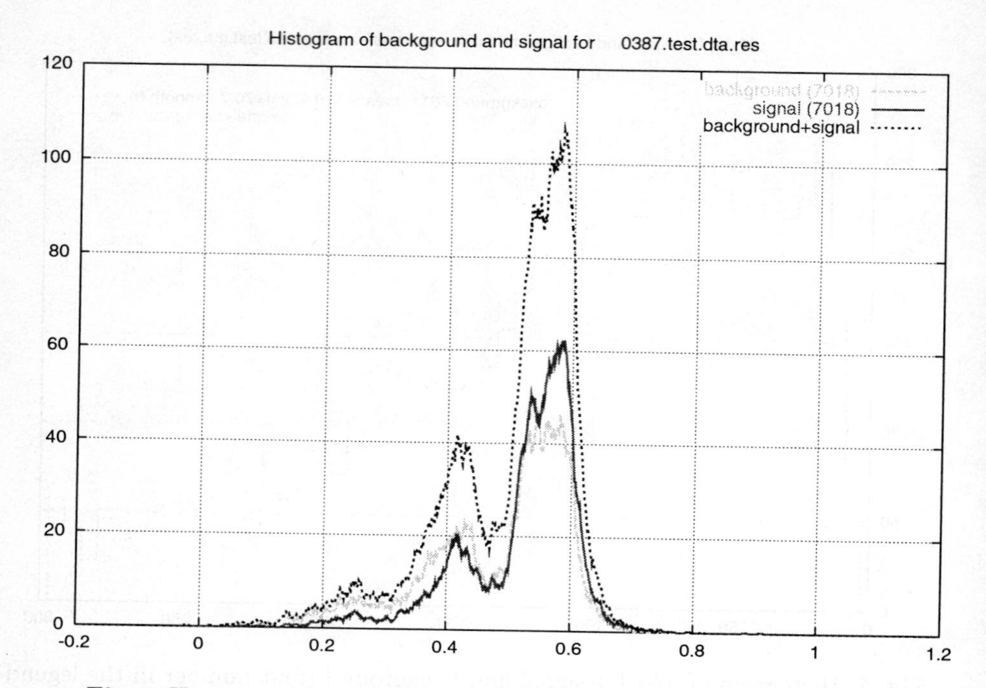

Fig. 5. Histogram of neural network output for signal and background.

It is evident that neural nets with switching units are able to partly separate signal with Higgs decay and background without Higgs decay. As we can seen on the plot mentioned, there is possible to choose an interval in which signal prevail over background, in case presented such interval should be $(0.45, 0.75)$. We call such interval "best signal window". On the other hand we can take interval, in which the signal is suppressed and background will be of dominant importance. In the case discussed now as this interval should serve $(0.20, 0.45)$. We call such interval consistently "best background window".

If Higgs decay is present than we can assume that plot of M_{bb} over all event whose are mapped by neural network into the best signal window will differs from the next one, based on events mapped into the best background window. Really, for our simulated data these plots differ, see figures 6 a) and 6 b). We can see visual differences between these two plots. Of course these plots should be different from the resemble plot on Fig. 4.

Hence our first experiments convinced us that chosen approach to separation of Higgs decay seem to be applicable and promise useful detection methods.

Finally we point out that developed separation method based on neural networks with switching units and genetic optimization is universal separation method which can be used for various pattern recognition problem. Perhaps someone can find this method too extensive, especially GA part, but no effec-

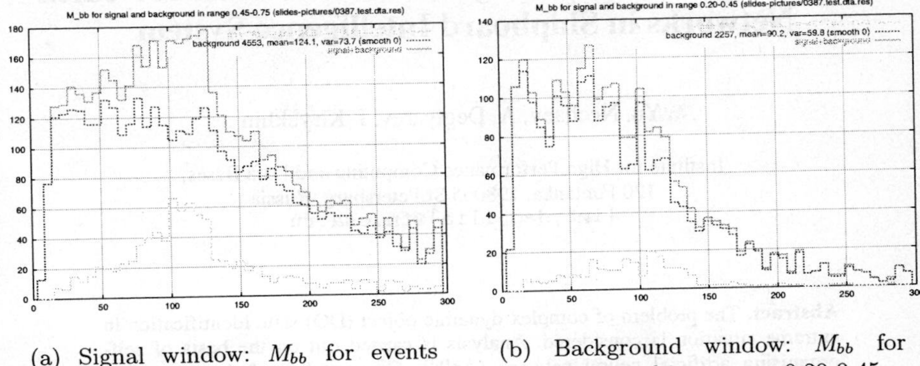

(a) Signal window: M_{bb} for events mapped into range 0.45-0.75.

(b) Background window: M_{bb} for events mapped into range 0.20-0.45.

Fig. 6. Signal and background windows.

tive method for neural net topology and parameter tunning is known to this time. In the next, we plan to implement further transfer function for functional nodes, paralleling of transfer function optimization via taking some parallel version of optimization routine (parallel LAPACK, for example) and applicate this separation tool to another pattern recognition problems.

References

1. F. Hakl and M. Jiřina. "Design of a neural net for level-two triggering in atlas nuclear experiments." *Neural Networks World*, vol. 6(6): pp. 951–973, 1996.
2. P. Bitzan, J. Šmejkalová, and M. Kučera, "Neural networks with switching units." *Neural Network World*, vol. 4, pp. 515–526, 1995.
3. M.Jiřina and M.Jiřina jr., "Neural network classifier based on growing hyperspheres" *Neural Network World*, vol. 3, pp. 417–428, 2000.
4. M. Sapinski, "The $t\bar{t}b\bar{b}$ background to the Higgs searches: Pythia versus CompHEP rates." Tech. rep. no. ATL-PHYS-2000-020, CERN Geneve, 2000.
5. John H. Holland. "Adaption in Natural and Artificial Systems." *A Bradford Book, The MIT Press*, 1992.
6. John R. Koza. "Genetic Programing." *A Bradford Book, The MIT Press*, 1992.
7. Lance Chambers (editors). "Practical Handbook of Genetic Algorithms." *CRC Press, Boca Racon*, 1995.
8. T. Sjöstrand. "PYTHIA 5.7 and JETSET 7.4 Physics and Manual." tech. report *CERN - TH . 7112/93*, dec 1993.

Complex Situation Recognition on the Basis of Neural Networks in Shipboard Intelligence System

Yu. Nechaev, A. Degtyarev, I. Kiryukhin

Institute for High Performance Computing and Data Bases,
120 Fontanka, 198005 St.Petersburg, Russia
{int,deg,ilia}@fn.csa.ru

Abstract. The problem of complex dynamic object (DO) state identification in extreme situation is considered. Analysis is carried out on the basis of self-organising artificial neural network (ANN). Compression of the measuring information about object dynamics is achieved by means of cognitive structures. Identification procedure is realised with the help of inference in intelligence system (IS) of unsinkability monitoring of ships and marine vehicles.

1 Introduction

One of the most complicated problems in estimation and forecast of DO behaviour in extreme conditions is situation identification. Critical obstacles we have especially for floating DO at holing and compartments flooding. In these conditions there are continuous modification of DO state and of factors defining its interaction with environment.

The authors use several approaches for this important practical problem solution [5,6]. Developed inferential mechanism includes various firmware procedures (statistical analysis, researches of phase portraits, mathematical modelling). However, in conditions of uncertainty and incompleteness of initial information such approach does not always ensure satisfactory results. Searching of more effective situation identification methods has resulted in using ANN theory [2,6-8,10]. Neural network approach is successfully applied by the authors in problems concerning with knowledge formalisation and multicriteria optimisation in IS of marine DO control.

Novelty of considered approach consists in the following:

- Processing of measuring information at input in ANN is realised by cognitive graphics methods;
- Improvement of identification effectiveness is achieved on the basis of alternatives analysis by comparison of results obtained on multilayer ANN models and Kohonen's self-organising maps;
- Results of identification are used as firmware procedure of fuzzy inferential mechanism.

In the paper ANN application is considered in framework of *competing computation technology* in parallel information processing on multiprocessor systems [7]. Principles of this technology allow to organise selection of the most effective

P.M.A. Sloot et al. (Eds.): ICCS 2002, LNCS 2331, pp. 564–573, 2002.

algorithm for measured information processing in real-time IS during alternatives analysis. As a result, reliability of assessments of characteristics of DO-environment interaction in complex situations is increased. Strategy of alternative selection in framework of competition principle is realised in the following way:

- generation of alternatives with the help of ANN basis and traditional algorithms of DO and environment characteristics control;
- alternatives analysis and choice of the best model;
- development of preferable alternative and result estimation.

Proposed conception is important both on stages of IS design and testing, and in operation. Comparison of different algorithms for obtaining dynamic characteristics of the control object and environment with modelling results based on ANN allows to estimate merits and demerits of used methods and to choose rational ways for information basis forming in production inference rules.

Another important advantage of developed technology consists in collecting data about ANN algorithms for solution of poor formalised problems. These data ensure formation of library of applied *neuromathematics* algorithms for realisation on *neurocomputers* which are very perspective high performance computational tools. The main advantage of ANN related with nonlinear information transformation allows to realise complicated algorithms of decision support for DO control in fuzzy environment. By collecting such information it is possible to organise learning process in such a way that the best solution will adopted in conditions of incompleteness and uncertainty of initial information.

2 Conceptual model

Let us consider method of automated synthesis of structure and determination of ANN parameters. Problem solution of DO state identification is carried out in accordance with realisation of parameters complex formed on the basis of information from measuring system. It is supposed that diagnosis areas in characteristics space are noncompact, i.e. nonempty intersection of the appropriate areas takes place. The concept and approach provide construction of ANN structure adequately describing characteristics of discovered situation. Sequence of information transformation determines solution of the following tasks:

- compression of information and its representation as convenient input for ANN perception;
- choice of ANN structure in conditions of continuous modification of object dynamics and environment;
- learning of ANN in accordance with sequence of extreme situations appropriating to classical flooding cases of the unsinkability theory.

Considered emergency conditions are represented by various areas of the diagnosis in attribute space. Correct recognition of situations by the developed classification rule can be reached on the basis of work [5]. ANN ability of self-learning and its parallelism allow to realise analysis and forecast of emergency situation in real time.

ANN is represented as "black box" at problem formalisation. Exterior information acts on input of such "black box". Desirable signal is realised on the output as vector of situation characteristics. As preliminary investigations have shown ANN is capable to completely transform its own behaviour at object dynamics and environment modification.

Algorithm of information transformation at realisation of inference mechanism is shown on the basis of production model [1,2,5,6]:

$$P_i : \langle if\ X_{i,q} \in \Phi_0 \ \&\ X^{*}_{j,k} \ \&\ X^{**}_{j,k} \in \Phi, then\ Y_i,\ else\ Z_i \rangle;$$
$$(i = 1,...,N;\ q = 1,...,S;\ j = 1,....,m;\ k = 1,...,n), \quad (1)$$

where P_i is the name of production; $X_{i,q}$ is the classified situation; $X^{*}_{j,k}$ is the vector of measurements; $X^{**}_{j,k}$ is the vector of modelling parameters; Φ_0 is the area of standard situations; Φ is the area of admissible values of situation characteristics; Y_i is the consequent of production; Z_i is the alternative choice; N is the number of productions; S is the number of standard situations classified by ANN; m, n are the numbers of parameters characterising vectors of measurements and modelling.

Thus various firmware procedures are used at interpretation of production kernel in antecedent of implication. Engineering process of IS construction assumes alternation of steps of situations modelling and decision making synthesis. Solution is achieved with the help of ANN and cognitive structures [11]. Check of extreme situation recognition reliability is carried out by methods of simulation modelling ensuring estimation and forecast of object (ship) dynamics with flooded compartment recognised by ANN:

$$F_{t,m} \in \{F_M, P(X)\}\ (t = \overline{1, T}), \quad (2)$$

where F_M is the operator mapping situations set {X} in behaviour set {Y} of local model F_m; t is considered time moment; $P(X)$ is the probability distribution describing set of input data acting from measuring system.

Adequacy condition is determined as

$$Y \in S\ \forall\ c \in C_m, \quad (3)$$

where S is the area bounding behaviour of model; C_m is the purposes of modelling.

Developed formalised knowledge system allows to supply monitoring and forecast of extreme situation development.

3 Cognitive structures

Solution of recognition problem by using cognitive paradigm [11] was carried out for extreme situation connected with estimation of damaged ship dynamics on waves. Such approach permits to obtain occurrence and development of oscillatory regime of damaged ship depending on character of flooding and level of external excitation. Searching of concrete mapping model was carried out with the help of cognitive spiral. It permits "to compress" initial information about non-linear rolling with a casual changing of periods. Construction of cognitive spiral on random function leads to periods "justification" and transformation of them to one value by means of affine compression or stretching of researched function intervals up to given size.

Cognitive spiral, as alternative to a phase portrait, contains more information, which it is possible to select visually orienting on image structure. Breadth of bands in spiral, its saturation by colour and frequency of changing of band of one colour by band of other colour can be referred to such information. Asymmetry of upper and lower parts of cognitive spiral and distribution of colour on angles close to 0 and 180 degrees can also play important role.

Considered recognition problem is connected with classification of "difficult to separate" typical cases of flooding among five classical situations defined at estimation of damaged ship dynamics [5,9]. These situations are characterised by variation of righting component of mathematical model of damaged ship rolling. Among them it is necessary to mark out second, third and fourth cases. The first case is trivial enough: lack of static heel at symmetric flooding and positive initial metacentric height. In other emergency states ship always has heel on a side. It allows to easily pick out first case without using complicated inference procedures. As regards to the fifth case it is usually considered as subset of fourth case. Similar pictures of oscillation are typical for this case.

Let us describe specific features of researched cases:

- *Second typical case.* Cognitive structures have intensive dark bands. These bands are divided by smaller in breadth light bands. Spiral is symmetric relative upper and lower parts and has steady light area at angles close to 0 and 180 degrees.
- *Third typical case.* Main difference of this cognitive structure is asymmetry of upper and lower parts of spiral. At angles close to 0 and 180 degree there no neutral light bands. Here there is no clear changing of bands, as in second case, dominance of light hues in lower part of spiral occurs.
- *Fourth typical case.* Cognitive spiral is symmetric and has very few light bands (actually there is alternation of more and less dark bands).

Examples of cognitive structures for cases 2 and 3 are shown in the fig.1.

Analysis of cognitive structures permits to mark out information for recognition. This problem is solved by determination of rational compromise between two factors. On the one hand, it is necessary to have maximum information for obtaining best classification possible. On the other hand, it is important to reduce information volume at entrance in ANN in connection with reduction of learning time and computer resources. Rational solution of this problem is obtained by selection of characteristic radial slits on some angles in cognitive spiral. Typical were slits at angles of 90 and 270 degrees representing frequency of bands change, their breadth and colour saturation. For qualitative comparison it is possible to use angles near 0 and 180 degrees, for example, ±10, 170 or 190 degrees.

Searching of rational ways of data transformation resulted in reviewing of no realisations, but of appropriate correlation functions or spectral densities constructed at quasistationarity intervals. As a result the constant number of input neurones is established, and ANN structure is determined. At that we have freedom in choice of realisation length depending on conditions of IS functioning reliability in extreme situations estimation. For recognition accuracy improvement and time reduction of ANN learning, moving average is excluded from used realisations. Correlation function for the "cleared" data is calculated.

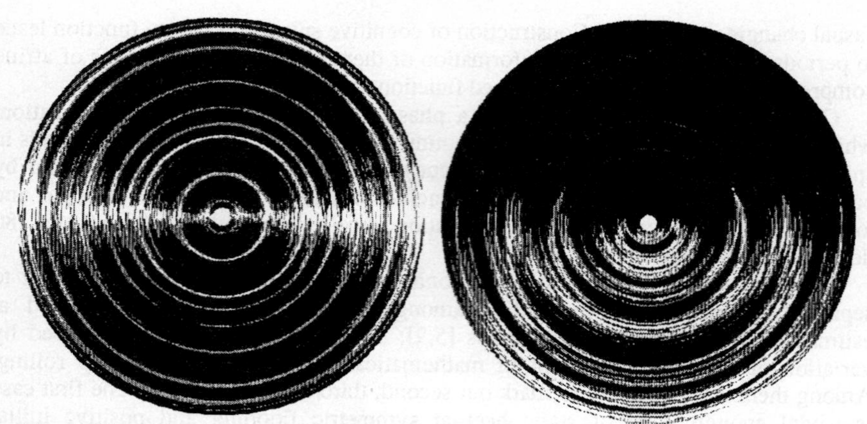

Fig. 1. Cognitive structure for flooding cases 2 (left image) and 3 (right image).

Comparison of spectral densities for considered flooding cases testifies about possibility to use them in classification problem. However correlation functions have great visual distinction in situations 2-4, therefore using of correlation functions are more preferable.

4 Multilayer neural network topology

ANN represents series connection of K layers of formal neurones in a view of cone-shaped configuration (fig.3). Each layer contains n_i $(i = 1,...,K)$ neurones, whose number regular by decreases with increasing of layer number. Each neuron or base processor element (BPE) is characterised by series connection of two components: linear converter (summator) of multidimensional input vector r_j into one-dimensional summator output vector S with weight coefficients W_{ij} $(i = 1,...,K;\ j = 1,...,n_i)$ and non-linear transformation of summator output S into output signal

$$q = f\left(\sum_{j=1}^{n} w_j r_j\right) = f(S). \tag{4}$$

Thus, multilayer ANN structure ensures complicated non-linear transformation of input vector $r = \{r_j\}$ depending on vectors of weight coefficients $W = \{w_{ij}\}$ and non-linear activation functions $f(\cdot)$.

Relation between input and output of ANN is determined by the following non-linear recurrent equation

$$\begin{aligned} Y = q^{(k)} &= f^{(k)}\Big(W_0^{(k)} + W_1^{(k)} f^{(k-1)}\Big(W_0^{(k-1)} + W_1^{(k-1)} f^{(k-2)}\Big(\ldots W_0^{(l)} + W_1^{(l)} f^{(l-1)} \times \\ &\Big(W_0^{(l-1)} + W_1^{(l-1)} f^{(l-2)} \times \Big(\ldots W_0^{(2)} + W_1^{(2)} f^{(1)}\Big(W_1^{(1)} + W_1^{(1)} q^{(0)}\Big)\ldots\Big)\Big)\ldots\Big)\Big)\Big) = F(r) \end{aligned} \tag{5}$$

where $F(r)$ is a non-linear function.

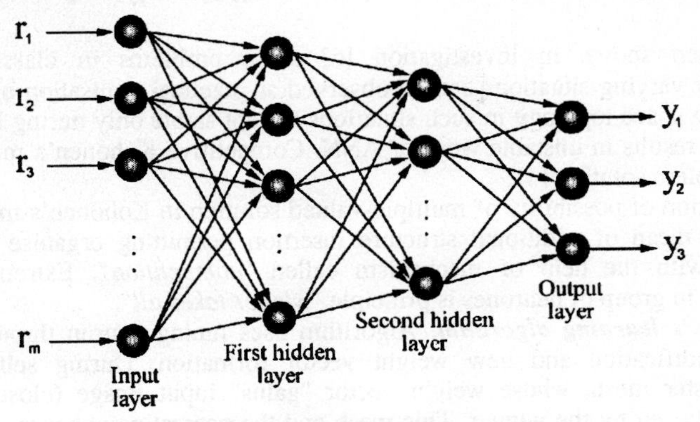

Fig. 2. Multilayer neural network

Both correlation functions and spectral densities of "cleared" slits of cognitive spirals are used as input information for recognition. In spite of so different input data ANN has identical structure. Input layer has the number of neurones corresponding to the number of points in given image. Output layer has 3 neurones (on one neuron in each classified case). Activation functions are chosen as follows: either sigmoid function is used for all three layer, or first two hidden layers have sigmoid function, and third, output layer, uses modified linear function.

Configuration and basic characteristics of ANN are shown in Table 1.

Table 1. ANN structure and basic characteristics

Characteristics	Values and descriptions
Input layer	16 neurones
First hidden layer	12 neurones, sigmoid activation function
Second hidden layer	8 neurones, sigmoid activation function
Output layer	3 neurones, modified linear activation function
The number of images	45
Learning coefficient	0.1
Learning accuracy, %	100
Number of learning steps	400

During ANN learning BPE weight coefficients are tuned, insuring solution of extreme situations recognition. "Back propagation" iterative procedure was used for determination of $W_{ij}(k)$ values:

$$W_{ij}^{(r)}(k+1) = W_{ij}^{(r)}(k) + \mu\left(-\nabla_{ij}^{(r)}(k)\right), \quad r = 1,2,\ldots \tag{6}$$

where k is the iteration number, which corresponds to input realisation $X^{o}(k)$ chosen by random way from population of input data; μ is the constant defining convergence of iterative procedure (6); $\nabla_{ij}(k)$ is the gradient computed by the formula

$$\nabla_{ij}^{r}(k) = \partial\varepsilon_{u}^{2}(k) / \partial W_{ij}^{(k)}(k)\,. \tag{7}$$

5 Competing model – Kohonen's self-organising map

As has been shown in investigation [6] some problems in classification of dynamically varying situations can be observed at practical realisation of multilayer ANN. Phase space topology in such situations is kept stable only during limited time intervals. It results in unstable work of ANN. Competitive Kohonen's model is used for this problem solution [4].

Elimination of possibility of multiple-valued solution in Kohonen's map (fig.3) is ensured by mean of additional structure insertion permitting organise recognition procedure with the help of mechanism called *"competition"*. Extreme form of competition in group of neurones is principle *"winner takes all"*.

Kohonen's learning algorithm. Algorithm uses tuning neuron through its own weights modification and new weight vector formation. During self-organising process cluster mesh, whose weight vector "gains" input image (closest to input image), is chosen by the winner. This mesh and the nearest neighbours change their own weights. At that learning velocity was accepted as slowly decreasing time function.

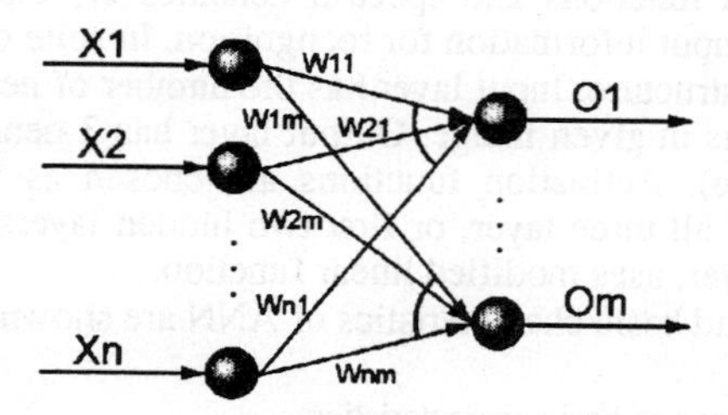

Fig. 3. Kohonen's self-organising map.

Searching of effective solutions has resulted in using of uncertainty at choice of weight coefficients, which are necessary for changing during learning. Standard learning method allows to change only gained neuron weights. In fuzzy approach all weights are varying in accordance with output neurones activation functions which are calculated for pattern given on ANN input. Such approach allows adjust weight matrix and divide space of objects into classes more precisely.

In considered classification problem uncertainty was used not only in learning, but also in inference. If we introduce normalisation of output neurones activity, we shall receive probabilities of acceptance of hypothesis about membership of image to concrete class. Analogy with fuzzy logic allows to consider Kohonen's self-organising map as a system for *object activation function* construction. Here competition for ANN organisation offered by Kohonen is retained, but it is less obvious at the fuzzy approach. Results of investigations show, that additional restrictions are imposed on initial conditions at learning with uncertainty elements. In particular, it is impossible to set identical columns in weight matrix. At lack of distinctions between weight coefficients their modification at the same activation of output neurones will happen quite equally. Therefore weight vectors could be tuned on the same class.

6 ANN weight coefficients adjustment by means of genetic algorithm

As competitive strategy of ANN learning (both multilayer ANN and Kohonen's self-organising map) genetic algorithm (GA) also has been used [3]. Let us note GA features applied for global optimum searching in space of weight coefficients:

- co-ordinates of weight coefficients were determined in range [–2;+2], and range was divided into 255 parts;
- chromosome coding was carried out with the help of Gray code;
- chromosome suitability was calculated through the goal function Q.

For multilayer ANN function Q looks like

$$Q = \sum_{k,j} (Y_j^k - D_j^k)^2 \rightarrow \min, \tag{8}$$

where Y_j^k is the signal from j-th neuron of output layer at the k-th learning example, D_k^j is the desired signal from j-th neuron of output layer.

For Kohonen's self-organising map goal function Q is

$$Q = \sum_k \left(\min_j \sum_i (W_i^j - X_i^k)^2 \right) \rightarrow \min, \tag{9}$$

where W^j is the vector adjusted to the gravity centre of j-th cluster, X^k is the k-th input vector.

Population size was selected a priori for practical reasons. As experiments have shown, GA demonstrates satisfactory results both in small and in large population sizes. With a view to decrease evolution time population size of 10 chromosomes was set up for our problem. In every evolution step one-point crossover, mutation, cyclic shift and selection proportionally suitability were realised. Probabilities of genetic operators are shown in Table 2.

Table 2. Genetic operators and there probabilities

Genetic operator	Probability
One-point crossover	1.00
Chromosome mutation	0.10
Bit inversion in chromosome mutation	0.10
Cyclic shift	0.05

Learning with the help of GA took more time than using of standard methods of networks learning. It is necessary to note that ANN with GA adjusted coefficients recognises cognitive spirals not worse than ANN, trained by standard method. GA advantages are:

- algorithm is not strongly linked to concrete activation function, as "back propagation" method (function can be non-differentiable and even discontinuous);
- searching in all weight coefficients space;

– this is stochastic searching.

Thus, GA is more universal algorithm of global optimum search, which, besides, is easily realised on parallel computing architecture.

7 Results of experiments

Investigation of possibility to use obtained results and efficiency of ANN was carried out. For these purposes practical classification problem of real IS was considered. Research included generation of real sea waves and damaged ship - environment interaction modelling with the help of specially developed tool for IS testing. This tool is complemented by subroutines of cognitive structures construction and processing and ANN learning. During research qualitative comparison of classification results for "short" and "long" realisations of non-linear stochastic processes was also carried out. These processes characterise damaged ship rolling on irregular waves.

Interesting results were obtained at the analysis of activation function influence on speed of ANN weights convergence. So, for example, semilinear (modified linear) activation functions

$$Y_i = \begin{cases} 0, \ S_j < 0; \\ kS_j, \ 0 \le S_j \le 1/k; \\ 1, \ S_j > 1/k. \end{cases}$$

could not be always used on all layers. At the same time, such function proves to be suitable for output layer where hypothesis about type of flooding is finally formed. Here semilinear function is limited and quickly reaches maximal values. Sigmoid function fits well for all other network layers (input and hidden). Using of such function on output layer only increases learning time and practically does not improve classification accuracy.

Research of realisation length influence is important for reduction of learning time at network adaptation in case when extreme situation development occurs during short time interval. In this case it is necessary to estimate situation very quickly and to take safety measures for ship survivability.

In using Kohonen's self-organising maps the same correlation functions were given on neural network input as on multilayer ANN input. It allowed to compare results and to estimate efficiency of used neural networks. As results of the research show, Kohonen's self-organising maps in some cases have certain advantages as compared to multilayer networks. These advantages provide certain split of situations and reference recognised object only for one concrete class. Besides, unsupervised learning has also other advantages. So, for example, in the second typical flooding case ship dynamics on irregular waves in conditions of limited excitations could be close to the third typical case. At supervised learning such situation results in occurrence of learning mistake: some rolling realisations of damaged ship could be incorrectly classified. But, if we use unsupervised learning, such variant joins class of realisations of the third typical case. As a result it is not possible to avoid learning mistakes. Way out from such situation is joint classification on the basis of considered ANN. It allows to essentially increase efficiency of recognition process.

8 Conclusions

Concept of on-board IS design as a system of parallel operation provides ANN using not only *as general method for analysis and dynamic measuring data interpretation in DO control systems, but also as competing technology for data processing in complex situations analysis*. As a result of alternatives analysis technology is preferable if it ensures more reliable assessment of researched situation in accordance with adopted discriminating rule. Just this technology will be developed in future during solution of current problem of analysis and interpretation of dynamic measuring data.

Researches carried out show great opportunities of using ANN in problems of extreme situations identification when DO–environment interaction conditions are continuously changing. Quality of ANN functioning is achieved by means of variation of operational principles and formation of criteria, which ensure flexibility and ability to external conditions adaptation. As a result perspectives of development of information technologies for important practical problems solution are opening. In this paper we have discussed application of such new information technologies to ensuring ships' unsinkability.

The research is supported by grant of RFBR N 00-07-90227.

References

1. Averkin A.A. Soft computing is base of new information technologies // Proc. of 5-th national conference on artificial intelligence with international participation, Kazan, 1996, vol.2, pp.237-239. (in Russian)
2. Bogdanov A., Degtyarev A., Nechaev Yu. Fuzzy logic basis in high performance decision support systems. Computational Science – ICCS 2001, LNCS 2074, Springer, 2001, pp.966-975
3. Goldberg D. E. Genetic algorithms in search, optimisation, and machine learning. Reading, MA: Addison-Wesley. 1989.
4. Kohonen T. Self-Organizing Maps. Springer-Verlag, Heidelberg, 1995
5. Nechaev Yu.I., Degtyarev A.B., Boukhanovsky A.V. Identification of extreme situation in fuzzy condition. // Proc. of international conference on fuzzy computing and measurement SCM-98. St. Petersburg, 1998, vol.1. pp.85-88. (in Russian)
6. Nechaev Yu.I., Siek Yu.L., Vasyunin D.A. Neural network technology in the intelligence system of marine technics. // Proc. of 6th national conference on artificial intelligence with international participation, Puschino, 1998, vol.2, pp.361-368. (in Russian)
7. Nechaev Yu.I. Neural network technologies in real time intelligence systems. //Proc. of 4th National conference "Neuroinformatics 2002", Moscow, 2002, v.1, pp.1-55 (in Russian)
8. Wassermam F. Neurocomputer technics. Theory and practice. – Moscow: Mir, 1992.
9. Ship Theory Handbook. Ed. by Y.I.Voitkounski In 3 volumes, Vol.2. Statics of Ship, Ship Motions, Leningrad. "Sudostroenie", 1985. (in Russian)
10. Zadeh L. Fuzzy logic, neural networks and soft computing. //Communication of the ACM, 1994, vol.37, N3, pp.77-84.
11. Zenkin A.A. Cognitive computer graphics. – Moscow: Nauka, 1991. (in Russian)

Dynamic Model of the Machining Process on the Basis of Neural Networks: from Simulation to Real Time Application

*Rodolfo E. Haber[1,3], R. H. Haber[2], A. Alique[1], S. Ros[1], J.R. Alique[1]

[1]Instituto de Automática Industrial (CSIC).
km. 22,800 N-III, La Poveda. 28500. Madrid.
SPAIN.
rhaber@iai.csic.es
[2]Departamento de Control Automático.
Universidad de Oriente. 90400.
CUBA
[3] School of Computer Science and Engineering.
Universidad Autónoma de Madrid.
Ciudad Universitaria de Cantoblanco
Ctra. de Colmenar Viejo,km 15. 28049 - Madrid
SPAIN
Rodolfo.Haber@ii.uam.es

Abstract. Nowadays, the modeling of complex manufacturing tasks is a key issue. In this work, as a case study is selected the application of a dynamic model to predict cutting force in machining processes. A model created using Artificial Neural Networks (ANN), able to predict the process output is introduced in order to deal with the characteristics of such an ill-defined process. This model describes the dynamic response of the output before changes in the process input command (feed rate) and process parameters (depth of cut). Experimental tests are made in a professional machining centre, with different cutting conditions, on real time data. The model provides sufficiently accurate prediction of cutting force, since the process-dependent specific dynamic properties are adequately reflected.

1 Introduction

New trends in modeling nonlinear processes have demonstrated that Artificial Neural Network (ANN) based algorithms are maybe the most suitable Artificial Intelligent technique to analyze the sensory information and depict process behavior. ANN represents an outstanding approach to model complex processes due to the feasibility for hardware implementation, real-time running, and a few prior assumptions for modeling [1]. Some results show that ANN can yield a more accurate process model than using the regression method [2]. In spite of its excellent interpolation capability, there are difficulties related with the poor extrapolation accuracy of the typical neural

* Corresponding author

P.M.A. Sloot et al. (Eds.): ICCS 2002, LNCS 2331, pp. 574–583, 2002.

networks and it is necessary to properly train models, requiring experimentation with a wide range of possible working conditions [3].

Among the industrial sectors, today's manufacturing industry has growing up demanding productivity and profitability requirements. Such demands can be satisfied only if the production systems are highly automated and extremely flexible. One of the main activities the manufacturing industry has to deal with is machining. In machining, as in any production process, an optimal performance is always a desirable feature. However, most machining processes such as milling, drilling, grinding and turning exhibit nonlinear and non-stationary behavior that make hard to perform optimization tasks.

The cutting force model of machining process has been extensively studied both analytically and empirically because of its invaluable importance for the assessment of cutting tool deflection, wear, breakage, vibration, and their effects on the quality of the elaborated part. The spectrum of available *conventional* methods for modeling is very wide. However, the different types of models developed up until only a few years have not lived up to expectations. Theoretical models are deduced by making certain approximations, which limits their validity and they may be cumbersome to handle. The empirical models thus developed far are valid only for the experimental conditions under which they were developed, which do not always coincide with industry's actual working conditions. Stochastic models cannot always make adequate prediction beforehand due to the vast variability of their estimates and the numerous variables involved in the complex processes of machining.

Presented in this paper is an intelligent model of the milling process. In order to deal with nonlinear process characteristics, a Neural Network Output Error (NNOE) model, able to predict online the resultant cutting force under actual cutting conditions, is proposed. In section 2 a brief presentation of neural networks that focuses on the type of ANN used, including training algorithm is given. In section 3, a short description of the machining process is shown. In section 4 the experimental set-up, the design considerations, the network topology for the best training result obtained and the experimental results are presented. Finally, the authors conclude on the model suitability and some suggestions for its further improvement and future works.

2 Artificial Neural Networks and Dynamic Process Models

Among all modelling structures, the so-called Output Error (OE) model is one of the most widely used. In this configuration noise is assumed to corrupt the process additively at the output. OE models are often more realistic models and therefore they often perform better than other configurations.

$$\hat{y}(t) = g(\boldsymbol{\varphi}(t), \boldsymbol{\theta}). \tag{1}$$

where $\boldsymbol{\varphi}(t) = [\hat{y}(t-1|\boldsymbol{\theta}), \ldots, \hat{y}(t-n_A|\boldsymbol{\theta}), u(t-1), \ldots, u(t-n_B)]^T$ is a regression vector, n_A is the number of past predictions used for determining the predictions, n_B is the number of past inputs, $\boldsymbol{\theta}$ is the vector containing the weights and $g(\cdot)$ is the function performed by the neural network.

One of the most widely used ANN paradigms, for its suitability for modelling and control applications, is the so-called Multilayer Perceptron (MLP). The class of MLP considered here consists in only one hidden layer with hyperbolic tangent activation function H, and a linear activation function, L, at the output. Such configuration is advantageous considering a foreseeable real-time implementation of the network.

$$\hat{y}_i(\mathbf{w},\mathbf{W}) = L_i\left[\sum_{j=1}^{Q} W_{ij} H_j\left(\sum_{k=1}^{M} w_{jk}u_k + w_{j0}\right) + W_{i0}\right]. \tag{2}$$

where Q is the number of output neurons, M is the number of neurons in the hidden layer, **u** are the inputs, and $\hat{y}_i$ is the output of the network. The weights are specified by the matrices **w** (input-to-hidden layer weights) and **W** (hidden-to-output layer weights). Both matrices are included in $\boldsymbol{\theta}$.

The identification problem can be viewed as the determination of the mapping from the set of data $z^N = [\mathbf{u}\ \mathbf{y}]^T$ (training set) to the set of possible weights ($\hat{\boldsymbol{\theta}}$) so that the network can produce a prediction $\hat{y}(t)$ as close as possible to the actual output $y(t)$ [4].

$$z^N \rightarrow \hat{\boldsymbol{\theta}}. \tag{3}$$

Using a prediction error identification method

$$E(\boldsymbol{\theta}, z^N) = \frac{1}{2N}\sum_{t=1}^{N}\left(y(t) - \hat{y}(t|\boldsymbol{\theta})\right)^T\left(y(t) - \hat{y}(t|\boldsymbol{\theta})\right). \tag{4}$$

The weights are calculated as

$$\hat{\boldsymbol{\theta}} = argmin\left(E(\boldsymbol{\theta}, z^N)\right). \tag{5}$$

As the training algorithm a version of the Levenberg-Marquardt method was selected [5].

$$\boldsymbol{\theta}^{(i+1)} = \boldsymbol{\theta}^{(i)} + \mu^{(i)} \cdot \mathbf{S_D}^{(i)}. \tag{6}$$

where the search direction is calculated from

$$\mathbf{S_D}^{(i)} = -\left[\mathbf{R}\left(\boldsymbol{\theta}^{(i)}\right) + \lambda^{(i)}\mathbf{I}\right]^{-1}\mathbf{d}\left(\boldsymbol{\theta}^{(i)}\right). \tag{7}$$

where **R** is the Hessian approximation, λ a positive scalar, μ the step size for iterations, **d** the gradient of the predictions computed with respect to the weights, and **I** the identity matrix. The size of the elements of the diagonal matrix added to **R** is adjusted according to the size of the ratio between actual and predicted decrease. For a step to be accepted the ratio

$$\rho^{(i)} = \frac{E(\boldsymbol{\theta}^{(i)}, z^{N}) - E(\boldsymbol{\theta}^{(i)} + \mathbf{S_D}^{(i)}, z^{N})}{E(\boldsymbol{\theta}^{(i)}, z^{N}) - \Lambda(\boldsymbol{\theta}^{(i)} + \mathbf{S_D})}. \quad (8)$$

must exceed some small positive number, where

$$\Lambda(\boldsymbol{\theta}^{(i)} + \mathbf{S_D}) = \left\| y(t) - \hat{y}(t, \boldsymbol{\theta}^{(i)}) - \mathbf{S_D}^{T} \left[\frac{\partial \hat{y}(t, \boldsymbol{\theta})}{\partial \boldsymbol{\theta}} \right]_{\boldsymbol{\theta} = \hat{\boldsymbol{\theta}}} \right\|_2^2. \quad (9)$$

$$\Lambda(\boldsymbol{\theta}^{(i)} + \mathbf{S_D}) = E(\boldsymbol{\theta}^{(i)}, z^{N}) + \mathbf{S_D}^{T} \mathbf{d}(\boldsymbol{\theta}^{(i)}) + \tfrac{1}{2} \mathbf{S_D}^{T} \mathbf{R}(\boldsymbol{\theta}^{(i)}) \mathbf{S_D}. \quad (10)$$

The algorithm adjusts λ according to whether $E(\boldsymbol{\theta}^{(i)}, z^{N})$ is increasing or decreasing.

After this brief explanation, becomes obvious that the attention is focused on determining a nonlinear OE model, i.e., a dynamic system, by training a two layer neural network with a Levenberg-Marquardt method.

3 The Machining Process

For the case study, among the various machining operations, is selected the milling process (see figure 1). This choice obeys a pessimistic criterion, since milling is one of the most complex of machining operations [6].

Fig. 1. Overall View of a Typical Machining Centre

Among the enormous quantity of variables and parameters involved in the machining process, the most relevant factors (see fig. 2a), in terms of control tasks are:

The spatial position of the cutting tool, considering the Cartesian axes (x_p, y_p, z_p),
Spindle speed (s) [*rpm*],
Relative feed speed between tool and worktable (f, feed rate) [*mm/min*],
Cutting power invested in removing metal chips from the workpiece (P_c)[*kW*],

Cutting force exerted during the removal of metal chips (F, cutting force) [N],
Radial depth of cut (a, cutting depth) [mm],
Diameter of the cutting tool (d) [mm].

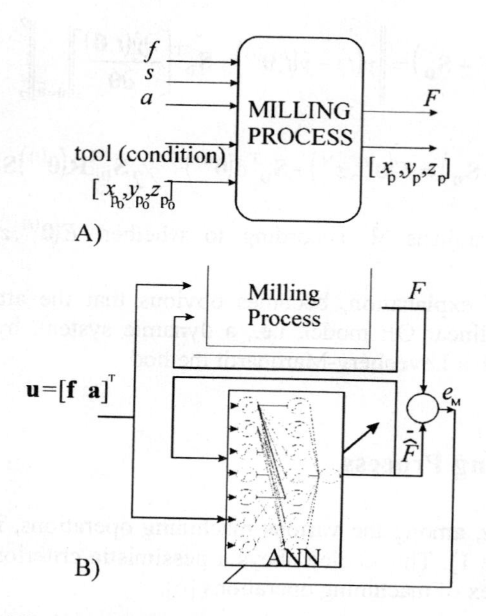

Fig. 2. a) Milling process representation b) Scheme for identification

On the basis of the previous system approach and aiming at intelligent supervision, the milling process can be formally described by a discrete nonlinear relationship

$$\hat{F}(t) = G(\mathbf{F}, \mathbf{f}, \mathbf{a}). \tag{11}$$

where G is an unknown function to identify and $\mathbf{f}, \mathbf{a}$ and $\mathbf{F}$, are the inputs and output respectively defined as $\mathbf{F} = [F(t-1) \;\; \cdots \;\; F(t-n)]$, $\mathbf{f} = [f(t-1), \;\; \cdots \;\; ,f(t-m)]$, $\mathbf{a} = [a(t-1) \;\; \cdots \;\; a(t-m)]$, t, is the discrete time instant and $n, m \in Z$.

If the model is simulated by means of an ANN parallel-identification scheme, the resulting estimated output is

$$\hat{F}(t) = g(\hat{\mathbf{F}}, \mathbf{f}, \mathbf{a}). \tag{12}$$

where g represents the ANN input-output mapping, $\hat{\mathbf{F}} = [\hat{F}(t-1), \;\; \cdots \;\; ,\hat{F}(t-n)]$ and $\hat{F}(t)$ is the one-step prediction of the model.

4 Experimental Set-up

The experimental tests are conducted on a 5.8kW-4 axes milling machine equipped with CNC, which is interfaced with a personal computer by an RS-232 communication link. The tailor-made architecture is illustrated in figure 3.

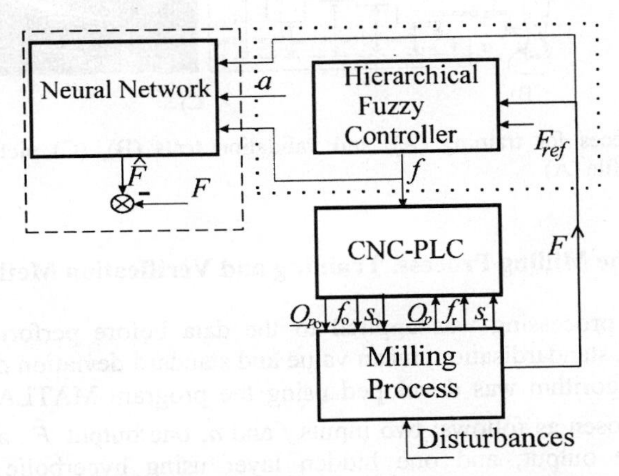

Fig. 3. Scheme for control and modeling of milling process

A personal computer is added in order to carry two important tasks. Firstly, to evaluate the measured values of the cutting force F and to perform the generation of f values in the block named hierarchical fuzzy controller. Details concerning hierarchical control system can be found in [7]. Secondly, and the main goal of this work, is to implement a neural network model (12). The model objective is to predict in real-time, one step ahead, the resultant mean cutting force $\hat{F}$. The input data for the model identification are obtained from the feed rate command signal f, the depth of cut a and the cutting force F.

For the training phase of the ANN, only new milling cutters 25 mm in diameter are used. Two workpieces with several changes in a are chosen for training the ANN (see fig. 4A) and online running of the model (see fig. 5B). Slot milling operation is supposed to be done in one direction only (see fig. 4C).

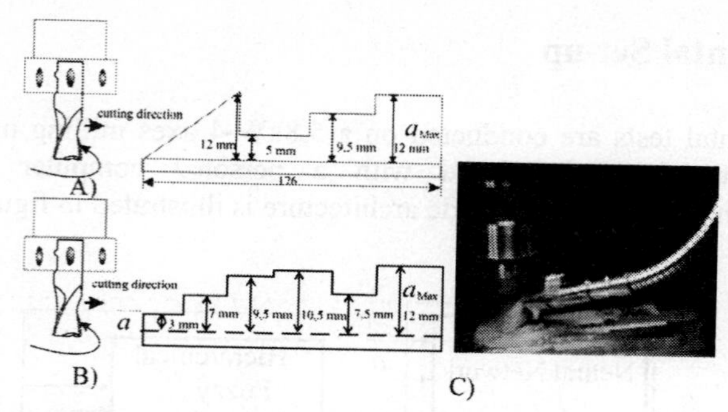

Fig. 4. Workpieces for training (A), and validation tests (B), (C) picture of slot milling operation on profile (A)

4.1 Model of the Milling Process. Training and Verification Methodology

A preliminary processing was applied to the data before performing the training procedure (e.g., standardisation, mean value and standard deviation of the signals) [8]. The training algorithm was developed using the program MATLAB. The topology was initially chosen as follows: two inputs f and a, one output $\hat{F}$, a linear activation function at the output, and one hidden layer using hyperbolic tangent for the activation function. The type of model was selected starting from the *a priori* knowledge of the milling process and the types of models considered in previous works (see fig. 2a). An ANN with 6 neurons in the input, 12 neurons in the hidden layer and one in the output layer, was selected.

Modifying (12) for a second order output error model, the one step prediction is evaluated with the previous model outputs

$$\hat{F}(t) = g\left(\hat{F}(t-1),\ \hat{F}(t-2),\ f(t-1),\ f(t-2),\ a(t-1),\ a(t-2)\right). \tag{13}$$

The initial values of the weights were randomly chosen. The initial cutting conditions used were f_0=100 mm/min., s_0=1000 rpm., a_0=6 mm. At the end of the training stage a pruning algorithm was performed in order to optimise the size of the network, removing the superfluous weights [9]. In order to validate the model, data analysis (mean squared value analysis, auto correlation analysis and transient data study) were done. The robustness property of the model, considering changes in cutting conditions (material hardness, tool wear, and disturbances in depth of cut), is also examined. Additionally, real-time performance of the model before these same situations was investigated. The comparison among different models analysed on the basis of final prediction error (FPE) is depicted in Table 1.

Table 1. Neural Network Output Error (NNOE) models.

Models	Final Prediction Error (%) Training	Verification
$a(t-1)$, $f(t-2)$, $f(t-1)$, $\hat{F}(t-1)$, $\hat{F}(t-2)$ → $\hat{F}(t)$	14.3	46.1
$a(t-1)$, $f(t-1)$, $\hat{F}(t-1)$, $\hat{F}(t-2)$ → $\hat{F}(t)$	18.3	50.4
$f(t-2)$, $f(t-1)$, $\hat{F}(t-1)$, $\hat{F}(t-2)$ → $\hat{F}(t)$	11.3	23.5
$a(t-2)$, $a(t-1)$, $f(t-2)$, $f(t-1)$, $\hat{F}(t-1)$, $\hat{F}(t-2)$ → $\hat{F}(t)$	13.2	16.3
$a(t-2)$, $a(t-1)$, $f(t-2)$, $f(t-1)$, $\hat{F}(t-1)$, $\hat{F}(t-2)$ → $\hat{F}(t)$	8.7	9.4

In order to evaluate the behaviour of the model considering actual working conditions, tailor-made software for simulation and real time supervision of machine tools was developed. The application for Windows NT was programmed in Visual C++. The libraries of controllers (e.g., different types of fuzzy controllers) and process models (e.g., neural networks) were programmed in C++ and compiled in two dynamic link libraries (DLL).

For the sake of clarity, only the results of two simulations and two real-time applications are shown in figure 5. The vector of initial conditions used was $\hat{F}(t=1) = g(100\text{N}, 75\text{ N}, 9\text{ mm/min}, 0\text{ mm/min}, 3\text{mm}, 0)$.

In fig. 5 the horizontal dot line means the set point of cutting force control system (not addressed in this work, but the software includes this option). The vertical line represents the actual (simulation or real-time values) values shown at the bottom of the dialog in text boxes. For each case considered, the chosen profile is depicted at the bottom of the picture. The feed rate command signal is shown in the middle of the graph. The behaviour of predicted and measured cutting forces is shown at the top of the picture. For simulations, only one signal is depicted, which corresponds to the output of the process model.

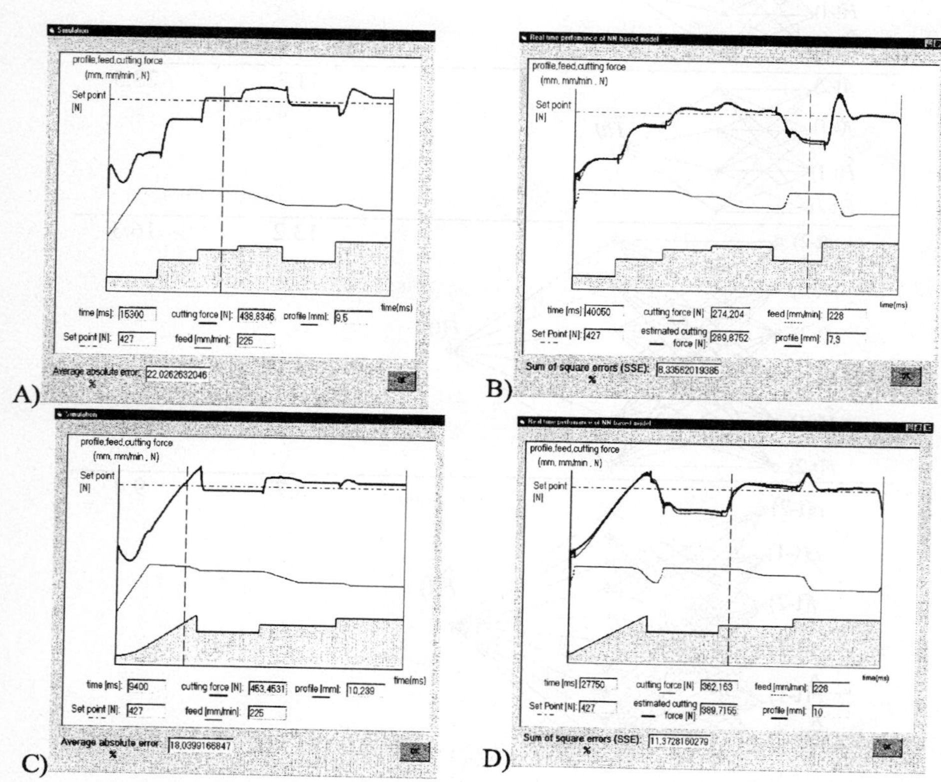

Fig. 5. Simulation and real-time running using different workpiece

The simulation serves also for evaluating the performance of the intelligent control system (i.e., accuracy, closed-loop stability, etc.) which is an important step to design closed-loop control systems under varying milling process dynamics (see figure 5A and 5C). On the other hand, the dynamic behaviour of the cutting force provides information about tool and process conditions (see figure 5B and 5D). Peaks and undesirable oscillations in cutting force can be the result of inappropriate cutting conditions as well as the change in tool condition. Good properties of the model can be inferred from these results, as well as its generalisation capability. This dynamic neural network-based model provides sufficiently accurate prediction of cutting force, since the process-dependent specific dynamic properties of milling operation are adequately reflected.

Conclusion

Various models of the milling process were obtained using the parallel identification scheme. The real time test of ANN-based models demonstrates how, through simple topologies, quite precise models can be obtained. These models can play an important role in the design and evaluation of controller performance under variations in the milling process dynamics. The results can be used to develop complex adaptive controllers and new monitoring systems.

The strength of the proposed approach lies in the solution provided for modeling. The neural network is able to describe the nonlinear characteristics of the milling process, showing superior learning capability, a suitable noise suppression and inclusion of non-measurable information (e.g., workpiece hardness, material). Drawbacks however are the need of neural network retraining and the building of a wider library of models. In spite of that a sufficient exciting of signals avoids a high number of retraining stages.

References

1. Haykin S. Neural Networks: A comprehensive foundation. 2nd edn. IEEE Press (1999).
2. Das S., Chattopadhyay A.B., Murthy A.S.R. Force parameters for on- line tool wear estimation - a neural network approach. Neural Networks 9(9) (1996) 1639-1645.
3. Dimla D.E., Lister P.M., Leighton N.J. Neural network solutions to the tool condition monitoring problem in metal cutting – a critical review of methods. International Journal of Machine Tools and Manufacturing 37(9) (1997) 1219-1241.
4. Ljung L., System Identification: Theory for the user: 2nd. Edn. Prentice Hall, (1999).
5. Fletcher R., Practical Methods of Optimization. 2nd. Edition, John Wiley&Sons, (2000).
6. King R.J. Handbook of High Speed Machining Technology. Mc.Graw Hill, New York, (1985).
7. Haber R.E., Peres C.R., et al. Towards intelligent machining: hierarchical fuzzy control for end milling process: IEEE Transactions on Control Systems Technology 2(6) (1998) 188-199.
8. Bendat J.S., Piersol A. G. Random Data, Analysis and Measurement Procedures. 3rd. Edition, Wiley & Sons, (2000).
9. B. Hassibi, et al. Optimal brain surgeon: Extensions, streamlining and performance comparisons, Advances in Neural Information Processing Systems 6 (1994) 263-271.

Incremental Structure Learning of Three-Layered Gaussian RBF Networks

David Coufal[1]

Institute of Computer Science, Academy of Sciences of the Czech Republic,
Pod Vodárenskou věží 2, 182 07 Praha 8, Czech Republic
coufal@cs.cas.cz

Abstract. In the paper a new incremental algorithm for structure learning of three-layered RBF neural network is proposed. The algorithm is intended to be used for solving high-dimensional learning tasks since they makes troubles to standard structure learning techniques.

1 Introduction

It is well known that supervised learning of RBF neural networks generally consists of two subtasks named as *structure learning* and *parameters learning*. Within structure learning a number of network neurons is determined together with initial setting of network parameters. Parameters learning is then only a suitable readjustment of initially set parameters to network's operation matches given training input/output set as close as possible.

Structure learning is a harder part of a learning process. This fact is well documented by relevant literature offering a variety of different approaches to this task but without such a degree of unification that it is available for parameters learning where we almost always deal with an instance of an optimization problem. In spite of this variety, however, several classes of structure learning algorithms can be recognized. We review in short common of them in the third section of the paper.

As we will see from this review each of presented algorithms suffers from some disadvantage which decreases its applicability. As a reaction to this situation we formulate a new algorithm minimizing presented limitations. The algorithm itself is described within the fourth and the fifth section of the paper. In the sixth section there are presented results of several experiments regarding the algorithm. The paper concludes by the seventh section.

As it is indicated by the title of the paper the new algorithm is considered to be applied on structure learning of three-layered feedforward RBF networks. Especially, we will consider networks with n-dimensional input and only one-dimensional output. Regarding type of radial basis function we employ Gaussian one but other types of functions can be used as well. In the following section we state an explicit description of a network architecture we are interested in.

P.M.A. Sloot et al. (Eds.): ICCS 2002, LNCS 2331, pp. 584–593, 2002.

2 Three-layered Gaussian RBF neural network

It is well known that a three-layered feedforward network's architecture is given by serial interconnection of an input, a hidden and an output layer, see Fig. 2. For our case an input layer is considered to consist of n neurons, the hidden layer of m neurons and the output layer only of one neuron.

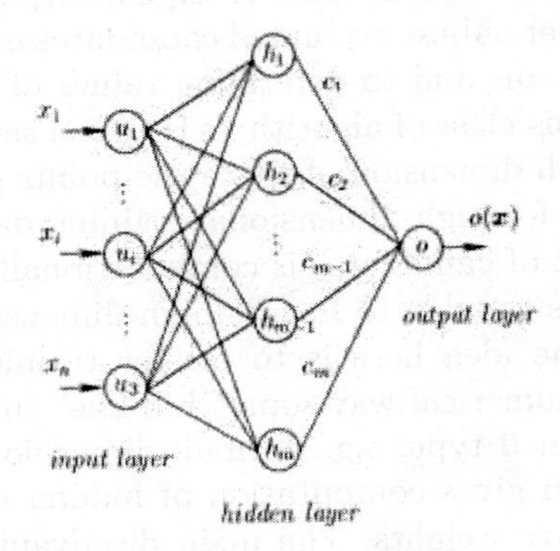

Fig. 1. A three-layered feedforward RBF neural network.

Input neurons are considered to be only transmitting ones. Hidden neurons are considered to be represented by multidimensional Gaussians, hence to act as

$$h_j(\boldsymbol{x}) = \exp\left[-\sum_{i=1}^{n}\frac{(x_i - a_{ji})^2}{2b_j^2}\right], \tag{1}$$

where $\boldsymbol{x}$ is an input $\boldsymbol{x} = (x_1, \ldots, x_n)$, $\boldsymbol{a}_j$ is a central point, $\boldsymbol{a}_j = (a_{j1}, \ldots, a_{jn})$, and $b_j \neq 0$ is a width parameter. Connections from hidden layer to output layer are endowed by weights $c_1, \ldots, c_m$. An output neuron then forms a linear combination of outputs of hidden neurons.

Therefore, mathematically, a network's computation can be seen as mapping from $\mathcal{R}^n$ to $\mathcal{R}$ given by a linear combination of m multidimensional Gaussians

$$o(\boldsymbol{x}) = \sum_{j=1}^{m} c_j \cdot h_j(\boldsymbol{x}) = \sum_{j=1}^{m} c_j \cdot \exp\left[-\sum_{i=1}^{n}\frac{(x_i - a_{ji})^2}{2b_j^2}\right]. \tag{2}$$

Clearly, a network just described has $3m$ parameters. These are given by triplets $\boldsymbol{a}_j, b_j, c_j,\ j = 1, \ldots m$, where $\boldsymbol{a}_j \in \mathcal{R}^n$, $b_j \in \mathcal{R}$, $b_j \neq 0$ and $c_j \in \mathcal{R}$.

3 Structure learning of RBF networks

In this section we shortly review several approaches currently used for structure learning of RBF neural networks.

The most simple approach is the random one. This is to fix a priori a number of hidden neurons and then set their parameters in a random way. Apparently, there is no systematicness in this approach so it would be generally rejected.

A better class of approaches is a class of decremental algorithms [1]. The common strategy employed here consists of the following consecutive steps. Firstly, a broad set of candidates for hidden neurons is created and each candidate is evaluated for its suitability with respect to input/output training set. Secondly, an a priori chosen number of best evaluated candidates is taken to represent computation of hidden neurons and to determine values of respective weights. The main disadvantage of this class of algorithms is that a set of candidates typically exponentially grows with dimension of space the points of training set are taken from. This implies that for high-dimensional training data these algorithms are inapplicable because set of candidates is computationally intractable.

Common approaches capable to handle high-dimensional data are based on fuzzy clustering [2]. The idea here is to cluster training set and found clusters, representing in a numerical way some "hill like" function, transform to an analytic form of predefined type, e.g., to multidimensional Gaussians. An analytic representation then gives computation of hidden neurons and also determines values of respective weights. The main disadvantage of fuzzy clustering approaches is that an a priori chosen number of clusters has to be stated in advance.

The third class of algorithms forms incremental algorithms. These algorithms adds neurons sequentially to match as close as possible training set [3]. A general idea here is to locally fit numerical function presented by training set in each step. If there is a point which current network not “incorporate” sufficiently a new neuron is added. Typically here is mixed partial parameter learning with a structure learning. These algorithms are capable to treat high-dimensional data.

In this paper we present new incremental algorithm based on ideas issuing from neuro-fuzzy systems area.

4 Basic idea of algorithm

An inspiration to algorithm comes from the area of neuro-fuzzy systems [1] where three-layered RBF neural networks are used to represent fuzzy inference systems. Under this representation hidden neurons of a network are seen as particular rules of a fuzzy system. A motivation for such a representation is driven by an effort for an incorporation of neural networks' learning ability into process of creation of fuzzy systems from data. On the other hand some ideas coming from fuzzy computing can be natively applied in the area of neural networks.

One of these ideas actually gave birth to a just presented algorithm. As Kosko showed theoretically [4] particular rules of fuzzy system covers extrema of function the system is performing. With respect to fuzzy view of neural network this means that hidden neurons should in some way represent extrema of learned function. Actually, this is a main idea of algorithm which we further develop.

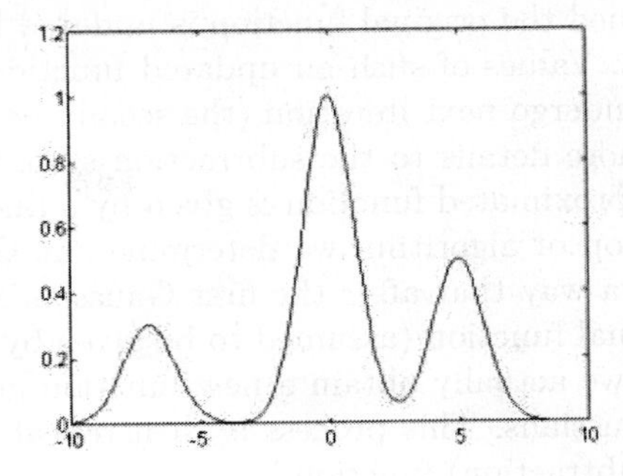

Fig. 2. A sum of three Gaussians.

We start with this one-dimensional example, see Fig. 2. A function presented here is required to be approximated by a linear combination of three Gaussians. To state explicitly parameters of these Gaussians and coefficients of the combination we can proceed in the following way. We put centers of Gaussians at points where local maxima of function are reached. Consequently, to fit function values at these points we have to multiply each Gaussian by a factor just given by the value of respective maxima. Hence maxima values gives us parameters of the linear combination. Finally, regarding width parameters, these can be set on base of a local fitting of Gaussians to approximated function. In fact the procedure presented in this example forms the skeleton of our structure learning algorithm. We now formalize this procedure in a more precise way and we also generalize it on a case of common learning data.

Regarding a generalization we have to point out that generally a function we try to approximate can also take negative values so central points has to be localized at local minima of function as well. But this situation can be transformed to only maxima case by considering in centers localization step not the original function but its absolute value. That is, we set centers at points where local maxima of absolute value of approximated function are reached.

Now, we aim on process of maxima localization. Visually, in Fig. 2., we are able to localize maxima in one instant but algorithmically this has to be formulated as iterative process assuring that linear combination of localized Gaussians gives good representation of approximated function. To fulfil this a cyclic form of algorithm is adopted where in the one step three particular actions are taken - a point of maxima of absolute value of approximated function is localized, width parameter is fitted and values of approximated function are updated. Let us demonstrate these steps on the first loop of the algorithm.

In the first loop a maximum of absolute value of original function is localized. The point at which this maximum is reached gives center a_1 of the first Gaussian. Parameter of linear combination c_1 is given by the value of the original function (not its absolute value) at this point. Then local fitting is taken, see next section for details, to set up width parameter b_1. Now, after parameters of the first

Gaussian are determined the original function is updated by subtraction of this first Gaussian from it. Values of such an updated function are then taken as a new function which undergo next iteration (the second loop) of the algorithm.

Remark here in more details to the subtraction step. It comes from an assumption that the approximated function is given by a linear combination of m Gaussians. In each loop of algorithm we determine one Gaussian of this sum. This is made in such a way that after the first Gaussian's localization we subtract it from the original function (assumed to be given by a linear combination of m Gaussians) and we actually obtain a new function given by a linear combination of $m-1$ Gaussians. This process is then repeated with the values of a new (obtained by subtraction) function.

Of course the task is now when we should stop the algorithm. If our assumption would be a perfectly true and sampled data were taken from a function just given by a linear combination of m Gaussians then after m steps subtraction yields zero function (note here explicitly that maximum of absolute value of zero function is zero) but this is not a real case. Therefore we stop the algorithm when value of maximum found is lower than some τ near to zero. Regarding setting of terminal parameter τ, its value cannot be stated in an absolute form because we have to cope with generally different ranges of values of approximated functions. From this reason we state its value in a relative way as some low percentage (say $\tau = 0.1$) of the first maximum found. This assures that linear combination of Gaussians localized by the algorithm differs maximally for $100\tau\%$ of the first maximum found from approximated data.

This finishes detail linguistic description of the algorithm. In the following section we give the procedural transcription of the algorithm.

5 The algorithm

In Table 1 there is given procedural transcription of our structure learning algorithm. As it was stated purpose of the algorithm is to perform supervised structure learning task of three-layered Gaussian RBF network based on training set given as $\mathcal{T} = \{(\boldsymbol{x}_k, t_k)\}$, $\boldsymbol{x}_k \in \mathcal{R}^n$, $t_k \in \mathcal{R}$, $k = 1, \ldots, N$. With respect to Table 1 we have to explain several things.

In step 01 there is made only formal renotation of original training set to its updates can be referred to.

In step 02 index k_1^* of point at which at which first local maximum is reached is localized. Respective value $t_{k_1^*}$ then gives parameter of linear combination c_1.

In step 03 value of terminal parameter τ is set, e.g., $\tau = 0.1$. The explanation of η value is given further. Index j refers to particular loop of the algorithm.

Steps 04-12 forms the main loop of algorithm. The loop is performed until found maximum is below τ percent of the first maximum found. In step 05 central point of jth Gaussian is set. Note here that generally we work with the n-dimensional inputs hence center is given by a point $\boldsymbol{a} \in \mathcal{R}^n$. In steps 06 and 07 respective width parameter is set. We will deal with this topic in more details now.

```
01. denote T = {(x_k, t_k)} as T^1 = {(x_k, t_k^1)};
02. denote k_1^* = argmax_k, {|t_k^1|}; set c_1 = t_{k_1^*}^1;
03. set τ ∈ (0,1); set η = 1.001; set j = 1;
04. while |c_j| > τ|c_1| do
05.    set a_j = x_{k_j^*};
06.    minimize E_j(b) given by (3) w.r.t. b;
07.    state found optimal b as b_j;
08.    update T^j to T^{j+1} = {(x_k, t_k^{j+1})}, where
       t_k^{j+1} = t_k^j − η c_j g(x_k, a_j, b_j);
09.    denote k_{j+1}^* = argmax_k, {|t_k^{j+1}|};
10.    set c_{j+1} = t_{k_{j+1}^*}^{j+1};
11.    j = j + 1;
12. end
13. m = j − 1;
```

Table 1. Incremental structure learning algorithm.

As it was already mentioned width parameter should be set on base of local fitting of just identified Gaussian to approximated function. Apparently, this can be quantified in standard sense of least squares as minimization of error function $E(b)$ given as

$$E(b) = \sum_{k=1}^{N} (t_k^j - \eta c_j g(\boldsymbol{a}_j, \boldsymbol{x}_k, b)). \tag{3}$$

Regarding η parameter consider for now that $\eta = 1$.

To minimize error function $E(b)$ we can use various algorithms such as golden section search, bisection method and so on [5]. But what is important that to optimization be effective we have to provide optimization algorithms by so called bracketing triplet [5]. In our case this is a triplet of points $b_{left} < b_{mid} < b_{right}$ such that $E(b_{left}) > E(b_{mid}) < E(b_{right})$. In the following text we show that in case of our algorithm such a triplet can always be stated.

To state bracketing triplet for error function (3) with respect to b we start by discussion of its properties. The first remark is that it is a function defined on intervals $(-\infty, 0) \cup (0, +\infty)$ and it is an even function which is given by the fact that b occurs in Gaussians, and therefore in $E(b)$, only in its second power.

The second observation is that although originally for $b = 0$ error E is not defined it can be defined here. This is given by the fact that at point $b = 0$ error function has discontinuity of the first order, hence we can continually define $E_j(0)$ by its limit at this point, i.e.,

$$E_j(0) = \lim_{b \to 0} E_j(b) = \sum_{k_p} (t_{k_p}^j)^2 + \sum_{k_c} (t_{k_c}^j - \eta c_j)^2. \tag{4}$$

In this formula k_p are indices of proper points of $\mathcal{T}^j$ which are the ones with $d^2(\boldsymbol{x}_{k_p}, \boldsymbol{a}_j) > 0$. Indices k_c are indices of points coinciding with central point of Gaussian, i.e., the ones with $d^2(\boldsymbol{x}_{k_c}, \boldsymbol{a}_j) = 0$, where $d^2(\boldsymbol{x}, \boldsymbol{a}) = \sum_i (x_i - a_i)^2$.

In our algorithm we will consider error function enhanced on value of $b = 0$. Let us now state the following assertion which proof is given in Appendix.

Lemma 1. *For error function (3) there exists $b_{max} > 0$ such that inequality $E_j(2b_{max}) > E_j(b_{max})$ holds if there is at least one non-coinciding point in $\mathcal{T}^j$. This b_{max} is given as*

$$b_{max} = \sqrt{\frac{d^2_{max}}{2\ln(\eta)}},$$

where $0 < d^2_{max} = \max_k\{d^2_k\}$, $d^2_k = \sum_i (x_{ki} - a_{ji})^2$.

On base of this lemma we can set bracketing triplet in the following way. Firstly, we compute values $E_j(0)$ and $E_j(b_{max})$ and we have three cases possible.

If $E(0) < E(b_{max})$ we have $E(-b_{max}) > E(0) < E(b_{max})$ which gives bracketing triplet as $b_{left} = -b_{max}$, $b_{mid} = 0$, $b_{right} = b_{max}$.

If $E(0) > E(b_{max})$ then because we have $E(b_{max}) < E(2b_{max})$ we can set triplet as follows $b_{left} = 0$, $b_{mid} = b_{max}$, $b_{right} = 2b_{max}$.

In the case of equality $E(0) = E(b_{max})$ we have $E(b_{max}) < E(2b_{max})$ hence $E(0) < E(2b_{max})$ which is in fact the first case. Therefore we can set triplet as $b_{left} = -2b_{max}$, $b_{mid} = 0$, $b_{right} = 2b_{max}$.

Having bracketing triplet set minimization of (3) can be performed by some one-dimensional optimization method [5]. But there are two minor complication. Theoretically, output of minimization can be $b = 0$, in this case we set b as very close to zero, e.g. $b = 0.001$. The second thing is that formula (1) has good meaning only for $\eta > 1$. To minimize impact of η on sum of squares error (3) we will consider η value very close the one, we use $\eta = 1.001$. In fact above lemma justifies employment of η parameter in error function (3).

After explanation of 06 and 07 steps of algorithm we can proceed to step 08 which states only subtraction update of approximated function.

In steps 09 and 10 we find localization of a new maxima and we set new parameter of linear combination. If found maximum fulfils condition in step 04 then loop is repeated. If not then we stop algorithm with set of m localized Gaussians. This finishes explanation of algorithm.

6 Experiments

To demonstrate practically behavior of presented algorithm we performed the following experiment. Other experiments cannot be referred here because of lack of space. In the experiment we used sampled sinusoid function $y = sin(x)$. Training set was formed by sampling sinusoid on interval $[0, 8]$ with step 0.5. Hence training set $\mathcal{T} = (x_k, t_k)$ was given as $x_k = \{0, 0.5, \ldots, 0.75, 8\}$ and respective t_k as $t_k = sin(x_k)$.

We used three structure learning algorithms - a decremental one, the one based on FCM clustering method and our incremental algorithm. In case of decremental algorithm we generates 100 candidates regulary spread on space

$[0,8] \times [0,8] \subset \mathcal{R}^2$ and we chose 3 best firing. In case of FCM clustering we set number of clusters also to be 3. For incremental clustering we used value of terminal parameter $\tau = 0.2$ which gives us by three Gaussians found. Hence result in all cases gives neural network with three hidden neurons. In Table 2 we see computational error of the network after learning. The error is meant in sense of least squares, i.e., $E = \sum_k (y_k - o(x_k))^2$.

	decremental (100/3)	*fuzzy clustering* (3)	*incremental* ($\tau = 0.2$)
$y = sin(x)$	28.2 (3)	16.2 (3)	1.3 (3)

Table 2. Errors of structure learning.

From the table we see that our algorithm was able to adaptively set number of neurons (Gaussians) as 3. Output error of structure learnt network is due to adaptive fitting step less than for other two approaches hence our algorithm produce network which can be better handled by parameters learning step.

7 Conclusion

In the paper we propose new structure learning algorithm for three-layered Gaussian RBF networks. This algorithm does not suffer from curse of dimensionality and it behaves well with comparison with other commonly used algorithms. Presented algorithm does not require an a priori specification of number of neurons identified, however, this number is driven by the value of terminal parameter τ. In the future work we aim on rejection of this parameter.

Acknolegment This work was supported by grant GACR 201/00/1489.

8 Appendix

Proof of Lemma 1: We start by notation remark. Denoting $d^2 = \sum_i (x_i - a_i)^2$ we have Gaussian written as $g(d^2, b)$. Especially, for some given $\boldsymbol{x}_k$, we have $d_k^2 = \sum_i (x_{ki} - a_{ji})^2$ and Gaussian written as $g(d_k^2, b)$.

The second remark is that sometimes not all arguments for Gaussian will be written explicitly. This means that for example $g_k(b)$ means Gaussian with given values $\boldsymbol{a}_j$, $\boldsymbol{x}_k$ known from context and we are aimed only at parameter b which is usually considered as variable.

To proceed, we remain two facts valid for Gaussian and one other fact for inequalities. The first fact valid for Gaussians is that

$$\text{For fixed } d^2 > 0 \text{ and for } 0 < b_1 < b_2 \text{ inequality } g(b_1) < g(b_2) \text{ holds.} \tag{5}$$

The second fact is that

$$\text{For fixed } b \text{ and for } 0 < d_1^2 < d_2^2 \text{ inequality } g(d_1^2) > g(d_2^2) \text{ holds.} \tag{6}$$

The last fact we remind is that

For $0 \leq x_1 < x_2$ raising to second power retains strict inequality i.e., $x_1^2 < x_2^2$. (7)

In the following text we omit index j in $E_j(b)$, i.e., we will write only $E(b)$. Note that overall error is given by two parts $E(b) = E_c + E_p(b)$. Constant term is given by coinciding points $\boldsymbol{x}_{k_c}$, $d^2(\boldsymbol{x}_{k_c}, \boldsymbol{a}) = 0$, in form $E_c = \sum_{k_c}(t_{k_c} - \eta c)^2$. The second term is given by non-coinciding (proper) points $\boldsymbol{x}_{k_p}$, $d^2(\boldsymbol{x}_{k_p}, \boldsymbol{a}) > 0$, as $E_p(b) = \sum_{k_p}(t_{k_p} - \eta c g_{k_p}(b))^2$. Only this second term can be affected by b parameter setting.

Now we can approach to proof itself. Let b_{max} be given in such a way that for all k_p and for some $\eta > 0$

$$\eta|c|g_k(b_{max}) \geq |t_k| \tag{8}$$

inequality holds. We show that this assumption implies $E(2b_{max}) > E(b_{max})$.

According to (5) we have $g_{k_p}(2b_{max}) > g_{k_p}(b_{max})$ for all k_p. Hence

$$g_{k_p}(2b_{max}) > g_{k_p}(b_{max}), \tag{9}$$

$$\eta|c|g_{k_p}(2b_{max}) > \eta|c|g_{k_p}(b_{max}), \tag{10}$$

$$\eta|c|g_{k_p}(2b_{max}) - |t_{k_p}| > \eta|c|g_{k_p}(b_{max}) - |t_{k_p}|. \tag{11}$$

Since from our assumption (8) we have $(\eta|c|g_{k_p}(b_{max}) - |t_{k_p}|) \geq 0$ we can the last equation rewrite according to (7) as

$$(\eta|c|g_{k_p}(2b_{max}) - |t_{k_p}|)^2 > (\eta|c|g_{k_p}(b_{max}) - |t_{k_p}|)^2. \tag{12}$$

Left side of (12) can be written as

$$\eta^2c^2g_{k_p}^2(2b_{max}) - 2\eta|c|g_{k_p}(2b_{max})|t_{k_p}| + t_{k_p}^2 \tag{13}$$

which is equal to

$$(\eta c g_{k_p}(2b_{max}) - t_{k_p})^2 + 2\eta c g_{k_p}(2b_{max})t_{k_p} - 2\eta|c|g_{k_p}(2b_{max})|t_{k_p}|. \tag{14}$$

Similarly, the right side of (12) can be rewritten as

$$(\eta c g_{k_p}(b_{max}) - t_{k_p})^2 + 2\eta c g_{k_p}(b_{max})t_{k_p} - 2\eta|c|g_{k_p}(b_{max})|t_{k_p}|. \tag{15}$$

Let $E_{k_p}(2b_{max}) = (\eta c g_{k_p}(2b_{max}) - t_{k_p})^2$ and $E_{k_p}(b_{max}) = (\eta c g_{k_p}(b_{max}) - t_{k_p})^2$ then we have (12) in form

$$E_{k_p}(2b_{max}) - E_{k_p}(b_{max}) >$$

$$2\eta|c|g_{k_p}(2b_{max})|t_{k_p}| - 2\eta c g_{k_p}(2b_{max})t_{k_p} + 2\eta c g_{k_p}(b_{max})t_{k_p} - 2\eta|c|g_{k_p}(b_{max})|t_{k_p}| \tag{16}$$

Term (16) can be rewritten as

$$2\eta g_{k_p}(2b_{max})(|c||t_{k_p}| - ct_{k_p}) + 2\eta g_{k_p}(b_{max})(ct_{k_p} - |c||t_{k_p}|) \tag{17}$$

which is

$$2\eta(|ct_{k_p}| - ct_{k_p})(g_{k_p}(2b_{max}) - g_{k_p}(b_{max})). \quad (18)$$

Since $|x| - x \geq 0$ for all $x \in \mathcal{R}$ and $(g_{k_p}(2b_{max}) - g_{k_p}(b_{max})) > 0$ according to (5) we have term (18) ≥ 0. That is, for all k_p holds $E_{k_p}(2b_{max}) - E_{k_p}(b_{max}) > 0$ which gives, summing through k_p, $E_p(2b_{max}) - E_p(b_{max}) > 0$. Since we assumed that there is at least one proper point we have with respect to overall error $E(b) = E_c + E_p(b)$

$$E(2b_{max}) > E(b_{max}). \quad (19)$$

Now, we aim on task how to set b_{max} to condition (8) holds. Since for c inequality $|c| \geq |t_k|$ holds for all k, see point 02 or 09 of Table 1, a setting of b_{max} in such a way that for all k

$$\eta|c|g_k(b_{max}) \geq |c| \quad (20)$$

solves the problem. Clearly (20) can be written as

$$g_k(b_{max}) \geq 1/\eta. \quad (21)$$

Considering $0 < d^2_{\max} = \max_k\{d_k^2\}$, $d_k^2 \leq d^2_{max}$, we have according to (6) for all k, $g_k(b_{max}) = g(d_k^2, b_{max}) \geq g(d^2_{max}, b_{max})$. Due to this inequality inequality (21) is valid when b_{max} is set in such a way that

$$g(d^2_{\max}, b_{max}) = 1/\eta. \quad (22)$$

This gives for b_{max} expression

$$\frac{d^2_{max}}{2b^2_{max}} = -\ln(\frac{1}{\eta}) \quad \text{and therefore} \quad b_{max} = \sqrt{\frac{d^2_{max}}{2\ln(\eta)}}.\square \quad (23)$$

References

1. Nauck D., Klawonn F., Kruse R.: Foundations of Neuro-Fuzzy Systems. John Wiley & Sons, 1997
2. Höppner F., et al.: Fuzzy cluster analysis. John Wiley & Sons, 1999
3. Cho B.C., Wang B.H.,: Radial basis function based adaptive fuzzy systems and their applications to system identification and prediction. Fuzzy sets and systems, **83**, no.2, (1996) 325–339
4. Kosko B.: Optimal fuzzy rules cover extrema. Int. J. Intell Syst., **10**, no.2, (1995) 249-255
5. Press W.H., Teukolsky S.A., Vetterling W.T, Flannery B.P., Numerical Recipes in C, The Art of Scientific Computing, Second Edition, Cambridge University Press, 1992; *internet version is available at* http://www.nr.com

Hybrid Learning of RBF Networks

Roman Neruda* and Petra Kudová

Institute of Computer Science, Academy of Sciences of the Czech Republic, P.O. Box 5, 18207 Prague, Czech Republic
roman@cs.cas.cz

Abstract. Three different learning methods for RBF networks and their combinations are presented. Standard gradient learning, three-step algoritm with unsupervised part, and evolutionary algorithm are introduced. Their perfromance is compared on two benchmark problems: Two spirals and Iris plants. The results show that three-step learning is usually the fastest, while gradient learning achieves better precission. The combination of these two approaches gives best results.

1 Introduction

By an *RBF unit* we mean a neuron with multiple real inputs $\vec{x} = (x_1, \ldots, x_n)$ and one output y. Each unit is determined by an n-dimensional vector $\vec{c}$ which is called *center*. It can have an additional parameter b usually called *width*.

The output y is computed as:

$$y = \varphi(\xi); \quad \xi = \frac{\| \vec{x} - \vec{c} \|}{b} \tag{1}$$

where $\varphi : \mathbb{R} \to \mathbb{R}$ is a suitable activation function, typically Gaussian $\varphi(z) = e^{-z^2}$.

For evaluating $\frac{\|\vec{x}-\vec{c}\|}{d}$, the Euclidean norm is usually used. In this paper we consider a general weighted norm instead of the Euclidean norm. A weighted norm is determined by a $n \times n$ matrix $\mathbf{C}$ and is defined as

$$\| \vec{x} \|_C^2 = (\mathbf{C}\vec{x})^T(\mathbf{C}\vec{x}) = \vec{x}^T\mathbf{C}^T\mathbf{C}\vec{x}. \tag{2}$$

It can be seen that the Euclidean norm is a special case of a weighted norm determined by an identity matrix. In further text we will use the symbol $\mathbf{\Sigma}^{-1}$ instead of $\mathbf{C}^T\mathbf{C}$.

In order to use a weighted norm each RBF unit has another additional parameter matrix $\mathbf{C}$.

An *RBF network* is a standard 3-layer feedforward network with the first layer consisting of n input units, a hidden layer consisting of h RBF units and an output layer of m linear units. Thus, the network computes the following function $\vec{f} : \mathbb{R}^n \to \mathbb{R}^m$:

$$f_s(\vec{x}) = \sum_{j=1}^{h} w_{js}\varphi\left(\frac{\| \vec{x} - \vec{c}_j \|_{C_j}}{b_j}\right), \tag{3}$$

* This work has been partially supported by GACR under grants 201/00/1489 and 201/02/0428.

P.M.A. Sloot et al. (Eds.): ICCS 2002, LNCS 2331, pp. 594–603, 2002.

where $w_{ji} \in \mathbb{R}$ and f_s is the output of the s-th RBF unit.

Denote $T = \{(\vec{x}(t), \vec{d}(t); t = 1, \dots, k\}$ a *training set* — a set of examples of network inputs $\vec{x}(t) \in \mathbb{R}^n$ and desired outputs $\vec{d}(t) \in \mathbb{R}^m$. For every training example we can compute the actual network output $\vec{f}(\vec{x}(t))$ and error $e_j(t)$ of each of the output units:

$$e_j(t) = d_j(t) - f_j(t).$$

The instantaneous error $\mathcal{E}(t)$ of the whole network is then:

$$\mathcal{E}(t) = \frac{1}{2} \sum_{j=1}^{p} e_j^2(t). \tag{4}$$

The goal of learning an RBF net is to minimize an error function

$$E = \sum_{t=1}^{k} \mathcal{E}(t). \tag{5}$$

1.1 Three step learning

The gradient learning described in previous section unifies all parameters by treating them in the same way. Now we introduce a learning method taking advantage of the well defined meaning of RBF network parameters (cf. [1], [2]).

There are three categories of RBF network parameters, so we can divide the learning into three consequent steps and customize the method of each step for the appropriate parameter.

The first step consists of determining the hidden unit centers. The positions of centers should reflect the density of data points and thus various clustering or vector quantization techniques can be used. Using a genetic algorithm during the first step will be discussed in 1.2.

The second phase sets the additional hidden unit parameters if there are any. There can be a parameter called width or a weighted norm matrix. These parameters determine the size and the shape of the area controlled by the unit. Suitable parameter values can be found by gradient minimization of function

$$E(b_1, \cdots, b_h; \mathbf{\Sigma_1^{-1}}, \cdots, \mathbf{\Sigma_h^{-1}}) = \tfrac{1}{2} \sum_{r=1}^{h} \left[\sum_{s=1}^{h} \varphi(\xi_{sr}) \, \xi_{sr}^2 - P \right]^2 \tag{6}$$

$$\xi_{sr} = \frac{\| c_s - c_r \|_{C_r}}{b_r}$$

where P is the overlap parameter controlling the overlap between areas of importance belonging to particular units.

In case of units with widths we can get around the minimization using simple heuristics. The often used one called the q-neighbours rule simply set the width proportionally to the average distance of q nearest neighbouring units.

The third step is a usual supervised learning known from multilayer perceptron networks reduced to a linear regression task. The only parameters to be set are the weights between the hidden and the output layer which represent the coefficients of linear combinations of RBF units outputs. Our goal is to minimize the overall error function:

$$E = \frac{1}{2} \sum_{t=1}^{k} \sum_{s=1}^{m} (d_s^{(t)} - f_s^{(t)})^2 \quad . \tag{7}$$

It can be achieved using gradient minimization or assuming the partial derivative $\frac{\partial E}{\partial w_{ij}}$ equal to zero and finding the solution in terms of linear optimalization using any of various linear least squares methods.

$$\sum_{t=1}^{k} (d_r^{(t)} y_q(\vec{x}^{(t)})) - \sum_{j=1}^{h} w_{jr} \sum_{t=1}^{k} \left(y_j(\vec{x}^{(t)}) y_q(\vec{x}^{(t)}) \right) = 0 \quad , \tag{8}$$

where $q = 1, \ldots, h$ and $r = 1, \ldots, m$.

It is true, however, that the success of this learning step depends on the previous steps.

1.2 Evolutionary learning

The third learning method is based on using a genetic algorithm. It is a stochastic optimization method inspired by evolution, using principals as selection, crossover and mutation.

A genetic algorithm works with a *population* of *individuals*. An individual (see fig. 5) represents some feasible values for all parameters of an RBF net being learned. Each individual is associated with the value of the error function of a corresponding network.

Starting with a population of random individuals new populations are produced using operators of *selection*, *mutation* and *crossover*. The *selection* guarantees that the possibility of being chosen to the new population is the higher the smaller is the error function of the corresponding network. The *crossover* compose a pair of new individuals combining parts of two old individuals. The *mutation* brings some random changes into the population. We iterate until population contains an individual with an error small enough.

Genetic algorithm can be combined with previous methods. Specifically, the determination of centers in the three-step method can be done by means of genetic algorithm. Than an individual codes only values of centers and its error is computed as

$$E_{vq} = \frac{1}{k} \sum_{t=1}^{k} \| \vec{x}^t - \vec{c}_c^{\,2} \|, \quad c = \text{argmin}_{i=1,..h} \{ \| \vec{x}^t - \vec{c}_i \| \}, \tag{9}$$

where $\vec{x}^t$ is training sample a $\vec{c}_i$ is the center of ith unit.

We implemented also canonical version of genetic learning described in [3].

2 Experiments

In the following sections results from our experiments will be presented. The first is a classification task, called *Two Spirals*. We will demonstrate an advantage of using a weighted norms. The second task, also a clasification – the known *Iris Plants*, compare all three methods described in previous sections and shows the advantage of combining two of them, specifically the three step method and the gradient learning.

All experiments were run on the Linux cluster. Each computation was run on an individual node with a Celeron 533 MHz processor and 384 MB memory.

2.1 Two spirals

The task of the first experiment, *Two Spirals*, is to discriminate between two sets of training points which lie on two distinct spirals in the 2D plane. The training set contains 372 training samples, each 2 input values (2D coordinates) and 1 output value (classification – either value 0.0 or value 1.0).

Considering the character of the training data we expect that a rather high number of RBF units will be needed. We used a network with 150 RBF units and both the Euclidean norm and a weighted norm. This network was trained using the gradient learning and the three step learning. The genetic algorithm isn't suitable because of the higher number of RBF units.

Gradient learning Knowing that a result of the gradient learning is dependent on the initial setup of parameters, the gradient learning was run five times using the Euclidean norm and five times using weighted norms and we consider the average, the worst and the best computation.

All computations were stopped after 5 000 iterations, the average time of 100 iteration was 361 seconds for an RBF net with weighted norms and 115 seconds for an RBF net with the Euclidean norms. In Figure 1 you can see the fall of the error function for the average computation using the Euclidean norm and for the average computation using weighted norms. The average error after 5000 iterations was 0.0167 for Euclidean norm, and 0.0088 for the weighted norm. Table 1 compares the time and the number of iterations needed for the fall of the error function under a given ε with an RBF net using the Euclidean norm and an RBF net using a weighted norm.

However a computation using weighted norms is slower than using the Euclidean norms, fewer iterations are needed to reach a given error and in the end a better solution is obtained.

Three step learning The three step learning was the second method applied on *Two Spirals*. Since the training samples are distributed on two spirals, we used several vector quantization methods to distribute the centers of RBF units. The Lloyd algorithm and the k-means clustering are known vector quantization methods. The genetic algorithm is our application of a common genetic algorithm to vector quantization.

The resulting value of the error function is comparable for all methods (see Table 2), but in case of the genetic algorithm there is the least number of unused units. However, the genetic algorithm has much higher time requirements.

The second step was realized by a gradient minimization of the error function (see section 1.1), 200 iterations (17 seconds for Euclidean norms, 90 seconds for weighted norms) were needed. For the determination of weights a linear least squares method was used (16 seconds for Euclidean norms, 90 seconds for weighted norms). The errors of the RBF nets learned by the three step learning are 1.101 for Euclidean norm, and 0.051 for the weighted norm.

We used two different methods to learn the *Two Spirals problem*. In both of them we saw the difference between the RBF net using Euclidean norms and the RBF net using weighted norms. In both of them the RBF net using weighted norm has a smaller error. We can interpret a use of a weighted norm as a transformation of a radial field of an RBF unit to an oval one. Then covering an input space by ovals is easier than using circles.

2.2 Iris

In the second experiment we used a well-known data set *Iris Plants*. It contains 3 classes of 50 instances each, where each class refers to a type of an iris plant. One class is linearly separable form the others, the other are not linearly separable from each other.

We used a net with three output neurons, one neuron for each class. The class number is then coded by three binary values, value 1 on the position corresponding with the number of class and zeros on the others. So each training sample consists of 4 input values describing some features of a plant and 3 output values coding its class.

We split *The Iris Plants* data set into two parts. The first containing 120 instances (40 per class) is used as a training set, the second containing other 30 instances (chosen randomly) is used for testing.

We applied all three methods (the gradient learning,the three step method and the genetic learning) on an RBF net with 3, 6 and 9 hidden units, all with weighted norms.

Gradient learning The gradient algorithm was run five times and the average, the minimum and the maximum computation was picked up. Figure 3 compares the fall of the average gradient algorithm error function. The number of iterations needed to reach a given error is shown in Table 3. Table 4 shows the error of the learned RBF net on the training set and the testing set with the number of misclassified samples.

Three step learning The three step learning consisted of a vector quantization using both the Lloyd algorithm and the genetic algorithm, the gradient minimization in the second step and the linear least squares.

Figures 3 and 4 show the fall of the vector quantization error function for the Lloyd algorithm and the genetic learning. 100 iterations of the genetic algorithm needs 3 seconds, 1000 iterations of the Lloyd algorithm need 2 seconds. Table 5 you see the resulting error (the average, the minimum, the maximum of five computations). 10 iterations of the Lloyd algorithm or 2000 iterations of the genetic algorithm were needed. The genetic algorithm is better than the Lloyd algorithm, which is highly dependent on its random initialization. However, the time requirements of the genetic algorithm are much higher.

In the second step a gradient minimization was used (1000 iterations, 1s) and in the third step the linear least squares (1s). In Table 6 you see the resulting errors and the numbers of misclassified samples.

Because of the resulting error of the three step learning is much worse than the one of the gradient learning, we decided to add a fourth step. Specifically we used the RBF net learned by the three step learning as the initialization for the gradient learning. Then the fourth step consists of some iterations of gradient learning, we practiced 5 000 iterations. The fall of the error function is shown on Figure 4, the number of iterations and the time needed to reach the given error is in Table 7. The review of the resulting errors is in Table 8.

Although the results of the three step learning were not the best, its time requirements are very low and so it can be used successfully as initialization of the gradient learning.

Genetic learning We ran the genetic algorithm five times and consider the average, the minimum and the maximum computation. All computations worked with a population of 50 individuals, an elite of 2 individuals and the mutation rate 0.2, the crossover rate 0.7. The average time of 100 iterations was 32.8 s using 3 units, 114 s using 6 units and 149 s using 9 units. In Figure 5 see the fall of the error function, Table 9 shows the number of iterations and time needed to reach a given error. The review of resulting errors is in Table 10.

The genetic algorithm is a little bit worse than previous methods. Its great disadvantage are its time requirements.

All three methods described in section 1 were demonstrated on the *Iris Plants* task. The genetic learning ended with a higher error and the highest time requirements. The gradient learning converged to the lowest error. The three step method has the very lowest time requirements. We showed that the best way is to apply the three step learning followed by the gradient learning.

References

1. K. Hlaváčková and R. Neruda. Radial basis function networks. *Neural Network World*, 3(1):93–101, 1993.
2. J. Moody and C. Darken. Learning with localized receptive fields. In D. Touretzky, G. Hinton, and T. Sejnowski, editors, *Proceedings of the 1988 Connectionist Models Summer School*, San Mateo, CA, 1989. Morgan Kaufmann.
3. R. Neruda. *Functional Equivalenece and Gentic Learning of RBF Networks*. PhD thesis, Institute of Computer Science, Academy of Sciences of the Czech Republic, Prague, Czech Republic, 1998.
4. Kudová P. Neuronové sítě typu RBF pro analýzu dat. Master's thesis, Charles University, Prague, Faculty of Mathematics and Physics, 2001.

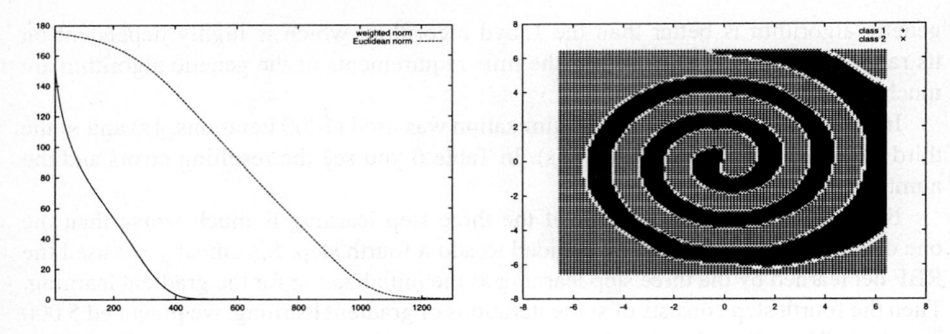

Fig. 1. Two spirals: a) Te gradient learning error function. b) The network output partitioning the intput space.

Table 1. Two spirals: The average number of iterations and time to reach a given ε.

	Euclidean norm		Weighted norm	
ε	iterations	time	iterations	time
10	1011	19 min 23 s	353	21 min 17 s
1	1192	22 min 51 s	441	26 min 35 s
0.5	1272	24 min 24 s	470	28 min 20 s
0.1	1929	37 min 0 s	689	41 min 32 s
0.01	–	–	3829	3 hour 50 min 49 s

Table 2. Two spirals: Vector quantization, 1st step.

	error	number of passes through the trainset	time
Lloyd alg.	0.1296	8	1 s
K-means clustering	0.1066	50	9 s
K-means cl. with local memory	0.0940	200	1 min 6 s
Genetic algorithm	0.1183	50 × 2 500	4hours 31 min

Table 3. Iris: Average number of iterations and time to reach a given ε.

	3 units		6 units		9 units	
ε	iterations	time	iterations	time	iterations	time
100	1	< 1s	1	< 1s	1	< 1s
50	5	1s	2	1s	176	38s
10	1832	1 min 49s	141	14s	427	1 min 33s
5	1833	1 min 49s	695	1 min 9s	445	1 min 37s
3	—	—	—	—	1380	5 min 3s

Table 4. Iris: The error divided by the number of samples and the number of misclassified samples.

	Error on trainset			Error on test set		
	average	minimum	maximum	average	minimum	maximum
3 units	0.029 (0)	0.026 (0)	0.034 (1)	0.092 (2)	0.089 (2)	0.099 (2)
6 units	0.037 (0)	0.017 (0)	0.065 (4)	0.100 (2)	0.087 (2)	0.123 (2)
9 units	0.020 (0)	0.010 (0)	0.025 (0)	0.106 (2)	0.090 (2)	0.123 (2)

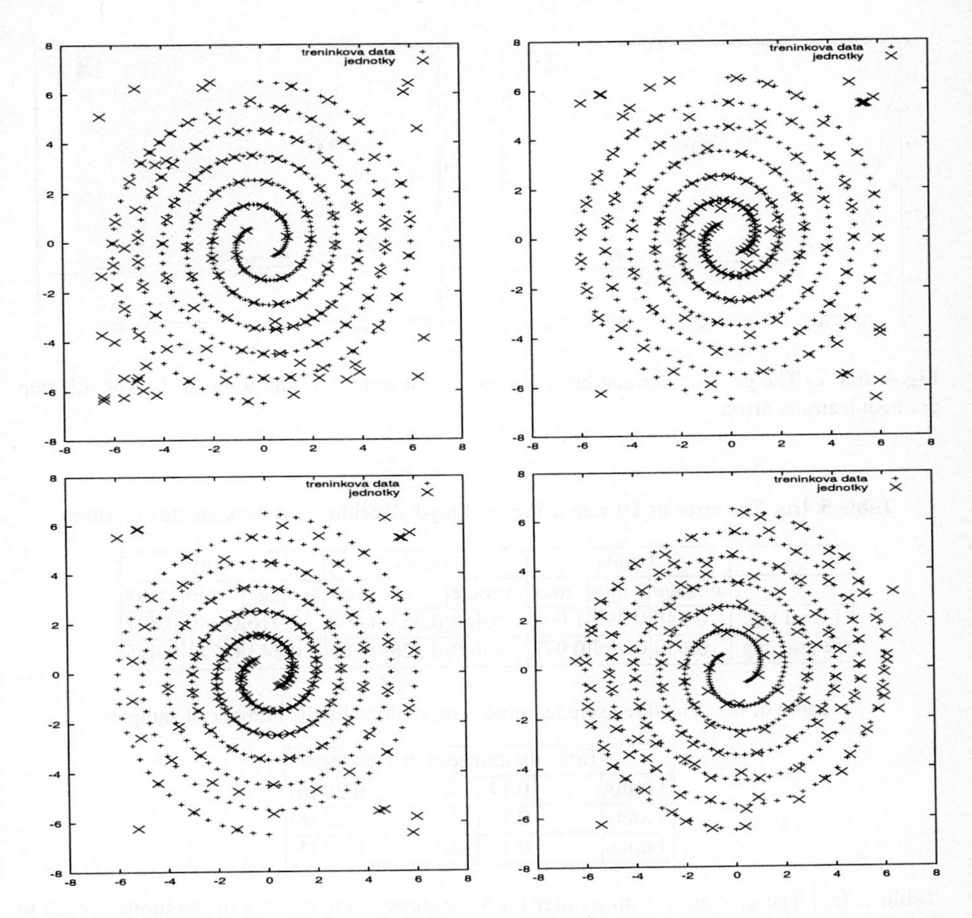

Fig. 2. Lloyd algorithm. K-means clustering. K-means clustering with local memory. Genetic algortihm.

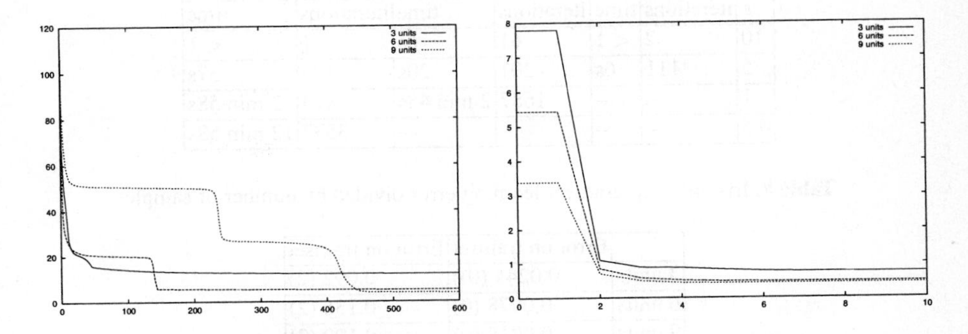

Fig. 3. Iris: a) The gradient learning error function using an RBF net with 3, 6 and 9 units. b) The VQ learning error function – 1st step (3, 6 and 9 units)

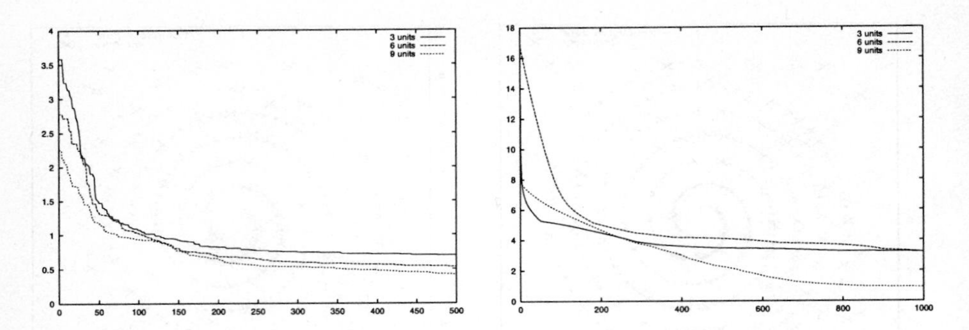

Fig. 4. Iris: a) The genetic learning error function – 1st step (3, 6 and 9 units). b) The 4th step gradient learning error.

Table 5. Iris: The error of 1st step using the Lloyd algorithm and the Genetic algorithm

	3 units			6 units			9 units		
	average	min	max	average	min	max	average	min	max
Lloyd alg.	0.748	0.497	1.000	0.460	0.355	0.498	0.416	0.340	0.497
Genetic alg.	0.650	0.499	0.977	0.449	0.326	0.666	0.343	0.243	0.425

Table 6. Iris: The three step learning error divided by the number of samples.

	Error on trainset	Error on testset
3 units	0.14 (13)	0.20 (6)
6 units	0.14 (18)	0.18 (6)
9 units	0.12 (14)	0.17 (5)

Table 7. Iris: The gradient learning (after the three steps) – the number of iterations needed to reach a given ε

	3 units		6 units		9 units	
ε	iterations	time	iterations	time	iterations	time
10	2	< 1	61	6s	1	< 1
5	111	6s	201	20s	171	37s
1	–	–	1687	2 min 48s	813	2 min 58s
0.5	–	–	–	—	3537	12 min 58s

Table 8. Iris: 4th step gradient learning error divided by number of samples

	Error on trainset	Error on trainset
3 units	0.0244 (0))	0.092 (2)
6 units	0.0078 (0)	0.136 (2)
9 units	0.0018 (0)	0.109 (2)

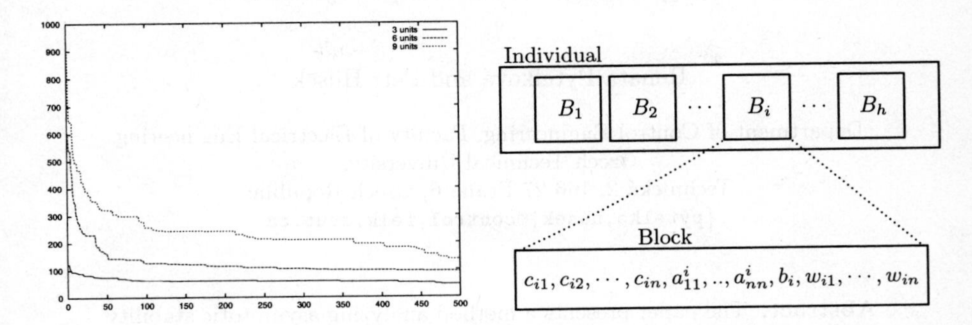

Fig. 5. a) The genetic learning error function using an RBF net with 3, 6 and 9 units.b) An individual representing an RBF net.

Table 9. The average number of iterations and time to reach a given ε.

	3 units		6 units		9 units	
ε	iterations	time	iterations	time	iterations	time
100	5	1s	317	6 min 4s	963	24 min 4s
50	468	2 min 29s	5443	1hour 46 min 14s	7515	3hours 7 min 52s
40	2325	12 min 24s	7835	2hours 30 min 10s	15990	6hours 39 min 45s
30	8859	47 min 14s	15925	5hours 5 min 13s	39819	16hours 35 min 28s
20	—	—	52753	16hours 51 min 5s	—	—

Table 10. The error divided by number of samples, number of misclassified samples.

	Error on trainset			Error on test set		
	average	minimum	maximum	average	minimum	maximum
3 units	0.249 (11)	0.134 (5)	0.361 (17)	0.225 (3)	0.182 (1)	0.292 (6)
6 units	0.161 (6)	0.089 (0)	0.284 (8)	0.272 (3)	0.165 (2)	0.429 (5)
9 units	0.242 (10)	0.205 (6)	0.335 (16)	0.399 (4)	0.223 (1)	0.662 (7)

Stability Analysis of Discrete-Time Takagi-Sugeno Fuzzy Systems

Renata Pytelková and Petr Hušek

Department of Control Engineering, Faculty of Electrical Engineering
Czech Technical University,
Technická 2, 166 27 Praha 6, Czech Republic
{pytelko,husek}@control.felk.cvut.cz

Abstract. The paper presents a method analyzing asymptotic stability of discrete-time Takagi-Sugeno fuzzy systems. It is supposed that the plant is described by the Takagi-Sugeno fuzzy system with linear state-space submodels in the consequents of rules and the controller by the Takagi-Sugeno fuzzy system with linear state feedback submodels. These fuzzy systems can be considered as linear time-varying systems, where coefficients are varying in given intervals. The method tests stability by checking only products of vertex (extreme) matrices.

1 Introduction

Stability is one of the most important issues in analysis of control systems. In the case of fuzzy systems no particular stability theory exists. In fact the fuzzy system is a kind of nonlinear system with the disadvantage of having a complex analytical description.

There exist several particular methods designed for some special case of fuzzy systems. These methods are based on methods used in nonlinear control theory, for example small gain theorem and its related techniques (the circle criterion, conicity methods etc.), qualitative methods (stability indices) and the Lyapunov stability theory, among others. These methods usually put some constraints on the controller or to the model of the controlled system.

Special case of fuzzy systems are so-called Takagi-Sugeno fuzzy systems ([4]). In fuzzy control theory Takagi-Sugeno fuzzy systems with linear state-space or input-output submodels in the consequents of rules are used. The main advantage is, that this approach (sometimes called model based fuzzy approach) provides to combine together the knowledge of linear systems and fuzzy logic and use it for stability analysis. There exist two main approaches of stability analysis of such systems.

The first one is based on finding a common Lyapunov function for all subsystems used in consequents of rules. Problem of finding of this function is solved by LMI. This method states sufficient condition of stability and more details concerning this method can be found for example in [5]. However this algorithm is unable to decide about stability in many practical situations due to its big conservatism.

P.M.A. Sloot et al. (Eds.): ICCS 2002, LNCS 2331, pp. 604–612, 2002.

Another approach can be found in [3]. In this paper it was shown, that problem of stability analysis of Takagi-Sugeno fuzzy systems with linear submodels in consequents of rules can be transformed to robust stability analysis of a polynomial with polynomic structure of its coefficients. Stability is tested by the Modified Jury or the Modified Routh criterion. It states sufficient and necessary stability conditions for slowly varying systems.

In this paper the new method for stability analysis of discrete-time Takagi-Sugeno fuzzy systems, where the plant is described by the Takagi-Sugeno fuzzy system with linear state-space submodels in the consequents of rules and the controller by the Takagi-Sugeno fuzzy system with linear state feedback submodels.

The paper is organized as follows. In section 2 the Takagi-Sugeno fuzzy model of a plant and the PDC controller is described. In section 3 the computation procedure of vertex extreme matrices of the closed-loop Takagi-Sugeno fuzzy system is introduced. Section 4 is devoted to the stability analysis method. Illustrative example is shown in the section 5 and some concluding remarks are mentioned in section 6.

2 Takagi-Sugeno Fuzzy Systems

In this section Takagi-Sugeno fuzzy systems with linear submodels in the consequents of rules will be introduced. Linear state-space submodels are supposed.

The Takagi-Sugeno fuzzy system with the state-space submodels is consider in the following form:

$$\begin{aligned} R^i &: \text{IF } z_1(t) \text{ is } M_1^i \text{ and } \ldots \text{ and } z_g(t) \text{ is } M_g^i \\ &\text{THEN } \mathbf{x}(t+1) = \mathbf{A}^i\mathbf{x}(t) + \mathbf{B}^i\mathbf{u}(t) \end{aligned} \tag{1}$$

for $i = 1, 2, \ldots, R$ where $\mathbf{z}(t) = [z_1(t), \ldots, z_g(t)]$ is a vector of variables of the premise (some measurable system variables), M_j^i are fuzzy sets, $\mathbf{x}(t) \in \Re^n$ is the state vector, $\mathbf{u}(t) \in \Re^m$ is the input vector, $\mathbf{A}^i \in \Re^{n\times n}$ are linearized system matrices and $\mathbf{B}^i \in \Re^{n\times m}$ are linearized input matrices.

The total output of the fuzzy system is inferred as follows:

$$\begin{aligned} \mathbf{x}(t+1) &= \frac{\sum_{i=1}^{R} w_i(\mathbf{z}(t))\{\mathbf{A}^i\mathbf{x}(t) + \mathbf{B}^i\mathbf{u}(t)\}}{\sum_{i=1}^{R} w_i(\mathbf{z}(t))} \\ &= \sum_{i=1}^{R} h_i(\mathbf{z}(t))\{\mathbf{A}^i\mathbf{x}(t) + \mathbf{B}^i\mathbf{u}(t)\} \end{aligned} \tag{2}$$

where

$$w_i(\mathbf{z}(t)) = \prod_{j=1}^{g} M_j^i(z_j(t)) \tag{3}$$

$M_j^i(z_j(t))$ is the grade of membership of $z_j(t)$ in M_j^i. It is assumed that $w_i(\mathbf{z}(t)) \geq 0$, for $i = 1, 2, \ldots, R$ and $\sum_{i=1}^{R} w_i(\mathbf{z}(t)) > 0$ for all t. Therefore

$$\sum_{i=1}^{R} h_i(\mathbf{z}(t)) = 1 \tag{4}$$

The PDC controller of the following form is considered:

$$\begin{aligned} R^i : \text{IF } z_1(t) \text{ is } M_1^i \text{ and } \ldots \text{ and } z_g(t) \text{ is } M_g^i \\ \text{THEN } \mathbf{u}(t) = -\mathbf{K}^i\mathbf{x}(t) \end{aligned} \tag{5}$$

The total output of this fuzzy controller is

$$\begin{aligned} \mathbf{u}(t+1) &= \frac{-\sum_{i=1}^{R} w_i(\mathbf{z}(t))\mathbf{K}^i\mathbf{x}(t)}{\sum_{i=1}^{R} w_i(\mathbf{z}(t))} \\ &= -\sum_{i=1}^{R} h_i(\mathbf{z}(t))\mathbf{K}^i\mathbf{x}(t) \end{aligned} \tag{6}$$

By substituting (6) into (2) the total state space model of the fuzzy closed-loop system is obtained:

$$\mathbf{x}(t+1) = \sum_{i=1}^{R}\sum_{j=1}^{R} h_i(\mathbf{z}(t))h_j(\mathbf{z}(t))\{\mathbf{A}^i - \mathbf{B}^i\mathbf{K}^j\}\mathbf{x}(t) \tag{7}$$

3 Computation of Vertex Extreme Matrices

The presented stability analysis method is based on stability analysis of linear time-varying systems, where coefficients are varying in given intervals. The method tests stability by checking only products of vertex extreme matrices.

Let's define the system matrix of the closed-loop system obtained from (7) as follows:

$$\mathbf{G}(\mathbf{h}(\mathbf{z}(t))) = \sum_{i=1}^{R}\sum_{j=1}^{R} h_i(\mathbf{z}(t))h_j(\mathbf{z}(t))\{\mathbf{A}^i - \mathbf{B}^i\mathbf{K}^j\} \tag{8}$$

By substituting $h_R(\mathbf{z}(t)) = 1 - \sum_{i=1}^{R-1} h_i(\mathbf{z}(t))$ into (8), $\mathbf{G}(\mathbf{h}(\mathbf{z}(t)))$ can be expressed as:

$$\begin{aligned} \mathbf{G}(\mathbf{h}(.)) = \sum_{i=1}^{R-1}\sum_{j=1}^{R-1} h_i(.)h_j(.)\big(\mathbf{B}^j\mathbf{K}^R + \mathbf{B}^R\mathbf{K}^j - \mathbf{B}^i\mathbf{K}^j - \mathbf{B}^R\mathbf{K}^R\big) \qquad (9) \\ + \sum_{j=1}^{R-1} h_j(.)\big(\mathbf{A}^j - \mathbf{B}^j\mathbf{K}^R - \mathbf{A}^R - \mathbf{B}^R\mathbf{K}^j + 2\mathbf{B}^R\mathbf{K}^R\big) + \mathbf{A}^R - \mathbf{B}^R\mathbf{K}^R \end{aligned}$$

Each element $g_{ij}(\mathbf{h}(\mathbf{z}(t)))$ of the matrix $\mathbf{G}(\mathbf{h}(\mathbf{z}(t)))$ is a polynomic function of the second order of the time-varying parameter $\mathbf{h}(\mathbf{z}(t))$ and can be rewritten to the following form:

$$g_{ij}(\mathbf{h}) = \mathbf{h}^T \mathbf{C}^{(ij)} \mathbf{h} + \left(\mathbf{d}^{(ij)}\right)^T \mathbf{h} + v^{(ij)} \tag{10}$$
$$\mathbf{C}^{(ij)} = \left[c^{(ij)}\right] \in \Re^{R-1,R-1}, \mathbf{d}^{(ij)} \in \Re^{R-1}, v^{(ij)} \in \Re$$

and

$$\mathbf{h} \in H \subset \Re^{R-1} \tag{11}$$
$$H = \left\{\mathbf{h} \in H \subset \Re^{R-1} : h_k \in [0,1],\ \sum_{k=1}^{R-1} h_k \leq 1,\ k = 1, \ldots, R-1\right\}$$

where $\mathbf{h}$ denotes $(\mathbf{h}(\mathbf{z}(t)))$.

The system matrix $\mathbf{G}(\mathbf{h})$ can be overbounded by the interval matrix $\mathbf{G}^I$ in the following form

$$\mathbf{G}^I = \begin{bmatrix} [g_{11}^{\min}, g_{11}^{\max}] & \cdots & [g_{1n}^{\min}, g_{1n}^{\max}] \\ \vdots & & \vdots \\ [g_{n1}^{\min}, g_{n1}^{\max}] & \cdots & [g_{nn}^{\min}, g_{nn}^{\max}] \end{bmatrix} \tag{12}$$

The problem of finding extreme values of $g_{ij}(\mathbf{h})$ on a box H is a task of mathematical programming ([1]). General formulation of a task of mathematical programming is as follows.

Let us consider the problem of minimization of a function $f_0(\mathbf{x})$, where the constraints are given in the form of inequalities

$$\min\left\{f_0(\mathbf{x}) \mid f_j(\mathbf{x}) \leq b_j,\ j = 1, \ldots, m\right\} \tag{13}$$

Necessary conditions of extreme values can be determined by the following theorem.

Definition 1. *Let a point ${}^0\mathbf{x}$ satisfy all constraints of (13). Let $J({}^0\mathbf{x})$ be the set of indices, for which the corresponding constraints are active (e.g. inequality changes to equality):*

$$J({}^0\mathbf{x}) = \left\{j \mid f_j({}^0\mathbf{x}) = b_j\right\} \tag{14}$$

The point ${}^0\mathbf{x}$ is said to be a regular point of the set X given by constraints in (13), if the gradients $\nabla f_j({}^0\mathbf{x})$ are linearly independent $\forall j \in J({}^0\mathbf{x})$.

Theorem 1 (Kuhn-Tucker). *Let ${}^*\mathbf{x}$ be a regular point of a set X and a function $f_0(\mathbf{x})$ has in some neighbourhood of ${}^*\mathbf{x}$ continuous first partial derivatives. If the function $f_0(\mathbf{x})$ has in the point ${}^*\mathbf{x}$ the local minimum on X, then there exists a (Lagrange) vector ${}^*\lambda \in \Re^m$ that*

$$\nabla f_0({}^*\mathbf{x}) + \sum_{r=1}^{m} {}^*\lambda_r \nabla f_r({}^*\mathbf{x}) = 0$$
$${}^*\lambda_j\left(f_j({}^*\mathbf{x}) - b_j\right) = 0 \tag{15}$$
$${}^*\lambda_j \geq 0$$

hold $\forall j = 1, \ldots, m$.

Remark 1. For maximization of a function $f_0(\mathbf{x})$ the last inequality of (15) is replaced by $^*\lambda_j \leq 0$.

To apply Theorem 1 for solving our problem it is necessary to check whether the preconditions of this theorem are satisfied. As $g_{ij}(\mathbf{h})$ is a polynomic function of the second order, its first partial derivatives are continuous $\forall \mathbf{h} \in H$ and the second assumption is satisfied. In our case

$$\begin{aligned} f_0(\mathbf{h}) &= g_{ij}(\mathbf{h}) \\ f_j(\mathbf{h}) &= (-1)^{j+1} h_i \\ b_j &= 0 \text{ for } j \text{ even} \\ b_j &= 1 \text{ for } j \text{ odd} \\ f_{2R-1}(\mathbf{h}) &= \sum_{k=1}^{R-1} h_k \\ b_{2R-1} &= 1 \end{aligned} \tag{16}$$

where $i = 1, \ldots, R-1,\ j = 1, \ldots, 2(R-1),\ j = 2i-1, 2i$.

Then

$$\begin{aligned} \nabla f_j(\mathbf{h}) &= (-1)^{j+1} \mathbf{e}^{(i)},\ \forall \mathbf{h} \in H,\ j = 1, \ldots, 2(R-1) \\ i &= \tfrac{j+1}{2} \text{ for } j \text{ odd } i = \tfrac{j}{2} \text{ for } j \text{ even} \\ \nabla f_{2R-1}(\mathbf{h}) &= [1, \ldots, 1]^T \end{aligned} \tag{17}$$

where $e^{(i)} = [0, \ldots, 0, 1, 0, \ldots, 0]^T$ with 1 on the i-th position. Because for some $\mathbf{h} \in H$ only even or only odd constraints (or none of them) can be active $\forall i = 1, \ldots, R-1$, $\nabla f_j(\mathbf{h})$, are linearly independent $\forall \mathbf{h} \in H,\ j \in J(\mathbf{h})$. It means that all points $\mathbf{h} \in H$ are regular ones.

According to Theorem 1 it is necessary to determine the gradient $\nabla g_{ij}(\mathbf{h})$. The components are

$$\nabla g_{ij}(\mathbf{h}) = \left[\frac{\partial g_{ij}(\mathbf{h})}{\partial h_1} \quad \cdots \quad \frac{\partial g_{ij}(\mathbf{h})}{\partial h_{R-1}} \right]^T \tag{18}$$

follows from (10)

$$\frac{\partial g_{ij}(\mathbf{h})}{\partial h_l} = 2c_{ll}^{(ij)} \cdot h_l + \sum_{\substack{r=1 \\ r \neq l}}^{R-1} \left(c_{lr}^{(ij)} + c_{rl}^{(ij)} \right) \cdot h_r + d_l^{(ij)} \tag{19}$$

$$i, j, l = 1, \ldots, R-1$$

After substituting (16), (17), (18) and (19) to (15) the following system of equations and inequalities is obtained:

$$\begin{bmatrix} 2c_{11}^{(ij)} & c_{12}^{(ij)} + c_{21}^{(ij)} & \cdots & c_{1,R-1}^{(ij)} + c_{R-1,1}^{(ij)} \\ c_{21}^{(ij)} + c_{12}^{(ij)} & 2c_{22}^{(ij)} & \cdots & c_{2,R-1}^{(ij)} + c_{R-1,2}^{(ij)} \\ \vdots & \vdots & \ddots & \vdots \\ c_{R-1,1}^{(ij)} + c_{1,R-1}^{(ij)} & c_{R-1,2}^{(ij)} + c_{2,R-1}^{(ij)} & \cdots & 2c_{R-1,R-1}^{(ij)} \end{bmatrix} \cdot \begin{bmatrix} h_1 \\ h_2 \\ \vdots \\ h_{R-1} \end{bmatrix} +$$

$$+\begin{bmatrix} 1 & -1 & 0 & 0 & 0 & \cdots & 0 & 0 & 1 \\ 0 & 0 & 1 & -1 & 0 & \cdots & 0 & 0 & 1 \\ \vdots & \vdots & \vdots & \vdots & \vdots & \ddots & \vdots & \vdots & \vdots \\ 0 & 0 & 0 & 0 & 0 & \cdots & 1 & -1 & 1 \end{bmatrix} \cdot \begin{bmatrix} \lambda_1 \\ \lambda_2 \\ \vdots \\ \lambda_{2R-1} \end{bmatrix} = \begin{bmatrix} d_1^{(ij)} \\ d_2^{(ij)} \\ \vdots \\ d_{R-1}^{(ij)} \end{bmatrix} \quad (20)$$

$$\begin{aligned} \lambda_1(h_1 - 1) &= 0 \\ \lambda_2 h_1 &= 0 \\ \lambda_3(h_2 - 1) &= 0 \\ \lambda_4 h_2 &= 0 \\ &\vdots \\ \lambda_{2\cdot R-3}(h_{R-1} - 1) &= 0 \\ \lambda_{2\cdot(R-1)} h_{R-1} &= 0 \\ \lambda_{2\cdot R-1}\Big(\sum_{k=1}^{R-1} h_k - 1\Big) &= 0 \end{aligned} \quad (21)$$

$\lambda_1, \ldots, \lambda_{2\cdot R-1} \geq 0$ for minimization
$\lambda_1, \ldots, \lambda_{2\cdot R-1} \leq 0$ for maximization

The important fact is that the equation (20) is linear. The computational procedure of solving (20), (21) runs as follows. At first all solutions of (21) (nonlinear) are determined. This corresponds to determining of all the parts of the box H - the interior and all the parts of the boundary of H (all manifolds with the dimension i, $i = 0, \ldots, R-2$ containing only points on the boundary of H). Each solution of (21) corresponds to $2R-1$ linear equations (from (21) it follows that at least one of $\lambda_{2i-1}, \lambda_{2i} \forall i = 1, \ldots, R-1$ has to equal zero; if $\lambda_{2i-1} = 0$ then either $\lambda_{2i} = 0$ or $h_i = 0$, if $\lambda_{2i} = 0$ then either $\lambda_{2i-1} = 0$ or $h_i = 1$, $\forall i = 1, \ldots, R-1$; λ_{2R-1} can be zero anytime). These $2R-1$ equations together with $R-1$ equations of (20) form $3R-2$ linearly independent linear equations for $3R-2$ unknown variables. It means that there exists a unique solution $({}^*\lambda, {}^*\mathbf{h})$ (for each solution of (21)) of (20) and (21). As the number of manifolds with the dimension i, $i = 0, \ldots, R-1$ containing only points on the boundary of H is $\binom{R}{i}$, the total number n_s of all solutions of (20) and (21) is given by

$$n_s = \sum_{i=0}^{R-1} \binom{R}{i} = 2^R - 1 \quad (22)$$

In the next step is checked whether ${}^*\lambda_j^{(l)} \geq 0$ $({}^*\lambda_j^{(l)} \leq 0)$, $\forall j = 1, \ldots, 2(R-1)$ and ${}^*\mathbf{h}^{(l)} \in H$ for each $l = 1, \ldots, n_s$. Denote by $L_{\min}(L_{\max})$ the set of l for which these conditions are satisfied.

$$\begin{aligned} L_{\min} &= \{l : \ {}^*\mathbf{h}^{(l)} \in H, \ {}^*\lambda_j^{(l)} \geq 0, \ \forall j = 1, \ldots, 2(R-1)\} \\ L_{\max} &= \{l : \ {}^*\mathbf{h}^{(l)} \in H, \ {}^*\lambda_j^{(l)} \leq 0, \ \forall j = 1, \ldots, 2(R-1)\} \end{aligned} \quad (23)$$

Then

$$g_{ij}^{\min}(\mathbf{h}) = \min_{l \in L_{\min}} \left(g_{ij}({}^*\mathbf{h}^{(l)})\right)$$
$$g_{ij}^{\max}(\mathbf{h}) = \max_{l \in L_{\max}} \left(g_{ij}({}^*\mathbf{h}^{(l)})\right) \tag{24}$$

Now let's define the set of extreme matrices $\mathbf{G}^E \subset \mathbf{G}^I$ such that

$$\mathbf{G}^E = \left\{[g_{ij}^E] \mid g_{ij}^E = g_{ij}^{\min} \text{ or } g_{ij}^{\max};\ i, j = 1, \dots, R\right\} \tag{25}$$

Obviously number of elements of $\mathbf{G}^E$ is 2^p, where $p = n^2$ element of $\mathbf{G}^E$ will be refered as $\mathbf{G}_j^E$ with $j \in \{1, \dots, 2^p\}$.

Let denote the set of all products of extreme matrices of length k by

$$\overline{\mathbf{P}}_k^E = \{\mathbf{P}_k^E(i_1, \dots, i_k) = \mathbf{G}_{i_1}^E \cdot \dots \cdot \mathbf{G}_{i_k}^E | (i_1, \dots, i_k) \in \{1, \dots, 2^p\}^k\} \tag{26}$$

where $\{1, \dots, 2^p\}^k$ is the k-th cross product of $\{1, \dots, 2^p\}$. Therefore the number of elements of $\overline{\mathbf{P}}_k^E$ is 2^{pk}.

4 Stability Analysis

Let us define ([2]) the following theorem and lemma:

Theorem 2. $\|\prod_{n=1}^k \mathbf{G}(n)\| < 1$ *for all choices of* $\mathbf{G}(n) \in \mathbf{G}^I$, $n = 1, \dots, k$, *iff* $\|\mathbf{P}_k^E(i_1, \dots, i_k)\| < 1$ *for all* $\mathbf{P}_k^E(i_1, \dots, i_k) \in \overline{\mathbf{P}}_k^E$.

The symbol $\| \ \|$ denotes the induced matrix 1-norm or the ∞-norm.

Lemma 1. *Let* $\mathbf{G}(n) \in \mathbf{G}^I$. *Then the time-variant Takagi-Sugeno fuzzy system (7) is asymptotically stable if there exists a finite* k *such that* $\|\prod_{n=1}^k \mathbf{G}(n)\| < 1, \forall \mathbf{G}(n) \in \mathbf{G}^I$, $n = 1, \dots, k$.

The theorem together with the lemma can be used to formulate the following corollary:

Corollary 1. *The time-variant Takagi-Sugeno fuzzy system (7) is asymptotically stable if there exists a finite* k *such that* $\|\mathbf{P}_k^E(i_1, \dots, i_k)\| < 1, \forall (i_1, \dots, i_k) \in \{1, \dots, 2^p\}^k$

Proofs of the theorem, lemma and corollary can be found in [2].

Stability test algorithm

1. Compute the interval matrix $\mathbf{G}^I$
2. Set $k = 1$, compute the set of extreme matrices $\mathbf{G}^E$
3. Generate all 2^{pk} products $\mathbf{P}_k^E(i_1, \dots, i_k)$ and compute the norm of all products.
4. If all norms are smaller than 1, the system is stable. In other case set $k = k+1$ and repeat step 3.

5 Example

Define the Takagi-Sugeno fuzzy system described by two following rules:

$$R^1 : \text{IF } x_1(t) \text{ is } M_1^1 \text{ THEN } \mathbf{x}(t+1) = \mathbf{A}^1\mathbf{x}(t) + \mathbf{B}^1\mathbf{u}(t)$$
$$R^2 : \text{IF } x_1(t) \text{ is } M_1^2 \text{ THEN } \mathbf{x}(t+1) = \mathbf{A}^2\mathbf{x}(t) + \mathbf{B}^2\mathbf{u}(t)$$

and define controller described by the following rules:

$$R^1 : \text{IF } x_1(t) \text{ is } M_1^1 \text{ THEN } \mathbf{u}(t) = \mathbf{K}^1\mathbf{x}(t)$$
$$R^2 : \text{IF } x_1(t) \text{ is } M_1^2 \text{ THEN } \mathbf{u}(t) = \mathbf{K}^2\mathbf{x}(t)$$

where

$$\mathbf{A}^1 = \begin{bmatrix} -0.2 & 0.4 \\ 0.5 & 0.5 \end{bmatrix} \mathbf{B}^1 = \begin{bmatrix} 1 \\ 1 \end{bmatrix} \mathbf{K}^1 = \begin{bmatrix} 0.4 & 0.3 \end{bmatrix}$$
$$\mathbf{A}^2 = \begin{bmatrix} -0.4 & 0.2 \\ 0 & 0.5 \end{bmatrix} \mathbf{B}^2 = \begin{bmatrix} 1 \\ 1 \end{bmatrix} \mathbf{K}^2 = \begin{bmatrix} -0.2 & 0.1 \end{bmatrix}$$

The interval matrix $\mathbf{G}^I$ is

$$\mathbf{G}^I = \begin{bmatrix} [-0.6, -0.2] & [0.1, 0.1] \\ [0.1, 0.2] & [0.2, 0.4] \end{bmatrix}$$

In the following table the extreme product norms are shown:

Table 1. Extreme product norms

k	$\max \lVert \cdot \rVert_1$	$\max \lVert \cdot \rVert_\infty$
1	0.47	0.42

Table 1 shows that the Takagi-Sugeno system is stable.

6 Conclusions

In the paper a new method for stability analysis of discrete-time Takagi-Sugeno fuzzy systems is presented. Linear state space submodels are supposed in rule consequents of the model of the plant and the controller is described by the linear state feedback submodels (so called PDC controller). The method states a sufficient condition of stability and it is based on the stability analysis of time-varying systems, where the system matrix is varying in given interval. This condition allows to test stability by checking only products of extreme matrices. This condition is simple to implement, but the computational complexity increases exponentially with the the dimension n.

Acknowledgements

This work has been supported by the research program No. J04/98:212300013 "Decision Making and Control for Manufacturing" of the Czech Technical University in Prague (sponsored by the Ministry of Education of the Czech Republic).

References

1. Kuhn H. W., A. Tucker.: Nonlinear Programming, in Proceedings of the 2nd Berkeley Symposium on Mathematical Statistics and Probability, University of California Press, Berkeley, (1951)
2. Bauer P.H., K. Premaratne, J. Durán (1993).: A Necessary and Sufficient Condition for Robust Asymptotic Stability of Time-Variand Discrete Systems, IEEE Transactions on Automatic Control, Vol. **38**, No. **9** (1993) 1427–1430
3. Pytelková R., P. Hušek.: Stability of Slowly Varying Takagi-Sugeno Fuzzy Systems, in Proceedings of IFSA/NAFIPS World Congress, Vancouver, Canada, (2001)
4. Takagi T., M. Sugeno.: Fuzzy Identification of System and Its Applications to Modelling and Control, IEEE Transactions on Systems, Man & Cybernetics, Vol. **15** (1985) 116–132
5. Tanaka K., T. Ikeda, H.O. Wang.: Robust Stabilization of a Class of Uncertain Nonlinear Systems via Fuzzy Control: Quadratic Stabilizability, H_∞ Control Theory, and Linear Matrix Inequalities, IEEE Transactions on Fuzzy Systems, Vol. **4**, No. **1** (1996) 1–13.
6. Hušek, R. Dvořáková.: Robust Stability of Polynomials with Polynomic Structure of Coefficients, in Proceedings of 8th IEEE Mediterranean Conference on Control & Automation, Patras, Greece, (2000)

Fuzzy Control System Using Nonlinear Friction Observer for the Mobile Robot

Woo-Young Lee[1], Il-Seon Lim[2], Uk-Youl Huh[3]

[1] Dept. of Electrical Engineering, Inha University, Incheon 402-751, Korea
woollung@hotmail.com
[2] Dept. of Electrical Engineering, Inha University, Incheon 402-751, Korea
islim@kopo.ac.kr
[3] Dept. of Electrical Engineering, Inha University, Incheon 402-751, Korea
uyhuh@inha.ac.kr

Abstract. In this paper, a fuzzy control system with the nonlinear friction observer is presented. This observer is used to compensate the nonlinear friction that makes the position accuracy in a wheeled mobile robot be limited. Adding the nonlinear friction observer in the system, the tracking accuracy of the robot is improved without any breakaway when it traces the reference trajectory. And experimental results show that the fuzzy controller with the observer has better performance even though it has difference between the right wheel inertia and the left wheel inertia.

1 Introduction

Fuzzy logic control is based on human intelligence and an expert's knowledge and performs well for the system that has nonlinear characteristic. And fuzzy logic control is robust on the change of system parameters and copes with the disturbance that has bad effects on the system [1][2]. Because of these merits, fuzzy logic control is used to control a mobile robot nowadays. A mobile robot uses the sensor information and makes path planning. After that, it carries out path tracking. In the process, the path consists of straight routines and curves. However, due to contour error and direction error, the robot stays easily out of the way. Those errors are caused by some unknown disturbance like nonlinear friction disturbance. In mobile robot, nonlinear friction is the factor which makes stick-slip induced by stiction and Stribeck effect [3]. To reduce these errors, we add a nonlinear friction observer to fuzzy controller. It is necessary to control the robot with high performance and accuracy by proper modeling and friction compensation. In designing the fuzzy controller with an observer, we separate friction torque characteristic into two parts as a positive part and a negative part due to nonlinear characteristic. According to analysis using this observer, we can compensate the errors with the observer that may cause breakaway when the robot moves. Simulation results show that fuzzy controller using the observer has good performance even if the robot has different inertia between two wheels.

P.M.A. Sloot et al. (Eds.): ICCS 2002, LNCS 2331, pp. 613–621, 2002.

2 Nonlinear Friction Observer

It is very important to deal with uncertainty for the system speed and position in the mobile robot. Nonlinear friction causes stick-slip in the very low speed movement and is the original nonlinear factor, which makes tracking error in the path tracking control process due to stiction (breakaway force) and Stribeck effect [3]. In this paper we design the observer to estimate and compensate the friction torque in the 2-wheel mobile robot movement. Using the observer, it is possible to prevent the robot from breakaway and the navigation performance is improved.

2.1 Nonlinear Friction Torque Modeling

The friction with uncertain characteristic is changed by the speed and lubrication condition. The friction modeling [6][7] is described in equation (1).

$$T_f(\omega(t)) = T_v(\omega(t)) + T_c \operatorname{sgn}(\omega(t)) + T_e \qquad (1)$$

T_c is coulomb friction torque. Normally the moving speed is proportional to the friction torque and the positive slope characteristic of the relation between friction torque and speed is shown in region B of Figure 1. Consequently, the friction torque is getting bigger in the very low speed scope. As it moves faster, the friction torque is getting smaller and this effect is called "Stribeck effect". Region A of Figure 1 shows the negative slope characteristic of the relation between friction torque and speed. T_v is nonlinear viscous friction caused by stiction (breakaway force), which appears when the mobile robot starts. T_e is static friction caused by Stribeck effect in the low speed scope and this negative characteristic describes the exponential decrement [4][5]. In Fig. 1, the friction torque modeling type is shown by the characteristic curve between the friction torque and moving speed. In this curve, the friction curve has the positive slope or the negative slope as the speed is changed. And the friction direction is changed with the opposite sign symmetrically as the sign of the speed is changed [6][7].

$$T_f(\omega(t)) = \operatorname{sgn}(\omega) K \left(\alpha_1 |\omega| + \beta_1 + C_1 e^{-(|\omega|/C_2)^{C_3}}\right) \qquad (2)$$

$$\text{Where, } \begin{cases} \operatorname{sgn}(\omega) = 1 & at\ \omega(t) > 0 \\ \operatorname{sgn}(\omega) = -1 & at\ \omega(t) < 0 \end{cases}$$

In equation (2), the parameter α_1 is the torque coefficient of the viscous friction at region B and β_1 is the coulomb friction torque coefficient when the robot starts. C_1, C_2 and C_3 are the negative friction torque coefficients caused by Stribeck effect at region A. K is the gain coefficient for the entire friction. The friction torque function included in system is a nonlinear function of the speed. Depending on the motor direction, the function type is decided and the performance on the speed control is affected. Especially when the low speed and the high-speed navigation are repeated or when the forward and backward navigation are repeated, the control signal and the

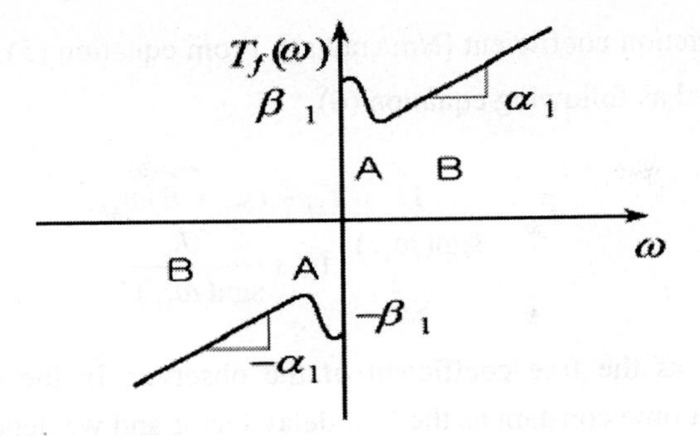

Fig. 1. Model of friction torque

navigation performance are disturbed. The friction torque is also shown as the disturbance that has a bad effect on the system control. In this paper, we consider the relationship between the tracking performance and breakaway in the planned path when the friction torques of the two wheels become unbalanced by stick and slip which are caused by inertia. When the robot turns left in the curve path after the straight movement, the speed of the right wheel increases and the speed of the left wheel decreases. In that case, the right wheel would keep the straight movement by inertia and the friction torque works to the contrary. Thus, stick that restrains the moving speed for the command of the torque increment happens. On the contrary, slip that would increase the speed for the decrement command happens.

2.2 Friction Torque Observer

The dynamic equation of the mechanical system is shown in equation (3).

$$T = J\frac{d\omega}{dt} + \tau_f \qquad (3)$$

After converting equation (3) into Laplace Transformation, we get the equation (4).

$$sJ\omega_M = \tau_M - \tau_f \qquad (4)$$

Where, J is the moment inertia of a system, τ_M is the motor torque, s is the Laplace operation. Assuming that the friction torque τ_f consists of the viscous friction torque and the coulombs friction in equation (4), equation (5) is obtained.

$$\begin{cases} sJ\omega_M = \tau_M - (B\omega_M + \tau_{fc}) \\ \tau_{fc} = \text{sgn}(\omega_M)T_{fc} \\ \text{sgn}(\omega_M) = 1 \quad (\omega_M > 0) \\ \text{sgn}(\omega_M) = 0 \quad (\omega_M < 0) \end{cases} \qquad (5)$$

Where, B is the friction coefficient [Nm/(rad/s)]. From equation (5), let $d\tau_{fc}/dt = 0$ and $\hat{T}_{fc}$ is estimated as following equation (6).

$$\hat{T}_{fc} = \frac{1}{\mathrm{sgn}(\omega_M)} \frac{\tau_M - (sJ + B)\omega_M}{1 + s\dfrac{J}{\mathrm{sgn}(\omega_M)L}} \qquad (6)$$

Where, L is used as the free coefficient of the observer. In the denominator, the coefficient of s is a time constant as the first delay factor and we denote this as K_τ.

$$K_\tau = \frac{J}{\mathrm{sgn}(\omega_M)L} \qquad (7)$$

The sign of K_τ depends on the change of $\mathrm{sgn}(\omega_M)$. For the observer to be stable, it is necessary that the roots of characteristic equation are at the left half plane in the s-plane and choose L to make the condition K_τ>0. The feed forward compensated torque is decided as following equation (8).

$$\hat{\tau}_{fc} = \mathrm{sgn}(\omega_M)\hat{T}_{fc} \qquad (8)$$

The speed control block using the proposed friction torque observer is shown in Fig 2. Where, Q*(s)* is defined in equation (9) and the motor torque τ_M is calculated in equation (10).

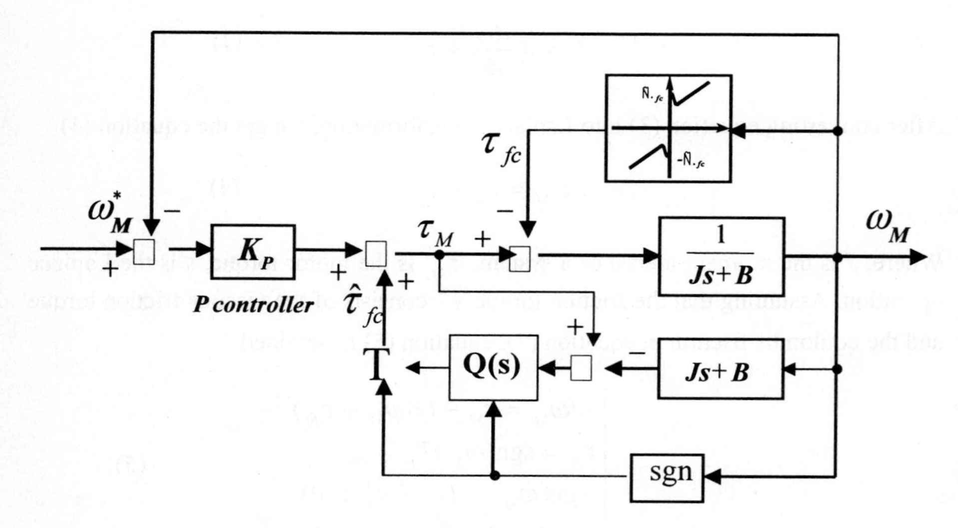

Fig. 2. The proposed friction torque observer

$$Q(s) = \frac{1}{\text{sgn}(\omega_M)} \frac{1}{1 + s\frac{1}{\text{sgn}(\omega_M)L}} \tag{9}$$

$$\tau_M = K_p(\overset{*}{\omega}_M - \omega_M) + \hat{\tau}_{fc} \tag{10}$$

Where, K_p is the proportional speed gain and $\overset{*}{\omega}_M$ is the speed command. The proposed friction torque observer is a kind of the adaptive control and the free coefficient of the observer L is changed as the sign of the moving speed is changed.

3 Fuzzy Control System Modeling

The performance of fuzzy controller is related to fuzzification, rule base, and I/O linguistic parameters. Especially, premise and consequent linguistic values are very important factors in designing the fuzzy controller. And rule bases are decided after a lot of trials and errors. First of all, the direction error e_θ that is the angle between the robot's proceeding direction and the reference path and contour error e_c that is the distance of the perpendicular line from the robot's weight center to the reference path are defined as the premise parameter for path tracking control [8]. And the consequent parameters v_d and v_s are defined in equation (11).

$$v_d = v_R - v_L \tag{11}$$

$$v_s = \frac{v_L + v_R}{2} \tag{12}$$

Where, v_L and v_R are the left and right speed input values respectively. For the fuzzification method, the triangular membership functions are used.

Table 1. Fuzzy rule tables for the nonlinear friction observer

v_d		e_θ NL	NS	Z	PS	PL
	NL	PL	PL	PM	Z	NS
	NS	PL	PM	PS	NS	NM
e_c	Z	PM	PS	Z	NS	NM
	PS	PM	PS	NS	NM	NL
	PL	PS	Z	NM	NL	NL

v_s		e_θ NL	NS	Z	PS	PL
	NL	M	S	M	M	Z
	NS	S	S	M	M	S
e_c	Z	S	S	L	S	S
	PS	S	M	M	S	S
	PL	Z	M	M	S	M

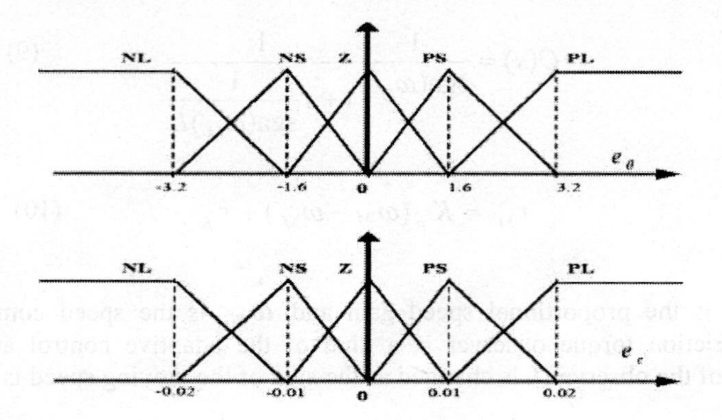

Fig. 3. Premise variable e_θ [deg] and e [m]

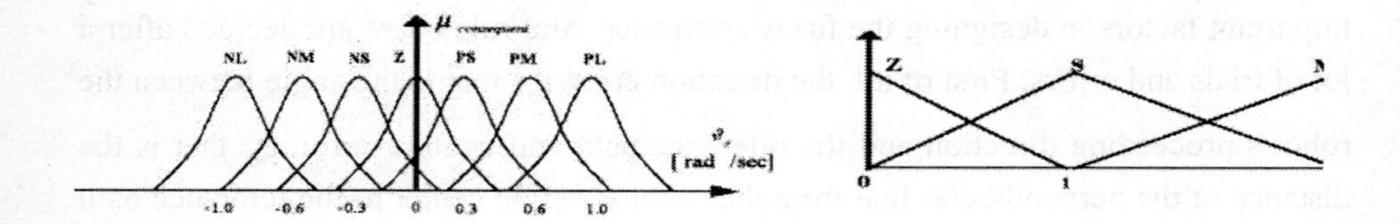

Fig. 4. Consequent variable v_d [rad/sec] and v_s [rad/sec]

Each fuzzy rule base for the nonlinear friction observer consists of 25 rules that are shown in table1. These values in premise variables are decided with considering the distance among sensors. To apply the designed controller, we should make the membership function of premise variables and consequent variables. They are shown in Fig. 3 and Fig. 4. For the direction angle and the moving speed, each rule base is made. That is, even if the inputs are the direction error and the contour error, the outputs are the direction angle and the moving speed with two 2⊑1 structures. For the fuzzy inference, Mamdani's product implication is used, and as t-norm and s-norm operations, algebraic product and max are used respectively. We use a lookup table to reduce calculation amount and time. After the sensor information is checked with this lookup table the control output is generated.

4 Simulation

In this paper, path tracking is simulated using Matlab program. We give the reference

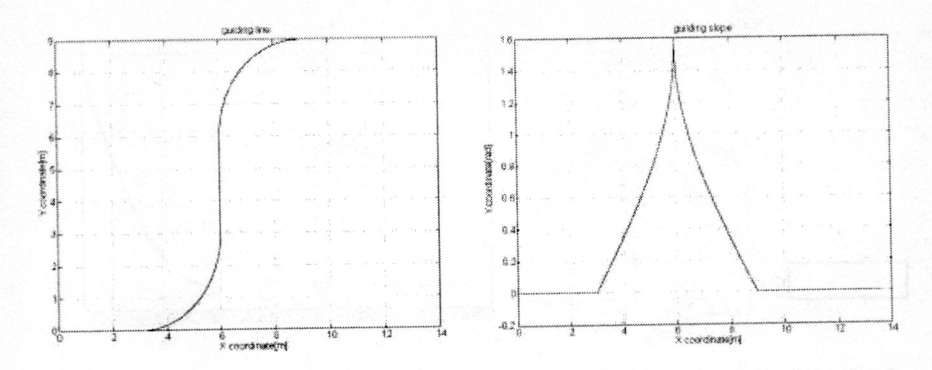

Fig. 5. The left figure shows the reference trajectory and the tangent slope of the trajectory is shown in the left figure

path and make the mobile robot follow that path. And the reference path is given like Fig. 5. We initialize the start position (0,0) and assume that the initial direction is the same as the moving direction.

4.1 Path Tracking without Observer

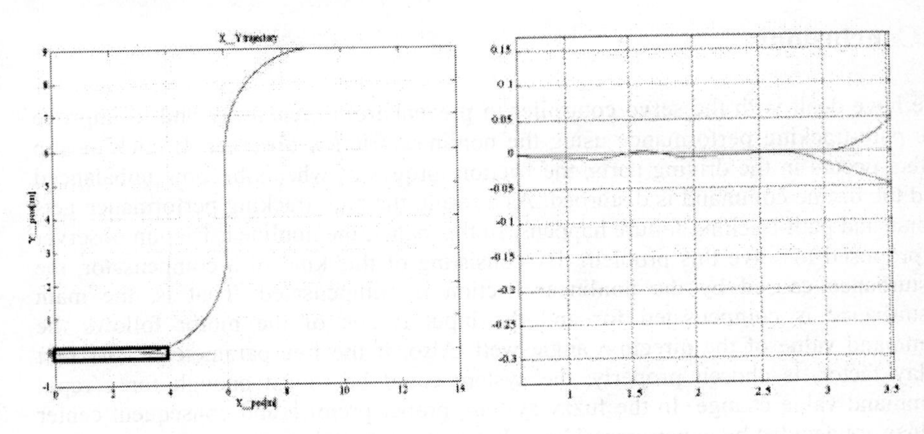

Fig. 6. The left result shows that the robot is out of the reference without using the observer and the right graph is a magnified one

The simulation result in Fig. 6 shows that when there is no difference between the left wheel inertia and the right wheel inertia or when the robot moves in the straight routines, it trace very well without the observer. But if there is some difference, the robot is out of the way as shown in Fig. 6 when it traces curves because the heavy slip effect caused by the nonlinear friction on the right wheel disturbs control commands.

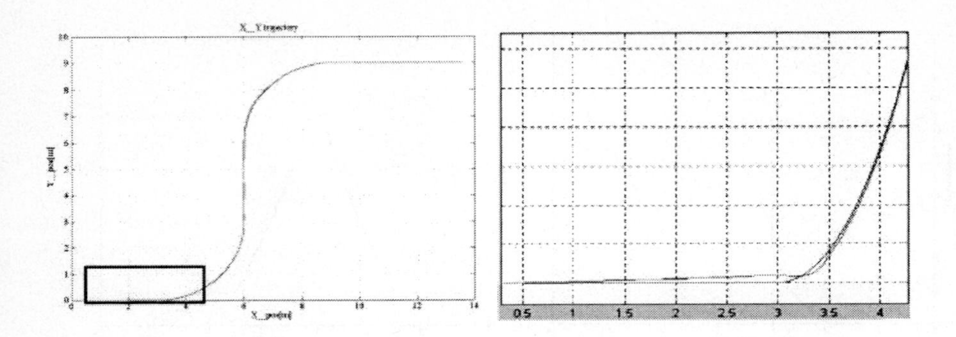

Fig. 7. The left result shows that the robot tracks the reference very well using the observer and the right graph is a magnified one

4.2 Path Tracking with Observer

In Fig. 7, the robot tracks both the straight path and the curve well. Especially when the robot follows the curve, the observer estimates the friction torque even though there is some difference between the left and the right wheel inertia. After that, the controller compensates the error caused by the nonlinear friction properly.

5 Conclusions

We have dealt with the servo controller to prevent from breakaway and to improve the path-tracking performance using the nonlinear friction observer. If stick or slip effect occurs in the driving parts, the friction torques of wheels become unbalanced and the torque command is disturbed. As a result, the path tracking performance gets worse and path-tracking failure happens. In this paper, the nonlinear friction observer is proposed to solve this problem. By consisting of this kind of a compensator, the disturbance caused by the nonlinear friction is compensated. That is, the main disturbance is compensated for and the input torque of the motor follows the command value of the direction angle well. Also, if the free parameter L, the first delay factor, is chosen properly, the system could be robust enough for a rapid command value change. In the fuzzy system, proper premise and consequent center values are decided by experimental knowledge and simulation. The system, therefore, would become stable and could have enhanced performance.

References

1. Sungchul Jee, "Fuzzy logic controls for CNC machine tools", a dissertation submitted in partial fulfillment of the requirements for the degree of Doctor of Philosophy in the University of Michigan, pp. 101-111

2. Kevin M. Passino, Stephen Yurkovich, "Fuzzy Control", Addison-Wesley, 1998.

3. B. Armstrong-Helouvry, P. Dupont, and C.C. de Wit, "A survey of models, analysis

tools and compensation methods for the control of machines with friction." Automatica, vol. 30, no 7, pp. 1083-1138, 1994

4. C. C. de Wit, H. Olsson, K. J. Astorm, and P. Lischinsky. "The new model for control of systems with friction." IEEE Trans. Automat. Contr., vol. 40, pp. 419-425, Mar. 1995.

5. M. Iwasaki and N. Matsui, "Observer-based nonlinear friction compensation in servo drive system." in Proc., IEEE 4th Int. Workshop AMC. 1996, pp. 344-348.

6. Makoto Iwasaki, "Disturbance-Observer-Based Nonlinear Friction Compensation in Table Drive System." IEEE/ASME Trans. on Mechatronics. vol. 4, No.1, March 1999.

7. P. E. Dupont, "Avoiding Stick-Slip through PD Control." IEEE Trans. on Automatic Control, vol. 39, no .5, pp. 1094 -1096, May. 1994.

8. L.Feng, "Cross-Coupling Motion Controller for Mobile Robots" IEEE Control Systems Magazine, Volume: 13 Issue: 6, pp. 35 –43 Dec. 1993.

Efficient Implementation of Operators on Semi-Unstructured Grids

Christoph Pflaum, David Seider
Institut für Angewandte Mathematik und Statistik,
Universität Würzburg
pflaum@mathematik.uni-wuerzburg.de,
seider@mathematik.uni-wuerzburg.de

Abstract. Semi-unstructured grids consist of a large structured grid in the interior of the domain and a small unstructured grid near the boundary. It is explained how to implement differential operators and multigrid operators in an efficient way on such grids. Numerical results for solving linear elasticity by finite elements are presented for grids with more than 10^8 grid points.

1 Introduction

One main problem in the discretization of partial differential equation in 3D is the large storage requirement. Storing the stiffness matrix of a FE-discretization of a system of PDE's on unstructured grids is very expensive even in case of relative small grids. To avoid this problem, one prefers fixed structured grids, since these kind of grids lead to fixed stencils independent of the grid point. Another interesting property of structured grids is that several numerical algorithms can be implemented more efficient on structured grids than on unstructured grids (see [7]). On the other hand, fixed structured grids do not have the geometric flexibility of unstructured grids (see [3], [10], [8], and [9]). Therefore, we apply semi-unstructured grids (see [6] and [5]). These grids consist of a large structured grid in the interior of the domain and a small unstructured grid, which is only contained in boundary cells. Since semi-unstructured grids can be generated automatically for general domains in 3D, they have the geometric flexibility of unstructured grids. In this paper, we explain how to implement discrete differential equations in an efficient way on semi-unstructured grids. It will be shown that algorithms based on semi-unstructured grids require much less storage than on unstructured grids. Therefore, these grids are nearly as efficient as structured grids.

2 Semi-Unstructured Grids

Assume that $\Omega \subset \mathbb{R}^3$, is a bounded domain with a piecewise smooth boundary. Semi-unstructured grids are based on the following infinite structured grids with

P.M.A. Sloot et al. (Eds.): ICCS 2002, LNCS 2331, pp. 622–631, 2002.

meshsize $h > 0$:

$$\Omega_h^\infty := \{(ih, jh, kh) \mid i, j, k \in \mathbb{Z}\} \quad .$$

The set of cells of this grid is

$$\mathcal{Z}_h^\infty := \{[ih, (i+1)h] \times [jh, (j+1)h] \times [kh, (k+1)h] \mid i, j, k \in \mathbb{Z}\} .$$

Theses cells are called *interior*, *exterior*, or *boundary cells.* The classification in these three types of cells has to be done carefully (see [6]). Otherwise, it is very difficult to construct a semi-unstructured grid. Let us denote the set of *boundary cells* by $\mathcal{Z}_h^{\text{boundary}} \subset \mathcal{Z}_h^\infty$ and the set of *interior cells* by $\mathcal{Z}_h^{\text{interior}} \subset \mathcal{Z}_h^\infty$.

To obtain a finite element mesh, we subdivide every interior and every boundary cell by tetrahedra, such that suitable properties are satisfied (see [6]). One of these properties is that the interior angles ϕ of every tetrahedron are smaller than a fixed maximal angle $\phi_{max} < 180°$, which does not depend on the meshsize h and the shape of the domain. This means

$$\phi \leq \phi_{max} < 180° \tag{1}$$

for every interior angle. A suitable subdivision of the boundary cells is described in [5]. Here, we briefly explain the subdivision of the interior cells. To this end, let us mark the corners of an interior cell Z as in Figure 2. Then, this cell is subdivided as follows:

Regular Subdivision

```
Tet. 1: WNT, WND, WST, EST
Tet. 2: EST, WND, WST, ESD
Tet. 3: WND, WSD, WST, ESD
Tet. 4: EST, WND, ESD, END
Tet. 5: ENT, WNT, EST, END
Tet. 6: WNT, WND, EST, END
```

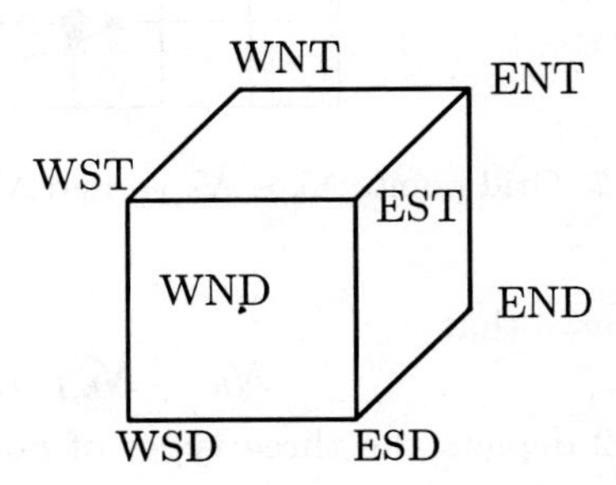

Fig. 1. Subdivision of an Interior Cell Z.

The subdivision τ_h of the interior cells and boundary cells by tetrahedra leads to a discretization domain

$$\Omega_{dis,h} = \bigcup_{\Lambda \in \tau_h} \overline{\Lambda}$$

and a set of nodal points

$$\mathcal{N}_h = \bigcup_{\Lambda \in \tau_h} \mathcal{E}(\Lambda),$$

where $\mathcal{E}(\Lambda)$ is the set of corners of a tetrahedron Λ.

We have to distinguish three types of nodal points:
I. The regular interior points:

$$\mathcal{N}_{h,I} := \{P \in \mathcal{N}_h \mid P \text{ is corner point of eight interior cells.}\}.$$

II. The boundary points and the points contained in boundary cells:

$$\mathcal{N}_{h,II} := \{P \in \mathcal{N}_h \mid P \text{ is not adjacent or contained in an interior cell.}\}.$$

III. The regular points near the boundary:

$$\mathcal{N}_{h,III} := \{P \in \mathcal{N}_h \mid P \text{ is corner point of at least one interior cell and one boundary cell.}\}.$$

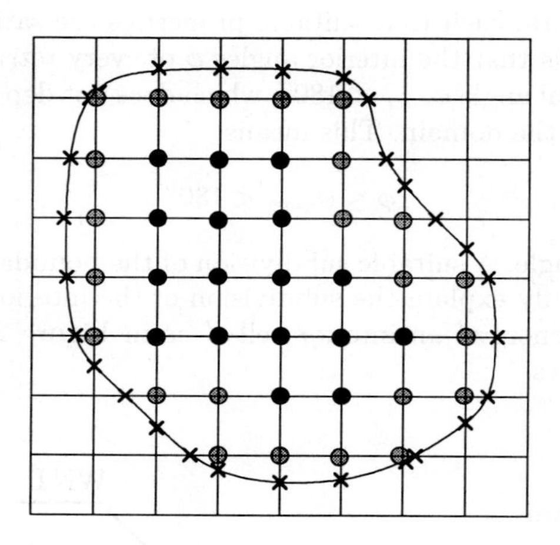

Fig. 2. Grid points $\mathcal{N}_{h,I}$, $\mathcal{N}_{h,II}$, and $\mathcal{N}_{h,III}$ for a semi-unstructured grid in 2D.

Observe that

$$\mathcal{N}_h := \mathcal{N}_{h,I} \cup \mathcal{N}_{h,II} \cup \mathcal{N}_{h,III}.$$

Figure 2 depicts the three types of nodal points for semi-unstructured grid in 2D.

For the discretization of partial differential equations, we apply the finite element space with linear elements on τ_h:

$$V_h := \left\{ v \in H^1(\Omega) \;\middle|\; v\big|_{\Lambda} \text{ is linear on every tetrahedron } \Lambda \in \tau_h \right\}.$$

3 Implementation of Differential Operators with Constant Coefficients

To avoid a large storage requirement, we want to implement operators like discrete differential operators without storing the stiffness matrix or the local stiffness matrices. This always can be done on a finite element grid by a recalculation of the local stiffness matrices when they are need. But such an implementation is very time consuming. On structured grids one can get a more efficient implementation by fixed stencils.

In this section, we explain how to obtain an efficient implementation of discrete differential operators with constant coefficients on semi-unstructured grids. For reasons of simplicity, we restrict ourselves to the discrete differential operator corresponding to the bilinear form

$$(u,v) \to a(u,v) = \int_{\Omega} \nabla u \, \nabla v \, d\mu. \tag{2}$$

To obtain an efficient implementation of the discrete differential operator corresponding to this bilinear form, one has to distinguish three cases:

I. Implementation at regular interior points $M \in \mathcal{N}_{h,I}$.
The eight cells next to the regular interior points $M \in \mathcal{N}_{h,I}$ are interior cells. Therefore, we can implement the operator corresponding to (2) by the following fixed stencil

$$\begin{aligned}
&(20.0 * u_M \\
&-4.0 * (u_T + u_D + u_E + u_W) \\
&-2.0 * (u_N + u_S) \\
&-(u_{ND} + u_{WN} - u_{WT} - u_{ED} + u_{ST} + u_{ES} \\
&-u_{EST} - u_{WND})) * h/3.0
\end{aligned}$$

Here, E means the next point in the east direction: $E = (1,0,0) + M$. The other directions W, N, S, D, T are defined analogously.

II. Implementation at points $P_1 \in \mathcal{N}_{h,II}$.
To calculate a discrete differential operator at a point $P_1 \in \mathcal{N}_{h,II}$, we need the local stiffness matrix of every tetrahedron $T \in \mathcal{T}_h$ adjacent to P_1. These tetrahedra are contained in boundary cells, since $P_1 \in \mathcal{N}_{h,II}$. We do not want to store the local stiffness matrix of T completely and we do not want to recalculate them each time they are needed. Therefore, we use the following approach, which stores 13 values for each tetrahedron $T \in \mathcal{T}_h$ contained in a boundary cell. Using these 13 values one can calculate the local stiffness matrix easily for different bilinear forms. To explain the definition of the 13 values $\xi_0, \xi_{x,1} \ldots, \xi_{x,4}, \xi_{y,1} \ldots, \xi_{y,4}, \xi_{z,1} \ldots, \xi_{z,4}$, let $T \in \mathcal{T}_h$ be a tetrahedron contained in a boundary cell and let us denote P_1, P_2, P_3, P_4 the corners of T. Furthermore, let v_p be the nodal basis function at the corners $p = P_1, P_2, P_3, P_4$ of T and

$$M = \frac{1}{4}(P_1 + P_2 + P_3 + P_4).$$

Then, let

$$\begin{aligned}
\xi_0 &:= \mathrm{vol}(T)/6, \\
\xi_{x,i} &:= \xi_0 \cdot \frac{\partial v_{P_i}}{\partial x}(M), \quad \text{for } i = 1,2,3,4,
\end{aligned}$$

$$\xi_{y,i} := \xi_0 \cdot \frac{\partial v_{P_i}}{\partial y}(M), \quad \text{for } i = 1, 2, 3, 4,$$

$$\xi_{z,i} := \xi_0 \cdot \frac{\partial v_{P_i}}{\partial z}(M), \quad \text{for } i = 1, 2, 3, 4.$$

Using these values, one can calculate the stiffness matrix corresponding to the bilinear form $a(u, v)$. For example, the entry $a(v_{P_1}, v_{P_2})$ of the local stiffness matrix is

$$\int_\Omega \nabla v_{P_1} \nabla v_{P_2} \, d\mu = (\xi_{x,1}\xi_{x,2} + \xi_{y,1}\xi_{y,2} + \xi_{z,1}\xi_{z,2})/(6\xi_0).$$

III. Implementation at points $P_1 \in \mathcal{N}_{h,III}$.
The cells adjacent to a point $P_1 \in \mathcal{N}_{h,III}$ are boundary cells and interior cells. Therefore, we have to distinguish to cases:

a) Let c be a boundary cell adjacent to P_1. Then, the local stiffness matrix of c can be calculated using the local stiffness matrices of the tetrahedra $T \in \tau_h$ contained in c.
b) Let c be an interior cell adjacent to P_1. Then, the local stiffness matrix corresponding to (2) is a fixed 8×8 matrix, which does not depend on c. Therefore, storing this matrix is very inexpensive.

4 Implementation of Operators with Variable Coefficients

Now, let us explain the implementation of discrete differential operators with variable coefficients.

For reasons of simplicity, let us restrict ourselves to the bilinear form

$$(u, v) \to a(u, v) = \int_\Omega b \nabla u \nabla v \, d\mu, \tag{3}$$

where $b \in \mathcal{C}(\Omega)$ and $\min_{p \in \Omega}(b(p)) > 0$. It is well-known, that, in general, one cannot directly implement the differential operator corresponding to a. Therefore, one has to approximate a by a bilinear form a_h. The standard way to approximate a is to apply a numerical integration with one Gauss integration point for every $T \in \tau_h$. Here, we use another approach. To explain this approach, let us classify the tetrahedra of τ_h in the following way:

$$\tau_{h,i} := \{T \in \tau_h \mid T \text{ is contained in an interior cell}\}$$
$$\tau_{h,b} := \{T \in \tau_h \mid T \text{ is contained in a boundary cell}\}.$$

Obviously, it is $\tau_h = \tau_{h,i} \cup \tau_{h,b}$. Let us define an interpolation operator I_h as follows:

$$I_h : \mathcal{C}(\Omega) \to L^\infty(\Omega)$$
$$I_h(b)(p) := b(p) \quad \text{if } p \text{ is the middle point of an interior cell.}$$
$$I_h(b)(p) := b(p) \quad \text{if } p \text{ is the middle point of a tetrahedron } T \in \tau_{h,b}.$$
$$I_h(b) \quad \text{is constant on every interior cell.}$$
$$I_h(b) \quad \text{is constant an every tetrahedron } T \in \tau_{h,b}.$$

Now, let the approximation of a be the following bilinear form

$$(u,v) \to a_h(u,v) = \int_\Omega I_h(b)\, \nabla u\, \nabla v\, d\mu$$

By the Bramble-Hilbert Lemma and by some calculations one can show (see [2] or [1]),

$$\begin{aligned} |a(u,v) - a_h(u,v)| &\le Ch\|b\|_{C^1}\|u\|_{H^1}\|v\|_{H^1} \\ |a(u,v) - a_h(u,v)| &\le Ch^2\|b\|_{C^2}\|u\|_{H^2}\|v\|_{H^2} \\ |v|^2_{H^1} &\le Ca_h(v,v). \end{aligned}$$

Using these inequalities one can prove an $O(h^{2-\delta})$ convergence with respect to the L^2-norm for the finite element solution corresponding to the bilinear form (3) for every $\delta > 0$ (see [6] and [4]) . The advantage of the above construction of a_h is that one can implement the differential operator corresponding to a_h similar to the construction in section 3.

I. Implementation at regular interior points $M \in \mathcal{N}_{h,I}$.
The discrete differential operator corresponding to (3) can be implemented by the following stencil at points $M \in \mathcal{N}_{h,I}$

$$\begin{aligned}
&(b_{cellWSD} * (3.0 * u_M - u_W - u_S - u_D)+\\
&b_{cellESD} * (5.0 * u_M - 3.0 * u_D - u_E - u_S - u_{ES} + u_{ED})+\\
&b_{cellWND} * (7.0 * u_M - 3.0 * (u_W + u_D) - u_N - u_{ND} - u_{WN} + 2.0 * u_{WND})+\\
&b_{cellEND} * (5.0 * u_M - 3.0 * u_E - u_N - u_D - u_{ND} + u_{ED})+\\
&b_{cellWST} * (5.0 * u_M - 3.0 * u_W - u_T - u_S - u_{ST} + u_{WT})+\\
&b_{cellEST} * (7.0 * u_M - 3.0 * (u_E + u_T) - u_S - u_{ST} - u_{ES} + 2.0 * u_{EST})+\\
&b_{cellWNT} * (5.0 * u_M - 3.0 * u_T - u_N - u_W - u_{WN} + u_{WT})+\\
&b_{cellENT} * (3.0 * u_M - u_E - u_N - u_T)\\
&) * h/6.0.
\end{aligned}$$

Here, $cellWSD$ means the middle point of the next cell point of M in the west-south-down direction.

II. and III. Implementation at points $P_1 \in \mathcal{N}_{h,II} \cup \mathcal{N}_{h,III}$.
The discrete differential operator at points $P_1 \in \mathcal{N}_{h,II} \cup \mathcal{N}_{h,III}$ can be implemented for variable coefficients similar to the case of constant coefficients. To explain this, let $\mathcal{M}_{const}$ be the local stiffness matrix for a constant coefficient 1.0 at a tetrahedron $T \in \tau_{h,b}$ or an interior cell c. Then, $I_h(b)$ is a constant value β on this tetrahedron T or interior cell c, respectively. Therefore, the local stiffness matrix $\mathcal{M}_{var}$ in case of variable coefficients is

$$\mathcal{M}_{var} = \beta \mathcal{M}_{const}.$$

The above implementation of discrete differential operators for variable coefficients is not much more time and storage consuming than for constant coefficients. To explain this, one has to study the implementation at regular interior points $P_1 \in \mathcal{N}_{h,I}$, since the computational time at these points dominates the total computational time on semi-unstructured grids. Counting the number of operations shows that the above discrete differential operator for variable coefficients requires $71/19 \approx 3.7$ more operations than in case of constant coefficients. Since, the computational time often is strongly influenced by problems with a small cache in the computer architecture, the computational time will increase by a factor smaller than 3.7 (see numerical results in section 6).

5 Implementation of Multigrid Operators and Coarse Grid Operators

For reasons of simplicity, let us assume that $\Omega \subset]0,1[^3$ and that $h = 2^{-n}$. A multigrid algorithm is based on a sequence of fine and coarse grids

$$\mathcal{N}_h = \mathcal{N}_n, \mathcal{N}_{n-1}, \cdots, \mathcal{N}_2, \mathcal{N}_1.$$

To define these grids, we recursively define sets of coarse grid cells:

$$\begin{aligned}
\mathcal{Z}_n^{\text{boundary}} &:= \{c \in \mathcal{Z}_{2^{-n}}^{\infty} \mid c \text{ is a boundary cell on the finest grid.}\} \\
\mathcal{Z}_k^{\text{boundary}} &:= \{c \in \mathcal{Z}_{2^{-k}}^{\infty} \mid \text{there exists a cell } c_b \in \mathcal{Z}_{k+1}^{\text{boundary}} \text{ such that } c_b \subset c\} \\
\mathcal{Z}_n^{\text{interior}} &:= \{c \in \mathcal{Z}_{2^{-n}}^{\infty} \mid c \text{ is an interior cell on the finest grid.}\} \\
\mathcal{Z}_k^{\text{interior}} &:= \{c \in \mathcal{Z}_{2^{-k}}^{\infty} \mid \text{there exist 8 cells } c_i \in \mathcal{Z}_{k+1}^{\text{interior}} \text{ such that } c_i \subset c\}
\end{aligned}$$

Furthermore, we define the maximal and minimal coarse grid

$$\begin{aligned}
\mathcal{N}_{k,\max} &:= \{p \mid p \text{ is a corner of a cell } c \in \mathcal{Z}_k^{\text{interior}} \cup \mathcal{Z}_k^{\text{boundary}}\}, \\
\mathcal{N}_{k,\min} &:= \{p \mid p \text{ is a corner of exactly 8 cells } c \in \mathcal{Z}_k^{\text{interior}} \cup \mathcal{Z}_k^{\text{boundary}}\}.
\end{aligned}$$

The coarse grid $\mathcal{N}_k$ has to be constructed such that the following inclusions are satisfied:

$$\mathcal{N}_{k,\min} \subset \mathcal{N}_k \subset \mathcal{N}_{k,\max}.$$

Here, we do not want to explain in detail how to choose $\mathcal{N}_k$. But let us mention, that the construction of $\mathcal{N}_k$ depends on the boundary conditions. The coarse grid $\mathcal{N}_k$ is the union of two disjoint types of grids:

$$\begin{aligned}
\mathcal{N}_k &= \mathcal{N}_{I,k} \cup \mathcal{N}_{II,k}, \\
\mathcal{N}_{I,k} &= \{p \mid p \text{ is a corner of exactly 8 cells } c \in \mathcal{Z}_k^{\text{interior}}\}.
\end{aligned}$$

To obtain a multigrid algorithm, we need a function space V_k on the coarse grid $\mathcal{N}_k$ such that

$$\begin{aligned}
V_1 \subset V_2 \subset \ldots \subset V_n \\
\dim(V_k) = |\mathcal{N}_k|.
\end{aligned}$$

On the interior coarse cells, this function space is given in a natural way by finite elements. On the boundary coarse grid cells one has to construct the function space V_k such that the boundary conditions of the fine grid space are preserved. A detailed description of the construction of V_k will be given in a subsequent paper. The coarse spaces V_k lead to natural restriction and prolongation operators R_k and P_k and to coarse grid differential operators L_k. The relation between these operators is

$$L_k = R_k L_{k+1} P_{k+1}.$$

At the grid points $\mathcal{N}_{I,k}$, the restriction and prolongation operators are operators with fixed stencils.

For the implementation of the coarse grid differential operators, we distinguish to cases:

I. Implementation at regular interior points $P \in \mathcal{N}_{I,k}$.

- If the fine grid operator L_h is an operator with constant coefficients, then the corresponding coarse grid differential operator L_k at interior points $P \in \mathcal{N}_{I,k}$ is

  $$L_k(P) = L_{2^{-k}}(P).$$

  Therefore, we do not need additional storage to evaluate this operator.
- If the fine grid operator L_h is an operator with variable coefficients, then one has to restrict the local stiffness matrices and one has to store the local stiffness matrices on all coarse cells. A differential operator can be evaluated using the local coarse grid stiffness matrices.

II. Implementation at points near the boundary $P \in \mathcal{N}_{II,k}$.

- If the fine grid operator L_h is an operator with constant coefficients, then the local stiffness matrices only at the coarse cells $\mathcal{Z}_k^{\text{boundary}}$ have to be stored.
- If the fine grid operator L_h is an operator with variable coefficients, then the local stiffness matrices at all coarse cells $\mathcal{Z}_k^{\text{boundary}} \cup \mathcal{Z}_k^{\text{interior}}$ have to be stored.

To store a local stiffness matrix, one has to store 8×8 values. Since these local stiffness matrices are stored only on coarse grids, storing the local stiffness matrices costs only a small percentage of the total storage.

6 Numerical results

In this section, we present two kinds of numerical results.

Numerical result 1: Differential operator with variable coefficients. Table 1 shows the computational time for the evaluation of one discrete Laplace operator on a Sun Ultra 1 workstation. t_c is the computational time of the operator with a constant coefficient corresponding to the following bilinear form

$$a(u, v) = \int_\Omega \nabla u \nabla v \, d\mu.$$

t_v is the computational time of the operator with a variable coefficient $\alpha(x, y, z)$ corresponding to the bilinear form

$$a(u, v) = \int_{\Omega} \nabla u \nabla v \ \alpha(x, y, z) \ d\mu.$$

One can see that the computational time increases only by a factor less than 2 in case of a variable coefficient.

Table 1. Parallel solution of linear elasticity equation.

grid size	t_c in sec	t_v in sec	t_v/t_c in sec
21 247	0.23	0.31	1.33
154 177	1.24	1.98	1.59
1 164 700	8.92	16.1	1.81

Numerical result 2: Multigrid for linear elasticity with traction free (Neumann) boundary conditions. We want to solve the linear elasticity equations with traction free boundary conditions. The solver is a cg-solver with a multigrid algorithm as a preconditioner. The multigrid algorithm is a stable solver even for traction free boundary conditions, since the coarse grid space is constructed in such a way that it contains the rigid body modes. Table 2 shows the computational time for one cg-iteration with multigrid preconditioning with respect to the number of grid points. The number of unknowns is the number of grid points multiplied by 3. The calculations were done on ASCI Pacific Blue.

Table 2. Parallel solution of linear elasticity equation.

processors	time in sec	grid size	number of unknowns
600	120	121 227 509	363 682 527
600	26	15 356 509	46 069 527
88	18	1 973 996	5 921 988
98	3.8	259 609	778 827
88	1.1	35 504	106 512
12	19.6	259 609	778 827
4	8.9	35 504	106 512

Acknowledgment. The author Christoph Pflaum would like to thank the Center for Applied Scientific Computing for supporting his research during his stay at Lawrence Livermore National Laboratory.

References

1. S. C. Brenner and L. R. Scott. *The Mathematical Theory of Finite Element Methods.* Springer, New York, Berlin, Heidelberg, 1994.
2. Ch. Großmann and H.-G. Roos. *Numerik partieller Differentialgleichungen.* Teubner, Stuttgart, 1992.
3. W.D. Henshaw. Automatic grid generation. *Acta Numerica*, 5:121–148, 1996.
4. C. Pflaum. Discretization of second order elliptic differential equations on sparse grids. In C. Bandle, J. Bemelmans, M. Chipot, J. Saint Jean Paulin, and I. Shafrir, editors, *Progress in partial differential equations, Pont-á-Mousson 1994, vol. 2, calculus of variations, applications and computations*, Pitman Research Notes in Mathematics Series. Longman, June 1994.
5. C. Pflaum. The maximum angle condition of semi-unstructured grids. In *Proceedings of the Conference: Finite Element Methods, Three-dimensional Problems (Jyväskylä, June 2000)*, Math. Sci. Appl., pages 229–242. GAKUTO Internat. Series, Tokio, 2001. to appear.
6. C. Pflaum. Semi-unstructured grids. *Computing*, 67(2):141–166, 2001.
7. L. Stals, U. Rde, C. Wei, and H. Hellwagner. Data local iterative methods for the efficient solution of partial differential equations. In *In proceedings of the eighth biennial computational techniques and applications conference, Adelaide, Australia*, 1997.
8. J.F. Thompson. *Numerical Grid Generation.* Amsterdam: North–Holland, 1982.
9. C. Yeker and I. Zeid. Automatic three-dimensional finite element mesh generation via modified ray casting. *Int. J. Numer. Methods Eng.*, 38:2573–2601, 1995.
10. M. A. Yerry and M. S. Shephard. Automatic three-dimensional mesh generation by the modified-octree technique. *Int. J. Numer. Methods Eng.*, 20:1965–1990, 1984.

hypre: A Library of High Performance Preconditioners

Robert D. Falgout and Ulrike Meier Yang

Center for Applied Scientific Computing, Lawrence Livermore National Laboratory, P.O.Box 808, L-560 Livermore, CA 94551

Abstract. *hypre* is a software library for the solution of large, sparse linear systems on massively parallel computers. Its emphasis is on modern powerful and scalable preconditioners. *hypre* provides various conceptual interfaces to enable application users to access the library in the way they naturally think about their problems. This paper presents the conceptual interfaces in *hypre*. An overview of the preconditioners that are available in *hypre* is given, including some numerical results that show the efficiency of the library.

1 Introduction

The increasing demands of computationally challenging applications and the advance of larger more powerful computers with more complicated architectures have necessitated the development of new solvers and preconditioners. Since the implementation of these methods is quite complex, the use of high performance libraries with the newest efficient solvers and preconditioners becomes more important for promulgating their use into applications with relative ease.

hypre has been designed with the primary goal of providing users with advanced scalable parallel preconditioners. Issues of robustness, ease of use, flexibility and interoperability have also been very important. It can be used both as a solver package and as a framework for algorithm development. Its object model is more general and flexible than the current generation of solver libraries [7].

hypre also provides several of the most commonly used solvers, such as conjugate gradient for symmetric systems or GMRES for nonsymmetric systems to be used in conjunction with the preconditioners.

Design innovations have been made to enable application users access to the library in the way that they naturally think about their problems. For example, applications developers that use structured grids, typically think of their problems in terms of stencils or grids. *hypre*'s users do not have to learn complicated sparse matrix structures; instead *hypre* does the work of building these data structures through various conceptual interfaces. The conceptual interfaces currently implemented include stencil-based structured/semi-structured interfaces, a finite-element based unstructured interface, and a traditional linear-algebra based interface.

P.M.A. Sloot et al. (Eds.): ICCS 2002, LNCS 2331, pp. 632–641, 2002.

The first part of this paper describes these interfaces and the motivations behind their design. The second part gives an overview of the preconditioners that are currently in the library with brief descriptions of the algorithms and some highlights of their performance characteristics. Since space is limited, it is not possible to describe the algorithms in detail, but various references are included for those who are interested in further information. The paper concludes with some remarks on additional software and improvements of already existing codes that are planned to be included in *hypre* in the future.

2 Conceptual Interfaces

Each application to be implemented lends itself to natural ways of thinking of the problem. If the application uses structured grids, a natural way of formulating it would be in terms of grids and stencils, whereas for an application that uses unstructured grids and finite elements it is more natural to access the preconditioners and solvers via elements and element stiffness matrices. Consequently the provision of various interfaces facilitates the use of the library.

Conceptual interfaces also decrease the coding burden for users. The most common interface used in libraries today is a linear-algebraic one. This interface requires that the user compute the mapping of their discretization to row-column entries in a matrix. This code can be quite complex, e.g. consider the problem of ordering the equations and unknowns on the composite grids used in structured adaptive mesh refinement (SAMR) codes. The use of a conceptual interface merely requires the user to input the information that defines the problem to be solved, leaving the forming of the actual linear system as a library implementation detail hidden from the user.

Another reason for conceptual interfaces, maybe the most compelling one, is that they provide access to a large array of powerful scalable linear solvers that need the extra information beyond just the matrix. For example, geometric multigrid (GMG) can not be used through a linear-algebraic interface, since it is formulated in terms of grids.

Similarly, in many cases, these interfaces allow the use of other data storage schemes with less memory overhead and provide for more efficient computational kernels.

Fig. 1 illustrates the idea of conceptual interfaces. On the left are specific interfaces with algorithms and data structures that take advantage of more specific information. On the right are more general interfaces, algorithms and data structures. Note that the more specific interfaces also give users access to general solvers like algebraic multigrid (AMG) or incomplete LU factorization (ILU). The top row shows various concepts: structured grids, composite grids, unstructured grids or just plain matrices. In the second row, various solvers/ preconditioners are listed. Each of those requires different information from the user, which is provided through the conceptual interfaces. Geometric multigrid, e.g., needs a structured grid and can only be used with the left most interface, AMGe [2], an algebraic multigrid method, needs finite element information, whereas

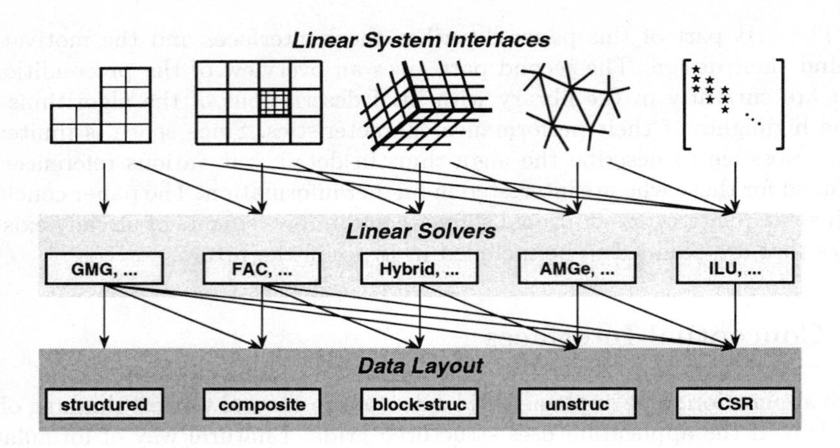

Fig. 1. Graphic illustrating the notion of conceptual interfaces.

general solvers can be used with any interface. The bottom row contains a list of data layouts or matrix/vector storage schemes that can be used for the implementation of the various algorithms. The relationship between linear solver and storage scheme is similar to that of interface and linear solver.

hypre currently supports four conceptual interfaces: a structured-grid system interface, a semi-structured-grid system interface, a finite-element interface and a linear-algebraic interface.

Note that *hypre* does not partition the problem, but builds the internal parallel data structures (often quite complicated) according to the partitioning of the application that the user provides.

2.1 Structured-Grid System Interface (Struct)

This interface is appropriate for scalar applications whose grids consists of unions of logically rectangular grids with a fixed stencil pattern of nonzeros at each grid point. It also enables users access to *hypre*'s most efficient scalable solvers for scalar structured-grid applications, such as the geometric multigrid methods SMG and PFMG. See also Sections 3.1 and 3.2. The user defines the stencil and the grid; the right hand side and the matrix are then defined in terms of the stencil and the grid.

2.2 Semi-Structured-Grid System Interface (SStruct)

This interface is appropriate for applications whose grids are mostly structured, but with some unstructured features, e.g. block structured grids (such as shown in Fig. 2), composite grids in structured adaptative mesh refinement (AMR) applications, and overset grids. It additionally allows for more general PDEs than the Struct interface, such as multiple variables (system PDEs) or multiple variable types (e.g. cell centered, face centered, etc.). The user needs to define stencils, grids, a graph that connects the various components of the final grid, the right hand side and the matrix.

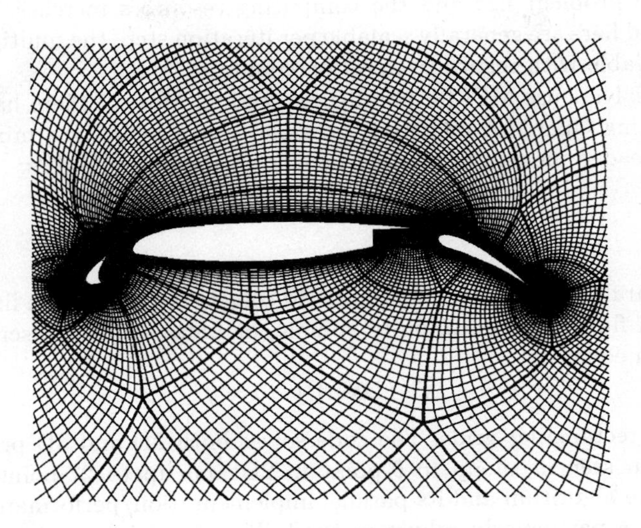

Fig. 2. An example block-structured grid, distributed across many processors.

2.3 Finite Element Interface (FEI)

This is appropriate for users who form their systems from a finite element discretization. The interface mirrors typical finite element data structures, including element stiffness matrices. Though this interface is provided in *hypre* , its definition was determined elsewhere [8]. This interface requires the definition of the element stiffness matrices and element connectivities. The mapping to the data structure of the underlying solver is then performed by the interface.

2.4 Linear-Algebraic System Interface (IJ)

This is the traditional linear-algebraic interface. The user needs to define the right hand side and the matrix in the general linear-algebraic sense, i.e. in terms of row and column indices. This interface provides access only to the most general data structures and solvers and as such should only be used when none of the grid-based interfaces is applicable.

3 Preconditioners

This section gives an overview of the preconditioners currently available in *hypre* via the conceptual interfaces. *hypre* also provides solvers to be used in conjunction with the preconditioners such as Jacobi, conjugate gradient and GMRES.

Great efforts have been made to generate highly efficient codes. Of particular concern has been the scalability of the solvers. Roughly speaking, a method is scalable if the time required to produce the solution remains essentially constant as both the problem size and the computing resources increase. All methods implemented here are generally scalable per iteration step, the multigrid methods are also scalable with regard to iteration count.

All the solvers use MPI for parallel processing. Most of them have also been threaded using OpenMP, making it possible to run *hypre* in a mixed message-passing/threaded mode, of potential benefit on clusters of SMPs.

3.1 SMG

SMG is a parallel semicoarsening multigrid solver targeted at the linear systems arising from finite difference, finite volume, or finite element discretizations of the diffusion equation

$$\nabla \cdot (D \nabla u) + \sigma u = f \tag{1}$$

on logically rectangular grids. The code solves both 2D and 3D problems with discretization stencils of up to 9-point in 2D and up to 27-point in 3D. For details on the algorithm and its parallel implementation/performance see [21, 3, 10]. SMG is a particularly robust method. The algorithm semicoarsens in the z-direction and uses plane smoothing. The xy plane solves are effected by one V-cycle of the 2D SMG algorithm, which semicoarsens in the y-direction and uses line smoothing

3.2 PFMG

PFMG is a parallel semicoarsening multigrid solver similar to SMG. It is described in detail in [1, 10]. PFMG uses simple pointwise smoothing instead of plane smoothing. As a result, it is less robust than SMG, but more efficient per V-cycle. The largest run with PFMG as a preconditioner for conjugate gradient was applied to a problem with 1 billion unknowns on 3150 processors of the ASCI Red computer and took only 54 seconds. Recently we added a PFMG solver for systems of PDEs available through the semi-structured interface.

3.3 *BoomerAMG*

BoomerAMG is a parallel implemenation of algebraic multigrid. It requires only the linear system. *BoomerAMG* uses two types of parallel coarsening strategies. The first one, refered to as RS-based coarsening, is based on the highly sequential coarsening strategy used in classical AMG [20]. To obtain parallelism, each processor coarsens independently, followed by various strategies for dealing with the processor boundaries. Obviously, this approach depends on the number of processors and on the distribution of the domain across processors. The second type of coarsening, called CLJP-coarsening [9], is based on parallel maximum independent set algorithms [19, 16] and generates a processor independent coarsening. CLJP-coarsening has proven to be more efficient for truly unstructured grids, whereas RS-based coarsenings lead to better results on structured problems. For more detailed information on the implementation of the CLJP coarsening scheme see [11]. For a general description of the coarsening schemes and the interpolation used within *BoomerAMG* as well as various numerical results, see [12].

BoomerAMG provides classical pointwise smoothers, such as weighted Jacobi relaxation, a hybrid Gauß-Seidel/ Jacobi relaxation scheme and its symmetric variant. It also provides more expensive smoothers, such as overlapping Schwarz smoothers, as well as access to other methods in *hypre* such as ParaSails, PILUT and Euclid. These smoothers have shown to be effective for certain problems for which pointwise smoothers have failed, such as elasticity problems [22].

BoomerAMG can also be used for solving systems of PDEs if given the additional information on the multiple variables per points. The function or 'unknown' approach coarsens each physical variable separately and interpolates only within variables of the same type. By exploiting the system nature of the problem, this approach often leads to significantly improved performance, lower memory usage and better scalability. See Table 1 which contains results for a structured 2-dimensional elasticity problem on the unit square, run on the ASCI Blue Pacific computer.

Table 1. Test results for a 2-dimensional model elasticity problem

grid size	# of procs.	scalar *BoomerAMG* time (# of its.)	systems *BoomerAMG* time (# of its)
80×80	1	42.4(58)	4.1 (8)
160×160	4	130.4(112)	6.3 (9)
320×320	16	317.5(232)	8.6(10)
640×640	64	1238.2(684)	14.4(13)

Table 2 contains results for a 3-dimensional elasticity problem on a thin plate with a circular hole in its center. The problem has 215,055 variables and was run on 16 processors of the ASCI White computer. The results show that for this problem *BoomerAMG* as a solver is not sufficient, but it does make an effective preconditioner.

3.4 ParaSails

ParaSails is a parallel implementation of a sparse approximate inverse preconditioner. It approximates the inverse of A by a sparse matrix M by minimizing the Frobenius norm of $I - AM$. It uses graph theory to predict good sparsity patterns for M. ParaSails has been shown to be an efficient preconditioner for many problems, particularly since the minimization of the Frobenius norm of $I - AM$ can be decomposed into minimization problems for the individual rows of $I - AM$,

Table 2. Test results for an elasticity problem

Solvers	# of its.	total time in secs.
scaled CG	1665	34.8
ParaSails-CG	483	26.6
scalar *BoomerAMG*	n.c.	-
scalar *BoomerAMG*-CG	53	28.9
systems *BoomerAMG*	78	40.6
systems *BoomerAMG*-CG	19	12.3

leading to a highly parallel algorithm. A detailed description of the algorithm can be found in [4] and implementation details in [5]. Particular emphasis has been placed on a highly efficient implementation that incorporates special, more efficient treatment of symmetric positive definite matrices and load balancing. The end result is a code that has a very scalable setup phase and iteration steps. See Table 3, which shows test results for ParaSails applied to the 3-dimensional constant coefficient anisotropic diffusion problem $0.1u_{xx} + u_{yy} + 10u_{zz} = 1$ with Dirichlet boundary conditions. The local problem size is $60 \times 60 \times 60$. Unlike multigrid, convergence is not linearly scalable, and the number of iterations will increase as the problem size increases. However, ParaSails is a general purpose solver and can work well on problems where multigrid does not.

Table 3. Scalability of ParaSails with increasing problem size (216,000 per proc.)

# of procs	# of its.	setup time	solve time	time per it.
1	107	12.1	75.3	0.70
8	204	13.8	247.9	1.22
64	399	15.4	536.6	1.34
216	595	15.8	856.4	1.44
512	790	17.4	1278.8	1.62
1000	979	17.1	1710.7	1.75

3.5 PILUT

PILUT is a parallel preconditioner based on Saad's dual-threshold incomplete factorization algorithm. It uses a thresholding drop strategy as well as a mechanism to control the maximum size of the ILU factors. It uses the Schur-complement approach to generate parallelism. The original code was written by Karypis and Kumar for the T3D [18]. This version differs from the original version in that it uses MPI and more coarse-grain parallelism.

3.6 Euclid

Euclid is a scalable implementation of the Parallel ILU algorithm. It is best thought of as an "extensible ILU preconditioning framework", i.e. Euclid can support many variants of ILU(k) and ILUT preconditionings. Currently it supports Block Jacobi ILU(k) and Parallel ILU(k) methods. Parallelism is obtained via local and global reorderings of the matrix coefficients. A detailed description of the algorithms can be found in [14,15].

Euclid has been shown to be very scalable with regard to setup time and triangular solves. Fig. 3 shows results for a 5 point 2D convection diffusion problem with 256 × 256 unknowns per processor.

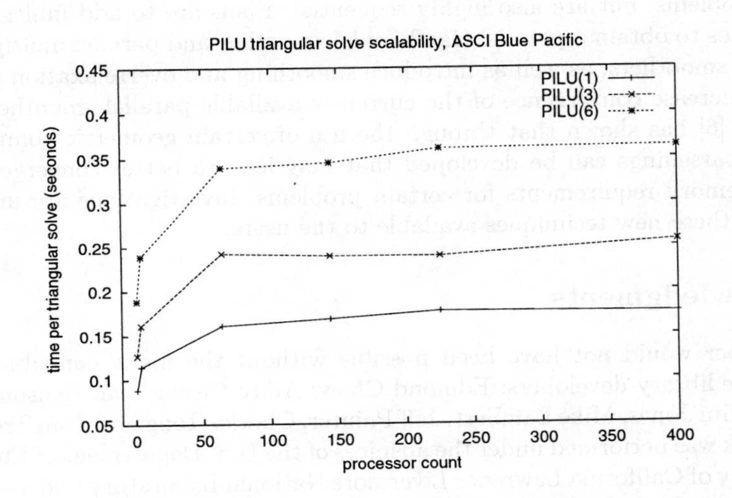

Fig. 3. Some scalability results for Euclid

4 Additional Information

The *hypre* library can be downloaded by visiting the *hypre* home page at the URL `http://www.llnl.gov/CASC/hypre`. It can be built by typing `configure` followed by `make`. There are several options that can be used with configure. For information on how to use those, one needs to type `configure --help`. Although *hypre* is written in C, it can also be called from Fortran. More specific information on *hypre* and how to use it can be found in the users manual and the reference manual, which are also available at the same URL.

5 Conclusions and Future Work

Overall, *hypre* contains a variety of highly efficient preconditioners and solvers, available via user-friendly conceptual interfaces. Nevertheless, it is a project in progress. As new research leads to better and more efficient algorithms, new preconditioners will be added and old preconditioners will be improved.

On the list of new codes to be made available shortly is AMGe, an algebraic multigrid method based on the use of local finite element stiffness matrices [2, 17]. This method has proven to be more robust and to converge faster than classical AMG for some problems, e.g. elasticity problems. This code will be available directly through the FEI interface.

Various improvements are planned for *BoomerAMG*. Classical Gauß-Seidel relaxation as well as multiplicative Schwarz smoothers are some of the numerically most efficient methods, i.e. they lead to good convergence for AMG for some problems, but are also highly sequential. Plans are to add multi-coloring techniques to obtain a parallel Gauß-Seidel smoother and parallel multiplicative Schwarz smoothers, as well as introduce smoothing and overrelaxation parameters to increase convergence of the currently available parallel smoothers. New research [6] has shown that through the use of certain geometric components, better coarsenings can be developed that may lead to better convergence and lower memory requirements for certain problems. Investigations are underway to make these new techniques available to the users.

Acknowledgments

This paper would not have been possible without the many contributions of the *hypre* library developers: Edmond Chow, Andy Cleary, Van Henson, David Hysom, Jim Jones, Mike Lambert, Jeff Painter, Charles Tong and Tom Treadway. This work was performed under the auspices of the U.S. Department of Energy by University of California Lawrence Livermore National Laboratory under contract No. W-7405-Eng-48.

References

1. Ashby, S., Falgout, R.: A parallel multigrid preconditioned conjugate gradient algorithm for groundwater flow simulations. Nuclear Science and Engineering **124** (1996) 145–159
2. Brezina, M., Cleary, A., Falgout, R., Henson, V., Jones, J., Manteuffel, T., McCormick, S, Ruge, J.: Algebraic multigrid based on element interpolation (AMGe). SIAM J. Sci. Comput. **22** (2000) 1570–1592
3. Brown, P., Falgout, R., Jones, J.: Semicoarsening multigrid on distributed memory machines. SIAM J. Sci. Comput. **21** (2000) 1823-1834
4. Chow, E.: A priori sparsity patterns for parallel sparse approximate inverse preconditioners. SIAM J. Sci. Comput. **21** (2000) 1804-1822
5. Chow, E.: Parallel implementation and practical use of sparse approximate inverses with a priori sparsity patterns. Int'l J. High Perf. Comput. Appl. **15** (2001) 56–74
6. Chow, E.: An unstructured multigrid method based on geometric smoothness. submitted to Num. Lin. Alg. Appl. Also available as Lawrence Livermore National Laboratory technical report UCRL-JC-145075 (2001)
7. Chow, E., Cleary, A., Falgout, R.: Design of the *hypre* preconditioner library. In Henderson, M., Anderson, C., Lyons, S., eds: Proc. of the SIAM Workshop on Object Oriented Methods for Inter-operable Scientific and Engineering Computing (1998) SIAM Press

8. Clay, R. et al.: An annotated reference guide to the Finite Element Interface (FEI) specification, version 1.0. Technical Report SAND99-8229, Sandia National Laboratories, Livermore, CA (1999)
9. Cleary, A., Falgout, R., Henson, V., Jones, J.: Coarse-grid selection for parallel algebraic multigrid. in Proc. of the 5th Intern. Sympos. on Solving Irregularly Structured Problems in Parallel, Lecture Notes in Computer Science **1457** (1998) 104–115
10. Falgout, R., Jones, J.: Multigrid on massively parallel architectures. In Dick, E., Riemslagh, K., and Vierendeels, J., eds: Multigrid Methods VI, Lecture Notes in Computational Science and Engineering, vol. **14** (2000) 101–107, Berlin. Springer
11. Gallivan, K., Yang, U. M.: Efficiency issues in parallel coarsening schemes. LLNL technical report (2001)
12. Henson, V. E., Yang, U. M.: *BoomerAMG*: a parallel algebraic multigrid solver and preconditioner. To appear in Applied Numerical Mathemaitics. Also available as LLNL technical report UCRL-JC-133948 (2000)
13. Henson, V.E., Vassilevski, P.: Element-free AMGe: General algorithms for computing interpolation weights in AMG. to appear in SIAM J. Sci. Comput. Also available as LLNL technical report UCRL-JC-139098
14. Hysom, D., Pothen, A.: Efficient parallel computation of ILU(k) preconditioners. SC99, ACM (1999), CDROM, ISBN #1-58113-091-0, ACM Order #415990, IEEE Computer Society Press Order # RS00197
15. Hysom, D., Pothen, A.: A scalable parallel algorithm for incomplete factor preconditioning. SIAM J. Sci. Comput. **22** (2001) 2194–2215
16. Jones, M., Plassman, P.; A parallel graph coloring heuristic. SIAM J. Sci. Comput. **14** (1993) 654–669
17. Jones, J., Vassilevski, P.: AMGe based on element agglomeration. to appear in SIAM J. Sci. Comput. Also available as LLNL technical report UCRL-JC-135441
18. Karypis, G., Kumar, V.: Parallel threshold-based ILU factorization. Technical Report 061 (1998) University of Minnesota, Department of Computer Science/ Army HPC Research Center, Minneapolis, MN
19. Luby, M.: A simple parallel algorithm for the maximal independent set problem. SIAM J. on Computing **15** (1986) 1036–1053
20. Ruge, J., Stüben, K.: Algebraic Multigrid (AMG). in McCormick, S., ed. Multigrid Methods, Frontiers in Applied Mathematics vol. **3** (1987) 73–130, SIAM, Philadelphia
21. Schaffer, S.: A semi-coarsening multigrid method for elliptic partial differential equations with highly discontinuous and anisotropic coefficients. SIAM J. Sci. Comput. **20** (1998) 228–242
22. Yang, U. M.: On the use of Schwarz smoothing in AMG. 10th Copper Mt. Conf. Multigrid Meth.. Also available as LLNL technical report UCRL-VG-142120 (2001)

Data Layout Optimizations for Variable Coefficient Multigrid*

Markus Kowarschik[1], Ulrich Rüde[1] and Christian Weiß[2]

[1] Lehrstuhl für Systemsimulation (Informatik 10), Institut für Informatik, Universität Erlangen–Nürnberg, Germany
[2] Lehrstuhl für Rechnertechnik und Rechnerorganisation (LRR-TUM), Fakultät für Informatik, Technische Universität München, Germany

Abstract. Efficient program execution can only be achieved if the codes respect the hierarchical memory design of the underlying architectures; programs must exploit caches to avoid high latencies involved with main memory accesses. However, iterative methods like multigrid are characterized by successive sweeps over data sets, which are commonly too large to fit in cache.
This paper is based on our previous work on data access transformations for multigrid methods for constant coefficient problems. However, the case of variable coefficients, which we consider here, requires more complex data structures.
We focus on data layout techniques to enhance the cache efficiency of multigrid codes for variable coefficient problems on regular meshes. We provide performance results which illustrate the effectiveness of our layout optimizations in conjunction with data access transformations.

1 Introduction

There is no doubt about the fact that the speed of computer processors has been increasing and will even continue to increase much faster than the speed of main memory components. As a general consequence, current memory chips based on DRAM technology cannot provide the data to the CPUs as fast as necessary. This memory bottleneck often results in significant idle periods of the processors and thus in very poor code performance compared to the theoretically available peak performances.

To mitigate this effect modern computer architectures use cache memories in order to store data that are frequently used by the CPU (one to three levels of cache are common). Caches are usually based on SRAM chips which, on the one hand, are much faster than DRAM components, but, on the other hand, have rather small capacities for both technical and economical reasons [7].

From a theoretical point of view multigrid methods are among the most efficient algorithms for the solution of large systems of linear equations. They

* This research is being supported in part by the *Deutsche Forschungsgemeinschaft* (German Science Foundation), projects Ru 422/7–1,2,3.

P.M.A. Sloot et al. (Eds.): ICCS 2002, LNCS 2331, pp. 642–651, 2002.

belong to the class of iterative schemes. This means that the underlying data set, which in general is very large, must be processed repeatedly. Efficient execution can only be achieved if the algorithm respects the hierarchical structure of the memory subsystem including main memory, caches and the processor registers, especially by the order of memory accesses [14]. Unfortunately, today's compilers are still far away from automatically applying cache optimizations to codes such complex as multigrid. Therefore much of this optimization effort is left to the programmer.

Semantics–maintaining cache optimization techniques for constant coefficient problems on structured grids have been studied extensively in our *DiME*[1] project [9, 16]. Our previous work primarily focuses on data access transformation which improve temporal locality. With multigrid methods for variable coefficient problems a reasonable layout of the data structures, which implies both high spatial locality and low cache interference, becomes more important. Thus this paper focuses on data layout optimization techniques for variable coefficient multigrid on structured meshes. We investigate and demonstrate the effectiveness of the data access transformations in conjunction with our data layout optimization techniques. Of course, it is not always appropriate or even possible to use such regular grids. Complex geometries, for instance, may require the use of irregular meshes. Thus our techniques must be seen as efficient building blocks which motivate the use of regular grid structures whenever this appears reasonable.

First considerations of data locality optimizations for iterative methods have been published by Douglas [3] and Rüde [14]. Their ideas initiated our *DiME* project [9, 16] as well as other research [1, 15]. All techniques are mainly based on data access transformation techniques like loop fusion and tiling for multigrid methods on structured grids. More recent work [4, 8] also focuses on techniques for multigrid on unstructured meshes. Keyes et al. have applied data layout optimization and data access transformation techniques to other iterative methods [6]. Genius et al. have proposed an automatable method to guide array merging for stencil–based codes based on a meeting graph method [5]. Tseng et al. have recently demonstrated how tile size and padding size selection can be automated for computational kernels in three dimensions [13].

This paper is organized as follows. In Section 2 we consider data structures and data layout strategies for stencil–based computations on arrays. Furthermore we explain array padding as an additional data layout optimization technique. Section 3 discusses performance results for data access optimizations in conjunction with various data layouts on several machines. Finally Section 4 summarizes our results and draws some final conclusions.

2 Data Layout Optimizations

As our optimization target we choose a multigrid V–cycle correction algorithm, which is based on a 5–point discretization of the differential operator, and assume Dirichlet boundaries. Consequently each inner node is connected to four

[1] **D**ata–**l**ocal **i**terative **ME**thods for the efficient solution of PDEs

neighboring nodes. We use a red/black Gauss–Seidel smoother, full–weighting to restrict the fine–grid residuals, and linear interpolation to prolongate the coarse–grid corrections.

2.1 Data Storage Schemes

The linear system of equations is written as $Au = f$, the number of equations is denoted by n. The linear equation for a single inner grid point $i, 1 \leq i \leq n$ reads as: $so_i u_{so(i)} + we_i u_{we(i)} + ce_i u_i + ea_i u_{ea(i)} + no_i u_{no(i)} = f_i$. In the case of a constant coefficient problem an iterative method only needs to store five floating–point values besides the unknown vector u and the right–hand side f.

For variable coefficient problems, however, five coefficients must be stored for each grid point. Hence, the memory required for the coefficients outnumbers the storage requirements for the vectors u and f. There is a variety of data layouts for storing the unknown vector u, the right–hand side f, and the bands of the matrix A. In the following we will investigate three different schemes.

- *Equation–oriented storage scheme:* For each equation the solution, the right–hand side and the coefficients are stored adjacently as shown in Figure 1. This data layout is motivated by the structure of the linear equations.
- *Band–wise storage scheme:* The vectors u and f are kept in separate arrays. Furthermore, the bands of A are stored in separate arrays as well. This rather intuitive data layout is illustrated in Figure 2.
- *Access–oriented storage scheme:* The vector u is stored in a separate array. For each grid point i, the right–hand side f_i and the five corresponding coefficients so_i, we_i, ce_i, ea_i, and no_i are stored adjacently, as illustrated in Figure 3.

While the access–oriented storage scheme does not seem intuitive, it is motivated by the architecture of cache memories. Whenever an equation is being relaxed, its five coefficients and its right–hand side are needed. Therefore it is reasonable to lump these values in memory such that cache lines contain data which are needed simultaneously. This *array merging* technique [7] thus enhances spacial locality.

In the following, we will investigate the performance of a variable coefficient multigrid code which is written in C and uses double precision floating–point numbers. In our standard version one red/black Gauss–Seidel iteration is implemented as a first sweep over all red nodes and a second sweep over all black nodes. Cache–aware smoothing methods will be discussed in Section 3. Figure 4 shows the resulting MFLOPS rates for the three data layouts on different architectures. The finest grid comprises 1025 nodes in each dimension[2]. We use Poisson's equation as our model problem.

[2] Our experiments have been performed on a Compaq XP 1000 (A21264, 500 MHz, Compaq Tru64 UNIX V4.0E, Compaq cc V5.9), a Digital PWS 500au (A21164, 500 MHz, Compaq Tru64 UNIX V4.0D, Compaq cc V5.6), and a Linux PC (AMD Athlon, 700 MHz, gcc V2.35). On all platforms the compilers were configured to perform a large set of compiler optimizations.

It is obvious that the access–oriented storage scheme leads to the best performance on each platform. This validates our above considerations concerning its locality behavior. Moreover, except for the Compaq XP 1000, the equation–oriented technique yields higher execution speeds than the band–wise storage scheme as long as array padding is not introduced. This is due to the fact that the band–wise layout is highly sensitive to *cache thrashing*, see Section 2.2.

It is remarkable that, for example, the access–oriented data layout yields about 60 MFLOPS on the Compaq XP 1000 machine. This corresponds to 6% of the theoretically available peak performance of approximately 1 GFLOPS. The results for the A21164–based Digital PWS 500au and for the Athlon–based PC are even worse, since — according to the vendors — these three machines provide the same theoretical peak performances.

2.2 Array Padding

The performance of numerically intensive codes often suffers from *cache conflict misses*. These misses occur as soon as the associativity of the cache is not large enough. As a consequence, data that are frequently used may evict each other from the cache [12], causing cache thrashing. This effect is very likely in the case of stencil–based computations where the relative distances between array entries remain constant in the course of the passes through the data set. It is particularly severe as soon as the grid dimensions are chosen to be powers of 2, which is often the case for multigrid codes. In many cases *array padding* can help to avoid cache thrashing: the introduction of additional array entries, which are never accessed during the computation, changes the relative distances of the array elements and therefore eliminates cache conflict misses.

The automatic introduction of array padding to eliminate cache conflict misses is an essential part of today's compiler research [12]. However, current techniques to determine padding sizes are based on heuristics which do not lead to optimal results in many cases. It is thus a common approach to run large test suites in order to determine appropriate padding sizes [17].

Figure 5 shows the performance of a multigrid code based on band–wise data storage scheme for a variety of intra– and inter–array paddings on a Digital PWS 500au. The finest grid comprises 1025 nodes in each dimension. If no padding is applied (this situation corresponds to the origin $(0,0)$ of this graph), poor performance results due to severe cache thrashing effects between the arrays holding bands of the matrix [11].

Our experiments have shown that the application of array padding hardly influences the execution times obtained for the equation–oriented storage scheme. This can be explained by the inherent inefficiency of this data layout: whenever an unknown is relaxed, the approximations corresponding to its four neighboring grid nodes are needed. Since, for each unknown, the approximative solution, the coefficients and the right–hand side are lumped in memory (Figure 1), most of the data which are loaded into the cache are not used immediately. This is particularly true for the coefficients and the right–hand side corresponding to the southern neighbor of the current grid point. It is likely that these data are

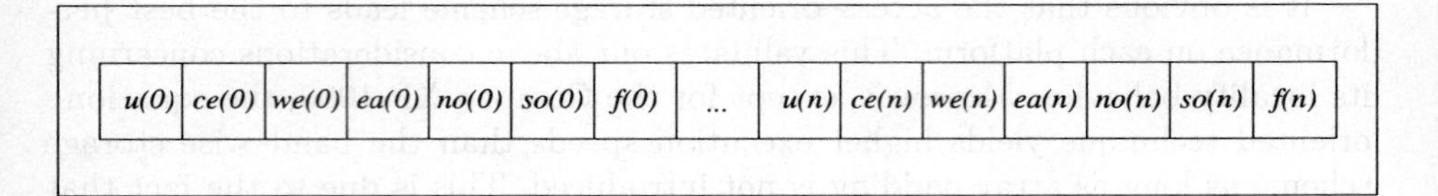

Fig. 1. Equation–oriented storage scheme.

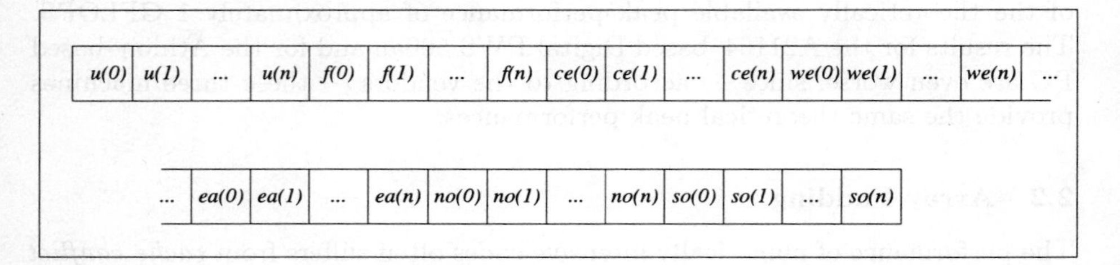

Fig. 2. Band–wise storage scheme.

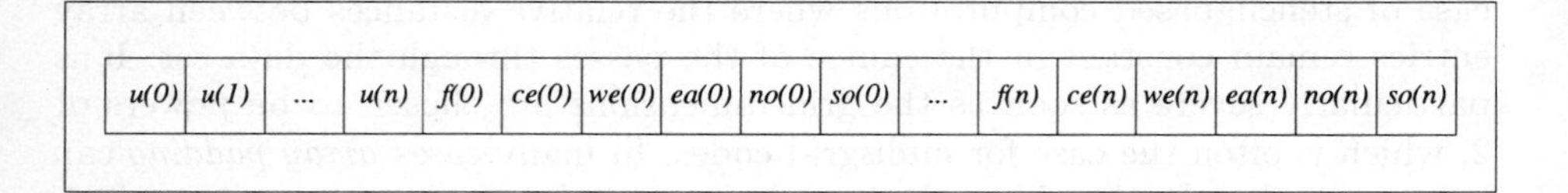

Fig. 3. Access–oriented storage scheme.

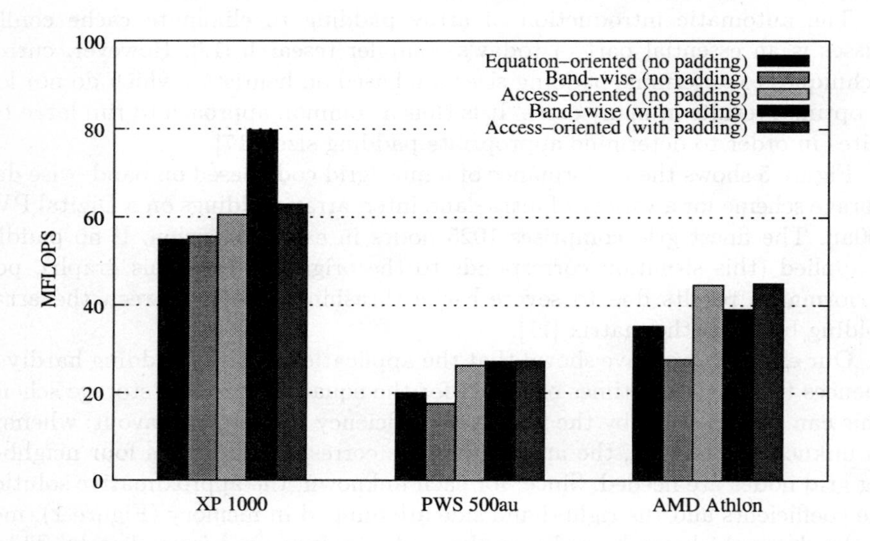

Fig. 4. CPU times for the multigrid codes based on different data layouts with and without array padding.

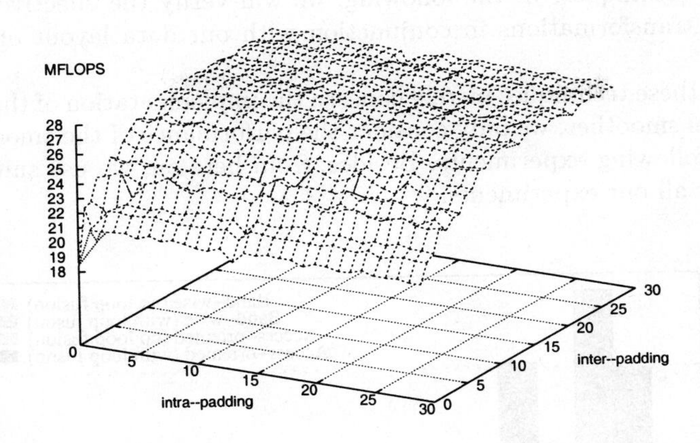

Fig. 5. MFLOPS rates for a multigrid code with different padding sizes using the band–wise storage scheme on a Digital PWS 500au .

evicted from the cache before they will be reused in the course of the next iteration. Consequently, the equation–oriented data layout poorly exploits spatial locality and will therefore no longer be considered here.

On all machines under consideration — the introduction of appropriate array paddings implies lower execution times if the band–wise storage scheme is used. The sensitivity of the code efficiency on the padding sizes mainly depends on the cache characteristics of the underlying machine; e.g., on the degrees of associativity. Detailed profiling experiments exhibit that, particularly for the Digital PWS 500au, the L1 miss rate and the L2 miss rate are reduced by more than 40% and 30%, respectively, as soon as padding is applied suitably to the band–wise data layout.

The third observation is that the performance of the multigrid code which employs the access–oriented storage scheme is always better or at least close to the performance for the band–wise data layout and, moreover, rather insensitive to array padding. Measurements using *PCL* [2] reveal that the cache miss rates almost remain constant. Therefore, as long as neither the programmer nor the compiler introduce array padding, this must be regarded as an advantage of the access–oriented storage scheme.

3 Data Access Optimizations

Data access transformations have been shown [9, 16] to be able to accelerate the red/black Gauss–Seidel smoother for constant coefficient problems by a multiple. Since the smoother is by far the most time–consuming part of a multigrid method this also leads to a significant speedup of the whole algorithm. The optimization techniques described extensively in [16] include the *fusion*, *1D blocking*, and *2D blocking* techniques. In the following, we will verify the effectiveness of the data access transformations in conjunction with our data layout optimization techniques.

Since all these techniques merely concern the implementation of the red/black Gauss–Seidel smoother, we only consider the performance of the smoothing routine in the following experiments. Besides, from here on, we use suitable array paddings for all our experiments.

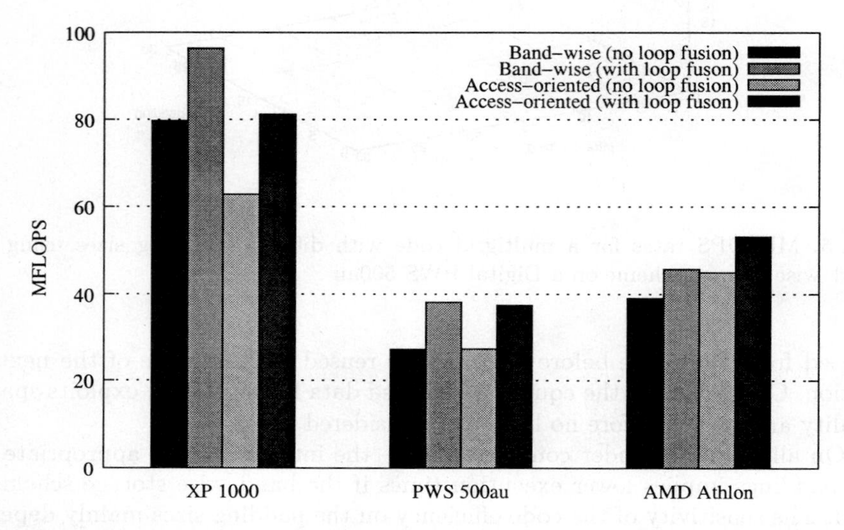

Fig. 6. MFLOPS rates for the smoothing routine based on different data layouts with and without loop fusion.

Figure 6 shows MFLOPS rates for the red/black Gauss–Seidel smoother on a square grid with 1025 nodes in each dimension. Again, we consider the efficiency of our codes on various platforms, with and without introducing the loop fusion technique. Both the band–wise storage scheme and the access–oriented data layout are taken into account. The efficiency for both Alpha–based machines still benefits from the application of loop fusion, whereas the performance gain on the Athlon–based PC is only marginal. This is due to the fact that the L2 cache of this processor has a capacity of 512 KB, which turns out to be too small to keep a sufficient number of neighboring grid lines, each of which containing 1025 nodes.

The same argument applies in the case of the 1D blocking. The application of the 1D blocking technique does not significantly enhance the performance of our smoothing routine further on the Athlon–based PC (Figure 7). However, both Alpha–based architectures have an additional off–chip cache of 4 MB, and, as a consequence, they benefit from blocking two ($m = 2$) or even four ($m = 4$) Gauss–Seidel iterations into one single pass through the grid.

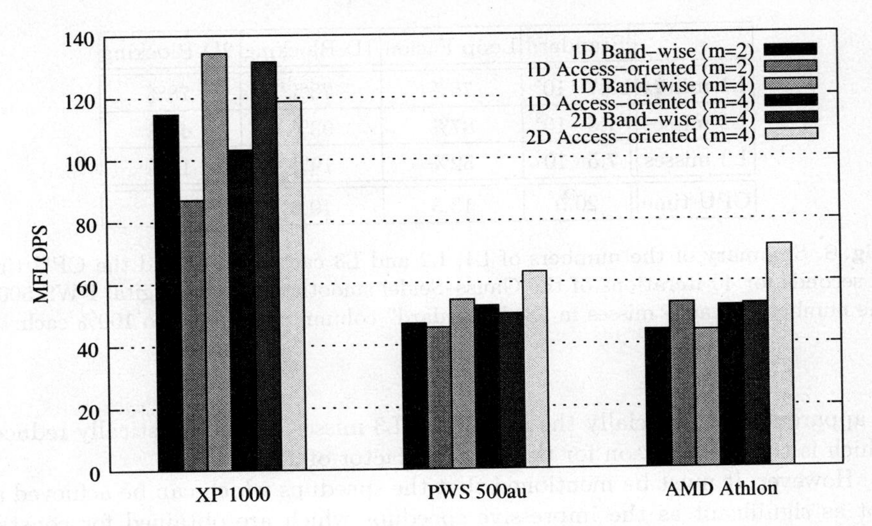

Fig. 7. MFLOPS rates for the smoothing routine based on different data layouts with 1D blocking ($m = 2, 4$) and 2D blocking ($m = 4$).

The situation, however, is different in the case of the 2D blocking technique. Figure 7 shows the performance after applying the 2D blocking technique to the red/black Gauss–Seidel smoother of our multigrid code. Four ($m = 4$) Gauss–Seidel iterations have been blocked into a single sweep over the grid, thus enhancing the reuse of cache contents and reducing the number of cache misses. The most important observation is that not only for both Alpha–based machines with the large off–chip caches, but also for the PC with only two levels of smaller caches, the MFLOPS rates can drastically be increased.

Consider for instance the MFLOPS rates for the AMD Athlon machine in Figure 6, which have been obtained by introducing the loop fusion technique. This comparison shows that, if the access–oriented storage scheme is used, the application of the 2D blocking technique can raise the MFLOPS rate by another 70%.

Varying the grid sizes reveals that, for smaller grids, 1D blocking leads to better performances than 2D blocking. The reason for this is that, if the grid lines are small enough, a sufficient number of them can be kept in cache, and 1D blocking causes efficient reuse of data in cache. If, however, the grid lines are getting larger, not enough of them can be stored in cache, and thus the additional overhead caused by the 2D blocking approach is over–compensated by the performance gain due to a higher cache reuse.

Figure 8 summarizes the influence of our optimizations on the cache behavior and the resulting execution times of our red/black Gauss–Seidel smoothers on the A21164–based Digital PWS 500au, again using a square grid with 1025 nodes in each dimension. The results for the optimization techniques loop fusion, 1D blocking and 2D blocking are based on the use of appropriate array paddings. It

	Standard	Loop Fusion	1D Blocking	2D Blocking
L1 misses	$3.7 \cdot 10^8$	76%	73%	78%
L2 misses	$1.5 \cdot 10^8$	87%	93%	41%
L3 misses	$7.5 \cdot 10^7$	52%	19%	16%
CPU time	20.0	13.3	10.0	8.0

Fig. 8. Summary of the numbers of L1, L2 and L3 cache misses and the CPU times in seconds for 40 iterations of the Gauss–Seidel smoothers on the Digital PWS 500au, the numbers of cache misses in the "Standard" column correspond to 100% each.

is apparent that especially the number of L3 misses can be drastically reduced, which is the main reason for the speedup factor of 2.5.

However, it must be mentioned that the speedups which can be achieved are not as significant as the impressive speedups which are obtained for constant coefficient codes, see e.g. [16]. This is due to the higher memory traffic required by variable coefficient codes. Nevertheless, our techniques yield speedup factors of 2 to 3.

4 Conclusions

In order to achieve efficient code execution it is inevitable to respect the hierarchical memory designs of today's computer architectures. We have presented optimization techniques which can enhance the performance of stencil–based computations on array data structures. Both data layout transformations and data access transformations can help to enhance the temporal and spatial locality of numerically intensive codes and thus their cache performance. We have shown that the choice of a suitable data layout — including the introduction of appropriate array padding — is crucial for efficient execution. This has been demonstrated for a variety of platforms.

Our research clearly illustrates the inherent performance penalties caused by the enormous gap between CPU speed — in terms of MFLOPS rates — and the speed of main memory components — in terms of access latency and memory bandwidth — and the resulting high potential for optimization.

Our experiments motivate the research on new numerical algorithms which can exploit deep memory hierarchies more efficiently than conventional iterative schemes. Therefore, our future work will focus on the development, performance analysis and optimization of patch–adaptive multigrid methods [10], which are characterized by a high inherent potential of data locality.

References

1. F. Bassetti, K. Davis, and D. Quinlan, *Temporal Locality Optimizations for Stencil Operations within Parallel Object–Oriented Scientific Frameworks on*

Cache–Based Architectures, in Proc. of the International Conf. on Parallel and Distributed Computing and Systems, Las Vegas, Nevada, USA, Oct. 1998, pp. 145–153.

2. R. BERRENDORF, *PCL — The Performance Counter Library: A Common Interface to Access Hardware Performance Counters on Microprocessors (Version 2.0)*, Forschungszentrum Juelich GmbH, Germany, `http://www.fz-juelich.de/zam/PCL`, Sept. 2000.
3. C. DOUGLAS, *Caching in With Multigrid Algorithms: Problems in Two Dimensions*, Parallel Algorithms and Applications, 9 (1996), pp. 195–204.
4. C. DOUGLAS, J. HU, M. KOWARSCHIK, U. RÜDE, AND C. WEISS, *Cache Optimization for Structured and Unstructured Grid Multigrid*, Electronic Transactions on Numerical Analysis, 10 (2000), pp. 21–40.
5. D. GENIUS AND S. LELAIT, *A Case for Array Merging in Memory Hierarchies*, in Proceedings of the 9th Workshop on Compilers for Parallel Computers (CPC'01), Edinburgh, Scotland, June 2001.
6. W. GROPP, D. KAUSHIK, D. KEYES, AND B. SMITH, *High Performance Parallel Implicit CFD*, Parallel Computing, 27 (2001), pp. 337–362.
7. J. L. HENNESSY AND D. A. PATTERSON, *Computer Architecture — A Quantitative Approach*, Morgan Kaufmann Publishers, second ed., 1996.
8. J. HU, *Cache Based Multigrid on Unstructured Grids in Two and Three Dimensions*, PhD thesis, Department of Mathematics, University of Kentucky, 2000.
9. M. KOWARSCHIK, U. RÜDE, C. WEISS, AND W. KARL, *Cache–Aware Multigrid Methods for Solving Poisson's Equation in Two Dimensions*, Computing, 64 (2000), pp. 381–399.
10. H. LÖTZBEYER AND U. RÜDE, *Patch–Adaptive Multilevel Iteration*, BIT, 37 (1997), pp. 739–758.
11. H. PFÄNDER, *Cache-optimierte Mehrgitterverfahren mit variablen Koeffizienten auf strukturierten Gittern*, Master's thesis, Department of Computer Science, University of Erlangen–Nuremberg, Germany, 2000.
12. G. RIVERA AND C.-W. TSENG, *Data Transformations for Eliminating Conflict Misses*, in Proceedings of the 1998 ACM SIGPLAN Conference on Programming Language Design and Implementation (PLDI'98), Montreal, Canada, June 1998.
13. G. RIVERA AND C.-W. TSENG, *Tiling Optimizations for 3D Scientific Computation*, in Proceedings of the ACM/IEEE SC00 Conference, Dallas, Texas, USA, Nov. 2000.
14. U. RÜDE, *Iterative Algorithms on High Performance Architectures*, in Proceedings of the EuroPar97 Conference, Lecture Notes in Computer Science, Springer, Aug. 1997, pp. 26–29.
15. S. SELLAPPA AND S. CHATTERJEE, *Cache–Efficient Multigrid Algorithms*, in Proceedings of the 2001 International Conference on Computational Science (ICCS 2001), vol. 2073 and 2074 of Lecture Notes in Computer Science, San Francisco, California, USA, May 2001, Springer, pp. 107–116.
16. C. WEISS, W. KARL, M. KOWARSCHIK, AND U. RÜDE, *Memory Characteristics of Iterative Methods*, in Proceedings of the ACM/IEEE SC99 Conference, Portland, Oregon, Nov. 1999.
17. R. C. WHALEY AND J. DONGARRA, *Automatically Tuned Linear Algebra Software*, in Proceedings of the ACM/IEEE SC98 Conference, Orlando, Florida, USA, Nov. 1998.

gridlib: Flexible and Efficient Grid Management for Simulation and Visualization*

Frank Hülsemann[1], Peter Kipfer[2], Ulrich Rüde[1], and Günther Greiner[2]

[1] System Simulation Group of the Computer Science Department, Friedrich-Alexander University Erlangen-Nuremberg, Germany, frank.huelsemann@cs.fau.de,
[2] Computer Graphics Group of the Computer Science Department, Friedrich-Alexander University Erlangen-Nuremberg, Germany, kipfer@cs.fau.de

Abstract. This paper describes the *gridlib* project, a unified grid management framework for simulation and visualization. Both, adaptive PDE-solvers and interactive visualization toolkits, have to manage dynamic grids. The *gridlib* meets the similar but not identical demands on grid management from the two sides, visualization and simulation. One immediate advantage of working on a common grid is the fact that the visualization has direct access to the simulation results, which eliminates the need for any form of data conversion. Furthermore, the *gridlib* provides support for unstructured grids, the re-use of existing solvers, the appropriate use of hardware in the visualization pipeline, grid adaptation and hierarchical hybrid grids. The present paper shows how these features have been included in the *gridlib* design to combine run-time efficiency with the flexibility necessary to ensure wide applicability. The functionality provided the *gridlib* helps to speed up program development for simulation and visualization alike.

1 Introduction

This article gives an overview of the *gridlib*[1] grid management project, its aims and the corresponding design choices [5], [6], [7], [8]. The *gridlib* combines grid manipulation requirements of mesh based PDE-solvers and visualization techniques into one single framework (library). Thus it offers developers of simulation programs a common interface for the computational and graphical parts of a project.

For interactive computer graphics, the efficient manipulation of grids and the data attached has always been important. In the numerical PDE community, it is the development of adaptive h-refinement algorithms in several space dimensions that led to recognising grid management as a worthwhile task in its own right.

* This project is funded by a KONWIHR grant of the Bavarian High Performance Computing Initiative.

[1] This is a temporary name. Choices for the final name of the whole project are currently being considered.

P.M.A. Sloot et al. (Eds.): ICCS 2002, LNCS 2331, pp. 652–661, 2002.

Despite the shared need to administer dynamically changing grids, there seems to be little joint work. Many PDE-packages, such as deal.II[2] or Overture[3] for example, include tools for the visualization of the results. However, these graphics components are usually tied to the solver part of the package and as such, they are too specific to be widely applicable. Likewise, although the numerous visualization libraries available, such as AVS[4] or VTK[5] for example, obviously display gridded data, they delegate the work of integrating the visualization into the solver to the solver developers. This assumes that an integration is possible at all, which is not obvious, given that some toolkits modify the submitted data for optimisation purposes.

The *gridlib* is a joint effort of three groups to exploit the synergy offered by one common grid management. The development is shared mainly between a visualization- and a simulation group, while the third, from computational fluid dynamics, provides valuable input from the users' perspective. Although the overall task of grid management is shared, the two communities, simulation and visualization, put different emphasis on the features of a grid administration software. The high performance computing community has demonstrated time and again that it is willing to put runtime efficiency (as measured in MFLOPS) above all other considerations. Visualization is a much more interactive process, which has to be able to respond to the choices of a user with very low latency. Consequently, visualization requirements result in higher emphasis on flexibility than is the norm (traditionally) in the HPC context, willing to trade CPU performance and memory usage for interactivity. This paper shows how the *gridlib* meets the demands from both sides. After an overview of the *gridlib* system architecture in Sect. 2, the topic of flexibility is discussed in Sect. 3. This is followed by the efficiency considerations in Sect. 4, before the main points of the paper are summed up in the conclusion in Sect. 5.

2 System Architecture of the gridlib

The *gridlib* is a framework library for the integration of simulation and visualization on adaptive, unstructured grids. Its infrastructure serves two main purposes. First, it supports developers of new simulation applications by providing subsystems for I/O, grid administration and grid modification, visualization and solver integration. Second, its parametrised storage classes allow (in principle) the re-use of any existing solvers, even those only available in binary format. For the special group of solvers that do not perform grid management themselves, the *gridlib* can provide plug-and-play functionality.

This high level of integration is achieved by three abstraction levels:

---

[2] deal.II homepage: `http://gaia.iwr.uni-heidelberg.de/~deal/`
[3] Overture homepage: `http://www.llnl.gov/CASC/Overture/overview.html`
[4] AVS homepage: `http://www.avs.com`
[5] VTK homepage: `http://public.kitware.com/VTK`

1. The lowest level offers an interface to describe the storage layout. This is the part of the library that has to be adapted when integrating an existing solver.
2. The level above implements abstraction of the geometric element type. Relying on the storage abstraction, it provides object oriented element implementations for the higher abstraction levels.
3. The highest level offers the interface to operations on the whole grid. It employs object oriented design patterns like functors for frequently needed operations.

3 Flexibility

The *gridlib* intends to be widely applicable. From a simulation perspective, this implies that the user should be able to choose the grid type and the solver that are appropriate for the application. For the visualization tasks, the *gridlib* must not assume the existence of any dedicated graphics hardware. However, if dedicated hardware like a visualization server is available, the user should be able to decide whether to use it or not. The following subsections illustrate how these aims have been achieved in the *gridlib* design.

3.1 Unstructured Grids

The scientific community remains divided as to what type of grid to use when solving PDEs. As a consequence, there are numerous different grid types around, ranging from (block-)structured over hybrid up to unstructured grids, each of them with their advantages and problems and their proponents. A grid software that intends to be widely applicable cannot exclude any of these grid types. Thus, the *gridlib* supports completely unstructured grids[6], which include all other more specialised grid types. Furthermore, the *gridlib* does not make any assumptions about the mesh topology nor the geometrical shape of the elements involved. Currently supported are tetrahedra, prisms, pyramids, octahedra and hexahedra. The *gridlib* is designed in such a way that other shapes can be added easily using object oriented techniques.

3.2 Integrating Existing Solvers

As mentioned before, the *gridlib* supports the re-use of existing solvers, even those only available in binary form. To this effect, the *gridlib* provides the grid data in the format required by a given solver. For example, this could imply storing the grid data in a particular data file format or arranging certain arrays in main memory to be passed as arguments in a function call. Clearly, for this approach to work, the input and output formats of the solver have to be known. In this case, the integration involves the following steps:

[6] One repeated argument against the use of unstructured grids in the scientific computing community is their alleged performance disadvantage. We will return to this point in Sect. 4.2.

1. Implementation of the storage format for the element abstraction.
2. Creation of an object oriented interface, which can be inherited from a provided, virtual interface. This step effectively "wraps" a potentially procedural solver into an object oriented environment.
3. Link the solver together with the *gridlib*.

Note that in many cases, the second step can be performed automatically by the compiler through the object-oriented template patterns already provided by the *gridlib*. If the source code of the solver can be modified, the first two steps can be combined, which results in the native *gridlib* storage format to be used throughout.

3.3 Visualization Pipeline

In the *gridlib*, the visualization is based on a attributed triangle mesh which in turn is derived from the original data or a reduced set of it. By working directly on the grid data as provided by the grid administration component of the library, the visualization subsystem can exploit grid hierarchies, topological and geometrical features of the grid and the algorithms for grid manipulation. This approach provides a common foundation for all visualization methods and ensures the re-usability of the algorithmic components.

In the visualization pipeline, the data is represented in the following formats:

1. As simulation results on the compute grid
2. As data on a modified grid (reduced, progressive, changed element types, ...)
3. As visualization geometries (isosurfaces, stream lines, ...)
4. As bitmap or video (stored in a file or displayed immediately)

These stages can be distributed across several machines. In the context of large scale simulations, a common distribution of tasks involves a *compute node* for the first step, a *visualization server* for the second and third, and lastly, the user's workstation for the forth. For a given project, these three functions, compute server, visualization server and front end workstation, have to be assigned to the available hardware. The *gridlib* makes provisions for different configurations that help the user to adequately match the given hardware to the tasks. The following factors influence the visualization pipeline:

1. Availability and performance of an interactive mode on the compute node. This is often an issue on batch-operated super computers.
2. Bandwidth and latency of the involved networks.
3. Availability and performance of a dedicated visualization server.
4. Storage capacity and (graphics-) performance of the front end workstation.

Given the concrete configuration, it is the user who can decide how to trade-off requirements for interaction with those for visualization quality. Conceptually, the *gridlib* supports different scenarios:

- Remote rendering on the compute node. Being based on the complete set of high resolution simulation results, this approach yields the maximum visualization quality. However, on batch-operated machines, no form of interaction is possible.
- Postprocessing of the simulation results on the compute node and subsequent transfer of a reduced data set to the visualization server or front end. Once the simulation results are available, this strategy offers the maximum of interaction in displaying the results but places high demands the servers and the networks, as even reduced data sets can still be large in absolute terms.
- Local rendering of remotely generated visualization geometries. The user experiences (subjective) fast response times but can only work on a given number of data sets. This approach allows high visualization quality but requires fast networks and high storage facilities.
- Two stage rendering. First, the user determines the visualization parameters (view point, cut plane, ...) on a reduced quality set, then transfers these parameters to the compute node, where they will be used for remote rendering at maximum quality.

Supporting all these scenarios is ongoing work. Several components have already been implemented. Progressive mesh techniques allow to trade visualization quality for faster response time (resolution on demand), see [5]. Slice- and isosurfaces geometries can be computed and displayed via various rendering options, see [8]. The most generally applicable renderer is a software-only implementation, which is useful on machines without dedicated graphics hardware. It can be run transparently in parallel on any multiprocessor machine with MPI support. Figure 1 illustrates the data flow for the parallel software renderer. The alternative is tuned for hardware accelerated OpenGL environments. Thus the *gridlib* lets the user choose a compromise between visualization quality and interaction.

4 Efficiency

This section introduces the two main features of the *gridlib* that are useful in the development of high performance solvers, for which maximum runtime efficiency is important. These two features are the provision of grid adaptation techniques and the concept of hierarchical hybrid grids.

4.1 Grid Adaptation

Adaptive h-refinement techniques, usually based on error estimators, have attracted considerable interest in the numerical PDE community over the last twenty years, see, for instance, [2], [1]. For many applications, these techniques are well-established and reliable error-estimators are available [1], [9], [3]. By providing functions for the uniform, adaptive or progressive subdivision and coarsening of the mesh, the *gridlib* is a well-suited platform for the implementation of

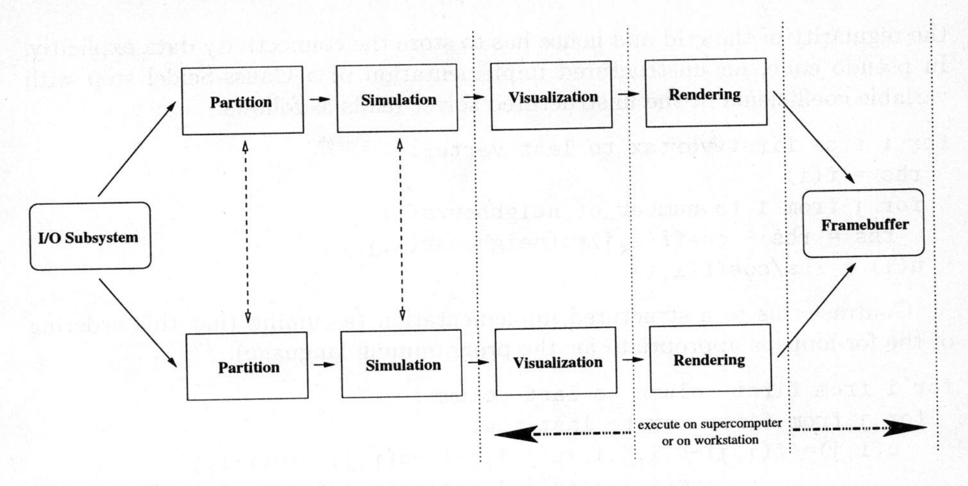

Fig. 1. Data flow for the parallel software renderer: The renderer processes the distributed simulation results concurrently before merging the individual parts together into the final picuture. The diagram emphasises the various stages in the visualization pipeline that can be assigned to the available hardware.

h-refinement algorithms. The user need only specify a criterion that marks the grid cells to be subdivided. The *gridlib*'s refinement algorithm performs the subdivision and ensures that the resulting grid is consistent and that hanging nodes are avoided (red-green refinement). For subdividing tetrahedra, the algorithm of Bey [4] has been chosen because of the small number of congruency classes it generates.

Provided that the user contributes a sharp error estimator, the *gridlib* features make it easy to generate solution adapted unstructured grids. Such grids are the essential tool to improve the accuracy of the solution for a given number of grid cells.

4.2 E!ciency Limits of Unstructured Grids and What to Do about It

It is important to note that adaptive refinement of unstructured grids (alone) cannot overcome the problem of low MFLOPS performance when compared to (block-)structured approaches.

The performance problem of solvers on unstructured grids results from the fact that the connectivity information is not available at compile time. Hence the resulting program, although very flexible, requires some form of book-keeping at run time. In structured codes, the connectivity is known at compile time and can be exploited to express neighbourhood relations through simple index arithmetic.

The following, deliberately simple example illustrates the difference between the two approaches. Given the unit square, which is discretised into square cells of side length h using bi-linear elements. An unstructured solver "does not see" the regularity of the grid and hence has to store the connectivity data explicitly. In pseudo code, an unstructured implementation of a Gauss-Seidel step with variable coefficients in the unstructured solver reads as follows:

```
for i from first vertex to last vertex:
  rhs = f(i)
  for j from 1 to number_of_neighbours(i)
    rhs = rhs - coeff(i,j)*u(neighbour(i,j))
  u(i) = rhs/coeff(i,i)
```

Contrast this to a structured implementation (assuming that this ordering of the for-loops is appropriate for the programming language):

```
for i from first column to last column:
  for j from first row to last row:
    u(i,j)=(f(i,j)-c(i,j,1)*u(i-1,j-1)-c(i,j,2)*u(i-1,j)
                  -c(i,j,3)*u(i-1,j+1)-c(i,j,4)*u(i+1,j-1)
                  -c(i,j,5)*u(i+1,j)  -c(i,j,6)*u(i+1,j+1)
                  -c(i,j,7)*u(i,j+1)  -c(i,j,8)*u(i,j-1))/c(i,i)
```

The work as measured in floating point operations is the same in both implementations, but their run-time performance differs significantly as the second version, being much more explicit, lends itself much better to compiler optimisation than the first one. On one node (8 CPUs) of a Hitachi SR8000 at the Leibniz Computing Centre in Munich, the MFLOPS rate of the (straightforward) structured version is a factor of 20 higher than the one of the similarly straightforwardly implemented unstructured algorithm.

The *gridlib* introduces the concept of hierarchical hybrid grids to overcome the performance penalty usually associated with unstructured grids while retaining their geometric flexibility.

The main idea behind the hierarchical hybrid grids is to deal with geometric flexibility and computing performance on different grid levels. The coarse grid levels are in general unstructured and ensure the geometric flexibility of the approach. The coarse grids are nested in the sense that the finer ones are generated through uniform or adaptive refinement from the coarser ones. The finest unstructured grid is assumed to resolve the problem domain adequately and is therefore referred to as the geometry grid. The fine grids, on which the computations are to be carried out, are generated through regularly subdividing the individual cells of the geometry grid. Figure 2 illustrates the concept.

As shown above, it is essential for high floating point performance that the implementation of the computing algorithms takes the regular structure of the compute grid within each cell of the geometry grid into account. Given that the compute grid is only patchwise regular, some fraction of the computations still require unstructured implementations. Obviously, the finer the compute grid, the more the overall floating point performance is dominated by the contribution from the structured parts.

The following discussion confirms this expectation for a vertex based algorithm like Gauss-Seidel. Let N_c be the number of vertices in the (unstructured)

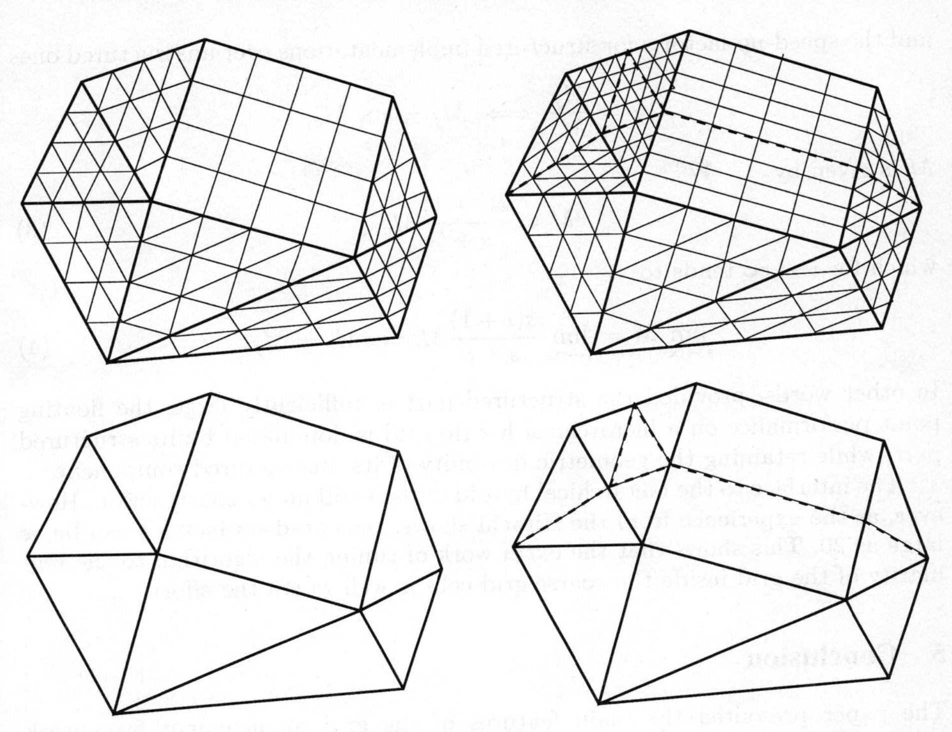

Fig. 2. Bottom left: coarsest base grid, bottom right: geometry grid after one unstructured refinement step, top row: compute grids after two regular subdivision steps of the respective coarse grids below

geometry grid and N_f be the number of vertices in the structured refinements. The unstructured algorithm achieves M_c MFLOPS while the structured part runs at M_f MFLOPS. Under the assumption that N_{op}, the number of floating point operations per vertex, is the same for both grid types (as it was in the Gauss-Seidel example above), then the execution time of one Gauss-Seidel iteration over the compute grid is given by

$$\frac{N_c \times N_{op}}{M_c} + \frac{N_f \times N_{op}}{M_f}. \tag{1}$$

Dividing the total number of operations, $N_{op} \times (N_c + N_f)$, by this execution time, one finds the MFLOPS value for the whole grid, M say, to be

$$M = \frac{(N_c + N_f) \times M_c M_f}{N_c M_f + N_f M_c}. \tag{2}$$

Introducing the fine to coarse ratio i

$$i = \frac{N_f}{N_c} \iff N_f = i \times N_c$$

and the speed-up factor s for structured implementations over unstructured ones

$$s = \frac{M_f}{M_c} \iff M_f = s \times M_c,$$

M is given by

$$M = \frac{s(i+1)}{s+i} M_c, \tag{3}$$

which for $i \to \infty$ tends to

$$\lim_{i\to\infty} M = \lim_{i\to\infty} \frac{s(i+1)}{s+i} M_c = sM_c = M_f. \tag{4}$$

In other words, provided the structured part is sufficiently large, the floating point performance on a hierarchical hybrid grid is dominated by its structured part, while retaining the geometric flexibility of its unstructured component.

The interface to the hierarchical hybrid grids is still under construction. However, as the experience from the Hitachi shows, the speed-up factor s can be as large as 20. This shows that the extra work of tuning the algorithm to the regularity of the grid inside the coarse grid cells is well worth the effort.

5 Conclusion

The paper presented the main features of the grid management framework *gridlib*. It combines the grid management requirements of both, visualization and simulation developers, into a single framework. By providing subsystems for frequently needed tasks in PDE solvers, such as I/O, adaptive grid refinement and, of course, visualization, the *gridlib* helps to speed up the development of such programs.

The article described the main features of the *gridlib* from the two perspectives of flexibility and (run-time) efficiency. Through its support of unstructured grids and numerous cell geometries, the *gridlib* is widely applicable. In case a particular cell geometry is not already included, the object-oriented design of the *gridlib* ensures that the user can add the required object easily. It was shown how existing solvers, that do not include any grid management, can be combined with the *gridlib*, so that these solvers, too, can benefit from the visualization facilities of the framework. For the visualization of large scale simulations, the *gridlib* supports different hardware scenarios, from which the user can choose to meet the project-specific requirements concerning visualization quality and interactivity. Its provision of algorithms for the consistent, adaptive subdivision of unstructured grids in three space dimensions makes the *gridlib* an ideal platform for implementing and experimenting with adaptive h-refinement methods. To close the gap in MFLOPS performance between unstructured and structured grids, the *gridlib* introduces the concept of hierarchical hybrid grids. This approach employs a hierarchy of two different grid types on the different levels to combine the advantages of the unstructured grids (geometric flexibility) with those of structured ones (high floating point performance). The coarse levels are made up of nested, unstructured grids. The patchwise structured grids on the finer levels are constructed through repeated *regular* subdivision of the cells of the finest, unstructured grid. Adapting the algorithm to take the grid structure into account increased the floating point performance of a Gauss-Seidel iteration inside the patches on a Hitachi SR8000 by a factor of twenty. The promise of the approach is therefore evident. However, more work in specifying user interfaces for the hierarchical hybrid grids among other things has to be done.

6 Acknowledgements

The authors wish to thank Dr. Brenner from the fluid dynamics group of Erlangen University for helpful discussions, U. Labsik, G. Soza from the Computer Graphics Group at Erlangen University for their input concerning geometric modelling and mesh adaptivity, S. Meinlschmidt from the same group for his work on the various options in the visualization pipeline, M. Kowarschik from the System Simulation Group, also at Erlangen University, for his insights in exploiting grid regularity for iterative methods. As mentioned in the introduction, this project is funded by a KONWIHR grant of the Bavarian High Performance Computing Initiative, which provided the access to the Hitachi SR8000.

References

1. Ainsworth, M., Oden, J.T.: A posteriori error estimation in finite element analysis. Comp. Methods Appl. Mech. Engrg. **142** (1997) 1–88
2. Babuska, I., Rheinboldt, W. C.: Error estimates for adaptive finite element computations. SIAM J. Numer. Anal. **15** (1978), 736–754
3. Becker, R., Rannacher, R.: Weighted A posteriori error control in FE methods. IWR Preprint 96-1, Heidelberg, 1996
4. Bey, J.: Tetrahedral Grid Refinement. Computing **55** (1995), 355–378
5. Labsik, U., Kipfer, P., Meinlschmidt, S., Greiner, G.: Progressive Isosurface Extraction from Tetrahedral Meshes. Pacific Graphics 2001, Tokio, 2001
6. Labsik, U., Kipfer, P., Greiner, G.: Visalizing the Structure and Quality Properties of Tetrahedral Meshes. Technical Report 2/00, Computer Graphics Group, University Erlangen, 2000
7. Greiner, G., Kipfer, P., Labsik, U., Tremel, U.: An Object Oriented Approach for High Performance Simulation and Visualization on Adaptive Hybrid Grids. SIAM CSE Conference 2000, Washington, 2000
8. Kipfer, P., Greiner, G.: Parallel rendering within the integrated simulation and visualization framework "gridlib". VMV 2001, Stuttgart, 2001
9. Süli, E.: A posteriori error analysis and adaptivity for finite element approximations of hyperbolic problems. In: Kröner, D., Ohlberger, M., Rohde C. (Eds.): An Introduction to Recent Developments in Theory and Numerics for Conservation Laws. Lecture Notes in Computational Science and Engineering **5**, 123–194 Springer-Verlag, 1998

Space Tree Structures for PDE Software

Michael Bader[1], Hans-Joachim Bungartz[2], Anton Frank[3], and Ralf Mundani[2]

[1] Dept. of Informatics, TU München, D-80290 München, Germany
[2] IPVR, Universität Stuttgart, D-70565 Stuttgart, Germany
[3] 4Soft GmbH, D-80336 München, Germany

Abstract. In this paper, we study the potential of space trees (boundary extended octrees for an arbitrary number of dimensions) in the context of software for the numerical solution of PDEs. The main advantage of the approach presented is the fact that the underlying geometry's resolution can be decoupled from the computational grid's resolution, although both are organized within the same data structure. This allows us to solve the PDE on a quite coarse orthogonal grid at an accuracy corresponding to a much finer resolution. We show how fast (multigrid) solvers based on the nested dissection principle can be directly implemented on a space tree. Furthermore, we discuss the use of this hierarchical concept as the common data basis for the partitioned solution of coupled problems like fluid-structure interactions, e. g., and we address its suitability for an integration of simulation software.

1 Introduction

In today's numerical simulations involving the resolution of both time and space, we are often confronted with complicated or even changing geometries. Together with increasing accuracy requirements, this fact is responsible for some kind of dilemma: On the one hand, orthogonal or Cartesian grids are simpler with respect to mesh generation, organization, and changes, but need very or even too high levels of refinement in order to resolve geometric details in a sufficient way. Unstructured grids, on the other hand, are clearly better suited for that, but entail costly (re-) meshing procedures and an often significant overhead for grid organization.

For the Cartesian world, one possibility to get out of this dilemma is to decouple the resolutions of the *geometric* grid (used for geometry representation and discretization) and of the *computational* grid (used for the (iterative) solution process). This seems to be justified by the fact that orthogonal grids come along with an $O(h)$ error concerning geometry, but are able to produce $O(h^2)$ discretization errors for standard second order differential operators. Hence, for a balance of error terms, it is definitely not necessary to iterate over all those tiny cells needed to resolve geometry, if some way is found to collect geometric details from very fine cells for the discrete equations of a surrounding coarser cell. Space trees provide this possibility, and they do it within the same data structure. This aspect is important, since the accumulation of geometric details

P.M.A. Sloot et al. (Eds.): ICCS 2002, LNCS 2331, pp. 662–671, 2002.

is not some kind of a "once-and-for-all" process, but there will be situations where the fine world has to be revisited (in case of adaptive refinement of the computational grid, for example).

In the following, we first summarize the principal ideas and properties of the hybrid space tree concept. Then, we demonstrate how space trees can be directly used for grid organization and representation in a PDE context by implementing a nested-dissection-based fast iterative solver. Afterwards, the use of space trees as the common geometry representation in a software environment for the partitioned solution of coupled problems and their potential for an embedding of PDE software into a broader context (integration of CAD and numerical simulation, for example) are discussed. Finally, we give some conclusions and an outlook over future work in this field.

2 Space Trees

Space trees [3, 6] are generalized quad- or octrees [5, 9–11]. The first generalization refers to the now arbitrary number d of dimensions. A second generalization is their ability to associate data not only with d-dimensional cells (i. e. volumes), but also with the $d-1$-dimensional hypersurfaces representing the boundary of a cell (and so on, recursively) – which is important if we think of boundary value problems. Hence, the basis is a successive spatial partitioning of a square (cube) in four (eight) attributed congruent subsquares (subcubes). Among the fields of application for such hierarchical structures, there are image processing, geometric modelling, computer graphics, data mining, and many more. For our purposes, the most interesting features of space trees are the reduced complexity (the boundary of an object determines its storage requirements, see the left part of Fig. 1), the availability of efficient algorithms for set operations, neighbour

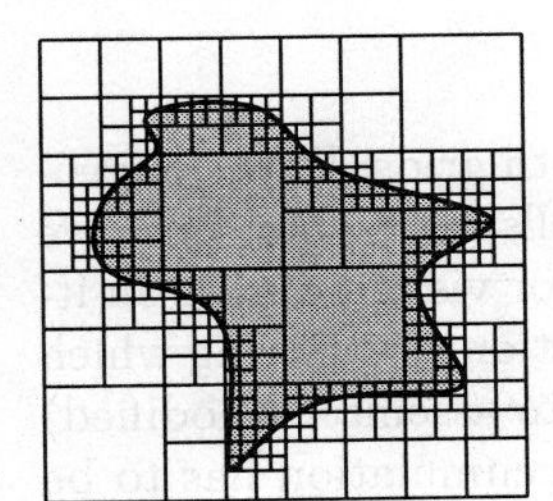

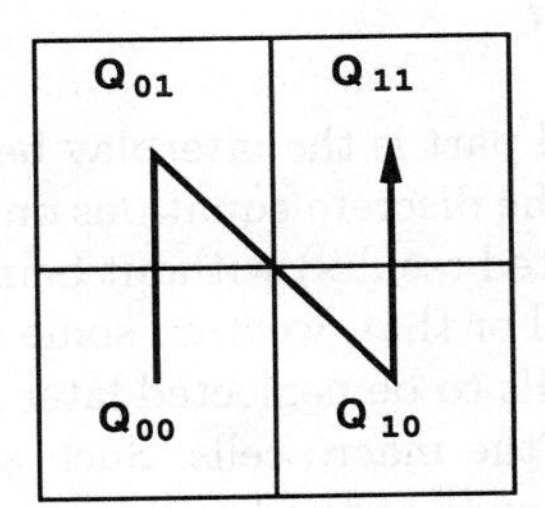

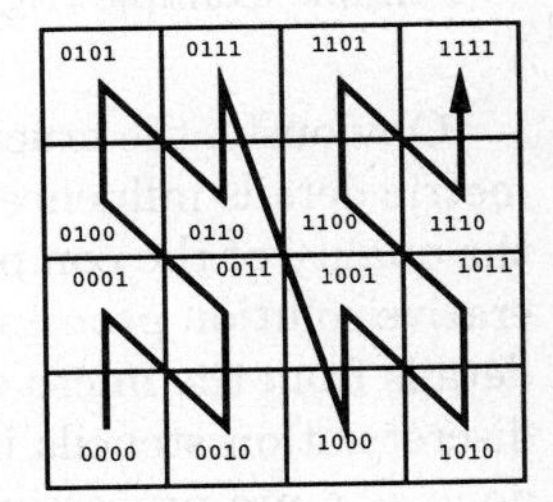

Fig. 1. Quadtree: successive refinement (left) and relations to Morton ordering and Lebesgue's space-filling curve (centre and right; helpful for a simple and efficient parallelization of computations on very heterogeneous grids)

detection, or movement, the inherent information for data sequentialization (see the right part of Fig. 1; this is especially important for a simple parallelization), and, finally, the possibility of compact implementations via bit coding, as long as we are only interested in the merely geometric information. For the remainder, the discussion is restricted to 2 D without loss of generality.

For some given geometry, which may be a technical object in a CAD representation or a measured or analytically described domain, the first step is to generate a space tree to hold the complete geometric information, but nothing else. This means that we have to choose a very high characteristic resolution h_g (the resolution of the input data, for example), but that we can use a compact implementation like a bit-coded linearized tree. Note that, if nevertheless too big for main memory, the space tree does not have to be kept there at one time as a whole. As a next step, the *computational* grid with characteristic mesh width h_c, i. e. the set of cells which will be related to degrees of freedom later, has to be built up. This will typically be a small subset of the above space tree, but can also contain finer cells than the geometric grid (in subdomains without any geometric details and, hence, coarse geometric cells, for example). Concerning the concrete implementation, think of a standard tree structure with several floating point variables in the nodes, which is explicitly generated and kept in main memory. Hence, following the maxim that the problem's physics and not its geometry should determine the level of detail during computations, we now have a hybrid or decoupled overall representation within the space tree concept. Figure 2 illustrates the relations between geometric and computational grid.

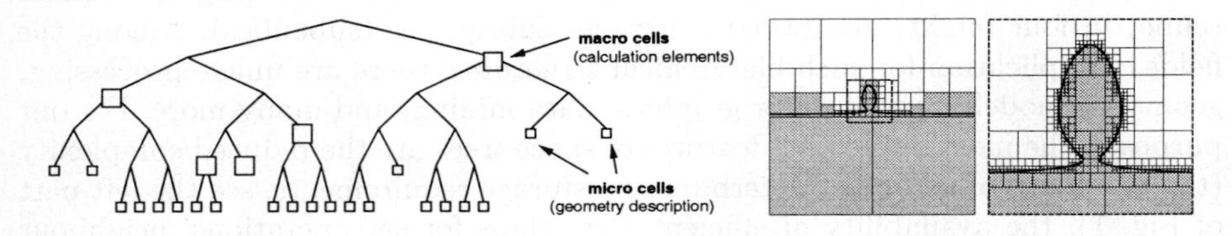

Fig. 2. Hybrid concept: macro and micro layer in the data structure (left) and applied to a simple example (right)

Obviously, the crucial part is the interplay between both grids: How can geometric details influence the discrete equations on coarse cells (and, thus, improve the quality of the computed results) without being stored or visited during an iterative solution process? For that, we need some accumulation step during which details from the micro cells to be neglected later are used to assemble (modified) discretization stencils in the macro cells. Such a global accumulation has to be done as a pre-processing once at the beginning, and it may have to be repeated partially each time the computational grid changes due to local adaptive refinement. Apart from these accumulation steps, the micro layer is not needed for the solution of the discretized equations.

The accumulation starts at the leaves of the micro layer with some atomic stencils or element matrices which, of course, depend on the given PDE, on the chosen discretization scheme (finite differences or elements, e. g.), and on the local geometry (a cell's corner outside the problem's domain will influence the atom and, hence, the following). Next, four neighbouring atoms are assembled

– in the sense of both stencils and matrices. Figure 3 shows such an atom and the assembly for the simple case of the standard finite difference approach for the Laplacian with all involved points lying within the domain. Since we do not

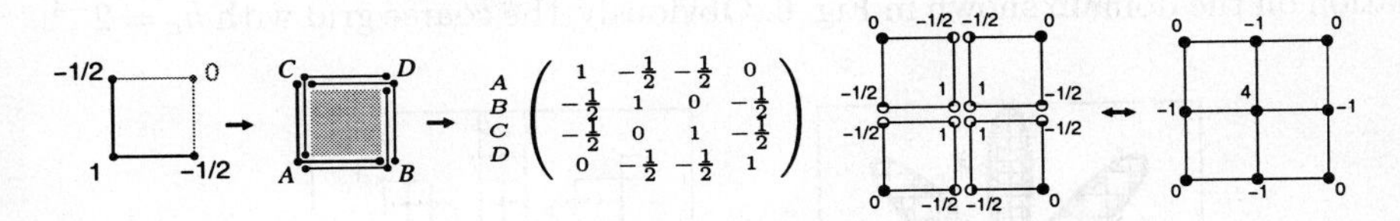

Fig. 3. Atoms (left, associated with cells or elements) as stencils or matrices (centre) and their assembly (right)

want to keep the fine grid points as degrees of freedom for the computations on the macro layer, a hierarchical transformation separating coarse and fine points is applied. The fine points are eliminated as in a nested dissection process (see Sect. 3) and even explicitly removed. Now, we have again atoms – related to larger cells than before, but again with four grid points involved. The whole process is illustrated in Fig. 4.

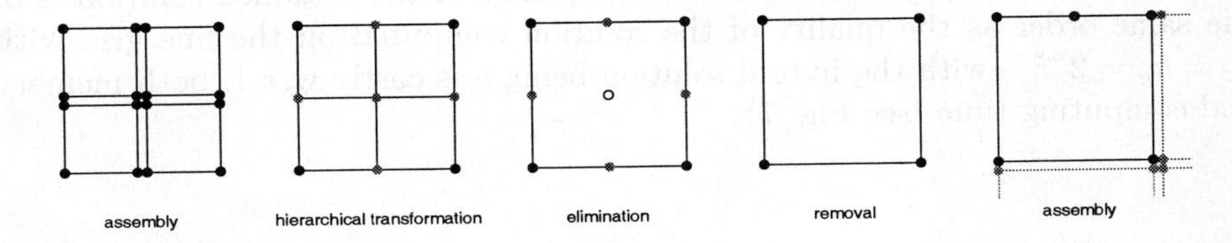

Fig. 4. Accumulating geometric detail information: assembly, hierarchical transformation, elimination, removal, and next assembly

This completes the description of the geometry accumulation process. Now,

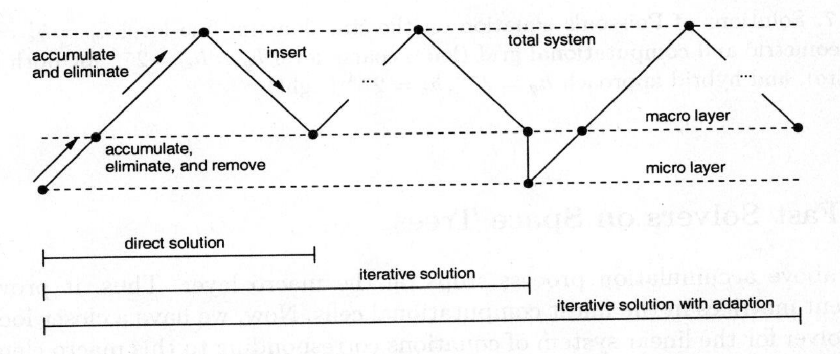

Fig. 5. Accumulation of geometric details and following direct or iterative solution on the computational grid

the system (with the local stencils or matrices derived above) must be solved on the cells of the macro layer. For that, principally, a direct or an iterative algorithm can be used (see Fig. 5). The design of suitable solvers will be studied in the next section. Here, we just present some numerical results for a Poisson equation on the domain shown in Fig. 6. Obviously, the coarse grid with $h_c = 2^{-4}$

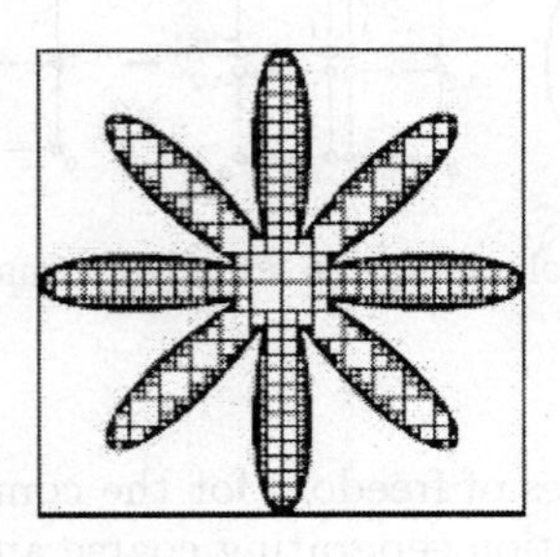
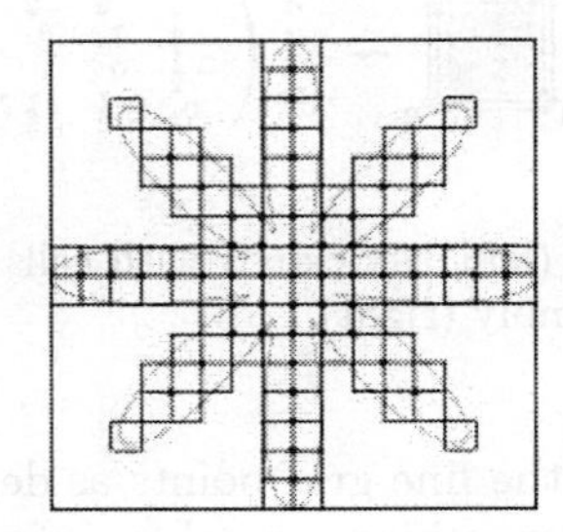

Fig. 6. 2 D star domain: used geometric grid (adaptive, finest occurring h is $h_g = 2^{-8}$; left) and used computational grid (regular, $h_c = 2^{-4}$; right)

is not able to produce reasonable results without the collected micro details. With our hybrid approach, however, the quality of the obtained solution is of the same order as the quality of the solution computed on the fine grid with $h_c = h_g = 2^{-8}$ – with the hybrid solution being less costly w. r. t. both memory and computing time (see Fig. 7).

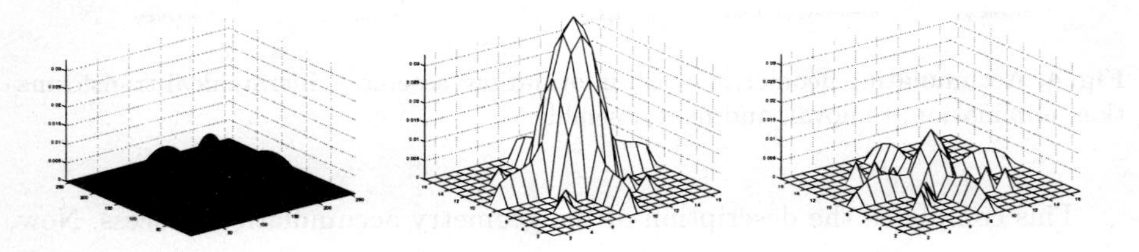

Fig. 7. Solutions of Poisson's equation on the star domain: fine level $h_g = h_c = 2^{-8}$ for geometric and computational grid (left), coarse level $h_g = h_c = 2^{-4}$ for both grids (centre), and hybrid approach $h_g = 2^{-8}, h_c = 2^{-4}$ (right)

3 Fast Solvers on Space Trees

The above accumulation process stops on the macro layer. Thus, it provides element matrices in the finest computational cells. Now, we have a closer look at the solver for the linear system of equations corresponding to this macro element information. Since the space tree idea is based on a recursive substructuring of the domain, a nested dissection approach [7] turns out to be a quite natural choice. In order to make nested dissection's successive bisections of the domain consistent with the space tree subdivision into 2^d successor nodes, we have to perform and combine d alternate bisections (one for each dimension). Hence, in the following, we can restrict the presentation to the standard case of a bisection in 2 D.

On each node of the space tree between its root and the macro leaves where geometry accumulation has provided element stencils or matrices, we define local sets $\mathcal{I}$ and $\mathcal{E}$ of unknowns related to grid points inside or on the boundary of the local cell (subdomain), resp. Due to the recursive bottom-up elimination of the nested dissection scheme, $\mathcal{I}$ is restricted to points on the so-called *separator*, the line separating the two subdomains of the local cell. For the precise definition of the iterative solver, we introduce a further set $\mathcal{G}$ of *coarse grid unknowns*, which will form the coarse grid in the sense of a multilevel solver. If the local system of equations is restricted to the unknowns in $\mathcal{G}$ (which is actually what is done during the geometry setup), we should get a system that describes physics on this coarser level sufficiently well. Figure 8 illustrates this classification.

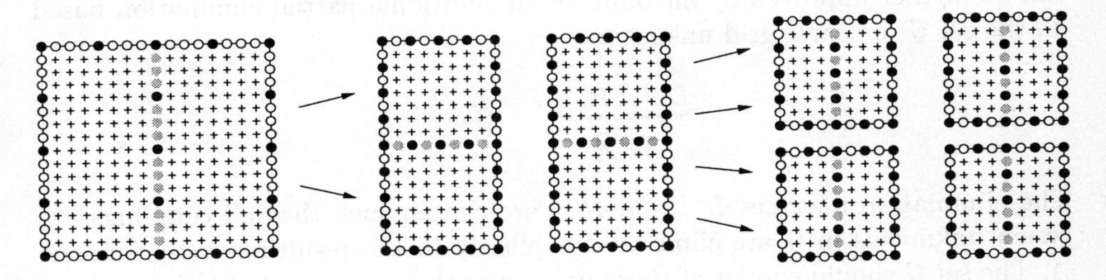

Fig. 8. Recursive substructuring by alternate bisection: Unknowns in $\mathcal{E}$ are painted in white, unknowns in $\mathcal{I}$ in grey. If the unknowns are also in $\mathcal{G}$, they are painted as black nodes. The little crosses indicate unknowns that are no longer present on the local subdomain due to their elimination on the child domains.

Starting at the leaves of the macro layer, we perform a block elimination on each local system of equations based on the partitioning of the unknowns:

$$\underbrace{\begin{pmatrix} \mathrm{Id} & -A_{\mathcal{E}\mathcal{I}}A_{\mathcal{I}\mathcal{I}}^{-1} \\ 0 & \mathrm{Id} \end{pmatrix}}_{=:L^{-1}} \underbrace{\begin{pmatrix} A_{\mathcal{E}\mathcal{E}} & A_{\mathcal{E}\mathcal{I}} \\ A_{\mathcal{I}\mathcal{E}} & A_{\mathcal{I}\mathcal{I}} \end{pmatrix}}_{=A} \underbrace{\begin{pmatrix} \mathrm{Id} & 0 \\ -A_{\mathcal{I}\mathcal{I}}^{-1}A_{\mathcal{I}\mathcal{E}} & \mathrm{Id} \end{pmatrix}}_{=:R^{-1}} = \underbrace{\begin{pmatrix} \widehat{A}_{\mathcal{E}\mathcal{E}} & 0 \\ 0 & A_{\mathcal{I}\mathcal{I}} \end{pmatrix}}_{=:\widehat{A}}, \quad (1)$$

where $\widehat{A}_{\mathcal{E}\mathcal{E}} := A_{\mathcal{E}\mathcal{E}} - A_{\mathcal{E}\mathcal{I}} \cdot A_{\mathcal{I}\mathcal{I}}^{-1} \cdot A_{\mathcal{I}\mathcal{E}}$ is the so-called *Schur complement.* Thus, the full information from $\mathcal{I}$ is preserved in $\widehat{A}_{\mathcal{E}\mathcal{E}}$. The submatrix $\widetilde{A}_{\mathcal{E}\mathcal{E}}$ is then transferred to the father who collects and assembles the local systems of his two sons and proceeds recursively with block elimination and assembly until the root of the space tree, i. e. the cell representing the overall domain of our problem, is reached. After this bottom-up assembly of equations, we start from the root and use the unknowns in the local $\mathcal{E}$ (available from the boundary conditions or from the respective father node) to compute the unknowns in the local $\mathcal{I}$ on every subdomain in a top-down traversal.

So far, the approach leads to a direct solver quite similar to the original nested dissection method from [7]. Likewise, its computing time grows like $\mathcal{O}(N^{3/2})$ with the number of unknowns N (2D). Since this, of course, is too expensive, the block elimination (1) should be replaced by some iterative scheme (a suitable preconditioner, e. g.). Then, the single top-down solution pass is replaced by a sequence of top-down (compute approximations based on the current residuals) and bottom-up (collect updated residuals from the leaves to the root) traversals, see also Fig. 5.

In our solver, we use the transformation to hierarchical bases or generating systems [8] as a preconditioner,

$$\underbrace{H^T A H}_{=:\,\bar{A}}\,\bar{x} = \underbrace{H^T b}_{=:\,\bar{b}}, \tag{2}$$

the latter leading to a true multigrid method. In both cases, the preconditioning can be further improved by introducing an additional partial elimination based on the set $\mathcal{G}$ of coarse grid unknowns:

$$\underbrace{L^{-1}\bar{A}\,R^{-1}}_{=:\,\widetilde{A}}\,\widetilde{x} = \underbrace{L^{-1}b}_{=:\,\widetilde{b}}. \tag{3}$$

The elimination matrices L^{-1} and R^{-1} are chosen such that all couplings between unknowns in $\mathcal{G}$ are eliminated explicitly in the resulting system matrix $\widetilde{A}$. The set $\mathcal{G}$ should consist of those unknowns that are expected to be strongly coupled. Hence, a good heuristics for choosing $\mathcal{G}$ can often be derived from the underlying physics of the respective PDE.

For Poisson equations, it is usually sufficient to choose just the unknowns on the corners of the subdomains. This is consistent with the accumulation process from Sect. 2 and leads to a multilevel method close to standard multigrid with uniformly coarsened grids. For other PDEs, however, this simple choice may no longer be appropriate. For convection diffusion equations, for example, especially in the case of increasing strength of convection, additional unknowns on the boundary of the local subdomain should be used for $\mathcal{G}$. This corresponds very much to using semi-coarsening in classical multigrid methods. In [2], we present a method that increases the number of coarse grid unknowns proportionally to the square root of the number of local unknowns in the cell. This approach balances the influence of both diffusion and convection in each cell. If convection is not too strong (mesh Péclet number bounded on the finest grid), the resulting algorithm has a complexity of $\mathcal{O}(N)$ with respect to both computing time and memory requirements [2].

Figure 9 (right) shows the convergence rates for the Poisson equation on the star domain from Fig. 6 and for a convection diffusion problem with circular convection field and varying strength of convection (centre). The streamlines of the convection field are given in the leftmost picture. Homogeneous Dirichlet

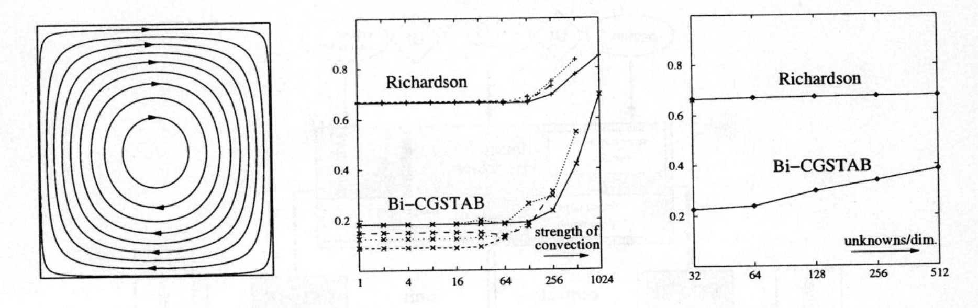

Fig. 9. Convergence rates for the convection diffusion problem (centre) with corresponding convection field (left) and for the Poisson equation on the star domain

boundary conditions were used in both examples. The results indicate that the convergence rates are generally independent of the number of unknowns and also constant for increasing strength of convection. However, this holds only up to a certain upper bound which depends on the mesh size of the finest grid. At least to some extent, the rates are also independent of the geometry of the problem's domain (see [1] for a more detailed discussion). In both examples, using the method as a preconditioner for Bi-CGSTAB can further improve the overall performance of the algorithm.

With the presented iterative solver, we are now able to efficiently solve a PDE on complicated domains with simple Cartesian grids. Due to the logical separation of the resolution of geometry and computations, there are no drawbacks concerning accuracy compared to more complicated unstructured grids. The further potential of space trees is studied in the next section.

4 Coupled Problems, Software Integration

Coupled or multi-physics problems involve more than one physical effect, where none of these effects can be solved separately. One of the most prominent examples is the interaction of a fluid with a structure – think of a tent-roof construction exposed to wind, or of an elastic flap in a valve opened and closed by some fluid. The numerical simulation of such interactions, which is not in the centre of interest here, is a challenging task, since expertise and model equations from different disciplines have to be integrated, and since the problem's geometry is not stationary (see [12], e. g.). In order to profit from existing models and code for the single phenomena and, hence, to reduce necessary new developments to the mere coupling of the different parts, the *partitioned solution* strategy based on a more or less sophisticated alternate application of single-effect solvers is very widespread. Figure 10 shows the basic structure of a possible modular framework that has been developed for the partitioned solution of problems with fluid-structure interactions [4, 6]. Input, output, the two solvers, and the coupling itself are strictly separated. The direct interplay of different codes entails the necessity to switch from one geometry representation of the common domain

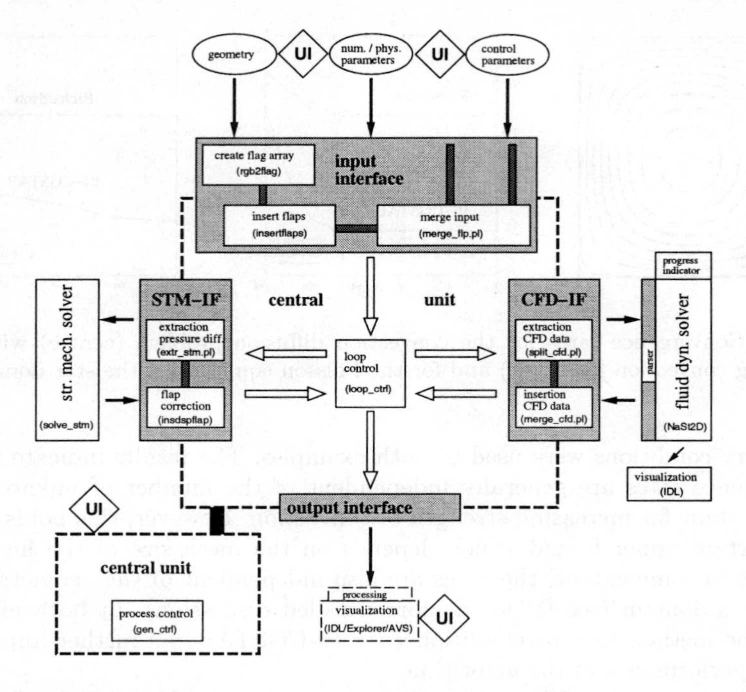

Fig. 10. Modular software environment for fluid-structure interactions (see [6])

to another. In our approach, we use space trees as the central data format. That is, if one of the solvers is to be exchanged, only a transformation between this solver's internal geometry representation and the central space tree has to be provided, which supports modularity. Again, the space tree allows us to keep geometric information at a very high resolution and to easily process changes of the geometry. Figure 11 shows the application of our software environment for coupled problems to the flow through a valve-driven micropump.

Finally, note that space trees are well-suited as general data basis for the integration of CAD, simulation, and visualization software [4]. For example, the bit-coded tree keeps the whole geometric information and allows for efficient CAD operations (affine transforms, collision tests, and so on).

5 Concluding Remarks

Space trees combines the simplicity, flexibility, and universality of Cartesian grids with the superior approximation properties of unstructured grids. In this paper, their main features as well as a fast linear solver and some attractive fields of application have been studied. Future work will cover the use of space trees for more complicated PDEs like the Navier-Stokes equations.

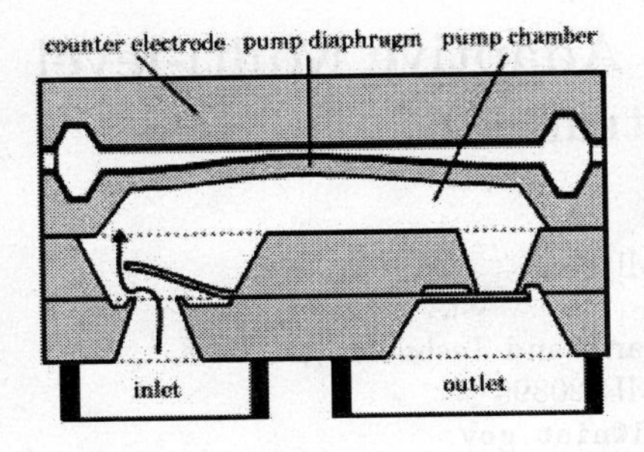

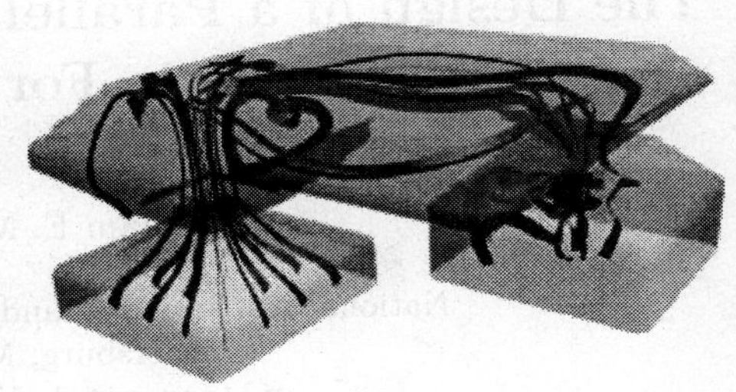

Fig. 11. Cross-section through a valve-driven micropump (left) and visualization of the simulated fluid-structure interaction at its valves (right; pump in octree representation and stream bands)

References

1. M. Bader. *Robuste, parallele Mehrgitterverfahren für die Konvektions-Diffusions-Gleichung.* PhD thesis, TU München, 2001.
2. M. Bader and C. Zenger. A robust and parallel multigrid method for convection diffusion equations. *ETNA*, 2002 (accepted).
3. P. Breitling, H.-J. Bungartz, and A. Frank. Hierarchical concepts for improved interfaces between modelling, simulation, and visualization. In B. Girod, H. Niemann, and H.-P. Seidel, editors, *Vision, Modelling, and Visualization '99*, pages 269–276. Infix, St. Augustin, Bonn, 1999.
4. H.-J. Bungartz, A. Frank, F. Meier, T. Neunhoeffer, and S. Schulte. Efficient treatment of complicated geometries and moving interfaces for CFD problems. In H.-J. Bungartz, F. Durst, and C. Zenger, editors, *High Performance Scientific and Engineering Computing*, volume 8 of *LNCSE*, pages 113–123. Springer-Verlag, Heidelberg, 1999.
5. R. A. Finkel and J. L. Bentley. Quad trees: a data stucture for retrieval on composite keys. *Acta Informatica*, 4:1–9, 1974.
6. A. Frank. *Organisationsprinzipien zur Integration von geometrischer Modellierung, numerischer Simulation und Visualisierung.* PhD thesis, TU München, Herbert Utz Verlag, München, 2000.
7. A. George. Nested dissection of a regular finite element mesh. *SIAM Journal on Numerical Analysis*, 10, 1973.
8. M. Griebel. *Multilevelmethoden als Iterationsverfahren auf Erzeugendensystemen.* Teubner Skripten zur Numerik, Stuttgart, 1994.
9. C. L. Jackins and S. L. Tanimoto. Oct-trees and their use in representing 3 D objects. *Computer Graphics and Image Processing*, 14:249–270, 1980.
10. D. Meagher. Geometric modelling using octree encoding. *Computer Graphics and Image Processing*, 19:129–147, 1982.
11. H. Samet. *The Design and Analysis of Spatial Data Structures.* Addison-Wesley, Reading, 1989.
12. S. Schulte. *Modulare und hierarchische Simulation gekoppelter Probleme.* PhD thesis, TU München, VDI Reihe 20 Nr. 271. VDI Verlag, Düsseldorf, 1998.

The Design of a Parallel Adaptive Multi-level Code in Fortran 90 *

William F. Mitchell

National Institute of Standards and Technology,
Gaithersburg, MD 20899
william.mitchell@nist.gov
http://math.nist.gov/~WMitchell

Abstract. Software for the solution of partial differential equations using adaptive refinement, multi-level solvers and parallel processing is complicated and requires careful design. This paper describes the design of such a code, PHAML. PHAML is written in Fortran 90 and makes extensive use of advanced Fortran 90 features, such as modules, optional arguments and dynamic memory, to provide a clean object-oriented design with a simple user interface.

1 Overview

Software for the solution of partial differential equations (PDEs) using adaptive refinement, multi-level solvers and parallel processing is complicated and requires careful design. This paper describes the design of such a code, PHAML (Parallel Hierarchical Adaptive Multi-Level). PHAML is written in Fortran 90 and makes extensive use of advanced Fortran 90 features, such as modules, optional arguments and dynamic memory, to provide a clean object-oriented design with a simple user interface.

The primary subroutine of PHAML solves a scalar, linear, second order elliptic PDE in two dimensions. More complicated PDE problems can be solved by multiple calls to the primary subroutine. This includes systems of equations, nonlinear equations, parabolic equations, etc. PHAML also provides for the solution of eigenvalue problems, like the linear Schrödinger equation.

The underlying numerical methods in PHAML are those used in the popular scalar PDE code MGGHAT [11], and are described in [9,10]. PHAML is a finite element program that uses newest node bisection of triangles for adaptive refinement/derefinement [9] and a hierarchical multigrid algorithm [10] for solution of the linear system. The multigrid method can be used either as a solver or as a preconditioner for conjugate gradients [2] or stabalized bi-conjugate gradients [2].

Several other software packages are optionally used by PHAML to increase its capability. Operation in parallel requires the use of either MPI [5] or PVM [4]. Visualization is provided through OpenGL [17], GLUT [6] and f90gl [12].

P.M.A. Sloot et al. (Eds.): ICCS 2002, LNCS 2331, pp. 672–680, 2002.

PHAML contains one method for partitioning the grid for load balancing, but additional methods are available through Zoltan [3]. Eigenvalue problems require the use of ARPACK [8]. Some operations use BLAS [7] and LAPACK [1]; performance may be improved by using a vendor implementation rather than source code included with PHAML.

PHAML uses the concept of data encapsulation from object-oriented program design. Using Fortran 90 modules with public and private attributes, the user can only manipulate top level data types using only the functions provided by PHAML. This not only removes the burden of knowing the details of the data structures from the user, but it provides for the evolution of those data structures without any changes to the user code, improving upward compatibility of revisions of PHAML.

Simplicity of the user interface is also improved by using the optional argument and dynamic memory features of Fortran 90. The argument list for the primary subroutine is rather long, but nearly all the arguments are optional. The user only provides the arguments for which a specific value is desired; all other arguments assume a reasonable default value. Finally, the use of dynamic memory (allocatable arrays) removes the burden of allocating workspace of the correct size, which is often associated with FORTRAN 77 libraries.

PHAML can run as either a sequential or parallel program. As a parallel program it is designed for distributed memory parallel computers, using message passing to communicate between the processors. It uses a master/slave model of computation with a collection of compute processes that perform most of the work, graphics processes that provide visualization, and a master process that coordinates the other processes. The executable programs can be configured as three separate programs for the three types of processes, or as a single program used by all of the processes. The single program approach is used when all processes are launched at once from the command line, but limits some of multiple-PDE capabilities of PHAML. In the multiple program approach, the master process is launched from the command line and spawns the other processes.

A simplified version of the algorithm of the primary subroutine is

```
initialize coarse grid
repeat
   if (predictive) then load balance
   refine/derefine
   if (not predictive) then load balance
   solve linear system
until termination criterion is met
```

Note there is option of performing *predictive load balancing* before the refinement of the grid occurs, or load balancing the grid after refinement. The numerical methods have been modified for parallel execution using the *full domain partition* approach [13–15]. This approach minimizes the frequency of communication between the processors to just a couple messages for each instance of refinement, load balancing or multigrid, which reduces the amount of time spent on communication, especially in high-latency, low-bandwidth environments like a cluster. Load balancing is performed by partitioning the grid and redistributing the data so that each process owns the data associated with a partition. The k-way refinement-tree partitioning method (RTK) [16] is used by default.

2 Modules

The Fortran 90 module is fundamental to good software design in Fortran 90. A module is a program unit that can contain variables, defined constants (a.k.a. parameters), type definitions, subroutines, and other entities. Each entity in a module can be private to the module or public. The entities that are public are available to any program unit that explicitly uses the module. Modules have many uses. For example, modules can contain global data to replace the old-style common blocks, contain interface blocks for an external library to provide strong type checking, or group together a collection of related subroutines.

One of the most important uses of modules is to provide data encapsulation, similar to a class in C++. The module contains one or more user defined types (a.k.a. structures) and functions that operate on those types. The type itself is made public, so that other program units can declare variables to be of that type, but the internals of the type are declared private, so that nothing outside the module can operate on the individual components of the type. For example

```
public hash_key
type hash_key
  private
  integer :: key
end type hash_key
```

Some of the functions in the module are made public to provide the means of operating on variables of the public type.

PHAML is organized into several modules, each of which contains either the structures and operations for some aspect of a parallel adaptive multi-level code, or global entities. The primary modules are:

`phaml`: This is the only module used directly by the user's code. It makes public the entities that the user needs, and contains all the functions that are directly callable by the user.

`linear_system`: This contains all data structures and operations related to creating and solving the linear system.

`grid`: This contains all operations on the grid, including refinement and redistribution.

`grid_type`: This contains the grid data type. It is separate from module `grid` because other modules also need access to this structure. For example, module `linear_system` needs to use the grid to create the linear system.

`load_balance`: This contains subroutines for partitioning the grid.

`message_passing`: This contains subroutines for communication between processes. It acts as an interface between a message passing library and the rest of the PHAML code.

`hash`: This contains operations on a hash table, which translates global IDs (known to all processes) to local IDs (known only to one process).

`global`: This contains global data which can be used by any program unit.

3 Data Structures

PHAML defines many data structures, most of which are private to a module or have private internals. A complete description of even the main structures is beyond the scope of this paper, and would be fruitless since they continue to evolve as the capabilities of PHAML expand. This section illustrates the flavor of the data structures through a snapshot of the current state of a few of them.

The only data type available to the user is `phaml_solution_type`, defined as

```
type phaml_solution_type
   private
   type(grids_type) :: grids
   type(proc_info) :: procs
   integer :: outunit, errunit, pde_id
   character(len=HOSTLEN) :: graphics_host
   logical :: i_draw_grid, master_draws_grid, &
              i_draw_reftree, master_draws_reftree, &
              still_sequential
end type phaml_solution_type
```

This structure contains all the information that one processor knows about the computed solution, grid and processor configuration for a PDE. See Sect. 4 for the operations that can be performed on a variable of this type. The first component contains the grid information. `type(grids_type)` contains the grid(s) corresponding to one or more partitions of the global grid. It allows for more than one partition to be assigned to a processor for possible future expansion to shared memory parallelism and/or cache-aware implementations. `type(proc_info)` contains information about the processor configuration for message passing. It is defined in module `message_passing` with private internals, and it's components depend on the message passing library in use. For example, the PVM version contains, among other things, the PVM task ids, while the MPI version contains the MPI communicators.

A slightly reduced version of the grid data type is

```
type grid_type
   type(element_type), pointer :: element(:)
   type(node_type), pointer :: node(:)
   type(hash_table) :: elem_hash, node_hash
   integer :: next_free_elem, next_free_node
```

```
   integer, pointer :: head_level_elem(:), head_level_node(:)
   integer :: partition
   integer :: nelem, nelem_leaf, nelem_leaf_own, nnode, &
              nnode_own, nlev
end type grid_type
```

The first two components are arrays containing the data for each element and node of the grid. These are allocatable arrays (the pointer attribute is used because Fortran 90 does not have allocatable structure components, but does allow a pointer to an array to be allocated), which allows them to grow as the grid is refined. The next two components are the hash tables, which are used for converting global IDs to local IDs. A global ID is a unique identifier for every element and node that may be created through refinement of the initial grid, and is computable by every processor. Global IDs are used for communication about grid entities between processors. The local ID is the index into the array for the element or node component. The next four components provide linked lists to pass through the elements or nodes of each refinement level. `partition` indicates which partition of the global grid is contained in this variable. Finally, the remaining scalars indicate the size of the grid and how much of it is owned by this partition.

Examining one level further, the data type for a node is given by

```
type node_type
   type(hash_key) :: gid
   type(point) :: coord
   real :: solution
   integer :: type, assoc_elem, next, previous
end type node_type
```

4 User Interface

The user interface to PHAML consists of two parts: 1) external subroutines written by the user to define the PDE problem, and 2) the PHAML subroutines that operate on a `phaml_solution_type` variable to solve the PDE problem.

The user must provide two external subroutines, `pdecoef` and `bcond`, to define the differential equation and boundary conditions, respectively. For problems involving the solution of more than one PDE, multiple interdependent PDEs can be defined in these subroutines, using the global variable `my_pde_id` to determine which one should be evaluated. An example of subroutine `pdecoef` is

```
subroutine pdecoef(x,y,p,q,r,f)

! pde is
!   -( p(x,y)*u ) -( q(x,y)*u ) +r(x,y)*u = f(x,y)
!              x x            y y
```

```
real, intent(in) :: x(:),y(:)
real, intent(out), optional :: p(:),q(:),r(:),f(:)

if (present(p)) p = 1.0
if (present(q)) q = 1.0
if (present(r)) r = 0.0
if (present(f)) f = x**2 + y**2

end subroutine pdecoef
```

Note that the arguments are arrays. This allows PHAML to call the subroutine with many quadrature points to reduce the overhead of calling it many times with one point. But, with Fortran 90's array syntax, in most cases the assignment can be done as a whole array assignment and look the same as the corresponding code for scalars. Also, the return arguments are optional, so the user must check for their existence with the intrinsic subroutine `present`. This allows PHAML to avoid unnecessary computation of coefficients that it does not intend to use at that time.

The user must also provide a subroutine to define the initial grid, which also defines the polygonal domain. At the time of this writing, an example is provided for rectangular domains, but it is difficult for a user to write the subroutine for more complicated domains. It is hoped that in the future PHAML will interface to a public domain grid generation code to define the initial grid.

Optionally, the user may also provide a subroutine with the true solution of the differential equation, if known, for computing norms of the error.

The user provides a main program for the master process. This program uses module `phaml`, and calls the public PHAML subroutines to perform the desired operations. The simplest program is

```
program user_main_example
use phaml
type(phaml_solution_type) :: pde
call create(pde)
call solve_pde(pde)
call destroy(pde)
end program
```

At the time of this writing there are nine public subroutines in module `phaml`. It is not the purpose of this paper to be a user's guide, so only a brief description of the function of the routines is given, except for the primary subroutine where some of the arguments are discussed.

`create, destroy`: These two subroutines are similar to a constructor and destructor in C++. Any variable of type `phaml_solution_type` must be passed to `create` before any other subroutine. Subroutine `create` allocates memory, initializes components, and spawns the slave and graphics processes. Subroutine `destroy` should be called to free memory and terminate spawned processes.

`solve_pde`: This is the primary subroutine, discussed below.

`evaluate`: This subroutine is used to evaluate the computed solution at a given array of points.

`connect`: With multiple PDEs, each one has its own collection of slave processes (see Sect. 5). For interdependent PDEs, these processes must be able to communicate with each other. This subroutine informs two `phaml_solution_type` variables about each other, and how to communicate with each other.

`store, restore`: These routines provide the capability of saving all the data in a `phaml_solution_type` variable to disk, and restoring it from disk at a later time.

`popen, pclose`: These routines provide parallel open and close statements, so that each process opens an I/O unit with a unique, but similarly named, file. This is used for the files in `store` and `restore`, and the `output_unit` and `error_unit` arguments to subroutine `create`.

Subroutine `solve_pde` is the primary public subroutine of PHAML. All the work of grid refinement and equation solution occurs under this subroutine. At the time of this writing it has 43 arguments to provide flexibility in the numerical methods and amount of printed and graphical output. All of these arguments are optional with reasonable default values, so the calling sequence need not be more complicated than necessary. Usually a user would provide them as keyword arguments (the name of the argument is given along with the value), which improves readability of the user's code. For example

```
call solve_pde(pde, &
               max_node = 20000, &
               draw_grid_when = PHASES, &
               partition_method = ZOLTAN_RCB, &
               mg_cycles = 2)
```

For many of the arguments, the acceptable values are given by defined constants (for example, `PHASES` and `ZOLTAN_RCB` above) which are public entities in module `phaml`.

Some of the arguments to `solve_pde` are:

`max_elem, max_node, max_lev, max_refsolveloop`: These are used as termination criterion.

`init_form`: This indicates how much initialization to do. `NEW_GRID` starts from the coarse grid, `USE_AS_IS` starts the refinement from an existing grid from a previous call, and `SOLVE_ONLY` does not change the grid, it just solves the PDE on the existing grid from a previous call.

`print_grid_when, print_grid_who`: These determine how often summary information about the grid should be printed, and whether it should be printed by the `MASTER, SLAVES` or `EVERYONE`. There are also similar arguments for printing of the error (if the true solution is known), time used, and header and trailer information.

`uniform`, `overlap`, `sequential_node`, `inc_factor`, `error_estimator`, `refterm`, `derefine`: These arguments control the adaptive refinement algorithm.

`partition_method, predictive`: These arguments control the load balancing algorithm.

`solver`, `preconditioner`, `mg_cycles`, `mg_prerelax`, `mg_postrelax`, `iterations`: These arguments control the linear system solver algorithm.

5 Parallelism

PHAML uses a master/slave model of parallel computation on distributed memory parallel computers or clusters of workstations/PCs. The user works only with the master process, which spawns the slave processes. PHAML also provides for sequential execution and for spawnless parallel execution, but this section assumes the spawning form of the program.

The parallelism in PHAML is hidden from the user. One conceptualization is that the computational processes are part of a `phaml_solution_type` object, and hidden like all the other data in the object. In fact, one of the components of the `phaml_solution_type` structure is a structure that contains information about the parallel processes. The user works only with the master process. When the master process calls subroutine `create` to initialize a `phaml_solution_type` variable, the slave processes are spawned and the `procs` component of the `phaml_solution_type` variable is initialized with whatever information is required for the processes to communicate. If another `phaml_solution_type` variable is initialized, a different set of slave processes are spawned to work on this one. When the master process calls any of the other public subroutines in module `phaml`, it sends a message to the slaves with a request to perform the desired operation. When the operation is complete, the slave waits for another request from the master. When subroutine `destroy` is called from the master process, the slave processes are terminated.

PHAML was written to be portable not only across different architectures and compilers, but also across different message passing means. All communication between processes is isolated in one module, `message_passing`. This module contains the data structures to maintain the information needed about the parallel configuration, and all operations that PHAML uses for communication, such as `comm_init` (initialization), `phaml_send`, `phaml_recv`, `phaml_global_max`, etc. Thus to introduce a new means of message passing, one need only write a new version of module `message_passing` that adheres to the defined API. PHAML contains versions for PVM, MPI 1 (without spawning), MPI 2 (with spawning), and a dummy version for sequential programs.

6 Conclusion

This paper described the software design of PHAML, a parallel program for the solution of partial differential equations using finite elements, adaptive refinement and a multi-level solver. The program is written in Fortran 90 and makes heavy use of modules for program organization and data encapsulation. The user interface is small and makes use of optional and keyword arguments to keep the calling sequence short and readable. The parallelism is hidden from the user, and portable across different message passing libraries.

The PHAML software has been placed in the public domain, and is available at *URL to be determined.*

References

1. Anderson, E., Bai, Z., Bischof, C., Demmel, J., Dongarra, J., Du Croz, J., Greenbaum, A., Hammarling, S., McKenney, A., Ostrouchov, S., Sorensen, D.: *LAPACK Users' Guide*, SIAM, Philadelphia, 1982
2. Barrett, R., Berry, M., Chan, T. F., Demmel, J., Donato, J., Dongarra, J., Eijkhout, V., Pozo, R., Romine, C., Van der Vorst, H.: *Templates for the Solution of Linear Systems: Building Blocks for Iterative Methods*, SIAM, Philadelphia, 1994
3. Boman, E., Devine, K., Hendrickson, B., Mitchell, W. F., St. John, M., Vaughan, C.: Zoltan: A dynamic load-balancing library for parallel applications, user's guide, Sandia Technical Report SAND99-1377 (2000)
4. Geist, A., Beguelin, A., Dongarra, J., Jiang, W., Manchek, R., Snderam, V.: *PVM: Parallel Virtual Machine. A Users' Guide and Tutorial for Networked Parallel Computing*, MIT Press, Cambridge, 1994
5. Gropp, W., Huss-Lederman, S., Lumsdaine, A., Lusk, E., Nitzberg, B., Saphir, W., Snir, M.: *MPI: The Complete Reference*, MIT Press, Cambridge, MA, 1998
6. Kilgard, M.: The OpenGL Utility Toolkit (GLUT) programming interface API version 3, http://www.opengl.org (1996)
7. Lawson, C. L., Hanson, R. J., Kincaid, D., Krogh, F. T.: Basic Linear Algebra Subprograms for FORTRAN usage, ACM Trans. Math. Soft. **5** (1979) 308–323
8. Lehoucq, R. B., Sorensen, D. C., Yang, C.: *ARPACK Users' Guide*, SIAM, Philadelphia, 1998
9. Mitchell, W. F.: Adaptive refinement for arbitrary finite element spaces with hierarchical bases, J. Comp. Appl. Math. **36** (1991) 65–78
10. Mitchell, W. F.: Optimal multilevel iterative methods for adaptive grids, SIAM J. Sci. Statist. Comput. **13** (1992) 146–167
11. Mitchell, W. F.: MGGHAT user's guide version 1.1, NISTIR 5948 (1997)
12. Mitchell, W. F.: A Fortran 90 interface for OpenGL: Revised January 1998, NISTIR 6134 (1998)
13. Mitchell, W. F.: The full domain partition approach to distributing adaptive grids, Appl. Num. Math. **26** (1998) 265–275
14. Mitchell, W. F.: The full domain partition approach to parallel adaptive refinement, in *Grid Generation and Adaptive Algorithms, IMA Volumes in Mathematics and it Applications* **113** Springer-Verlag (1998) 151–162
15. Mitchell, W. F.: A parallel multigrid method using the full domain partition, Elect. Trans. Num. Anal. **6** (1998) 224–233
16. Mitchell, W. F.: The refinement-tree partition for parallel solution of partial differential equations, NIST J. Res. **103** (1998) 405–414
17. Woo, M, Neider, J., Davis, T., Shreiner, D.: *The OpenGL Programming Guide*, Addison-Wesley, 1999

OpenMP versus MPI for PDE Solvers Based on Regular Sparse Numerical Operators *

Markus Nordén, Sverker Holmgren, and Michael Thuné

Uppsala University, Information Technology, Dept. of Scientific Computing, Box 120
SE-751 04 Uppsala, Sweden
{markusn, sverker, michael}@tdb.uu.se

Abstract. Two parallel programming models represented by OpenMP and MPI are compared for PDE solvers based on regular sparse numerical operators. As a typical representative of such an application, the Euler equations for fluid flow are considered.
The comparison of programming models is made with regard to UMA, NUMA, and self optimizing NUMA (NUMA-opt) computer architectures. By NUMA-opt, we mean NUMA systems extended with self optimizations algorithms, in order to reduce the non-uniformity of the memory access time.
The main conclusions of the study are: (1) that OpenMP is a viable alternative to MPI on UMA and NUMA-opt architectures, (2) that OpenMP is not competitive on NUMA platforms, unless special care is taken to get an initial data placement that matches the algorithm, and (3) that for OpenMP to be competitive in the NUMA-opt case, it is *not* necessary to extend the OpenMP model with additional data distribution directives, *nor* to include user-level access to the page migration library.

Keywords: OpenMP; MPI; UMA; NUMA; Optimization; PDE; Euler; Stencil

1 Introduction

Large scale simulations requiring high performance computers are of importance in many application areas. Often, as for example in fluid dynamics, electromagnetics, and acoustics, the simulations are based on PDE solvers, i.e., computer programs for the numerical solution of partial differential equations (PDE). In the present study, we consider parallel PDE solvers involving regular sparse operators. Such operators typically occur in the case of finite difference or finite volume methods on structured grids, either with explicit time-marching, or with *implicit* time-marching where the resulting algebraic systems are solved using an iterative method.

In the present article, we compare two programming models for PDE solver applications: the shared name space model and the message passing model. The

* The work presented here was carried out within the framework of the Parallel and Scientific Computing Institute. Funding was provided by Sun Microsystems and the Swedish Agency for Innovation Systems.

P.M.A. Sloot et al. (Eds.): ICCS 2002, LNCS 2331, pp. 681–690, 2002.

question we pose is: *will recent advances in computer architecture make the shared name space model competitive for simulations involving regular sparse numerical operators*? The tentative answer we arrive at is *"yes"*.

We also consider an additional issue, with regard to the shared name space model, viz., whether it requires explicit data distribution directives. Here, our experiments indicate that the answer is *"no"*.

The state-of-the-art for parallel programming of large scale parallel PDE solvers is to use the message passing model, which assumes a local name space in each processor. The existence of a default standard for this model, the Message Passing Interface (MPI) [5], has contributed to its strong position. However, even more important has been its ability to scale to large numbers of processors [6]. Moreover, many major massively parallel computer systems available on the market present, at least partly, a local name space view of the physically distributed memory, which corresponds to the assumptions of the message passing model.

However, with recent advances in SMP server technology has come a renewed and intensified interest in the shared name space programming model. There is now a de facto standard also for this model: OpenMP [1]. However, it is still an open question how well OpenMP will scale beyond the single SMP server case. Will OpenMP be a viable model also for clusters of SMPs, the kind of computer architecture that is currently dominating at the high end?

Clusters of SMPs typically provide non-uniform memory access (NUMA) to the processors. One approach to OpenMP programming in a NUMA setting is to extend the model with directives for data distribution, in the same spirit as in High-Performance Fortran. By directing the initial data placement explicitly, the same way as an MPI programmer would need to do, the user would be able to ensure that the different OpenMP threads get reasonably close to their data. This argument was put forward in, e.g., [4].

Another, more orthodox approach, was taken by Nikolopoulos et al. [2, 3], who claim that data distribution directives should *not* be added to OpenMP, since that would contradict fundamental design goals for the OpenMP standard, such as platform-independence and ease of programming. Moreover, they claim that directives are *not necessary* for performance, provided that the OpenMP implementation is supported by a dynamic page migration mechanism. They have developed a user-level page migration library, and demonstrate that the introduction of explicit calls to the page migration library into the OpenMP code enables OpenMP programs without distribution directives to execute with reasonable performance on both structured and non-structured scientific computing applications [2, 3].

Our contribution is in the same spirit, and goes a step further, in that we execute our experiments on a *self optimizing* NUMA (NUMA-opt) architecture, and rely exclusively on its built-in page migration and replication mechanisms. That is, no modifications are made to the original OpenMP code. The platform we use is the experimental Orange (previously known as Wildfire) architecture from Sun Microsystems [7]. It can be configured, in pure NUMA mode (no page migration and replication), and alternatively in various self optimization modes (only migration, only replication, or both). Moreover, each node of the system is an SMP, i.e., exhibits UMA behavior. Thus, using one and the same platform, we have been able to experiment with a variety of computer architecture types under *ceteris paribus* conditions.

Our Orange system consists of two 16-processor nodes, with UltraSparc II processors (i.e., not of the most recent generation), *but* with a sophisticated self optimization mechanism. Due to the latter, we claim that the Orange system can be regarded as a prototype for the *kind* of parallel computer platforms that we will see in the future. For that reason, we find it interesting to study the issue of OpenMP versus MPI for this particular platform.

The results of our study are in the same direction as those of Nikolopoulos et al. Actually, our results give even stronger support for OpenMP, since they do not presume user-level control of the page migration mechanisms. Moreover, our results are in agreement with those of Noordergraaf and van der Pas [10], who considered data distribution issues for the standard five-point stencil for the Laplace equation on a Sun Orange system. Our study can be regarded as a generalization of theirs to operators for non-scalar and non-linear PDEs, and also including a comparison to using a message passing programming model.

2 The Stencil Operator

The experiments reported below are based on a stencil which comes from a finite difference discretization of the nonlinear Euler equations in 3D, describing compressible flow. The application of this stencil operator at a certain grid point requires the value of the operand grid function at 13 grid points. This corresponds to 52 floating point numbers, since the grid function has four components.

Moreover, we assume that the physical structured grid is *curvilinear*, whereas the computations are carried out on a rectangular computational grid. This introduces the need for a mapping from the computational to the physical grid. Information about this mapping has to be available as well, and is stored in a 3×3-matrix that is unique for each grid point, which means nine more floating point numbers. In all, 61 floating point numbers have to be read from memory and approximately 250 arithmetic operations have to be performed at each grid point in every iteration.

The serial performance of our stencil implementation is close to 100 Mflop/s. This is in good agreement with the expectations according to the STREAM benchmark [9]. (See [11] for further discussion of the serial implementation.)

3 Computer System Configurations

On the Sun Orange computer system used here, there are a number of configurations to choose between. First of all, there are two self optimization mechanisms, page migration and replication, that can be turned on and off independently in the operating system.

Table 1. The computer system configurations used in the parallel experiments

Configuration	Thread scheduling	Memory allocation	Page migration	Page replication	Architecture type
Configuration 1	One node	One node	Off	Off	UMA
Configuration 2	Default	One node	On	On	NUMA-opt
Configuration 3	Default	Matching	On	On	NUMA-opt
Configuration 4	Balanced	One node	On	On	NUMA-opt
Configuration 5	Balanced	Matching	On	On	NUMA-opt
Configuration 6	Balanced	One node	Off	Off	NUMA
Configuration 7	Balanced	Matching	Off	Off	NUMA
Configuration 8	Balanced	One node	Off	On	NUMA-opt
Configuration 9	Balanced	One node	On	Off	NUMA-opt

Using these mechanisms it is possible to configure the Orange systems so as to represent a variety of architecture types. First, using only one server of the system gives a UMA architecture. Secondly, using both servers, but turning off the self optimization mechanisms, gives a NUMA. Finally, self optimizing NUMA systems with various degrees of self optimization can be studied by turning on the page migration and/or replication.

For the investigation of how OpenMP performs in different environments, we are interested in the variation not only in architecture type, but also in thread placement and data placement. This variation can also be achieved in the Orange system.

Table 1 summarizes the different Orange system configurations that were used in the parallel experiments reported below. With Configuration 1 we only use the resources in one node, i.e. we are running our program on an SMP server and the number of threads is limited to 16.

Configuration 2 and 3 both represent the default Orange system settings, with all self optimization turned on. The difference is that for the former all the data are initially located in one node, whereas for the latter they are distributed in a way that matches the threads already from the beginning.

For Configuration 4–9 the load of the SMP nodes is balanced, in that the threads are scheduled evenly between the nodes. The configurations differ, however, in the way that data are initialized and which self optimization mechanisms are used. Configuration 6 and 7, with no self optimization, represent pure NUMA systems.

We used the Sun Forte 6.2 (early access, update 2) compiler. It conforms to the OpenMP standard, with no additional data distribution directives. The configurations with matching data distribution were obtained by adding code for *initializing* data (according to the first-touch principle) in such a way that it was placed were it was most frequently needed. In this way, the same effect was obtained *as if* OpenMP had been extended with data distribution directives.

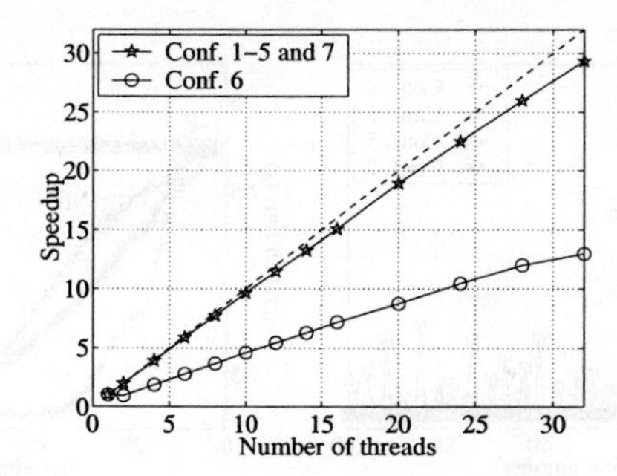

Fig. 1. Speedup per iteration for an OpenMP solver for the nonlinear Euler equations in 3D. The speedup was measured with respect to the time needed to carry out one iteration, once the system is fully adapted. Different curves correspond to different configurations of the parallel computer platform, see Table 1

4 OpenMP and the Effect of Self Optimization

The first series of experiments studies the performance of OpenMP for all the configurations discussed above. In particular we are interested in evaluating the effect of self optimization (in the computer system) on the performance of parallel programs based on OpenMP.

The reason for introducing self optimization is to allow the system to scale well even when more than one SMP node is used. Therefore, speedup is a good measure of how successful the optimization is. However, it is important to let the system adapt to the algorithm before any speedup measurements are done. Consequently, we have measured speedup with respect to the time needed to carry out one iteration once the system is fully adapted, i.e., after a number of iterations have been performed, see below.

The time-per-iteration speedup results for our application are shown in Figure 1. As can be seen, all configurations scale well, except for Configuration 6. The latter corresponds to a NUMA scenario, with no self optimization, and where data are distributed unfavorably. The configurations that rely on the self optimization of the system show identical speedup as the configurations that rely on hand-tuning. That is, after the initial adaption phase, the self optimization mechanisms introduce no further performance penalty.

Next, we study how long it takes for the system to adapt. The adaption phase should only take a fraction of the total execution time of the program, otherwise the self optimization is not very useful.

We have measured the time per iteration for the different configurations. As can be seen in Figure 2, the adaption phase takes approximately 40–60 iterations for our program. This is fast enough, since the execution of a PDE solver usually involves several hundreds or even thousands of such iterations.

Consequently, the conclusion of the speedup and adaption-time experiments, taken together, is that the self optimization in the Orange system serves its purpose well for the kind of application we are considering.

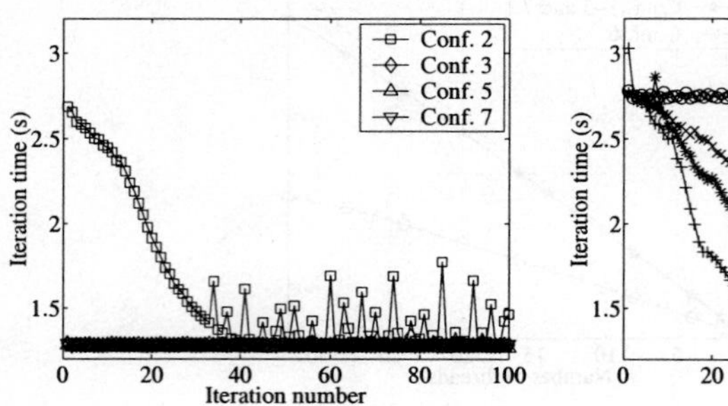

(a) Different thread scheduling, data distribution and optimization

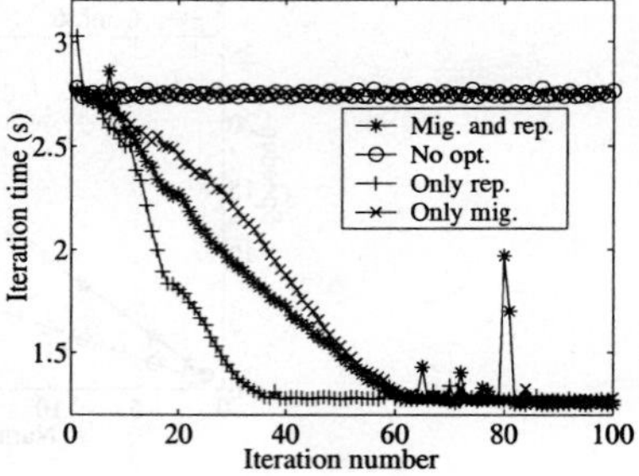

(b) Different self optimization mechanisms

Fig. 2. Time per iteration of our test program for different configurations of the computer system. The graphs refer to the 24 processor case, and similar results were obtained for other numbers of processors. After 40–60 iterations the system has adapted to the memory access pattern of the algorithm. This overhead is negligible in comparison to the typical total number of iterations

With regard to OpenMP, the conclusion is that additional data distribution directives are *not* needed for PDE solvers based on regular sparse numerical operators. This holds provided that the computer system is equipped with efficient self optimization algorithms, as is the case with the Orange system prototype used in our experiments.

In Table 2 we also show how many memory pages are migrated and replicated when using the different self optimization techniques. All the data were initialized to reside on one node, and the thread allocation is balanced over the nodes.

When using both migration and replication, approximately half of the data are migrated to the node where they are used. There are also some memory pages that are replicated, probably those that are used to store data on the border between the two halves of the grid, and therefore are accessed by threads in both nodes.

When using only replication, approximately half of the data are replicated to the other node and when only migration is allowed, half of the data are migrated.

With regard to the optimization modes, the conclusion of our experiments is that migration is sufficient for the kind of numerical operators we consider here. Combined replication and migration does not lead to faster adaption. The third alternative, replication only, gives slightly faster adaption, but at the expense of significant memory overhead.

Table 2. Iteration time and the number of pages that are migrated and replicated using the different optimization mechanisms. The threads are scheduled evenly between the nodes but data initially resides in just one of the nodes. In these experiments 24 threads were used and in all 164820 memory pages were used

Optimization	Iter. time	# Migrs	# Repls
Mig. and rep. (4)	1.249	79179	149
Only rep. (8)	1.267	N/A	79325
Only mig. (9)	1.264	79327	N/A

4.1 OpenMP versus MPI

We now proceed to comparing OpenMP and MPI. We have chosen to use balanced process/thread scheduling for both the MPI and OpenMP versions of the program. Every process of the MPI program has its own address space and therefore matching allocation is the only possibility. It should also be mentioned that since the processes have their own memory, there will be no normal memory pages that are shared by different processes. Consequently, a program that uses MPI will probably not benefit from migration or replication. This is also confirmed by experiments, where we do not see any effects of self optimization on the times for individual iterations as we did in the previous section.[1]

The experiments below also include a hybrid version of the program, which uses both MPI and OpenMP. There, we have chosen OpenMP for the parallelization within the SMP nodes and MPI for the communication between the nodes.

Now to the results. We have already seen that the OpenMP version scales well for both the UMA and self optimizing NUMA architectures. The results for Configuration 1 and the different NUMA-opt configurations were virtually identical. On the other hand, the speedup figures for the NUMA type of architecture (Configuration 6) were less satisfactory.

Turning to MPI, that programming model is not aware of the differences between the three architecture types, as discussed above. The same holds for the hybrid version, since it uses one MPI process for each node, and OpenMP threads within each such process. Consequently, the execution time was virtually the same for all MPI cases, regardless of architecture type, and similarly for the hybrid OpenMP/MPI cases.

[1] Accesses are made to the same address by different processes when we use MPI communication routines. This communication is normally performed so that one process writes the data to an address that is shared, and another process subsequently reads from the same address. Since the memory access pattern for that memory page is that one process always writes, after which another process reads, neither migration nor replication would improve performance. The reason is that in the case of migration the page would always be remotely located, as seen from one of the processes, and in the case of replication every new cache line that is to be read would result in a remote access since it has been updated on the other node since it was fetched last time.

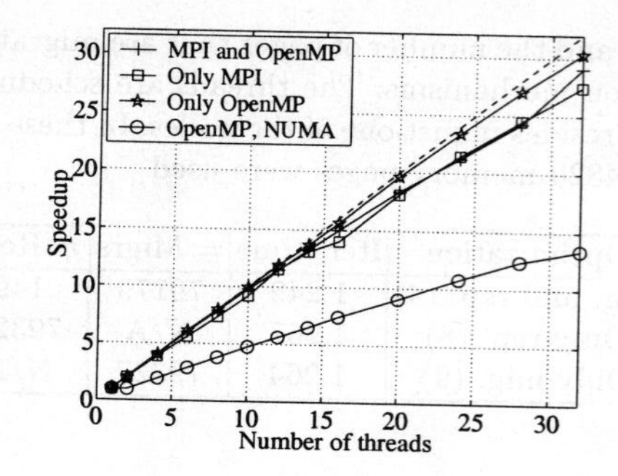

Fig. 3. Speedup for different versions of the non-linear Euler solver

Figure 3 shows the results in terms of time-per-iteration speedup. The MPI and hybrid versions give the same performance for all three architecture types. For OpenMP, the NUMA architecture gives significantly lower performance. However, for the UMA and NUMA-opt configurations OpenMP is competitive, and even somewhat better than the other alternatives.

Most likely, the MPI and hybrid versions scale less well because of time needed for buffering data during the communication. The reason why the hybrid version does not scale better than the pure MPI version is that even though there are fewer processes in the hybrid version than in the MPI version when the same number of processors are used, the amount of data to be exchanged is still the same for each process. The communication has to be done serially within the processes. Therefore, while one of the threads of an MPI process is busy sending or receiving data, the other threads of that process will be idle, waiting for the communication to take place.

5 Conclusions and Future Work

The main conclusions of our study are:

1. OpenMP is competitive with MPI on UMA and self optimizing NUMA architectures.
2. OpenMP is not competitive on pure (i.e., non-optimizing) NUMA platforms, unless special care is taken to get an initial data placement that matches the algorithm.
3. For OpenMP to be competitive in the self optimizing NUMA case, it is *not* necessary to extend the OpenMP model with additional data distribution directives, *nor* to include user-level access to the page migration library.

Clearly, there are limitations to the validity of these conclusions:

– They refer to applications involving regular sparse numerical operators. Such operators exhibit a very regular memory access pattern with only local communication, therefore it should be quite easy for the system to adapt to the algorithm. Further investigations are needed before the conclusions can be extended to applications with a highly irregular memory access pattern. However, the results reported by Nikolopoulos et al. [3], for OpenMP extended with user-level calls to a page migration library, give hope that our conclusions *will* in fact generalize to such applications.

- The Orange system used in this study has only two nodes. A massively parallel platform with a large number of nodes would be more challenging for the self optimization algorithms. We expect such systems to appear in the future, and we conjecture that it will be possible to generalize the migration and replication algorithms of the Orange system in such a way that the OpenMP model will be competitive on them as well. However, this remains to be proven.

For the near future, the really large scale computations will be carried out on massively parallel clusters of SMP (or heterogeneous clusters in a "grid" setting), with a local name space for each node. Then MPI, or the hybrid OpenMP/MPI model are the only alternatives. In fact, the results reported in [6] indicate that for some applications, the hybrid model is to prefer for large numbers of processors.

Our results for the NUMA case show that even for an SMP cluster equipped with an operating system that presents a shared name space view of the entire cluster, the MPI and hybrid models are still the best alternatives, in comparison with standard OpenMP. The data placement required for OpenMP to be competitive indicates the need for additional data distribution directives. On the other hand, since many platforms use the first-touch principle, an alternative way to achieve such data placement is via a straightforward initialization loop. Consequently, in our opinion, adding data distribution directives to OpenMP, in order to address the NUMA type of architecture, would not be worth its prize in terms of contradicting the design goals of OpenMP.

In the long term perspective, our results speak in favor of efficiently self optimizing NUMA systems, in combination with standard OpenMP, i.e., with no additional data distribution directives. As mentioned, we conjecture that self optimization algorithms of the type found in the Orange system can be generalized to work efficiently also for massively parallel NUMA systems. If this turns out to be true, programming those systems with standard OpenMP will allow for rapid implementation of portable parallel codes.

The work reported here is part of a larger project, "High-Performance Applications on Various Architectures" (HAVA). Other subprojects of HAVA consider other kinds of applications, for example pseudospectral solvers [8, 12], and solvers based on unstructured grids. The next phases of the present subproject will be to consider first a finite difference based *multi-grid solver* for the Euler equations, and then structured adaptive mesh refinement for the same application. The latter, in particular, provides additional challenges, for self optimization algorithms *as well as* for user-provided load balancing algorithms.

References

1. OpenMP Architechture Review Board. OpenMP Specifications.
2. D. S. Nikolopoulos et al. Is Data Distribution Necessary in OpenMP? In *SC2000 Proceedings.* IEEE, 2000.
3. D. S. Nikolopoulos et al. Scaling Irregular Parallel Codes with Minimal Programming Effort. In *SC2001 Proceedings.* IEEE, 2001.
4. J. Bircsak et al. Extending OpenMP for NUMA Machines. In *SC2000 Proceedings.* IEEE, 2000.
5. Message Passing Interface Forum. MPI Documents.
6. W. D. Gropp et al. High-performance parallel implicit CFD. *Parallel Computing,* 27:337–362, 2001.
7. E. Hagersten and M. Koster. WildFire: A Scalable Path for SMPs. In *Proceedings of the 5th IEEE Symposium on High-Performance Computer Architecture (HPCA),* 1999.
8. S. Holmgren and D. Wallin. Performance of a pseudo-spectral PDE solver on a self-optimizing NUMA architecture. In *Proc. of Euro-Par 2001.* Springer-Verlag, 2001.
9. J. D. McCalpin. Sustainable Memory Bandwidth in Current High Performance Computers. Technical report, Advanced Systems Division, Silicon Graphics, Inc., 1995.
10. L. Noordergraaf and R. van der Pas. Performance Experiences on Sun's WildFire Prototype. In *SC99 Proceedings.* IEEE, 1999.
11. M. Nordén, S. Holmgren, and M. Thuné. OpenMP versus MPI for PDE solvers based on regular sparse numerical operators. Report in preparation.
12. D. Wallin. Performance of a high-accuracy PDE solver on a self-optimizing NUMA architecture. Master's thesis, Uppsala University School of Engineering, 2001. Report No. UPTEC F 01 017.

High-Level Scientific Programming with Python

Konrad Hinsen[1]

Centre de Biophysique Moléculaire (UPR 4301 CNRS), Rue Charles Sadron, 45071 Orléans Cedex 2, France
hinsen@cnrs-orleans.fr
WWW home page: http://dirac.cnrs-orleans.fr/

Abstract. Scientific computing is usually associated with compiled languages for maximum efficiency. However, in a typical application program, only a small part of the code is time-critical and requires the efficiency of a compiled language. It is often advantageous to use interpreted high-level languages for the remaining tasks, adopting a mixed-language approach. This will be demonstrated for Python, an interpreted object-oriented high-level language that is particularly well suited for scientific computing. Special emphasis will be put on the use of Python in parallel programming using the BSP model.

1 Introduction

Scientific computing has some specific requirements that influence the choice of programming tools. The most outstanding property of scientific computing is its explorative nature: although some standard methods are used over and over again, they are used in different combinations every time, and often it is necessary to add custom algorithms and programs to a collection of well-established standard code. Although the literature on scientific computing leaves the impression that all that matters are efficient number crunching and visualization methods, the day-to-day work of a computational scientist involves a lot of interfacing, file format conversion, bookkeeping, and similar tasks, often made difficult by bad user interface design and lack of documentation. These lengthy and unattractive tasks often discourage scientists to pursue a potentially interesting idea. Good programming tools can thus make an important contribution to good computational science.

High-level languages can help in several ways. At the simplest level, they can be used to write all tools that are not time critical, such as simple analysis programs, file format converters, etc. As a general rule, high-level languages are much better suited for I/O- and text-oriented tasks than the standard programming languages used in scientific computing (Fortran, C, C++). However, as this article will show, they can be useful in number-crunching applications as well, making them easier to develop and to use. The key to these applications is mixed-language programming, i.e. combining a high-level and a low-level language in order to get the best of both worlds.

P.M.A. Sloot et al. (Eds.): ICCS 2002, LNCS 2331, pp. 691–700, 2002.

To avoid misunderstandings, an explanation of the term "high-level" is in order. Most of all, it implies no judgement of quality. High-level languages are by definition those whose constructs and data types are close to natural-language specifications of algorithms, as opposed to low-level languages, whose constructs and data types reflect the hardware level. With high-level languages, the emphasis is on development convenience, whereas low-level languages are designed to facilitate the generation of efficient code by a compiler. Characteristic features of high-level languages are interactivity, dynamic data structures, automatic memory management, clear error messages, convenient file handling, libraries for common data management tasks, support for the rapid development of graphical user interfaces, etc. These features reduce the development and testing time significantly, but also incur a larger runtime overhead leading to longer execution times.

Note that what is called "high-level" in this article is often referred to as "very high level"; different authors use different scales. Some authors refer to these languages as "scripting languages", which however seems too limiting, as scripting is merely one of their applications.

The high-level language that is used as an example in this article is Python [1], a language that is becoming increasingly popular in the scientific community. Although other suitable languages exist and the choice always involves personal preferences, Python has some unique features that make it particularly attractive: a clean syntax, a simple but powerful object model, a flexible interface to compiled languages, automatic interface generators for C/C++ and Fortran, and a large library of reusable code, both general and scientific. Of particular importance is Numerical Python [5], a library that implements fast array operations and associated numerical operations. Many numerical algorithms can be expressed in terms of array operations and implemented very efficiently using Numerical Python. Moreover, Numerical Python arrays are used at the interface between Python and low-level languages, because their internal data layout is exactly that of a C array.

2 Scripting and Computational Steering

A typical situation in computational science is the following: an existing program contains all the revelant methods, but its user interface is cumbersome, I/O facilities not sufficient, and interfacing with other programs could be easier. Another common case is the existence of a library of computational algorithms which is used by relatively simple application programs that are constantly modified. In this case, modification and testing of the applications often take a significant amount of time.

A good solution in both situations is the use of a high-level language for *scripting*, which is sometimes called *computational steering* in the context of scientific computing. The user of a Python-scripted application/library writes simple Python programs that make heave use of calls to the application/library, but can also use other Python modules, e.g. for I/O. The advantage of scripting is increased flexibility, e.g. being able to use variables and loops to define calculation paramters, and shorter development and testing times for specific application programs. It is also rather easy to add a graphical user interface.

In case of an existing monolithic program, the first step would be to isolate the computational parts of the code and turn them into a (highly specialized) library; much of the user interface and I/O code would be discarded. This library then has to be provided with a Python interface; in most cases this task can be handled by an automatic interface generator such as SWIG [2] for C/C++ and Pyfort [3] or F2PY [4] for Fortran.

Another possibility, preferred when scripting is not the standard user interface, is *embedding* Python into an existing application. This option is limited to C and C++ programs. A program with an embedded Python interpreter can ask the interpreter to execute code, run a script, etc. A typical case would be a program with a graphical user interface that also offers scripting for advanced users. The difference to the straightforward scripting approach described above is that the application program is in charge of script execution, it can decide if and when to run Python code.

An advantage of the scripting approach is that it is easy to realize. Existing code can be used without extensive modifications, and users need only learn the basics of Python in order to be able to profit from scripting. On the other hand, the benefit is mostly limited to the users, developers work much like before. The majority of the code is still written in a low-level language, and the design of the low-level code, especially its data structures, determine the design of the scripting layer.

3 High-Level Design

The complementary approach to scripting is to design an application or library for the high-level language, switching to low-level code only for specific time-critical parts. The roles of the two languages are thus inversed, the low-level code is written specifically to fit into the high-level design. The developer can profit from all of the advantages of high-level languages to reduce development and testing time, and – assuming a good programming style – the code becomes more compact and much more readable. However, this approach makes it less straightforward to integrate existing low-level code, unless it takes the form of a library with a well-designed interface. High-level design also requires a good knowledge of Python and object-oriented techniques.

It must be stressed that the result of this approach is very different from simple scripting, and that the practical advantages are significant. In the course of time, a computational scientist can build up a library of problem-specific code, written by himself or obtained from others, that uses the same scientific concepts and abstractions as natural language: numbers, vectors, functions, operators, atoms, molecules, flow fields, differential equation solvers, graphical representations, etc. In low-level code, with or without scripting, everything would have to be expressed in terms of numbers and arrays plus functions working on these data.

To give a simple example, suppose you have a large compressed text file containing one number per line and you want to plot a histogram of that number set. In Python this can be written as follows:

```
from Scientific.IO.TextFile import TextFile
from Scientific.Statistics.Histogram import Histogram
from Gnuplot import plot

data = []
for line in TextFile('data.gz'):
    data.append(float(line))
plot(Histogram(data, 100))
```

The class TextFile presents a simple abstraction of a text file to the user: a sequence of lines that can be iterated over. Internally it handles many details: it can deal with standard as well as compressed files, and it accepts URLs instead of filenames, in which case it automatically downloads the file from a remote server, stores it temporarily, and deletes the local copy when it has been read completely. The user need not know how any of this is accomplished, for him a text file always remains just a sequence of lines.

The class Histogram provides a similar abstraction for histograms. You give it the data points and the number of bins, and that is all you need to know. Of course the classes TextFile and Histogram must be written first, but only once by one person, they can then be used by anyone anywhere, even interactively, without the need to know how they works internally.

It is often said that object-oriented low-level languages, such as C++, can be used in the same way. However, the higher implementation effort is often discouraging and one settles for the simplest solution that will do the job at hand, even if that means starting from scratch for the next project. Moreover, the code would have to be recompiled for each application, whereas the Python code can be used interactively in a calculator-style fashion. Python with a problem-specific library can be used as a "numerical workbench" for explorative computing.

As a general rule, code reuse works much better in Python than in low-level languages, whose strict typing rules make it difficult to design sufficiently flexible interfaces. With the exception of libraries designed by expert programmers with the explicit goal of generality (e.g. LAPACK), scientific code in low-level languages is almost never directly reusable. In Python, reusability is much easier to achieve, and the weak type compatibility rules, combined with independent name spaces for modules, ensure that even libraries designed completely independently work well together.

4 Parallel Computing

As an example of the use of high-level design in a traditional heavy-duty computing field, this section shows how Python can be used to facilitate the development of parallel programs.

Most textbooks on parallel computing focus on algorithmic aspects, mentioning implementation issues only in passing. However, the implementation of parallel algorithms is far from trivial, since a real-life application uses several different algorithms and requires a significant amount of bookkeeping and I/O. Although many computational scientists envisage parallelization at some time, few ever get beyond simple test programs, because development and debugging become too cumbersome.

A major reason for the difficulty of parallel programming is the low-level nature of the most popular parallel communications library, the Message Passing Interface (MPI). MPI has a large number of functions that permit the optimization of many communication patterns, but it lacks easy-to-use high-level abstractions. Most importantly, it places the responsibility for synchronization fully on the programmer, who spends a lot of time analyzing deadlocks. Moreover, MPI does not offer much support for transferring complex data structures.

A much simpler and more convenient parallel programming model is the Bulk Synchronous Parallel (BSP) model [6]. In this model, computation and communication steps alternate, and each communication step involves a synchronization of all processors, effectively making deadlocks impossible. Another advantage of bundling communication in a special step is the possibility for an underlying communications library to optimize data exchange for a given machine, e.g. by combining messages sent to the same processor. The analysis of algorithms is also facilitated, making it possible to predict the performance of a given algorithm on a given parallel machine based on only three empirical parameters. The Python implementation of BSP (which is part of the Scientific Python package [7]) adds the possibility of exchanging almost arbitrary Python objects between processors, thus providing a true high-level approach to parallelization.

4.1 Overview

An important difference for readers familiar with MPI programming is that a Python BSP program should be read as a program for a parallel machine made up of N processors and *not* as a program for one processor that communicates with $N-1$ others. A Python BSP program has two levels, local (any one processor) and global (all processors), whereas a typical message-passing program uses only the local level. In message-passing programs, communication is specified in terms of local send and receive operations. In a BSP program, communication operations are synchronized and global, i.e. all processors participate.

The two levels are reflected by the existence of two kinds of objects: local and global objects. Local objects are standard Python objects, they exist on a single processor. Global objects exist on the parallel machine as a whole. They have a local value on each processor, which may or may not be the same everywhere.

For example, a global object "processor id" would have a local value equal to the respective processor number. Global objects also often represent data sets of which each processor stores a part, the local value is then the part of the data that one processor is responsible for.

The same distinction applies to functions. Standard Python functions are local functions: their arguments are local objects, and their return values are local objects as well. Global functions take global objects as arguments and return global objects. A global function is defined by one or more local functions that act on the local values of the global objects. In most cases, the local function is the same on all processors, but it is also common to have a different function on one processor, usually number 0, e.g. for I/O operations.

Finally, classes can be local or global as well. Standard Python classes are local classes, their instances are local objects, and their methods act like local functions. Global classes define global objects, and their methods act like global functions. A global class is defined in terms of a local class that describes its local values.

Communication operations are defined as methods on global objects. An immediate consequence is that no communication is possible within local functions or methods of local classes, in accordance with the basic principle of the BSP model: local computation and communication occur in alternating phases. It is, however, possible to implement global classes that are not simply global versions of some local class, and that can use communication operations within their methods. They are typically used to implement distributed data structures with non-trivial communication requirements. The design and implementation of such classes requires more care, but they allow the complete encapsulation of both the calculation and the communication steps, making them very easy to use. An example within the Python BSP package is a class that represents distributed netCDF files, and which ensures automatically that each processor handles a roughly equal share of the total data. From a user's point of view, this class has a programming interface almost identical to that of the standard Python netCDF module.

4.2 Standard Global Classes

The simplest and most frequent global objects are those which simply mirror the functionality of their local values and add communication operations. They are represented by the classes `ParConstant`, `ParData`, and `ParSequence`, all of which are subclasses of `ParValue`. The three classes differ in how their local representations are specified.

`ParConstant` defines a constant, i.e. its local representation is the same on all processors. Example:

```
zero = ParConstant(0)
```

has a local representation of 0 on all processors.

`ParData` defines the local representation as a function of the processor number and the total number of processors. Example:

```
pid = ParData(lambda pid, nprocs: pid)
```

has an integer (the processor number) as local representation

`ParSequence` distributes its argument (which must be a Python sequence) over the processors as evenly as possible. Example:

```
integers = ParSequence(range(10))
```

divides the ten integers among the processors. With two processors, number 0 receives the local representation `[0, 1, 2, 3, 4]` and number 1 receives `[5, 6, 7, 8, 9]`. With three processors, number 0 receives `[0, 1, 2, 3]`, number 1 receives `[4, 5, 6, 7]`, and number 2 receives `[8, 9]`.

All these classes define the standard arithmetic operations, which are thus automatically parallelized. They also support indexing and attribute extraction transparently.

Global functions are created using the class `ParFunction` when the local representation is the same local function on all processors (the most common case). Another frequent case is to have a different function on processor 0, e.g. for I/O operations. This is arranged by the class `ParRootFunction`.

4.3 A Simple Example

The first example illustrates how to deal with the simplest common case: some computation has to be repeated on different input values, and all the computations are independent. The input values are thus distributed among the processors, each processor calculates its share, and in the end all the results are communicated to one processor that takes care of output. In the following example, the input values are the first 100 integers, and the computation consists of squaring them.

```
from Scientific.BSP import ParSequence, ParFunction, \
                           ParRootFunction

import operator

# The local computation function.
def square(numbers):
    return [x*x for x in numbers]

# The global computation function.
global_square = ParFunction(square)

# The local output function
def output(result):
    print result

# The global output function - active on processor 0 only.
global_output = ParRootFunction(output)
```

```
# A list of numbers distributed over the processors.
items = ParSequence(range(100))

# Computation.
results = global_square(items)

# Collect results on processor 0.
all_results = results.reduce(operator.add, [])

# Output from processor 0.
global_output(all_results)
```

The local computation function is a straightforward Python function: it takes a list of numbers, and returns a list of results. The call to `ParFunction` then generates a corresponding global function. `ParSequence` takes care of distributing the input items over the processors, and the call to `global_square` does all the computation. Before processor 0 can output the results, it has to collect them from all other processors. This is handled by the method `reduce`, which works much like the Python function of the same name, except that it performs the reduction over all processors instead of over the elements of a sequence. The arguments to `reduce` are the reduction operation (addition in this case) and the initial value, which is an empty list here because we are adding up lists.

This program works correctly independently of the number of processors it is run with, which can even be higher than the number of input values. However, the program is not necessarily efficient for any number of processors, and the result is not necessarily the same, as the order in which the local result lists are added up by the reduction operation is not specified. If an identical order is required, the processes have to send their processor ID along with the data, and the receiving processor must sort the incoming data according to processor ID before performing the reduction.

One possibly critical aspect of this program is that each processor needs to store all the data initially, before selecting the part that it actually works on. When working with big data objects, it might not be feasible to have each processor store more than the data for one iteration at the same time. This case can be handled with synchronized iterations, in which each processor handles one data item per step and then synchronizes with the others in order to exchange data.

4.4 Systolic Algorithms

The next example presents another frequent situation in parallel programming, a systolic algorithm. It is used when some computation has to be done between all possible pairs of data items distributed over the processors. In the example, a list of items (letters) is distributed over the processors, and the computational task is to find all pairs of letters (in a real application, a more complex computation would of course be required).

The principle of a systolic algorithm is simple: each data chunk is passed from one processor to the next, until after $N-1$ iterations each processor has seen all data. The new features that are illustrated by this example are general communication and accumulation of data in a loop.

```
from Scientific.BSP import ParData, ParSequence, \
    ParAccumulator,  ParFunction, ParRootFunction, \
    numberOfProcessors

import operator

# Local and global computation functions.
def makepairs(sequence1, sequence2):
    pairs = []
    for item1 in sequence1:
        for item2 in sequence2:
            pairs.append((item1, item2))
    return pairs
global_makepairs = ParFunction(makepairs)

# Local and global output functions.
def output(result):
    print result
global_output = ParRootFunction(output)

# A list of data items distributed over the processors.
my_items = ParSequence('abcdef')

# The number of the neighbour to the right (circular).
neighbour_pid = ParData(lambda pid, nprocs: [(pid+1)%nprocs])

# Loop to construct all pairs.
pairs = ParAccumulator(operator.add, [])
pairs.addValue(global_makepairs(my_items, my_items))
other_items = my_items
for i in range(numberOfProcessors-1):
    other_items = other_items.put(neighbour_pid)[0]
    pairs.addValue(global_makepairs(my_items, other_items))

# Collect results on processor 0.
all_pairs = pairs.calculateTotal()

# Output results from processor 0.
global_output(all_pairs)
```

The essential communication step is in the line

```
other_items = other_items.put(neighbour_pid)[0]
```

The method `put` is the most basic communication operation. It takes a list of destination processors (a global object) as its argument; in this example, that list contains exactly one element, the number of the successor. Each processor sends its local representation to all the destination processors.

In the example, each processor receives exactly one data object, which is extracted from the list by a standard indexing operation. The result of the line quoted above thus is the replacement of the local value of `other_items` by the local value that was stored in the preceding processor. After repeating this $N-1$ times, each processor has seen all the data.

It is instructive to analyze how the systolic loop would be implemented using the popular MPI library in a low-level language. First, either the "items" have to represented by arrays, or appropriate MPI data types need to be defined. Then each processor must send its own data and receive the data from its neighbour. If standard send and receive calls are used, the programmer must take care not to use the same order (send/receive or receive/send) on all processor, as this creates a risk of deadlock. MPI provides a special combined send+receive operation for such cases. The programmer must also allocate a sufficiently large receive buffer, which implies knowing the size of the incoming data. All this bookkeeping overhead would easily exceed the code size of the whole Python program, the programmer would have to take care that it doesn't cause run-time overhead when executing on a single processor, and even more effort would be required to make the program work serially without MPI being present.

5 Conclusion

This goal of this article has been to show how a high-level language can be very useful in application domains that are generally considered to be reserved to efficient, compiled low-level languages. The major point to keep in mind is the problem-oriented nature of high-level programming, and its much better possibilities for reusing existing code and for sharing code with other computational scientists.

References

1. The Python Web site: http://www.python.org/
2. The SWIG Web site: http://www.swig.org/
3. The Pyfort Web site: http://pyfortran.sourceforge.net/
4. The F2PY Web site: http://cens.ioc.ee/projects/f2py2e/
5. The Numerical Python Web site: http://numpy.sourceforge.net/
6. D.B. Skillicorn, J.M.D. Hill and W.F. McColl, "Questions and Answers about BSP", Technical Report PRG-TR-15-96, Oxford University Computing Laboratory, available from http://www.bsp-worldwide.org/implmnts/oxtool.htm
7. The Scientific Python Web site: http://dirac.cnrs-orleans.fr/ScientificPython/

Using CORBA Middleware in Finite Element Software

J. Lindemann, O. Dahlblom and G. Sandberg

Division of Structural Mechanics, Lund University
`strucmech@byggmek.lth.se`

Abstract. Distributed middleware technologies, such as CORBA can enable finite element software to be used in a more flexible way. Adding functionality is possible without the need for recompiling client code. Transfer of data can be done directly, without the need for intermediate input and output files. The CORBA software components can be easily configured and distributed tranparently over the network. A sample structural mechanics code, implemented in C++ is used to illustrate these concepts. Some future directions, such as placing CORBA enabled finite element software on HPC centres are also discussed.

1 Introduction

A complex hardware product often consists of many exchangeable components. As long as a component fits into the product, the internal implementation can differ. Software components are analogous to hardware components. Components in programs can be exchanged without the need for recompilation, as long as the component interface is unchanged. The use of components in software development has increased during the last few years. The reason for this is the need to reduce the size of the client programs. When the first client/server systems appeared, the client software were often large programs. Most of the processing was done in the client program and the database server was used as data storage. The problem with these systems was the cost of installing and maintaining the client software. New systems developed today often use a thin client with little or no data processing capabilities. Instead of calling the database servers directly, they use a set of components placed on central servers for data processing. These components then access the database servers. The advantage of this approach is that the components can be placed on powerful systems, reducing the amount of processing needed at the client. This approach has been successfully applied to database applications. It is of interest to apply this technique to analysis software as well. Using the technique of distributed computing, clients can use components as if they were located on the same machine, making it possible to create integrated programs with transparent access to computational resources, such as available workstations on the network or resources at High Performance Computing (HPC) centres. This would make high performance computing more available to a wider user group.

P.M.A. Sloot et al. (Eds.): ICCS 2002, LNCS 2331, pp. 701–710, 2002.

The present work describes structural analysis software, where the computational parts of analysis codes can be placed as components on remote servers. Before describing the structural analysis code, a brief overview of client/server architecture will be given.

2 Client/server architecture

Three-tier and n-tier applications emerged from the need to shield the client program from changes at the server side by placing a layer between the client and the server. The history of the client/server architecture is described by Schussel [15]. For a more detailed description over the client/server architecture, see Orfali and Harkey [12]. The logical three-tier or n-tier model divides an application into three or more logical components. Each component is responsible for a well-defined task. In a database application there would be a presentation layer for displaying data and modifying data, a logic and rules layer and a database layer responsible for storing the data.

The components of the logical model can be grouped together in different configurations to form a physical model. One of the most interesting combination of the logical model is when the three logical services are placed as separate applications on different computers, forming a physical three-tier application. This implementation enables developers to have a greater flexibility in the choice between different hardware and software configurations.

3 Distributed computing

Distributed computing is defined as a type of computing in which different components and objects comprising an application can be located on different computers connected to a network; for an overview see [10].

Currently, there are today three coexisting technologies for distributed object computing DCOM [2], RMI [14] and CORBA [1]. Microsoft's distributed COM (DCOM) extends the Component Object Model to be used over the network. RMI or Remote Method Invokation [14] is a distributed technology based on the Java language. CORBA is the Object Management Group's [1] specification for interoperability and interaction between objects and applications. Objects and applications can be placed on any platform and accessed from any platform. CORBA is a specification, and therefore platform-independent.

This paper describes an implementation in CORBA. In a previous paper [7] a DCOM based implementation has been studied.

4 CORBA

4.1 Concepts and Terminology

To describe a CORBA based implementation, it is important to understand some terminology and concepts of a CORBA implementation. A more thorough description can be found in Henning and Vinoski [6]. Some of the more important concepts and terminology is shown below.

- A *client* is an entity that invokes a request on a CORBA object.
- A *CORBA object* is a "virtual" entity capable of being located by an ORB and having client requests invoked on it.
- A *server* is an application with one or more CORBA objects.
- An *object reference* is a handle used to identify, locate and address a CORBA object. Object references is the only way for a client to access CORBA objects.
- A *servant* is a programming language entity that implements one or more CORBA objects.

Communication in CORBA is done by a client invoking requests on a CORBA object through either a statically linked stub in the client application or through the dynamic invocation interface (DII). The requests are dispatched to the local ORB which in turn dispatches these requests to an ORB on the remote machine. The remote ORB then dispatches the request to an object adaptor, which then directs the request to the servant implementation code.

4.2 Interface definition language

To access a CORBA object the client must know which methods and properties it contains. This description is called an interface. To describe such interfaces CORBA uses the Interface Definition Language (IDL). In this language the object interfaces are described. Using a separate language for describing the objects makes CORBA language neutral. This enables CORBA applications to be implemented in a variety of different languages. To implement CORBA clients and objects the IDL definition is compiled using an IDL compiler. This compiler takes the interface definition and generates the implementation code for both client and server, in the desired implementation language.

The following code shows an example of a simple IDL interface, declaring an interface to an `Echo` object. In this case the object echoes the string word back to the calling client.

```
interface Echo {
    string Shout(in string word);
}
```

Compiling this example using a C++ IDL compiler, will generate a header file and an implementation source file for accessing the object described from a C++ based application and the skeleton code for implementing the servant object in C++.

4.3 Name service

One of the biggest benefits of CORBA is location transparency. Information about server location is often not included in the client application. This makes it easy to configure a client server setup. A client only needs an object reference to connect to an object. Object references are unique identifiers, which also include information about the location of objects. To connect to objects the client needs a way of retrieving an object reference. Before the introduction of CORBA 2.3, object references were often transferred using files over a network file system or using a non-standard method of name lookup. In CORBA 2.3 a name service was introduced. The name server stores object references in a human readable form. When a server is started, it creates an entry in the name server for the object reference. The client then queries the server by name to receive the object reference. By using a name server, client/server configuration can be done transparently. Name server location is the only thing that has to be configured for the servers and the clients. Clients and servers get the location of the name server by specifying special command line options.

4.4 Object creation and destruction

Before request to an object can be made, the object implementation (servant) must be instantiated and activated. In CORBA this is done by the object adaptor. Earlier CORBA specifications only included a limited basic object adapter (BOA). To enhance the functionality of this object adaptor many ORB vendors added non-standard extensions. The consequence of this was that the server side of a CORBA application became ORB dependent. With CORBA 2.3 this limitation was removed by the introduction of the Portable Object Adaptor (POA).

Different types of policies for the creation and destruction of objects can be specified using lifetime policies for the portable object adaptor (POA) in CORBA. Figure 1 illustrates the typical lifetime of a CORBA object. The default

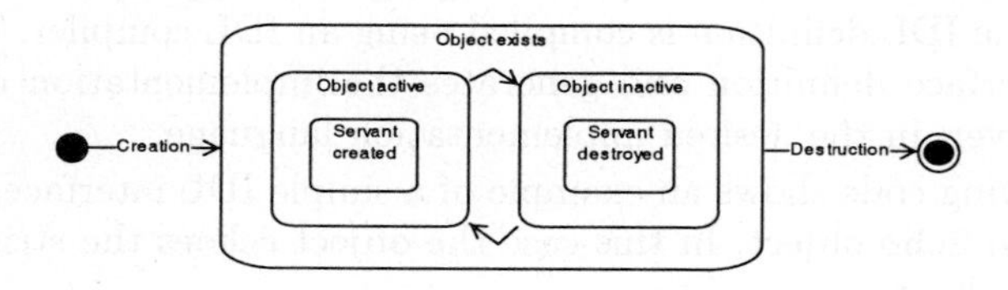

Fig. 1. Object creation and destruction

policy is `TRANSIENT`. In this policy the object can not be reactivated, when it has been deactivated. The object reference of an `TRANSIENT` object is only valid when the object is active. The `PERSISTENT` lifetime policy enables objects to be activated and deactivated multiple times. This requires that the object servants are able to store their state in a persistent form between the activations.

Because CORBA is a distributed technology, the creation of objects must be handled in a different way than it is handled when creating local objects.

In a CORBA system, objects are created by special factory objects. These factory objects can be seen as the equivalent of an object constructor in the C++ language.

The destruction of a CORBA object is not done by the factory, instead a special method is declared in the object interface for removing the object. If the factory was responsible for destroying the object, the client referencing the object would also have to reference the factory when destroying the object. This can be quite complex if the object reference has been passed from object to object. The process of creating and destroying is discussed in detail in Henning and Vinoski [6].

5 Finite element CORBA implementation

The educational software ForcePAD [4] was modified to use a CORBA based finite element solver. The ForcePAD application is an intuitive tool for visualising the behaviour of structures subjected to loading and boundary conditions. ForcePAD uses a bitmap canvas on which the user can draw the finite element model using standard drawing tools. When the calculation is executed the bitmap image is transferred to a finite element grid, which is then solved. The main window is shown in Figure 2. The application consists of four components divided

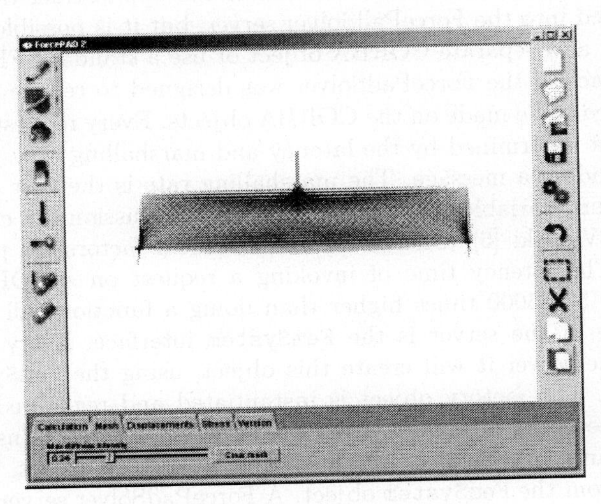

Fig. 2. Sample CORBA application

into three layers, as shown in figure 3. The user interface is responsible for interactively defining the problem. The ForcePadSolver component contains the interfaces used to describe the finite element model used in the application. The name server components handles the location of available CORBA ForcePAD-Solver components in the network. The FE solver components are responsible for executing the calculations. By providing the functionality of the application

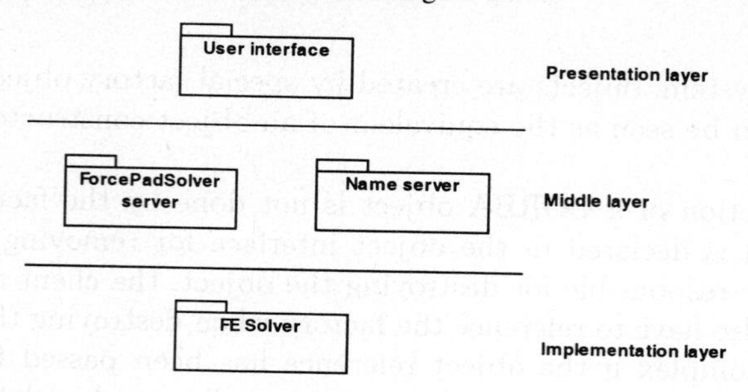

Fig. 3. Application components

in a component based form the application can be configured and maintained in a more flexible way.

5.1 ForcePadSolver server

The middle layer of the application is implemented in a single server. The ORB used in the implementation is ORBacus [11], which is a commercial ORB available with source for multiple platforms including Microsoft Windows and many Unix dialects. It can also be used without cost for non-commercial use. The FE solver is implemented in C++ using the newmat09 [8] library, which is freely available with source code. In this version of the application, the FE solver is statically linked into the ForcePadSolver server, but it is possible to implement the FE Solver as a separate CORBA object or use a standard FE code.

The interface of the ForcePadSolver was designed to reduce the number of requests needed to be made on the CORBA objects. Every request on a CORBA object has cost determined by the latency and marshalling rate. The latency is the cost of sending a message. The marshalling rate is the cost of sending the input and return variables. For a more detailed discussion see chapter 22.3 in Henning and Vinoski [6]. One of the most critical factors for performance is the latency. The latency time of invoking a request on a CORBA object is approximately 500-5000 times higher than doing a function call in C++. The main interface in the server is the `FemSystem` interface. Every time a client connects to the server it will create this object, using the `FemSystemFactory` factory object. The factory object is instantiated and registered in the name server when the server is started. The `FemSystem` object, when instantiated will create an instance of a `FemSolver` object and a `FemGrid` object. These objects are returned from the `FemSystem` object. A ForcePadSolver server can hold one instance of `FemSystem` objects for each client connected to the server, as shown in figure 4.

The code below shows how a `FemSystem` object is created from C++ using the `FemSystemFactory` object.

```
femSystemFactory = ... Get from name server ...
femSystem = femSystemFactory->create();
femGrid = femSystem->getFemGrid();
femSolver = femSystem->getFemSolver();
```

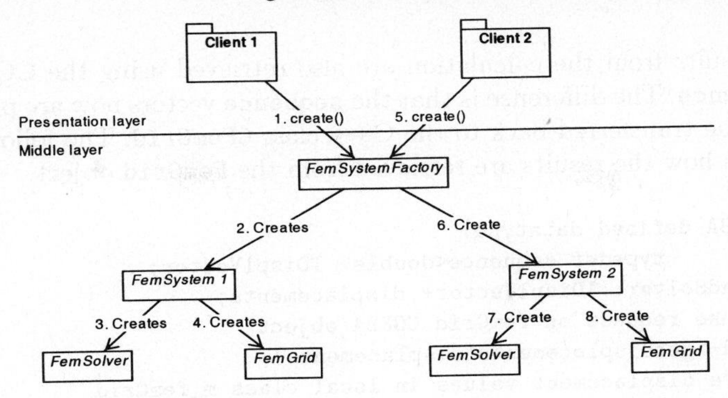

Fig. 4. Object creation using the FemSystemFactory object

The `FemGrid` object defines the finite element model and the `FemCalc` is used to control the calculation of the finite element model.

To reduce the marshalling times for the FE model, data will mainly be transferred using the CORBA data type `sequence`. This data type is a dynamic array of a specified type. The following code illustrates a typical data transfer from the client to a CORBA object in the ForcePAD client application.

```
// CORBA defined datatype:
//     typedef sequence<double> TStiffnessVector;
ForcePadSolver::TStiffnessVector stiffnessVector(nStiffness);
stiffnessVector.length(nStiffness);
// Transfer internal fem model to stiffnessVector
l = 0; float value;
for (i=0; i<rows; i++)
    for (j=0; j<cols; j++)
        for (k=0; k<2; k++) {
            value = m_femGrid->getGridValue(i, j, k);
            stiffnessVector[l++] = (double)value;
        }
// Invoke request on femGrid CORBA object
femGrid->setStiffness(stiffnessVector);
```

When all input data has been transferred to the CORBA object `FemGrid`, the finite element model can be solved. The execution of the finite element solver is controlled by the `FemCalc` object. The following code from the client application shows how the calculation is initiated:

```
femSolver->execute();
error = femSolver->getLastError();
```

In the ForcePadSolver server the `execute()` method is implemented as a blocking call. This means that the execution of the client application will wait until the server is finished. To solve this, the `execute()` could be implemented as an asynchronous method call in CORBA. Additional methods for monitoring the execution would have to be added to the interface as well.

The results from the calculation are also retrieved using the CORBA data type `sequence`. The difference is that the `sequence` vectors now are preallocated and must be transferred back to the C++ class `CFemGrid`. The following client code shows how the results are retrieved from the `FemGrid` object.

```
// CORBA defined datatype:
//          typedef sequence<double> TDisplVector;
ForcePadSolver::TDisplVector* displacements;
// Invoke request on femGrid CORBA object
femGrid->getDisplacements(displacements);
// Store displacement values in local class m_femGrid
m_femGrid->setDisplacementSize(displacements->length()); for (i=0;
i<displacements->length(); i++)
    m_femGrid->setDisplacement(i+1, (*displacements)[i]);
// We are responsible for deleting the return values
delete displacements;
```

The lifetime policy used in the ForcePadSolver server is `TRANSIENT`. A calculation in ForcePAD does not execute over several days, so the policy `PERSISTENT` will not be necessary in this case, it is better suited for applications executing over several days. The client applications can then connect and disconnect to object during the execution.

5.2 Server implementation

The ForcePAD solver server is implemented as a C++ console application using the ORBacus [11] ORB. A skeleton implementation for the server is generated using a special switch in the ORBacus IDL compiler.

To handle object creation and destruction automatically, each servant is also derived from the `RefCountServantBase` base class. This class implements a reference counting scheme which automatically destroys the object servant when there are no connections to the object. Depending on the implementation, more complex schemes of object creation and destruction can be implemented, see [6] for more details.

The process of executing a calculation starts with a request to the `FemSolver` method `execute()`. The `FemSolver` reads the input model from the `FemGrid` object and assembles the finite element model. The solver from the `newmat09` [8] is then called. When the solution is found the results are stored back in the `FemGrid` object. The results are now available to the client application.

5.3 Client/server configurations

The easiest configuration of the finite element system is to install the client application together with the ForcePADSolver server and the finite element solver on a single computer, see Figure 5. This configuration is typically used to do calculations that fit into the memory of the local machine.

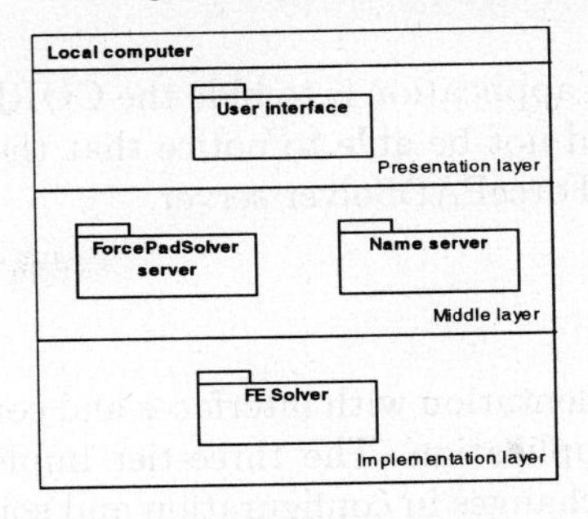

Fig. 5. Local configuration

In the first distributed configuration, the middlelayer and implementation are moved to a separate computer. This configuration requires the server to be able to run a CORBA ORB. If the server running the finite element solver does not support running an ORB, the middlelayer can be placed on a separate computer. Execution of the finite element solver can then be done using `rexec`, `rsh` or `ssh` utilities. Figure 6 show two of the possible configurations. Many more configurations are possible. By providing location transparency, the CORBA objects can be configured in almost any way without needing to recompile the clients and the servers.

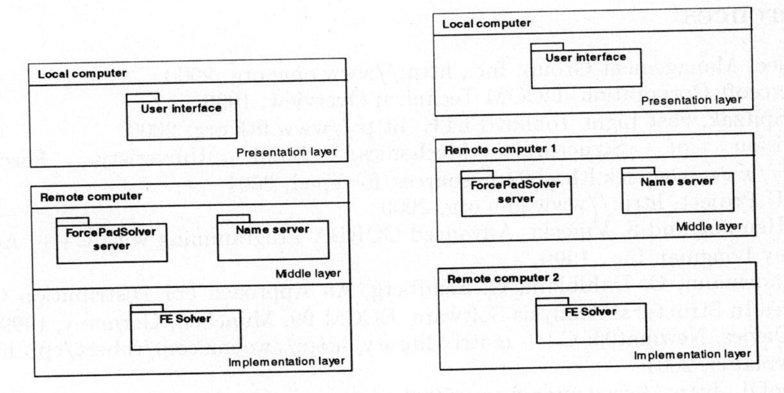

Fig. 6. Remote configuration 1 and 2

5.4 Client application

To create a platform independent application, ForcePAD uses the fast light toolkit (FLTK) [3]. FLTK [3] is a lightweight user interface toolkit written in C++. The toolkit can be used on Windows 98/NT/2000 and most Unix dialects with good performance. The 2D graphics in ForcePAD is implemented using OpenGL [9].

One goal of the client application is to hide the CORBA implementation from the user. The user should not be able to notice that the client is using CORBA for interfacing with the ForcePADSolver server.

6 Conclusion

Using a three-tier implementation with interfaces and components, creates a very flexible finite element application. The three-tier implementation protects the client applications from changes in configuration and solver design. Components are easily configurable and maintainable, reducing the need for further development. By using interfaces when communicating with components, the need to recompile client software when a new functionality is introduced in the solver components is reduced. Interfaces can also be published enabling other software to use the finite element application in an effective way. The CORBA specifications also enable new ways of using software. Client software can easily distribute calculations over available workstations. High Performance Computing (HPC) centres would be able to host a set of applications as CORBA objects. From a web site, users can register themselves as users and download client applications that connect to the objects. This would make high performance computing more available to a wider user group.

References

1. Object Management Group, Inc., http://www.omg.org, 2000
2. Microsoft Corporation, DCOM Technical Overview, 1996
3. B. Spitzak, Fast Light Toolkit FLTK, http://www.fltk.org, 2000
4. Division of Structural Mechanics, Lund Univeristy, ForcePAD, http://www.byggmek.lth.se/bmresources/forcepad, 2001
5. GNU Project, http://www.gnu.org, 2000
6. M. Henning and S. Vinoski, Advanced CORBA Programming with C++, Addison Wesley Longman Inc., 1999
7. J.Lindemann, O. Dahlblom, G. Sandberg, An Approach For Distribution Of Resources In Structural Analysis Software, ECCM 99, München, Germany, 1999
8. R. Davies, Newmat09: C++ matrix library, http://webnz.com/robert/cpp_lib.htm #newmat09, 2001
9. OpenGL, http://www.opengl.org, 2000
10. The Open Group, http://www.opengroup.org/dce, 2000
11. Object Oriented Concepts Inc., ORBacus 4.0, http://www.ooc.com/ob, 2000
12. R. Orfali and D. Harkey, Client/server programming with Java and CORBA. - 2nd ed., John Wiley and Sons Inc., 1998
13. G. Reilly, Developing Active Server Components with ATL, Microsoft Corporation, 1997
14. Sun Microsystems Inc., JavaTM Remote Method Invocation, http://java.sun.com/j2se/1.3/docs/guide/rmi/index.html, 2001
15. G. Schussel, Client/Server: Past, Present and Future, http://www.dciexpo.com/geos/dbsejava.htm, 1996
16. S. Williams and C. Kindel, Microsoft Corporation, The Component Object Model: A Technical Overview, 1994

On Software Support for Finite Difference Schemes Based on Index Notation

Krister Åhlander and Kurt Otto

Department of Scientific Computing, Uppsala University,
Box 120, 751 04 Uppsala, Sweden,
{krister,kurt}@tdb.uu.se

Abstract. A formulation of finite difference schemes based on the index notation of tensor algebra is advocated. Finite difference operators on regular grids may be described as sparse, banded, "tensors". Especially for 3D, it is claimed that index notation better corresponds to the inherent problem structure than does conventional matrix notation. The transition from mathematical index notation to implementation is discussed. Software support for index notation that obeys the Einstein summation convention has been implemented in the C++ package EinSum. The extension of EinSum to support typical data structures of finite difference schemes is outlined. A combination of general index notation software and special-purpose routines for instance for fast transforms is envisioned.

1 Introduction

Basic finite difference schemes for partial differential equations on structured grids are, in principle, simple to implement. If we oversimplify, the process is as follows. The derivative at a certain point of the grid is approximated with a particular finite difference stencil. For example, if we consider the Laplace equation in 2D, a second-order finite difference scheme yields the well-known five-point stencil. If curvilinear grids are used, the stencil weights vary at every point of the grid. A relation between grid functions (discrete approximations of fields) is thus obtained. This relation is usually expressed as a matrix vector multiplication, where each row of the matrix corresponds to a stencil at a specific grid point. The resulting matrix is a band matrix, whose structure can be exploited in order to develop efficient algorithms. Since linear algebra is such a well-developed area of research, both with respect to mathematical aspects as well as to efficient computer programs, the restructuring of the original formulation into a matrix vector multiplication has several advantages.

However, the original formulation relates two grid functions through a stencil at each point of the grid. This data structure can be described as a "grid stencil". Such a grid stencil may of course be implemented as a sparse band matrix, but in this process the conceptual picture of the grid stencil is obscured. This potential risk of missing information makes it more difficult to express finite

P.M.A. Sloot et al. (Eds.): ICCS 2002, LNCS 2331, pp. 711–718, 2002.

difference schemes mathematically, particularly in 2D and 3D. It also makes the implementation of finite difference schemes more awkward.

In this paper, we discuss an alternative approach to finite difference schemes, which better takes into account the inherent structure of the problem. This formulation is based on index notation, including the Einstein summation convention. We also study software design based on this formulation. Particularly, we discuss the extension of EinSum to support typical data structures of finite differences.

EinSum is a C++ package for index notation. It allows index notation expressions to be written as plain code, thereby avoiding a notational gap between index notation and implementation code. The EinSum package is well suited for tensor algebra. It has been presented in more detail in [1], and its support of tensors with general symmetries is described in [2]. In the present paper, we focus on an extension to general sub and super diagonals (bands) in multi-dimensional arrays (tensors).

The remainder of this paper is outlined as follows. In Section 2, we exemplify index notation for finite differences. In Section 3, we discuss software support for index notation. Finally, in Section 4, we discuss future work and the relation between different software implementation strategies. We find that general index notation support is useful, for instance for expressing the finite difference operator. To address performance, it should be combined with special-purpose software for performance critical tensor operations.

2 Index notation and finite differences

2.1 Introductory examples

Band matrices arise in finite difference schemes, for instance for solving PDEs in one space dimension. In higher dimensions, a more appropriate data structure is a "band tensor". For instance, the well known five-point stencil D can be interpreted as a $(2;2)$ "tensor", i.e., a multi-dimensional array with 2 upper and 2 lower indices. It operates on a gridfunction x with two upper indices, i.e., a $(2;0)$ tensor, to produce another $(2;0)$ tensor y. In index notation, where the Einstein summation convention is adopted, this operation reads

$$y^{i,j} \leftarrow D^{i,j}_{k,\ell} x^{k,\ell}. \tag{1}$$

Here, the summation convention implies summation over k and ℓ, since they are repeated indices (see Section 3).

The nonzero elements of D are:

$$D^{i,j}_{k,\ell} = -4, \; i = k \text{ and } j = \ell,$$
$$D^{i,j}_{k,\ell} = 1, \; i = k \text{ and } j = \ell \pm 1,$$
$$D^{i,j}_{k,\ell} = 1, \; i = k \pm 1 \text{ and } j = \ell.$$

Using the conventions (see Section 3) that summation is always suppressed on the left-hand side in assignments, and that free indices vary over their range, we may assign the nonzero elements of D as

$$\begin{array}{ll} D^{i,j}_{i,j} \leftarrow -4, & \\ D^{i,j}_{i,j+1} \leftarrow 1, & D^{i,j}_{i,j-1} \leftarrow 1, \\ D^{i,j}_{i+1,j} \leftarrow 1, & D^{i,j}_{i-1,j} \leftarrow 1. \end{array} \tag{2}$$

Another way of forming the five-point stencil is as follows. Let the nonzero elements of Δ_+ and Δ_- be

$$[\Delta_+]^i_{i+1} \leftarrow 1, \ [\Delta_+]^i_i \leftarrow -1,$$
$$[\Delta_-]^i_i \leftarrow 1, \ [\Delta_-]^i_{i-1} \leftarrow -1.$$

The five-point stencil is then given by

$$D^{i,j}_{k,\ell} \leftarrow [\Delta_+]^i_m [\Delta_-]^m_k + [\Delta_+]^j_n [\Delta_-]^n_\ell. \tag{3}$$

For curvilinear grids, the band tensors are no longer constant along the diagonals. As an illustration, consider a two-dimensional transformation $(x, y) = (x(\xi,\eta), y(\xi,\eta))$ and let ξ_x denote $\partial\xi/\partial x$ etc. By the chain rule,

$$\frac{\partial}{\partial x} = \xi_x \frac{\partial}{\partial \xi} + \eta_x \frac{\partial}{\partial \eta}.$$

Using the centred difference $\Delta_0 = \frac{1}{2}(\Delta_+ + \Delta_-)$ the corresponding differentiation operator Δ_x may be expressed as:

$$[\Delta_x]^{i,j}_{k,\ell} \leftarrow {\xi_x}^{i,j} [\Delta_0]^i_k + {\eta_x}^{i,j} [\Delta_0]^j_\ell. \tag{4}$$

In summary, finite difference operators may often conveniently be expressed as band tensors. Moreover, index notation provides a useful way to manipulate with the operators as well as the "grid functions".

2.2 Band tensors for the Helmholtz equation

In this subsection, we illustrate the use of index notation as an efficient tool for realistic problems. This section is based mainly upon results in [8–10].

Consider the Helmholtz equation,

$$u_{xx} + u_{yy} + \kappa^2 u = g,$$

for a 2D duct, see Figure 1. Using an orthogonal transformation, the equations may be rewritten for computational coordinates (ξ, η):

$$(a u_\xi)_\xi + (a^{-1} u_\eta)_\eta + e u = g.$$

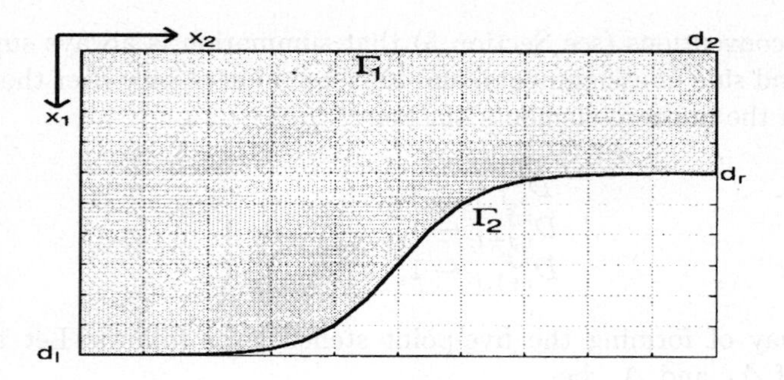

Fig. 1. The Helmholtz equation for the following duct is an application of interest in underwater acoustics.

Here, a and e are coefficients that depend on the transformation. A fourth-order Numerov scheme may be used (see [9]), to express the finite difference scheme in index notation as

$$B^{i,j}_{k,\ell} u^{k,\ell} = g^{i,j} . \tag{5}$$

In the interior, B is a band tensor with nine diagonals.

The system may be solved iteratively using a Krylov subspace method with a normal block preconditioner. Here, we sketch the process of forming the preconditioner, using index notation. The theory is developed in [8], using the considerably more complex matrix notation.

Choose a unitary $(1;1)$ tensor Q with components Q^i_j. Consider partially diagonal $(2;2)$ tensors Λ such that

$$\Lambda^{i,j}_{k,\ell} = 0, \ i \neq k.$$

Compose a $(2;2)$ preconditioner $M^{i,j}_{k,\ell} \leftarrow Q^i_\alpha \Lambda^{\alpha,j}_{\alpha,\ell} [Q^*]^\alpha_k$.

The choice of Λ that minimizes the Frobenius norm of $(B - M)$ is given by

$$\Lambda^{i,j}_{i,\ell} \leftarrow [Q^*]^i_\alpha \, B^{\alpha,j}_{\beta,\ell} Q^\beta_i . \tag{6}$$

(Note that summation over i is suppressed due to the range and the assignment conventions, see Section 3.)

Equation (5) may now be solved efficiently using M as a preconditioner. The preconditioning step can be performed cheaply, using fast transforms [10].

3 Software support

The conclusion of the previous section is that index notation and tensor-like data structures are convenient for expressing finite difference methods. Consequently, we investigate software support for such operations. This allows for a smooth transition from the mathematical formulation to the implementation of finite difference methods. In this section, we focus on index notation support in the EinSum package. The design of EinSum has been presented in more detail in [1, 2].

Firstly, EinSum grammar is based on the following well-known conventions (see, e.g., [11]):

1. The summation convention: If an index is repeated, both as an upper index and as a lower index, summation is understood.
2. The range convention: A free index is understood as a loop index, ranging over all its values.

EinSum is originally developed for supporting common tensor algebra operations. For example, a contraction of a (3; 1) tensor D may be assigned to a (2; 0) tensor C as

$$C^{i,j} \leftarrow D^{i,k,j}_{k} .$$

In EinSum, the corresponding assignment is straightforward. First, some indices are declared, with an approriate range N according to the dimension.

```
EinIndex I(1,N), J(1,N), K(1,N), L(1,N);
```

Next, the tensors are declared with references to appropriate index spaces. Index spaces (without symmetry properties) may be described by "juxtaposing" indices, using the `operator|`. See [2] for more details on index space construction. Below, `D` is associated with an index space of rank 4, whose elements range from $(1,1,1,1)\ldots(N,N,N,N)$ and `C` with an index space of rank 2, whose elements range from $(1,1)\ldots(N,N)$. Furthermore, `D` and `C` are declared as (3; 1) and (2; 0) tensors, respectively.

```
EinTensor<double> D(I|J|K|L,3,1), C(I|J,2,0);
```

After data initialization, the assignment may now be carried out:

```
C(I|J) = D(I|K|J,K);
```

EinSum requires tensors to be indexed according to their rank. For `C`, as a tensor with no lower indices, `operator()` accepts one argument. `D` must be indexed with two arguments, "multi-indices" of rank three and one, respectively. Again, `operator|` is used for juxtaposition. When the assignment is carried out, EinSum interprets the expression and recognizes that there is an implicit loop over indices `I` and `J`, as well as an implicit summation over `L`. See [1] for more details on the interpretation of the EinSum notation. This includes more complex cases, and also the introduction of a third convention:

3. The assignment convention: Summation is suppressed in the left-hand side of assignments.

This convention yields a natural and intuitive notation for assignments along diagonals. Consider, for instance, the assignments

$$A^i_i \leftarrow B^i_i,\ A^i_i \leftarrow B^j_j,\ A^i_j \leftarrow B^i_j .$$

According to the conventions, all these assignments have different meaning. The first statement assigns the diagonal values of B to the corresponding diagonal elements of A. The second statement assigns the trace of B, i.e., B^j_j. The third statement assigns all elements of B to all elements of A, a "block assignment". All three statements are readily coded in EinSum:

```
A(I,I) = B(I,I);  A(I,I) = B(J,J);  A(I,J) = B(I,J);
```

Compare with for instance Matlab or Fortran notation, or other libraries with some kind of index abstraction, for instance [5, 6, 12]. In these languages or libraries, only the block assignment is directly supported. We argue that the notation supported in EinSum is more powerful, in particular concerning the ability to address diagonals for tensors of higher order.

In the present contribution, we extend these ideas so that we are also able to treat generalized sub and super diagonals, as motivated for instance by equation (2). A crucial point in the implementation is the identification of indices. To this end, each index has a unique ID, automatically generated. In order to handle sub and super diagonals, we need to slightly modify this technique. For instance, i and $i+1$ are two expressions which refer to the same ID, but whose ranges differ. It is achieved by overloading `operator+` to take an integer value and return a new index object with the same ID as the original, but with the range properly adjusted. This makes it possible to initiate the five-point stencil according to equation (2) as follows:

```
EinTensor D(I|J|K|L,2,2);

D(   I|J, I|J ) = -4;
D( I+1|J, I|J ) =  1;  D( I-1|J, I|J ) =  1;
D( I|J+1, I|J ) =  1;  D( I|J-1, I|J ) =  1;
```

Of course, equation (3) provides another way of creating D (assuming that Δ_+ and Δ_- are properly initiated):

```
D(I|J,K|L) = DeltaPlus(I,M)*DeltaMinus(M,K) +
             DeltaPlus(J,N)*DeltaMinus(N,L);
```

Application of the difference operator, see equation (1), may be carried out using the already implemented interpretation mechanisms of EinSum. It reads:

```
Y(I|J) = D(I|J,K|L)*X(K|L);
```

Generalized tensor diagonals are also present when simulating the Helmholtz equation. The difference operator B is a banded tensor, and the computation of Λ, see equation (6), can be written as one line of EinSum notation:

```
Lambda(I|J,I|L) = Qstar(I,Alpha)*B(Alpha|J,Beta|L)*Q(Beta,I);
```

The summation is over α and β, in accordance with the conventions.

4 Discussions and future work

We have discussed index notation as an alternative to matrix formulation in order to express finite difference algorithms for simulating PDEs. For structured grids, we find that index notation better corresponds to the data structures of the problem. These data structures are multi-dimensional arrays and may be compared with tensors. For finite differences, the sparsity pattern of typical stencils generates band tensors.

Support of tensor notation has already been developed in the EinSum package. In the present paper, we stress that it can be used also for expressing finite difference schemes. In particular, the support for an intuitive notation for generalized sub and super diagonals has been presented. The current implementation does not utilize the sparsity patterns of the band tensors, though. This may be addressed firstly by extending the index space hierarchy of EinSum. As explained in [2], index spaces contain information for instance about structurally implied zero elements. It should not be difficult to develop a new index space with relevant information for banded tensors. The idea of banded index spaces is actually similar to the stencil structures found in for instance PETSc, [4]. An advantage with banded index spaces is the potential for tailoring them according to the particular stencil.

It is, however, more difficult to exploit the sparsity pattern of band tensors with respect to algorithms, when a notation as general as index notation is used. In addition, for some operations other performance improving techniques may be available, which are difficult to cater for in a general package. For example, as part of the preconditioning step for the Helmholtz equation, the following unitary transformations are carried out (see [10]):

$$v^{i,j} \leftarrow Q^i_k y^{k,j} .$$

This is computed according to a fast transform algorithm, using special-purpose software.

As often is the case in scientific computing, there is a trade-off between general packages and specialized routines. Our aim is to utilize the best of both worlds. We believe that the simulation of the Helmholtz equation (5), for example, can benefit from a general index notation library, particularly in the assembly process of the finite difference operator B. Also, as was noted in [7], preconditioners for this and similar problems would gain from a general library with index notation support. However, regarding the number-crunching, highly optimized routines should be used. Even though these may be less general, their design should still be based upon the data structures of the problem, see [3]. For finite differences, we believe that a general band tensor abstraction is such a data structure.

References

1. K. Åhlander. Einstein summation for multi-dimensional arrays. In E. Munthe-Kaas et al., editors, *Norsk Informatikk Konferanse – NIK'2000*, pages 67–78. Tapir,

Norway, 2000.
2. K. Åhlander. Supporting tensor symmetries in EinSum. Technical Report 212, Dept. of Informatics, University of Bergen, Bergen, Norway, June 2001. To appear in Computers and Mathematics with Applications.
3. K. Åhlander, M. Haveraaen, and H. Munthe-Kaas. On the role of mathematical abstractions for scientific computing. In R. Boisvert and P. Tang, editors, *The Architecture of Scientific Software*, pages 145–158. Kluwer Academic Publishers, Boston, 2001.
4. S. Balay, W. D. Gropp, L. C. McInnes, and B. F. Smith. Efficient management of parallelism in object-oriented numerical software libraries. In E. Arge, A. M. Bruaset, and H. P. Langtangen, editors, *Modern Software Tools for Scientific Computing*, pages 163–202. Birkhäuser, 1997.
5. J.C. Cummings et al. Rapid application development and enhanced code interoperability using the POOMA framework. In S. L. L. M. E. Henderson and C. R. Anderson, editors, *Object-oriented Methods for Interoperable Scientific and Engineering Computing*, Philadelphia, 1999. SIAM. ch. 29.
6. M. Lemke and D. Quinlan. P++, a parallel C++ array class library for architecture-independent development of structured grid applications. *ACM SIGPLAN Notices*, 28(1):21–23, 1993.
7. E. Mossberg, K. Otto, and M. Thuné. Object-oriented software tools for the construction of preconditioners. *Sci. Programming*, 6:285–295, 1997.
8. K. Otto. A unifying framework for preconditioners based on fast transforms. Report 187, Dept. of Scientific Computing, Uppsala Univ., Uppsala, Sweden, 1996.
9. K. Otto. Iterative solution of the Helmholtz equation by a fourth-order method. *Boll. Geof. Teor. Appl.*, 40 suppl.:104–105, 1999.
10. K. Otto and E. Larsson. Iterative solution of the Helmholtz equation by a second-order method. *SIAM J. Matrix Anal. Appl.*, 21:209–229, 1999.
11. J.G. Papastavridis. *Tensor calculus and analytical dynamics.* Library of engineering mathematics. CRC Press LLC, 1999.
12. T. Veldhuizen. Arrays in Blitz++. In *Proceedings of the 2nd International Scientific Computing in Object-Oriented Parallel Environments (ISCOPE'98)*, Lecture Notes in Computer Science. Springer-Verlag, 1998.

A Component-based Architecture for Parallel Multi-Physics PDE Simulation

Steven G. Parker[1]

Scientific Computing and Imaging Institute
School of Computing
50 S. Central Campus Dr., Room 3490
University of Utah
Salt Lake City Utah 84112, USA,
sparker@cs.utah.edu,
WWW home page: http://www.cs.utah.edu/~sparker

Abstract. We describe the Uintah Computational Framework (UCF), a set of software components and libraries that facilitate the simulation of Partial Differential Equations (PDEs) on Structured AMR (SAMR) grids using hundreds to thousands of processors. The UCF uses a non-traditional approach to achieving parallelism, employing an abstract taskgraph representation to describe computation and communication. This representation has a number of advantages that affect the performance of the resulting simulation. We demonstrate performance of the system on a solid mechanics algorithm, two different Computational Fluid-Dynamics (CFD) algorithms, and coupled CFD/solids algorithms. We illustrate the performance of the UCF for jobs requiring up to 2000 processors.

1 Introduction

Computational scientists continue to push the capabilities of current computer hardware to the limits in order to simulate complex real-world phenomena. These simulations necessitate the use of ever increasing computational resources. Furthermore, the software written to model real-world scientific and engineering problems is typically very complex. Grid generation, non-linear and linear solvers, visualization systems, and parallel runtime systems all combine to provide a very powerful environment for solving complex scientific and engineering problems. Such complexities are further compounded when multiple simulation codes are combined to simulate the interaction of multiple phenomena.

The Uintah Computational Framework (UCF) is a set of software components and libraries that facilitate the simulation of Partial Differential Equations (PDEs) on Structured AMR (SAMR) grids using hundreds to thousands of processors. The UCF uses a non-traditional approach to achieving parallelism, employing an abstract taskgraph representation to describe computation and communication. This representation has a number of advantages, including efficient fine-grained coupling of multi-physics components, flexible load balancing mechanisms, and a separation of application concerns from parallelism concerns.

P.M.A. Sloot et al. (Eds.): ICCS 2002, LNCS 2331, pp. 719–734, 2002.

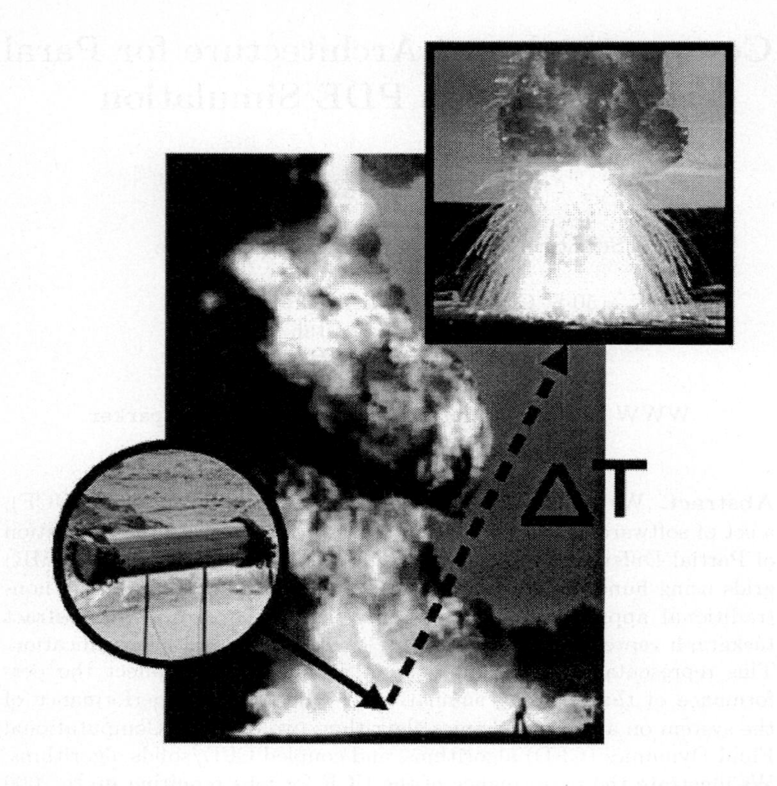

Fig. 1. A typical C-SAFE problem involving hydrocarbon fires and explosions of energetic materials.

In this paper, we will describe the taskgraph concept, along with a number of the implementation details. We will discuss the advantages and disadvantages of this methodology, and will present results from UCF simulations.

2 Overview

The system described here involves several connections between a number of different projects, including C-SAFE, CCA, SCIRun and Uintah. In order to adequately describe how they all interact, we first give an overview of the individual pieces in this section. The remainder of the paper will discuss how those pieces fit together to achieve large-scale parallel simulations.

2.1 C-SAFE

In 1997, the University of Utah created an alliance with the DOE Accelerated Strategic Computing Initiative (ASCI) to form the Center for the Simulation of Accidental Fires and Explosions (C-SAFE) [1, 2]. C-SAFE focuses specifically on providing state-of-the-art, science-based tools for the numerical simulation of accidental fires and explosions, especially within the context of handling and storage of highly flammable materials. The primary objective of C-SAFE is to provide a software system comprising a PSE in which fundamental chemistry and engineering physics are fully coupled with non-linear solvers, optimization, computational steering, visualization and experimental data verification. The availability of simulations using this system will help to better evaluate the risks and safety issues associated with fires and explosions. Our goal is to integrate and deliver a system that is validated and documented for practical application to accidents involving both hydrocarbon and energetic materials. Efforts of this nature requires expertise from a wide variety of academic disciplines. A typical C-SAFE problem is shown in Figure 2.1.

2.2 CCA

The CCA forum [3, 4] was established by a group of researchers from several Department of Energy national laboratories, and several universities to address the need for a software component architecture that fulfilled the needs of high-performance computing. The CCA architecture aims to provide higher performance, explicit support for multi-dimensional arrays, and explicit support for parallelism. Uintah, described below, is one implementation of the CCA architecture which we use as a research vehicle for experimenting with these ideas and for exercising their efficacy on a complex scientific application, such as C-SAFE simulations.

2.3 SCIRun

SCIRun[1] is a scientific problem solving environment that allows the interactive construction and steering of large-scale scientific computations [5–7]. A scientific application is constructed by connecting computational elements (modules) to form a program (network). This program may contain several computational elements as well as several visualization elements, all of which work together in orchestrating a solution to a scientific problem. Geometric inputs and computational parameters may be changed interactively, and the results of these changes provide immediate feedback to the investigator. SCIRun is designed to facilitate large-scale scientific computation and visualization on a wide range of machines from the desktop to large supercomputers.

2.4 Uintah

C-SAFE's Uintah Problem Solving Environment [8, 9] is a massively parallel, component-based, PSE designed to simulate large-scale scientific problems, while

[1] Pronounced "ski-run." SCIRun derives its name from the Scientific Computing and Imaging (SCI) Institute at the University of Utah.

allowing the scientist to interactively visualize, steer, and verify simulation results. Uintah is a derivative of the SCIRun PSE and adds support for the more powerful CCA component model and support for distributed-memory parallel computers.

3 Component Architecture

Due to space constraints, we will only briefly describe the overall Uintah architecture here. Further details can be found in [4, 8, 9], as well as in future publications from C-SAFE.

The Uintah Component Architecture is based on the CCA component architecture. Since the CCA component architecture is an evolving standard, we focused on a subset of the overall standard in order to focus on C-SAFE simulations. In particular, we created a C++ only implementation and ignored the multi-language features that were still under development.

The primary feature of the CCA is its port model. The port model, also called gPorts (for generalized ports), is the mechanism by which components communicate. In C++, this looks like little more than an abstract class on which the caller can perform method invocations. Components *provide* a set of interfaces, which other components can *use*. At component creation time, the component declares a set of Provides and Uses ports. An external entity, called a builder, connects the provides port of one component to the uses port of another component. For the simulations described here, this connection is provided by a stand-alone main program. However, in general, the builder can be a script, a graphical user interface, or even another component.

The CCA architecture utilizes the Scientific Interface Definition Language (SIDL) [10]. SIDL is the mechanism by which component interfaces are described. A SIDL compiler generates code for inter-language operation, as well as, for remote method invocations in a distributed memory environment. Although Uintah contains support for such distributed memory operations, the component interactions described here all occur within a single address space. Operation in a distributed-memory parallel environment occurs with MPI calls within individual components.

4 Uintah Computational Framework

The Uintah Computational Framework (UCF) is implemented in the context of the Uintah PSE Environment (PSE). The Uintah PSE Architecture is very general, facilitating a wide range of computational and visualization applications. On top of the Uintah architecture, we have designed a set of components and supporting libraries that are targeted toward the solution of PDEs on massively parallel architectures (hundreds to thousands of processors).

4.1 Overview

The UCF employs a non-traditional approach to achieving parallelism. Instead of explicit MPI calls placed throughout the program, applications are cast in terms of a *taskgraph*, a construct that describes the data dependencies between various steps in the problem.

The UCF exposes flexibility in dynamic application structure by adopting an execution model based on software-based "macro" dataflow. Computations are expressed as directed acyclic graphs of *tasks*, each of which produces some output and consumes some input (which is in turn the output of some previous task). These inputs and outputs are specified for each patch in a structured AMR grid. Tasks form a UCF data structure called the *taskgraph*, which represents imminent computation. Associated with each task is a C++ method, which is used to perform the actual computation. UCF data structures are compatible with Fortran arrays, so that the application writer can use Fortran subroutines to provide numeric kernels on each patch.

Each execution of a taskgraph integrates a single timestep, or a single non-linear iteration, or some other coarse algorithm step. Tasks "communicate" with each other through an entity called the *DataWarehouse*. The DataWarehouse is accessed through a simple name-based dictionary mechanism, and provides each task with the illusion that all memory is global. If the tasks correctly describe their data dependencies, then the data stored in the DataWarehouse will match the data (variable and region of space) needed by the task. In other words, the DataWarehouse is an abstraction of a global single-assignment memory, with automatic data lifetime management and storage reclamation. Values stored in the DataWarehouse are typically array-structured.

Communication is scheduled by a local scheduling algorithm that approximates the true globally optimal communication schedule. Because of the flexibility of single-assignment semantics, the UCF is free to execute tasks close to data or move data to minimize future communication.

The UCF storage abstraction is sufficiently high-level that it can be efficiently mapped onto both message-passing and shared-memory communication mechanisms. Threads sharing a memory can access their input data directly; single-assignment dataflow semantics eliminate the need for any locking of values. Threads running in disjoint address spaces communicate by message-passing protocol, and the UCF is free to optimize such communication by message aggregation. Tasks need not be aware of the transports used to deliver their inputs and thus UCF has complete flexibility in control and data placement to optimize communication both between address spaces or within a single shared-memory SMP node. Latency in requesting data from the DataWarehouse is not an issue; the correct data is deposited into the DataWarehouse before the task is executed.

Consider the taskgraph in Figure 4.1. Ovals represent tasks, each of which are a simple array program and easily treated by traditional compiler array optimizations. Edges represent named values stored by UCF. Solid edges have values defined at each material point (Particle Data) and dashed edges have values defined at each grid vertex (Grid Data). Variables denoted with a prime (') have

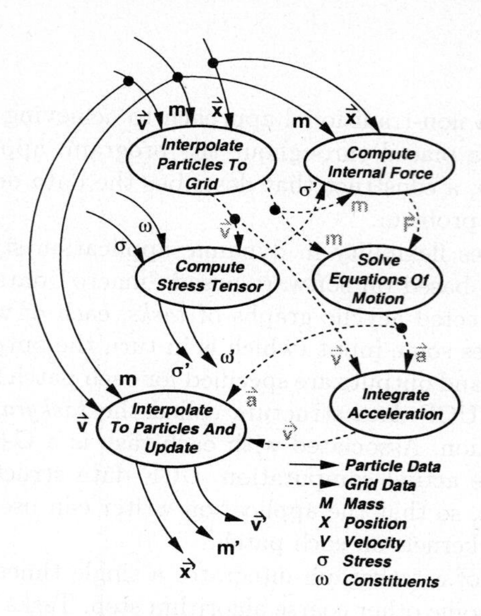

Fig. 2. An Example UCF Taskgraph

been updated during the time step. The figure shows a portion of the actual Uintah Material Point Method (MPM) [11–13] taskgraph concerned with advancing Newtonian material point motion on a single patch for a single timestep.

The idea of the dataflow graph as an organizing structure for execution is well known. The SMARTS [14] dataflow engine that underlies the POOMA [15] toolkit shares goals and philosophy with the UCF. SISAL compilers [16] used dataflow concepts at a much finer granularity to structure code generation and execution. Dataflow is a simple, natural and efficient way of exposing parallelism and managing computation, and is an intuitive way of reasoning about parallelism. What distinguishes implementations of dataflow ideas is that each caters to a particular higher-level presentation. SMARTS caters to POOMA's C++ implementation and stylistic template-based presentation. The SISAL compiler was of course developed to support the SISAL language. UCF is implemented to support a presentation catering to C++ and Fortran based mixed particle/grid algorithms on a structured adaptive mesh. The primary algorithms of importance to C-SAFE are the Material Point Method (MPM), and Eulerian CFD algorithms.

4.2 Taskgraph Advantages and Disadvantages

This dataflow-based representation of parallel computation fits very well with the SAMR grid, and with the nature of the computations that C-SAFE is performing. In particular, we decided to use this approach in order to accommodate a number of important needs.

First, the taskgraph helps accommodate flexible multi-physics integration needs. In particular, integration of a particle-based solid mechanics algorithm

(MPM) with a state-of-the-art CFD code was a primary C-SAFE goal. However, scientists still wanted to execute the CFD algorithm by itself, or the MPM algorithm by itself or even with a different CFD algorithm. The taskgraph facilitates this integration by allowing each application component (MPM and CFD in this example) to describe their tasks independently. The scheduler connects these tasks where data is exchanged between the different algorithm phases. In this fashion, a fine-grained interleaving of these different algorithms is obtained.

Second, the taskgraph can accommodate a wide range of unforeseen workloads. In C-SAFE simulations, load imbalances arise from a variety of situations: with particle-based methods particles may exist only in small portions of the domain; or ODE-based reaction solvers may be more costly in some regions of space than in others. Using the taskgraph, the UCF can map patches to other processors to minimize overall load imbalance. Communication is performed automatically, and the system has the information necessary to predict whether data motion is likely to pay off. These features would be more difficult to implement if each scientist were burdened with these complexities when writing simulation components.

Third, the taskgraph helps manage the complexity of a mixed threads/MPI programming model. Many modern supercomputing architectures employ a number of SMP nodes connected together by a fast interconnect. These architectures are often programmed with a flat MPI model, ignoring the fact that 2-128 of the nodes actually share a single memory. Using the UCF, tasks can be mapped to threads in order to achieve multi-threaded execution within a node. The semantics of the DataWarehouse enable true sharing of the data, eliminating explicit communication and data redundancy between neighboring SMP processors. Once again, this would not be possible if simulation scientists were burdened with these additional complexities.

Fourth, the taskgraph can accommodate a mix of static and dynamic load balancing mechanisms. In particular, we are developing a mechanism that uses fine-grained dynamic load balance within an SMP node, and a relatively static coarse-grained mechanism between address spaces.

Fifth, the taskgraph facilitates development of simulation *components* that allow pieces of the simulation to evolve independently. In many respects, this is the most important advantage for a large interdiscplinary project such as C-SAFE. Since C-SAFE is a research project, we need to accommodate the fact that most of the simulation components are still under development. The component-based architecture allows pieces of the system to be implemented in a basic form at first, and then to evolve as the technologies mature. Most importantly, the UCF allows the aspects of parallelism (schedulers, load-balancers, parallel I/O, and so forth) to evolve independently of the simulation components. This allows the computer science effort to focus on these problems without waiting for the completion of the scientific applications or vice-versa.

However, in addition to these advantages there are some disadvantages to the approach that we chose. First, creating an optimal schedule for the taskgraph is known to be an NP-hard problem. However, we have found that using

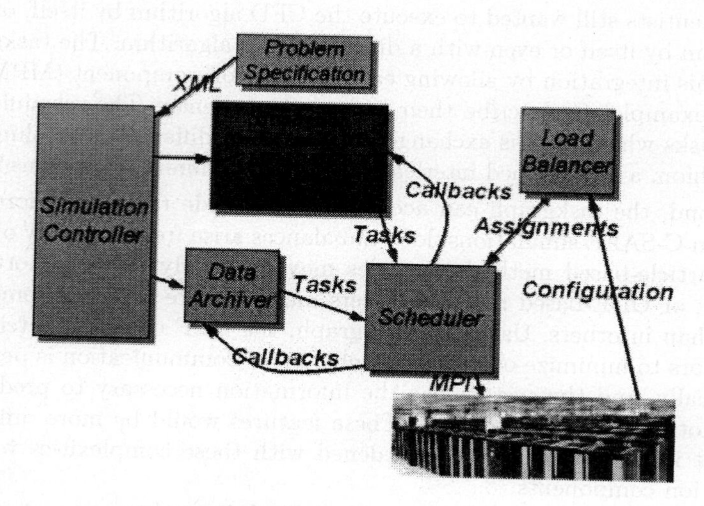

Fig. 3. UCF Simulation Components

some simple heuristics, and exploiting the regularity in the problem, that we can obtain respectable performance using straightforward scalable algorithms. Further refinements in these algorithms will receive more attention in the near future. Second, creation of the schedule can be costly. We take advantage of the fact that the schedule does not need to be recomputed for each execution. It does need to be recomputed if the algorithm changes, or if the grid changes. In addition, the schedule may need to be recomputed periodically to maintain an optimal load-balance. Third, the taskgraph requires a mental shift for parallel application programmers. Nevertheless, we found that the programmers were able to easily take a description of their algorithm and cast that into a set of tasks. The application was then able to run in parallel, even on hundreds of processors. For our applications, this benefit outweighed the cost of casting the algorithms in the dataflow execution model.

For C-SAFE, the advantages far outweighed the disadvantages. We note that for a typical purely structured-grid computation, the taskgraph may be overkill. For a purely unstructured-grid computation, the granularity would be too small and data dependencies would be more complex. The SAMR grids employed by C-SAFE seem to have just the right granularity for this approach to be successful.

4.3 Components Involved

Figure 4.3 shows the main components involved in a typical C-SAFE simulation. The SimulationController is the component in charge of the simulation. It will manage restart files if necessary, and control the integration through time. First, it reads the specification of the problem from an XML input file. After setting up the initial grid, it passes the description to the simulation component. The

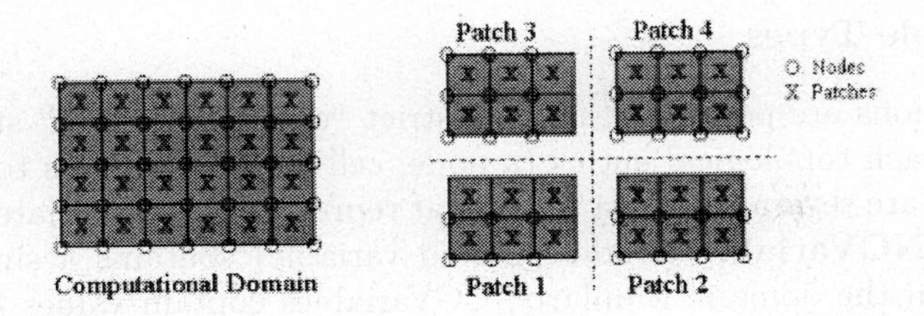

Fig. 4. A simple computational domain and a 4-patch decomposition of that domain

simulation component can be a number of different things, including one of two different CFD algorithms, the MPM algorithm, or a coupled MPM-CFD algorithm. The simulation components define a set of tasks to the scheduler. In addition, the DataArchiver component describes a set of output tasks to the scheduler. These tasks will save a specified set of variables to disk. Once all tasks are known to the scheduler, the load-balancer component uses the configuration of the machine (including processor counts, communication topologies, etc.) to assign tasks to processing resources. The scheduler uses MPI to communicate the data to the right processor at the right time and then executes callbacks into the simulation or DataArchiver components to perform the actual work. This process continues until the taskgraph is fully executed. The the execution process is repeated to integrate further timesteps.

Each of these components run concurrently on each processor. The components communicate with their counterparts on other processors using MPI. However, the scheduler is typically the only component that needs to communicate with other processors.

4.4 Definitions

Consider Figure 4.4. We define several terms which we use in discussing Structured AMR grids:

- **Patch:** A contiguous rectangular region of index space and a corresponding region of simulated physical space. The domain on the right of Figure 4.4 is the same as the domain on the left, except that is has been decomposed into two patches.
- **Cell:** A single coordinate in the integer index space, also corresponding to the smallest unit in simulated physical space. A variable centered at the cells in the simulation would have a value corresponding to each of the **X**'s in Figure 4.4.
- **Node:** An entity at the corners of each of the cells. A variable centered at the nodes in the simulation would have a value corresponding to each of the **O**'s in Figure 4.4.
- **Face:** The faces join two cells. The UCF represents values on X, Y, and Z faces separately.
- **Ghost cell:** Cells (or nodes) that are associated with a neighboring patch, but are copied locally to fulfill data dependencies from out side of the patch.

4.5 Variable Types

UCF simulations are performed using a strict "owner computes" strategy. This means that each topological entity (a node, cell or face) belongs to exactly one patch. There are several variable types that represent data associated with these entities. An **NCVariable** (Node-centered variable) contains a single value at each **Node** in the domain. Similarly, CCVariables contain values for each cell, and XFCVariables, YFCVariables and ZFCVariables are Face-centered values for the faces corresponding to the X, Y and Z axes. Each of these variables types are C++ template classes, therefore a node/cell/face-centered value can be any arbitrary type. Typically values are a double-precision number representing pressure, temperature, or some other scalar throughout the field, but values may also be a more complex entity, such as a vector representing velocity or a tensor representing a stress.

In addition to the topological based variables described above, there is one additional variable type: ParticleVariable. This variable contains values associated with each particle in the domain. A special particle variable contains the position of the particle. Other particle variables are defined by the simulations, and in the case of the MPM algorithm include quantities like temperature, acceleration, stress and so forth. For the purposes of the discussions below, particles can be considered a fancy type of cell-centered variable, since each particle is associated with a single cell. It is important however, to point out that explicit lists of particles within a cell are not maintained. We have found it more efficient to determine particle/cell associations as they are needed instead of paying the high cost of maintaining lists of particles for each cell.

4.6 X around Y

Tasks describe data requirements in terms of their computations on Node, Cell and Face-centered quantities. A task that computes a cell-centered quantity from the values on surrounding nodes would establish a requirement for 1 layer of nodes around the cells within a patch. This is termed "nodes around cells" in UCF terminology. As shown in Figure 4.6, a layer of *ghost nodes* would be copied from neighboring patches to the top and right edges of the lower-left patch. In a four-processor simulation this copy would involve MPI messages from each of the other three processors. It is important to note the asymmetry in this process; data is often not required from all 26 (in 3D) neighbors to satisfy a computation. Symmetry comes when a subsequent step uses "cells around nodes" to satisfy another data dependency.

In this fashion, each task in the algorithm specifies a set of data requirements. Similarly, each task specifies a set of data which it will compute, but in this case no ghost cells are necessary (or allowed). These "computes & requires" lists for each task are collected to create the full taskgraph.

A task could specify that it requires data from the entire computational domain. However, for typical scalable algorithms, the tasks ask for only one (or possibly two) layers of data outside of the patch.

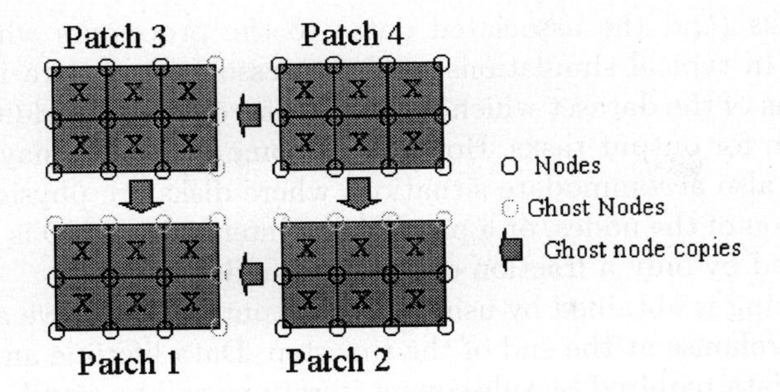

Fig. 5. Communication of ghost nodes in a simple 4-patch domain

4.7 Compilation

Using the data dependencies described above, the UCF scheduler compiles a coarse-representation of the taskgraph. This compilation process proceeds simultaneously on each processor without communication between processors. This representation contains all of the data dependency information, but contains only a single vertex (graph bubble) for each task. The full taskgraph will contain a vertex for each patch/task combination. However, this full representation is not fully instantiated on each processor.

Initially, a topological sort is performed on the coarse taskgraph. The scheduler then creates a *detailed taskgraph* for the subregion of the total graph which covers the neighborhood of a particular processor. It will instantiate graph vertices for the task/patch combinations that are assigned to the processor, and will instantiate graph edges which connect to or from those vertices. Finally, it will create vertices that connect to those edges. In this fashion, the detailed taskgraph contains the vertices for tasks owned by this processor and for those tasks on other processors with which it will communicate. The detailed taskgraph uses a very compact representation since the number of detailed tasks can be significant. However, there are $O(1)$ detailed tasks on each processor for scalable simulations.

After instantiating the detailed tasks, the UCF scheduler performs a set of analysis functions on the resulting taskgraph. It ensures that the application programmers have used every variable that it computes, and that they do not expect variables that are never produced. It also ensures that the types of variables match between the computes and requires, including the type of the underlying templated data. Finally, the scheduler analyzes the lifetimes of each variable used throughout the execution of the taskgraph. Variables that hold intermediate quantities are scheduled for deletion when no more tasks require them.

4.8 I/O and Checkpointing

Data output is scheduled using the taskgraph just like any other computation. Constraints specified with the task allow the load balancing components to direct those tasks (and the associated data) to the processors where data I/O should occur. In typical simulations, each processor writes data independently for the portions of the dataset which it owns. This requires no additional parallel communication for output tasks. However, in some cases this may not be ideal. The UCF can also accommodate situations where disks are physically attached to only a portion of the nodes, or a parallel filesystem where I/O is more efficient when performed by only a fraction of the total nodes.

Checkpointing is obtained by using these output tasks to save all of the data in the data warehouse at the end of the timestep. Data lifetime analysis ensures that only the data required by subsequent iterations will be saved. If the simulation components have been correctly written to store all of their data in the data warehouse, restart is a trivial process. During restart, the components process the XML specification of the problem that was saved with the datasets, and then the UCF creates input tasks that load the data warehouse from the checkpoint files. If necessary, data redistribution is performed automatically during the first execution of the taskgraph. In a similar fashion, changing the number of processors is possible. The current implementation does not redistribute data among the patches when the number of processors are changed. Patch redistribution is a useful component even beyond changing the processor count, and will be implemented in the future.

4.9 Legacy MPI Compatibility

In order to accommodate software packages that were not written using the UCF execution model, we allow tasks to be specially flagged as "using MPI". These tasks will be gang-scheduled on all processors simultaneously, and will be associated with all of the patches assigned to each processor. In this fashion, UCF applications can use available MPI libraries, such as PETSc [17] and hypre [18].

4.10 Execution

On a single processor, execution of the taskgraph is simple. The tasks are simply executed in the topologically sorted order. This is valuable for debugging, since multi-patch problems can be tested and debugged on a single processor. In most cases, if the multi-patch problem passes the taskgraph analysis and executes correctly on a single processor, then it will execute correctly in parallel.

In a multi-processor machine the execution processes is more complex. In an MPI-only implementation, there are a number of ways to utilize MPI functionality to overlap communication and I/O. We describe one way that is currently implemented in the UCF.

We process each detailed task in a topologically sorted order. For each task, the scheduler posts non-blocking receives (using MPI_Irecv) for each of the data dependencies. Subsequently we call MPI_Waitall to wait for the data to be sent from neighboring processors. After all data has arrived, we execute the task. When the task is finished, we call MPI_Isend to initiate data transfer to any dependent tasks. Periodic calls to MPI_Waitsome for these posted sends ensure that resources are cleaned up when the sends actually complete.

The mixed MPI/thread execution is somewhat different. Initially, non-blocking MPI_Irecvs are posted for *all* of the tasks assigned to the processor. Then each thread will concurrently call MPI_Waitsome and will block for internal data dependencies (i.e. from other tasks) until the data dependencies for any task are complete. That task is executed and data that it produces is sent out. The thread then goes back and tries to complete a next task. This implements a completely asynchronous scheduling algorithm. Preliminary results for this scheduler indicate that a performance improvement of approximately 2X is obtainable. However, thread-safety issues in vendor MPI implementations have slowed this effort.

It can be seen that dramatically different communication styles can be employed by simply changing out the scheduler component. The application components are completely insulated from these variations. This is a very important aspect that allows the Computer Science team members to focus on the best way to utilize the communication software and hardware on the machine without requiring sweeping changes in the application. Each scheduler implementation consists of less than 1000 lines of code, so it is relatively easy to write one that will take advantage of the properties of the communication hardware available on a machine. Often, the only difficult part is getting the correct information from the vendor in order to determine the best strategy for communicating data.

4.11 Load Balancing

The Load Balancer component is responsible for assigning each detailed task to one processor. To date, we have implemented only simple static load-balancing mechanisms. However, the UCF was designed to allow very sophisticated load-balance analysis algorithms. In particular, a cost-model associated with each task will allow an optimization process to determine the optimal assignment of tasks to processing resources. Cost models associated with the communication architecture of the underlying machine are also available. One interesting aspect of the load-balance problem is that integrated performance analysis in the UCF will allow the cost-models to be corrected at run-time to provide the most accurate cost information possible to the optimization process.

The mixed thread/MPI scheduler described above implements a dynamic load-balancing mechanism (i.e. a work queue) within an SMP node, and uses a static load-balancing mechanism between nodes. We feel that this is a powerful combination that we will pursue further.

Careful readers will pick up on the fact that the creation of detailed tasks require knowledge of processor assignment. However, sophisticated load-balance components may require this detailed information before they can optimize the task/processor assignments. We use a two-phase approach where tasks are assigned arbitrarily, then an optimization is performed and the final assignments are made to the tasks. Subsequent load-balance iterations use the previous approximation as a starting point for the optimization process.

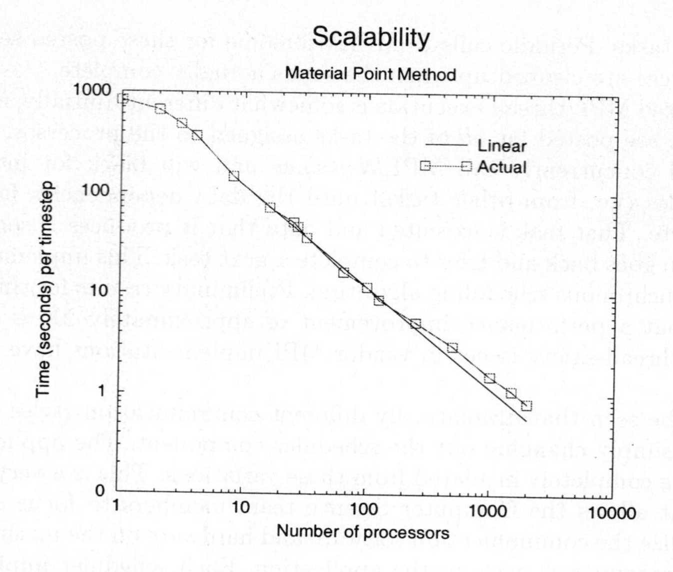

Fig. 6. Parallel Performance of a typical C-SAFE problem. This is a 16 million particle Material Point Method (MPM) computation, running on up to 2000 processors of the Los Alamos National Laboratory Nirvana machine. The straight line represents ideal performance, while the boxes show data points of performance actually obtained. There is a slight slowdown as the computation spans more than one 128 processor Origin 2000, but continues on linearly to the maximum number of processors on the machine.

5 Applications and Results

The system described here has been used to implement a variety of simulations. Two different CFD algorithms, one MPM (solid mechanics) algorithm, and two different coupled CFD-MPM algorithms have been implemented to date. Each of these simulations run quite well in parallel. Figure 5 shows the scalability of the MPM application up to 2000 processors of the ASCI Nirvana machine at Los Alamos National Laboratory. Nirvana is a cluster of 16 SGI Origin 2000 ccNUMA machines, each consisting of 128 MIPS R10K processors running at 250 Mhz. The machines are connected with a 6.4 Gigabit GSN (Gigabyte System Network). It should be noted that the problem shown is relatively small, with timesteps completing in much less than one second for the large processor configurations. One of the CFD components (Arches) has similar scaling properties to MPM; the others we have not had the opportunity to push to large numbers of processors. Figure 7 shows results from 1000 processor simulations using the Arches CFD component and the MPM component.

6 Conclusions and Future Work

The past few years of this project have focused on developing a flexible framework in which to accomplish scalable parallel simulations. Now that we have the basic

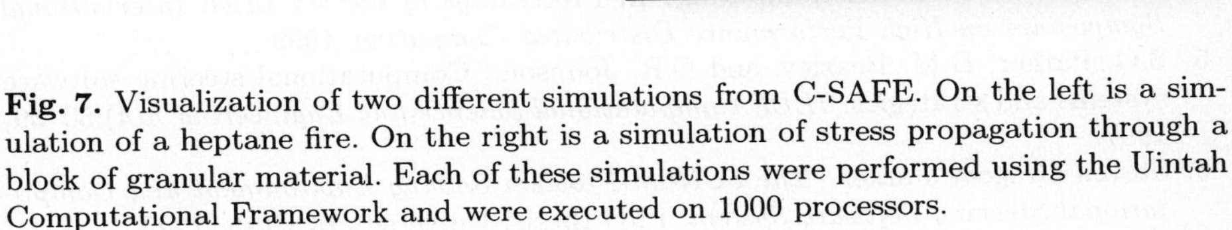

Fig. 7. Visualization of two different simulations from C-SAFE. On the left is a simulation of a heptane fire. On the right is a simulation of stress propagation through a block of granular material. Each of these simulations were performed using the Uintah Computational Framework and were executed on 1000 processors.

infrastructure in place, there are a number of research projects worthy of pursuit, including: development of feedback-based load balancing components that utilize task performance information collected at runtime to optimize resource allocations; further development and tuning of the mixed thread/MPI scheduler; investigation of and development of optimization algorithms that operate on an incomplete representation of the global graph; and further development of the UCF's AMR capabilities.

We have presented a powerful way of building complex PDE simulation software. We have demonstrated that the UCF system can obtain good performance on up to 2000 processors. Using the UCF, C-SAFE scientists have been able to simulate complex physical phenomena that was not achievable using smaller-scale versions of the code.

We have described only a small fraction of a very complex system. Other interesting aspects, such as the template implementations of the data warehouse, the augmented run-time typing system, and other features will be described in more detail in the future.

7 Acknowledgments

This work was supported by the DOE ASCI ASAP Program. James Bigler, J. Davison de St. Germain, and Wayne Witzel are very valuable members of the C-SAFE PSE team that developed this software. C-SAFE visualization images were provided by Kurt Zimmerman and Wing Yee. Datasets were created by Scott Bardenhagen, Jim Guilkey, Seshadri Kumar, Rajesh Rawat, and John Schmidt. The DOE ASCI ASAP program also provided computing time for the simulations shown.

References

1. Center for the Simulation of Accidental Fires and Explosions - Annual Report, Year 2. http://www.csafe.utah.edu/documents.
2. T. C. Henderson, P. A. McMurtry, P. J. Smith, G. A. Voth, C. A. Wight, and D. W. Pershing. Simulating accidental fires and explosions. *Comp. Sci. Eng.*, 2:64–76, 1994.
3. Common Component Architecture Forum. http://z.ca.sandia.gov/ cca-forum.
4. R. Armstrong, D. Gannon, A. Geist, K. Keahey, S. Kohn, L. McInnes, S. Parker, and B. Smolinski. Toward a Common Component Architecture for High-Performance Scientific Computing. In *Proceedings of the 8th IEEE International Symposium on High Performance Distributed Computing*, 1999.
5. S.G. Parker, D.M. Beazley, and C.R. Johnson. Computational steering software systems and strategies. *IEEE Computational Science and Engineering*, 4(4):50–59, 1997.
6. Steven Gregory Parker. *The SCIRun Problem Solving Environment and Computational Steering Software System.* PhD thesis, University of Utah, 1999.
7. S.G. Parker and C.R. Johnson. SCIRun: A scientific programming environment for computational steering. In *Supercomputing '95.* IEEE Press, 1995.
8. J. Davison de St. Germain, John McCorquodale, Steven G. Parker, and Christopher R. Johnson. Uintah: A Massively Parallel Problem Solving Environment. In *Proceedings of the Ninth IEEE International Symposium on High Performance and Distributed Computing*, August 2000.
9. J. McCorquodale, S. Parker, J. Davison, and C. Johnson. The utah parallelism infrastructure: A performance evaluation. In *2001 High Performance Computing Symposium (HPC'01), Advanced Simulation Technologies Conference*, 2001.
10. A. Cleary, S. Kohn, S.G. Smith, and B. Smolinski. Language Interoperability Mechanisms for High-Performance Scientific Applications. In *Proceedings of the SIAM Workshop on Object-Oriented Methods for Interoperable Scientific and Engineering Computing*, October 1998.
11. D. Sulsky, Z. Chen, and H. L. Schreyer. A Particle Method for History Dependent Materials. *Comp . Methods Appl. Mech. Engrg*, 118, 1994.
12. S.G Bardenhagen, J.E. Guilkey, K.M Roessig, J.U. Brackbill, W.M. Witzel, and J.C. Foster. An improved contact algorithm for the material point method and application to stress propagation in granular material. *Computer Modeling in Engineering and Sciences*, 2(4):509–522, 2001.
13. J.E. Guilkey and J.A. Weiss. Implicit time integration with the Material Point Method (submitted). *International Journal for Numerical Methods in Engineering*, 2002.
14. S. Vajracharya, S. Karmesin, P. Beckman, J. Crotinger, A. Malony, S. Shende, R. Oldehoeft, and S. Smith. Smarts: Exploiting temporal locality and parallelism through vertical execution, 1999.
15. S. Atlas, S. Banerjee, J.C. Cummings, P.J. Hinker, M. Srikant, J.V.W. Reynders, and M. Tholburn. POOMA: A high-performance distributed simulation environment for scientific applications. In *Supercomputing '95 Proceedings*, December 1995.
16. J. T. Feo, D. C. Cann, and R. R. Oldehoeft. A report on the sisal language project. *Journal of Parallel and Distributed Computing*, 10(4):349–366, 1990.
17. S. Balay, W. Gropp, L. McInnes, and B. Smith. Petsc home page, 1999.
18. hypre: high performance preconditioners. http://www.llnl.gov/casc/hypre/.

Using Design Patterns and XML to Construct an Extensible Finite Element System*

J. Barr von Oehsen[1], Christopher L. Cox[2], Eric C. Cyr[2], and Brian A. Malloy[3]

[1] CAEFF, Clemson University, Clemson, SC 29634, USA
vonoehse@ces.clemson.edu

[2] Department of Mathematical Sciences, Clemson University
Clemson, SC 29634, USA
clcox@ces.clemson.edu

[3] Department of Computer Science, Clemson University
Clemson, SC 29634, USA
{ecyr,malloy}@cs.clemson.edu

Abstract. We present an object-oriented finite element software package which employs XML and design patterns in order to solve a problem which is fundamental to the simulation of viscoelastic flows. The XML format increases flexibility in data handling, while design patterns provide a high-level organizational structure. The application problem and the finite element methodology are described, along with the opportunities presented for incorporation of the software tools. Numerical experiments comparing results with and without XML and design patterns are also presented.

1 Introduction

The goal of the Center for Advanced Engineering Fibers and Films is to develop computational models which predict final fiber and film properties based on processing conditions. A primary goal of the CAEFF modeling group is to develop a widely-applicable finite element software package for modeling of viscoelastic flows that is easily maintainable and fully modular, so that the code can be readily customized to the user's choice of material parameters, governing equations, solution domain and computer platform. Most efforts to improve finite element codes have concentrated on the linear or nonlinear system solvers, because normally this stage dominates the overall CPU time. For applications such as ours, it is also worthwhile to make the assembly phase as flexible as possible with efficiency also a major consideration. This flexibility will be especially important when the code is ported to a parallel environment, where both assembly and solution phases are distributed over many processors. Ease of handling data in a variety of formats is a necessary aspect of the software package, because the target code will link the simulation routines with an experimental database and advanced visualization software.

* This work was supported primarily by the ERC program of the National Science Foundation under Award Number EEC-9731680.

P.M.A. Sloot et al. (Eds.): ICCS 2002, LNCS 2331, pp. 735–744, 2002.

In this paper, we present the design and implementation of a prototype system which will undergird a viscoelastic flow simulation package. Our model includes several design patterns that assure ease in maintenance and modification of the code. Design patterns help developers resolve software architecture bottlenecks and provide a common vocabulary for building a simulation package. Those familiar with the patterns used to develop a system, will find the code easier to read and maintain. The end-user, likely one who is not familiar with object-oriented programming, will see modules identified according to application.

In the next section we provide background about XML and the design patterns that we use in our system. In Section 3, we describe our finite element solution to a prototype viscous flow problem that we consider, and in Section 4, we provide an overview of the design of our system and the flow of data through the system. In Section 6 we describe our use of XML to present our output to any viewer for which an accompanying XSLT style sheet is provided. In Section 7 we describe the results of comparing our patterned approach with a non-patterned approach also written in C++. Finally, in Section 8, we draw conclusions and describe our ongoing work.

2 Background

In this section, we provide background about XML and design patterns, including an overview of the Factory Method and Singleton patterns that facilitate our modeling of a finite element solution of partial differential equations[13, 4].

2.1 XML

The extensible markup language, XML, has become one of the hottest concepts in computer science, web authoring and web programming. Like its HTML counterpart, XML is derived from the standard generalized markup language, SGML. However, HTML is a markup language used for displaying information content; XML, on the other hand, is a markup language for creating markup languages. HTML is a markup language for marking documents using tags for headings and paragraphs, whereas XML enables the creation of new tags for marking anything, such as mathematical formulas, molecular structure of chemicals, music scores and any other document. HTML limits the user to a fixed collection of tags, used primarily to describe the content that is displayed in a browser.

An XML document consists of a list of element types (tags), together with their attributes. The relationships of these elements, to each other, can be specified by an optional XML Schema. XML Schemas express shared vocabularies and allow machines to carry out rules made by people. They provide a means for defining the structure, content and semantics of XML documents. A schema is not required for a document but is recommended for document conformity. By combining an XML document with its corresponding XML Schema, an XML parser can determine the content and structure of an XML document. We have created our own fiber and film XML Schema - which we hope will become the industry standard.

For web authoring, HTML has emerged as the technology of choice for describing the content of an HTML document; cascading style sheets, CSS, has emerged as the technology of choice for describing the form of an HTML document. Similarly, the extensible style language, XSL, was developed as a technology for describing the form of an XML document. An XSL style sheet provides the rules for displaying or organizing an XML document's data. XSL also provides elements that define rules for describing how to transform one XML document into another XML document. For example, an XML document can be transformed into an HTML document. The facet of XSL that addresses the problem of transforming XML documents is called XSL transformations, or XSLT.

2.2 Design Patterns

A fundamental aspect to any science or engineering discipline is a common vocabulary that practitioners can use to express concepts, and a language for relating the concepts. The development of a catalog of design patterns provides such a vocabulary, together with a body of literature to help software developers resolve recurring problems encountered in software development. Patterns describe design practices that capture experience in a way that enables others to reuse this experience. The primary focus is not so much on technology as it is on creating documentation of sound software engineering design practices.

The *Factory Method* is a creational pattern that describes an extensible approach to the construction of objects. This pattern is typically used to create instances of objects described by a class hierarchy. The Factory Method eliminates the need for binding application-specific code about the construction of the objects into the system[4]; instead, the construction is incorporated into a function, or method, that abstracts the details of construction. This method typically accepts a parameter that encodes information about the type of object to create and then the Factory Method, after interpreting the information, constructs the object. If the Factory Method is a class constructor, than this pattern is sometimes referred to as the *Virtual Constructor Pattern*[4].

The *Singleton Pattern* is another creational pattern that permits the developer to restrict creation to a single instance, while providing global access to this instance. To do this, the Singleton class constructor is placed either in the private or protected section, and a *construction function* is placed in the public section of the Singleton class. Instantiation of the Singleton class must be made through a call to the *construction function*, which maintains a count of the number of instantances previously created, as well as a pointer to the single instance. If the Singleton class has already been instantiated, than the pointer to this instantiation is returned; otherwise, an instance of the Singleton class is constructed and the resulting pointer returned.

3 A Finite Element Solution of a Viscous Flow Problem

A mathematical model for the flow of a viscoelastic fluid consists of the standard momentum, mass, and energy balance equations, plus a constitutive equation representing the manner in which the stress depends on the velocity gradient. A common form of these equations for isothermal flow, with inertial terms dropped, is [2]

$$\rho\frac{\partial \mathbf{v}}{\partial t} - \nabla\cdot\boldsymbol{\tau} + \nabla p = \mathbf{f} \tag{1}$$

$$\nabla\cdot\mathbf{v} = 0 \tag{2}$$

$$\boldsymbol{\tau}_p + \lambda_1\boldsymbol{\tau}_{p(1)} - \alpha\frac{\lambda_1}{\eta_p}\{\boldsymbol{\tau}_p\cdot\boldsymbol{\tau}_p\} = -\eta_p\dot{\gamma}, \tag{3}$$

In (1), ρ is the fluid density, $\mathbf{v}$ is the velocity vector, p is pressure, $\boldsymbol{\tau}$ is the extra stress tensor, and $\mathbf{f}$ is the forcing term vector.

The extra stress has been split according to $\boldsymbol{\tau} = \boldsymbol{\tau}_n + \boldsymbol{\tau}_p$ where $\boldsymbol{\tau}_n$, the Newtonian part, is a constant multiple of $\dot{\gamma} = \nabla\mathbf{v} + (\nabla\mathbf{v})^T$. The polymeric part, $\boldsymbol{\tau}_p$, satisfies a nonlinear differential or integral equation. For example, (3) is known as the Giesekus model. In this equation λ_1, α and η_p are fitted parameters and $\boldsymbol{\tau}_{p(1)} = \frac{d\boldsymbol{\tau}_p}{dt} - (\nabla\mathbf{v})^T\cdot\boldsymbol{\tau}_p - \boldsymbol{\tau}_p\cdot\nabla\mathbf{v}$.

Several finite element formulations for approximating the solution of (1)-(3) appear in the literature. Realistic simulations,in three dimensions, require the solution of systems containing hundreds of thousands or even several million unknowns. Therefore iterative methods are normally chosen over direct methods. One such iterative approach is the θ-method [10], which will be briefly described. For the general problem of finding the solution $\mathbf{U}$ for the system

$$M\frac{d\mathbf{U}}{dt} + A(\mathbf{U}) = 0 \tag{4}$$

with initial condition $\mathbf{U}(0) = \mathbf{U}_0$, the θ-method is based on splitting the nonlinear operator A into a sum of simpler (linear) operators A_1 and A_2. For equations (1)-(3), $\mathbf{U} = [\mathbf{v}\ \ p\ \ \boldsymbol{\tau}_p]^T$, or more precisely the coefficients in the finite element approximation for each function. The operator A is the nonlinear integral/differential operator which arises in the variational formulation of (1)-(3). One complete θ-method update for (4) on time interval $[t,\ \ t+\Delta t]$ consists of three steps. For the operator A associated with the finite element approximation of (1)-(3), a splitting can be chosen so that the second step is effectively a transport equation for $\boldsymbol{\tau}_p$, while the first and third steps dominate the computational work and have the form

$$\begin{aligned} \beta\mathbf{v} - \nu\Delta\mathbf{v} + \nabla p &= \mathbf{f} \\ \nabla\cdot\mathbf{v} &= 0 \end{aligned} \tag{5}$$

which is the Stokes Problem varied by the addition of the $\beta\mathbf{v}$ term.

Because the solution of the Stokes Problem is fundamental to the θ-method for solving (1)-(3), and because the theme of this paper is software design for finite element approximation, we will concentrate on the Stokes Problem, i.e. (5) with $\beta = 0$), with simple Dirichlet boundary condition $\mathbf{v} = 0$ and computational domain the unit square in the first quadrant, with interior denoted Ω and boundary Γ.

The finite element solution is based on the variational formulation, i.e. find $(\mathbf{v},\ p)$ in an appropriate product space of functions (in this case $H_0^1(\Omega) \times L_2(\Omega)$) so that

$$\begin{aligned} \nu \int_\Omega (\nabla \mathbf{v} : \nabla \mathbf{w} - p \nabla \cdot \mathbf{w}) &= \int_\Omega \mathbf{f} \cdot \mathbf{w} \\ - \int_\Omega q \nabla \cdot \mathbf{v} &= 0 \end{aligned} \tag{6}$$

for all $(\mathbf{w},\ q)$ in a similar product space [5]. In (6), $\boldsymbol{\sigma} : \boldsymbol{\tau} = \sigma_{ij}\tau_{ij}$, with summation on repeated indices.

A suitable choice of finite dimensional subspace-pair for the approximate solution of (6) is the Taylor-Hood element on a triangular mesh [12], comprised of continuous piecewise quadratic functions for $\mathbf{v}$ and $\mathbf{w}$, and continuous piecewise linears for p and q. For uniqueness, an additional condition must be imposed on the pressure space. We impose the condition $\int_\Omega p = 0$.

The finite element solution is computed in three steps: mesh generation, assembly, and solution of the matrix system. For mesh generation we use the object-oriented package QMG [9]. An efficient, modular formulation of the assembly stage is especially important for this code because it must eventually be applicable to a range of fluid characterizations (i.e. (3) will change significantly) in various physical settings (e.g. a variety of boundary conditions). Assembly on each triangular element is carried out with the standard technique of mapping to a canonical element, using isoparametric elements to allow for curved boundaries [5]. The use of this mapping has a significant impact on data structures, allowing the use of *singletons* for basis functions and quadrature rules. For this test problem the linear system was solved using banded Gaussian elimination with partial pivoting. A benchmark analytical solution was used to check convergence with respect to mesh refinement, in order to assure that the code was working properly.

4 System Overview

Figure 1 summarizes our system to provide a finite element solution to a viscoelastic flow problem. The two circles on the left of the diagram illustrate input to the system: the top circle on the left represents the mesh, and the bottom circle on the left represents a polymer read from an existing database. Both inputs to our system are in XML format, using a schema that we have developed for viscoelastic flow. Our input is accepted by the `Finite Element Factory`, represented by the square on the left of Figure 1. The `Finite Element Factory` builds the necessary objects, used in the assembly of the linear system.

The `Assembly Routine` is represented by a circle in Figure 1, and uses methods in two other factories to facilitate assembly. These methods are part of our `Basis Function Factory` and `Quadrature Factory`, represented by the

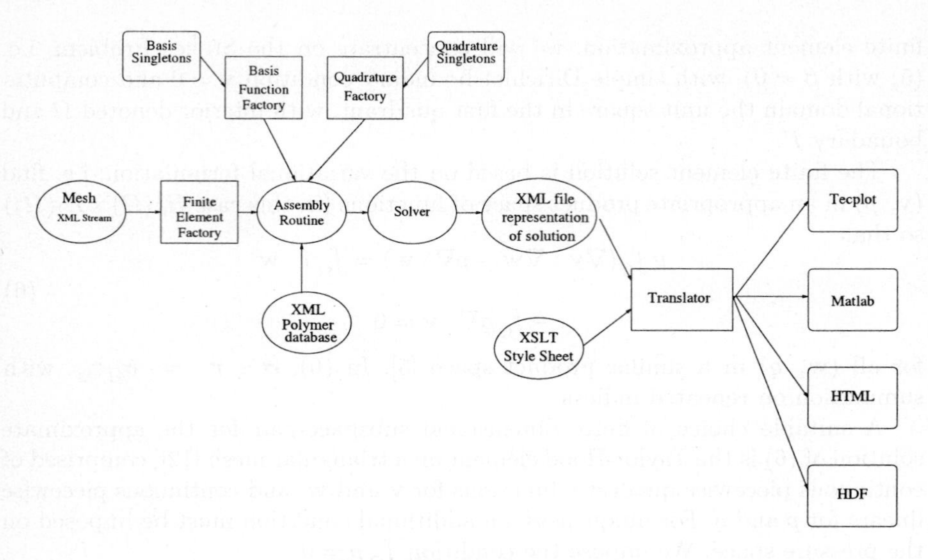

Fig. 1. *System Overview.*

squares at the top of the figure. Once assembled, the mesh is input to the `solver`. The output of the solver is an XML representation of the solution, which together with an `XSLT Style Sheet`, is input to our `Translator`. Using an appropriate XSLT style sheet, the output can be formatted for any viewing tool, such as *Tecplot*, *Matlab*, [1, 6], or any of the other viewing tools illustrated on the right side of Figure 1.

In the next section, we provide details about the `Finite Element Factory` and `Singleton`, that is a pivotal class hierarchy in our design.

5 Finite Element Factory and Singleton Patterns

The creation of objects (in our case 1D, 2D, and 3D elements, basis functions, and quadratures) occur through common factories (finite element factory, basis element factory, quadrature factory) rather than allowing the creation of these objects to be distributed throughout the system. The advantage of this approach is that if adding a new object to our design is necessary, we need only modify the factory - not the entire system. For example, if we need to add another type of basis function (which we will do when we include 3D models) or constitutive equation to our system, the factory pattern allows this flexibility without having to modify the whole code. Not only does this design allow for easy maintenance, but enables us to make our code modular.

The singleton pattern allows the creation of one and only one instance of an object. This design works particularly well with our system since we are pushing all of our finite element analysis onto a canonical element. There is no need to have multiple copies of a particular basis function object stored in memory if

```
( 1)  < ?xml version="1.0"? >
( 2)  < FINITE_ELEMENT
( 3)     xmlns : xsi = "http : //www.w3.org/2001/XMLSchema − instance"
( 4)     xsi : noNamespaceSchemaLocation = "finite_element.xsd" >
( 5)       < NODES NUMBER_OF_NODES = "1681" >
( 6)       < NODE GLOBAL_NODE_NUMBER = "0" >
( 7)       < VECTOR DIM = "2" >
( 8)       < COORD > 0 < /COORD >
( 9)       < COORD > 0 < /COORD >
(10)       < /VECTOR >
(11)       < VELOCITY >
(12)       < VECTOR DIM = "2" >
(13)       < CRD > 0 < /CRD >
(14)       < CRD > 0 < /CRD >
(15)       < /VECTOR >
(16)       < /VELOCITY >
(17)       < PRESSURE >-104.249< /PRESSURE >
(18)       < /NODE >
(19)       < /NODES >
(20)       < ELEMENTS NUMBER_OF_ELEMENTS = "800" >
(21)       < ELEMENT GEOMETRY = "TRIANGLE_SIX_POINT"
(22)         GLOBAL_ELEMENT_NUMBER = "0" >
(23)       < VECTOR DIM = "6" >
(24)       < COORD > 0 < /COORD >
(25)       < COORD > 2 < /COORD >
(26)       < COORD > 84 < /COORD >
(27)       < COORD > 1 < /COORD >
(28)       < COORD > 43 < /COORD >
(29)       < COORD > 42 < /COORD >
(30)       < /VECTOR >
(31)       < /ELEMENT >
(32)       < /ELEMENTS >
(33)    < /FINITE_ELEMENT >
```

Fig. 2: *XML Example.* This figure illustrates our use of XML to describe the data structures that we use in our modeling of a viscous flow problem. Our approach can be applied to most mathematical models where data needs some form of visualization.

all instances are exactly the same. Thus, the singleton pattern allows us to save on the memory used and also increase the overall performance of the assembly routine. Moreover, we allow global access to the singleton object exclusively through a static member function.

6 XML Representation of the Data

Figure 2 illustrates our use of XML to describe the finite element solution of a viscous flow problem. The finite element XML Schema includes XML tags to describe nodes and elements used in finite element analysis. These tags are represented by `NODES` and `ELEMENTS` on lines 5 and 20 in the figure. The children of `NODES` are one or more `NODE` elements, which are composed of coordinate vectors, velocity vectors and pressure, illustrated on lines 6 through 18 in the figure. We have abbreviated the XML representation but in the actual system these XML elements completely describe our finite element solution. Unfortunately, the solution is almost meaningless if one can not use other tools to view the results. Here lies the problem - different tools require different file formats. It is precisely this reason that flexibility in translation is absolutely essential.

Depending on the solution of the linear system and the visualization tool that we use, we sometimes need to create more than one formatted file. By exploiting the XML tools made available by the Apache XML project (Xerces and Xalan), the Mozilla Organization (Rhino - JavaScript for Java), and the Simple API for

XML (SAX) we have designed our system so that most decisions of this type are transparent to the user[7, 3, 8, 11]. For example, we accomplish the transparency of creating multiple formatted files by first parsing the XML file to check tags and then set certain values dependent upon the parsed outcome. Once completed, our system then reads in a given XSLT style sheet and substitutes our set values into parameterized conditional statements. The translation only takes place where the conditional statements are true (see Figure 3).

```
( 1)  < xsl:when test="$param='velocity_pressure'" >
( 2)  variables="x","y","u","v","p"
( 3)  < /xsl:when >
( 4)  < xsl:when test="$param='no_pressure'" >
( 5)  variables="x","y","u","v"
( 6)  < /xsl:when >
( 7)  < xsl:when test="$param='no_velocity'" >
( 8)  variables="x","y","p"
( 9)  < /xsl:when >
```

Fig. 3: *Parameterized Conditional Statements.* This figure summarizes how one uses parameterized conditional statements within an XSLT style sheet. If the variable "param" is set to "velocity_pressure," then only the text on line (2) will be used in the translation. Likewise for the other conditional statements.

As shown in Figure 1, once the system of equations has been solved, the results are output into an XML file. The oval to the right of the solver represents these results. The XML representation of the solution, together with an XSLT style sheet, becomes input to a translator, represented by the rectangle toward the right of Figure 1. An XSLT style sheet is tailored to a particular viewing tool and is used to translate the XML representation into a file for that particular tool. Finally, the translated file is used as input to the tool and can be viewed by the user in a variety of file formats.

Using XML has several advantages: (1) ease of maintenance and less code bloat (we do not have to hard code different formats into our system), (2) the user does not need to wait for a software upgrade just so that he or she can use their favorite visualization/computer algebra system tool, (3) writing an XSLT style sheet is not very difficult, (4) the user can save output in multiple formats, and (5) if the user decides later that he or she would rather use a different file format after the fact, then (as long as the user has the proper style sheet) getting a new translation can easily be managed without having to run the finite element code from the start.

7 Results

In this section, we present the results of our object-oriented design and implementation, written in C++, that uses the Factory Method and Singleton patterns. To show the effects of our use of the patterns, we compare our patterned approach to an object-oriented approach without patterns (Fortran legacy code rewritten to be object oriented), also coded in C++.

All of the experiments in this section were executed on a Dell Optiplex GX1 workstation, equipped with a 500 MHz PIII processor and 768 Megabytes of memory. Our operating system is Red Hat Linux 7.1. The C++ compiler is gcc version 3.0.2.

Number of Unknowns	Experiments with Assembly Routine (sec)			
	C++ No Patterns		Patterned C++	
	mean	std dev	mean	std dev
5,477	27.76	0.077	8.63	0.01
22,202	113.82	0.210	34.82	0.12
89,402	697.50	6.670	382.43	27.93

Table 1: **Results of Experiments.**

This table illustrates the results of our experiments with two versions of the assembly routine that builds the linear system. Each column reports the results of 50 executions.

Table 1 illustrates the results of our experiments with two versions of the assembly routine that builds the linear system. There are three rows of data in the table, representing the number of unknowns to be solved within the linear system. There are three columns of data in the table, including headings for the experiments with the C++ code that did not use patterns, `C++ No Patterns`, and a heading for the C++ code that used patterns, `Patterned C++`. For each approach, we report the `mean` and `standard deviation` for fifty executions of the assembly routine, making note of the number of unknowns for each experiment. For example, for the experiment with 5,477 unknowns, the C++ code without patterns required an average of 27.76 seconds, and the C++ with patterns required an average of 8.63 seconds.

The difference between the C++ code without patterns and the C++ code with patterns is that the unpatterned code distributes object creation throughout the application; in the patterned code, object creation is localized in the factory method for the respective class framework. For example, the factory method for the finite element framework accepts a dimension parameter and then builds the elements and sets up the bookkeeping necessary for solving a particular problem, simplifying code maintenance.

In comparing the results of the C++ code without patterns to the C++ code with patterns, we see a clear performance advantage to the organization and modularity provided by the patterns. For example, for the test case with 89,402 unknowns, the C++ code with no patterns required on average 697.50 seconds, while the patterned C++ code required on average only 382.43 seconds. This performance benefit is due, in part, to our use of the singleton pattern, which obviates the need to repeatedly create and destroy objects during execution. Instead, we create a singleton of each of the objects and then use this singleton when it is needed.

8 Conclusions and Future Work

In this paper, we have described an object-oriented finite element software package that incorporates XML formatting and design patterns to provide ease of maintenance and extensibility. Numerical experiments indicate no additional cost in CPU time with the new design. Rather, the CPU time for the assembly stage of the finite element process decreased when the code was modified to include XML and design patterns. In the immediate future, we will benchmark the code against a conventional C-language code and Fortran. We are also extending the code to simulate non-Newtonian and viscoelastic flows through complex geometries. An effort is underway to develop a software database of experimental results that will provide input parameters for the code. The technique presented in this paper will be especially useful for each of these next phases.

References

1. Amtec Engineering Incorporated. *Tecplot, Version 9.0.* http://www.amtec.com/, 2001.
2. F. P. T. Baaijens. Mixed finite element methods for viscoelastic flow analysis: a review. *J. Non-Newtonian Fluid Mech.*, 79:361–385, 1998.
3. Fluent. *Fluent Flow Modeling Software.* http://www.fluent.com/, 2001.
4. E. Gamma, R. Helm, R. Johnson, and J. Vlissides. *Design Patterns: Elements of Reusable Object-Oriented Software.* Addison-Wesley, 1995.
5. C. Johnson. Numerical solution of partial differential equations by the finite element method. *Cambridge University Press*, 1992.
6. Mathworks Incorporated. *Matlab, Version 6.1.* http://www.mathworks.com/, 2001.
7. Oasis. *SAX developed collaboratively by the members of the XML-DEV mailing list.* http://www.oasis-open.org/, 2001.
8. Mozilla Organization. *Rhino: JavaScript for Java.* http://www.mozilla.org/, 2001.
9. QMG. 2.0 mesh generation software. *www.cs.cornell.edu/home/vavasis/qmg2.0/qmg2_0_home.html*, 1999.
10. P. Saramito. A new θ-scheme algorithm and imcompressible fem for viscoelastic fluid flows. *Math. Modelling Numer. Anal.*, 28(1):1–34, 1994.
11. The XML Apache Organization. *The Apache XML Project.* http://xml.apache.org/, 2001.
12. F. Thomasset. Implementation of finite element methods for navier-stokes equations. 1981.
13. J. B. von Oehsen, R. C. Jenkins, C. L. Cox, and B. A. Malloy. Exploiting xml to provide a uniform interface for graphical representation of finite element analysis. *Proceedings of the International Conference on Computer and Information Systems*, pages 181–185, Oct 2001.

GrAL- The Grid Algorithms Library

Guntram Berti

C&C Research Laboratories, NEC Europe Ltd.
Rathausallee 10, 53757 St. Augustin, Germany
berti@ccrl-nece.de

Abstract. Dedicated library support for mesh-level geometry components, central to numerical PDE solution, is scarce. We claim that the situation is due to the inadequacy of traditional design techniques for complex and variable data representations typical for meshes. As a solution, we introduce an approach based on generic programming, implemented in the C++ library GrAL, whose algorithms can be executed on any mesh representation. We present the core design of GrAL and highlight some of its generic components. Finally, we discuss some practical issues of generic libraries, in particular efficiency and usability.

1 Introduction

Software for the numerical solution of PDEs generally involves a lot of components, which can be partitioned into several layers. At the bottom, there are container data structures and corresponding algorithms, such as sorting and searching. Here, we are concerned with the next higher level, namely representations for geometric structures (meshes), and associated algorithms.

On top of these geometric components operate numerical discretization such as FEM or FV algorithms. Often, we also need algebraic components, namely, matrix and vector representations implementing the vector-space operations, and algorithms for solving linear or non-linear systems, which again may access the grid data structures, e. g. preconditioners or multi-grid algorithms. In addition, handling boundary conditions may involve complex geometric calculations, for instance for contact problems. Besides the numerical algorithms, components for pre- and postprocessing also operate directly on grids in a multitude of ways.

All this means that the mesh layer is central to numerical PDE solution. To the author's knowledge, however, there is no library dedicated to meshes available from which we can, say, take a mesh contact detection algorithm for inclusion into a PDE solver.

We feel that a primary cause for this apparent lack is that traditional ways of library design cannot cope with the variability of geometric data structures, because they either introduce a tight coupling between representational and algorithmic code, or rely on a conversion of data structures.

In this paper, we introduce the Grid Algorithms Library GrAL (available at [1]), which overcomes these difficulties by defining an abstract interface for

P.M.A. Sloot et al. (Eds.): ICCS 2002, LNCS 2331, pp. 745–754, 2002.

grids. Using a generic programming approach in C++, GrAL achieves a complete decoupling of algorithms and data structures.

After briefly analyzing the problems with traditional library design, we give an overview over the core design of GrAL, and highlight some of GrAL's generic components, which includes support for parallel PDE solution. Then, we discuss some practical aspects by which generic libraries differ from other approaches. Finally, future options are outlined.

2 Problems with Traditional Grid-related Components

If we consider grids used for numerical simulation, we see a wide variety of different types: Cartesian and curvilinear mapped structured grids, multi-block and semi-structured [15], unstructured pure-simplex or general cell grids, to hybrid or chimera-type grids. Taking into account the virtually unlimited possibilities for representing these grids, it becomes clear that no "standardization" approach on the representation level can ever be successful. It is also clear that the chance for using algorithms implemented for one kind of grid data structure in a different context are practically zero. So, basically, traditional approaches to create reusable grid components are limited to the following:

1. Use a grid API.
2. Use a file coupling
3. Always use the same standard data structure,
4. implement the algorithm ad-hoc for a given data structure

In the API approach, the grid has to be copied into the data structures predefined by the library routine, and possibly vice versa, which might be not trivial. Second, copying can be grossly inefficient if the algorithm itself is fast or operates only locally (e. g. point location). Memory might become a bottleneck, especially if the grid uses an optimized data structure with implicit connectivity. File coupling evidently has the same – worsened – difficulties.

In sum, these approaches are viable only if a substantial amount of work is done on the grids, justifying both the overhead of copying and the programming work for converting the data structures. This is commonly the case for instance for mesh generation or visualization, where file coupling is the rule.

Using standard data structures works only for cases with limited representational choice, such as structured grids, see [6, 8]. The most prominent example for this approach (albeit not for grids) probably is LAPACK, which prescribes a set of representations for dense matrices. Thus, in spite of its deficiencies, option 4 all to often is the only choice left.

The solution we offer here has none of these drawbacks. It uses the common underlying mathematical structure of grids to define an abstract interface for them, which is powerful enough that a large class of algorithms can be implemented *generically* on top of this interface. This approach reverses the flow of information implicit in a classical API: Whereas in the case of an API, we have to learn the interface of the library component; in our approach, we present the internals of our data representation in a structured way to the library component.

The solution is inspired by the C++ Standard Template Library STL [10]. Similar approaches are BGL [16] for graphs, CGAL [7] for general computational geometry, and VIGRAL [9] for image processing.

3 Design and Core Components of GrAL

There are some essential requirements which our library should fulfill:

1. Complete decoupling of algorithms from data structures: Algorithms shall be usable with *any* data structure providing sufficient functionality
2. Constant reuse cost: Shall be able to use any algorithm, by creating a thin adaptation layer user data *once*
3. Shall maintain high performance with respect to direct implementation

Thus, the process of creating an adequate interface is a compromise between expressiveness (we want to access the essential properties of grids), minimality (we don't want to create new interfaces for each new algorithm) and efficiency (we want algorithms to run almost as fast as direct implementations).

Now, having a unified interface does neither mean that we are restricted to some least common denominator, nor does it place unrealistic requirements on the functionality of data structures. Concerning the first point, note that *specialization* is an integral part of generic programming. For example, if we implement a generic search algorithm / data structure, we realize that it can be implemented much more efficiently for Cartesian grids. Thus, we can create a special implementation for that case, and when the generic search component is used, the compiler possesses enough information to decide which version to use.

Similarly, there is no need for a concrete grid component to support the complete interface – in fact, it is in general not possible. For instance, if our data structure does not know about cell neighbor relationships, we cannot support the corresponding concept of the interface. Again, generic components can specialize according to the supported subset of the interface, or the missing functionality can be added on-the-fly.

3.1 The Kernel Interface – A Very Quick Tour

The following gives a bird's-eye view only. Details can be found [2, 3], and the formal interface specifications are continually updated in [1].

The definition of a grid corresponding to the interface given below is slightly more general than that originally given in [2]. Restrictions for meeting requirements of algorithms are made on a case-by-case basis. We distinguish between a combinatorial grid (abstract complex) and its geometric embedding, a distinction which is preserved in the interface definition. Also, the notion of mappings defined on grids finds its equivalent in the concept of a grid function.

Definition 1 (Abstract complex) *An abstract finite complex* $\mathcal{C}$ *of dimension* d *is a set of* elements e, *together with a mapping* $\dim : \mathcal{C} \mapsto \{0, \ldots, d\} \subset \mathbb{N}$, *(*$\dim(e)$ *is called the dimension of* e*), and a partial order* $<$ (side-of *relation) with* $e_1 < e_2 \Rightarrow \dim(e_1) < \dim(e_2)$. *Elements are named according to table 1. A* morphism *between abstract complexes* $\mathcal{C}_1, \mathcal{C}_2$ *is a mapping* $\Phi : \mathcal{C}_1 \mapsto \mathcal{C}_2$ *with* $e < f \Rightarrow \Phi(e) < \Phi(f)$.

An abstract complex is a purely combinatorial entity, also known as *poset*. We need the notion of a geometric complex, too:

Definition 2 (Geometric realization of an abstract complex) *A geometric realization* Γ *of an abstract complex* $\mathcal{C}$ *is a Hausdorff space* $\|\mathcal{C}\|$ *and a mapping*

$$\Gamma : \mathcal{C} \mapsto \Gamma(\mathcal{C}) = \|\mathcal{C}\| = \bigcup_{e \in \mathcal{C}} \Gamma(e) \quad \textit{with}$$

$$e_1 < e_2 \Leftrightarrow \Gamma(e_1) \subset \partial\Gamma(e_2) \quad \textit{and} \quad \partial\Gamma(e_2) = \bigcup_{e_1 < e_2} \Gamma(e_1) \quad \forall e_1, e_2 \in \mathcal{C}$$

Combinatorial Grid Interface The combinatorial layer is concerned only with abstract complexes. The *elements* of the grid are its "atoms" and named according to their dimension or codimension, see table 1. A minimal representation of an element of a fixed grid is called *element handle*, which may be simply an integer. Handles are useful e. g. for subranges.

At a very basic level, a grid is a set of sequences: A sequence of its vertices, of its edges, and so on. We can model this property by introducing *grid sequence iterators* (table 1), which just have the standard (STL) iterator interface.

Table 1. Combinatorial grid entities

Element	dim	codim	Sequence Iterator
Vertex	0	d	VertexIterator
Edge	1	d-1	EdgeIterator
Facet	d-1	1	FacetIterator
Cell	d	0	CellIterator

In order to access the incidence relationship, we need *incidence iterators* (table 2). These allow for example to access the sequence of all vertices of a cell (VertexOnCellIterator), see fig. 1. The number of different incidence iterators is $d(d-1)$, where d is the grid dimension.

A similar concept are *adjacency iterators*, which relate elements of the same dimension. We define them only for vertices and cells, because there is no "natural" definition for the intermediate dimensions, and they seem to be hardly used.

Table 2. The full set of incidence and adjacency (A) iterators in 3D

VertexOnVertexIt (A)	VertexOnEdgeIt	VertexOnFacetIt	VertexOnCellIt
EdgeOnVertexIt		EdgeOnFacetIt	EdgeOnCellIt
FacetOnVertexIt	FacetOnEdgeIt		FacetOnCellIt
CellOnVertexIt	CellOnEdgeIt	CellOnFacetIt	CellOnCellIt (A)

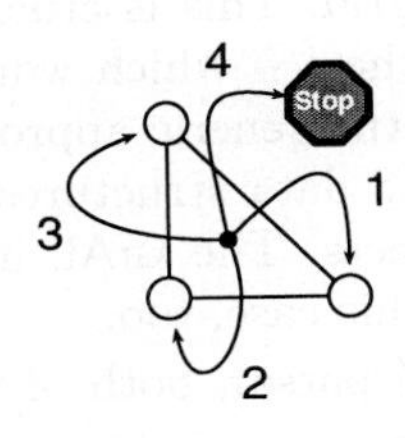

Fig. 1. Action of a VertexOnCellIterator (*Incidence iterator*)

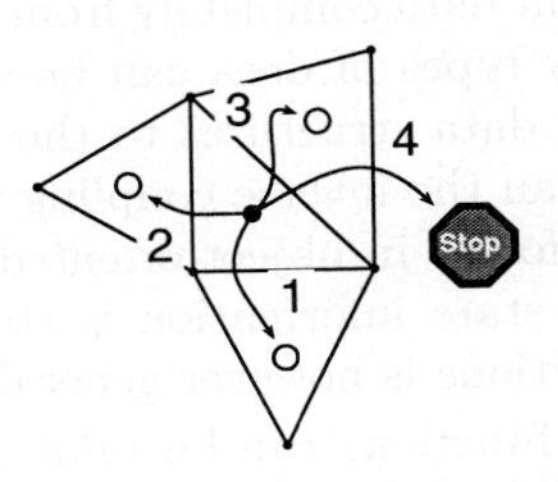

Fig. 2. Action of a CellOnCellIterator (*Adjacency iterator*)

As already mentioned, it is not required to implement all types of elements or iterators. Also, even if the kernel interface for an element type is supported, it does not need to be stored explicitly. The best example here is a Cartesian grid, where everything is given implicitly.

More Combinatorial Functionality: Switch Operator and Archetype Incidence iterators suffice for implementing a surprisingly large class of algorithms. However, there is no ordering relationship between different incidence iterators, for example in 2D, vertices and edges incident to a cell can be ordered independently.

If we need such relationships, we can use the *switch* operator, which allows e. g. to traverse a connected component of a grid's boundary (see [2] for details). Also, the boundary of a cell (its *archetype*) can be accessed as a grid of dimension $d-1$. The interface for switch and archetypes is still experimental.

Grid Geometries Grid geometries represent geometric realizations (embeddings) of combinatorial grids. Thus, they map combinatorial to geometric entities: Vertices to points, edges to arcs, and so on. The grid geometry interface is open for additional properties or measures, for example lengths of edges, thus entailing better encapsulation of geometric decision: If edge lengths are computed in client code under the implicit assumption of linear segments, it would fail to profit from pre-calculated edge lengths, or would break for, say, isoparametric elements of higher order.

The separation from the combinatorial grid layer has a number of practical advantages. It allows to reuse the same combinatorial grid data structure with different embeddings, for example 2D domain and 3D surface grids. We can also use different geometries for the same grid simultaneously, for instance use straight edges for FEM computation, and curved edges on the boundary for grid refinement.

Grid Functions Grid functions allow to store and access data on grid elements of any dimension. It is possible to decouple the storage of this application-dependent data completely from the combinatorial grid data structure, such that arbitrary types of data can be stored on a given grid. This is crucial to avoid coupling data structures to the algorithms using them – which would be even worse than the inverse coupling we overcame with the generic approach. Yet, it is often found in object-oriented approaches to grid data structures, where for instance state information is stored in vertex objects. The GrAL interface for grid functions is however general enough to cover this case, too.

Grid functions can be *total* (dense) or *partial* (sparse), both of which have generic default implementations, see below and [4].

Mutating Primitives A large class of grid algorithms important for PDE solutions does need only read access to the grids; for these, the interface components presented so far are sufficient.

However, in some cases we need to change the grid: Obviously, this is the case if we read the grid from some file, or copy it from another grid. Also, for grid refinement, coarsening or optimization, it has to be changed.

In search of a general solution which allows efficient implementations for a large class of data structures, we found that in virtually all cases we investigated it proved sufficient to use *coarse grained* mutating primitives (in contrast to *atomic* primitives like Euler operators [12]). We can do with just three of them: Copying grids, enlarging (gluing) grids, and cutting something off a grid. These primitives maintain a grid morphism between the source and the copy, in order to allow the transfer of additional information, such as grid functions. Mutating primitives are discussed in more detail in [2] and [3].

The copy primitive can be seen as a generalized constructor, and it can be used to implement transparent file I/O or data structure conversion, necessary for using traditional libraries using API or file coupling. To this end, a pair of input/output adapters which both have a minimal grid interface is implemented for each file format. Reading the file is achieved by copying from the input adapter, and writing is equivalent to copying to the output adapter. An example in GrAL is the output adapter for GMV [14].

A generic copy operation also poses interesting challenges: For instance, in 3D grids, numbering of local vertices often differs between applications e. g. for hex cells. In order to copy from one numbering to another, we need to calculate a grid isomorphism between the two hex representation (or more precisely, their *archetypes*, see above), which is performed by a GrAL algorithm.

4 Components of GrAL

A full, up-to-date catalogue of components can be found on the GrAL web-site [1]. Here, we present a small selection of basic components.

Mesh Data Structures Several grid data structures are implemented in GrAL: Cell complexes with general cells in 2D and 3D, Cartesian grids in 2D and 3D, and a simple triangulation data structure in 2D. As the focus of GrAL currently is more on algorithms than on data structures, these serve merely as examples and 'working horses' – typically, the generic components of GrAL will be instantiated for user-provided data structures.

Generic Iterators In order to ease the creation of a GrAL adaptation layer for user-defined data structures, a number of generic iterator classes are defined in GrAL, for example edges and facets for 3D grids. A consistent framework aiming for minimizing adaptation work is of high priority.

Closure Iterators When using grid subranges, for instance subsets of grid cells, we often need to access all elements of a given dimension incident to them, without duplication. This is what the generic closure iterators do.

Incidence Calculation Typically, grid data structures provide only a limited amount of incident information; e. g. in a basic FEM implementation, there is often no need for cell neighbor information. In case this information is needed at a later point, it can be calculated by a generic GrAL algorithm.

Grid Functions As already mentioned, GrAL contains generic implementations for both total and partial grid functions. Depending on the storage characteristics of elements, either array-based or hash-table-based implementations are selected.

Distributed Grids For parallel PDE solution, grids and grid functions have to be distributed and augmented by additional information. The necessary data structures and algorithms for creating distributed grids and grid functions (using arbitrary overlaps) are generic GrAL components and can be wrapped around any sequential grid representation, see [2] or [5] for details.

5 Using Generic Libraries

Generic and traditional libraries have some marked differences, for example concerning generality, granularity, efficiency, ease-of-use, and tool support. We will discuss some of these issues.

5.1 Generality & Granularity

The desire for a higher level of generality has been the driving force behind the development of GrAL. In the case of grids, a similar degree of generality is simply not achievable for traditional libraries.

The fact that no copying is needed makes much smaller-grained components practical, for example, we can provide iterators traversing boundary components, or local search algorithms. Thus, the number of components exposed to the user increases, as does the flexibility of the components themselves. This creates serious challenges with respect to their documentation.

5.2 Efficiency

The overhead of generic components with respect to code specialized to concrete data structures is called *abstraction penalty.* Measuring this penalty is practical only for small pieces of code – in more complex cases it is often just to tedious to create an equivalent ad-hoc low-level component. Such measurements are also highly dependent on the compilers used. In some cases, the penalty can be removed completely, in others, an overhead of 50% or more can be measured, see [2]. It has to be kept in mind that the loops presented there do essentially measure the pure overhead and thus give a sort of worst-case result.

Coming back to section 2, we have to keep in mind that this comparison is with respect to ad-hoc implementations for a given data structure. Comparing with API or file coupling approaches, the performance of generic components is in general much better, and memory bottlenecks can be avoided.

All this means that our generic grid components are usable very well for high-performance applications. First, their performance is in general quite good, even compared with direct implementations. Second, if there really is a hot spot, we have always the possibility of transparently specializing the generic version *after profiling.* In general, there tend to be only few such hot spots.

However, we have to use an optimizing compiler; non-optimized generic code will in general run an order of magnitude slower.

5.3 Ease-of-use, Documentation and Testing

A very important advantage of the generic approach is that it makes it practical to provide sufficiently small, focused and self-contained components, which can be tested separately and automatically. Such unit tests are currently being implemented successively for all GrAL components, and run on a nightly basis.

By using different types for instantiating and running generic components, it can also be checked much better than in the non-generic case that they do not depend on some arbitrary property of their arguments, which could break later. Whereas in non-generic code assumptions on the functionality of arguments are *implicit* – they just rely on what is provided – generic components can (or should) make only *explicit* assumptions on their argument types.

These assumptions (or requirements) have to be documented thoroughly; often encountered sets of requirements can be captured by definition of *concepts*, a technique put forward by the SGI STL documentation [18], and also used for GrAL. Such concepts ultimately lead to a domain-specific language, which is an invaluable aid for reasoning and communicating. So, besides the runtime pre- and postconditions for the runtime arguments, documentation of generic components involves in addition the description of the compile-time type arguments.

How to enforce the requirements (constraints) imposed by a generic component is a controversial issue. C++ offers no built-in ways of doing so, unlike Eiffel [13]. However, there are means of checking constraints near the library entry points by so-called *concept checks* [17]. Such checks, accompanied with a meaningful message in case of failure, are of great help to a user and can compensate somewhat for the additional source of errors introduced by the additional degrees of freedom. An important task for future work will be to provide a layered user interface for heavily parameterized generic components, offering a path from minimal parameterization to the most general versions.

All in all, tool support for generic programming is still inadequate, resulting in nuisances such as long compile times and incomprehensible error messages.

A serious use of GrAL typically requires the user to create an adaptation layer for his data structure. Although not difficult, this requires a prior understanding of the underlying abstractions, and blocks quick success. Thus, enhancing the support for adaptation layers is a key task.

6 Discussion

The generic approach presented here provides for the first time universally usable mesh-based components for direct incorporation into software for PDE solution, *independently* of the underlying data structures. The effort for reusing GrAL components is restricted to creating a thin interface adaptation layer once, and thus independent of the number of components used. By creating wrappers for traditional mesh-based libraries in GrAL, these can be used with no extra effort.

It has to be acknowledged that using generic libraries still faces difficulties, some of a more technical nature, related to insufficient tool support, others due to the raised level of abstraction. The latter point poses an initial hurdle; however, in the long run it substantially contributes to better understanding.

Efficiency of generic components is a design criterion and turns out to be quite satisfactory, although the overhead cannot be eliminated in all cases. In view of the inefficiencies of the traditional approaches (due to the necessary copying), and the fine-grained opportunities for specialization and tuning, the approach of GrAL should be seen as a step ahead also in terms of efficiency.

Future work will on the one hand concentrate on easing the use of GrAL by enhancing support for adapting user mesh data structures. Plans for further extension of GrAL include components for mesh optimization and checking, partitioning and visualization. Also, coupling GrAL to other generic libraries of interest for PDE solution, such as MTL [11] or BGL [16] will help to assess the viability of the generic approach.

References

1. Berti, G.: GrAL – the Grid Algorithms Library. http://www.math.tu-cottbus.de/~berti/gral (2001)
2. Berti, G.: Generic software components for Scientific Computing. PhD thesis, Faculty of mathematics, computer science, and natural science, BTU Cottbus, Germany (2000)
3. Berti, G.: A generic toolbox for the grid craftsman. In Hackbusch, W., Langer, U., eds.: Proceedings of the 17th GAMM Seminar on Construction of Grid Generation Algorithms, Online proceedings at http://www.mis.mpg.de/conferences/gamm/2001/ (2001)
4. Berti, G.: Generic components for grid data structures and algorithms with C++. In: First Workshop on C++ Template Programming, Erfurt, Germany. (2000)
5. Berti, G.: A calculus for stencils on arbitrary grids with applications to parallel PDE solution. In Sonar, T., Thomas, I., eds.: Proceedings of the GAMM Workshop on Discrete Modelling and discrete Algorithms in Continuum Mechanics, Logos Verlag Berlin (2001) 37–46
6. Brown, D.L., Quinlan, D.J., Henshaw, W.: Overture - object-oriented tools for solving CFD and combustion problems in complex moving geometries. http://www.llnl.gov/CASC/Overture/ (1999)
7. The CGAL Consortium: The CGAL home page – Computational Geometry Algorithms Library. http://www.cgal.org (1999)
8. Karmesin, S., et al.: POOMA: Parallel Object-Oriented Methods and Applications. http://www.acl.lanl.gov/PoomaFramework/ (1999)
9. Köthe, U.: VIGRA homepage. http://kogs-www.informatik.uni-hamburg.de/~koethe/vigra/ (2000)
10. Lee, M., Stepanov, A.A.: The standard template library. Technical report, Hewlett-Packard Laboratories (1995)
11. Lumsdaine, A., Siek, J.: The Matrix Template Library (MTL). http://www.lsc.nd.edu/research/mtl/ (1999)
12. Mäntylä, M.J.: Computational topology: a study of topological manipulations and interrogations in computer graphics and geometric modeling. Acta Polytech. Scand. Math. Comput. Sci. Ser. **37** (1983) 1–46
13. Meyer, B.: Eiffel: The Language. Object-Oriented Series. Prentice Hall, New York, NY (1992)
14. Ortega, F.: General mesh viewer (GMV) homepage. http://www-xdiv.lanl.gov/XCM/gmv/GMVHome.html (1996)
15. Pflaum, C.: Semi-unstructured grids. Computing **67** (2001) 141–166
16. Siek, J., Lee, L.Q., Lumsdaine, A.: BGL – the Boost Graph Library. http://www.boost.org/libs/graph/doc/table_of_contents.html (2000)
17. Siek, J., Lumsdaine, A.: Concept checking: Binding parametric polymorphism in C++. In: First Workshop on C++ Template Programming, Erfurt, Germany. (2000)
18. Silicon Graphics Inc.: SGI Standard Template Library Programmer's Guide. http://www.sgi.com/tech/stl (since 1996)

A Software Strategy Towards Putting Domain Decomposition at the Centre of a Mesh-Based Simulation Process

Peter Chow and Clifford Addison

Fujitsu European Centre for Information Technology Limited, Hayes Park Central, Hayes End Road, Hayes, Middlesex UB4 8FE, UK
{chow, addison}@fecit.co.uk

Abstract. In the mesh-based computer aided engineering sphere the most manpower intensive stages in the simulation process chain are the model creation and mesh generation. Together, they can account for 60 to 90 per cent of the total modelling time for complex industrial 3D models. In component meshing and gluing (CMG) we take a novel approach, using domain decomposition technology, to try to reduce the total modelling time further within the existing process without incurring significant re-engineering costs. The thinking is to off-load a sizeable portion of the workload performed by manpower in the creation and meshing stages to the downstream stage, the analysis, performed by computing horsepower. In this paper we describe the concept and software strategy for such an approach.

1 Introduction

In mesh-based simulations the most time consuming part of the whole process – consuming some 60 to 90 per cent of total modelling time – is the stage for geometry creation and mesh generation. This is the case with almost all of the mesh-based simulation software of today. In computer aided engineering (CAE) space there is much development towards a computer aided design (CAD) data exchange standard to enabled direct importing of geometry data created in CAD into the stage of model geometry creation. Combined with the development in tools for repairing flawed CAD data (flawed for meshing, not for CAD) for mesh generation, and together with advances in automatic meshing, this co-ordination has simplified and speeded up the processes of geometry creation and mesh generation considerably. Large and complex geometry problems still require a high degree of user interactions and time; thus geometry creation and mesh generation remains the most time intense stage in the whole process. The problem is often made more acute by the physics involved, which requires a particular degree of mesh density and alignment, and geometric features of contrasting scales.

Figure 1 shows a common mesh-based CAE design and optimisation process found in the manufacturing industries. Like most manufacturing processes the cost of the

P.M.A. Sloot et al. (Eds.): ICCS 2002, LNCS 2331, pp. 755–763, 2002.

change needs to be weighed against the associated benefits. Naturally, only cases where the benefits outweigh the cost of changing will be considered otherwise it is pointless. In the sections below we describe the concept of the approach to try further streamline the simulation process – geometry, meshing, analysis, and visualisation – and a description of the software strategy and data structures.

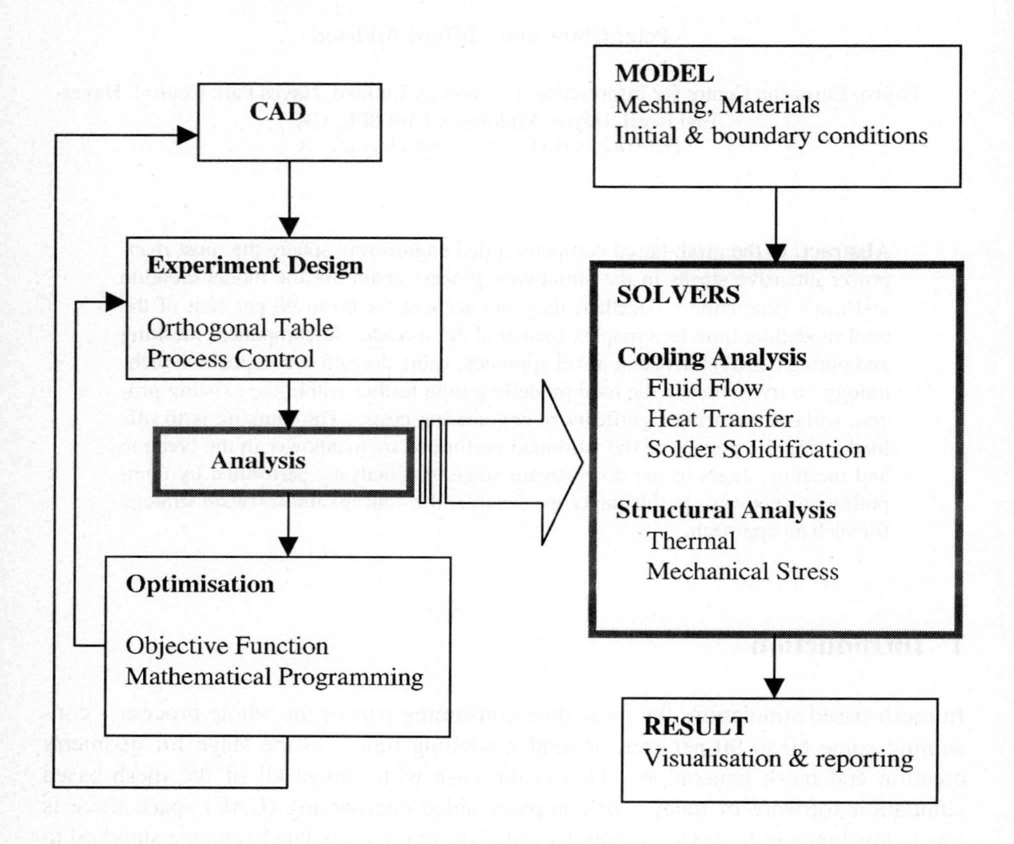

Fig. 1. A CAE mesh-based simulation process chain

2 CMG Concept

The concept of the novel approach, which we refer to as component meshing and gluing (CMG), is to create and solve individual parts (components) that make up the model. The solution is arrived at by the exchange of boundary conditions between the components. The components are created and meshed individually in the same way as existing models, for example doors, wheels etc. for automobile parts. The fastener here (the glue) is the domain decomposition technology [1], which ties the components together at the analysis solver stage. Figure 2 shows the modification and the new elements necessary to convert an existing mesh-based simulation process to the CMG concept. The new elements are 1) find the interface region between two meshed components and generate a surface mesh for the interface with prescribed points at the model building stage. The methods for doing these tasks are commonly found in solid modelling and mesh generation technologies. And 2) incorporate the interface regions and conditions into the existing solution procedure as boundary conditions at the solver stage.

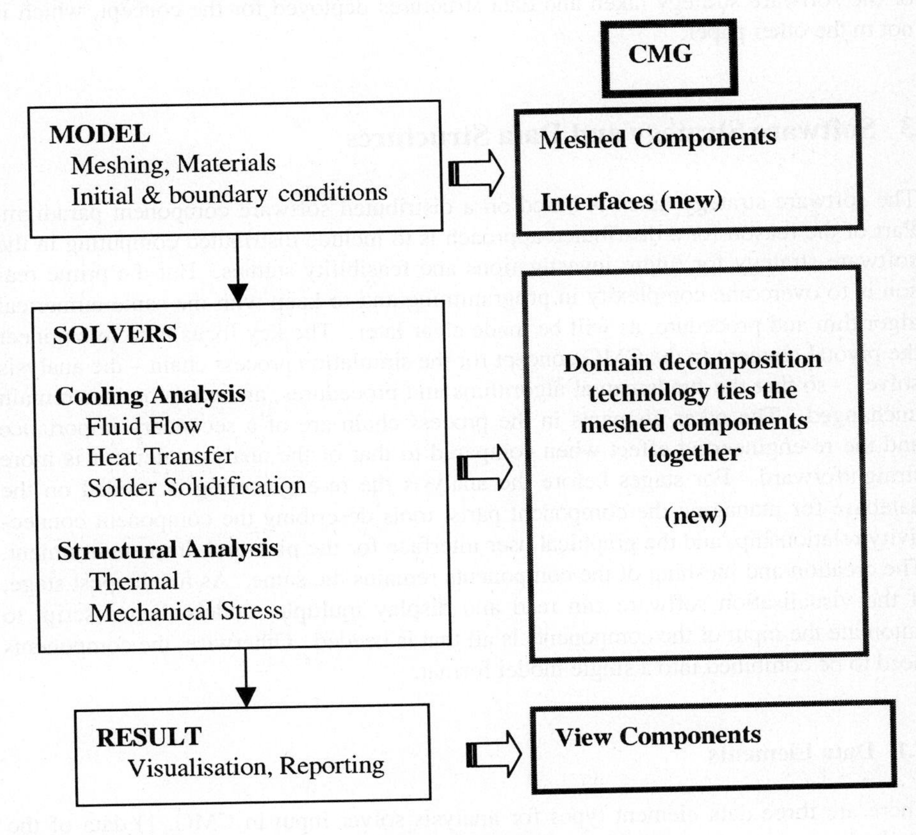

Fig. 2. CMG modification for mesh-based simulation process

With the entire model made-up of "decoupled" meshed components, the compute time at the solver stage will be longer than otherwise. How much longer than a coupled mesh model will depend on the numerical technology employed under the umbrella of the domain decomposition methods. Therefore, if the decoupled mesh solver procedure is not too expensive then for certain types of applications where meshed components and even models can be reused, the CMG concept could be of value. The thinking behind the CMG concept is that the geometry creation and mesh generation in the process can be converted into a "pick-and-drop" stage, picking from a database of meshed components and dropping into the model folder with component connectivity information to process in the analysis solver stage. With this, it is possible that model building could be a fraction of the time it takes to do then now and offset the extra compute time in the decoupled mesh solver to have a shorten overall modelling time. Recent results obtained by the present authors, given in a recent paper [2], on 2D models shows potential for shortening total modelling time on non-linear problems. The paper [2] provides detailed description of the numerical methods and techniques, which will not be repeated here. Instead the pages below provide a description of the software strategy taken and data structures deployed for the concept, which is not in the other paper.

3 Software Strategy and Data Structures

The software strategy taken is based on a distributed software component paradigm. Part of the reason for a distributed approach is to include distributed computing in the software strategy for future investigations and feasibility studies. But the prime reason is to overcome complexity in programming and to keep with the same numerical algorithm and procedure, as will be made clear later. The key focus is to re-engineer the pivotal element in the CMG concept for the simulation process chain – the analysis solver – so that the fundamental algorithms and procedures, and core routines remain unchanged. The other elements in the process chain are of a secondary importance and the re-engineering effect when compared to that of the analysis element is more straightforward. For stages before the analysis the re-engineering is needed on the database for managing the component parts, tools describing the component connectivity relationship, and the graphical user interface for the pick-and-drop environment. The creation and meshing of the components remains the same. As for the post stage, if the visualisation software can read and display multiple models then a script to automate the input of the components is all that is needed. Otherwise, the components need to be combined into a single model format.

3.1 Data Elements

There are three data element types for analysis solver input in CMG, 1) data of the meshed components, 2) data of the interface components, and 3) the model data. The data element in the meshed components can be the same or similar to common standard meshed data formats. The data element in the interface component has the interface mesh and the one-to-one mesh element connectivity between the interface and neighbouring meshed components (see Figure 3). This data element is explicitly created 1) for external interrogation and 2) to simplify the complexity in programming. Finally there is the data element that describes and co-ordinates the components in the model. The model data consists of lists of meshed and interface components and program control parameters. This is similar to instruction or command files in many CAE packages.

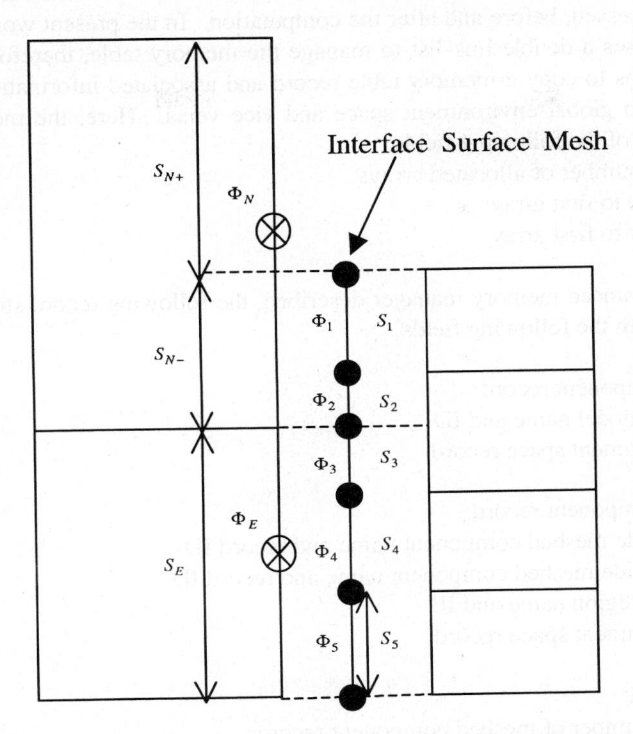

Fig. 3. Interface surface mesh and neighbouring meshed components

With a single instruction set (i.e. the single solver code) performing on multiple data sets (data sets of meshed and interface components) data management is more challenging than with the single data set paradigm employed in conventional solver codes, for example, handling naming conflicts and array indexing between data sets. Some of these challenges are also found in multi-block and/or multi-grid codes. In keeping with the single instruction and single data (SISD) paradigm, and to minimise re-engineering of the solver code, each data element must have its own memory data-base so to remove the majority of the above mentioned conflicts and complexity.

3.2 Advanced Memory Manger

To allow each data element to have its own memory database, within the single code model, an advanced version of the memory manager is required. This is based on the SISD memory manager, where allocation and de-allocation of memory is by the memory manger in the form of a memory table to keep track of all the allocated arrays. Treating the table as an explicit record and creating a record for each of the data elements, one can derive a multiple memory table version of the memory manager. This means virtually no revision to the memory manger are required, apart from the extra routines needed to load and save the memory table record each time the data element is processed, before and after the computation. In the present work the memory manager uses a double link-list to manage the memory table, therefore these are simply functions to copy a memory table record and associated information from the data element to global environment space and vice versa. Here, the memory table record consists of the following fields:

- ☐ Total number of allocated arrays
- ☐ Pointer to first array
- ☐ Pointer to first array

With the advanced memory manager described, the following record structures can be deployed with the following fields.

Meshed Component record:

- ☐ Mesh model name and ID
- ☐ Environment space record

Interface Component record:

- ☐ Left-side meshed component name and record ID
- ☐ Right-side meshed component name and record ID
- ☐ Union region name and ID
- ☐ Environment space record

Model record:

- ☐ Total number of meshed component records
- ☐ Total number of interface component records
- ☐ Meshed component records
- ☐ Interface component records
- ☐ Environment space record

Environment Space record:

- ☐ Memory table record
- ☐ (plus other software modules that requires its own environment settings, e.g. input/output and libraries)

Global Environment Space record:

- ☐ Model record
- ☐ Current loaded meshed component record ID
- ☐ Current loaded interface component record ID
- ☐ Environment space record

3.3 Revised Solution Procedure

The revised solution procedure with CMG is:

Input:
- ☐ Model data
- ☐ Meshed component data
- ☐ Interface component data

Solution Processing:
- ☐ Initialisation

Transient and/or non-linear iteration loop
- ☐ Solve unknown in each meshed components
- ☐ Solve unknown in each interface components

Repeat loop if residual is too high

Output:
- ☐ Model data
- ☐ Meshed component data
- ☐ Interface component data

The stages remain virtually the same with the exception of processing two different types of multiple data sets, meshed and interface data elements. Inside the bullet items, the data element is processed individually with a load and save of the environment space record and component record ID, respectively before and after processing. The pseudo code is (and repeated for each type of data elements):

```
For I = 1 to N-components {
    Load environment space record for component(I)
    Process data element
    Save environment space record for component(I)
}
```

4 Coupling (Gluing) of Individual Meshed Components

In the current implementation the coupling is by information exchanges between the meshed and interface components. The base solver is a finite-volume unstructured mesh method using a cell-centred approach with polyhedral control volumes [3], [4]. For this solver it is best, for minimal changes, to insert the code for data exchanges in the routine that constructs the system matrix for the system of equations. First, there is a call to revise the boundary conditions for boundaries on the interface regions and to update the boundary values, all before the beginning of the routine. In the non-matching grid case, boundary facets could partly be inside the interface region and partly outside (see Figure 3) then a weighted-average value for the boundary facet is sought for. Equation 1 is one example of evaluating a weighted-average quantity. And similarly, Equation 2 follows the same form of the weighted-average quantity for facets totally inside the interface region. Note: the start and end limits in the summation in both the equations are used to make clear the quantities to be sum corresponding to Figure 3. In the generic case the summation is over the apply region. The symbol S is the surface area.

$$\overline{\Phi_N} = \frac{\sum_{i=1}^{2} \Phi_i S_i + \Phi_N S_{N+}}{S_{N+} + S_{N-}} \tag{1}$$

$$\overline{\Phi_E} = \frac{\sum_{i=3}^{5} \Phi_i S_i}{S_E} \tag{2}$$

Secondly, there could be several unknowns to be solved. In such a situation the unknown variables to be solved for need to be declared to the interface module. Otherwise, the interface module couldn't be expected to provide the right interface quantities. For this reason a variable identifier record is constructed for this purposes, and it has the following fields:

- □ Name of variable
- □ Name of variable's gradient
- □ Name of diffusion coefficient
- □ Name of convection coefficient

The names or IDs will depend on the memory manager, in our case we opted for the character string on the memory table that is associated with the allocated array for readability and flexibility in programming. Any calls to the interface region routines will require the variable identifier record. The call to declare the variable to be solved can be positioned anywhere after the allocation of the arrays for the variable and before the looping structures. For our implementation it has been placed in the initialisation stage of the solution procedure.

Apart from the addition of load and save calls for each data element in the solution procedure, calls to declare unknowns, calls to revise boundary conditions and to update boundary quantities, the core of the solver is untouched. The new mechanism for the evaluation of interface quantities and management of data exchanges is encapsulated in the new module for interface handling. The procedure for data exchange inside the interface module is then something like:

1. Assign spaces for incoming data from meshed components.
2. Make a copy of present global environment space record.
3. Load the left-side meshed component's environment space record into the global environment space.

4. Find and retrieve the data required in the variable identifier record and store in assigned spaces.
5. Save the global environment space record into the left meshed component space.
6. Repeat steps 3 to 5 for the right-side meshed component.
7. Load back the global environment space record made in step 2.
8. Perform the task needed for interface region and release the assigned spaces.

5 Summary

A description of the CMG concept and software strategy used to implement such an approach has been provided. The software strategy taken and data structures deployed provide a straightforward mechanism to convert existing analysis solvers with finite-volume cell-centred methods, with minimal changes, and provide a framework for the adaptation of the CMG concept onto a simulation process chain. The study so far suggests the re-engineering costs shouldn't be costly as it reuses all existing software components in the process chain. And in a recent publication [2] by the same authors the CMG concept shows potential on non-linear problems in complex geometry and merits further investigations.

References

1. Domain Decomposition Methods in Sciences and Engineering, proceedings of international conference series on domain decomposition methods, published by DDM.org, URL: www.ddm.org
2. Chow, P., Addison, C.: Putting domain decomposition at the heart of a mesh-based simulation process. To appear in the LMS workshop special issue of the International Journal for Numerical Methods in Fluids
3. Chow, P., Cross, M., Pericleous, K.: A natural extension of the conventional finite volume method into polygonal unstructured meshes for CFD application. Applied Mathematical Modelling, vol. 20, (1996) 170-183
4. Ferziger, J.H., Peric, M.: Computational Methods for Fluid Dynamics. Springer-Verlag, Berlin (1996)

A Software Framework for Mixed Finite Element Programming

Hans Petter Langtangen and Kent-Andre Mardal

Simula Research Laboratory
and
Dept. of Informatics, University of Oslo
{hpl,kent-and}@simula.no

Abstract. Many code developers think the implementation of mixed finite elements on general unstructured meshes is challenging. This assertion is reflected in the many efforts on constructing stabilized finite element methods (e.g. for flow problems), where standard elements can be used, and the fact that there are very few libraries offering a mixed finite element programming environment. In this paper we described a software framework for mixed finite element programming, which makes the use of mixed methods as easy as standard finite element methods. The various abstractions used in this framework will be explained. The tools are made available in the generic Diffpack software.

1 Introduction

Mixed finite elements refers to solving a system of partial differential equations (PDEs) using different finite elements for the different unknown functions. For example, when solving the Navier-Stokes equations for incompressible viscous fluid flow, a standard finite element method, where the pressure and velocity components are approximated using the same elements, will normally not work well. A stable and robust discretization can be achieved when the pressure and velocities are approximated by different elements, one example being quadratic elements for the velocities and linear elements for the pressure. Similarly, when solving the Poisson equation $-\nabla \cdot (k\nabla u) = f$ as a system of equations,

$$\nabla \cdot \boldsymbol{q} = f, \quad \boldsymbol{q} = -k\nabla u,$$

stability of the discretization and optimal convergence rate is ensured by using different elements for u and $\boldsymbol{q}$. The working choices of combination of elements are dictated by the Babuska-Brezzi condition [1, 2, 11]. The great advantage of the system-formulation of Poisson's equation is the possibility to approximate, at a given cost, $\boldsymbol{q}$ with higher accuracy than when applying the conventional finite element method to $-\nabla \cdot (k\nabla u) = f$ and then numerically differentiate u to get $\boldsymbol{q} = -k\nabla u$. In problems where $\boldsymbol{q}$ is of primary interest (e.g., porous media flow, potential flow, torsion problems) the use of mixed finite elements is hence desirable.

P.M.A. Sloot et al. (Eds.): ICCS 2002, LNCS 2331, pp. 764–773, 2002.

Programming with mixed finite elements needs considerable more bookkeeping than programming with conventional finite elements. This is one reason why there have been numerous attempts to stabilize numerical methods to avoid mixed finite elements. One common approach is to add stabilization terms to the PDEs. This introduces new parameters in the problem that need to be determined experimentally. The advantage of mixed finite elements is that the numerical problem is stable without parameters to be tuned.

Another disadvantage of mixed finite elements is that the resulting linear systems are often indefinite. A typical linear system can be written

$$\begin{bmatrix} \boldsymbol{A} & \boldsymbol{B} \\ \boldsymbol{B}^T & \boldsymbol{0} \end{bmatrix} \begin{bmatrix} \boldsymbol{v} \\ \boldsymbol{p} \end{bmatrix} = \begin{bmatrix} \boldsymbol{f} \\ \boldsymbol{g} \end{bmatrix}. \tag{1}$$

Finding efficient iterative methods and preconditioners for linear systems with this type of coefficient matrices has been a difficult problem. The indefiniteness may cause breakdown of iterative methods like preconditioned Krylov solvers, such as GMRES. GMRES handles indefinite and nonsymmetric matrices, but the preconditioner needs to be constructed properly such that the eigenvalues are not close to zero. A promising remedy for this problem seems to be block preconditioning, which will be described from an implementional point of view later (the mathematical properties are considered in [10]).

A variety of alternative solution algorithms exists (see e.g., [5, 9, 11] for an overview of methods for the Navier-Stokes equations.). Of particular interest is the basic iteration on the pressure Schur complement, considered thoroughly in [11]. It basically consist in eliminating $\boldsymbol{v}$ from (1) to obtain an equation for the pressure,

$$\boldsymbol{B}^T \boldsymbol{A}^{-1} \boldsymbol{B} \boldsymbol{p} = \boldsymbol{B}^T \boldsymbol{A}^{-1} \boldsymbol{f} - \boldsymbol{g} \tag{2}$$

This equation is then solved iteratively with the basic iteration and a preconditioner for $\boldsymbol{B}^T \boldsymbol{A}^{-1} \boldsymbol{B}$. Recent research indicates that there is a very close connection between different methods like operator splitting and proper preconditioners for the fully coupled mixed system [5, 11]. The principle idea for both methods is to explore the block form of the matrix and construct special preconditioners for $\boldsymbol{A}$ and $\boldsymbol{B}^T \boldsymbol{A}^{-1} \boldsymbol{B}$. Numerical experiments have shown promising results in particular test cases for both operator splitting algorithms and strategies for the fully coupled system, but the robustness and efficiency in general are still unknown issues.

In addition, it is an open question whether the increased accuracy that can be obtained with mixed finite elements also imply increased computational time because of the lack of efficient linear solvers. If this is the case, conventional finite elements with more degrees of freedom can perhaps deliver the same accuracy at the same CPU cost. Hence it is desirable to have a framework were all methods can be tested in each given application.

Mixed finite elements and associated iterative solution strategies can now be used to construct accurate, efficient, and robust numerical methods for many problems, especially in fluid dynamics applications governed by the Navier-Stokes equations or the Poisson equation. The major obstacle to increased use of mixed finite elements is that the computer programming on general unstructured grids is difficult. In addition, the theory of mixed elements is very abstract, and it is difficult to see how it should be implemented and what is actually gained in practice. In this paper we address software abstractions that make mixed finite element as easy to program with as conventional finite elements. Furthermore, we address software abstractions for easy programming of preconditioners utilizing the block structure of the coefficient matrices. Special emphasis is paid to constructing the software abstractions on basis of the abstractions already present in conventional finite element programs. This makes it possible to add mixed finite element support "on top" of existing codes, at least in principle.

We have implemented the suggested mixed finite element programming tools in the software system Diffpack [3, 4, 6, 7]. Diffpack is a collection of C++ classes organized in libraries. Originally, Diffpack was designed and implemented for conventional finite element methods. The mixed finite element support was added many years later. Writing an application for solving a system of PDEs is typically a matter of

- defining and reading problem-specific parameters,
- defining a finite element grid and associated scalar and vector fields over the grid representing each unknown field in the system of PDEs,
- defining essential boundary conditions,
- defining the integrand in the weak formulation of the problem,
- specifying linear system storage and associated (iterative) solver,
- dumping results (computed fields) to databases for visualization.

An application code is normally just a few pages long. All the numerics that can be shared among different applications take place in generic libraries. The advantage of such a design is that the application codes become short, since the libraries provide the major parts of the the code. Moreover, the applications are more reliable and easier to debug, because a well-tested numerical kernel is reused in many different problems.

Due to the space limitations of this paper we concentrate the exposition on principal ideas rather than on Diffpack-specific implementational details.

2 Software Abstractions for Mixed Finite Elements

Conventional finite elements are associated with a grid, which defines the geometry of the domain and elements as well as the basis functions (also called trial functions or shape functions). In particular, the geometry of the elements is described by a mapping of a reference element onto a possible deformed element in the physical coordinates. When the mapped basis functions coincide with the geometry functions, the element is referred to as *isoparametric*.

In problems solved by mixed finite elements, there will be at least one element that is not isoparametric. For example, in a fluid flow problem with quadratic triangular elements for the velocity components and linear triangular elements for the pressure, the mapping can be based on the quadratic basis functions, allowing curved sides in the elements. The velocity elements are then isoparametric, but the pressure elements are not since different basis functions are used for the mapping and for approximating the unknown. The next subsection describes how to deal with a common grid for all unknowns, but allow different unknowns to have different sets of basis functions.

2.1 Grids for Basis Functions

We apply a standard finite element grid to describe the geometry of all elements. The basis functions associated with conventional isoparametric elements in this grid are used in the mapping of the reference elements onto elements in the physical coordinate system. These basis functions are referred to as *geometry functions*.

Each primary unknown over this grid has elements whose geometries are described by the geometry functions, but the unknown itself is approximated by a sum involving *basis functions*. Different unknowns may have different basis functions over an element, but a common element geometry. We therefore need a way to hold information about the basis functions associated with an unknown. This information is provided by a kind of grid overlay, called *basis function grid*. The basis function grid holds the nodes associated with the definition of the basis functions and a numbering of the corresponding degrees of freedom. That is, we need to store the coordinates of each basis function node and which nodes that belong to which elements (connectivity information). We also need to mark basis function nodes on the boundary. In case the element is isoparametric, no additional information about the basis function nodes need to be stored since all necessary information is available in the underlying geometry grid. In our example with quadratic triangular elements for the velocity components and linear triangular elements for the pressure, the velocities apply standard isoparametric finite elements, whereas the pressure needs a basis function grid. In a triangular element for the pressure, there are three basis function nodes, one at each corner of the element. (The geometry functions are defined in terms of six nodes.)

If the code is organized in terms of classes and objects, it is natural to have a special class for conventional finite element grids. We then introduce the basis function grid as a new class containing (i) a pointer to a conventional grid class for holding the element geometry information, and (ii) data structures for holding the basis function nodes (coordinates, connectivity, boundary markers). All information about the basis function nodes is provided through member functions in the class. In case the element is isoparametric, these member functions look up the requested information in the underlying geometry grid, otherwise they access the additional data structures.

Each unknown in the system of PDEs is represented by a field class, which holds a pointer to a basis function grid and an array of degrees of freedom, plus interpolation and other types of convenient functionality. In this way we achieve a clean design. Given a field object, we can access element information from its basis function grid, which behaves similarly to a conventional grid.

2.2 Programming with Mixed Finite Elements

We have described how the information about the mixed finite elements is stored in a basis function grid associated with each unknown. When it comes to programming with mixed finite elements, we need to evaluate formulas in the integrand in the weak formulation. An example of an integrand in a mixed finite element problem may be written as $L_i \nabla N_j$, where L_i and N_i are different basis functions. At an integration point in an element we therefore need to evaluate the basis functions and their derivatives (with respect to physical coordinates) associated with every unknown function in the PDE system.

In a conventional finite element program, we will normally collect all information about the basis functions and the isoparametric mapping in a class, here called the `FiniteElement` class. Any object of this class can evaluate the basis functions and their derivatives at a given point, map local coordinates in the reference domain to the corresponding coordinates in the physical coordinate system, calculate the Jacobian of the mapping and the derivatives of the basis functions with respect to physical coordinates. With this information at hand, it is easy to evaluate expressions in the integrand in the weak form.

A mixed finite element method has in principle one type of finite element for each unknown in the system of PDEs. Any type of finite element can be represented by a `FiniteElement` object. Hence, to program with mixed finite elements, it is natural to introduce a new class, called `MxFiniteElement`, which holds a list of `FiniteElement` objects, one for each unknown in the PDE system. With the `MxFiniteElement` class coded in C++, we can then evaluate an expression like $L_i \nabla N_j$ with the following (hopefully) self-explanatory and easily readable code:

```
// given an MxFiniteElement mfe object
// and
// enum unknowns { P=0, U=1, V=2 }
// enum coordinates { x=1, y=2 }

mfe(P).N(i) * mfe(U).dN(j,x)  // L_i*dN_j/dx
mfe(P).N(i) * mfe(U).dN(j,y)  // L_i*dN_j/dx
```

The indexing `mfe(P)`, here meaning `mfe(0)`, just returns access to a conventional `FiniteElement` object that holds the information about a pressure element. An `MxFiniteElement` object is initialized from a list of all the fields associated with all the unknowns in the system of PDEs. Each field provides a basis function grid from which the various `FiniteElement` objects can be constructed. Clearly, programming with `MxFiniteElement` is as easy as programming with a conventional `FiniteElement` object. Figure 1 displays an overview of the type of classes we have dealt with so far.

The generic finite element "engine" needs to loop over the (geometry) elements and integration points, and for each integration point call a problem-specific function for sampling the integrand in the weak form at this point. An `MxFiniteElement` object must be prepared for each (geometry) element and all its `FiniteElement` objects must be initialized, i.e., all basis functions and their derivatives, plus the Jacobian information, must be evaluated at the current integration point. This `MxFiniteElement` object is then sent to the problem-specific function. All the application programmer has to do is to write code as outlined above, using the `MxFiniteElement` object. Any programmer who is familiar with conventional finite element method in this software system will be quickly up and going with mixed finite elements in the same system since the interfaces are similar. In particular, mixed finite elements on unstructured, possible locally refined grids, represent no extra difficulty.

Our impression is that many code developers have struggled with mixed finite element in general on unstructured meshes. An indicator for this assertion is that most papers investigating various "exotic" mixed finite elements involve numerical examples on uniformly partitioned rectangular or box-shaped domains. With our suggested support for mixed finite element programming, it is straightforward to test a new element in any desired geometry.

2.3 Adding New Mixed Elements to the Element Library

How is a new mixed finite element actually implemented in the element library? Every element is implemented as a subclass of the `ElmDef` base class. This subclass is supposed to provide member functions for evaluating the geometry functions, the basis functions, their derivatives, location of geometry and basis function nodes, among other things. All information is given in the local reference coordinate system for the element. A range of conventional and mixed finite elements have been implemented in Diffpack. Some of these are: 9-node quadrilateral geometry with bilinear basis functions, 9-node quadrilateral geometry with a constant basis function, 6-node triangular geometry with linear basis functions, 6-node triangular geometry with a constant basis function, 4-node quadrilateral geometry with a constant basis function, and the similar elements in 3D. Among the non-conforming elements we find the Rannacher-Turek and the Crouzeix-Raviart elements (4-node quadrilateral and 3-node triangular geometry with basis function nodes on the middle of each side and continuity at this mid-point only), the MINI element (which is a standard triangular linear element and a cubic bobble function that is zero on the element boundary), the Raviart-Thomas elements (with vector basis functions with nodes at the mid-point of each side and continuity only in the normal direction on the mid-point), the Mardal-Tai-Winther [8] element (which is a non-conforming Stokes element that also works for the mixed Poisson problem, it is a cubic polynomial with 9 nodes and is continuous in the normal direction and has continuous middle value in the tangential direction). The Raviart-Thomas and the Mardal-Tai-Winther elements require special care. They are vector elements and since the standard geometry mapping does not preserve the continuity in the normal direction at each node, this mapping cannot be used. A special vector mapping needs to be used (see [1]); this mapping can also be described as a matrix mapping of the reference vector element. Since this mapping cannot be described through a `FiniteElement` object for scalar elements, `MxFiniteElement` also has the additional functionality for vector elements. This is realized such that `MxFiniteElement` has a pointer to a `MxMapping` object, which is the implementation of this special mapping. The `MxMapping` object is initialized and used behind the curtain if it is necessary.

Hence, the application coder does not need to worry about these issues, but needs to change the code slightly. The above example then reads,

```
mfe.N(P,i) * mfe.dN(U,j,x)   // L_i*dN_j/dx
mfe.N(P,i) * mfe.dN(U,j,y)   // L_i*dN_j/dx
```

For elements that do not require this special mapping `mfe.N(P,i)` and `mfe.dN(U,j,x)` are identical to `mfe(P).N(i)` and `mfe(U).dN(j,x)`, respectively.

2.4 Modification of Finite Element Algorithms

We have tried to explain how the mixed finite element software abstractions are built "on top" of conventional finite element abstractions. However, the generic algorithms for looping over elements and integration points and assembling element-level contributions must be rewritten for mixed finite elements in a strongly typed language such as C++, because the mixed finite element classes have other names than the conventional finite element classes (i.e., the original algorithms work with `FiniteElement` objects, whereas the mixed versions work with `MxFiniteElement`). This could be avoided by parameterizing classes in the generic finite element "engine", using either templates or object orientation to hide the difference between (e.g.) `FiniteElement` and `MxFiniteElement`. We have considered this approach, but found it cleaner to duplicate finite element algorithms, mainly because the implementation of these algorithms is quite compact and because "too parameterized" template implementations might be difficult to understand and modify for a user.

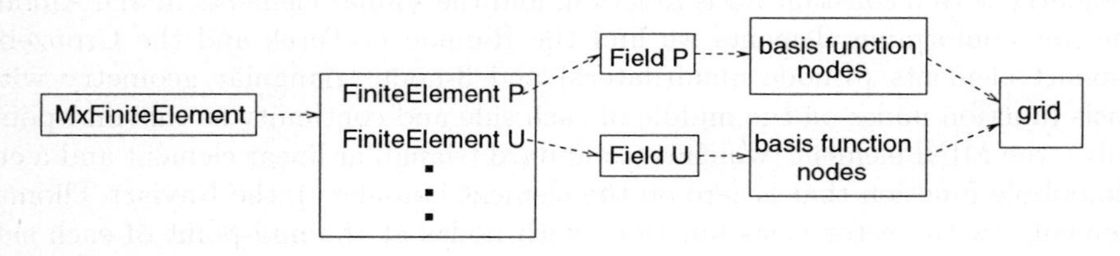

Fig. 1. Relation between central classes and objects related to the mixed elements. Dashed arrows represent pointers.

3 Block Matrices and Block Preconditioning

The initial version of Diffpack did not have support for block matrices. However, the standard matrix in Diffpack, called `Matrix` (a base class for a wide range of matrix formats), was wrapped in a class `LinEqMatrix` for use in linear systems. Several years later, the `LinEqMatrix` class was extended to handle matrices of `Matrix` pointers, i.e., block matrices. Similarly, the initial version of Diffpack wrapped ordinary vectors in a class `LinEqVector` for use in linear systems, and

`LinEqVector` could easily be extended to a collection of block vectors. By adding a relatively small piece of block matrix support in `LinEqMatrix` and `LinEqVector`, the whole Diffpack system could easily exploit the block form of linear system. The coding effort was limited to implementing

– a data-structure that has a matrix of pointers to matrices,
– block-matrix-block-vector product,
– various printing, initialization and data access functionality.

Similarly, a block preconditioner,

$$\begin{bmatrix} \boldsymbol{M} & \boldsymbol{0} \\ \boldsymbol{0} & \boldsymbol{L} \end{bmatrix} \tag{3}$$

is implemented as matrix of preconditioners. The needed functionality is similar to what is needed for block matrices

– a data-structure that has a matrix of pointers to preconditioners,
– block-matrix-block-vector product of preconditioners applied to vectors,
– various printing, initialization and data access functionality.

The preconditioned system reads

$$\begin{bmatrix} \boldsymbol{M} & \boldsymbol{0} \\ \boldsymbol{0} & \boldsymbol{L} \end{bmatrix} \begin{bmatrix} \boldsymbol{A} & \boldsymbol{B}^T \\ \boldsymbol{B} & \boldsymbol{0} \end{bmatrix} \begin{bmatrix} \boldsymbol{v} \\ \boldsymbol{p} \end{bmatrix} = \begin{bmatrix} \boldsymbol{M} & \boldsymbol{0} \\ \boldsymbol{0} & \boldsymbol{L} \end{bmatrix} \begin{bmatrix} \boldsymbol{f} \\ \boldsymbol{g} \end{bmatrix} \tag{4}$$

Hence, we see that when using iterative methods, block matrix and block preconditioning support can be implemented relatively easy. All that is needed is to include a block-matrix-block-vector product of simpler matrices and preconditioners. These sub matrices and preconditioners are usually already implemented and tested extensively.

These outlined modifications are localized and fairly small thanks to the use abstract data types and object-oriented programming (in our case through C++ class hierarchies). A lesson that we have repeatedly learned during the Diffpack development over the last decade, is that a flexible object-based/object-oriented design allows a package to be extended far beyond its initial application area *with ease and reliability* (the less code that needs to be the touch, the fewer bugs will show up later).

A lot of effort has been put into developing fast iterative methods for the indefinite problem (1). However, the knowledge of the efficiency and robustness of these methods have not yet reached the maturity level of, e.g., elliptic problems. It therefore seems natural to assume that several numerical solution strategies should be tested to find the optimal approach in each particular test case. Most solvers around use operator splitting techniques to solve (1). They are usually closely related to a basic iteration on the pressure Schur complement (2) [11]. We therefore explain briefly the basics of these methods before we see how they can be reused in the above framework. The basic steps in these solvers are,

1. Solve $\boldsymbol{A}\boldsymbol{u}^n = \boldsymbol{l}$,
2. Solve $\boldsymbol{N}\boldsymbol{p}^n = \boldsymbol{m}$, where $\boldsymbol{N} \sim \boldsymbol{B}\boldsymbol{A}^{-1}\boldsymbol{B}$.

Here, the right-hand sides $\boldsymbol{l}$ and $\boldsymbol{m}$ depend on which algorithm that is used as the outer iteration, but they are related to the residual. The preconditioners or solvers for $\boldsymbol{A}$ and $\boldsymbol{N}$ are usually iterative solvers (like multilevel methods) that are used with some kind of convergence criterion. These solvers are, in our case and in general, solvers that have been developed, debugged and tested thoroughly, and the point is that they can often be reused without modifications. In the basic iteration on the pressure Schur complement described in [11], $\boldsymbol{A}$ have to be solved exactly, that is, $\boldsymbol{A}^{-1}$ is usually not inverted, but computed with an iterative method with sufficient accuracy to avoid affecting the outer iteration. One the other hand, $\boldsymbol{N}$ is a preconditioner and influences the efficiency of the solution algorithm only and whether it converges or not. There are also, usually, some parameters that need to be chosen and tuned numerically for optimal performance and avoiding divergence. Notice that the steps 1-2 are present in most solvers and consumes often a considerable amount of code lines. Wrapping 1-2 in suitable objects and including these objects in the sketched framework may extend existing codes with a range of solution strategies.

Alternatively, one could solve the coupled system (4) directly. The disadvantage is that the preconditioned system is still indefinite, in contrast to the pressure Schur complement, and breakdown may occur. However, if $\boldsymbol{M}$ and $\boldsymbol{L}$ are constructed properly, then the block preconditioner described in (4) bound the eigenvalue spectrum away from zero and breakdown is no longer a problem [10]. Furthermore, the components in the preconditioner (3) should be exactly the same as the ones used in the operator splitting methods,

$$\boldsymbol{M} \sim \boldsymbol{A}^{-1}, \quad \boldsymbol{L} \sim \boldsymbol{N}^{-1}.$$

Therefore, the software developed for operator splitting schemes can be reused, basically as they are, but with an outer loop that is more robust (no need to tune parameters).

One other advantage with this approach is that $\boldsymbol{M}$ can be a cheap approximation of $\boldsymbol{A}^{-1}$, e.g, one multigrid sweep. In this case there is no need to choose an inner stop criterion that is strict enough to preserve the accuracy of the outer iteration. Since $\boldsymbol{A}^{-1}$ is usually computed by an iterative method and $\boldsymbol{L} = \boldsymbol{N}^{-1}$ already is a preconditioner, it is straightforward to utilize these solvers as preconditioners. Krylov solvers are particularly interesting as outer iterations, because there is no need to tune parameters, they "always" converge, but the convergence is sensitive to the condition number of the preconditioned system. Therefore, it is not easy to determine which approach is best in a particular application. A framework that makes the application tester capable of choosing various algorithms is desirable.

4 Concluding Remarks

In this paper we have outlined how an existing finite element code can be extended to handle mixed finite elements and associated block preconditioning in an easy way. Basic objects in the conventional code, such as grids, finite elements, and scalar/vector fields, are re-used as building blocks in new objects for mixed finite elements for each unknown in the PDE system and grids with basis function information. This allows the application programming interface to the library to remain (almost) the same when mixed elements are added.

Mixed finite elements lead to indefinite matrix problems, and promising efficient solution methods exploit the block nature of the system. We have outlined the software support for block-oriented linear algebra (matrices, vectors, preconditioners) and mentioned that the the "block management" is merely a top layer in a package. That is, exploiting the block structure requires just some managing software at a much higher abstraction level than the number crunching loops. Much of the software pieces needed in block-oriented algorithms are classical numerical strategies already available in many codes.

References

1. F. Brezzi and M. Fortin. *Mixed and Hybrid Finite Element Methods.* Springer, 1991.
2. P. M. Gresho and R. L. Sani. *Incompressible Flow and the Finite Element Method.* Wiley, 1998.
3. Internet. Diffpack software package. *http://www.diffpack.com.*
4. H. P. Langtangen. *Computational Partial Differential Equations – Numerical Methods and Diffpack Programming.* Lecture Notes in Computational Science and Engineering. Springer, 1999.
5. H. P. Langtangen, K.-A. Mardal, and R. Winther. Numerical methods for incompressible viscous flow. *Advances in Water Resources*, 2002. submitted.
6. K.-A. Mardal and H. P. Langtangen. Mixed finite elements. In H. P. Langtangen and A. Tveito, editors, *Computational Partial Differential Equations using Diffpack - Advanced Topics.* Springer, 2002. In preparation.
7. K.-A. Mardal, J. Sundnes, A. Tveito, and H. P. Langtangen. Block preconditioning for systems of PDEs. In H. P. Langtangen and A. Tveito, editors, *Computational Partial Differential Equations using Diffpack - Advanced Topics.* Springer, 2002. In preparation.
8. K.-A. Mardal, X.-C. Tai, and R. Winther. Robust finite elements for Darcy-Stokes flow. *(Submitted)*, 2001.
9. R. Rannacher. Finite element methods for the incompressible Navier-Stokes equation. 1999. *http://www.iwr.uni-heidelberg.de/sfb359/Preprints1999.html.*
10. T. Rusten and R. Winther. A preconditioned iterative method for saddlepoint problems. *SIAM J. Matrix Anal.*, 1992.
11. S. Turek. *Efficient Solvers for Incompressible Flow Problems.* Springer, 1999.

Fast, Adaptively Refined Computational Elements in 3D

Craig C. Douglas[1,2], Jonathan Hu[3], Jaideep Ray[3], Danny Thorne[1], and Ray Tuminaro[3]

[1] University of Kentucky, Department of Computer Science, 325 McVey Hall, Lexington, KY 40506-0045, USA
{douglas,thorne}@ccs.uky.edu

[2] Yale University, Department of Computer Science, P.O. Box 208285 New Haven, CT 06520-8285, USA
douglas-craig@cs.yale.edu

[3] Sandia National Laboratory, Livermore, CA 94550, USA
{jhu,jairay,tuminaro}@california.sandia.gov

Abstract. We describe a multilevel adaptive grid refinement package designed to provide a high performance, serial or parallel patch class for use in PDE solvers. We provide a high level description algorithmically with mathematical motivation. The C++ code uses cache aware data structures and automatically load balances.

1 Introduction

In this paper, we use adaptively refined [2, 3, 15] multilevel [5, 11, 12] procedures to solve problems of the form

$$\begin{cases} \mathcal{L}(\phi) = \rho \text{ in } \Omega, \\ \mathcal{B}(\phi) = \gamma \text{ on } \partial\Omega, \end{cases} \tag{1}$$

subject to standard conditions that ensure ellipticity and well posedness [1]. The procedure is derived from the adaptive grid refinement process, not from the multigrid procedure.

This paper assumes the reader is familiar with how to discretize and solve a partial differential equation. For remedial information, see [8, 13, 16].

In §2, mathematical formalities are provided. In §3, we define a simple problem and then use it in later sections to motivate the definitions and methods. In §4, we define a multilevel adaptive mesh refinement method that is both complicated and directly implemented in C++ classes. In §5, we provide some implementation details. In §6, we draw some conclusions.

2 Mathematical Formalities

The basic algorithms are geometrically inspired. Definitions that assume nesting are defined from a domain (or subdomain), not a grid, perspective. This is utterly common in adaptive grid refinement literature, but is less so in multigrid literature.

P.M.A. Sloot et al. (Eds.): ICCS 2002, LNCS 2331, pp. 774–783, 2002.

We begin by assuming that Ω is overlaid by a union of tensor product meshes $\Lambda^{1,j}$, $j = 1, \ldots, n_1$,

$$G^1 = \bigcup_{j=1}^{n_1} \Lambda^{1,j}, \quad \text{where } G^1 \subset \Omega^1 = \Omega.$$

Normally $n_1 = 1$, but the method works fine for $n_1 > 1$. This is referred to as the level 1, or coarsest, grid and there will be operators defined on it later.

We may have many patches that have been determined through an adaptive grid refinement process. The set of local grid patches corresponding to $\ell - 1$ refinements ($1 < \ell \leq lmax$) are denoted

$$G^\ell = \bigcup_{j=1}^{n_\ell} \Lambda^{\ell,j}$$

where the $\Lambda^{\ell,j}$ are also tensor product meshes and have been obtained by adaptively refining the $\Lambda^{\ell-1,j}$ meshes. We define the domains Ω^ℓ and $\Omega^{\ell,j}$ as the minimum domains that include G^ℓ and $\Lambda^{\ell,j}$, respectively. Normally, Ω^ℓ will be a union of disconnected subdomains (one subdomain corresponding to each level ℓ patch).

Note that since our code is really an adaptive grid refinement code with a multigrid procedure added as an afterthought, $lmax$ can change (increase or decrease) during the course of solving an actual problem.

We assume there are projection and refinement operators defined, P and R, respectively. The notation is standard adaptive mesh refinement notation, but is different from multigrid notation (where the symbols are reversed, sadly). The operators are used interchangeably with either domains Ω^ℓ or grids G^ℓ. In terms of superscripts of domains or grids, P *projects* from "fine to coarse," i.e., $\ell \to \ell - 1$ and R *refines* from "coarse to fine," i.e., $\ell \to \ell + 1$.

We require strict nesting of grids from a geometric viewpoint:

$$G^{lmax} \subseteq G^{lmax-1} \subseteq \cdots \subseteq \cdots G^1.$$

The domain version of this requirement can be written as

$$R(P(\Omega^{\ell+1})) \subseteq \Omega^{\ell+1} \quad \text{and} \quad P(\Omega^{\ell+1}) \subset \Omega^\ell$$

and

$$\Omega = \bigcup_{\ell=1}^{lmax} (\Omega^\ell - P(\Omega^{\ell+1})), \quad \text{where } \Omega^{lmax+1} = \phi.$$

Note that interpolation can be used to extend the method to nonnested grids quite easily, however.

The use of tensor product meshes allows for fairly straightforward finite difference/finite volume type stencils to define the discrete operator. At internal patch boundaries, however, some care must be taken. The general idea is to define ghost points near internal patch boundaries so that locally equispaced unknowns are available for use with a a regular stencil. In essence, simple B-splines [4] are used along the boundaries to produce the needed ghost point values. The key point is that the resulting discretization enforces C^1 continuity along the patch boundaries.

C^1 continuity is different than what is normally required in the multigrid literature (which normally only imposes C^0 continuity). For serial computing, C^0 continuity is usually sufficient. However, for problems with severe fronts or near discontinuities in the solution, C^0 continuity is not always sufficient or desirable.

For parallel multigrid with a local point relaxation smoother, only requiring C^0 continuity requires a special procedure for data that is next to processor imposed subdomain boundaries. Otherwise the method is normally globally stable, but inconsistent. Hence the method does not necessarily converge through failing the well known numerical analysis theorem *stability+consistency=convergence.* C^1 continuity imposes a process that guarantees consistency without disturbing stability. However, it imposes a C^1 numerical solution, surprisely, which turns out not to be much of a constraint for some very difficult problems.

The boundaries of the domains, $\{\partial\Omega^\ell\}$, are required to meet the following condition:

$$\partial\Omega^{\ell+1} \cap \partial\Omega^\ell \subset \partial\Omega$$

only. This merely ensures that the C^1 continuity procedure that we use is well defined and of the right approximation order near patch boundaries in the interior of Ω [11].

3 A Simple Example

For (1), assume we are solving Poisson's equation on the unit cube, with a cell centered finite volume method with uniform mesh spacing h_ℓ on a level ℓ. Note that our code supports variable coefficients as well, however.

Let $h_\ell = 2^{1-\ell}h_1$. Let $h = h_\ell$ and $\phi = \phi^\ell$ unless otherwise noted, and let $N = 1/h$ (the number of grid points in each dimension).

In the interior of Ω^ℓ, we have the standard seven point stencil:

$$(\Delta\phi)_{i,j,k} = (\phi_{i+1,j,k} + \phi_{i-1,j,k} + \phi_{i,j+1,k} + \phi_{i,j-1,k} + \\ \phi_{i,j,k+1} + \phi_{i,j,k-1} - 6\phi_{i,j,k})h^{-2}.$$

On the physical boundaries, we must incorporate boundary values that match flux conditions necessary to ensure a C^1 continuity. Assume the ϕ's with fractional indices are supplied boundary values. For instance, on the side boundary in the negative x-direction ($i = 0$), we have

$$(\Delta\phi)_{0,j,k} = (\frac{4}{3}\phi_{1,j,k} + \frac{8}{3}\phi_{-\frac{1}{2},j,k} + \phi_{i,j+1,k} + \phi_{i,j-1,k} + \\ \phi_{i,j,k+1} + \phi_{i,j,k-1} - 8\phi_{0,j,k})h^{-2}.$$

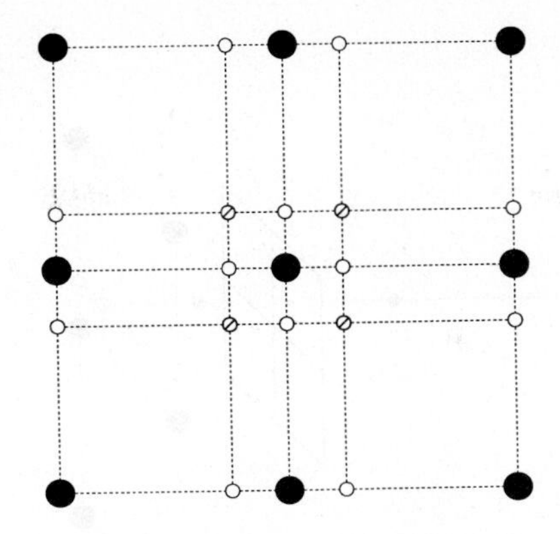

Fig. 1. 3D Interpolation of Ghost Points, Step 1.

At the corner boundary where $(i,j,k) = (0,N,N)$, we have

$$(\Delta\phi)_{0,N,N} = (\frac{4}{3}\phi_{1,j,k} + \frac{8}{3}\phi_{-\frac{1}{2},j,k} + \frac{8}{3}\phi_{i,N+\frac{1}{2},k} + \frac{4}{3}\phi_{i,N-1,k} + \frac{8}{3}\phi_{i,j,N+\frac{1}{2}} + \frac{4}{3}\phi_{i,j,N-1} - 12\phi_{0,N,N})h^{-2}.$$

The rest of the boundary discretizations can be produced similarly.

Near the interface between a coarse grid, Ω^ℓ, and a fine grid, $\Omega^{\ell+1}$, stencils must be defined on both the coarse and fine grid. The general idea is to use a flux differencing form of the equations $\Delta\phi^\ell = \nabla\cdot f^\ell$, where $f^\ell = \nabla\phi^\ell$:

$$(\Delta\phi^\ell)_{i,j,k} = \frac{1}{h_\ell}\left(f^\ell_{i+\frac{1}{2},j,k} - f^\ell_{i-\frac{1}{2},j,k} + f^\ell_{i,j+\frac{1}{2},k} - f^\ell_{i,j-\frac{1}{2},k} + f^\ell_{k+\frac{1}{2},j,k} - f^\ell_{k-\frac{1}{2},j,k}\right).$$

Normally, these fluxes are defined as

$$f^\ell_{i+\frac{1}{2},j,k} = (\phi^\ell_{i+1,j,k} - \phi^\ell_{i,j,k})h_\ell^{-1}, \text{ and } f^\ell_{i-\frac{1}{2},j,k} = (\phi^\ell_{i,j,k} - \phi^\ell_{i-1,j,k})h_\ell^{-1},$$

and similarly in the y and z directions. However, some of the $\phi^\ell_{i,j,k}$'s may not be available. To approximate a flux across a coarse edge that spans several fine grid edges (of a neighboring patch), each fine grid edge flux is first approximated using ghost values defined via interpolation or extrapolation. The individual fine grid fluxes are then summed to obtain the flux across the coarse edge. The interpolation/extrapolation procedure guarantees C^1 continuity near a coarse-fine interface in the interior of Ω.

As a concrete example, consider Figs. 1 and 2. Only one dimensional, quadratic interpolation and extrapolation is used. First (Fig. 1), the coarse element centers, represented by the large ●s, are used to calculate values at the ○ points, which are used in turn to calculate the values at the ⊘ points. Second (Fig. 2), we use these ⊘ points and two existing fine element centers, represented by the small •s, to get ghost point values at the ⊗ points, which are at the centers of where fine elements would be.

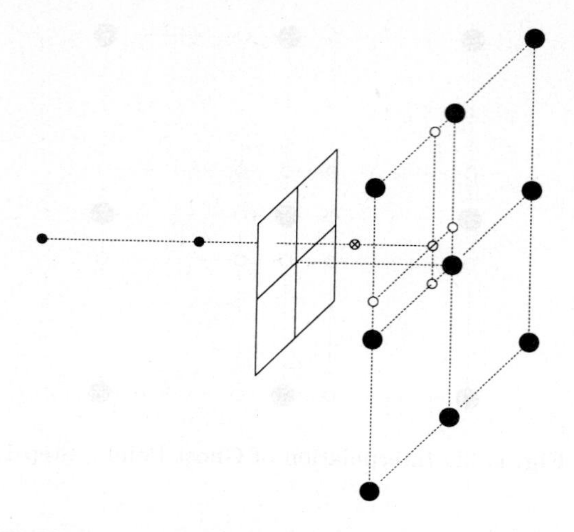

Fig. 2. 3D Interpolation of Ghost Points, Step 2.

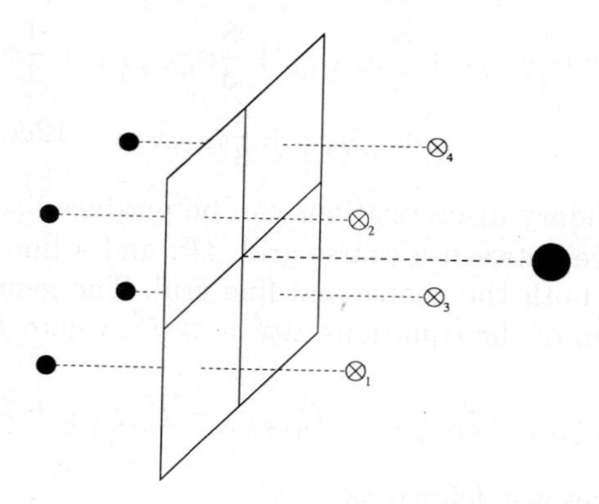

Fig. 3. 3D Flux Matching.

These $\otimes$ points are used to approximate the average of the fluxes across the four fine grid cell walls at the coarse-fine interface. For instance,

$$f^{\ell}_{i-\frac{1}{2},j,k} = \tfrac{1}{4}\left(\tfrac{1}{h^{\ell+1}}\left(\delta_1+\delta_2+\delta_3+\delta_4\right)\right),$$

where

$$\begin{aligned}\delta_1 &= u^{\ell+1}_{\otimes_1} - u^{\ell+1}_{2(i-1),2j-1,2k-1}, \quad \delta_2 = u^{\ell+1}_{\otimes_2} - u^{\ell+1}_{2(i-1),2j,2k-1},\\ \delta_3 &= u^{\ell+1}_{\otimes_3} - u^{\ell+1}_{2(i-1),2j-1,2k}, \text{ and } \delta_4 = u^{\ell+1}_{\otimes_4} - u^{\ell+1}_{2(i-1),2j,2k},\end{aligned}$$

as illustrated in Fig. 3, where the coarse-fine interface is in the negative x-direction from the coarse grid point at which the operator is being applied.

There can be up to three coarse-fine interfaces at a given coarse cell. In the case of multiple coarse-fine interfaces, an analogous flux matching procedure is applied at each interface. A minor difference is that some of the $\oslash$ points must be extrapolated from the coarse grid data.

More than three coarse-fine interfaces at one coarse cell is a case that is avoided by the mesh refinement algorithm. More generally, isolated coarse grid cells, squeezed between fine grid regions, is a case that is avoided by the mesh refinement algorithm. We must have three adjacent (and colinear) coarse grid cells, including the one at which the flux matching is being applied, in order to do the quadratic interpolation of ghost points. When coarse-fine interfaces abut the physical boundary, boundary values may be used in the interpolation procedure to compute the $\oslash$ points.

4 Multilevel Adaptive Mesh Refinement

In this section, we define the operators, vectors, and algorithms needed to solve (1) numerically on a composite grid. Much of the material follows [11].

For solving (1) on level ℓ, we need to store data and information about

$$G^\ell, \ \phi^\ell, \text{ and } \rho^\ell.$$

However, only data and information that cannot be associated with a finer level $\ell+\epsilon$, $\epsilon>0$, is stored on level ℓ. Hence, only the part of level ℓ associated with $\Omega^\ell - P(\Omega^{\ell+1})$ is stored. Data is stored for grid points only on the finest grid that that a grid point exists. This is the opposite storage strategy from the one used for hierarchical basis multigrid implementations [18].

We define two discrete versions of $\mathcal{L}$: one that is defined on $\Omega^\ell - P(\Omega^{\ell+1})$ and another that is defined assuming that no finer level exists. The difference is both subtle and confusing. We take great care to motivate the difference between the two and, in particular, how the patch boundary computations differ.

First, we define $\mathcal{L}^\ell(\phi)$ on $\Omega^\ell - P(\Omega^{\ell+1})$. In the interior of Ω^ℓ, this is the standard discretization. By the $\Omega^\ell/\Omega^{\ell+1}$ interface, we extrapolate ghost points, then do the standard discrete operator. We always use either the physical boundary conditions or the flux matching condition across the boundary.

Second, we define $\mathcal{L}^{nf,\ell}(\epsilon^\ell,\epsilon^{\ell-1})$ on Ω^ℓ. This is the standard discretization on G^ℓ without any regard for the existence of finer levels. On G^1, we use the standard discretization across all of Ω. On G^ℓ, $\ell>1$, we only use coarse grid data to extrapolate ghost point information.

Multigrid methods always have at least one solver (called a smoother or rougher) and sometimes more than one of each. In the multilevel algorithm that we will define shortly, we need an operator $S^\ell(\epsilon^\ell, R^\ell, h_\ell)$ on G^ℓ.

The operator S^ℓ is typically a damped Gauss-Seidel iteration using either the natural, red-black, or a multi-color ordering. First we compute S^ℓ pointwise on level ℓ without regard for any other level:

$$\epsilon^\ell_{ijk} \leftarrow \epsilon^\ell_{ijk} + \lambda\left[\mathcal{L}^{nf,\ell}(\epsilon^\ell,0) - R^\ell_{ijk}\right].$$

All coarse grid boundary components are set to zero above: $\epsilon^{\ell-1}=0$, i.e., Dirichlet type boundary conditions are imposed. The damping factor for a Gauss-Seidel operator S^ℓ can be chosen to be

$$\lambda_{interior} = \frac{h_\ell^2}{6} \quad \text{and} \quad \lambda_{boundary} = \frac{5\lambda_{interior}}{6}.$$

These damping factors correspond to the recipricol of the discrete operator diagonal at the interior and boundary.

We need a method for computing a composite residual on coarser levels. On the finest level,

$$R^{lmax} = \rho^{lmax} - \mathcal{L}^{lmax}(\phi),$$

where ϕ is defined over all composite grids. On any level $\ell < lmax$,

$$R^\ell = \begin{cases} \text{Average}(R^{\ell+1} - \mathcal{L}^{nf,\ell+1}(\epsilon^{\ell+1}, \epsilon^\ell)) \text{ in } P(\Omega^{\ell+1}) \\ \rho^\ell - \mathcal{L}^\ell(\phi) \text{ in } \Omega^\ell - P(\Omega^{\ell+1}) \end{cases}$$

where Average() is a standard nine point weighted restriction from the finer level to the coarser level [5].

Lastly, we define a composite grid version of a multigrid μ cycle, where μ determines how many correction cycles to do from a given level $\ell > 1$.

Algorithm MGCycle(ℓ, μ)
If ($\ell = lmax$) then $R^\ell \leftarrow \phi^\ell - \mathcal{L}^{nf,\ell}(\phi^\ell, \phi^{\ell-1})$
If ($\ell = 1$) then
  solve $\mathcal{L}^1 e^1 = R^1$ on G^1
Otherwise repeat μ times :
  $\phi^{\ell,save} \leftarrow \phi^\ell$
  $e^\ell, e^{\ell-1} \leftarrow 0$
  $e^\ell \leftarrow S^\ell(e^\ell, R^\ell, h^\ell)$
  $\phi^\ell \leftarrow e^\ell$
  $R^{\ell-1} \leftarrow$ Average($R^\ell - \mathcal{L}^{nf,\ell}(e^\ell, e^{\ell-1})$ on $P(\Omega^\ell)$)
  MGCycle($\ell - 1, \mu$)
  $e^\ell \leftarrow e^\ell +$ Interpolate($e^{\ell-1}$)
  $R^\ell \leftarrow R^\ell - \mathcal{L}^{nf,\ell}(e^\ell, e^{\ell-1})$
  $\bar{e}^\ell \leftarrow 0 \quad \bar{e}^\ell \leftarrow S^\ell(\bar{e}^\ell, R^\ell, h^\ell)$
  $e^\ell \leftarrow e^\ell + \bar{e}^\ell$
  $\phi^\ell \leftarrow \phi^{\ell,save} + e^\ell$

Choosing $\mu = 1$ is the multigrid V cycle. Choosing $\mu = 2$ is the multigrid W cycle. If we double the number of iterations of the smoother or rougher each time we move to a coarser grid, we get a variable V or W cycle.

5 Implementation Details

We have implemented a set of C++ classes to solve elliptic PDEs. Our main usage is in solving nonlinear elliptic and parabolic PDEs for complex problems, e.g., combustion simulation.

To implement multigrid on adaptively refined meshes, we use two main types of objects: grids and grid functions. In this section we describe our grids and grid functions and methods that are defined for each.

The classes store data and operate on it based on the following: discrete operators, solvers, residuals, and a composite multilevel algorithm. The form implemented follows the descriptions in §4.

A grid is, first of all, a bounding box. It contains the beginning and ending coordinates of a brick shaped grid stored as integer coordinates of the beginning and ending grid points from the discretization. We also store the real coordinates of the corresponding region from the domain. In addition, a grid includes the mesh spacing in each direction, the number of grid points in each direction, and the identity of the processor that it belongs to.

Multiple grids are combined to form a grid level. A grid level is characterized by its mesh spacing, and the mesh spacings between adjacent levels differ by a factor, usually two or four, called the refinement factor. Multiple grid levels are combined to form a grid hierarchy.

Finer grids can be thought of as either nested within or hovering above a coarser grid. In either case, they can be called child grids of the coarser grid. Grids maintain pointers to their children and parent. A grid has at most one parent.

In the beginning, a grid hierarchy usually consists of only one grid level with only one grid. Through the refinement and load balancing processes this may develop into an elaborate hierarchy of many grid levels with many grids designed to efficiently resolve the features of the problem.

Grid functions store the values associated with each grid point on a grid. A grid function is initialized with a reference to the corresponding grid. The grid functions of all the grids in the hierarchy are combined into a composite grid function for which methods are written for applying operations over all the grid functions of the grid hierarchy as a whole. All the grid functions exist on all the processors, but they only allocate space for data on the processor that their corresponding grid belongs to.

We derive these grid functions from an array class that has been highly optimized for cache performance and robustness. Grid function data can be accessed using Fortran-like syntax that allows for memory management to be done at a very low level. This allows us to experiment with different cache optimizations on different processors easily.

We are exploring a number of storage schemes for optimizing cache effects. For serial computers, modifying algorithms and data structures for standard multigrid has been investigated for a variety of problems [7, 9, 10, 17]. For adaptively refined grids on parallel processors, there are many more possibilities than in the serial case. We are implementing many possibilities in order to see which ones work well on which parallel architectures.

After approximating the solution on the grid hierarchy, there may be points at which the solution is not sufficiently accurate according to some error approximation such as Richardson extrapolation [11]. Clustering is the process of grouping these points into conveniently sized regions. We use clustering routines based on ones in GrACE (Grid Adaptive Computation Engine) [14], which is an object-oriented C++ package for managing adaptively refined structured meshes. It provides an infrastructure for storage of and computation on adaptively refined meshes.

Bigger regions are more convenient for doing computations. There is overhead associated with applying operations to each grid, so minimizing the number of grids by making the grids bigger means less overhead. Also, more work has to be done at the interface between a coarse and fine grid, so it is better to use bigger grids which have a greater proportion of interior nodes.

On the other hand, smaller regions are more convenient for load balancing and getting an entire patch to fit into cache. It is hard to distribute a few big grids evenly over many processors. Smaller regions are desirable since they can more efficiently capture the groups of points that need refinement. Since the refinement regions are brick shaped and the groups of points needing refinement are most likely not, it is probable that the refinement region will include points that do not need refinement. Using smaller grids makes it possible to minimize the number of points included that do not need refinement.

Refinement is the process of building finer grids over the regions computed by the clustering process. For each such region, a new grid is created and incorporated into the hierarchy. The mesh spacing for the new grid is equal to the mesh spacing for its parent grid divided by the refinement factor. The new grid creates a pointer to its parent grid, and the parent grid creates a pointer to its new child grid.

Load balancing is the process of determining a distribution of the grids over more than one processor in a way that will balance the amount of work each processor has to do. We use Zoltan [6] for load balancing.

The load balancing procedure assigns each grid to a processor. The grid functions must then distribute their data accordingly. They might need to move their data to a different processor and they might need to interpolate their data to the grid function of a new grid that was created in the refinement stage. Grid functions and composite grid functions are equipped with methods for handling both of these tasks: message passing and interpolation.

The components of the solution procedure are the $\mathcal{L}$ operators, the smoother, and the transfer operators, projection and interpolation. and implement what was described in §4.

We provide the user with an object that is composed of the grid hierarchy object and the composite grid function object. The code uses methods for initializing the base grid and grid function(s) and then uses a solve method. We try to minimize the programming required of the user while guaranteeing robustness.

6 Conclusions

There are a number of strategies for implementing adaptively refined mesh solvers for elliptic problems in 3D. Besides what type of meshes are chosen, there are nonobvious caching techniques that must be implemented and evaluated in as portable a manner as possible. By portable, a small set of easily determined parameters must be left to the user to choose, preferably in as automatic a manner as possible. The library described in this paper is providing a useful testbed for such an evaluation, besides being useful software.

References

1. Agmon, S.: *Lectures on Elliptic Boundary Value Problems.* Van Nostrand Reinhold, New York, 1965.
2. Berger, M.J., and Colella, P.: Local adaptive mesh refinement for shock hydrodynamics. *J. Comput. Phys. 82* (1989), 64–84.
3. Berger, M.J., and Oliger, J.: An adaptive mesh refinement for hyperbolic partial differential equations. *J. Comput. Phys. 53* (1984), 484–512.
4. Boor, C.: *A Practical Guide to Splines.* Springer-Verlag, New York, 1978.
5. Briggs, W.L., Henson, V.E., and McCormick, S.F.: *A Multigrid Tutorial.* SIAM Books, Philadelphia, 2000. Second edition.
6. Devine, K., Hendrickson, B., Boman, E., St.John, M., and Vaughan, C.: Design of dynamic load-balancing tools for parallel applications. In *Proc. International Conference on Supercomputing* (Santa Fe, 2000).
7. Douglas, C.C.: Caching in with multigrid algorithms: problems in two dimensions. *Paral. Alg. Appl. 9* (1996), 195–204.
8. Douglas, C.C., Haase, G., and Langer, U.: A tutorial on elliptic pde's and parallel solution methods. http://www.mgnet.org/~douglas/ccd-preprints.html, 2002.
9. Douglas, C.C., Hu, J., Kowarschik, M., Rüde, U., and Weiss, C.: Cache optimization for structured and unstructured grid multigrid. *Elect. Trans. Numer. Anal. 10* (2000), 21–40.
10. Hu, J.: *Cache Based Multigrid on Unstructured Grids in Two and Three Dimensions.* PhD thesis, University of Kentucky, Department of Mathematics, Lexington, KY, 2000.
11. Martin, D., and Cartwright, K.: Solving Poisson's equation using adaptive mesh refinement. http://seesar.lbl.gov/anag/staff/martin/tar/AMR.ps, 1996.
12. McCormick, S.F.: The fast adaptive composite (FAC) method for elliptic equations. *Math. Comp. 46* (1986), 439–456.
13. Morton, K.W., and Mayers, D.F.: *Numerical Solution of Partial Differential Equations.* Cambridge University Press, Cambridge, 1994.
14. Parashar, M.: GrACE. http://www.caip.rutgers.edu/~parashar/TASSL/Projects/GrACE, 2001.
15. Rüde, U.: *Mathematical and Computational Techniques for Multilevel Adaptive Methods*, vol. 13 of *Frontiers in Applied Mathematics.* SIAM, Philadelphia, 1993.
16. Varga, R.S.: *Matrix Iterative Analysis.* Prentice–Hall, Englewood Cliffs, NJ, 1962.
17. Weiss et al, C.: Dimepack. http://wwwbode.cs.tum.edu/Par/arch/cache.
18. Yserentant, H.: On the multi–level splitting of finite element spaces. *Numer. Math. 49* (1986), 379–412.

Preconditioning Methods for Linear Systems with Saddle Point Matrices

Owe Axelsson[1] and Maya Neytcheva[2]

[1] Department of Mathematics, University of Nijmegen, Toernooiveld 1
6525 ED Nijmegen, The Netherlands, axelsson@sci.kun.nl
[2] Department of Scientific Computing, Uppsala University
Box 120, 75104 Uppsala, Sweden, maya@tdb.uu.se

Abstract. Preconditioning methods for matrices on saddle point form, as typically arising in constrained optimization problems, are surveyed. Special consideration is given to two methods: a nearly symmetric block incomplete factorization preconditioning method and an indefinite matrix preconditioner. Both methods result in eigenvalues with positive real parts and small or zero imaginary parts. The behaviour of the methods are illustrated on solving a regularized Stokes problem. ...

1 Introduction

Constrained optimization problems involve at least two variables, one of which is the so-called Lagrange multiplier. In the context of mixed variable or constrained finite element methods, the Lagrange multiplier acts as a natural physical variable. The arising systems are normally symmetric and indefinite, but have a block matrix structure which can be utilized. Various types of preconditioners for such systems have been proposed through the years. They include block diagonal preconditioners and block incomplete factorization preconditioners where the iteration matrix is symmetrizable but still indefinite, requiring some generalized iterative solution algorithm such as a minimum residual conjugate gradient method. Reduced matrix approaches, where a Schur complement system for one of the variables is solved, have also been proposed. The latter system is normally symmetric and positive definite and a symmetric positive definite preconditioner can be chosen. If one uses inner iterations, perturbations occur so one may be forced to use a (truncated) version of a generalized conjugate gradient method, such as GCG or GMRES.

Finally, indefinite preconditioners have been proposed. For the simplest type of those, the corresponding preconditioned matrix has only positive eigenvalues but has, in general, a deficient eigenvector space, i.e., the set of eigenvectors do not span the whole space. This may cause problems in the convergence of the iterative solution method.

In this paper, the above methods are surveyed. Two methods are given particular attention and it is shown how to avoid the eigenvector deficiency problem referred to above. An extension of this approach to a regularized system is

P.M.A. Sloot et al. (Eds.): ICCS 2002, LNCS 2331, pp. 784–793, 2002.

presented. Regularization is needed for systems not satisfying the so called LBB-stability condition. The regularization corresponds to the use of an augmented Lagrangian method in constrained optimization.

The remainder of the paper contains the following parts. Section 2 presents preconditioning methods for matrices on (regularized) saddle point form. Section 3 contains some numerical tests for the regularized Stokes solver in 2D.

2 Solving indefinite matrix problems by iteration; preconditioners for indefinite matrix problems

We consider here indefinite matrices of the form

$$\mathcal{A} = \begin{bmatrix} M & B^T \\ B & 0 \end{bmatrix} \quad \text{and linear systems} \quad \begin{bmatrix} M & B^T \\ B & 0 \end{bmatrix} \begin{bmatrix} \mathbf{u} \\ \mathbf{x} \end{bmatrix} = \begin{bmatrix} \mathbf{a} \\ \mathbf{b} \end{bmatrix}, \tag{1}$$

where B has full row rank ($= m$) and M, of order ($n \times n$), is symmetric. The block matrix M can be indefinite in general, but we assume that M is positive definite on $ker(B)$, which implies that $\varepsilon M + B^T B, 0 < \varepsilon < \varepsilon_0$, is positive definite for some sufficiently small ε_0. Such problems occur in equality constrained optimization problems, for instance.

If M is indefinite, we may replace the system in (1) by

$$\begin{bmatrix} M + \frac{1}{\varepsilon} B^T B & B^T \\ B & 0 \end{bmatrix} \begin{bmatrix} \mathbf{u} \\ \mathbf{x} \end{bmatrix} = \begin{bmatrix} \mathbf{a} + \frac{1}{\varepsilon} B^T \mathbf{b} \\ \mathbf{b} \end{bmatrix}, \tag{2}$$

which is equivalent to the latter in the sense that both systems have the same solution. In particular, since $M + \frac{1}{\varepsilon} B^T B$ is positive definite, it follows readily that a solution exists. Due to the above equivalence we can equally well assume that M in (1) is positive definite from the onset.

Solving indefinite matrix problems by iteration requires special care, because the iterative method may diverge, or a breakdown (division by zero), or a near breakdown (division by a number which is small) can occur. As is well-known, division by small numbers usually causes large relative round-off errors.

We shall now consider various approaches to solve (1).

2.1 Schur complement approaches

One way to solve (1) is via the Schur complement $S = -BM^{-1}B^T$. We then eliminate the first component ($\mathbf{u}$) to get

$$BM^{-1}B^T\mathbf{x} = BM^{-1}\mathbf{a} - \mathbf{b}, \tag{3}$$

which is solved first, followed by $M\mathbf{u} = \mathbf{a} - B^T\mathbf{x}$. Here $-S$ is symmetric and also positive definite since B has full rank. Clearly, in an iterative method there is no need to form the matrix S explicitly, as it is only the action of the matrix which is needed. In each iteration step one such action is performed. However, each action requires the solution of the "inner" system with the matrix M which, in general, for large scale problems is solved by iteration as well. Hence, a coupled inner-outer iteration method must be used. The major expense in such an iterative scheme is normally related to the actions of M^{-1}, and the accuracy with which the systems with M are solved plays thereby an important role.

When M and/or $BM^{-1}B^T$ are ill-conditioned, it can be efficient to use the corresponding reformulation of the problem in (2). There holds (see e.g. [1]) the following lemma.

Lemma 1. *Let $M_\varepsilon = \varepsilon M + B^T B$, where M is positive definite and B has full rank. Then*
$BM_\varepsilon^{-1} = (\varepsilon I + BM^{-1}B^T)BM^{-1}$, $BM_\varepsilon^{-1}B^T = I - \varepsilon(\varepsilon I + BM^{-1}B^T)^{-1}$,
and the eigenvalues of $BM_\varepsilon^{-1}B^T$ are contained in the interval $\left[\frac{\lambda_1}{\varepsilon+\lambda_1}, 1\right)$, where λ_1 is the smallest eigenvalue of $BM^{-1}B^T$.

Lemma 1 shows that, under the stated assumptions, we can make the matrix $BM_\varepsilon^{-1}B^T$ arbitrarily well-conditioned by choosing ε sufficiently small. Hence, we can get an arbitrarily fast rate of convergence, leading to just a few iteration steps, when solving the reduced system of (2), i.e.,

$$BM_\varepsilon^{-1}B^T\mathbf{x} = BM_\varepsilon^{-1}(\mathbf{a} + \frac{1}{\varepsilon}B^T\mathbf{b}) - \frac{1}{\varepsilon}\mathbf{b}, \tag{4}$$

by an iterative solution method such as the conjugate gradient (CG) method, with no need to use a preconditioning.

Remark 1. There is an alternative derivation of (2). The system (2) is namely the saddle point matrix for the augmented Lagrangian for the optimization problem $\min\left\{\frac{1}{2}\mathbf{u}^T M\mathbf{u} - \mathbf{a}^T\mathbf{u}\right\}$, subject to the constraint $B\mathbf{u} = \mathbf{b}$. If $\mathbf{x}$ is the Lagrangian multiplier, the augmented Lagrangian takes the form
$\inf_{\mathbf{x}} \sup_{\mathbf{u}} \left\{\frac{1}{2}\mathbf{u}^T M\mathbf{u} - \mathbf{a}^T\mathbf{u} + \mathbf{x}^T(B\mathbf{u} - \mathbf{b}) + \frac{1}{2\varepsilon}(B\mathbf{u} - \mathbf{b})^T(B\mathbf{u} - \mathbf{b})\right\}$, where $\varepsilon > 0$. The idea with the augmented Lagrangian is to improve the conditioning of the matrix M or, if M is indefinite, let ε be sufficiently small so that $M + \frac{1}{\varepsilon}B^T B$ is positive definite.

As we have seen, the reduced system gets increasingly well-conditioned by choosing ε smaller. On the other hand, as ε decreases, $M_\varepsilon = \varepsilon M + B^T B$ becomes increasingly ill-conditioned, which may require more computational efforts for solving the inner systems with M_ε. The optimal choice of ε for smallest total computational complexity must be be a balance between the rate of convergence of the global method to solve (4) and the cost of solving systems with M_ε.

Using the relations in Lemma 1 an elementary computation shows that (4) can be rewritten in the form

$$\mathbf{u} - \varepsilon(\varepsilon I + BM^{-1}B^T)^{-1}(\mathbf{u} + \frac{1}{\varepsilon}(BM^{-1}\mathbf{a} - \mathbf{b})) = \mathbf{0}, \tag{5}$$

which can be used when computing the residuals in the iterative (CG) method.

Assuming that $M^{-1}\mathbf{a}$ is computed initially before the iterations, the latter method requires the action of the matrix $\varepsilon I + BM^{-1}B^T$, which can be somewhat better conditioned for a proper choice of ε than $BM^{-1}B^T$, when λ_1 is small, as its smallest eigenvalue is $\lambda_1 + \varepsilon$ while its largest eigenvalue is negligibly increased. Each action of this matrix requires essentially the same amount of computation as the action of $BM^{-1}B^T$.

2.2 Block Gauss-Seidel preconditioners

We consider now preconditioners for the whole, unreduced, system (1).

A simple, but still efficient preconditioner has the form $\mathcal{A}_1 = \begin{bmatrix} D_1 & 0 \\ B & -D_2 \end{bmatrix}$, where D_1 and D_2 are symmetric positive definite preconditioners to M and $BD_1^{-1}B^T$, respectively. An elementary computation reveals that the preconditioned matrix is

$$\begin{bmatrix} D_1 & 0 \\ B & -D_2 \end{bmatrix}^{-1} \begin{bmatrix} M & B^T \\ B & 0 \end{bmatrix} = \begin{bmatrix} I_1 & D_1^{-1}B^T \\ 0 & D_2^{-1}BD_1^{-1}B^T \end{bmatrix} + \begin{bmatrix} D_1^{-1}(M-D_1) & 0 \\ D_2^{-1}BD_1^{-1}(M-D_1) & 0 \end{bmatrix}.$$

It follows that if $D_1 = M$ then the eigenvalues of the preconditioned matrix are real and equal to the unit number (with multiplicity at least n) and the eigenvalues of $D_2^{-1}BD_1^{-1}B^T$, respectively, which latter are positive. By choosing D_2 sufficiently close to $BD_1^{-1}B^T$ we can hence cluster the eigenvalues around the unit number.

If D_1 is not equal to M we get complex eigenvalues in general, but if D_1 is sufficiently close to M the eigenvalues have positive real parts, close to the unit number and small imaginary parts.

The arising preconditioned system can be solved with a minimal residual iterative method. Depending on the expense in solving systems with D_1 and D_2, this method can work quite efficiently.

Remark 2. Preconditioners of the above form are also suitable for cases when M itself is nonsymmetric, as for the linearized Navier-Stokes equations, see e.g. [9, 7, 8]. As pointed out in [8], if D_2 takes the form $D_2 = (BB^T)(BMB^T)^{-1}(BB^T)$, then under a certain additional condition, all eigenvalues of $\mathcal{A}^{-1}A$ equal the unit number and no Jordan block has order higher than two.

Preconditioners of the form $(BB^T)(BMB^T)^{-1}(BB^T)$ to $BM^{-1}B^T$ have been considered previously in [4] for nonlinear diffusion problems, where it enabled for a cheap update of the nonlinear matrix M.

2.3 Congruence transformations

In order to avoid the strongly unsymmetric matrix in the former method, we can use a symmetric form of the preconditioning matrix. Consider first

$$\widetilde{\mathcal{A}} = \begin{bmatrix} L^{-1} & 0 \\ 0 & N^{-1} \end{bmatrix} \begin{bmatrix} M & B^T \\ B & 0 \end{bmatrix} \begin{bmatrix} L^{-T} & 0 \\ 0 & N^{-T} \end{bmatrix} = \begin{bmatrix} L^{-1}ML^{-T} & L^{-1}B^TN^{-T} \\ N^{-1}BL^{-T} & 0 \end{bmatrix}, \quad (6)$$

where L, N are nonsingular matrices, the choice of which will be discussed below. The negative Schur complement of $\widetilde{\mathcal{A}}$ equals $N^{-1}BM^{-1}B^TN^{-T}$, which does not depend on L. Therefore, we can choose N independently of L, so that the outer matrix $N^{-1}\left(BM^{-1}B^T\right)N^{-1}$ is well-conditioned. By choosing LL^T as an accurate preconditioner of M, such as an incomplete Cholesky factorization, we can solve the arising inner systems with $L^{-1}ML^{-T}$ efficiently.

The matrix $\widetilde{\mathcal{A}}$ in (6) is still on saddle point form. The next congruence transformation reduces $\widetilde{\mathcal{A}}$ to block-diagonal form. Consider then first the more general matrix $A = \begin{bmatrix} A_{11} & A_{12} \\ A_{21} & A_{22} \end{bmatrix}$, where A_{11} is symmetric positive definite (s.p.d.), $A_{21}^T = A_{12}$ and A_{22} is symmetric. Then, if $S = A_{22} - A_{21}A_{11}^{-1}A_{12}$,

$$\begin{bmatrix} I_1 & 0 \\ -A_{21}A_{11}^{-1} & I_2 \end{bmatrix} \begin{bmatrix} A_{11} & A_{12} \\ A_{21} & A_{22} \end{bmatrix} \begin{bmatrix} I_1 & -A_{11}^{-1}A_{12} \\ 0 & I_2 \end{bmatrix} = \begin{bmatrix} A_{11} & 0 \\ 0 & S \end{bmatrix}. \tag{7}$$

Hence, this transformation reduces A to a block-diagonal form with symmetric block matrices. For a saddle point matrix, where $A_{22} = 0$, we have $S = -A_{21}A_{11}^{-1}A_{12}$, so S is negative (semi)definite.

In this method one must find a preconditioner to S. In some problems such as arising in certain partial differential equations like the Stokes problem, S may be quite well-conditioned and it may therefore suffice to precondition it by a diagonal matrix. However, systems with A_{11} must still be solved.

We consider next applying a transformation to a matrix with positive definite symmetric part, where we have changed the sign in the lower block rows in one of the factors (as in [3] among others). We can apply the alternative transformation

$$\begin{bmatrix} I_1 & 0 \\ A_{21}A_{11}^{-1} & -I_2 \end{bmatrix} \begin{bmatrix} A_{11} & A_{12} \\ A_{21} & A_{22} \end{bmatrix} \begin{bmatrix} I_1 & -A_{11}^{-1}A_{12} \\ 0 & I_2 \end{bmatrix} = \begin{bmatrix} A_{11} & 0 \\ 0 & -S \end{bmatrix}, \tag{8}$$

in which case, for a saddle point problem, $-S$ is positive (semi)definite and all eigenvalues are nonnegative (positive if $-S$ is positive definite, which latter holds if A_{21} has full rank).

When applying transformation (8) we must solve systems with A_{11} and in the corresponding iterative method we must solve the Schur complement system, which again involves the action of A_{11}^{-1}. Clearly, such a method is not viable, as we might better have solved just the reduced system for the component $\mathbf{x}$ first and then computed the other component ($\mathbf{u}$). Therefore, when applied for the saddle point matrix (1), we combine the transformations in (7) and (8), i.e., we use the left and right transformations $\begin{bmatrix} I_1 & 0 \\ \widetilde{B}\widetilde{M}^{-1} & -I_2 \end{bmatrix} \begin{bmatrix} L^{-1} & 0 \\ 0 & N^{-1} \end{bmatrix}$ and $\begin{bmatrix} L^{-T} & 0 \\ 0 & N^{-T} \end{bmatrix} \begin{bmatrix} I_1 & \widetilde{M}^{-1}\widetilde{B}^T \\ 0 & I_2 \end{bmatrix}$, where $\widetilde{M} = L^{-1}ML^{-T}$ and $\widetilde{B} = N^{-1}BL^{-T}$. The transformed matrix takes then the form

$$\begin{bmatrix} L^{-1} & 0 \\ \widetilde{B}\widetilde{M}^{-1}L^{-1} & -N^{-1} \end{bmatrix} \begin{bmatrix} M & B^T \\ B & 0 \end{bmatrix} \begin{bmatrix} L^{-T} & -L^{-T}\widetilde{M}^{-1}\widetilde{B}^T \\ 0 & N^{-T} \end{bmatrix} = \begin{bmatrix} \widetilde{M} & 0 \\ 0 & \widetilde{B}\widetilde{M}^{-1}\widetilde{B}^T \end{bmatrix},$$

which is block-diagonal. However, this matrix transformation still requires the action of $\widetilde{M}^{-1} = L^T M^{-1} L$. We replace therefore finally $\widetilde{M}$ with a diagonal matrix $\widetilde{D}$ and and elementary computation shows that the left and right preconditioned matrix takes then the form

$$\begin{bmatrix} L^{-1} & 0 \\ \widetilde{B}\widetilde{D}^{-1}L^{-1} & -N^{-1} \end{bmatrix} \begin{bmatrix} M & B^T \\ B & 0 \end{bmatrix} \begin{bmatrix} L^{-T} & -L^{-T}\widetilde{D}^{-1}\widetilde{B}^T \\ 0 & N^{-T} \end{bmatrix} =$$
$$\begin{bmatrix} \widetilde{M} & (I_1 - \widetilde{M}\widetilde{D}^{-1})\widetilde{B}^T \\ \widetilde{B}(\widetilde{D}^{-1}\widetilde{M} - I_1) & 2\widetilde{B}\widetilde{D}^{-1}\widetilde{B}^T - \widetilde{B}\widetilde{D}^{-1}\widetilde{M}\widetilde{D}^{-1}\widetilde{B}^T \end{bmatrix} \tag{9}$$

This preconditioner involves actions of L^{-1}, L^{-T} and N^{-1}, N^{-T}, as well as of $\widetilde{B}\widetilde{D}^{-1}$ and $\widetilde{D}^{-1}\widetilde{B}^T$, which latter involves additional actions of L^{-1} and $^{-T}$.

If LL^T is a sufficiently close approximation to M, then $\widetilde{M}$ is nearly diagonal and we may choose $\widetilde{D}$ as a sufficiently close approximation of $\widetilde{M}$ and therefore the eigenvalues of the symmetric part of (9) will be positive. Furthermore, the skew-symmetric part gets arbitrarily small. Hence, the transformed matrix has eigenvalues with positive real part, which cluster around the unit number as the imaginary parts of the eigenvalues are small. Unfortunately, this preconditioner involves much computation - two actions of L^{-1} and L^{-T} and one action of N^{-1} and N^{-T}.

The final preconditioner we present here involves less computational effort and gives positive eigenvalues as well, even though it is based on an indefinite matrix.

2.4 A preconditioner on saddle point form

We consider now finally preconditioners of the same, indefinite, form as the given matrix. For the analysis of the first preconditioner we shall use the next lemma.

Lemma 2. *Let B, C, E be real matrices of order $n \times m$, $m \times m$ and $n \times n$ respectively, where B has full rank ($= m$), C is positive definite and E is symmetric. Then the eigenvalues of the generalized eigenvalue problem*

$$\gamma \begin{bmatrix} I & B^T \\ B & -C \end{bmatrix} \begin{bmatrix} \mathbf{u} \\ \mathbf{x} \end{bmatrix} = \begin{bmatrix} E & 0 \\ 0 & 0 \end{bmatrix} \begin{bmatrix} \mathbf{u} \\ \mathbf{x} \end{bmatrix}, \quad |\mathbf{u}| + |\mathbf{x}| \neq \mathbf{0} \tag{10}$$

where $\mathbf{u} \in \mathbb{C}^n$ and $\mathbf{x} \in \mathbb{C}^m$, satisfy

(a) $\gamma = \frac{\mathbf{u}^* E \mathbf{u}}{\mathbf{u}^*(I + B^T C^{-1} B)\mathbf{u}} \neq \mathbf{0}$, if $E\mathbf{u} \neq \mathbf{0}$ and $\gamma = 0$, if and only if $\mathbf{u} = \mathbf{0}, \mathbf{x} \neq \mathbf{0}$,
(b) the dimension of the eigenvector space corresponding to the zero eigenvalue is $m + q$, where $q = dim\{ker(E)\}$;
(c) the nonzero eigenvalues are contained in the interval $\lambda_{min}(E) \leq \gamma \leq \lambda_{max}(E)$.

Consider now the generalized eigenvalue problem (10), where $C = 0$. Here it holds $\gamma(\mathbf{u} + B^T\mathbf{x}) = E\mathbf{u}$ and $\gamma B\mathbf{u} = \mathbf{0}$. Thus, in this case, at least one of $\gamma = 0$

or $Bu = \mathbf{0}$ must hold. If $Bu = \mathbf{0}$ but $\gamma \neq 0$ then $\gamma BB^T\mathbf{x} = BE\mathbf{u}$ (i.e., there holds here $E\mathbf{u} \neq \mathbf{0}$) and $\gamma\mathbf{u}^*\mathbf{u} = \mathbf{u}^*E\mathbf{u}$, i.e., $\mathbf{x} = \frac{1}{\gamma}(BB^T)^{-1}BE\mathbf{u}$, where $\gamma = \frac{\mathbf{u}^*E\mathbf{u}}{\mathbf{u}^*\mathbf{u}}$. Further, $\gamma\mathbf{u} + B^T(BB^T)^{-1}BE\mathbf{u} = E\mathbf{u}$ or $\gamma\mathbf{u} = (I - B^T(BB^T)^{-1}B)E\mathbf{u}$. Hence, for any $\mathbf{u} \in ker(B), \mathbf{u} \neq \mathbf{0}$ and eigenvalue $\gamma \neq 0$, there holds that $[\mathbf{u}, \frac{1}{\gamma}(BB^T)^{-1}BE\mathbf{u}]^t, \gamma = \mathbf{u}^*E\mathbf{u}/\mathbf{u}^*\mathbf{u}$ is an eigenvector for this γ.

If $\gamma = 0$, then it must hold $E\mathbf{u} = \mathbf{0}$. If $E\mathbf{u} = \mathbf{0}$ then $\gamma = 0$ is an eigenvalue for such a vector $\mathbf{u}$ and any $\mathbf{x}$. Hence, there holds that $[0, \mathbf{x}^{(i)}]^T, i = 1, 2, \cdots, m$ and $[\mathbf{u}^{(j)}, \mathbf{0}]^T, j = 1, 2, \cdots, q$ are eigenvectors for $\gamma = 0$, where $\{\mathbf{x}^{(i)}\}_1^m$ span $\mathbb{R}^m$ and $E\mathbf{u}^{(j)} = \mathbf{0}, \mathbf{u}^{(j)} \neq \mathbf{0}$. Hence, there are $m + q$ linearly independent eigenvectors for $\gamma = 0$.

For some of the eigenvectors $\mathbf{u}^{(j)}$ it may hold $\mathbf{u}^{(j)} \in ker(B)$. If $ker(E) \cap ker(B) = 0$ then the algebraic multiplicity of $\gamma = 0$ is $n + m - (n - m) = 2m$. In this case the index of eigenvector deficiency is $2m - (m + q) = m - q$.

If, however, $ker(E) \cap ker(B) \neq 0$, then the algebraic multiplicity is increased and the index of eigenvector deficiency is correspondingly increased.

If one can construct D so that $E = D^{\frac{1}{2}}MD^{-\frac{1}{2}} - I$ has m zero eigenvalues, then there would be no eigenvector deficiency. It is clear, however, that this requirement is not viable for most applications.

Consider now the matrix $\mathcal{A} = \begin{bmatrix} M & B^T \\ B & -C \end{bmatrix}$. Let D be a symmetric and positive definite preconditioner to M and let $\begin{bmatrix} D & B^T \\ B & -C \end{bmatrix}$ be a preconditioner to $\mathcal{A}$. For the generalized eigenvalue problem $\lambda \begin{bmatrix} D & B^T \\ B & -C \end{bmatrix} \begin{bmatrix} \mathbf{u} \\ \mathbf{x} \end{bmatrix} = \mathcal{A} \begin{bmatrix} \mathbf{u} \\ \mathbf{x} \end{bmatrix}$ we have $\gamma \begin{bmatrix} D & B^T \\ B & -C \end{bmatrix} \begin{bmatrix} \mathbf{u} \\ \mathbf{x} \end{bmatrix} = \begin{bmatrix} M - D & 0 \\ 0 & 0 \end{bmatrix} \begin{bmatrix} \mathbf{u} \\ \mathbf{x} \end{bmatrix}$, where $\gamma = \lambda - 1$, or $\gamma \begin{bmatrix} I & \widetilde{B}^T \\ \widetilde{B} & -C \end{bmatrix} \begin{bmatrix} \widetilde{\mathbf{u}} \\ \mathbf{x} \end{bmatrix} = \begin{bmatrix} \widetilde{E} & 0 \\ 0 & 0 \end{bmatrix} \begin{bmatrix} \widetilde{\mathbf{u}} \\ \mathbf{x} \end{bmatrix}$, where $\widetilde{B} = BD^{-\frac{1}{2}}, \widetilde{E} = D^{-\frac{1}{2}}MD^{-\frac{1}{2}} - I$ and $\widetilde{\mathbf{u}} = D^{-\frac{1}{2}}\mathbf{u}$.

This problem has the same form as the generalized eigenvalue problem in Lemma 2. Hence, the previous analysis of the eigenvalues and eigenvectors is applicable.

In this paper we propose the following preconditioner on regularized form,

$$\begin{bmatrix} D_1 & B^T \\ B & -C \end{bmatrix}, \tag{11}$$

where D_1 is a preconditioner to M. The systems with this preconditioner will be solved via the Schur complement $C + BD_1^{-1}B^T$.

A preconditioner for the Schur complement system must be found. In applications such as for Stokes problem, where D_1 is a sufficiently accurate preconditioner to M, it can be chosen as a diagonal matrix D_2. The computational effort in solving systems with the preconditioner (using the conjugate gradient method) is therefore not big. Furthermore, as we have already commented on, the number of outer iterations if using a GCG method will be few.

Remark 3. Indefinite preconditioners for the saddle point problem (10) have been proposed previously in [3] and [10]. However, it was not pointed out that the eigenvector space for the unit eigenvalue, $\lambda = 1$, is deficient in general. This was, however, done in [11]. The eigenvector deficiency may cause problems in the iterative solution algorithm. For instance, the rate of convergence of minimal residual iterative methods are based on the expansion of the initial residual using the eigenvectors as basis vectors. This approach is inapplicable when the space is deficient. Alternatively, see e.g. [5], one can use estimates, based on the Jordan canonical form but this shows also a possible long delay in convergence. Other preconditioners have been proposed in [11] where one may avoid the eigenvalue deficiency problem. As we have seen, there is no eigenvalue deficiency if one uses the regularized form of the preconditioner.

2.5 An indefinite preconditioner on factorized form

Following [3], let now the preconditioner be given on factorized form

$$\mathcal{A}_0 = \begin{bmatrix} D & B^T \\ B & -R \end{bmatrix} = \begin{bmatrix} D & 0 \\ B & L \end{bmatrix} \begin{bmatrix} I_1 & D^{-1}B^T \\ 0 & -L^T \end{bmatrix}, \tag{12}$$

where $R = LL^T - BD^{-1}B^T$. Here D is a preconditioner to M and LL^T is a preconditioner to $C + BD^{-1}B^T$. We will use $\mathcal{A}_0$ as a preconditioner to $\mathcal{A} = \begin{bmatrix} M & B^T \\ B & -C \end{bmatrix}$. Depending on the choice of L and D, this preconditioner can be positive definite or indefinite.

For its analysis, we consider then the generalized eigenvalue problem $\lambda \mathcal{A}_0 \begin{bmatrix} \mathbf{u} \\ \mathbf{x} \end{bmatrix} = \mathcal{A} \begin{bmatrix} \mathbf{u} \\ \mathbf{x} \end{bmatrix}$ or $\gamma \begin{bmatrix} D & B^T \\ B & -R \end{bmatrix} \begin{bmatrix} \mathbf{u} \\ \mathbf{x} \end{bmatrix} = \begin{bmatrix} M - D & 0 \\ 0 & R - C \end{bmatrix} \begin{bmatrix} \mathbf{u} \\ \mathbf{x} \end{bmatrix}$, where $\gamma = \lambda - 1$. We find then $\gamma \begin{bmatrix} \mathbf{u} \\ \mathbf{x} \end{bmatrix} = \begin{bmatrix} D^{-1} - D^{-1}B^TS^{-1}BD^{-1} & D^{-1}B^TS^{-1} \\ S^{-1}BD^{-1} & -S^{-1} \end{bmatrix} \begin{bmatrix} M - D & 0 \\ 0 & R - C \end{bmatrix} \begin{bmatrix} \mathbf{u} \\ \mathbf{x} \end{bmatrix}$, where $S = R + BD^{-1}B^T = LL^T$, or

$$\gamma \begin{bmatrix} \widetilde{\mathbf{u}} \\ \widetilde{\mathbf{x}} \end{bmatrix} = \begin{bmatrix} I_1 - \widetilde{B}^T\widetilde{B} & \widetilde{B}^T \\ \widetilde{B} & -I_2 \end{bmatrix} \begin{bmatrix} \widetilde{M} - I_1 & 0 \\ 0 & \widetilde{R} - \widetilde{C} \end{bmatrix} \begin{bmatrix} \widetilde{\mathbf{u}} \\ \widetilde{\mathbf{x}} \end{bmatrix}, \tag{13}$$

where $\widetilde{\mathbf{u}} = D^{\frac{1}{2}}\mathbf{u}$, $\widetilde{\mathbf{x}} = L^T\mathbf{x}$, $\widetilde{B} = L^{-1}BD^{\frac{1}{2}}$, $\widetilde{M} = D^{-\frac{1}{2}}MD^{-\frac{1}{2}}$, $\widetilde{R} = L^{-1}RL^{-T} = I_2 - \widetilde{B}\widetilde{B}^T$, $\widetilde{C} = L^{-1}CL^{-T}$.

To analyse this further we make first the assumption that $\widetilde{M} \geq I_1$ and $\widetilde{R} \geq \widetilde{C}$, i.e., $I_2 \geq \widetilde{C} + \widetilde{B}\widetilde{B}^T$. This holds if D and LL^T are proper preconditioners to M and $C + BD^{-1}B^T$, respectively.

It follows then from (13), that if $\widehat{\mathbf{u}} = (\widetilde{M} - I_1)^{\frac{1}{2}}\widetilde{\mathbf{u}}$ and $\widehat{\mathbf{x}} = (\widetilde{R} - \widetilde{C})^{\frac{1}{2}}\widetilde{\mathbf{x}}$,

$$\gamma \begin{bmatrix} \widehat{\mathbf{u}} \\ \widehat{\mathbf{x}} \end{bmatrix} = \begin{bmatrix} (\widetilde{M} - I_1)^{\frac{1}{2}}(I_1 - \widetilde{B}^T\widetilde{B})(\widetilde{M} - I_1)^{\frac{1}{2}} & (\widetilde{M} - I_1)^{\frac{1}{2}}\widetilde{B}^T(\widetilde{R} - \widetilde{C})^{\frac{1}{2}} \\ (\widetilde{R} - \widetilde{C})^{\frac{1}{2}}\widetilde{B}(\widetilde{M} - I_1)^{\frac{1}{2}} & -(\widetilde{R} - \widetilde{C}) \end{bmatrix} \begin{bmatrix} \widehat{\mathbf{u}} \\ \widehat{\mathbf{x}} \end{bmatrix}. \tag{14}$$

This matrix is symmetric but indefinite. Its eigenvalues are hence real and the absolute values of them can be controlled by choosing sufficiently accurate preconditioners D to M and LL^T to $C+BD^{-1}B^T$. In this way we may get $\gamma > -1$, so $\lambda = \gamma + 1 > 0$. On the other hand, if $\widetilde{R} \leq \widetilde{C}$, then (13) can be transformed to

$$\gamma \begin{bmatrix} \widehat{\mathbf{u}} \\ \widehat{\mathbf{x}} \end{bmatrix} = \begin{bmatrix} (\widetilde{M}-I_1)^{\frac{1}{2}}(I_1 - \widetilde{B}^T\widetilde{B})(\widetilde{M}-I_1)^{\frac{1}{2}} & -(\widetilde{M}-I_1)^{\frac{1}{2}}\widetilde{B}^T(\widetilde{C}-\widetilde{R})^{\frac{1}{2}} \\ (\widetilde{C}-\widetilde{R})^{\frac{1}{2}}\widetilde{B}(\widetilde{M}-I_1)^{\frac{1}{2}} & (\widetilde{C}-\widetilde{R}) \end{bmatrix} \begin{bmatrix} \widehat{\mathbf{u}} \\ \widehat{\mathbf{x}} \end{bmatrix},$$

where now $\widehat{\mathbf{x}} = (\widetilde{C}-\widetilde{R})^{\frac{1}{2}}\mathbf{x}$. In this case, the preconditioned matrix, after transformation, is nonsymmetric, but with positive definite symmetric part. Hence the eigenvalues may be complex but there holds that $Re(\lambda) \geq 1$. Further, as before, the eigenvalues cluster around the unit number when D is sufficiently close to M and LL^T is sufficiently close to $C+BD^{-1}B^T$.

Finally, if $\widetilde{M}-I_1$ and/or $\widetilde{R}-\widetilde{C}$ are indefinite, then the eigenvalues may be complex but we can estimate the absolute value of γ by simple norm inequalities and clustering around the unit number occurs as before.

3 Numerical tests

We choose to illustrate the behaviour of the proposed indefinite system preconditioners on the stationary Stokes problem, described in detail in [2].

$$\begin{aligned} -\Delta u + p_x &= f_1(x,y) \text{ in } \Omega, \\ -\Delta v + p_y &= f_2(x,y) \text{ in } \Omega, \\ u_x + v_y &= 0 \text{ in } \Omega, \\ u(x,y)|_{\partial\Omega} = g_1(x,y), &\quad v(x,y)|_{\partial\Omega} = g_2(x,y). \end{aligned} \tag{15}$$

Problem 1. Let $\Omega = (0,1)^2$. The functions f_1 and f_2, and the boundary conditions are computed so that the exact solution of (15) is $u(x,y) = x^3 + x^2 - 2xy + x$, $v(x,y) = -3x^2y + y^2 - 2xy - y$ and $p(x,y) = x^2 + y^2$.

We use the regularized method described above, namely, we solve a system with a matrix $\mathcal{A} = \begin{bmatrix} M & B^T \\ B & -\sigma C \end{bmatrix}$, and the choice of the regularization parameter σ is broadly discussed in [2].

Problem 1 is solved using the preconditioned GCG-MR method (cf. [1]) and a preconditioner $\widetilde{\mathcal{A}}$ to $\mathcal{A}$ of the form (11). D_1 in our experiments is the AMLI-preconditioner constructed for the diagonal blocks of M. During each GCG-MR iteration, systems with the preconditioner are solved via its Schur complement $S_{\widetilde{\mathcal{A}}} = -(\sigma C + B\,\mathrm{AMLI}[M]\,B^T)$. Relative stopping criteria are used for both GCG-MR and for the CG methods. $\|\mathbf{r}^{(k)}\|/\|\mathbf{r}^{(0)}\|$ is checked to be less than 10^{-6} and 10^{-4}, respectively. The results of this experiment are shown in Table 1. The systems with $S_{\widetilde{\mathcal{A}}}$ can be solved by a unpreconditioned CG method since for the particular regularized formulation its condition number is independent of the discretization parameter h (column 4 in Table 1). One can further improve the total complexity of the method by using a preconditioned CG to solve $S_{\widetilde{\mathcal{A}}}$ (column 5 in Table 1).

The numerical tests are performed in `Matlab`.

Table 1. Method (11), Problem 1

h	$size(\mathcal{A})$	GCG-MR iter.	Aver. CG iter.	Aver. PCG iter.
0.0667	768 (3×256)	7	24	14
0.0323	3072 (3×1024)	7	25	16
0.0159	12288 (3×4096)	8	25	18
0.0039	196608 (3×65536)	8	25	18

References

1. Axelsson O.: *Iterative Solution Methods*, Cambridge University Press, Cambridge, 1994.
2. Axelsson O., Barker V.A., Neytcheva M., Polman B.: Solving the Stokes problem on a massively parallel computer. *Mathematical Modelling and Analysis*, 4 (2000), 1-22.
3. Axelsson O., Gustafsson I.: An iterative solver for a mixed variable variational formulation of the (first) biharmonic problem. *Computer methods in Applied Mechanics and Engineering*, 20 (1979), 9-16.
4. Axelsson O., Gustafsson I.: An efficient finite element method for nonlinear diffusion problems. *Bulletin Greek Mathematical Society*, 22 (1991), 45-61.
5. Axelsson O., Makarov M.: On a generalized conjugate orthogonal residual method, *Numerical Linear Algebra with Applications*, 2 (1995), 467-480.
6. Axelsson O., Vassilevski P.S.: Variable-step multilevel preconditioning methods. I. Selfadjoint and positive definite elliptic problems. *Numer. Linear Algebra Appl.* 1 (1994), 75-101.
7. Braess D.: *Finite elements. Theory, fast solvers, and applications in solid mechanics.* Cambridge University Press, Cambridge, 2001. (Second edition)
8. Elman H.C.: Preconditioning for the steady-state Navier-Stokes equations with low viscosity. *SIAM Journal on Scientific Computing*, 20 (1999), 1299-1316.
9. H.C. Elman and D. Silvester, Fast nonsymmetric iterations and preconditioning for Navier-Stokes equations. *SIAM Journal on Scientific Computing*, 17 (1996), 33-46.
10. Ewing R.E., Lazarov R., Lu P., Vassilevski P.: Preconditioning indefinite systems arising from mixed finite element discretization of second order elliptic problems. In Axelsson O., Kolotilina L. (eds.): *Lecture Notes in Mathematics* No. 1457, Springer-Verlag, Berlin, 1990.
11. Lukšan L., Vlček J.: Indefinitely preconditioned inexact Newton method for large sparse equality constrained non-linear programming problems. *Numerical Linear Algebra with Applications*, 5(1998), 219-247.

Mixed-hybrid FEM Discrete Fracture Network Model of the Fracture Flow

Jiří Maryška[1], Otto Severýn[1], and Martin Vohralík[2]

[1] Technical University of Liberec, Faculty of Mechatronics
Hlkova 6, 461 17 Liberec 1, Czech Republic
[2] Czech Technical University, FNPE,
Trojanova 13, 120 00 Praha 2, Czech Republic

Abstract. A stochastic discrete fracture network model of Darcy's underground water flow in disrupted rock massifs is introduced. Mixed finite element method and hybridization of appropriate lowest order Raviart–Thomas approximation is used for the special conditions of the flow through connected system of 2-D polygons placed in 3-D. Model problem is tested.

1 Introduction

We consider a steady saturated Darcy's law governed flow of an incompressible fluid through a system of 2-D polygons placed in the 3-D space and connected under certain conditions into one network. This may simulate underground water flow through natural geological disruptions of a rock massif, fractures, e.g. for the purposes of finding of suitable nuclear waste repositories. Note that intersection of three or more triangles through one edge in the discretization is possible owing to the special geometrical situation, see Fig. 1. We study the existence and uniqueness of weak and discrete mixed solutions, and finally use the hybridization of the lowest order Raviart-Thomas mixed approximation, see [3], [4] respectively. For technical details of the following, see [5].

2 Mathematical-physical Formulation

We suppose that we have

$$\mathcal{S} = \Big\{ \bigcup_{\ell \in L} \overline{\alpha_\ell} \setminus \partial \mathcal{S} \Big\}, \tag{1}$$

where α_ℓ is an opened 2-D polygon placed in a 3-D Euclidean space; we call $\overline{\alpha_\ell}$ as a fracture. We denote as L the index set of fractures, $|L|$ is the number (finite) of considered fractures. We suppose that all closures of these polygons are connected into one "fracture network", the connection is possible only through an edge, not a point. Moreover, we require that if $\overline{\alpha_i} \bigcap \overline{\alpha_j} \neq \emptyset$ then $\overline{\alpha_i} \bigcap \overline{\alpha_j} \subset \partial\alpha_i \bigcap \partial\alpha_j$, i.e. the connection is possible only through fracture boundaries, cf.

P.M.A. Sloot et al. (Eds.): ICCS 2002, LNCS 2331, pp. 794–803, 2002.

Fig. 1 (we state this requirement in order to be able to define correct function spaces).

Let us have a 2-D orthogonal coordinate system in each polygon α_ℓ. We are looking for the fracture flow velocity $\mathbf{u}$ (2-D vector in each α_ℓ), which is the solution of the following problem

$$\mathbf{u} = -\mathbf{K}\Big(\nabla p + \nabla z\Big) \quad \text{in} \quad \mathcal{S}\,, \tag{2}$$

$$\nabla \cdot \mathbf{u} = q \quad \text{in} \quad \mathcal{S}\,, \tag{3}$$

$$p = p_D \quad \text{in} \quad \Lambda_D, \qquad \mathbf{u}\cdot\mathbf{n} = u_N \quad \text{in} \quad \Lambda_N\,, \tag{4}$$

where all variables are expressed in appropriate local coordinates of α_ℓ and also the differentiation is always expressed towards these local coordinates. The equation 2 is Darcy's law, 3 is the mass balance equation and 4 is the expression of appropriate boundary conditions. The variable p denotes the modified fluid pressure p ($p = \frac{\mathrm{p}}{\varrho g}$), g is the gravitational acceleration constant, ϱ is the fluid density, q represents stationary sources/sinks density and z is the elevation, positive upward taken vertical 3-D coordinate expressed in appropriate local coordinates. We require the second rank tensor $\mathbf{K}$ to be symmetric and uniformly positive definite on each α_ℓ. Λ_D is a part of $\partial\mathcal{S}$ where Dirichlet's type boundary conditions is given and similary Λ_N is a part of $\partial\mathcal{S}$ where Neumann's type boundary condition is given. Of course $\partial\mathcal{S} = \overline{\Lambda_D \cup \Lambda_n}$ holds.

3 Function Spaces

We start from $L^2(\alpha_\ell)$, $\|u\|_{0,\alpha_\ell} = (\int_{\alpha_\ell} u^2\, dS)^{\frac{1}{2}}$ and $\mathbf{L}^2(\alpha_\ell) = L^2(\alpha_\ell) \times L^2(\alpha_\ell)$ in order to introduce

$$L^2(\mathcal{S}) \equiv \prod_{\ell\in L} L^2(\alpha_\ell)\,, \ \mathbf{L}^2(\mathcal{S}) \equiv L^2(\mathcal{S}) \times L^2(\mathcal{S})\,. \tag{5}$$

We begin with classical Sobolev space $H^1(\alpha_\ell)$ of scalar functions with square integrable weak derivatives, $H^1(\alpha_\ell) = \{\varphi \in L^2(\alpha_\ell);\ \nabla\varphi \in \mathbf{L}^2(\alpha_\ell)\}$, $\|\varphi\|_{1,\alpha_\ell} = (\int_{\alpha_\ell}[\varphi^2 + \nabla\varphi\cdot\nabla\varphi]\, dS)^{\frac{1}{2}}$, so as to introduce

$$\begin{aligned} H^1(\mathcal{S}) \equiv \{v \in L^2(\mathcal{S})\,;\ & v|_{\alpha_\ell} \in H^1(\alpha_\ell) \quad \forall \ell \in L\,, \\ (v|_{\alpha_i})|_f = (v|_{\alpha_j})|_f \quad & \forall f = \overline{\alpha_i} \textstyle\bigcap \overline{\alpha_j}\,,\, i,j \in L\}\,. \end{aligned} \tag{6}$$

We note that this is possible even for the investigated geometrical situation. We then have the spaces $H^{\frac{1}{2}}(\partial\mathcal{S})$ and $H^{-\frac{1}{2}}(\partial\mathcal{S})$ and the surjective continuous trace operator $\gamma\,:\, H^1(\mathcal{S}) \to H^{\frac{1}{2}}(\partial\mathcal{S})$ as in the standard planar case.

We denote as $\mathbf{H}(div, \alpha_\ell)$ the Hilbert space of vector functions with square integrable weak divergences, $\mathbf{H}(div,\alpha_\ell) = \{\mathbf{v} \in \mathbf{L}^2(\alpha_\ell)\,;\ \nabla\cdot\mathbf{v} \in L^2(\alpha_\ell)\}$, $\|\mathbf{u}\|_{\mathbf{H}(div,\alpha_\ell)} = (\|\mathbf{u}\|^2_{0,\alpha_\ell} + \|\nabla\cdot\mathbf{u}\|^2_{0,\alpha_\ell})^{\frac{1}{2}}$. We can define now

$$\begin{aligned} \mathbf{H}(div,\mathcal{S}) \equiv \{\mathbf{v} \in \mathbf{L}^2(\mathcal{S})\,;\ \mathbf{v}|_{\alpha_\ell} \in \mathbf{H}(div,\alpha_\ell) \quad \forall\ell \in L\,, \\ \textstyle\sum_{i\in I_f} \langle \mathbf{v}|_{\alpha_i}\cdot\mathbf{n}_i, \varphi_i\rangle = 0\} \end{aligned} \tag{7}$$

$$\forall f \text{ such that } |I_f| \geq 2\,,\ I_f = \{i \in L\,;\ f \subset \partial\alpha_i\}\,,\ \forall\varphi_i \in H^1_{\partial\alpha_i\setminus f}\,.$$

Again, such "local" definition is necessary, since we do not deal with a standard planar case. It naturally expresses the continuity of the normal trace of functions from $\mathbf{H}(div, \mathcal{S})$ even for the given geometrical situation. We have the surjective continuous normal trace operator $\zeta \; : \; \mathbf{u} \in \mathbf{H}(div, \mathcal{S}) \rightarrow \mathbf{u} \cdot \mathbf{n} \in H^{-\frac{1}{2}}(\partial \mathcal{S})$ as in the standard planar case. We further define the space $\mathbf{H}_{0,N}(div, \mathcal{S}) = \{\mathbf{u} \in \mathbf{H}(div, \mathcal{S}) \; ; \; \langle \mathbf{u} \cdot \mathbf{n}, \varphi \rangle_{\partial \mathcal{S}} = 0 \quad \forall \varphi \in H^1_D(\mathcal{S})\}$. (where $H^1_D(\mathcal{S}) = \{\varphi \in H^1(\mathcal{S}) \; ; \; \gamma\varphi = 0 \text{ on } \Lambda_D \}$) Naturally, the norms on the spaces defined by 5, 6, 7 are given as

$$\| \; \|^2_{\cdot,\mathcal{S}} = \sum_{\ell=1}^{|L|} \| \; \|^2_{\cdot,\alpha_\ell} \, . \tag{8}$$

Remark 31 *Note that definitions 5, 6, 7 are essential. The system $\mathcal{S}$, however consisting of plane polygons, is not planar by oneself. Moreover, one edge can be common to three or more polygons α_ℓ creating the system $\mathcal{S}$.*

4 Weak Mixed Solution

Let us denote $\mathbf{A} = \mathbf{K}^{-1}$ on each α_ℓ, characterizing the medium resistance. Let us now consider such $\tilde{\mathbf{u}}$ that $\tilde{\mathbf{u}} \cdot \mathbf{n} = u_N$ on Λ_N in appropriate sense.

Definition 41 *As a weak mixed solution of the steady saturated fracture flow problem described by 2 – 4, we understand a function $\mathbf{u} = \mathbf{u}_0 + \tilde{\mathbf{u}}$, $\mathbf{u}_0 \in \mathbf{H}_{0,N}(div, \mathcal{S})$, and $p \in L^2(\mathcal{S})$ satisfying*

$$\begin{aligned}(\mathbf{A}\mathbf{u}_0, \mathbf{v})_{0,\mathcal{S}} - (\nabla \cdot \mathbf{v}, p)_{0,\mathcal{S}} = -\langle \mathbf{v} \cdot \mathbf{n}, p_D \rangle_{\Lambda_D} + (\nabla \cdot \mathbf{v}, z)_{0,\mathcal{S}} - \\ -\langle \mathbf{v} \cdot \mathbf{n}, z \rangle_{\partial \mathcal{S}} - (\mathbf{A}\tilde{\mathbf{u}}, \mathbf{v})_{0,\mathcal{S}} \qquad \forall \mathbf{v} \in \mathbf{H}_{0,N}(div, \mathcal{S}) \, ,\end{aligned} \tag{9}$$

$$-(\nabla \cdot \mathbf{u}_0, \phi)_{0,\mathcal{S}} = -(q, \phi)_{0,\mathcal{S}} + (\nabla \cdot \tilde{\mathbf{u}}, \phi)_{0,\mathcal{S}} \qquad \forall \phi \in L^2(\mathcal{S}) \, . \tag{10}$$

Our requirements are $A_{ij} \in L^\infty(\mathcal{S})$, $q \in L_2(\mathcal{S})$, $p_D \in H^{\frac{1}{2}}(\Lambda_D)$ and $u_N \in H^{-\frac{1}{2}}(\Lambda_N)$.

Theorem 41 *The problem (9), (10) has a unique solution.*

Proofs of this and all following theorems and lemmas can be found in [5].

5 Mixed Finite Element Approximation

Let us suppose a triangulation $\mathcal{T}_h$ of the system $\mathcal{S}$ from now on. We define an index set J_h to number the elements of the triangulation, $|J_h|$ denotes the number of elements. We define a 3-dimensional space $\mathbf{RT}^0(e)$ of vector functions linear on a given element e with the basis $\mathbf{v}^e_i$, $i \in \{1, 2, 3\}$, where

$$\mathbf{v}^e_1 = k^e_1 \begin{bmatrix} x - \alpha^e_{11} \\ y - \alpha^e_{12} \end{bmatrix}, \quad \mathbf{v}^e_2 = k^e_2 \begin{bmatrix} x - \alpha^e_{21} \\ y - \alpha^e_{22} \end{bmatrix}, \quad \mathbf{v}^e_3 = k^e_3 \begin{bmatrix} x - \alpha^e_{31} \\ y - \alpha^e_{32} \end{bmatrix}.$$

Concerning its dual basis, we state classically N_j^e, $j = 1, 2, 3$, $N_j^e(\mathbf{u}_h) = \int_{f_j^e} \mathbf{u}_h \cdot \mathbf{n}_j^e \, dl$, with each functional N_j^e expressing the flux through one edge for $\mathbf{u}_h \in \mathbf{RT}^0(e)$; we have $N_j^e(\mathbf{v}_i^e) = \delta_{ij}$ after appropriate choice of $\alpha_{11}^e - \alpha_{32}^e$, $k_1^e - k_3^e$. The local interpolation operator is then given by

$$\pi_e(\mathbf{u}) = \sum_{i=1}^{3} N_i^e(\mathbf{u})\mathbf{v}_i^e \quad \forall\, \mathbf{u} \in (H^1(e))^2 \,. \tag{11}$$

We start from the Raviart–Thomas space $\mathbf{RT}_{-1}^0(\mathcal{T}_h)$ of on each element linear vector functions without any continuity requirements,

$$\mathbf{RT}_{-1}^0(\mathcal{T}_h) \equiv \{\mathbf{v} \in \mathbf{L}^2(\mathcal{S})\,;\ \mathbf{v}|_e \in \mathbf{RT}^0(e) \quad \forall e \in \mathcal{T}_h\}\,,$$

to define the "continuity assuring" space $\mathbf{RT}_0^0(\mathcal{T}_h)$ by

$$\begin{aligned}\mathbf{RT}_0^0(\mathcal{T}_h) \equiv \{\mathbf{v} \in \mathbf{RT}_{-1}^0(\mathcal{T}_h)\,;\ &\textstyle\sum_{i\in I_f} \mathbf{v}|_{e_i} \cdot \mathbf{n}_{f,\partial e_i} = 0\, \forall f \text{ such that} \\ |I_f| \geq 2\,,\ I_f = \{i \in J_h\,;\ f \subset \partial e_i\} &= \mathbf{RT}_{-1}^0(\mathcal{T}_h) \cap \mathbf{H}(div, \mathcal{S})\,.\end{aligned}$$

We set furthermore

$$\mathbf{RT}_{0,N}^0(\mathcal{T}_h) \equiv \{\mathbf{v} \in \mathbf{RT}_0^0(\mathcal{T}_h)\,;\ \mathbf{v}\cdot\mathbf{n} = 0 \text{ in } \Lambda_N\} = \mathbf{RT}_{-1}^0(\mathcal{T}_h) \cap \mathbf{H}_{0,N}(div, \mathcal{S})$$

and

$$M_{-1}^0(\mathcal{T}_h) \equiv \{\phi \in L^2(\mathcal{S})\,;\ \phi|_e \in M^0(e) \quad \forall e \in \mathcal{T}_h\}\,,$$

where $M^0(e)$ is the space of scalar functions constant on a given element e. Looking for the basis, appropriate dual basis, and global interpolation operator for $\mathbf{RT}_0^0(\mathcal{T}_h)$, we have the following definitions and lemmas:

We set $\mathcal{N}_h = \{N_1, N_2, \ldots, N_{I_{\mathcal{N}_h}}\}$ as the dual basis of $\mathbf{RT}_0^0(\mathcal{T}_h)$, where for each border edge f, we have one functional N_f defined by $N_f(\mathbf{u}_h) = \int_f \mathbf{u}_h|_e \cdot \mathbf{n}_{\partial e}\, dl$, and for each inner edge f common to elements $e_1, e_2, \ldots, e_{I_f}$, we have $|I_f| - 1$ functionals given by

$$N_{f,j}(\mathbf{u}_h) = \frac{1}{|I_f|}\int_f \mathbf{u}_h|_{e_1} \cdot \mathbf{n}_{\partial e_1} dl - \frac{1}{|I_f|}\int_f \mathbf{u}_h|_{e_{j+1}} \cdot \mathbf{n}_{\partial e_{j+1}} dl, \quad j = 1, \ldots, |I_f| - 1\,.$$

Lemma 1. *For all* $\mathbf{u}_h \in \mathbf{RT}_0^0(\mathcal{T}_h)$, *from* $N_j(\mathbf{u}_h) = 0\ \forall\ j = 1, \ldots, I_{\mathcal{N}_h}$ *follows that* $\mathbf{u}_h = 0$.

We set $\mathcal{V}_h = \{\mathbf{v}_1, \mathbf{v}_2, \ldots, \mathbf{v}_{I_{\mathcal{N}_h}}\}$, where for each border edge f, we have one base function $\mathbf{v}_f$ defined by $\mathbf{v}_f = \mathbf{v}_f^e$ with $\mathbf{v}_f^e$ being the local base function appropriate to the element e and its edge f, and for each inner edge f common to elements $e_1, e_2, \ldots, e_{I_f}$, we have $|I_f| - 1$ base functions given by

$$\mathbf{v}_{f,i} = \sum_{k=1,\ k\neq i+1}^{|I_f|} \mathbf{v}_f^{e_k} - (I_f - 1)\mathbf{v}_f^{e_{i+1}} \quad , \quad i = 1, \ldots, |I_f| - 1\,.$$

Lemma 2. *For the bases $\mathcal{N}_h$ and $\mathcal{V}_h$, $N_j(\mathbf{v}_i) = \delta_{ij}$, $i,j = 1,\ldots,I_{\mathcal{N}_h}$ holds.*

We introduce first a space smoother than $\mathbf{H}(div,\mathcal{S})$, corresponding to the classical $(H^1(\mathcal{S}))^2$,

$$\begin{aligned}\mathbf{H}(grad,\mathcal{S}) = \{v \in \mathbf{L}^2(\mathcal{S})\,;\, \mathbf{v}|_{\alpha_\ell} \in (H^1(\alpha_\ell))^2 \quad \forall \ell \in L\,,\\ \textstyle\sum_{i\in I_f} \mathbf{v}|_{\alpha_i}\cdot \mathbf{n}_{f,\partial\alpha_i} = 0\\ \forall f \text{ such that } |I_f| \geq 2\,,\, I_f = \{i \in L\,;\, f \subset \partial\alpha_i\}\,,\end{aligned}$$

in order to set the global interpolation operator

$$\pi_h(\mathbf{u}) = \sum_{i=1}^{I_{\mathcal{N}_h}} N_i(\mathbf{u})\mathbf{v}_i \quad \forall\, \mathbf{u} \in \mathbf{H}(grad,\mathcal{S})\,. \tag{12}$$

Lemma 3. *Concerning the local and global interpolation operators given by 11, 12 respectively, we have their equality on each element, i.e.*

$$\pi_h(\mathbf{u})|_e = \pi_e(\mathbf{u}|_e) \quad \forall\, e \in \mathcal{T}_h,\ \forall\, \mathbf{u} \in \mathbf{H}(grad,\mathcal{S})\,.$$

Lemma 4. *Even for the considered special function spaces and their finite dimensional subspaces, we have*

$$\begin{array}{ccc} \mathbf{H}(grad,\mathcal{S}) & \xrightarrow{\text{div}} & L^2(\mathcal{S}) \\ \Big\downarrow \pi_h & & \Big\downarrow P_h \\ \mathbf{RT}_0^0(\mathcal{T}_h) & \xrightarrow{\text{div}} & M_{-1}^0(\mathcal{T}_h) \end{array}\,, \tag{13}$$

i.e. the commutativity diagram property, where π_h is the global interpolation operator defined in (12), and P_h is the $L^2(\mathcal{S})$-orthogonal projection onto $M_{-1}^0(\mathcal{T}_h)$.

Definition 51 *As the lowest order Raviart–Thomas mixed approximation of the the problem (9), (10), we understand functions $\mathbf{u}_{0,h} \in \mathbf{RT}_{0,N}^0(\mathcal{T}_h)$ and $p_h \in M_{-1}^0(\mathcal{T}_h)$ satisfying*

$$\begin{aligned}(\mathbf{A}\mathbf{u}_{0,h},\mathbf{v}_h)_{0,\mathcal{S}} - (\nabla\cdot\mathbf{v}_h, p_h)_{0,\mathcal{S}} = -\langle \mathbf{v}_h\cdot\mathbf{n}, p_D\rangle_{\Lambda_D} + (\nabla\cdot\mathbf{v}_h, z)_{0,\mathcal{S}} - \\ -\langle \mathbf{v}_h\cdot\mathbf{n}, z\rangle_{\partial\mathcal{S}} - (\mathbf{A}\tilde{\mathbf{u}},\mathbf{v}_h)_{0,\mathcal{S}} \qquad \forall\,\mathbf{v}_h \in \mathbf{RT}_{0,N}^0(\mathcal{T}_h)\,,\end{aligned} \tag{14}$$

$$-(\nabla\cdot\mathbf{u}_{0,h},\phi_h)_{0,\mathcal{S}} = -(q,\phi_h)_{0,\mathcal{S}} + (\nabla\cdot\tilde{\mathbf{u}},\phi_h)_{0,\mathcal{S}} \qquad \forall\phi_h \in M_{-1}^0(\mathcal{T}_h)\,. \tag{15}$$

Theorem 51 *The problem (14), (15) has a unique solution.*

6 Error Estimates and Hybridization of the Mixed Method

If the solution $(\mathbf{u}_0, p)$ of (9), (10) is such that $(\mathbf{u}_0,p) \in \mathbf{H}(grad,\mathcal{S}) \times H^1(\mathcal{S})$ and $\nabla\cdot\mathbf{u}_0 \in H^1(\mathcal{S})$ and if $(\mathbf{u}_{0,h}, p_h)$ is the solution of (14), (15), then

$$\|\mathbf{u}_0 - \mathbf{u}_{0,h}\|_{\mathbf{H}(div,\mathcal{S})} + \|p - p_h\|_{0,\mathcal{S}} \leq Ch(|p|_{1,\mathcal{S}} + |\mathbf{u}_0|_{1,\mathcal{S}} + |\nabla\cdot\mathbf{u}_0|_{1,\mathcal{S}})\,,$$

where the constant C does not depend on h and $|\varphi|_{1,\mathcal{S}} = \|\nabla\varphi\|_{0,\mathcal{S}}$, $|\mathbf{u}|^2_{1,\mathcal{S}} = \sum_{i=1}^{2} |\mathbf{u}_i|^2_{1,\mathcal{S}}$ (see [4], Theorem 13.2).

Intending to hybridize the mixed approximation, we define two sets of edges,

$$\Lambda_h = \cup_{e\in\mathcal{T}_h} \partial e \quad , \quad \Lambda_{h,D} = \cup_{e\in\mathcal{T}_h} \partial e - \Lambda_D \, .$$

If $f \in \Lambda_h$, we define first the space $M^0(f)$ of functions constant on this edge and finally

$$M^0_{-1}(\Lambda_{h,D}) \equiv \{\mu_h : \Lambda_h \to R;\ \mu_h|_f \in M^0(f) \quad \forall f \in \Lambda_h ,$$
$$\mu_h|_f = 0 \quad \forall f \in \Lambda_D\} \, .$$

It now follows immediately that if $\mathbf{v}_h \in \mathbf{RT}^0_{-1}(\mathcal{T}_h)$, then $\mathbf{v}_h \in \mathbf{RT}^0_{0,N}(\mathcal{T}_h)$ if and only if

$$\sum_{e\in\mathcal{T}_h} \langle \mathbf{v}_h \cdot \mathbf{n}, \lambda_h \rangle_{\partial e\cap\Lambda_{h,D}} = 0 \quad \forall \lambda_h \in M^0_{-1}(\Lambda_{h,D}) \, ,$$

which allows us to state the hybrid version of the lowest order Raviart–Thomas mixed method:

Definition 61 *As the lowest order Raviart–Thomas mixed-hybrid approximation of the the problem (9), (10), we understand functions* $\mathbf{u}_{0,h} \in \mathbf{RT}^0_{-1}(\mathcal{T}_h)$, $p_h \in M^0_{-1}(\mathcal{T}_h)$ *and* $\lambda_h \in M^0_{-1}(\Lambda_{h,D})$ *satisfying*

$$\sum_{e\in\mathcal{T}_h} \{(\mathbf{A}\mathbf{u}_{0,h}, \mathbf{v}_h)_{0,e} - (\nabla\cdot\mathbf{v}_h, p_h)_{0,e} + \langle \mathbf{v}_h\cdot\mathbf{n}, \lambda_h\rangle_{\partial e\cap\Lambda_{h,D}}\} =$$
$$= \sum_{e\in\mathcal{T}_h} \{-\langle \mathbf{v}_h\cdot\mathbf{n}, p_D\rangle_{\partial e\cap\Lambda_D} + (\nabla\cdot\mathbf{v}_h, z)_{0,e} - \langle \mathbf{v}_h\cdot\mathbf{n}, z\rangle_{\partial e} - (\mathbf{A}\tilde{\mathbf{u}}, \mathbf{v}_h)_{0,e}\}$$
$$\forall \mathbf{v}_h \in \mathbf{RT}^0_{-1}(\mathcal{T}_h) \, , \tag{16}$$

$$-\sum_{e\in\mathcal{T}_h} (\nabla\cdot\mathbf{u}_{0,h}, \phi_h)_{0,e} = -\sum_{e\in\mathcal{T}_h} \{(q, \phi_h)_{0,e} - (\nabla\cdot\tilde{\mathbf{u}}, \phi_h)_{0,e}\}$$
$$\forall \phi_h \in M^0_{-1}(\mathcal{T}_h) \, , \tag{17}$$

$$\sum_{e\in\mathcal{T}_h} \langle \mathbf{u}_{0,h}\cdot\mathbf{n}, \mu_h\rangle_{\partial e\cap\Lambda_{h,D}} = \sum_{e\in\mathcal{T}_h} \{\langle u_N, \mu_h\rangle_{\partial e\cap\Lambda_N} - \langle \tilde{\mathbf{u}}\cdot\mathbf{n}, \mu_h\rangle_{\partial e\cap\Lambda_{h,D}}\}$$
$$\forall \mu_h \in M^0_{-1}(\Lambda_{h,D}) \, . \tag{18}$$

Due to the previously mentioned, the triple $\mathbf{u}_{0,h}$, p_h, λ_h surely exist and is unique, $\mathbf{u}_{0,h}$ and p_h are moreover at the same time the unique solutions of (14), (15); moreover, the multiplier λ_h is an approximation of the trace of p on all edges from $\Lambda_{h,D}$. Consequently, all error estimates are valid also for the mixed-hybrid solution triple $\mathbf{u}_{0,h}, p_h, \lambda_h$. Thus, we have the following theorem:

Theorem 61 *The problem (16) – (18) has a unique solution.*

7 Model Problem

We consider a simple model problem with the system $\mathcal{S}$ viewed in Figure 1,

$$\mathcal{S} = \overline{\alpha_1} \bigcup \overline{\alpha_2} \bigcup \overline{\alpha_3} \bigcup \overline{\alpha_4} \setminus \partial \mathcal{S} ,$$

$$\mathbf{u} = -\Big(\nabla p + \nabla z\Big) \quad \text{in} \quad \mathcal{S} ,$$
$$\nabla \cdot \mathbf{u} = 0 \quad \text{in} \quad \mathcal{S} ,$$

$$\begin{aligned} p = 0 \quad &\text{in} \quad \Lambda_1 \quad , \quad p = 0 \quad \text{in} \quad \Lambda_2 \\ \mathbf{u} \cdot \mathbf{n} = 0 \quad &\text{in} \quad \Lambda_3 \quad , \quad \mathbf{u} \cdot \mathbf{n} = 0 \quad \text{in} \quad \Lambda_4 \\ p = \sin\Big(\tfrac{\pi x_1}{2X}\Big) \sinh\Big(\tfrac{\pi (A+B)}{2X}\Big) + S \cdot A \quad &\text{in} \quad \Lambda_5 \quad , \quad p = S \cdot y_1 \quad \text{in} \quad \Lambda_6 \end{aligned} \tag{19}$$

$$\begin{aligned} p = 0 \quad &\text{in} \quad \Lambda_7 \quad , \quad p = 0 \quad \text{in} \quad \Lambda_8 \\ \mathbf{u} \cdot \mathbf{n} = 0 \quad &\text{in} \quad \Lambda_9 \quad , \quad \mathbf{u} \cdot \mathbf{n} = 0 \quad \text{in} \quad \Lambda_{10} \\ p = \sin\Big(\tfrac{\pi x_4}{2X}\Big) \sinh\Big(\tfrac{\pi (B+B)}{2X}\Big) \quad &\text{in} \quad \Lambda_{11} \quad , \quad p = 0 \quad \text{in} \quad \Lambda_{12} \,. \end{aligned}$$

The exact solutions in α_1 can be easily found as

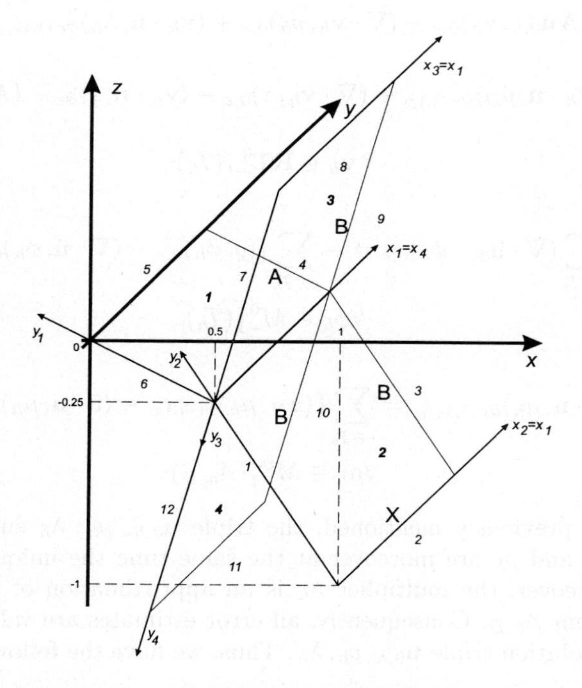

Fig. 1. Considered Domain for the Model Problem

$$p_{\alpha_1} = \sin\Big(\frac{\pi x_1}{2X}\Big)\sinh\Big(\frac{\pi(y_1+B)}{2X}\Big) + S\cdot y_1\,,$$

$$\mathbf{u}_{\alpha_1} = \Big(-\frac{\pi}{2X}\cos\Big(\frac{\pi x_1}{2X}\Big)\sinh\Big(\frac{\pi(y_1+B)}{2X}\Big),$$

$$-\frac{\pi}{2X}\sin\Big(\frac{\pi x_1}{2X}\Big)\cosh\Big(\frac{\pi(y_1+B)}{2X}\Big) - S - \nabla z^y_{\alpha_1}\Big)\,,$$

where $\nabla z_{\alpha_1} = (0, \nabla z^y_{\alpha_1})$, $S + \nabla z^y_{\alpha_1} = \nabla z^y_{\alpha_2}$.

The following table gives pressure, velocity, and pressure trace approximation errors in the first fracture α_1. There is the expected $O(h)$ convergence in pressure and velocity, but only $O(h^{\frac{1}{2}})$ in pressure trace in $\|\cdot\|_{0,\Lambda_{h,D}}$ norm. All the computations were done in double precision on a personal computer, the resulting symmetric indefinite systems of linear equations were solved by the solver GI8 of the Institute of Computer Science, Academy of Sciences of the Czech Republic, see [2]. This is based on the sequential elimination onto a system with Schur's complement and subsequent solution of this system by the specially preconditioned conjugate gradients method. The solver accuracy was set to 10^{-8}.

Table 1. Pressure, Velocity, and Pressure Trace Errors in α_1 for the Model Problem

N	triangles	$\|p-p_h\|_{0,S}$	$\|\mathbf{u}-\mathbf{u}_h\|_{\mathbf{H}(div,\mathcal{T}_h)}$	$\|\lambda-\lambda_h\|_{0,\Lambda_{h,D}}$
2	8×4	0.4445	1.2247	1.4973
4	32×4	0.2212	0.6263	1.0562
8	128×4	0.1102	0.3150	0.7509
16	512×4	0.0550	0.1577	0.5332
32	2048×4	0.0275	0.0789	0.3779
64	8192×4	0.0138	0.0394	0.2676
128	32768×4	0.0069	0.0197	0.1893
256	131072×4	0.0034	0.0099	0.1339

8 Example of real-world problem

Results of model problem presented in previous section prooved correctness of mathematical model as well as correctness of its numerical implementation. Therefore we have tried to use the model for solving a real-world problem.

The problem was based on measurements in the boreholes PTP–3 and PTP–4 situated in Krun Hory mountains, Czech Republic. Results of these measurements have given us data for creating computer approximation of fractured environment in rock massif and, consequently, for its discretizing to FEM/FVM mesh. Then, boundary contition was set and calculation has been started.

Example of results of such calculation is shown at figure 2. Mesh presented of this figure covers volume 5x10x10 meter. It consists of approx. 200 fractures and 3000 triangle elements.

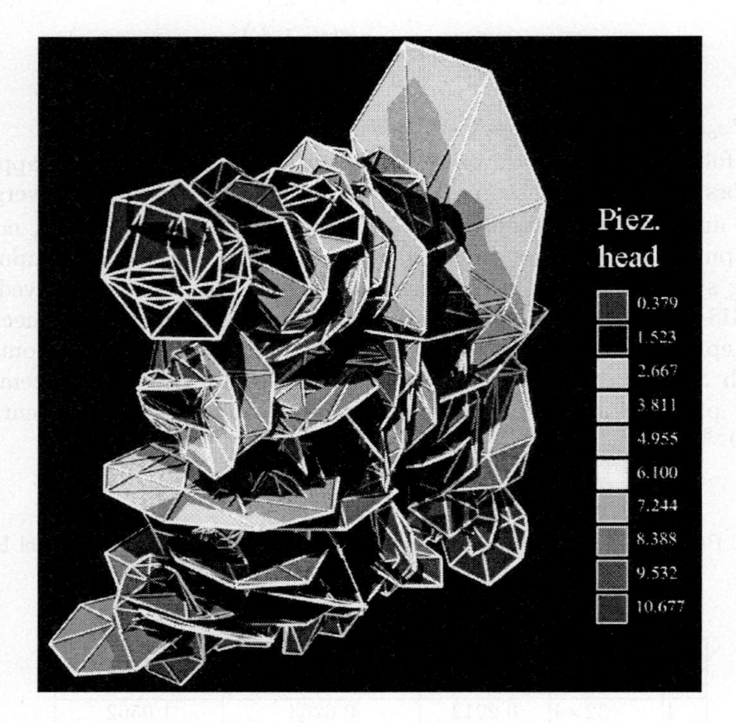

Fig. 2. Example of real-world problem calculation

9 Conclusion

Mathematical model of groundwater flow was described. This model is based on assumption, that flow in particular fracture can be approximated by Darcy's law. Mixed-hybride approximation of solution of problem was introduced and error estimation for such approximation was derived. Practical tests prooved correctness of presented approach.

References

1. QUARTERONI A., VALLI A.: *Numerical Approximation of Partial Differential Equations*, Springer-Verlag Berlin Heidelberg, Berlin, 1994.
2. MARYŠKA J., ROZLOŽNÍK M. TŮMA M.: *Schur Complement Reduction in the Mixed-hybrid Finite Element Approximation of Darcy's Law: Rounding Error Analysis*,

Technical Report TR-98-06, Swiss Center for Scientific Computing, Swiss Federal Institute of Technology, Zurich, Switzerland, pp. 1-15., June 1998.
3. RAVIART P.A., THOMAS J.M.: *A Mixed Finite Element Method for Second-order Elliptic Problems*, in: GALLIGANI I., MAGENES E.: *Mathematical Aspects of Finite Element Methods*, Lecture Notes in Mathematics 606, pp. 292–315, Springer, Berlin, 1977.
4. ROBERTS J.E., THOMAS J.-M.: *Mixed and Hybrid Methods* in: CIARLET P.G., LIONS J.L.: *Handbook of Numerical Analysis, vol. II, Finite Element Methods (Part 1)*, pp. 523–639, Elsevier Science Publishers B.V. (North-Holland), Amsterdam, 1991.
5. VOHRALÍK M.: *Existence- and Error Analysis of the Mixed-hybrid Model of the Fracture Flow*, Technical Report MATH-NM-06-2001, Dresden University of Technology, Dresden, 2001.

Parallel Realization of Difference Schemes of Filtration Problem in a Multilayer System

Miron Pavluš[1,2] and Edik Hayryan[2]

[1] Technical University of Košice, Department of Mathematics,
Vysokoškolská 4, 042 00 Košice, Slovakia
Miron.Pavlus@tuke.sk

[2] Joint Institute of Nuclear Research, Laboratory of Information Technology,
141 980 Dubna, Moscow Region, Russia
{Pavlus, Ayrjan}@cv.jinr.ru

Abstract. Early suggested difference schemes with splitting according to the physical processes are used and applied for a plane filtration problem in a multilayer system. Parallel algorithm connected with the solving of the filtration problem (one water-carrying layer on one processor) is constructed. Program realization on the multiprocessor system $SPP2000$ is discussed. Some results of sequential and parallel programs are compared with regard to the used time $MPI\_Wtime()$ procedure.

1 Introduction

Modeling of the water filtration in practical applications of hydro-geology dynamics is based very often on the using of models of plane filtration [1],[2],[3],[4] with homogenization of filtration flows in the direction of the layer thickness. A typical situation occurs when horizontal water-carrying layers are alternated by the horizontal weakly permeable ground layers. Mathematical models of filtration in a multilayer system were constructed provided that a longitudinal flow of water takes place mainly in the water-carrying layers and cross flow of water takes place through separated layers (Mityaev-Girinski model [5]). These mathematical models can be valid by the homogenization theory [6] and represent a complex system.

Various methods [7] were used for numerical solution of the plane filtration problems. The simplest of them consists of using of common difference schemes with weights [8]. For example, the implicit scheme for a parabolic constrained system of equations for the piezometric head in each water-carrying layer can be used. After some suitable iteration process based on the idea of determining of piezometric head in each separate water-carrying layer can be applied. Difference schemes with splitting in space variables were constructed in [7]. But realization of such approach for a multilayer system even for one dimensional problems (the dependence only on one longitudinal variable) is constrained with using of three diagonal matrix inversion that, certainly, substantially complicates computational schemes.

P.M.A. Sloot et al. (Eds.): ICCS 2002, LNCS 2331, pp. 804–812, 2002.

In this paper according to [9] we formulate a problem of the plane filtration in the multilayer systems and we use, suggested in [9], difference schemes with splitting that are free of mentioned inconvenience. The used difference schemes are splitted according to the physical processes when two processes – filtration along the water-carrying layers and flows between the layers are separated. Moreover, the schemes are additive and absolutely stable.
The formulation of the problem and difference schemes for its solution are presented in Section 2. Parallel algorithms for solution of the used difference schemes of filtration problems in multilayer systems are constructed in Section 3. In Section 3 we also suggest the most natural approach that consists of the solution when the filtration problem in each water-carrying layer is processed by one processor. Program realization of such approach is discussed for the multiprocessor system $SPP2000$ in Section 3 as well.

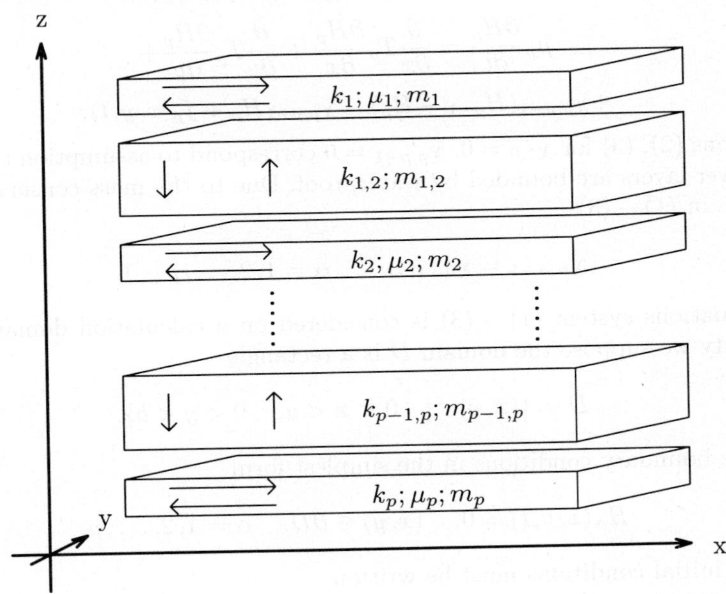

Fig. 1 A Multilayer System of Water-carrying and Weakly Permeable Layers

2 Formulation of the Problem and Difference Schemes

Let us consider, according to [9], the plane filtration in a multilayer system which consists (see Fig. 1) of p water-carrying layers that are separated by weakly permeable layers. Let us suppose all layers are horizontal. We denote the hydrodynamical piezometric head $H_\alpha(x, y, t)$ in the point (x, y) of the layer α, $\alpha = 1, 2, \ldots, p$ at the moment t. Let $k_\alpha = k_\alpha(x, y)$ be filtration coefficients, m_α – thickness of a water-carrying layer α, $T_\alpha = k_\alpha m_\alpha$ – conductivity of layer and $\mu_\alpha = \mu_\alpha(x, y)$ – elastic capacity of layer. Correspondingly, $k_{\alpha,\beta} = k_{\alpha,\beta}(x, y)$, $\beta = \alpha+1, \alpha-1$ are filtration coefficients and $m_{\alpha,\beta}$ – thickness of a weakly permeable layer, $\chi_{\alpha,\beta} = k_{\alpha,\beta}/m_{\alpha,\beta}$ – coefficients of the cross flow. If the conditions of the cross flow ($k_\alpha/m_\alpha << k_{\alpha,\beta}/m_{\alpha,\beta}$) hold then we use for description of filtration in each water-carrying layer α, $\alpha = 2, 3, \ldots, p-1$ the following equation

$$\mu_\alpha \frac{\partial H_\alpha}{\partial t} = \frac{\partial}{\partial x} T_\alpha \frac{\partial H_\alpha}{\partial x} + \frac{\partial}{\partial y} T_\alpha \frac{\partial H_\alpha}{\partial y} + \tag{1}$$

$$+\chi_{\alpha,\alpha+1}(H_{\alpha+1} - H_\alpha) + \chi_{\alpha,\alpha-1}(H_{\alpha-1} - H_\alpha) + f_\alpha(x, y, t),$$

where $f_\alpha(x, y, t)$ is a water source and/or a water sink. We have for lower and upper layers

$$\mu_1 \frac{\partial H_1}{\partial t} = \frac{\partial}{\partial x} T_1 \frac{\partial H_1}{\partial x} + \frac{\partial}{\partial y} T_1 \frac{\partial H_1}{\partial y} + \tag{2}$$

$$+\chi_{1,2}(H_2 - H_1) - \chi_{1,0} H_1 + f_1(x, y, t),$$

$$\mu_p \frac{\partial H_p}{\partial t} = \frac{\partial}{\partial x} T_p \frac{\partial H_p}{\partial x} + \frac{\partial}{\partial y} T_p \frac{\partial H_p}{\partial y} + \tag{3}$$

$$+\chi_{p,p-1}(H_{p-1} - H_p) - \chi_{p,p+1} H_p + f_p(x, y, t),$$

Equations (2), (3) for $\chi_{1,0} = 0$, $\chi_{p,p+1} = 0$ correspond to assumption that upper and lower layers are bounded by waterproof. Due to the mass conservation law we have in (1) – (3)

$$\chi_{\alpha,\alpha+1} = \chi_{\alpha+1,\alpha}, \qquad \alpha = 1, 2, \ldots, p-1. \tag{4}$$

The equations system (1) – (3) is considered on a calculation domain D. For simplicity we suppose the domain D is a rectangle

$$D = \{(x, y) \mid 0 < x < a, \quad 0 < y < b\}.$$

We take boundary conditions in the simplest form

$$H_\alpha(x, y, t) = 0, \quad (x, y) \in \partial D, \quad \alpha = 1, 2, \ldots, p. \tag{5}$$

Finally, initial conditions must be written

$$H_\alpha(x, y, 0) = H_\alpha^0(x, y), \quad (x, y) \in D, \quad \alpha = 1, 2, \ldots, p. \tag{6}$$

We rewrite the imposed initial boundary value problem (1) – (6) in the brief form. Let us introduce some denotations. Let the vector

$$U = U(x, y, t) = \{H_1, H_2, \ldots, H_p\}$$

be a solution of the problem (1) – (6), and

$$F = F(x, y, t) = \{f_1, f_2, \ldots, f_p\}.$$

Let us define the diagonal matrices M and θ

$$M = \{\mu_\alpha(x, y)\delta_{\alpha,\beta}\}, \quad \alpha, \beta = 1, 2, \ldots, p, \tag{7}$$

$$\theta = \{T_\alpha(x,y)\delta_{\alpha,\beta}\}, \quad \alpha, \beta = 1, 2, \ldots, p,$$

where $\delta_{\alpha,\beta}$ be the Kronecker's symbol. Now, let K is a three diagonal matrix that corresponds to cross flow through separating layers

$$K = \{\kappa_{\alpha,\beta}\}, \quad \alpha, \beta = 1, 2, \ldots, p, \tag{8}$$

in which connection

$$\kappa_{\alpha,\alpha-1} = -\chi_{\alpha,\alpha-1}, \quad \kappa_{\alpha,\alpha+1} = -\chi_{\alpha,\alpha+1}, \quad \kappa_{\alpha,\alpha} = \chi_{\alpha,\alpha-1} + \chi_{\alpha,\alpha+1}. \tag{9}$$

Finally, we define

$$LU = -\frac{\partial}{\partial x}\theta\frac{\partial U}{\partial x} - \frac{\partial}{\partial y}\theta\frac{\partial U}{\partial y}. \tag{10}$$

Taking into account (7) – (10) we write the system of equations (1) – (3) in the form

$$M\frac{\partial U}{\partial t} + LU + KU = F(x,y,t), \quad (x,y) \in D. \tag{11}$$

Using analogical denotations the boundary and initial conditions (5), (6) have the following form

$$U(x,y,t) = 0, \quad (x,y) \in \partial D, \tag{12}$$

$$U(x,y,0) = U^0(x,y), \quad (x,y) \in D. \tag{13}$$

Next, we introduce difference schemes with weights for solution of the brief formulated problem (11) – (13). Let us introduce the equidistant net

$$\omega_\tau = \{t \mid t = t_n = n\tau, \quad n = 0, 1, \ldots\}$$

with respect to time variable the time step of which is $\tau > 0$. We also examine the approximated equations (11) – (13) with respect to the space variables discretization. We introduce an equidistant rectangle net in the rectangular domain D with discretizing parameters h_x and h_y

$$\overline{\omega}_h = \omega_h + \gamma_h = \{(x,y) | (x,y) = (x_i, y_j),$$

$$x_i = ih_x, i = 0, 1, \ldots, N_x, \quad y_j = jh_y, j = 0, 1, \ldots, N_y,$$

$$N_x h_x = a, N_y h_y = b\},$$

where ω_h means the set of inner nodes and γ_h means the set of boundary nodes of defined net. We denote V^n the approximate solution of the problem (11) – (13) at the moment t_n. Note, the constructed operators M, L and K are constant (do not depend on n). Now, the used difference schemes are the following

$$M\frac{V^{n+1/4} - V^n}{\tau} + L_{hx}[\sigma_1 V^{n+1/4} + (1-\sigma_1)V^n] = F^n, \tag{14}$$

$$M\frac{V^{n+1/2} - V^{n+1/4}}{\tau} + L_{hy}[\sigma_1 V^{n+1/2} + (1-\sigma_1)V^{n+1/4}] = F^n, \tag{15}$$

$$M\frac{V^{n+3/4}-V^{n+1/2}}{\tau}+K^{+}[\sigma_2 V^{n+3/4}+(1-\sigma_2)V^{n+1/2}]=F^n, \qquad (16)$$

$$M\frac{V^{n+1}-V^{n+3/4}}{\tau}+K^{-}[\sigma_2 V^{n+1}+(1-\sigma_2)V^{n+3/4}]=F^n, \qquad (17)$$

where

$$L_h = L_{hx}+L_{hy}, \quad L_{hx}V_h=-(\theta_1 V_{h\bar{x}})_x, \quad L_{hy}V_h=-(\theta_2 V_{h\bar{y}})_y,$$

$$K=K^{+}+K^{-}, \quad (K^{+})^{*}=K^{-},$$

i.e. the three diagonal matrix K is splitted to lower and upper triangular matrices. Generally speaking the existence of the inverse operator K^{-1} doesn't take place in our case (see equations (1)-(3) and condition (4)). We observe a more pleasant situation when upper and lower layers are not bounded by waterproof but also by weakly permeable layers i.e. $\chi_{1,0}$ and $\chi_{p,p+1}$ are positive. The schemes (14) – (17) were constructed in [9] and are additive. It was proven in [9] that if $F^n = 0$, $\sigma_1 \geq 0.5$ and $\sigma_2 = 1$ then the schemes are stable with regard to the initial condition $V^0 = U^0$. The stability according to the right hand side F^n can be also established. Moreover, the schemes are economical (the number of arithmetical operations related to one node when moving to a next time layer doesn't depend on the general number of nodes). The schemes possess the following important property. The transfer of water in the water-carrying layers are described by the first two time quarter-steps (see (14), (15)) and by the last two time quarter-steps (see (16), (17)) the transfer between weakly permeable layers are described. In this context the schemes (14) – (17) of sum approximation can be treated as schemes of splitting according to the physical processes.

3 Parallel Realization of Difference Schemes

Let us consider the difference scheme (14) – (17) of splitting of the problem according to the physical processes. On the first time half-step ((14), (15)) we consider the water transfer in the horizontal water-carrying layers and on the second time half-step ((16), (17)) we consider the flow of the water in the vertical direction through the weakly permeable layers. The equations (14), (15) represent systems of linear algebraic equations with symmetric three diagonal matrices on the each water-carrying layer α. Equations (16), (17) tie together unknowns on the water-carrying layers and represent systems of linear algebraic equations with upper and lower triangle two diagonal matrices, respectively.

The most natural parallel realization of the difference scheme consists of the solution of the filtration problem on one processor for each water-carrying layer. In this case one time step of the difference scheme can be realized by the following algorithm.

1. Each fixed processor α has values of the piezometric head v_α^n, $\alpha = 1, 2, \ldots, p$ on the net ω_h on the water-carrying layer α e.i. processor α contains α-th component of the vector V^n.

2. According to (14) we use $N_y - 1$ of the Gauss elimination in the direction x on each water-carrying layer. As a result we receive values $V^{n+1/4}$.
3. According to (15) we use $N_x - 1$ of the Gauss elimination in the direction y on each water-carrying layer. As a result we receive values $V^{n+1/2}$.
In cases 2. and 3. the Gauss eliminations are realized in parallel, need no data exchange and represent well known ADI method [10], [11].
4. We solve the system (16) with the upper triangle matrix from the bottom to the top (from the layer p to the first layer) for all nodes ω_h. As a result we receive values $V^{n+3/4}$.
5. We solve the system (17) with the lower triangle matrix from the top to the bottom (from the first layer to the layer p) for all nodes ω_h. As a result we receive values V^{n+1}.

However, an effective parallel realization of the steps 4. and 5. requires data exchange between processors. For this reason we divide the net ω_h on p equal parts choosing the value N_x in the form $N_x = pk_x + 1$ and we introduce auxiliary nets (see Fig. 2)

$$\omega_\alpha = \{(x_i, y_j) | x_i = i.h_x, i = (\alpha - 1)k_x + 1, (\alpha - 1)k_x + 2, \ldots, \alpha k_x;$$
$$y_j = j.h_y, j = 1, 2, \ldots, N_y - 1\},$$

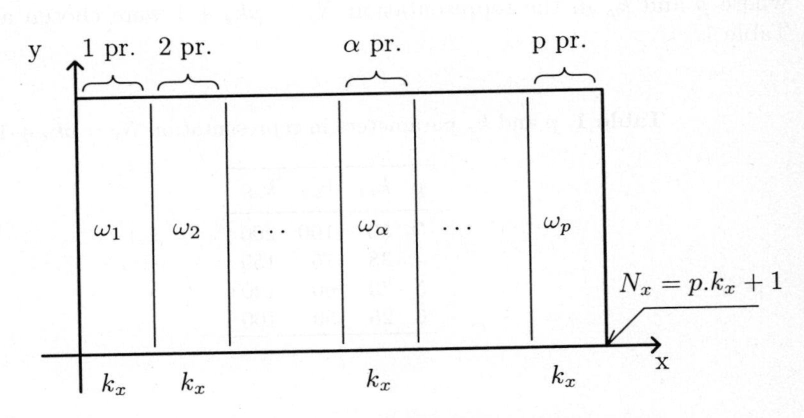

Fig. 2 Distribution of the Net Points between **p** Processors

We prescribe to processor α the processing of unknowns on the net ω_α. The processor α before the 4th step

- receives values $v_1^{n+1/2}, \ldots, v_{\alpha-1}^{n+1/2}, v_{\alpha+1}^{n+1/2}, \ldots, v_p^{n+1/2}$, on the nets ω_α from processors $1, \ldots, \alpha - 1, \alpha + 1, \ldots, p$ and puts them in stead of unknowns $v_\alpha^{n+1/2}$ on the nets ω_β, $\beta = 1, \ldots, \alpha - 1, \alpha + 1, \ldots, p$
- transmits values $v_\alpha^{n+1/2}$ from nets ω_β, $\beta = 1, \ldots, \alpha - 1, \alpha + 1, \ldots, p$ to processor β.

For finishing one time step of difference scheme it remains "to return" obtained values V^{n+1} to correspondent processors.

Two codes for one and for p processors $p = 3, 4, 5, 6$ were written for multiprocessor system $SPP2000$ using FORTRAN'77 with application of MPI (Message Passing Interface) [12]. The code for one processor (sequential program) processes sequentially p water-carrying layers. The following coefficients of equations, initial and boundary conditions, parameters of difference scheme were chosen

$$a = b = 1, \quad \tau = 0.1, \quad \sigma_1 = 0.5, \quad \sigma_2 = 1,$$

$$f_\alpha(x, y, t) = 0, \quad V_\alpha^0(x, y) = \alpha x(a - x)y(b - y)/p, \quad \mu_\alpha(x, y) = \alpha/p,$$

$$\theta_{1\alpha}(x, y) = \alpha(x - \frac{h_x}{2})y/p, \quad \theta_{2\alpha}(x, y) = \alpha x(y - \frac{h_y}{2})/p, \quad \chi_{\alpha,\alpha-1} = \chi_{\alpha,\alpha+1} = 5,$$

$$\alpha = 1, 2, \ldots, p.$$

Calculations were carried out for three following nets

$$(N_x, N_y) = \{(151, 100); (301, 200); (601, 400)\},$$

where p and k_x in the representation $N_x = pk_x + 1$ were chosen according to Table 1.

Table 1. p and k_x parameters in representation $N_x = pk_x + 1$

p	k_{x1}	k_{x2}	k_{x3}
3	50	100	200
4	38	75	150
5	30	60	120
6	25	50	100

The Fig.3 presents the dependence of work time of processors on the number of layers. A "plus" sign denotes the mean time of a processor work if the parallel code was applied. A "minus" sign denotes time of a single processor for comparison with the sequential code. Both cases correspond to p water-carrying layers $p = 3, 4, 5, 6$. We remind that in parallel code the number p also denotes the number of used processors. The $MPI\_Wtime()$ procedure for time counting out was used. The values of time are in seconds. The Fig. 3 shows that calculation time of parallel code inessentially increases when the number p of water-carrying layers increases. This is a good evidence of parallelism of algorithm.

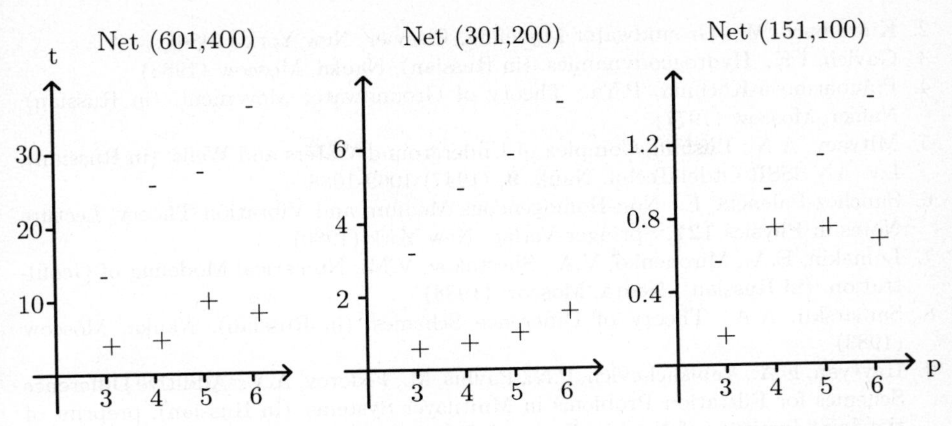

Fig. 3 The Dependence of Processors Work Time on the Number of Layers respectively on the Number of Processors

4 Concluding Remarks

At present the existence theorem was not proved neither for exact nor for weak solution of the formulated problem (11) – (13) but the proof of existence follows from the estimations given in [9].
Present paper deals with only parallel realization of the specific difference scheme. The used ADI method (equations (14)-(15)) can be replaced by other methods like FEM or a suitable modification of relaxation method. That means the suggested parllelism does not depend on the Gauss eliminations in steps 2. and 3. of parallel realization of difference schemes. A main advantage of the scheme consists of the following. It reflects physical processes – a longitudinal flow of water takes place mainly in the water-carrying layers and cross flow of water takes place through separated layers. The parallel realization conserves this property and can be used for modeling, for example, a water or naphtha movement in the multilayer system. This realization speeds up the calculation process and opens a new way for the multilayer systems modeling.

5 Acknowledgement

Authors wish to express thanks to Russian Fond of Fundamental Research, Grant 99-01-01101, with the support of which this paper was executed. Authors also thank Mr. Sapozhnikov A.P. and Mrs. Sapozhnikova T.F. for help with *SPP*2000. We thank Austrian Center in Košice for the conference ICCS-2002 presentation support. We also thank anonymous reviewers for their helpful comments.

References

1. Fried, J.J.: Groundwater Pollution. Elsevier, New York (1975)

2. Kinzelbach, W.: Groundwater Modeling. Elsevier, New York (1986)
3. Gavich, I.K.: Hydrogeodynamics. (in Russian), Nauka, Moscow (1988)
4. Polubarinova-Kochina, P.Ya.: Theory of Groundwater Movement. (in Russian), Nauka, Moscow (1977)
5. Mityaev, A.N.: Pushing Complex of Underground Waters and Wells. (in Russian), Izv. AN SSSR Otdel Techn. Nauk, **9**, (1947) 1069-1088
6. Sanchez-Palencia, E.: Non-Homogeneous Medium and Vibration Theory. Lecture Notes in Physics **127**, Springer-Verlag, New York (1980)
7. Lomakin, E.A., Mironenko, V.A., Shestakov, V.M.: Numerical Modeling of Geofiltration. (in Russian), Nedra, Moscow (1988)
8. Samarskij, A.A.: Theory of Difference Schemes. (in Russian), Nauka, Moscow (1983)
9. Hayryan, E.A., Vabishchevich, P.N., Pavluš, M., Fedorov, A.V.: Additive Difference Schemes for Filtration Problems in Multilayer Systems. (in Russian), preprint of the Joint Institute of Nuclear Research P11-2000-51, Dubna (2000)
10. Peaceman, D.W., Rachford H.H.: The Numerical Solution of Parabolic and Elliptic Differential Equation, J.Soc.Indus.Appl.Math. **3**, (1955) 28-42
11. Young D.M.: Iterative Solutions of Large Linear Systems, Academic Press, New York (1971)
12. Snir, M., Otto, S.W., Huss-Lederman, S., Walker, D.W., Dongarra, J.: MPI. The Complete Reference. The MIT Press, Massachusetts Institute of Technology, Cambridge (USA) (1997)

Stokes Problem for the Generalized Navier-Stokes Equations

Andrei Bourchtein, Ludmila Bourchtein

FAPERGS Science Foundation/Pelotas State University, Department of Mathematics
Campus Universitario da UFPel, 96010-900, Brazil
E-mail: burstein@conex.com.br

Abstract. The generalized Navier-Stokes equations for incompressible viscous flows through isotropic granular porous medium are considered. First Stokes problem is solved applying Laplace transform with respect to time variable and evaluating the inverse transform integrals by the residue calculus. The derived analytical solution includes the classic one as a particular case, that is, it can be obtained from the generalized solution by putting the porosity parameter equal to 1. The use of the derived exact solutions for benchmarking purposes is described.

1. Introduction

Due to nonlinearity of the Navier-Stokes equations only a small number of exact solutions have been found. The most recent reviews of analytical solutions of the Navier-Stokes equations and its classification were given by Wang [7,8]. These solutions are important because they represent some fundamental fluid flows and serve for checking the accuracy of approximate methods, in particular, numerical schemes. We consider one model of laminar flows through granular porous medium which can be represented in the form of the generalized Navier-Stokes equations. It is useful to generalize some known analytical solutions of the Navier-Stokes equations to the case of the considered model. Couette-Poiseuille, Hagen-Poiseuille and Beltrami generalized solutions were obtained by Bourchtein et al. [1]. Here we derive the Stokes generalized solution and show its possible application for benchmarking.

2. Differential problem

The generalized Navier-Stokes equations applied to the description of incompressible viscous laminar flows through a rigid isotropic granular porous medium have been developed by DuPlessis and Masliyah [3]. The advantages of their model are its applicability to granular porous media over the entire porosity range and simple adaptability to numerical simulations. These equations have the following form:

P.M.A. Sloot et al. (Eds.): ICCS 2002, LNCS 2331, pp. 813–819, 2002.

$$\rho \mathbf{V}_t + \rho(\mathbf{V}\cdot\nabla)\frac{\mathbf{V}}{n} = -n\nabla p + \mu\nabla^2\mathbf{V} - \mu F\mathbf{V} - n\rho\mathbf{g}, \ \nabla\cdot\mathbf{V} = 0 \tag{1}$$

Here, the common denotations are used for fluid and porous medium characteristics: $\mathbf{V}$ is a fluid velocity vector, p is a pressure; g is the gravitational force. The fluid is specified by definition of density ρ and dynamic viscosity μ. The characteristics of porous medium are porosity n and porosity function F. Porosity n is defined as a ratio of volume of the void space to the bulk volume of a porous medium and it changes from zero to one ($n \in (0,1]$). Function F represents an additional drag force, which describes influence of porous medium on flow. This function depends on porosity n only, it is continuous, decreasing and positive on interval $(0,1]$. F becomes infinite as n approaches zero and its limit is equal to 0 as n approaches 1. Comparing this system with the usual incompressible Navier-Stokes equations, one can see that the latter is the particular case of model (1) when $n = 1$.

Considering the primitive system in 4D domain $[0,T]\times\overline{\Omega}$, where $\overline{\Omega} = \Omega + \partial\Omega$ is a 3D space bounded domain with boundary $\Gamma = \partial\Omega$, we have to specify the initial and boundary conditions to define the unique solution:

$$\mathbf{V} = \mathbf{V}_0 \text{ on } \overline{\Omega} \text{ at } t=0 \tag{2}$$

(initial condition, $\mathbf{V}_0$ is the given function of spatial variables),

$$\mathbf{V} = \mathbf{V}_\Gamma \text{ on } \Gamma = \partial\Omega \text{ for all } t \in [0,T] \tag{3}$$

(no-slip boundary condition, $\mathbf{V}_\Gamma$ is the given function of the time variable and two spatial variables). These conditions have to subject to some constraints specified in [2,5]. We assume that the last constraints are satisfied by the appropriate choice of initial and boundary conditions, so we will not mention these conditions anymore.

3. Solution of Stokes problem

We consider the problem of unsteady flow of a semi-infinite fluid, which is caused by a plate moving in its own plane. There are two independent variables in this problem: the time t and the spatial variable y (a figure can be found in [9]). Using the simplification conditions $v \equiv 0, w \equiv 0$, primitive system (1) can be reduced to the following initial boundary value problem involving velocity component u:

$$u_t = \nu u_{yy} - \nu F u, \ u(t,0) = A, \ u(t,+\infty) = 0, \ u(0,y) = 0. \tag{4}$$

Here $\nu = \mu/\rho$ is the kinematic viscosity. In the classic case (when $F = 0$) the solution of this problem can be found by the similarity method, defining the only independent variable of the problem and reducing the primitive bivariate partial differential equation to an ordinary one [7,9]. Unfortunately, an introduction of additional porous drag force eliminates this way of derivation of solution, because there are two dimensionless combinations of primitive variables in this case. So we have been forced to choose another method. Due to linearity and parabolicity of the u-component equation (4) the Laplace transform with respect to variable t

$$U(s,y)=L[u(t,y)]\equiv\int_0^\infty u(t,y)e^{-st}dt \tag{5}$$

is suitable to be applied. This transform reduces (4) to an ordinary differential problem:

$$U_{yy}=\frac{s}{\nu}U+FU,\ U(s,0)=\frac{A}{s},\ U(s,+\infty)=0. \tag{6}$$

A solution of (6) is

$$U=\frac{A}{s}e^{-\sqrt{\frac{s}{\nu}+F}y}. \tag{7}$$

Applying inverse transform we obtain solution of (4) in the form

$$u(t,y)=L^{-1}[U(s,y)]\equiv\frac{1}{2\pi}\int_{c-i\infty}^{c+i\infty}U(s,y)e^{st}ds=\frac{1}{2\pi}\int_{c-i\infty}^{c+i\infty}\frac{A}{s}e^{-\sqrt{\frac{s}{\nu}+F}y}e^{st}ds. \tag{8}$$

To evaluate the integral on the right hand side of (8) we change variable s and parameter y:

$$z=s+\nu F,\ d=\frac{y}{\nu} \tag{9}$$

and we get

$$\frac{1}{2\pi}\int_{c-i\infty}^{c+i\infty}\frac{A}{s}e^{-\sqrt{\frac{\rho}{\mu}s+F}y}e^{st}ds=\frac{1}{2\pi}\int_{c+\nu F-i\infty}^{c+\nu F+i\infty}\frac{A}{z-\nu F}e^{-\sqrt{zd}}e^{zt}e^{-\nu Ft}dz=\frac{Ae^{-\nu Ft}}{2\pi}\int_{c_1-i\infty}^{c_1+i\infty}g(z)dz, \tag{10}$$

where

$$g(z)=\frac{1}{z-\nu F}e^{-\sqrt{zd}}e^{zt},\ c_1=c+\nu F. \tag{11}$$

Function $\sqrt{z}$ can be made a single-valued regular function of z, if a cut along the negative real axis is applied to the z-plane. In this case, function $e^{-\sqrt{zd}}$ will also be a single-valued regular function. We choose the branch of $\sqrt{z}$ for which $\sqrt{1}=1$.

Let us consider a closed contour $C=C_R \cup C_+ \cup C_- \cup C_r \cup C_l$ shown in Fig.1, where C_R is the part of the circular arc with radius $R>c_1$ lying on the left of line $\operatorname{Re} z=c_1$, C_r is a circular arc with radius $r<c_1$, C_+ and C_- are the parts of upper and lower edges of the cut along the negative real axis, connecting the circular arcs with radius R and r, C_l is the part of line $\operatorname{Re} z=c_1$ inside the circle with radius R.

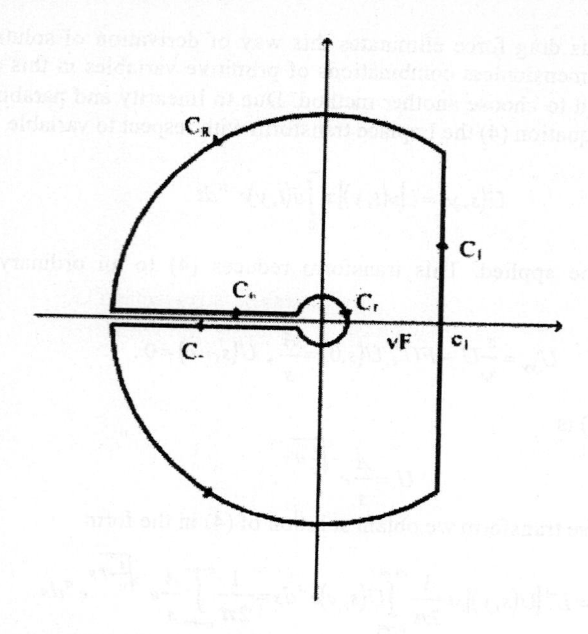

Figure 1. Closed contour C and its partition $C = C_R \cup C_+ \cup C_- \cup C_r \cup C_l$.

The chosen branch of the function $g(z)$ is regular in the simply-connected domain D bounded by closed contour C except at a point $z = \nu F$ where it has a simple pole. Therefore, by the residue theorem,

$$\int_C g(z)dz = 2\pi i \cdot \operatorname*{res}_{z=\nu F} g(z) = 2\pi i \cdot e^{-\sqrt{z}d} e^{zt}\Big|_{z=\nu F} = e^{-\sqrt{\nu F}d} e^{\nu Ft} \tag{12}$$

On the other hand, we can represent the integral along C as the sum of the integrals along respective parts of contour C (see Fig.1):

$$\int_C g(z)dz = \int_{C_R} + \int_{C_+} + \int_{C_-} + \int_{C_r} + \int_{C_l} g(z)dz \tag{13}$$

Formulas (12) and (13) are true for any radius R and r such that $R > c_1 > r$.

Let us evaluate each integral in (13) separately, as R approaches ∞ and r approaches 0:

1) The chosen single-valued branch of function $\sqrt{z}$ has value $\sqrt{R}e^{i\varphi/2}$ at the points of C_R. Then

$$\left|\left.\frac{e^{-\sqrt{z}d}}{z - \nu F}\right|_{C_R}\right| = \left|\frac{e^{-\sqrt{R}d(\cos\varphi/2 + i\sin\varphi/2)}}{Re^{i\varphi} - \nu F}\right| \le \frac{e^{-\sqrt{R}d\cos\varphi/2}}{R - \nu F} \xrightarrow[R\to\infty]{} 0 \tag{14}$$

because

$$-\frac{\pi}{2} \leq \frac{\varphi}{2} \leq \frac{\pi}{2}, \ \cos\frac{\varphi}{2} \geq 0, \ d > 0. \tag{15}$$

Therefore, by Jordan's lemma [6],

$$\int_{C_R} g(z)dz \xrightarrow[R\to\infty]{} 0. \tag{16}$$

2,3) Now we evaluate together the second and third integrals in (13)

$$\int_{C_+} g(z)dz + \int_{C_-} g(z)dz = \int_R^r \frac{1}{-a-\sqrt{F}} e^{-i\sqrt{ad}} e^{-at}(-da) + \int_r^R \frac{1}{-a-\sqrt{F}} e^{i\sqrt{ad}} e^{-at}(-da) = \tag{17}$$

$$= 2i\int_r^R \frac{\sin\sqrt{ad}}{a+\sqrt{F}} e^{-at} da \xrightarrow[R\to\infty,\, r\to 0]{} 2i\int_0^{+\infty} \frac{\sin\sqrt{ad}}{a+\sqrt{F}} e^{-at} da.$$

Making the substitution $a = x^2$, we obtain

$$\int_0^{+\infty} \frac{\sin\sqrt{ad}}{a+\sqrt{F}} e^{-at} da = 2\int_0^{+\infty} \frac{\sin xd}{x^2+\sqrt{F}} e^{-tx^2} x dx. \tag{18}$$

The last integral can be found in [4]:

$$2\int_0^{+\infty} \frac{\sin xd}{x^2+\sqrt{F}} e^{-tx^2} x dx = -\frac{\pi}{2} e^{t\sqrt{F}} \left[2\sinh\sqrt{\sqrt{F}}d + e^{-\sqrt{\sqrt{F}}d} \Phi\left(\sqrt{t\sqrt{F}} - \frac{d}{2\sqrt{t}}\right) - e^{\sqrt{\sqrt{F}}d} \Phi\left(\sqrt{t\sqrt{F}} + \frac{d}{2\sqrt{t}}\right) \right] \tag{19}$$

where

$$\Phi(x) = \frac{2}{\sqrt{\pi}} \int_0^x e^{-t^2} dt \tag{20}$$

is the probability integral.

4) Since the chosen branch of $g(z)$ is regular in neighborhood of the point $z = 0$ cutting along negative real axis, there exists a number M such that $|g(z)| \leq M$ for all z in this neighborhood. Therefore

$$\left| \int_{C_r} g(z)dz \right| \leq \int_{C_r} |g(z)||dz| \leq M \cdot 2\pi r \xrightarrow[r\to 0]{} 0 \tag{21}$$

5) It is evident that

$$\int_{C_l} g(z)dz \xrightarrow[R\to\infty]{} \int_{c_1-i\infty}^{c_1+i\infty} g(z)dz. \tag{22}$$

Finally, putting the right hand sides of (12) and (13) to be equal, calculating limits as R approaches ∞ and r approaches 0 and using the results (16), (19), (21), (22), we obtain

$$2i\pi e^{-\sqrt{F}d}e^{Ft} = -i\pi e^{\sqrt{F}}\left[2\sinh\sqrt{F}d + e^{\sqrt{F}d}\Phi\left(\sqrt{tF}-\frac{d}{2\sqrt{t}}\right) - e^{\sqrt{F}d}\Phi\left(\sqrt{tF}+\frac{d}{2\sqrt{t}}\right)\right] + \int_{q-i\infty}^{q+i\infty} g(z)dz \tag{23}$$

Therefore, returning to (8), we can express the solution of (4) in the following form

$$u(t,y) = \frac{Ae^{-Ft}}{2\pi}\left\{2i\pi e^{-y\sqrt{F/v}}e^{Ft} + i\pi e^{F}\left[2\sinh y\sqrt{F/v} + e^{-y\sqrt{F/v}}\Phi\left(\sqrt{tF}-\frac{y}{2\sqrt{tv}}\right) - e^{y\sqrt{F/v}}\Phi\left(\sqrt{tF}+\frac{y}{2\sqrt{tv}}\right)\right]\right\} \tag{24}$$

or, simplifying,

$$u(t,y) = A\cosh y\sqrt{F/v} + \frac{A}{2}\left[e^{-y\sqrt{F/v}}\Phi\left(\sqrt{tF}-\frac{y}{2\sqrt{tv}}\right) - e^{y\sqrt{F/v}}\Phi\left(\sqrt{tF}+\frac{y}{2\sqrt{tv}}\right)\right]. \tag{25}$$

It is evident that the initial condition is satisfied in limit form as t approaches 0 because

$$\Phi\left(\sqrt{tvF}-\frac{y}{2\sqrt{tv}}\right)\xrightarrow[t\to+0]{}-1,\quad \Phi\left(\sqrt{tvF}+\frac{y}{2\sqrt{tv}}\right)\xrightarrow[t\to+0]{}1. \tag{26}$$

To transform this generalized solution to classic one it is sufficient to set $F=0$ ($n=1$).

There are detailed tables and different programs for calculating the probability integral values, so the derived solution can be evaluated with any order of precision.

The application of this solution to benchmarking is simple. It is sufficient to introduce bounded domain with respective boundary conditions. For example, considering equation (1) in rectangular parallelepiped and using Cartesian coordinates, we can complete it by initial conditions (2) and boundary conditions (3) specified as follows:

$$\begin{gathered} v_{t=0}=v_0=0,\ w_{t=0}=w_0=0,\ u_{t=0}=u_0=0; \\ v_\Gamma=0,\ w_\Gamma=0,\ u_{z=0}=u_{z=c}=0,\ u_{x=0}=u_{x=a}=u(t,y),\ u_{y=0}=A,\ u_{y=b}=u(t,b). \end{gathered} \tag{27}$$

4. Conclusion

First Stokes problem for generalized Navier-Stokes equations is considered. Laplace transform is applied to reduce this problem to the set of ordinary differential problems which permit simple solutions. Inverse transform is calculated using some results of the theory of complex variable functions. A possible application of obtained exact solution to benchmarking is discussed.

Acknowledgments

We are grateful to the Brazilian foundation FAPERGS which supported this work by a grant 01/60053.9.

References

1. Bourchtein, A., Bourchtein, L., Lukaszczyk, J.P.: Numerical simulation of incompressible flows through granular porous media. Applied Numerical Mathematics **40** (2002) 291-306
2. Constantin, P., and Foias, C. Navier-Stokes equations. University of Chicago Press, Chicago (1989)
3. DuPlessis, P.J., and Masliyah, J.H.: Flow through isotropic granular porous media. Transport in Porous Media, **6** (1991) 207-221
4. Gradshteyn, I.S., Ryzhik, I.M.: Tables of integrals, series and products. Academic Press, Boston (1994)
5. Gresho, P.M.: Incompressible fluid dynamics: some fundamental formulation issues. Annual Reviews of Fluid Mechanics, **23** (1991) 413-453
6. Shilov, G.E.: Elementary real and complex analysis. Dover Publications, New York (1996)
7. Wang, C.Y.: Exact solutions of the unsteady Navier-Stokes equations. Applied Mechanics Review, **42** (1989) 269-282
8. Wang, C.Y.: Exact solutions of the steady-state Navier-Stokes equations. Annual Reviews of Fluid Mechanics, **23** (1991) 159-177
9. Yih, C.-S.: Fluid mechanics. McGraw-Hill, New York (1977)

Domain Decomposition Algorithm for Solving Contact of Elastic Bodies

Josef Daněk

University of West Bohemia, Univerzitní 22, 306 14 Pilsen, Czech Republic
danek@kma.zcu.cz

Abstract. A nonoverlapping domain decomposition algorithm of Neumann-Neumann type for solving variational inequalities arising from the elliptic boundary value problems in two dimensions with unilateral boundary condition is presented. First, the linear auxiliary problem, where the inequality condition is replaced by the equality condition, is solved. In the second step, the solution of the auxiliary problem is used in a successive approximations method. In these solvers, a preconditioned conjugate gradient method with Neumann-Neumann preconditioner is used for solving the interface problems, while local problems within each subdomain are solved by direct solvers. A convergence of the iterative method and results of computational test are reported.

1 Equilibrium of a system of bodies in contact

We consider a system of elastic bodies decomposed into subdomains each of which occupies, in reference configuration, a domain Ω_i^M in $\mathbb{R}^2$, $i = 1, \ldots, I_M$, $M = 1, \ldots, \mathcal{J}$, with sufficiently smooth boundary $\partial\Omega_i^M$. Suppose that boundary $\bigcup_{M=1}^{\mathcal{J}} \partial\Omega^M$ consists of four disjoint parts Γ_u, Γ_τ, Γ_c and Γ_o and that the displacements $u : \Gamma_u \to \mathbb{R}^2$ and forces $P : \Gamma_\tau \to \mathbb{R}^2$ are given. The part Γ_c denote the part of boundary that may get into unilateral contact with some other subdomain and the part Γ_o denote the part of boundary on that is prescribed the condition of the bilateral contact.

We shall look for the displacements that satisfy the conditions of equilibrium in the set $K = \{v \in V |\, v_n^k + v_n^l \leq 0 \text{ on } \Gamma_c\}$ of all kinematically admissible displacements $v \in V$, $V = \{v \in \mathcal{H}^1(\Omega) |\, v = u_0 \text{ on } \Gamma_u, v_n = 0 \text{ on } \Gamma_o\}$, $\mathcal{H}^1(\Omega) = [H^1(\Omega_1^1)]^2 \times \ldots \times [H^1(\Omega_{I_\mathcal{J}}^\mathcal{J})]^2$ is standard Sobolev space. The displacement $u \in K$ of the system of bodies in equilibrium then minimizes the energy functional $\mathcal{L}(v) = \frac{1}{2}a(v, v) - L(v)$:

$$\mathcal{L}(u) \leq \mathcal{L}(v) \text{ for any } v \in K. \tag{1}$$

Conditions that guarantee existence and uniqueness of the solution may be expressed in terms of coercivity of $\mathcal{L}$ and may be found, for example, in [1].

We define $\Gamma_i^M = \partial\Omega_i^M \setminus \partial\Omega^M$ and the interface $\Gamma = \bigcup_{M=1}^{\mathcal{J}} \bigcup_{i=1}^{I_M} \Gamma_i^M$. Let us introduce

$$T^M = \{j \in \{1, \ldots, I_M\} : \quad \Gamma_c \cap \partial\Omega_j^M = \emptyset\}, \qquad M = 1, \ldots, \mathcal{J}. \tag{2}$$

P.M.A. Sloot et al. (Eds.): ICCS 2002, LNCS 2331, pp. 820–829, 2002.

The number of a separate subset Γ_c is P_c, i.e. $\Gamma_c = \bigcup_{j=1}^{P_c} \Gamma_{cj}$. Further, we denote

$$\Omega^{*j} = \{x \in \bigcup_{M=1}^{\mathcal{J}} \bigcup_{i=1}^{I_M} \Omega_i^M : \quad \partial\Omega_i^M \cap \Gamma_{cj} \neq \emptyset\}, \qquad j = 1, \dots, P_c, \tag{3}$$

$$\vartheta^j = \{[i, M] : \partial\Omega_i^M \cap \Gamma_{cj} \neq \emptyset\}, \qquad j = 1, \dots, P_c, \tag{4}$$

i.e. $\Omega^{*j} = \bigcup_{[i,M]\in\vartheta^j} \Omega_i^M$, $j = 1, \dots, P_c$. We suppose that $\Gamma \cap \Gamma_c = \emptyset$ then $V_\Gamma = \gamma K|_\Gamma = \gamma V|_\Gamma$ for trace operator $\gamma : [H^1(\Omega_i^M)]^2 \to [L^2(\partial\Omega_i^M)]^2$. We suppose that $\gamma^{-1} : V_\Gamma \to V$ is arbitrary linear inverse mapping for which

$$\sum_{M=1}^{\mathcal{J}} (\gamma^{-1}\bar{v}^M)_n = 0 \quad \forall \bar{v} \in V_\Gamma \quad \text{on } \Gamma_c. \tag{5}$$

After denoting restrictions $\bar{R}_i^M : V_\Gamma \to \Gamma_i^M$, $L_i^M : L^M \to \Omega_i^M$, $a_i^M(.,.) : a^M(.,.) \to \Omega_i^M$, $V(\Omega_i^M) : V(\Omega^M) \to \Omega_i^M$ and introduction

$$V^0(\Omega_i^M) = \{v \in V \,|\, v = 0 \text{ on } (\bigcup_{M=1}^{\mathcal{J}} \Omega^M) \setminus \Omega_i^M\},$$

we can formulate the theorem 1.

Theorem 1. *A function $u \in K$ is the solution of the global problem (1) if and only if the function u satisfies:*

1.

$$\sum_{M=1}^{\mathcal{J}} \sum_{i=1}^{I_M} (a_i^M(u_i^M(\bar{u}), \gamma^{-1}\bar{w}) - L_i^M(\gamma^{-1}\bar{w})) = 0 \quad \forall \bar{w} \in V_\Gamma, \ \bar{u} \in V_\Gamma, \tag{6}$$

for the trace $\bar{u} = \gamma u|_\Gamma$ on the interface Γ.

2. Its rescriction $u_i^M(\bar{u}) \equiv u|_{\Omega_i^M}$ satisfies following conditions:

a)

$$\begin{aligned} a_i^M(u_i^M(\bar{u}), \phi_i^M) &= L_i^M(\phi_i^M) \quad \forall \phi_i^M \in V^0(\Omega_i^M), \\ u_i^M(\bar{u}) &\in V(\Omega_i^M), \ \gamma u_i^M(\bar{u})|_{\Gamma_i^M} = \bar{R}_i^M \bar{u}, \end{aligned} \tag{7}$$

for $i \in T^M$, $M = 1, \dots, \mathcal{J}$,

b)

$$\begin{aligned} \sum_{[i,M]\in\vartheta^j} a_i^M(u_i^M(\bar{u}), \phi_i^M) &\geq \sum_{[i,M]\in\vartheta^j} L_i^M(\phi_i^M) \\ &\forall \phi \equiv (\phi_i^M, [i, M] \in \vartheta^j), \ \phi_i^M \in V^0(\Omega_i^M) \end{aligned} \tag{8}$$

such that

$$u + \phi \in K;$$

$$\gamma u_i^M(\bar{u})|_{\Gamma_i^M} = \bar{R}_i^M \bar{u} \quad \textit{for} \quad [i, M] \in \vartheta^j,$$

for $j = 1, \dots, P_c$.

2 The Schur complements

We now want to write the interface problem (6) in operator form. For this purpose, we first introduce additional notation. We introduce the local trace spaces

$$V_i^M = \{\gamma v|_{\Gamma_i^M} \mid v \in K\} = \{\gamma v|_{\Gamma_i^M} \mid v \in V\} \tag{9}$$

and the extension $Tr_{iM}^{-1} : V_i^M \to V(\Omega_i^M)$ defined by

$$\begin{aligned}
&\gamma(Tr_{iM}^{-1}\bar{u}_i^M)|_{\Gamma_i^M} = \bar{u}_i^M, \qquad i = 1, \dots, I_M,\ M = 1, \dots, \mathcal{J}, \\
&a_i^M(Tr_{iM}^{-1}\bar{u}_i^M, v_i^M) = 0 \qquad \forall v_i^M \in V^0(\Omega_i^M),\ Tr_{iM}^{-1}\bar{u}_i^M \in V(\Omega_i^M), \\
&\qquad\qquad\qquad\qquad\qquad \text{for } i \in T^M,\ M = 1, \dots, \mathcal{J}.
\end{aligned} \tag{10}$$

For subdomains Ω^{*j}, $j = 1, \dots, P_c$, we completed definition Tr_{iM}^{-1} with boundary condition

$$\sum_{[i,M]\in\vartheta^j} (Tr_{iM}^{-1}\bar{u}_i^M)_n = 0 \quad \text{on } \Gamma_{cj}, \ \text{for } j = 1, \dots, P_c,$$

i.e.

$$\begin{aligned}
&\sum_{[i,M]\in\vartheta^j} a_i^M(Tr_{iM}^{-1}\bar{u}_i^M, v_i^M) = 0 \quad \forall (v_i^M,\ [i,M] \in \vartheta^j) : \ v_i^M \in V^0(\Omega_i^M), \\
&\qquad\qquad \text{so that} \quad \sum_{[i,M]\in\vartheta^j} (v_i^M)_n = 0 \text{ on } \Gamma_{cj}, \quad j = 1, \dots, P_c.
\end{aligned} \tag{11}$$

Definition 1. *The local Schur complement, for $i \in T^M$, $M = 1, \dots, \mathcal{J}$, is operator $S_i^M : V_i^M \to (V_i^M)^*$ defined by*

$$\langle S_i^M \bar{u}_i^M, \bar{v}_i^M \rangle = a_i^M(Tr_{iM}^{-1}\bar{u}_i^M, Tr_{iM}^{-1}\bar{v}_i^M) \quad \forall \bar{u}_i^M, \bar{v}_i^M \in V_i^M. \tag{12}$$

In matrix form, we have

$$S_i^M \bar{U}_i^M = (\bar{A}_{iM} - B_{iM}^T \mathring{A}_{iM}{}^{-1} B_{iM})\, \bar{U}_i^M, \tag{13}$$

where we decompose the degrees of freedom U_i of u_i into internal degrees of freedom $\mathring{U}_i^M$ and interface degrees of freedom $\bar{U}_i^M$:

$$U_i^M = \left[\mathring{U}_i^M,\ \bar{U}_i^M\right]^T.$$

With this decomposition, the matrix representation of $a_i^M(.,.)$ on $H^1(\Omega_i^M)$ take the form

$$A_{iM} = \begin{bmatrix} \mathring{A}_{iM} & B_{iM} \\ B_{iM}^T & \bar{A}_{iM} \end{bmatrix}. \tag{14}$$

Definition 2. *The combined local Schur complement, for subdomains* Ω^{*j}, $j = 1,\dots,P_c$, *is operator*

$$S_{*j}: (V_i^M, [i,M] \in \vartheta^j) \to (V_i^M, [i,M] \in \vartheta^j)^*, \quad j = 1,\dots,P_c,$$

defined by

$$\langle S_{*j}(\bar{u}_i^M, [i,M] \in \vartheta^j), (\bar{v}_i^M, [i,M] \in \vartheta^j)\rangle = \sum_{[i,M]\in\vartheta^j} a_i^M(u_i^M(\bar{u}_i^M), Tr_{iM}^{-1}\bar{v}_i^M)$$
$$\forall (\bar{v}_i^M, [i,M] \in \vartheta^j) \in (V_i^M, [i,M] \in \vartheta^j), \tag{15}$$

where $u_i^M(\bar{u}_i^M)$ *is the solution of the problem (8) and* $\bar{R}_i^M \bar{u} \equiv \bar{u}_i^M$, $[i,M] \in \vartheta^j$.

Lemma 1. *The condition (6) for the function* $\bar{u}$ *on interface* Γ *is equivalent to the condition (16):*

$$\sum_{M=1}^{\mathcal{J}} \sum_{i\in T^M} \langle S_i^M \bar{u}_i^M, \bar{w}_i^M\rangle + \sum_{j=1}^{P_c} \langle S_{*j}(\bar{u}_i^M, [i,M] \in \vartheta^j), (\bar{w}_i^M, [i,M] \in \vartheta^j)\rangle =$$
$$= \sum_{M=1}^{\mathcal{J}} \sum_{i\in I_M} L_i^M(Tr_{iM}^{-1}\bar{w}_i^M), \quad \forall \bar{w} \in V_\Gamma, \text{ where } \bar{w}_i^M = \bar{R}_i^M \bar{w},\ \bar{u}_i^M = \bar{R}_i^M \bar{u}, \tag{16}$$

by using the local Schur complements.

We rewrite the condition (16) in the form

$$S_0\bar{U} + S_{KON}\bar{U} = F, \tag{17}$$

where

$$S_0 = \sum_{M=1}^{\mathcal{J}} \sum_{i\in T^M} (\bar{R}_i^M)^T S_i^M \bar{R}_i^M, \quad S_{KON} = \sum_{j=1}^{P_c} \bar{R}_{*j}^T S_{*j} \bar{R}_{*j},$$

$$F = \sum_{M=1}^{\mathcal{J}} \sum_{i\in I_M} (\bar{R}_i^M)^T (Tr_{iM}^{-1})^T L_i^M$$

and

$$\bar{R}_{*j}\bar{u} = (\bar{R}_i^M \bar{u}, [i,M] \in \vartheta^j)^T, \quad \bar{u} \in V_\Gamma, \quad \forall j = 1,\dots,P_c.$$

By reason that operator S_{KON} is nonlinear, we solve the equation (17) successive aproximations method. We choose the solution of the auxiliary linear problem as an initial aproximation $\bar{U}^0$. In the auxiliary problem we replace the set K by

$$K^0 = \{v \in V | \sum_{[i,M]\in\vartheta^j} (v_i^M)_n = 0 \quad \text{on } \Gamma_{cj}\}$$

and we obtain

$$u_0 = \arg\min_{v\in K^0} \mathcal{L}(v),$$
$$\bar{U}^0 = \gamma u^0|_\Gamma.$$

Now we come back to the equation (17) and we compute $\bar{U}^k$as the solution of the linear problem

$$S_0\bar{U}^k = F - S_{KON}\bar{U}^{k-1}, \qquad k = 1,2,\dots \tag{18}$$

3 The linearized problem

We solve the variational equation

$$u^0 \in K^0, \quad D\mathcal{L}(u^0, v) = 0 \quad \forall v \in K^0. \tag{19}$$

For problem (19) we can describe the analogy of theorem 1 with one different in case 2b) where inequality is replaced by equality on K^0. For solution of this variational equality we define combined local Schur complement S^0_{*j}, $j = 1, \ldots, P_c$ same as in definition 2.

Definition 3. *We define a global Schur complement:*

$$S = \sum_{j=1}^{P_c} \bar{R}^T_{*j} S^0_{*j} \bar{R}_{*j} + \sum_{M=1}^{\mathcal{J}} \sum_{i \in T^M} (\bar{R}^M_i)^T S^M_i \bar{R}^M_i \tag{20}$$

and the condition (6) on the interface Γ has form

$$S\bar{U} = F \tag{21}$$

in dual space $(V_\Gamma)^*$.

The equation (21) we solve by a conjugate gradient method with Neumann-Neumann preconditioner. This method does not require the explicit construction of the local Schur complement matrix S^M_i but does require an efficient preconditioner $\mathcal{M}^{-1}$. Its inverse $(S^M_i)^{-1}$, resp. $(S^0_{*j})^{-1}$ simply consists in associating to the generalized derivative $g \in (V^M_i)^*$ the trace $\gamma\phi^M_i$ on Γ^M_i of the solution ϕ^M_i of the corresponding Neumann problem.

Definition 4. *We define an injection*

$$\begin{aligned} &D^M_i : \quad V^M_i \to V_\Gamma, \quad i \in T^M, \quad M = 1, \ldots, \mathcal{J}, \\ &D_{*j} : \quad (V^M_i, [i, M] \in \vartheta^j) \to V_\Gamma, \ D_{*j} = (D^M_i, [i, M] \in \vartheta^j), \quad j = 1, \ldots, P_c, \end{aligned} \tag{22}$$

such that on each interface degree of freedom is

$$D^M_i \bar{v}(P_l) = \bar{v}(P_k) \frac{\varrho^M_i}{\varrho_T}, \quad i = 1, 2, \ldots I_M, \ M = 1, \ldots, \mathcal{J}, \tag{23}$$

if the lth degree of freedom of V_Γ corresponds to the kth degree of freedom of V^M_i and

$$D^M_i \bar{v}(P_l) = 0, \quad \text{if not.} \tag{24}$$

Here ϱ^M_i is a local measure of the stiffness of subdomain Ω^M_i (for example an average Young modulus on Ω^M_i) and

$$\varrho_T = \sum_{P_l \in \bar{\Omega}^M_j} \varrho^M_j \tag{25}$$

is the sum of ϱ^M_j on all subdomains na Ω^M_j containing P_l.

The Neumann-Neumann precoditioner supposes that the solution of each local Neumann problem is uniquely defined, whereas rigid body motions are possible. This weakness can be fixed by replacing $(S_i^M)^{-1}$, resp. $(S_{*j}^0)^{-1}$ by a regularized inverse $(\tilde{S}_i^M)^{-1}$, resp. $(\tilde{S}_{*j}^0)^{-1}$. We introduce on each subdomain Ω_i^M, resp. Ω^{*j} a small local coarse space Z_i^M, resp. Z^{*j} containing all rigid body motion.

The general trick to upgrade the original preconditioner then consists in adding to the initial local contribution ϕ_i^M, resp. ϕ_j a "bad" $z_i^M \in Z_i^M$, resp. $z_j \in Z^{*j}$ which is chosen in order to get the smallest difference $(\mathcal{M}^{-1} - S^{-1})$.

We suppose that L satisfies the invariance property

$$\langle L, D_i^M \gamma z_i^M \rangle = 0 \qquad \forall z_i^M \in Z_i^M, \quad i \in T^M, \quad M = 1, \ldots, \mathcal{J}, \tag{26}$$

$$\langle L, D_{*j} \gamma z_j \rangle \equiv \sum_{[i,M] \in \vartheta^j} \langle L, D_i^M \gamma z_i^M \rangle = 0 \qquad \forall z_j \in Z^{*j}, \quad j = 1, \ldots, P_c. \tag{27}$$

We introduce a closed orthogonal complement space $Q(\Omega_i^M)$ of Z_i^M in $V(\Omega_i^M)$ and a closed orthogonal complement space $Q(\Omega^{*j})$ of Z^{*j} in $\hat{V}_j$ where

$$\hat{V}_j = \{(v_i^M, [i,M] \in \vartheta^j) | v_i^M \in V(\Omega_i^M), \quad \sum_{[i,M] \in \vartheta^j} (v_i^M)_n = 0 \text{ on } \Gamma_{cj}\}.$$

Let then $\phi_i^{0M} \in Q(\Omega_i^M)$ be the particular solution of the variational problem defined by

$$a_i^M(\phi_i^{0M}, v_i^M) = \langle L, D_i^M (\gamma v_i^M)|_{\Gamma_i^M} \rangle \qquad \forall v_i^M \in V(\Omega_i^M) \tag{28}$$

and $\phi_{*j}^0 = (\phi_i^{0M}, [i,M] \in \vartheta^j) \in Q(\Omega^{*j})$ be the particular solution of the variational problem defined by

$$\sum_{[i,M] \in \vartheta^j} a_i^M(\phi_i^{0M}, v_i^M) = \sum_{[i,M] \in \vartheta^j} \langle L, D_i^M (\gamma v_i^M)|_{\Gamma_i^M} \rangle \qquad \forall v_j \in \hat{V}_j. \tag{29}$$

Equations (28), (29) are well posed varitional problems set on $Q(\Omega_i^M)$, $Q(\Omega^{*j})$.

Definition 5. *We define Neumann-Neumann preconditioner $\mathcal{M}^{-1}(z^0)$ by*

$$\mathcal{M}^{-1}(z^0)L = \sum_{M=1}^{\mathcal{J}} \sum_{i=1}^{I_M} D_i^M \gamma(\phi_i^{0M} + z_i^{0M})|_{\Gamma_i^M}, \tag{30}$$

with the solution z_i^{0M} of the minimization problem

$$z^0 = \arg\min_{z \in \Pi Z} \underbrace{\langle S(\mathcal{M}^{-1}(z) - S^{-1})L, (\mathcal{M}^{-1}(z) - S^{-1})L \rangle}_{J(z)}, \tag{31}$$

$$\Pi Z \equiv \Big(\bigotimes_{i \in T^M, M=1,\ldots,\mathcal{J}} (Z_i^M) \Big) \times \Big(\bigotimes_{j=1,\ldots,P_c} (Z^{*j}) \Big).$$

4 Successive approximations method

Now we solve, by the successive approximations method, the equation (18). We must effectively compute the solution $\bar{U}^k$ of the linear problem

$$S_0\bar{U}^k = b^k, \tag{32}$$

with

$$S_0 = \sum_{M=1}^{\mathcal{J}} \sum_{i \in T^M} (\bar{R}_i^M)^T S_i^M \bar{R}_i^M, \qquad b^k = F - S_{KON}\bar{U}^{k-1},$$

$$F = \sum_{M=1}^{\mathcal{J}} \sum_{i \in I_M} (\bar{R}_i^M)^T (Tr_{iM}^{-1})^T L_i^M, \qquad S_{KON} = \sum_{j=1}^{P_c} \bar{R}_{*j}^T S_{*j} \bar{R}_{*j}.$$

The equation (32) we solve by a preconditioned conjugate gradient method.

Definition 6. *We define an injection*

$$D_i^M : \quad V_i^M \to V_\Gamma, \quad i \in T^M, \quad M = 1, \ldots, \mathcal{J}, \tag{33}$$

such that on each interface degree of freedom is

$$D_i^M \bar{v}(P_l) = \bar{v}(P_k), \quad \text{if } P_k \in \Gamma_i^M \subset \partial\Omega^{*j} \text{ for any } j \in \{1, \ldots, P_c\}, \tag{34}$$

$$D_i^M \bar{v}(P_l) = \bar{v}(P_k) \frac{\varrho_i^M}{\varrho_T}, \quad \text{for } P_k \in \Gamma_i^M \not\subset \partial\Omega^{*j} \;\; \forall j = 1, \ldots, P_c, \tag{35}$$

if the lth degree of freedom of V_Γ corresponds to the kth degree of freedom of V_i^M and

$$D_i^M \bar{v}(P_l) = 0, \quad \text{if not.} \tag{36}$$

Let $\phi_i^{0M} \in Q(\Omega_i^M)$ be the particular solution of the variational problem defined by (28). Similarly to the linearized problem we define a preconditioner.

Definition 7. *We define Neumann-Neumann preconditioner $\mathcal{M}_o^{-1}$ by*

$$\mathcal{M}_o^{-1}(z^0)L = \sum_{M=1}^{\mathcal{J}} \sum_{i \in T^M} D_i^M \gamma(\phi_i^{0M} + z_i^{0M})|_{\Gamma_i^M}, \tag{37}$$

with the solution z_i^{0M} of the minimization problem

$$z^0 = \arg \min_{z \in \Pi_o Z} \langle S_0(\mathcal{M}_o^{-1}(z) - S_0^{-1})L, (\mathcal{M}_o^{-1}(z) - S_0^{-1})L \rangle, \tag{38}$$

$$\Pi_o Z \equiv \bigotimes_{i \in T^M, M=1,\ldots,\mathcal{J}} (Z_i^M).$$

We introduce the coarse trace space

$$V_{oH} = \sum_{M=1}^{\mathcal{J}} \sum_{i \in T^M} D_i^M \gamma Z_i^M, \tag{39}$$

a set $V_{oH}^{\perp} \subset (V_\Gamma)^*$ given by

$$L \in V_{oH}^{\perp} \Leftrightarrow \langle L, z \rangle = 0 \qquad \forall z \in V_{oH}.$$

A convergence theorem requires to introduce some definitions. Let Θ be an ortogonal complement of V_{oH} in V_Γ. We introduce seminorms

$$|\bar{R}_{*j}\bar{v}|_{a_{*j}} = \sqrt{\sum_{[i,M] \in \vartheta^j} a_i^M (Tr_{iM}^{-1} \bar{R}_i^M \bar{v}, Tr_{iM}^{-1} \bar{R}_i^M \bar{v})}, \qquad j = 1, \ldots, P_c.$$

Lemma 2. *The expression*

$$\|\bar{u}\|_Q^2 = \langle S_0 \bar{u}, \bar{u} \rangle$$

is a norm on Θ where

$$Q = \bigotimes_{i,M:\ i \in T^M;\ M=1,\ldots,\mathcal{J}} Q(\Omega_i^M).$$

Definition 8. *Let $\mathcal{T} : \Theta \to \Theta$ be a mapping defined by*

$$\langle S_0(\mathcal{T}\bar{y}), \bar{v} \rangle = \langle F - S_{KON}(\bar{y}), \bar{v} \rangle \qquad \forall \bar{v} \in \Theta. \tag{40}$$

Theorem 2. *Assume that there exists a constant $\lambda < \frac{1}{\sqrt{2P_c}}$ such that the following condition hold:*

$$|\bar{R}_{*j}\bar{u}|_{a_{*j}} \leq \lambda \|\bar{u}\|_Q, \quad \forall \bar{u} \in \Theta, \quad \forall j \in \{1, \ldots, P_c\}. \tag{41}$$

Then the mapping $\mathcal{T}$ is the contraction on Θ. If $\bar{U}^0 \in \Theta$ then the sequence of the iterations $\bar{U}^k$, computed by (32), are convergent and the limit is a fixed point $\bar{U}$ of the mapping $\mathcal{T}$. The following error estimate holds

$$\|\bar{U}^k - \bar{U}\|_Q \leq \frac{(2\lambda^2 P_c)^k}{1 - 2\lambda^2 P_c} \|\bar{U}^0 - \mathcal{T}\bar{U}^0\|_Q.$$

5 Numerical experiments

In this section, we illustrate the practical behavior of our algorithm on solution of the geomechanical model problem describing loaded tunnel which is crossing by the deep fault and based on the geomechanical theory and models having connection with radioactive waste repositories (see [3]). The introduced algorithm has been implemented in MPI version 1.2.0 by using FORTRAN 77 compiler. A geometry of the problem is in Fig. 1.

Material parameters: 2 regions with Young's modulus $E = 0.52^{10}$[Pa] and Poisson's ratio $\nu = 0, 18$.

Boundary conditions: Prescribed displacement $(-2, 5 \times 10^{-2}, 0)$ [m] on 3-4. Pressure $0, 5 \times 10^7$[Pa] on 1-4. Bilateral contact boundary: 1-2 and 2-3. Unilateral contact boundary: 5-6 and 7-8.

Discretization statistics: 12 subdomains, 5501 nodes, 9676 elements, 10428 unknowns, 89 unilateral contact conditions, 466 interface elements.

Convergence statistics: 19 iterations of the PCG algorithm for the auxiliary problem 14 iterations of the successive approximations method, total 38 iterations of the PCG algorithm for the original problem.

Fig. 2 represents detail of deformations and Fig. 3 demonstrates detail of principal stresses in model problem.

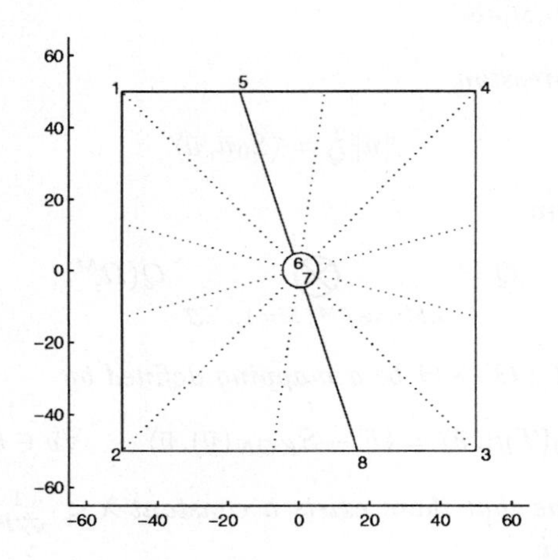

Fig. 1. A geometry of the problem

References

1. Hlaváček, I., Haslinger, J., Nečas, J., Lovíšek, J.: *Solution of Variational Inequalities in Mechanics.* Appl. Math. Sci. 66, Springer-Verlag, New York, 1988
2. Hlaváček, I.: *Domain Decomposition Applied to a Unilateral Contact of Elastic Bodies.* Technical report V-785 ICS AS CR, Prague 1999
3. Nedoma, J.: *Numerical Modelling in Applied Geodynamics.* Wiley, Chichester, 1998
4. Le Tallec, P.: *Domain Decomposition Methods in Computational Mechanics.* Computational Mechanics Advances 1 (pp. 121 - 220), North-Holland, 1994
5. Le Tallec, P., Vidrascu, M.: *Generalized Neumann-Neumann Preconditioners for Iterative Substructuring.* Proceedings of the 9th International Conference on Domain Decomposition Method, pp. 413-425, 1998
6. Daněk, J.: *A solving unilateral contact problems with small range of contact in 2D by using parallel computers.* Ph.D. Thesis, UWB Pilsen, 2001.
7. Daněk, J.: *Domain decomposition method for contact problems with small range contact.* J. Mathematics and Computers in Simulation, Proceedings of the conference Modelling 2001, UWB Pilsen, 2001.

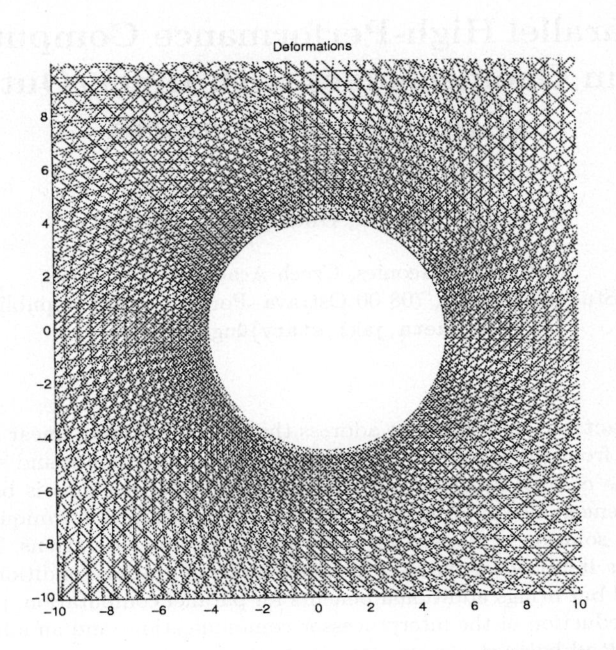

Fig. 2. A detail of deformations in model problem (enlarge factor=10)

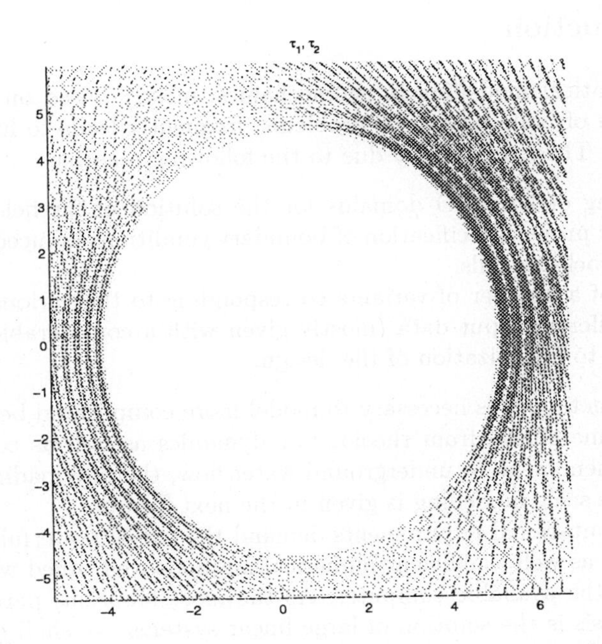

Fig. 3. A detail of principal stresses in neighbourhood of the tunnel. Figure show that maximal pressure is along right-hand side of the tunnel.

Parallel High-Performance Computing in Geomechanics with Inner/Outer Iterative Procedures

Radim Blaheta, Ondřej Jakl, Jiří Starý

Institute of Geonics, Czech Academy of Sciences,
Studentská 1768, 708 00 Ostrava - Poruba, Czech Republic
{blaheta,jakl,stary}@ugn.cas.cz

Abstract. In this paper we address the solution of large linear systems arising from the mathematical modelling in geomechanics and show an example of such modelling. The solution of linear systems is based on displacement decomposition or domain decomposition techniques with inexact solution of the arising subproblems by inner iterations. The use of inner iterations requires a generalization of the preconditioned CG method but brings additional benefits for parallel computation, possibility of reduction of the interprocessor communications and an additional tool of load balance.

1 Introduction

The mathematical modelling in geomechanics, mostly based on the finite element solution of boundary value problems, frequently leads to high computing requirements. These are mainly due to the following reasons:

– considering of large 3D domains for the solution of far field problems as well as for proper specification of boundary conditions induced by the virgin stress in rocks or soils,
– solution of a number of variants corresponding to the various construction stages, different input data (mostly given with a considerable uncertainty) as well as to optimization of the design.

Moreover, sometime it is necessary to model more complicated behaviour of geomaterials, time effects from rheology or dynamics as well as coupling of the mechanical phenomena to underground water flow, thermal loading etc. An example of large scale modelling is given in the next Section.

High computational requirements demand the use of powerful parallel computer systems as well as efficient numerical methods well tuned with the solved problem and the available computer. As the most laborious part of the finite element analysis is the solution of large linear systems, we shall concentrate in this paper to this topic.

For the solution of large linear systems, we suggest here to use decomposition techniques, which lead to a straightforward parallelization. Namely, we

P.M.A. Sloot et al. (Eds.): ICCS 2002, LNCS 2331, pp. 830–839, 2002.

shall consider displacement decomposition and overlapping domain decomposition, see Section 3. For the solution of the arising subproblems, we suggest to use inner iterations, which also allow to optimize the computer time consisting from both computation and interprocessor communication and balance the load of processors. The inner iterations disturb the standard properties of the preconditioners, which should be treated in the outer iterations, see Section 4. The results from testing the described methods on a benchmark problem, which is derived from the modelling described in Section 2, are reported in the last Section.

2 An example of large scale modelling

As an example of large scale geomechanical problem which requires application of high-performance computing, we shall describe the assessment of the development of the stress fields during the mining at the uranium ore deposit Rožná in the Czech Republic. The deposit is situated in metamorphic sedimentary - effusive rocks with uranium mineralization of hydrothermal origin situated in zones of veins arising from longitudinal faults with inclination 45° - 70° to the West, see Figure 1. For the modelling, a 3D domain is selected, which has the

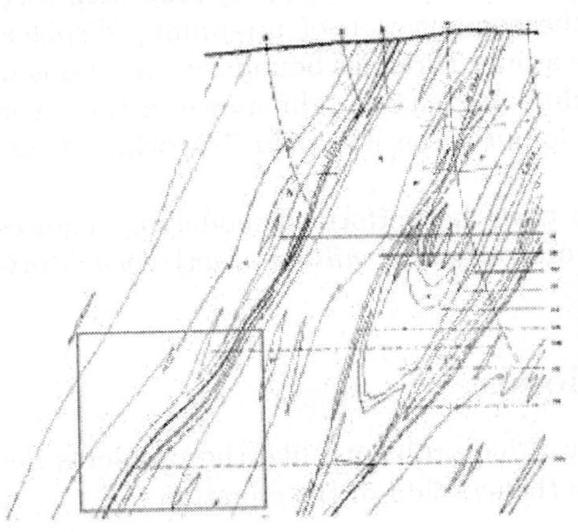

Fig. 1. (left) East-West cross-section of the deposit with 4th zone and the modelled area in the left down corner.

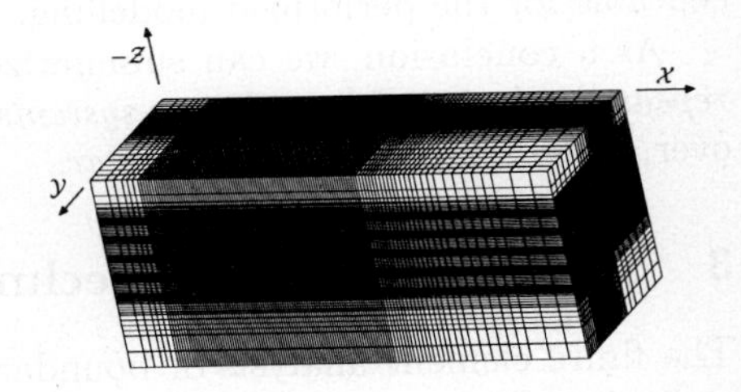

Fig. 2. FE mesh: 124×137×76 nodes.

dimensions 1430×550×600 meters. For the finite element modelling, this domain is divided into hexahedral bricks, which are subsequently divided, each brick into six tetrahedra. The finite element analysis with linear tetrahedral elements uses 3 873 264 degrees of freedom. The exploited mesh can be seen from Figure 2. The loading is given by the weight of the material and the action of the surrounding rocks existing under the pre-mining stress state. Due to the performed measurements of this pre-mining stress, the modelling was performed for two extreme cases: the isotropic pre-mining stress $\sigma_{xx} = \sigma_{yy} = \sigma_{zz}$, $\sigma_{xy} = \sigma_{yz} = \sigma_{zx} = 0$ and the anisotropic one $\sigma_{xx} = \sigma_{yy} = 1.25\,\sigma_{zz}$, $\sigma_{xy} = 0.75\,\sigma_{zz}$, $\sigma_{yz} = \sigma_{zx} = 0$. The vertical stress σ_{zz} is given by the weight of the rocks, $\sigma_{zz} = -\gamma\,h$ where γ is the averaged specific weight of the rocks, h is the depth under surface. For a future modelling, an identification procedure was suggested in [7], which allows to reduce the uncertainties concerning this pre-mining stress state.

The modelling itself consists in solving a sequence of boundary value problems with different material distribution, which corresponds to four selected stages of mining. In these stages, we solve the equilibrium equations with the use of a proper linear constitutive relations. Such approach requires a estimate of the effective elastic moduli for the rocks, ore as well as for the damaged rocks (the goaf), which appear in the vicinity of the mine openings. The estimate of the extent of the zone of damaged material and the estimate of the material constants for the goaf is another important source of uncertainty, see e.g. the sensitivity analysis performed in [6].

As the pre-mining stress is not compatible with any apriori known displacements, the modelling sequence consists from solving four pure Neumann boundary value problems. In this case, the use of different weights of intact and damaged materials leads to some global disbalance, which can be overcome by finding generalized solution of the arising singular systems. Note also that stresses, which are of our primary interest, are unique.

The above sequence of Neumann problems (Neumann sequence) can be solved by iterative solvers. But more advantageous may be to use another sequence of problems (Dirichlet sequence), which consists from one auxiliary step with solving the pure Neumannn problem for the computation of pre-mining displacements and using these displacements as a pure Dirichlet boundary conditions in the subsequent modelling steps. Note that the maximal difference in the stress computed by these two sequences was found to be less than 7 %, which is acceptable for the performed modelling.

As a conclusion, we can summarize that the performed modelling requires *repeated solution* of *large linear systems* of nearly *four million of unknown*. Moreover, these systems can be *singular*.

3 Two decomposition techniques

The finite element analysis of boundary value problems, like the problems described in the previous Section, requires the solution of the equation

$$Au = b, \tag{1}$$

which can be interpreted as the operator equation in the finite element space V or the corresponding linear system in the space of algebraic vectors R^n, where n is large. For 3D analysis n is typically in the range 10^5 - 10^7.

The solution of (1) is usually done by the conjugate gradient (CG) method with a proper preconditioning, see e.g. [1]. The construction of preconditioning can be based on a decomposition of V,

$$V = V_1 + \ldots + V_p\ , \tag{2}$$

where V_k are subspaces of V, which are not necessarily linearly independent. Of course, the corresponding decomposition of the algebraic space R^n will be used for implementation of the preconditioners.

The general scheme for the construction of space decomposition preconditioners is the following. Let

$$R^n \leftrightarrow V, \quad R^{n_k} \leftrightarrow V_k , \tag{3}$$

$$I_k : R^{n_k} \rightarrow R^n \quad \text{be the prolongation given by the inclusion } V_k \subset V , \tag{4}$$

$$R_k : R^n \rightarrow R^{n_k} \quad \text{be the restriction given by } R_k = I_k^T . \tag{5}$$

Let A be the $n \times n$ stiffness matrix and $A_k = R_k A I_k$ be the matrices corresponding to the subproblems on the subspaces. Note that if A is symmetric positive definite then A_k has the same properties.

Now, the preconditioner can be defined by the following algorithm for the computation of the pseudoresidual g from the residual r. It can be interpreted as a linear mapping $G : r \mapsto g$.

Algorithm 1

$g = 0$
for $k = 1, \ldots, p$ **do**
$\quad g \leftarrow g + I_k A_k^{-1} R_k z_k$
end

The operations

$$w_k = A_k^{-1} v_k$$

can be implemented as

$$w_k = S_k(v_k) ,$$

where $S_k(v_k)$ results from solving the subproblem $A_k w_k = v_k$ by inner iterations. The inner iterations can be given again by the CG method stopped by the condition

$$\| v_k - S_k(v_k) \| \leq \varepsilon_0 \| v_k \| .$$

In the case of inexact solution of the subproblems, the mapping G become nonlinear. It may require some measures to be implemented in outer iterations, see the next Section.

For more details about general space decomposition preconditioners, see e.g. [13] and the references therein. Note that in this paper, we consider only the additive preconditioners, which lead to a straightforward parallelization.

We shall use two particular decomposition techniques, the displacement decomposition (see [3], [5], [8]) and the overlapping domain decomposition (see e.g. [12] for the description). Possibly, we can also use a combination of these techniques, cf. [9].

The *displacement decomposition* concerns the solution of the elasticity problem in a domain $\Omega \subset R^d$ $(d = 2, 3)$ by the finite element method with Lagrangian finite elements. In this case, the algebraic vectors from R^n represent nodal displacements. In a special, so called separate displacement component ordering of the degrees of freedom, the vectors $v \in R^n$ have block form $v = (v_1, \cdots, v_d)^T$, where v_k represents nodal displacement in $k-th$ coordinate direction of the coordinate system exploited for description of Ω. Then R^{n_k} correspond to the blocks,

$$I_k: \ v_k \mapsto (v_1, \ldots, v_d)^T \ , \quad v_l = 0 \text{ for } l \neq k \ .$$

This decomposition allows to construct the preconditioners, which have been studied in [3], [5], [2], [8].

In this paper, the *domain decomposition* concerns again the solution of elasticity problems in the domain Ω, which is divided into finite elements $E \in \mathcal{T}_h$. The decomposition starts from decomposition of Ω into non-overlapping subdomains $\hat{\Omega}_k$, which are subsequently extended into overlapping subdomains Ω_k, $\Omega = \bigcup_{k=1}^{p} \Omega_k$. We assume that each $\hat{\Omega}_k$, Ω_k can be represented as a union of some elements from the global division $\mathcal{T}_h$. Then the division of Ω induces a decomposition of the finite element space V with the subspaces V_k,

$$V_k = \{v \in V : v = 0 \text{ in } \Omega \setminus \Omega_k\} \ .$$

In all spaces V, $V_1, \ldots, V_p$, we can use the same finite element basis functions. The isomorphism with R^n, $R^{n_1}, \ldots, R^{n_p}$ then allows simple construction of the prolongation represented by a Boolean matrix,

$$I_k = [c_{ij}], \ 1 \leq i \leq n, \ 1 \leq j \leq n_k$$

where $c_{ij} = 1$ if the degrees of freedom i and j correspond to the same finite element basis function, otherwise $c_{ij} = 0$.

The efficiency of the domain decomposition preconditioner improves with the increasing overlap, but deteriorates with the increasing number of subdomains p, see e.g. [12] for the explanation. This drawback can be remowed and overall efficiency can be improved by using extended decomposition with an additional subspace V_0 corresponding to discretization of the global problem with the aid of a coarser finite element division $\mathcal{T}_H$. If $\mathcal{T}_h$ is a refinement of $\mathcal{T}_H$, then I_0 is simply defined by the inclusion $V_0 \subset V$. If $\mathcal{T}_H$ and $\mathcal{T}_h$ are not nested, then we have to use more complicated interpolation I_0, which may be relatively costly to create and perform. For this reason, we consider also another choice of V_0 constructed from V by aggregation. This construction was introduced for multigrid methods [4], its use for the overlapping Schwarz method is analyzed e.g. in [11].

Let $\{\phi_i : i = 1, \ldots, n\}$ is the FE basis of V_h and let $\{1, \ldots, n\} = \bigcup_{k=1}^{N} J_k$ where J_k are disjoint sets. Then the aggregation space can be defined as $V_0 = \operatorname{span}\{\Psi_k : k = 1, \ldots, N\}$, $\Psi_k = \sum_{i \in J_k} \phi_i$. In this case, we can again construct a Boolean prolongation $I_0 : R^N \to R^n$, which also allows a cheap construction of the matrix $A_0 = (I_0)^T A I_0$.

4 GPCG method

The use of inner iterations for solving the subproblems leads to generally nonlinear preconditioner G, which approximates the linear space decomposition preconditioner defined by the same decomposition and exact solution of subproblems. Such nonlinear preconditioner can be implemented within the standard

CG method, but the arising preconditioned CG may be not efficient or even fail to converge. Therefore, we shall describe here a simple generalization of the CG method, which is convenient to use in this case. For more details, see [1], Chapter 12, and [8].

Let $\langle u, v\rangle = u^T v$ denotes the standard inner product in R^n, $J(v) = \frac{1}{2}\langle Av, v\rangle - \langle b, v\rangle$ be the energy functional, whose minimization is equivalent to the solution of the system (1). Then the GPCG iterations with a general preconditioner G use the following steps.

GPCG [m] iteration, $1 \le m \le \infty$.

Given $u^i, v^i, \ldots, v^{i+1-m_{i+1}}$, where $m_i = \min\{i, m\}$, compute:

1. $\alpha_i \in R: \; J(u^i + \alpha_i v^i) = \min_{\alpha \in R} J(u^i + \alpha v^i)$
2. $u^{i+1} = u^i + \alpha_i v^i \;, \; r^{i+1} = r^i - \alpha_i A v^i$
3. $g^{i+1} = G(r^{i+1})$
4. $v^{i+1} = g^{i+1} + \sum_{k=1}^{m_{i+1}} \beta_{i+1}^{(k)} v^{i+1-k} \;, \; \langle Av^{i+1}, v^{i+1-k}\rangle = 0 \quad \forall k = 1, \ldots, m_{i+1}.$

The following Theorem is important for the implementation of GPCG.

Theorem 1.

1. $\alpha_i = \dfrac{\langle r^i, v^i\rangle}{\langle Av^i, v^i\rangle} = \dfrac{\langle r^i, g^i\rangle}{\langle Av^i, v^i\rangle}$

2. $\beta_{i+1}^{(k)} = -\dfrac{\langle g^{i+1}, Av^{i+1-k}\rangle}{\langle Av^{i+1-k}, v^{i+1-k}\rangle} = \dfrac{\langle g^{i+1}, r^{i+2-k}\rangle - \langle g^{i+1}, r^{i+1-k}\rangle}{\langle g^{i+1-k}, r^{i+1-k}\rangle}$

3. the GPCG[m] *algorithm will not break down if* $r \neq 0$ *implies*

$$\langle r, g\rangle = \langle r, G(r)\rangle \neq 0.$$

The correctness and convergence of GPCG[m], $1 \le m \le \infty$, in the case of the additive space decomposition preconditioner with inexact subproblem solvers can be shown by using the following Theorem, for the proof see [8].

Theorem 2.
Let G be an approximation to a SPD matrix B,

$$\| G(r) - B^{-1} r \|_B \le \varepsilon_0 \| B^{-1} r \|_B$$

and let γ_1, γ_2 be two positive constants such that

$$\gamma_1 \langle Bv, v\rangle \le \langle Av, v\rangle \le \gamma_2 \langle Bv, v\rangle \quad \forall v \in R^n \,.$$

Then for $m \ge 1$ and arbitrary $\varepsilon_0 \in \langle 0, 1)$ the GPCG[m] *method does not break down and converges,*

$$\| u^{i+1} - A^{-1} b \|_A \le \sqrt{1 - (1 - \varepsilon_o^2)\kappa^{-1}} \; \| u^i - A^{-1} b \|_A \,, \qquad \kappa = \gamma_2 / \gamma_1 \,.$$

5 Parallel implementation and inner/outer iterations

The use of additive space decomposition preconditioners leads to very natural parallelization of the computation of the pseudoresiduals, when p processors compute simultaneously the contributions from p subspaces. Moreover, the described decomposition techniques induce also (nonoverlapping) decomposition of the global matrix and global vectors, which can be used for parallelization of the further steps in outer iterations.

There are two additional roles of inner iterations if the parallel implementation is used. Firstly, an increase of the accuracy of inner iterations makes the outer iteration more costly but reduces the number of outer iterations. This may be very advantageous if we use parallel system with relatively slow communication rate. This effect will be demonstrated in the next Section. Secondly, variable number of iterations for different subproblems can be used for better balancing the load of processors and speed-up the computations.

6 Performance of the methods

For numerical testing, we shall use the linear system arising from the stress computation concerning the last stage of the modelling described in Section 2 as a benchmark. In this benchmark, we use the pure Dirichlet boundary conditions with the prescribed displacement computed in a preliminary stage.

We present here results from computing on a multi-processor computer SUN HPC 6500 with processors UltraSPARC II/400, 1 GB RAM, a simple cluster of PC's with Intel Pentium III/500, 384 MB RAM and Ethernet 100 interconnection (CLUSTER-1), and more powerful cluster of PC's with AMD Athlon/1400, 768 MB RAM and Ethernet 100 interconnection (CLUSTER-2). Note that the numerical experiments on clusters of PC's are still in progress.

Solver	Number of iterations	SUN Time [s]	CLUSTER-1 Time [s]	CLUSTER-2 Time [s]
$PCG - G_F$	83	668	2 653	420
$PCG - G_I$	11	653	918	326
$GPCG[1] - G_I$	8	557	787	274

Table 1.

Table 1 shows the numbers of iterations and computer times from solving the benchmark problem with the aid of the additive displacement decomposition preconditioner. Parallel computations are performed on 3 processors from the described computer systems. Zero initial guess and stopping by relative accuracy $\varepsilon = 10^{-4}$ is used. PCG means the use of the preconditioned conjugate gradient method with orthogonalization of the new direction to the previous one by using

$$v^{i+1} = g^{i+1} + \beta_i v^i$$

with standard formula

$$\beta_i = \frac{\langle g^{i+1}, r^{i+1} \rangle}{\langle g^i, r^i \rangle} ,$$

which was introduced already in the pioneering paper [10]. On the other hand, GPCG[1] uses the modified formula, see Theorem 1 (2.). G_F means inexact solution of the subproblems by replacing the subproblem matrices by their incomplete factorization, G_I uses inner CG iterations with the same incomplete factorization as inner preconditioner. The relative accuracy for the inner iterations is $\varepsilon_0 = 10^{-1}$. By numerical experiments, it was found that this value is nearly optimal.

From Table 1 we can see that more exact solution of the subproblems by inner iterations substantially decrease the number of outer iterations. This fact has a big influence when the computations are performed on a cluster with a slow interprocessor communications.

The displacement decomposition works well and in case of 3 or 4 processor computations, it is more efficient then the domain decomposition with the same number of subproblems. But the domain decomposition is scalable and this allows to get better times by using more processors, see Figure 3.

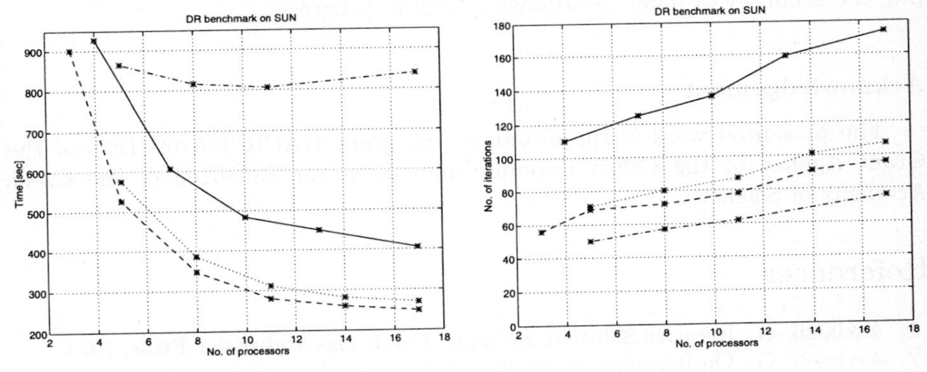

Fig. 3.

This Figure contains diagrams with numbers of outer iterations and computer time for the solution of the benchmark problem with the additive preconditioner based on domain decomposition with minimal overlap and no coarse grid problem (solid lines) and with coarse grid problem created by aggregation - grouping of 3×3×3 nodes (dot-dashed), 6×6×6 nodes (dashed) and 9×9×6 nodes (dotted lines). The subproblems corresponding to subdomains are solved by replacing the subproblem matrices by incomplete factorization, the subproblem corresponding to the coarse grid problem is solved with relative accuracy $\varepsilon_0 = 10^{-1}$ by inner PCG iterations with the same incomplete factorization preconditioning. The number of processors is equal to the number of subproblems.

From this Figure, we can see that aggregation helps, but relatively rough aggregation was required for decrease of the computing times. Finer aggregation reduced the number of outer iterations but lead to a worse load balance of the processors. In this respect, we will try to further optimize the performance as well as test the method on clusters of PC's.

Note also that the displacement decomposition and domain decomposition can be combined, see [9].

7 Concluding remarks

In this paper, we consider inner/outer iteration processes, which arise from using space decomposition preconditioning with inexact solution of the subproblems by inner iterations.

The space decomposition is used for parallelization of the computations. Moreover, the inner/outer iterations can be used for balancing the ratio between the amount of communications (the increase of the number of inner iterations can decrease the number of outer iterations and the amount of communications among the processors). Different numbers of inner iterations for different subproblems can keep all the processors busy and speed up the solution process.

Two special decomposition techniques for solving elasticity problems have been used, the displacement decomposition and domain decomposition. For outer iterations, we used recently developed generalized preconditioned CG method. Further numerical experiments with balancing parallel computations by controlling the accuracy of inner iterations are still in progress.

Acknowledgement

The presented work is supported by the grant GAČR 105/99/1229 of the Grant Agency of the Czech Republic and the grant S3086102 of the Czech Academy of Sciences.

References

1. Axelsson, O.: Iterative Solution Methods. Cambridge University Press, 1994
2. Axelsson, O.: On iterative solvers in structural mechanics; separate displacement orderings and mixed variable methods. Mathematics and Computers in Simulation 50 (1999), 11-30
3. Axelsson, O., Gustafsson, I.: Iterative methods for the solution of the Navier's equations of elasticity. Computer Meth. Appl. Mech. Engng. 15 (1978), 241-258
4. Blaheta, R.: A multilevel method with overcorrection by aggregation for solving discrete elliptic problems. J. Comp. Appl. Math. 24 (1988), 227-239
5. Blaheta, R.: Displacement Decomposition - Incomplete Factorization Preconditioning Techniques for Linear Elasticity Problems. Num. Lin. Alg. Appl. 1 (1994), 107-128
6. Blaheta, R., Kohut, R.: Uncertainty and sensitivity analysis in problems of geomechanics. Report DAM 2000/1, IGAS Ostrava, 2000, 16pp. (in Czech)
7. Blaheta, R.: Identification methods for application in geomechanics and materials science. Report DAM 2000/2, IGAS Ostrava, 2000, 29pp.
8. R. Blaheta: *GPCG - generalized preconditioned CG method and its use with nonlinear and nonsymmetric displacement decomposition preconditioners.* Report DAM 2001/2, Institute of Geonics AS CR, Ostrava, 2001 (submitted).
9. Blaheta, R., Byczanski, P., Jakl, O., Starý, J.: Space decomposition preconditioners and their application in geomechanics. Math. Computer Simulation, submitted
10. Hestenes, M.R., Stiefel, E.: Methods of conjugate gradients for solving linear systems. Journal of Research of the National Bureau of Standards 49 (1952), 409-436

11. Jenkins, E.W., Kelley, C.T., Miller, C.T., Kees, C.E.: An aggregation-based domain decomposition preconditionerfor groundwater flow. SIAM J. Sci. Comput. 23 (2001), 430-441
12. Smith, B.F., Bjørstad, P.E., Gropp, W.D.: Domain Decomposition Parallel Multilevel Methods for Elliptic Partial Differential Equations. Cambridge University Press, 1996
13. Xu, J.: The method of subspace corrections. J. Comp. Appl. Math. 128 (2001), 335-362

Reliable Solution of a Unilateral Frictionless Contact Problem in Quasi-Coupled Thermo-Elasticity with Uncertain Input Data

Ivan Hlaváček[1] and Jiří Nedoma[1]

Institute of Computer Science, Academy of Sciences of the Czech Republic
Pod Vodárenskou věží 2, 182 07 Prague 8, Czech Republic

Abstract. A unilateral contact problem without friction in quasi-coupled thermo-elasticity and with uncertain input data is analysed. The worst scenario method is used to find the most "dangerous" admissible input data.

1 Introduction

In this contribution we deal with contact problems without friction (see [4], [5], [6]) in quasi-coupled thermo-elasticity considering uncertain input data representing extension of problems solved in [5] and [6]. By uncertain input data we mean physical coefficients, right-hand sides, etc., which cannot be determined uniquely but only in some intervals determined by the measurements. The reliable solution is defined as the worst among a set of possible solutions, and the degree of badness is measured by a criterion-functional (see [1]). The main aim of our contribution will be to find maximal values of this functional. We prove the solvability of the corresponding maximization (worst scenario) problems.

2 Formulation of the Problem

Let us assume a union Ω of bounded domains Ω^ι, $\iota = 1, \ldots, s$, with Lipschitz boundaries $\partial\Omega^\iota$, occupied by elastic bodies such that $\Omega = \bigcup_{\iota=1}^{s} \Omega^\iota \subset R^2$. Let the boundary $\partial\Omega = \cup_{\iota=1}^{s} \partial\Omega^\iota$ consist of three disjoint parts Γ_τ, Γ_u and Γ_c, such that $\partial\Omega = \overline{\Gamma}_\tau \cup \overline{\Gamma}_u \cup \overline{\Gamma}_c$, $\Gamma_c = \bigcup_{k,l} \Gamma^{kl}$, $\Gamma^{kl} = \partial\Omega^k \cap \partial\Omega^l$, $1 \le k,\, l \le s$, for $k \ne l$, and $\overline{\Gamma}_\tau$, $\overline{\Gamma}_u$, $\overline{\Gamma}_c$ denotes the closures in $\partial\Omega$.

Let the heat sources W^ι, the prescribed temperature T_1, the body forces $\mathbf{F}$, the surface forces $\mathbf{P}$, displacements $\mathbf{u}_0$, elastic coefficients c_{ijkl}, coefficients of thermal expansion β_{ij} and the reference temperature T_0 be given. Throughout the paper we use the summation convention, i.e. a repeated index implies summation from 1 to 2. Furthermore, $\mathbf{n}^k = (n_i^k)$, $i = 1, 2, 1 \le k \le s$, denotes the unit normal with respect to $\partial\Omega^k$, $\mathbf{n}^k = -\mathbf{n}^l$ on Γ^{kl}. Assume that κ^ι and C^ι are positive definite symmetric matrix functions,

P.M.A. Sloot et al. (Eds.): ICCS 2002, LNCS 2331, pp. 840–851, 2002.

$$0 < \kappa_0^\iota \le \kappa_{ij}^\iota \zeta_i \zeta_j |\zeta|^{-2} \le \kappa_1^\iota < +\infty \text{ for a.a. } \mathbf{x} \in \Omega^\iota,\ \zeta \in R^2\,,$$
$$0 < c_0^\iota \le c_{ijkl}^\iota \xi_{ij} \xi_{kl} |\boldsymbol{\xi}|^{-2} \le c_1^\iota < +\infty \text{ for a.a. } \mathbf{x} \in \Omega^\iota,\ \boldsymbol{\xi} \in R^4,\ \xi_{ij} = \xi_{ij}\,,$$

where κ_0^ι, κ_1^ι, c_0^ι, c_1^ι are constants independent of $\mathbf{x} \in \boldsymbol{\Omega}^\iota$. Let $\kappa_{ij}^\iota \in L^\infty(\Omega^\iota)$, $W^\iota \in L^2(\Omega^\iota)$, $T_1^\iota \in H^1(\Omega^\iota)$, $T_1^k = T_1^l$ on $\bigcup\limits_{k,l} \Gamma^{kl}$, $c_{ijkl}^\iota \in L^\infty(\Omega^\iota)$, $F_i^\iota \in L^2(\Omega^\iota)$, $P_i \in L^2(\Gamma_\tau)$, $\beta_{ij}^\iota \in L^\infty(\Omega^\iota)$, $\mathbf{u}_0^\iota \in [H^1(\Omega^\iota)]^2$.

We will deal with the following problem:

Problem ($\mathcal{P}$): Find a pair of functions $(T, \mathbf{u})$ satisfying

$$\frac{\partial}{\partial x_i}\left(\kappa_{ij}^\iota \frac{\partial T^\iota}{\partial x_j}\right) + W^\iota = 0,\quad \frac{\partial}{\partial x_j}\tau_{ij}(\mathbf{u}^\iota, T^\iota) + F_i^\iota = 0 \ \text{ in } \Omega^\iota,\ 1 \le \iota \le s,\ i = 1,2 \tag{1}$$

$$\tau_{ij}(\mathbf{u}^\iota, T^\iota) = c_{ijkl}^\iota e_{kl}(\mathbf{u}^\iota) - \beta_{ij}^\iota (T^\iota - T_0^\iota) \ \text{ in } \Omega^\iota,\ 1 \le \iota \le s,\ i = 1,2 \tag{2}$$

$$\kappa_{ij} \frac{\partial T}{\partial x_j} n_i = 0,\quad \mathbf{u} = \mathbf{u}_0 \ \text{ on } \Gamma_u\,, \tag{3}$$

$$T = T_1,\quad \tau_{ij}(\mathbf{u}, T) n_j = P_i \ \text{ on } \Gamma_\tau\,, \tag{4}$$

$$T^k = T^l,\quad \left(\kappa_{ij} \frac{\partial T}{\partial x_j} n_i\right)^k - \left(\kappa_{ij} \frac{\partial T}{\partial x_j} n_i\right)^l = 0 \ \text{ on } \bigcup_{k,l} \Gamma^{kl},\ 1 \le k,l \le s\,, \tag{5}$$

$$u_n^k - u_n^l \le 0,\quad \tau_n^k \le 0,\quad (u_n^k - u_n^l)\tau_n^k = 0 \ \text{ on } \bigcup_{k,l} \Gamma^{kl},\ 1 \le k,l \le s\,, \tag{6}$$

$$\tau_t^k = -\tau_t^l = 0 \ \text{ on } \bigcup_{k,l} \Gamma^{kl},\ 1 \le k,l \le s\,, \tag{7}$$

where $e_{ij}(\mathbf{u}) = \frac{1}{2}\left(\frac{\partial u_i}{\partial x_j} + \frac{\partial u_j}{\partial x_i}\right)$, $u_n^k = u_i^k n_i^k$, $u_n^l = u_i^l n_i^l = -u_i^k n_i^k$ (no sum over k or l), $u_t^k = (u_{ti}^k)$, $u_{ti}^k = u_i^k - u_n^k n_i^k$, $u_t^l = (u_{ti}^l)$, $u_{ti}^l = u_i^l - u_n^l n_i^l$, $i = 1,2$, $\tau_n^k = \tau_{ij}^k n_i^k n_j^k$, $\tau_t^k = (\tau_{ti}^k)$, $\tau_{ti}^k = \tau_{ij}^k n_j^k - \tau_n^k n_i^k$, $\tau_n^l = \tau_{ij}^l n_i^l n_j^l$, $\tau_t^l = (\tau_{ti}^l)$, $\tau_{ti}^l = \tau_{ij}^l n_j^l - \tau_n^l n_i^l$.

Since the stress and strain tensors and coefficient of thermal expansion are symmetric then the entries of any symmetric 3×3 matrices $\{\tau_{ij}\}$ can be rewritten in the vector notation $\{\tau_j\}$, $j = 1,2,3$ and similarly the symmetric matrices $\{e_{ij}\}$, $\{\beta_{ij}\}$ by vectors $\{e_j\}$, $\{\beta_j\}$. Then (2) can be rewritten as

$$\tau_i(\mathbf{u}^\iota, T^\iota) = \sum_{j=1}^{3} A_{ij}^\iota e_j(\mathbf{u}^\iota) - \beta_i^\iota (T^\iota - T_0^\iota) \ \text{ in } \Omega^\iota,\ 1 \le \iota \le s,\ 1 \le i,j \le 3, \tag{8}$$

where A^ι is a symmetric 3×3 matrix, $A_{ik}^\iota \in L^\infty(\Omega^\iota)$, $\iota = 1, \dots, s$. Since $\tau_{ij} e_{ij} = \sum\limits_{i=1}^{2} \tau_i e_i + 2\tau_3 e_3$, we can write

$$c^{\iota}_{ijkl}e_{ij}e_{kl} = \sum_{i,j=1}^{3} B^{\iota}_{ij}e_i e_j \,,$$

where B^{ι} is a symmetric 3×3 matrix such that $B^{\iota}_{ij} = A^{\iota}_{ij}$ for $i,j = 1,2$, $B^{\iota}_{ij} = \frac{3}{2}A^{\iota}_{ij}$ for $i = 1,2$, $j = 3$ and $B^{\iota}_{ij} = 2A^{\iota}_{ij}$ for $i,j = 3$.

In what follows, we denote

$$W_1 = \sqcap^{s}_{\iota=1} H^1(\Omega^{\iota}), \; \|w\|_{W_1} = \left(\sum_{\iota \le s} \|w^{\iota}\|^2_{1,\Omega^{\iota}} \right)^{\frac{1}{2}},$$

$$W = \sqcap^{s}_{\iota=1} [H^1(\Omega^{\iota})]^2, \; \|v\|_W = \left(\sum_{\iota \le s} \sum_{i \le 2} \|v^{\iota}_i\|^2_{1,\Omega^{\iota}} \right)^{\frac{1}{2}},$$

$$V_1 = \left\{ z | z \in W_1, \; z = 0 \text{ on } \Gamma_{\tau}, \; z^k = z^l \text{ on } \bigcup_{k,l} \Gamma^{kl} \right\},$$

$$V = \{\mathbf{v} | \mathbf{v} \in W, \; \mathbf{v} = 0 \text{ on } \Gamma_u\}, \quad K = \left\{ \mathbf{v} | \mathbf{v} \in V, \; v^k_n - v^l_n \le 0 \text{ on } \bigcup_{k,l} \Gamma^{kl} \right\}.$$

Definition 1. We say that the pair of functions T and $\mathbf{u}$ is a weak solution of problem $(\mathcal{P})$, if $T - T_1 \in V_1$,

$$b(T,z) = s(z) \quad \forall z \in V_1 \,, \tag{9}$$

$$\mathbf{u} - \mathbf{u}_0 \in K \,,$$

$$a(\mathbf{u}, \mathbf{v} - \mathbf{u}) \ge S(\mathbf{v} - \mathbf{u}, T) \quad \forall \mathbf{v} \in \mathbf{u}_0 + K \,, \tag{10}$$

where

$$b(T,z) = \sum_{\iota=1}^{s} \int_{\Omega^{\iota}} \kappa^{\iota}_{ij} \frac{\partial T^{\iota}}{\partial x_i} \frac{\partial z^{\iota}}{\partial x_j} d\mathbf{x}, \quad s(z) = \sum_{\iota=1}^{s} \int_{\Omega^{\iota}} W^{\iota} z^{\iota} d\mathbf{x} \,, \tag{11}$$

$$a(\mathbf{u}, \mathbf{v}) = \sum_{\iota=1}^{s} \int_{\Omega^{\iota}} \sum_{i,j=1}^{3} B^{\iota}_{ij} e_i(\mathbf{u}^{\iota}) e_j(\mathbf{v}^{\iota}) d\mathbf{x} \,, \tag{12}$$

$$S(\mathbf{v}, T) = \sum_{\iota=1}^{s} \int_{\Omega^{\iota}} F^{\iota}_i v^{\iota}_i d\mathbf{x} + \int_{\Gamma_{\tau}} P_i v_i ds - \sum_{\iota=1}^{s} \int_{\Omega^{\iota}} \beta^{\iota}_i (T^{\iota} - T^{\iota}_0) v^{\iota}_i d\mathbf{x} \,. \tag{13}$$

Remark 1. In $S(\mathbf{v}, T)$ we insert the weak solution T of (9). Moreover, we assume that $\mathbf{u}_0$ satisfies

$$u^k_{0n} - u^l_{0n} = 0 \quad \text{on } \bigcup_{k,l} \Gamma^{kl} \,. \tag{14}$$

3 Worst Scenario Method for Uncertain Input Data

Let us assume that the input data

$$A = \{B^\iota, \kappa^\iota, F_i^\iota, W^\iota, \beta_i^\iota, P_i, u_{0i}, T_1, \iota = 1, \dots, s, \ i = 1, 2\}$$

are uncertain, and belong to some sets of admissible data, i.e.

$$A \in U_{ad} \Leftrightarrow B^\iota \in U_{ad}^{B^\iota}, \ \kappa^\iota \in U_{ad}^{\kappa^\iota}, \ F_i^\iota \in U_{ad}^{F_i^\iota}, \ W^\iota \in U_{ad}^{W^\iota}, \ \beta_i^\iota \in U_{ad}^{\beta_i^\iota},$$
$$P \in U_{ad}^{P_i}, \ u_{0i} \in U_{ad}^{u_{0i}}, \ T_1 \in U_{ad}^{T_1}.$$

We will assume that all the bodies Ω^ι are piecewise homogeneous, so that partitions of $\overline{\Omega}^\iota$ exist such that

$$\overline{\Omega}^\iota = \bigcup_{j=1}^{r^\iota} \overline{\Omega}_j^\iota, \ \Omega_j^\iota \cap \Omega_k^\iota = \oslash \ \text{for } j \neq k, \ 1 \leq \iota \leq s, \tag{15}$$

$$\Gamma^{kl} = \bigcup_{q=1}^{Q_{kl}} \overline{\Gamma}_q^{kl}, \ \Gamma_q^{kl} \cap \Gamma_p^{kl} = \oslash \ \text{for } q \neq p, \ \forall k, l. \tag{16}$$

Let the data B^ι, κ^ι, $\mathbf{F}^\iota$, W^ι, β^ι be piecewise constant with respect to the corresponding partitioning (15) and let us denote

$$\Gamma_u^\iota = \Gamma_u \cap \partial\Omega^\iota, \ \iota = 1, \dots, s \ \text{and} \ \Gamma_\tau^\iota = \Gamma_\tau \cap \partial\Omega^\iota, \ \iota \leq s. \tag{17}$$

Further, we define the sets of admissible matrices:

$$U_{ad}^{B^\iota} = \{3 \times 3 \text{ symmetric matrices } B^\iota : \underline{B}_{ik}^\iota(j) \leq B_{ik|\Omega_j^\iota}^\iota = \text{const.} \leq \overline{B}_{ik}^\iota(j),$$
$$j \leq r^\iota, \ i, k = 1, \dots, 3\} \tag{18}$$

where $\underline{B}^\iota(j)$ and $\overline{B}^\iota(j)$ are given 3×3 symmetric matrices, $\iota = 1, \dots, s$, and let there exist positive constants $c_B^\iota(j)$ such that

$$\lambda_{\min}\left(\frac{1}{2}(\underline{B}^\iota(j) + \overline{B}^\iota(j))\right) - \rho\left(\frac{1}{2}(\overline{B}^\iota(j) - \underline{B}^\iota(j))\right) \equiv c_B^\iota(j)$$
$$\text{for } j = 1, \dots, r^\iota, \ \iota = 1, \dots, s, \tag{19}$$

where $\lambda_{\min}$ and ρ denotes the minimal eigenvalue and the spectral radius, respectively,

$$U_{ad}^{\kappa^\iota} = \{2 \times 2 \text{ symmetric matrices } \kappa^\iota : \underline{\kappa}_{ik}^\iota(j) \leq \kappa_{ik|\Omega_j^\iota}^\iota = \text{const.} \leq \overline{\kappa}_{ik}^\iota(j),$$
$$j \leq r^\iota, \ i, k \leq 2\} \tag{20}$$

where $\underline{\kappa}^\iota(j)$ and $\overline{\kappa}^\iota(j)$ are given 2×2 symmetric matrices, $j = 1, \dots, r^\iota$, $\iota = 1, \dots, s$, and let there exist positive constants $c_B^\iota(j)$ such that

$$\lambda_{\min}\left(\frac{1}{2}(\underline{\kappa}^\iota(j) + \overline{\kappa}^\iota(j))\right) - \rho\left(\frac{1}{2}(\overline{\kappa}^\iota(j) - \underline{\kappa}^\iota(j))\right) \equiv c_\kappa^\iota(j) \text{ for } j \le r^\iota,\ \iota \le s\,, \tag{21}$$

where $\lambda_{\min}$ and ρ denotes the minimal eigenvalue and the spectral radius, respectively. If (18) and (19) are satisfied, then the matrices $B^\iota(j) \equiv B^\iota_{|\Omega_j^\iota}$ are positive definite for any $B^\iota \in U_{ad}^{B^\iota}$, $\iota = 1, \dots, s$ and any $j \le r^\iota$ (see [8]) and the matrices $\kappa^\iota(j) = \kappa^\iota_{|\Omega_j^\iota}$ are positive definite for any $\kappa^\iota \in U_{ad}^{\kappa^\iota}$, $\iota \le s$, $j \le r^\iota$.

Furthermore, we define

$$U_{ad}^{F_i^\iota} = \{f \in L^\infty(\Omega) : \underline{F}_i^\iota(j) \le f_{|\Omega_j^\iota} = \text{const.} \le \overline{F}_i^\iota(j),\ j \le r^\iota\}\,, \tag{22}$$

for $i \le 2$, $\iota \le s$, where $\underline{F}_i^\iota(j)$ and $\overline{F}_i^\iota(j)$ are given constants;

$$U_{ad}^{W^\iota} = \{w \in L^\infty(\Omega) : \underline{W}^\iota(j) \le w_{|\Omega_j^\iota} = \text{const.} \le \overline{W}^\iota(j),\ j \le r^\iota\}\,, \tag{23}$$

for $\iota \le s$, where $\underline{W}^\iota(j)$ and $\overline{W}^\iota(j)$ are given constants;

$$U_{ad}^{T_1} = \{T \in L^\infty(\Gamma_\tau) : \underline{T}_1(\iota) \le T_{|\Gamma_\tau^\iota} = \text{const.} \le \overline{T}_1(\iota),\ \iota \le s\}\,, \tag{24}$$

where $\underline{T}_1(\iota)$ and $\overline{T}_1(\iota)$ are given constants;

$$U_{ad}^{u_{0i}} = \{u \in L^\infty(\Gamma_u) : \underline{u}_{0i}(\iota) \le u_{|\Gamma_u^\iota} = \text{const.} \le \overline{u}_{0i}(\iota),\ \iota \le s\}\,, \tag{25}$$

where $\underline{u}_{0i}(\iota)$ and $\overline{u}_{0i}(\iota)$, $i = 1, 2$, are given constants;

$$U_{ad}^{P_i} = \{p \in L^\infty(\Gamma_\tau) : \underline{P}_i(\iota) \le p_{|\Gamma_\tau^\iota} = \text{const.} \le \overline{P}_i(\iota),\ \iota \le s\}\,, \tag{26}$$

where $\underline{P}_i(\iota)$ and $\overline{P}_i(\iota)$, $i = 1, 2$, are given constants;

$$U_{ad}^{\beta_i^\iota} = \{b \in L^\infty(\Omega) : \underline{\beta}_i^\iota(j) \le b_{|\Omega_j^\iota} = \text{const.} \le \overline{\beta}_i^\iota(j),\ j \le r^\iota\}\,, \tag{27}$$

for $i \le 3$, $\iota \le s$, where $\underline{\beta}_i^\iota(j)$ and $\overline{\beta}_i^\iota(j)$ are given constants.

Finally, we define the set of admissible data by

$$\begin{aligned} U_{ad} = \sqcap_{\iota \le s} U_{ad}^{B^\iota} \times \sqcap_{\iota \le s} U_{ad}^{\kappa^\iota} \times \sqcap_{\iota \le s, i \le 2} U_{ad}^{F_i^\iota} \times \sqcap_{\iota \le s} U_{ad}^{W^\iota} \times \\ \times \sqcap_{\iota \le s, i \le 2} U_{ad}^{\beta_i^\iota} \times \sqcap_{i \le 2} U_{ad}^{P_i} \times \sqcap_{i \le 2} U_{ad}^{u_{0i}} \times \sqcap_{\iota \le s} U_{ad}^{T_1}\,. \end{aligned} \tag{28}$$

Further, instead of $b(T, z)$, $a(\mathbf{u}, \mathbf{v})$, $s(z)$, $S(\mathbf{v}, T)$ we will write $b(A; T, z)$, $a(A; \mathbf{u}, \mathbf{v})$, $s(A; z)$, $S(A; \mathbf{v}, T)$ for any $A \in U_{ad}$.

The next results are parallel to those of [3] for the general case with friction.

Lemma 1. There exist positive constants c_i, $i = 0, 1, \dots, 5$ independent of $A \in U_{ad}$, such that

$$
\begin{aligned}
b(A; z, z) &\geq c_0 \|z\|^2_{W^1} \quad \forall z \in V_1 \,, && (29)\\
|b(A; z, y)| &\leq c_1 \|z\|_{W^1} \|y\|_{W^1} \quad \forall z, y \in W_1 \,, && (30)\\
a(A; \mathbf{v}, \mathbf{v}) &\geq c_2 \|\mathbf{v}\|^2_W \quad \forall \mathbf{v} \in V \,, && (31)\\
|a(A; \mathbf{v}, \mathbf{w})| &\leq c_3 \|\mathbf{v}\|_W \|\mathbf{w}\|_W \quad \forall \mathbf{v}, \mathbf{w} \in W \,, && (32)\\
|s(A; z)| &\leq c_4 \|z\|_{0,\Omega} \quad \forall z \in V_1 \,, && (33)\\
|S(A; \mathbf{v}, T)| &\leq c_5 (\|\mathbf{v}\|_{0,\Omega} + \|\mathbf{v}\|_{0,\Gamma_\tau} + \|T - T_0\|_{0,\Omega} \|\mathbf{v}\|_W) \quad \forall \mathbf{v}, \mathbf{w} \in W \,. && (34)
\end{aligned}
$$

Proposition 1. There exists a unique weak solution $(T(A), \mathbf{u}(A))$ of the problem $(\mathcal{P})$ for any $A \in U_{ad}$. Moreover, $\|T(A)\|_{W_1} \leq c$, where c is independent of A.

To find the most "dangerous" input data A in the set U_{ad}, we will introduce a criterion, i.e. defined a functional, which depends on the solution $(T(A), \mathbf{u}(A))$ of the problem $(\mathcal{P})$. Such criteria can be as follows:

Let $G_r \subset \bigcup_{\iota \leq s} \Omega^\iota$, $r = 1, \dots, \bar{r}$, be subdomains adjacent to the boundaries $\partial\Omega^\iota$. Then we define

$$\Phi_1(T) = \max_{r \leq \bar{r}} \varphi_r(T) = \max_{r \leq \bar{r}} \left[(\text{meas}_2\, G_r)^{-1} \int_{G_r} T d\mathbf{x} \right] ; \tag{35}$$

let $G'_r \subset \Gamma_u$, $r \leq \bar{r}$ and

$$\Phi_2(T) = \max_{r \leq \bar{r}} \psi_r(T) = \max_{r \leq \bar{r}} \left[(\text{meas}_1\, G'_r)^{-1} \int_{G'_r} T ds \right] ; \tag{36}$$

and

$$\Phi_3(\mathbf{u}) = \max_{r \leq \bar{r}} \chi_r(\mathbf{u}) = \max_{r \leq \bar{r}} \left[(\text{meas}_2\, G_r)^{-1} \int_{G_r} u_i n_i(X_r) d\mathbf{x} \right] ; \tag{37}$$

where $\mathbf{n}(X_r)$ is the unit outward normal at a fixed point $X_r \in \partial\Omega^\iota \cap \partial G_r$ (if $G_r \subset \Omega^\iota$) to the boundary $\partial\Omega^\iota$;

$$\Phi_4(\mathbf{u}) = \max_{r \leq \bar{r}} \chi'_r(\mathbf{u}) = \max_{r \leq \bar{r}} \left[(\text{meas}_1\, G'_r)^{-1} \int_{G'_r} u_i n_i(X_r) ds \right] ; \tag{38}$$

where $G'_r = \bigcup_{\iota \leq s} \partial\Omega^\iota \backslash \Gamma_u$. Since the weak solution $\mathbf{u}(A)$ of our problem (10) depends on $T(A)$, then $\mathbf{u}(A) = \mathbf{u}(A; T(A))$ and instead of $\Phi_i(\mathbf{u})$ we write $\Phi_i(A; \mathbf{u}, T)$. Thus we may define

$$\Phi_5(A; u, T) = \max_{r \leq \bar{r}} \omega_r(A; u, T) = \max_{r \leq \bar{r}} \left[(\text{meas}_2\, G_r)^{-1} \int_{G_r} I_2^2(\tau(A; \mathbf{u}, T)) d\mathbf{x} \right] ; \tag{39}$$

here $I_2(\tau) = \left(\sum_{i,j=1}^{3} \tau_{ij}^D \tau_{ij}^D \right)^{\frac{1}{2}}$ is the intensity of shear stress, where $\tau_{ij}^D = \tau_{ij} - \frac{1}{3}\tau_{kk}\delta_{ij}$ and $\tau(A; \mathbf{u}, T)$ is defined by (2). Finally, we may choose

$$\Phi_6(A; u, T) = \max_{r \le \bar{r}} \mu_r(A; u, T) = \max_{r \le \bar{r}} \left[(\text{meas}_2\, G_r)^{-1} \int_{G_r} (-\tau_n(A; \mathbf{u}, T)) d\mathbf{x} \right]; \tag{40}$$

where G_r is a small subdomain adjacent to Γ_c.

Now we formulate the worst scenario problems as follows: find

$$A^{0i} = \arg \max_{A \in U_{ad}} \Phi_i(T(A)), \quad i = 1, 2 \tag{41}$$

and

$$A^{0i} = \arg \max_{A \in U_{ad}} \Phi_i(\mathbf{u}(A), T(A)), \quad i = 3, 4, 5, 6, \tag{42}$$

where $(T(A), \mathbf{u}(A))$ is weak solution of the problem $(\mathcal{P})$.

4 Stability of Weak Solutions

To prove the solvability of worst scenario problems (41), (42), we have to study the mapping $A \mapsto T(A)$, $A \mapsto \mathbf{u}(A, T(A))$. We introduce the decomposition of $A \in U_{ad}$ as $A = \{A', A''\}$, where

$$A' = \{\sqcap_{\iota \le s} \sqcap_{j \le r^\iota} \kappa^\iota(j), \sqcap_{\iota \le s} \sqcap_{j \le r^\iota} W^\iota(j), \sqcap_{\iota \le s} T_1^\iota\}, \; A' \in R^{p_1}, \; p_1 = 4 \sum_{\iota \le s} r^\iota + s,$$

and

$$A'' = \{\sqcap_{\iota \le s} \sqcap_{j \le r^\iota} B^\iota(j), \sqcap_{\iota \le s} \sqcap_{j \le r^\iota} \mathbf{F}^\iota(j), \sqcap_{\iota \le s} \mathbf{P}^\iota, \sqcap_{\iota \le s} \mathbf{u}_0^\iota, \sqcap_{\iota \le s} \sqcap_{j \le r^\iota} \beta^\iota(j)\},$$

$$A'' \in R^{p_2}, \; p_2 = \left(\sum_{\iota \le s} r^\iota \right) [9 + 2(1 + 2s)].$$

We are going to show the continuity of the mappings $A' \mapsto T(A')$, $A \mapsto \mathbf{u}(A, T(A'))$ for $A' \in U'_{ad} = \sqcap_{\iota \le s} U_{ad}^{\kappa^\iota} \times \sqcap_{\iota \le s} U_{ad}^{W^\iota} \times U_{ad}^{T_1^\iota}$ and $A'' \in U''_{ad} = \sqcap_{\iota \le s} U_{ad}^{B^\iota} \times \sqcap_{\iota \le s, i \le 2} U_{ad}^{F_i^\iota} \times \sqcap_{\iota \le s, i \le 2} U_{ad}^{\beta_i^\iota} \times \sqcap_{i \le 2} U_{ad}^{P_i} \times \sqcap_{i \le 2} U_{ad}^{u_{0i}}$, respectively. Since the problem discussed is quasi-coupled, we will prove the following theorems and lemma:

Theorem 1. Let $A' \in U'_{ad}$, $A'_n \to A'$ in R^{p_1} as $n \to \infty$. Then

$$T(A'_n) \to T(A) \quad \text{in } W_1 .$$

Sketch of the proof: Since

$$b(A; z, z) \geq \left(\min_{\iota \leq s, j \leq r^{\iota}} c^{\iota}_{K}(j) \right) \sum_{\iota \leq s} \int_{\Omega^{\iota}} |\operatorname{grad} z^{\iota}|^2 \, d\mathbf{x} , \tag{43}$$

for $T_n := T(A'_n)$ we obtain $\|T_n\|_{W_1} \leq c$ for all n. Then a $T \in W_1$ and a subsequence $\{T_m\} \subset \{T_n\}$ exist such that

$$T_m \rightharpoonup T \quad \text{weakly in } W_1 . \tag{44}$$

By definition

$$b(A'_m; T_m, z) = s(A'_m; z) \quad \forall z \in V_1, \ \forall m . \tag{45}$$

Since

$$|b(A'_m; T_m, z) - b(A'; T, z)| \to 0 \,, \text{ as } m \to \infty \,,$$
$$|s(A'_m; z) - b(A'; z)| \to 0 \,, \text{ as } m \to \infty \,,$$

we prove that

$$b(A'_m; T_m, z) \to b(A'; T, z) \text{ as } m \to \infty \,, \tag{46}$$
$$s(A'_m; z) \to s(A'; z) \text{ as } m \to \infty \,. \tag{47}$$

Then we pass to the limit with $m \to \infty$ in (45). Using (46), (47) we prove that $T = T(A')$ is a weak solution of thermal part of the problem. Since it is unique, the whole sequence $\{T_n\}$ tends $T(A')$ weakly in W_1. □

Remark 2. It can be proved that $T_m \to T$ converges also strongly in W_1.

Lemma 2. If $A''_n \in U_{ad}$, $A''_n \to A''$ in R^{p_2}, and $\mathbf{u}_n \to u$ weakly in W, then

$$a(A''_n; \mathbf{u}_n, \mathbf{v}) \to a(A''; \mathbf{u}, \mathbf{v}) \quad \forall \mathbf{v} \in W \,, \tag{48}$$
$$S(A''_n; \mathbf{u}_n, T) \to S(A''; \mathbf{u}, T) \quad \forall T \in W_1 \,. \tag{49}$$

Sketch of the proof: The proof follows from the fact that

$$|a(A''_n; \mathbf{u}_n, \mathbf{v}) - a(A''; \mathbf{u}, \mathbf{v})| \to 0 \quad \text{for } n \to \infty \,,$$
$$|S(A''_n; \mathbf{u}_n, T) - S(A''; \mathbf{u}, T)| \to 0 \quad \text{for } n \to \infty \,.$$

□

Theorem 2. Let $A_n \in U_{ad}$, $A_n \to A$ in $U \equiv R^{p_2}$. Then

$$\mathbf{u}(A_n) \to \mathbf{u}(A) \quad \text{in } W\,. \tag{50}$$

Sketch of the proof: Let us denote $\mathbf{u}_n := \mathbf{u}(A_n)$, $\mathbf{u} := \mathbf{u}(A)$, $\mathbf{u}_{0n} := \mathbf{u}_0(A_n)$, $\mathbf{u}_0 := \mathbf{u}_0(A)$, $T_n := T(A_n)$, $T := T(A)$. Inserting $\mathbf{u} := \mathbf{u}_0 + \mathbf{w}(A)$, $\mathbf{w}(A) \in K$, $\mathbf{u}_n := \mathbf{u}_{0n} + \mathbf{w}_n(A)$, $\mathbf{w}_n(A) \in K$, $\mathbf{v} := \mathbf{u}_0 + \mathbf{w}$ or $\mathbf{v} := \mathbf{u}_{0n} + \mathbf{w}$, $\mathbf{w} \in K$ into the variational inequality (10), we obtain

$$a(A_n; \mathbf{w}_n, \mathbf{w} - \mathbf{w}_n) \geq S(A_n; \mathbf{w} - \mathbf{w}_n, T_n) - a(A_n; \mathbf{u}_{0n}, \mathbf{w} - \mathbf{w}_n)\,. \tag{51}$$

Hence, putting $\mathbf{w} = 0$, using Lemma 1, Theorem 1, definition of $U_{ad}^{u_{0i}}$, after some modifications we find that

$$c_0 \|w_n\|_W^2 \leq c_7 \|w_n\|_W + c_8\,.$$

As a consequence, $\mathbf{w}_n$ are bounded in W and there exists a subsequence $\{\mathbf{w}_k\}$ and a function $\boldsymbol{\omega} \in W$ such that

$$\mathbf{w}_k \rightharpoonup \boldsymbol{\omega} \quad \text{weakly in } W, \text{ as } k \to \infty\,. \tag{52}$$

It can be shown that $\omega = \mathbf{w}(A)$. Thus, since $\boldsymbol{\omega} \in K$ and since $a(A_k; \mathbf{w}_k - \boldsymbol{\omega}, w_k - \boldsymbol{\omega}) \geq 0$, after some modification and using Lemma 2, we obtain $\liminf a(A_k; \mathbf{w}_k, \mathbf{w}_k - \boldsymbol{\omega}) \geq \lim a(A_k; \boldsymbol{\omega}, \boldsymbol{w}_k - \boldsymbol{\omega}) = 0$. Inserting $\mathbf{w} := \boldsymbol{\omega}$ into (51) we arrive at

$$a(A_k; w_k, \omega - w_k) \geq S(A_k; \omega - w_k, T_k) - a(A_k; u_{0k}, \omega - w_k)$$

and

$$\limsup a(A_k; \mathbf{w}_k, \mathbf{w}_k - \boldsymbol{\omega}) \leq \limsup S(A_k; \mathbf{w}_k - \boldsymbol{\omega}, T_k) + \limsup a(A_k; \mathbf{u}_{0k}, \boldsymbol{\omega} - \mathbf{w}_k)\,.$$

For any $A \in U_{ad}$, $T \in W_1$ we can show that $\lim S(A_k; \mathbf{w}_k - \boldsymbol{\omega}, T_k) = 0$ and $\lim a(A_k; \mathbf{w}_k, \mathbf{w}_k - \boldsymbol{\omega}) = 0$ as $\limsup a(A_k; \mathbf{w}_k, \mathbf{w}_k - \boldsymbol{\omega}) \leq 0$, from which it follows that $\lim a(A_k; \mathbf{w}_k, \mathbf{w}_k - \boldsymbol{\omega}) = 0$. It can be shown that $|a(A_k; \mathbf{w}_k, \mathbf{w} - \mathbf{w}_k) - a(A; \boldsymbol{\omega}, \mathbf{w} - \boldsymbol{\omega})| \to 0$; then

$$\lim a(A_k; w_k, w - w_k) = a(A; \omega, w - \omega)$$

and since $|S(A_k; \mathbf{w} - \mathbf{w}_k, T_k) - S(A; \mathbf{w} - \boldsymbol{\omega}, T)| \to 0$, then

$$\lim S(A_k; w - w_k, T_k) = S(A; w - \omega, T)\,.$$

Moreover, we have $|a(A_k; \mathbf{w} - \mathbf{w}_k, \mathbf{u}_{0k}) - a(A; \mathbf{w} - \boldsymbol{\omega}, \mathbf{u}_{0k})| \to 0$, where Lemma 1, Lemma 2 and the convergence $\mathbf{u}_{0k} \to \mathbf{u}_0$ in W were used. Thus

$$\lim a(A_k; \mathbf{w} - \mathbf{w}_k, \mathbf{u}_{0k}) = a(A; \mathbf{w} - \boldsymbol{\omega}, \mathbf{u}_0)\,.$$

Passing to the limit with $k \to \infty$, we obtain

$$a(A;\boldsymbol{\omega},\mathbf{w}-\boldsymbol{\omega}) \geq S(A;\mathbf{w}-\boldsymbol{\omega},T) - a(A;\mathbf{w}-\boldsymbol{\omega},u_0)\,. \tag{53}$$

Since the variational inequality (10) has a unique solution, $\boldsymbol{\omega} = \mathbf{w}(A)$ follows from (53) and moreover, the whole sequence $\{\mathbf{w}(A_n)\}$ tends to $\mathbf{w}(A)$ weakly in W.

Furthermore, the strong convergence can also be proved.

5 Existence of a Solution of the Worst Scenario Problem

To prove the existence of a solution of the worst scenario problem, we will use the following lemma.

Lemma 3.

(i) Let $\Phi_i(T)$, $i = 1, 2$, be defined by (35), (36) and let $T_n \to T$ in W_1, as $n \to \infty$. Then

$$\lim_{n\to\infty} \Phi_i(T_n) = \Phi_i(T), \quad i = 1, 2\,. \tag{54}$$

(ii) Let $\Phi_i(\mathbf{u})$, $i = 3, 4$, be defined by (37), (38) and let $\mathbf{u}_n \to \mathbf{u}$ in W, as $n \to \infty$. Then

$$\lim_{n\to\infty} \Phi_i(\mathbf{u}_n) = \Phi_i(\mathbf{u}), \quad i = 3, 4\,. \tag{55}$$

(iii) Let $\Phi_i(A;\mathbf{u},\mathbf{T})$, $i = 5, 6$, be defined by (39), (40) and let $A_n \to A$ in U, $A_n \in U_{ad}$, $\mathbf{u}_n \to \mathbf{u}$ in W and $T_n \to T$ in $L^2(\Omega)$, as $n \to \infty$. Then

$$\lim_{n\to\infty} \Phi_i(A_n,\mathbf{u}_n,T_n) = \Phi_i(A,\mathbf{u},T), \quad i = 5, 6 \tag{56}$$

The proof is a modification of that of [3].

As the main result of the paper we present the following theorem:

Theorem 3. There exists at least one solution of the worst scenario problems (41), (42), $i = 1, \ldots 6$.
The proof is a modification of that of [3].

6 Conclusion

Mathematical models connected with the safety of construction and of operation of the radioactive waste repositories involve input data (thermal conductivity and elastic coefficients, body and surface forces, thermal sources, coefficients of thermal expansion, boundary values, coefficient of friction on contact boundaries, etc.) which cannot be determined uniquely, but only in some intervals, given by the accuracy of measurements and the approximate solutions of identification problems. The notation "reliable solution" denotes the worst case among a set of possible solutions where the degree of badness is measured by a criterion functional. For the safety of the radioactive waste repositories we seek the maximal value of this functional, which depends on the solution of the mathematical model. Then for the computations of such problems (some mean values of temperatures, displacements, intensity of shear stresses, principal stresses, stress tensor components, normal and tangential components of the displacement or stress vector on the contact boundaries, etc.) we have to formulate a corresponding maximization (worst scenario) problem. Then methods and algorithms known from "optimal design" can be used.

To construct a model of structures under the influence of critical conditions the influence of global tectonics onto a local area, where the critical structure is built as well as the influence of the resulting local geomechanical processes on a critical structure must be taken into account ([6]). Problems of this kind with uncertain input data are problems with high level radioactive waste repositories. In the case of the high level radioactive waste repositories the effects of geodynamical processes in the sense of plate tectonics must be taken into consideration, namely in regions near tectonic areas (e.g. the Japan island arc, the Central and South Europe, etc), but also in the platform regions (as in Sweden, Canada, etc.). Another example is represented by modelling an interaction between a tunnel wall and a rock massif in the radioactive waste repository tunnels or by modelling of a tunnel crossing by an active deep fault(s), respectively.

Acknowledgements

The first author thankfully acknowledges the support of the Grant Agency of the Czech Republic under the grant 201/01/1200. Both authors acknowledge the support under the grant COPERNICUS-HIPERGEOS II INCO-KIT-977006 and the grant of the Ministry of Education, Youth and Sports of the Czech Republic No OK-407.

The paper is prepared for the workshop on "Numerical Models in Geomechanics" of the conference ICCS'2002, Amsterdam, April 2002 and will be published in the Springer Lecture Notes in Computer Science.

References

1. Hlaváček, I.: Reliable Solution of a Signorini Contact Problem with Friction, Considering Uncertain Data. Numer. Linear Algebra Appl. **6** (1999) 411–434
2. Hlaváček, I., Nedoma, J.: On a Solution of a Generalized Semi-Coercive Contact Problem in Thermo-Elasticity. Mathematics and Computers in Simulation (in print) (2001)
3. Hlaváček, I., Nedoma, J.: Reliable Solution of a Unilateral Contact Problem with Friction and Uncertain Data in Thermo-Elasticity (to apperar) (2002)
4. Nečas, J., Hlaváček, I.: Mathematical Theory of Elastic and Elasto-Plastic Bodies: An Introduction. Elsevier, Amsterdam (1981)

5. Nedoma, J.: On the Signorini Problem with Friction in Linear Thermo-Elasticity. The quasi-coupled 2D-case. Appl.Math. **32** (3) (1987) 186–199
6. Nedoma, J.: Numerical Modelling in Applied Geodynamics. John Wiley&Sons, Chichester, New York, Weinheim, Brisbane, Singapore, Toronto (1998)
7. Nedoma, J., Haslinger, J., Hlaváček, I.: The Problem of an Obducting Lithospheric Plate in the Aleutian Arc System. A Finite Element Analysis in the Frictionless Case. Math. Comput. Model. **12** (1989) 61–75
8. Rohn, J.: Positive Definiteness and Stability of Interval Matrices. SIAM J. Matrix Anal. Appl. **15** (1994) 175–184

Computational Engineering Programs at the University of Erlangen-Nuremberg

Ulrich Ruede

Lehrstuhl für Simulation,
Institut für Informatik
Universität Erlangen
WWW home page: `http://www10.informatik.uni-erlangen.de/ ruede`
`ruede@cs.fau.de`

Abstract. The Computational Engineering Program at the University of Erlangen was initiated by the Department of Computer Science as a two year graduate program leading to a Master's degree in 1997. In 1999, a three year undergraduate program was added. As a peculiarity, the Masters program is taught in English and is thus open to international students who do not speak German.
The curriculum consists of approximately equal parts taken from computer science, (applied) mathematics, and an engineering discipline of choice. Within the current program, this includes material sciences, mechanical engineering, fluid dynamics, and several choices from electrical engineering, such as automatic control, information technology, microelectronics, semiconductor technology, or sensor technology.

1 Organisational Background

Computational Engineering (CE) at the University of Erlangen was initiated by the Department of Computer Science in 1997 as a two year postgraduate program leading to a Masters degree. Originally the program was funded by the *Deutsche Akademische Austauschdienst (DAAD)* within a project to attract more international students to Germany. Thus all core courses of the Master's program (but not the undergraduate program) are taught in English and are thus open to international students without knowledge of German.

The CE program is presently offered within Erlangen's School of Engineering (Technische Fakultät), which includes the Departments of Computer Science, Electrical, Chemical, and Mechanical Engineering, and the Department of Material Sciences. The mathematics courses of the curriculum are being taught by the Department of Applied Mathematics in the School of Sciences.

The Bachelor-Master structure of academic programs is still in an experimental stage in Germany. The traditional German Diploma degree still dominates and both faculty and (prospective) students view the new Bachelor or Master programs with considerable scepticism. Despite this, the CE program is currently accepting approximately 50 new undergraduate students and 40 graduate students annually.

P.M.A. Sloot et al. (Eds.): ICCS 2002, LNCS 2331, pp. 852–857, 2002.

Formal Ph.D. programs are traditionally not offered within the German university system. However, sufficiently well qualified Master graduates are routinely being offered Ph.D. studentships and research assistantships by the faculty participating in the CE program.

2 Undergraduate Curriculum

The undergraduate curriculum consists of an approximately equal number of credits in computer science, (applied) mathematics, and an engineering discipline of choice.

Required courses consist of the traditional German four semester sequence in *Engineering Mathematics* which contains elements from calculus, analysis, linear algebra, and ordinary and partial differential equations(PDEs). Starting with the second year, students are required to take a two semester sequence in *Numerical Analysis*.

The core computer science courses are taken from Erlangen's standard computer science curriculum. CE students must take the two semester sequence *Algorithmik* that teaches the standard algorithms of computer science, data structures, programming languages, compilers, plus some elements of theoretical computer science. Additionally, students are required to take two semesters of *Organisation and Technology of Computers* and *Systems Programming* which include basic computer architecture and operating systems.

All students must choose a *Technical Application Field*. Presently, the program offers specialisation in

- Mechanical Engineering,
- Micro Electronics,
- Information Technology,
- Automatic Control,
- Thermo- and Fluid Dynamics,
- Material Sciences, and
- Sensor Technology.

Depending on the particular choice of the technical application, each student must take a sequence of 10 to 20 credits of application specific required courses. With specialisation in, for example, *Thermo- and Fluid Dynamics* the required courses are

- Elementary Physics,
- Thermo Dynamics (two semesters),
- Heat- and Mass Transfer, and
- Fluid Dynamics (two semesters).

Additional requirements are a 12 week internship in industry and participation in a seminar.

All courses listed above are taken from existing degree curricula. A new course, specifically designed for the CE program, is *Scientific Computing* (6 credits). This is the core required course in the second year and is aimed at teaching the specific techniques of computational science and engineering on an elementary to moderate level.

The specific emphasis of the present course is on practical and algorithmic aspects, rather than theory. Currently the contents are

- cellular automata (including lattice gases),
- solution of dynamical systems (initial and boundary value problems),
- basic visualisation techniques,
- elementary numerical PDEs, and
- iterative solution techniques.

Besides the core curriculum of required courses, students can and must select additional 10-20 credits from an exhaustive list of optional courses that can be taken either from the student's application field, Computer Science, or Applied Mathematics. Though any course from the conventional degree programs of the participating departments can be chosen, students are intensively advised and guided individually to enable them to find suitable combinations of courses. Any selection should primarily provide application oriented and practical skills. Typical courses for such a specialisation include

- Computer Graphics and Visualisation,
- High Performance and Parallel Programming,
- Pattern Recognition,
- Computer Networking,
- Advanced Numerical Methods (including numerics of PDEs),
- Finite Elements, and
- Optimisation.

Since the traditional German education system puts strong emphasis on independent, project oriented work, the final requirement is a project of three months duration and includes writing a short thesis.

3 Masters Curriculum

Graduate students are selected on a competitive bases from Mathematics, Computer Science, or a technical discipline. In a first *orientation semester*, students are required to take courses in

- Usage of Computer Systems and Operating Systems,
- Advanced Computer Programming and Data Structures,
- Computer Architecture,
- Numerical Mathematics,
- Scientific Computing, and
- an overview course according to the application field of choice.

These courses are designed to provide new students, who may come from quite different educational backgrounds, with a uniform basis of knowledge in basic CE techniques.

For sufficiently well qualified students, their participation in individual courses of the orientation semester, or even the complete orientation semester, can be waived. This will be routinely done for students coming from a computational science or engineering undergraduate program that conforms with our curriculum.

The main part of the Masters curriculum consists of a total of 38 credits of elective courses which should be taken within two semesters. A minimum of 10 credits must be taken from a technical application where the list of choices for the application field is the same as in the undergraduate curriculum (though only a subset is offered in English).

Sixteen credits must be earned in courses of advanced computer science and 10 credits in interdisciplinary courses which may be taken from either the technical application, Mathematics, or Computer Science. Required courses depend on the application field. As example, the required courses for a specialisation in Thermo- and Fluid Dynamics are:

- Numerical Fluid Mechanics,
- Applied Thermodynamics,
- Numerical Methods for PDEs, and
- Special Topics in Scientific Computing,

while courses like Visualisation, High Performance and Parallel Computing, Non-Newtonian Fluids, or Turbulence, are recommended optional courses. Participating in a seminar is also required in the Masters program.

As in traditional German Diploma degree programs, a strong emphasis is put on project oriented, independent work. A full six month period is therefore set aside for work on a Master thesis. The thesis topic may again be chosen from either the student's technical application, Mathematics, or Computer Science. An excellent thesis is usually considered as the primary prerequisite to being offered a Ph.D. studentship.

4 CE Specific Courses

As described above, the majority of courses in the Erlangen CE curriculum are taken from conventional programs. One notable exception is the *Scientific Computing* class that is mandatory in the undergraduate program and is part of the graduate student orientation semester.

The remaining courses of the graduate orientation semester are also specifically taught for the CE program. Thus, for example, the computer architecture class for CE can put more emphasis on high performance and parallel computing than the standard classes taught for Computer Science students, where computer architecture would be treated in more generality and breadth. Similarly, the numerical analysis class for the graduate CE program is specifically tuned for the audience and puts more emphasis on algorithms and program development than on analysis and theory.

At the more advanced level, we have also begun to create new classes with the goal to better bridge the gap between the disciplines. These courses are interdisciplinary and are taught jointly by faculty from the different departments.

One such course is *Numerical Simulation of Fluids* which is taught jointly by Chemical Engineering and Computer Science faculty. Using [NFL] as the basic text, the course guides students to develop an incompressible Navier-Stokes solver and to adapt and apply their code to a nontrivial application scenario. The first half of the course has weekly assignments that result in a core 2D fluid simulator. The method is based on a staggered grid finite difference discretisation and explicit time stepping for the velocities. When the students have completed the core program, they form project teams of up to three to develop some advanced extensions of their choice. Typical projects include extending the 2D solver for 3D flows, handling free surfaces, adding heat transport and simulating buoyancy driven flows, advanced flow visualisation, or parallelizing the pressure correction step. For more details, see [NFLW].

A second course, which follows a similar pattern is offered jointly by Electrical Engineering and Computer Science. Here the goal is to develop a finite element program for simulating electrical fields. The emphasis is on code development using modern object oriented techniques, data structures for handling unstructured grids, and solution by iterative techniques.

5 Future Development

The University of Erlangen's CE program is one of the first such efforts in Europe. Within the past five years, the program has grown to respectable breadth and is attracting a substantial number of students from Germany and internationally.

However, the program is still clearly in the experimental phase and will continue to be adapted and extended. The discussion about what makes CE different from a traditional engineering discipline, (applied) mathematics, or computer science is still continuing. We have learned that composing a curriculum by simply combining existing courses from the participating departments can only be a beginning. Both graduate students and undergraduates need systematic courses to integrate the different components and to teach them truly interdisciplinary work.

The SIAM white paper on CSE [SIAM-CSE] defines the field as being primarily oriented at developing *new and improved methods* for problems and applications. The novelty of the application is secondary. However, engineers or scientists from the potential CSE application fields are primarily interested in solving new and interesting *problems*, even if they can be dealt with using standard computational techniques and tools. For them, developing new algorithms and software is secondary.

This conflict of interest is also visible in our current curriculum, where some application choices are primarily oriented at developing new methods and exploiting new computing technology, while others are primarily oriented at teaching a broad range of existing methods and tools within their application field. From our experience, this is a core issue that will continue to be a source of interesting discussions.

References

[ER-CE] Erlangen Computational Engineering WWW-Site:
http://www10.informatik.uni-erlangen.de/CE

[Guide01] Erlangen CE Study Guide (2001 edition):
http://www10.informatik.uni-erlangen.de/CE/officialdocs/ce_study_guide.ps.gz

[NFL] Michael Griebel, Thomas Dornseifer, and Tilman Neunhoeffer: *Numerical Simulation in Fluid Dynamics: A Practical Introduction*, SIAM, 1997.

[NFLW] Numerical Simulation in Fluids Web Site:
http://www10.informatik.uni-erlangen.de/teaching/NuSiF/Project.html

[SIAM-CSE] SIAM Computational Science and Engineering WWW-Site:
http://www.siam.org/cse/index.htm

Teaching Mathematical Modeling: Art or Science?

Wolfgang Wiechert

University of Siegen, FOMAAS,
Paul-Bonatz-Str. 9-11, D-57068 Siegen, Germany.
E-Mail address: wiechert@simtec.mb.uni-siegen.de,
WWW home page: http://www.simtec.mb.uni-siegen.de

Abstract. Modeling and simulation skills are two core competencies of computational science and thus should be a central part of any curriculum. While there is a well-founded theory of simulation algorithms today the teaching of modeling skills bears some intrinsic problems. The reason is that modeling is still partly an art and partly a science. As an important consequence for university education the didactic concepts for teaching modeling must be quite different from those for teaching simulation algorithms. Some experiences made with a two term course on 'Modeling and Simulation' at the University of Siegen are summarized.

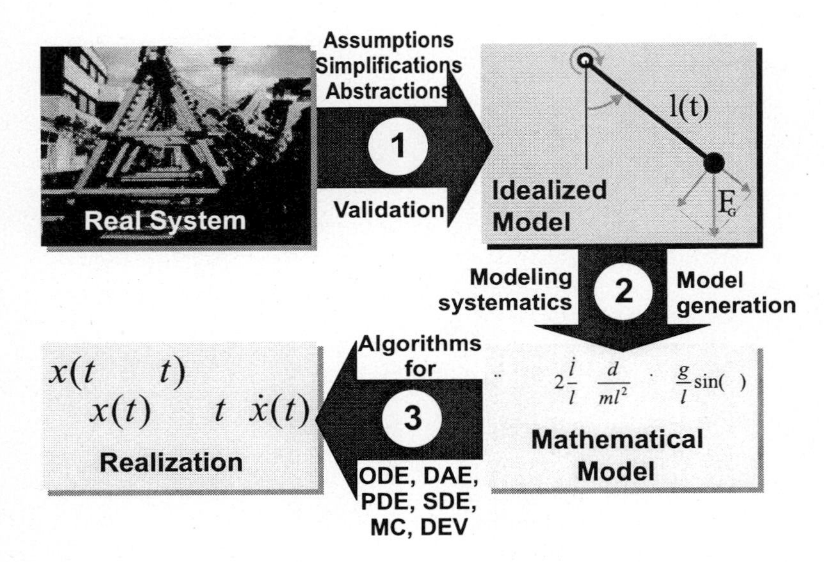

Fig. 1. General steps in modeling and simulation. The example simulation project is concerned with the modeling and simulation of a boat swing. The goal is to perform a 360° looping. Details are given in the main text.

P.M.A. Sloot et al. (Eds.): ICCS 2002, LNCS 2331, pp. 858–862, 2002.

1 General Modeling and Simulation Methodology

As a widely accepted approach to the modeling and simulation of a real system the overall task is divided into the following three steps, although these steps are not always clearly separated in practice (cf. Figure 1):

Step 1: From reality to an idealized model. In the first step (frequently also called 'physical modeling' [1]) the real system is replaced by an idealized model. This model is characterized by a well-defined semantics which means that once the model has been constructed there is nothing more to discuss about the physical laws that have to be used in order to describe the system. From a didactic viewpoint there are two different aspects to consider:
1. General problem of modeling reality and possible pitfalls of modeling.
2. Concepts, methods and tools for model validation.

Step 2: From the idealized model to a mathematical model. By a mathematical model a formal representation using mathematical concepts like differential equations, probability distributions or state machines is meant. The (possibly automatic) generation of the mathematical model from the idealized model essentially is an algorithmic task and is very well understood in many application disciplines (like multibody systems or electrical circuits). The following aspects are relevant in education:
1. General systematics of modeling.
2. Types of mathematical models (ODE, DAE, PDE, DES etc.).
3. Algorithms for automatic model generation.

Step 3: From the mathematical model to a computer realization. In general mathematical models represent a system in a descriptive or implicit way. A computer realization of a mathematical model means to compute a set of (possibly time and state dependent) variables that approximatively solve the model equations. This is the field of simulation algorithms which can be roughly subdivided into:
1. Numerical algorithms for AEs, ODEs, DAEs, PDEs, etc.
2. Random number generation, stochastic algorithms, MC methods etc.
3. Algorithms for discrete event systems, automata, Petri nets, etc.

2 Teaching Mathematical Modeling

This contribution further concentrates on the aspect of teaching mathematical modeling. It is not easy to answer the question of how to do this in the most efficient way. In the three step scheme given above the modeling activities are spread over Steps 1 and 2 while Step 3 is best described as the 'simulation' step in a narrower sense. Basically three aspects of modeling have to be integrated into a course on computational science:

Finding the right idealization of reality: This is the most difficult part in the modeling process. It is an important point that once the idealized model has been created (Step 1) we are no more confronted with reality but with models. In particular any mistakes made in the early phase of model building can never be corrected later on even by the highest precision numerical algorithms. These interconnections must be made very clear to the students. Otherwise there is a high risk to get lost in the difficult but always well-defined mathematical problems of designing powerful simulation algorithms while loosing the connection to the real system and the goal of the simulation project. Unfortunately, many textbooks start with the idealized model (as e.g. the mathematical or physical pendulum) instead of reality.

Building well-structured models: Although all the basic modeling decisions are made in the first step this does not imply that the mathematical model built in Step 2 is also well-structured (i.e. easy to maintain, suited for automatic model generation, well prepared for simulation algorithms). In recent years there has been great progress in the field of modeling systematics and several highly developed modeling methodologies for different model types exist [1–4]. All these methodologies are mathematically well formalized and algorithmically specified. Thus – in contrary to the first step – there is a sound scientific platform for teaching these aspects of modeling.

Model validation: Finally it should always kept in mind that modeling is an iterative procedure guided by the attempt to validate a given model. Thus model validation always accompanies the whole process of modeling and simulation. Unfortunately, the concept of 'model validity' is not at all well-defined and leads into deep methodological problems. Some well-founded concepts for model validation have been developed in a statistical framework [5] but all these methods generally rely on even more assumptions on the real system (like e.g. a known error distribution).

In summary, building the idealized model in Step 1 is still rather an art than a science. A lot of experience and specific know how about the application domain is required. On the other hand the building and automatic generation of well-structured models from a physical model (Step 2) has a sound scientific background. Likewise most of the aspects related to simulation algorithms (Step 3) are covered in well written textbooks.

3 Three approaches to modeling

The big challenge of university education in modeling and simulation is how to teach the art of modeling (i.e. Step 1) in an efficient way. From the experiences gathered with the lectures on 'Modeling and Simulation' at the University of Siegen the general problem of modeling can be best understood by studying 'fruitful' examples and making practical experiences with simple modeling problems. The main difficulty herein is to avoid an overload of technical problems that are more related to simulation than to modeling. Three types of examples can be distinguished. The first one is best suited for lectures, the second one for practical exercises and the last one for a seminar:

Extensive example study: The lecture starts with a simple yet not oversimplified example of a typical simulation project. It illustrates all steps of the modeling and simulation process without requiring any prior simulation knowledge from the students. The goal is to design a swing boat which can perform a 360° looping (Figure 1). To this end the right dimensions of the swing boat must be determined, a strategy to accelerate the swing must be found and the resulting forces on the person must be computed [6]. Essentially this ends up with the classical mathematical pendulum with time dependent length. However, a lot of simplifying assumptions must be made for this reduction and some of these assumptions (e.g. on friction in the bearing) can be validated experimentally. This study serves well to give a first impression of the general relation between modeling and simulation activities in a realistic project.

Prototypical examples for certain modeling problems: These are strongly reduced examples which cover only a certain modeling problem (like e.g. friction or contacts in mechanical systems). The examples are particularly well chosen if they seduce the students to make a typical mistake. Such examples are treated for example in the first 7 units of the simulation laboratory at the University of Siegen [6] or in the EUROSIM comparisons [7]. Another source of fruitful examples turned out to be the modeling and simulation of physical toys like the woodpecker toy that illustrates the stick slip effect [8]. This is also a typical problem for the last exercise (unit 8) of the simulation laboratory where a small project has to be managed in about three days with a final presentation.

Multiscale studies of one specific system: The choice of model type usually depends on the time or space scale on which reality has to be understood. Examples by which several different scales and hence different modeling approaches can be studied are mechanical springs (Hook's law, non-linear springs, valve springs with contacts, strain inside a spring, fracture mechanics, molecular basis of elasticity), chemical reaction systems (reaction kinetics, stochastic particle system, molecular dynamics, Schrödinger equations) or manufacturing systems (strategic simulation, continuous material flow, discrete material flow, machine modeling, virtual reality).

4 Conclusions

Computational science involves skills from different scientific disciplines (numerics, statistics, computer science,...) and a profound knowledge of the application domain (physics, chemistry, biology, engineering,...). At the University of Siegen a general two term graduate lecture on 'Modeling and Simulation' serves as a link between the mathematical disciplines and the application domains. It demonstrates the different steps in the modeling and simulation process and clarifies the role that each single discipline plays in the whole concert. The interdisciplinary lectures are suited for engineers, mathematicians, computer scientists and natural scientists.

Due to its broad scope the mathematical demand of the introductory lecture is rather low and only the basic ideas of simulation algorithms are explained.

Those students who want to get a deeper insight into a certain topic can attend other lectures on numerics, finite elements, computational fluid dynamics, parallel computing, software engineering etc. which are given in different departments of the University of Siegen. Thus without currently having established a dedicated masters course on computational science it is already possible for students of different disciplines to put their focus on it.

An organizational frame for computational science at the University of Siegen is given by the Research Centre for Multidisciplinary Analysis and Applied Systems Optimization (FOMAAS) which is an interdisciplinary union of scientists from the engineering, mathematics, and computer science departments [9]. FOMAAS works on the analysis, simulation and optimization of complex systems. Moreover the author is responsible for the coordination of educational activities in the German/Austrian/Swiss simulation organization ASIM. More details about the simulation courses at the University of Siegen and curricular aspects can be taken from [6,10,11].

References

1. Tiller, M.: Introduction to physical modeling with Modelica, Kluwer Academic, 2001
2. Cellier, F.E.: Continuous System Modeling, Springer (1991).
3. Farlow, S.J.: Partial Differential Equations for Scientists and Engineers, Wiley (1993).
4. Fishman, G.S.: Discrete-Event Simulation: Modeling, Programming and Analysis, Springer (2001).
5. Burnham, K.P., Anderson, D.R.: Model Selection and Inference - A Practical Information-Theoretic Approach, Springer (1998).
6. Wiechert, W.: Lectures on Simulation: Lecture Notes and Materials, http://www.simtec.mb.uni-siegen.de
7. ARGESIM, Arbeitsgemeinschaft Simulation, EUROSIM Comparisons, TU Wien. http://www.argesim.org.
8. Ucke, C.: Literature data base on physical toys, http://fluorine.e20.physik.tu-muenchen.de/~cucke/
9. FOMAAS (Research Centre for Multidisciplinary Analysis and Applied System Optimization): University of Siegen, http://www.fomaas.uni-siegen.de.
10. Wiechert, W.: Multimediale Lehre im Fach Simulationstechnik. Pp. 39-53 in: Großmann, K.; Wiemer, H., SIM2000 - Simulation im Maschinenbau, scan-factory, Dresden (2000).
11. Wiechert, W.: Eine Checkliste für den Aufbau einer Simulationstechnik Vorlesung. Pp. 151-156 in: Panreck, K.; Dörrscheid, F. 15. Symposium Simulationstechnik, Paderborn, SCS Publishing House (2001).

CSE Program at ETH Zurich: Are We Doing the Right Thing?

Rolf Jeltsch
Seminar for Applied Mathematics
ETH Zurich
CH-8092 Zurich
jeltsch@math.ethz.ch

Kaspar Nipp
Seminar for Applied Mathematics
ETH Zurich
CH-8092 Zurich
nipp@math.ethz.ch

1 Introduction

Computational techniques, next to theory and experiments, have become a major and widespread tool to solve large problems in Science and Engineering. Especially in places where large computer power was available, such as in North America, Japan and Europe, the need was recognized to start to educate persons in the new discipline Computational Science and Engineering (CSE). A first survey of the state of affairs in this new subject is given in the report 'Graduate Education in Computational Science and Engineering' of the SIAM working Group [3].

At the Swiss Federal Institute of Technology (ETH) in Zurich, W. Gander (Institute for Scientific Computing, Computer Science), M. Gutknecht (Swiss Center for Scientific Computing until 1998, since 1999 Seminar for Applied Mathematics, Mathematics) together with R. Jeltsch (Seminar for Applied Mathematics, Mathematics) started the discussion of implementing CSE education at ETH in 1995. The first discussion was to find ways to add a CSE direction in each engineering and each science curriculum. This turned out not to be practical at that time, due mostly to the regulations amongst the curricula differing a lot. For example, this would have meant that one would have to teach several courses in Numerical Analysis with a different number of teaching hours per week, each for a very small number of students. Moreover, in the diploma degree received by a student, CSE would then only have been mentioned as a minor, if at all. The Executive Committee of ETH Zurich strengthened the group at the end of 1995 by making it a formal committee and adding two more members, W. van Gunsteren (Computational Chemistry, Chemistry) and K. Nipp (Seminar for Applied Mathematics, Mathematics).

In this article we describe the program developed by this committee. It leads to a Diploma (M.S.). The first students started in October 1997. Since then

P.M.A. Sloot et al. (Eds.): ICCS 2002, LNCS 2331, pp. 863–871, 2002.

the curriculum has undergone minor modifications and we present here the one which will be in effect for students starting this year in fall.

The key elements of this curriculum are as follows. The overall study time to reach the Diploma in CSE should not be longer than the other curricula leading to a diploma at ETH, i.e., 4.5 years (9 Semesters). The curriculum should build upon other classical studies in engineering and sciences. A student would first study two years in one of these subjects before changing to the CSE curriculum which then would last only for 2.5 years (5 semesters). Depending on the knowledge a student has when entering the program, a certain number of supplementary courses have to be taken in order to bring all students to an approximately equivalent level of education.

In the present curriculum in CSE at ETH 'computational' is understood as mainly numerical computing, rather than logical computing or data bases. Hence a large portion of the curriculum is concerned with differential equations, ordinary and partial, optimization in all variations. In a sense we consider CSE a new expression superseding 'scientific computing' by being more applied. See for a comment on this in the interview of B. Engquist by O. Nevanlinna, [2].

In Section 2 we describe the curriculum in detail while in Section 3 we briefly mention the organizational structure. In Section 4 the experiences from the past are described and future development is touched upon. We finish with some conclusions in the last section.

2 Curriculum at ETH

2.1 Design principles

a) Aim of curriculum

When designing a new curriculum one should think of what a student should know, which abilities she or he should have acquired, when finishing successfully the program. Since ETH had no previous program in CSE we had the luxury to create the curriculum we felt would be best.

The major objectives and design principles are:

- The emphasis is on Science and Engineering and not just computing for the sake of computing, i.e., we want that the student knows what she or he is computing.
- Computing has to be an essential tool in any field taught.
- A student should make an in depth study in at least one or maybe two fields.
- A student should obtain a broad view and should have knowledge of many applications.
- A student should acquire the ability to work in a team.

- A student should have the ability to work together with people from different backgrounds (e.g., he or she must be able to talk to the scientists and or engineers who know the application area and he or she must be able to talk to mathematicians and computer scientists because one can not expect of the person educated in CSE to know that much mathematics and computer science that he or she can solve all problems by himself or herself).
- A student should have the ability to enter quickly in an ongoing research project of a team, make his contribution and transmit it successfully to the team.
- A student should acquire communication abilities.
- The overall study time should not be longer as with other students at ETH, i.e., 4 years and an additional 4 months for the diploma thesis.

b) ETH boundary conditions

Some of the boundary conditions at ETH are time dependent, i.e., as time goes on they might change and then the CSE program will have to change too. Here are some of the conditions and their effect on the curriculum.

- Curricula at ETH in Zurich draw first semester students mostly from the German and Italian speaking part of Switzerland. The number of students therefore for a curriculum in CSE starting in the first semester is small and hence not very cost effective. Due to this, it was decided to start the curriculum only after two years of basic studies in an appropriate science or engineering curriculum. This has the additional side effect that the students come from different fields and learn on a daily basis do communicate with persons from other fields.
- Since we did not expect many students and ETH had already many courses which could be modified, such that these became excellent for our CSE curriculum, we wanted to use as many as possible existing courses. It is clear that as ETH acquires more lecturers in CSE and more departments add scientists who use computation as a major tool, the number of courses taught for the CSE students alone would increase. This has already happened. In the first year, only three courses were taught due to the CSE curriculum. In the next academic year this number will at least be doubled.

2.2 Curriculum

a) Overview

The above mentioned design principles have lead to the following curriculum which should be covered in four semesters. The hours mentioned in the second column are the number of hours per week in a semester added up over the four semesters. For example, the total of 83 hours means that each semester a student has approximately 21 hours of courses per week.

Overview of the Curriculum

	Hours of work per week per semester	short description
Core Courses	37	methodology in mathematics, computer science, important application areas, case studies
Field of Specialization	10	work in one computational application area, specialization
Elective Courses	12	must be CSE related
Two Term Papers	12	application oriented, work in a team, computational, at least 180 hours of work
Supplementary Courses	12	to complete foundations for CSE education
Total	83	

In addition to this every student at ETH must acquire 8 credits from the department of Humanities, Social and Political Sciences during the four years of studies. The duration of the diploma thesis is four months and it is usually done after the four semesters of study.

b) Core Courses

The nine core courses consist of eight courses covering numerics of differential equations, parallel computing, optimization techniques, computational statistics, methods of computer-oriented quantum mechanics and statistical mechanics, software engineering, visualization and the so called 'case studies'. In these first eight core courses, a student learns CSE related methodologies which enables her or him to use these as problem solving tools. Clearly these eight courses are mandatory and they are offered each year. The course 'case studies' is offered each semester and it is mandatory for a student to participate. The case studies have two functions. Lecturers from ETH and specialists from outside institutions, from industry or academia, demonstrate in two lectures, each 45 minutes, one area of application from their own fields of work, from the modelling to the solution of the problem with the help of the computer. A list of topics treated in the winter semester 2001/2002 is added below. From this list one sees that students have to do presentations to improve their communication abilities. There are two types of student presentations. Each semester, every student has to give a ten minute presentation of a short scientific article. He or she can choose an article from literature after consultation with the course leader or can choose from a set of predetermined articles. Each presentation is followed by a question and answer session and a discussion of the communication skills. The second type of student presentation is the oral communication of the result of a term paper. The presentation lasts usually 20 - 30 minutes.

In the case studies, the student gets a broad overview and learns many applications. In addition his or her communication abilities are improved.

Case Studies	Winter semester 2001/2002
25. 10. 01	Student presentations
1. 11. 01	E. Kissling, Geophysics, ETH Seismic Tomography
8. 11. 01	Xavier Daura, Computational Chemistry, ETH Computer Simulation in Biochemistry
15. 11. 01	L. Scapozza, Pharmaceutical Biochemistry, ETH Computational Science within Life Sciences
22. 11. 01	F. Kappel, University of Graz Modeling of the Cardio Circulatory System
29. 11. 01	Student presentations
6. 12. 01	W. Benz, Institute of physics, University of Bern Formation and Evolution of Planetary Systems
13. 12. 01	Student presentations
20. 12. 01	Student presentations
10. 1. 02	A. Troxler, Seminar for Applied Mathematics, ETH Aerodynamic Design of Gas Turbines
17. 1. 02	W. Flury, Esoc / ESA, Darmstadt Optimization of Inter-planetary Trajectories Using Solar Electric Propulsion and Gravity
24. 1. 02	Jochen Maurer, Time-steps, Affoltern am Albis Pricing and Risk Analysis of Energy Portfolios
31. 2. 02	Student presentations
7. 2. 02	Student presentations

c) Field of Specialization

The student has to follow courses in one of the following fields of Specialization: Atmospheric Physics, Computational Chemistry, Computational Fluid Dynamics, Control Theory, Robotics, Theoretical Physics. A field of specialization typically comprises three courses totalling approximately ten hours per week during a semester. Typically one or two of these courses are not computational but provide the student with the scientific or engineering background to be able to understand what one wants to compute in the remaining courses. Typically a student would do one term paper in his or her field of specialization. It is at this point that a student learns to go in depth in one field.

The list of possible fields of specialization depends on the research groups which compute at ETH. To be able to have a field of specialization, such a research group has to guarantee that these three courses are taught regularly every year. This means that the group has to consist of at least two lecturers.

The list is flexible, i.e., as new professors come to ETH new fields can be listed. For example in the near future computational astrophysics and computational biology might be added.

d) Elective Courses

During their studies a student must attend four elective courses totalling at least 12 hours per week. These courses have to be computational or provide the students with even more scientific or engineering background in her or his field of specialization. A student can use these courses in different ways:

- to deepen the topic treated in a core course.
- to deepen the understanding and knowledge in the chosen field of specialization.
- to get a second field of specialization.
- to broaden the students view by attending additional courses in other areas.

A list of such courses is presented in the Lecture Catalogue. Other choices might be approved by the Studies Advisor, however. The courses often concern research fields where the size of the group is not large enough to sustain a course program for a field of specialization. Moreover, these courses might change every year.

e) Term Papers

It is here where a student learns to get integrated in a team. He or she is assigned to a running research group which is modelling a problem or which is developing a large code. The time spent is at least 180 hours, which means that during a whole semester of 14 weeks she or he works 3 hours, four afternoons per week. In the first few weeks, members of the group introduce the student to the code and the problem. Then the actual work begins. A term paper is produced in the end and the result is presented to the group and in the 'case studies'.

f) Supplementary Courses

As students enter from different backgrounds, they have to do supplementary courses fundamental to CSE. The aim is that upon finishing their studies they all have the same knowledge on the fundamentals and the core courses. How are these supplementary courses determined? A virtual ideal curriculum for the non-existing first two years of studies was determined by the CSE Executive Committee. When a student enters after four semesters of basic education, the courses actually taken are compared to the virtual ideal curriculum. Then it is determined which of the missing lectures he or she still has to follow and which exams have to be taken. For the two-year curricula of 7 departments of the ETH, the supplementary courses to be taken by the student following the CSE program are predetermined. For all others, especially the ones coming from outside of ETH, the Studies Advisor determines the supplementary courses.

g) Diploma Thesis

The time to complete a diploma thesis is four months. Generally it is done right after the final examination following 2 years of studies in CSE, the topics being from similar fields as the term papers.

3 Organizational structure

The general problem with a curriculum such as CSE is the following. Since the focus is on science and engineering, it is preferred to have the professors remaining in the department where their scientific or engineering interest lies. Hence the computational fluid dynamics person will stay in mechanical engineering, the applied mathematician in mathematics, the computational chemistry person in chemistry. It is believed that in this way the continuity of high quality research is better maintained. The curriculum involves in this way many departments and this makes things complicated.

At ETH a curriculum has to be attached to one department and hence the question arose to which one the CSE curriculum should be attached, e.g., to Computer Science or Mathematics. Fortunately, ETH has a long tradition that the Physics and the Mathematics Department do their curricula by a common body, the so called 'Unterrichtskonferenz' (Conference for teaching). All lecturers in mathematics and physics are members of this body. In addition, if a topic in CSE arises, a certain representation of the CSE lecturers are invited too. This representation consists of the Studies Advisor, the CSE Executive Committee, the coordinators for the fields of specialization and the lecturers of the core courses as well as the courses of the fields of specialization. The most active part of the CSE structure is the small CSE Executive Committee, which has only six members (the Studies Advisor, five members from different departments, one of them acting as secretary of the committee). This construction is very good in the sense that it reduces the number of meetings and the number of persons having to sit in meetings. However this structure will have to evolve in time too.

4 Experiences and future development

What are the experiences from the past years? On the positive side we have to mention that the program attracted extremely good students. The grades in their Diploma examination have been on average higher than in traditional curricula. In addition, they were all very fine personalities. On the downside one has to mention that the number is still small. While in 1997 11 entered the program, 2 of which withdrew, last year 15 entered. What are the reasons for this? The program is difficult for the students, because we use mostly existing courses from other curricula and this means that conflicts in schedule can not be avoided. Hence, often a student learns partly from the professor's manuscript because he or she can not attend all lectures. However, as ETH hires more and more computationally oriented scientists, we will have the staff to make more courses taught for CSE students only. In addition, the number of supplementary courses which have to be taken and the additional examinations which have to be passed turned out to make the studies even harder. This number will be reduced therefore, starting this coming autumn. For the above reasons, the students often feel overloaded. Since there have not been many courses for the CSE students alone, some felt that they were missing an identity. Again, this problem will be eased once more professors in this area are hired. The students did appreciate that they had close contact and much advice from the Secretary of the CSE Executive Committee and from the Studies Advisor. The low number of students is also probably due to the fact that CSE is still not a very common curriculum, it is also very difficult to reach people in the middle of their studies and to motivate them to change.

On the positive side one can mention that last year 10 students did their diploma thesis and thus finished successfully the program, see [1], p. 11. Of these, 4 chose Computational Fluid Dynamics as their specialization, 4 chose robotics, 1 chose theoretical physics and 1 atmospheric physics. In addition, contrary to expectation most students wanted to go on and do a Ph.D.

On the positive side for ETH, the fact is that this additional curriculum does not cost very much. 15'000 CHF each year are used for the course 'case studies' and some other items. The senior scientist who is the secretary of the CSE Executive Committee uses about 60 % of his time for running the curriculum. In addition, colleagues have been positively motivated to review and renew the contents of their courses, and new courses were created. The most interesting one was the course in computational physics. This was needed to make physics a Field of Specialization. Now, there are much more physics students in this course, than students from CSE. We expect this to happen in other new courses too.

What are the changes for the future? As ETH will change from the Diploma system to a Bachelor and Master system, the CSE curriculum will have to undergo change too. We expect this to happen in autumn 2003. With this change, ETH will abolish the 2nd year examination. With the present system it was natural to move from a classical curriculum to the CSE curriculum right after this second year examination. Whether one will still be able to change after two years of studies or not, will have to be discussed.

In another move ETH decided last year to create three professorships in CSE. One has been filled last year by a professor who used to be in computational fluid dynamics. Currently this increased manpower is used to strengthen the CSE part in different curricula, such as in mechanical engineering, in computer science and in biology. In addition, an interdepartmental Center for CSE is planned.

5 Conclusions

At ETH a CSE program to obtain a Diploma (M.S.) was successfully started in 1997. It builds mainly on existing courses, but is flexible enough so as to accommodate the change in the faculty, which definitely moves to more scientists using computation as a major tool in their research. The focus is on science and engineering and not on computing for the sake of it. A small number of students have entered and successfully finished the program with very high marks. It is planned that most courses will be taught in English as soon as possible. A Bachelor and a Master program are considered. The curriculum in CSE has been very interesting and changes had to be implemented in a flexible and fast way. This will definitely remain so in the coming years.

For more information on the curriculum see the web page www.cse.ethz.ch. Reference [1] can be ordered from the Seminar for Applied Mathematics, ETH Zurich, CH-8092 Zurich.

References

1. CSE Computational Science and Engineering, Annual Report 2000/2001, R. Jeltsch, K. Nipp, W. van Gunsteren, edits., ETH Zurich, 2001
2. O. Nevanlinna, Interview with Bjoern Engquist, interviewer O. Nevanlinna, EMS-Newsletter June 2001, p 15
3. SIAM Working Group on CSE Education, Graduate Education in Computational Science and Engineering, SIAM Rev.43, (2001), 163-177

An Online Environment Supporting High Quality Education in Computational Science

Luis Anido, Juan Santos,
Manuel Caeiro, and Judith Rodríguez

Grupo de Ingeniería de Sistemas Telemáticos
ETSI Telecomunicación
University of Vigo, SPAIN
{lanido, jsgago, mcaeiro, jestevez}@det.uvigo.es

Abstract. Education in Computational Science is a demanding task, especially when online support is required. From the learner's point of view, one of the most challenging issues is to get used to the supporting tools, especially with the computing devices employed (e.g. DSPs). For this, students need to understand concepts from the Computer Architecture area. This paper presents a WWW-based virtual laboratory that provides learners with a suitable environment where Computer Architecture concepts, needed to face Computational Science education, can be acquired. Our contribution can be described as an open distributed platform to provide practical training. Services offered support the seamless integration of simulators written in Java, and include features like student tracking, collaborative tools, messaging, task and project management, or virtual file repositories. A CORBA-based distributed architecture to access real computer systems, available in labs at University facilities, is also described.

1 Introduction

Computer Architecture has some specific properties that makes it a challenging matter when trying to provide adequate supporting tools for e-learning. While the implementation of the first phases of the learning process in this field, typically through theoretical lectures, poses no major problems today given the help provided by available authoring systems and course tools and resources, remote laboratory settings are far more demanding. Student interaction with virtual equipment (i. e. simulators) or remote devices like Digital Signal Processors (DSPs) and complex computing equipment should be guaranteed to adequately support virtual presence, which demands efficient transmission protocols. Note that, to acquire the skills needed to adequately interact with Computational Science related equipment, hands-on experience is a must. Additionally, to provide an adequate learning environment, interactions among students and lecturers should be also supported, which poses a need for distributed communication services.

With respect to the target equipment themselves, typically virtual devices or remotely accessed laboratory premises, their implementation and configuration

P.M.A. Sloot et al. (Eds.): ICCS 2002, LNCS 2331, pp. 872–881, 2002.

is not a simple task either. They should be robust beyond their standard, non pedagogical versions, and they should support educational-oriented features like guided operation, improved fault tolerance or activity logging for further study by lecturers.

In this paper we describe our approach to virtual laboratories in the field of Computational Science, whose main objective is to make students gain acquittance with the complex hardware and software tools used in the field. Our contribution can be described as an open distributed platform to support practical training. Services are offered to easily integrate third-party developed educational simulators. More specifically, they support the seamless integration of simulators written in Java, and include features like student tracking, collaborative tools, messaging, task and project management, or virtual file repositories.

Additionally, we propose a CORBA-based distributed architecture to access real computer systems, available in labs at University facilities, which can be seen as a complementary approach to that based on simulation.

All these approaches can be easily integrated to provide a true practical distance learning experience.

2 Teaching Computer Architecture

Distance practical training in Computer Architectures is not straightforward. Hands-on practice is a must to adequately grasp the fundamental concepts in this field. Skills on machine and assembly programming, microprocessor architectures and low-level computer communications can only be adequately obtained when the students test their own developments and see how computers evolve as directed by them. For this, we have to develop an adequate environment to offer practical training, taking into account lessons learnt from previous experiences. Current approaches to practical training over the Internet can be classified into two groups [1]: (1) those based on educational simulations and (2) settings that provide access to real laboratory equipment. In our case, for the former approach students would be provided with simulators of computer systems to study the simulated behaviour of the real equipment. In the latter case, students would control real computer systems, placed in academic institutions, via the Internet.

In our case, distance learning in this field is intended to provide students the needed skills to take the most of available equipment in Computational Science labs (e.g. DSPs, Single Board Computers, etc.). As a general rule, simulation is adequate for first- and second- year students. There are many educational simulators of many different architectures available, and for this target group a simulator provides additional pedagogical advantages, like a safe, controlled environment, ability to easily reconstruct student activities to analyze errors, or guided simulation.

On the other side, access to the real thing is targeted to more advanced students. For them, practice with real computer systems is a natural step ahead after introductory courses on the matter.

3 SimulNet: Simulators over the Network

SimulNet [2] provides students with a teleteaching environment where the "learning-by-doing" paradigm is possible. Unlike other distance teaching systems whose aim is to achieve a virtual classroom, SimulNet provides a virtual laboratory to put theoretical knowledge into practice. Because SimulNet is a 100% pure Java system, our labware can be run on any computer and operating system. Our approach is based on the simulation of the actual laboratory tools that are delivered through the Internet (Java applets) or by CD-ROM technology (Java applications). Although SimulNet can be used in a remote access way, Java allows us to provide always the highest level of interactivity, which is an essential feature in any distance education system. In addition, SimulNet also provides a set of communication and tutoring tools for learners and instructors, providing a full cooperative learning atmosphere. We believe distance education should not mean to study alone and, therefore, we made an extra effort to provide an environment where students and teachers feel as is they were in a virtual lecture room.

3.1 Architecture

The implementation of SimulNet is based on currently available Internet technologies, especially in the Java computing. Thanks to the Java computing technology we have achieved several important features in SimulNet:

- Platform independence. SimulNet client and server applications may be run on any computer whatever its operating system or architecture is.
- Java eases the development of WWW-based applications. SimulNet provides a WWW client (based on Java applets) and a stand alone client (based on Java applications) delivered by CD-ROM, see Fig. 1.

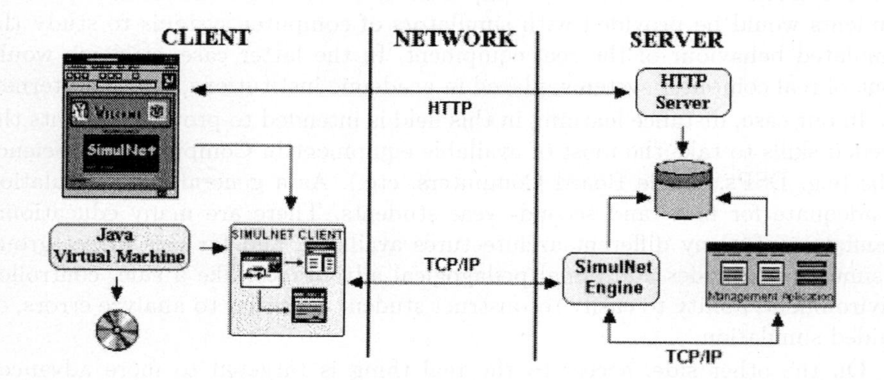

Fig. 1. Simplez Structural Model

The simulators are interactive. There is no network overhead as the simulators run on the student's own computer. This is an essential feature to provide to trainees the same feeling as if they indeed were in the laboratory using the real training tools.

Users access SimulNet using a standard WWW browser, which presents an HTML document provided by the HTTP server at the server side. This document contains an embedded Java applet that would start and stop any other Java application provided by the server side through the Internet. In this way, no additional software is required apart from the browser itself.

Alternatively, students can use a fully independent Java application delivered by CD-ROM. In this case, the SimulNet simulators could be used in a standalone way. At the same time, the user could connect the Java application to the server side to benefit from the virtual laboratory advantages: communication channel, trainees' traces, etc.

3.2 Collaborative Learning

SimulNet provides several communication tools to support cooperative learning. All of them were developed from scratch by our team, and they can be easily included in new simulators thanks to the SimulNet API. In this way, students are provided with an easy-to-use set of collaborative facilities to ease learning on Computational Science or any other field. On the one hand, there is no need to set up or use additional software at client computers, which may be difficult for trainees. The needed software is automatically downloaded from the network and can be run wherever the user may be, regardless of the particular hardware platform or operating system. The developed communication tools are (c.f. Fig. 2): Mail Tool, Bulletin Board, Talk, Multi-Talk, Whiteboard and Project Management Tool.

3.3 First steps in Online Computational Science Education: acquiring experience with supporting and related concepts

Our platform provides a set of tools to acquire the needed experience with computational devices. For this, we have used pedagogical computers that are good enough to explain and practice with the fundamental concepts needed to manage more elaborated artifacts used in Computational Science Education.

We have developed simulators of the computers described in [3], from the simplest one to the most complex. Thus, students are guided in the right way, starting from a simple assembler language and ending with the complex world of microinstructions, firmware and datapaths. In the following subsections we introduce the main features of three pedagogical computers: Simplez, Simplez+i4 and Algoritmez.

All of our simulators have several common features: an editor, an assembler and the simulator of the computer itself. Programs written by students in assembler code using the editor are transformed into object code that can be run by the computer model. This task is done by the assembler. It also offers information

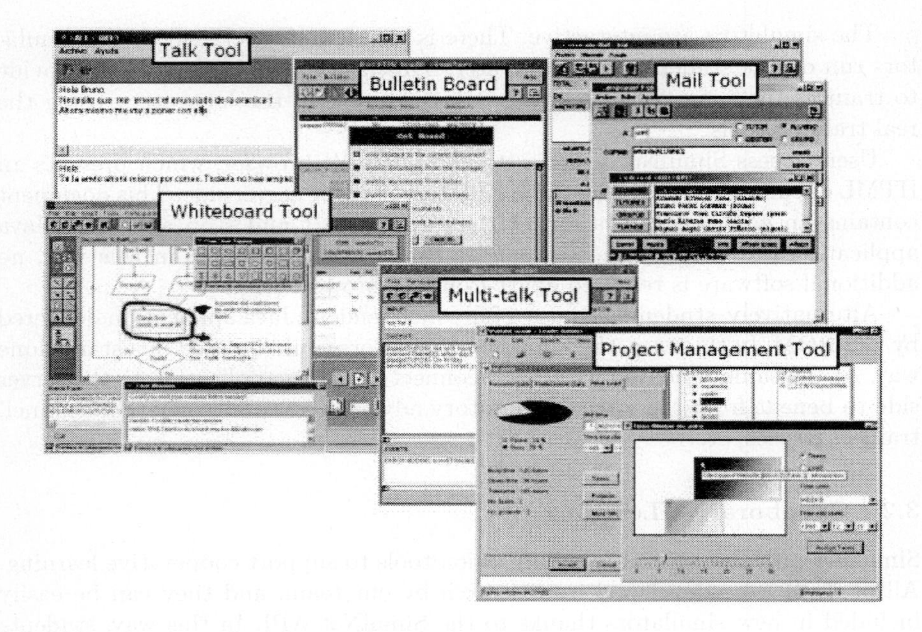

Fig. 2. SimulNet Communication Tools

about different labels and constants used by the student. The simulator of the computer model executes the object code, which can also be displayed through the memory viewer. It has several execution modes: complete run, trace, step by step. Students can also use breakpoints or modify memory or registers at any time. All of these are common features the student must be familiar with in order to be able to manage himself properly in Computational Science labs.

Simplez Simplez has 512 memory locations, 1 register, 8 instructions, I/O mechanism via memory mapped I/O and 1 addressing mode (direct addressing mode). With this architecture, trainees are shown how to implement basic algorithms and how difficult could be to implement the more complex ones. The Simplez simulator embodies an editor, an assembler, the processor simulator itself, a code viewer and the Simplez monitor.

Simplez+i4 Simplez+i4 is a more complex computer. It is the next step in our students learning process. This computer is based on Simplez but adding three different addressing modes: indirect (the first i), indexed (the second i) and indexed+indirect (the third i). It also includes a simple interrupt mechanism (the fourth i) to implement communication with two peripherals (keyboard and monitor). Its conventional machine level incorporates an index register, 4096 memory locations and a more elaborated format for the eight instruction set. After students have acquired the fundamental concepts of Simplez operation, they are able to be introduced to new ones with a higher level of difficulty such as computer interruptions and addressing modes.

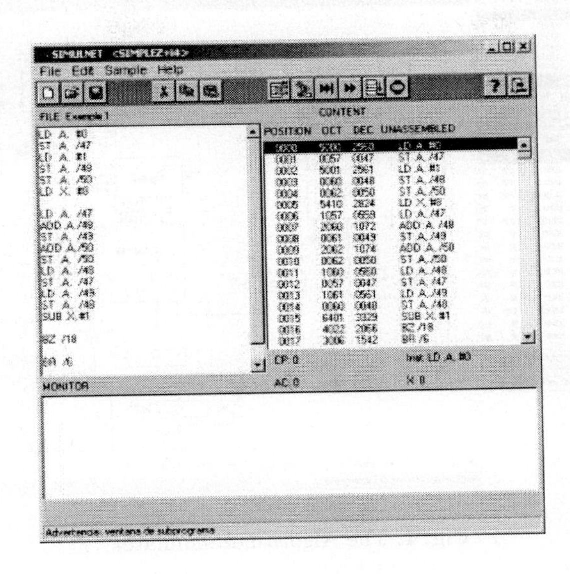

Fig. 3. The Simplez+i4 simulator

Algoritmez Algoritmez is a pedagogical computer closer to commercial ones. Its structural model presents a 64K local memory (including a stack), 256 I/O ports, 16 different registers (two of them used as the program counter and the stack pointer), a status register with several flags and 54 different instructions (related to the arithmetic and shift units, access to stack and memory locations, I/O and, of course, flow control). Through the functional model that we set up over this structural model, students gain further insight into the actual behavior of computers.

Furthermore, the simulator of Algoritmez gives the possibility of microprograming. The implementation of a computer model at micromachine level can be internally microprogrammed allowing designers to easily change its contents and, thus, adapting the system to different instruction sets or data paths. Students are allowed to use the Algoritmez's data path and microprogram the instruction set. In this way, a higher degree of abstraction is offered and the student is not restricted to a fixed model. We usually redefine many of the characteristics of our pedagogical computer to let students interact with different (virtual) machines that are emulated using the own Algoritmez (simulated) hardware.

We include a datapath viewer, see Fig. 4, that shows the connections among the different parts of the computer and let the user to check the signals that are being generated by the control unit, how data flows through the buses and

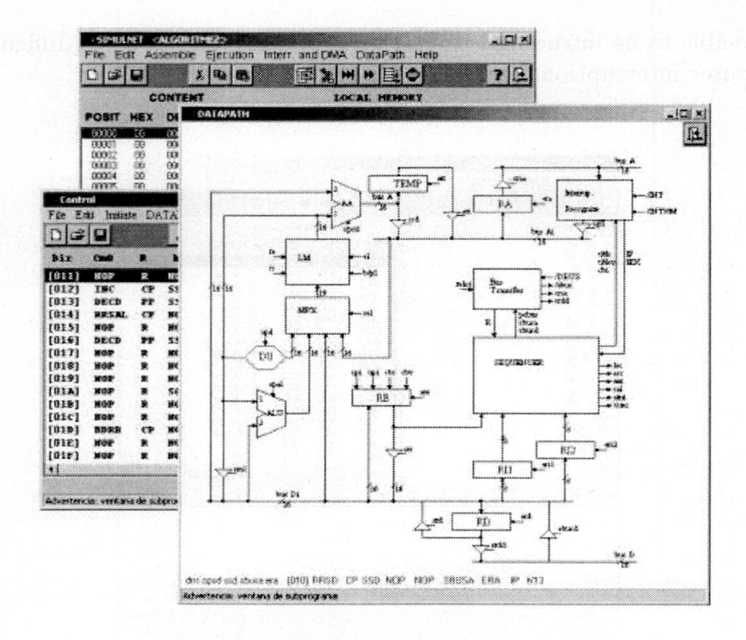

Fig. 4. The Algoritmez simulator

arrive to registers, memory locations, etc. This machine is complex enough to let students understand the more elaborated concepts behind the computational resources used in most Computational Science labs.

3.4 Tutoring

The most experienced teachers provide information about what actions performed on the simulator should be considered as important from a pedagogical point of view. These action will be reflected in students' traces. Whenever a student performs a given task, a report is sent to the tutor responsible for monitoring his or her actions (c.f. Fig. 5). So, without being in the same room, the teacher is able to follow students' performance and, if necessary, to teach how to do the training practice via several available communication tools, see section 3.2.

4 Accessing to real equipment

Although it is possible to use simulation to teach many practical skills to students, there exist several situations where the use of the real equipment is compulsory: either the development of a simulator from scratch is not feasible or real industry equipment is too complex to simulate. In this case, in order to design a distance education environment, we need to manage real instruments

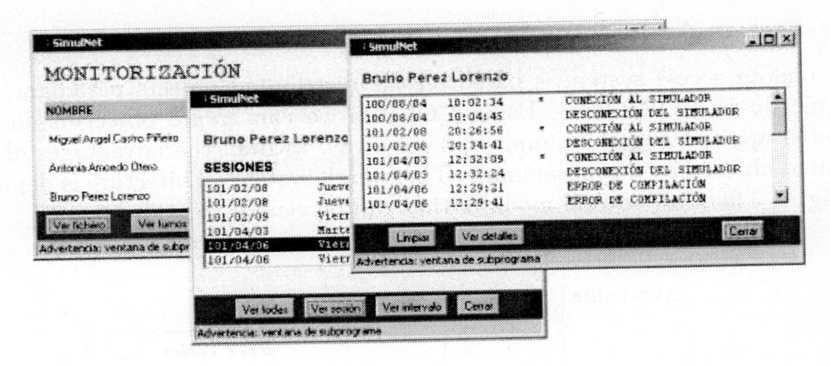

Fig. 5. Monitoring students' behaviour

and equipment remotely. We have developed a system that is centered in this remote operation context: we have to carry out the experiment as if students were in the actual laboratory, and we need to provide them with the output and results of every action, command or modification, as the experiment takes place.

In our introductory laboratory to Computational Science, students are provided with a Java/CORBA-based environment [4] where real DSP devices can be controlled remotely via the Internet. Apart from student accessibility advantages, this solution generates important savings for the institution responsible for maintaining laboratory facilities, both in equipment and staff.

4.1 Description of the environment

In order to acquire the needed experience in DSP programming, students have full access to the real computer where Flite Electronic's DSP25 Cards [6] are used. These cards includes a TMS320C25 DSP chip programmed by students to put theoretical concepts into practice.

DSP programming learning would not be feasible without a proper theoretical introduction. For this, we have included as part of our environment an online course. After this introduction, learners are supposed to be able to deal with the real DSP using the virtual lab.

In this virtual lab learners access DSP25 cards using a WWW browser. This client provides a complete development environment, since it handles all applications used in the conventional laboratory (editor, assembler, debugger, etc.) First practices are typically quite simple as its aim is just to familiarise with the working environment [7]. Last steps consist of developing a small project cooperatively among a group of partners (SimulNet provides both communication and task-based learning supporting tools). Typical practices would be the development of digital filters. Eventually, in the very last phases they need to use the conventional lab facilities where they test the programmes developed using the analog instruments available at the lab (signal generators, oscilloscopes, etc.)

4.2 System Architecture

Our remote access system is based on the distributed objects paradigm, and specifically in CORBA [4]. Using CORBA, we can create object-based distributed applications in a simple and easy way, with all the advantages of distributed object-based programming. The overall system architecture is depicted in Fig. 6. There are several modules that can be clearly identified:

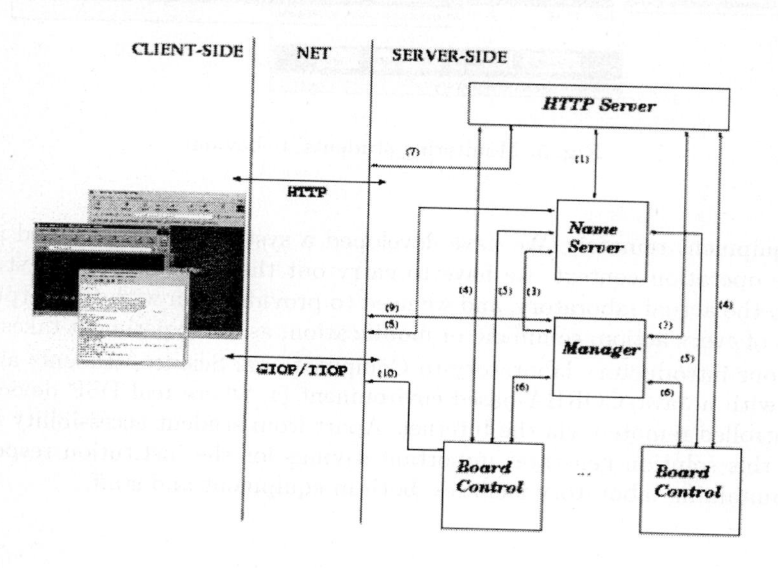

Fig. 6. System architecture

- Client. It is the application that wants to establish remote access to a DSP25 board. It must negotiate with server-side processes to get a free board, and with board-control processes to actually access the board and use additional features.
- Name server. It acts as a bridge used by clients and board-control processes to access the system.
- System manager process. It is responsible for two different tasks. The first one is registering new boards in the system. The second is guaranteeing that, in case there is an available board, any authenticated client can use it.
- Board-control processes. Each of them controls a DSP25 board and other resources accessible by the client in the server file system (files, I/O, execution of programs, etc.).
- HTTP server. It is used as a main door to the system, used to download WWW laboratory pages and also to access the name server.

With this configuration, we can obtain savings up to 84% comparing to the real lab [5]. A reduction of the system availability allows this savings. We use the fact that in a virtual laboratory, with no schedules at all, students will not access the system simultaneously. Although there might be some access rejections at peak time, our experience demonstrates savings are worth enough as the rejection rate is low.

5 Conclusions

We have presented our experience in designing practical training support for remote learning in the field of Computational Science. The systems described allow students to gain expertise in the use of complex computing devices, like the ones they will encounter in laboratories in the field. Two approaches have been discussed: training through educational simulators of real systems, and remote access to the systems themselves, including additional supporting services to provide pedagogical added value.

We think that this distance learning is particularly adequate in this case, because our main objective is to provide previous hands-on experience to students that will have to interact with real computing equipment in regular courses in Computational Science. In this sense, e-learning serves as a complement to the target audience, where lectures can be taken independently of time constraints or physical location. These complementary remote courses will enable students to get the most from the corresponding regular courses.

References

1. Anido, L., Llamas, M., Fernández, M.J.: Internet-based Learning by Doing. IEEE Transactions on Education, Vol. 44, No. 2, CD-ROM Folder 09. ISSN 0018-9359 (2001)
2. Llamas, M., Anido, L., Fernández, M. J.: Simulators over the Network. IEEE Transactions on Education, Vol. 44, No. 2, CD-ROM Folder 09. ISSN 0018-9359 (2001)
3. Fernández, G.: Conceptos básicos de Arquitectura y Sistemas Operativos. Curso de Ordenadores. Sistemas y Servicios de Comunicación, S.L. ISBN 84-605-0522-7.
4. Harkey, D., Orfali, R.: Client/Server Programming with Java and CORBA, 2nd Edition. John Wiley & Sons. ISBN 047124578X.
5. Castaño, F.J., Anido, L., Vales, J., Fernández, M.J., Llamas, M., Rodríguez, P.S., Pousada, J.M.: Internet-based Access to Real Equipment at Computer Architecture Laboratories using the Java/CORBA Paradigm. Computers & Education, Vol. 36, No. 2, pp. 151-170, Elsevier Science. ISSN 0360-1315 (2001)
6. Flite Electronics Ltd. web page at http://www.flite.co.uk/index.html
7. Fuchiwaki, Y., Usuki, N., Arai, T., Murahara, Y.: *The DSP Experiments for Under Graduate Students.* ICASSP (IEEE) 6:3526-3529 (2000)

Computing, Ethics and Social Responsibility: Developing Ethically Responsible Computer Users for the 21st Century

Mildred D. Lintner

Department of Computer Science. Eastern Michigan University,
511 Pray-Harrold Hall, Ypsilanti, Michigan 48197 USA
mlintner@online.emich.edu
http://www.emich.edu/compsci/faculty/f_mlintn.html

Abstract. Computer technology has proven itself a double-edged gift, alternately improving and threatening our lives and environment. In the past year alone we encountered computer viruses, interruptions of power, invasions of privacy, cyber-pornography, and many thefts. Our obligation to students is to include in their education the means to understand the relationship between the most powerful current influence on our lives and the ethical values necessary to use that influence responsibly. This paper outlines educational topics and methodologies for helping both beginning and pre-professional students apply ethically responsible values to using computer technology in the 21st century.

1 Introduction: A Double Edged Gift

Throughout mankind's history, every powerful discovery — fire, gunpowder, medicine, nuclear power, petroleum — has proven to have two sides: Each has the potential either to improve or to threaten our lives and environments immeasurably. For each discovery, we have had to learn how to use it to best advantage to benefit mankind. We have also discovered, sometimes very painfully indeed, how easily these gifts can be misused, causing instead indescribable devastation.

In the twenty-first century, mankind has once again been entrusted with a powerful gift — computer technology. Like fire, computing has awesome beneficial powers, enabling all manner of human endeavor. It also supports devastating technological threats. Computer viruses, invasions of privacy, threats to intellectual property rights, interruptions of internet service, theft of property or identity — these and other misuses of technology fill our news media and threaten our everyday lives and activities.

Unfortunately, the technological threats we hear about or experience are the tip of an ever-growing iceberg. As we have moved into the 21st century, computer technology has pervaded every venue of human endeavor. Computer-related technology now shapes the sociological, economical and political values of our world. As a result we can no longer afford to teach and learn only how to make computers work. We must also develop discipline from within to control use of computer technology. Our obli-

P.M.A. Sloot et al. (Eds.): ICCS 2002, LNCS 2331, pp. 882–887, 2002.

gation to students is to provide them with the tools to harness this powerful new influence, as well as the ethical values needed to use that influence responsibly

2 Target Audience

Who should study the relationship between computing and ethical values? The easy answer is *everyone*. Computers pervade business, industry, education and personal interconnectivity. Few human activities today are performed without the intervention and/or support of computer technology. Thus, the need to understand the dangers and threats that accompany computer convenience and technology is universal.

Nevertheless, three distinct groups of students emerge as needing instruction in computing ethics: general education freshmen, pre-service teachers and computing pre-professionals.

2.1 General Education Freshmen

American colleges and universities require that beginning students, whatever their eventual specialization, take a series of *general education* courses. These are designed to insure that students meet at least minimal standards in essential subjects — written and oral communication skills, mathematics and computer fluency, among others.

First year college students exhibit a broad range of computer experiences and skill-levels. Some have had high school courses in keyboarding, beginning programming (in BASIC) or using standard productivity tools. Others are self-educated experts at interactive online games and chat rooms. Whatever computer skills these students bring with them, most approach their required general education computer course with little motivation. They feel they either know enough about computers, or can live without them. For these students, content material concerning ethical use of computers, if included at all, is usually presented in one unit at the very end of the course.

2.2 Pre-Service Teachers

Most United States teacher certification programs require that pre-service teachers pass state-administered qualifying examinations covering both subject content and instructional methodology. In computer science, examination areas usually include computer fluency, societal impact, programming, computer organization and architecture, professional studies and instructional technology. While few states include specific course requirements in ethical use of computing, most qualifying examinations include some questions involving ethical values. The breadth of the teaching curriculum provides many opportunities for pre-service teachers to investigate computing ethics as they relate to societal impact and computer use in schools and other public places.

2.3 Computing Pre-Professionals.

Computational engineers, programmers, scientists and administrators acquire special knowledge and skills enabling them to exert great power over computer and information technology. With this power comes a variety of responsibilities, to employers, to clients, to other members of the profession, and to public safety and well-being. Thus an ethical approach to computing is an essential part of the education of a computing pre-professional.

Curricula for preparing computer professionals vary greatly, dependent on career goals and specific disciplines. While some programs do include a course in professional ethics, most do not, and provide little opportunity for adding such a course. An alternative strategy for developing ethical values in computing pre-professionals is to include ethics in many contexts and projects across the curriculum.

3 Objectives

Computing ethics might be taught as a single unit of a general education computer course, as one or more discrete courses in a professional curriculum, or incorporated into the content and activities of many courses across a computing discipline. Whichever academic format is used, the content and instructional methodology of a computing ethics program should guide students in attaining the following:

A Basic Foundation in Philosophical Ethics. Before students can discuss ethical concepts and evaluate computer use based on ethical concepts, they must accumulate some background knowledge in common. Minimal preparation in philosophical ethics will provide essential definitions and understandings, including the basics of several different ethical value systems that have existed throughout history.

Some Intellectual Criteria for Ethical Examination and Judgement. Students need to understand several different sets of measurements commonly used for making ethical decisions, and to be able to apply them appropriately to various technological situations.

An Understanding of Mankind's Various Uses of Double Edged Gifts. Computer technology is not the first great invention that can be used for both good and evil. Understanding earlier inventions will provide a firm base for applying ethical criteria to technological situations.

An Understanding of Everyday Ethical Issues Involving Computing. Examination of situations involving college students, young computing professionals, teenagers, parents with young children and everyday workers will help students internalize both the need for ethical values and their applications to familiar experiences.

A Responsible Attitude Toward Use of Computers and the Internet. Most students do not connect the reported horrors of Internet crime with their own uses of the Web's easy accessibility to ideas and products. Developing a responsible stand is especially important to pre-service teachers.

Effective Application of Critical Thinking Skills to Information and Communication Technology. Young computing professionals must be able to envision the societal impact of the products and services they will develop during their working careers.

A Code of Values Defining Personal and Commercial Uses of Computers and Related Technologies. This involves the development of random concepts concerning ethics and computer use into a consistent and cohesive guide for technology use.

Effective Communication about Abstract Ideas and Values.. This involves learning to organize ideas and support materials into a cohesive ethical stand, and to use information and communication technology appropriately to support such a stand.

4 Content Outline

The following content outline is suggested to support and help students attain the stated objectives. It is based on a course outline used twice so far: first with pre-professional computer programmers, and once with general education freshmen. Within the next year these offerings will be repeated, and an additional class for computer science pre-service teachers will be offered.

- Definitions of ethics and ethical use of powerful tools.
- Historical perspective
  - Weapons and gun laws
  - The pen and laws of libel
  - The absence of a cyber-ethic
- Developing a computer ethic
- Computing and Piracy
  - Issues of theft
  - Ownership (of data, goods and identity)
  - Intellectual property rights.
  - Software piracy
  - Cyber-plagiarism
  - Images, sounds and multimedia
- Computing and Protection
  - Security
  - Passwords
  - Encryption
  - Identity
  - firewalls
- Censorship vs. Freedom of speech
  - Hate-mongering
  - Pornography

- Appropriate use
- Computing and Privacy
    - Benign invasions
        - Electronic supervision
        - Quality control
    - Vicious invasions
- Consumer Protection
    - Consumer profiling
    - Data mining
    - Supervisory monitoring
    - Secure servers
    - Safe practices
- Computers and Damage Control
    - Hackers and crackers
    - Viruses and other dangerous invaders
    - Service interruptions
- Conclusion

5 Examples of Typical Assignments

The study of computer ethics provides opportunity for a broad variety of assignments and projects. Here are a few examples of those used in conjunction with the outline listed above. They should, of course, be customized to both the skills level and eventual goals of the students. For all assignments, student need to prepare reports of their work. Reports can be written, using good verbal expression, or oral, using slides or other visuals, clear spoken communication and good organization.

5.1 Ethics Scenario Evaluation

For this assignment, the students are assigned to examine fairly detailed scenarios of computer-related situations. The scenarios can be either hypothetical or taken from real life. Several students may be assigned to the same scenario, but they seem to accomplish most if each works alone. In the first part of the assignment, each student must analyze the scenario without making any judgements, seeking specific information, such as the following: Who are the stakeholders in the situation? What are the benefits and the costs involved for each stakeholder? What actions were taken in the scenario? What alternative actions were possible?

The second part of the assignment requires the students to make decisions on ethical issues as they pertain to everyday computer tasks. Students use their own personal values as basis for all decisions. It is important for this part of the study for students to be judgmental: Was the *right* (or best) action taken? What would you have done if you were one of the stakeholders? Apply your own code of ethics to this scenario. Who 'won'? Who 'lost'?

5.2 Science Fiction Evaluation

This assignment is best done as a group discussion project. Students read a science fiction book or short story, or see a sci-fi movie in which technology plays a major role. The technology presented does not have to be either futuristic or imaginary. Books by Jules Verne, Isaac Asimov or George Orwell or films based on their work would be appropriate.

Students begin by summarizing the story. They then list the ethical, technical and social issues addressed by the book or film. Finally they discuss the author's view of the technology used in the story - does the author consider technology good, evil, neutral, or in control? Students are then asked to compare the view of technology in their science fiction work to what they know of current technological reality - How is today's technology presented to the public? Is it good or evil? How does that compare with the author or film maker's viewpoint?

5.3 Societal Impact Analysis

This final example was designed specifically for computer pre-professionals. In it, students study an existing computer system or a project they have worked on and completed themselves. The system being studied should be real and functioning. The project's purpose is to understand the societal impact of the system on the population it serves.

Students analyze the working system, to determine the uses made of the system, its functional impacts on the organization and the ethical concerns of its stakeholders. Students must be able to talk to all types of stakeholders interacting with the system to determine their reactions to the workings of the system - do they consider the computer system to be a positive or negative influence on the service provided? Would using the system differently have a beneficial or destructive impact on the organization? Has the system impacted numbers or types of jobs in the organization, number or efficiency of clients served? Which stakeholders have reaped the greatest benefits (or losses) because of the system.

References

1. Hester, D. Micah, Ford, Paul J.: Computers and Ethics in the Cyberage. Prentice Hall Inc., Upper Saddle River, New Jersey (2001)
2. Johnson, Deborah G.: Computer Ethics. 3rd edition Prentice Hall, Inc., Upper Saddle River, New Jersey (2001)

Teaching Parallel Programming Using Both High-Level and Low-Level Languages

Yi Pan

Georgia State University, Atlanta, GA 30303, USA
pan@cs.gsu.edu

Abstract. We discuss the use of both high-level and low-level languages in the teaching of senior undergraduate and junior graduate classes in parallel and distributed computing. We briefly introduce several language standards and discuss why we have chosen to use OpenMP and MPI in our parallel computing class. Major features of OpenMP are briefly introduced and advantages of using OpenMP over message passing methods are discussed. We also include a brief enumeration of some of the drawbacks of using OpenMP and how these drawbacks are being addressed by supplementing OpenMP with additional MPI codes and projects. Several projects given in our class are also described in this paper.

1 Introduction

Parallel computing, the method of having many small tasks solve one large problem, has emerged as a key enabling technology in modern computing. The past several years have witnessed an ever-increasing acceptance and adoption of parallel processing, both for high-performance scientific computing and for more "general-purpose" applications. The trend was a result of the demand for higher performance, lower cost, and sustained productivity. The acceptance has been facilitated by two major developments: massively parallel processors and the widespread use of clusters of workstations.

In the last ten years, courses on parallel computing and programming have been developed and offered in many institutions as a recognition of the growing significance of this topic in computer science [1],[7],[8],[10]. Parallel computation curricula are still in their infancy, however, and there is a clear need for communication and cooperation among the faculty who teach such courses.

Georgia State University (GSU), like many institutions in the world, has offered a parallel programming course at the graduate and Senior undergraduate level for several years. It is not a required course for computer science majors, but a course designated to accomplish computer science hours. It is also a course used to obtain a Yamacraw Certificate. Yamacraw Training at GSU was created in response to the Governor's initiative to establish Georgia as a world leader in highbandwidth communications design. High-tech industry is increasingly perceived as a critical component of tomorrow's economy.

Our department offers a curriculum to prepare students for careers in Yamacraw target areas, and Parallel and Distributed Computing is one of the

P.M.A. Sloot et al. (Eds.): ICCS 2002, LNCS 2331, pp. 888–897, 2002.

courses in the curriculum. Graduate students from other departments may also take the course in order to use parallel computing in their research.

Low-level languages and tools that have been used at GSU for the course includes Parallel Virtual Machine (PVM) and the Message Passing Interface (MPI) on an SGI Origin 2000 shared memory multiprocessor system. As we all know, the message passing paradigm has several disadvantages: the cost of producing a message passing code may be between 5 and 10 times that of its serial counterpart, the length of the code grows significantly, and it is much less readable and less maintainable than the sequential version. Most importantly, the code produced using the message passing paradigm usually uses much more memory than the corresponding code produced using high level parallel languages since a lot of buffer space is needed in the message passing paradigm. For these reasons, it is widely agreed that a higher level programming paradigm is essential if parallel systems are to be widely adopted. Most schools teaching the course use low-level message passing standards such as MPI or PVM and have not yet adopted OpenMP [1], [7], [8], [10]. To catch up with the industrial trend, we decided to teach the shared-memory parallel programming model beside the message passing parallel programming model. This paper describes experience in using OpenMP as well as MPI to teach a parallel programming course at Georgia State University.

2 About OpenMP

The rapid and widespread acceptance of shared-memory multiprocessor architectures has created a pressing demand for an efficient way to program these systems. At the same time, developers of technical and scientific applications in industry and in government laboratories find they need to parallelize huge volumes of code in a portable fashion.

The OpenMP Application Program Interface (API) supports multi-platform shared-memory parallel programming in C/C++ and Fortran on all architectures, including Unix platforms and Windows NT platforms. Jointly defined by a group of major computer hardware and software vendors, OpenMP is a portable, scalable model that gives shared-memory parallel programmers a simple and flexible interface for developing parallel applications for platforms ranging from the desktop to the supercomputer, [2]. It consists of a set of compiler directives and library routines that extend FORTRAN, C, and C++ codes for shared-memory parallelism.

OpenMP's programming model uses fork-join parallelism: the master thread spawns a team of threads as needed. Parallelism is added incrementally: i.e. the sequential program evolves into a parallel program. Hence, we do not have to parallelize the whole program at once. OpenMP is usually used to parallelize loops. A user finds his most time consuming loops in his code, and splits them up between threads. In the following, we give some simple examples to demonstrate the major features of OpenMP.

Below is a typical example of a big loop in a sequential C code:

```
void main()
{
  double A[100000];
  for (int i=0;i<100000;i++) {
   big_task(A[i]);
  }
}
```

In order to parallelize the above code in OpenMP, users just need to insert some OpenMP directives to tell the compiler how to parallelize the loop.

A short hand notation that combines the Parallel and work-sharing construct is shown below:

```
void main()
{
     double Res[100000];
#pragma omp parallel for
     for(int i=0;i<100000;i++)
     {
          big_task(Res[i]);
     }
}
```

The OpenMP work-sharing construct basically splits up loop iterations among the threads in a team to achieve parallel efficiency. By default, there is a barrier at the end of the "omp for". We can use the "nowait" clause to turn off the barrier.

Of course, there are many different OpenMP constructs available for us to choose. The most difficult aspect of parallelizing a code using OpenMP is the choice of OpenMP constructs, and where these should be inserted in the sequential code. Smart choices will generate efficient parallel codes, while bad choices of OpenMP directives may even generate a parallel code with worse performance than its original sequential code due to communication overheads.

When parallelizing a loop in OpenMP, we may also use the schedule clause to perform different scheduling policies which effects how loop iterations are mapped onto threads. There are four scheduling policies available in OpenMP. The static scheduling method deals-out blocks of iterations of size "chunk" to each thread. In the dynamic scheduling method, each thread grabs "chunk" iterations off a queue until all iterations have been handled. In the guided scheduling policy, threads dynamically grab blocks of iterations. The size of the block starts large and shrinks down to size "chunk" as the calculation proceeds. Finally, in the runtime scheduling method, schedule and chunk size are taken from the `OMP_SCHEDULE` environment variable and hence are determined at runtime.

The section work-sharing construct gives a different structured block to each thread. This way, task parallelism can be implemented easily if each section has a task (procedure call). The following code shows that three tasks are parallelized using the OpenMP section work-sharing construct.

```
#pragma omp parallel
#pragma omp sections
{
        task1();
#pragma omp section
        task2();
#pragma omp section
        task3();
}
```

Another important clause is the reduction clause, which effects the way variables are shared. The format is `reduction (op : list)`, where `op` can be any general operation such as $+$, max, etc. The variables in each "list" must be shared in the enclosing parallel region. Local copies are reduced to a single global copy at the end of the construct. For example, here is an example for global sum and the final result is stored in the variable res.

```
#include <omp.h> #define NUM_THREADS 2 void main ()
{       int i;
        double ZZ, func(), res=0.0;
        omp_set_num_threads(NUM_THREADS)
#pragma omp parallel for reduction(+:res) private(ZZ)
        for (i=0; i< 1000; i++)
        {
                ZZ = func(I);
                res = res + ZZ;

        }
}
```

Programming in a shared memory environment is generally easier than in a distributed memory environment and thus saves labor costs. However, programming using message passing in a distributed memory environment usually produces more efficient parallel code. This is much like the relationship between assembly languages and high level languages. Assembly codes usually run faster and are more compact than codes produced by high-level programming languages and are often used in real-time or embedded systems where both time and memory space are limited, and labor costs are not the primary consideration. Besides producing efficient codes, assembly languages are also useful when students learn basic concepts about computer organization, arithmetic operation, machine languages, addressing, instruction cycles, etc. When we need to implement a large complicated program, high-level languages such as C, C++, or Java are more frequently used. Similarly, students can learn a lot of concepts such as scalability, broadcast, one-to-one communication, performance, communication overhead, speedup, etc, through low-level languages such as MPI or PVM. These concepts are hard to obtain through high-level parallel programming languages due to the fact that many details are hidden in the language constructs. However, students can implement a relatively large parallel program using a high-level parallel language such as OpenMP or HPF easily within a short period of time.

We believe that the future of high performance computing heavily depends on high level parallel programming languages such as OpenMP due to the increasingly high labor costs and the scarcity of good parallel programmers. High level parallel programming languages are one way to make parallel computer systems popular and available to non-computer scientists and engineers. Hence, teaching students how to use high level parallel programming languages as well as the low level message passing paradigm is an important task for teaching parallel programming.

3 Why OpenMP and MPI

There are currently four major standards for programming parallel systems that were developed in open forums: High Performance Fortran (HPF) [6], OpenMP [2], PVM [3] and MPI [5].

HPF relies on advanced compiler technology to expedite the development of data-parallel programs [6]. Thus, although it is based on Fortran, HPF is a new language, and hence requires the construction of new compilers. As a consequence each implementation of HPF is, to a great extent, hardware specific, and until recently there were very few complete HPF implementations. Furthermore most of the current implementations are proprietary and quite expensive. HPF has been written for the express purpose of writing data-parallel programs, and, as a consequence, it is not well-suited for dealing with irregular data-structures or control-parallel programs.

The Parallel Virtual Machine (PVM) system uses the message-passing model to allow programmers to exploit distributed computing across a wide variety of computer types, including multiprocessor systems [3]. A key concept in PVM is that it makes a collection of computers appear as one large virtual machine, hence its name. The PVM computing model is simple yet very general, and accommodates a wide variety of application program structures. The programming interface is deliberately straightforward, thus permitting simple program structures to be implemented in an intuitive manner. The user writes his application as a collection of cooperating tasks. Tasks access PVM resources through a library of standard interface routines. These routines allow the initiation and termination of tasks across the network as well as communication and synchronization between tasks. The PVM message-passing primitives are oriented towards heterogeneous operation, involving strongly typed constructs for buffering and transmission. Communication constructs include those for sending and receiving data structures as well as high-level primitives such as broadcast, barrier synchronization, and global sum.

MPI specifies a library of extensions to C and Fortran that can be used to write message passing programs [5]. So an implementation of MPI can make use of existing compilers, and it is possible to develop more-or-less portable MPI libraries. Thus, unlike HPF, it is relatively easy to find an MPI library that will run on existing hardware. All of these implementations can be freely downloaded from the internet. Message passing is a completely general method for parallel programming. Indeed, the generality and ready availability of MPI have made it one of the most widely used systems for parallel programming. Compared with the PVM library, MPI has recently become more popular.

OpenMP has emerged as the standard for shared memory parallel programming. For the first time, it is possible to write parallel programs which are portable across the majority of shared memory parallel computers. OpenMP is a portable, scalable model that gives shared-memory parallel programmers a simple and flexible interface for developing parallel applications for platforms ranging from the desktop to the supercomputer.

The most important reason for our adoption of OpenMP in a parallel programming class is that students can parallelize some realistic code (not toy problems) within a short period of time due to the ease of programming that it offers. Students can also experiment with different scheduling schemes such as static or dynamic loop scheduling policies within a short period of time, which would be impossible otherwise using MPI. Using OpenMP also has the advantage that task parallelism can be easily implemented by just inserting several OpenMP directives. By combining loop parallelism and task parallelism, better performance and higher scalability can be achieved. On the other hand, task parallelism is difficult to implement using MPI or HPF.

Another reason for our selection of OpenMP in our class is that we have an SGI Origin 2000 shared memory multiprocessor system in our department. A shared memory programming model fits in well.

Besides the above reasons, OpenMP has the following benefits for parallel programming compared with message passing models such as MPI: a) A user just needs to add some directives to the sequential code to instruct the compiler how to parallelize the code. Hence, it has unprecedented programming ease, making threading faster and more cost-effective than ever before. b) The directives are treated as comments when running a single processor. Hence, a single-source solution can be used for both serial and threaded applications, lowering code maintenance costs. c) Parallelism is portable across Windows NT and Unix platforms. d) The correctness of the results generated using OpenMP can be verified easily which dramatically lowers development and debugging costs.

Hence, our strategy is to teach students the basic concepts in parallel programming such as scalability, broadcast, one-to-one communication, performance, communication overhead, speedup, etc, through a low-level parallel programming language, and teach other concepts such as various scheduling policies and task parallelism through a high-level parallel programming language. Since MPI and OpenMP are the most widely used languages in the two categories, these are selected to teach parallel programming.

4 Some Pitfalls with OpenMP

Because OpenMP is a high level parallel language, many details are hidden from a programmer. The good thing is that students can learn quickly and start to program immediately after learning some techniques. The pitfall is that students cannot clearly see the communications involved in a parallel program. Our approach to overcome this problem is to supplement OpenMP projects with some simple MPI programs.

Students first learn the basics of parallel programs in a distributed memory environment. They start to parallelize a sequential code using simple MPI constructs such as `MPI_Bcast`, `MPI_Reduce`, `MPI_Send`, and `MPI_Recv`. Through several small projects, they learn the concepts of one-to-one communication, multicast, broadcast, reduction, synchronization, and concurrency.

Later, when they use OpenMP to parallelize a program, they already have a deep understanding of communication structure, communication overhead, scalability and performance issues.

The second shortcoming with OpenMP is that it does not provide memory allocation schemes for arrays and other data structures since OpenMP is designed for shared memory machines. Again, this relieves the students from complicated memory allocation decisions, allowing concentration on loop and task parallelism. This is good for the ease of programming, but students do not know the details of array allocation schemes such as BLOCK or CYCLIC schemes commonly used in distributed memory environments. Since the memory on the SGI Origin 2000 is not physically shared, SGI provides data distribution directives to allow users to specify how data is placed on processors. If no data distribution directives are used, then data are automatically distributed via the "first touch" mechanism [4] which places the data on the processor where it is first used. Because different allocation schemes may affect the performance of a program greatly, SGI data distribution directives are required in the final project to show the performance improvement.

For example, the following data distribution directive distributes the 4D array H on dimension 2:

```
!$sgi distribute_reshape H(*,BLOCK,*,*)
```

Students are required to try several data distribution schemes, to observe the running times and to comment on the timing results as described below. In this way, the relationship between memory allocation schemes and performance is demonstrated.

Due to these pitfalls with OpenMP, students would not learn all the concepts and the whole picture in parallel programming using OpenMP alone. Our strategy is to supplement OpenMP with explanation on several typical MPI codes and small projects using the MPI standard. Then, students experiment with various scheduling policies and complicated parallelization methods in OpenMP. In this way, students experience various parallel schemes and techniques in a short period of time. This would be very hard to achieve if only the MPI or PVM programming model were used in teaching parallel programming because of the time demands for implementation and parallelization of large codes in MPI or PVM. The following section details the strategy of using both OpenMP and MPI in the class.

5 Using MPI and OpenMP in Projects

The parallel programming class at GSU is a semester-long class for upper-level undergraduates and beginning graduate students. In it tutorials on MPI and OpenMP from the Ohio Supercomputing Center were used as supplements to a parallel algorithms textbook [9]. The code presented in the lectures uses both C and Fortran.

The course begins with an overview of parallel computing and continues with a brief introduction to parallel computing models such as various PRAMs, shared memory models and distributed memory models. The concepts of data parallelism and pipelining are also introduced at that time. The next block of lectures forms a transition into a more or less standard parallel algorithms course. First serial and parallel versions for a very simple computation – e.g., prefix sums and prime finding, are discussed. In the course of analyzing the performance of these algorithms, the concepts of speedup, scalability and efficiency are developed. The deterioration of the performance of the parallel algorithm as the number of processes is increased leads naturally to a discussion of Amdahl's Law and scalability.

The course work consists of two tests, a final exam, five programming projects and a research paper. Since the course's emphasis is on parallel programming, projects are an important part of the course. The purpose of the first project is simply to acquaint students with the system and programming environment of the Origin 2000. In this they write a simple addition code, and measure the parallel times using different numbers of processors.

In the second project, the students implement an MPI code to calculate π using Simpson's Rule instead of the rectangle rule discussed in class, where students are exposed to various MPI communication functions. For timing measurements and precision, they need to test the code using several different numbers of subintervals to see the effect on the precision of results and different number of processors on the execution times.

In the third project, students implement the parallel game of life. Through the assignment, students learn various domain decomposition strategies. All the above projects are implemented in MPI.

Once students understand the communication mechanisms of parallel computing systems, and communication overhead within a parallel code, it is time to introduce OpenMP. After briefly discussing the use of OpenMP and illustration of OpenMP through several examples, students are asked in the fourth project to initialize a huge array A so that each element has its index as its value. A second real array B which contains the running average of array A is then created. The loops are parallelized with all four scheduling schemes available in OpenMP (static, dynamic, guided, and runtime) and the running times are measured with different scheduling policies and different chunk sizes. Students write up their observations on the timings using the four different scheduling policies and explain why the performance differs in these cases.

In the fifth project, students learn how to parallelize a real research Fortran code in OpenMP. The project contains several parts which includes the parallelization of the major loops in the code in OpenMP using both the loop and task parallelism of OpenMP. Thus the best scheduling policy, best chunk size for the policy, and both loop and task parallelism are obtained in the above two steps, while array mapping is done automatically by the OpenMP compiler. For the final step, the arrays are distributed manually using SGI array distribution directives because array distribution directives are not available in OpenMP. The purpose is for students to understand the effect of array distribution on the runtime performance. Students are also required to write a short report to summarize the results obtained. Through these steps, students learn how to parallelize a real code in a step-by-step fashion.

Students are also required to write a research paper or a survey paper on a chosen topic in parallel processing. The purpose is for the students to apply the knowledge learned in the course to an application. Some students have implemented algorithms using MPI and/or OpenMP using various strategies and compared the performance of their implementations with the results published in the literature. At the end of the term, students need to present their findings and submit a paper.

The outcome of the course is very good. Based on student evaluations and comments on the course, most students feel that they learn a lot in the course. Some of the students have already applied the knowledge learned in the course to research projects supported by the NSF and Air Force. One student implemented a parallel program for Cholesky factorization using both MPI and OpenMP, and did a lot of testing using various scheduling and partition strategies. He also did a comprehensive comparison among the different implementations, and wrote an excellent research paper at the end of the course. The paper is being revised and potentially could be published in a conference. This would have been impossible if only MPI had been taught in the course.

6 Conclusion

As OpenMP becomes more popular for parallel programming because of its many advantages over message passing programming models, it is important to introduce OpenMP in a parallel programming course. However, OpenMP also has some shortcomings for teaching parallel programming concepts. Our strategy is to use MPI to convey the basic concepts of parallel programming and to use OpenMP to tackle more complicated problems such as various scheduling policies and combined loop and task parallelism. It seems that the strategy is well received by the students.

References

1. F.C. Berry. An undergraduate parallel processing laboratory, IEEE Trans. Educations, vol. 38, pp. 306-311, (Nov. 1995)
2. Rohit Chandra, Ramesh Menon, Leo Dagum, David Kohr, Dror Maydan, and Jeff McDonald, Parallel Programming in OpenMP, Morgan Kaufmann Publishers, 300 pages, (October 2000)
3. J. Dongarra, P. Kacsuk, N. Podhorszki(Editors): Recent Advances in Parallel Virtual Machine and Message Passing Interface: 7th European PVM/MPI Users' Group Meeting, Balatonfuered, Hungary, (September 2000)
4. J. Fier. Performance Tuning Optimization for Origin 2000 and Onyx 2. Silicon Graphics, (1996), http://techpubs.sgi.com
5. W. Gropp, E. Lusk, A. Skjellum. Using MPI : portable parallel programming with the message- passing interface, MIT Press, Cambridge, Mass., (1994)
6. C. H. Koelbel. The High performance Fortran handbook, MIT Press, Cambridge, Mass., (1994)
7. R. Miller. The status of parallel processing education, Computer, vol. 27, no. 8, pp. 40-43, (Aug. 1994)
8. C. H. Nevison. Parallel computing in the undergraduate curriculum, Computer, vol. 28, no. 12, pp. 51-53, (Dec. 1995)
9. M. J. Quinn. Parallel Computing - Theory and Practice, McGraw-Hill, INC., (1994)
10. B. Wilkinson and M. Allen. A state-wide senior parallel programming course, IEEE Trans. Educations, vol. 42, no. 3, pp. 167-173, (1999)

Computational Science In High School Curricula: The ORESPICS Approach

P. Mori and L. Ricci

Dipartimento di Informatica,Università di Pisa
Corso Italia 40, 56125-Pisa (Italy)
{mori,ricci}@di.unipi.it

Abstract. This paper presents a new approach for the introduction of computational science into high level school curricula. The approach is based on the definition of an ad hoc environment, including a programming language, suitable for this target of age. The language includes a set of simple constructs supporting both concurrency and the management of a graphical interface. The solutions of some classical problems are shown.

1 Introduction

Computational science is a new interdisciplinary research area which applies concepts and techniques from mathematics and computer science to solve real life problems. Computational science has introduced a new methodology to investigate real life problems. Instead of defining a theory of a physical phenomenon and verify it through a set of experiments, a computational model of the phenomenon that can be simulated through the computer is defined. In this way, the phenomenon can be monitored directly through the computer. This methodology is currently supported by sophisticated environments resulting from the recent advances in computer technology.

Several university curricula include computational science courses. Furthermore, a set of proposals for the introduction of computational science into the high school curricula have also been presented [4, 8]. The goal of these proposals is to increase the interest of a larger number of students in scientific disciplines: the rationale is that students are more interested in learning mathematical concepts if these can be applied to real life problems. Furthermore, the use of the computer makes the learning even more appealing. Yet, the introduction of computational science into the high school curricula and/or in the undergraduate courses requires to settle some issues. First of all, the basic mathematical skills to support a first training in computational science are to be defined. These skills should support the development of models for a minimal, yet significative, kernel of applications. These applications should be characterized either by simple mathematical models or by a complex one that may be simplified without loosing its connection with real life. A further critical issue is a software environment suitable for young students. This issue is one of the most challenging because most current tools to develop computational software have been defined for expert users only. As a matter of fact, most applications are developed through *FORTRAN* or *C* extended with a set of libraries supporting concurrency and visualization of scientific data. These

P.M.A. Sloot et al. (Eds.): ICCS 2002, LNCS 2331, pp. 898–907, 2002.

libraries are often tied to a specific operating system and the user needs some knowledge of this system as well.

We believe that a more friendly environment, including a programming language, should be developed specifically for these introductory courses. In this way, all the constructs are integrated in the same language, rather than spread across several libraries. The language should preserve the main features of existing ones, like concurrency and graphical interface support. On the other hand, the set of constructs should be reduced to a minimal kernel.

The didactic language should be *concurrent* because concurrency is a powerful tool to simplify the description of the applications. Several phenomena can be modeled as a set of concurrent, interacting entities. Consider, for instance, the simulation of the dynamics of a fluid or of a gas which can be modeled as a set of interacting molecules. An example in shown in sect. 3. Furthermore, most scientific applications are developed on highly parallel systems because of their high computational needs and the software development for these systems usually requires the knowledge of a concurrent language. Hence, the basic concepts of concurrency should be acquired as soon as possible. However, libraries such as *MPI*, *PVM* [11, 13], or *OpenMp* are not suitable, because of their complexity. As a matter of fact these libraries include several semantic equivalent primitives differing, for instance, only because of their implementation. A didactic environment should introduce a single construct for each different concept of the language. The didactic language should support a simple *graphical interface* as well, so that the student can monitor the behaviour of the concurrent activities directly on the screen. Complex visualization techniques based on sophisticated mathematical techniques, like rendering or textures, are not required in an introductory didactic environment. However, the teacher can exploit the basic mechanisms of the language to implement more sophisticated visualization techniques.

The remainder of this paper presents the *ORESPICS* environment, the new didactic environment we propose to support the teaching of computational science in high schools. The environment includes a new language, *ORESPICS-PL* that integrates a minimal set of graphical primitives with a minimal set of concurrent constructs. The graphical primitives are mostly taken from the Logo language [2], while the concurrent part of the language is based on the *message passing paradigm*. The environment and the language are fully described in [5, 6]. Sect. 2 briefly recalls the main constructs of the language and describes the ORESPICS environment. To describe how *ORESPICS* can be exploited in an introductory course, sect. 3 and 4 show some simple, yet significative problems and their *ORESPICS* solutions. Sect. 5 presents some related work. Sect. 6 draws some preliminary conclusions.

2 The ORESPICS Language and Development Environment

The sequential part of Orespics-PL includes traditional imperative constructs (repeat, while, if,...), the turtle primitives of the Logo language [2] to control the agents' movements and a set of primitives to modify the external aspect of an agent and to create sounds. The language supports all the elementary data types (integer, boolean,..) and the only data structure is the list.

An Orespics-PL program includes a set of *agents*, interacting through messages exchange. It is possible both to pair each agent with a different code and to define a *breed* of agents characterized by the same behaviour. Each agent belonging to a breed may be identified by a set of indexes. In this way, a *SPMD* programming style can be exploited.

Agent interact through a minimal, but complete set of communication primitives. Two basic kinds of communication modes, corresponding respectively to *synchronous* and *buffered* communication modes of MPI, are available. The corresponding sends are:

SendAndWait *msg* **to** *agent*

SendAndnoWait *msg* **to** *agent*

Buffer management for buffered communications is delegated to the run time support.

Orespics-PL does not define other communication modes. This is consistent with the choice to include constructs corresponding only to the *basic mechanisms* of the message passing paradigm. Other communication modes in fact, like blocking, ready or persistent modes of MPI, can be considered *optimized versions* of the previous ones, to enhance the performance of parallel programs.
The receive construct:

WaitAndReceive msg **from** agent

waits until a message is received from the selected agent. Orespics-PL also defines an *asymmetric version* of the receive:

WaitAndReceiveAny msg **from** agent

In this case, the receiver selects, according to a *non deterministic strategy*, one of the messages sent by any active agent of the microworld. Furthermore, the function

inmessage(*agent*)

allows polling of incoming messages, without actually receiving them. *Inmessage(A)* returns true if there is at least an incoming message from agent *A*. The message can be received through a **WaitAndReceive** command. The function *Inmessage(Any)* tests the presence of messages incoming from any agent. The set of collective communications includes the synchronization barrier:

Waitagents()

and two versions of the broadcast send, respectively synchronous and asynchronous:

SendAllAndWait()
SendAllAndnoWait()

Each agent involved in a broadcast communication executes a different primitive, a broadcast send, or a receive. A single primitive with distinct semantics relating to the agent executing it, could be confusing. Other collective communication, like MPI scatter, gather or reduce, can be emulated through point to point or broadcast communications. Since Orespics-PL provides a mechanism to define macros, the students can develop their own implementation of these primitives.

Collective communication involving *subsets* of agents, can be defined in Orespics-PL by associating a set of properties with each agent. The simplest kind of property is its *breed*. The breed of an agent is defined during the initialization phase and can be exploited in the communication commands to restrict the set of senders/receivers in a communication. Each agent belonging to a breed can be further identified by a set of properties. The value of these properties may be statically initialized in the declarative part of the agent's code, through the *property construct*. Afterwards it can be dynamically updated. The property mechanism is fully described in [6] and it will be exploited in the applications of section 3.

The Orespics environment supports the development of Orespics-PL programs. The user may define the appearance and the kind of each agent and pair a set of animations and/or sound with it. Furthermore, the user defines the execution environment, i.e. a microworld where the agents moves and interact. The kind of an agent A defines if there is a single instance of A or if it belongs to a breed and, in this case, how many instances of A should be created. Even if Orespics associates a default icon with each agent, the user can change this icon or choose a new one from a predefined set of images. Furthermore, several images can be defined for each agent. We will see an example of this in sect. 4. Each image is uniquely identified by its name, and each agent, at run time, can select its image through the *setimage* command. An agent may be paired with an animation as well. This is realized by selecting a set of frames which can be displayed cyclically or from the first to the last one and on the other way round, continuously. It is also possible to associate a sound with any agent. Each sound is uniquely identified and can be selected by an agent through the *setsound* command.

The execution environment is initialized by choosing the background image and music. At the microworld initialization, the standard icons of all the agents are displayed. The user defines the initial position of each agent by simply dragging and dropping its icon, whilst the initial position of the agents belonging to a breed is automatically decided by the system, but it can be updated through the *ORESPICS-PL* positioning commands. The whole environment is based on a friendly interface, based on a set of windows. A detailed description of the environment is presented in [6].

3 A Cellular Automata

This section and the following one show how $ORESPICS$ should be exploited in introductory computational science courses. The first example, presented in this section, shows a cellular automata simulating the dynamic behaviour of a gas. Sect. 4 discusses the solution of searching and optimization problems.

The definition of models for fluid or gas dynamics is an active area in computational science. Computational fluid dynamics describes physical phenomena through partial differential equation, like the *Navier-Stokes* ones, whose solution requires non trivial mathematical techniques which are generally acquired in advanced mathematical course. Nevertheless, simpler models result from solving these equations through *finite differencing methods* that introduce a set of discrete approximations and these models are closer to the real phenomena. These models can be exploited to present basic concepts of computational fluid dynamics in introductory courses. As an example of a simple, yet realistic, problem consider the diffusion of heat on a square metal sheet. The temperature at an inner point can be computes as the average of the temperatures of the four neighboring points. In $ORESPICS$, it is rather simple to define a data parallel concurrent program, where the agents corresponds to the points of the sheets: the program could display the temperature of the sheet by pairing distinct temperature values with distinct colors.

```
Agent Particle_{i,j}
    property mypos
    setimage Particle
    initial-positioning()
    repeat
        Waitagents()
        \* Movement
        forward 1
        mypos ← pos()
        mydir ← heading()
        Waitagents()
        \* Directions Echange
        SendAllAndNoWait mydir toBreed (Particles) withProp (position=mypos)
        Waitagents()
        \* Conflict Resolution
        count ← 0
        turn ← false
        while inmessage() fromBreed(Particles)
            WaitAndReceive dir fromBreed(Particles)withProp (position=mypos)
            count ← count +1
            if abs(mydir-dir)= 180° then turn ← true
        endwhile
        if (count=1 and turn) then left 90°
    forever
```

Fig. 1. The cellular automata

Another class of computational models for molecular dynamics is that of *lattice gas automata* [12, 7] that model a fluid as a system of particles moving on the edges of a lattice, according to a set of rules. In general, these models assume that at most one particle enters a given node of the lattice, from a given direction. The particles move at discrete time steps, at a constant speed. Particles entering the same node at the same time step may collide: the rules to solve collisions guarantee the conservation of the total number of particles and of the angular moment. Several lattice gas automata have been proposed. The simplest one, based on the $HPPmodel$ [7], exploits a square lattice and it considers only a simple collision rule: when exactly two particles enters the same lattice vertex from opposite directions, a collision is detected and the particles change their direction by turning left 90° . In all other case, for instances when the directions of two particles are orthogonal or when more that two particles find themselves at the same vertex, the particles do not change their direction.

In Fig. 1, we show an *ORESPICS* agent implementing a single particle: all the particles are characterized by the same behaviour.

Each agent is characterized by a property, defining its coordinates on the screen: the value of the property is dynamically updated whenever the agents moves. Initially, each agent places itself at a lattice node and chooses a direction. The initial positioning must

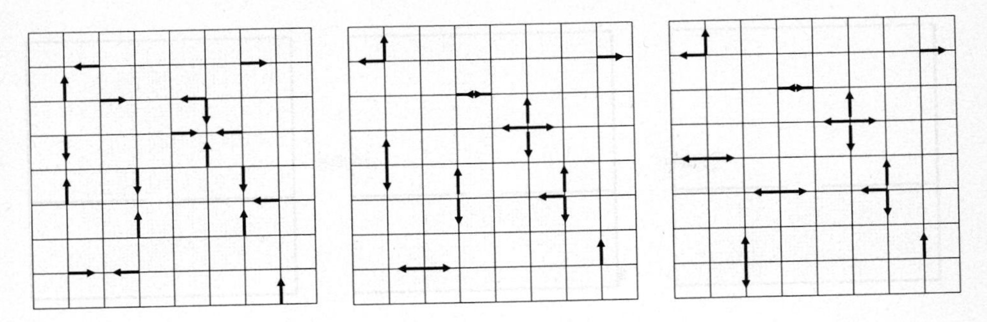

Fig. 2. Evolution of the cellular automata

guarantee that at most a particle is positioned at a vertex with a given direction. After the initialization phase, each agent iteratively executes three distinct phases. In the movement phase, it moves one step along its direction. In the second phase, all particles lying in the same node exchange their directions. The last step implements the interactions among particles: each particles collects all the incoming messages and detects any possible collision. If a collision is detected, each particle involved in the collision changes its direction, by a 90° left turn. These phases are separated by *synchronization barriers*, implemented through the *Waitagents()* primitive, to guarantee that each phase is initiated by any agent only when all others agents have completed the previous phase. For instance, the second barrier guarantees that each particle at node *N* starts collecting the messages only after any particle at *N* has sent its direction. In this way, all messages will be received.

ORESPICS supports a straightforward implementation of both the concurrent behaviour of the automata and the graphical interface. As far as concerns concurrency, the property mechanism is exploited in the second phase, to select the proper set of receivers, i.e. the set of particles lying at the same vertex. The graphical interface exploits the *LOGO* turtle graphics to show the evolution of the automata. We recall that, in *LOGO*, each turtle is characterized by its position, i.e. its coordinate on the screen and by its heading which is measured in degrees clockwise from North. Since two particles may collide if and only if their headings are directed against each other, the collision may be detected by checking if the absolute value of the difference of the headings' values is 180°. Furthermore, each particle involved in a collision, simply turns towards its own left through 90°.

Fig. 3 shows how the evolution of the system in the various phases can be monitored in *ORESPICS*. The left snapshot shows the initial situation, the central one the situation after the movement of the particles, the left one the situation after conflict resolution.

To model a realistic situation, some constraints have to be added to this simple model. For instance, consider a gas constrained in a container. The container is divided into two parts, separated by a wall with a hole. Initially, the gas particles are confined in the bottom part of the container, then they start flowing through the hole to the upper part till an equilibrium is reached. The behaviour of the particles bumping against the walls of the container is modeled by adding a new rule to the automata: a particle

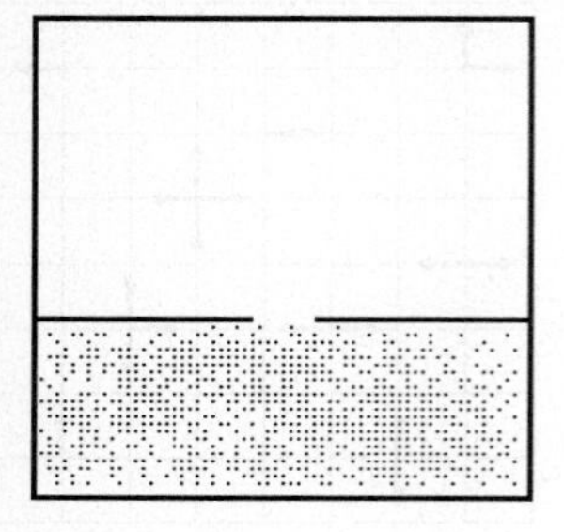
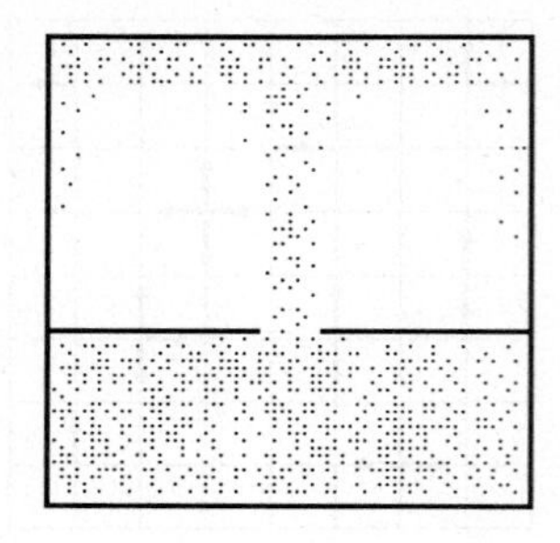
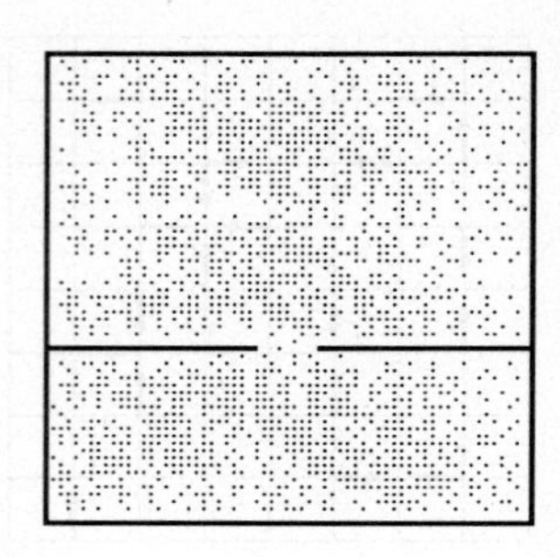

Fig. 3. Time evolution of a HPP gas

bumping against a wall, bounces back from where it came. The student can monitor the evolution of the system as shown in Fig. 3.

The code shown in Fig. 1 can be easily modified to implement more complex lattice models. For instance, in the *FHH model* [12] the particles moves along the edges of an hexagonal lattice. Even if a larger number of conflict situations have to be considered, each conflict can be simply implemented by changing the direction of a particle through $ORESPICS$ graphical commands.

4 Searching and Optimization Problems

Another class of computational science problems are search and optimization ones. The heuristic techniques usually exploited in this class, e.g. branch and bound, hill climbing, simulated annealing, genetic algorithm, often present a simple mathematical formulation and can be applied to real life problems. Hence, these problems are suitable for our target.

This section shows how the hill climbing search technique can be introduced through a real life problem. The problem is proposed in [3] as follows:

The recently discovered planetoid, Geometrica, has a most unusual surface. By all available observation, the surface can be modeled by the function $h(\theta, \rho)$*:*

$$h = 35000sin(3\theta)sin(2\rho) + 9700cos(10\theta)cos(20\rho) - 800sin(25\theta + 0.03\pi) + 550cos(\rho + 0.2\pi)$$

where h is the height above or below sea level, θ *is the angle in the equatorial plane (defines longitude on earth), and* ρ *is the angle in the polar plane (defines latitude on Earth). A space-ship has landed on Geometrica. The main goal of the astronauts is to find the* (θ, ρ) *position of the highest point above the sea level on Geometrica surface.*

To reach the topmost point of Geometrica, an astronaut may adopt an *hill climbing* strategy and move always uphill. Obviously, this does not guarantee that the highest point will be reached, because the astronaut can be stucked at the top of a low hill. To increase the probability of reaching the top of Geometrica, the national minister for space

```
Agent Hiker_i
    WaitAndReceive (x_min,x_max) from Master
    x ← random(x_min,x_max)
    goto (x, h(θ̄, x))
    setimage Astronaut
    Δ ← ε
    stop ← false
    repeat
        h ← h(θ̄, x)
        if h(θ̄, x + Δ) > h then
            pendown
            x ← (x + Δ)
            goto (x, h(θ̄, x))
        else
            if h(θ̄, x − Δ) > h then
                pendown
                x ← (x − Δ)
                goto (x, h(θ̄, x))
            else
                setimage Flag
                SendAndnoWait h to Master
                stop ← true
            endif
        endif
    until stop
```

Fig. 4. Hill climbing

missions engages a large number of astronauts: each astronaut should start climbing at a different position, chosen randomly. The chief of the mission remains on space-ship and collects the results from the hikers, thus determining the global maximum.

It is worth noticing that this example could be exploited also to introduce *Monte Carlo numerical techniques* because the basis of these techniques is the use of a large set of randomly generated values used to define different, independent computations.

To simplify the $ORESPICS$'s implementation of the previous problem, we consider an equivalent $2D$ problem, by fixing the parameter $\theta = \overline{\theta}$ in the h function. Furthermore, each hiker performs a single exploration in its area. The resulting implementation is shown in Fig. 4. The code of the master is not shown because it is very simple. It partitions the area to be searched among the different astronauts, collects the results, and computes the maximum height. This corresponds to a static assignment of the tasks to the astronauts. Each hiker receives the coordinates of its area and puts itself in a position of the area chosen randomly. Then, it tries to move uphill: if this is not possible, it puts a flag on the top of the hill, to show it has been visited. This is implemented through the *setimage* command which changes the aspect of the hiker. At this point, the hiker can stop (as in Fig. 4) or continue the exploration choosing a new starting point.

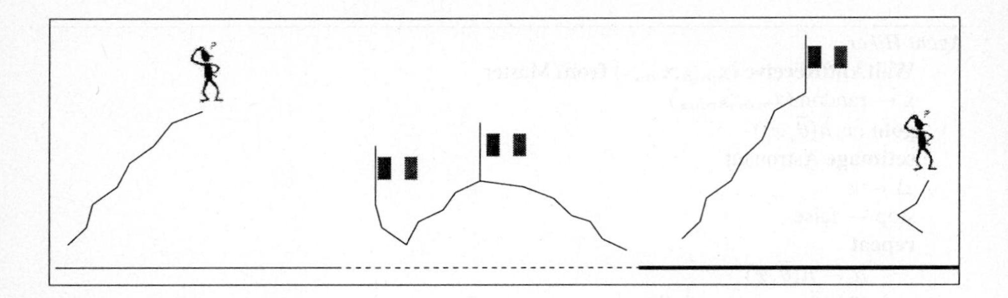

Fig. 5. Hill Climbing

The evolution of the search is shown in Fig. 5. Segments representing areas assigned to distinct hikers are represented through different line styles. We can note that some astronauts may have a longer way than others to reach their local peak, or some astronauts may climb faster because they are younger. For instance, in Fig. 5 the hiker assigned to the central area has completed its exploration, while the others are still climbing. This can be solved through a dynamic assignment. The master partitions the area into smaller segments and initially assign a segment to each astronaut. When an astronaut reaches a local peak, it asks for a new area. When no more areas to search are left, the master sends a termination message to each astronaut.

5 Related Work

Several proposals to introduce computational science in high level schools have been proposed. In [8] a set of proposals for the introduction of computational science education in high school curricula is presented. This paper discusses also how the introduction of supercomputers and high-performance computing methodology can be instrumental in getting the attention of the teenagers and attracting them to science. A presentation of more recent proposals can be found in [4].

Like *ORESPICS*, Starlogo [9] is a programming environment which is based on an extension of *LOGO*. This language has been proposed to program the behavior of decentralized systems. A student may program and control the behavior of hundreds of turtles. The world of the turtles is alive: it is composed of hundreds of patches that may be thought of as programmable entities but without movement. Turtles move parallel to one another and use the patches to exchange messages. Since the underlying concurrency paradigm is the shared memory one, this completely differentiates Starlogo from Orespics. The main goal of the Starlogo is the analysis and the simulation of the decentralized systems of the world, in contrast with more traditional models based on centralized ones. It helps users to realize that the individuals of a population may organize themselves without the presence of a centralized point of control.

Recently, several visual environments [1, 10] have been defined to support the development of parallel programs. These proposals do not define a language designed for didactic purposes, but provide support for editing and monitoring the execution of parallel programs written in C with calls to the PVM or MPI library. No particular support is provided to program real life situations: the user has to link some classical graphical library to the C program.

6 Conclusions

In this paper, we have presented *ORESPICS*, a programming environment supporting the learning of computational science in high school curricula. We are currently experimenting the system with a group of students and the first results are satisfactory. Problems from different areas, i.e. cellular automata programming, genetic programming, simulated annealing, have been programmed through *ORESPICS*. The system has also been adopted to introduce some classical computational science algorithms, like algorithms from matrix algebra, or graph algorithms. As an example, we have defined a set of animations to introduce systolic algorithms for matrix manipulation, like matrix multiplication, transposition and transitive closure computation. Currently, we are improving the system in several directions. A richer set of functionalities to monitor the execution of the programs will be defined. Furthermore, we are defining a library, including a set of complex visualization techniques through *ORESPICS* basic constructs. Finally, we plan to extend the language with constructs to support the shared memory paradigm as well.

References

1. A.Beguelin, J.Dongarra, A.Geist, and V.Sunderam. Visualization and debugging in a heterogeneous environment. *IEEE Computer*, 26(6), June 1993.
2. B.Harvey. *Computer Science Logo style*. MIT press, 1997.
3. B.Wilkinson and M.Allen. *Parallel Programming techniques and applications using networked workstations and parallel computers*. Prentice Hall, 1999.
4. C.Swanson. Computational science education. In *www.sgi.com/education/whitepaper.dir/*.
5. G.Capretti, M.R.Lagana', and L.Ricci. Learning concurrent programming: a constructionist approach. *Parallel Computing Tecnologies, PaCT*, 662:200–206, September 1999.
6. G.Capretti, M.R.Lagana', L.Ricci, P.Castellucci, and S.Puri. Orespics: a friendly environment to learn cluster programming. *IEEE/ACM International Symposium on Cluster Computing and the Grid, CCGRID 2001*, pages 498–505, May 2001.
7. J.Hardy, Y.Pomeau, and O.de Pazzis. Time evolution of two-dimensional model system. Invariant states and time correlation functions. *Jour.Math.Phys.*, 14:1746–1759, 1973.
8. M.Cohen, M.Foster, D.Kratzer, P.Malone, and A.Solem. Get high school students hooked on science with a challange. In *ACM 23 Tech. Symp. on Computer Science Education*, pages 240–245, 1992.
9. M.Resnick. *Turtles, termites and traffic jam: exploration in massively paralle micro-world*. MIT Press, 1990.
10. P.Kacsuk and al. A graphical development and debugging environment for parallel programming. *Parallel Computing Journal*, 22(13):747–770, February 1997.
11. P.Pacheco. *Parallel programming with MPI*. Morgan Kauffmann, 1997.
12. U.Frish, B.Harlacher, and Y.Pomeau. Lattice-gas automata for the navier-stokes equation. *Physical Review Letters*, 56(14):1505–1508, 1986.
13. V.Sunderam. PVM: a framework for parallel distributed computing. *Concurrency;Practice and experience*, 2(4):315–339, 1990.

Parallel Approaches to the Integration of the Differential Equations for Reactive Scattering

Valentina Piermarini, Leonardo Pacifici, Stefano Crocchianti and Antonio Laganà

Dipartimento di Chimica, Università di Perugia, Via Elce di Sotto, 8, 06123 Perugia, Italy

Abstract. Parallel restructuring of computational codes devoted to the calculation of reactive cross sections for atom-diatom reactions plays an important role in exploiting the potentialities of concurrent platforms. Our reactive scattering codes have been parallelized on different platforms using MPI and performances have been evaluated to figure out the most efficient organization models. The same codes have been used for testing new parallel environments and related coordination languages.

1 Introduction

Properties of elementary chemical reactions [1] and related virtual reality constructions [2] can be determined accurately by using rigorous quantum methods to solve the Schrödinger equation. After separating the electronic motion, the motion of the nuclei can be described by a set of time dependent Schrödinger equations:

$$\hat{H}_n(\{\mathbf{W}\})\Psi_n(\{\mathbf{W}\},t) = i\hbar\frac{\partial}{\partial t}\Psi_n(\{\mathbf{W}\},t) \tag{1}$$

where t is time, $\{\mathbf{W}\}$ is the set of nuclear coordinates, $\hat{H}_n$ is the Hamilton operator and $\Psi_n(\{\mathbf{W}\},t)$ is the time dependent nuclear wavefunction (from now on the index n will be dropped by considering only the case of the ground electronic state). The solution of this differential equation delivers the information necessary to rationalize the dynamics of the reaction considered and to evaluate its measurable properties. Such a solution can be obtained either by choosing time as a continuity variable and integrating directly the time dependent formulation of the Schrödinger equation or by factorizing the time dependence in the system wavefunction and integrating the related stationary formulation. As a case study we consider here the atom diatom systems for which is:

$$\hat{H}(\mathbf{R}_\tau,\mathbf{r}_\tau)\Psi(\mathbf{R}_\tau,\mathbf{r}_\tau,t) = \left[-\frac{\hbar^2}{2\mu}(\nabla^2_{\mathbf{R}_\tau}+\nabla^2_{\mathbf{r}_\tau}) + V(R_\tau,r_\tau,\Theta_\tau)\right]\Psi(\mathbf{R}_\tau,\mathbf{r}_\tau,t) \tag{2}$$

where $\mathbf{R}_\tau$ and $\mathbf{r}_\tau$ are the mass scaled atom-diatom Jacobi coordinates of arrangement τ, $\Psi(\mathbf{R}_\tau,\mathbf{r}_\tau,t)$ is the nuclear wavefunction and $V(R_\tau,r_\tau,\Theta_\tau)$ is the potential expressed in terms of R_τ and r_τ (the moduli of $\mathbf{R}_\tau$ and $\mathbf{r}_\tau$) and of the

P.M.A. Sloot et al. (Eds.): ICCS 2002, LNCS 2331, pp. 908–917, 2002.

angle Θ formed between them. In practice, by exploiting the properties of the rotations of the rigid bodies the dimensionality of the problem can be further reduced by formulating also the hamiltonian and the wavefunctions in terms of R_τ, r_τ and Θ (or the corresponding hyperspherical coordinates, like the APH3D ones ρ, θ and χ [3]) and the three Euler angles α, β and γ.

2 Time dependent versus time independent approaches

Time dependent approaches to the solution of the Schrödinger equation are conceptually quite simple since they set the system wavefunction in its initial configuration and then let it evolve in time under the effect of the Hamiltonian operator. The scheme of the time dependent code (TIDEP) is shown below. The outcome of the calculation depends on the initial conditions chosen. In the scheme the C coefficients represent the matrix elements which are manipulated

```
                              SECTION I
Read input data
Set inizialization and initial conditions
Calculate auxiliary variables
                              SECTION II
LOOP on t
    Perform the propagation step
    Store the C coefficients
END LOOP on t
                              SECTION III
Calculate final quantities from C coefficients and print the output
```

Fig. 1. Scheme of the TIDEP program

in SECTION III in order to obtain dynamical information. Time independent approaches to the solution of the Schrödinger equation have a more complex structure. They are typical irregular problems. The related hyperspherical coordinate computational procedure, APH3D, is articulated into several computer programs. The first of these programs is ABM. ABM calculates the eigenvalues and the eigenfunctions Φ of the hyperangular (in θ and χ) part of the hamiltonian for all the sectors in which the hyperradius ρ has been partitioned. The code then assembles the coupling matrix for the propagation on ρ to be performed at a fixed value of the total energy by the second program LOGDER. These two programs are the most time and memory consuming components of the computational procedure. The schemes of the ABM and LOGDER programs are given in Fig. 2 and Fig. 3, respectively.

```
                                    SECTION I
Input general data
Calculate quantities of common use
                                    SECTION II
LOOP on sectors
   Calculate the value of ρ at the sector midpoint
   LOOP on Λ
      Calculate eigenvalues and surface functions
      Store on disk eigenvalues
      IF(not first ρ)
         THEN
            Calculate overlaps with previous sector functions
            Store on disk the overlap matrix
      END IF
   END the Λ loop
   LOOP on Λ
      Calculate the coupling matrix
   END the Λ loop
   Store on disk the coupling matrix
END the sector loop
```

Fig. 2. Scheme of the ABM program.

```
                                    SECTION I
Input general data
Read from disk data stored by ABM
Calculate quantities of common use
                                    SECTION II
LOOP on N_E energies
   Embed the energy dependence into the coupling matrix
   LOOP over sectors
      Single step propagate the fixed J and p logarithmic derivative matrix
   END the sector loop
   Calculate and store the final logarithmic derivative matrix elements on disk
END the energy loop
```

Fig. 3. Scheme of the LOGDER program.

As is apparent from the schemes of Fig. 1 and Fig. 3 a critical point of both TIDEP and LOGDER calculations is the fact that the natural computational grain is the fixed J (J is the total angular momentum quantum number whose projections on the quantization axis are the $J+1$ values of Λ) and p (parity) integration of the scattering equations.

3 Data structure and Data streams

The flux diagram of ABM is given in Fig. 4. In the figure the nodes represent

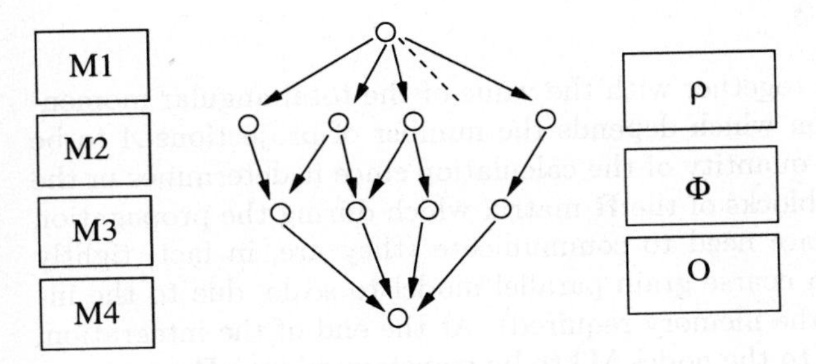

Fig. 4. ABM application graph

the computational moduli while the arches represent fluxes of data. The name of the moduli are given in the left hand side boxes while the names of the matrices containing the key information are given in the right hand side boxes. In particular, the first relevant data are the surface functions Φ, in which the fixed J and p partial wave has been expanded and that are calculated at the midpoint value of ρ for each sector. The other relevant data are the overlap matrix **O** and the coupling matrix **D** that are stored on disk for further use by LOGDER. The flux diagram indicates that the M1 module reads the input data and performs preliminary calculations of general use including the partitioning of ρ into sectors. The M2 module calculates then the surface functions Φ and sends them to module M3 to calculate the overlap matrix between the surface functions of adjacent sectors at the common border. The calculation of these overlap integrals introduces an order dependence that can be solved in different ways. Additional calculations are performed by the sequential module M4 in order to calculate the coupling matrix **D** out of the vector eigenvalues and overlap matrices. The **D** matrix is then stored on disk for a subsequent use by the LOGDER program. The flux diagram of the second program, LOGDER, which carries out the propagation of the solutions of the scattering equations at a fixed value of energy E, is given in Fig. 5. For this program the most important data are the coupling matrix **D** and the solution matrix **R**, either at the final integration point (**Rf**) or at its asymptotic limit (**Ra**). The flux diagram indicates that the M1 module after reading the input data (and in particular the coupling matrix **D**) and carrying out the preliminary calculations of general variables, passes the **D**

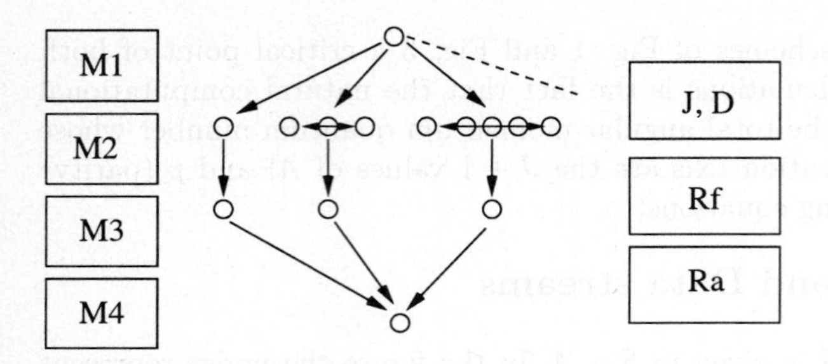

Fig. 5. LOGDER application graph

matrix to the next module together with the value of the total angular momentum quantum number J, on which depends the number of projections Λ to be considered. This is the key quantity of the calculation since it determines in the module M2 the number of blocks of the **R** matrix which during the propagation along the reaction coordinate need to communicate (they are, in fact, tightly coupled and do not allow a coarse grain parallel model to scale, due to the increase with the J value of the memory required). At the end of the integration, the **Rf** matrices are passed to the nodes M3 to be transformed into **Ra** matrices which are then passed to module M4 to be assembled and stored for further use. Despite the several conceptual differences, the same graph applies to TIDEP as also apparent from the schemes of Figs. 1 and 3.

4 Message passing parallelization

The parallel organization of the computational moduli as well as the management of the fluxes of data described above, have been implemented using MPI. The model used for the ABM program is a dynamic task-farm in which the calculation of the surface functions Φ of each sector is assigned to the first available node by a dynamic allocation controlled by the master. The surface functions are then used for calculating the **O** matrix of the overlap integrals between surface functions of adjacent sectors at the common border and for the construction of the coupling matrix **D**. In this case shared memory would allow an access to data of different sectors in the sequence dictated by the order in which overlaps between functions of adjacent sectors need to be calculated. Using the message passing paradigm of MPI this can be obtained by the (repeated) direct calculation of the surface functions when needed, either by solving again the Schrödinger equation for the previous sector, or in a simpler way, by combining the primitive functions using the already calculated eigenvectors. This choice, together with a dynamic distribution of the work, greatly reduces the communication time and minimizes the work unbalance. Accordingly, the master reads the input data and distributes them to the slaves as soon as they are available and ready to start a new job. The MPI structure of the ABM master is shown in Fig. 6.

On their side, the slaves receive at first the input data and afterwards the ρ value

```
MASTER PROCESS
Read input data
Send input data to the slaves
LOOP on sector index
      Calculate the value of ρ at the sector midpoint
      Call MPI_SEND(ρ)
END LOOP on sector index
```

Fig. 6. Pseudo MPI code for the master process in ABM

associated with the sector for which they have to calculate the surface functions Φ. The slaves (see in Fig. 7 the related MPI structure), after calculating the surface functions, write on disk the eigenvectors, which are the coefficients of the surface functions in the given basis set. When the sector is not the first one,

```
SLAVE PROCESS
    Receive input data
10 Call MPI_RECV(ρ)
        LOOP on Λ
            Construct the primitive basis set at the given value of ρ
            Solve the angular Schrödinger equation by expanding in the primitives
            Store on disk eigenvalues and eigenvectors
            IF(not first sector) then
                Construct the basis set at the previous value of ρ
                Read from disk related eigenvectors
                Compute overlap integrals at the common edge of the two sectors
            END IF
        END LOOP on Λ
    Calculate the coupling matrix
    Store on disk the coupling matrix for use from LOGDER
    GOTO 10
```

Fig. 7. Pseudo MPI code for the slave process in ABM

the process has to be repeated (or corresponding eigenvectors have to be read from disk). This allows the calculation of surface functions of the previous sector, and the evaluation of the overlap integrals between them. Then, the coupling matrix is calculated and stored on disk, together with the overlap matrix, for use by the LOGDER program. This parallel model has been implemented on the CRAY T3E of EPCC (Edinburgh, UK) using up to 128 processors. The individual processor computing time never exceed more than 10% the average one and the speedup was never smaller than 70 % the ideal value [4].

A similar task farm model was adopted for LOGDER and TIDEP. However a similar excellent performance was obtained only for the lowest J=0 runs, due to the difficulty of handling shared memory with MPI. The only possibility of being able to deal with this class of irregularity in a portable way (the CRAY machine has its own shared memory environment) is to turn into abstract level coordination languages.

5 The coordination language parallelization

A very promising coordination language being developed by prof. Vanneschi and collaborators at the Department of Informatics of Pisa is ASSIST [5]. It sprouts out of the skeleton based language SkIE (Skeleton-based Integrated Environment) [6]. SkIE defines some moduli (called skeletons) structured as parallel model prototypes that can be used as building blocks of a parallel application. ASSIST introduces a more general and flexible module, called parmod, whose structure can be defined by the user by making reference to a set of virtual processors (VP) as well as a tool allowing the use of external objects (variables, routines, etc.) and emulating the shared memory. The use of these concepts has allowed us to develop several constructs to implement ABM and it has made possible also a portable parallelization of LOGDER and TIDEP for an arbitrary value of J. As for ABM, see Fig. 8, the first model is straightforwardly derived by

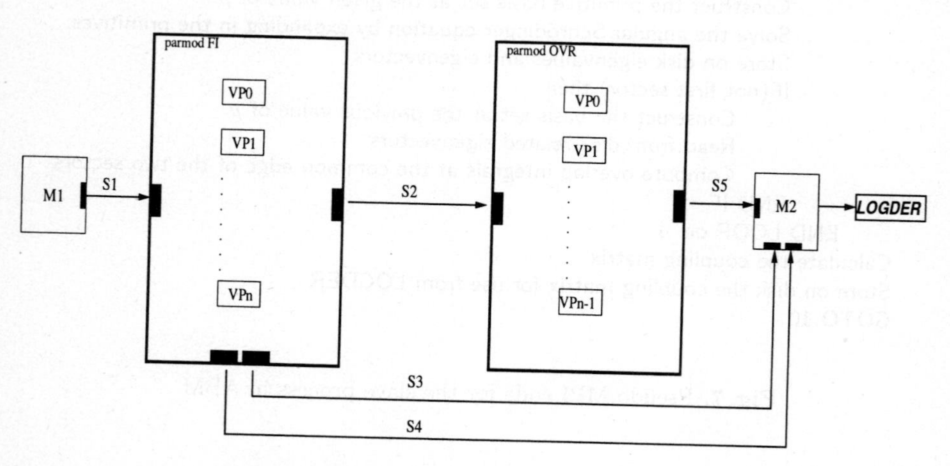

Fig. 8. ASSIST scheme of the first model for ABM

a message passing implementation of the sequential version and is characterized by four moduli: a sequential (M1) and three parallel (FI, OV and M2) ones. M1 executes some preliminary calculations and sends the value of the reaction coordinate of each sector to parmod FI, using the output stream S1 (indicated in the figure as an arrow connecting M1 to F1). The VPs of parmod F1 are arranged in a monodimensional array and calculate the surface functions by solving the Schrödinger equation on the hyperangles. Eigenvalues and components of the coupling matrix built out of the surface functions of the same sectors are sent to parmod M2 by the output streams S3 and S4. Surface functions are sent to the parmod OV through the output stream S2 to calculate the overlap matrix **O** between surface functions of adjacent sectors. The parmod OV, that has the same topology as FI, after receiving the values of the surface functions of the involved sectors at the quadrature points needed for the evaluation of the overlap, builds the matrix **O** to be used for mapping the **R** matrix from one sector to the next one and sends it to parmod M2 by the output stream S5. The second model is quite similar to the first one (see Fig. 9). In fact it is still constituted by 4 moduli one of which is sequential and three are parallel. The main difference between the two models is that in the latter one the transfer of the values of the surface functions has been replaced by a transfer of the eigenvectors and the parameters of the primitive functions through the output streams S2 and S3, as it has been done by using MPI. This enables the next module, parmod OV, to calculate values of surface functions of the previous sector on the quadrature grid points of the current sector and to evaluate the overlap integrals by minimizing communications. The transfer of the coupling matrix terms and eigenvalues is still performed directly using the parallel module M2 (using the output streams S4 and S5) which collects also the output of parmod OV thanks to the output stream S6. The third model is characterized by the fact that there is only

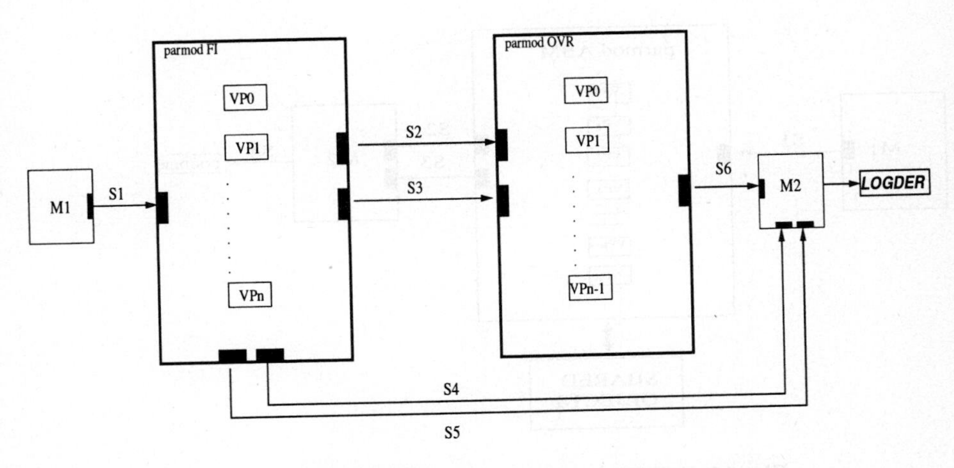

Fig. 9. ASSIST scheme of the second model for ABM

one parmod module placed between the two sequential ones, as can be seen in Fig. 10. While M1 works exactly in the same way as in the first two models, the parmod ABM does not have anymore the structure of a monodimensional array. In fact, in this case, the VPs of the parmod are organized in a 2xN matrix in which the virtual processors of the first row elaborate the data coming from the

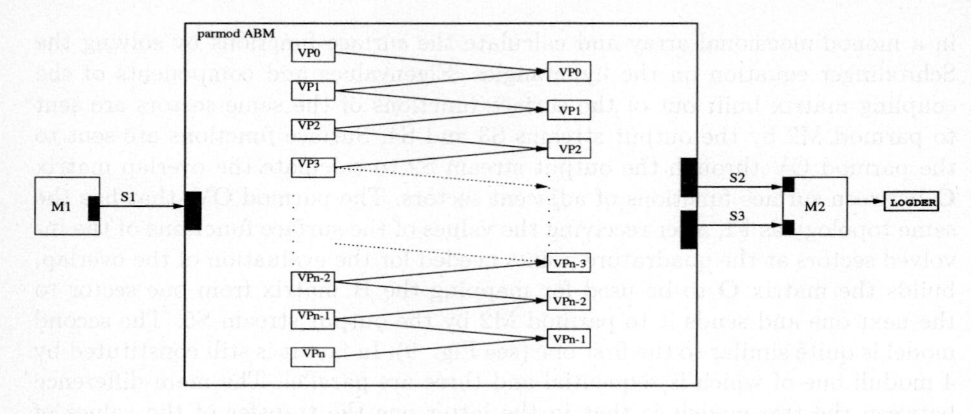

Fig. 10. ASSIST scheme of the third model for ABM

input stream S1 and calculate the surface functions. On the other hand, virtual processors of the second row calculate the overlap integrals on the data coming from the processors of the first row. The transfer of the coupling matrix and of the eigenvalues terms derived from the surface functions of the same sector to module M2 is performed through the output stream S2, while the transfer of the coupling matrix is performed through the output stream S3. The fourth

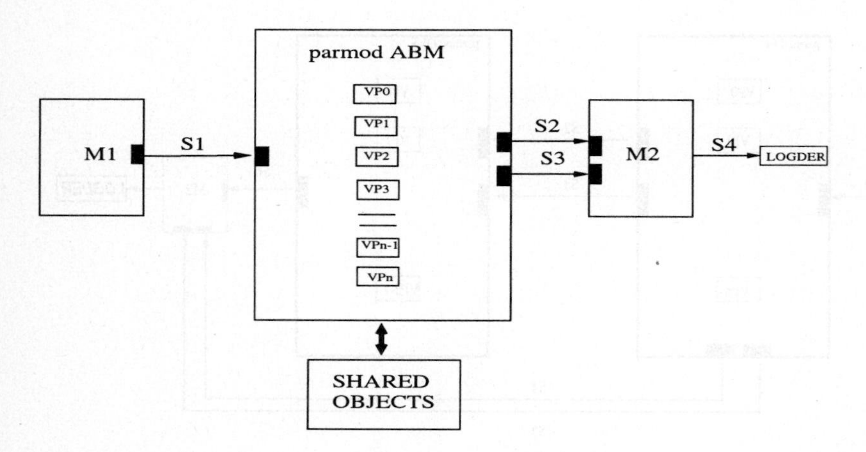

Fig. 11. ASSIST scheme of the fourth model for ABM

model is the simplest one and the closer to the sequential version of the code. It is characterized by a single parmod sandwiched in a PIPE between two sequential modules M1 and M2. The structure of the model is, therefore, identical to that of the third one, yet the array structure of the VPs is monodimensional. This is due to the use of shared objects (see Fig. 11) to handle the sequentiality

associated with the calculation of overlap integrals between surface functions of adjacent sectors. Data transfers are ensured by streams of output like S1 (from the sequential module M1 to parmod ABM) and S2 and S3 (from parmod ABM to the sequential module M2). Again, S4 sends the coupling matrix **D** to the LOGDER program.
A similar extremely simple approach can be adopted for LOGDER and TIDEP. As shown by Fig. 12, the sequential module M1 arranges the **D** and **O** matrices

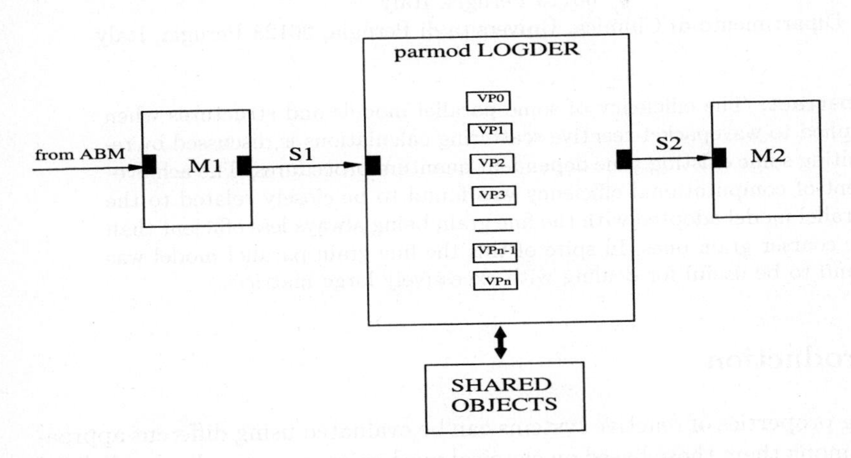

Fig. 12. ASSIST scheme of LOGDER

(read from disk) for use by the pipeline of VPs given in the central parmod LOGDER. Then the resulting **Rf** matrix final elements are converted into **Ra** matrix (asymptotic) and passed to subsequent analysis program.

References

1. Laganà, A., Riganelli, A.: Computational Reaction and Molecular Dynamics: from Simple Systems and Rigorous Methods to Large Systems and Approximate Methods. Lecture Notes in Chemistry, **75** (2000) 1-12
2. Laganà, A., Riganelli, A.: Chimica e Informatica: Dal laboratorio alla realtà virtuale molecolare (Morlacchi, Perugia) (2001) ISBN 88-87716-46-3
3. Parker, G. A., Pack, R. T, Laganà, A., Archer, B. J., Kress, J.D. and Bačic, Z.: Exact quantum results for reactive scattering using hyperspherical (APH) coordinates. Supercomputer Algorithms for reactivity, Dynamics and Kinetics of Small Molecules, A. Laganà Ed. (Kluwer, Dordrecht) (1989) 105-129
4. Laganà, A., Crocchianti, S., Bolloni, A.: Quantum isotopic effects and reaction mechanism: the Li + HF reaction. Phys. Chem. Chem. Phys. **2** (2000) 535
5. Ciullo, P., Danelutto, M., Vaglini, L., Vanneschi, M., Guerri, D., Lettere, M. Progetto ASI-PQE2000, Workpackage 1, Ambiente ASSIST: modello di programmazione e linguaggio di coordinamento ASSIST-CL (versione 1.0) (2001)
6. Vanneschi, M.: Parallel Paradigms for Scientific Computing. Lecture Notes in Chemistry, **75** 1 (2000) 168-181

Fine Grain Parallelism for Discrete Variable Approaches to Wavepacket Calculations

Daniele Bellucci[1], Sergio Tasso[1] and Antonio Laganà[2]

1 - Dipartimento di Matematica e Informatica, Università di Perugia, 06123 Perugia, Italy
2 - Dipartimento di Chimica, Università di Perugia, 06123 Perugia, Italy

Abstract. The efficiency of some parallel models and structures when applied to wavepacket reactive scattering calculations is discussed by revisiting some existing time dependent quantum procedures. The achievement of computational efficiency was found to be closely related to the parallel model adopted with the fine grain being always less efficient than the coarser grain ones. In spite of this the fine grain parallel model was found to be useful for dealing with excessively large matrices.

1 Introduction

Scattering properties of reactive systems can be evaluated using different approaches[1]. Among them those based on classical mechanics are naturally parallelized at large grain since each trajectory calculation is an independent computational task. On the contrary, approaches based on quantum mechanics are difficult to parallelize because of the spread nature of the wavefunction describing the quantum system. The spread nature of the quantum representation of the system shows up in the use of a large basis set or of a large pointwhise representation of the wavefunction.

Quantum wavepacket methods differ from time-independent quantum methods in that they integrate in time the time-dependent Schrödinger equation starting from a known quantum state of the reactants.

In the numerical procedure we use (TIDEP), only the real part of the wavepacket is propagated [2]. For the generic atom-diatom collinear (two mathematical dimensions) reactions A + BC(v, j) the initial wavepa-cket is set up by expressing the wavefunction in terms of the initial diatomic molecule BC wavefunction and its analysis is performed at a cut corresponding to a large fixed B-C vibrational coordinate [2]. To start the propagation, the initial wavepacket (the vibrational, v, and rotational, j, wavefunction of the diatomic reactant times a normalized Gaussian function and a phase factor giving a relative momentum towards the interaction region[2]) is placed in the entrance channel. The method can be implemented in a way that only the real component of wavepacket can be explicitly propagated [2] and a collocation method can be used. Accordingly, the potential and the wavefunction are represented by their values on a regular grid that must be large enough to contain the initial wavepacket, the region where the analysis line is drawn, and the interaction region. The grid must also be fine

P.M.A. Sloot et al. (Eds.): ICCS 2002, LNCS 2331, pp. 918–925, 2002.

enough to accurately describe the structure of the wavefunction. The real part of the wavepacket is propagated in time until it has mainly been absorbed near the edge of the grid. This reduces the calculation to a continuous manipulation of a certain number of multidimensional matrices.

As already pointed out elsewhere[3], a coarse grain model is better suited to parallelize TIDEP. Coarse grain parallel implementations of the code and related advantages and disadvantages are discussed in section 2. Finer grain parallelization models are, under certain respects, more interesting. They act, in fact, at the innermost level of the matrices implying a highly repetitive execution of some operations that is a favourable case for parallelism. In addition, the bigger dimensionality of the matrices used in these approaches makes more likely the possibility that the memory limits of the machine used are hit. In section 3, we discuss several details of a fine grain parallel implementation of TIDEP.

2 The coarse grain parallelization of TIDEP

The parallelization tool used in our work is MPI. In TIDEP, calculations are performed for a given range of energy, at a fixed value of the vibrotational quantum number (vj) of the reactant diatom and at a single value of the total angular momentum quantum number J and parity p. Therefore, the coarsest grain level of parallelism that can be adopted is the one distributing the calculation for a given quartet of v, j, J, p quantum numbers. In this case, due to the increasingly (with J) different characteristics of the various tasks, the best choice is to adopt a task farm organization dynamically assigning the computational workload [3]. This very coarse grain approach was fruitfully implemented on a cluster of powerful workstations. In this case, however, the highest J calculations require increasingly longer times to run, make the check for convergency with partial waves difficult and the imbalance of the load rapidly growing.

A next lower level of parallelization of TIDEP is the one based on the combined distribution of fixed J, p and Λ calculations. As it has been already pointed out above, while it is natural to distribute fixed J calculations (these calculations are fully decoupled) the decoupling of Λ is not natural since one has to introduce physical constraints of the centrifugal sudden type (*i.e.* the projection of J on the z axis of the body fixed frame is kept constant during the collision). This allows to perform separately the step-propagation of the wavepacket for blocks of fixed Λ values and recombine the various contributions only at the end of each propagation step. This is a key feature of the adopted computational scheme since it allows a decomposition of the domain of the wavepacket in J blocks of size equivalent to that of block $J = 0$. This converts a request for increasing the node memory proportionally to J into a request for increasing the number of nodes proportionally to J while keeping the node memory constant.

To carry out the calculations the $O(^1D)+HCl$ atom diatom reaction [4, 5] was taken as a case study. Accordingly, the mass values were chosen to be 15.9949 amu, 1.00783 amu and 34.96885 amu (for O, H and Cl, respectively). The energy range covered by the calculation was approximately 1 eV, the initial vibrotational state used for the test was $v = 0$ and $j = 0$. The potential energy surface used for the calculations is described in ref. [4, 5] where other details are also given. Two sizes ((a) 127×119 points, (b) 251×143 points) were used for the dimension of the R' and r' matrices while the angular part was expanded in a set of basis functions (80 in both (a) and (b) cases). Time propagation was carried out for about 40000 steps to properly diffuse the wavepacket. Production runs took about 3 weeks on a single node of a Sylicon Graphics PowerChallenge supercomputer at $J = 0$. A first version of the parallel code was run [5] on the Cray T3E at the EPCC of Edinburgh (UK) for the simplest case of $J = 0$ and $J = 1$ in which only three pairs of J and Λ values are considered and only three nodes are used. Measured speedups are 2.6 for the propagation grid (a) and 2.5 for the propagation grid (b).

The model was then generalized to higher J values. In this generalized model node zero was exclusively devoted to act as a master and the I/O was decentralized. At the same time, the feature of carrying out fixed J calculations in pairs by associating one high J value with its complement to the maximum value of J was adopted in order to keep the number of processors used constant. To evaluate the performance of this model, the calculations were carried out on the Origin 3800 at Cineca (Bologna, I) using the same set of parameters adopted for the grid (a) test described above yet reducing the basis set expansion for the angular part to 10. Table 1 shows the value of the percentual increase of the node computing time with respect to that of the node carrying out the $J = 0$.

Table 1. Percentual time increment (with respect to $J = 0$)

J	1	2	3	4	5	6	7	8	9	10	11	12	13
% time	0.9	1.8	3.2	6.4	6.4	8.2	15.5	17.3	16.4	19.1	20.0	21.8	22.7

As clearly shown by the results reported in the Table 1 the computing time per node (averaged over the various values of Λ) increases with J up to about 20%. This indicates that, although one has to pay an extra price to increase the maximum allowed value of J, for this parallel model the increase of communication time associated with an increase in the number of allowed Λ only marginally reduces the advantage of having distributed the calculations over several computing nodes.

3 The fine grain parallelization of TIDEP

The key feature of TIDEP is the iterated use of a time propagator which is characterized by a determined and recursive structure of matrix operations, such as the fast Fourier transform, which could allow a re-use of the resources. This requires, however, that the matrix operations (multiplications, transpositions, Fourier transforms) of the algorithmic sequence are performed in a proper way.

At fine grain level, this means to focus the parallelization work on the routines BLAS DCOPY and DAXPY. In fact TIDEP calls these routines more than hundred thousands times per propagation. If use is made of propagation techniques involving a continuous transformation between coordinate and momentum space, the use of the BLAS routines is accompanied by the use of a Fast Fourier Transform routine that makes the computational burden even heavier.

An alternative approach is that based on the Discrete Variable Representation (DVR) method. This is based on the reiterated application of operations like

$$\mathbf{H} = \mathbf{A} \cdot \mathbf{C} + \mathbf{C} \cdot \mathbf{B}^{\mathrm{T}} + \mathbf{V} \odot \mathbf{C} \tag{1}$$

where $\mathbf{A}$ and $\mathbf{B}$ are the matrix representations of the two terms (one for each dimension) of the Laplacian operator, $\mathbf{C}$ is the collocation matrix of the wavefunction, $\mathbf{V}$ is the matrix representation of the potential operator (accordingly $\mathbf{V} \odot \mathbf{C}$ is the direct product of the single component $\mathbf{V}$ matrix with $\mathbf{C}$).

```
LOOP of iv from 1 to nv
  LOOP of ir from 1 to nr
    h(ir, iv) = 0
  END loop of ir
END loop of iv
LOOP of iv from 1 to nv
  LOOP of i from 1 to nr
    LOOP of ip from 1 to nr
      h(i, iv) = h(i, iv) + a(i, ip) · c(ip, iv)
    END loop of ip
  END loop of i
END loop of iv
LOOP of i from 1 to nr
  LOOP of iv from 1 to nv
    LOOP of ivp from 1 to nv
      h(i, iv) = h(i, iv) + c(i, ivp) · b(iv, ivp)
    END loop of ivp
  END loop of iv
END loop of i
LOOP of iv from 1 to nv
  LOOP of i from 1 to nr
    h(i, iv) = h(i, iv) + v(i, iv) · c(i, iv)
  END loop of i
END loop of iv
```

Fig. 1. Pseudo code for the section of the *av* routine associated with eq. (1)

In the reduced dimensionality version of TIDEP used for the parallelization, above matrix operations are performed inside the routine *av*. Inside *av*, two matrix times vector and one vector times vector operations are performed according to the computational scheme given in Fig. 1.

When all the matrices involved are distributed per (groups of) rows among a certain number of nodes, all the operations sketched above imply a quite significant amount of communication to allow the nodes have the updated version of the matrices involved.

As already mentioned, the fine grain approach has the advantage of allowing an increase of the size of the involved matrices beyond the capacity of the node memory. In this approach, in fact, the request of memory is drastically reduced by partitioning the space (and momentum) domain. The choice we made was to partition the representation domain by rows and to adopt a management of the memory that takes into account the hierarchy of the memory including the I/O levels.

Accordingly, out of eq. (1) one obtains

$$Row(i,H) = \sum_{k=1}^{nr} A(i,k) \cdot Row(k,C) + Row(i,C) \cdot B^T + Row(i,V) \odot Row(i,C)$$

whose algorithmic structure is given in Fig. 2.

In this algorithm the matrix C is always handled by rows. The parallel model adopted is a task farm that performs the calculation for the first two operations of the right hand side of eq. (1) at worker level and leaves the third one with the master. Each worker has access to a local (unshared) secondary memory in which the elements of the proper partitions of A, B and C are stored. In the startup phase the C matrix is distributed to the workers (this avoids possible subsequent I/O conflicts). In the same phase the first row of A is distributed by the master that, after reading sequentially the rows of C, forwards the pairs $\langle A(1,k), Row(k,C)\rangle$ to the workers using a *roundrobin* policy. This implies the use of a buffer of nr elements in which, at each time, the C row is stored. The dimension of the buffer is determined by the number of workers (M). Accordingly, the ith worker is assigned the pairs $\langle A(1,k), Row(k,C)\rangle$ with $k \equiv i \bmod M$.

Rows of matrix A and B are stored in the same local secondary memory. Each worker after receiving the row vector of C performs its multiplication

$$\sum_{k \in \mathcal{D}_i} Row(k,C) \cdot A(1,k) \tag{2}$$

where $\mathcal{D}_i = \{x \in N | x \equiv i \bmod M\}$.

The product $Row(1,C) \cdot B^T$ is then computed by multiplying inside each node the first row of C by the related partition of rows of B (avoiding so far the transposition). These terms are then summed to the quantity (2). The sum of the vectors computed by the workers and the master determines the first row of H. This is performed via a *reduce* called by all the farm processes in which the master deals with $Row(1,V) \odot Row(1,C)$ and the workers the computed vectors. The sum is saved into the logical space of the master that stores it into the secondary memory space assigned to the matrix H.

```
{Let nr = nv}
LOOP of i from 1 to nr
  ReadFromFile Row(i, A)
  ReadFromFile Row(i, V)
  ReadFromFile Row(i, C)
  LOOP of j from 1 to nv
    Temp(j) = 0
  END loop of j
  LOOP of k from 1 to nr
    ReadFromFile Row(k, C)
    LOOP of j from 1 to nv
      Temp(j) = Temp(j) + A(i, k) · C(k, j)
    END loop of j
  END loop of k
  LOOP of j from 1 to nv
    Temp(j) = Temp(j) + V(i, j) · C(i, j)
  END loop of j
  LOOP of w from 1 to nr
    ReadFromFile Row(w, B)
    LOOP of j from 1 to nv
      Temp(w) = Temp(w) + C(i, j) · B(w, j)
    END loop of j
  END loop of w
  LOOP of j from 1 to nv
    H(i, j) = Temp(j)
  END loop of j
  WriteToFile Row(i, H)
END loop of i
```

Fig. 2. Sequential version of the section of the *av* routine associated with eq. (1)

To minimize the worker *idle* time, the master *broadcasts* to all workers the pair $\langle Row(2, A), Row(2, C)\rangle$ before entering the state of waiting for the completion of the *reduce*. This allows the worker to immediately resume their calculations after executing the *reduce*. This has the effect of overlapping (and therefore masking) the time needed for the completion of the *broadcast* through the computing time of the workers. It is worth noting that the master performs the *broadcast* while no other process of the farm attempts an access to the communication channel. As a result, the performance is not affected by possible network access conflicts.

Then each worker W_i can access at the same time its own block of C rows stored in the startup phase with no conflicts and alike in the previous startup phase performs the sum of scalar products by taking from the vector $Row(2, A)$ the proper elements of index i modulus the number of workers (M). In a similar way, the C rows of index greater than 1 are generated.

Test runs have been performed on a Beowulf made of 8 monoprocessor (Pentium III 800 MHz) nodes having 512 MB of central memory using square matrices of size 512, 768 and 1024. As shown by Table 2, the elapsed times measured for the parallel version are definitely smaller than those of the serial version. As a result, speedups are significant and the advance apparent especially if a comparison is made between with the previous version for which the elapsed time of a five processor parallel run would hardly break even with that of a single processor sequential run[3]). It is particularly worth noting also that, in the investigated range of matrix sizes, the speedup is constant.

Table 2. Elapsed times and speedups

Matrix size	Seq. time/s	Par. time/s	Speedup
512x512	1549.1	231.1	6.7
768x768	5193.2	768.3	6.8
1024x1024	12271.0	1814.2	6.8

4 Conclusions

The need for pushing the parallelization of wavepacket reactive scattering codes to a fine level in order to deal with matrices of large dimensions associated with the solution of problems of high dimensionality has been discussed. The code considered by us for parallelization makes use of a collocation method and a discrete variable technique. Then the domain is decomposed in a way that the sequence of matrix operations can be performed by minimizing the time needed for the reorganization of the matrices during the operations and by overlapping communication to execution. The study has shown that in this way it is not only possible to deal with systems whose collocation matrices are too large to be accomodated in the local memory of the nodes but it is also possible to achieve a significant parallel speedup.

5 Acknowledgments

This research has been financially supported by MIUR, ASI and CNR (Italy) and COST (European Union).

References

1. Laganà, A., Innovative computing and detailed properties of elementary reactions using time dependent approaches, Computer Physics Communications, **116** (1999) 1-16.

2. Balint-Kurti, G.G., Time dependent quantum approaches to chemical reactions, Lecture Notes in Computer Science, **75** (2000) 74 - 88.
3. V. Piermarini, L. Pacifici, S. Crocchianti, A. Laganà, Parallel models for reactive scattering calculations, Lecture Notes in Computer Science **2110** (2001) 194 - 203.
4. V. Piermarini, G. Balint-Kurti, S. K. Gray, G.F. Gogtas, M.L. Hernandez, A. Laganà, and M.L. Hernandez, Wave Packet Calculation of Cross Sections, Product State Distributions, and Branching Ratios for the $O(^1D)$+ HCl Reaction, J. Phys. Chem. A **105(24)** (2001) 5743-5750.
5. V. Piermarini, A. Laganà, G. Balint-Kurti, State and orientation selected reactivity of $O(^1D)$+ HCl from wavepacket calculations, Phys. Chem. Chem. Phys., **3** (2001) 4515-4521.

A Molecular Dynamics Study of the Benzene...Ar2 Complexes

A. Riganelli, M. Memelli and A. Laganà

Dipartimento di Chimica, Università di Perugia, Via Elce di Sotto, 8,
06123 Perugia, Italy

Abstract. A simulation of benzene..Ar_2 clusters has been performed using molecular dynamics software. Details on development tools, interaction formulation and calculation parameters are given. Estimates of macroscopic properties as well as elements for understanding the dynamics of the systems are discussed.

1 Introduction

The purpose of this work is twofold. On one side we illustrate our efforts to use and develop object oriented tools and applications to be compared and complemented with the more traditional procedural programming techniques are illustrated. On the other side our efforts to develop computational chemistry applications aimed at producing realistic simulations of complex chemical processes in a priori fashion are discussed. To this end the formulation of the interaction and of the dynamics, the design of the computational algorithms, the averaging over unobserved parameters and the production of visualization tools are developed having in mind the objective of creating an environment enhancing chemical intuition and insight. The context of these efforts is Simbex [1]. Simbex is a Problem solving environment for the simulation of molecular processes at microscopic level using a priori means and aimed at reproducing quantities measured by experimental apparatuses (and in particular molecular beams) that is being assembled at our Laboratory.

The specific process considered here is the formation of benzene-Ar_2 clusters both by considering the C_6H_6 molecule frozen at its equilibrium geometry and by allowing it to deform. Van der Waals clusters formed by the benzene molecule with rare gas atoms have been the focus of a considerable number of experimental and theoretical studies (see for example [2, 3]). These studies were the basis of our investigation.

The computational tools used for the theoretical investigation of these systems are those of Molecular Dynamics (MD) or Monte Carlo (MC)approaches. In the MD approach, the one used in this work, after constructing a proper analytical formulation of the molecular interaction the system is represented as an appropriate set of particles, positions and momenta that is allowed to evolve in time according to the laws of classical mechanics.

P.M.A. Sloot et al. (Eds.): ICCS 2002, LNCS 2331, pp. 926–931, 2002.

In the second section of the paper we describe the object oriented computational procedure. In the third section we describe the construction of the intermolecular interaction. In the fourth section we describe the MD techniques used. In the fifth section we examine the results obtained.

2 The computational procedure

The computational procedure used for the calculations was based entirely on MMTK [4]. MMTK is a set of libraries particularly designed to assemble computational procedures of Molecular Modeling. The novelty of this computational procedure lies in the that it is totally written in Python. Python is a high-level object-oriented general-purpose programming language whose Scientific Python subset [5] has modules suited for integrating differential equations, for interpolating numerical functions, for manipulating geometrical figures, for managing PDB files, for interfacing MPI libraries, etc.. In Python we have integrated C/C++ and Fortran routines as shared libraries (in particular routines from ref. [6]).

This has allowed us to build our own computational procedure by importing from Python almost all modules and writing a few new ones for which we exploited the inheritance properties of MMTK modules. In particular, use has been made of modules for the construction of the chemical universe and chemical objects (atoms, groups, molecules, molecular complexes), for the definition of the force field (Amber and Lennard Jones), for the integration of the trajectories (initial, final and instant positions, integrators, optimizers) and for the analysis of the results (visualization, averaging, integrators).

Due to the particular simplicity of the chemical system under investigation only a few atomic chemical objects were defined and the subset of operations that was considered was also limited. The same is true for the various phases of the MD treatment. The initial conditions of the system were set by defining the universe of initial velocities. This was, eventually, scaled using *thermostat* and *barostat* objects to impose a given distribution of temperature or pressure. In certain runs bond constraints were also imposed. Trajectory data such as atomic positions, velocities, energetic contributions, gradients, etc., were saved using netCDF libraries [7] which allow an economical binary storage of the information for further use in the analysis. The approach used for the energy minimization were based on the conjugate gradient algorithm.

The calculations were performed on a 900Mhz PC with 128 Mb RAM memory and taking each run few ours CPU time.

3 Inter- and intra-molecular potential

Intermolecular interaction of the van der Waals molecule are particularly difficult to characterize. The small value of the binding energy makes the treatments based upon ab initio calculation machinery quite inefficient. At the same time, experimental studies based upon crossed molecular beams are also difficult to perform. For this reason, as is often the case, we addressed ourselves to model

Diatom	σ	ϵ
Ar-Ar	3.35	99.09
Ar-C	3.42	40.20
Ar-Ar	3.21	33.00

Table 1. Lennard-Jones parameters used in this work. σ is given in $\mathring{A}$ and ϵ in cm^{-1}.

formulations. In particular, here we deal both with different expansions and model formulations.

The first intermolecular potential used for the simulation of the benzene-Ar system is a pair-additive one obtained by summing all the two body terms of the interaction of the C and H atoms of the benzene molecule with the Argon atoms plus that between the two Ar atoms. For these two body interactions, as usual, a Lennard-Jones 12-6 model potential [8] was adopted

$$V^{LJ}(r) = 4\epsilon \left[\left(\frac{\sigma}{r} \right)^{12} - \left(\frac{\sigma}{r} \right)^{6} \right] \quad (1)$$

In eq. 1 r is the internuclear distance between the two atoms, ϵ is the energy minimum of the diatomic interaction, σ is the corresponding equilibrium value. As for the intramolecular potential we have either assumed that the benzene is frozen at its equilibrium position (though free of translating and rotating) or allowed the molecule to deform according to a force field of the Amber type [9]. The parameters for the Ar-Ar, Ar-C and Ar-H interactions are given in Table 1 and were taken from Ref. [10]

An extension of the computational procedure to other potentials is being considered using a polynomial in the bond order (BO) coordinates (that has been already extensively used for the study of triatomic and tetratomic systems [11]). Another potential that is also being considered is the product of a Morse-Switching-van der Waals radial term times the spherical harmonics describing the angular dependence of the interaction. Related parameters have been otpimized to the reproduction of crossed molecular beam data [12]. A conjugate-gradient technique was used to locate the minima on the potential energy surface. Figure 1 shows these minima corresponding to $R = 3.49\,\mathring{A}$ for the (1|1) isomer and $R = 3.70\,\mathring{A}$, $r_1 = 3.50\,\mathring{A}$ and $r_2 = 3.10\,\mathring{A}$ for the (2|0) isomer. In Table 2 these values and the corresponding point groups are given.

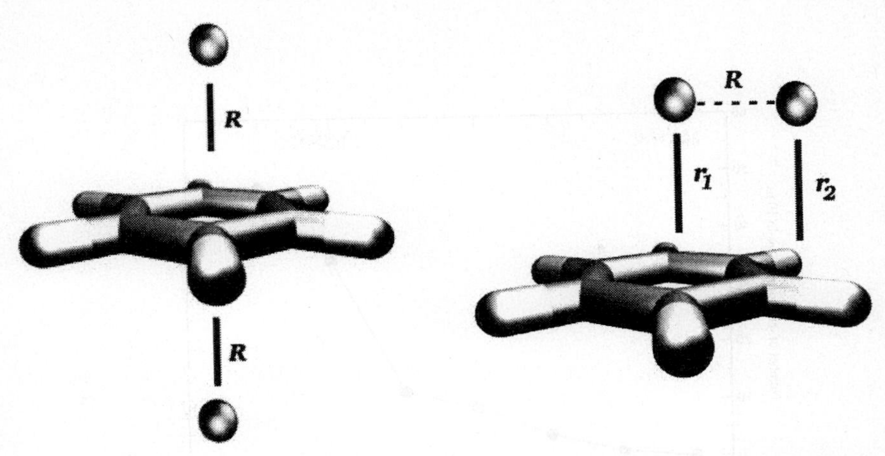

Fig. 1. Location of the energy minima for the benzene-Ar_2 cluster isomer (1|1) (left hand side panel) and isomer (2|0) (right hand side panel).

n	Label	PG	Energy
1	(1|0)	$C_{6}v$	-356.0
2	(1|1)	$D_{6}h$	-711.0
	(2|0)	C_s	-665.0

Table 2. Energies (in millihartree) and point groups (PG) for the minima of benzene-Ar_n clusters. The meaning of the labels is described in the text.

4 Calculations and results

MD simulations were performed using a microcanonical ensemble (NVE). The duration of these simulations was of 100 ns. A timestep of 0.5 fs was adopted to obtain an energy conservation of about 2 cm^{-1}. In the case in which the benzene was kept rigid the timestep was set at 2 fs. Initial positions were chosen so as to start with the isomer (1|1) at the energy global minimum. For this configuration the initial velocities were generated and the system was let to run for about 10^5 steps in order to allow thermalization and then NVE (microcanonical ensemble) conditions were imposed. When this dynamical balancing process turned out not to be successful velocities were scaled and the balancing process started again. When the balancing process was successful the simulation was run. At the end of the simulation the system is analyzed and relevant parameters calculated.

For our investigation, a temperature range of 10 K starting from 25 K was chosen. The temperature was increased in steps of 1 K up to 35 K.

A first interesting parameter of the process is the number of internconversion (NINT) from isomer (1|1) to the isomer (2|0) observed at the various temperatures. The value of NINT is plotted as a function of the temperature in Fig. 2.

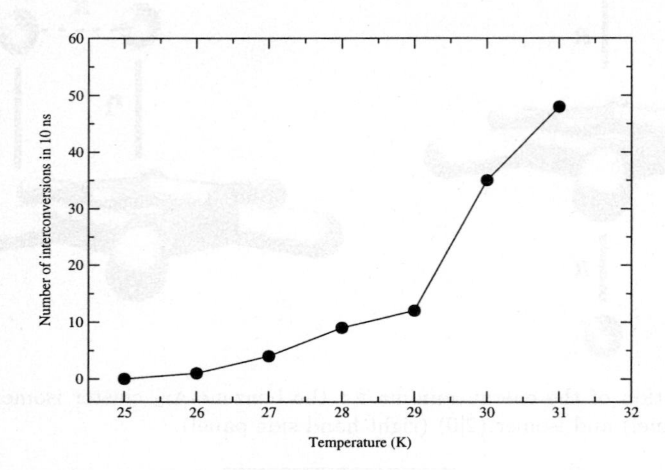

Fig. 2. Number of interconversions as a function of temperature

The figure shows, as expected, an increase of the number of interconversions as the temperature increases. However, there is a sharp variation of the slope of the curve around T = 29 K. These values were determined for a simulation time of 10 ns. For each temperature the lifetime of the complex was monitored during the trajectory. The complex dissociates when the distance between the two Argon atoms and the center of mass of benzene molecule is larger than 10 Å. We found that at temperatures larger than 31 K the lifetime of the complex decreases suddenly. In the 32-40 K temperature range the complex dissociates after 2 ns.

Another relevant property is the relative population (RP) determined as the fraction of time spent by the system in a given configuration with respect to the total time. The value of RP slightly decreases with temperature for isomer (1|1) while it slightly increases for isomer (2|0), the ratio of the relative percentages being 57/43 at 25 K and 62/38 at 31 K.

5 Conclusions

The ability of object oriented programming in allowing the construction of conceptually simple computational procedures has been proven to be succesful. In our case, use was made of the object oriented Scientific Python library. The simulator of molecular processes, that was built in this way, allows the reproduction of the signal of molecular beam experiments starting from first principles. The application was shown to be addressable both to the question of building suitable formulations of the interaction of rarified gas systems and to the understanding of the mechanisms leading to the formation of Ar-benzene clusters.

6 Acknowledgments

This research has been financially supported by MIUR, ASI and CNR (Italy) and COST (Europe).

References

1. Gervasi, O., Cicoria, D., Laganà, A., Baraglia, R., Animazione e calcolo parallelo per lo studio delle reazioni elementari. Pixel. **10** (1994) 19-26. Gervasi, O. and Laganà, A., A distributed computing approach to the simulation of elementary reactions, Lecture Notes in Comp. Science, in press, this issue.
2. Vacek, J., Konvicka, K., Hobza, P., A molecular dynamics study of benzene...Ar_2 complex. Application of the nonempirical ab initio and empirical Lennard-Jones 6-12 potentials. Chem. Phys. Lett. **220** (1994) 85-92
3. Dullweber, A. Hodges, M. P., Wales, D. J., Structure, dynamics and thermodynamics of benzene-Ar_n clusters ($1 \leq n \leq 8$ and $n = 19$), J. Chem. Phys. **106** (1997) 1530-1544
4. Hinsen, K., The Molecular Modeling Toolkit: A New Approach to Molecular Simulations, J. Comp. Chem. **21**, (2000) 79-85
5. `http://starship.python.net/crew/hinsen/scientific.html`
6. Press, W. H., Teukolski, S. A., Vetterling, W. T., Flannery, B. P., Numerical Recipes in Fortran: the Art of Scientific Computing, Cambridge University Press, (1992)
7. `http://www.unidata.ucar.edu/packages/netcdf/`
8. Stone, A., The Theory of Intermolecular Forces. Oxford University Press Oxford (1996)
9. `http://www.amber.ucsf.edu/amber/amber.html`
10. Wales, D. J., Popelier, P. L. A., Stone, J., Potential energy surface of van der Waals complexes of water and hydrogen halides modeled using distributed multiples., J. Chem. Phys. **102** (1995) 5551-5565
11. Laganà, A., Ochoa de Aspuru, G. and Garcia, E., The largest angle generalization of the rotating bond order potential: three different atom reactions. J. Chem. Phys. **108** (1998) 3886-3896
12. Pirani, F., Cappelletti, D., Bartolomei, M., Aquilanti, V., Scotoni, M., Vescovi, M., Ascenzi, D., Bassi, D., Orientation of benzene in supersonic expansions probed by IR-laser absorption and by molecular beam scattering. Phys. Rev. Letter **86** (2001) 5035 - 5038

Beyond Traditional Effective Intermolecular Potentials and Pairwise Interactions in Molecular Simulation

Gianluca Marcelli,[1] B. D. Todd,[2] Richard J. Sadus[2]

[1] Department of Chemistry, Imperial College of Science, Technology and Medicine, University of London, London, UK

[2] Centre for Molecular Simulation and School of Information Technology, Swinburne University of Technology, PO Box 218 Hawthorn, Victoria 3122, Australia
http://www.it.swin.edu.au/centres/cms
RSadus@swin.edu.au

Abstract. Molecular simulation methods such as Monte Carlo simulation and both equilibrium and nonequilibrium molecular dynamics are powerful computational techniques that allow the exact calculation of molecular properties with minimal approximations. The main approximations are the choice of intermolecular potential and the number of particles involved in each interaction. Typically, only pairwise interactions are counted using a simple effective intermolecular potential such as the Lennard-Jones potential. The use of accurate two-body potentials and calculations that explicitly include three or more body interactions are rare because of the large increase in computational cost involved. Here, we report recent progress in the use of both genuine two-body potentials and calculations involving three-body interactions. We show that in some cases, the contribution of three-body interactions can be accurately estimated from two-body interactions without the increase in computational cost involved in explicitly accounting for three-body interactions. As an example of the benefit of including three-body interactions, the improvement in the prediction of vapour-liquid equilibria is examined.

1 Introduction

Molecular simulation [1] is a generic term that encompasses both Monte Carlo (MC) and molecular dynamics (MD) algorithms. The appeal of molecular simulation is that it provides a means of predicting the properties of matter by evaluating the underlying intermolecular interactions. Unlike other computational techniques, the calculations are exact and require very few simplifying assumptions or approximations. The main assumptions are the choice of intermolecular potential and how many atoms or molecules are involved in each interaction. The heart of a molecular simulation is the evaluation of intermolecular energy (MC simulation) or forces (MD simulation). For example, in the absence of external influences, the potential energy of N interacting particles can be obtained from:

$$E_{pot} = \sum_{i}\sum_{j>i} u_2(r_i, r_j) + \sum_{i}\sum_{j>i}\sum_{k>j>i} u_3(r_i, r_j, r_k) + \dots\dots , \tag{1}$$

P.M.A. Sloot et al. (Eds.): ICCS 2002, LNCS 2331, pp. 932–941, 2002.

where the summations are performed for all distinct particle interactions, u_2 is the potential between pairs of particles, u_3 is the potential between particle triplets etc. An analogous expression can be written in terms of force, which is simply the derivative of the potential with respect to intermolecular separation.

The point of truncation of Eq. (1) determines the overall order of the algorithm. Including pair, three-body and four-body interactions result in algorithms of ON^2, ON^3 and ON^4, respectively. It should be noted that computation saving strategies [2] have been developed that mean that these theoretical limits are rarely approached in practice. Nonetheless, the large increase in computing time involved when three or more body interactions are involved has meant that, until recently, it has only been computationally feasible to calculate pair interactions. In a molecular simulation Eq. (1) is typically truncated after the first term and the two-body potential is replaced by an effective potential:

$$E_{pot} = \sum_{i} \sum_{j>i} u_{eff}(r_i, r_j). \tag{2}$$

Therefore, only pairwise interactions are calculated and the effects of three or more body interactions are crudely incorporated in the effective intermolecular potential.

Many intermolecular potentials have been proposed [1,3], but molecular simulations are overwhelming performed using the Lennard-Jones potential:

$$u_{eff}(r_i, r_j) = 4\varepsilon_{ij}\left[\left(\frac{\sigma_{ij}}{r_{ij}}\right)^{12} - \left(\frac{\sigma_{ij}}{r_{ij}}\right)^{6}\right] \tag{3}$$

where ε_{ij} is the potential minimum and σ_{ij} is the characteristic diameter between particles i and j. The use of the Lennard-Jones potential is not confined only to atoms. It is also widely used to calculate the site-site interactions of molecules and it is the non-bonded contribution of molecular force fields such as AMBER [4] and CHARMM [5].

The use of effective intermolecular potentials and confining the calculations to pairwise interactions makes molecular simulation computationally feasible for a diverse range of molecules. Effective intermolecular potentials generally yield good results. However, the use of effective intermolecular potentials also means that the effects of three-body interactions remain hidden. It has recently become computationally feasible to perform molecular simulations involving three-body interactions. Here, we examine the consequences of these calculations and show how the contribution of three-body interactions can be obtained from two-body intermolecular potentials.

2 Simulation Details

2.1 Simulation Algorithms

The simulations discussed in this work are the result of implementing two different algorithms. Simulations of vapour-liquid coexistence equilibria were obtained using the Gibbs Ensemble algorithm [6] as detailed elsewhere [7]. Nonequilibrium molecular dynamics (NEMD) calculations used the SLLOD algorithm [8] as discussed recently [9].

The simulations typically involved 500 particles and conventional periodic boundary conditions and long-range corrections were applied. It should be noted that simulations involving three-body interactions require additional care to maintain the spatial invariance of three particles with respect to periodic boundary conditions [10].

2.2 Intermolecular Potentials

A feature of the simulations discussed here is that Eq. (1) is truncated after the three-body term. Therefore, expressions for both u_2 and u_3 are required. Several accurate two-body potentials for noble gases are available in the literature [1]. In addition, some recent work has also been reported on ab initio potentials [11,12]. However, the focus of our examination is the Barker-Fisher-Watts (BFW) potential [13] which has the following functional form:

$$u_2 = \varepsilon\left[\sum_{i=0}^{5} A_i (x-1)^i \exp\left[\alpha(1-x)\right] - \sum_{j=0}^{2} \frac{C_{2j+6}}{\delta + x^{2j+6}}\right]. \tag{4}$$

In eq. (4), $x = r/r_m$, where r_m is the intermolecular separation at which the potential passes through a minimum and the other parameters are obtained by fitting the potential to experimental data for molecular beam scattering, second virial coefficients, and long-range interaction coefficients. The contribution from repulsion has an exponential-dependence on intermolecular separation and the contribution to dispersion of the C_6, C_8 and C_{10} coefficients are included.

The main contribution to three-body dispersion can be obtained from the triple-dipole term determined by Axilrod and Teller [14]:

$$u_3 = \frac{\nu\left(1 + 3\cos\theta_i \cos\theta_j \cos\theta_k\right)}{\left(r_{ij} r_{ik} r_{jk}\right)^3} \tag{5}$$

where ν is the non-additive coefficient, and the angles and intermolecular separations refer to a triangular configuration of atoms.

3 Results and Discussion

At the outset it should be noted that calculations of the effect of three-body interactions have been performed in the past. For example Barker et al. [13] estimated the three-body energy of argon, Monson et al. [15] investigated three-body interactions in diatomic molecules, Rittger [16] analyzed thermodynamic data for the contribution of three-body interaction and Sadus and Prausnitz [17] estimated the three-body contribution to energy on the vapour-liquid coexistence curve. Typically, in these early reports the contributions of three-body interactions were periodically estimated during the course of an otherwise conventional pairwise simulation. However, recently [9-12] three-body interactions have actually been used to determine the outcome of the simulation by contributing to the acceptance criterion for each and every step of a Markov chain in a MC simulation or contributing to every force evaluation in a MD simulation.

The atomic noble gases have provided the focus of most of the work on three-body interactions because they are free from the additional complexities of molecular shape and polar interactions. The contributions to three-body dispersion for atomic systems are documented [18,19]. In addition to the triple-dipole term (Eq. (5)) contributions can be envisaged from dipole-dipole-quadrupole (DDQ), dipole-quadrupole-quadrupole (DQQ), triple-quadrupole (QQQ), dipole-dipole-octapole (DDQ) and fourth-order triple-dipole (DDD4) interactions.

The phase behaviour of fluids is likely to be very susceptible to the nature of intermolecular interactions. Simulations [1] with the Lennard-Jones potential indicate good agreement for the vapour-liquid phase envelope except at high temperatures resulting in an overestimate of the critical point. In Fig. 1, a comparison is made between experiment and simulations of the vapour-liquid phase envelope of argon in the reduced temperature ($T^* = kT/\varepsilon$, k is Boltzmann's constant) – reduced density ($\rho^* = \rho\sigma^3$) projection. It is apparent from Fig. 1 that the BFW potential alone cannot be used for accurate predictions. However, when the BFW potential is used in conjunction with contributions from three-body terms (DDD + DDQ + DDQQ + QQQ + DDD4), very good agreement with experiment is obtained. In particular, the improved agreement for the liquid phase densities can be unambiguously attributed to the importance of three-body interactions. A similar conclusion can be made for the other noble gases [10].

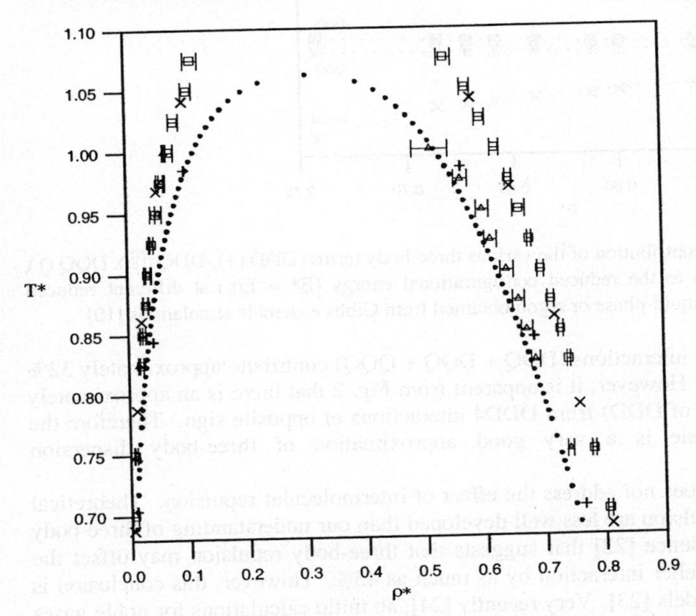

Fig. 1. Comparison of experiment (•) with Gibbs ensemble simulations using the BFW potential (Eq. (4)) (⊡), the Aziz-Slaman potential (×, [20]), the Aziz-Slaman + Axilrod-Teller (+, [20]) and the BFW + three-body (DDD + DDQ + DQQ + QQQ + DDD4) potentials (Δ) for the vapour-liquid coexistence of argon [10]

Anta et al. [20] also reported good agreement with experiment for the Aziz-Slaman [21] two-body potential in combination with only the Axilrod-Teller term (see Fig. 1). The signs of the various contributions to three-body interactions are different and it was believed [22] that a large degree of cancellation occurs. Figure 2, illustrates the contribution to energy of the various contributions at different densities along the liquid side of the coexistence curve of argon [10].

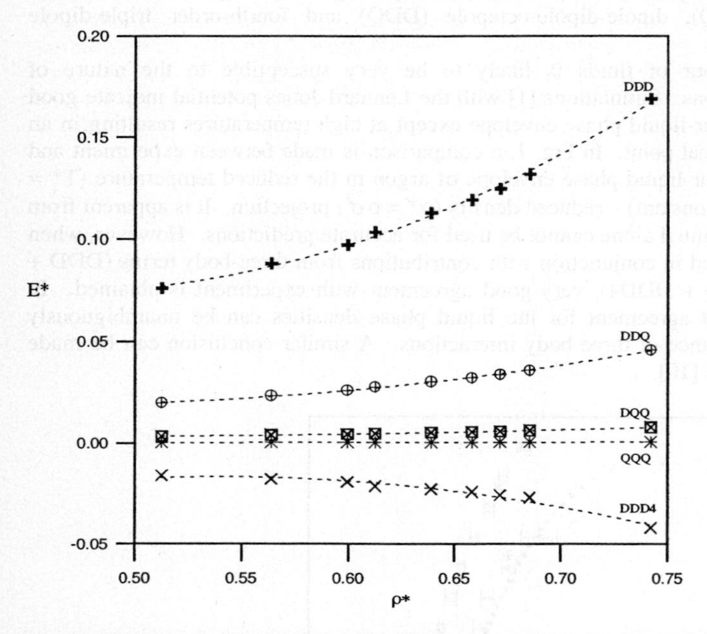

Fig. 2. Comparison of the contribution of the various three-body terms (DDD (+), DDQ (☒), DQQ ()), QQQ (□) and DDD4 (□)) to the reduced configurational energy ($E^* = E/\varepsilon$) at different reduced densities ($\rho^* = \rho\sigma^3$) of the liquid-phase or argon obtained from Gibbs ensemble simulations [10]

The third order multipole interactions (DDQ + DQQ + QQQ) contribute approximately 32% of the triple-dipole term. However, it is apparent from Fig. 2 that there is an approximately equal contribution (26 % of DDD) from DDD4 interactions of opposite sign. Therefore the Axilrod-Teller term alone is a very good approximation of three-body dispersion interactions.

The above analysis does not address the effect of intermolecular repulsion. Theoretical models of three-body repulsion are less well developed than our understanding of three-body dispersion. There is evidence [22] that suggests that three-body repulsion may offset the contribution of Axilrod-Teller interaction by as much as 45%. However, this conclusion is based on approximate models [23]. Very recently [24], ab initio calculations for noble gases have been reported which explicitly account for three-body repulsion. It was observed that the effect of repulsion is also offset by other influences, leaving the Axilrod-Teller term alone as a very good approximation of the overall three-body interaction.

Irrespective of the fact that the Axilrod-Teller term alone is a very good approximation of three-body simulations, using it for routine simulations is computationally prohibitive. For example on a vector computer such as the NEC SX-5, the two-body calculations of the phase coexistence typically require 1 CPU hour per point to obtain meaningful results. In contrast, the same calculations involving three-body interactions require 12 CPU hours. Furthermore, to achieve the figure of 12 CPU hours, the code must be optimized to take advantage of the vector architecture as detailed elsewhere [25]. These impediments are magnified further if we attempt to incorporate three-body interactions for molecular systems. Another consequence of this computational impediment is that genuine two-body potentials have little role in molecular simulation because they alone cannot be used to accurately predict fluid properties. To make use of two-body potentials, a computationally expedient means of obtaining the three-body contribution is required.

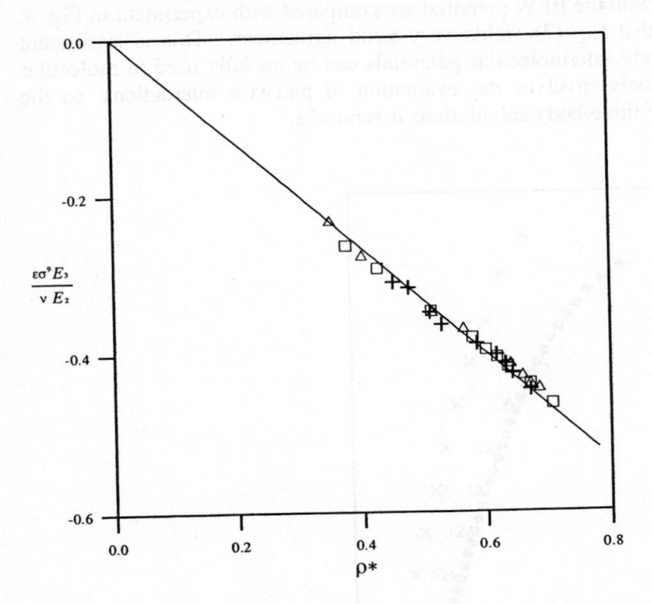

Fig. 3. The ratio of three-body and two-body energies obtained from molecular simulation at different reduced densities. Results are shown for argon (Δ), krypton (+) and xenon (□). The line through the points was obtained from Eq. (6) [26]

Responding to this need, Marcelli and Sadus [26] analyzed molecular simulation data and obtained the following simple relationship between the two-body (E_2) and three-body (E_3) configurational energies of a fluid:

$$E_3 = -\frac{2\nu\rho E_2}{3\varepsilon\sigma^6}, \tag{6}$$

where ν is the non-additive coefficient, ε is the characteristic depth of the pair-potential, σ is the characteristic molecular diameter used in the pair-potential and $\rho = N/V$ is the number density obtained by dividing the number of molecules (N) by the volume (V). As illustrated in Fig. 3, this simple relationship predicts the three-body energy with an average absolute deviation of 2%.

A useful consequence of Eq. (6) is that it can be used to derive an effective potential from any two-body potential via the relationship:

$$u_{eff} = u_2\left(1 - \frac{2\nu\rho}{3\varepsilon\sigma^6}\right). \tag{7}$$

The results of Gibbs ensemble simulations [26] for the vapour-liquid equilibria of argon using Eq. (7) in conjunction with the BFW potential are compared with experiment in Fig. 4. It is apparent from Fig. 4 that Eq. (7) yields very good agreement. This is significant because it means that two-body intermolecular potentials can be usefully used in molecular simulations. Equation (7) only involves the evaluation of pairwise interactions, so the computational impediment of three-body calculations is removed.

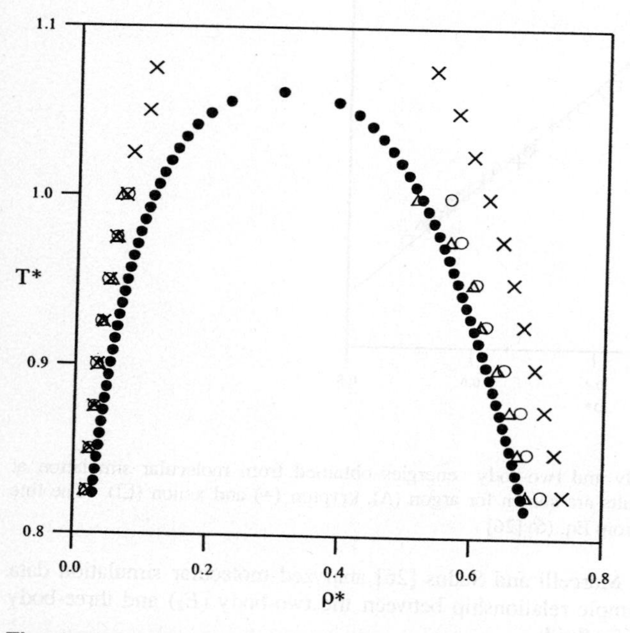

Fig. 4. Comparison of Gibbs ensemble calculations with experiment (●) for the vapour-liquid equilibria of argon in the reduced temperature-density projection. Results are shown for the BFW potential (×), the BFW + Axilrod-Teller potential (Δ) and Eq. (7) using the BFW potential (O) [26]

Although Eq. (7) was determined from Monte Carlo Gibbs ensemble simulation, it can also be used in MD. The forces required in MD are simply obtained from the first derivative of Eq. (7) with respect to intermolecular pair separation. Recently [27] it has been established that Eq. (7) can be used in NEMD [8,9] simulations of transport phenomena such as shear viscosity. As illustrated in Fig. 5, it was shown [27] that the ratio of two-body and three-body energies is largely independent of the reduced strain rate ($\dot{\gamma}^* = \dot{\gamma}\,\sigma\sqrt{m/\varepsilon}$, m is the mass).

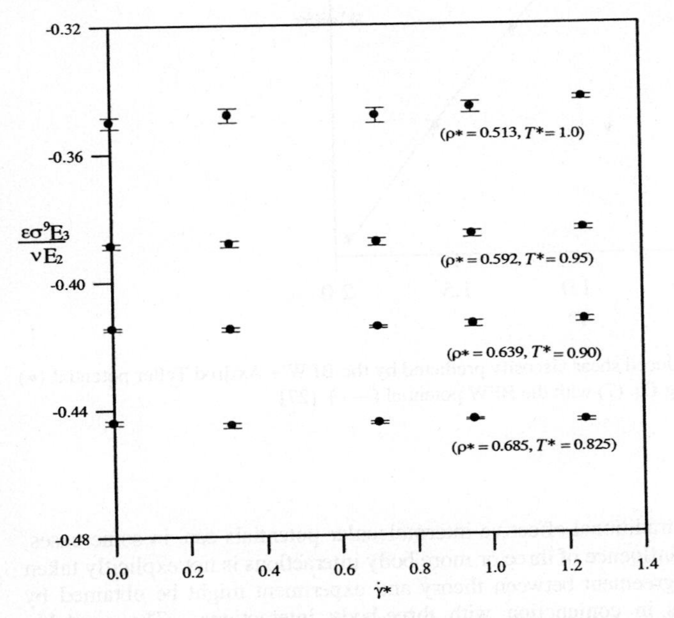

Fig. 5. The ratio of three-body to two-body energies for argon obtained from NEMD simulations at different strain rates and state points [27]

Furthermore, Eq. (6) can be used to predict the three-body energy obtained from NEMD simulations to average absolute deviation of 2.3%. This suggests that Eq. (7) can be used in an NEMD simulation to predict shear viscosities with a similar degree of accuracy as full two-body + three-body simulations. Some results [27] for the reduced shear viscosity ($\eta^* = \eta\sigma^2/\sqrt{m\varepsilon}$) are illustrated in Fig. 6, which confirm the usefulness of Eq. (7).

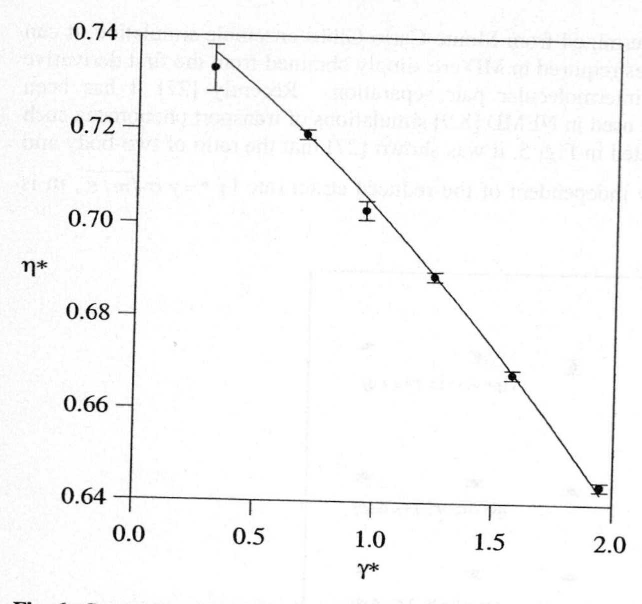

Fig. 6. Comparison of the reduced shear viscosity predicted by the BFW + Axilrod Teller potential (•) with NEMD simulations using Eq. (7) with the BFW potential (⟶) [27]

4 Conclusions

The results obtained from traditional effective intermolecular potentials can, in some cases, be misleading because the influence of three or more body interactions is not explicitly taken into account. Improved agreement between theory and experiment might be obtained by using two body potentials in conjunction with three-body interactions. The available evidence suggests that the Axilrod-Teller term alone is a good description of three-body interactions. There is also evidence that there is a simple relationship between two-body and three-body interaction energies, which can be used to formulate an effective potential from two-body potentials. This means that genuine two-body potentials can be used to accurately predict the properties of fluids.

References

1. Sadus, R. J.: Molecular Simulation of Fluids: Theory, Algorithms and Object-Orientation. Elsevier, Amsterdam (1999).
2. Frenkel, D., Smit, B.: Understanding Molecular Simulation: From Algorithms to Applications. Academic Press, San Diego (1996).
3. Stone, A. J.: The Theory of Intermolecular Forces. Clarendon Press, Oxford (1996).
4. Weiner, S. J., Kollman, P. A., Case, D. A., Singh, U. C., Ghio, C., Alagona, G., Profgeta Jr, S., Weiner, P.: A New Force Field for Molecular Mechanical Simulation of Nucleic Acids and Proteins. J. Am. Soc. **106** (1984) 765-784.

5. Brooks, B. R., Bruccoleri, R. E., Olafson, B. D., States, D. J., Swaminathan, S., Karplus, M.: CHARMM: A Program for Macromolecular Energy, Minimization, and Dynamics Calculations. J. Comput. Chem. **4** (1983) 187-217.
6. Panagiotopoulos, A. Z.: Direct Determination of Phase Coexistence Properties of Fluids by Monte Carlo Simulation in a New Ensemble. Mol. Phys. **61** (1987) 813-826.
7. Panagiotopoulos, A. Z.: Direct Determination of Fluid Phase Equilibria by Simulation in the Gibbs Ensemble: A Review. Mol. Sim. **9** (1992) 1-23.
8. Evans, D. J. and Morriss, P.: Statistical Mechanics of Nonequilibrium Liquids. Academic Press, London (1990).
9. Marcelli, G., Todd, B. D., Sadus, R. J.: Analytic Dependence of the Pressure and Energy of an Atomic Fluid Under Shear. Phys. Rev. E. **63** (2001) 021204.
10. Marcelli, G., Sadus, R. J.: Molecular Simulation of the Phase Behavior of Noble Gases Using Two-Body and Three-Body Intermolecular Potentials. J. Chem. Phys. **111** (1999) 1533-1540.
11. Vogt, P.S., Liapine, R., Kirchner, B., Dyson, A. J., Huber, H., Marcelli, G., Sadus, R. J.: Molecular Simulation of the Vapour-Liquid Phase Coexistence of Neon and Argon using Ab Initio Potentials. Phys. Chem. Chem. Phys. **3** (2001) 1297-1302.
12. Leonhard, K., Deiters, U.K.: Monte Carlo Simulations of Neon and Argon Using Ab Initio Potentials. Mol. Phys. **98** (2000) 1603-1616.
13. Barker, J. A., Fisher, R. A., Watts, R. O.: Liquid Argon: Monte Carlo and Molecular Dynamics Calculations. Mol. Phys. **21** (1971) 657-673.
14. Axilrod, B. M., Teller, E: Interaction of the van der Waals' Type Between Three Atoms. J. Chem. Phys. **11** (1943) 299-300.
15. Monson, A. P., Rigby, M., Steele, W. A.: Non-Additive Energy Effects in Molecular Liquids. Mol. Phys. **49** (1983) 893-898.
16. Rittger, E.: Can Three-Atom Potentials be Determined from Thermodynamic Data? Mol. Phys. **69** (1990) 867-894.
17. Sadus, R. J., Prausnitz, J. M.: Three-Body Interactions in Fluids from Molecular Simulation: Vapor-Liquid Phase Coexistence of Argon. J. Chem. Phys. **104** (1996) 4784-4787.
18. Bell, R. J.: Multipolar Expansion for the Non-Additive Third-Order Interaction Energy of Three Atoms. J. Phys. B **3** (1970) 751-762.
19. Bell, R. J. and Zucker, I. J. in Klein, M. L. and Venables, J. A. (Eds): Rare Gas Solids, Vol. 1., Academic Press, London (1976).
20. Anta, J. A., Lomba, E., Lombardero, M.: Influence of Three-Body Forces on the Gas-Liquid Coexistence of Simple Fluids: the Phase Equilibrium of Argon. Phys. Rev. E **55** (1997) 2707-2712.
21. Aziz, R. A., Slaman, M. J.: The Argon and Krypton Interatomic Potentials Revisited. Mol. Phys. **58**, (1986) 679-697.
22. Maitland, G. C., Rigby, M., Smith, E. B., Wakeham, W. A.: Intermolecular Forces: Their origin and Determination, Clarendon Press, Oxford (1981).
23. Sherwood, A. E., de Rocco, A. G., Mason, E. A.: Nonadditivity of Intermolecular Forces: Effects on the Third Virial Coefficient. J. Chem. Phys. **44** (1966) 2984-2994.
24. Bukowski, R., Szalewicz, K.: Complete Ab Initio Three-Body Nonadditive Potential in Monte Carlo Simulations of Vapor-Liquid Equilibria and Pure Phases of Argon. J. Chem. Phys. **114** 9518-9531 (2001).
25. Marcelli G.: The Role of Three-Body Interactions on the Equilibrium and Non-Equilibrium Properties of Fluids from Molecular Simulation. PhD Thesis, Swinburne University of Technology (2001), http://www.it.swin.edu.au/staff/rsadus/cmsPage/.
26. Marcelli, G., Sadus, R. J.: A Link Between the Two-Body and Three-Body Interaction Energies of Fluids from Molecular Simulation. J. Chem. Phys. **112** (2000) 6382-6385.
27. Marcelli, G., Todd, B. D., Sadus, R. J.: On the Relationship Between Two-Body and Three-Body Interactions from Nonequilibrium Molecular Dynamics Simulation. J. Chem. Phys. **115** (2001) 9410-9413.

Density Functional Studies of Halonium Ions of Ethylene and Cyclopentene

Michael P. Sigalas and Vasilios I. Teberekidis

Laboratory of Applied Quantum Chemistry, Department of Chemistry, Aristotle University of Thessaloniki, 540 06 Thessaloniki, Greece. E-mail: sigalas@chem.auth.gr

A computational study of a variety of $C_2H_4X^+$, $C_5H_8X^+$, $C_5H_{8-n}(OH)_nX^+$ (n=1, 2), where X= Cl and Br, has been carried out. The potential energy surfaces of all molecules under investigation have been scanned and the equilibrium geometries and their harmonic vibrational frequencies have been calculated at the Becke3LYP/6-311++G(d,p) level of theory. The bonding in bridged halonium ions is discussed in terms of donor – acceptor interaction between ethylene and halogen orbitals in the parent ethylenehalonium ion. The relative energies, the equilibrium geometries and the proton and carbon NMR chemical shifts calculated are in good agreement with existing experimental and theoretical data.

1 Introduction

Organic halogen cations have gained increasing significance both as reaction intermediates and preparative reagents. They are related to oxonium ions in reactivity but they offer greater selectivity. They can be divided into two main categories namely acyclic (open-chain) halonium ions and cyclic halonium ions.[1] In 1937, Roberts and Kimball [2] proposed a cyclic bromonium ion intermediate to explain the stereoselective bromination reactions with alkenes, whereas in 1965, the chloronium ion analogue was found by Fahey et al. [3, 4]

A series of *ab initio* calculations have been reported for the $C_2H_4X^+$ [X=F, Cl, Br] cation. [5-10] In all these calculations the *trans*-1-bromoethyl cation, **1**, is less stable than the corresponding bridged bromonium ion, **2**, whereas the *cis*-1-bromoethyl cation, **3**, is a transition state. For X=F or Cl structure **1** is more stable than **2**, with **3** being also a transition state.

1 **2** **3**

Ab initio and semiempirical calculations have been carried out in more complicated systems like $C_4H_8X^+$, [8] and $C_6H_{10}X^+$. [10] Except from a brief *ab initio* study of Damrauer et al., [10] of $C_5H_8Br^+$ and a semiempirical study of $C_5H_7(OH)Br^+$ [11] there is no systematic study of the potential energy surface for halonium ions of substituted or non substituted cyclopentene.

In this work we present a detailed study of the conformational space of halonium ions of ethylene $C_2H_4X^+$ and cyclopentenes like $C_5H_8X^+$ and $C_5H_{8-n}(OH)_nX^+$ (n=1, 2), where X= Cl and Br at the Becke3LYP/6-311++G(d,p) level of theory. The relative energies, the equilibrium geometries and the proton and carbon NMR chemical shifts calculated are discussed in relation to existing experimental and theoretical data.

P.M.A. Sloot et al. (Eds.): ICCS 2002, LNCS 2331, pp. 942–949, 2002.

2 Computational details

The electronic structure and geometry of the halonium ions studied were computed within the density functional theory, using gradient corrected functionals, at the Becke3LYP [12] computational level. The basis set used was 6-311G++(d,p) [13,14]. Full geometry optimizations were carried out without symmetry constraints. Frequency calculations after each geometry optimization ensured that the calculated structure is either a real minimum or a transition state in the potential energy surface of the molecule. The ^{13}C and ^{1}H NMR shielding constants of the B3LYP/6-311++G(d,p) optimized structures were calculated with the gauge-independant atomic orbital (GIAO) method [15] at the B3LYP/6-311+G(2d,p) level and were converted to the chemical shifts by calculating at the same level of theory the ^{13}C and ^{1}H shieldings in TMS. All calculations were performed using the Gaussian98 package. [16]

3 Results and discussion

3.1 $C_2H_4X^+$ (X=Cl, Br)

An assessment of the computational level and basis set necessary to achieve reasonable energy comparisons for the cyclopentyl cations was made by reexamining previous *ab initio* works on the $C_2H_4X^+$ (X=Cl, Br) system. The agreement of the relative energies and geometries of the species calculated at Becke3LYP/6-311++G(d,p) level with those found at the CISD, [7] QCISD and MP2 [10] levels of theory suggest that the energy differences depend more on the quality of the basis set used than on correlation effects.

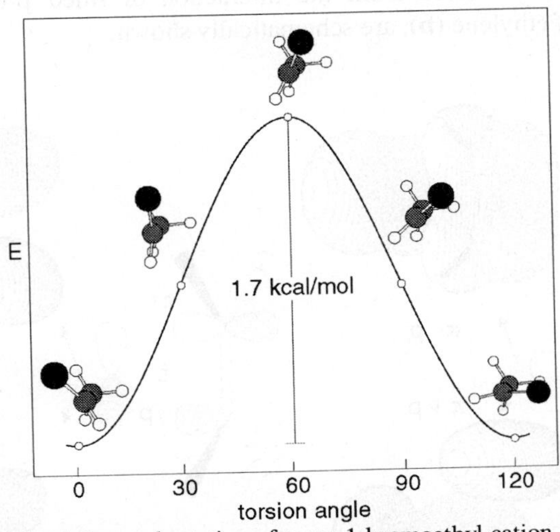

Fig. 1. Internal rotation of *trans*-1-bromoethyl cation, **1**

For X=Br, the bridged bromonium ion, **2**, is more stable than *trans*-1-bromoethyl cation, **1**, by 0.4 kcal/mol. The *cis*-1-bromoethyl cation, **3**, with energy 2.1 kcal/mol above **2**, is a transition state in the maximum of the potential energy path related to the internal rotation of **1**, as shown in Figure 1. Each point in this path has been partially optimized keeping the torsion angle fixed. For X=Cl structure **1** is the global

minimum. The bridged cation, **2**, and the transition state , **3**, are located 6.4 and 1.7 kcal/mol higher respectively.
The bromonium, **2**, C-C bond length was calculated equal to 1.450 Å between the usual values of 1.34 Å for C=C and 1.54 Å for C-C. This distance is 1.449 Å and 1.442 Å for 2-bromoethyl cations **1** and **3** respectively. The C-X bond length is larger for **2** than for **1** or **3** for both X=Cl and Br. For example, in the bromonium ion the C-Br distance of 2.053 Å is a bit longer than a typical single bond length of 1.94 Å. [17] The C-Br distance from the X-ray determination of a substituted ethylenebromonium ion with a Br3- counterion (formed from bromination of adamantylidene-adamantane) [18] is 2.155 Å, which is 0.1 Å longer than our calculated value of for the parent cation. The C-Br bond length was calculated equal to 1.794 Å and 1.791 Å for 2-bromoethyl cations **1** and **3** respectively. The ethylene part of the bromonium ion, **2**, is near planar as the sum of the C-C-H, H-C-H, and C-C-H angles were computed equal to 357.3° and 357.2° for X=Cl or Br respectively. This sum was calculated equal to 357.3° for X=Br at the density functional level with effective core potentials [19] and 356.6° and 356.6° for X=Cl and Br respectively at the MP2 level. [10]
Considerable discussion has been done in whether the three membered ring in bridged halonium ions is a σ-complex or a π-complex. The relationship between π-complexes and true 3-membered rings has been discussed by Dewar [20] and Cremer, [21] whereas Schaefer [7] has stated that there is no sharp boundary between the two. Indeed an examination of the orbitals calculated for bromonium ion revealed that both interactions are present. In figure 2 the shapes of the bonding and antibonding orbitals derived from the interaction of the filled ethylene π-orbital and vacant p-orbital of Br (**a**), as well as these derived from the interaction of filled p-orbital of Br and vacant π^*-orbital of ethylene (**b**), are schematically shown.

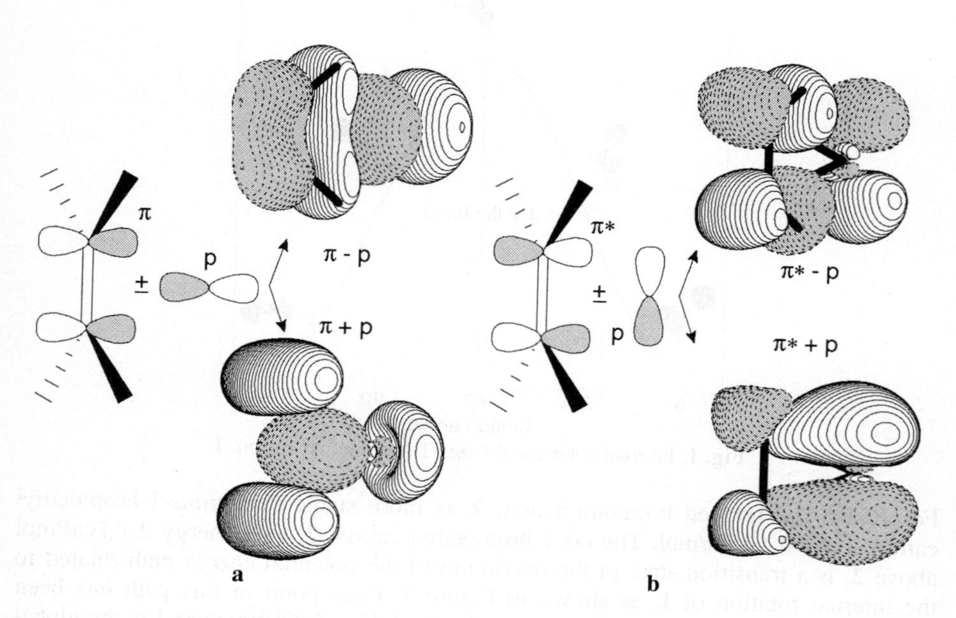

Fig. 2. Orbital interactions in bridged ethylene bromonium ion

3.2 $C_5H_8X^+$ (X=Cl, Br)

We have studied the three possible chloro and bromocyclopentyl cations: namely, the 1-halocyclopentylium (**4a,b**), the 1,2-bridged (**5a,b**), and the 1,3-bridged (**6a,b**) cations with geometry optimizations at the Becke3LYP/6-311++G(d,p) level.

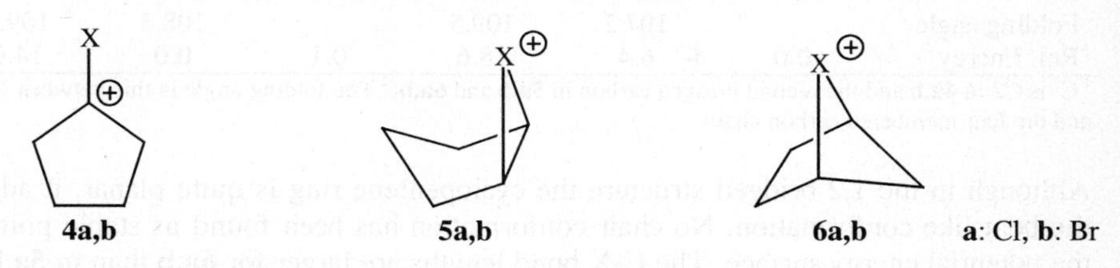

The optimized structures are shown in Fig.3, whereas the relative energies and selected optimized geometrical parameters in Table 1. Frequencies calculations have shown that all structures are minima on the potential energy surfaces.

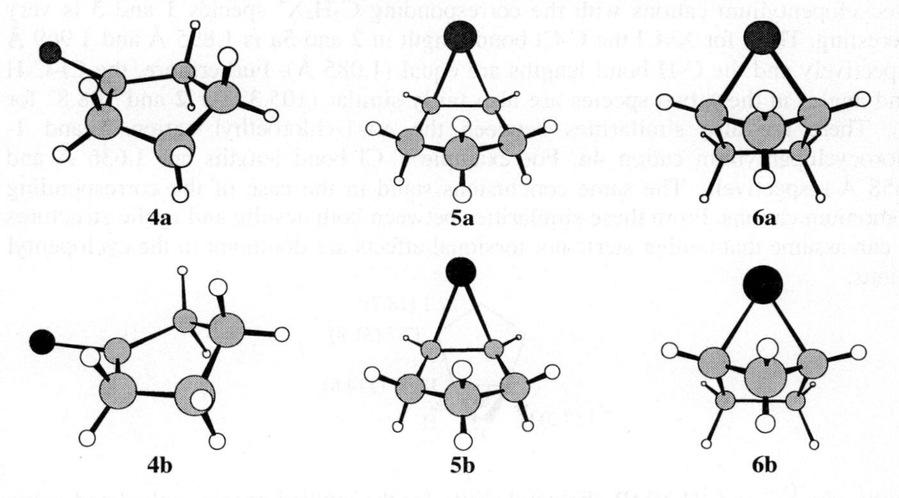

Fig. 3. Optimized structures of $C_5H_8X^+$ cations

The most stable $C_5H_8Cl^+$ cation is the 1-chlorocyclopentylium cation (**4a**) being 6.4 kcal/mol lower in energy than the 1,2-bridged chlorocyclopentylium (**5a**). In the bromonium cations the energy order is reversed with **5b** being 0.1 kcal/mol more stable than **4b**. Apparently the larger and less electronegative bromine atom stabilizes more effectively the bicyclic bridged structure than chlorine. These computations are consistent with the observations of Olah and co-workers. [22,23] Thus, although they have achieved to prepare **5b** from *trans*-1,2-dibromocyclopentane, in a similar experiment with *trans*-1,2-dichlorocyclopentane they obtained, instead of **5a**, only the **4a**. The 1,3-bridged structures **6a,b** are more higher in energy due to high strain energy.

Table 1. Calculated energies (kcal/mol) and geometrical parameters (Å, °) of $C_5H_8X^+$ cations

	4a	**5a**	**6a**	**4b**	**5b**	**6b**
X	Cl	Cl	Cl	Br	Br	Br
C-C' [1]		1.462			1.458	
C-X	1.658	1.969	2.027	1.818	2.123	2.177
X-C-C'	123.9	68.0	57.8	124.0	69.9	59.8
Folding angle [2]		107.2	109.5		108.3	109.7
Rel. Energy	0.0	6.4	18.6	0.1	0.0	14.0

[1] C' is C2 in **4a,b** and the second bridged carbon in **5a,b** and **6a,b**. [1] The folding angle is this between XCC' and the four membered carbon chain.

Although in the 1,2 bridged structure the cyclopentene ring is quite planar, it adopts the boat like conformation. No chair conformation has been found as stable point in the potential energy surface. The C-X bond lengths are larger for **6a,b** than in **5a,b** by near 0.5 Å and the folding angle of the XCC' bridge with the rest of the molecule is between 107-110°.

The comparison of the bridged 1,2-halonium cyclopentylium and 1-halocyclopentylium cations with the corresponding $C_2H_4X^+$ species **1** and **3** is very interesting. Thus, for X=Cl the C-Cl bond length in **2** and **5a** is 1.895 Å and 1.969 Å respectively and the C-H bond lengths are equal (1.085 Å). Furhermore, the Cl-C-H bond angles in these two species are also fairly similar (105.3° for **2** and 108.8° for **5a**). There are also similarities between the *cis*-1-chloroethyl cation **3** and 1-chlorocyclopentylium cation **4a**. For example C-Cl bond lengths are 1.636 Å and 1.658 Å respectively. The same conclusions stand in the case of the corresponding bromonium cations. From these similarities between both acyclic and cyclic structures we can assume that neither steric nor torsional effects are dominant in the cyclopentyl cations.

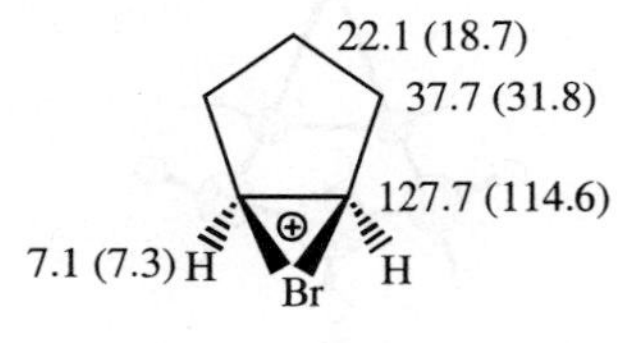

7

Finally, the ^{13}C and ^{1}H NMR chemical shifts for the studied species calculated using the GIAO method are in very good agreement with the existing experimental data. The ^{13}C chemical shifts of the carbon atoms and the proton shifts for the two equivalent olefin-type protons for the bridged 1,2-bromonium cyclopentylium are given in 7, along with the experimental values [23] in parentheses.

3.3 $C_5H_7(OH)X^+$ (X=Cl, Br)

The potential energy surface for the chloro and bromo hydroxycyclopentyl cations has been scanned in an energy window of about 20.0 kcal/mol at the Becke3LYP/6-311++G(d,p) level. The optimized structures found and their relative energies are shown in Fig. 4. All structures are real minima since no imaginary frequencies were calculated.

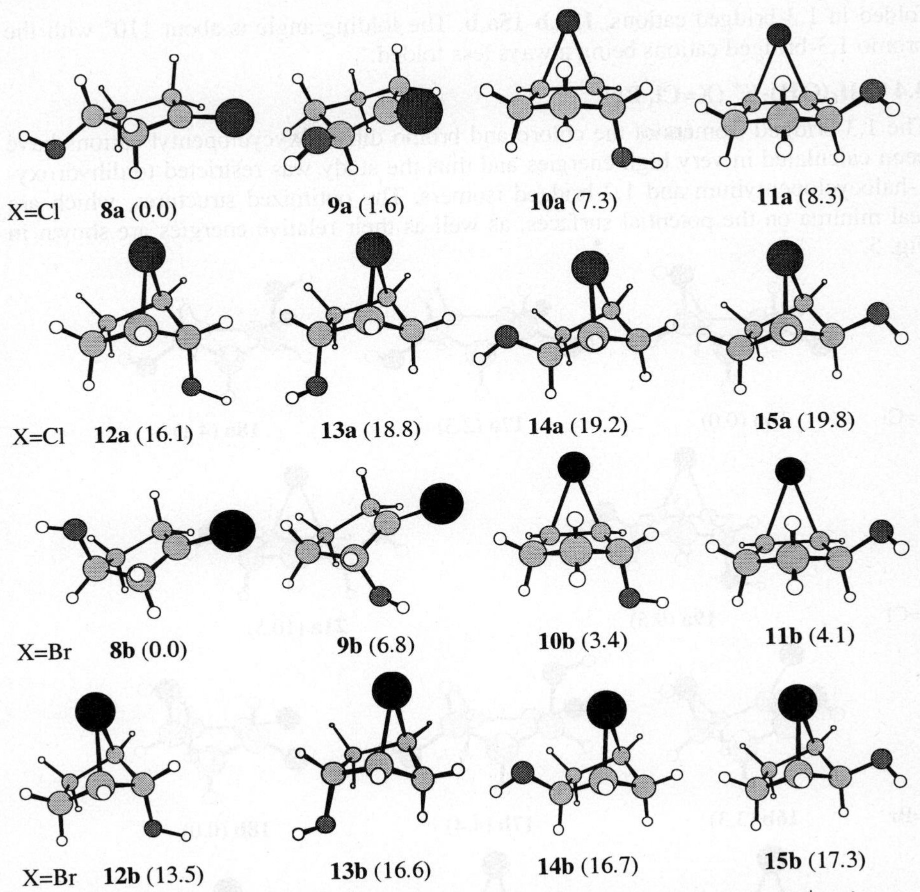

Fig. 4. Optimized structures and relative energies (kcal/mol) of $C_5H_7(OH)X^+$ cations

In contrast to what has been found in the parent halonium cations of ethylene and unsubstituted cyclopentene, the 3-hydroxy-1-halocyclopentyliums, **8a,b**, are the most stable isomer for both chlorine and bromine. However, the tendency of bromine to stabilize the 1,2-bridged structure is present and in this system. Thus, the two bridged 1,2-bridged 3-hydroxybromocyclopentylium, **10b** and **11b**, are only 3.4 and 4.1 kcal/mol higher from 3-hydroxy-1-bromocyclopentylium and much bellow the 2-hydroxy-1-bromocyclopentylium isomer. In the case of chlorine both two hydroxy-1-chlorocyclopentylium isomers are more stable than the two 1,2-bridged structures. In both cases the 1,2-bridged structure with halogen and hydroxyl in *anti* position are more stable by about 1 kcal/mol. All 1,3-bridged isomers are an order of magnitude higher in energy with bromine derivatives being less destabilized.

The presence of hydroxyl does not affect the calculated overall geometry of the isomers. For example the C-X bond length in hydroxy-1-halocyclopentyliums, **8a,b** and **9a,b**, are equal with those of 1-halocyclopentylium, **4a,b**, (C-Br = 1.818 Å and C-Cl = 1.650 Å). The cyclopentene ring is nearly flat in **8a,b-11a,b**, whereas it is folded in 1,3-bridged cations, **12a,b-15a,b**. The folding angle is about 110° with the bromo 1,3-bridged cations being always less folded.

3.4 $C_5H_6(OH)_2X^+$ (X=Cl, Br)

The 1,3-bridged isomers of the chloro and bromo dihydroxycyclopentyl cations have been calculated in very high energies and thus the study was restricted to dihydroxy-1-halocyclopentylium and 1,2-bridged isomers. The optimized structures, which are real minima on the potential surfaces, as well as their relative energies are shown in Fig. 5.

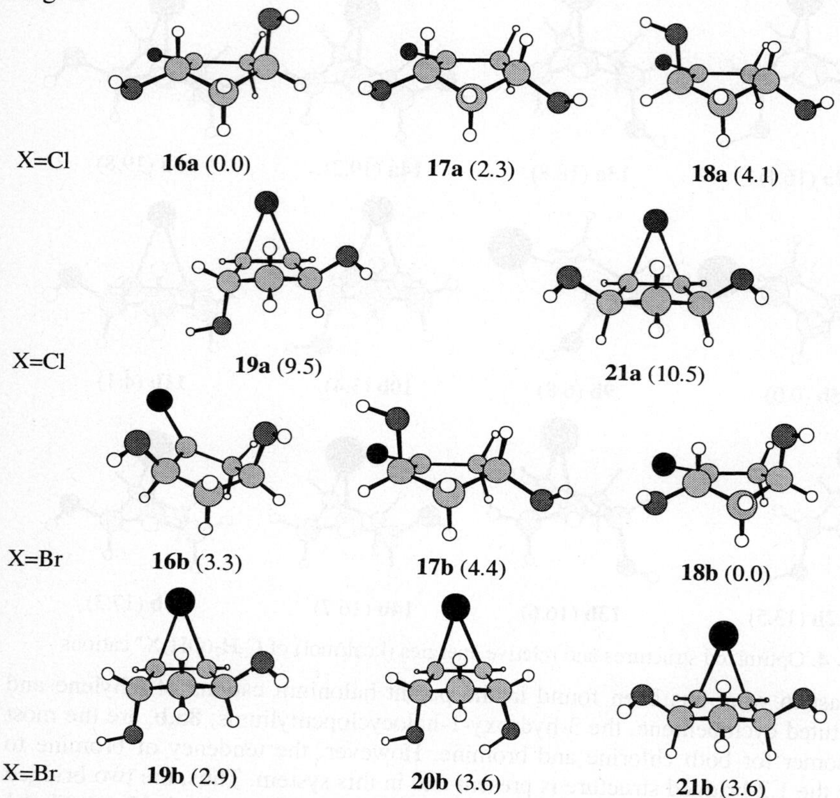

Fig. 5. Optimized structures and relative energies (kcal/mol) of $C_5H_6(OH)_2X^+$ cations

As in the case of the hydroxycyclopentene derivatives a dihydroxy-1-halocyclopentylium is the most stable isomers for both chlorine (**16a**) and bromine, (**18b**), but the presence of a the second hydroxyl seems to decrease the energy gap between the 1-halocyclopentylium the 1,2-bridged isomers. Once again, the bromine atom stabilizes the 1,2-bridged structures more than chlorine. The optimized geometrical parameters are very similar to those for the corresponding cyclopentene and hydroxysyclopentene derivatives. Finally, the cyclopentene ring is nearly flat in all structures.

References

1. Olah, G. A.: Halonium Ions. Wiley Interscience, New York, 1975
2. Roberts, I., Kimball, G. E.: J. Am. Chem. Soc. 59 (1937) 947
3. Fahey, R. C., Schubert, C. J. Am. Chem. Soc. 63 (1965) 5172
4. Fahey, R. C. J. Am. Chem. Soc. 88 (1966) 4681
5. Ford, G. A., Raghuveer, K. S. Tetrahedron 44 (1988) 7489
6. Reynolds, C. H. J. Chem. Soc. Chem. Commun. (1990) 1533
7. Hamilton, T. P., Schaefer, H. F., III. J. Am. Chem. Soc. 112 (1990) 8260
8. Reynolds, C. H. J. Am. Chem. Soc. 1992, 114, 8676
9. Rodriquez, C. F., Bohme, D. K., Hopkinson, A. C. J. Am. Chem. Soc. 115 (1993) 3263
10. Damrauer, R., Leavell, M. D., Hadad, C. M. J. Org. Chem. 63 (1998) 9476
11. Nichols, J., Tralka, T., Goering, B. K., Wilcox, C. F., Ganem B. Tetrahedron 52 (1996) 3355
12. A. D. Becke, A. D. J. Chem. Phys. 98 (1993) 5648
13. Raghavachari, K., Pople, J. A. , Replogle, E. S., Head-Gordon, M., J. Phys. Chem. 94 (1990) 5579
14. Frisch, M. J., Pople, J. A., Binkley, J. S. J. Chem. Phys. 80 (1984) 3265
15. Wolinski, K., Hilton, J. F., Pulay, P. J. Am. Chem. Soc. 112 (1990) 8251
16. Frisch, M. J., Trucks, G. W., Schlegel, H. B., Scuseria, G. E., Robb, M. A., Cheeseman, J. R., Zakrzewski, V. G., Montgomery, J. A., Stratmann, R. E., Burant, J. C., Dapprich, S., Millam, J. M., Daniels, A. D., Kudin, K. N., Strain, M. C., Farkas, O., Tomasi, J., Barone, V., Cossi, M., Cammi, R., Mennucci, B., Pomelli, C., Adamo, C., Clifford, S., Ochterski, J., Petersson, G. A., Ayala, P. Y., Cui, Q., Morokuma, K., Malick, D. K., Rabuck, A. D., Raghavachari, K., Foresman, J. B., Cioslowski, J., Ortiz, J. V., Stefanov, B. B., Liu, G., Liashenko, A., Piskorz, P., Komaromi, I., Gomperts, R., Martin, R. L., Fox, D. J., Keith, T., Al-Laham, M. A., Peng, C. Y., Nanayakkara, A., Gonzalez, C., Challacombe, M., Gill, P. M. W., Johnson, B. G., Chen, W., Wong, M., W. Andres, J. L., Head-Gordon, M., Replogle E. S., Pople, J. A. Gaussian 98, Revision A.1, Gaussian Inc., Pittsburgh PA, 1998
17. Huheey, J. E.: Inorganic Chemistry. Harper and Row, New York, 1983
18. Slebocka-Tilk, H., Ball, R. G., Brown, R. S. J. Am. Chem. Soc. 107 (1985) 4504
19. Koerner, T., Brown, R. S., Gainsforth, J. L., Klobukowski, M. J. Am. Chem. Soc. 120 (1998) 5628
20. Dewar, M. J. S., Ford, G. P. J. Am. Chem. Soc. 101 (1979) 783
21. Cremer, D., Kraka, E. J. Am. Chem. Soc. 107 (1985) 3800
22. Prahash, G. K., Aniszfeld, R., Hashimoto, T., Bausch, J. W., Olah, G. A. J. Am. Chem. Soc. 111 (1989) 8726
23. Olah, G. A., Liang, G., Staral J. J. Org. Chem. 96 (1974) 8112

Methodological Problems in the Calculations on Amorphous Hydrogenated Silicon, a-Si:H

Alexander F. Sax and Thomas Krüger

Institut für Chemie, Karl-Franzens-Universität Graz, Strassoldogasse 10, A-8010 Graz, Austria
alexander.sax@uni-graz.at, thomas.krueger@uni-graz.at

Abstract. Large silicon clusters (up to 230 atoms) with vacancies of different size were used to model the electronic structure of defects in amorphous silicon. We used a mechanical embedding technique, where a small core of atoms around a vacancy is surrounded by a larger number of bulk atoms. The electronic structure of the core was calculated with DFT methods, the bulk was treated with a semiempirical quantum chemical method. With this hybrid technique we investigated the structure of the cluster and ground and excited electronic states.

1 Introduction

Amorphous hydrogenated silicon, a-Si:H, is a cheap and technically versatile material used for example in solar cells. Its use is limited by the Staebler-Wronski effect [1], [2] which describes the degradation of a-Si:H manifesting in a considerable loss in photo-conductivity and dark conductivity after several hours of light exposure. This conductivity loss can, however, be fully reversed by thermal annealing in the dark. It is well known that a-Si:H has a large number of radical centers in the bulk due to unpaired electrons, called native dangling bonds. The existence of these dangling bonds in the bulk is one reason for the amorphous character of a-Si:H because not all silicon atoms are fourfold coordinated. Note, that missing covalent bonds lead immediately to geometry reorganization in the surrounding of the radical center and, thus, to deviations from the regular crystalline structure. The existence of dangling bonds in the amorphous material is also one explanation of the Staebler-Wronski effect, however, not the native dangling bonds are made responsible but so called light-induced or metastable dangling bonds. This explanation is based on the observation that after light exposure the number of dangling bonds, i.e. the spin density, increases by one to two orders of magnitude, after thermal annealing the spin density drops to the value typical for native dangling bonds. The metastable dangling bonds are produced from excitation of defect precursors.

Small angle X ray scattering investigations [3], [4], [5] show up to 5×10^{19} microvoids per cm^3 in a-Si:H. The microvoids are on average spherical with a mean radius of 3.3 to 4.3Å, corresponding to 16 to 25 missing atoms, the distribution is, however, rather broad. The large internal surfaces are made by

P.M.A. Sloot et al. (Eds.): ICCS 2002, LNCS 2331, pp. 950–955, 2002.

silicon atoms with dangling bonds which can form new bonds. As a result we expect a strong reorganization of the internal surface: The building of new bonds results in forces acting on old bonds. This leads to deviations from the standard bond length, bond angle and dihedral angles in crystalline silicon. It depends on the size and form of the void how strongly the reorganized structure deviates from the structure of crystalline silicon. One objective of our investigation was, therefore, to find out whether or not regularities in the structural reorganization exist.

Stretched bonds and deformed bond angles in the bulk as well as newly formed bonds influence the electronic spectrum of the system, so the defects in a-Si:H may be found at these microvoids. Moreover, it depends on the number of dangling bonds and the geometry of the void how many unpaired electrons couple to form new bonds and how many of them remain uncoupled. Such weakly coupled electrons can be easily excited and, thus, geometric arrangements leading to such electronic structures can be thought of being a defect precursor. Therefore, a detailed investigation of the electronic structure of a-Si:H was the second objective of this study.

2 Methodological Problems

The reliability of investigations using embedding techniques depends strongly on the correct size of the embedded core and the method chosen for its description as well as on the proper treatment of the surrounding bulk. In the core we must correctly describe the formation of new bonds or the reorganization of existing bonds which demands methods that account for electron correlation. Moreover, it is necessary to decide on how many silicon atoms define the core, only the atoms with dangling bonds, that is the atoms forming the internal surface of the void, or some more layers. Since the number of core atoms can become rather large density functional methods are certainly best suited for such investigations. The description of the bulk demands a method that allows for an elastic response of the bulk on forces that result from bond formation in the core. The bulk must only prevent the core atoms from collapsing into the void. The methods used for the bulk can, therefore, be at a much lower level than the method for the core. Whenever by the embedding the boundary between core and bulk region cuts through covalent bonds between core and bulk atoms, the proper treatment of this boundary region is crucial for the use of the embedding method.[6] Whenever the high and the low level methods describe the potential curves of the cut Si-Si bonds differently, i.e. the minima and the curvatures do not match, the convergence of the geometry optimization can be extremely slow or can completely fail and the final geometry parameters can be unreliable.

Molecular mechanics methods are frequently used as low level method but to fulfill the above mentioned requirement to match with the high level method re-parametrization of the force field parameters is often necessary. This is true for both, force fields used in solid state physics as well as force fields used in chemistry. Because the time for our project was limited we had to find a combination of existing methods that could be applied to our problem. We finally decided for a combination of the density functional BP86[7],[8] as the high level method and the semiempirical method AM1[9],[10] as the low level method. We used the ONIOM[11] embedding scheme which is implemented in the Gaussian98 software package.[12] The 6-31G* basis set was used in all calculations. Justification for our choice of the density functional method was given in [13]. We used a semiempirical quantum chemical method instead of a force field because the Gaussian98 suite did not correctly handle the cluster with vacancies when the implemented force fields were combined with the density functional method, but this problem did not exist for semiempirical methods. We chose AM1 because with this method geometry optimization of large cluster converged, other semiempirical methods converged for small cluster but failed completely for the large ones.

The ability of the bulk to prevent a vacancy from collapsing into the void depends on its size. To find out how large the bulk must be we had to make test calculations on clusters of increasing size and of different form. For vacancies with one missing atom a cluster of 121 silicon atoms gave converged geometry data. For vacancies with more missing atoms cluster of more than 200 silicon atoms gave trustworthy results, sometimes even smaller clusters could be used.[14] Cluster size is the limiting factor in the applicability of this methodological approach because for a cluster with more than about 250 silicon atoms and vacancies in it the AM1 method shows serious convergence problems.

3 Results

3.1 The structure of vacancies

Whenever a single silicon atom is removed from a silicon crystal a simple vacancy (monovacancy) is created with four radical silicon atoms in tetrahedral arrangement. From the four atomic orbitals at the radical centers we can build four molecular orbitals, the one with the lowest orbital energy is totally symmetric and the remaining three orbitals form the basis for a threefold degenerate representation. Placing of four electrons in these four molecular orbitals results in a partial occupation of the degenerate orbitals yielding a degenerate electronic state which gives rise to a Jahn-Teller distortion of the nuclear frame. The local symmetry is lowered from T_d to C_2 and the distances between two pairs of silicon atoms shrink from 3.82Å to about 2.45Å.[14] We get, thus, two new Si-Si single bonds which are about 5% longer than the Si-Si single bonds in crystalline silicon. Because the symmetry is reduced to C_2 or C_s the new bonds are not orthogonal to each other but twisted. This causes a deformation of the bond angles and dihedral angles in the surrounding of the vacancy. Due to Jahn-Teller distortion every vacancy contributes, thus, to a reduction of the crystalline character and to an increase of the amorphous character of the material.

Removing two adjacent atoms gives a vacancy with six radical centers in D_{3h} arrangement, two sets of three radical centers with local C_3 symmetry (equilateral triangle). For each set we get three molecular orbitals, one totally symmetric and one doubly degenerate. Putting three electrons in these three orbitals yields again a degenerate electronic state and, thus, gives rise to Jahn-Teller distortion. The local C_3 symmetry is lowered to C_{2v} and the radical centers can arrange in two possible structures: an acute angled and an obtuse angled isosceles triangle. We find in our calculations the obtuse angled structure with two newly formed Si-Si single bonds of about 2.5Å. The bond angle between them is about 90°. Bonding interaction of the electrons in the two sets reduces the distances between the apex silicon atoms in the local triangles from 5.9Å in the crystal to 4.7Å. The Si-Si distances between the other silicon atoms shrink from 4.5Å in the crystal to 4.3Å. The apex atoms in the obtuse angles triangles have three old Si-Si single bonds to their neighbors and two new Si-Si single bonds to former radical centers, they are five-fold coordinated.[14] Again we find strong deviation from the crystalline structure in the neighborhood of such vacancies.

Vacancies with three[14] and four[15] missing silicon atoms show a much greater variety of structures. When three or four adjacent silicon forming a "linear" chain are removed the vacancy has the form of a tube. We find then always the formation of three or four new Si-Si single bonds which make the tube shrink. At both ends of the tube we find a single radical center. When four atoms forming a "circular" chain are removed a bond is formed that reminds at a twisted and stretched disilene, i.e. a twisted Si-Si double bond. When four silicon atoms that form a pyramid are removed we get the first vacancy which is rather hollow even after geometrical rearrangement.

These vacancies are clearly the limit of systems that can be treated with this methodological approach. Investigations of larger vacancies need certainly larger bulks and, therefore, methods that can handle them properly.

We calculated also systems with hydrogen placed in the vacancies. Preliminary results show clearly that formation of Si-H bonds leads to large shifts in the Si-Si bonds, therefore, the geometries can show great structural differences compared with the vacancies without hydrogen. Hydrogen atoms that form strong Si-H bonds at the internal surface of vacancies or microvoids could, thus, help to optimally separate the radical centers.

3.2 The electronic structure of vacancies

Light induced dangling bonds are thought to result from the electronic excitation of weak bonds.[14] So we calculated not only the geometric structure of the vacancy in its electronic ground state but also the lowest excited states. In a monovacancy the two newly formed Si-Si single bonds are so strong that the excitation energy is similar to that of a normal Si-Si bond in crystalline silicon. In vacancies with two or more missing atoms we obtain rather low lying excited singlet states which lie within the band gap of a-Si:H which is about 1.7 eV. These excited states result from the excitation of weak "bonds" like the weakly coupled electrons from the ends of the tubes mentioned above. The larger the distance between the radical center becomes the easier it is to excite an electron. Whether or not a low lying excited state can indeed be excited by light depends, however, on the oscillator strength of this excitation. Indeed, only few transitions to low lying excited states have a considerable oscillator strength so that the transition can lead to metastable dangling bonds. With our methods it is very laborious to find out whether or not such excited states have local minima. The existence of a local minimum is, however, a necessary condition for a stable defect structure.

Use of density functional methods does not pose big problems for most ground state investigations even when the ground state can be deferred from a large number of dangling bonds. The calculation of the excited states that have strong multiconfigurational character is, however, rather tricky. Unfortunately, calculation of excited states with density functional methods can not yet that routinely be done as the calculation of grounds states. Due to computer time and convergency reasons we have calculated the excitation spectra by conventional singly excited CI based on a Hartree-Fock wave function. This mixing of methods is, however, undesirable for a consistent investigation of a large series.

4 Summary and Outlook

With the described hybrid technique we are able to investigate structure and electronic spectrum of vacancies. We showed that formation of new bonds between dangling bonds yields a strong geometric reorganization of the cluster which is partially responsible for the amorphous character of the material. We could also show that such vacancies lead to electronic structures with low lying excited states. This technique has two main limits:

- The number of atoms in the cluster is limited to about 250 atoms, treatment of larger cluster is prevented by convergence problems caused by the semiempirical method as well as the drastically increasing computer time.
- Geometry optimization in excited states is not that routinely possible with the standard density functional methods as it is for ground states.

Therefore, our future work will focus on the selection and reparametrization of a force field for silicon that can be used in combination with density functional methods. Only then we have the preposition to enlarge the bulk and calculate the structure of larger vacancies or microvoids with 10 to 20 missing atoms.

Methodic developments in the density functional methods that allow a more efficient geometry optimization in excited electronic states would be highly welcomed.

5 Acknowledgement

This work was supported by grant No. S7910-CHE from the Austrian FWF (Fonds zur Förderung der wissenschaftlichen Forschung) within the scope of the joint German-Austrian silicon research focus (Siliciumschwerpunkt).

References

1. Staebler, D. L., Wronski, C. R., Appl. Phys. Lett. **31** (1977) 292
2. Staebler, D. L., Wronski, C. R. J., Appl. Phys. **51** (1980) 3262
3. Williamson, D.L., Mahan, A. H., Nelson, B. P., Crandall, R. S., Appl. Phys. Lett. **55** (1989) 783
4. Mahan, A. H., Williamson, D.L., Nelson, B. P., Crandall, R. S., Phys. Rev. B **40** (1989) 12024
5. Remes, Z., Vanecek, M., Mahan, Crandall, R. S., Phys. Rev. B **56** (1997) R12710
6. Sauer, J., Sierka, M. J., Comput. Chem. **21** (2000) 1470
7. Becke, A. D., Phys. Rev. A **38** (1988) 3098
8. Perdew, J. P., Phys. Rev. B **33** (1986) 8822
9. Dewar, M. J. S., Zoebisch, E. G., Healy, E. F., J. Am. Chem. Soc. **107** (1985) 3902
10. Dewar, M. J. S., Reynolds, C. H., J. Comput. Chem. **2** (1986) 140
11. Dapprich, S., Komaromi, I., Byun, K. S., Morokuma, K., Frisch, M. J., J. Mol. Struct. (Theochem) **461-462** (1999) 1
12. M. J. Frisch, et al., Gaussian 98 (Revision A.7), Gaussian, Inc., Pittsburgh PA, 1998
13. Krüger, T., Sax, A. F., J. Comput. Chem. **22** (2001) 151
14. Krüger, T., Sax, A. F., Phys. Rev. B, **64** (2001) 5201
15. Krüger, T., Sax, A. F., Physica B, in press.

Towards a GRID based Portal for an a priori Molecular Simulation of Chemical Reactivity

Osvaldo Gervasi[1], Antonio Laganà[2], and Matteo Lobbiani[1]

[1] Department of Mathematics and Informatics, University of Perugia, via Vanvitelli, 1, I-06123 Perugia, Italy
osvaldo@unipg.it
http://www.unipg.it/~osvaldo

[2] Department of Chemistry, University of Perugia, via Elce di Sotto, 8, I-06123 Perugia, Italy
lag@unipg.it
http://www.chm.unipg.it/chimgen/mb/theo1/text/people/lag/lag.html

Abstract. The prototype of an Internet Portal devoted to the Simulation of Chemical reactivity has been implemented using an engine running in parallel. The application makes use of PVM, and it has been structured to be ported on a GRID environment using MPI.

1 Introduction

This paper illustrates the development and the implementation of a Problem Solving Environment (PSE)[1] for the Simulation of Chemical Reactive Processes. The application is based on an Internet Portal connected to a computer grid[2, 3], updating a set of visualization facilities and to monitor in real-time the evolution of the simulation.

As a prototype PSE we consider here an a priori Simulator of Molecular Beam Experiments, **SIMBEX**, for atom-diatom reactions[4]. Crossed Molecular Beams are a crucial experiment providing a stringent test for the understanding of molecular interactions and the rationalization of chemical processes[5]. Their a priori simulation is a high demanding computational procedure that for its progress relies on the advance in computing technologies. For this reason **SIMBEX** has been specifically designed for distributed computing platforms.

The rapid evolution of networking technologies is, in fact, making it feasible to run complex computational applications on platforms articulated as a geographically dispersed large clusters of heterogeneous computers ranging from versatile workstations to extremely powerful parallel machines (Computing Grid). This opens the perspective of carrying out realistic simulations of complex chemical systems by properly coordinating the various computational tasks distributed over the network. Such an approach challenges also the exploitation of a remote cooperative use of software, hardware and intellectual resources belonging to a cluster of various research Centers and Laboratories. On this ground Metalaboratories devoted to various complex computational applications in chemistry are

P.M.A. Sloot et al. (Eds.): ICCS 2002, LNCS 2331, pp. 956–965, 2002.

being established in Europe[1] dealing with different targets in complex computational chemistry[8].

Among this, **SIMBEX** is an application aimed at designing a distributed version of the Crossed Molecular Beam Simulator prototype reported in the literature a few years ago [9, 10].

The basic articulation of the Web structure of **SIMBEX** consists of a client, a back-end and a middleware component. The client level consists of a Web browser connected to the Portal: the web pages drive the user to the selection of the application, to the collection of the input data and to the recollection and presentation of the results. The authentication of the user is taken care by the initial Web page. Then the user is offered a list of applications to run on the back-end system by activating the related hyperlinks. After the configuration, the simulation starts and the user can follow the quantities of interest in Virtual Monitors that are Java Applets downloaded from the Web server to the client. The back-end presently consists of a cluster of workstations distributed among the Department of Chemistry in Perugia (Italy), where crossed beam experiments are run, the Computer Center of the University of Perugia (Italy), that also shares with the cluster a section of its IBM SP2, and the Department of Physical Chemistry of the University of the Basque Country in Vitoria (Spain). An extension of the cluster to other Laboratories is under way.

The middleware layer consists of a static Web server (Apache), a Java Web Server (Tomcat) and a daemon called `SimGate`. The Web server deals with the client and with the Java server handling the requests of the users. `SimGate` is devoted to handle the communication with the applets, freeing the farmer of such task.

The paper is articulated as follows: In section 2 the architecture of the Internet portal is discussed. In section 3, the articulation of related computational procedures is analysed in order to single out models and templates for their distributed use. In section 4, the particular case of a prototype atom diatom reaction is discussed for illustrative purposes.

2 The Architecture of the Internet Portal

To allow the user to access the Problem Solving Environment a Java2 enabled Web browser is used. The problems related to the management of the distributed environment, and the communications between the various components are solved at the Server Web level.

The product has been developed using *Freesoftware* components. In particular use has been made of the Apache Web Server, the Tomcat Java Web Server, Java2, MySQL and PVM, powered by Linux RedHat.

[1] To incentive the gathering of research Laboratories having complementary expertises on common research projects, the European Community has launched within the COST (Collaboration in Science and Technology) in Chemistry[6] initiative the Action D23[7].

Access to the Portal requires an account. We have used MySQL to save the data of the users and to handle the authentication phase. Because of the multiuser environment, multiple requests to the web server are dealt using the parameter *SessionID*.

After the user has defined the configuration of the simulation, the Java Servlets start `ABCtraj`, the parallel application, as a Farmer process running on the Web server and the Worker programs running on the PVM distributed nodes. The communication between the Servlets and `ABCtraj` occurs trough a Unix Socket. The Servlets start up also a daemon called `SimGate` devoted to the communication between the Java Applets related to the Virtual Monitors and the server Web for the on-line update of the Virtual Monitors. The communications between `SimGate` and the applets occurs through a stateless protocol that makes use of a TCP/IP socket using a service port chosen between a set of reserved values. During the simulation the applet asks for updates to `SimGate`, that returns the values received by `ABCtraj`. Different contemporary simulations are easily handled by the multithread nature of Java.

The communication between the Farmer and `SimGate` has the following syntax:
`UPD MON` $[0-11]$ `ID` $SessionID$ `DATA -` $a_1, a_2, ..., a_{20}$
with a_i being floating point numbers. The $SessionID$ parameter allows `SimGate` to control that the flow originates from the right Farmer (the one associated to $SessionID$). `SimGate` answers with a success or an error message.

The protocol between `SimGate` and the applet is slightly different: the applet asks for new data, specifying its own $Monitor\ number$ and $SessionID$:
`GET MON` $[0-11]$ `ID` $SessionID$
to which `SimGate` answers sending the data or an error message. The amount of data exchanged between the server web and the applets related to the Virtual Monitors is estimated in 0.2K bytes per virtual Monitor for each update process.

3 The Parallel Application ABCtraj

As already mentioned in Section 2, the computational engine of the Problem Solving Environment is the `ABCtraj` parallel application, that allows the study of atom-diatom reactivity using a Quasiclassical trajectory approach[11].

`ABCtraj` has been structured as a task farm. The Farmer section of the code is sketched in Fig.1. The Worker portion of the code is sketched in Fig.2.

The Farmer receives the data from the Servlets, performs the initialization and calculates a seed for each trajectory to be computed. The seed will be used by the Worker to generate the string of pseudorandom numbers needed by each trajectory in a deterministic way.

The Farmer enrolls all Workers in the PVM[12] environment and sends to the Worker a bulk of common data to be used for the entire simulation. It also sends then to each Worker a trajectory number, the related random seed and waits for one of the Workers to complete its work unit. When a Worker has finished its

```
                    FARMER code
Receive from Servlets the input data via Unix Socket
Initialize the PVM environment, enrolling the worker program to Workers
Calculate a seed for each trajectory
Send initial data to all Workers

WHILE all_Workers_complete_work
      Waits for a Worker to complete its work unit
      Send to the same Worker a new work unit
      Update SimGate
END WHILE

Write out final histograms
Shutdown the PVM environment
Exit
```

Fig. 1. Scheme of the FARMER portion of the trajectory code for atom-diatom systems.

work unit, the Farmer receives the data and updates `SimGate` and sends to the Worker a new work-unit, until the last trajectory has been calculated.

The final section of the code carries out the remaining (non iterative) calculations relevant to the evaluation and the print out of rate coefficients, cross sections and product distributions for the different reaction channels.

After the Worker receives a trajectory number and the related random seed it starts the integration of the trajectory step by step to follow the evolution in time of positions and momenta. When the trajectory ends, the Worker sends the results to the Farmer and waits for a new work unit to be assigned. If no more trajectories have to be run ($trajectory\ number = 0$) statistics manipulations of the trajectory results are performed to evaluate reaction probabilities and cross sections, product energy and space distributions.

4 The atom diatom H + I Cl reaction case study

To compare with an already investigated system, we discuss here the case of the atom-diatom reaction $H + ICl \rightarrow HI + Cl, HCl + I$. This is a simple heavy heavy light system for which all parameters have been already given in ref.[11] where a simulation environment based on a Graphical user Interface (GUI) developed in X-Windows and Motif environments was discussed.

In Fig. 3 is shown the entry point of the Portal, from wich the researcher has two main possibilities: start a new simulation or analize the Virtual Monitors of a simulation already carried out in the past.

Before starting the simulation, the user must *login* into the PSE. As already mentioned this step is necessary to control who is using the PSE. However it is

```
                              WORKER code
Receive preliminar data from Farmer

Set initializations
Calculate auxiliary variables

WHILE not_last_trajectory
    Receive number of trajectory and related seed for
        random number generation from Farmer

    Generate the subset of pseudorandom numbers characterizing the trajectory
    Calculate the corresponding initial conditions
        LOOP on time
            IF (asymptote is not reached)
                THEN perform integration step
                ELSE exit time loop
            ENDIF
        END the time loop
    Calculate product properties
    Update statistical indexes
    Send to the Farmer the trajectory results to update the Virtual Monitors
END WHILE

Leave PVM
Exit
```

Fig. 2. Scheme of the WORKER portion of the trajectory code for atom-diatom systems.

also necessary to allow the user to build a customized environment. From this page the user can select the type of application he wishes to run (presently, only `ABCtraj` is available). The user can also select the type of Database to be used. The default Database contains all Chemical Systems known by the Portal and available on the various sites of the grid, on Databanks of the Web and on the user's Database that contains the data and the systems already defined by the user. The Chemical System that will be used for the simulation is selected from a selection box built from the directories available on the Database chosen. After this selection, the researcher should choose one of the files listed in the directory to define the configuration of the simulation. In Fig. 4 is shown how to tune some parameters of the configuration. After the configuration phase, the application `ABCtraj` and the parallel environment are activated and the simulation starts.

The user is also enabled to access from the Web the Virtual Monitors he likes (at the moment only the Angular Distribution and the Opacity Function Monitors are activated) from the configurations he wants to study. When the hyperlink of a selected Virtual Monitor is accessed, a Java Applet is downloaded from the HTTP server to the researcher's client and the data of the simulation are shown and updated dinamically. In Fig. 5 an example of the $H+ICl \rightarrow HI+Cl$ Angular Distribution Virtual Monitor produced while the simulation was running is shown. The production of this or other Virtual Monitor at the experimental site supplies useful indications to the experimentalists on how modify measurement conditions.

5 Conclusions

In this paper we have discussed a prototype implementation of an Internet Portal for the distributed simulation of crossed beam processes. The system considered (the atom-diatom reaction $H + ICl \rightarrow HI + Cl, HCl + I$ for which a previous study has been made and the same parameters have been used) has made it possible to carry out a comparison with results obtained using an older version of the simulator. This implementation has shown that **SIMBEX** is a suitable test bed for grid approaches in which computing resources and complementary know how have to be gathered togheter. In particular the grid implementation of **SIMBEX** allows theorists to work on the application in variuos places and experimentalists to start their simulations at the site where the experiment is being carried out. The structure of the simulation is such that the calculations can be spread over a wide network of computers to run concurrently. This allows the simultaneous dealing of a large number of events and a real time clock of the simulation.

Extensions of the procedure to other phases of the simulation are in progress as well as more complex systems and a richer set of interfaces and monitors.

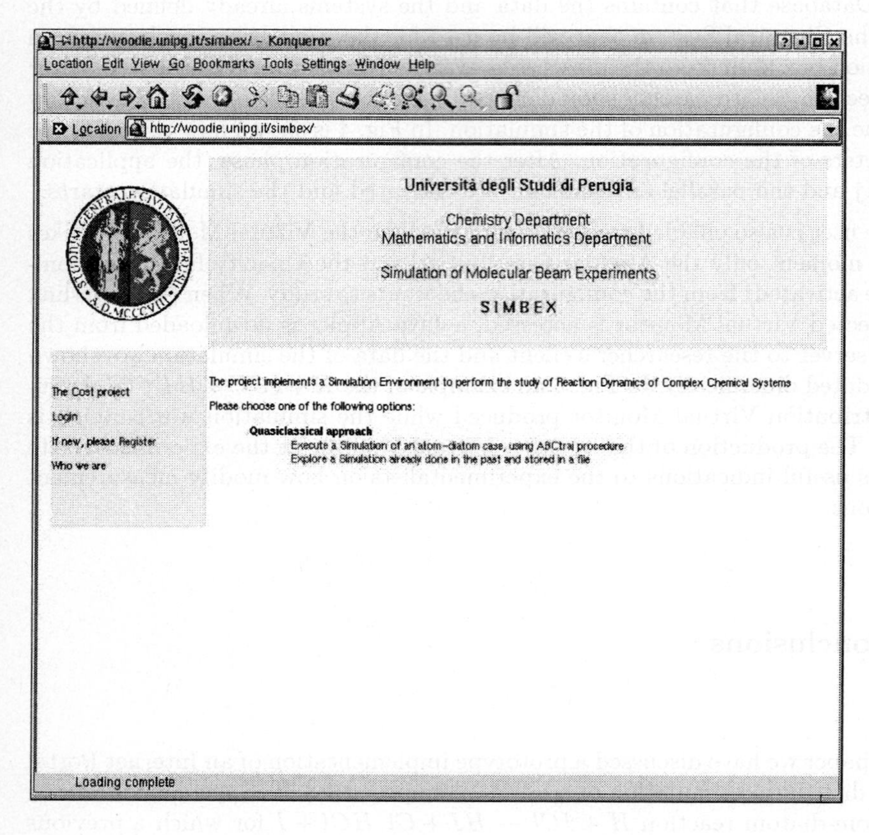

Fig. 3. The Portal entry point of **SIMBEX**

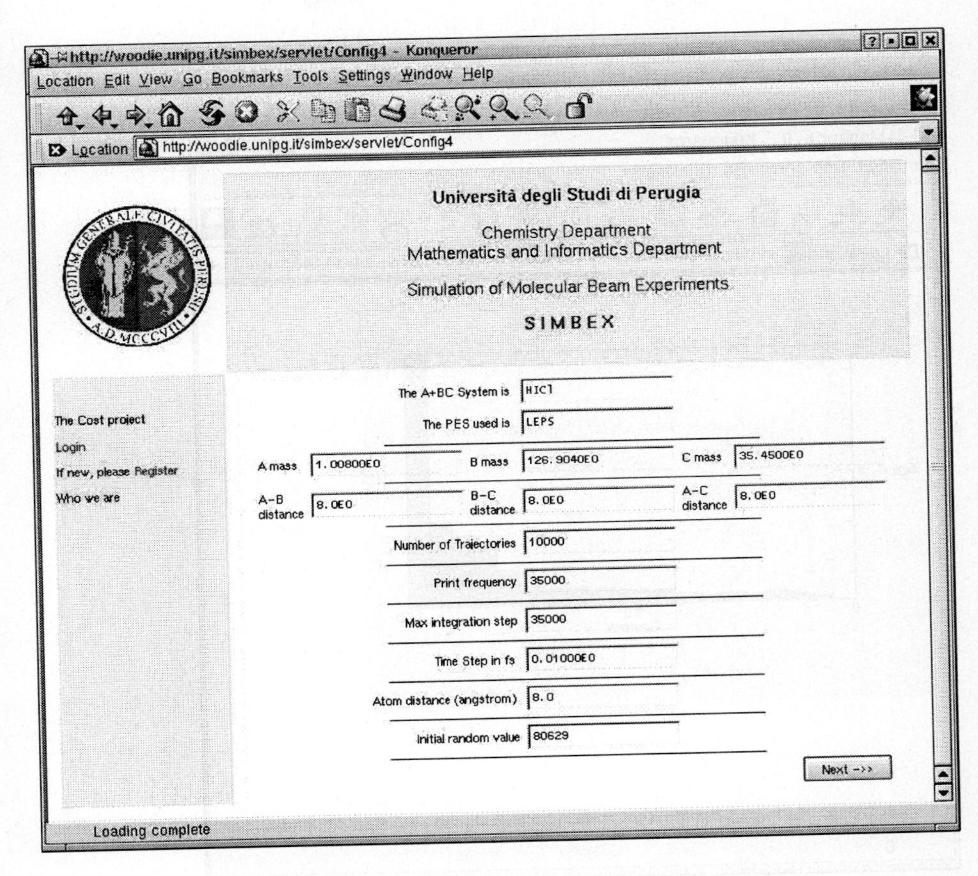

Fig. 4. Definition of the configuration of the simulation, by tuning the parameters related to the Chemical System considered.

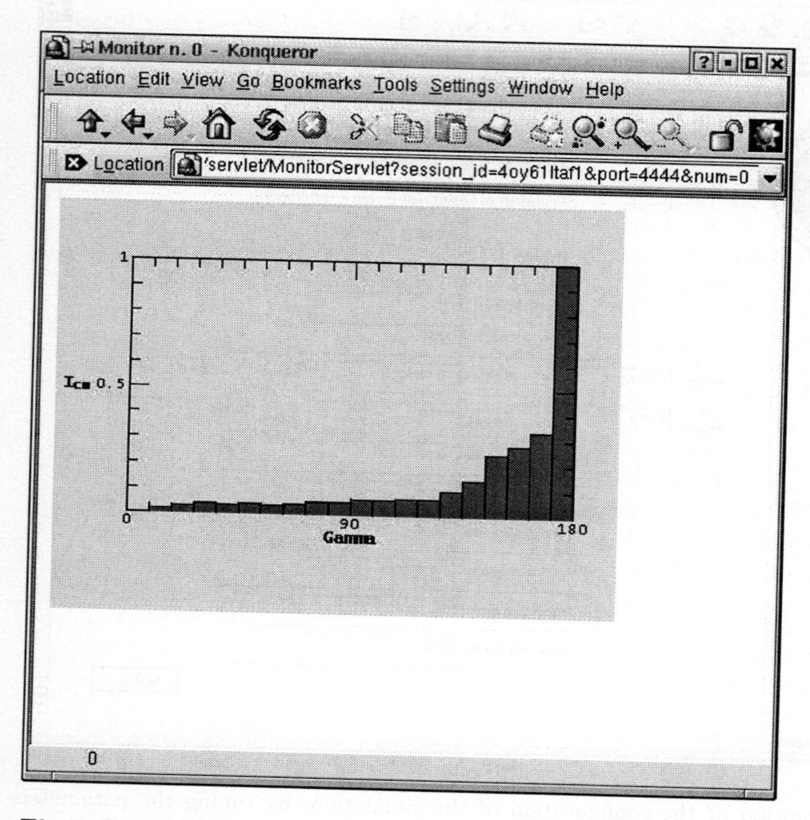

Fig. 5. Example of Virtual Monitor of the Angular Distribution for $H+ICl \rightarrow HI+Cl$ reaction, taken while the simulation is running

6 Acknowledgments

This research has been financially supported by MIUR, ASI and CNR (Italy) and COST (European Union).

References

1. Gallopoulos, S., Houstis, E., Rice, J., Computer as Thinker/Doer: Problem-Solving Environments for Computational Science, IEEE Computational Science and Engineering, Summer (1994).
2. Foster, I. and Kesselman, C. Eds., The Grid: Blueprint for a Future Computing Infrastructure, Morgan Kaufmann Publishers, USA (1999).
3. Baker, M., Buyya, R., Laforenza, D., The Grid: International Efforts in Global Computing, SSGRR2000, L'Aquila, Italy, July (2000).
4. Gervasi, O., COST in Chemistry Action N.D23, Project 003/2001, *SIMBEX: a Metalaboratory for the a priori Simulation of Crossed Molecular Beam Experiments*
5. Casavecchia, P., Chemical Reaction Dynamics with Molecular Beams, Rep. Prog. Phys. **63** (2000), 355-414. J. Chem. Phys., **73** (1980), 2833-2850.
6. `http://www.unil.ch/cost/chem`
7. Laganà, A., METACHEM: Metalaboratories for cooperative innovative computational chemical applications, METACHEM workshop, Brussels, November (1999) (see also [4]).
8. Gavaghan, H., Nature, **406** (2000), 809-811.
9. Gervasi, O., Cicoria, D., Laganà, A., Baraglia, R., Pixel, **10** (1994) 19-26.
10. Gervasi, O, Cicoria, D., Laganà, A., Baraglia, R., Animation and parallel Computing for the Study of Elementary Reactions, Scientific Visualization '95, R. Scateni Ed., CRS4, Cagliari, Italy, World Scientific (1995), 69-78
11. Laganà, A., Gervasi, O., Baraglia, R., and Laforenza, D., From parallel to Distributed Computing for Reactive Scattering Calculations, Int. J. Quantum Chem.: Quantum Chem. Symp., **28** (1994), 85-102.
12. Beguelin, A., Dongarra, J., Geist, A., Manchek, R., and Sunderam, V., A user's guide to PVM Parallel Virtual Machine,Oak Ridge National Laboratory, Tennessee, 1992.
    Geist, A., Beguelin, A., Dongarra, J., Jiang, W., Manchek, R., and Sunderam, V., PVM: Parallel Virtual Machine A Users' Guide and Tutorial for Networked Parallel Computing, MIT Press, Scientific and Engineering Computation, Janusz Kowalik, Editor, Massachusetts Institute of Technology, 1994.
    (`http://www.netlib.org/pvm3/book/pvm-book.html`);
    `http://www.epm.ornl.gov/pvm/pvm_home.html`;
    `http://www.netlib.org/pvm3`

The Enterprise Resource Planning (ERP) System and Spatial Information Integration in Tourism Industry---- Mount Emei for Example

YAN Lei[1], WANG Jing-bo[2], MA Yuan-an[2], DOU Jing[1]

[1] Peking University Beijing 100871, P.R. China
lyan@pku.edu.cn, doujing@263.net
[2] Governmental Committee of Emei Tour Sites, Emei City, Sichuan 614200, P.R. China
jbwang747@yahu.com

Abstract. An integrated information system (ERP system) of tourism industry is proposed in this paper for the management of the planning of Mt. Emei and the relative industries. The authors also demonstrate the advantages of ERP solution, including its construction and functional realization in detail. The fusion of much spatial information in the ERP system is discussed and a spatial integrated information scheme is proposed.

1 Overview

Tourism is a service industry. As to the development of computer technology and Tourism in these years, cyber-systems are used to manage the tourism and its corporations. Due to tourism as an opening, complicated and huge system [1], the research about it should covered with synthetical and integrated methods. Also GIS (Geography Information System) could be used in the inosculation of spatial information and other ones [2][3].

The method of ERP (Enterprise Resource Planning) system to manage and research the tourism system is proposed ---- Mt. Emei for example. The artifices of GIS are utilized in the integration of spatial information. Tour sites of Mt. Emei are managed by the Governmental Committee, including hotels, tourism agencies, etc.

2 The Application of ERP in Tourism Industry

ERP is an advanced operation system in industry [4]. It divides an enterprise into several subsystems. These subsystems are interdependent and cooperated each other. The workflow of the corporation is looked upon a chain of closely supplied links. Not only the manufacturing corporation but also the service one could be managed by the ERP system.

P.M.A. Sloot et al. (Eds.): ICCS 2002, LNCS 2331, pp. 966–974, 2002.

As to the tourism, ERP could be used to manage every taches of the traveling service chain, including management of finance, human resource, service, maintenance, projects, stocking, investment, risk, decision-making, payoff analyzing, region planning, region business, and intelligent traffic. The mainframe of Mt. Emei Information System (ERP) is shown in Fig.1.

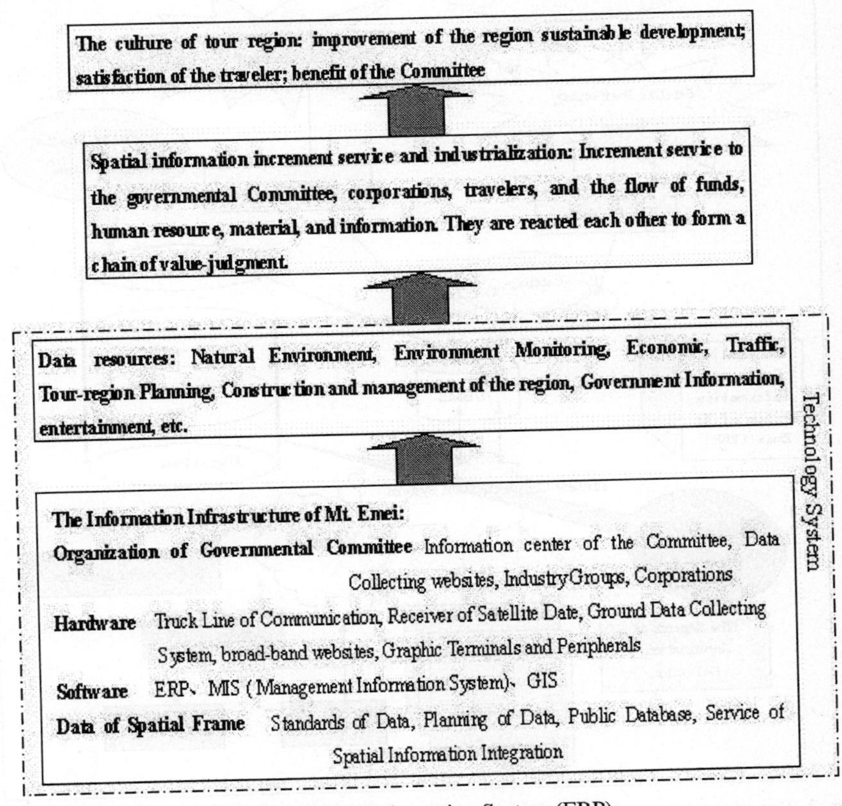

Fig. 1. The Concept Frame of Mt. Emei Information System (ERP)

2 1 The General Frame of Integrated Information System (ERP) on Mount Emei

The general frame is composed of a system for data obtaining and updating, database, data processing, and information announcement.

For the ERP, the multimedia database is its base and the function modules are its structure. And the laters can be divided into the application of government, corporation and business according to the attribute of the users. The general frame of the system is showing in Fig.2. Here RS means Remote Sensing, GPS does Global Positioning System, PDA does Personal Digital Assistant.

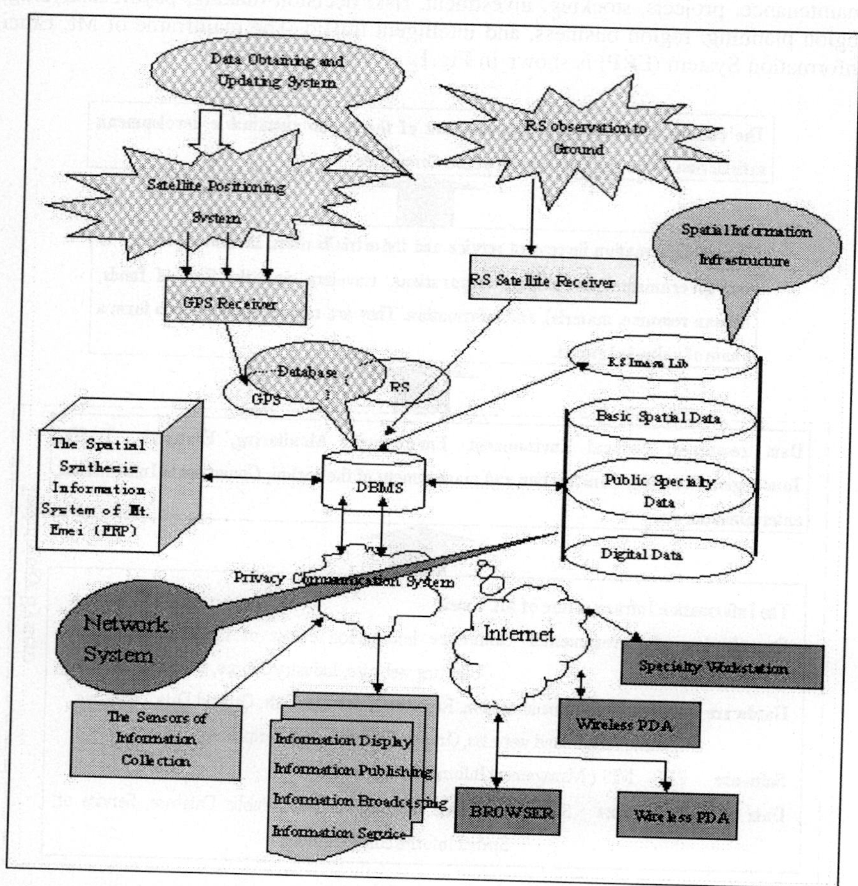

Fig. 2. The general structure of Synthesis Information System on Mt. Emei

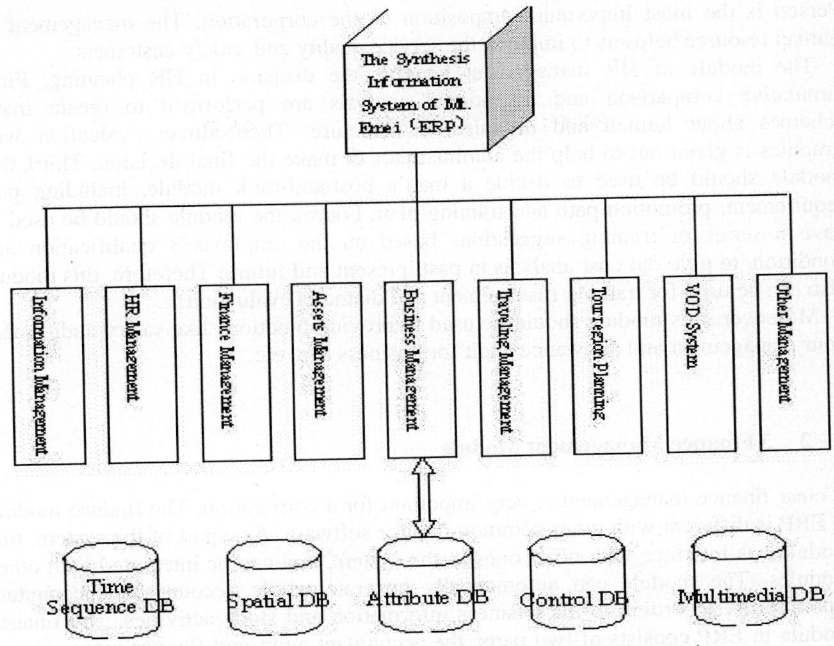

Fig. 3. The Organization Frame of Synthesis Information System of Mt. Emei

2 2 System Functions of Synthesis Information System(ERP) on Mount Emei

The organization frame of Synthesis Information System (ERP) of Mt. Emei is shown in Fig.3. Here VOD means Video-On-Demand; HR does Human Resource; DB does Date-Base.

2 2 1 Information Management Module of Public Data

The maintenance and management of basic data is provided in this module, including the materials about maps, departments, projects, personals, finance, etc. All basic data can be introduced into other modules.

The system setting and maintenance are also provided, including the function modules about the setting of system information and database, application component setup, re-organization for entertainment module, user or computer management, etc.

2 2 2 The Module of Human Resource Management

Person is the most important composition of the corporation. The management of human resource helps us to improve the service quality and satisfy customers.

The module of HR management benefits the decision in HR planning. First, simulative comparison and operational analysis are performed to create many schemes about human and organization structure. Then, direct evaluation with graphics is given out to help the administrator to make the final decision. Third, this module should be used to decide a man's post-and-rank module, including post requirement, promotion path and training plan. Fourth, the module should be used to have a series of training suggestions based on the employee's qualification and condition, to give out cost analysis in past, present and future. Therefore, this module also can be used for training management and diathesis evaluation.

Moreover, this module should be used to provide functions like salary audit, man-hour management and allowance audit for business trip, etc.

2 2 3 Finance Management Module

A clear finance management is very important for a corporation. The finance module in ERP is different with other common finance software. As a part of the system, this module has interface with other ones in the system, and can be integrated with other modules. The module can automatically generate whole account and accountant report forms according to the business information and stock activities. The finance module in ERP consists of two parts: the accountant audit and finance management. The former functionality is to record, audit and analyze the capital flow and results in the corporation activities. The later one is to analyze the former's data and give out corresponding prediction, management and control. After the ERP works, the budget of every department is included in the capital management of the corporation. Administrator can check the capital, selling costs, incomed or outgone money. The ERP can also be integrated with other finance software and set the standardization of the finance management.

2 2 4 Asset Management Module

As an integrated information system, ERP can also be used for asset management. This column includes purchase, usage and maintenance of the facilities. For example, as vital assets, vehicles can be analyzed for its transportation power, route and consume. And we can know vehicles' condition and make sure the tourist's security.

2 2 5 Business Management Module

The ERP will integrate most of the sales' business in the corporation, and realize market sales, finance prediction, dispatch of the product power and its resource. It is the main assistant platform for sale management and decision in the corporation. The ERP will help the corporation to streamline the flow of order-stock- (production)-sale, to refine the finance management, to realize the resource share for the material, capital and information flows.

2 2 6 Tourism Management Module

As a tourism corporation, its management is very important. The management module includes all the information about the tourism services. This information comes from the different department of Mt. Emei. The module also consists of many decision-making and analysis tools. These tools will help the tour manager to provide comprehensive, detail and thoughtful services to customers, develop the more and better planning, and to improve the tourist's services for his consume activities. This module is related to tourists and the business department of Mt. Emei.

The module functions of tourism management include:

(1) Ticket incoming management: strict ticket management to improve the incoming of the ticket sales and to facilitate tourist management.
(2) Support to remote users: Direct illustration of the panorama of Mt. Emei through integration of video, sound, 3D and text.
(3) Tourism market analysis: Information query for users in menu or map form.

2 2 7 Planning Management Module on Mount Emei

The incorporation of GIS into ERP will help the planning stuff of Mt. Emei get different materials, productivity, the basis for the leaders to understand, grasp, analyze present situation of Mt. Emei and to have decision-making.

2 2 8 VOD Video On Demand System

The VOD is based on Open Video system, which is a browser-server module. It has a multi-layer structure, and consists of video server, WWW server and database server. VOD system can be provided to the tourists with VOD services.

3 Integration of Tourism Management and Spatial Compositive Information

The ERP of Mt. Emei is a highly integrated system, which involves many domains. Shown in Figure 4, the information project for the governmental committee of Mt. Emei is based on this system.

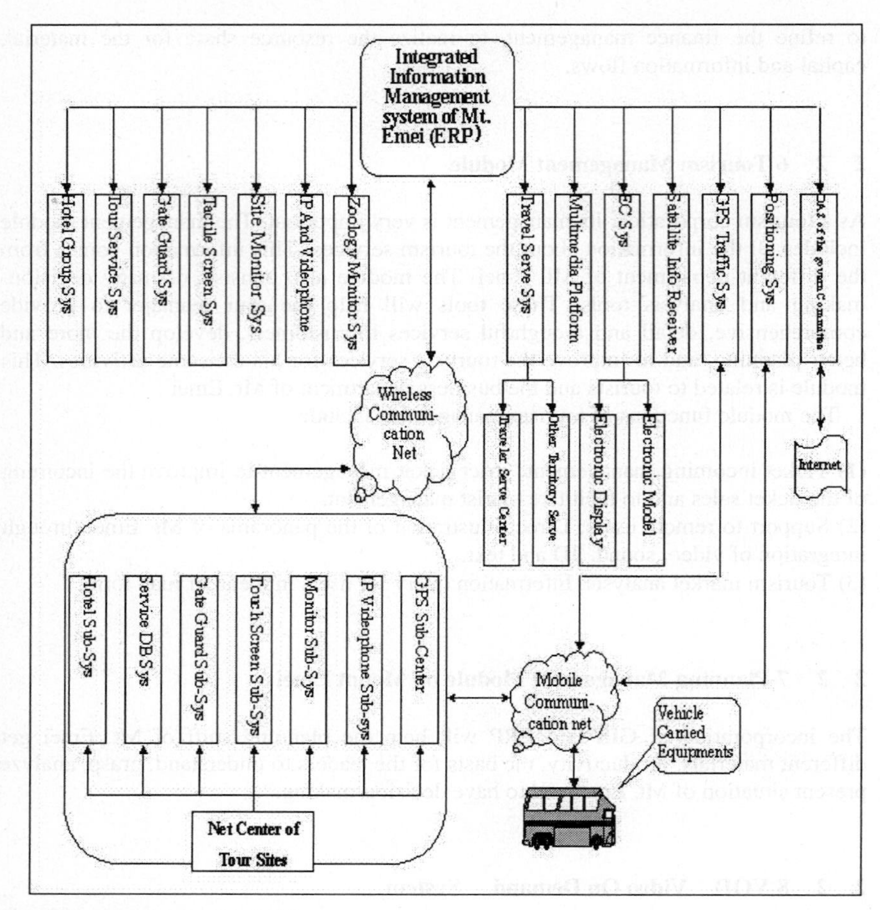

Fig. 4. System Structure of ERP

Here EC means Electronic Commercial, OAS does Office Automation System.

Many models in the system will produce all kinds of spatial information:

(1) Vehicle Management System: Control of the vehicle operation and schedules by GPS technology to ensure tourist security.
(2) Surveillant System: large scale of surveillance for tour spots and hotels.
(3) Tiros Receiver System: reception of Tiros signals to give an alarm for large scale of calamity, such as flood, fire, earthquake and so on.
(4) Positioning and Searching System: Definition of positioning and searching services by wireless technology to ensure the tourists security.
(5) Entironment Monitor System: Surveil lance of the entironment to support the sustainable development of Mt. Emei.

All above systems should be integrated, and can be efficiently merged with each other via GIS platform in the ERP system. It will benefit information interchange in the corporation and analysis/decision support of information management. Furthermore, this kind of spatial integrated information can be intuitively reflected in other systems.

4 Some Information Sub-system Realization

All the information system is very large and complex. Its realization is divided into five stages. Stage one is the informatics accidence of tourism service; Stage two is the governmental monitor and upgrade of the tourism brand; Stage three is the construction of technology platform of tourism service; Stage four is the integration service of tour information; Stage five is general information platform of service. Currently, Stage one and two have been finished. The main functional module/subsystem includes:

(1) Electrical sand table: Present Mt. Emei in the microform by high technologies like sound, light and electronic to the tourists.
(2) Tactile screen: Spread the touch screen systems throughout Mt. Emei; propagate all the tour spots in the form of multimedia.
(3) Management system of the Governmental Committee: Manage all daily works of the governmental committee of Mt. Emei.
(4) Tourist center: As a comprehensive and integrated service system, provide the tourists with query, reference, simulation experience, shopping service and so on.
(5) Scatter tourist service center: Provide scatter tourist with an easy, comfort and personalized tourism service.
(6) Emei online website: Provide tourist services like tour sites, cultural, tour path introduction, as well as other facilities.

Generally, when all the planed modules are finished, they will be interacted each other on the ERP system. At the same time, the excellent scalability, interface design of the ERP system and the component based on software development will ensure the constant upgrade of the overall system and the improvement of the function modules.

5 Conclusions

Due to the uncertain factor for service industry, tourism system is a very complex huge system. An ERP system for the tour service industry is proposed in this paper for tourism planning and the management of Tourism Corporation. The system integrates much spatial information by the GIS platform, provides scientific foundation and reduces uncertainties for decision and planning. Some realized

Subsystems have worked well and have been proven that the general designs of the authors are successful and efficient.

6 References

1. S.L.Zhang, Z.J.Zou, Tourism Planning Synthesis Integrated Methodology Dealing with the Open, Complicated and Large System, Human Geography, Vol.16, No.1, Feb. 2001, P11 15
2. C.Q.Li, etc, Research based on GIS System to the National Tourism Beauty Spots. Science of Surveying and Mapping, Vol 26 No.2 Jun., 2001, P35 38
3. L.Wu, L.Yu, J.Zhang, X.J.Ma, S.Y.Wei, Y.Tian, 2001, Geography Information System Principle, Methodology and Application, Science Press
4. L.Yan, 1998, Base of sustainable development System and Structure Control on Resource, Environment & Ecology, Huaxia Publishing House

3D Visualization of Large Digital Elevation Model (DEM) Data Set

Min Sun[1], Yong Xue[2,3], Ai-Nai Ma[1] and Shan-Jun Mao[1]

[1]Institute of Remote Sensing and Geographic Information System, Peking University, Beijing 100871, China
{Sm6902@263.net, maainai@pku.edu.cn}

[2]Laboratory of Remote Sensing Information Sciences, Institute of Remote Sensing Applications, Chinese Academy of Science, PO Box 9718, Beijing 100101, China
{yxue@irsa.irsa.ac.cn}

[3]School of Informatics and Multimedia Technology, University of North London, 166-220 Holloway Road, London N7 8DB, UK
{y.xue@unl.ac.uk}

Abstract. With the rapid development of Earth observation technologies, raster data has become the main Geographic Information System (GIS) data source. GIS should be able to manage and process huge raster data sets. In this paper, the problem of 3-dimensional visualization of huge DEM data sets and its corresponding images data sets in GIS was addressed. A real-time delamination method based on raster characteristics of DEM and image data has been developed. A simple geometry algorithm was used to implement dynamic Level of Division (LOD) delamination of DEM and image data and to realize real-time 3-dimensional visualization of huge DEM and image data set based on RDBMS management.

1 Introduction

In recent years, Earth observation technologies such as remote sensing, SAR, airborne laser scan, and Photogrammetry, have got a rapid progress. The amount of the data captured by these technologies is increasing in geometric series. Raster data would become the main GIS data source. Many research works on huge raster data management have been done, e.g., Fang *et al.* [2] developed a GeoStar software to manage multi-resolution image data using pyramid style in file system. The software can be used to seamlessly manage multi-scale and multi-resolution image data. Nebiker [6] proved that high-performance relational database management systems could be used to manage conceptually unlimited amounts of raster data. But how to perform data operation on the base of huge data management is still a difficult question, e.g. 3-dimensional visualization of huge DEM data and its image data.

Most 3-dimensional DEM data visualization methods and theories are used for single DEM data block. They are emphasized on how to simplify and efficiently manage the terrain data in order to realize visualization in high speed with high precision. Many algorithms introduced in order to simplify the data were usually

P.M.A. Sloot et al. (Eds.): ICCS 2002, LNCS 2331, pp. 975–983, 2002.

established on the data structure which were used to manage the terrain data, e.g., the data structure based on quadtree. These methods pretreated DEM data before visualization. It not only effects real-time rendering speed, but also limits visual data area, therefore, its difficult to realize visualization of huge DEM data set in real time.

Some typical works on 3-dimensional visualization of DEM data are: Cignoni *et al.* [1] expressed multi-resolution DEM data using hierarchical TIN structure. This method needs great calculation and it is difficult to realize real-time visualization. Lindstrom [5] put forward a real-time, continuous LOD rendering method. The basic idea of this method is to manage DEM data using quadtree data structure. When the DEM data block is huge, the quadtree data structure itself would need a huge memory space (e.g.: 20480×20480 grid DEM data, about 400Mb, if 4×4 grid form a node, each node occupies 16 bits, then it would cast 400Mb memory space). Despite the fact that this method could simplify a huge amount of data, it still needs a large amount of data in order to render details in high precision. The algorithm have an great effect on rendering speed, so it is difficult to deal with huge DEM data sets. Hoppe [3] also put forward an algorithm that could be used to dynamically calculate LODs based on viewpoint. But this method is still difficult to deal with huge DEM data real-time rendering. All these works are based on TIN structure. They need great calculation in real-time rendering and pretreatment, and also high hardwires configure. They are suitable to run on graphic workstations.

Comparing with TIN structure, regular grid structure is much simple and DEM usually was expressed using regular grid in practice. A common style is point matrix of raster structure, saving in image format. Research works on 3D visualization of DEM data based on regular grid are: Kofler [4] developed a method combining LOD and R-tree structure. This method used R-tree instead of quadtree. It would be more difficult for this method used to realize huge DEM data sets 3D visualization. Wang *et al.* [7] put forward a method based on quadtree structure and simplify the data depending viewpoint position. As this method used quadtree structure to manage DEM data, it used a lot of memory and it is difficult for huge DEM data sets visualization.

So far, there is no one method which could be used to solve the problem of huge DEM data sets real-time visualization properly.

2 Real-time LOD delamination based on regular grid

2.1 Regular grid division

Regular grid structure is easy to process and DEM data formats are mainly using regular grid. The basis of our method for huge DEM data sets real-time visualization is to express DEM data using regular grid. DEM regular grid format is in 2-dimensional matrix. As its length and width is known, one dimension array, e.g., $p[n]$, is used normally. If a plane's original point position is (0,0), the length and width of DEM are s and t, the height value of a random point (i, j) is $p[\, i*t + j]$. This concise structure of regular grid actually manages DEM data very good. There is no need to manage it using extra structure.

According to LOD idea, terrain areas in far viewpoint would not need same precision as that in near viewpoint. So we could use viewpoint distance to divide grid LOD delamination in order to improve visualization speed.

In Figure1, e is the viewpoint position, a is observation angle, b is the angle between sightline and vertical direction and p is the center point of view area on DEM. Then the dash-line circle marked area needs high rendering precision LOD. For the convenience of calculation, we use the square instead of this circle. The different LOD areas are expressed by nested square area. Based on this simple thinking, we could easily get LOD division of any random time and random viewpoint position.

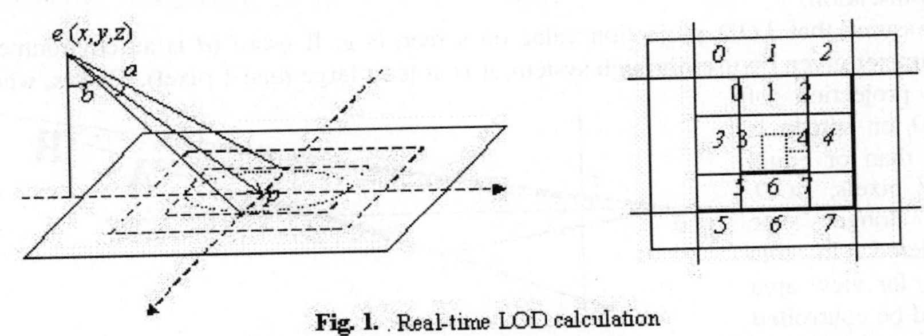

Fig. 1. . Real-time LOD calculation

Right picture in Figure 1 shows the principle of LOD repetition calculation. Any LOD area could be consider as combination of 8 blocks. Its size could be calculated using formula $D = m\times2^n$, m expresses the size of first LOD area where the viewpoint is in. The value of m could change with the distance from e to p.

Regular grid data usually results in interposition from discrete points on plane. It possess a huge number of abundance data. Adjacent points have close height values, so we could simplify the DEM and its corresponding image data by resample method to get LOD real-time rendering data. For the first LOD area around viewpoint p, resample one in two points on the base of original DEM and image data, for second LOD area resample one in four points, for n LOD area resample interval use 2^n(n = 1, 2, …).

Figure 2 shows the above resample process. If we divided according $m\times2^n$, then the actual rendering DEM grid number tm in real-time is:

$$tm = (2m/2)^2 + (4 \times[(m\times2^{n-1}/2^n)^2 + (2m + m\times2^{n-1}) \times m\times2^{n-1}/2^n])$$
$$= m^2 + m^2 (3 + 2^{n+1})$$
$$= (3n + 2^{n+2}) m^2.$$

The actual DEM size is:

$$[2\times(m + \ldots + 2^{n-1}m)] \times [2\times(m + \ldots + 2^{n-1}m)]$$
$$= (2^{n+1} - 2)^2m^2$$

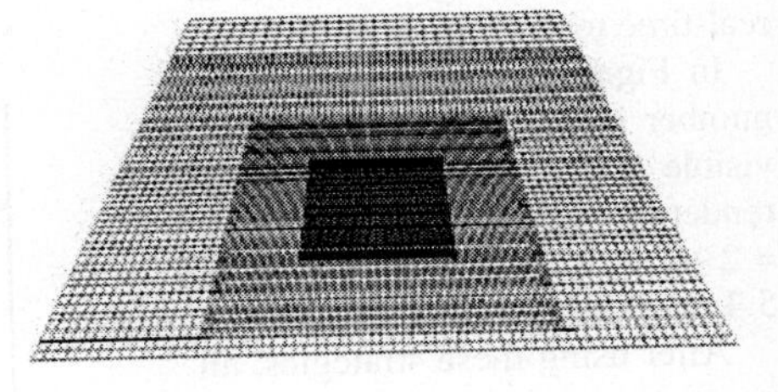

Fig. 2. Real time LOD level division

if we let m = 64 n = 5, then real-time rendering grid number is: 143×64^2 = 585,728.

While DEM size is: $(2^6 - 2)^2 \times 64^2 = (62 \times 64)^2 = 3968 \times 3968$

The actual DEM grid number is 15,745,024, the simplify process discards 96.3% original data.

2.2 Further simplify strategies

For personal computer environment, it is still difficult to render in real-time although more than ten millions grids have been simplified to less than one million. As above simplification does not consider the observation angle, the distance from viewpoint to observation point and view area, these parameters could be used to do further simplification.

Assume that LOD_i projection value on screen is g. If $g<=d$ (d is an environment parameter given by user through system, it is at least large than 1 pixel). That is, when the projection of LOD_i on screen is less than or equal to d pixels, LOD_i no longer be rendered. In this way, far view area could be controlled in real-time rendering. In Figure 3, AB is the width of LOD_i, and its projection value on screen is$\hat{o}$. When$\hat{o}$is too small to be noticed, there is no need to render AB. Therefore, the fact of this strategy is to controlled DEM visual area so as to decrease DEM rendering data in real time.

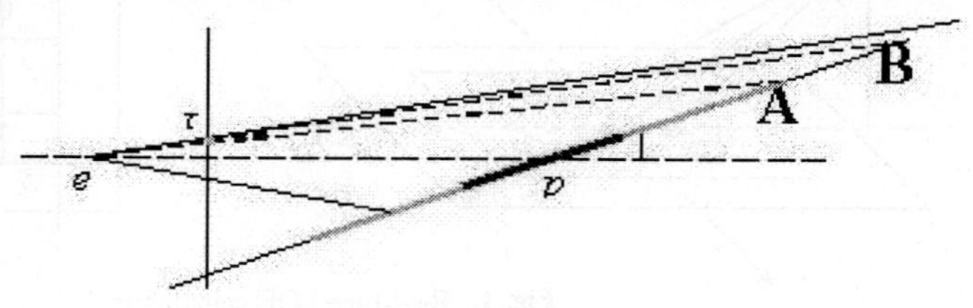

Fig. 3. The projected DEM area on the screen from telescopic view

Also, assume that the present screen width is r and LOD_i projection width on screen is f. If $r > f$, LOD_i will be replaced by LOD_{i+1}, contrarily LOD_{i+1} is replaced by LOD_i. In such LOD exchange process, firstly the system does not keep so much DEM blocks which are read from the database. Secondly the width of view area could cut off at least 50% grid number in real-time rendering.

In Figure 4, LOD_{i+1} rendering number is less than $2/3m^2$. When visible LOD number is n, total rendering grid numbers are: $S = 2 + 2\ n/3\ \ m^2$, when $n = 5$, $S = 5.3\ m^2$, if $m = 64$, than $S = 21{,}708$.

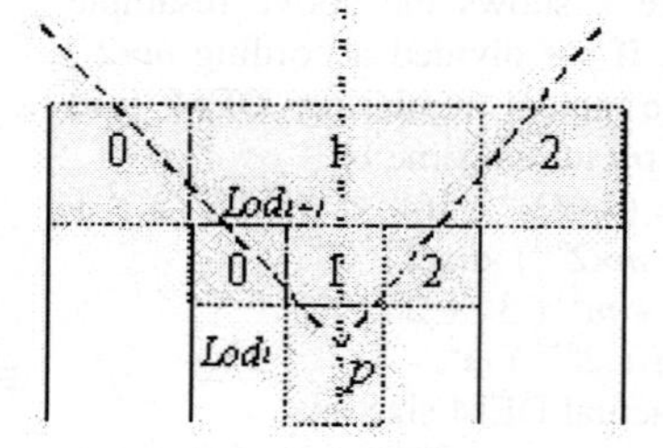

Fig. 4. LOD blocks cutting out by view field

After using these strategies, all necessary rendering grid numbers have been cut down to feasible degree for animation. But for real-time animation this number should not be bigger than 10,000. Therefore, we need further processing which will be addressed in next section.

2.3 Seam problem between two LODs

During regular grid division process, different levels use different resample intervals. If we don't adapt further special processing, a seam would appear between two LODs. As seams appear on the sides of two adjacent LOD areas, and resample interval ratio of these two LOD areas is 2, we could change drawing one square to three triangles. Four sides need special processing respectively. Figure 5 shows the processing.

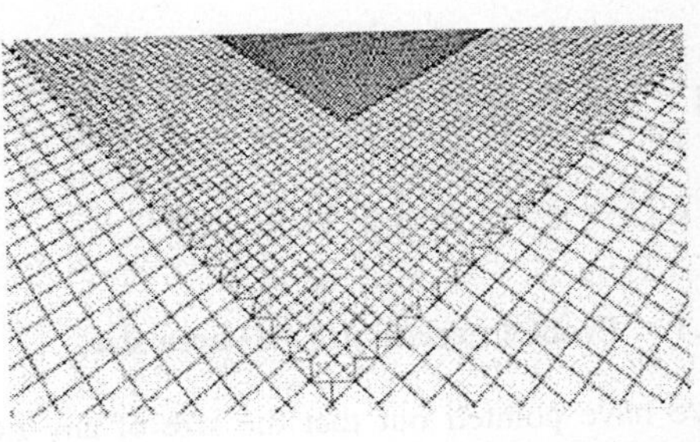

Fig. 5. Filling gaps between two levels

3 Huge DEM data sets management and scheduling

Before discussing visualization, we firstly need to manage the huge DEM and its corresponding image data sets. As many research works on huge raster data management have been done, we use following management method in our paper:

- Delaminate DEM and image data according its resolution and precision
- Each level has same resolution and precision
- Precision or resolution for different level increases principally in 2^n;

In order to visualize DEM data in real-time, DEM and image data has to be divided into blocks. Blocks size should be (2^n, 2^m). (1024, 1024) is the ideal size. DEM division and image division must be kept consistent.

Divided blocks are managed in DBMS using BLOB type according to their levels. One block has a record such as: serial number, row and column number, position in original raster data, corresponding spatial data and the block's raster data matrix;

An index file is established to save raster data according to their level structures. This index file is read into memory in real-time visualization process. The file manages reading from and releasing to each block data.

In real-time visualization process, it should permit observer to view terrain area both partially and fully. So the data scheduling need to switch DEM and image data between blocks and levels. The details are:

- Using view parameters calculate LOD areas and select visible DEM level. Then look up DEM and image data blocks in database and if blocks spatial area intersect with LOD areas, then read these blocks;

- Judge each data block in memory whether it is still visible. If not, release memory occupied by this data block;
- As data blocks reading and releasing would affect DEM visualization continuity, it is needed to set a buffer with suitable size. In addition, visualization process and data access process should be processed in each program thread.

4 Discussion

Two key points need to be discussed.

4.1 Problem relate to LOD area size

We have pointed out that the size of any LOD_n could be expressed using $2^n m$. As resample interval of LOD_n is 2^n, rendering grid numbers of LOD_n are decided by the value of m. At the same time, according to above scheduling, data blocks that need to be read into memory are determined by the value of m. For personal computer environment, the block numbers in memory need especially to be controlled carefully. From our experiment processes, we have examined m value and found that 64 is an ideal value for m. When resample interval value is 2^n $n = 2, 4, 8,\ldots$, the sequence values of LOD area are 256, 512, 1024, 2048, ... When 5 LODs are used in visualization, three data levels are rightly need to be read at each time. If each level need eight blocks (DEM data blocks and its corresponding image data blocks), real-time rendering to three level data needs 24 data blocks to be read. Through cut off by view field (at least 50%), real-time rendering grid number are (64×64 + 64×32 + 64×32 +64×32 +64×32) = 12288.

4.2 Problem of continuity of frame speed and visualization effect

With the observer position changing, view field should also be changed; therefore, switches between LODs are needed. As different resolution levels in database are equal to dynamic delamination and dynamic delamination is calculated at real-time, LOD switch is the exchanges between two different resolution data levels. It is actually data reading and releasing process. One switch process might need reading eight data blocks including DEM and image data. This would greatly affect continuity of frame speed. In order to solve this problem, a buffer must be used to read these data blocks that would be rendered in recent frames. For high fly speed, this question is still there.

Figure 6 shows four visualization effects rendered using different resample interval at same place. It is found that visualization effects for resample interval value 8 and 1 are same for same image data. If we use 8 as the highest resample interval value, then value 12288 is reduced to 3072. This value is really a widely acceptable value. Therefore, users can select between visual speed and effect from the system.

One obvious defect of our method is that many distortions would appear during animation process in sides of LOD areas. Such distortion is effected by terrain undulation. But distortion is not too much as it appears in far view areas, and it could be implemented by adding plants and fog effect.

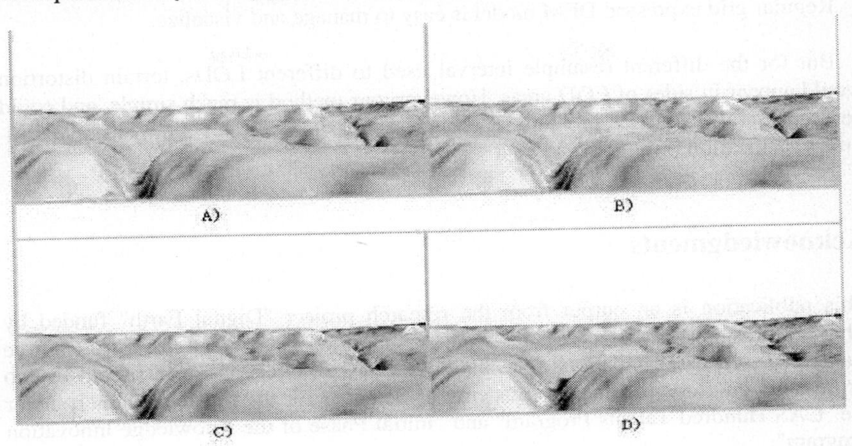

Fig. 6. Visualization effects using different sample intervals (Resample interval of A, B, C, D are 1, 2, 4, 8 respectively)

5 Experiment

As it is difficult to obtain a huge DEM data set we used a DEM data with block size 2048×4096 and its corresponding image block, which is downloaded from "ftp://ftp.research.microsoft.com/users/hhoppe/data/terrain/". We copied one DEM block (8Mb) and its corresponding image block (24Mb) into 64 copies. All DEM and image data blocks were divided into subblocks of size 1024×1024 (about 3Mb), and stored into DBMS database. Its spatial position assigned according to blocks arrangement and blocks position in original data. We used resample interval sequence 2^n (n = 4, 8, 16 …). The hardware configures of our PC are Window2000 professional, OpenGL 1.1, PIII CPU 667, RAM 128M, display card ELSA Erazor III LT, display memory 32M. Real-time rendering speed is 20fps. Figure 7 shows five pictures rendering by the experiment system, which were copied from five different viewpoints.

6 Conclusion

In this paper, we developed a simple 3-dimensional visualization method. Comparing with existing methods, it has following merits:

- Much less calculation is used in real-time visualization process, so the visual process was not affected by the visualization calculation.

- In addition to DEM and images data, few extra memories are used for visualization process. This is very useful for fly on terrain with a huge DEM data set;
- Regular grid expressed DEM model is easy to manage and visualize.

But for the different resample interval used to different LODs, terrain distortion would appear in sides of LOD areas. However, our method is much simple, and could be used to solve practice problem. It provides a feasible way for 3-dimensional real-time visualization of huge DEM data sets.

Acknowledgments

This publication is an output from the research project "Digital Earth" funded by Chinese Academy of Sciences and is partly supported by a grant from the Chinese National Sciences Foundations (No. 59904001). Also, Dr. Yong Xue should like to express his gratitude to Chinese Academy of Sciences for the financial support under the "CAS Hundred Talents Program" and "Initial Phase of the Knowledge Innovation Program".

Reference

1 Cignoni P., Puppo E., Scopigno R.: Representation and visualization of terrain surface at variable resolution, The Visual Computer, 13 (1997) 199-217

2 Fang T., Li Deren, Gong Jianya, Pi Minghong: Development and Implementation of Multiresolution and Seamless Image Database System GeoImageDB, Jorunal of WUHAN Technical University of Surveying and Mapping, 24 (1999) p222

3 Hoppe H.: Smooth view-dependent level-of-detail control and its application to terrain rendering, IEEE Visualization (1998), 35-42.

4 Kofler M.: R-trees for Visualizing and Organizing Huge 3D GIS Databases, [Ph.D. Dissertation], Graz Technischen Universität Graz, (1998)

5 Lindstrom P., Koller D., Ribarsky W., Hodges L.F., Faust N.: Real-Time, Continuous Level of Detail Rendering of Height Fields, [In] Proceedings of ACM SIGGRAPH 96, (*http://www.cc.gatech.edu/ ~lindstro*), (1996), 109-118.

6 Nebiker, S.: Spatial Raster Data Management for Geo-Information Systems–A Database Perspective, PhD Dissertation, Swiss Federal Institute of Technology Zurich (1997).

7 Wang H.W., Dong S.H.: A view-Dependent Dynamic Multiresolution Terrain Model, Journal of Computer Aided Design and Computer Graphics, 12 (2000) 575-579

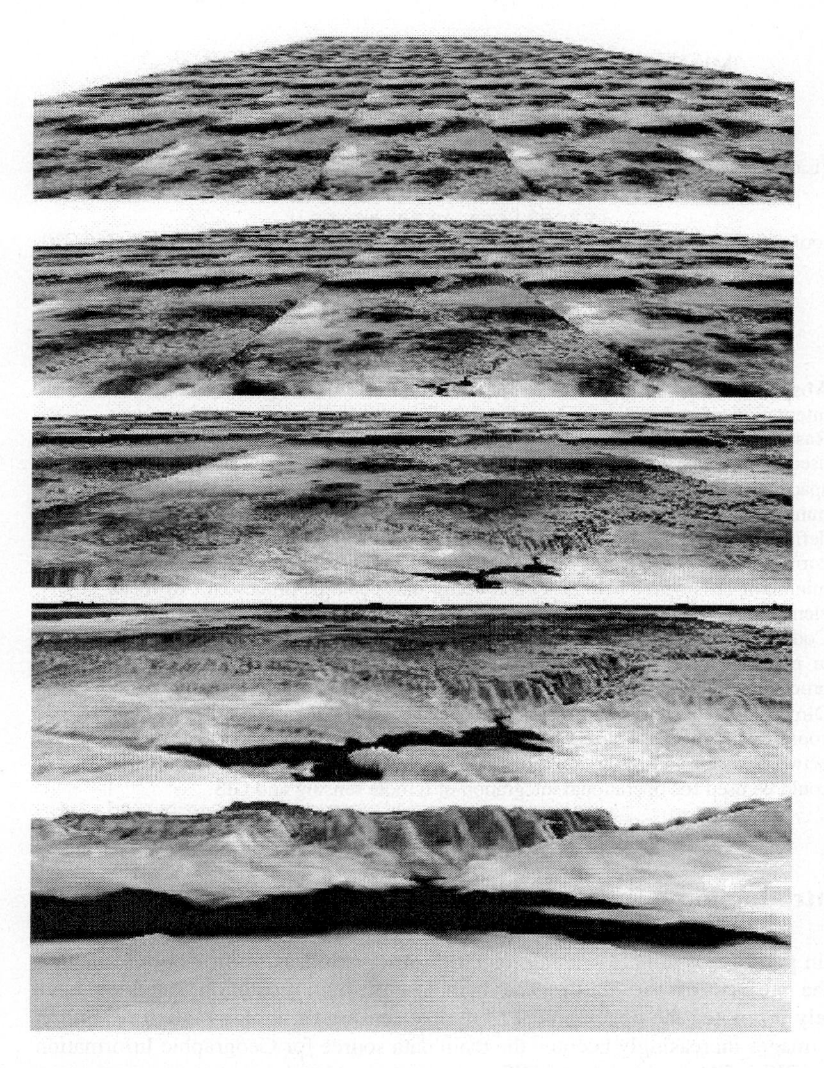

Fig. 7. 3D visualization of a terrain landscape using the algorithm developed in the paper

Dynamic Vector and Raster Integrated Data Model Based on Code-Points

Min Sun[1], Yong Xue[2,3], Ai-Nai Ma[1] and Shan-Jun Mao[1]

[1]The Institute of GIS & RS of Peking University, Beijing, 100871 China
{Sm6902@263.net, maainai@pku.edu.cn}
[2]Laboratory of Remote Sensing Information Sciences, Institute of Remote Sensing Applications, Chinese Academy of Sciences, Beijing, 100101, China
{yxue@irsa.irsa.ac.cn}
[3]School of Informatics and Multimedia Technology, University of North London, 166-220 Holloway Road, London N7 8DB, UK
{y.xue@unl.ac.uk}

Abstract. With the rapid development of remote sensing technology, the integration of raster data and vector data becomes more and more important. Raster data models are used in tessellation spaces and vector data models are used in discrete spaces respectively. The relationships between tessellation space and discrete space have to be established for integrated data models. The minimum cells where both raster and vector data could be processed have to be defined. As it is very easy to establish relationships between vector points and corresponding raster cells, we defined those raster cells as Code-Points, the minimum cells where both raster and vector data could be processed. All vector elements such as lines, faces and bodies are composed directly or indirectly of Code-Points. This can be done by using interpolation algorithms to Code-Points in real-time. We have developed an integrated data model based on above procedures. In addition, we have developed a geometric primitive library for 3-Dimensional objects in order to improve the processing speed. This library could be hardware realized as a graphic accelerator card. If the conversion between vector and raster could be done in real time, the integrated data model could be used for operational integration of remote sensing and GIS.

1. Introduction

The main purpose of remote sensing, or Earth observation, is to provide information about the surface of the Earth. The launches of high-resolution satellites have effectively promoted the development of remote sensing technology. High-resolution satellite images increasingly become the main data source for Geographic Information Systems (GISs). The integration of GIS and remote sensing becomes more important. The fundament of the integration of remote sensing and GIS is the integrated vector and raster data models. Many related research works in this field have been done, e.g.: Ballard [1] developed a Strip Tree Model, dividing vector lines into several segments, and expressed vector lines using binary tree data structure; Randal [6] put forward to

P.M.A. Sloot et al. (Eds.): ICCS 2002, LNCS 2331, pp. 984–993, 2002.

express vector data using quadtree hierarchical structure; Gong [2] developed a vector and raster integrated data model using multilevel grid and quardtree data structure.

The real world is very complex and it is very difficult to build a common 3D data model. Molannar [4] developed a 3DFS data model. Pilouk *et al.* [5] developed a tetrahedral based data model. Li [3] developed a CSG based data model and Tempfle [8] developed a 3D topological data model for 3D city. All these data models are vector data models.

If vector data of a face element is expressed by quadtree, the quardtree expressed data layer must be repartitioned in quardtree when vector data changed. Because of such defects, there are still no operational GISs which could integrate vector and raster data dynamically. The fundamental problem of the fully integrated system is to choose appropriate data structures, those problems are remained unsolved [9].

In the paper, we developed a new method, which is based on the code-point, a minimum unit which both raster data and vector data can be processed directly. First, we discuss its main idea, followed by spatial coding algorithm. Integrated data model is explained in Section 4. After the discussion of data structure in Section 5, the conclusion has been given.

2. Basic Idea of the Integrated Data Model

As we know, a vector object is expressed using discrete points in a discrete space (vector space), and a raster object is expressed using continuous cells in a tessellation space (raster space). In order to integrate vector and raster data, a unified space should be setup and the minimum cells should be identified and defined on which both raster data and vector could be disposed, where two different kinds of data could be processed directly without any conversions.

As discrete space could be considered as a special case of tessellation space, tessellation space could be used for the integration of vector and raster data. Vector point is the basic element in vector space, and raster cell is the minimize cell in raster space. When a raster space is expressed using Octree data structure, raster cell could be expressed using a code. If this code is a vector point, we call it a Code-Point (CP). The Code-Point is the minimum cell in the integrated vector and raster data model developed in the paper. In tessellation space, an object is expressed using a set of raster cells. But for an object in a vector space, it becomes much more difficult as a vector object, especially a solid object, usually combined by lines or faces. In order to represent a vector object in Code-Points, the object must be converted from its present expression to Code-Point expression. Basically, this conversion is same as the conversion from vector to raster. Normally, the conversion of a vector object using raster cells would not only increase the data abundance, but also not be convenient to update data. In our vector and raster integrated data model, we use algorithms to conduct the conversion from vector data to raster data in real-time.

First, raster spaces are represented in octree data structure. Octree is an effective data structure to represent a 3D tessellation space. A vector point corresponds to a raster cell, i.e. an octree node. As it needs a large mount of computer memory to establish an octree data structure, we use a code-point to represent the relationship between a vector point and a raster cell (an octree node).

Interpolation algorithms will be used to represent vector elements in Code-Points in tessellation space. For example, a line could be expressed by a linear interpolation of Code-Points. However, it will be much more difficult and complex for a face and a solid. In the following section, we developed some methods to solve this problem. Thus, the key point for the integrated data model is the real-time raster interpolation algorithms. Base on the following several reasons, we consider that the real-time interpolation algorithm is feasible:

- Real-time interpolation could avoid data redundancy, and it is easy to update data;
- It could be much more easier to manage and deal the vector objects if the raster form is interpolated from Code-Points in real-time;
- In the interactive graphic mode, the frequent calculation usually occurs in small area. The amount of calculation could be reduced;
- To build a library for primitive geometric objects and related conversion algorithms which could be performed by hardware, would reduce the calculation complexity significantly and improve the processing speed.
- The new developments of data processing capability of the present computer hardware provide good opportunities for the fast calculation in practice.

Figure 1 shows the representation of a line object and a face object based on Code-Points. The raster forms of the line and the face are not continuous. Interpolation process is needed. Firstly, we discuss the algorithm of Space Coding (Octree-ID).

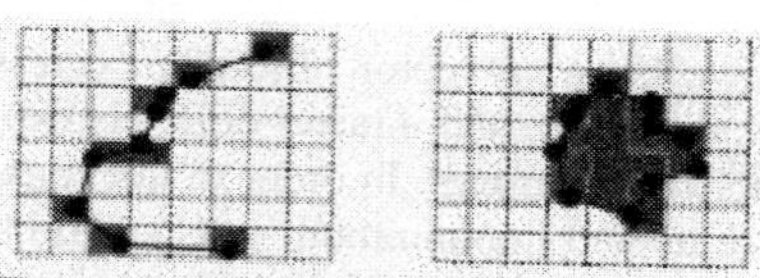

Fig. 1. The expression of line object and face object in the integrated data model

3. Space coding algorithm

The establishment of an octree structure for a raster space would use a large amount of RAM space. In order to save the memory space, we only calculate corresponding raster octree codes for each vector point. Here the code means Morton code of octree node. The main idea of the algorithm is to calculate Morton codes for octree nodes that correspond to the vector points according their coordinates. The algorithm frame is as following:

Algorithm_Calculate_Octree-id (X, Y, Z → Octree-ID):

/* *Assuming MinX, MaxX, MinY, MaxY, MinZ, MaxZ are the maximum values and the minimum values in x direction, y direction and z direction, respectively. And assuming Min_Dx, Min_Dy and Min_Dz are the least cell values in x direction, y direction and z direction, respectively. The three values could be equal. Then we have* */

$Wx = MaxX - MinX;$

$Wy = MaxY - MinY;$

```
Wz = MaxZ – MinZ;
/* The code in x direction is {0 ,1}, in y direction needs add {0   2}, in z direction
needs add {0, 4}    as well.*/
/* Define the point structure     */
  typedef point {
  float x;  float y;  float z;
  char pCode[10]; }
/* Calculate the minimal interval values in x y and z direction*/
  Min_Width = Min( Min_Dx,  Min_Dy,  Min_Dz);
/* Calculate the length of Octree codes */
  t =f(int[Wx / Min_Width]) );
/* Calculate Octree-Ids */
  for(int i = 0; i < t ; dx++)
  {for(j = 0; j < pt_num; j++)
     {if(p[j].x > (MinX + Wx * (c))) // c = 1/2 + 1/4 + 1/8 + 1/16..... 1/(2^t)
            p[j].pCode[i] = 1;
     else   p[j].pCode[i] = 0;
            if(p[j].y > (MinY + Wy * (c)))
     p[j].pCode[i] += 2;
            if(p[j].z > (MinZ + Wz * (c)))
     p[j].pCode[i] += 4;
            }}
```

The length of Octree-ID is t. The larger t is, the smaller the octree cell is. It also means that the space is partitioned more deeply. The value of t depends on real situations. However, the value can't be too small. Otherwise, it needs much more interpolation calculations. In the next, we discuss the establishment of integrated data model in detail.

4. Integrated data model

4.1 Vector element expression

Normally, vector elements in a vector model are points, lines, faces and bodies. A line is composed of points, a face is composed of lines and a solid is composed of faces. In a raster space, all kinds of elements are expressed in raster cells. Vector points and lines are relatively simple. In the paper, we mainly discuss face elements and solid elements. When a face element is a spatial plan or a combination of a set of spatial plans, it could be expressed by its sidelines. If it is a spatial curve surface, it should be expressed by regular grid or TIN in order to simplify the procession.

Spatial solid objects could be divided into homogeneous objects and heterogeneous objects according to its internality. And it also could be divided into regular objects and irregular objects (here regular objects are spatial objects expressed by regular geometry primitives such as cuboids, prisms, spheres, cylinders and cones. It also includes complex spatial objects expressed by the combination of these regular geometry primitives. All primitives that could be expressed by parameters could also be called regular geometry primitives). For a homogeneous object, it could be expressed by its surfaces. However, for a heterogeneous object, it can't be simply expressed by its surfaces, as its internal needs to be expressed as well. Molenaar's 3DFS data model is good to express a homogeneous object, and is also good to express a regular object in some degrees [4]. The tetrahedral based data model developed by Pilouk *et al.* [5] is good to express a heterogeneous object, and it is also good to express an irregular object. But it is not good to express a regular object. The regular object tetrahedral partition expression increases unnecessary calculations in some degrees. Considering the spatial regularity and its internal uniformity of solid elements, we divided solid objects into the following categories, and expressed them using different ways:

- Regular homogeneous objects: expressed using regular geometry primitives or their combinations;
- Irregular homogeneous objects: its whole solid could be represented by its surface. It could be expressed by the combinations of several face objects in order to simplify the raster procession. Surface is expressed using regular grids;
- Regular heterogeneous objects: expressed using regular geometry primitives and tetrahedrons;
- Irregular heterogeneous objects: expressed using tetrahedral partition.

Regular geometric primitives and tetrahedrons obviously become the basic geometric primitives to express solid objects. Therefore, it is necessary to establish a library for geometric primitives.

4.2 Establishment of Geometry Primitives Library

It is same as libraries in AutoCAD and 3DMAX, the following geometry primitives in our integrated data model are included in the geometric primitives library: Rectangle, Triangle, Cuboid, cylinder, cone, prism and sphere, Frustum of a pyramid, Tetrahedron. In addition, any geometry primitives that have analytical forms could be considered as regular primitives, and could be included in the geometric primitive library.

In order to express more complex objects, the library should be an open graphic library. It will allow users to add more geometric primitives in order to express some special objects. Besides, any new geometric primitive added to the library should be associated with an interpolation algorithm for vector and raster data integration. There should also be some basic operational functions, such as rotate, pan and zoom, in the library.

4.3 Raster interpolation Algorithm

In order to do the integration, vector elements have to be converted to raster cells. In our integrated data model, we will use raster interpolation methods with Code-Points. Now we discuss the interpolation methods for line, face and solid objects, respectively.

Line Object: a line object composes of a set of points. Assuming that two points are linked with a straight line, a line interpolation algorithm could be illustrated in Figure. 2, that is to calculate all raster cells along the line AB.

A vector line L = $\{p_{1\ x,y,z}, p_{2\ x,y,z}, \ldots, p_{n\ x,y,z}\}$ in raster space should be L = $\{p_{1(pcode)}, p_{2(pcode)}, \ldots, p_{n(pcode)}\}$. The vector line L should be continuous when the length of Octree-ID is equal to 1. The deviation of $p_{1(pcode)}, p_{2(pcode)}, \ldots, p_{n(pcode)}$ is positive proportional to the length of Octree-ID.

We assume that any two Code-Points are connected with a straight line. The interpolation between two code-points is to calculate raster cells passed by the line segment. Each cell is expressed by an octal Morton code. Figure 2 shows, for a 2-dimensional plane, the four corners of the cell where Code-Point A is are A_1, A_2, A_3, A_4, respectively. And the slopes of the four segments A_1A, A_2A, A_3A, A_4A are l_1, l_2, l_3, l_4, respectively. The slope of the segment AB is l. Assume $x = B_x - A_x$ and $y = B_y - A_y$. Following the positive and negative values of x and y, the next cell could be located by comparing the value between l and l_1, l_2, l_3, l_4, respectively. For 3-Dimensional space, many more judgment conditions need to be added. This algorithm is similar to the linear scan conversion in computer graphics. A hardware chip could be designed to perform this process.

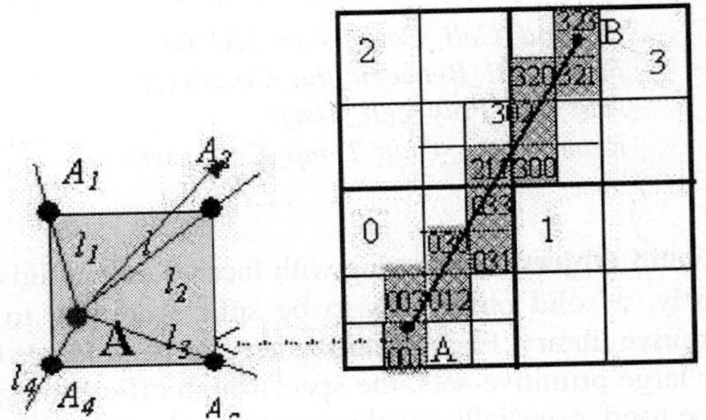

Fig. 2. Interpolation of a line using raster cells

Face object: In 2D plane, a face usually is composed of lines. The interpolation of a face could be processed on the base of linear interpolation algorithms. But in 3D space, a face object is usually very complex. For the convenience of raster interpolation, we only consider to express face objects using regular grid and TIN. Now, the question is the interpolation of spatial rectangles and spatial triangles.

The interpolation algorithm for rectangles could be considered in the way shown in Figure 3. A spatial rectangle has projections on three planes in 3D coordinate system respectively. The

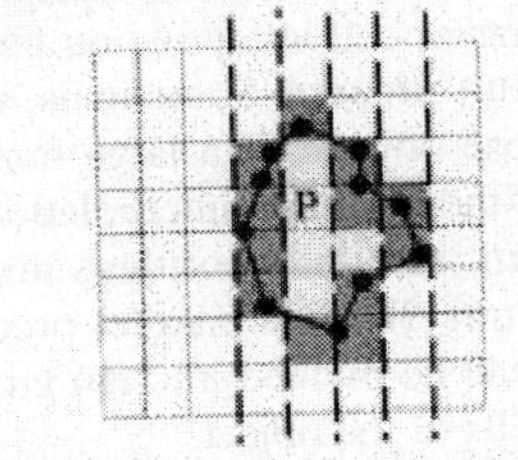

Fig. 3. Raster cell Interpolation of a polygon

projections are parallelograms on planes. For the interpolation algorithm for 2D plane area (see Fig. 3), it is simple to calculate the raster cell sets in three directions of x y and z. From the calculation of their intersections, we could get a raster cell set of the spatial rectangles. Repeating the same way to all spatial rectangles that are used to express the face objects, all raster cells of a face object could be calculated.

To write a 2D face object interpolation algorithm, we assume that one polygon is surround by n lines. Because the polygon on the plane is a close area, there are at least two same X direction Code-Points (or X values) in Y direction. All other cells in Y direction between two cells must belong to the polygon. From above algorithm, we could find out all raster cells of the polygon.

```
Algorithm_VtoR_Poly(P)
  { // Let a polygon is P = {L1, L2,…,Ln};
    // All raster cells are stored in the table "Poly_Cell_List"
    for(i = 0; i < n; i++)
    {
      Algorithm_VtoR_Line(Li);
      Add_Line_Cell_To_Poly_Cell_List( );}
      Temp_cell_List = Poly_cell_List;
   While(! IsEmpty(Temp_Cell_List) )
   {
     if( Find_Cell_Couple( )= TRUE)
     Find_Cell_Between_the_Couple();
     Add_cell(Poly_cell_List);
     Remove_the_Cell( Temp_Cell_List);
   }
  }
```

Solid Object: Comparing with face objects, solid objects are much more complex. Firstly, a solid object has to be split according to the elements in the geometric primitive library. Each primitive need to be processed with corresponding algorithms. For large primitive sets, the special high efficient algorithms for each primitive have to be used, especially for the operation algorithms of these primitives, such as rotate, pan and zoom etc. As our aim is the real-time raster interpolation for solid objects, the algorithms for every geometry primitives could not be too difficult to run in real time.

For geometric primitives expressed analytically, such as cuboids, cylinders and cones, etc., the centerlines or center points could be used for calculation. Frustums of a quadrangle or a polygon may be partitioned into tetrahedrons, and then processed using tetrahedrons. Frustum of a right circular cone could be calculated analytically as well. The raster cell interpolation for irregular and heterogeneous solid is possibly the most difficult process. Sometimes, these are hundreds of tetrahedrons are needed to express those objects. If a large number of such objects have to be processed at the same time, the real-time interpolation becomes much more difficult. In order to solve this problem, parallel algorithms are needed by grouping the tetrahedrons according to their Octree_ID. If a parallel process has ten thread processes, all tetrahedrons of the object can be divided into ten groups. Delete repeat cells in the results, we can get all raster cells of the object.

In order to process raster interpolation much fast, the whole geometric primitives library can be hardware realized, including those interpolation algorithms of each primitives. We call this hardware "raster interpolation card ". The vector solid object can be split using primitives, and can be grouped according to their types. Tetrahedrons have to be grouped separately because there are usually a large number of them. All grouped data will be sent into "raster interpolation card " for processing. There are many calculation cells for different primitives types on cards. These cells are parallel to each other. The repeat raster cells have to be removed from results. The rest of raster cells will be integrated before they are used to express raster solid objects and face objects.

4.4 Integrated Data Model

From above analysis, we could conclude an integrated data model (Figure 4). The data model includes a raster space and a vector space at the same time. Objects are expressed using raster cell sets in the raster space and expressed as same as traditional methods in the vector space. A geometric primitive library was established. Face objects and solid objects are expressed using geometric primitives from the library. Polygon is introduced as a primitive for face objects in 2-Dimensional plane. Point elements are converted from vector to raster space using spatial code-points. But lines are converted using linear interpolations and face and solid objects are converted using the geometric primitive library.

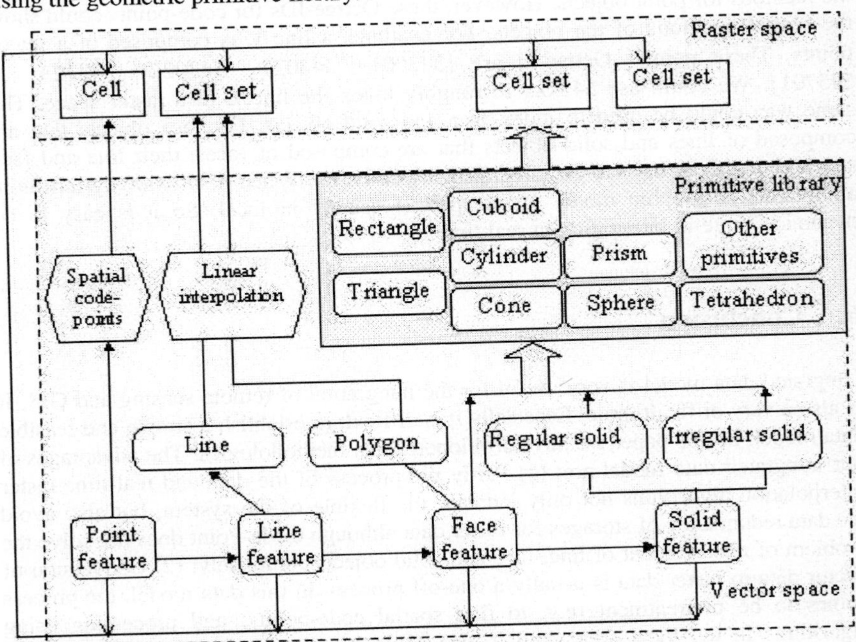

Fig. 4. Diagram of Raster and vector integrated data model

5. Data Structure

The foundations of the integrated data model are Code-Points and raster interpolation algorithms for vector objects. Although we have encapsulated geometric primitives that are used to express face objects and solid objects into a primitive library, the data model does not show object organizations. Hence, we need to discuss the data structure of the data model.

Table structure is the popular data structure in traditional 2-Dimensional GIS. The utilization of Oracle's Spatial Cartage module in GIS realizes the integrated management of vector spatial data and attribute database on RDBMS. Spatial Cartage uses quardtree index to manage spatial objects. In our integrated data model, octree is used to divide 3-dimensional space. Octree-IDs are used to manage vector points. But for other vector elements, such lines, face and solid elements, it is difficult to manage them. We develop a data structure based on list structure to manage those vector elements.

In order to manage point elements, we define a structure *Point_Code {*point, Octree-id}* and use a list to manage it. Structure *Point_Code* is sorted according to code-point values in the list. As it's easy to insert, delete and find in the list, this structure is efficient to manage point objects. For line, face and solid objects, each object may have a large amount of space code-points. It cannot be organized same as the methods for point objects. However, these Octree-IDs for code-points could show the spatial positions of the objects. For example, a line L is composed of a set of points. These points' Octree-id are {543001 543021 543022 543123, …, 543701}. We could use 543xxx to roughly index the line L in a raster space. The same way could be used to index face and solid objects. For face objects that are composed of lines and solid objects that are composed of faces, their line and face space code-points have to be integrated respectively). When a software system is in an interactive graphic mode, visible part is usually in local. So it is easy to do dynamic Octree-id integration in real-time.

6. Conclusion

Integrated data model is very useful for the integration of remote sensing and GIS. It is also a very difficult issue. Especially it is difficult to establish a simple and feasible data model. In this paper we have developed some methodologies. The advantages of our integrated data model are: (1) Partly pre-process of the data and real-time raster interpolation algorithms not only increase the flexible of the system, but also avoid the data redundancy of storages for raster data although Code-Point does not solve the problem of management of line, face and solid objects completely; (2) Conversion of vector data to raster data is usually a one-off process. In this data model, the process comes to be pretreatment (e.g. to find spatial code-points) and processing using algorithms in the local area. And it also could be processed in parallel based on Octree-ids; (3) It could be used to manage spatial objects efficiently although the index based on space code Octree-id is not as easy as octree index. It avoids the infeasibility caused by the excess occupation of RAM space from octree structure itself; (4) The data model keeps the original vector characteristics, such as high precision and fewer amounts of data. And it also does not increase the complex of object expression and data operation. In addition, this integrated data model is compatible to existing vector data models in 3-Dimensional GIS systems.

Topological relationships are not included in this data model. Although expression and procession of spatial topological relationships distinguish GIS from other graphics processing systems especially CAD system, there is no big difference between GIS and CAD when the expression forms turn to be processed by algorithms. The question becomes whether we must express topological relationships. In fact, very limited topological relationships can be expressed in current 2-Dimensional and 3-Dimensional GIS data models. It is obviously a very difficult task if we try to calculate spatial relationships for real spatial objects from several simple topological relationships expressed in data models. Further more, expressions of topological relationships usually limit the flexibility of the data model and introduce the complexity for data updating. We propose that topological relationships should also be calculated in real-time.

Acknowledgments

This publication is an output from the research projects "CAS Hundred Talents Program", "Initial Phase of the Knowledge Innovation Program" and "Digital Earth" funded by Chinese Academy of Sciences (CAS) and is partly supported by a grant from the Chinese National Sciences Foundations (No. 59904001).

References

1. Ballard, D. H.: Strip Trees Hierarchical Representation for Curves, CACM, 24 (1981) 310-321.
2. Gong, J. Y.: 1993, The Research on Integrated Spatial Data Structure. Wuhan Technical University of Surverying and Mapping, PhD thesis, 1993 (In Chinese)
3. Li. R. X.: Data Structure and Application Issues in 3-D Geographical Information Systems. *Geomatics*, 48 (1994) 209-224.
4. Molenaar, M.: A Formal Data Structure For Three Dimensional Vector Maps, Proceedings of the 4th International Symposium on Spatial Data Handling, Zurich. 1990, pp830-843.
5. Pilouk M., Tempfli K., Molenaar M.: A Tetrahedron-Based 3D Vector Data Model for Geoinformation, Advanced Geographic Data Modeling. Netherlands Geodetic Commission, Publications on Geodesy, 40 (1994) 129-140.
6. Randal C., Samet H.: A Consistent Hierarchical Representation for Vector Data, Siggraph, 20 (1986) 197-206.
7. Sun, J.G., and Yang, C.G.: 1997, Computer Graphics (Tsinghua University Press, Beijing) (In Chinese)
8. Tempfle, K.: 3D topographic mapping for urban GIS. *ITC Journal*, 3/4 (1998) 181-190.
9. Wilkinson, G. G.: A review of current issues in the integration of GIS and remote sensing data. International Journal of Geographic Information System, 10 (1996) 85-101.

K-Order Neighbor: the Efficient Implementation Strategy for Restricting Cascaded Update in Realm[1]

Yong Zhang[1], Lizhu Zhou[1], Jun Chen[2], RenLiang Zhao[2]

[1] Department of Computer Science and Technology, Tsinghua University
Beijing, P.R.China, 100084
zhangy97@mails.tsinghua.edu.cn
[2] National Geomatics Center of China
Beijing, P.R.China, 100044
chenjun@nsdi.gov.cn

Abstract. A realm is a planar graph over a finite resolution grid that has been proposed as a means of overcoming problems of numerical robustness and topological correctness in spatial database. One of the main problems of realm is cascaded update. Furthermore, cascaded update causes heavy storage and complex management of transaction. Virtual realm partially resolves the problem of space overhead by computing the portion of realm dynamically. K-order neighbor is a concept commonly used in Delaunary triangulation network. We use K-order neighbor in the Voronoi diagram of realm objects to restrict cascaded update. Two main update operations – point insertion and segment insertion are discussed. In point insertion, the distortion caused by cascaded update is restricted to 1-order neighbor of the point. In segment insertion, two end points of the segment are treated specially. This strategy can be used in both stored realm and virtual realm.

1 Introduction

A realm [3] [4] is a planar graph over a finite resolution grid that has been proposed as a means of overcoming problems of numerical robustness and topological correctness in spatial database. These problems arise from the finite accuracy of number in computer. In realm based spatial database, the intersections between spatial objects are explicitly represented in insertion/update, can be modified slightly if necessary.

One of the main problems of realm is cascaded update [6] [7], that is, the update of some spatial object can modify the value of spatial objects previously stored. Furthermore, cascaded update causes heavy storage and complex transactions.

ROSE algebra [4] [5] is a collection of spatial data types (such as points, lines and regions) and operations. It supports a wide range of operations on spatial data type,

[1] This paper is supported by Natural Science Foundation of China (NSFC) under the grant number 69833010.

P.M.A. Sloot et al. (Eds.): ICCS 2002, LNCS 2331, pp. 994–1003, 2002.

and has been designed in the way that all these operations are closed. In this paper, we call the value of spatial data type as spatial object.

In the implementation in [3] [4], realm is stored explicitly (called as stored realm). Stored realm is organized as a separated layer and has spatial index. Virtual realm [8] stored realm objects within spatial objects. It partially resolves the heavy storage problem. However, both approaches do not resolve the problem of cascaded update.

K-order neighbor [9] is a concept commonly used in Delaunary triangulation network. We use K-order neighbor in Voronoi diagram of realm objects to restrict cascaded update. Two main update operations – point insertion and segment insertion are discussed. In point insertion, the distortion caused by cascaded update is restricted to 1-order neighbor of the point. In segment insertion, two end points of the segment are treated specially. This strategy can be used in both stored realm and virtual realm.

This paper is organized as follows: in section 2, we describe redrawing in data update, and concepts of stored realm and virtual realm; in section 3, K-order neighbor is described; in section 4, we apply K-order neighbor into realm; section 5 gives a comparison; and the last section is conclusion.

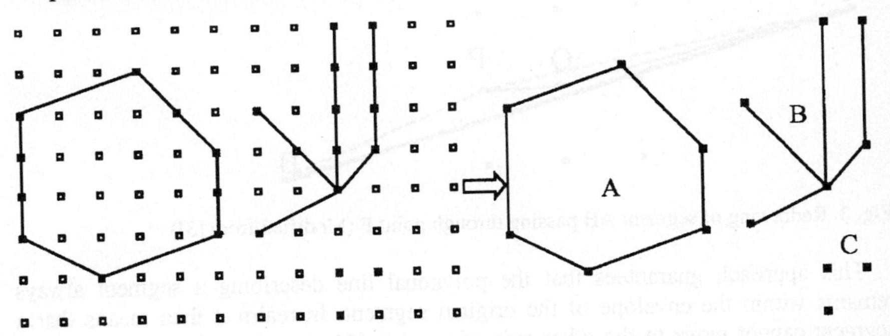

Fig. 1. The example of realm **Fig. 2.** Spatial objects built from realm

2 Basic Concepts in Realm

Fig. 1 shows an example of realm that is a set of points and non-intersecting segments in the finite resolution grid. (Here we call the point in realm as point, and the segment in realm as segment.) In applications, all spatial objects take these points and segments as elements. Fig. 2 shows a regions object A, a lines object B and a points object C, all these objects can be constructed from the points and segments in Fig. 1.

One of the main problems of realm is cascaded update. A point or segment inserted can modify the points and segments in database (to guarantee numerical robustness and topological correctness), that is, data update is cascaded. During this procedure, many segments have to be redrawn. So many segments are generated. At the same time, too many segments needed redrawing make the strategy of locking very complex; therefore it is difficult to manage the transactions.

In the following, we explain redrawing in data update, and the basic concepts of stored realm and virtual realm.

2.1 Redrawing in Data Update

Redrawing [1] of segment is the source of cascaded update. After redrawing, one segment is divided into several segments. The idea is to define for a segment s an envelope E(s) roughly as the collection of points that are immediately above, below, or on s. An intersection point between s and other segment may lead to a requirement that s should pass through some point P on its envelope. This requirement is then fulfilled by redrawing s by some polygonal line within the envelope rather than by simply connecting P with the start and end points of s. Fig. 3 shows that P lies on the envelope of AB; after redrawing segment AB is divided into AQ, QP and PB rather than AP and PB.

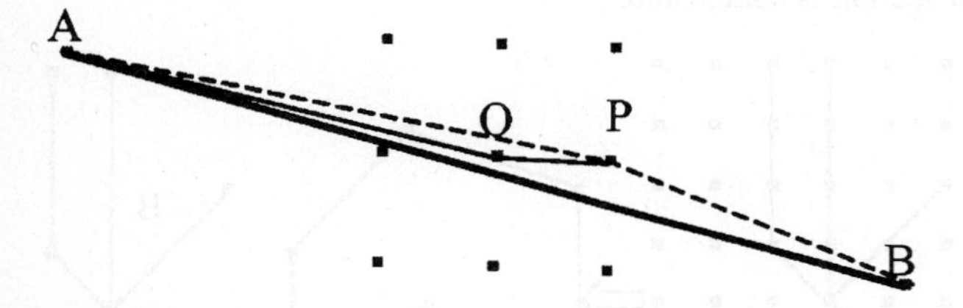

Fig. 3. Redrawing of segment AB passing through point P (Modified from [3])

This approach guarantees that the polygonal line describing a segment always remains within the envelope of the original segment. In realm it then means that a segment cannot move to the other side of a point. [3] extended envelope to "proper envelope" that is the subset of envelope points that are not end points of segments (denoted as $\overline{E}(s)$ for segment s). [3] also added an integrity rule for points and segments that are very close to each other:

No (R-)point lies on the proper envelope of any (R-)segment.

In the worst case, the number of redrawn segments of one segment is logarithmic in the size of the grid. At the same time, if a point or segment is inserted into an area with a high concentration of points and segments, there may be many segments needed redrawing. Therefore, we must decrease the segments needed redrawing as few as possible.

2.2 Stored Realm and Virtual Realm

In stored realm [3] [5], there exist bi-links between spatial objects and realm objects. The realm objects compose a single layer and have spatial index; the spatial objects compose another layer and also have spatial index. The bi-links between realm objects and spatial objects are redundancy, because we can use the pointers in one direction to get the pointers in the contrary direction by traveling.

The links from realm objects to spatial objects are only needed in data update. In stored realm, we firstly operate on the realm layer, and then propagate the changed realm objects to the corresponding spatial objects. There is another implementation approach – virtual realm [8]. In this approach, the realm objects are stored within spatial objects. During data update, we firstly find the spatial objects influenced, and then operate on the set of the realm objects corresponding to these spatial objects. Both approaches can get the same result, but they do not resolve the problem of cascaded update.

3 K-Order Neighbor

K-order neighbor [9] is a concept commonly used in Delaunary triangulation network. Here we define K-order neighbor using Voronoi diagram. Voronoi diagram is the partition of the space based on the neighbor relation.

Spatial neighbor is the degree of the distance between two spatial objects; it is a fuzzy spatial relation. Voronoi diagram provide a clear definition of spatial neighbor [2]: If the Voronoi regions of two objects have the common boundary, then they are defined as spatial neighbor (Fig. 4(a)).

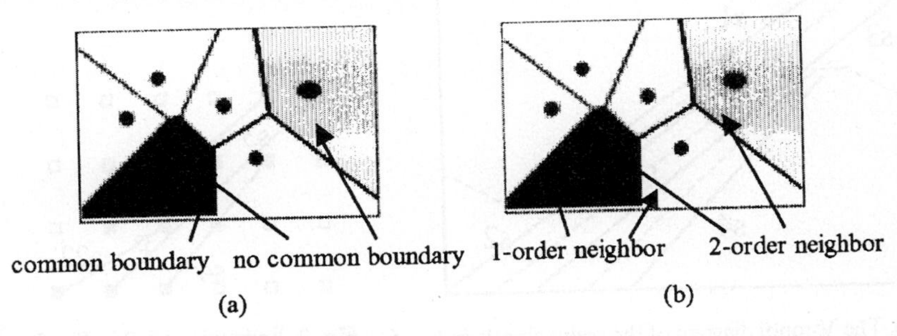

Fig. 4. Spatial neighbor and K-order neighbor

This definition only describes the relation between two objects whose Voronoi regions have common boundary, but does not consider two objects that do not have common Voronoi boundary. K-order neighbor is the extension of spatial neighbor. It is can be used to describe the relation between two objects whose Voronoi regions do not have common boundary. We give the definition using Voronoi diagram (Fig.4(b)):

(1) If the Voronoi regions of P and Q have common boundary, then P is 1-order neighbor of Q, and Q is 1-order neighbor of P.

(2) If the Voronoi region of P has the common boundary with the Voronoi region of one of the 1-order neighbors of Q, and P is not 1-order neighbor of Q, then P is 2-order neighbor of Q.

……

(k) If the Voronoi region of P has the common boundary with the Voronoi region of one of the (k-1)-order neighbors of Q, and P is not (k-1)-order neighbor of Q, then P is k-order neighbor of Q.

4 Application of K-Order Neighbor in Realm

As Fig. 5 Shows, a point is inserted into a realm with a high concentration of segments; point p lies on the proper envelopes of s1, s2, s3 and s4. Using algorithms in [3] and [8], s1-s4 are all needed redrawing. After redrawing, four segments are divided into 10 segments. Moreover, all these segments intersect at the same point, which changes the original topological relations (separated) between them.

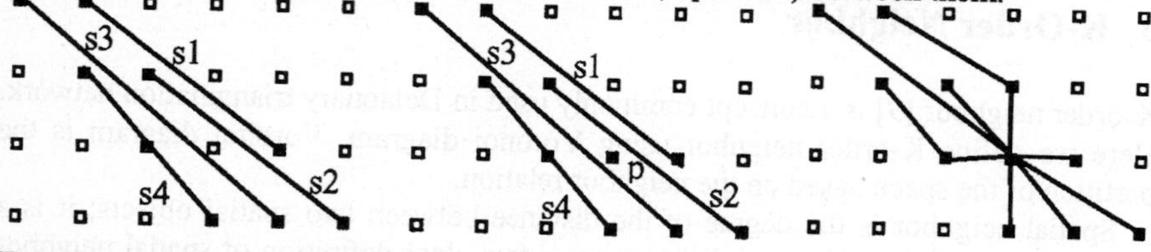

Fig. 5. Insert a point into a realm with a high concentration of segments

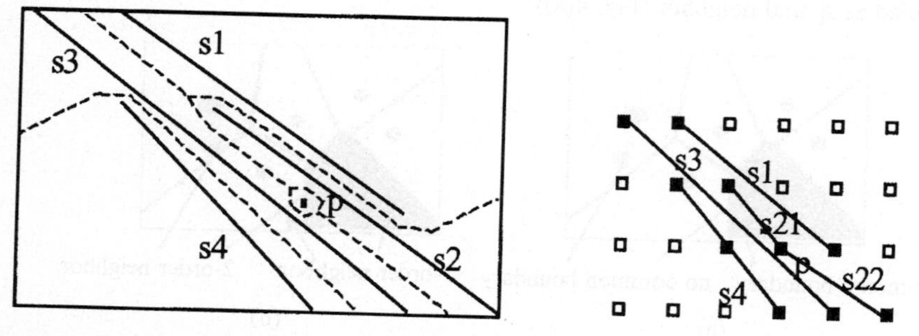

Fig. 6. The Voronoi diagram of the realm objects in Fig. 5 **Fig. 7.** Redrawing of s2 in Fig. 5

Fig. 6 shows the Voronoi diagram of p being inserted into realm. (Notice we use more precision here.) From Fig. 6 we can find that point p is 1-order neighbor of s2 and s3, and 2-order neighbor of s1 and s4. We want to restrict the influence of point p to the scope of its 1-order neighbor, that is, although s1 and s4 are also in its envelope, they are not the "nearest", so we do not need to redraw them.

Let us analyze the point insertion in Fig. 5 step by step. Because point p lies on the proper envelope of s2, we firstly redraw s2, and then s2 is divided into two segments s21 and s22 (As Fig.7 shows). Then we can find that point p has become the end point of s21 and s22, and the proper envelopes of s1, s3 and s4 have also been changed, so point p does not lie on the proper envelopes of them. Hence we do not need to redraw s1, s3 and s4. Therefore, the distortion is restricted to the scope of 1-order neighbor of point p. In practice, we do it as follows: firstly computing the local Voronoi diagram containing point p, and then redrawing the 1-order neighbor segment of p firstly found.

There are three kinds of approaches to get the Voronoi diagram: (1) generate the Voronoi diagram dynamically; (2) store the Voronoi diagram in database; (3) select some not easily changed spatial objects, and store the Voronoi diagram of them in database, and then generate detail Voronoi diagram dynamically. The generating of Voronoi diagram is very slow, and the boundaries generated is in the size of the actual objects, so the first and second approaches are not practical, here we use the third approach.

In the above, what we considered is only the insertion of isolated point. In the algorithms of segment insertion in [3] [8], two end points are firstly inserted into realm as isolated points, and then the segment is inserted. In fact, the insertions of the end points of a segment are different from the insertion of isolated point greatly. As in Fig. 5, if point p is one of end point of a segment to be inserted, then it does not lie on any proper envelope of the segments. Here we must ensure that the insertions of two end points of a segment and the insertion itself are included in a transaction. If the end points of the segment are inserted, but the insertion of the segment fails, then we have to remove the end points from realm and recover realm to the state before insertion.

Overall, using our approach, there are two cases to decrease the segments needed redrawing:

(a) The inserted point lies on the proper envelopes of many envelops, but only one segment needs to be redrawn;

(b) The inserted point is one of the end points of a segment to be inserted, then no segment needs to be redrawn or only one segment needs to be redrawn.

Our approach can be used in both stored realm and virtual realm. In the following, we use stored realm to explain the algorithms in our approach.

4.1 Algorithm of Point Insertion

The algorithm presented in Fig. 8 for inserting a point into a realm is similar to that given in [3]. In point insertion, we have to deal with four cases:

(1) The point is in the realm (line 13).

(2) The point is new, and does not lie on any envelope (line 15).

(3) The point is in some segment; then we separate the segment into two new segments (line 17).

(4) The point lies on one or more proper envelopes (but not in any segment). Here are two conditions:

(a) The point is an end point of a segment to be inserted, and then we do nothing (line 19).

(b) The point is an isolated point, and then we select the segment firstly found in 1-order neighbor of the point to be redrawn (line 20-23).

00 Algorithm: InsertPoint(R, p, flag, R', r, SP)

```
01   Input: R: realm, and R = P ∪ S, P is the point set, and S is the segment set
02          p: point
03          flag: the type of point, 0: isolated point, 1: the end point of some segment
04   Output: R': modified realm
05          r: realm object identifier corresponding to p
06          SP: the set of identifiers of influenced spatial objects
07
08   Step 1: Initialization
09   SP:=∅;
10
11   Step2: Find the segment to be redrawn
12   If ∃q∈P: p=q; (one such point at most)
13   Then r:=roid(q); R' := R; return;
14   Else if ∀s∈S: p∉ E(s) (not exists, and not lie on any proper envelope)
15       Then r:=roid(p); R'=R∪{(p,r,∅)}; return;
16       Else if ∃ s∈S: p in S
17           Then Insert a hook h=<p,p> on s;
18           Else if flag = 1
19               Then r:=roid(p); R'=R∪{(p,r,∅)}; return;
20               Else r:=roid(p);
21                   Generate the Voronoi diagram near p;
22                   Search the 1-order neighbor of p in the Voronoi diagram, s
     is first segment found;
23                   Insert a hook h = <base(p,s),p> on s;
24
25   Step3: Redraw the segment with hook
26   Redraw segment s according to [1];
27   Let {s1, ..., sn} is the set of segments after redrawing, such that si=(qi-1,qi),
     i∈{1,...,n};
28
29   Step 4: Update realm
30   R' := R\{s, roid(s), scids(s)};
31   (Insert the end points and segments in the set of segments of the redrawings, if
     they do not already exist in the realm)
32   for each i in 0..n do
33           if not ExistsPoint(qi) then R' := R' ∪ {( qi, roid(qi), ∅)};
34   for each i in 1..n do
```

```
35              if not ExistsSegment(s_i) then R':= R' ∪ {(s_i, roid(s_i), ∅)};
36      SP := {(sc, {(s_i, roid(s_i))|i∈{1,...,n}}) | sc ∈ scids(s)};
37
38      End InsertPoint
```

Fig. 8. Algorithm of point insertion

4.2 Algorithm of Segment Insertion

```
Begin Transaction
    InsertPoint(R, p, 1, R', r, SP);
    InsertPoint(R, q, 1, R', r, SP);
    InsertSegment(R, s, R', RD, SP) (see [3] for detail, we omit parameter "ok");
Commit Transaction
```

Fig. 9. The algorithm of segment insertion

Our algorithm of segment insertion is similar to those in stored realm and virtual realm. The process of inserting a segment s = (p, q) requires three steps: (1) insert p into realm; (2) insert q into realm; (3) insert s into realm. Because in point insertion, we distinguish if a point is an end point of a segment to be inserted, these three steps must be completed in one transaction (Fig. 9). Otherwise, the integrity rule of proper envelope will be violated.

5 Performance Analysis

Taking stored realm as an example, this section presents an informal comparison of the performance of point insertions using K-order neighbor and not using it.

If the distribution of spatial objects is dense, the advantage of our approach is very clear: the segments needed redrawing decrease greatly. Therefore, fewer segments are generated after redrawings. At the same time, the restriction of distortion provides a good foundation for the simplicity of transaction management.

Table 1 summarizes that tasks while inserting a point using K-order neighbor and not. The column of difference indicates whether K-order neighbor increase (+) or decrease (-) the cost of performing input-output (I/O) or processing (CPU) tasks.

Table 1. The influence of using K-order neighbor on point insertion in stored realm (modified from [8])

Stored realm	Stored realm using K-order neighbor	Difference
Retrieve required nodes of spatial index (many entries)	Retrieve required nodes of spatial index (many entries)	-I/O, -CPU

Retrieve segments and points from MBR of inserted segment (possible many)	Retrieve segments and points from MBR of inserted segment (possible many)	-I/O
	Retrieve the corresponding part of basic Voronoi diagram	+I/O
	Compute the local detail Voronoi diagram	+CPU
	Compute the modification of Voronoi diagram	+CPU
	Use Voronoi diagram to search the neighbors	+CPU
Compute changes in the realm	Compute changes in the realm	-CPU
Delete changed segments from disk	Delete changed segments from disk	- I/O
Write new segments into disk	Write new segments into disk	- I/O
Compute changes to spatial index	Compute changes to spatial index	-CPU
Write changed index to disk	Write changed index to disk	- I/O
	Write changed basic Voronoi diagram to disk (maybe not necessary)	+I/O
Retrieve the spatial objects related to the changed segments	Retrieve the spatial objects related to the changed segments	- I/O
Replace the changed segments in spatial objects	Replace the changed segments in spatial objects	- I/O
Write changed spatial objects to disk	Write changed spatial objects to disk	- I/O
	Delete the local detail Voronoi diagram	+CPU

The results presented in the table can be summarized with respect to the effect on I/O costs and CPU time:

I/O: (a) There are two factors to increase I/O activities, the reason is that we store a basic Voronoi diagram in database; (b) There are eight factors to decrease I/O activities, the reason is that the total number of points and segments in database is fewer.

CPU: (a) There are four factors to increase CPU time, the reason is that the processes related Voronoi diagram are added; (b) There are three factors to decrease CPU time, the reason is that the total number of points and segments in database is fewer.

Overall, there are two main reasons that influence I/O costs and CPU time: Voronoi diagram and the realm objects in database. It is hard to say that because of using K-order neighbor, I/O costs and CPU time are saved. The saving of I/O costs and CPU times depends on the distribution of spatial objects (i.e., the applications). However, we can conclude that using K-order neighbor decreases the realm objects and simplifies the transaction management. Furthermore, the stored Voronoi diagram can speed many ROSE algebra operations such as *inside, intersection, closest* and so on.

6 Conclusion

This work is being carried out in the context of a project of 3D spatial database based on realm. We find that in some applications (especially the distribution of segments is dense), in spite of using stored realm or virtual realm [3] [8], point insertions and segment insertions bring out boring cascaded update. In these conditions, many segments have to be redrawn, which result many segments to occupy large storage space and make the management of transactions very complex. So we want to find an approach to restrict cascaded update.

K-order neighbor [9] is a concept commonly used in Delaunary triangulation network. We use Voronoi diagram to describe this concept. K-order neighbor restricts cascaded update efficiently. Presently we have implemented the algorithm of K-order neighbor and used it in the data update of realm.

References

1. Greene, D., Yao, F.: Finite-Resolution Computational Geometry. Proc. 27th IEEE Symp. on Foundations of Computer Science (1986) 143-152
2. Gold, C.M.: The Meaning of "Neighbour". In Frank, A. U., Campari, I. And formentini, U. (Eds) Theories of Spatial -Temporal Reasoning in Geographic Space. Lecture Notes in Computer Science 639, Berlin: Springer-Verlag (1992) 220-235
3. Guting, R.H., Schneider, M.: Realms: A Foundation for Spatial Data Type in Database Systems. In D. Abel and B.C. Ooi, editors, Proc. 3rd Int. Conf. on Large Spatial Databases (SSD), Lecture Notes in Computer Science, Springer Verlag, (1993) 14-35
4. Guting, R.H., Ridder, T., Schneider, M.: Implementation of the ROSE Algebra: Efficient Algorithms for Realm-Based Spatial Data Type. 4th Int. Symp. on Advances in Spatial Databases (SSD), LNCS951, Springer Verlag (1995) 216-239
5. Guting, R.H., Schneider, M.: Realm-Based Spatial Data Types: The Rose Algebra. VLDB Journal, Vol.4 (1995) 100-143
6. Cotelo Lema, J. A., Guting, R.H.: Dual Grid: A New Approach for Robust Spatial Algebra Implementation. FernUniversity Hagen, Informatik-Report 268 (2000)
7. Cotelo Lema, J. A.: An Analysis of Consistency Properties in Existing Spatial and Spatiotemporal Data Models. Advances in Databases and Information Systems, 5th East – European Conference ADBIS' 2001, Research Communications, A. Caplinskas, J. Eder (Eds.): Vol. 1 (2001)
8. Muller, V., Paton, N.W., Fernandesyy, A.A.A., Dinn, A., Williams, M.H.: Virtual Realms: An Efficient Implementation Strategy for Finite Resolution Spatial Data Types, In 7th International Symposium on Spatial Data Handling - SDH'96, Amsterdam (1996)
9. Zhang, C.P., Murayama, Y.J.: Testing local spatial autocorrelation using k-order neighbours. Int. J. Geographical Information Science, Vol.14, No.7, (2000) 681-692

A Hierarchical Raster Method for Computing Voronoi Diagrams Based on Quadtrees

Renliang ZHAO [1,3], Zhilin LI [2], Jun CHEN [3], C.M. Gold[2] and Yong ZHANG[4]

[1]Department of Cartography and GIS, Central South University, Changsha, China, 410083
{ zhaorl@sina.com }
[2]Department of Land Surveying and Geo-Informatics
The Hong Kong Polytechnic University, Kowloon, Hong Kong
{ lszlli@polyu.edu.hk }
[3]National Geometrics Center of China,
No. 1 Baishengcun, Zizhuyuan, Beijing, China, 100044
{ chenjun@nsdi.gov.cn }
[4]Department of Computer Science and Technology, Tsinghua University ,Beijing, 100084

Abstract. Voronoi diagram is a basic data structure in geometry. It has been increasingly attracting the investigation into diverse applications since it was introduced into GIS field. Most current methods for computing Voronoi diagrams are implemented in vector mode. However, the vector-based methods are good only for points and difficult for complex objects. At the same time, most current raster methods are implemented only in a uniformed-grid raster mode. There is a lack of hierarchical method implemented in a hierarchical space such as quadtrees. In this paper such a hierarchical method is described for computing generalized Voronoi diagrams by means of hierarchical distance transform and hierarchical morphological operators based on the quadtree structure. Three different solutions are described and illustrated with experiments for different applications. Furthermore, the errors caused by this method are analyzed and are reduced by constructing the dynamical hierarchical distance structure elements.

1 Introduction

A Voronoi diagram is one of fundamental geometric structure, actually describes the spatial influent region for each a generator and each point in the influent region associated with a generator is closer to the generator than the others [1], [14], seen as Fig. 1. Voronoi diagrams have been applied widely in various areas since they were originally used to estimate regional rainfall averages in 1911 [1], [20]. In GIS area, Voronoi diagrams are also taken as one useful tool and have been increasingly attracting the investigation into diverse applications of Voronoi methods [3],[5],[6], [8],[10], [15], [18], [25].

Generally, Voronoi diagrams can be implemented in vector space and also in raster space. But most current methods are vector-based, for example, the classic divide and conquer method, incremental method, sweepline method [1],[9],[15] [20].

P.M.A. Sloot et al. (Eds.): ICCS 2002, LNCS 2331, pp. 1004–1013, 2002.

However, such vector-based methods are good only for discrete spatial points, difficult for complex objects such as line and area objects.

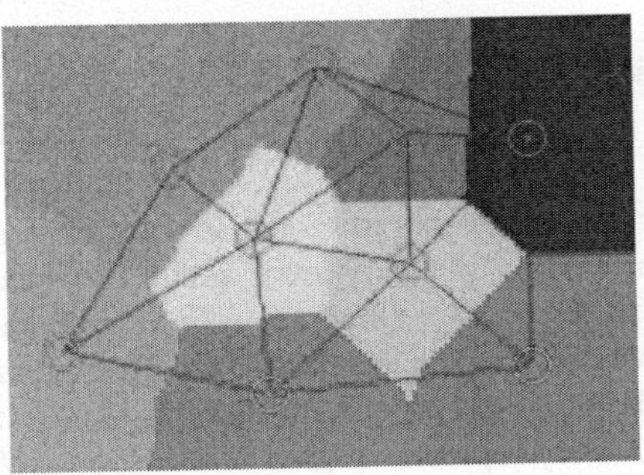

Fig. 1. A Voronoi diagram for points

At the same time, it has been found that Voronoi diagrams can be implemented very well in a raster space. Compared with vector-based methods, raster methods are performed faster and more simply [2],[13],[15],[20]. However, these methods are implemented only in a uniformed grid space. Compared with a uniformed grid structure, a hierarchical structure of space like a quadtree often occupies less space and costs less execution time [23]. In fact, hierarchical methods like that of quadtree have been popular and proved efficient in many areas of GIS including spatial modeling, spatial query, spatial analysis, spatial reasoning and so on[7],[21].

But so far there are a few efforts related to the hierarchical implementation of Voronoi diagrams, despite the fact that the quadtree data structure was used very early in the implementation of Voronoi diagrams. For instance, in Ohya's method, quadtrees are actually only used as an index of buckets to reduce the time of the initial guess of inserted new generator points [17],[19]. However, this method works only for points in a vector space.

In another research closely related to the computation of Voronoi diagrams based on framed-quadtrees, Chen et.al. (1997) adopted a modified quadtree structure to compute Euclidean shortest paths of robotics in raster (or image) space [4]. But the "quadtree structure" there is not the ordinary quadtree, it must be modified into the framed-quadtree whose each leaf node has a border array of square cells with the size of smallest cell before the method can work.

In this paper, a hierarchical method is presented for implementing Voronoi diagrams, directly based on the ordinary quadtree structure as a very good structure representing multi-resolution data or for space saving. In this method, generalized Voronoi diagrams are performed with hierarchical distance transform and mathematical morphological operators in a raster space represented with the standard qudatrees. In the following Section 2, some related operators for the hierarchical computation of quadtrees are introduced. In Section 3 Voronoi diagrams are implemented hierarchically based on these operators, and three solutions are described and illustrated with experiments for different applications. In Section 4 the errors caused by this method are analyzed and a corresponding improvement solution is given by means of constructing the dynamical hierarchical distance structure elements. Conclusions are given in Section 5.

2 Quadtrees and Hierarchical Distance Computation

As well known, quadtree is one of popular data structure representing spatial data in a variable multi-resolution way in many areas related to space such as GIS, image processing and computer graphics. Quadtree is a hierachical data structure and is based on the successive subdivision of space into four equal-sized quadrants[22], seen as Fig. 2. In such a quadtree, there are three types of nodes: black nodes, grey nodes and white nodes. They represent region of one spatial object, mixed regions of two or more spatial objects and free space respectively. Especially, if a quadtree only records its leaf nodes according to a certain location code, such quadtree is called linear quadtree, while the above quadtree is called regular quadtrees. Different kinds of quadtrees could be suitable for different applications.

The reduction in storage and execution time is derived from the fact that for many images, the number of nodes in their quadtree representation is significantly less than the number of pixels. In terms of the definition of Voronoi diagrams, the formation of Voronoi diagrams is based on the distance. So the most important operation is the computation and propagation of distance in a hierarchical space. This operation can be accomplished with a local method or global method.

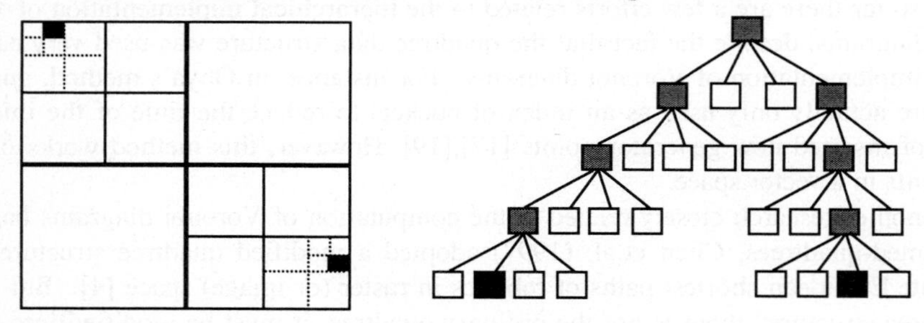

Fig. 2. A regular quadtree

The local method often uses small masks (for instance, the mask consisting of neighbouring pixels with 3 columns and 3 rows) to spread distance and obtain global distance, where the small masks are often associated with definitions of local distance such as city block and Chamfer distance. This is actually the distance transform using small neighbourhoods in image processing [2]. The high efficiency is one of main reasons why this method is popular in image processing. But it will be also costly if the propagation of distance operation covers all pixels of each level. The local method can be straightly generalized to linear quadtrees, shown in Fig. 3(a). The main difference is the distance transform is measured in multiples of the minimum sized quadrant.

The global method is different the local method. In the global method, the global distance can be directly calculated with an equation of distance in a coordinate system in which each pixel is assigned to a coordinate such as the number of row and column. For instance, the Euclidean distance D between two pixels $P_1(i_1,j_1)$ and $P_2(i_2,j_2)$ can be calculated with the following equation.

$$D(P_1, P_2) = \sqrt{(i_2 - i_1)^2 + (j_2 - j_1)^2} \tag{1}$$

The global method can get the distance between any pixels in the whole space more accurately and flexibly than the local method, but the cost will be very high while all of distances between any pixels need.

Therefore, in order to keep the balance between efficiency and precision, it is an alternative choice to integrate the local method and global method, i.e., the distance operation can be accomplished by combining the direct calculation of global distance and the distance transform with small masks. In this mixed method, one can firstly get the global distance between some connected pixels and a spatial object, and then make distance transform outside the spatial object at the beginning of these connected pixels with the global distance. (Seen in Fig. 3(b)) Derived from the above procedure, the efficiency and precision of this integrated method are between the local method and global method for distance calculation.

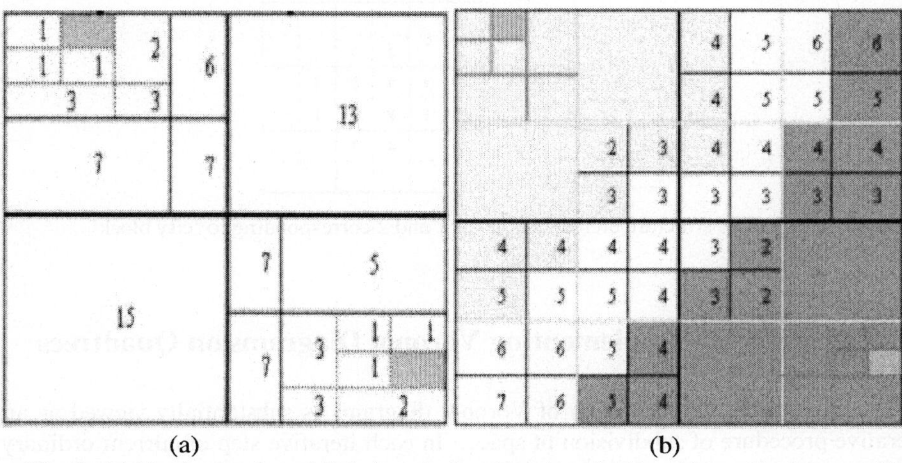

(a) (b)

Fig. 3. (a) The integration of global method and local method in a hierarchical space; (b) Distance transform applied to linear quadtrees

The mixed method can play a better role in the distance calculation of the hierarchical space since it is more flexible than ordinary distance transform. As known, it is derived that some regions of a certain level do not have to continue to be processed in next levels in a hierarchical space. In this case, the flexibility of mixed method makes it possible to omit the distance calculation over the regions unnecessary to be processed. In Fig. 3(b), for instance, it is supposed that the white regions are unnecessary to be continued to process, others are to be continued. With the mixed method, only a part of pixels (with bold lines in Fig. 3(b)) are involved in the global distance calculation, the following distance propagation begins with these pixels with the global distance as seen in Fig. 3(b).

In addition to the above approximate method for the distance calculation on linear quadtrees, such distance calculation between free pixels and spatial objects can be also obtained using the dilation operator in mathematical morphology. The dilation operator is one of two basic operators in mathematical morphology, represented as follows [24]:

$$A \oplus B = \cup_{b \in B} A_b \tag{2}$$

Where A is the original space, and B is the structure element.

The morphological dilation operator has been not only successfully applied to the distance calculation in a uniform grid space [13],[15], but also used for the distance calculation in a hierarchical structure such as the pyramid [16]. Here, it is attempted to implement a new method for the distance calculation using morphological dilation operator in regular quadtrees or linear quadtrees, called hierarchical morphological methods based on quadtrees. The key of a hierarchical dilation operator is substantially made up of the sequence of hierarchical structure elements in quadtrees corresponding to the ordinary structure element. Each hierarchical structure element belongs to a level of quadrees. Fig. 4 shows the structure elements at level 1 and 2 corresponding to 'city block'.

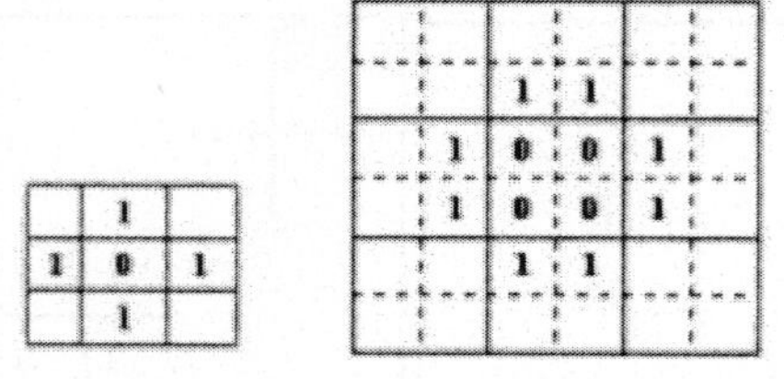

Fig. 4. The structure elements at level 1 and 2 corresponding to 'city block'

3 Hierarchically Implementing Voronoi Diagrams on Quadtrees

In raster space, the formation of Voronoi diagrams is substantially viewed as an iterative procedure of subdivision of space. In each iterative step of current ordinary raster methods, each pixel will be determined to belong a certain spatial object according to the distance calculation values. However, it is unnecessary to determine all pixels because possible changed pixels are only those pixels located in boundaries between regions of different spatial objects or between free regions and regions of spatial objects. The application of quadtrees can avoid those redundant operations and finding those possible changed pixels efficiently. The quadtree based hierarchical implementation of Voronoi diagrams is actually also the procedure of continuous subdivision and reconstruction of the quadtree. Due to different ways of hierarchical processing, the hierarchical method for Voronoi diagrams is also implemented in different routines. Different routine of the implementation will use different distance calculation.

3.1 Mixed Distance Calculation Routine

For regular quadtrees, it is a good way to implement Voronoi diagrams from top to bottom and level by level. In this way, the distance calculation can be accomplished with the mixed method in Section 2. The hierarchical implementation can consist of the following steps.

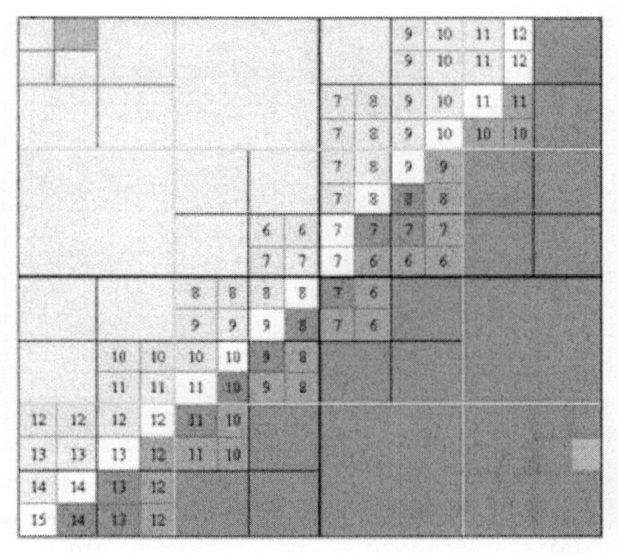

Fig. 5. The hierarchical implementation for Voronoi diagrams using the mixed distance calculation based routine (Total pixels for distance transform 16+40+84=140, but total pixel will be 256 using non hierarchical methods)

(1) Use global distance calculation method to compute the distance between white nodes and black nodes or grey nodes at the third highest level of the quadtree, then assign those white nodes to the corresponding black nodes or grey nodes
(2) Search for those nodes belonging to each black node or grey node including only one spatial object and not adjacent to other nodes belonging other spatial objects, label them as unused nodes because they are unnecessary to continue to be processed, and label those nodes adjacent to other nodes as boundary nodes;
(3) Subdivide the grey nodes and boundary nodes into less nodes at the next lower level, computing the global distance between less boundary nodes and black nodes or grey nodes including only one spatial objects;
(4) Perform the distance transform using the local method for distance calculation in the regions except nodes labeling 'unused', assign those white nodes to the corresponding black nodes or grey nodes;
(5) Repeat step (2)-(4) until the bottom level.

Fig. 5 gives an example of the above routine.

3.2 Linear Quadtrees Routine

While the raster space is organized into linear quadtrees, it will be very efficient to perform the hierarchical computation of Voronoi diagrams in a way of mixed levels. The feature of this routine is that local distance transform is straightly applied to all levels. Based on the technique, the hierarchical implementation for Voronoi diagrams is described as follows.

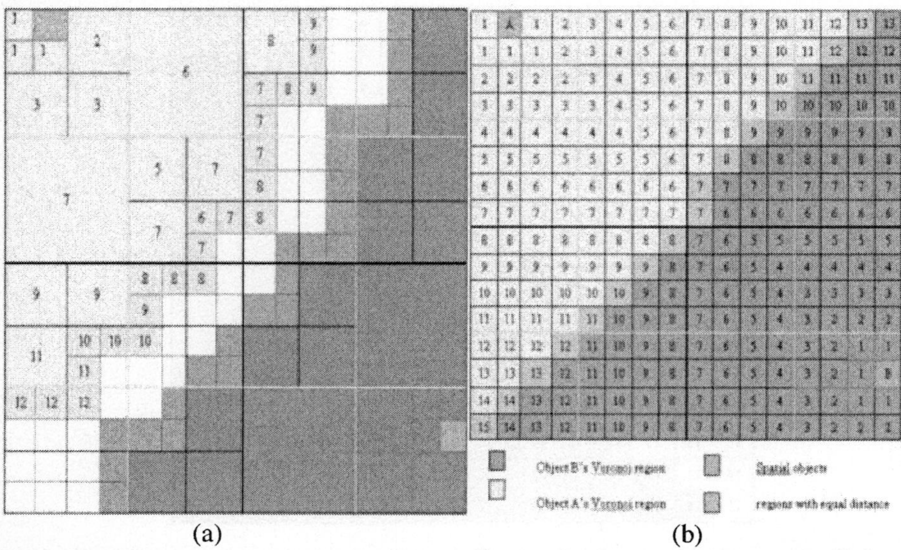

Fig. 6. (a) The hierarchical implementation for Voronoi diagrams using linear quadtrees based routine, (b) Hierarchical morphological method for Voronoi diagrams

(1) Perform the distance transform on linear quadtrees, make each white nodes associated with a certain spatial object;
(2) Search for those white nodes which belong to each black node and are not adjacent to other white nodes belonging other spatial objects, label them as unused nodes because they are unnecessary to continue to be processed, and label those nodes adjacent to other nodes as boundary nodes;
(3) Subdivide the larger boundary nodes into less nodes at the next lower level,
(4) Perform the distance transform in the regions except nodes labeling 'unused', make each white nodes associated with a certain spatial object;
(5) Repeat (2)-(3) till no white node need subdivision.

An example using this routine is shown in Fig. 6(a).

3.3 Morphological Routine

When original raster data are represented in the uniform grid format, it is necessary to reorganize them into quadtrees firstly for hierarchically computing Voronoi diagrams. At the same time, the hierarchical distance calculation based on morphological method is actually performed at the bottom level of qudtree i.e., the original uniform grids. So in this case, it is suitable for implementing the Voronoi diagrams hierarchically using morphological dilation operator from the each higher level to the bottom level. It can be represented as follows.

(1) Renew to organize the original raster data into a quadtree and construct a sequence of hierarchical structure elements corresponding to the definition of distance in raster space;
(2) For each black node, perform the hierarchical dilation operation stated in the previous section;
(3) Make the distance transform only those pixels dilated by hierarchical operator, assign corresponding spatial object to each pixel of those;
(4) Performing the union operation of quadtrees, merger those dilated pixels and update the current quadtree;
(5) Repeat (2)-(4) till no distance value distance.

Fig. 6(b) illustrates the result of this procedure.

4 Comparison and Improvement

The hierarchical implementation of Voronoi diagrams can be realized in three routines. The first routine based on mixed distance calculation method can be suitable for regular quadtrees. The second routine directly using distance transform on linear quadtrees is more suitable for raster data in the format of linear quadtrees. The third routine based on hierarchical morphological operators can be performed in uniformed grid space with the combination of linear quadtrees. More important, the three routines are also different in the efficiency and precision.

From the viewpoint of efficiency, the best routine is the second routine because the distance transform is directly implemented on all levels and it cost less time to get an appropriate distance of all leaf nodes. The time cost time of the third routine is more than the others since it involves many the operation of union quadtrees in each iterative procedure of the implementation of Voronoi diagrams.

In precision, it is derived that the third routine is the best because all operations are actually performed at the bottom level and the result is the same as that of the implementation of non hierarchical methods. In the other two routines, the distance calculation is performed at various levels and causes more error than the third one. However, as pointed out in the literature [15], ordinary distance transforms increase error varying with the growth of distance. This results in larger error of Voronoi diagrams via ordinary distance transforms. In order to improve the precision, an dynamical distance transform hierarchical method is presented here by constructing a set of hierarchical structure of elements close to a circle.

The key hierarchical structure of elements can be constructed with the method introduced in Section 2, but the difference is that each structure element corresponding to a level here must consist of a set of elements whose combined effect of dilation is required to be a region very close a circle. Applying dynamically these hierarchical dilation operators, the third routine is improved.

5 Conclusions

A Voronoi diagram is a geometric subdivision of space that identifies the regions closet to each of its generator sites. Voronoi diagrams have been regarded as an alternative data model and the researches on the applications of Voronoi diagrams have been attracted increasingly in GIS area[5],[8],[9],[11],[12],[15],[25].

The quadtree is very good data structure representing a hierachical or multiscale spatial data. A quadtree based method for the hierarchical computation of Voronoi diagrams in raster space is described in this paper. It can be implemented in three different routines in order to adopt different conditions. Three approaches are different in efficiency and precision. The morphological routine is the best in the aspect of precision, and linear quadtree based routine is more efficient that others in general cases. One can select a suitable routine in his practical needs.

Acknowledgement. The work was substantially supported by a grant from the Research Grants Council of the Hong Kong Special Administrative Region (Project No. PolyU 5048/98E) and partially supported by a grant from the National Natural Science Foundation (No. 40025101).

References

1. Aurenhammer, Franz, 1991, Voronoi diagrams—A survey of a fundamental geometric data structure. *ACM Computing Surveys*, 23 (3), 345-405.
2. Borgefors, G., 1986, Distance transformations in digital images. Computer Vision, Graphics and Image Processing, 34, 344-371
3. Chakroun, H.; Benie, G.B; O'Neill, N. T., Desilets, J., 2000, Spatial analysis weighting algorithm using Voronoi diagrams. International Journal of geographical Information Science.14(4), 319-336
4. Chen D. Z., Szczerba R. J. and Uhran, J., 1997, A framed-quadtree approach for determining Euclidean shortest paths in a 2D environment. IEEE Transaction on Robotics and Automation, vol.13, pp668-681
5. Chen, Jun, Li, C., Li, Z. and Gold, C., 2001, A Voronoi-based 9-intersection model for spatial relations. International Journal of geographical Information Science, 15(3): 201-220
6. Chen, Jun, Zhao, R.L., and Li., Zhi-Lin, 2000, Describing spatial relations with a Voronoi diagram-based Nine Intersection model. In: SDH'2000, Forer, P., Yeh, A. G. O. and He, J. (eds.), pp4a.4-4a.14

7. David, M, Lauzon, J. P. and Cebrian, J. A., 1989, A review of qudtree-based strategies for interfacing coverage data with digital elevation models in grid form. Int. J. Geographical Information Systems, 3(1): 3-14.
8 .Edwards ,Geoffrey, 1993, The Voronoi Model and Cultural Space: Applications to the Social Sciences and Humanities In: Spatial information theory : a theoretical basis for GIS : European Conference, COSIT'93, Marciana Marina, Elba Island, Italy, September 19-22, Berlin ; New York : Springer-Verlag, Lecture Notes in Computer Science 716, pp202-214
9 .Gahegan, M. and Lee, I., 2000, Data structures and algorithms to support interactive spatial analysis using dynamic Voronoi diagrams, Computers, Environment and Urban Systems, 24: 509-537
10.Gold, C. M., 1994a, a review of the potential applications for Voronoi methods in Geomatics. In: the proceeding of the Canadian Conference of GIS, Ottawa,1647-1652
11.Gold, C.M., 1994b, Advantages of the Voronoi spatial model. In: Frederiksen, P. (ed.). Proceedings, Eurocarto XII; Copenhagen, Denmark, 1994. pp1-10.
12.Gold, C.M.; Remmele, P.R. and Roos, T., 1997, Voronoi methods in GIS. In: Van Kreveld, M., Nievergeld, J., Roos, T. and Widmeyer, P. (eds.), "Algorithmic Foundations of GIS. Lecture Notes in Computer Science No. 1340", Springer-Verlag, Berlin, Germany, pp. 21-35.
13.Kotroplulos, C., Pitas, I. and Maglara, M., 1993, Voornoi tessellation and Delauney triangulation using Euclidean disk growing in Z^2. IEEE, V29-V32.
14.Lee, D. T. and Drysdale, R.L., 1981, Generalization of Voronoi Diagram in the plane. SIAM Journal of Computing, **10**, 73-87.
15.Li, C., Chen, J. and Li, Z. L., 1999, Raster-based methods or the generation of Voronoi diagrams for spatial entities. International Journal of Geographical Information Science, **13** (3), 209-225.
16.Liang, E. and Lin, S.1998, A hierarchical approach to distance calculation using the spread function, International Journal of Geographical Information Science, 12(6), 515-535
17.Marston, R.E.; Shih, J.C. 1995, Modified quaternary tree bucketing for Voronoi diagrams with multi-scale generators Multiresolution Modelling and Analysis in Image Processing and Computer Vision, IEE Colloquium on , 1995 Page(s): 11/1 -11/6
18.Mioc, D.; Anton, F.; Gold, C.M. and Moulin, B., 1998, Spatio-temperal change representation and map updates in a dynamic Voronoi data structure. In: Poiker, T.K. and Chrisman, N.R. (eds.). Proceedings, 8th International Symposium on Spatial Data Handling; Vancouver, BC, 1998. pp. 441-452.
19.Ohya, T., Iri, M. and Murota, 1984, A fast Voronoi diagram with quadternay tree bucketing. Information Processing Letters, Vol. 18, pp. 227-231
20.Okabe, A., Boots, B. and Sugihara, K., 1992, Spatial Tessellations: Concepts and Applications of Voronoi diagrams, Chichester, England, New York, Wiley & Sons.
21.Papadias, D. and Theodoridis Y., 1997, Spatial relations, minimum bounding rectangles, and spatial data structures. International Journal of Geographical Information Science, 11(2), pp111-138
22.Samet, H., 1982, Neighbor finding techniques for images represented by quadtrees. Computer Graphics and Image Processing, Vol.18, pp.37-57.
23.Samet, H., 1989, Design and analysis of spatial data structures: quadtrees, octress, and other hierarchical methods, Reading MA.
24.Serra, Jean Paul., 1982, Image analysis and mathematical morphology London ; New York : Academic Press.
25.Wright, D.J., and Goodchild., M. F., 1997, Data from the deep: implications for GIS community. International Journal of Geographical Information Science, **11**, 523-528

The Dissection of Three-Dimensional Geographic Information Systems

Yong Xue[1,2], Min Sun[3], Yong Zhang[4] and RenLiang Zhao[5]

[1]LARSIS, Institute of Remote Sensing Applications, Chinese Academy of Sciences, Beijing, 100101, PR China
[2]School of Informatics and Multimedia Technology, University of North London, 166-220 Holloway Road, London N7 8DB, UK
{y.xue@unl.ac.uk}
[3]Institute of RS & GIS, Peking University, Beijing, 100871, PR China
[4]Institute of Computer technology, Tsinghua University, Beijing, 100871, PR China
[5]Institute of Surveying and Mapping, Central South University, Changsha, 410083, PR China

Abstract. In this paper we dissected the 3-dimensional Geographic Information Systems (3D GISs) and disambiguated some points such as (1) 3D GIS should be a component of GIS, it's a subsystem; (2) data modelling is not the main obstacle to its development; (3) it's no necessary and also very difficult for 3D GIS to replace GIS; (4) the main developing direction of 3D GIS should be special subsystems, that is, 3D GIS research must be based on relative application areas.

1. Introduction

We are living in a true 3-dimensional (3D) world. The environment and almost all artificial objects have 3-demensional sizes. However, as so far, it's difficult to establish a uniform data model to represent the environment and the objects. In GIS, objects are represented using its projections on 2D plane. Although this kind of representation could solve many problems, it is still very difficult to represent and process the geometry and attributes for third spatial dimension. For example, GIS can't represent real 3-dimensional nature and artificial phenomena in geosciences area, such as in geology, mining, petroleum, etc. Many research works about 3D GIS have been carried out in order to remedy such defects in solving true 3-dimensional problems. It has been found that it is almost impossible to establish a 3D GIS that has similar functions as those in current GISs, especially for the processing of complex 3-dimensional spatial relationships.

Data modelling of 3D objects is one of the most difficult problems. As data model is the fundament to the establishment of GIS, research works on 3D GIS have been mainly focused on data model. So far, there isn't a perfect data model found. The great difficulties for 3D GIS research could be summarized in three points:

P.M.A. Sloot et al. (Eds.): ICCS 2002, LNCS 2331, pp. 1014–1023, 2002.

1. Try to find a completely new 3-dimensional data model to substitute GIS vector data model. Third dimensional value Z not only changes from attribute data to spatial data, i.e.: (X, Y): Z → (X, Y, Z) but also a new volume element is added in addition to three elements such as point, line and face;
2. Try to represent various 3-dimensional objects in one data model;
3. Try to represent objects, especially complex real objects, in high precision.

Most research works on data modelling have been to establish a common 3-dimensional data model based on 3D GIS as a common platform for real world complex objects, including geology, mining, petroleum, cataclysm, ocean, atmosphere, and urban environments. However, no significant progresses have been made so far. In this paper, the concept and applications of 3D GIS will be addressed first. Data modelling will be studied and the further development of 3D GIS will be discussed at the end.

2. 3D GIS CONCEPT AND APPLICATIONS

2.1 Applications of 3D GIS

We could find GIS applications almost in any areas that relate to spatial data. Object scales might vary from 1:500 to 1:1,000,000, even larger or smaller. That's in such scale areas, we could find actual objects represented by GIS data model, e.g.: in 1:500 scale a polygon might represent a house, while in 1:1,000,000 scale and a polygon might represent a city area. But in 3D GIS, the case changes a lot. We could analyse this question from GIS data models.

GIS is mainly established on vector data model. In this model, line is composed of points and face is composed of lines.

- Point element $P(x, y, z)$ may represent any single object occupying one spatial point location, and also an object occupying 2-dimensional spatial areas, e.g.: streetlight, road crossing, station or even a whole city;
- Line element $L(P_1, P_2, P_3, \ldots)$ may represent any object that occupies 1-dimensional or 2-dimensional space, e.g.: pipelines, roads and rivers;
- Face element $S(L_1, L_2, L_3, \ldots)$ may represent any object that occupies 2-dimensional space, from single position to small area or even a very large space, e.g.: house, square, city and whole country terrain.

Therefore, GIS vector data model could represent random size objects in plane, and could be used in many spatial related applications. In 3D GIS, space is extended from 2D to 3D. Volume element has to be introduced in both vector model and raster model [14], [15], [16], [20].

If we consider volume element represented by components of point, line and face elements, i.e.: $V(R(P, L, S))$, then in theory, volume V could be used to represent any objects in Euclid space with random dimension (d <= 3) and random size. However, such representation is limited in practice. For example,

- *City environment:* in city environment, the main objects are artificial objects (including constructers, bridges, etc.). Their third dimensional sizes are usually in a certain range, e.g.: from 1 meter to 100 meters. If we use scale 1:10,000 to represent one constructer with 100m in its width, length and height, its map size is only 1cm×1cm×1cm. Such a size representation is much similar to its plane representation, and for many smaller objects, it already loses 3D representation significance. In addition, third information significance of 3D city object would change with scale. For example, it has significance to represent one balcony in 1: 1,000. But in 1: 10, 000 scale, representation of whole building in detail would lose significance. Therefore, to city environment, 3D GIS spatial data modelling should be limited to large scale, e.g. no large than 1:10,000.
- *Ocean and atmosphere environments:* these two environments are reverse to city environment. Information is captured usually in small scales, and artificial objects would lose significance. Information represented usually only in small scales would have significance (e.g.: smaller than 1: 100,000), e.g. temperature and humidity distributing; clouds and ocean flow excursion etc. At the same time, in such environments, medium distributing has continuity. It is difficult to model according individual object as in city environment. Therefore, in such environments, 3D GIS data modelling is not only limited to small scales, but also much different from city environment. Data modelling can't be established on individual object.
- *Mining, geology, cataclysm and petroleum environments*: in these environments, there are both nature objects and artificial objects. But if we review these environments, we could conclude two main cases. One related to resource exploring, which information capture is usually in large scales while with high precision. Another relates to macro geosciences research, such as: terra construction, cataclysm monitoring, etc., which information capture is mainly limited to small scales, similar to atmosphere and ocean environments. In present research works on 3D GIS, most works are related to resource exploring. Therefore, we could get a rough conclusion that in such environments, 3D GIS data modelling are mainly limited to large scales, but have great difference to city, atmosphere and ocean environments.

Now we can summarize a conclusion that 3D GIS data modelling not only has great differences in different environments, but also has great different spatial scale requirements in different environments. It is very difficult to find a kind of volume element in data modelling to meet many different object representations in various environments and various scales.

2.2 3D GIS Concept and its research contents

The current GISs should be called 2D GISs because they are developed from plane map. They could only disposal actual 2D spatial information and their data models are mainly based on 2D vector elements, such as point, line and face. Deunis et al. defined GIS as a software system that contains function to perform input, storage, editing, manipulation, analysis and display of geographically located data [4]. Here geographically located data are referred to 2D spatial data because in GIS data model, third dimensional data is disposed as attribute data.

So far, there is no formal definition for 3D GIS. Different people have different considerations. Fritsch pointed out that main research work of 3D GIS is the integration of height information into 2D GIS [6]. Kofler considered that one major difference between 3D GIS and 2D GIS is the amount of data to be processed, and the second fundamental difference is the user interface [11]. Pigot pointed out that similar to 2D variants, 3D GIS should be capable to perform metric (distance, length, area, volume, etc), logic (intersection, union, difference), generalization, buffering, network (shortest way) and merging operations [19]. Pfound pointed out that today, most applications and data structures for 3D GIS are optimised for visualization [18]. They usually omit topological information in order to get a better performance. Actually, these hints could be found in many papers. These hints imply that 3D GIS is a new system that should be established on the base of 2D GIS. From there, we could define that 3D GIS is a software system that contains function to perform input, storage, editing, manipulation, analysis and visualization of three-dimensional geographical information data in one certain application area.

2.3. Research content

The majority of research works on 3D GIS are 3D data modelling and 3D data structure [6], [15], [16], [19], [20]. A few is on 3D visualization [11] and others are on 3D web GIS [13][26]. From the scientific view, 3D GIS research content must be larger. On the other hand, if we consider the above definition, 3D GIS research content may focus on following points:

- 3D data modelling base on actual applications;
- 3D data structure base on actual applications;
- 3D visualization technologies for actual applications;
- Huge multimedia data management;
- 3D technologies on web for actual applications.

These contents have strong relations with application areas. The majority of 3D GIS research works are carried on with the other research works for actual application. Hence such considerations would meet the practice needs. While the mainstream of 3D GIS research works seldom consider situations in actual application, this is the main reason why we can't see one 3D GIS used in practice beside some special software used in certain areas like petroleum or city.

3. Data Modelling in 3D GIS

The final aim of data modelling is to truthfully represent the real world. However, because of limitation of theories and technologies, it is difficult to redivivus real world objects through data modelling in GIS. The data models could be classified into three levels or three types according to their representation precision:

Symbol model: in 2D GIS vector data model, point and line element could be considered as symbols on plane as a face is composed of lines. We can consider face is represented by line symbols because 2D GIS comes from map, and objects representation manners are not changed from map to GIS. 2D GIS vector data model can be considered as a symbol model and this model can represent objects location and its rough shape. Symbols are suitable to represent symbolic logic. This is the reason that 2D GIS vector data model could be easily used to represent spatial topology.

Basic model: a few spatial data, and predefined elements, such as points, lines, faces and bodies can be used to represent various objects, and could basically reflect object's shape, spatial extension and higher precision 3-dimensional spatial location. 3D GIS data models are mainly used for the representation of basic information of objects in real world. For example, tetrahedron models were used to represent irregular mine body [20] and CSG primitives were used to represent regular buildings [14]. In city environment, third dimensional spatial information can be added to construct 3D urban GIS models easily [7][9]. Today, the majority of data models available are belong to basic model category.

Refined model: a large number of spatial data are used in this kind of models. It is established mainly through interactive methods, and it is difference from above two kinds significantly. Many types of elements are used in this kind of data modelling, and these elements could be adjusted to meet modelling needs during modelling process. This kind of data model could reflect objects with high precision in spatial location, subtle spatial shape and vision characteristics, etc. It is mainly used in Virtual Reality GIS (VRGIS), especially in 3-dimensional landscape reconstruction based on GIS spatial data. As VR technology need very high precision representation of 3-dimensional objects (e.g.: construction, road bridges, trees, etc.), high precision are also requested in data capture and data modelling functions. Therefore, at the present time, this kind of data models could only be established using software packages like 3Dmax and AutoCAD. It is very difficult for GIS to perform this kind of data modelling.

Obviously, refined model needs very high precision representation. This may cause huge amount of data and modelling complexity. As different objects may have different characteristics, it is difficult to realize automatic data modelling in large quantities. Special modelling software packages must be used. Therefore, refined model is not suitable in 3D GIS and it is more suitable for VRGIS.

On the other hand, as *symbol model* can't be used in 3D GIS, *basic models* are usually used in 3D GIS data modelling. *Basic model* can still be classified into two types according its modelling purposes:

1. *Currency model:* the aim of this kind of models is to represent various objects in different application areas in real world. The majority of present data models are belong to this category. For example, Molenaar put forward a 3D FDS data structure [15]. Pilouk et al. put forward a tetrahedron based data model [20]. Pigot put forward a topological data model [19]. Li put forward a CSG based data model [14], and etc. Although these data models have different advantages, there are still some defects in different degrees as currency models to represent various complex 3-dimensional objects in real world. The realisation of truly 3-dimensional GIS solutions is not yet in view [3].
2. *Special model:* it is established for certain application area. Early in 1989, Peter pointed out that elements in geosciences have a lot of communities [17]. It could be possible to establish one uniform geosciences data model. Raper discussed 3D data modelling for geosciences [21]. He pointed out that 3D data models have strong links with underground data samples and parameters. Turer pointed out that an exact model of the subsurface is not possible, simplified versions may be derived from geological and engineering data developed from exploration procedures, including borings, well tests, seismic exploration, and stratigraphic or sedimentological descriptions of depositional systems [25]. In city environment, Zlatanova used faces and nodes as main elements combining R-tree to establish a 3-dimensional topological data model on Web [26]. Gruen et al. developed a TOBAGO system using point sets to establish 3D data model by interactive method [8]. Tempfle put forward a 3D topological urban data model [24]. Koehl put forward a data modelling method based on CSG level structure [10] and Sun et al. put forward a 3D spatial data model based on surface partition [23].

Because special models disposal objects limited in certain application areas, data modelling is relatively simple. It is also easy to implement in practice. For example, using 3D surface modelling method, ArcView 3D analyst module could realize modelling to construction [5].

4. Is there a perfect spatial data model for 3D GIS?

4.1. Perfect 3D GIS data model

Pfund pointed out in his research work that a perfect 3D GIS data model should be vector data structure, boundary representation and topology [18]. Peter pointed out that boundary representation (B-rep) is a suitable modelling method for geosciences elements representation and operation when he did research works on one Geoscientfic Resource Management System [17]. Pigot had summarized 3D GIS modelling research works [19]. He pointed out that 3D GIS data modelling began with Molenaar's 3D FDS data structure and many others research works are developed on the base of it. Pigot himself put forward a 3D spatial data model based on manifold topology. This model is similar to 3D data models developed by Pilouk et al. [20].

Actually, Molenaar's 3D FDS is an extension of 2D GIS vector data model and it is a B-rep data model. B-rep data model is a kind of perfect data model in some degree. The reasons are:

- The basic elements are point, line, face and body. Point and line are the most basic elements, so it is easy to combine a 3D GIS with a 2D GIS;
- Amount of data is not large, and it is easy to establish data model and to manage;
- Easy to establish topology;
- Easy to realize visualization, and easy to do data operations;

But B-rep method also has some defects. It is difficult to verify representation validity, to perform Boolean operations, and to require complex algorithms for model generation [17]. However, if we don't pay too attention to verification of representation validity, it doesn't effect today's 2D GIS so much. Boolean operations are not so urgent and complex algorithms become simple to today's computer hardware. So we think B-rep data model is a kind of perfect 3D GIS data model.

4.2. Questions about spatial topology relationships

This question may raise many arguments, especially in 3D GIS. Topology representation is considered as the typical feature of GIS, but why do we have to represent topology relationships? Spatial topology could be established dynamically. The essence of this thinking is to calculate topology in real-time (e.g.: Voronoi diagram is a good structure). Actually, current 2D GISs could only represent a few topology relationships, and topology problems on 2D plane are still not solved very well [1].

Topology relationships representation usually limits the establishment of data model, and it also increase the complexity of data updating and operation. The topology relationships in 3D GIS could be calculated in real-time. AutoCAD might give us a good example. There are no any topology concepts in AutoCAD, but it could be used very well in industry, such as machine assemble. Bernhardsen pointed out that the essential difference between GIS and CAD becomes more and more blur [1]. By adding some attribute data, many CAD system vendors could effectively compete GIS market. Kofler used a CAD model and R-tree combined method in Vienna walking system [11]. Stoter also introduced a CAD model into GIS in order to solve the problem of 3D cadastral registration [22]. Koehl put forward CSG level model. His main purpose was to solve CAD models representation [10]. In VRGIS, data modelling mainly comes from 3Dmax or CAD.

4.3. 3D GIS data modelling for geoscience

Data models based on tetrahedron (or 3-simplex) could preferably be used to represent various elements in geosciences. In geosciences, most objects are natural objects not artificial ones. Lattuada and Raper put forward that 3-dimensional tetrahedron network is suitable for modelling in these areas [12]. Recently, we also did some research works in terra area, we reconstruct 3D geological body on the base of simulate artesian well data using tetrahedron network, and also reconstruct complex geological bodies on the constraint tetrahedron network. Tetrahedron model is not suitable to represent regular bodies, especially some bodies with arc shape, such as cylinder, cone and etc. Extra partition calculation and redundancy data will be introduced if tetrahedron is used to represent regular objects.

In city environment, almost all objects are artificial objects. Besides, it is not necessary to represent inner attributes of city objects in GIS. As majority of city objects are regular ones, they could be represented by some primitives like rectangle, triangle, cuboids, cylinder, cone, etc. or their combinations [10][15][23]. Some CSG primitives could not be represented by R-rep, but they could be deal with parameter expressions.

5. 3D GIS Development Directions

Although the general 3D GIS system have not be progressed significantly because of the baffle in data modelling, the great progress have been got in Three-Dimensional City Model and Virtual Geographic Information System. These systems could provide very excellent visualization effects and interactive functions.

3D GIS should be designed as a subsystem of 2D GIS. Of course, it could run alone. The establishment of 3D GIS should be compatible with 2D GIS functions. It should have interface same as that for 2D GIS. And reducing 2D GIS functions would decrease much complexity in 3D GIS. On one hand, 3D GIS should have unique visualization interface as well as operation and analysis functions. Many systems are established on the base of GIS [9]. While in geology, mining, etc., although subsurface data is the principal, ground data is also important. For example, 2D data in GIS is usually the reference to underground works in resource mining process.

Therefore, It is a crucial point to set up a connect between 3D GIS and 2D GIS, we have several views as (1) Establish special SDE for 3D spatial data management; (2) Extend present 2D GIS data manage functions and make them compatible with 3D spatial information management; (3) Develop 3D GIS modules in component methods; (4) Establish data change interface between 2D to 3D.

Cambray et al. put forward a multidimensional geographical data model, through add some faces with different heights to solve this problem [2]. Their data model could only solve regular and protruding enveloped objects. Arc view 3D analyst is a well known 3D GIS module developed based on component. Haala et al. established a 3-dimensional city model by adding some extra information and changing 2D data to 3D data [9].

3D GIS has no great significance in many application areas, as in small scales space (e.g.: scales less then 1:10,000). There are great different spatial characteristics and requirements in city, geology, mining, ocean, atmosphere and other environments. It is impossible to establish a currency data model to these areas in theory. In fact, from the review of GIS software packages (e.g. Arc/Info, MapInfo, GeoMedia, AotoDesk), we would found that in these GIS software packages, 3D functions are limited to a few modules. For city environment, functions are limited to visualization and simple queries.

We conclude that 3D GIS development should be based on special application. The research should focus on application areas information representation and analysis. 3D spatial data is only a carrier. Data modelling should focus on representation integrality to special information. It should not simply emphasize precision and topology consistency of spatial data representation. At the time being, it is not necessary, and also very difficult to establish a currency 3D GIS platform.

ACKNOWLEDGMENTS

This publication is an output from the research project "Digital Earth" funded by Chinese Academy of Sciences. Also, Dr. Yong Xue should like to express his gratitude to Chinese Academy of Sciences for the financial support under the "CAS Hundred Talents Program" and "Initial Phase of the Knowledge Innovation Program".

Reference

1. Bernhardsen T. "*Geographic Information Systems An Introduction*". John Wiley &Sons Inc., New York, 1999.
2. Cambray B., Yeh T. S. "A Multidimensional (2D, 2.5D, 3D) GeoGraphical Data Model". *Proceedings of the International Conference on Management of Data* (COMAD'94), Bangalore, India, 1994, pp317-336.
3. Carosio A. "Three-Dimensional Synthetic Landscapes: Data Acquisition, Modelling and Visualisation". *Photogrammetry week'99*, 1999, pp293-302.
4. Deunis R. S., Arithur R. P., "Three-dimensional GIS for the earth sciences". In: Three dimensional applications in Geographical Information Systems, Taylor & Francis, Edit by Jonathan F. Raper, pp149-154, 1989.
5. Esri Company "ArcView 3D Analyst". http://www.esri.com
6. Fritsch D., "Three-Dimensional Geographic Information Systems -- Status and Prospects". *Proceeding of International Archives of Photogrammetry and Remote Sensing,* Vol. 31/4, ISPRS, Vienna, Austria, 1996, pp215-221.
7. Gruber M. "The CyberCity Concept from 2D GIS to the Hypermedia Database". *Proceedings of UM3'98,* International Workshop on Urban Multi-Media/3D Mapping, 1998, pp47-54.
8. Gruen A. "Tobago—A semi-automated approach for the generation of 3D building models". *ISPRS Journal of Photogrammetry & Remote Sensing,* 53, 108-118, 1998.
9. Haala N., Brenner C. and Anders K. H. "Generation of 3D City Models From Digital Surface Models and 2D GIS". *ISPRS,* Vol.32, Part 3-4W2, 3D Reconstruction and Modelling of Topographic Objects, Stuttgart, 1997, pp68-75.

10.Koehl M. "The Modelling Of Urban Landscapes." *Proc of International Archives of Photogrammetry and Remote Sensing*. Vol. XXXI, Part B4. Vienna, 1996, pp460-464.
11.Kofler M. "R-trees for Visualizing and Organizing Large 3D GIS Databases". *PhD. thesis*, Technischen Universität Graz, Graz, 1998.
12.Lattuada R., Raper J. "Applications of 3D Delaunay triangulation algorithms in geoscientific modelling". http://geog.bbk.ac.uk
13.Lee H.G., Kim K.H., Lee K., "Development of 3-Dimensional GIS Running on Internet". *IEEE*, pp1046-1049, 1998.
14.Li R.X., "Data Structure and Application Issues in 3-D Geographical Information Systems". *Geomatics,* 48(3): pp209-224, 1994.
15.Molenaar M., "A Formal Data Structure For Three Dimensional Vector Maps", *Proc 4th International Symposium on spatial data handling,* Zurich, 1990, pp830-843.
16.Oosterom P.V., *Reactive Data Structures for Geographic Information Systems,* Oxford University Press, Oxford, 1993.
17.Peter R. G., Andrew J. B. "Three dimensional representation in a Geosicientfic Resource Management System for the minerals industry." *Three-dimensional applications in Geographical Information Systems,* Taylor & Francis, Edit by Jonathan F. Raper, 1989, pp155-182.
18.Pfund M. "Topologic Data Structure for a 3D GIS". http://www.gis.ethz.ch/publications
19.Pigot S. "A topological model for a 3-dimensional spatial information system". *PhD. thesis*, University of Tasmania, 1998.
20.Pilouk M., Tempfli K., Molenaar M., "A Tetrahedron-Based 3D Vector Data Model for Geoinformation." In: *Advanced Geographic Data Modelling,* Netherlands Geodetic Commission, Publications on Geodesy, (40): pp129-140, 1994.
21.Raper J. "The 3-dimensional geoscientific mapping and modelling system: a conceptual design." *Three-dimensional applications in Geographical Information Systems*, Taylor & Francis, Edit by Jonathan F. Raper, 1989, pp11-19.
22.Stoter J. "Needs possibilities and constraints to develop a 3D cadastral registration system". http://www.geo.tudelft.nl/GISt/gist_e/index.htm
23.Sun M., Chen Jun, Zhang X. Z. "3DCM data modelling based on surface partition". *Acta Geodateica et Cartographica Sinica,* Vol.29, No.3, 257-265, 2000.
24.Tempfle K. "3D topographic mapping for urban GIS". ITC journal 1998-3/4, 181-190, 1998.
25.Turer A. K. "The role of three-dimensional geographic information systems in subsurface characterization for hydrogeological applications". *Three-dimensional applications in Geographical Information Systems,* Taylor & Francis, Edit by Jonathan F. Raper, 1989, pp115-127.
26.Zlatanova S., Tempfli K. "Modelling for 3D GIS: Spatial Analysis and Visualisation through web". *Proc of IAPRS*, Vol. XXXIII, Amsterdam, 2000, pp1257-1264.

Genetic Cryptoanalysis of Two Rounds TEA

Julio César Hernández[1], José María Sierra[1], Pedro Isasi[2] and Arturo Ribagorda[1]

[1] Computer Security Group, Computer Science Dept., Carlos III University, 28911 Leganés,Madrid, Spain
{jcesar, sierra, arturo}@inf.uc3m.es
[2] Artificial Intelligence Group, Computer Science Dept., Carlos III University, 28911 Leganés, Madrid, Spain
{isasi}@ia.uc3m.es

Abstract. Distinguishing the output of a cryptographic primitive such as a block cipher or a hash function from the output of a random mapping seriously affects the credibility of the primitive security, and defeats it for many cryptographic applications. However, this is usually a quite difficult task. In a previous work [1], a new cryptoanalytic technique was presented and proved useful in distinguishing a block cipher from a random permutation in a completely automatic way. This technique is based in the selection of the worst input patterns for the block cipher with the aid of genetic algorithms. The objective is to find which input patters generate a significant deviation of the observed output from the output we would expect from a random permutation. In [1], this technique was applied to the case of the block cipher TEA with 1 round. The much harder problem of breaking TEA with 2 rounds is successfully solved in this paper, where an efficient distinguisher is also presented.

1 Introduction

A cipher [2] is an algorithm characterised by a tuple (Π, Γ, K, $\in$, Δ) that verifies:

1. Π is the finite set of plaintexts
2. Γ is the finite set of ciphertexts
3. K is the finite set of possible keys
4. For every κ in K there is a ciphering function ε_κ in $\in$ and a deciphering function δ_κ in Δ that verifies $\delta_\kappa(\varepsilon_\kappa(\pi))=\pi$ for every plaintext π de Π.

The fundamental property here is number 4, that implies that the original plaintext can be recovered if the ciphering key is known. But this knowledge of the ciphering key must not only be sufficient for recovering the text, but also necessary. Any robust cipher must not allow the recovery of plaintext without the knowledge of the key used.

P.M.A. Sloot et al. (Eds.): ICCS 2002, LNCS 2331, pp. 1024–1031, 2002.

Block ciphers are a special class of ciphers, characterised for operating over bit blocks instead of over single bits as the stream ciphers do. This block length (typically 64, 128 or 256 bits) and the key length used to cipher (typically 64 or 128 bits) are fixed and fundamental for the block cipher's strength.

In an ideal block cipher, every key must define a random permutation of the input bits over the output bits. This is why the problem of deciding if a given block cipher is secure or not can be translated into the problem of deciding if the permutations it generates when fixing keys are or not random. Unfortunately, this problem is very hard. Randomness, although being a paramount matter in security, is quite hard to evaluate.

However, and independently of the techniques used, if we are able of distinguishing a cryptographic primitive from a random mapping in an efficient and statistically significant way, then this cryptographic primitive must be discarded for any cryptographic application, and immediately removed and replaced by a more robust algorithm. This is why the techniques for evaluating randomness have a paramount importance in cryptography.

This paper is organised as follows: Section I is a brief introduction to the concept of block ciphers and the importance of distinguishers in cryptoanalysis. Section II briefly introduces the block cipher TEA, which will be the objective of our attacks. Section III highlights the main ideas behind what we call genetic or poisoned cryptoanalysis. Section IV presents known results of this technique over TEA with 1 round, mostly for comparison porpoises, and new results over the harder TEA with 2 rounds. Section V presents some conclusions and Section VI finally mentions some selected references.

2 The TEA Algorithm

TEA stands for Tiny Encryption Algorithm. It is the name of a block cipher invented by David Wheeler and Roger Needham, members of the Computer Security Laboratory of Cambridge University. It was presented in the 1994 Fast Software Encryption Workshop [3].

TEA is a very fast block cipher that does not use predefined tables or s-boxes and does not need much initialisation time. It is a Feistel type algorithm, thus named because it divides its input in two halves and operates over them individually in each round and interchanges them at the end of every round. It works over 64 bit blocks and uses keys of 128 bits, which are large enough for today's security standards.
As they authors say, it has a security (with 8, 16 or 32 rounds) at least comparable with the Data Encryption Standard (DES), and it is quite faster. The block cipher TEA has some additional advantages: it is very robust (only one published attack [4]

in more than 7 years of existence, and it is one *academic* attack, that is, a theoretical attack with little or no practical implications) and it is very portable, simple and efficient as its compact code below shows:

```
void code(long* v, long* w, long* k){
unsigned long y=v[0],z=v[1],sum=0,delta=0x9e3779b9,n=8;
while (n-->0)   {
  sum += delta ;
  y +=  (z<<4)+k[0]  ^ z+sum  ^  (z>>5)+k[1]  ;
  z +=  (y<<4)+k[2]  ^ y+sum ^  (y>>5)+k[3]  ;    }
w[0]=y ; w[1]=z ; }
```

3 Cryptoanalysis with genetic algorithms

Some classic statistical tests [5] are based in observing the output of a block cipher algorithm while *feeding* it with inputs as random as possible. This is because the idea behind these test is to verify if the execution of the block cipher algorithm maintains the good randomness properties of the input, so this must be as random as possible. Our proposal is based in a complete different idea. Instead of using a random input, we use what can be called *poisoned* inputs.

We try to check if the fixing of some of the input bits simplifies the process of discovering correlations between the input and the output bits. If significant correlations exist, we would be able to infer information about some bits of the plaintext just from the output bits of the ciphertext. These correlations could also be used to distinguish the block cipher algorithm from a random permutation in a simple and efficient way.

It is important to mention that this fixing of some bits in the input must be reasonable in length to allow for enough different inputs to mathematically verify the effect it has over the output. If a statistically significant deviation from randomness is found, then we would have an easy method for distinguishing the primitive from a random mapping and the algorithm must be discarded for any cryptographic use.

But how to decide which bits to fix in the input and to which values is a very complex problem. It is, essentially, a search in a huge space with $2^{\Pi+K}$ elements. For TEA this space bitmask has 2^{192} possible values, so an exhaustive search is infeasible. For finding the best bitmasks in those huge spaces we propose the use of genetic algorithms in which individuals will codify bitmasks.

In our method, schematically represented in Figure 1, individuals codify bitmasks that will be used to perform a logical AND with every random input. In this way, by means of an AND with a given bitmask, we manage to fix some of the input bits to zero.

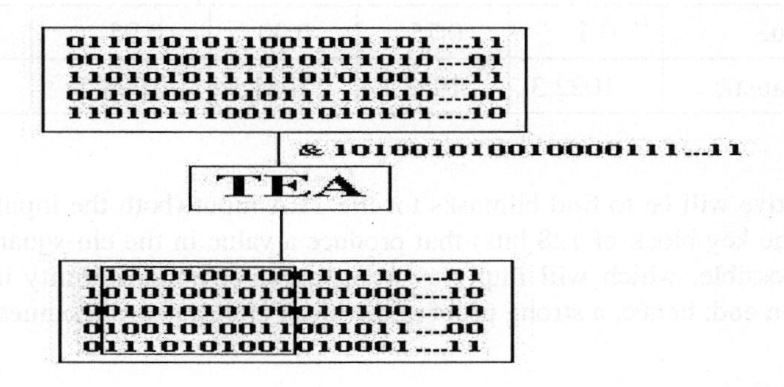

Figure 1: A schema of the procedure used for fixing some of the input bits to zero by performing an AND with a bitmask.

For every bitmask (individual) in a given population of the genetic algorithm, we will observe the deviation it produces in the observed output and we will decide if this deviation is or not statistically significant. Repeating this process, we will try to find the best individuals (bitmasks) that will we those which produce more significant deviations from the expected output.

We have used the implementation of the genetic algorithm of William M. Spears, from the Navy Center for Applied Research. After a number of preliminary tests, we determined that a 0.95 probability of *crossover* and a 0.05 probability of *mutation* were adequate for our problem and we decide to fix them to these values.

The *fitness* function we try to maximise is a chi-square statistic of the observed output. We decided to observe the distribution of the 10 least significant bits of the first output word of TEA because some authors, notably [6], have shown that block ciphers that use rotation as one of their round operations (as is the case of TEA) tend to show bad distributions in their lest significant bits. So we will measure the fitness of an individual by how the bitmask that it codifies affects the probability distribution of the 10 rightmost bits of the first output word of TEA.

These 10 bits can be interpreted as the binary representation of the integers between 0 and 1023, and their distribution should uniformly take all these values. For checking if this is the case, we will perform a statistical chi-square test, which is one of the most extended test in cryptoanalysis due to its high sensibility to little deviations.

The statistic distribution should correspond to a chi-square distribution with 1024-1=1023 degrees of freedom. The values for different percentiles of this distribution are shown in Table 1:

Table 1. Values of the Chi-Square distribution with 1023 D.o.F. for different percentils

p-value	0.5	0.75	0.90	0.95	0.99
X^2 Statistic	1022.3	1053.1	1081.3	1098.5	1131.1

Our objective will be to find bitmasks for the TEA input (both the input block of 64 bits and the key block of 128 bits) that produce a value in the chi-square statistic as high as possible, which will imply a strong proof of nonuniformity in the output distribution and, hence, a strong proof of the block cipher nonrandomness.

4 Results

Every mask (codified as individuals in the genetic algorithm population) was evaluated by performing an AND with 2^{13} random inputs, different for every individual and every generation. This makes convergence harder, but improves the generality of the results obtained because it makes overfitting nearly impossible. An example of a bitmask for TEA1 obtained using this approach is:

{01010000100010101011010000000000,01010101110000111000000000000000,01000011
11101000110010000000000,00101001101101010100000000000000,01000011111101010111
100010101011,01011101000110100011011111011001}

This bitmask has a length of 192 bits (192=64+128) and a weight of 73. This implies that, choosing input vectors at random and applying this mask over them could give us 2^{73} different inputs to the block cipher TEA1. This is a huge number, so this bitmask is useful and applicable. This would not occur if the weight of the bitmask were very low. It is clear that if two masks provoke the same deviation in the output we should prefer the heavier one because more ones in the bitmask imply more input bits that do not affect the behaviour of the observed output.

The chi-square statistic we are using as a *fitness* function can not increase indefinitely, but has a maximum, as we will show.

As we make 2^{13} tests for every individual and any of them can return a value between 0 and 1023 that we expect will be uniformingly distributed, the number of occurrences of every value must be close to $2^{13}/2^{10}=2^3=8$. The maximum value for the chi-square statistic under these assumptions will occur when the observed distribution is as far as uniform as possible, that is, when only one of these 1024 possible values really occurs and all the rest do not occur. In this case we say the distribution has collapsed in a single output value and the *fitness* will be

$$\chi^2 = \sum_{i=0}^{1023} (n_i - 8)^2 / 8 = \sum_{i=0}^{1022} (0-8)^2 / 8 + (2^{13} - 8)^2 / 8 = 8380416$$

This is exactly the case for the bitmask show before. It produces a collapse of all the 10 rightmost bits of the first output word of TEA1 into a single value.

For assuring the generality of this bitmask, it was tested with other sets of inputs not previously seen by the generic algorithm and in every case it produced a collapse of the output. So the bitmask shows that the 10 bits we observe do not depend uniformingly of every input bit (those positions that have a value of 1 in the bitmask do not affect the output bits we are observing), which is a property that block ciphers must have.

So, apart from showing that there are high correlations between the input and the output of TEA1, we can also use this result for constructing a distinguisher able of deciding if a given function is TEA1 or a random mapping using as input a single random vector. This distinguisher algorithm can be described as:

```
INPUT: F: Z_2^192->Z_2^64, a random mapping or TEA1
ALGORITHM:
Generate a random vector v of Z_2^192
Apply the mask m getting v'= v & m that can take 2^73
possible values
Compute F(v´)=w[0]w[1]
Compute r = w[0] & 1023
OUTPUT:
If r=441 then F is TEA1 else F is not TEA1
```

It is important to note that the algorithm for distinguishing TEA1 from a random mapping only needs one plaintext (input vector). This is a very unusual fact in cryptoanalysis, because most attacks and distinguishers need a large number of texts (numbers like 2^{47} are common) to work properly. This distinguisher is able of distinguishing TEA1 from a random mapping with an extremely low probability of false positives (only 1/1024=0.000977, less than a 0.1%) and a zero probability of false negatives.

The case of TEA with 2 rounds is quite harder. An additional round significantly improves the strength of the algorithm, so the same approach does not produce interesting results, even after many generations and different iterations of the genetic algorithm. Different fitness functions were tested for extending this method to TEA2, and most of then were discarded. We finally got results using as a fitness function not the chi-square statistic used in TEA1, but its fourth power. This has the effect of amplifying differences in the chi square statistic that had little influence in the previous fitness function. It also makes the selection procedure of the genetic algorithm, that in our implementation is proportional to fitness, closer to a selection proportional to rank. Using this approximation, we managed to obtain the following bitmask

{00110101101000111000001010110010,00110110000100000011000110001000,000101000
11110111111000011100000,01100011010001001101000111011010,0000000010010000110
1010000100001,00011110001111001111111000000000}

that produces chi-square values that vary around 1900. As this statistic value is extremely high (see Table 1) and the weigh of the mask is 77, this bitmask can be used to design a distinguisher for TEA2. The construction of the distinguisher, once we have the bitmask, is trivial. The algorithm will be:

INPUT: **F**: $Z_2^{192} \rightarrow Z_2^{64}$, a random mapping or TEA2

ALGORITHM:

Generate 2^{13} random vectors $\mathbf{v}_i$ of Z_2^{192}

Apply the mask $\mathbf{m}_2$ to every vector $\mathbf{v}_i$, getting $\mathbf{v}_i' = \mathbf{v}_i$

& $\mathbf{m}_2$ that can take 2^{77} possible values

Compute $\mathbf{F}(\mathbf{v}_i') = \mathbf{w}^i[0]\mathbf{w}^i_{[1]}$

Compute $\mathbf{r}_i = \mathbf{w}^i_{[0]}$ **& 1023**

Perform a chi-square test for checking if the observed distribution of $\mathbf{r}_i$ is consistent with the expected uniform distribution, calculating the corresponding chi-square statistic $\bullet^2$

OUTPUT: If $\bullet^2 > 1291.44$ then **F** is TEA2 else **F** is not TEA2

This is a quite interesting distinguisher in the sense that it will produce a very low ratio of false positives (a value greater than 1291.44 will only occur in a chi-square statistic with 1023 degrees of freedom one in 10^{-8} times) and also a very low probability of false negatives (the average of the statistics produced by this bit mask is around 1900 so a value of less than 1291.442306 is extremely unlikely).

It is also worthy to mention that this distinguisher uses 2^{13} input vectors, not at all a huge number but many more than the corresponding distinguisher does for TEA1 does. This is because we must perform the chi-square test and we need enough inputs for expecting at least 5 occurrences of all the 1024 possible outputs. Slightly increasing this minimum of 5 to 8 leads to the actual number $2^{3+10=13}$ of input vectors proposed.

4 Conclusions

In this work over the use of genetic algorithms in cryptoanalysis we have shown how a new technique which is useful to perform an automatic cryptoanalysis of certain cryptographic primitives (named poisoned or genetic cryptoanalysis) can be implemented with the aid of genetic algorithms. Although this fact was previously stated in [1], it was only proven to produce results over the very limited TEA1. By showing that this model is also able of finding strong correlations in variants of the block cipher TEA with more rounds (TEA2) we have finally provided enough evidence of the interest of this technique for cryptoanalysis.

References

1. Hernández J.C., Isasi P., and Ribagorda A.: An application of genetic algorithms to the cryptoanalysis of one round TEA. Proceedings of the 2002 Symposium on Artificial Intelligence and its Application. (To appear)

2. Douglas R. Stinson, Cryptography, Theory and Practice, CRC Press, 1995.

3. D. Wheeler, R. Needham: TEA, A Tiny Encryption Algorithm, Proceedings of the 1995 Fast Software Encryption Workshop. pp. 97-110 Springer-Verlag. 1995

4. John Kelsey, Bruce Schneier, David Wagner: Related-Key cryptoanalysis of 3-WAY, Biham-DES, CAST, DES-X, NewDES, RC2 and TEA, Proceedings of the ICICS'97 Conference, pp. 233-246, Springer-Verlag, 1997.

5. Juan Soto et. al., NIST Randomness Testing for Round 1 AES Candidates Proceedings of the Round 1 AES Candidates Conference, 1999

6. John Kelsey, Bruce Schneier, David Wagner: Mod n cryptoanalysis with applications against RC5P and M6, Proceedings of the 1999 Fast Software Encryption Workshop, pp. 139-155 Springer-Verlag, 1999.

Genetic Commerce – Intelligent Share Trading

Clive Vassell

Harrow Business School,
University of Westminster,
London
Email: vasselc@wmin.ac.uk,
URL: users.wmin.ac.uk/~vasselc

Abstract. In time, it seems feasible that genetic algorithms will help to achieve similar levels of productivity gains in many service domains as has been achieved in line and, more recently, batch manufacture. And e-commerce will embody the standards and guidelines necessary to enable managers to make the most of the possibilities offered by this development. It will help them to both satisfy and retain the custom of their clients; and make it possible to operate in a highly efficient and effective manner. The paper discusses the nature of these changes and it assesses some of their implications for organisations and their management. It introduces the concept of intelligent share trading; a future manifestation of these developments. And it talks about the important role artificial intelligence, particularly genetic algorithms, will play in these systems.

Electronic Commerce

The Internet looks set to become a key plank in the infrastructure needed to support a new era of business development. The promise of a network connecting all significant economic agents of the world (both human and software) and many of the devices on which they rely creates the possibility of a huge array of information services [1].

If the Internet is, in large measure, the infrastructure which facilitates this new era, e-commerce is the 'business protocol' which determines the standards and norms by which trade is conducted in this new context.

It covers such issues as electronic data interchange (EDI) between the various members of the virtual supply chain, the payment systems which are to be used and/or permitted, and the maintenance of the levels of security necessary to reassure customers and potential customers.

Just as importantly, it encapsulates the understanding gleamed about what works and what doesn't in the information age; the 'strategic protocol' which leads to satisfied and loyal customers as well as robust and profitable businesses.

P.M.A. Sloot et al. (Eds.): ICCS 2002, LNCS 2331, pp. 1032–1041, 2002.

It is, as yet, early days. We still have a considerable amount to learn; and we still have many tried and trusted approaches carried over from the preceding era which will need to be unlearned as they no longer apply.

A few indicators of the likely form of these protocols are beginning to emerge however. First of all, the customer is going to be better placed tomorrow than he or she is today [2]. The information age will make information available to the connected world in abundance; it will no longer be the preserve of the resourceful or powerful.

Information on available products, prices, relative performance, cost of production, methods of production and distribution, environmental friendliness, suitability for various tasks and/or users, and much, much more will be available to all those interested to look for it [3].

And looking for such information will become progressively easier. We suffer from information overload now not so much because there is too much information out there but rather because the information searching and filtering tools we have at present are not sufficiently sophisticated.

As these tools improve, so our ability to more effectively manage large amounts of information will increase as will our ability to be selective about the theme of the information presented, its format, and its level of detail.

It will become successively more difficult, indeed counterproductive, for large suppliers to misrepresent their products or services. Should they do so, an army of empowered consumers may abandon their offerings and might well petition (via the Net) industry watchdogs, MPs, consumer groups or TV programs or web sites, or any others they feel appropriate to take action against the offending firm.

Furthermore, for the foreseeable future there will remain a need for effective logistics [4]. Indeed e-commerce is likely to increase the need for this and yet it is an area that is often overlooked. Even information services require facilitating goods (equipment and consumables) and the more dispersed the service provision, the more carefully the supporting logistics will have to be planned and implemented.

Recently, data mining has become an area of considerable interest to organisations. Many large firms have huge data warehouses full of transaction data, management information, information on the external environment, and information on much else of potential value. Data mining approaches, including artificial intelligence, can be very useful in making sense of this mountain of data and using that understanding to improve decision making [5], [6], [7].

Artificial intelligence is being used in a wide range of applications. It is being used to better facilitate manufacturing [8], and to make intelligent agents more effective [9], and in a host of other applications and domains in between.

And, inevitably, it is being used in the financial arena. Neural networks and genetic algorithms are the preferred AI environments in this area. The vast amounts of data available and the complexity of the applications make them particularly well suited to the domain. And thus they are being used for a range of financial applications, including stock market prediction [10], [11].

These are but a few examples of the kind of insight of this new order which e-commerce will have to encapsulate. There will be many more which will emerge over the years ahead.

Organisations will have to be sensitive to the effectiveness of the initiatives they introduce and be ready to respond rapidly where and when required. This will be no easy task but if it is done well enough and soon enough, the rewards could be a place among the great and the good of the new order.

Intelligent Share Trading

So how might all this work in practice? Well, one important application of e-commerce is share trading. There are many organisations which provide a service that enables customers to buy and sell shares of their choice on one or more stock exchange(s).

Typically the user is required to select the shares he or she wishes to buy or sell and carry out the transaction. The system makes it possible for the user to do so simply and quickly, and typically provides up-to-date prices and some relevant news. But it does not make the choices and it does not automatically conduct transactions.

In principle, however, it would be perfectly feasible for such systems to be extended to include these two additional functions. The systems could use a selection strategy to choose which shares to buy and a deselection strategy to determine which ones to sell. It could then conduct the buy and sell transactions whenever the appropriate conditions applied.

The selection and deselection strategies could be specified by the user so that the system behaves in the way that the user would but does so automatically and responds to changed circumstances almost instantaneously. Alternatively, the systems could use artificial intelligence to determine appropriate selection and deselection criteria.

The latter is, in essence, an intelligent share trading system. It would automatically trade on its user's behalf according to criteria it has determined; the user simply needs to set the broad performance goals (such as strong growth or modest volatility), and periodically review the performance of the portfolio and the history of transactions.

The Constituents of an Intelligent Share Trading System

The main constituents of an intelligent share trading system would be the data collection module, the share trading module, the share selection module and the strategy optimisation module.

The data collection module would collect share price information and store it in a database ready for processing by the share selection and strategy optimisation modules. The selection module would apply the current selection and deselection strategy to determine whether any shares ought to be bought or sold at present. If so it would request that the share trading module conduct a trade. The trading module would then buy or sell the requested quantity of the required share.

The strategy optimisation module would run periodically and, using historical data, determine which investment/trading strategy would have given optimal results – in line with the broad performance objectives specified by the user. This would then become the strategy to be applied by the share selection module.

Genetic algorithms would be used to effect the optimisation. They are arguably one of the best tools for finding optimal or near optimal solutions given an infinite or very large range of potential solutions.

In fact the genetic algorithms would operate on two levels. On one level they would find the optimum strategy given a set of attributes; on the second level they would find the set of attributes which yield the best strategy given a larger universe of attributes.

(This would help to ensure that the system comes up with both the optimal set of share attributes on which to base many of the better strategies, and the optimum strategy itself).

The system would be designed to devise strategies which were robust or, in other words, optimised for say ten years (rather than 'over optimised' for a short period of time and/or a particular set of market circumstances). This should mean that the strategy selected would not need to change often unless the user changed the overall performance objectives.

Strategy Optimisation Module
Share Selection Module
Share Trading Module
Data Collection Module

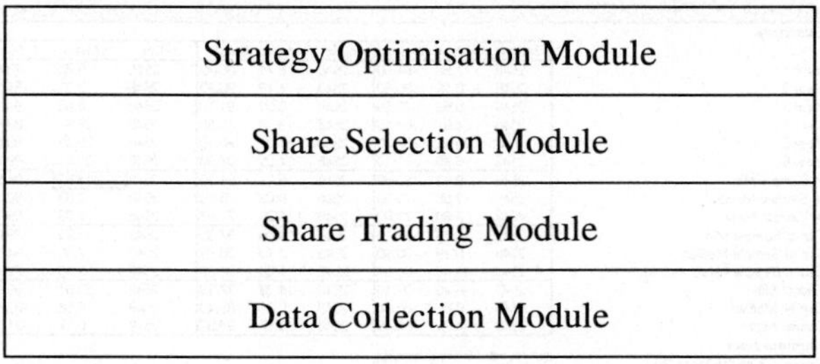

Fig. 1. The main components of an intelligent trading system

Genetic Algorithms and Optimisation

Genetic algorithms apply the principles of natural selection to finding optimum solutions to a problem. They operate on a population of potential solutions and apply the principles of selection, crossover and mutation to produce a new generation of candidate solutions.

The selection operation is used to choose the best candidates from the population by testing each solution against a fitness or target function. The crossover operator is used to produce a child solution from two parent solutions by combining elements of the chromosomes of one parent with elements of the chromosomes of the other. The mutation operator introduces an element of randomness into each population.

The combination of these three operators leads to new generations of solutions which tend to improve in performance in relation to previous generations but which will not simply converge on sub-optimal solutions, in much the same way as they help living organisms to thrive.

The Nature of the Optimisation

So what might the optimisation dialogue look like? It would have screen displays similar to some of the screens taken from one of the models I am currently using for my research in this area. I shall use this model to give an example of how this part of the intelligent share trading system might look and feel.

An important component of the Strategy Optimisation Module of the intelligent share trading system would be the performance summary screen. An example of how it might look is shown below.

The summary would indicate what the current share selection strategy was and how that strategy would have performed in the past. In fig. 2 no strategy has been selected. This is indicated by the fact that the chromosome values are all zero.

Microsoft Excel - Optimisation.xls

File Edit View Insert Format Tools Data Window Help

Summary

	3mCnt	3mMean	3mMin	6mCnt	6mMean	6mMin	1yCnt	1yMean	1yMin
Year 1	2548	1.56	-87.50	2548	5.71	-66.40	2548	9.83	-83.60
Year 2	2548	0.19	-90.90	2548	4.19	-90.90	2548	7.33	-94.30
Year 3	2548	-0.52	-79.50	2548	8.09	-91.10	2548	-8.63	-94.90
Year 4	2548	4.87	-83.10	2548	13.33	-79.90	2548	26.61	-83.50
Year 5	2548	25.61	-67.00	2548	31.97	-76.10	2548	25.45	-90.80
Year 6	2548	-6.46	-91.10	2548	-14.26	-97.30	2548	-23.61	-99.64
In Sample Min	2548	-0.52	-87.50	2548	5.71	-91.10	2548	-8.63	-94.90
In Sample Median	2548	1.56	-79.50	2548	8.09	-76.10	2548	9.83	-90.80
In Sample Mean	2548	8.88	-78.00	2548	15.26	-77.87	2548	8.89	-89.77
Out of Sample Min	2548	-6.46	-91.10	2548	-14.26	-97.30	2548	-23.61	-99.64
Out of Sample Median	2548	0.19	-90.90	2548	4.19	-90.90	2548	7.33	-94.30
Out of Sample Mean	2548	-0.46	-88.37	2548	1.09	-89.37	2548	3.45	-92.48
Overall Min	2548	-6.46	-91.10	2548	-14.26	-97.30	2548	-23.61	-99.64
Overall Median	2548	0.87	-85.30	2548	6.90	-85.40	2548	8.58	-92.55
Overall Mean	2548	4.21	-83.18	2548	8.17	-83.62	2548	6.17	-91.12

Chromosomes

Previous 3 Month Price Rise	Cppr3m	0	0-100
Previous 6 Month Price Rise	Cppr6m	0	0-100
Previous 1 Year Price Rise	Cppr1y	0	0-100
Price/Earnings Ratio	Cpe	0	0-100
Dividend Yield	Cyld	0	0-100
Estimated Capital (£100k)	Ccap*100	0	0-100
1 Year Earnings Per Share Growth %	Cepsgr	0	0-100
1 Year Profit Growth %	Cprofgr	0	0-100
1 Year Dividend Growth %	Cdivgr	0	0-100
1 Year Turnover Growth %	Ctogr	0	0-100

Evolver Log / Share Details / Summary / Results

Start | Microsoft Excel - Opti... | Screens.doc - Microsoft W... | 12:30

Fig. 2. The strategy performance summary screen

Microsoft Excel - Optimisation.xls

Summary

	3mCnt	3mMean	3mMin	6mCnt	6mMean	6mMin	1yCnt	1yMean	1yMin
Year 1	8	-8.76	-25.60	8	-5.46	-23.10	8	2.45	-32.40
Year 2	17	2.54	-8.00	17	8.88	-40.60	17	29.51	-40.60
Year 3	26	7.38	-23.90	26	36.08	-26.40	26	25.37	-54.00
Year 4	23	5.87	-17.90	23	13.98	-28.40	23	11.90	-47.30
Year 5	101	-0.45	-50.40	101	-6.04	-65.80	101	-1.36	-80.80
Year 6	46	-3.43	-37.50	46	-5.02	-38.20	46	-10.81	-72.22
In Sample Min	8	-8.76	-50.40	8	-6.04	-65.80	8	-1.36	-80.80
In Sample Median	26	-0.45	-25.60	26	-5.46	-26.40	26	2.45	-54.00
In Sample Mean	45	-0.61	-33.30	45	8.19	-38.43	45	8.82	-55.73
Out of Sample Min	17	-3.43	-37.50	17	-5.02	-40.60	17	-10.81	-72.22
Out of Sample Median	23	2.54	-17.90	23	8.88	-38.20	23	11.90	-47.30
Out of Sample Mean	29	1.66	-21.13	29	5.95	-35.73	29	10.20	-53.37
Overall Min	8	-8.76	-50.40	8	-6.04	-65.80	8	-10.81	-80.80
Overall Median	25	1.05	-24.75	25	1.93	-33.30	25	7.18	-50.65
Overall Mean	37	0.52	-27.22	37	7.07	-37.08	37	9.51	-54.55

Chromosomes

Previous 3 Month Price Rise	Cppr3m	0	0-100
Previous 6 Month Price Rise	Cppr6m	0	0-100
Previous 1 Year Price Rise	Cppr1y	10	0-100
Price/Earnings Ratio	Cpe	0	0-100
Dividend Yield	Cyld	5	0-100
Estimated Capital (£100k)	Ccap*100	0	0-100
1 Year Earnings Per Share Growth %	Cepsgr	0	0-100
1 Year Profit Growth %	Cprofgr	0	0-100
1 Year Dividend Growth %	Cdivgr	0	0-100
1 Year Turnover Growth %	Ctogr	0	0-100

Evolver Log / Share Details / Summary / Results

Fig. 3. The strategy performance summary screen with a strategy selected

In fig. 3, the selection strategy is:

> Select those shares where the share price has risen by 10% or more in the previous year and the previous year's dividend has been 5% or more of the (prevailing) share price.

It is important to understand that this selection strategy would have been applied at the start of each of the six years represented in the data set. And the size and performance of the resulting share portfolio for each of the six years for the subsequent three months, six months and one year are recorded at the top of the screen.

The three 'in sample' rows provide summaries of the in sample data (in this case years one, three and five). The three 'out of sample' rows provide summaries of the out of sample data (years two, four and six) and the three overall summary rows provide summaries of all the data (years one to six).

The target cell is also shown (the in sample minimum of the one year means). The target function is to maximise the value of this cell providing the associated count (the in sample minimum of the counts of the shares selected for each year) is at least ten.

The target function is defined in the screen below:

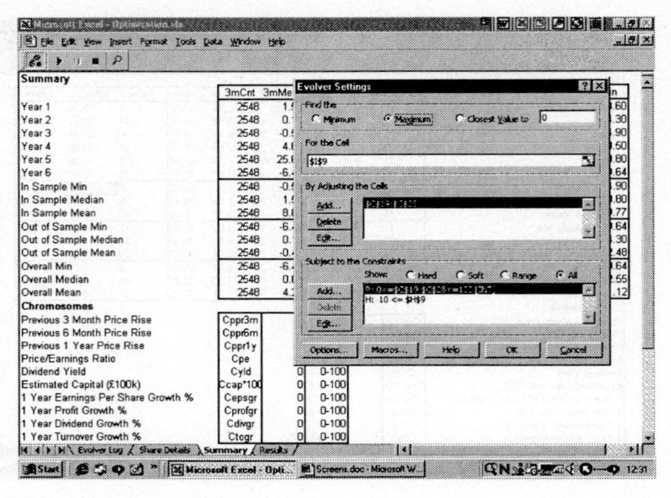

Fig. 4. Defining the *target function*

The Test attribute (or field) is used to determine which shares in the database meet the current share selection criteria. The contents of the Test attribute can perhaps best be explained if it is laid out in a logical format. This is done below. Here we can see how the genetic algorithm operates at the two levels spoken about earlier.

The Contents of the Test Attribute

```
=And (
Or (Cppr3m=0,     And (PriceRisePrev3m<>"",     PriceRisePrev3m>=Cppr3m) ),
Or (Cppr6m=0,     And (PriceRisePrev6m<>"",     PriceRisePrev6m>=Cppr6m) ),
Or (Cppr1y=0,     And (PriceRisePrev1y<>"",     PriceRisePrev1y>=Cppr1y) ),
Or (Cpe=0,        And (PERatio<>"",             PERatio<=Cpe) ),
Or (Cyld=0,       And (DividendYield<>"",       DividendYield>=Cyld) ),
Or (Ccap=0,       And (EstCapital<>"",          EstCapital>=Ccap) ),
Or (Cepsgr=0,     And (EPSgrowth<>"",           EPSgrowth>=Cepsgr) ),
Or (Cprofgr=0,    And (Profitgrowth<>"",        Profitgrowth>=Cprofgr) ),
Or (Cdivgr=0,     And (Dividendgrowth<>"",      Dividendgrowth>=Cdivgr) ),
Or (Ctogr=0,      And (Turnovergrowth<>"",      Turnovergrowth>=Ctogr) )
)
```

Where a chromosome has a value of zero, that chromosome plays no part in selecting shares. It is, in effect, switched off. Where it has a non-zero value, it is used in the selection process. Thus the genetic algorithm will, at any one time, be selecting which chromosomes are to be part of the selection criteria and which are not.

Where a chromosome has a non-zero value, the Test attribute selects those shares which have both a non-blank value in the appropriate attribute and the value of the attribute meets the criteria level of the associated chromosome.

In principle, this means we can have a large number of chromosomes but that, at any moment in time, the genetic algorithm will generally be using a few of them only. It means we should be able to test the effectiveness of a wide range of chromosome permutations while at the same time testing a range of selection strategies using each of those permutations. And we should be able to do this without having to build a whole suite of models.

(I think this principle may well reduce the likelihood of 'overfitting' in a relatively straightforward manner. Whether this particular application of the principle is likely to work remains to be seen.)

Conclusions

It should be noted that my main concern when trying to identify suitable strategies is robustness. In other words, I am interested in strategies which give good returns year in, year out rather than strategies which give high mean returns but are accompanied by high levels of volatility.

This is why my target function is the minimum one year mean (rather than the mean of all years). I am particularly interested in strategies which provide positive returns in all test years (both in sample and out of sample).

Perhaps in part because of the quite specific nature of my requirements, the results to date have not been entirely encouraging. I have not, so far, come across any strategy which meets these requirements (though there are one or two which show modest losses in one year only).

However the work continues. I plan to extend the range of candidate chromosomes and possibly introduce both maximum value and minimum value chromosomes to see whether this improves the performance of the best strategies.

While the results of the exercise are not yet terribly encouraging, there is some evidence to suggest that the search should prove fruitful in the end [12] and there is a prominent theoretical framework which is compatible with this kind of expectation [13], [14], [15].

Wider Implications

We are entering a new era in the history of commercial activity. The understanding which helps to crystallise this new era will be the body of knowledge which e-commerce will comprise. It will be the content of the MBAs of the online business schools of tomorrow and will enable executives to profitably steer their firms in the decades ahead.

The real winners however will probably not have learned much of what is critically important about e-commerce from business schools but rather by being the first to live them and learn from them in their own organisations [16]. And by being sufficiently capable as managers to capitalise extensively on the lead they so gain.

It is organisations like these that will fashion the new era; and it is those who study these organisations that will identify the e-commerce protocols associated with this novel form.

Intelligent share trading is an example of this kind of development. It results from a fusion of electronic commerce, artificial intelligence and online trading.

These systems are likely to prove profitable for their suppliers and their users alike. And the organisations who are the first to introduce robust and user friendly examples will in all probability do very well indeed.

And these systems will inevitably make extensive use of artificial intelligence. The huge amounts of historical data which can be called upon to facilitate understanding and the complexity of the application areas will make the use of data mining techniques and tools very valuable. And neural networks and genetic algorithms will likely prove extremely pertinent.

And genetic algorithms (and possibly hybrid approaches) will be of particular value in situations where it is necessary to both carry out appropriate actions on behalf of the users and explain to the users the underlying strategy behind those actions.

So important might this kind of artificial intelligence become that perhaps genetic commerce will be the most appropriate way to describe systems of the type outlined in this paper. Indeed its popularity might well signal the next phase of the incessant rise of the machine.

References

1. Aldrich Douglas F, 'Mastering the Digital Marketplace: Practical strategies for competitiveness in the new economy', John Wiley, 1999
2. Hagel John, Armstrong Arthur G, 'Net Gain: Expanding markets through virtual communities', Harvard Business School Press, 1997
3. Evans Philip, Wurster Thomas S, 'Getting Real About Virtual Commerce', Harvard Business Review, November-December 1999
4. Jones Dennis H, 'The New Logistics: Shaping the new economy', Ed: Don Tapscott, Blueprint to the Digital Economy: Creating wealth in the era of e-business, McGraw Hill, 1998
5. Gargano Michael L, Raggad Bel G, 'Data Mining – A Powerful Information Creating Tool', OCLC Systems & Services, Volume 15, Number 2, 1999
6. Lee Sang Jun, Siau Keng, 'A Review of Data Mining Techniques', Industrial management and Data Systems, Volume 101, Number 1, 2001
7. Bose Indranil, Mahapatra Radha K, 'Business Data Mining – A Machine Learning Perspective', Information & Management, Volume 39, Issue 3, December 2001
8. Burns Roland, 'Intelligent Manufacturing', Aircraft Engineering and Aerospace Technology: An International Journal, Volume 69, Number 5, 1997
9. Saci Emilie A, Cherruault Yves, The genicAgent: A Hybrid Approach for Multi-Agent Problem Solving', Kybernetes: The International Journal of Systems & Cybernetics, Volume 30, Number 1, 2001

10.Wittkemper Hans-Georg, Steiner Manfred, 'Using Neural Networks to Forecast the Systemic Risks of Stocks', European Journal of Operational Research, Volume 90, Issue 3, May 1996

11.Back Barbo, Laitinen Teija, Sere Kaisa, 'Neural Networks and Genetic Algorithms for Bankruptcy Predictions', Expert Systems With Applications, Volume 11, Issue 4, 1996

12.Bauer Richard J, 'Genetic Algorithms and Investment Strategies', John Wiley & Sons, February 1994

13.Fama Eugene F, French Kenneth R, 'The Cross-Section of Expected Stock Returns', The Journal of Finance, Volume 47, Number 2, June 1992

14.Fama Eugene F, French Kenneth R, 'Size and Book-to-Market Factors in Earnings and Returns', The Journal of Finance, Volume 50, Number 1, March 1995

15.Fama Eugene F, French Kenneth R, 'Multifactor Explanations of Asset Pricing Anomalies', The Journal of Finance, Volume 51, Number 1, March 1996

16.Senge Peter M, 'The Fifth Discipline: The art and practice of the learning organisation', Business (Century/Arrow), 1993

Efficient Memory Page Replacement on Web Server Clusters

Ji Yung Chung and Sungsoo Kim

Graduate School of Information and Communication
Ajou University, Wonchun-Dong, Paldal-Gu
Suwon, Kyunggi-Do, 442-749, Korea
{abback, sskim}@madang.ajou.ac.kr

Abstract. The concept of network memory was introduced for the efficient exploitation of main memory in a cluster. Network memory can be used to speed up applications that frequently access large amount of disk data. In this paper, we present a memory management algorithm that does not require prior knowledge of access patterns and that is practical to implement under the web server cluster. In addition, our scheme has a good user response time for various access distributions of web documents. Through a detailed simulation, we evaluate the performance of our memory management algorithms.

1 Introduction

With the growing popularity of the internet, services using the world wide web are increasing. However, the overall increase in traffic on the web causes a disproportionate increase in client requests to popular web sites. Performance and high availability are critical for web sites that receive large numbers of requests [1, 2, 3].

A cluster is a type of distributed processing system and consists of a collection of interconnected stand-alone computers working together. Cluster systems present not only a low cost but also a flexible alternative to fault tolerant computers for applications that require high throughput and high availability.

Processing power was once a dominant factor in the performance of initial cluster systems. However, as successive generations of hardware appeared, the processor decreased its impact on the overall performance of the system [4]. Now, memory bandwidth has replaced the role of the processor as a performance bottleneck. The impact of networking has also decreased with the 100Mbps ethernet. Thus, efficient memory management is very important for overall cluster system performance.

This work is supported in part by the Ministry of Information & Communication in Republic of Korea ("University Support Program<2001>" supervised by IITA).

This work is supported in part by the Ministry of Education of Korea (Brain Korea 21 Project supervised by Korea Research Foundation).

P.M.A. Sloot et al. (Eds.): ICCS 2002, LNCS 2331, pp. 1042–1050, 2002.

The concept of network memory was introduced for the efficient exploitation of main memory in a cluster. Network memory is the aggregate main memory in the cluster and can be used to speed up applications that frequently access large amounts of disk data.

This paper presents a memory management algorithm that always achieves a good user response time for various access distributions of web documents.

The remainder of the paper is organized as follows. Section 2 presents related work on cluster memory management. In section 3, we explain the clustered web server architecture that is considered in this paper. Section 4 explains our memory management algorithm and section 5 presents simulation results. Finally, section 6 summarizes with concluding remarks.

2 Related Work

Recently, some papers on the memory management of the cluster have studied the method of utilizing the idle client's memory [5, 6, 7]. The active client forwards cache entries that overflow its local cache directly to the idle node. The active client can then access this remote cache until the remote node becomes active. However, in these methods, a client's request must go to the disk if the requested block does not exist in the limited memory, even though another node has that block.

The Greedy Forwarding algorithm deals with the memory of the cluster system as a global resource, but the algorithm does not attempt to coordinate the contents of this memory [8]. The main problem with this policy is that global memory is under-utilized because of duplication.

Duplicate Elimination [9] takes the other extreme approach. Since it is inexpensive to fetch a duplicate page from remote memory, compared to a disk input/output (I/O), every duplicate page is eliminated before a single page. Each node maintains two LRU (Least Recently Used) lists, one for single pages and the other for duplicate pages. The advantage of Duplicate Elimination is that it has a high global hit rate because of the global memory management. However, a main drawback of Duplicate Elimination is that the local hit rate of some nodes reduces because of the smaller cache size, even if the global hit rate increases.

In order to adjust the duplication rate of the data page dynamically, the N-chance Forwarding algorithm forwards the last copy of the page from one server to a randomly chosen server, N times, before discarding it from global memory [8].

Also, the Hybrid algorithm dynamically controls the amount of duplication by comparing the expected cost of an LRU single page and an LRU duplicate page [9]. The expected cost is defined as the product of the latency to fetch the page back into memory and a weighting factor that gives a measure of the likelihood of the page being accessed next in memory. The Hybrid algorithm has a good response time on average, but it does not have a minimum response time for a special range of workload and node configuration.

In this paper, we present a memory management algorithm that does not require prior knowledge of access patterns and that is practical to implement under the web server cluster. Also, this method always has a good user response time for various access distributions of web documents.

3 Web Server Cluster Architecture

In order to handle millions of accesses, a general approach adopted by popular web sites is to preserve one virtual URL (Uniform Resource Locator) interface and use a distributed server architecture that is hidden from the user.

Thus, we consider the architecture of cluster system that consists of the load balancer and a set of document servers. Each of the document servers is a HTTP (Hyper Text Transfer Protocol) server.

Figure 1 presents the web server cluster architecture. In this architecture, the load balancer has a single, virtual IP (Internet Protocol) address and request routing among servers is transparent. Every request from the clients is delivered into the load balancer over the internet. After that, the load balancer redirects the request to a document server in a round-robin manner.

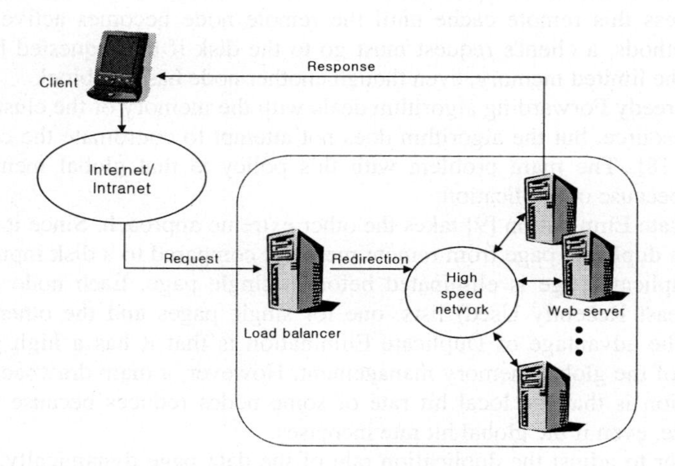

Fig. 1. Web server cluster architecture

Web documents that are serviced in the web server cluster are distributed on the disks of each node. Nodes that receive a user request are called the primary, and nodes that store the requested data on the disk are called the owner. This means that each node can be the primary or the owner as the case may be. The owner nodes maintain a directory in which they keep track of the copies of the data pages they own in global memory.

The only times a node has to be informed about status changes to the pages are when a page becomes a duplicate after being the only copy in global memory and when the page becomes the last copy in the node's memory after being a duplicate.

4 Memory Management of Web Server Cluster

Efficient memory management is the task of keeping useful data closer to the user in the memory hierarchy. Figure 2 shows the memory hierarchy that is considered in this paper.

The user request is divided into page-sized units and is serviced by the primary node. If the requested page exists in the primary node, it is serviced immediately. If the page is absent, the primary node requests the owner node for the page. If the page presents in the owner node, it is forwarded to the primary node. Otherwise, the owner node forwards the request to the other node that has the requested page. The node receiving the forwarded request sends the data directly to the primary node. However, if no node contains the page, the request is satisfied by the owner's disk.

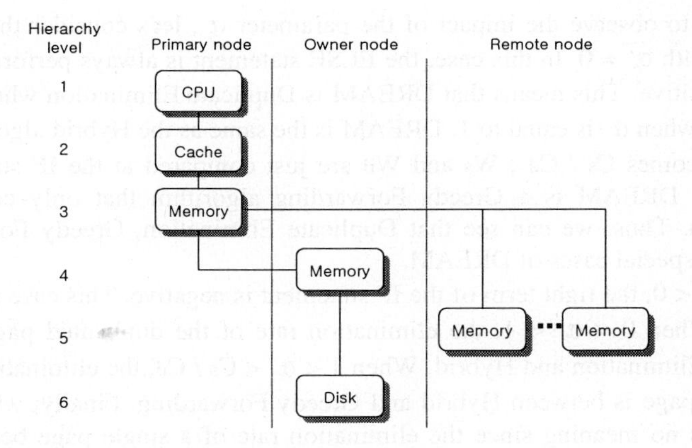

Fig. 2. Cluster memory hierarchy

When the owner node reads the page from disk, it retains a copy of the page in local memory. In order to improve global memory utilization, these pages are maintained on a FIFO (First In First Out) list and should be removed from the memory of the owner node as soon as possible.

The page replacement policy is to minimize the number of possible page faults so that effective memory access time can be reduced. The LRU algorithm is a popular policy and often results in a high hit ratio in a single server. This policy assumes that the costs of page faults are the same. However, this assumption is not valid in the cluster system, since the latency to fetch a page from disk and the latency to fetch a page from the remote node are different. Therefore, the elimination rate of the duplicated page in the global memory has to be higher than that of a single page.

Our proposed policy, DREAM (Duplication RatE Adjustment Method) for memory management of the web server cluster adjusts the duplication rate according to the various access distributions of web documents. The page replacement method of DREAM is as follows.

```
IF (Ws × Cs  < Wd × Cd × α)

          Replace the LRU single page

ELSE

          Replace the LRU duplicate page
```

Where W_s and W_d are the inverse of the elapsed time since the time of last access for the LRU single page and LRU duplicate page, respectively. C_s is the latency to fetch a page back into memory from disk and C_d is the latency to fetch a page back into local memory from the remote memory. α is a parameter for duplication rate adjustment. W_s, W_d, C_s and C_d are positive and C_d is lower than C_s. Thus C_s / C_d is always higher than 1.

In order to observe the impact of the parameter α, let's consider the behavior of DREAM with $\alpha = 0$. In this case, the ELSE statement is always performed since $W_s \times C_s$ is positive. This means that DREAM is Duplicate Elimination when α is equal to 0. Also, when α is equal to 1, DREAM is the same as the Hybrid algorithm.

If α becomes C_s / C_d, W_s and W_d are just compared at the IF statement. This means that DREAM is a Greedy Forwarding algorithm that only considers time information. Thus, we can see that Duplicate Elimination, Greedy Forwarding and Hybrid are special cases of DREAM.

When $\alpha < 0$, the right term of the IF statement is negative. This case is the same to α of 0. When $0 < \alpha < 1$, the elimination rate of the duplicated page is between Duplicate Elimination and Hybrid. When $1 < \alpha < C_s / C_d$, the elimination rate of the duplicated page is between Hybrid and Greedy Forwarding. Finally, when $\alpha > C_s / C_d$, there is no meaning since the elimination rate of a single page becomes higher than that of a duplicated page. Therefore, the valid range of parameter α is $0 \leq \alpha \leq C_s / C_d$. Figure 3 shows the relation for the elimination rate of duplicated page and parameter α.

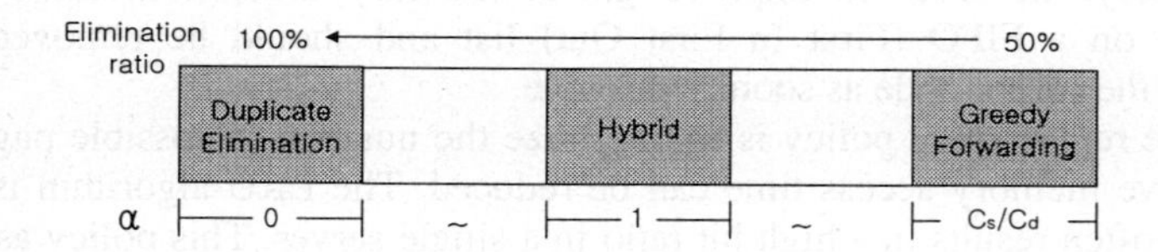

Fig. 3. The relation for the elimination rate of duplicated page and α

In general, the access frequencies of web documents closely resemble a Zifpian distribution [10]. In this distribution, the shape of the curve is decided by the skew parameter. When skew is 0 the access frequencies of web documents are uniform, and when the skew parameter is 2 the access frequencies of web documents are skewed.

The algorithm that has the best response time among Duplicate Elimination, Greedy Forwarding and Hybrid is decided by the skew parameter. DREAM has the best response time by adjusting parameter α according to the skew parameter. Also, it does not have an additional overhead for improving performance.

5 Performance Evaluation

As indicated by the title, this section is devoted to a description of the simulation results which we obtained using our memory management for web server cluster. Table 1 presents the simulation parameters and the Simscript Ⅱ.5 process oriented simulation package is used to model the system.

Table 1. Simulation parameters

Parameter	Value
Number of nodes	3
Memory per node	64MB
Number of files	384
File size	512KB
Page size	8KB
Message cost	1ms / page
Network bandwidth	15MB / sec
Disk bandwidth	10MB / sec

Figure 4 shows the response time as a function of skew. At low skew, Greedy Forwarding is worse than the other algorithms, but the response time decreases drastically as skew increases. To the contrary, Duplicate Elimination has a good response time at low skew, but the improvement of response time is little even if skew increases. Hybrid has a response time that is close to the minimum at both low and high skews, but it is not the best solution at all skews. DREAM eliminates duplicate pages in a similar way to Duplicate Elimination at low skew and it duplicates the hot pages in a similar way to Greedy Forwarding at high skew. Thus, it has always the best response time even though skew increases.

In Figure 5, the aggregate web document size that will be serviced is 2 times the size of global memory. In this case, Greedy Forwarding has the worst response time even though skew is high.

Also, the data size of Figure 6 is 0.5 times the size of global memory. In this case, Duplicate Elimination has the worst response time at every skew because each node has enough local memory. When the local memory is enough, the duplication of hot pages improves performance by eliminating network overhead.

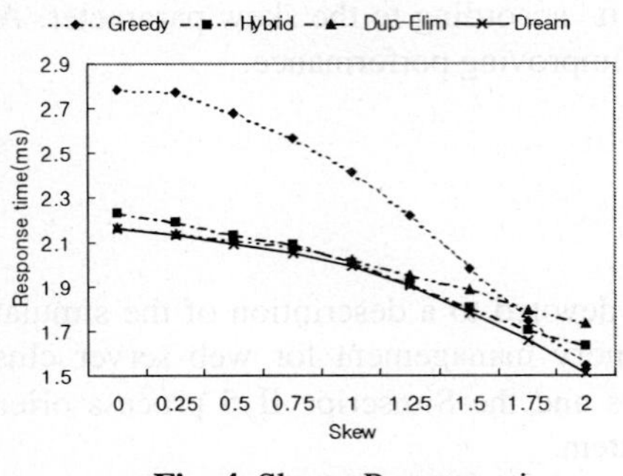

Fig. 4. Skew : Response time, (Number of files = 768)

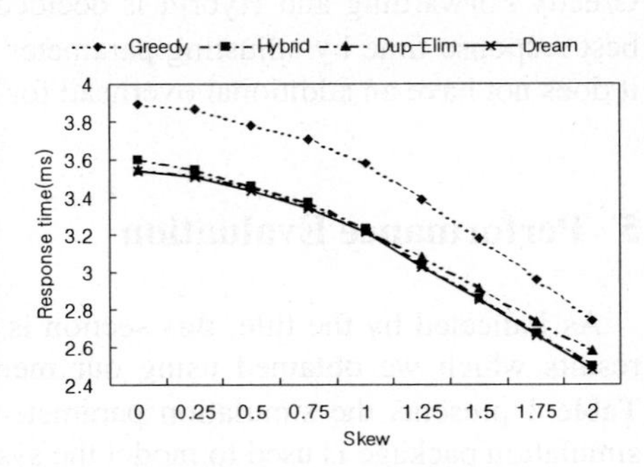

Fig. 5. Skew : Response time, (Number of files = 384)

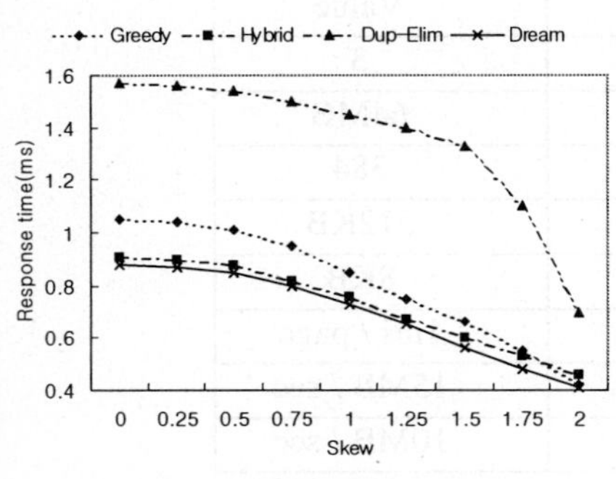

Fig. 6. Skew : Response time, (Number of files = 192)

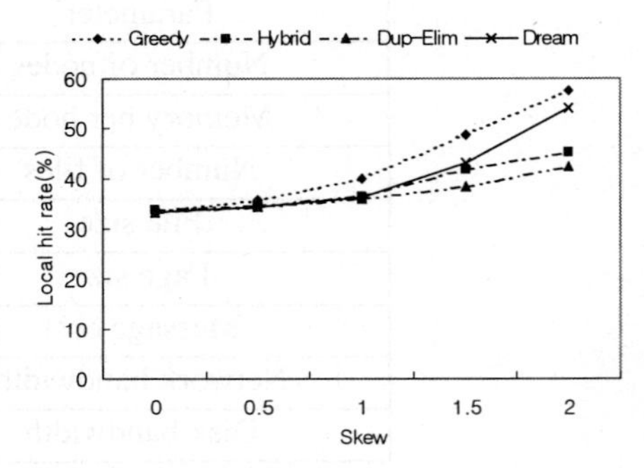

Fig. 7. Skew : Local hit rate

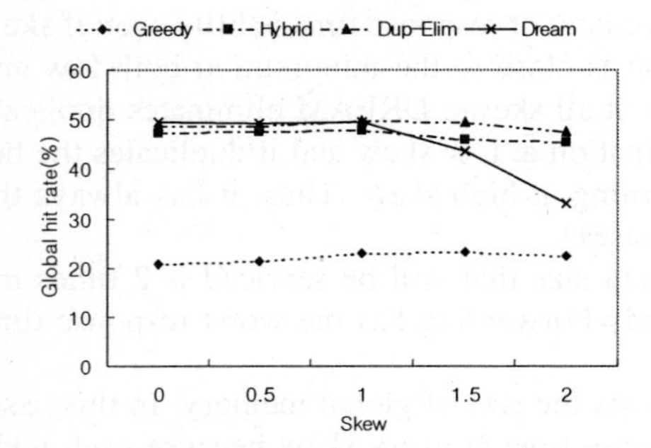

Fig. 8. Skew : Global hit rate

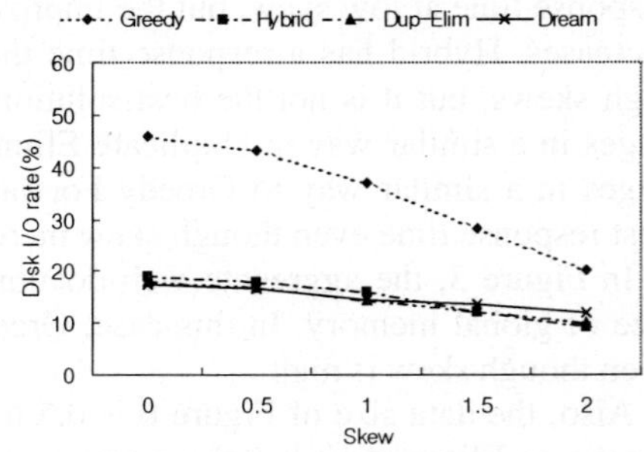

Fig. 9. Skew : Disk I/O rate

Hybrid has a good response time on the average, but it does not have the minimum response time at all skews. On the contrary, we can see that DREAM has a best response time by adjusting α according to the skew.

Figure 7 and Figure 8 show the average local hit rate and global memory hit rate, respectively. In Figure 8, global hit rate of Greedy Forwarding is lower than other algorithms because disk I/O rate is high at the low skew and local hit rate is low at the high skew.

In addition, Figure 9 shows the disk I/O rate. Greedy Forwarding performs frequent disk access at the low skew but has the best local memory hit rate at the high skew. Duplicate Elimination has the worst local memory hit rate but has a good global memory hit rate. DERAM has the best response time by optimizing the ratio of the local memory hit rate, global memory hit rate and disk access rate.

6 Conclusion

Cluster systems are emerging as a viable architecture for building high performance and high availability servers in application areas such as web services or information systems. In initial cluster systems, processing power was a dominant factor of the performance, but memory bandwidth has replaced the role of the processor as a performance bottleneck. Thus, efficient memory management is very important for the overall cluster system performance. In this paper, we proposed an efficient memory management algorithm under the web server cluster. In addition, simulation results show that DREAM always achieves a good user response time for various skew parameters.

References

1. Du, X., Zhang, X.: Memory Hierarchy Considerations for Cost-effective Cluster Computing. IEEE Transactions on Computer (2000) 915-933
2. Cardellini, V., Colajanni, M., Yu, P.S.: Dynamic Load Balancing on Web-server Systems IEEE Internet Computing (1999) 28-39
3. Zhu, H., Yang, T., Zheng, Q., Watson, D., Ibarra, O.H., Smith, T.: Adaptive Load Sharing for Clustered Digital Library Servers. Proceedings of the Seventh IEEE International Symposium on High Performance Distributed Computing (1998) 28-31
4. Buyya, R.: High Performance Cluster Computing: Architectures and Systems. Prentice-Hall (1999)
5. Feeley, M. et. al.: Implementing Global Memory Management in a Workstation Cluster. In Proceedings of the 15th ACM SOSP (1995)
6. Venkataraman, S., Livny, M., Naughton, J.: Impact of Data Placement on Memory Management for Multi-Server OODBMS. In Proceedings of the 11th IEEE ICDE (1995)
7. Koussih, S., Acharya, A., Setia, S.: Dodo: A User-Level System for Exploiting Idle Memory in Workstation Clusters. 8th IEEE International Symposium on High Performance Distributed Computing (1999)
8. Dahlin, M., Wang, R., Anderson, T., Patterson. D.: Cooperative Caching: Using Remote Client Memory to Improve File System Performance. In Proceedings of the First Symposium on Operating Systems Design and Implementation (1994)

9. Venkataraman, S., Livny, M., Naughton, J.: Memory Management for Scalable Web Data Servers. 13th International Conference on Data Engineering (1997)
10. Zipf, G.: Human Behavior and the Principle of Least Effort. Addison-Wesley (1949)
11. Guchi, M., and Kitsuregawa, M.: Using Available Remote Memory Dynamically for Parallel Data Mining Application on ATM-Connected PC Cluster. 14th International Parallel and Distributed Processing Symposium (2000)

Interval Weighted Load Balancing Method for Multiple Application Gateway Firewalls

B. K. Woo[1], D. S. Kim[1], S. S. Hong[1], K. H. Kim[1], and T. M. Chung[1]

[1]Real-Time Systems Laboratory,
School of Electrical and Computer Engineering,
SungKyunKwan University,
Chon-chon dong 300, Chang-an gu, Suwon, Kyung-ki do,
Republic of Korea
{bkwoo, dskim, sshong, byraven, tmchung}@rtlab.skku.ac.kr

Abstract. Firewalls are installed at network perimeters to secure organization's network as alternatives to general gateways. Because of potential performance problems on the gateways, load balancing technique has been applied. However, compared to general gateways, firewalls require more intelligent load balancing method to process massive network traffic because of their relatively complex operations. In this paper, we analyze the inherent problems of existing simple load balancing methods for firewalls and propose the interval weighted load balancing (IWLB) to enhance the processing of massive network traffics. The IWLB deals with network traffics in consideration of the characteristics of application protocols to achieve more effective load balancing. We observed that the IWLB outperforms other simple load balancing methods during our simulation. Therefore, we expect that the IWLB is suitable to balancing loads for multiple firewalls at a network perimeter.

1 Introduction

While the explosive growth of the Internet made it possible to exchange massive information, it caused some negative effects such as the increase of network traffics and security threats. Thus, organizations have a burden to deal with massive network traffic and protect their network from any malicious security threats.

The typical solution to counteract various security threats is deploying the firewall which applies policy-based access control to network traffic at a network perimeter [5, 6, 7]. However, a single firewall cannot operate properly when massive network traffics are applied but also become the dangerous security hole of networks [8]. Furthermore, as shown from DDoS attacks, such as the Trinoo, the Nimda worm, and so forth, using massive packet deliveries or service requests, these attacks degrade network services and performance and create serious operational failures, as well [12, 13].

To make firewall more robust and stable with massive network traffics, it is inevitable to install multiple firewalls to distribute network traffics by using various load

P.M.A. Sloot et al. (Eds.): ICCS 2002, LNCS 2331, pp. 1051–1060, 2002.

balancing methods. Thus, the performance and fairness of a load balancer is the critical factor to estimate the network performance in the environment above.

Although there are many existing load balancing methods based on different principles such as Round robin, Hashing, and Randomization, most of them have flaws to be deployed for firewalls because they do not consider the characteristics of application protocols as the parameter for load balancing.

In this paper, we propose the interval weighted load balancing (IWLB) method, which is designed to use the characteristics of application protocols as the parameter for load balancing decision, to overcome the critical problems of existing load balancing methods. Furthermore, the IWLB is able to enhance network performance and give more robustness in the environment that multiple firewalls are inevitable. With the analysis of simulation results, we observed that the IWLB outperforms other load balancing methods when it is deployed as the principle for the load balancing to distribute traffics.

This paper is organized as follows. In chapter 2, we issue the potential problems of existing load balancing methods when they are applied for multiple firewalls. In chapter 3, we introduce the principle and mechanism of the IWLB method. Our simulation model is described in chapter 4, and the analysis of simulation results and the comparison to other load balancing methods are presented in chapter 5. At last, we close this paper with conclusions in chapter 6.

2 Simple Load Balancing Methods

Under the environment that multiple firewalls are installed in a network for the load balancing purpose, it is desirable to build an appropriate load balancing principle to guarantee the performance of the individual firewall and network security. The most important factor for successful load balancing is to distribute service requests fairly to multiple firewalls.

However, the fairness must be defined differently in this environment because a firewall could discard incoming service requests in terms of its access control policy. That is, even though the load balancer of multiple firewalls distributes incoming service requests fairly, the active session distribution on multiple firewalls can be distributed unevenly. The existing load balancing methods show the limit to overcome this problem because they do not regard the characteristics of application protocols as the parameter for fair load balancing.

In the following sections, we issue the critical flaws of most widely used load balancing methods when they are deployed as the load balancer of multiple firewalls.

Round Robin. Round robin method distributes incoming service requests by simply allocating them to the next available firewall in the rotational manner. That is, this method does not consider the load or number of active sessions currently allocated to the individual firewall. Thus, it can cause the potential load imbalance and increase the skew when a specific application requires long service time [1].

Hashing. Hashing method distributes incoming service requests by hashing the information of service request packets such as source address, source port number, and so on [1]. However, when considering that the flooding attacks generally occur with the identical source information, Hashing method can be vulnerable to those attacks. Moreover, when deployed Hashing mechanism is exposed to attackers, the overall network can be plunged into the fatal situation. We strongly suggest not using Hashing method for the load balancer of multiple firewalls.

Randomization. Randomization method distributes requests to each node according to the value of pseudo random number [1]. In this method, the fine algorithm for random number generation is the key to the successful load balancing. Like other methods mentioned above, Randomization method does not consider the characteristics of application protocols or sessions. Therefore, it is hardly expected that this method is suitable for the load balancer of multiple firewalls.

3 The Interval Weighted Load Balancing

As mentioned in chapter 2, simple load balancing methods are not suitable for the fair load balancing of multiple application gateway firewalls, because they do not take the characteristics of application protocols in consideration. Thus, it is necessary to consider the characteristics of application protocols or sessions as the parameter in order to design more stable and efficient load balancer.

The proposed load balancing method named the interval weighted load balancing (IWLB) makes use of the weight value allocated to each application protocol. In the IWLB, the weight of an application protocol is defined as the interval in the order of firewalls in order to decide which firewall will process a current incoming request. That is, the IWLB decides the firewall for the current service request by adding the weight of the application protocol to the order of the previously selected firewall for the previous service request that has the same protocol as the current one. Since the initial value of the weight, based on the former research about the traffic pattern of application protocols [14], is assigned to each application protocol, the IWLB keep track of the weight value by calculating it periodically in a statistical manner.

To give a specific example how the IWLB obtains the weight value of each protocol, let's suppose the following situation. There are 8 firewalls and all of them can serve 4 different application protocols: HTTP, FTP, SMTP, and TELNET. And let's the weight of HTTP assigned by the IWLB is 3 at this moment. If 6 HTTP requests arrived at the IWLB load balancer sequentially, the order of firewalls to service these requests will be $Fw_1 \rightarrow Fw_4 \rightarrow Fw_7 \rightarrow Fw_2 \rightarrow Fw_5 \rightarrow Fw_8$. That is, the IWLB decides the firewall to service these requests by adding the weight value of HTTP to the previously selected firewall in a rotational manner. Fig. 1 depicts this example and difference from the conventional round robin method.

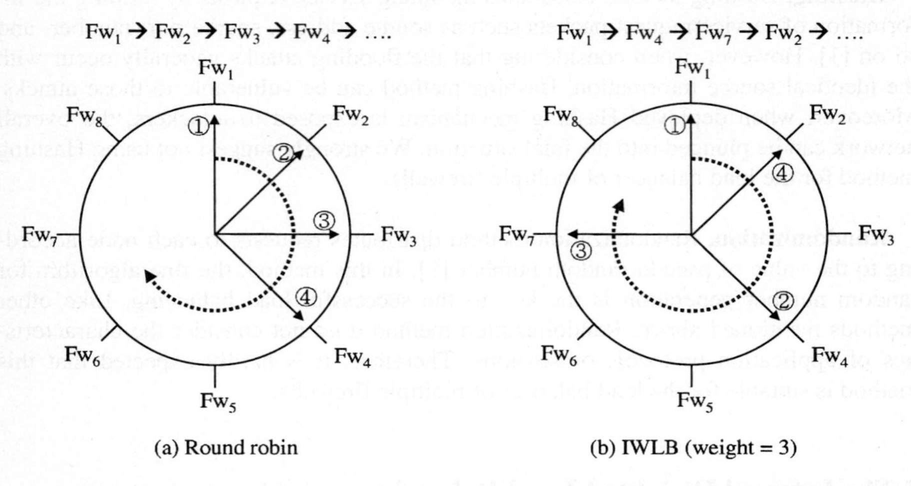

Fig. 1. The comparison of the selection sequence for Round robin and the IWLB

The IWLB uses the standard weight value to calculate the weight value of each application protocol, the standard weight value is set to 1 and assigned to the application protocol that has the largest average service time. After deciding the standard weight value, the weight values of other application protocols are decided as the ratio of their average service time to the average service time of the application protocol that has the standard weight value. Finally, the calculated ratio values must be rounded to a nearest integer value to be used as the weight values for application protocols.

Table 1 explains the rule that the IWLB generates the weight values of application protocols.

Table 1. The average service time and weight of application protocols

Application protocol	Average service time	Calculation rule	Weight
HTTP	970	$W_{http} = 3090/970 \doteqdot 3.18$	3
FTP	3090	W_{ftp} = 1 (standard weight)	1
SMTP	790	$W_{smtp} = 3090/790 \doteqdot 3.91$	4
TELNET	430	$W_{telne} = 3090/430 \doteqdot 7.18$	7

In Fig. 2, we depict how the IWLB balances the service requests with the example of Table 1. Load Distribution of each application protocol starts at the first firewall, i.e., Fw_1.

On the other hand, if the total number of firewalls is divisible by the weight value of a certain protocol, all of the incoming requests of the protocol would be assigned to the same firewall. To prevent this phenomenon, the weight for each application protocol must be the prime to the total number of firewalls.

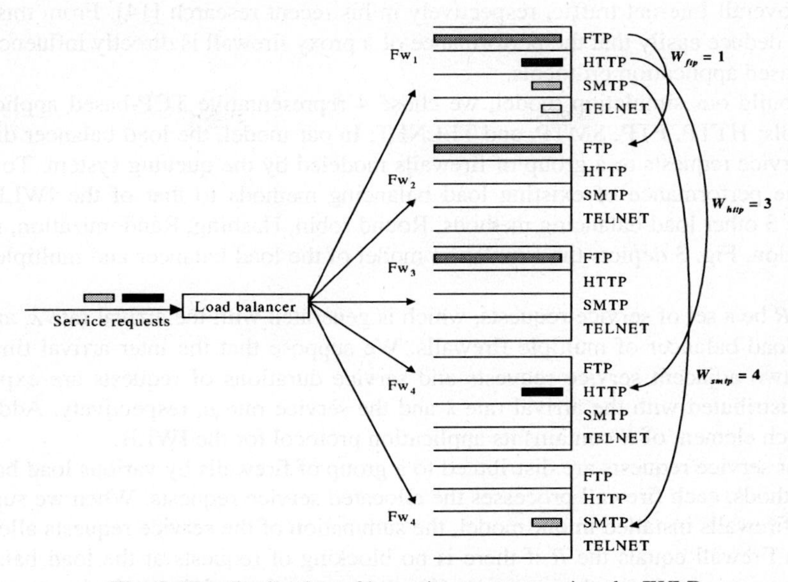

Fig. 2. The distribution of incoming requests using the IWLB

4 Modeling and Simulation

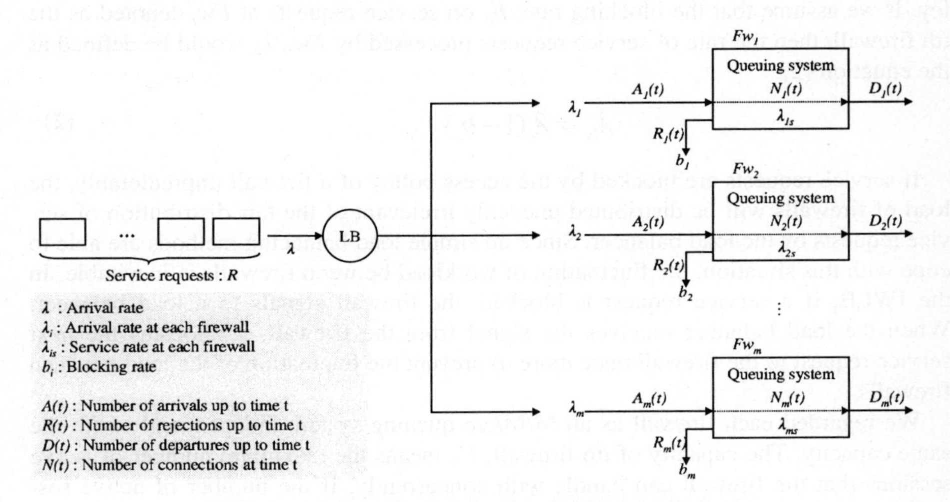

Fig. 3. The modeled load balancer and multiple firewalls

It is widely known that TCP-based application protocols occupy the large portion of the Internet traffic. McCreary announced that some TCP-based application protocols such as HTTP, FTP, SMTP, and TELNET occupy 42.52%, 2.59, 1.70%, and 0.13%, of the overall Internet traffic, respectively in his recent research [14]. From this fact, we can deduce easily that the performance of a proxy firewall is directly influenced by TCP-based application protocols.

To build our simulation model, we chose 4 representative TCP-based application protocols: HTTP, FTP, SMTP, and TELNET. In our model, the load balancer distributes service requests to a group of firewalls modeled by the queuing system. To compare the performance of existing load balancing methods to that of the IWLB, we include 3 other load balancing methods, Round robin, Hashing, Randomization, in our simulation. Fig. 3 depicts the simulation model of the load balancer and multiple firewalls.

Let R be a set of service requests, which is generated with the arrival rate λ, arrived at the load balancer of multiple firewalls. We suppose that the inter-arrival time between two adjacent service requests and service durations of requests are exponentially distributed with the arrival rate λ and the service rate μ, respectively. Additionally, each element of R contains its application protocol for the IWLB.

After service requests are distributed to a group of firewalls by various load balancing methods, each firewall processes the allocated service requests. When we suppose that m firewalls installed in our model, the summation of the service requests allocated to each firewall equals the R if there is no blocking of requests at the load balancer. Therefore, as shown in the equation (1), the summation of the arrival rate at each firewall equals the arrival rate at the load balancer.

$$\lambda = \sum_{i=1}^{m} \lambda_i \tag{1}$$

The service request allocated to a firewall can be blocked in terms of its access policy. If we assume that the blocking rate, b_i, on service requests at Fw_i denoted as the ith firewall, then the rate of service requests processed by Fw_i, λ_{is} would be defined as the equation (2)

$$\lambda_{is} = \lambda_i (1 - b_i) \tag{2}$$

If service requests are blocked by the access policy of a firewall unpredictably, the load of firewalls will be distributed unevenly irrelevant of the fair distribution of service requests by the load balancer. Since no simple load balancing methods are able to cope with this situation, the fluctuation of workload between firewalls is inevitable. In the IWLB, if a service request is blocked, the firewall signals to a load balancer. When the load balancer receives the signal from the firewall, it allocates the next service request to the firewall once more to prevent the fluctuation of the load between firewalls.

We regarded each firewall as an M/M/c/c queuing system independently with the same capacity. The capacity of ith firewall, C_i, means the maximum number of active sessions that the firewall can handle with concurrently. If the number of active sessions exceeds the C_i, then the allocated service request to firewall will be queued for a later service.

5 Simulation Result and Analysis

In the simulation, we assume that the load balancer interacts with 4 firewalls. We generated 200,000 service requests to monitor the fair distribution of service requests and applied 3 simple load balancing methods and the IWLB to the load balancer in our model. During the simulation, we supposed that the propagation delays from the load balancer to firewalls are ignorable.

For the analysis of simulation results, we monitored the summation of service time and the waiting time of service requests in the queue at each firewall every second. If we analyze these values between firewalls, we would judge whether the applied load balancing methods distributed the service requests optimally to firewalls.

Firstly, we compared the response time of the IWLB to that of other simple load balancing methods. **Fig. 4** depicts the maximum response time of Hashing, Randomization, Round robin, and the IWLB, respectively. We sampled the response times of each firewall every 1 second and select the maximum response time among the sampled value. The graph shows that the maximum response time of the IWLB is remarkably lower than others. Additionally, the fluctuation of its curve is relatively narrower than that of other simple methods. Note that we put the results of each load balancing method together in a graph for comparison.

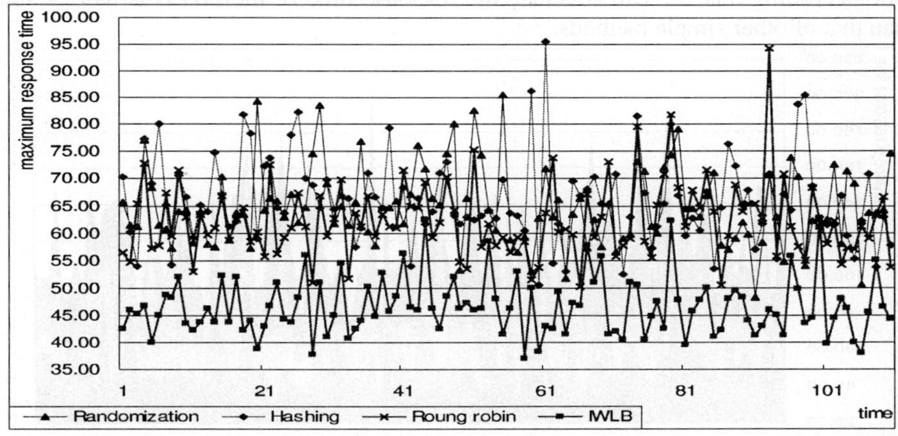

Fig. 4. Maximum response time

For more sophisticated comparison, we calculated the mean response time of each load balancing method and the mean response time of each method is depicted in **Fig. 5**. While the mean values of Round robin, Randomization, Hashing methods are not much different each other, those of the IWLB are quite different from them. The IWLB shows very lower mean response time during our simulation.

Now, from the two results of the IWLB, we can judge that the IWLB outperformed other simple load balancing methods.

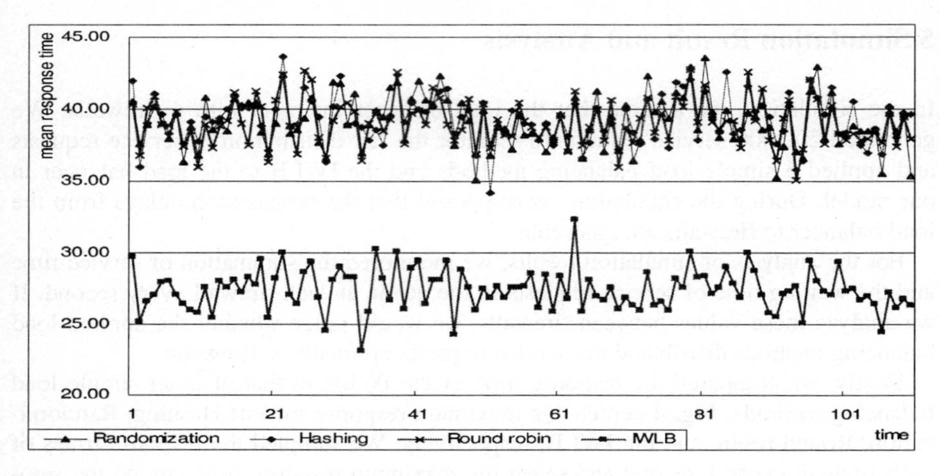

Fig. 5. Mean response time

In **Fig. 6**, we calculated the variance of the response time of each load balancing method. In this figure, we can see that the variance of the IWLB is lower than others, too. According this, we can conclude the response time of the IWLB is more stable than that of other simple methods.

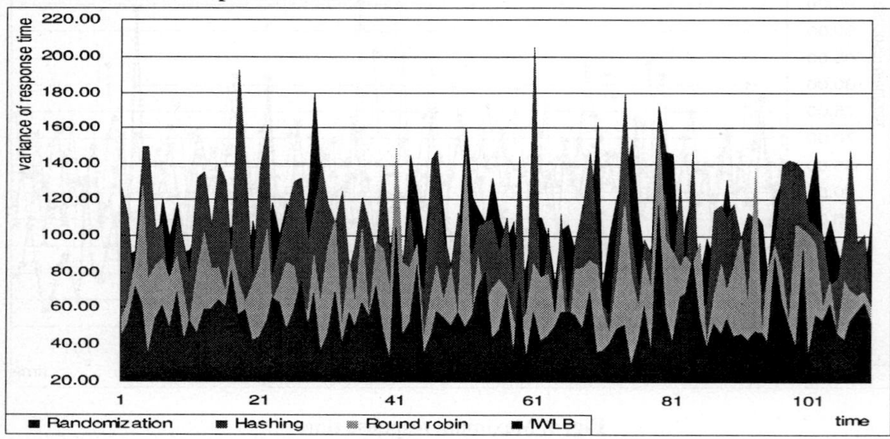

Fig. 6. The variance of response time

From the comparison, we can deduce two facts. One is that the distribution of service requests by simple load balancing methods causes the skewed load distribution among firewalls because they do not consider the blocking of requests in terms of the access policy of firewalls. The other is that the load balancing principle of the IWLB, considering the characteristics of application protocols, shows the positive effect for the fair traffic distribution.

As **Table 2** shows, we compared the mean and maximum buffer size of each firewall. In the case of the IWLB, the mean and maximum buffer sizes are greatly smaller than those of other simple methods. That is, the IWLB balances loads most fairly among simulated methods and can reduce the memory resource of firewalls. Note that the mean values of buffer size are calculated to the second decimal place.

Table 2. The comparison of the buffer size of firewalls

	Mean buffer size	Max buffer size
Round robin	8.13	40
Randomization	8.21	35
Hashing	8.12	39
IWLB	0.14	15

During the analysis of simulation results, we observed that the IWLB outperformed other simple load balancing methods in many aspects. Moreover, the simulation results explain that the IWLB proved that it is able to cope with massive traffic loads adequately. Consequently, it is strongly required to consider the weight of application protocols for fair load balancing to counteract massive service requests. Furthermore, when the load balancer interacts with firewalls, it should be able to manage the request blocking by the firewall to prevent the fluctuation of workloads between them. We expect that the IWLB meets to these requirements successfully.

6 Conclusion and Further Studies

Although many organizations are deploying firewalls for the purpose of network security, it is doubtable that they can process massive network traffics without performance degradation. To make it worse, considering the trends of preferring an application gateway firewall that performs more sophisticated operations, it is obvious the performance degradation will be more serious for massive network traffic. Because the performance degradation or malfunction of firewalls implies the critical security flaws, it is strongly required to make the firewall more tolerate against massive network traffic.

Several researches paid attention to deploying multiple firewalls and load balancers to counteract massive network traffic. It was unsuccessful to manage them because of the inherent drawbacks of existing simple load balancing methods for firewalls.

In this paper, we proposed the enhanced load balancing method, the IWLB, to manage overloaded network traffic efficiently. Since the IWLB makes use of the weight values of application protocols, calculated by the statistical traffic pattern of application protocols, for the load distribution, it is optimized for the load balancing method for multiple application gateway firewalls.

As shown in our simulation results, we observed that the IWLB outperformed other simple load balancing methods on the load distribution of massive TCP-based application service requests. From these results, we expect that the IWLB would be suitable for the load balancer for the network that deploying multiple application gateway firewalls is inevitable.

At this moment, we are planning to extend our evaluation model to various application protocols and performance measurement to various aspects such as resource usage, packet loss, and so on. Additionally, we will evaluate the scalability and tolerance of the IWLB when some firewalls are not functioning properly.

References

1. Rajkumar, B.: High Performance Cluster Computing: Architecture and Systems, Volume 1, Prentice Hall PTR, (1999)
2. Leon-Garcia, A.: Probability and Random Process for Electrical Engineering, 2nd Ed., Addison Wesley Publishing Company, Inc., (1994)
3. Molloy, K.M.: Fundamentals of Performance Modeling, Macmillan Publishing Company, (1989)
4. Law, M.A., Kelton, W.D.: Simulation Modeling & Analysis 2nd ed., McGraw-Hill Book Co., (1991)
5. Cheswick , R.W., Bellovin, M.S.: Firewalls and Internet Security : repelling the willy hacker, Addison Wesley, (1994)
6. Chapman, D.B., Zwicky, D.E.: Building Internet Firewalls, O Reilly & Associations, Inc., (1996)
7. Hare, C., Siyan, K.: Internet Firewalls and Network Security - 2nd ed., New Readers, (1996)
8. Kostic, C., Mancuso, M.: Firewall Performance Analysis Report, Computer Sciences Corporation, Secure Systems Center – Network Security Department, (1995)
9. Haeni, E. R.: Firewall Penetration Testing, The George Washington University, Cyberspace Policy Institute, (1997)

10 Test Final Report – Firewall Shootout Networkd+Interop, KeyLabs Inc., 28 May 1998.

11. Foundry ServerIron Firewall Load Balancing Guide, Foundry Networks, Inc., (2001)
12. Carnegie Mellon University, CCERT Advisory CA-2001-26 Nimda Worm, CERT/CC, http://www.cert.org/advisories/CA-2001-26.html, (2001)
13. Carnegie Mellon University, CERT Incident Note IN-99-07: Distributed Denial of Service Tools, CERT/CC, http://www.cert.org/incident_notes/IN-99-07.html, (1999)
14. McCreary, S., Claffy, K.: Trends in wide area IP traffic patterns - A view from Ames Internet Exchange, Proceedings of 13th ITC Specialist Seminar on Internet Traffic Measurement and Modeling, Monterey, CA. 18-20, (2000)

Modeling and Performance Evaluation of Multistage Interconnection Networks with Nonuniform Traffic Pattern*

Youngsong Mun[1] and Hyunseung Choo[2]

[1] School of Computing, Soongsil University, Seoul, KOREA
mun@computing.ssu.ac.kr
[2] School of Electrical and Computer Engineering
Sungkyunkwan University, Suwon 440-746, KOREA
choo@ece.skku.ac.kr

Abstract Even though there have been a number of studies about modeling MINs, almost all of them are for studying the MINs under uniform traffic which cannot reflect the realistic traffic pattern. In this paper, we propose an analytical method to evaluate the performance of ATM switch based on MINs under non-uniform traffic. Simulation results show that the proposed model is effective for predicting the performance of ATM switch under realistic nonuniform traffic. Also it shows that the detrimental effect of hot spot traffic on the network performance turns out to get more significant as the switch size increases.

1 Introduction

Since ATM has been adopted as a standard for broadband ISDN, many research efforts have been focused on the design of the next generation switching systems for ATM. The three main approaches employed for the design of an ATM switch are shared medium, shared memory, and space-division architecture [1]. In all these designs, the limitation on the switching size is the primary constraint in the implementation. To make a larger size ATM switch, thus, more than one system is interconnected in a multistage configuration [2].

Multistage interconnection networks (MINs) [3] constructed by connecting simple switching elements (SEs) in several stages have been recognized as an efficient interconnection structure for parallel computer systems and communication systems. There have been a number of studies investigating the performance of MINs in the literature [4-8]. However, almost all of these previous works are for studying the MINs under the uniform traffic pattern. Nonuniform traffic reflects the realistic traffic pattern of currently deployed integrated service network where a wide range of bandwidths needs to be accommodated. Therefore, the performance of the MINs under nonuniform traffic must be studied for obtaining efficient switch-based system. Even though

---

* This work was supported by Brain Korea 21 Project.

P.M.A. Sloot et al. (Eds.): ICCS 2002, LNCS 2331, pp. 1061–1070, 2002.

there have been some models considering nonuniform traffic patterns [5,7], they are not precise enough since the performance of the models has not been verified.

In this paper, we propose an analytical method to evaluate the performance of ATM switch based on MINs under nonuniform traffic. It is mainly achieved by properly reflecting the nonuniform dispatch probability in modeling the operation of each switch element. To evaluate the accuracy of the proposed model, comprehensive computer simulation is performed for two performance measures – throughput and delay. MINs of 6 and 10 stages with buffer modules holding single or multiple cells are considered for evaluation. As nonuniform traffic pattern, hot spot traffic of 3.5% and 7% are investigated. Comparison of the simulation data with the data obtained from the analytical model shows that the proposed model is effective for predicting the performance of ATM switch under realistic nonuniform traffic. The detrimental effect of hot spot traffic on the network performance turns out to get more significant as the switch size increases. For example, the throughput is about 0.3 for 6-stage switch with 3.5% hot spot traffic, while it becomes only about 0.03 for 10-stage switch.

2 The Proposed Model

2.1 Assumptions, Buffer States, and Definitions

In our models, 2×2 switching elements with the buffer modules of size m are used, and a network cycle consists of two phases. The sending buffer modules check the buffer space availability of the receiving buffer modules in the first phase. Based on the availability (and routing information) propagated backward from the last stage to the first stage, each buffer module sends a packet to its destination or enters into the blocked state in the second phase.

In each network cycle packets at the head of each buffer module (head packets) in an SE contend with each other if the destinations of them are same. Based on the status of the head packet, the state of a buffer module can be defined as follows. Figure 1 shows the state transition diagram of a buffer module in SEs.

- *State-* 0 : a buffer module is empty.
- *State-* n_k : a buffer module has k packets and the head packet moved into the current position in the previous network cycle.
- *State-* b_k : a buffer module has k packets and the head packet could not move forward due to the empty space of its destined buffer module in the previous network cycle.

The following variables are defined to develop our analytical model. Here $Q(ij)$ denotes the j-th buffer module in Stage-i. And its conjugate buffer module is represented as $Q(ij^c)$. Also t_b represents the time instance when a network cycle begins, while t_d represents the duration of a network cycle.

- m : the number of buffers in a buffer module.
- n : the number of switching stages. There are $n = \log_2 N$ stages for $N \times N$ MINs.

- $P_0(ij,t) / \overline{P(ij,k)}$: the probability that $Q(ij)$ is empty/not full at t_b .
- $P_{n_k}(ij,t)$: the probability that $Q(ij)$ is in *State*- n_k at t_b , where $1 \leq k \leq m$.
- $P_{b_k}(ij,t)$: the probability that $Q(ij)$ is in *State*- b_k at t_b , where $1 \leq k \leq m$.
- $SP_n(ij,t)$: $\sum_{k=1}^{m} P_{n_k}(ij,t)$
- $SP_b(ij,t)$: $\sum_{k=1}^{m} P_{b_k}(ij,t)$
- $P_b^u(ij,t) / P_b^l(ij,t)$: the probability that a head packet in $Q(ij)$ is a blocked one and destined to the upper/lower output port at t_b .

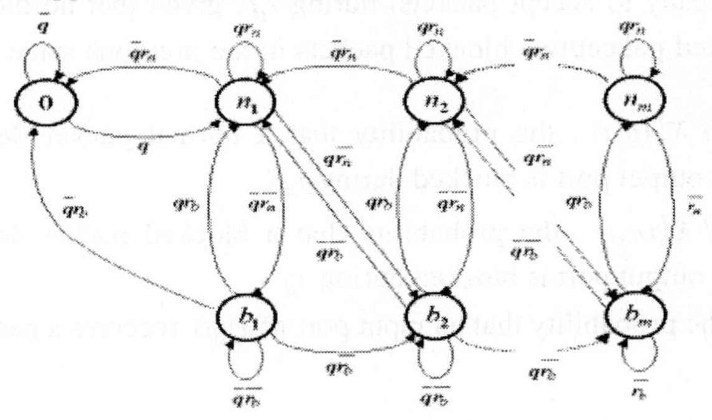

Figure 1. The state transition diagram of the proposed model.

- $r(ij) / r_x(ij,t)$: the probability that a normal/blocked head packet in $Q(ij)$ is destined to the upper output port.
- $q(ij,t)$: the probability that a packet is ready to come to the buffer module $Q(ij)$.
- $r_n(ij,t) / r_b(ij,t)$: the probability that a normal/blocked packet at the head of $Q(ij)$ is able to move forward during t_d .
- $r_n^u(ij,t) / r_n^l(ij,t)$: the probability that a normal packet at the head of $Q(ij)$ can get to the upper/lower output port during t_d .
- $r_b^u(ij,t) / r_b^l(ij,t)$: the probability that a blocked packet at the head of $Q(ij)$ can get to the upper/lower output port during t_d .
- $r_{nn}^u(ij,t) / r_{nn}^l(ij,t)$: the probability that a normal packet at the head of $Q(ij)$ can get to the upper/lower output port during t_d by considering $Q(ij^c)$ in either *State*- n or *State*- b . If $Q(ij^c)$ is in *State* - b , it is assumed that the blocked packet is destined to the lower/upper port (so no contention is necessary).
- $r_{nb}^u(ij,t) / r_{nb}^l(ij,t)$: the probability that a normal packet at the head of $Q(ij)$ is able to get to the upper/lower output port during t_d by winning the contention with a blocked packet at the head of $Q(ij^c)$.

- $r^u_{bn}(ij,t) / r^l_{bn}(ij,t)$: the probability that a blocked packet at the head of $Q(ij)$ is able to move forward to the upper/lower output port during t_d. Here it is assumed that $Q(ij^c)$ is empty or in the *State-* n.
- $r^u_{bb}(ij,t) / r^l_{bb}(ij,t)$: the probability that a blocked packet at the head of $Q(ij)$ is able to move forward to the upper/lower output port during t_d. Here it is assumed that $Q(ij^c)$ also has a blocked packet.
- $P^{na}(ij,t) / P^{ba}(ij,t) / P^{bba}(ij,t)$: the probability that a buffer space in $Q(ij)$ is avaible (ready to accept packets) during t_d, given that no blocked packet/only one blocked pakcet/two blocked packets in the previous stage is destined to that buffer.
- $X^u_n(ij,t) / X^l_n(ij,t)$: the probability that a normal packet destined to the upper/lower output port is blocked during t_d.
- $X^u_b(ij,t) / X^l_b(ij,t)$: the probability that a blocked packet destined to the upper/lower output port is blocked during t_d.
- $T(ij,t)$: the probability that an input port of $Q(ij)$ receives a packet.

2.2 Calculations of Required Measures

2.2.1 Obtaining $r_n(ij,t)$

A normal packet in an SE is always able to get to the desired output port when the other buffer module is empty or destined to a different port from it. When two normal packets compete, each packet has the equal probability to win the contention. The probability that a normal packet in $Q(ij)$ does not compete with a blocked packet in the other buffer module is $r(ij)\{1-r_x(ij^c,t)\} + \{1-r(ij)\}r_x(ij^c,t)$. Therefore, the probabilities $r^u_{nn}(ij,t)$ is as follows and $r^l_{nn}(ij,t)$ is obtained similarly.

$$r^u_{nn}(ij,t) = r(ij)P_0(ij^c,t) + [0.5r(ij)r(ij^c) + r(ij)\{1-r(ij^c)\}]SP_n(ij^c,t) + r(ij)\{1-r_x(ij^c,t)\}SP_b(ij^c,t) \tag{1}$$

$r^u_{nb}(ij,t)$ and $r^l_{nb}(ij,t)$ are the probabilities that a normal packet has the same destination as the blocked one in the other buffer module and wins the contention. Thus they are as follows:

$$r^u_{nb}(ij,t) = 0.5r(ij)r_x(ij^c,t)SP_b(ij^c,t), \tag{2}$$

The probability that a buffer module is not full ($\overline{P(ij,t)}$) is simply

$$\overline{P(ij,t)} = 1 - P_{n_m}(ij,t) - P_{b_m}(ij,t). \tag{3}$$

If the originating buffer module of a packet is in $State-b_i$ $(1 \le i \le m)$, then the destined buffer module must be in either $State-n_j$ $(1 \le j \le m)$ or $State-b_k$ $(2 \le k \le m)$. If it has received a packet in the previous network cycle, it can be in $State-n_j$ $(1 \le j \le m)$ or $State-b_k$ $(2 \le k \le m)$. If it does not have received a packet, it must be in $State-b_m$. Thus

$$P^{ba}(ij,t) = T(ij,t-1) \times A + \{1 - T(ij,t-1)\} \frac{P_{b_m}(ij,t) r_b(ij,t)}{P_{b_m}(ij,t)}. \tag{4}$$

Here $A = \dfrac{\sum_{k=1}^{m-1} P_{n_k}(ij,t) + \sum_{k=2}^{m-1} P_{b_k}(ij,t) + P_{n_m}(ij,t) r_n(ij,t) + P_{b_m}(ij,t) \times r_b(ij,t)}{1 - P_0(ij,t) - P_{b_1}(ij,t)}$.

The probabilities $r_n(ij,t)$ and $r_b(ij,t)$ will be discussed later in this section. $P^{na}(ij,t)$ is obtained similarly. If the destined buffer module has not received a packet, it must be in any state except $State\text{-}n_m$. Then

$$P^{na}(ij,t) = T(ij,t-1) \times A + \{1 - T(ij,t-1)\} \times B \tag{5}$$

Here $B = \dfrac{P_0(ij,t) + \sum_{k=1}^{m-1} P_{n_k}(ij,t) + \sum_{k=1}^{m-1} P_{b_k}(ij,t) + P_{b_m}(ij,t) \times r_b(ij,t)}{1 - P_{n_m}(ij,t)}$.

For a packet to move to the succeeding stage, it should be able to get to the desired output port and the destined buffer module should be available. Thus $r_n^u(ij,t)$ is as follows and $r_n^l(ij,t)$ is obtained similarly.

$$r_n^u(ij,t) = r_{nn}^u(ij,t) P^{na}((i+1),t) + r_{nb}^u(ij,t) P^{ba}((i+1),t) \tag{6}$$

So $r_n(ij,t)$ is

$$r_n(ij,t) = r_n^u(ij,t) + r_n^l(ij,t). \tag{7}$$

We can calculate $r_b(ij,t)$ using the similar method.

2.2.2 Obtaining $X_n^u(ij,t)$, $X_b^u(ij,t)$, $X_n^l(ij,t)$, $X_b^l(ij,t)$, and $r_x(ij,t)$

$X_n^u(ij,t)$ is the probability that a normal packet destined to the upper output port is blocked.

$$\begin{aligned} X_n^u(ij,t) = {} & r_{nn}^u(ij,t)\{1 - P^{na}((i+1),t)\} + r_{nb}^u(ij,t)\{1 - P^{ba}((i+1),t)\} \\ & + 0.5 r(ij) r(ij^c) SP_n(ij^c,t) + 0.5 r(ij) r_x(ij^c,t) SP_b(ij^c,t) \end{aligned}. \tag{8}$$

The first two terms in the equation above are the probabilities that the destination has no available space. The last two terms are for the case of lost contention. $X_b^u(ij,t)$ is the probability that a blocked packet destined to the upper output port is blocked again. We can calculate this probability easily by the approach emplyed in $X_n^u(ij,t)$.

$$X_b^u(ij,t) = r_{bn}^u(ij,t)\{1 - P^{ba}((i+1),t)\} + r_{bb}^u(ij,t)\{1 - P^{bba}((i+1),t)\} \\ + 0.5r_x(ij)r(ij^c)SP_n(ij^c,t) + 0.5r_x(ij)r_x(ij^c,t)SP_b(ij^c,t) \quad (9)$$

$X_n^l(ij,t)$ and $X_b^l(ij,t)$ are obtained similarly.

Also $r_x(ij,t)$, which is the probability that a blocked head packet is destined to the upper output port, is calculated as follows.

$$r_x(ij,t) = \frac{P_b^h(ij,t-1)}{P_b^h(ij,t-1) + P_b^l(ij,t-1)} \qquad (P_b^u(ij,t-1) + P_b^l(ij,t-1) \neq 0) \quad (10)$$

Here $P_b^u(ij,t)$ and $P_b^l(ij,t)$ are calculated as follows.

$$P_b^u(ij,t) = X_n^u(ij,t)SP_n(ij,t) + X_b^u(ij,t)SP_b(ij,t), \quad (11)$$

$$P_b^l(ij,t) = X_n^l(ij,t)P_n(ij,t) + X_b^l(ij,t)P_b(ij,t). \quad (12)$$

2.2.3 Obtaining $T(ij,t)$ and $q(ij,t)$

Due to its inherent connection property of MINs, the two buffer modules in an SE are connected to either upper or lower output ports of the SE of the previous stage. On the contrary, the buffer modules below it are connected to the lower output ports. We denote $T(ij,t)$ for the buffer modules connected to upper output ports as

$$T(ij,t) = SP_n((i-1)g,t)r_n^u((i-1)g,t) + SP_n((i-1)g^c,t)r_n^u((i-1)g^c,t) \\ + SP_b((i-1)g,t)r_b^u((i-1)g,t) + P_b((i-1)g^c,t)r_b^u((i-1)g^c,t) \quad (13)$$

The buffer modules which are connected to lower output ports of the previous stage are obtained similarly. $T(ij,t)$ $(1 \le i \le n)$ also has the following relation with $T(ij,t)$.

$$T(ij,t) = q(ij,t)[\overline{P(ij,t)} + P_{n_m}(ij,t)r_n(ij,t) + P_{b_m}(ij,t)r_b(ij,t)] \quad (14)$$

Finally, $q(ij,t)$ $(2 \le i \le n)$ is obtained.

$$q(ij,t) = \frac{T(ij,t)}{\overline{P(ij,t)} + P_{n_m}(ij,t)r_n(ij,t) + P_{b_m}(ij,t)r_b(ij,t)} \quad (15)$$

2.2.4 Calculating $r(ij,t)$

$r(ij)$ is calculated by using the transformation method proposed in [7]. It is a mapping scheme that transforms the given reference pattern into a set of $r(ij)$'s which reflect the steady state traffic flow in the network. For a steady state reference pattern, we represent it in terms of destination accessing probabilities A_j, the probability that a new packet generated by an inlet chooses the output port j as its destination. Then $r(ij)$ can be represented as the conditional probability that the sum of A_j 's which are connected to the upper output port of $Q(ij)$ given the sum of A_j 's of all possible destined output ports which are connected to the upper or lower output port of $Q(ij)$. For example, $r(ij)$'s in three stage MIN are described as follows.
For the last stage:

$$r(31) = r(32) = \frac{A_1}{A_1 + A_2}, \quad r(33) = r(34) = \frac{A_3}{A_3 + A_4},$$

$$r(35) = r(36) = \frac{A_5}{A_5 + A_6}, \quad r(37) = r(38) = \frac{A_7}{A_7 + A_8}.$$

For the second stage:

$$r(21) = r(22) = r(25) = r(26) = \frac{A_1 + A_2}{A_1 + A_2 + A_3 + A_4},$$

$$r(23) = r(24) = r(26) = r(27) = \frac{A_5 + A_6}{A_5 + A_6 + A_7 + A_8}.$$

For the first stage, all $r(1j)$ ($1 \le j \le N$) are same:

$$r(11) = r(12) = \cdots\cdots = (r18) = \frac{A_1 + A_2 + A_3 + A_4}{A_1 + A_2 + A_3 + A_4 + A_5 + A_6 + A_7 + A_8} = A_1 + A_2 + A_3 + A_4$$

2.3 Throughput and Delay

Normalized throughput of a MIN is defined to be the throughput of an output port of the last stage. If *Port-j* is the upper output port of an SE, the normalized throughput in this port is as follows.

$$\begin{aligned} TNET(j,t) = {} & SP_n(nj,t)r_n^u(nj,t) + SP_n(nj^c,t)r_n^u(nj^c,t) \\ & + SP_b(nj,t)r_b^u(nj,t) + SP_b(nj^c,t)r_b^u(nj^c,t) \end{aligned} \tag{16}$$

The delay occurred for a packet at the buffer module $Q(ij)$ in the steady state is calculated by using Little's formula.

$$D(ij) = \lim_{t \to \infty} \frac{\sum_{k=1}^{m} k\{P_{n_k}(ij,t) + P_{b_k}(ij,t)\}}{T(ij,t)} \tag{17}$$

As delay at each output port are different, the weight of it should be considered for obtaining the mean delay. Hence the mean delay is

$$D = \sum_{j=1}^{N} w_j D(j) \tag{18}$$

Here w_j – the weight of *Port-j* for the mean delay – is obtained by the rate of the normalized throughput of that port as follows.

$$w_j = \lim_{t \to \infty} \frac{TNET(j)}{\sum_{k=1}^{N} TNET(k,t)} \tag{19}$$

3 Verification of the Proposed Model

Correctness of our model in terms of network throughput and delay is verified by comparing them with the data obtained from computer simulation for various buffer sizes and traffic conditions. For the simulation, 95% confidence interval is used and the following approaches are employed for the computer simulation.

- Each inlet generates requests at the rate of the offered input traffic load.
- The destination of each packet follows the given hot spot nonuniform traffic pattern. Here each inlet makes a fraction h of their requests to a hot spot port, while the remaining $(1 - hP^{na}(ij,t))$ of their requests are distributed uniformly over all output ports including the hot spot port.
- If there is a contention between the packets in an SE, it is resolved randomly.
- The buffer operation is based on the FCFS principle.

Figure 2 shows the mean throughput and delay comparison of a 6-stage single buffered MIN with 7% the hot spot traffic. The offered traffic load varies form 0.1 to 1, and simulation data are obtained by averaging 10 runs. In each run, 1,000,000 iterations are taken to collect reliable data. The variations in the last 100,000 iterations are less than 0.1%. Figure 3 shows the comparison of the throughput of the hot spot port and other ports between the analytical model and computer simulation in this case. It reveals that the throughput of the hot spot port is more than two times higher than that of other ports since the access probability to the hot spot port is higher than others. Also Figures 4 and 5 show the comparison results of multiple buffer MINs. In case of uniform traffic, more buffer entries can increase the performance of MINs 10% to 20%. As identified here, in case of the nonuniform traffic, the increase in the throughput is as small as about 2% even though more buffers are added since blocking among the packets is more likely due to the nonuniform traffic. Similar result are shown in case of the 3.5% hot spot traffic.

The figures show that our models are effective for predicting the performance of MINs with realistic traffic. In case of the large sized MIN (1024×1024), the throughput of the hot spot port is always close to 1 since there always exists a packet to that port coming from a large number of input ports. However, those of other ports are as low as less than 0.03 since blocking is so severe.

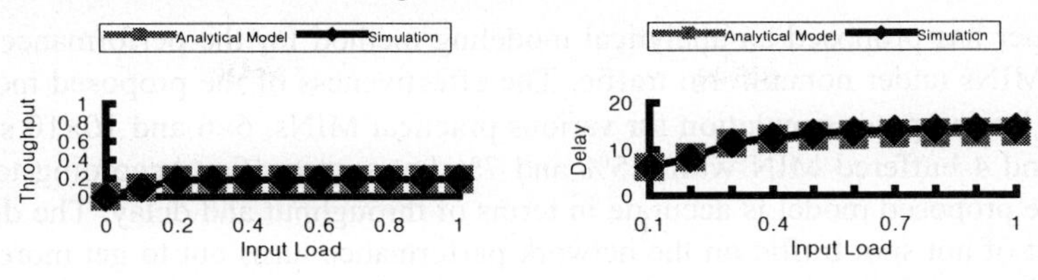

Figure 2. Comparison of the throughput and mean delay for 6-stage, single-buffered MIN delay with 7% hot spot traffic.

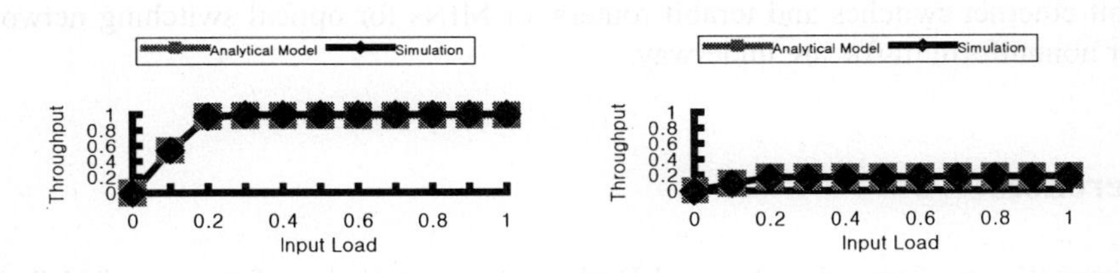

a) Hot spot port b) Other ports

Figure 3. Comparison of the throughput of hot spot port and other ports with 7% hot spot traffic.

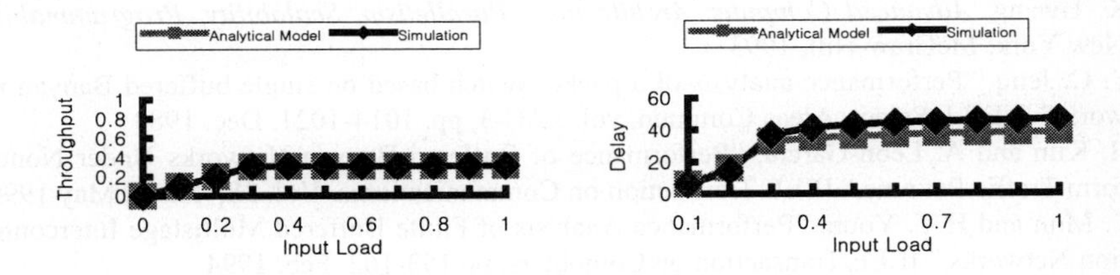

Figure 4. Comparison of the throughput and mean delay for 6-stage 4-buffered MIN with 7% hot spot traffic.

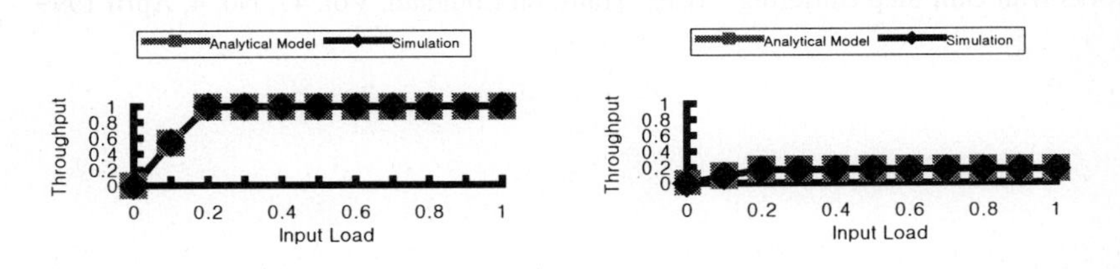

a) Hot spot port b) Other ports

Figure 5. Comparison of the throughput of hot spot port and other ports with 7% hot spot traffic.

4 Conclusion

This paper has proposed an analytical modeling method for the performance evaluation of MINs under nonuniform traffic. The effectiveness of the proposed model was verified by computer simulation for various practical MINs; 6×6 and 10×10 switches, single and 4-buffered MIN with 3.5% and 7% hot spot traffic. According to the results, the proposed model is accurate in terms of throughput and delay. The detrimental effect of hot spot traffic on the network performance turns out to get more significant as the switch size increases. For example, the throughput is about 0.3 for 6-stage switch with 3.5% hot spot traffic, while it becomes only about 0.03 for 10-stage switch. Therefore hot spot traffic needs to be avoided as much as possible for especially relatively large size switches. Performance analysis of other structures such as gigabit ethernet switches and terabit routers, or MINs for optical switching networks under nonuniform traffic are underway.

References

1. Hyoung-IL Lee, Seung-Woo Seo and Hyuk-jae Jang. "A High performance ATM Switch Based on the Augmented Composite Banyan Network," IEEE International Conference on Communications, Vol.1, pp.309-313, June 1998.
2. Muh-rong Yang and GnoKou Ma, "BATMAN : A New Architectural Design of a Very Large Next Generation Gigabit Switch," IEEE International Conference on Communications, Vol.2/3, pp.740-744, May 1997.
3. K. Hwang, *Advanced Computer Architecture: Parallelism, Scalability, Programmability.* New York: McGraw-Hill, 1993.
4. Y. C. Jenq, "Performance analysis of a packet switch based on single buffered Banyan network," IEEE J. Select. Areas Commun, vol. SAC-3, pp. 1014-1021, Dec. 1983.
5. H. Kim and A. Leon-Garcia, "Performance of Buffered Banyan Networks Under Nonuniform Traffic Patterns," IEEE Transaction on Communicationis, Vol. 38, No. 5, May 1990.
6. Y. Mun and H.Y. Youn, "Performance Analysis of Finite Buffered Multistage Interconnection Networks," IEEE Transaction on Computers, pp.153-162, Feb. 1994
7. T. Lin and L. Kleinrock, "Performance Analysis of Finite-Buffered Multistage Interconnection Networks with a General Traffic Pattern", ACM SIGMETRICS Conference on Measurement and Modeling of Computer Systems, San Diego, CA, pp. 68-78, May 21-24, 1991.
8. H.Y. Youn and H. Choo, "Performance Enhancement of Multistage Interconnection Networks with Unit Step Buffering," IEEE Trans. on Commun. Vol. 47, No. 4, April 1999.

Real-Time Performance Estimation for Dynamic, Distributed Real-Time Systems

Eui-Nam Huh[1], Lonnie R. Welch[2], and Y. Mun[3]

[1] Sahmyook University, Department of Computer Science
Chongyang, P.O.Box 118, Seoul, Korea
huh@syu.ac.kr
[2] Ohio University, School of Electrical Engineering & Computer Science
Athens, Ohio, USA
welch@ohiou.edu
[3] Soongsil University, School of Computer Science
Seoul, Korea
mun@computing.ssu.ac.kr

Abstract. The main contribution of this paper is accurate analysis of real-time performance for dynamic real-time applications. A wrong system performance analysis can lead to a catastrophe in a dynamic real-time system. In addition, real-time performance guarantee combined with efficient resource utilization is observed by experiments, while the previous worst-case approaches primarily focused on performance guarantee but resulted in typically poor utilization. The subsequent contribution is schedulability analysis for a feasible allocation of resource management on the Solaris operating system. This is accomplished with a mathematical model and by accurate response time prediction for a periodic, dynamic distributed real-time application.

1 Introduction

The rac-25 radiation machine killed cancer patients because of a software bug. Imagine that you have a car accident. After you have already been thrown into the windshield, the airbag inflates. This paper addresses the problem of certifying that software will respond to real-world events in a timely manner.

Use of real-time systems is being spread rapidly to many areas. Real-time services need to react to dynamic user requests. A dynamic real-time system used to offer services to the dynamic user requests should consider variable execution times and/or arrival rates of tasks during run-time.

One of the management problems of dynamic real-time systems that must be solved is to provide the quality of service (QoS) for a time-constrained task. The task commonly appears in systems such as air traffic control, robotics, automotive safety, mission control, and air defense. An example of this task is software that detects radar data, evaluates it and launches missiles if a hostile missile is detected. All of these dynamic real-time systems require consideration of variable execution times and/or arrival rates of tasks, as opposed to deterministic and stochastic real-time systems which have a priori known, fixed task execution times and arrival rates.

P.M.A. Sloot et al. (Eds.): ICCS 2002, LNCS 2331, pp. 1071–1079, 2002.

Resource allocation, mapping software (S/W) to hardware (H/W), for those systems is an essential component and needs to consider real-time constraints of the task and should analyze feasible resources. Thus, the feasible allocation of the available resources has to be carried out according to time-constrained requirements, failure of which might cause a catastrophe in hard real-time systems. To maintain the feasible allocation, shedulability analysis of the task is required. Prediction of the response time of the task compared to the time-constrained requirement is one of the schedulability analysis approaches. Furthermore, computing resources for those dynamic systems should be utilized efficiently. The distributed system is employed to provide scalable resources to the dynamic S/W systems that have various and unbounded execution times.

Generally, the real-time system is designed to analyze that a task can meet its time constraint before it is executed. The Rate Monotonic Analysis (RMA) introduced by Liu and Layland in [1] is used primarily to determine schedulability of an application by using a priori Worst-Case Execution Time (WCET) and the priority of the application. The priority of the application to be applied to RMA is dependent upon arrival patterns and rates. The application, which has a higher arrival rate of task or needs to be executed more frequently, has a higher priority level than any other applications.

However, as has been noted in [2], [3], [4], [5], and [6], resources are poorly utilized if the average case is significantly less than the worst case. Another drawback of RMA is that it cannot efficiently accommodate high-priority jobs that have relatively low rates. It must, however, be noted that RMA can be made to work in such cases, by transforming low-rate, high-priority jobs into high-rate jobs -- but this can be extremely wasteful in terms of resources.

It is stated in [6], [7] and [13] that accurately measuring the WCET is often difficult, and is sometimes impossible. Puschner and Burns in [8] consider WCET analysis to be hardware dependent, making it an expensive operation on distributed heterogeneous clusters.

The statistical RMA by Atlas and Bestavros in [9] considers tasks that have variable execution times and allocate resources to handle the expected case. The benefit of this approach is the efficient utilization of resources. However, there are shortcomings. Firstly, applications which have a wide variance in resource requirements cannot be characterized accurately by a time-invariant statistical distribution; and secondly, deadline violations occur when the expected case is less than the actual case. Similarly, real-time queuing theory by Lehoczky in [5] uses probabilistic event and data arrival rates for performing resource allocation analysis. On the average, this approach provides good utilization of resources. It must be noted, however, that applications which have a wide variance of resource requirements cannot be characterized accurately by a time-invariant statistical distribution. Called the dynamic real-time system by Welch and Masters in [10], there is a need for a new approach to the dynamic real-time system which would be more efficient than RMA and statistical RMA in assurance of the real-time QoS.

2 Problem Statement

This section uses mathematical notation to describe the problem of certifying that a real-time application meets (or does not meet) its real-time requirement on a host that is running other applications. The notation will be used in subsequent sections to concisely define certification methods. For convenience, the Data Flow Diagram (DFD) is used for description of problems.

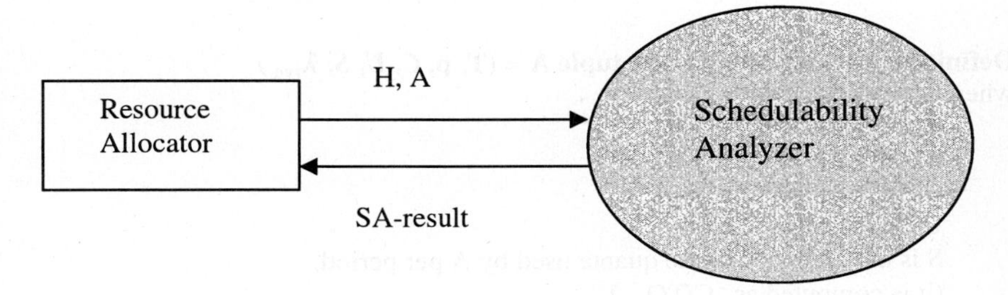

Fig. 1. Level 0 DFD of "Schedulabilty Analyzer System"

As shown in Fig.1, the Schedulability Analyzer certifies (or says that it cannot certify) that the application ('A') will meet its real-time requirement on the host (H). The application 'A' is represented as period, priority, execution time, segment and real-time requirement. The host 'H' is represented as its name, time quanta, and a list of applications. SA-result is returned to indicate the certification result, which consists of a boolean value and a predicted response time.

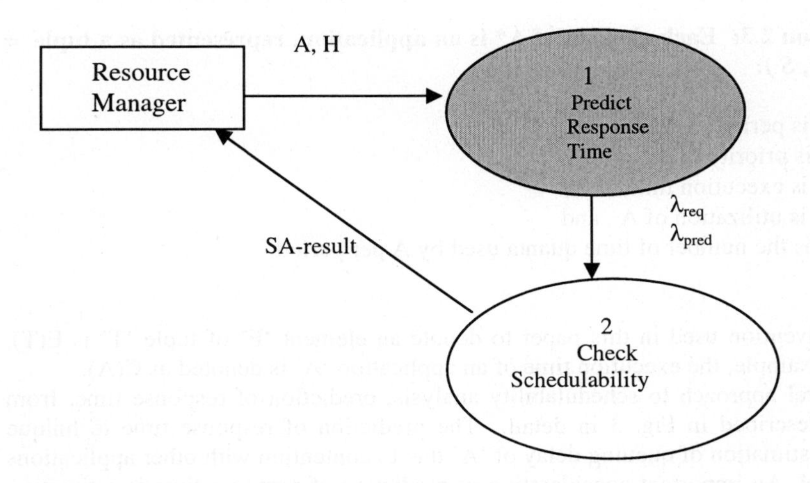

Fig. 2. Level 1 DFD of "Schedulability Analyzer"

The Schedulability Analyzer from Fig. 1 is depicted in Fig. 2. This paper presents a novel approach to schedulability analysis, which predicts response time (λ_{pred}) of 'A' and checks schedulability to certify that 'A' will meet its real-time requirement (λ_{req}) on H before allocation. 'Predict Response Time' reads application list ('A*') on H and computes λ_{pred} of 'A'. 'Check Schedulability' compares λ_{pred} with λ_{req} and returns 'true' if $\lambda_{pred} < \lambda_{req}$ and returns 'false' otherwise; the predicted response time; λ_{pred}, is also returned. 'A', H and the elements of A* are defined as tuples below.

Definition 2.1: An application tuple A = (T, p, *C*, *U*, *S*, λ_{req} *)*
where
T is period of A ,
p is priority of A,
C is execution time of A,
U is utilization of A ,
S is the number of time quanta used by A per period.
(it is computed as C/TQ .)
λ_{req} is the requirement of A. (note: we assume that $\lambda_{req} \leq$ T)

Definition 2.2: A host tuple H=(name, TQ, A*)
where
name is name of a host,
TQ is a vector defining the time quantum for each priority on a host
A* is a list of the applications:$\langle a_1, a_2,,, a_n \rangle$ that are currently allocated to H.

Definition 2.3: Each element of A* is an application, represented as a tuple = (T, p, *C*, *U*, *S*):
where
T is period of A ,
p is priority of A,
C is execution time of A,
U is utilization of A , and
S is the number of time quanta used by A per period.

The convention used in this paper to denote an element 'E' of tuple 'T' is E(T). Thus, for example, the execution time of an application 'A' is denoted as C(A).

The novel approach to schedulability analysis, prediction of response time, from Fig. 2 is described in Fig. 3 in detail. The prediction of response time technique considers estimation of queuing delay of 'A' due to contention with other applications ('A*') on H. An important consideration of prediction of response time is estimation of queuing delay due to the same priority (p) applications (D_{pred1}) as shown at 1.1 bubble in Fig.3. Queuing delay due to higher priority (p) applications (D_{pred2}) as shown as bubble 1.2 in Fig. 3 is also estimated as higher priority tasks always hold resources of H. Finally, Calculate response time of the application as shown as bubble

1.3 in Fig. 3 computes the estimated response time of 'A': $\lambda_{pred} = C + D_{pred1} + D_{pred2}$, where D_{pred1}: delay experienced by a due to waiting for $a_i \in$ 'A*' such that $p(a_i)=p(A)$, and D_{pred2}: delay experienced by waiting for $a_j \in$ 'A*' such that $p(a_j) > p(A)$.

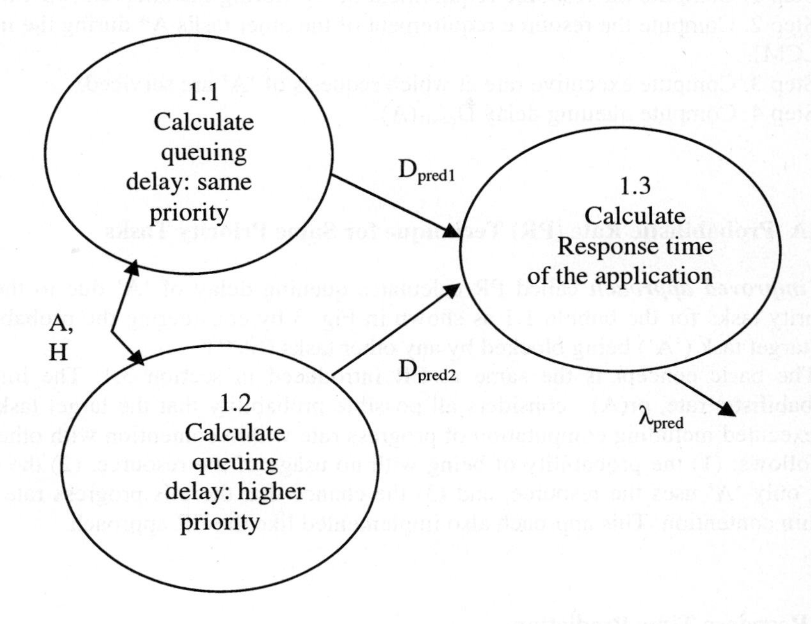

Fig. 3. Level 2 DFD of "Predict Response Time"

3 Real-Time Performance Analysis

This section presents approaches for performing estimation of real-time performance of dynamic real-time applications introduced as bubbles, 1.1, 1.2, and 1.3 as shown in Fig. 3. These approaches based on probabilistic techniques are extended to be applicable to time-sharing round-robin scheduler, which considers queuing delay due to the same priority tasks. As mentioned in section 1, the worst-case response time analysis approaches, [11] and stochastic approaches, [5] and [9] are not appropriate for response time analysis in dynamic real-time systems.

3.1 A Execution Rate (ER) Technique

This approach calculates queuing delay of 'A' due to the same priority tasks, and considers applications' periods to find total contention among them. Least Common Multiple (LCM) of periods of 'A' and 'A*' on H is used in this approach. To extend this approach to round-robin scheduler, the execution of ‘A’ is analyzed in the unit of time quantum (TQ) and segment (S). Following steps show how it is processed to compute $D_{pred1}(A)$ due to the same priority tasks.

Step 1. Compute the resource requirement of ‘A’ during the interval, [0,LCM].
Step 2. Compute the resource requirement of the other tasks A* during the interval, [0,LCM].
Step 3. Compute executive rate at which requests of ‘A’ are serviced.
Step 4. Compute queuing delay $D_{pred1}(A)$.

3.2 A Probabilistic Rate (PR) Technique for Same Priority Tasks

An ***improved approach*** called PR calculates queuing delay of ‘A’ due to the same priority tasks for the bubble 1.1 as shown in Fig. 3 by considering the probability of the target task (‘A’) being blocked by any other tasks (‘A*’).

The basic concept is the same as ER introduced in section 3.1. The improved probabilistic rate, pr(A), considers all possible probability that the target task could be executed including computation of progress rate within contention with other tasks as follows: (1) the probability of being with no usage of the resource, (2) the chance that only ‘A’ uses the resource, and (3) the chance that there is progress rate of ‘A’ within contention. This approach also implemented like the ER approach.

3.3 Response Time Prediction

This section simply combines C(A), $D_{pred1}(A)$, and $D_{pred2}(A)$ to compute the predicted response time of the task, $\lambda_{pred}(A)$ as shown as the bubble 1.3 in Fig. 3. That is, $\lambda_{pred}(A) = C(A) + D_{pred1}(A) + D_{pred2}(A)$.

4. Experiments and Summary

This section shows how the approaches accurately predict response time of an application using DynBench, path based s/w systems, introduced by Welch and Shirazi (see [12]).

The ‘filter1’ application (which filters noise from sensed data) in the first sensing path (denoted as “path 1”) as the task is examined on the variable size of workload scenario. Thirty samples of the response time of ‘filter1’ are collected and averaged, when workloads of the path 2 are changed dynamically and monotonically. The workload of sensing path 3 is fixed by 800, and that of sensing path 2 is suddenly increased to 1300 by adding 800 workloads at 1 minute 20 seconds; it drops by 600 at 2 minutes 40 seconds; it adds 700 again at 4 minutes, and drops 500 at 5 minutes 20 seconds. See Fig. 4.

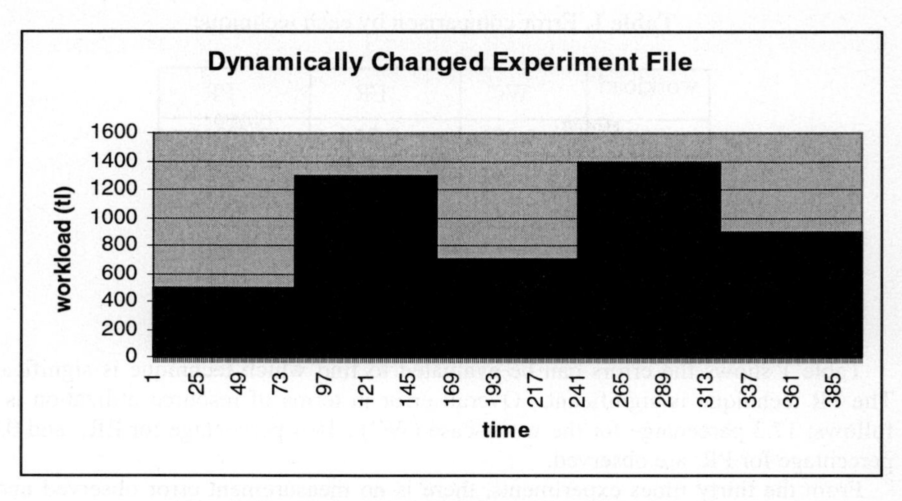

Fig. 4. Dynamically changed workload scenario

The response time, $\lambda_{pred}=D_{pred1}+D_{pred2}+C(\text{filter1})$, depicted in Fig. 5, is measured using the same approach for each contention case by ER and PR. And the worst-case analysis (denoted as WC) by Audsley in [11] is employed as well to see how the new technique is accurate.

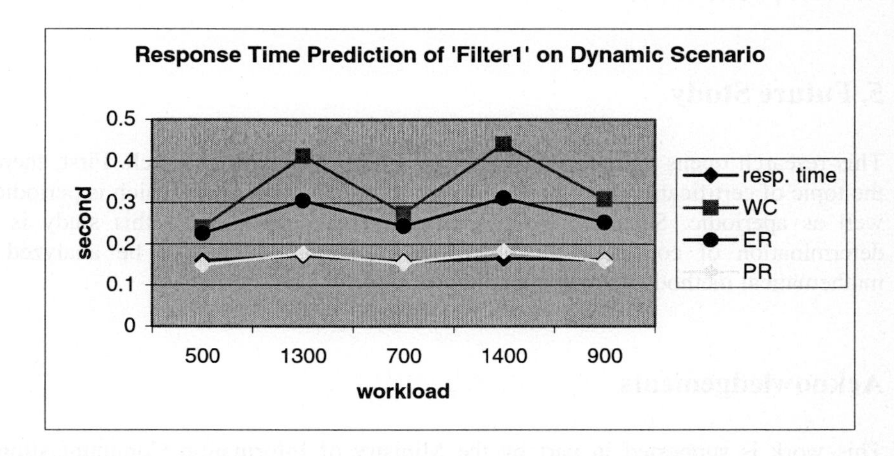

Fig. 5. Predicted Response time comparison

Table 1. Error comparison by each technique

workload	WC	ER	PR
500	0.081	0.066	0.011
1300	0.241	0.133	0.005
700	0.117	0.086	0.004
1400	0.277	0.146	0.018
900	0.147	0.091	0.004

Table 1 shows the errors can be evaluated to find which technique is significant. The PR technique is significant. Overall error in terms of resource utilization is as follows: 17.3 percenatge for the worst-case (WC) , 10.4 percentage for ER, and 0.08 percentage for PR are observed.

From the thirty times experiments, there is no measurement error observed under the 95% confidence interval. The PR, probabilistic contention analysis technique, can accurately predict response time of an application on a host, while the worst-case analysis poorly predicts response times. Therefore, these experiments for the dynamic real-time systems give strong analysis that the worst-case analysis poorly utilizes computational resources, and new approaches can predict response time accurately with the dynamic environment constraints using current utilization. The probabilistic response time prediction method for dynamic real-time systems rather than the worst-case is very useful in terms of resource utilization and certification of real-time performance.

5. Future Study

This research opens the possibility for yet future interesting research. First, there is the topic of certification for an event-driven real-time application which is periodic as well as aperiodic. Second, a new research area derived from this study is the determination of confidence levels of certification, which will be analyzed by mathematical methods with an upper bound, or error.

Acknowledgements

This work is supported in part by the Ministry of Information Communication of Korea, under the "Support Project of University Information Technology Research Center(ITRC)" supervised by KIPA".

References

1. Liu, C. L., and Layland , J.W.: Scheduling algorithms for multi-programming in a hard-real-time environment. Journal of the ACM, Vol. 20. (1973) 46-61
2. Ramamritham, J.A. Stankovic and Zhao, W.: Distributed scheduling of tasks with deadlines and resource requirements. IEEE Transactions on Computers, Vol. 38. (1989) 110-123
3. Haban, D. and Shin, K. G.: Applications of real-time monitoring for scheduling tasks with random execution times. IEEE Transactions on Software Engineering, Vol. 16. (1990) 1374-1389
4. Tia, T. S., Deng, Z., Shankar, M., Storch, M.,Sun, J., Wu, L.C., Liu, J.W.S.: Probabilistic performance guarantee for real-time tasks with varying computation times. Proceedings of the 1st IEEE Real-Time Technology and Applications Symposium. IEEE Computer Society Press (1995) 164-173
5. Lehoczky, J.P.: Real-Time Queueing Theory. Proceedings of IEEE Real-Time Systems Symposium. IEEE Computer Society Press (1996) 186-195
6. Abeni, L. and Buttazzo, G.: Integrating multimedia applications in hard real-time systems. Proceedings of the 19th IEEE Real-Time Systems Symposium. IEEE Computer Society Press (1998) 3-13
7. Stewart, D.B. and Khosla, P.K.: Mechanisms for detecting and handling timing errors. Communications of the ACM, Vol. 40. (1997) 87-93
8. Puschner, P., Burns, A.: A Review of Worst-Case Analysis. The International Journal of Time-Critical Computing Systems, Vol. 18. (2000) 115–128
9. Atlas, A., and Bestavros, A.: Statistical rate monotonic scheduling. Proceedings of the 19th IEEE Real-Time Systems Symposium. IEEE Computer Society Press (1998) 123-132
10. Welch, L.R and Masters, M.W.: Toward a Taxonomy for Real-Time Mission-Critical systems. Proceedings of the First International Workshop on Real-Time Mission-Critical Systems (1999)
11. Audsley, Neil C.: Deadline Monotonic Scheduling. Report YCS-90-146. Department of Computer Science, York University (1990)
12. Welch, L. R., Shirazi, B.: A Dynamic Real-Time Benchmark for Assessment of QoS and Resource Management Technology. IEEE Real-Time Application System (1999)
13. Burns, F., Koelmans, A., Yakovlev, A.: WCET Analysis of Superscalar Processors Using Simulation With Coloured Petrinets. The International Journal of Time-Critical Computing Systems, Vol. 18. (2000) 275–288

A Load Balancing Algorithm Using the Circulation of A Single Message Token

Jun Hwang, Woong J. Lee, Byong G. Lee, and Young S. Kim

School of Computer Science and Engineering, Seoul Women's University,
126 Nowon-Gu GongReung-Dong Seoul, Korea, 139-774
{hjun, wjlee, byongl, yskim}@swu.ac.kr

Abstract. We present an efficient load balancing algorithm which simplifies the system status information through the use of an information message (VISITOR) in distributed system. Using the proposed algorithm, information gathering for decision-making is possible with less number of messages than the one using existing load balancing algorithm. The proposed algorithm improves the performance of distributed systems over the existing algorithms, not only by exchanging fewer messages to gather the information for making decision on load balancing and migration, but also by automatically determining of node and when to migrate.

1 Introduction

Load Balancing Strategy (LBS) is an activity of process sharing performed by distributed processes. The activity is called as a global or distributed scheduling [14, 11] in that process allocation is targeted to all distributed processes while typical process scheduling such as FIFO, LRU, RR are called as a local scheduling. LBS minimizes response time and maximizes the throughput by considering a system as a whole virtual process. The performance of load balancing mainly depends on its algorithm, which can be categorized as following;

First, when a node has to inform its own load status to other nodes, or to select a node to migrate, load balancing algorithm takes necessary global information through message broadcasting and do process migration accordingly [16, 10]. Second, a node utilizes the global information only for figuring out the load status of each node. For other decision-making, such as load balancing and migration, is done by heuristic information [16]. Third, each node does not need the global information from every node but from adjacent node only, and the load balancing is done by the judgment of a local node [16, 7, 4].

These load balancing algorithms may cause the inundation of broadcasting messages in picking up of global information and in selecting a node for migration [1]. In this paper, proposed is an algorithm that performs

P.M.A. Sloot et al. (Eds.): ICCS 2002, LNCS 2331, pp. 1080–1089, 2002.

the global information retrieval and load balancing through a single information message by simplifying the way of acquiring the global information and of making decision for migration.

Chapter 2 describes a load balancing algorithm, e.g., VISITOR algorithm, and reviews its concept and characteristics. Chapter 3 introduces VISITOR (¥-) which enhances the VISITOR algorithm. In chapter 4, the simulation context and necessary parameters for VISITOR (II) are discussed, and the performance of the proposed algorithm is analyzed in comparison with VISITOR algorithm. Finally, Chapter 5 concludes the paper with a summary and future work.

2 Load Balancing Algorithm

2.1 The Concept of VISITOR Algorithm

Provided with the load information, each node compares its own load state with those of entire system. In case that load imbalance is detected by exceeding a neutral zone , process migration occurs from an overloaded node to an under-loaded node. According to which node requests the migration, different approaches can be used, e.g., 'Receiver-Initiative' or 'Sender-Initiative'.

The VISITOR message contains the global information that represents the load status and migration information of each node. This message is circulated along the nodes at constant time interval and supplies necessary information to each node.

2.2 VISITOR Algorithm

To find out whether the load status of each node is overload or under-load, an average load value must be used. The average load value influences so much to the performance and efficiency of load balancing algorithm [6, 15]. The tracing function H for measurement of average load value is defined as follows;

$$H^*(AVG_k) = AVG_k+1/m^*(LOCAL-AVG_k)+CD \quad ¡ \; ƒ \; ¡ \; ƒ(1)$$
$$= AVG_{k+1}$$
and $AVG_0 = 0$

where AVG_k : average load value on VISITOR message after k^{th} circulation

AVG_{k+1} : expected average load value after k+1 circulation

m : number of nodes in the distributed system

LOCAL : measured value of local load (number of processes in local queue)

CD : communication latency time

The function H is derived from the facts that the load value at a node influences to the global information at the ratio of 1/m. Except from the information that is used in process migration, a control message is also necessary to run VISITOR algorithm as following.

- VISITOR: circulates along all the nodes in the connection during load balancing and provides the load balancing information.
- ACCEPT: permits migration of overloaded node to under-loaded node.
- RECOMM: does not participate in the migration, but informs the overloaded nodes that other nodes are under-loaded states.

Figure 1 depicts the structure of VISITOR message. VISITOR message fills with the information such as average load value, the least loaded node, and the most loaded node, and circulates along logical ring consecutively. As the VISITOR message passes through nodes, each node keeps its local load value and figures out load status of its own as to the whole system.

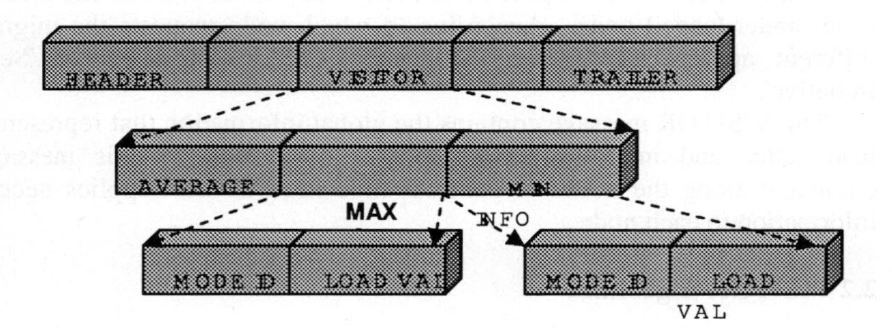

Fig. 1. The VISITOR message format

3 VISITOR(¥-) Algorithm

Because VISITOR is complicated algorithm, it is a burden for a distributed system to take responsibility of load balancing. To lessen the burden, VISITOR (¥-) algorithm is suggested by modifying original VISITOR algorithm. The difference between VISITOR and VISITOR (¥-) algorithm is that VISITOR (¥-) adopts 'Receiver-Initiative' method (VISITOR uses 'Sender/Receiver-Initiative' method). It is known that if the communication latency is short and the node is not overloaded, 'Receiver-Initiative' method performs better than 'Sender/Receiver-Initiative' method.

3.1 Detailed Description of VISITOR (¥–) Algorithm

In section 2.2, the heuristic function H was defined with equation (1). The function H is to calculate a new average load value with local load value and system-wise load value. However, with the function H, average load value is amplified without any control as the number of nodes and communication latency increases, thus resulting in high variance of average value. Thus, a function F that is complimentary of function H, is introduced.

$$F = AVG_K + ACE*1/m*(LOCAL-AVG_K) = AVG_{K+1}$$

and $AVG_0 = 0$,
$ACE = C/(LOCAL-AVG+C)$--(1)

where, LOCAL : number of process in local queue
m : number of nodes in distributed system
ACE: Adaptive Control Element (ACE) to reduce the error in tracing new average
C : communication Delay Factor*Constant

The ACF module influences the result of average load value. The ACF prevents the load value from influencing other nodes' values without any controlling over the value. That is, if the local load value in H is extremely high or low, its load value influences average load-tracing value of other nodes at the ratio of 1/m. This displacement occurs periodically and reacts sensitively if the state of entire distributed system changes. Therefore, a settlement is necessary in calculating the average load value of node. Adding a simple function called as ACF solves this problem. The influences of ACF on average load value should keep minimal, even as the difference increases [5]. In addition, VISITOR (II) includes message transmission information. The message format of VISITOR (¥–) is shown in figure 2.

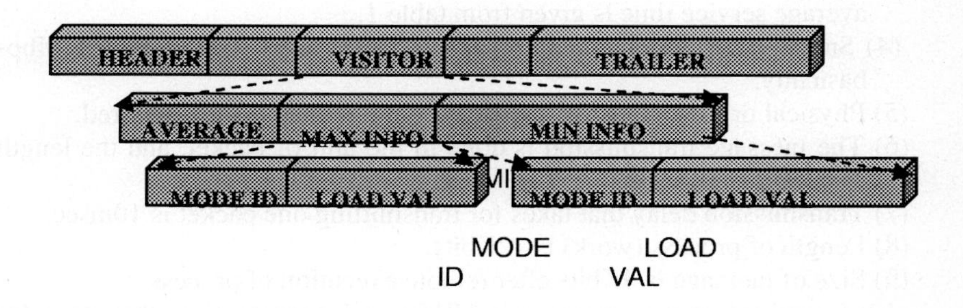

Fig. 2. VISITOR(¥–) message format

3.2 VISITOR(¥-) Algorithm

VISITOR(II) algorithm for load balancing follows;

CASE 1: In case of a VISITOR(¥-) message arriving,
- Calculate a new average load value through ACF and decide its own load status by averaging the load value.
- Execute the routine steps of load balancing.

CASE 2: In case of overload state,
- If there exist MIN information and its load value is less than average load value, choose a process from local queue.
- Migrate selected process to the under-loaded node.
- Eliminate the field information from MIN_INFO on VISITOR message.
- Pass the VISITOR(¥-) message to the next node.

CASE 3: In case of under-loaded state,
- If there exist MIN information and its load value is greater than average load value, insert its own node ID and load value information into MIN_INFO field.
- Send the VISITOR(¥-) message to the next node.

CASE 4: In case of optimal loaded state,
- Send the VISITOR(¥-) message to the next node.

CASE 5: In case of process arriving,
- Increase its own local load and put it into the local queue

4 The Simulation and Analysis

4.1 The Environment of Simulation

The simulation uses 5* (M/M/1) model with following assumptions.

(1) Fully-connected form of network topology is assumed.
(2)The node number of distributed system is 5 and VISITOR message circulates in a numerical order (clockwise or counterclockwise).
(3) Suppose that the load status of random node, average arrival rate and average service time is given from table 1.
(4) Suppose the communication capacity of network system be 0.1 Mbps basically.
(5) Physical or logical loss of message or any related error is ignored.
(6) The message transmission is done in the unit of packet, and the length of packet is 1,000bits.
(7) Transmission delay that takes for transmitting one packet is 10msec.
(8) Length of process (work) is 12Kbits.
(9) Size of message is 4Kbits after remote execution of process.
(10) Length of control message is 1Kbits and duration time that stays for transmission and collection of information in a node is 10msec.
(11) Length of VISITOR message is 1Kbits and duration time for staying and collecting of information is 10msec.

Table 1. The load status of each node

(unit: sec)

Load status	EU	UL	NL	OL	EO
Average arrival time	11.76	11.11	10.10	9.09	8.33
Average service time	10.0	10.0	10.0	10.0	10.0

EU (Extremely Under-loaded State), UL (Under-loaded State)
NL (Loaded Normally State), OL (Overloaded State)
EO (Extremely Overloaded State)

Communication Delay = (PACKET SIZE/0.1Mbps) * (PROCESS SIZE/1000)---(2)

The equation (2) calculates communication delay time independently with the simulation environment [12]. The total communication time is achieved by dividing the size of process with the length of packet (1000bits), and by multiplying it with communication latency time, which takes for transmitting one packet. Although above equation appears to be simple and insufficient, it still fully represents the characteristics of the communication system. The communication latency time is artificially calculated by equation (2) after accepting a message from a random node. After intended latency time is elapsed, the message is sent to a target node. This model is simulated in a fully connected network environment. Otherwise, more refinement should be done to the model.

4.2 The Performance of Average Load Value Tracing Function

The real load value of the entire system is calculated by dividing the total number of processes in waiting queue at random time with the number of nodes. The trace value of VISITOR message is calculated by function F and H (Figure 3). The status of each node is assumed as shown in table 2.

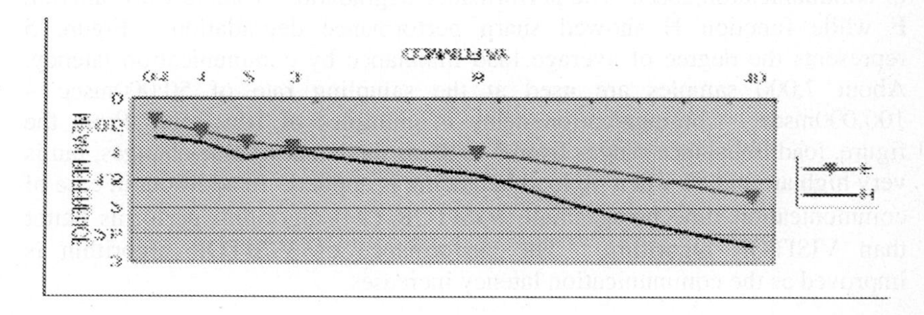

Fig. 3. The relation of real average load value (COMM.DELAY = 20msec) and tracing average load value by function F and H respectively

Table 2. The load status of each node

Node number	1	2	3	4	5
Load status	EU	OL	UL	EO	NL

To figure out the deviation between the real average load value and average tracing load value, the variation of standard deviation is used as depicted on figure 4. To get a accurate sample set, a sampling is done at a stable time of the system and the number of sample is about 3000. To get average deviation value, the difference of real value and traced value is calculated [3].

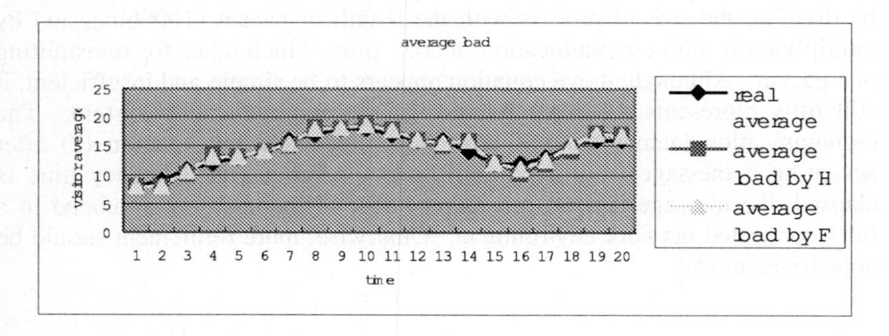

Fig. 4. The change of tracing value due to VISITOR passing delay

The actual average load value and sample mean of each trace function, F grows accordingly as communication latency increases. This means that the communication latency is proportional to error rate. Function F, compared to H, is appeared to be low in speed of performance decrease due to the increase of communication costs. The performance degradation is stable with function F while function H showed sharp performance degradation. Figure 5 represents the degree of average load imbalance by communication latency. About 7,000 samples are used at the sampling rate of 50,000msec ~ 100,000msec. Communication delay is multiples of 10m sec. From the figure, load imbalance ranges from 3 to 35 as the latency value changes, but is very high around 81 when no load balancing is applied. In addition, in case of communication time being small, VISITOR (¥-) algorithm performs better than VISITOR algorithm. The performance of VISITOR algorithm is improved as the communication latency increases.

4-3 Neutral Zone

Neutral zone is a handler for controlling load state of random node and regulating frequency of process migration. Therefore, suitable adjustment of Neutral zone is a key element for influencing the performance of load balancing algorithm. Neutral zone as shown in figure 6 has two neutral widths approximately and these widths represent optimal load state. If this zone gets narrower, the load difference between two nodes decreases picking up many load migration, and thus increases communication costs of load balancing. As the zone width gets larger, the frequency of migration becomes smaller and thus may allocate more time on process execution.

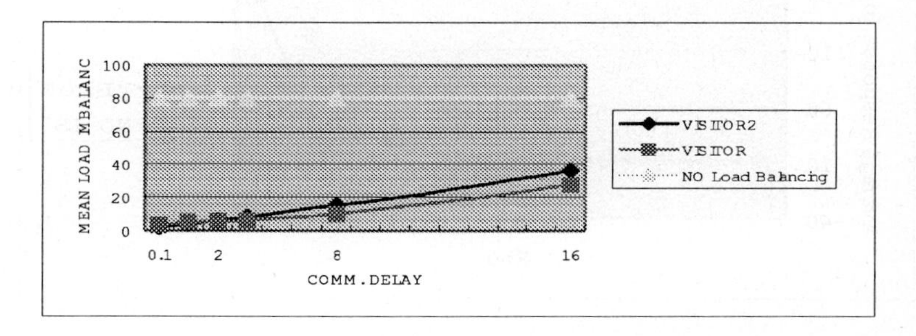

Fig. 5. The change of load imbalance due to communication latency

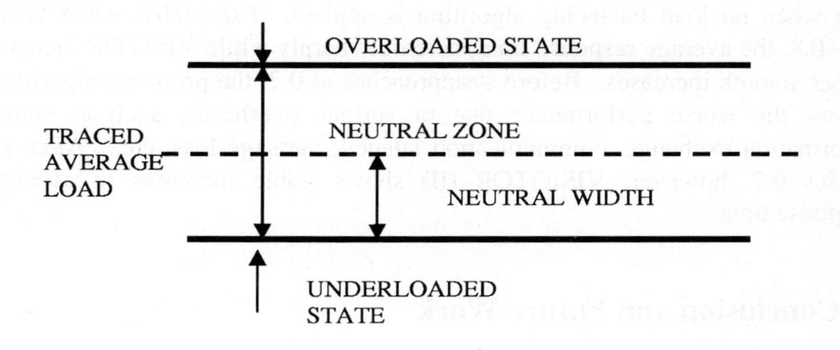

Fig. 6. The relation of neutral zone and load status

4.4 The Average Response Time

It is important to measure response time of random process for analyzing the system performance. The simulation uses equation (3) for measurement of average response time that represents the total length of visit from which the process arrives in waiting queue and until it leaves, including remote processing

MR(t) = ¢ †R(t)/p ---(3)

R(t) represents the response time of random process and p is the number of process. Defined in equation (4) is the average response time of total system.

ASRT(t) = ¢ †[¢ †R(t)/p]/m ------------------------------------(4)

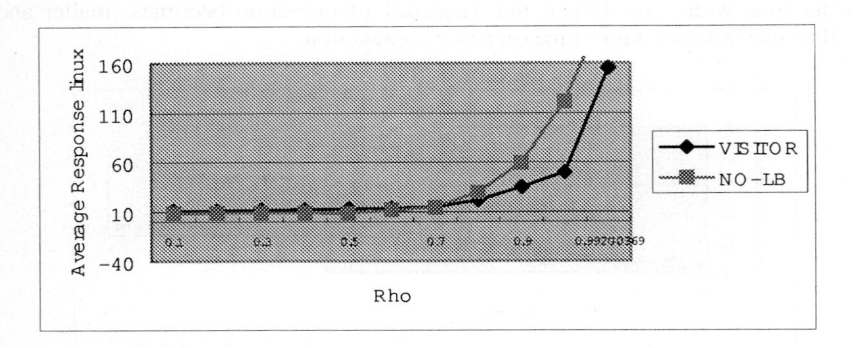

Fig. 7. The average arrival rate versus average service time

Figure 7 indicates that the result of average response time is better than the one when no load balancing algorithm is applied. Especially, when ¥æ is 0.7~0.8, the average response time increases sharply while VISITOR incurs a rather smooth increases. Before ¥æapproaches to 0.7, the proposed algorithm shows the worse performance due to various overheads, such as status information exchange, communication latency, message loss, etc.. After ¥æ passes 0.7, however, VISIOTOR (II) shows stable increases in average response time.

5 Conclusion and Future Work

In this paper, different methods of load balancing are presented to remove load imbalance, which occurs in typical distributed system. For this, two algorithms are introduced and analyzed with their performance comparison. With VISIOR (II) algorithm, the global information is achieved with less communication costs, compared with the one in broadcasting mechanism, and the accuracy of the information is proved. The performance, in terms of average response time, is improved about 50 ~ 80% when ¥æ is 0.9 ~ 0.99 than when no load balancing algorithm is applied. The communication cost for updating the status information is reduced remarkably because the global information is managed with simple message circulation and tracking functions.

"Victim Problem" [1] in VISITOR algorithm, however, can cause poor system performance as decision making at each nodes is combined. For future studies, cooperation and coordination mechanism to reduce the frequency of decision-making will be examined. Also, the recovery and rehabilitation mechanism for reducing the possibility of message loss in communication line will also be studied.

References

1. Alon, M., Barak, A., Manber, U., "On Dissemination Reliably without Broadcasting," IEEE The 7th International Conf. On Distributed/Computing Systems, pp. 74-81, Sep. 1987.
2. Casavant,T. and Singhal, M., Reading in Distributed Computing Systems, IEEE Computer Society Press, 1994.
3. Chao, L. L., *Statistics Methods and Analysis*, McGraw-Hill, 2nd Edition, pp. 188-192, 1974.
4. Eager, D. L., et al., "Adaptive Load Sharing in Homogeneous Distributed Systems," IEEE Trans. On Software, Vol. SE-12, No. 5, pp.662-675, May 1986.
5. Kness, C., Matkowsky B. J., Schuss, Z., Tier, C., "Two Parallel Queue with Dynamic Routing," IEEE Trans. on Comm., Vol. COM-34, No. 12, pp. 1170-1175, Dec. 1986.
6. Krueger, P., Livny, M., "A comparison of Preemptive and Non-Preemptive Load Distributing," The 8th International Conf. On DCS, pp. 123-130, 1988
7. Lin ,F.C.H., Keller, R. M., "The Gradient Model Load Balancing Method," IEEE Trans. on SE, Vol. SE-13, No. 1, Jan. 1987.
8. Ni , L. M., et al., "A Distributed Drafting Algorithm for Load Balancing," IEEE Trans. on Software Engineering, Vol. SE-11, No. 10, Oct. 1985.
9. Smith, R. G., "The Contract Net Protocol: High Level Communication and Control in a Distributed Problem Solver," IEEE Trans. on COMPUTERS, Vol. C-29, No. 12, Dec. 1980.
10. Stankovic, J., "A Perspective on Distributed Computer Systems," IEEE Trans. on Comp. Vol. C-33, No. 12, pp. 1102-1115, Dec. 1984.
11. Stankovic, J., "An Application of Bayesian Decision Theory to Decentralized Control of Job Scheduling," IEEE Trans. on Comp., Vol. C-34, No. 2, pp. 117-130, Feb. 1985.
12. Tay, Y. C. and Pang, HweeHwa, "Load Sharing in Distributed Multimedia-on-Demand Systems", IEEE Transactions on Knowledge and Data Engineering, Bol. 12, No. 3, May/June 2000.
13. Tonogai, D., "AI in Operating Systems: An Expert Scheduler," PROGRESS REPROT No. 88,12, Dec. 1988.
14. Zhou, S., Ferrari, D., "A Measurement Study of Load Balancing Performance," IEEE The 7th Conf. on DCS, pp. 490-497, Sep. 1987.
15. Zhou, S., Ferrai, D., "An Experimental Study of Load Balancing Performance," REPORT No. UCB/CSD/87/336, Berkeley, California, Jan. 1987.

A Collaborative Filtering System of Information on the Internet

DongSeop Lee[1], HyungIl Choi[2]

Soongsil University, 1-1 Sando-5Dong, DongJak-Gu, Seoul, Korea
[1] leeds@vision.soongsil.ac.kr,
[2] hic@computing.soongsil.ac

Abstract. In this paper we describe a collaborative filtering system for automatically recommending high-quality information to users with similar interests on arbitrarily narrow information domains. It asks a user to rate a *gauge set* of items. It then evaluates the user's ratings and suggests a *recommendation set* of items. We interpret the process of evaluation as an inference mechanism that maps a gauge set to a recommendation set. We accomplish the mapping with FAM (Fuzzy Associative Memory). We implemented the suggested system in a Web server and tested its performance in the domain of retrieval of technical papers, especially in the field of information technologies. The experimental results show that it may provide reliable recommendations.

1 Introduction

In this paper we describe a collaborative filtering system for automatically recommending high-quality information to users with similar interests on arbitrarily narrow information domains. Our system follows the same operational principle as the *Eigentaste* system [4]. It asks a user to rate a *gauge set* of items. It then evaluates the user's ratings and suggests a *recommendation set* of items. We interpret the process of evaluation as an inference mechanism that maps a gauge set to a recommendation set. We accomplish the mapping with FAM (Fuzzy Associative Memory).

FAM provides a framework that maps one family of fuzzy sets to another family of fuzzy sets [3]. This mapping can be viewed as a set of fuzzy rules that associate input fuzzy sets (gauge sets) with output fuzzy sets (recommendation sets). FAM also provides a Hebbian-style learning method that establishes the degree of association between an input and output [2]. This learning method is very simple and takes very little computation time.

Another aspect of collaborative filtering is how to form groups of users with similar tastes, so that the known preference of a group of users may be exploited to predict the unknown preference of a new user. This is a typical problem of clustering [6]. However, our approach does not require this type of explicit clustering, since the clustering is embedded in connection weights of FAM. In fact, FAM generates fuzzy rules that classify data into groups of classes. This grouping is supervised at the stage

P.M.A. Sloot et al. (Eds.): ICCS 2002, LNCS 2331, pp. 1090–1099, 2002.

of learning the connection weights. The details will be discussed in the next section[1].
Our collaborative filtering system consists of two main parts, a learning part and inferring part. The learning part operates off-line. It analyzes training data made up of input and output pairs in order to form fuzzy sets, where input data correspond to a set of rates on a gauge set of items and output data correspond to a set of rates on a recommendation set of items. Our system asks users to rate their preference on a continuous rating scale. To rate items, we may use a horizontal "rating bar" as in [4], where a user is supposed to click a mouse. The learning module then generates a correlation matrix that shows the degree of association between input and output fuzzy sets. The details will be discussed in section 3. The inferring part operates online. It presents a gauge set to a new user. It then processes the rates on the gauge set and draws a recommendation with the fuzzy rules built up in the learning part. Our system makes a conclusion in the form of induced preference rates on a recommendation set. The details will be discussed in section 2.

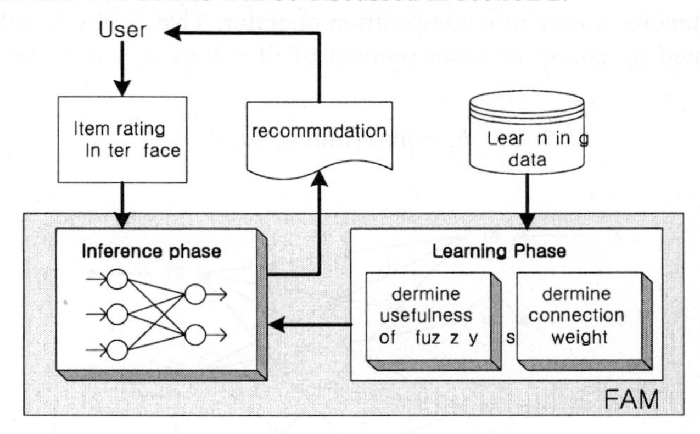

Fig. 1. System organization

2 Inferring Model With FAM

FAM can be viewed as the fusion of associative memory and fuzzy logic. It associates a family of fuzzy sets with another family of fuzzy sets. Figure 2 shows a basic structure of FAM. The antecedent term A_i in fuzzy association (A_i, B_j) denotes an input associant, the consequent term B_j denotes an output associant, and the synaptic weight w_{ij} denotes the degree of association between input and output associants. In our collaborative filtering system, the input associant corresponds to a fuzzy set of a gauge items, the output associant corresponds to a fuzzy set of recommendation items, and the synaptic weight corresponds to the degree of correlation between these two sets. As an example, one may consider fuzzy association ("*high preference of computer*", "*high preference of internet*") with the association degree of 0.8. There are several ways of interpreting the synaptic weight

w_{ij}. One popular interpretation considers it as a fuzzy Hebbian-style correlation coefficient in which the weight is encoded as the minimum of input and output associant values [3]. The details of this interpretation will be discussed in the third section.

If we now somehow encoded the set of synaptic weights, then FAM carries out forward recalling through max-min composition relation [3]. Suppose that a fuzzy set A_i is defined on the domain of an item x_i and $a_i = \mu_{A_i}(x_i)$ is the fit value of x_i to a membership function μ_{A_i}, and a fuzzy set B_j is defined on the domain of an item y_j and $b_j = \mu_{B_j}(y_j)$ is the fit value of y_j to a membership function μ_{Bj}. Then FAM exhibits forward recalling as in (1)

$$A \circ W = B \tag{1}$$

where $\circ$ denotes a max-min composition operator. That is, the recalled component b_j is computed by taking an inner product of fit values a_i's with the jth column of W.

$$b_j = \max_i \{\min(a_i, w_{ij})\} \tag{2}$$

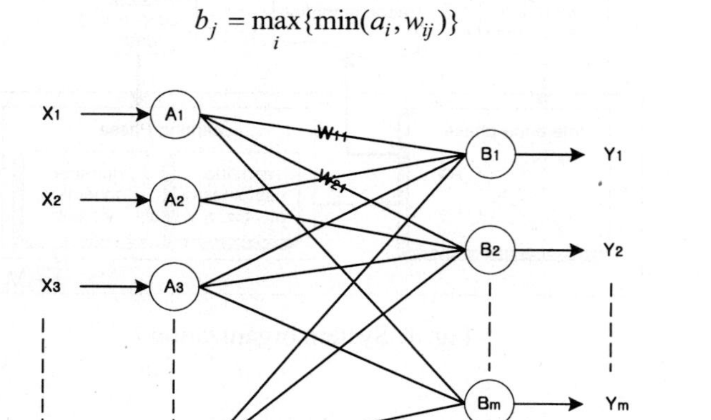

Fig. 2. Fuzzy associative memory

We suggest a fuzzy inference system that employs FAM for implementing fuzzy rules. These rules embody the task of grouping users with similar tastes in the form of association between gauge items and recommendation items. That is, we interpret a gauge items as an antecedent part of a fuzzy rule, a recommendation items as a consequent part, and a synaptic weight as the degree of reliability of the rule. When an antecedent part of a rule has several fuzzy terms, the corresponding input associant forms a conjunction of such fuzzy terms. Figure 3 shows the structure of our inferring model that consists of four layers. For the purpose of illustration, we assume that there are n gauge items x_i (i=1, ..., n) and m recommendation items y_j (j=1, ..., m), each furnishes p_i fuzzy sets, and each is represented by a fuzzy set B_j.

The input layer of Figure 3 just accepts input values that correspond to rates that a user assigns to a set of gauge items. Thus, the number of nodes in the input layer becomes *n*. The fuzzification layer contains membership functions of input items. Since there are *n* input items and each input item x_i produces p_i fuzzy sets, the total number of nodes in this layer becomes $\sum_{i=1}^{n} p_i$. The output of this layer then becomes the fit values of input rates to associated membership functions. The antecedent layer contains antecedent parts of fuzzy rules, which have the form of logical AND of individual fuzzy terms. This layer requires $N = \prod_{i=1}^{n} p_i$ nodes, since we allow every possible combination of fuzzy sets drawn one from each group of p_i fuzzy sets. Thus, each node in the antecedent layer has *n* incoming links. Each incoming link has a weight that represents the degree of usefulness of an associated fuzzy set. If links from some node of the fuzzification layer have a high value of weight, it means that the fuzzy set contained in the node is very useful in inferring a desired recommendation.

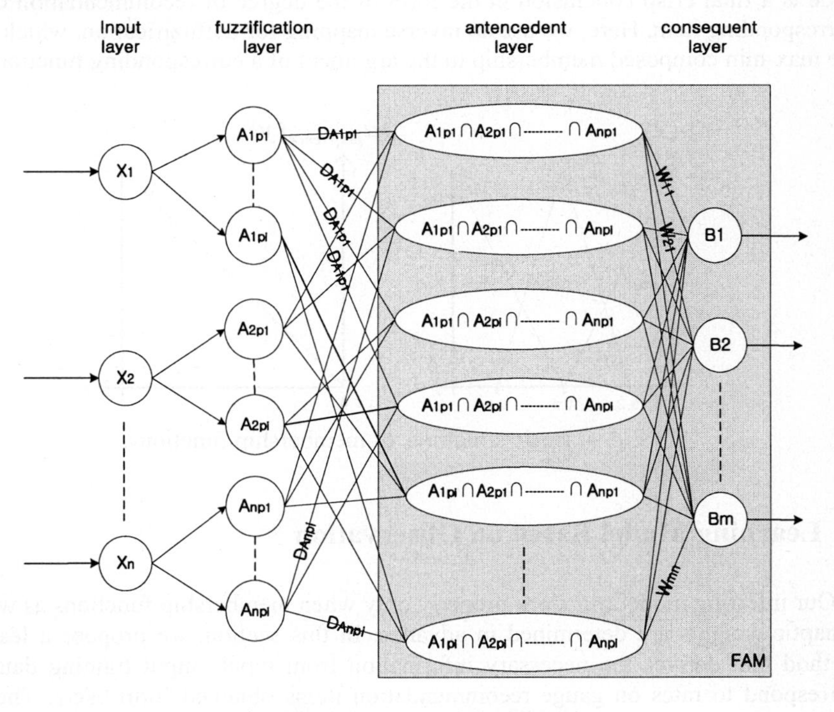

Fig. 3. Inference model with FAM

The details of how to determine the weights will be discussed in section 3. Each node of this layer just compares incoming weighted values and takes the minimum of them, since the truth value of an antecedent part of a rule is determined by taking a logical AND of individual truth values of participating fuzzy terms.

The consequent layer contains consequent parts of fuzzy rules. This layer contains m membership functions, each of which is to determine the preference of an individual recommendation item. We allow full connections between the antecedent layer and the consequent layer. But, each connection may have a different value of weight, which represents the degree of credibility of each connection. We basically follow the max-min compositional rule of inference. Thus, when N antecedent nodes $A_1, \cdots, A_N$ are connected to the jth consequent node B_j with weights w_{ij}'s, the output of the jth consequent node is a deffuzified value determined as in (3).

$$y^*_j = \mu^{-1}_{B_j}(\max_{1 \le i \le N}\{\min(w_{ij}, output(A_i))\}) \quad (3)$$

where $\mu_{B_j}(y)$ is a membership function contained in the jth consequent node, and $output(A_i)$ is an output of the ith antecedent node. The output of each consequent node is a final crisp conclusion in the form of the degree of recommendation on the corresponding item. Here, we use an inverse mapping for deffuzification, which maps the max-min composed membership to the argument of a corresponding function.

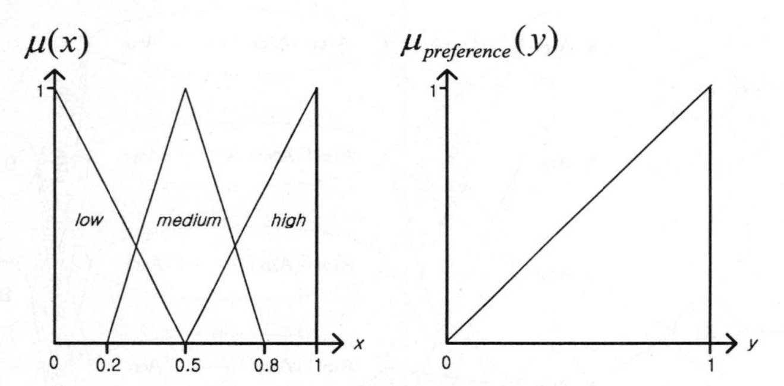

Fig. 4. Basic structures of membership functions

3 Learning Model Based on Observation

Our inferring model can work properly only when membership functions as well as synaptic weights are determined in advance. In this section, we propose a learning method that derives the necessary information from input-output training data that correspond to rates on gauge-recommendation items obtained form users. The first issue is how to determine the number of fuzzy sets for each item and corresponding membership functions. There must be some systematic way to divide the range of each input and output item into subranges and associate each subrange with a proper membership function. For this purpose, one may consider a tuning approach that exploits the distributions of data [2]. Here we take a simple approach and supplement it with the usefulness measure to be described later. Figure 4 shows the basic structures of our membership functions.

A user is supposed to rate his preference on an item by a degree between 0 and 1. A large value denotes a high preference and a small value denotes a low preference. For a gauge items, we divide the entire range into three parts and assign them with *low*, *medium*, *high* fuzzy sets. Their membership functions are defined as in (4).

For a recommendation items, we take an entire range for a fuzzy set "*preference*" and define its membership function to be monotonically increasing, $\mu_{preference}(y) = y$. This is to reduce the number of nodes in the consequent layer of our inferring model. But, one may introduce several fuzzy sets also for a recommendation items, if one wants to be more specific.

$$\mu_{low}(x) = \begin{cases} 1-2x, & 0 \le x < 0.5 \\ 0, & 0.5 \le x \le 1 \end{cases}$$

$$\mu_{medium}(x) = \begin{cases} (10x-2)/3, & 0.2 \le x < 0.5 \\ (-10x+8)/3, & 0.5 \le x \le 0.8 \end{cases} \tag{4}$$

$$\mu_{high}(x) = \begin{cases} 0, & 0 \le x < 0.5 \\ 2x-1, & 0.5 \le x \le 1 \end{cases}$$

To determine the usefulness of an input fuzzy set, we evaluate how the fuzzy set discriminates recommendation items. If an input fuzzy set shows such a characteristic that the training data that fall on the fuzzy set have their corresponding rates on recommendation items evenly distributed, we claim that the fuzzy set is not useful in terms of recommending some specific items. However, if their rates on recommendation items show a spiky type of distribution, we claim that the fuzzy set is useful in inferring recommendation. This criterion is equivalent to emphasizing a distinctive fuzzy set.

$$D_{A_l} = 1 + \sum_{j=1}^{M} P_{A_l}(j) \cdot \ln(P_{A_l}(j))$$

$$P_{A_l}(j) = \frac{\sum_{i=1}^{K} \mu_{A_l}(x_i) \cdot r_{B_j}(x_i)}{\sum_{j=1}^{M} \sum_{i=1}^{K} \mu_{A_l}(x_i) \cdot r_{B_j}(x_i)} \tag{5}$$

To evaluate the discriminating power, we use the measure of entropy. The entropy is a statistical measure of uncertainty, and it is good at revealing the dispersion of data. We compute degrees of usefulness of input fuzzy sets as in (5). In (5), x_i denotes ith input-output training data, $\mu_{A_l}(x_i)$ denotes the fit value of x_i to an input fuzzy set A_l, $r_{B_j}(x_i)$ denotes the rate of x_i on a recommendation item B_j, M is the number of recommendation items, and K is the total number of training data. Thus,

$P_{A_l}(j)$ denotes the probability density of rates on the jth recommendation item, which are associated with an input fuzzy set A_l.

Our inference model requires a predetermined correlation matrix that represents the degrees of associations between input and output fuzzy sets. We take a Hebbian-style learning approach to build up the correlation matrix. The Hebbian learning is an unsupervised learning model whose basic idea is that "the synaptic weight is increased if both an input and output are activated." In this way, the phenomena of habit and learning through repetition are often explained. Many artificial neural networks which take a Hebbian learning approach increase network weights according to the product of excitation levels of an input and output. In our fuzzy associative memory, input and output values are fit values to membership functions. Thus, we replace a product operation with a minimum operation and an addition operations with maximum operation. That is, when $a_i(n)$ is an input associant for the nth learning datum and $b_j(n)$ is an output associant for the nth learning datum, the change of weight is carried out as in (6).

$$w_{ii}(n) = (1-\eta) \cdot w_{ii}(n-1) + \eta \cdot (a_i(n) \quad AND \quad b_j(n)) \tag{6}$$

In (6), η is a positive learning rate that is not greater than 1. This learning rate controls the average size of weight changes. Our inference model allows full connection between the antecedent layer and the consequent layer. This full connection may present too many rules, causing too much computation time. The connection weight may be considered as the confidence measure of a related fuzzy rule, since our learning approach is based on Hebbian learning. We may then consider the rules with small connection weights to be unimportant in inferring a useful conclusion. Thus in practical situations, we may prune some of the connections whose weights are less than some predetermined threshold and reduce the number of rules.

4 Experimental Results and Conclusion

To confirm the effectiveness of our suggested model, we chose the domain of retrieval of technical papers, especially in the field of information technologies. We implemented our learning and inferring module using C++ in a Web server. We also developed Web interface module using *Java* that allows internet users to rate papers. The interface module presents papers and collects ratings as users click on a rating bar. In the learning phase, all the papers in both the gauge set and recommendation set are presented to users one by one. After a user rates each paper, another is presented. The collected ratings are then used to build up membership functions and connection weights of our FAM. In the testing phase, our interface module presents to a user the papers in the gauge set and asks for ratings on the papers. After all papers in the gauge set are rated, our system recommends to the user the papers in the recommendation set with induced preferences. Our system also collects user's ratings on each recommended paper, so that it may compare recommended ratings against user's ratings.

To evaluate the performance of our system, we performed three types of experiments: we first examined the usefulness measure of a fuzzy set in terms of its effect on accuracy of induced preferences. The second experiment was to examine the effect of the number of connections of FAM on the accuracy of induced preferences. We also compared the performance of our system against those of other systems. In our experiments, the learning rate η in (6) was set to 1.0 and the initial weights were set to 0. 100 users were involved in the learning phase, and also 100 different users were involved in the testing phase. The gauge set contains 10 different papers and the recommendation set contains 20 different papers, so the numbers of nodes in input layer and consequent layer are 10 and 20, respectively.

The table I lists the degrees of usefulness of input fuzzy sets. These values reflect how input fuzzy sets discriminate recommendation items. For illustration, A_{21} has the high value of 0.87, since the training data that fall on this fuzzy set have the most of their recommendation rates only in 4 items. In contrast, A_{62} has the small value of 0.17, since the training data that fall on this fuzzy set have their recommendation rates spreaded among 17 items.

Table I. Usefulness of each input fuzzy set

Fuzzy set	Usefulness	Fuzzy set	usefulness	Fuzzy set	Usefulness
A_{01}	0.78	A_{02}	0.85	A_{03}	0.32
A_{11}	0.76	A_{12}	0.44	A_{13}	0.15
A_{21}	0.87	A_{22}	0.92	A_{23}	0.86
A_{31}	0.05	A_{32}	0.46	A_{33}	0.65
A_{41}	0.84	A_{42}	0.65	A_{43}	0.82
A_{51}	0.65	A_{52}	0.14	A_{53}	0.09
A_{61}	0.86	A_{62}	0.17	A_{63}	0.60
A_{71}	0.92	A_{72}	0.82	A_{73}	0.14
A_{81}	0.57	A_{82}	0.65	A_{83}	0.79
A_{91}	0.87	A_{92}	0.88	A_{93}	0.88

$$OMAE = \frac{1}{M} \times \sum_{j=0}^{M} MAE(j)$$
$$MAE(j) = \frac{1}{N} \times \sum_{i=0}^{N} \left| r_{ij} - p_{ij} \right| \quad (7)$$

To evaluate the accuracy of the performance, we use the Overall Mean Absolute Error metric that is often used in the literature. In (7), p_{ij} is the prediction for how user i will rate item j and r_{ij} is the actual rate given by user i for item j, M is the number of items user i has rated, and N is the total number of users involved in the test.

In (7), *MAE(j)* represents the mean error for the *j*th recommendation item over the whole users, and *OMAE* represents the mean error over the whole items and users. Table II shows the performance of our system with and without the usefulness measure of input fuzzy set. As can be noted, the performance is improved by 0.07 on average with the help of the usefulness measure.

Table II. *MAE* of each recommendation item

Recommendation Item	MAE without usefulness measure	MAE with Usefulness measure	Recommendation item	MAE without usefulness measure	MAE with usefulness measure
Paper01	0.234	0.215	Paper11	0.154	0.142
Paper02	0.164	0.156	Paper12	0.189	0.185
Paper03	0.209	0.201	Paper13	0.220	0.215
Paper04	0.186	0.178	Paper14	0.237	0.231
Paper05	0.160	0.149	Paper15	0.205	0.196
Paper06	0.207	0.203	Paper16	0.156	0.154
Paper07	0.242	0.234	Paper17	0.151	0.147
Paper08	0.219	0.216	Paper18	0.214	0.206
Paper09	0.182	0.179	Paper19	0.178	0.171
Paper10	0.170	0.168	Paper20	0.203	0.194

We examined the effect of the number of synaptic connections on the performance of the system. We pruned out synaptic connections whose weights are less than some threshold *Th*, so that the number of connections is reduced and the system is simplified. For illustration, we have 25% reduced connections for *Th*=0.1, and 35% reduced connections for *Th*=0.2. We also compared the performance of our systems with those of other two systems: POP [5] and Eigentaste [4]. Table III shows the results. As can be noted, our system with *Th*=0.1 surpasses the others. It is interesting to note that the performance of our system does not drop drastically with increasing the threshold *Th*.

Table III. Comparison of performance

System	OMAE
Our system with th = 0.0	0.187
Our system with th = 0.1	0.192
Our system with th = 0.2	0.245
POP[9]	0.302
Eigentaste[6]	0.284

Acknowledgement

This work was supported by the Korea Science and Engineering Foundation (KOSEF) through the Advanced Information Technology Research Center(AITrc).

Reference

1. Dae-Sik Jang, Hyung-Il Choi, Fuzzy Inference system based on fuzzy associative memory, journal of Intelligent and fuzzy systems, Vol.5, 271-284. 1997.
2. Hideyuki T, Isao H, NN-Driven fuzzy reasoning, Int. J. Approximate Reasoning, 191-212. 1991.
3. Kosko B, Neural Networks and Fuzzy Systems," Prentice-Hall International. 1994.
4. Zimmermann HJ, Fuzzy Set Theory and Its Applications, KALA. 1987.
5. Ken Goldberg and Theresa Roeder and Dhruv Gupta and Chris Perkins, Eigentaste: A constant time Collaborative Filtering Algorithm, University of California, Berkeley, Electronics Research Laboratory Technical Report M00/41. 2000.
6. Jonathan Herlocker, Joseph Konstan, Al Borchers, and John Riedl. An algorithmic framework for performing collaborative filtering, In Proceedings of the SIGIR. ACM, August 1999.

Hierarchical Shot Clustering for Video Summarization

YoungSik Choi, Sun Jeong Kim, and Sangyoun Lee

Multimedia Technology Research Laboratory, Korea Telecom,
Seocho-Gu Woomyeon-dong 17
Seoul, Korea,
{choimail, sunjkim, leesy}@kt.co.kr

Abstract. Digital video is rapidly becoming a communication medium for education, entertainment, and a variety of multimedia applications. With the size of the video collections growing to thousnads of hours, efficient searching, browsing, and managing video information have become of increasing importance. In this paper, we propose a novel hierarchical shot clustering method for video summarization which can efficiently generate a set of representative shots and provide a quick and efficient access to a large volume of video content. The proposed method is based on the compatibility measure that can represent correlations among shots in a video sequence. Experimental results on real life video sequences show that the resulting summary can retain the essential content of the original video.

1 Introduction

With the recent advances in compression and communication technologies, vast amounts of video information are created, stored, and transmitted over networks for education, entertainment, and a host of multimedia application. Therefore, efficient searching, browsing, and managing video information have become of increasing importance. The MPEG group has recently begun a new standardization phase for efficient searching and managing multimedia content (MPEG-7). The MPEG-7 will specify the ways to represent multimedia information by means of descriptors and description schemes. The question of how to obtain these descriptors automatically is becoming a highly important research topic. Particularly, automatic video summarization is gaining the attention as a way to condense a large volume of video into smaller and comprehensible units, and allows quick and easy access to video content.

There are a few approaches to video summarization: (1) selecting and concatenating the "most representative" images or shots [3], [6]. (2) creating a "skim" video which represents a short synopsis [7]. In this paper, we address the problems with selecting the "representative" shots and propose a novel hierarchical shot clustering for video summarization.

Shot clustering has been frequently used for video summarization and segmentation. In

P.M.A. Sloot et al. (Eds.): ICCS 2002, LNCS 2331, pp. 1100–1107, 2002.

this approach, a video sequence is first segmented into shots and then shot clustering is applied to select the representative shots [3], [6] or to group the shots into scenes (story units) [4], [5]. Agglomerative hierarchical clustering with time constraint has been used for shot clustering [1], [2]. This approach can produce the tree-structured representation that is useful for video summarization. However, it is expensive, requiring insertions and deletions of the cluster dissimilarity matrix, and requiring a search for each closest cluster pairing. Window-based shot grouping has also been used [4], [5]. In this paradigm, incoming shots are compared with the shots in a given window in order to check if incoming shots may be included in the current segment. This method is computationally less expensive to group shots into scenes than agglomerative clustering. It is, however, difficult to have a compact hierarchical representation for video summarization.

To overcome the limitations in both approaches, we propose a novel shot clustering algorithm that can efficiently produce the representative shots and extract the hierarchical structure from a video sequence. The proposed clustering algorithm is based on the assumptions that the shots from a scene are more likely to be compatible with each other than those from other scenes, and that the shot highly compatible with other shots is more likely to be the representative shot of a scene. We define the compatibility measure to be the average value of the degrees to which a shot is similar to its neighboring shots within a given window. Using the compatibility measure, we develop an efficient clustering algorithm for video summarization.

2 Hierarchical Shot Clustering

2.1 Compatibility Measure

We define the compatibility measure as follows. Let $S = \{s_1, s_2, \ldots, s_k, \ldots, s_N\}$ denote a video sequence, where s_i is i-th shot and N is the total number of shots. The compatibility measure $C(s_i)$ of s_i within a given window is defined as:

$$C(s_i) = (1/\,N(i)) \textstyle\sum_{j\in N(i)} \mu_{ji}, \qquad \textbf{(1)}$$

where $N(i)$ is the set of the neighbors of s_i and μ_{ji} is a fuzzy membership function which determines the value of the degree to which s_i is similar to s_j. $N(i)$ may be $\{s_{i-2}, s_{i-1}, s_{i+1}, s_{i+2}\}$ if the window size is 4. The membership function can be defined as a monotonically decreasing function of the dissimilarity between s_i and s_j. In this Letter, we propose to use the following bell-shaped membership function:

$$\mu_{ji} = \exp(-D(s_i, s_j)/\beta_j). \qquad \textbf{(2)}$$

Note that in general, μ_{ij} is not equal to μ_{ji}. The $D(s_i, s_j)$ denote the dissimilarity between s_i and s_j. The β_j is a scaling factor and is determined for each s_j to consider the local context as follows.

$$\beta_j = (1/|N(j)|)\Sigma_{i\in N(j)}D(s_i, s_j), \tag{3}$$

where $|N(j)|$ is the number of neighbors of s_j. The dissimilarity $D(s_i, s_j)$ between s_i and s_j may be defined as a function of shot feature vectors. In this Letter, the shot feature vector is computed as the average of all frame feature vectors within a shot.

Fig. 1. Illustration of grouping and handling local minima

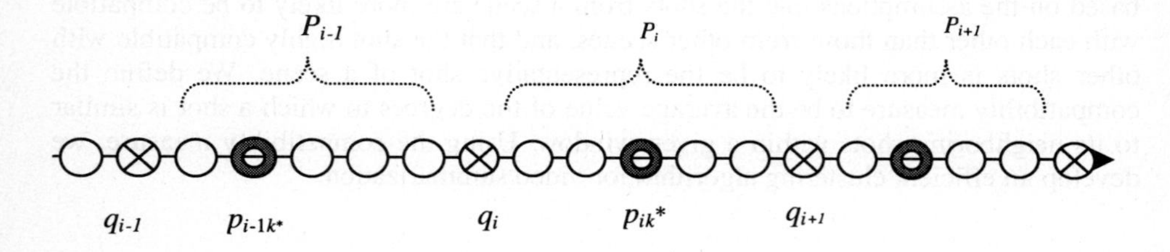

2.2 Clustering Algorithm:

The proposed compatibility measure can represent the correlations among shots in a video sequence. If the value of $C(s_i)$ is higher than the values of its neighbors, s_i is more compatible with its neighbors and is, therefore, more likely to be the representative of its neighbors. If the value of $C(s_i)$ is lower than its neighbors, s_i is less compatible with its neighbors and, therefore a scene boundary more likely exists around s_i. Taking these into accounts, we propose the following hierarchical clustering method.

(Step 1) *Initialization*: Let $S = \{s_1, s_2, \ldots, s_k, \ldots, s_N\}$ be a video sequence and $\{\boldsymbol{h}(s_1), \boldsymbol{h}(s_2), \ldots, \boldsymbol{h}(s_k), \ldots, \boldsymbol{h}(s_N)\}$ be the corresponding shot feature vectors, where $\boldsymbol{h}(s_k)$ is the feature vector of shot s_k. We initialize S as the set of initial clusters, P.

(Step 2) *Grouping and selecting a key shot*: For each cluster in P, we obtain the compatibility using equations (1), (2), and (3). We find the clusters with local minimum compatibility values. Let $\{q_1, q_2, \ldots, q_n\}$ denote the set of these clusters, where n is the total number of local minima (See Figure 1).
We group the clusters between clusters q_i and q_{i+1} into a new cluster P_i. Let $P_i = \{p_{i1}, \ldots, p_{ik}^*, \ldots, p_{in}\}$, where in is the total number of clusters between q_i and q_{i+1}, and p_{ik}^* is the cluster with the local maximum compatibility value between q_i and q_{i+1}. Note that only one p_{ik}^* exists between q_{i-1} and q_i. We select p_{ik}^* as the representative cluster of P_i.

If this is the first iteration, we select p_{ik}* as the key shot of P_i. Otherwise, we select p_{ik}*'s key shot as the key shot of P_i. Note that the representative clusters and the key shots are the same in the first iteration.

(Step 3) *Handling Local Minima*: We can consider $\{q_1, q_2, \ldots, q_n\}$ as outliers or boundary clusters. Therefore, we first check whether q_i is an outlier or a boundary cluster according to $C(q_i)$. If q_i is an outlier, q_i becomes a new cluster between P_{i-1} and P_i. Otherwise, we merge q_i into the closer cluster P_{i-1} or P_i with respect to the dissimilarity function defined in equation (4). The following describes this process.
For each local minimum cluster q_i starting from q_1, do the following.
(Step 3-a) If $C(q_i) < T_C$, then make q_i as a new cluster between P_{i-1} and P_i. T_C is the threshold value.
(Step 3-b) If $C(q_i) \geq T_C$, do the following.
If $A(q_i, P_{i-1}) < A(q_i, P_i)$, then add q_i into cluster P_{i-1}. Otherwise, add q_i into cluster P_i. $A(q_i, P_i)$ is the dissimilarity between clusters q_i and P_i, and defined as

$$A(q_i, P_i) = \min D(q_i, p_{ij}), \text{ where } p_{ij} \in P_i. \quad (4)$$

(Step 4) *Time constraint and terminating condition*: For each new cluster P_i, we check the time constraint with threshold value T_T. If the duration of cluster exceeds T_T, then we ungroup the cluster and make the clusters in P_i as new clusters between P_{i-1} and P_{i+1}. Note that this ungrouping is similar to clustering with time-constrained distance as in [1]. We terminate the clustering process if there is no change in clusters P_i.

(Step 5) *Update cluster feature vector*: We update the feature vectors of new clusters P_i such that the feature vectors of more compatible clusters may gain more weights than those of less compatible clusters. That is, we update the feature vector $\boldsymbol{h}(P_i)$ for each cluster P_i as the weighted average of feature vectors $\boldsymbol{h}(p_{ij})$ of clusters p_{ij} in P_i.

$$\boldsymbol{h}(P_i) = \Sigma C(p_{ij})\boldsymbol{h}(p_{ij}) / \Sigma C(p_{ij}),$$

where $p_{ij} \in P_i$. Set $P = \{P_1, \ldots, P_m\}$ and m is the number of new clusters P_i. Go to Step 2.

3 Video Summarization

The proposed clustering algorithm results in a hierarchical structure of a video sequence where each node corresponds to P_i. Each node has the representative cluster p_{ik}* and the key shot (See Step 2 in 2.2). The set of clusters P_i in the highest level can be considered as a partition of a video sequence (a video segmentation) and the set of the representative clusters p_{ik}* and key shots as a abridged version (a video summarization). Taking these into accounts, we present the following scheme for video summarization.

Suppose that our clustering method produces M clusters in the top level and we select the set of key shots in each cluster as a summary. Then, the length T_C of the summary becomes

$$T_C = \Sigma_{i=1..M} T(S_i),$$

where $T(S_i)$ is the length of key shot S_i from cluster P_i. Conversely, if a user requests a summary of length T_{req}, we need to determine the number of clusters in the top level. Assuming that $T(S_i)$ be the average shot length T_{avg} in a video sequence, the number of segments M becomes

$$M = T_{req}/T_{avg}.$$

Now, we need to adjust the threshold value T_T to produce M clusters in the top level. In the proposed clustering algorithm, the number of clusters M has an inverse relation to the threshold value T_T in Step 4 in Section 3. Therefore, we propose to set T_T as

M **5**

where T_{org} is the length of the original video and k is a constant greater than 1. Using equation (5), we can generate a summary as a user requests in terms of T_{req}, the length of the summary. Let M' be the actual number of clusters in the top level produced by using the threshold value in (5). Then, we can simply generate a summary by selecting the set of key shots from M' clusters.

4 Experimental Results

To test the proposed clustering and summarization method, we used a 55 minutes and a 45 minutes TV dramas. We segmented the 55 minutes drama into 399 shots and the 45 minutes drama into 291 shots using traditional histogram difference. We used mean color histogram for a shot feature vector as in [5].

We set the threshold values T_C and the window size for the compatibility computation as 0.1 and 4, respectively. With these values and T_T of 250 seconds, our clustering converged within 3-4 iterations. Figure 2 shows the compatibilities of the TV drama with window size 4. In this Figure, the peaks correspond to the representative shots and the valleys correspond to the possible scene boundary shots.

Fig. 2. Compatibilities versus shot numbers: (a), (b), and (c) compatibilities after 2, 1, and 0 iterations, respectively.

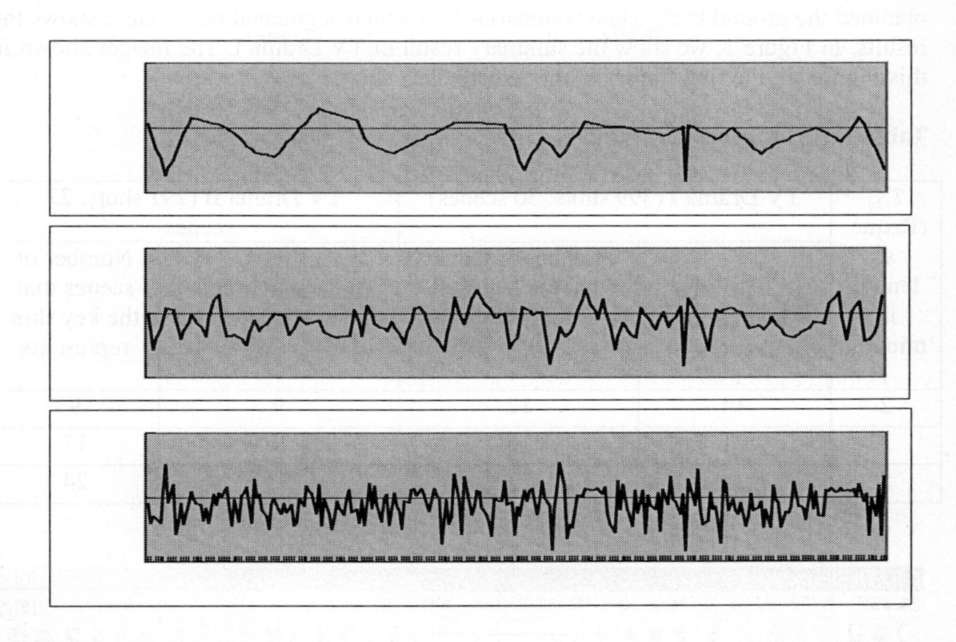

In order to test the effectiveness of equation (5), we varied T_{req} and compared the desired number of clusters with the actually generated number of clusters. Table 1 shows the results obtained from the two TV dramas with $k = 2$ in equation (5). The results show that the number of generated clusters is close to the desired number of clusters.

Table 1. Comparison of the desired and generated number of clusters with respect to T_{req}

T_{req} (Request length in minute)	TV Drama I (399 shots: 55 minutes)			TV Drama II (291 shots: 45 minutes)		
	C (Desired number of clusters)	T_T (Threshold value in second)	C' (Generated number of clusters)	C (Desired number of clusters)	T_T (Threshold value in second)	C' (Generated number of clusters)
2	14	480	14	12	450	9
3	21	320	21	19	284	22
4	28	240	31	25	216	33
5	35	192	31	32	169	36
10	71	95	72	64	84	71

We defined the number of key shots coming from the different scenes as the performance measure of the proposed summarization method. That is, the summary is better if it represents more scenes in the original video. For this experiment, we obtained the ground truth scene boundaries by manual segmentation. Table 2 shows the results. In Figure 3, we show the summary result of TV Drama I. The images shown in this Figure are the first frames of the selected key shots.

Table 2. Performance measure

T_{req} (Request length in minute)	TV Drama I (399 shots: 30 scenes)		TV Drama II (291 shots: 27 scenes)	
	C' (Generated number of clusters)	Number of scenes that the key shot represents	C' (Generated number of clusters)	Number of scenes that the key shot represents
2	14	12	9	9
3	21	15	22	17
5	31	23	33	24

5 Conclusions

In this paper, we presented a hierarchical clustering method based on the compatibility measure that can represent shot correlations. The proposed method can efficiently generate a set of representative shots and also extract the hierarchical structure of a video sequence. Experimental results show that our proposed summarization abridged the original video where compaction is up to 25:1 and still kept most of important scenes. This result is accredited to the clustering capability of extracting the hierarchical structure of a video sequence.

References

1. M Yeung and Boon-Lock Yeo, and Bede Liu "Extracting Story Units from Long Programs for Video Browsing and Navigation", Proceedings of IEEE International Conference on Multimedia Computing and Systems1996, pp. 296-305.
2. E. Venequ and et al, "From Video Shot Clustering to Sequence Segmentation", Proceedings of IEEE International Conference on Pattern Recognition 2000, pp. 254-257.
3. Shingo Uchihashi and, et al, "Video Magna: Generating Semantically Meaningful Video Summaries", Proceedings of ACM International Conference on Multimedia 1999, pp. 383-391
4. A. Hanjalic and et al, "Automated High-Level Movie Segmentation for Advanced video-Retrival Systems", IEEE Transactions On Circuits and Systems for Video Technology, Vol. 9, No. 4, June 1999, pp. 580-588.
5. Ong Lin and Hong-Jiang Zhang, "Automatic Video Scene Extraction by Shot Grouping", Proceedings of IEEE International Conference on Pattern Recognition 2000, pp. 39-42.
6. Nikolaos D. Doulamis and et al, "Efficient Summarization of Stereoscopic Video Sequences", IEEE Transactions on Circuits and Systems for Video Technology, Vol. 10, No. 4, June 2000, pp. 501-517.
7. Michael A. Smith and Takeo Kanade, "Video Skimming and Characterization through the Combination of Image and Language Understanding Techniques", Proceeding of IEEE, pp.775-781. 1997

On Detecting Unsteady Demand in Mobile Networking Environment *

V.V. Shakhov, H. Choo, and H.Y. Youn

School of Electrical and Computer Engineering
Sungkyunkwan University, Suwon 440-746, KOREA
{vova, choo, youn}@ece.skku.ac.kr

Abstract. One of the key issues in mobile communication system is how to predict the number of calls per each cell. It is an important parameter and usually assumed as random Poisson value. For effective management of cellular network, the average number of calls should be traced and the changes in the numbers need to be promptly detected. In this paper we propose an algorithm detecting the changes in the behavior of the users using the technique proposed for point-of-change problem based only on the number of call arrivals. Computer simulation reveals that the proposed method can effectively detect the discord, and the developed model is very accurate as showing mostly less than 1% differences.

1 Introduction

Recently information technology has been evolving into the direction leading to the convenience of users. Mobile and wireless communication have a significant role in this IT networking era due to the request of users. One of main elements of mobile networking, especially for network design, is the prediction mechanism for the number of calls per unit time in each cell [1]. The number is related to the number of frequency channels, the call blocking probability, and many other key issues of resource management in mobile computing. With steady-state behavior of mobile users, several techniques [1] can be employed for effective mobility management.

If the number of mobile users is simply modeled as a constant with a certain level of blocking probability, the required number of channels can be easily obtained. However, as we know, behavior of mobile users and their movement change dynamically. More specifically, the average and maximum number of calls can widely fluctuate. If they increase, the call blocking probability also increases and thus the quality of service drops. Meanwhile, if they decrease, less amount of resource will be required for maintaining the same quality of service. Hence, the number of calls is an important parameter in mobile network. For effective

* This work was supported in part by Brain Korea 21 and grant No. 2000-2-30300-004-3 from the Basic Research Program of Korea Science and Engineering Foundation. Dr. Choo is the corresponding author.

P.M.A. Sloot et al. (Eds.): ICCS 2002, LNCS 2331, pp. 1108–1117, 2002.

management of cellular network, the average number of calls should be traced and the changes in the numbers need to be promptly detected.

Sometimes it is enough to use a simple deterministic model for calculating the number of calls, where the parameters of the model are known. Usually a flow of call arrivals is represented by a stochastic model, and the number of calls is a random value decided depending on the behavior of users.

Several models for analyzing the movement of users were proposed in [2, 3]. The models require some additional information such as the speed of mobile users, the direction of movements, travel path, and so on. Since a lot of factors affect the system performance, an approximation technique is employed for simplifying the modeling. For example, assumption of uniform distribution for the speed of mobile users, a cellular network as a set of identical hexagons, etc. Note, however, that movement of users is irregular, and the assumptions employed for the sake of simplicity can cause unrealistic modeling for the average number of calls. In this paper, therefore, it is offered to detect the change in the behavior of the users using the point-of-change problem solution technique based only on the number of call arrivals.

The rest of the paper is organized as follows. Section 2 provides the basic notation and previous works. The main algorithm of discord detection under the Poisson distribution and an admissible lag is presented in Section 3. In Section 4, we present the results of numerous simulation experiments with the proposed algorithm. Section 5 is a brief conclusion.

2 Preliminaries

To calculate the required number of frequency channels or traffic load in a cell, we have to know the number of calls per unit time which is a random number. Let N^c be the required number of calls, N be the total number of users in a cell, L be the probability of a user to have a mobile terminal, and k be the probability of mobile users making network access. Then $N^c = NLk$. Here, N depends on the average distance between mobile users and k depends on the speed of mobile users [1].

Actually, the larger the sojourn time for a mobile user is, the higher the network access probability is. If traffic congestion occurs, the average distance between adjacent mobile terminals decreases, and thus the sojourn time increases. Hence, both N and N^c increase. In this situation, more channels are required for customers to have a certain level of quality of service. Refer to Figure 1. In this figure, N_i, k_i and N_i^c are the parameters discussed above before ($i = 1$) and after ($i = 2$) traffic congestion. Here we assume that each user carries a mobile terminal ($L = 1$) and two channels are assigned to a cell.

The number of calls can be predicted by mathematical modeling. In this section some models proposed for modeling the behavior of the users are discussed. Their approaches are usually based on a certain regularity of user movement. The models proposed in [2, 4] assume that the average speed of users and their directions are same in each cell. However, movements of customers are usually different in practice, and even the information on the direction and speed of the users are unavailable. In [5], the authors consider irregular movements of mobile users. For predicting the future location of mobile users a Markov chain model and database of the movement patterns are used. This scheme does not work if preliminary information such as pattern database is either absent or obsolete. On the other hand, an approach for predicting the behavior of the users based only on current location information is considered in [6]. This scheme is costly because no previous information is used while highly accurate data observation is necessary. A mobility prediction scheme based on neuro-fuzzy theory is offered in [3]. Here a unique reason can affect the behavior of a user, and thus all previous information and regulations can be out-of-date.

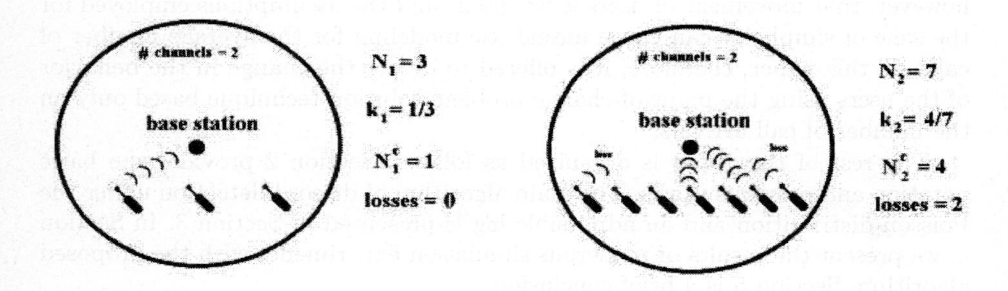

Fig. 1. The losses before and after traffic congestion.

We know that the movement of mobile users may be irregular, and thus prediction of their behavior is very difficult. However, we can observe and analyze the number of calls in a cell, which is considered as a random value. If the demand is steady we have a sequence of random values under one distribution. Otherwise, the distribution of the number of calls is changeable. Usually the following assumption is acceptable, where the number of calls ξ in a unit time is independent and identically distributed Poisson random value [7].

$$P(\xi = k) = \frac{\lambda^k}{k!} e^{-\lambda}, \quad k = 0, 1, \ldots. \tag{1}$$

The number of calls before traffic congestion has the Poisson distribution with parameter λ_1, and it has the same distribution with rate λ_2 after traffic congestion occurs. It is necessary to detect the demand change, in the strict sense, the moment when the distribution change occurs should be found. The detection problem is known as a point-of-change problem or a discord problem. It is a very popular problem [8, 9], and minimization of the time between the moment of discord and the moment of alarm is an important issue. As far as the channel resource is available, the quality of service is satisfied and the delay for the discord detection can be tolerated. In this paper, we focus on a point-of-change problem with acceptable delay, and propose a technique based on a point-of-change problem with admissible lag.

3 The Point-of-Change Problem with Poisson Distribution

3.1 Problem Statement and Discord Detection Algorithm

Let a sequence of independent random values ξ_i, $i = 1, 2, \ldots$, be under consideration. ξ_i denotes the number of calls in time unit i. Before discord, the distribution of ξ_i is modeled by the Poisson distribution of Eq. (1) with parameter λ_1. In unknown point of time t_D the intensity of flow of calls is changed. It means that the distribution parameter of ξ_i becomes different from λ_1. We denote the distribution parameter after the discord as λ_2. Without loss of generality, we assume that $\lambda_1 \leq \lambda_2$. The other case can be easily modeled with simple modification.

It is desirable and sometimes required to alarm system managers if a discord takes place in advanced systems. Let us designate the moment of alarm as t_A and T be an admissible lag. Then we know that a discord is detected if $t_D < t_A \leq t_D + T$. If $t_A < t_D$, it must be a false alarm.

Now we propose the following algorithm for discord detection.

Discord Detection Algorithm

Step 1. Calculation of the sum $S_T = \sum_{i=1}^{T} \xi_i$.

Let us denote the variate S_T has Poisson distribution [10].

Step 2. If $S_n > h$, then the system raises an alarm and go to Step 5. Otherwise, proceed to the next step. Here n takes the value $T, T+1, \ldots$. Note that h is a threshold value calculated prior to the beginning of the algorithm. It is considered in detail below.

Step 3. The following sum is computed

$$S_n = S_{n-1} + \xi_n - \xi_{n-T}, \text{where} \quad n = T+1, T+2, \ldots$$

It is obvious that the variate S_n also has the Poisson distribution because S_n is a sum of independent Poisson random values. Hence, if the distribution parameter of ξ_i is λ, the distribution parameter of S_n becomes $T\lambda$.

Step 4. Go to Step 2.

Step 5. End.

The value of h decides the tradeoff between the rate of discord detection and false alarm. For example, if $h = 0$, discord will be unambiguously detected, but a false alarm will be generated on each observation. If h value is high, false alarm will be seldom declared, but the probability of discord detection will be very low.

Let us discuss the basic concept of the proposed method. The random values $\xi_{n-T}, \ldots, \xi_n$ are under consideration. Assume that the probability of a false alarm is fixed if the following restriction is true.

$$P(S_n > h, n < t_D) \leq \alpha. \tag{2}$$

Here α is a constant representing the degree of false alarm. Before applying the method, it needs to be decided according to how often false alarm can be tolerated.

The second issue is the number of discord omissions. A discord is lost if $S_n < h$ and $n \geq t_D + T$. Hence, the probability of omission is

$$P(S_n < h, n = t_D + T). \tag{3}$$

For the quality of discord detection algorithm, it needs to minimize the probability of Eq. (3), while taking into account Eq. (2). It is achieved with an optimal choice of threshold value h_{opt}.

Theorem 1. *For the proposed method the optimal choice of threshold is*

$$h_{opt} = min\{h : \sum_{k:=0}^{h} \frac{(\lambda_1 T)^k}{k!} > e^{\lambda_1 T}(1-\alpha)\}. \tag{4}$$

The probability of omission is then

$$\sum_{k:=0}^{h_{opt}} \frac{(\lambda_2 T)^k}{k!} e^{-\lambda_2 T}.$$

Proof: Let us consider Step 2 of the algorithm. The decision rule is true, but the discord is not found ($n < t_D$). Hence, a false alarm takes place and S_n has the Poisson distribution with parameter $T\lambda_1$. It is clear that S_n is a positive integer. Hence, it can be assumed that h is a positive integer. We have

$$P(S_n > h, n < t_D) = 1 - P(S_n < h, n < t_D) = 1 - \sum_{k=0}^{h} \frac{(\lambda_1 T)^k}{k!} e^{-\lambda_1 T} \leq \alpha.$$

Thus, if the condition of Eq. (2) is true, then the threshold value belongs to a set

$$\{h : \sum_{k=0}^{h} \frac{(\lambda_1 T)^k}{k!} \geq (1-\alpha)e^{\lambda_1 T}\}. \tag{5}$$

Without loss of generality we assume that $t_D = 0$. An omission of point-of-change occurs if $S_n < h$ and $n \geq T$. The aim of the proposed method is to minimize the probability of Eq. (3) under the condition of Eq. (2). Hence, it needs to minimize the probability $P(S_T < h)$. h can take a value from the set of Eq. (5). Obviously, the discord omission probability decreases if the threshold decreases, but the probability of false alarm increases. So, an optimal value of threshold is the least possible value of the set of Eq. (5). Hence, we have Eq. (4).

At the same time

$$P(S_T \leq h_{opt}) = \sum_{k:=0}^{h_{opt}} P(S_T = k) = \sum_{k:=0}^{h_{opt}} \frac{(\lambda_2 T)^k}{k!} e^{-\lambda_2 T}.$$

This contradiction proves the theorem. □

The following theorem is needed for the sequel.

Theorem 2. *h_{opt} does not depend on λ_2. That is, the distribution after discord does not affect the optimal threshold value.*

The proof directly follows Eq. (4).

3.2 Varying Parameter after Discord

The rate of distribution after discord is a constant λ_2 as defined earlier. Sometimes it is possible to have unsteady rate such that the random value ξ_{t_D+1} has a Possion distribution with the rate $\lambda_2 - \delta_1$, ξ_{t_D+2} with the rate $\lambda_2 - \delta_2, \dots, \xi_T$ with the rate $\lambda_2 - \delta_T$. Here, we have $\delta_i < \lambda_2$. It is possible that $\delta_j = 0, \delta_{j+1} = 0, \dots \delta_T = 0; 0 \leq j \leq T$. In this case the quality of the proposed method is described by the following theorem.

Theorem 3. *The probability of discord omission is*

$$P(S_T < h_{opt}, t_D = 0) = \sum_{k:=0}^{h_{opt}} \frac{(\lambda_2 T - \Delta)^k}{k!} e^{-(\lambda_2 T - \Delta)}, \Delta = \sum_{i=1}^{T} \delta_i.$$

Here h_{opt} is calculated using Eq. (3).

Proof: By Theorem 2, h_{opt} is obtained by Eq. (3). Since $\xi_1, \dots, \xi_T$ are independent Poisson random values, the sum of those values, S_T, has the Poisson distribution too. The parameter of the distribution of S_T is equal to $T\lambda_2 - \Delta$. So,

$$P(S_T < h_{opt}, t_D = 0) = \sum_{k:=0}^{h_{opt}} \frac{(\lambda_2 T - \Delta)^k}{k!} e^{-(\lambda_2 T - \Delta)}.$$ □

4 Performance Evaluation

In this section we analyze the quality of the proposed method through computer simulation. For it, a pseudorandom generator for Poisson distribution is necessary, and the choice of best generator is discussed. Also, we compare the analytical data and simulation data, and verify that the degree of false alarm is limited. The proposed algorithm is tested for both the constant and varying parameters of the distribution. For each simulation experiment, 10^6 runs are averaged to show the sensitiveness of the algorithm to the lag and parameters of distributions.

4.1 Pseudorandom Generator for Poisson Distribution

For the performance evaluation of mobility management such as mobile location management and hand off schemes, simulation is frequently employed. Usually the ranges of parameter values of wireless network are quite large, and it results in complex calculation and large simulation time. Therefore, it is important to choose a fast pseudorandom number generator.

In this paper we use a sequential random values of Poisson distribution for modeling the number of calls per unit time. There exist many algorithms generating Poisson pseudorandom numbers [10–12]. They are compared taking into account the specific property of the target problem.

The methods in [10, 11] use the idea that if the time between arrivals is exponentially distributed, then the number of arrivals have the Poisson distribution. We also need to use a uniform pseudorandom number generator. The method [12] uses only one uniformly distributed numbers. In Table I below the result of the generators tested is given. It shows that the algorithm from [12] is the best for average and large values of the Poisson distribution parameter. Method (A) of Molloy and Knuth[10, 11] have an advantage for small value parameters.

Table I. The time for generating 1,000,000 pseudorandom numbers (nanosecond).

Distribution rate	Ermakov	Knuth; Molloy (A)	Knuth; Molloy (B)
0.0001	38	33	44
0.01	38	33	44
0.1	39	38	44
0.2	39	44	44
0.3	38	50	44
0.5	43	47	50
1	47	61	58
2	49	99	71
3	66	137	82
5	82	198	109
10	143	363	176
100	1027	3279	1373
300	3010	9694	4010
500	5141	16323	6712

4.2 Verification of Limited Degree of False Alarm

According to Eq. (2), the proposed method gives a small number of false alarm. In each test the method employs a random value T, and a counter is increased if a false alarm takes place. Let the degree of false alarm be equal to or less than 5% throughout the tests. An admissible lag is selected among different values.

Table II. The degree of false alarm.

λ_2	T	h_{opt}	**False alarm(%)**
0.01	20	5	2.0
1	10	15	5.0
2	10	28	3.4
3	10	39	4.8
3	20	73	4.3
3	50	170	4.9
3	5	22	3.3
3	3	14	4.0
3	1	6	3.5
10	10	117	4.2
50	10	537	4.8

Each row in Table II is a result of simulation experiment for different λ_2 values. At the beginning of each experiment, the optimal threshold is calculated based on **Theorem 1** and the value is used for all the runs. The number of false alarms is in the column for **false alarm** which is given in percentage. We see that the degree of false alarm is always lower than 5%.

4.3 The Efficiency of the Proposed Algorithm

Let us find out the quality of the proposed discord detection algorithm. Without loss of generality we assume that $t_D = 0$ for discord detection. So, we have a distribution with a discord at the beginning. The simulation method is the same as above and the degree of false alarm is equal to 5%.

The discord detection rates from the simulation and analytical model are shown in Table III. Notice that they are very close, which verifies the effectiveness of the proposed model. The proposed discord detection algorithm always detects a demand change for relatively large distribution parameter values even though λ_1 and λ_2 are close and the admissible lag is not large.

Let us now consider the case of non-constant rate after the point-of-change. Let a random value ξ_i, $i = 1, \ldots, T$, after the discord have the parameter of distribution $\lambda_1 + \delta i$, if $\lambda_1 + \delta i < \lambda_2$. Otherwise, the rate is equal to λ_2. In other words, the rate of distribution after the discord changes gradually with a constant step δ up to λ_2. The results for $\delta = 0.1$ are given in Table IV.

Table III. The efficiency of the algorithm.

λ_1	λ_2	T	h_{opt}	Simulation	Model
1	1.2	10	15	15.7	15.6
1	1.2	100	117	58.2	58.5
1	1.2	300	329	94.9	94.8
1	3	10	15	100	100
2	3	20	28	87.1	86.5
3	5	15	56	98.8	98.7
2	4	30	73	100	100
5	10	20	117	100	100
10	11	100	346	100	100
100	120	10	346	100	100

Table IV. Non-constant rate after discord.

λ_1	λ_2	T	Simulation	Model
1	2	10	48.5	48.3
1	2	20	91.1	88.3
10	11	10	12.1	12.2
10	11	20	29.7	26.8
1	3	10	88.4	88.9
100	120	10	100	100

We see that the quality of the proposed method in the case of varying distribution parameter is worse than that of the constant case. However, it is not true if T is relatively large. In Figure 2 the relationship between the quality and T is demonstrated for the case of $\lambda_1 = 2$, $\lambda_2 = 3$, and $\delta = 0.1$.

5 Conclusion

It is important to know the number of calls for moving users for effective management of cellular networks. The number depends on the behavior of customers, which is usually irregular and very difficult to model. The method for detecting the demand change in terms of the numbers has been presented in this paper. The point-of-change problem under admissible lag has been stated and analyzed by computer simulation with Poisson random value. The proposed algorithm for detecting the demand change shows high accuracy under heavy load, and the developed model can accurately predict the performance of the algorithm.

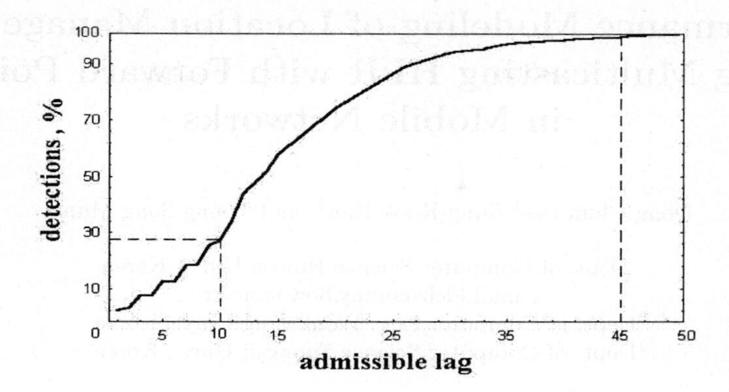

Fig. 2. The relationship between the detection rate(%) and admissible lag; $\lambda_1 = 2$, $\lambda_2 = 3$.

References

1. W. Lee, *Mobile Cellular Telecommunications: analog and digital systems,* Second Edition, New York: McGraw-Hill, 1995.
2. R. Thomas, H. Gilbert, and G. Mazziotto, "Influence of the Movement of Mobile Station on the Performance of the Radio Cellular Network," *Proceeding of the 3rd Nordic Seminar, Copenhagen,* September 1988.
3. C.Y. Park, Y.H. Han, C.S. Hwang, and Y.S. Jeong, "Simulation of a Mobility Prediction Scheme Based on Neuro-Fuzzy Theory in Mobile Computing," *Simulation,* v. 75, No 1, p. 6-17, 2000.
4. D. Hong and S. Rappaport, "Traffic Model and Performance Analysis for Cellular Mobile Radio Telephone Systems with Prioritized and Non-Prioritized Handoff Procedures," *IEEE Transactions on Vehicular Technology, v. 35, No 3,* pp. 77-91, August, 1986.
5. G. Liu and Jr. Maguire, " Class of Mobile Motion Prediction Algorithms for Wireless Mobile Computing and Communications," *Mobile Networks and Application,* v. 1, pp.113-121, 1996.
6. T. Lui, P. Bahl, and I. Chlamtac, "An Optimal Self-Learning Estimation for Predicting Inter-cell User Trajectory in Wireless Radio Networks," *IEEE 6th International Conference on Universal Personal Communications,* v. 2, pp.438-442, 1997.
7. Y. Fang, I. Chlamtac, and Y.-B. Lin, "Call completion probability for a PCS network", *IEEE 6th International Conference on Universal Personal Communications,* v. 2, pp. 567-571, 1997.
8. E.S. Page, *Continuous inspection schemes.* Biometrika, 1954, v.41, N2, p.100-114.
9. B. Yakir, "Dynamic sampling policy for detecting a change in distribution, with a probability bound on false alarm," *Ann. Statist,* v. 24, pp. 2199-2214, 1996.
10. M. Molloy, *Fundamentals of Performance Modeling,* New York : Macmillan Publishing Company, 1988.
11. D.E. Knuth, *The art of computer programming,* 3rd ed., v.2, pp. 137, Addison-Wesley, 1998.
12. S.M. Ermakov and G.A. Michaelov, *Statistics modelling,* oscow: Nauka, 1982 (in Russian).

Performance Modeling of Location Management Using Multicasting HLR with Forward Pointer in Mobile Networks

Dong Chun Lee[1] Sung-Kook Han[2], and Young Song Mun[3]

[1]Dept. of Computer Science Howon Univ., Korea
e-mail:ldch@sunny.howon.ac.kr
[2]Dept. of Computer Eng. Wonkwang Univ., Korea
[3]Dept. of Computer Science Soongsil Univ., Korea

Abstract. We propose a new location management called Multicasting HLR with Forward Pointer (MHFP) which exploits receiver side call locality in Mobile Networks (MN). When a call is established, Multicasting HLR (MH) records the caller's VLR ID according to the callee. Periodically, MH ranks the VLRs and determines which VLRs frequently make calls to the callee. During a location registration process, MH sends the terminal's location information to the determined VLRs. And also, when a mobile terminal frequently moves between two Registration Areas (RAs) or within a small area, the terminal's location information does not register to HLR but to VLR using the Forwarding Pointer (FP) with unit length and link both sides.

1 Introduction

For mobility management scheme in the MN, the standard commonly used in North America is the EIA / TIA Interim Standard 41 (IS-41), and in Europe the GSM[2]. And whenever a terminal crosses a RA or a call originates, the HLR should be updated or queried. Frequent DB accesses and message transfers may be cause the HLR bottleneck problem and then degrade the system performance. A number of related works have been reported to reduce overhead traffic of the HLR. In [7], [8], a Location Forwarding Strategy is proposed to reduce the signaling costs for location registration. A Local Anchoring Scheme is introduced in [1], [4]. Under these schemes, signaling traffic due to location registration is reduced by eliminating the need to report location changes to the HLR. Hierarchical database system architecture is introduced in [3]. These schemes can reduce both signaling traffics due to location registration and call tracking using the properties of call locality and local mobility. We propose a new location management scheme to reduce the location overhead traffic of HLR.

2 IS-41 Standard Scheme

The whole MN coverage area is divided into cells. Each mobile terminal within a cell communicates with the network through a Base Station (BS) which is

P.M.A. Sloot et al. (Eds.): ICCS 2002, LNCS 2331, pp. 1118–1127, 2002.

installed inside the cell. These cells are grouped together to form larger areas called RAs. All BSs, belonging to a given RA, are wired to a Mobile Switching Center (MSC). In this paper, we assume that the VLR is co-located with the MSC and a single HLR in the network [4]. In order to locate a terminal effectively when a call arrives, each terminal is required to report its location whenever it enters a new RA. We call this reporting process location registration. In order to track a call to the proper terminal, the HLR and the VLRs are queried to find the current RA, and all cells within the RA are paged to find it. Within the call tracking, we call the queries to the HLR and the VLRs as the Search. According to the IS-41 location strategy, the HLR always knows exactly the ID of the serving VLR of a mobile terminal. We outline the major steps of the IS-41 location registration. (For the details, refer to [4], [9].)

1. The mobile terminal sends a Registration Request (REGREQ) message to the new VLR.
2. The new VLR checks whether the terminal is already registered. If not, it sends a Registration Notification (REGNOT) message to the HLR.
3. The HLR sends a Registration Cancellation (REGCANC) message to the old VLR. The old VLR deletes the information of the terminal.

and the IS-41 call tracking is outlined as follows:

1. The VLR of caller is queried for the information of callee. If the callee is registered to the VLR, the SEARCH process is over and the call is established. If not, the VLR sends a Location Request (LOCREQ) message to the HLR.
2. The HLR finds out to which VLR the callee is registered, and sends a Routing Request (ROUTREQ) message to the VLR serving the callee. The VLR finds out the location information of the callee.
3. The serving MSC assigns a Temporary Local Directory Numbers (TLDN) and returns the digits to the VLR which sends it to the HLR.
4. The HLR sends the TLDN to the MSC of the caller.
5. The MSC of the caller establishes a call by using the TLDN to the MSC of the callee.
   Among the above 5 steps, the call tracking process is composed of step 1 and step 2.

3 MHFP Scheme

When a call is established, MH records the caller's VLR ID according to the callee. Periodically, MH ranks the VLRs and determine which VLRs frequently make calls to the callee. During a location registration process, MH sends the terminal's location information to the determined VLRs. And also, when a mobile terminal frequently moves between two Registration Areas (RAs) or within a small area, the terminal's location information does not register to HLR but to VLR using the Forwarding Pointer (FP) with unit length and with both side links. Fig. 1 shows one case of location registration when lenth of FP is longer than 1, and the VLR in the starting point of FP is not multicasting. And Fig. 2 also shows one case of call tracking when the receiving terminal has multicasting data, and the VLR has FP of the receiving terminal.

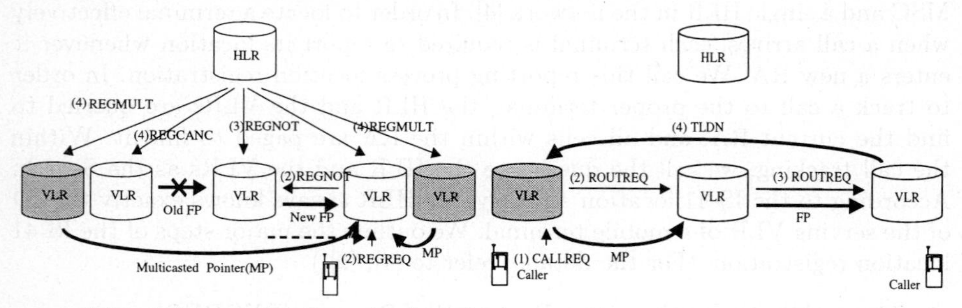

Fig. 1. Location registration(m=2).

Fig. 2. Call tracking.

Here REGMULT represents the registration multicasting and m represents the number of VLR to multicast the location information. We outline the major steps of location registration as follows (see Fig. 1):

- The REGREG message from a terminal is transferred to the VLR that messages a new RA. The VLR transfers the REGNOT message to previous VLR. Previous VLR performs the query by terminal number.
- If the FP of RT exists in the end point in query result,
  - The previous VLR of terminal transfers REGNOT message including FP information to the HLR. The previous VLR of terminal make new FP to current VLR of terminal. The HLR performs the query by number of received terminal.
  - If length of FP is longer than 1, and the VLR in the starting point of FP is not multicasting ,
    - The REGCANC message is transferred to the VLR which has previous FP of terminal. The REGMULT message is transferred to the multicasting VLRs.
  - If length of FP is longer than 1, and the VLR in the starting point of FP is multicasting ,
    - The REGMULT message is transferred to the multicasting VLRs.
  - Else if length of FP is less than 1,
    - Previous VLR of terminal makes new FP to current VLR of terminal.

And call tracking is outline as follows (see Fig.3):

- The sending terminal (ST) requests a call to the VLR. The VLR performs the query by number of RT.
- If the RT exists in query result,
  - The TLDN is assigned to the RT via the MSC.
- Else if the FP of RT exists in query result,
  - The ROUTREQ message is transferred to the VLR directed by the FP. The VLR directed by the FP assigns TLDN via the MSC and transfers to the VLR of ST.

- Else if the multicasting information of the RT exists in query result,
  - The ROUTREQ message is transferred to the VLR directed by multicasting information. The VLR directed by multicasting information performs the query by number of RT.
  - If RT has multicasting information and the VLR has in the RT in query result,
    - The TLDN is assigned to the RT via the MSC and is transferred to the VLR of sending terminal.
  - Else the RT has multicasting information, and the VLR has the FP of RT,
    - The ROUTREQ message is transferred to the VLR directed by the FP. The VLR directed by the FP assigns TLDN via the MSC and transfers to the VLR of ST.
- Else if the multicasting information of the RT doesn't exists in query result,
  - The VLR of ST transfers LOCREQ of message to the HLR. The HLR find the VLR of RT by query and transfers to ROUTREQ. The VLR which receive ROUTREQ message performs the query by number of RT.
  - If the RT must search to the HLR and the VLR directed by the HLR has the RT,
    - The TLDN is assigned to the RT via the MSC and is transferred to the HLR. The HLR transfer the TLDN to the VLR of ST.
  - Else the RT must search to the HLR and VLR directed by HLR has the FP of RT,
    - The ROUTREQ message is transferred to the VLR directed by the FP. The VLR directed by the FP assigns TLDN via the MSC and transfers to the VLR of ST. The HLR transfer The TLDN to the VLR of ST.

In selection of multicasting objects, m, Fig. 3 shows the conceptual structure of the CL field and RL field.

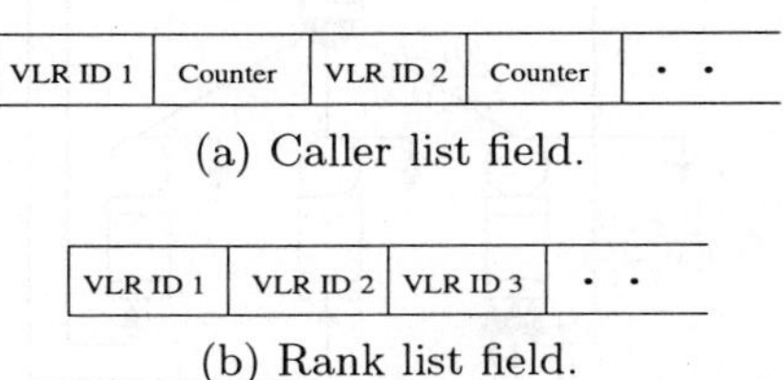

(a) Caller list field.

(b) Rank list field.

Fig. 3. Conceptual structure of the CL field and RL field.

Fig. 3 (a), the CL field consists of the pairs of the ID of frequently calling VLR and number of calls from it. Every established call adds a new pair or increases the number of existing calls. The CL fields are distributed over the HLR and multicast VLR recording the established calls, and merged into HLR periodically to construct the RL field. In Fig.3 (b), each element of RL field is an ID of VLR and the ID of VLR calling more frequently course before. The RL field exists only in the user profile of HLR while the CL field is dispersed over VLR and HLR.

4 Analytical Model

Form the user's point of view, the end-to-end service delay (location registration or call tracking) will be an important performance metric. To evaluate this end-to-end delay, we treat the database of MN as Jackson's network. The service time of each database operation is assumed to be a major delay, and we do not consider a link cost [5], [6]. We assume that there are n VLRs and one HLR in the system. The HLR is assumed to have an infinite buffer and single exponential server with the average service time $\frac{1}{\mu_h}$. Likewise, the VLR is assumed to have an infinite buffer and single exponential server with the average service time $\frac{1}{\mu_v}$. We assume that within a RA, the location registration occurs in a Poisson process with rate λ_u and the call origination occurs in a Poisson process with rate λ_c. With these assumptions, the MN using IS-41standard as a mobility management method becomes Jackson's network [10] as shown in Fig.4.
λ_{lr} and λ_{tt} represent the average arrival rate of REGCANC message and the average arrival rate of ROUTREQ message, respectively. λ_h represents the average arrival rate of messages to the HLR from other VLRs, and by the Burke's theorem [10] it is the same as the average departure rate of messages from the HLR. P_{vo} is the probability the departure message from the VLR leaves the system. P_{vh} is the probability the departure message from the VR enters the HLR. $P_{hv}$is the probability the departure message from the HLR enters one of n VLRs. From the definition of λ_{lr} and λ_{tt}, we know that these messages get out of the system after going through the VLR.

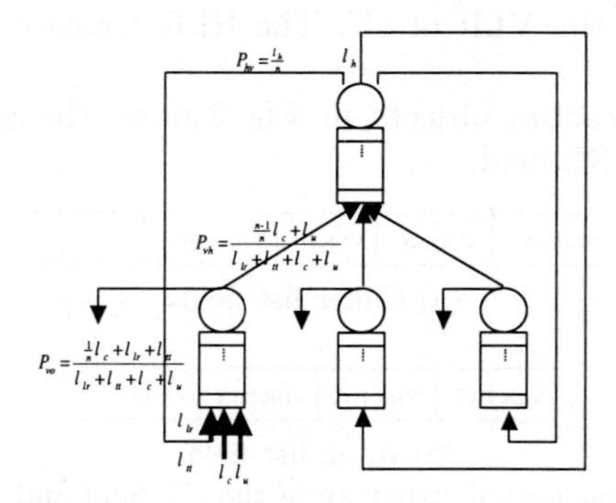

Fig. 4. Jackson's network modeling of the IS-41standard scheme.

λ_u enters the VLR in the form of REGREQ message, and after receiving services from the VLR it is delivered to the HLR in the form of REGNOT message. After receiving services from the VLR, $\frac{1}{n}\lambda_c$ get out of the system because the probability the callee is in the same RA with the caller is $\frac{1}{n}$. So, we have

$$P_{vo}=\frac{\frac{1}{n}\lambda_c+\lambda_{lr}+\lambda_{tt}}{\lambda_{lr}+\lambda_{tt}+\lambda_c+\lambda_{tt}}, \tag{1}$$

$$P_{vh}=\frac{\frac{n-1}{n}\lambda_c+\lambda_{tt}}{\lambda_{lr}+\lambda_{tt}+\lambda_c+\lambda_{tt}}, \tag{2}$$

$$P_{hv}=\frac{\lambda_h}{n}. \tag{3}$$

By the property of Jackson's network, we have

$$\lambda_h=n\times(\frac{n-1}{n}\lambda_c+\lambda_{tt}). \tag{4}$$

From $\frac{\lambda_h}{n}=\lambda_{lr}+\lambda_{tt}$, we have

$$\lambda_{lr}=\lambda_{tt}, \tag{5}$$

$$\lambda_{tt}=\frac{n_1}{n}\lambda_c. \tag{6}$$

Though Eqs. (5) and (6) could be directly inferred from the definitions of λ_c and λ_u, we lead these equations through Eqs. (1)-(4) to help understand how λ_c and λ_u are delivered to HLR and VLRs and how many messages arrive at these DBs on average using the property of Jackson's network. We will repeat these steps in the following Jackson's network modeling the proposed scheme.
Now, let W_v and W_h represent the average system time (queue plus service) in the VLR and the average system time in the HLR, respectively. By the Little's Rule [10], W_v and W_h becomes

$$W_v=\frac{1}{\mu_v-(\lambda_{lr}+\lambda_{tt}+\lambda_c+\lambda_{tt})}=\frac{1}{\mu_v-(\frac{2n-1}{n}\lambda_c+\lambda_{tt})}, \tag{7}$$

$$W_h=\mu_h-n(\frac{n-1}{n}\lambda_c+\lambda_{tt}). \tag{8}$$

Likewise, Fig.6 shows Jackson's network model of MHFP scheme. The focus of message flow aims at the quantity.

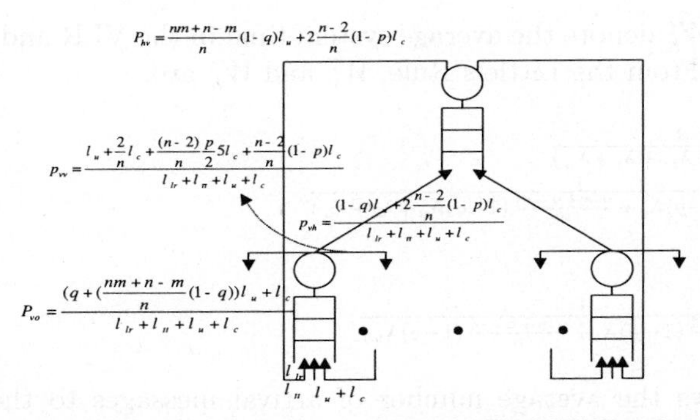

Fig. 5. Jackson's network modeling of the MHFP scheme.

The wide difference between Fig.4 and Fig. 5 is that the number of message out coming HLR. When call managed in HLR, passing one more device dealing with multicasting can solve this problem. Another difference is what communicates call between VLR passing HLR. It is because of obtaining location information of the receiving terminal by multicasting information, m. We need to calculate the probability that the departure message from the VLR is delivered to another VLR. We denote this probability by P_{vv}. Considering the definition of λ_{lr}, λ_{tt}, λ_c, λ_u as follows:

$$P_{vo}=\frac{(q+\frac{nm+n-m}{n}(1-q))\lambda_u+\lambda_c}{\lambda_u+\lambda_{lr}+\lambda_c+\lambda_{tt}}. \tag{9}$$

$$P_{vh}=\frac{(1-q)\lambda_u+2\frac{n-1}{n}(1-p)\lambda_c}{\lambda_u+\lambda_{lr}+\lambda_c+\lambda_{tt}}. \tag{10}$$

$$P_{vv}=\frac{\lambda_u+\frac{2}{n}\lambda_c+\frac{(n-2)}{n}\frac{p}{2}5\lambda_c+\frac{n-2}{n}(1-p)\lambda_c}{\lambda_u+\lambda_{lr}+\lambda_c+\lambda_{tt}}. \tag{11}$$

$$P_{hv}=\frac{nm+n-m}{n}(1-q)\lambda_u+2\frac{n-1}{n}(1-p)\lambda_c. \tag{12}$$

P represents probability that multicasting message search to VLR of receiving terminal ,and q is probability that the FP has less than 1. By the definition of λ_{tt} and λ_{lr}, we obtain

$$\lambda_{tt}=2\frac{n-1}{n}(1-p)\lambda_c, \tag{13}$$

$$\lambda_{lr}=\frac{nm+n-m}{n}(1-q)\lambda_u. \tag{14}$$

From the property of Jackson's network, we have

$$\lambda_h=n(2\frac{n-1}{n}(1-p)\lambda_c+\frac{nm+n-m}{n}(1-q)\lambda_u). \tag{15}$$

Let W_v' and W_h' denote the average system time in the VLR and in the HLR, respectively, From the Little's Rule, W_v' and W_h' are:

$$W_v'=\frac{1}{\mu_v-(\lambda_{tt}+\lambda_{lr}+\lambda_c+\lambda_u)}$$
$$=\frac{1}{\mu_v-(2\frac{n-1}{n}(1-p)\lambda_c+\frac{nm+n-m}{n}(1-q)\lambda_u+\lambda_c+\lambda_u)}. \tag{16}$$

$$W_h'=\frac{1}{\mu_v-(2\frac{n-1}{n}(1-p)\lambda_c+\frac{nm+n-m}{n}(1-q)\lambda_u)}. \tag{17}$$

Table 1 shows the average number of arrival messages to the VLR, and the average system time in the VLR when the proposed scheme is used as

Table 1. VLR comparisons between the MHFP and IS-41scheme

	Average number of arrival messages	Average system time in VLR
IS-41	$\frac{2n-1}{n}\lambda_c+2\lambda_u$	$\frac{1}{\mu_v-(\frac{2n-1}{n}\lambda_c+2\lambda_u)}$
MHFP	$(2\frac{n-1}{n}(1-p)\lambda_c$ $+\frac{nm+n-m}{n}(1-q)\lambda_u+\lambda_c+\lambda_u)$	$\frac{1}{\mu_v-(2\frac{n-1}{n}(1-p)\lambda_c+\frac{nm+n-m}{n}(1-q)\lambda_u+\lambda_c+\lambda_u)}$

Table 2. HLR comparisons between the MHFP and IS-41scheme

	Average number of arrival messages	Average system time in VLR
IS-41	$n(\frac{n-1}{n}\lambda_c+\lambda_u)$	$\frac{1}{\mu_h-n(\frac{n-1}{n}\lambda_c+\lambda_u)}$
MHFP	$n(2\frac{n-1}{n}(1-p)\lambda_c$ $+\frac{nm+n-m}{n}(1-q))$	$\frac{1}{\mu_h-n(2\frac{n-1}{n}(1-p)\lambda_c+\frac{nm+n-m}{n}(1-q)\lambda_u)}$

compared to those of IS-41scheme. And table 2 shows the average number of arrival messages to the HLR, and the average system time in the HLR. As shown in table 1 and table 2, the proposed scheme distributes messages from the HLR to the VLRs, and it also reduces the average system time in the HLR with the small increase of the average system time in the VLRs. Based on the delay times of the HLR and the VLR, we can calculate the mobility management costs for IS-41 standard and the proposed scheme as follows.

- The IS-41scheme for mobility management cost.
  (1) W_{IS-41L}(Location registration cost) $= W_v+W_h+W_v$.
  (2) W_{IS-41C}(Call tracking cost) $= \frac{1}{n}W_v+\frac{n-1}{n}(W_v+W_h+W_v)$.
  (3) W_{IS-41M}(Mobility management cost)
  $=\frac{\lambda_u}{\lambda_c+\lambda_u}\times \mathrm{W}_{IS-41L}+\frac{\lambda_c}{\lambda_c+\lambda_u}\times \mathrm{W}_{IS-41C}$.
- The proposed scheme for mobility management cost.
  (1) $W'_{ProposedL}$(Location registration cost)
  $=q\times 2W'_v+(1-q)(2W'_v+W'h)$.
  (2) $W'_{ProposedC}$(Call tracking cost)
  $=(\frac{1}{n})\times 4W'_v+\frac{n-2}{n}\times\frac{p}{2}\times 7W'_v+\frac{n-2}{n}\frac{(1-p)}{2}(7w'_v+4w'_h)$.
  (3) $W'_{ProposedM}$(Mobility management cost)
  $=\frac{\lambda_u}{\lambda_c+\lambda_u}\times W'_{ProposedL}+\frac{\lambda_c}{\lambda_c+\lambda_u}\times W'_{ProposedC}$.

5 Numerical Results

To get numerical results, we use the same value of system parameters as those in [2], n= 128, for example. From these parameters, the average occurrence rate of location registration in an RA, λ_u, is calculated as $\lambda_u =$

5.85/s. And the average call origination rate in an RA, λ_c, is calculated as $\lambda_c = 8.70/s$. We assume that the average service rates of HLR and VLR are $\mu_h = 2000/s, \mu_v = 1000/s$, and Probability p is 0.5.
In figures, X-coordinate shows probability q that the FP has less than 1 and Y-coordinate shows the ratio dividing the result value in IS-41scheme into the result value in proposed scheme. If the ratio value is 1, the IS-41scheme performance is equal to proposed scheme performance. If it has greater than 1, the performance of proposed scheme is superior to the performance of IS-41scheme. And if it has less than 1, the performance of IS-41scheme is superior to the performance of proposed scheme. The example of the lower graph can show m of multicasting factor. It is determined whether we multicast the location information of terminal into several VLRs.
In Fig.6, we can know that the results of performance are superior to IS-41scheme. As multicasting the location registration, this result becomes a matter of course. If m increases gradually, the level of performance is lower. In Fig.7, regardless of m, the graph is determined by the probability q, but m has no effect upon the value. Otherwise, a relation of between m and probability q is actually represented by an expression but also, it is facts that m has no effect upon the value. When m is 3, we may assume that probability q is 0.5. If probability q is 0, the number of the call tracking message is equal to the number of the location registration message.
Fig.8 shows message ratio in mobility management cost. If probability q is 0.5, and m is 3, the value is equal to 0.87 in short. It is fact that proposed scheme is 1.15 times as many messages as IS-41scheme.

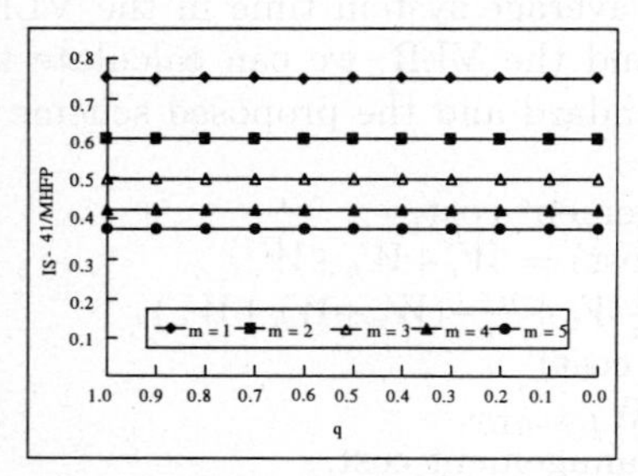

Fig. 6. Message number ratio in registration cost.

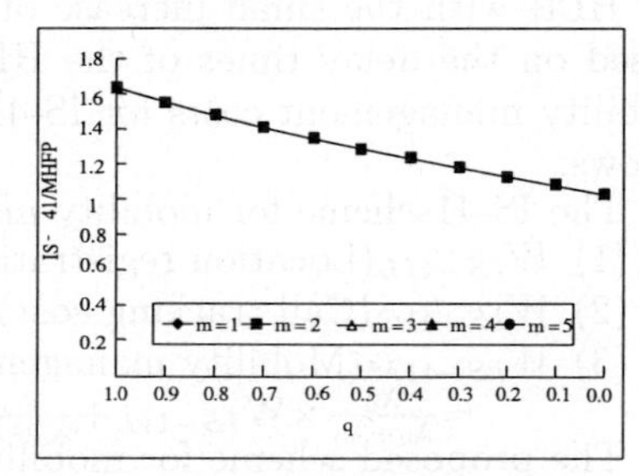

Fig. 7. Message number ratio in call tracking cost.

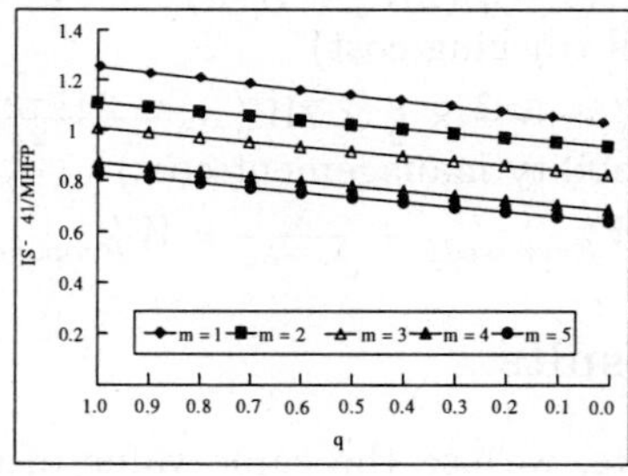

Fig. 8. Message number ratio in mobility management cost.

6 Conclusions

We propose MHFP scheme to reduce the location traffic of HLR. In numerical results, the proposed scheme has lower value in the number of call tracking message and the management delay time of HLR than the IS-41scheme. Especially, the management delay time of HLR has more performance than the IS-41scheme. The proposed scheme can be expected to have the prominent performance advancing for mobile management in the 3G mobile networks.

Acknowledgments

This work is supported in part by the ministry information Communication Communication of Korea, under the "Support Project of University information Technology Research Center(ITRC)" supervised by KIPA".

References

1. I. F. Akyildiz, J, McNair, J, Ho, H. Uzunalioglu, W. Wang, "Mobility Management in Current and Future Communication Networks," IEEE Network, Jul./Aug., 1998, pp. 39-49
2. R. Jain, Y. B. Lin, C. N. Lo., and S. Mohan, "A Caching Strategy to Reduce Network Impacts of PCS," IEEE Jour. on Selected Areas in Com., Vol. 12, No. 8, 1994, pp. 1434-1445
3. C. L. I, G. P. Plooini and R. D. Gitlin, "PCS Mobility Management using the Reverse Virtual All Setup Algorithm," IEEE/ACM Trans. Vehicle. Tech, 1994, pp. 1006-1010
4. T. Russel, Signaling System ♯7, McGraw-Hill, 1995
5. J. Z. Wang, "A Fully Distributed Location Registration Strategy for Universal Personal Communication Systems," IEEE Personal Com., First Quarter 1994, pp. 42-50995.
6. C. Eynard, M. Lenti, A. LOmbardo, O. Marengo, S. Palazzo, "Performance of Data Querying Operations in Universal Mobile Telecommunication System (UMTS)."Proceeding of IEEE INFOCOM '95, Apr. 1995, pp. 473-480
7. T. H. Cormen, C. E. Leiserson, R. L. Rivest, Introduction to Algorithms, MIT Press and McGraw-Hill, 1990
8. N. Shivakumar, J. Widom, "User Profile Replication for Faster Location Lookup in Mobile Environment," Proceeding of ACM MOBICOM, Nov. 1995, pp. 161-169
9. Y. -B. Lin, "Reducing Location Update Cost in a PCS Network," IEEE/ACM Trans. Networking Vol. 5, Feb. 1997, pp. 25-33
10. L. Kleinrock, " Queuing Systems", Vol. 1, Wiley-Interscience, 1975

Using Predictive Prefetching to Improve Location Awareness of Mobile Information Service

Gihwan Cho

Division of Electronic and Information Engineering, Chonbuk University,
664-14 Duckjin-Dong, Duckjin-Gu, Chonju, Chonbuk, 561-756, S. Korea
ghcho@dcs.chonbuk.ac.kr

Abstract. Mobile information services have to provide some degree of information adaptability onto the current service context, such as mobile user's location, network status. This paper deals with a predictive prefetching scheme for reducing the latency to get refreshed information appropriated on current location. It makes use of the velocity mobility model to exploit location knowledge about the terminal and/or user's mobility behavior. With considering the user's moving speed and direction, the prefetching zone has been proposed to effectively limit the prefetched information into the most likely future location context whilst to preserve the prefetching benefits. The proposed scheme has been evaluated with a simulator in the context adaptability point of view.

1 Introduction

With prosperity of hardware technology, computing devices are getting small and higher performance. Moreover, by cooperation of the wireless communication with hand-hold computer, computing users can access their own information anytime, anywhere, even they move around. Even if many existing computing services may be logically extended into this new coming computing paradigm, its inherent characteristics, such as user, terminal and/or service mobility, require new service methodologies. That is, mobile information services have to provide some degree of information adaptability onto the current service context, where context is defined as the characteristics of the user's situation like as current location, and the user supporting environment like as network status (bandwidth, broadcast etc.) and mobile terminal capabilities (size, color etc.). This feature is generally formalized as context-aware mobile information services [1].
In the user and/or terminal mobility environment, location context may be changed whilst the user is still interested in the corresponding answer. Therefore, it is required to be adaptively provided a response appropriate to the current location: a previously provided response must be refreshed, if the corresponding request is still active after a

This work was supported in part by University Research Program of Ministry of Information nd Communication in South Korea, and Center for Advanced Image and Information of Chonbuk National University in South Korea.

P.M.A. Sloot et al. (Eds.): ICCS 2002, LNCS 2331, pp. 1128–1136, 2002.

change of location that invalidates the old response [2]. Thus, the latency to get refreshed information appropriated on the current context is the most critical service parameter, especially in the frequently context changed environment. With the long latency, the user may suffer the long time duration provided with information not appropriate for the current location. To reduce the latency, a well-established technique, prefetching, can be adapted to fetch information in advance, usually based on what contexts are highly utilized [3]. Thus, it is very important to make use of some degree of knowledge to limit the amount of prefetched information; otherwise, unnecessarily prefetched information can overwhelm the benefit of latency reduction with communication and memory cost. Therefore, prefetching scheme is usually based on some sort of prediction strategy about the information that will be needed in the near future.

The predictive prefetching concept has been applied and evaluated to reduce the latency in retrieving a Web document on the wired environment, based on the user's navigation pattern [4]. It shows that the predictive prefetching reduces significantly in the average access time. The work in [3] provides the effectiveness of prefetching policies in the location-aware mobile information services. The work considered only the uniform mobility pattern based on the Markov model, and did not consider how to effectively confine the prefetching information to reduce communication cost. In the reference [5], a semantic based caching scheme has been presented to access location dependent data. The work is just centric to the cache information replacement strategy to utilize the semantic locality in terms of location.

In this paper, a predictive prefetching scheme is proposed by exploiting user movement knowledge to limit the prefetching to the most likely future contexts. Because the location is the most distinguished context for the context aware mobile information system, our discussion is limited in location context after this. The rest of the paper is organized as follows. In Section 2, a general architecture of mobile information service is presented, along with the problem definition. In Section 3, a velocity based mobility model is presented and a predictive prefetching scheme is formalized with the mobility model. Using a simulation, a numerical result is shown in Section 4. Finally a conclusion is given in Section 5.

2 Mobile Information Service

For mobile information service, a general architecture has been utilized to describe the location-aware feature [2], [3], [5], [6], [7]. The architecture consists of 3 logical components; mobile device that possibly moves around with a user, information server that would be located somewhere in Internet, and location-aware service manager that acts as intermediary between the mobile device and the information server. The manager performs a sequence of operations to retrieve information associated with the current location from the server, and to load it on the user device. To do this, it periodically monitors the current user location, and checks whether it belongs to the current information scope. If an out-of-scope condition is detected, the service manager restarts to retrieve the corresponding information on current location. A mobile user is assumed to equipped with a certain mechanism to obtain its current location,

such as the GPS or indoor infrared sensors.
For a given mobile information service, the set of possible location contexts is assumed to be divided into separated subsets, such as rooms, areas. Then a separate piece of information is associated to each subset, such as equipment available in a room, map of a given area. A unit area, which a user may move around with the same information piece, is divided into adjacent geographical portions. Each geographical portion has corresponding portion information that must be provided by the information service as response to a user's request within that geographical portion. Geographical portions are assumed to be equal shape represented as quadrangles. A discrete time model is used for a user's movement. At the end of each time slot, the user can remain in the same quadrangle, or move to an adjacent quadrangle through one of the shared edges. At a certain point of time, the choice of remaining inside or moving outside the quadrangle of a user is just depended on the current location and the moving direction and speed.

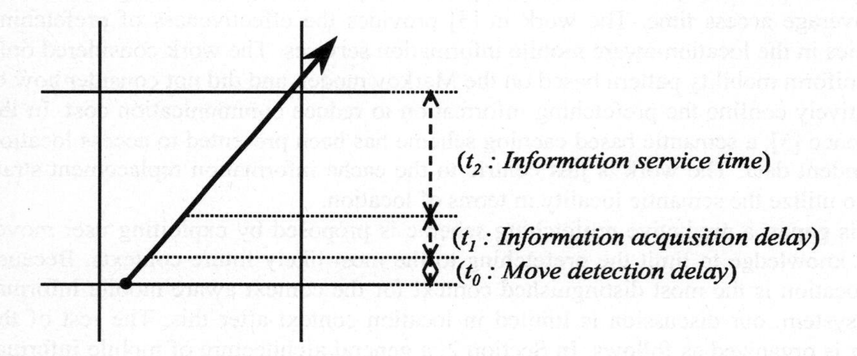

Fig. 1. Mobile information service delay on a given geographical portion

Fig. 1. shows the delay to get corresponding information for a given geographical portion (the quadrangle stands for a geographical portion). On crossing a geographical portion, the latency can be classified with two time factors; the delay needed to detect that the provided portion information is out of its current context, so t_0 : move detection delay, and the delay needed to retrieve and start loading the new information for current geographical portion, so t_1: information acquisition delay. By reducing two delays, a mobile user can be blessed enough with the current context information, so t_2: information service time. The effect of the first factor can be improved by reducing the time interval between two consecutive location checks. The most common technique to improve the second factor is to prefetch information that will be required in the near future. With the practical requirements such as communication and storage limitation of mobile device, the amount of prefetched information must be minimized as much as possible, whilst the prefetching amount reduction must not hurt the prefetching benefits.

3 Mobility Model and Predictive Prefetching

At a given location, the user's future location can be relatively represented by the moving direction and speed. Therefore, the moving behavior of mobile users can be now modeled from 3 parameters, current location, moving speed and direction.

3.1 Velocity based Mobility Model

To model both the direction and speed of a moving user, a velocity V is defined as a vector $< V_x, V_y >$, where V_x and V_y are integers. A pair of values, V_x and V_y, is the projections of its speed vector on x axis and y axis respectively (in practically, each axis can stand for the east or the north respectively). The signs of V_x and V_y, and the ratio of the absolute value of V_x and that of V_y determine the directions of the movements. The model assumes that a user moves around in a 2-dmenstional space at a speed that keeps constant during any observed time period and may change after a given unit time. So, at a certain point of time, when a user's current location has been given, the location in the future can be specified by considering the movement velocity. This concept has been mostly borrowed from [5].

3.2 Predictive Prefetching Strategy

Generally, a natural prefetching strategy is to prefetch the portion information corresponding to a set of geographical portions that are within a certain degree of outbound distance d from the current location. These geographical portions consist of a *prefetching zone* (shortly PZ). To form a PZ, one possible strategy is to include all the quadrangles within outbound distance d fraction (so, radius d) of rings 0, 1, 2, …n [3]. No remote loading is necessary unless the user moves around within its current PZ. When the user enters a portion outside its current PZ, the geographical portion becomes the new starting position to build a new PZ. The new PZ may include the portion information for quadrangles within distance d ring.

Here, our concern is to exploit some degree of prediction about the user behavior, in order to be prefetched only the information for most likely future contexts. When a user crosses a PZ, the moving speed is applied to estimate the distance d, and the moving direction is utilized to decide the width w of PZ. Now, a PZ shapes as a rectangle [7], rather than a ring. This is based on that, if a user moves with high speed, it would be very reasonable to set the distance higher than that of low speed movement. Similarly, if a user's behavior tends to move with high degree of obliquity, it would be very reasonable to set the width wider than that of low degree of obliquity.

Our strategy to predictive prefetching is as follows. The distance d of a rectangle PZ is to get the upper bound integer of the user's speed, that is, the squared root of the sum of squares of two speed values V_x and V_y, as $\left\lceil \sqrt{V_x^2 + V_y^2} \right\rceil$. Thus, the signs of speed values should be ignored because the distance considers the speed characteristic only. In addition, if one of speed values is zero, it must be replaced with the other non-zero value. Similarly, the width w of a rectangle PZ can be obtained with the upper bound integer of the mean value of the sum of two speed values, V_x and V_y, that is $\left\lceil \frac{|V_x|+|V_y|}{2} \right\rceil$. Note that the signs of speed values should be ignored because the direction reflects the degree of obliquity rather than that of dimension. Differently with the distance, zero valued speed will be ignored in the width. Fig. 2. shows an example for the predictive prefetching reconstruction idea for three different moves. The dark rectangles depict the new PZ that is constructed on each crossing of the PZ. For each moving with the velocity <2, 1>, <3, 5> and <5, 0>, the size of PZs is calculated as the distance $d = 3$, 6 and 8, the width $w = 2$, 4 and 3 respectively.

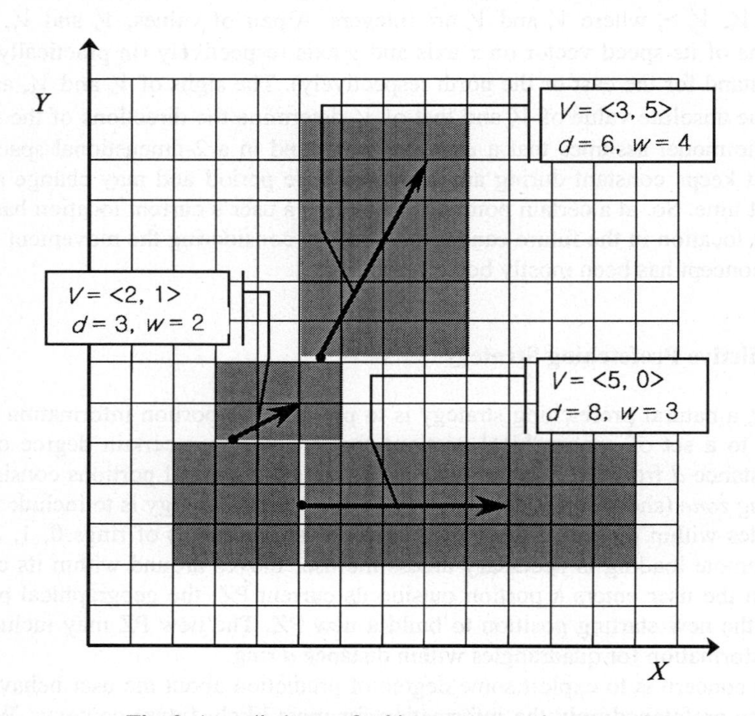

Fig. 2. A predictive prefetching reconstruction example

As shown in the Fig. 2., the gap between the PZs is inevitable. The location aware manager may start to build a new PZ and to prefetch the new PZ, just after it finds out the fact of crossing the current PZ, so the move detection delay in Fig. 1. Then, until the device is refreshed with information for the new PZ, so the information acquisition delay in Fig. 1., the mobile user therefore does not have any information appropriate to its current context. Due to the mobility feature, the user is in the out-of-scope condition of an effective information service. This situation can be improved with defining the edge geographical portions of a PZ into the *crossing zone*; it should be also defined with considering its moving speed and direction. When a user meets a geographical portion in the crossing zone, it starts to prefetch the new PZ in advance.

4 Numerical Result

In order to provide the effectiveness of proposed scheme, a comparison has been made between the ring PZ approach and the rectangle PZ one, through a simulation study. For simulation runs, a set of system parameters has been borrowed from other works, as shown in Table 1. Some numerical result has been measured for the amount of prefetched portion information, the ratio of utilization of the prefetched portion information, and the service time for end user. The first is significant to reduce the communication cost, the second stands for how much the prefetched portion information is made use of real information service, and the last is meaningful to configure the user's satisfaction for the service. The simulator has been implemented in C language using an event-based simulator CSIM [8].

Table 1. System parameters for simulation

Parameter	Description	Value
Wired bandwidth	bandwidth of the wired line	800 kbps - 1.2 Mbps
Wireles bandwidth	bandwidth of the wireless medium	80 kbps - 120 kbps
Move detection time	time to detect the fact a user accrosses a gegraphycal portion	100 ms
Residentiary time	mean time for a user staying in a gegraphycal portion	10 sec. (exponential distribution)
Portion size	information size which stands for a gegraphycal portion	1 MByte
Packet size	size of a unit data packet which is passing through the network	1440 byte

4.1 User Move Scenario

In the velocity based mobility model, a user moves around in a 2-dmenstional portion's geographical space at a speed that keeps constant during any given unit time period. Then it changes its speed and direction, and moves constantly within the unit time. Therefore, the model is much significant when a user's moving pattern is uniform. Practically, this is analogous to many real life cases, such as driving a car on highways or walking a street. A typical user move scenario has been utilized for the simulation. A user moves around 20 by 20 geographical portions, as shaped in Fig. 2. It starts from the position (10, 10), and moves 20 times, sequenced as (2, -4), (3, 2), (-3, 4), (-5, 0), (2, -5), (-3, -3), (5, -2), (0, 3), (4, -2), (-1, 3), (3, 3), (0, 5), (-2, 5), (-3, 0), (-5, -2), (5, -2), (-5, 0), (-2, 4), (-3, -6), (4, -5).

4.2 Simulation Experiments

With the given user mobility scenario, the number of prefetched portion information has been evaluated to show the effectiveness of prefetching strategies by comparing three different cases; prefetching with ring PZ, prefetching with rectangle PZ, prefetching is not applied. The same simulation parameters are used for each case, whilst different strategies are used to build a PZ. For the prefetching with ring PZ, radius of the ring is a critical simulation parameter. In order to provide some degree of fairness with that of prefetching with rectangle PZ, the speed value $\sqrt{V_x^2 + V_y^2}$ is adapted for calculating the radius of ring PZ.

Fig. 3. shows the number of prefetched portion information for each cases. The prefetching with ring PZ has relatively large variation depending on the speed value. Thus, some of portion information would be overlapped between that of the previous PZ and that of the current PZ. After excluding these overlapped portions, 1,105 portions are totally prefetched for the ring PZ strategy, and mean number of prefetched portion information is 55.25 for each move. In the rectangle PZ scheme, the number of prefetched portion information is stable; total number of prefetched portion information is 335, mean number of prefetched portion information is 16.75. In the case of prefetching is not applied, each portion information has to be acquired for every geographical portions in turn, on the path of the user moves through.

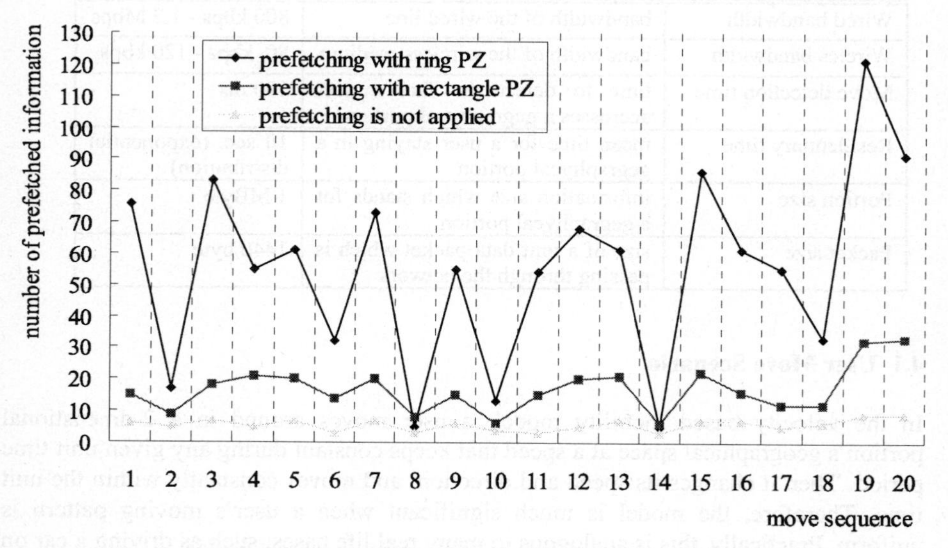

Fig. 3. The number of prefetched portion information

Now, it is meaningful to figure out how much the prefetched portion information would be actually participated for the user's location-aware service. Fig. 4. shows the utilization ratio of prefetched portion information for two prefetching strategies, so the ring PZ and the rectangle PZ. Note that the utilization ratio is just depended on the user's mobility pattern. With the given user move scenario, the utilization of rectangle PZ is much better than that of ring PZ; the former shows that over 30% of prefetched portion information is utilized for real service, whilst the later figures around 10%. We believe that the predictive prefetching with considering the user's move behavior plays great role in building a prefetching method for mobile information service.

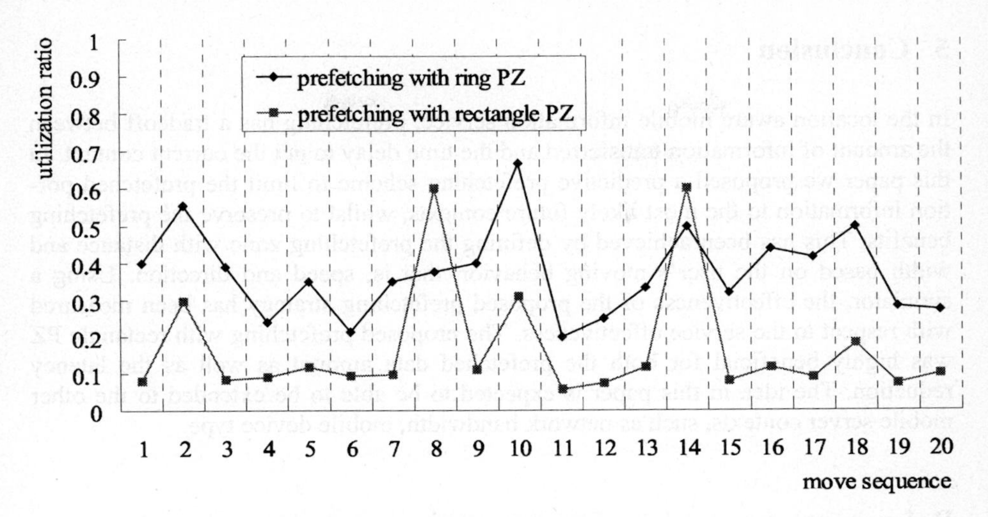

Fig. 4. Utilization ratio of the prefetched portion information

Fig. 5. shows the service time which has been spent to provide the corresponding information on each move. In the case of the prefetching is not applied, the time can be obtained by adding the service delay for getting each of portion information that the user has visited. So, mean service time is 570 sec. for each PZ. With the rectangle PZ strategy, the service delay is applied for only when a user crosses the current PZ. Once the service delay has been given on the portion, the other portion information within current PZ would be delivered in concurrently during the first portion information is serviced to end user. So the delay is the same for getting one of portion information on crossing PZ, and can be reduced as about 95 sec.

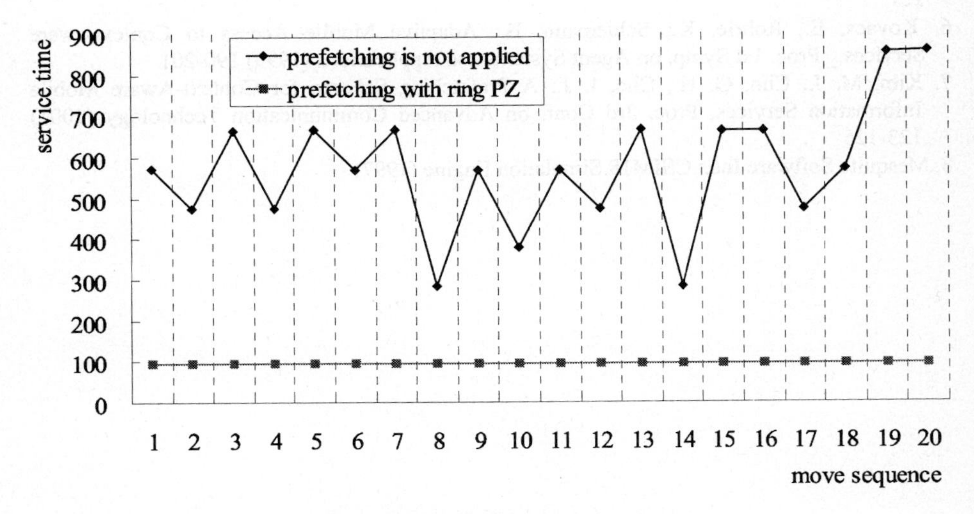

Fig. 5. Information service Time

5 Conclusion

In the location aware mobile information service, prefetching has a tradeoff between the amount of information transferred and the time delay to get the current context. In this paper we proposed a predictive prefetching scheme to limit the prefetched portion information to the most likely future contexts, whilst to preserve the prefetching benefits. This has been achieved by defining the prefetching zone with distance and width based on the user's moving behavior, that is, speed and direction. Using a simulator, the effectiveness of the proposed prefetching strategy has been measured with respect to the service effectiveness. The proposed prefetching with rectangle PZ was highly beneficial for both the prefetched data amount as well as the latency reduction. The idea in this paper is expected to be able to be extended to the other mobile server contexts, such as network bandwidth, mobile device type.

References

1. Brown, P. J., Bovey, J. D., Chen, X.: Context-aware Applications: From the Laboratory to the Marketplace. IEEE Personal Communications, Vol. 4, No. 5 (1997) 55-65
2. Couderc, P., Kermarrec, A. -M.: Improving Level of Service for Mobile Users Using Context-Awareness. Proc. 18th IEEE Symp. on Reliable Distributed Systems (1999) 24-33
3. Persone, V. N., Grassi, V., Morlupi, A.: Modeling and Evaluation of Prefetching Polices for Context-Aware Information Services. Proc. ACM Conf. on Mobile Computing and Networking (1998) 55-65
4. Padmanabhan, V. N., Mogul, J. C.: Using Predictive Prefetching to improve World Wide Web Latency._ACM Computer Communication Review, Vol. 26, No. 3 (1998) 22-36
5. Ren, Q., Dunham, M. H.: Using Semantic Caching to Manage Location Dependent Data in Mobile Computing. Proc. ACM Conf. on Mobile Computing and Networking (2000) 210-221
6. Kovacs, E., Rohrle, K., Schiemann, B.: Adaptive Mobile Access to Context-aware Services._ Proc. 1st Symp. on Agent Systems and Applications (1999) 190-201
7. Kim, M. J., Cho, G. H., Cho, I. J.: A Prefetching Scheme for Context-Aware Mobile Information Services, Proc. 3rd Conf. on Advanced Communication Technology (2001) 123-125
8. Mesquite Software Inc.: CSIM18 Simulation Engine (1997)

Dynamic and Stochastic Properties of Molecular Systems: From Simple Liquids to Enzymes

Igor V. Morozov[1,2], Guenri E. Norman[2,3], and Vladimir V. Stegailov[3,2]

[1] Department of Physics, Chair of General Physics and Wave Processes, Lomonosov Moscow State University, Vorob'evy Gory, 119899, Moscow, Russia, bogous@orc.ru

[2] Institute for High Energy Densities, Associated Institute for High Temperatures, Russian Academy of Sciences (IHED-IVTAN), Izhorskaya 13/19, Moscow 127412, Russia, henry_n@orc.ru

[3] Moscow Institute of Physics and Technology, Institutskii per. 9, 141700, Dolgoprudnyi, Russia stega@nusun.jinr.ru

Abstract. Molecular dynamics method (MDM) supplies to the solution of fundamental contradiction between macroscopic irreversibility and microscopic reversibility with data which help to reveal the origin of stochastization in many-particle systems. The relation between dynamic memory time t_m, fluctuation of energy dE and K-entropy (Lyapunov exponent) is treated. MDM is a method which retains Newtonian dynamics only at the times less than t_m and carries out a statistical averaging over initial conditions along the trajectory run. Meaning of t_m for real systems is related to the quantum uncertainty, which is always finite for any classical system and influence upon particle trajectories in a coarse-graining manner. Relaxation of kinetic energy to equilibrium state was studied by MDM for non-equilibrium strongly coupled plasmas. Two stages of relaxation were observed: initial fast non-Boltzmann oscillatory stage and further relatively slow Boltzmann relaxation. Violation of the microscopic reversibility principle in some enzymatic reactions is discussed.

Monte Carlo method is referred to stochastic methods of molecular simulation whereas molecular dynamics method (MDM) is usually called a dynamic method. The objective of the present paper is to show that MDM possesses both dynamic and stochastic features. Moreover if MDM had no hidden stochastic features MDM would not probably be able to achieve well-known successive results. Another objective is to present examples of MDM simulation of non-equilibrium relaxation when stochastic features influence the dynamics.

The topic is related to the occurrence of the irreversibility in the case of the classical molecular systems which has been discussing since Boltzmann-Zermelo debate. The present state was given at the round table "Microscopic origin of macroscopic irreversibility" at the XX International conference on statistical physics (Paris, 1998) [1,2], see also [3].

P.M.A. Sloot et al. (Eds.): ICCS 2002, LNCS 2331, pp. 1137–1146, 2002.

1 Molecular Dynamics Method

The idea of MDM is very simple: all possible classical systems and media are simulated by a set of N moving atoms and/or molecules, which interact with each other (e.g., see [4–11]). The numerical integration of the corresponding system of Newton equations

$$m_i \frac{d\mathbf{v}_i(t)}{dt} = \mathbf{F}_i[r(t)], \quad \frac{d\mathbf{r}_i(t)}{dt} = \mathbf{v}_i(t) \tag{1}$$

results in determination of the trajectories of all particles $\{\mathbf{r}, \mathbf{v}\}$. Here, m_i, $\mathbf{v}_i$, $\mathbf{r}_i$, and $\mathbf{F}_i$ are the mass, velocity, and coordinate of the ith particle and the force acting on this particle, respectively ($i = 1, \ldots, N$); the $\mathbf{v}_i$ and $\mathbf{r}_i$ values explicitly depend only on time t; $\mathbf{F}_i$ depends only on the coordinates of particles; $\mathbf{r}(t)$ is the set of the coordinates of all particles, $\mathbf{r}(t) = \{\mathbf{r}_1(t), \mathbf{r}_2(t), \ldots, \mathbf{r}_N(t)\}$; $\mathbf{v}(t)$ is defined similarly; and

$$\mathbf{F}_i = -\frac{\partial}{\partial \mathbf{r}_i(t)} U(\mathbf{r}_1, \mathbf{r}_2, \ldots, \mathbf{r}_N), \tag{2}$$

where U is the potential energy. Function U (forces $\mathbf{F}$) is assumed to be given in MDM. The total energy E of the system is the sum of the kinetic T and potential U energies,

$$E = T + U, \qquad T = \sum_{i=1}^{N} \frac{m v_i^2}{2} \tag{3}$$

Set (1) is exponentially unstable for a system of more than two particles (e.g., see [3–15]). The parameter that determines the degree of instability, that is, the rate of divergence of initially close phase trajectories, is Lyapunov exponent or K-entropy K. It can be determined by several ways. As example for a given U function and particles of the same mass m and for identical initial conditions corresponding to the kth point on an equilibrium molecular-dynamical trajectory, solutions $\{\mathbf{r}(t), \mathbf{v}(t)\}$ to system (1) are found in steps Δt and trajectories $\{\mathbf{r}'(t), \mathbf{v}'(t)\}$ are calculated in steps $\Delta t'$. Averaged differences of the coordinates (velocities) of the first and second trajectories are determined at coinciding time moments,

$$\langle \Delta v^2(t) \rangle = \frac{1}{NI} \sum_{j,k}^{N,I} \left(v_{jk}(t) - v'_{jk}(t) \right)^2, \tag{4}$$

$$\langle \Delta r^2(t) \rangle = \frac{1}{NI} \sum_{j,k}^{N,I} \left(r_{jk}(t) - r'_{jk}(t) \right)^2, \tag{5}$$

To improve accuracy, averaging over $k = 1, \ldots, I$ is also performed. In some transient time t_l the differences become exponentially increasing with the same value of K,

$$\langle \Delta v^2(t) \rangle = A \exp(Kt),$$
$$\langle \Delta r^2(t) \rangle = B \exp(Kt) \quad \text{at} \quad t_l < t < t_m. \tag{6}$$

The values of A and B are determined by the difference of Δt and $\Delta t'$. An example for the three-dimensional Lennard-Jones system is given in Fig. 1. We use reduced units in which $m = \epsilon = \sigma = 1$, and time is measured in $(m\sigma^2/\epsilon)^{1/2}$ units.

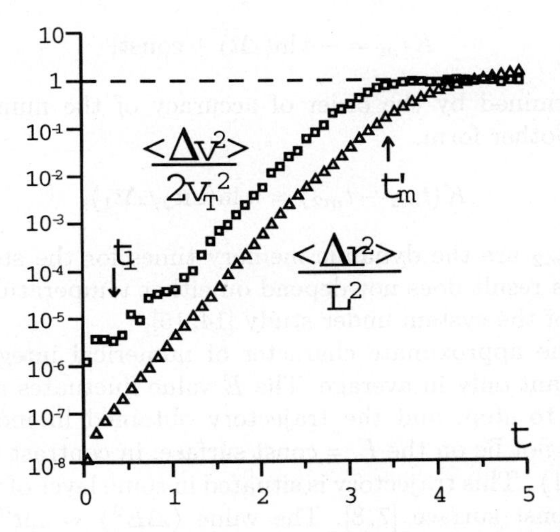

Fig. 1. Normalized averaged differences of velocities Δv^2 and coordinates Δr^2 at coinciding time moments along two trajectories calculated for identical initial conditions in steps $\Delta t = 0.001$ and $\Delta t' = 0.0001$; L is the calculation cell edge length, $N = 64$, $r = 0.5$, and $T = 0.44$.

The exponential increase of $\langle \Delta v^2(t) \rangle$ is limited by the finite value of the thermal velocity of particles. Thus after the time t'_m

$$t'_m \approx K^{-1} \ln (6kT/Am), \tag{7}$$

where T is the temperature, saturation is reached. t'_m is a dynamic memory time [14–17]. For $t > t'_m$

$$\langle \Delta v^2(t) \rangle = 2 \langle v^2 \rangle = 6kT/m, \tag{8}$$
$$\langle \Delta r^2(t) \rangle = 6D(t - t'_m) + \langle \Delta r^2(t'_m) \rangle, \tag{9}$$

where $3kT/m$ is the square of thermal velocity v_T and D is the diffusion coefficient. Estimates show that $\langle \Delta r^2(t_m) \rangle = r_{av}$, where $r_{av} = (\sqrt{2}n\sigma)^{-1}$ is the mean path of particles between collisions.

The dynamic memory times are determined by calculating at the same Δt value and different $\Delta t'$ values of $\Delta t/2$, $\Delta t/5$, $\Delta t/10$, etc. The limiting value when $\Delta t'/\Delta t \to 0$ is the dynamic memory time t_m for a given system and the selected numerical integration step Δt [14, 15]. During numerical integration, the system completely "forgets" its initial conditions in time t_m, and the calculated molecular-dynamical trajectory completely ceases to correlate with the initial hypothetical Newtonian trajectory. In other words it determines the time interval during which the behavior of a dynamic system can be predicted from initial conditions and deterministic equations of motion [16, 18].

The calculated dependencies of Kt_m on the numerical integration step Δt can be presented in the form

$$Kt_m = -n \ln(\Delta t) + \text{const}, \tag{10}$$

where n is determined by the order of accuracy of the numerical integration scheme, or, in another form,

$$K(t_{m1} - t_{m2}) = n \ln(\Delta t_2/\Delta t_1), \tag{11}$$

where t_{m1} and t_{m2} are the dynamic memory times for the steps Δt_1 and Δt_2, respectively. This result does not depend on either temperature, density, or the special features of the system under study [14, 15].

Because of the approximate character of numerical integration, energy E [Eq. (3)] is constant only in average. The E value fluctuates about the average value from step to step, and the trajectory obtained in molecular dynamics calculations does not lie on the $E =$ const surface, in contrast to exact solutions to Newton Eqs. (1) . This trajectory is situated in some layer of thickness $\Delta E > 0$ near the $E =$ const surface [7, 8]. The value $\langle \Delta E^2 \rangle \sim \Delta t^n$ depends on the accuracy and the scheme of numerical integration [7, 8, 22–25]. Therefore

$$Kt_m = -\ln\left(\langle \Delta E^2 \rangle\right) + \text{const} \tag{12}$$

Equation (12) relates the K-entropy and the dynamic memory time to the noise level in the dynamic system. This equation corresponds to the concepts developed in [16, 18].

It follows from (10)-(12) that t_m grows no faster than logarithmically as the accuracy of numerical integration increases. The available computation facilities allow ΔE to be decreased by 5 orders of magnitude even with the use of refined numerical schemes [23–25]. This would only increase t_m two times. Estimates of dynamic memory times showed that, in real systems, t_m lies in the picosecond range. It means that t_m is much less than MDM run. So MDM is a method which retains Newtonian dynamics only at the times less than t_m and carries out a statistical averaging over initial conditions along the trajectory run.

The K-values were calculated by MDM for systems of neutral particles [3, 7–11, ?,20], two-component [15] and one-component [17] plasmas and primitive polymer model [21]. The values of K turn out to be the same for both velocities and coordinates deviations. It is also seen that the K values for electrons and ions are close to each other at the initial stage of divergence. At $t = t_{me}$ the quantity $\langle \Delta v^2(t) \rangle$ for electrons reaches its saturation value and, therefore, at $t > t_{me}$ only ion trajectories continue to diverge exponentially with another value of K-entropy depending on M/m as $K_i \sim (M/m)^{-1/2}$. The ratio t_{me}/t_{mi} is a fixed value. The dependence of t_{mi} on the electron-ion mass ratio also fits the square root law.

System of 10 polymer molecules with atom-atom interaction potential and periodic boundary conditions was studied in [21]. Each model molecule consisted of 6 atoms with constant interatomic distances and variable angles ϕ between links. Divergence of velocities $\Delta v^2(t)$ and coordinates $\Delta r^2(t)$ for both atoms and molecule center-of-masses as well as angles $\Delta\phi^2(t)$ were calculated. All the five dependencies follow the exponential law before saturation. All the five exponents turned out to be equal to each other, as for electrons and ions in plasmas. One can expect that it is a general conclusion for systems with different degrees of freedom.

2 Stochastisity of dynamic systems

Kravtsov et al. [16, 18] considered the measuring noise, fluctuation forces and uncertainty in knowledge of differential deterministic equations of the system as the reasons why t_m has a finite value. It is a characteristic of a simulation model in [7, 14, 15, 17]. The time t_m was related to the concept of quasi-classical trajectory, which takes into account small but finite quantum effects in classical systems: broadening of particle wave packets and diffraction effects at scattering [14, 15, 26, 27], to weak inelastic processes [18]. Quasi-classical trajectories themselves are irreversible. The intrinsic irreversibility in quantum mechanics originates from the measurement procedure which is probabilistic by definition.

Our premise coincide with the Karl Popper's conviction foundation stone that "nontrivial probabilistic conclusions can only be derived (and thus explained) with the help of probabilistic premises" [28]. The probabilistic premise we use is the quantum nature of any motion which is used to be considered as a deterministic classical one. The idea was inspired by an old remark of John von Neumann [29] and Landau [30] that any irreversibility might be related to the probabilistic character of measurement procedure in quantum mechanics.

Estimates of dynamic memory times were obtained for molecular dynamics numerical schemes. Since t_m values very weakly (logarithmically) depend on the noise level, it allowed us to extend qualitative conclusions to real systems of atoms, in which the finiteness of the dynamic memory time is caused by quantum uncertainty.

Though the primary source of the stochastic noise is probabilistic character of measurement procedure there are other factors which remarkably increase the noise value and permit to forget about quantum uncertainty at simulation. For example it is water molecule background that creates the stochastic noise in electrolytes. One is able to add Langevin forces into (1) and apply MDM to study their influence on dynamic properties of Coulomb system [31]. The dependence of the dynamic memory time on the value of Langevin force is presented in Fig. 2. It is seen that collisions of ions with water molecules does not change essentially the value of t_m.

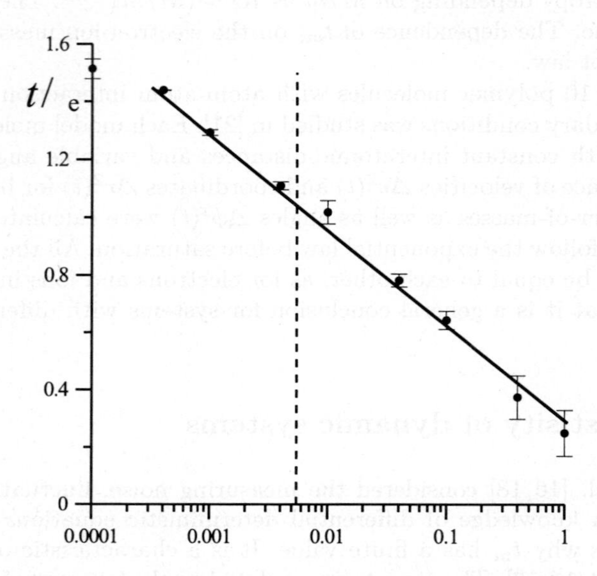

Fig. 2. Dynamic memory time for different values of Langevin force. The dashed line corresponds the level of the Langevin force which acts on the ions in the water solution.

3 Boltzmann and Non-Boltzmann Relaxation

Boltzmann equation is a fundamental equation of kinetic theory of gases; there are numerous attempts to modify the equation and extend it to dense media as well [32, 33]. Kinetic theory of gases deals only with probabilities of collision results. It is an initial assumption at the derivation of Boltzmann equation. Another fundamental assumption is Stosszahlansatz, which means that molecules are statistically independent. Molecular chaos hypothesis a base of the kinetic theory, i.e. it is implied that molecule motion is stochastized. However it is apparent that dynamic motion precedes to stochastic processes [13, 32, 33]. It is supposed that dynamic motion governed by intermolecular interactions defines the values of collision cross-sections but does not influence time dependence of kinetic processes. One can expect that Boltzmann description of kinetic processes is valid only for the times greater than t_m.

MDM can be a powerful tool for studying non-Boltzmann relaxation phenomena in more or less dense media. Some non-equilibrium processes have been already studied with MDM for example in [14, 19, 34–38].

MDM was applied in [39] to the study of electron and ion kinetic energy relaxation in strongly coupled plasmas. Two-component fully ionized system of $2N$ single-charged particles with masses m (electrons) and M (ions) was considered . It is assumed that the particles of the same charge interact via the Coulomb potential, whereas the interaction between particles with different charges was described by the effective pair potential ("pseudo-potential"). The nonideality was characterized by a parameter $\gamma = e^2 n^{-1/3}/kT$, where $n = n_e + n_i$ is the total density of charged particles. The values of γ were taken in the interval from 0.2 to 3. The details of the plasma model and numerical integration scheme are presented in [15].

The following procedure was used to prepare the ensemble of nonequilibrium plasma states. Equilibrium trajectory was generated by MD for a given value of γ. Then a set of $I = 50 - 200$ statistically independent configurations were taken from this run with the velocities of electrons and ions dropped to zero. Thus the ensemble of initial states of nonequilibrium plasma was obtained. MD simulations were carried out for every of these initial states and the results were averaged over the ensemble.

An example of relaxation of average kinetic energy $T(t) = \frac{1}{NI} \sum_{j,k}^{N,I} v_{jk}^2(t)$ is presented in Fig. 3a for the total relaxation period . The values of T for both electrons and ions are normalized by the final equilibrium value. The nonideality parameter γ_f in final equilibrium state differs from initial γ value. The time is measured in periods of plasma oscillations $\tau_e = 2\pi/\Omega_e$. Fig. 3a reveals Boltzmann character for time $t > 5\tau_e$. It is evident from the insertion where the difference between the electron and ion kinetic energies is presented in the semi-logarithmic scale. The character of this long time relaxation agrees with earlier results [34, 36].

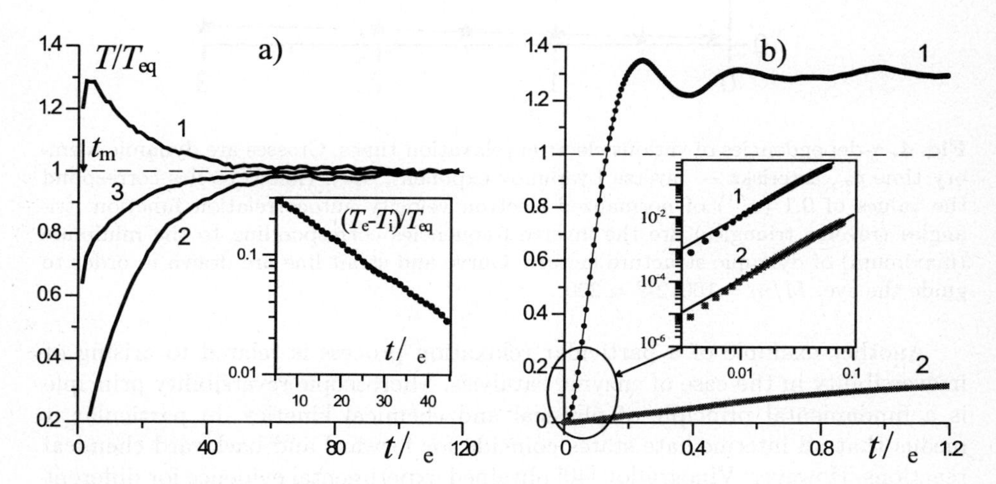

Fig. 3. The kinetic energy for electrons (1), ions (2) and average value (3): a) at Boltzmann relaxation stage; b) for times less than the dynamic memory time. The equilibrium dynamic memory time is given by a vertical arrow. $\gamma = 1, \gamma_f = 3.3, M/m = 100, 2N = 200$.

At the times less than $0.1\tau_e$ both electron and ion kinetic energies increase according to the quadratic fit (Fig. 3b). Then the electron kinetic energy passes through the maximum and undergoes several oscillations damping at $t \approx t_m$ while the ion kinetic energy increases monotonously. The possibility of two stages of relaxation was noted in [36].

The relative importance of non-Boltzmann relaxation stage for different nonideality parameters can be derived from Fig. 3. It is seen (as an example from the velocity autocorrelation function decay time) that it decreases with decrease of plasma nonideality. The calculation shows as well that the oscillatory character of non-Boltzmann relaxation vanishes when the nonideality parameter becomes less than $\gamma = 0.5$.

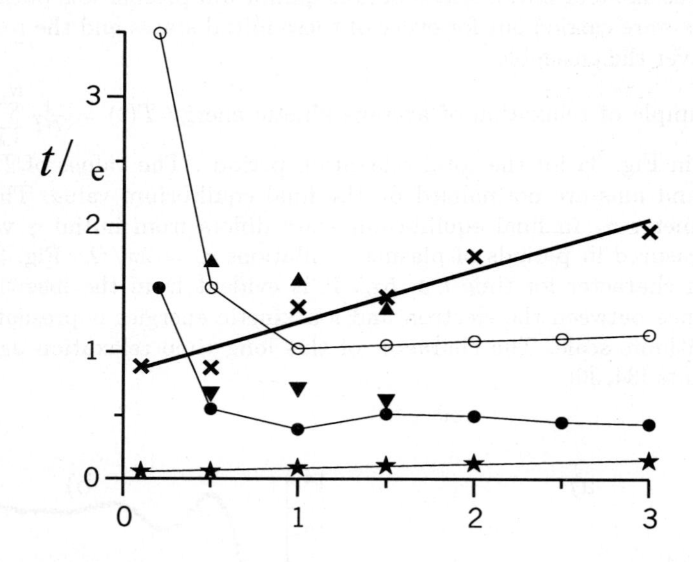

Fig. 4. γ-dependencies of various electron relaxation times. Crosses are dynamic memory time t_m, asterisks — inverse Lyapunov exponent, open (filled) circles correspond the values of 0.1 (e^{-1}) of normalized electron velocity autocorrelation function, triangles (reverse triangles) are the inverse frequencies corresponding to the minimum (maximum) of dynamic structure facture. Curve and strait line are drawn in order to guide the eye. $M/m = 100, 2N = 200$.

Another example of a particular relaxation process is related to arising of irreversibility in the case of enzyme catalysis. Microscopic reversibility principle is a fundamental principle of physical and chemical kinetics. In particular it means that all intermediate states coincide for forward and backward chemical reactions. However, Vinogradov [40] obtained experimental evidence for different pathways for direct and reverse enzymatic reactions in the case of hydrolysis and synthesis of ATP and some other mitochondrial molecular machines and supposed that "the one-way traffic" principle is realized at the level of a single enzyme.

Since microscopic reversibility principle follows from the time reversibility of fundamental dynamic equations, the occurrence of the irreversibility in the case of enzyme catalysis might be similar to the case of the classical molecular systems. The latter problem has been existing as intermediate between physics and philosophy. Hypothesis [40] switches it to experiment and applied science.

If there are two pathways along the hypersuface of potential energy between initial and final states there should be at least two bifurcation points. It is not Maxwell demon but Lyapunov instability, stochastic terms and asymmetry of complicated potential relief with developed system of relatively hard valence bonds that define the local choice of reaction pathway in the bifurcation point [41]. The physical sense of stochastic terms is related to thermal fluctuations of the relief and noise produced by collisions with water molecules while the main features of the relief do not depend on time essentially.

Molecular simulation example [42, 43] for a primitive model confirms this conclusion. The local choice is determined by the local parameters. The situation is equivalent to the statement that there is no thermodynamic equilibrium in the area around the bifurcation point and the theory of transient state is not valid here.

The work is supported by RFBR (grants 00-02-16310a, 01-02-06382mas, 01-02-06384mas).

References

1. Prigogine, I.: Physica A **263** (1999) 528–539
2. Lebowitz, J.L.: Physica A **263** (1999) 516–527
3. Hoover, W.G.: Time Reversibility, Computer Simulation and Chaos. World Scientific, Singapore (1999)
4. Ciccotti, G., Hoover W.G. (eds.): Molecular-Dynamics Simulation of Statistical Mechanical Systems. Proc. Int. School of Physics "Enrico Fermi", Course 97. North-Holland, Amsterdam, (1986)
5. Allen, M.P., Tildesley D.J.: Computer Simulation of Liquids. Clarendon, Oxford (1987)
6. van Gunsteren W.F.: In: Truhler, D. (ed.): Mathematical Frontiers in Computational Chemical Physics. Springer, New York (1988), 136–151.
7. Valuev A.A., Norman G.E., Podlipchuk V.Yu.: In: Samarskii, A.A., Kalitkin N.N. (eds.): Mathematical Modelling. Nauka, Moscow, (1989) 5–40 (in Russian)
8. Norman, G.E., Podlipchuk, V.Yu., Valuev, A.A.: J. Moscow Phys. Soc. Institute of Physics Publishing, UK **2** (1992) 7–21
9. Hoover, W.G.: Computational Statistical Mechanics. Elsevier, Amsterdam (1991)
10. Rapaport, D.C.: The Art of Molecular Dynamics Simulations, Parag. 3.8, 5.5.1. Cambridge University Press, Cambridge (1995)
11. Frenkel, D., Smith, B.: Understanding Molecular Simulations, Parag. 4.3.4. Akademic Press, London (1996)
12. Stoddard, S.D., Ford, J.: Phys. Rev. A **8** (1973) 1504–1513
13. Zaslavsky, G.M.: Stochastisity of dynamic systems. Nauka, Moscow (1984); Harwood, Chur (1985)
14. Norman, G.E., Stegailov, V.V.: Zh. Eksp. Theor. Phys. **119**, (2001) 1011–1020; J. of Experim. and Theor. Physics **92** (2001) 879–886

15. Morozov, I.V., Norman, G.E., Valuev, A.A.: Phys. Rev. E **63** 036405 (2001) 1–9
16. Kravtsov, Yu.A.: In: Kravtsov, Yu.A. (ed.): Limits of Predictability. Springer, Berlin (1993) 173–204
17. Ueshima, Y., Nishihara, K., Barnett, D.M., Tajima, T., Furukawa, H.: Phys. Rev. E **55** (1997) 3439-3449
18. Gertsenshtein, M.E., Kravtsov, Yu.A.: Zh. Eksp. Theor. Phys. **118** (2000) 761–763; J. of Experim. and Theor. Physics **91** (2000) 658–660
19. Hoover, W.G., Posch, H.A.: Phys. Rev. A **38** (1998) 473–480
20. Kwon, K.-H., Park, B.-Y.: J. Chem. Phys. **107** (1997) 5171-5179
21. Norman, G.E., Yaroshchuk, A.I.: (to appear)
22. Norman, G.E., Podlipchuk, V.Yu., Valuev, A.A.: Mol. Simul. **9** (1993) 417–424
23. Rowlands, G.J.: Computational Physics **97** (1991) 235–239
24. Lopez-Marcos, M.A., Sanz-Serna, J.M., Diaz, J.C.: J. Comput. Appl. Math. **67** (1996) 173–179
25. Lopez-Marcos, M.A., Sanz-Serna, J.M., Skeel, R.D.: SIAM J. Sci. Comput. **18** (1997) 223–230
26. Kaklyugin, A.S., Norman, G.E.: Zh. Ross. Khem. Ob-va im. D.I. Mendeleeva **44**(3) (2000) 7–20 (in Russian)
27. Kaklyugin, A.S., Norman, G.E.: J. Moscow Phys. Soc. Allerton Press, USA **5** (1995) 167-180
28. Popper, K.: Unended Quest. An Intellectual Autobiography. Fontana/Collins, Glasgow (1978)
29. von Neumann, J.: Z. Phys. **57** (1929) 30–37
30. Landau, L.D., Lifshitz, E.M.: Course of Theoretical Physics, Vol. 5, Statistical Physics, Part 1. Nauka, Moscow (1995); Pergamon, Oxford (1980);
Quantum Mechanics: Non-Relativistic Theory, Parag. 8, Vol. 3. 4th ed. Nauka, Moscow, (1989); 3rd ed. Pergamon, New York, (1977)
31. Ebeling, W., Morozov, I.V., Norman, G.E.: (to appear)
32. Balescu, R.: Equilibrium and Nonequilibrium Statistical Mechanics. London Wiley, New York (1975).
33. Zubarev, D.N., Morozov, V.G., Roepke, G.: Statistical Mechanics of Nonequilibrium Processes. Akademie-Verlag, Berlin (1996)
34. Hansen, J.P., McDonald, I.R.: Phys. Lett. **97A** (1983) 42–45
35. Norman, G.E., Valuev, A.A.: In: Kalman, G., Rommel, M., Blagoev, K. (eds.): Strongly Coupled Coulomb Systems. Plenum Press, New York (1998) 103–116
36. Norman, G.E., Valuev, A.A., Valuev, I.A.: J. de Physique (France) **10**(Pr5) (2000) 255–258
37. Hoover, W.G., Kum, O., Posch, H.A.: Phys. Rev. E **53** (1996) 2123–2132
38. Dellago, C., Hoover, W.G.: Phys. Rev. E **62** (2000) 6275–6281
39. Morozov, I.V., Norman, G.E.: (to appear)
40. Vinogradov, A.D.: J. Exper. Biology **203** (2000) 41–49; Biochim. Biophys. Acta **1364** (1998) 169–185
41. Kaklyugin, A.S., Norman, G.E.: Zhurnal Ross. Khem. Ob-va im D.I. Mendeleeva (Mendeleev Chemistry Journal) **45**(1) (2001) 3–8 (in Russian)
42. Norman, G.E., Stegailov, V.V.: ibid. **45**(1) (2001) 9–11
43. Norman, G.E., Stegailov, V.V.: Comp. Phys. Comm. (to appear)

Determinism and Chaos in Decay of Metastable States

Vladimir V. Stegailov

Moscow Institute of Physics and Technology, Institutskii per. 9, 141700, Dolgoprudnyi, Russia
stega@nu.jinr.ru

Abstract. The problem of numerical investigation of metastable states decay is described in this work on example of melting of the superheated solid crystal simulated within the framework of molecular dynamics method. Its application in the case of non-equilibrium processes has certain difficulties, including the averaging procedure. In this work an original technique of averaging over the ensemble of configuration is presented. The question of the instability of the phase space trajectories of many-particle system (i.e. chaotic character of motion) and its consequences for simulation are also discussed.

1 Introduction

Melting is still not completely understood phenomenon. There exists the question of estimation of the maximum possible degree of crystal superheating. This problem attracts certain interest in connection with the experiments dealing with intensive ultrafast energy contributions where specific conditions for superheating are realized [1]. Superheated crystal is metastable, therefore it melts in some finite time. It is obvious that the decay of ordered phase happens sooner at higher degrees of superheating. The utmost superheating therefore should be characterized by the values of life time comparable to the period of crystal node oscillations. At this time scale it is possible to apply molecular dynamics (MD) method of numerical simulation to investigate melting at high degrees of superheating on the microscopic level [2, 3]. In this work I dwell on the peculiarities of application of the MD method in study of this non-equilibrium process on example of direct calculation of nucleation rate and melting front propagation velocity.

2 Model and Calculation Technique

The model under consideration is the fcc-lattice of particles interacting via homogeneous potential $U = \varepsilon(\frac{\sigma}{r})^m$. Calculations were made for $m = 12$. To model an initially surface-free and defect-free ideal crystal periodic boundary conditions were used. For numerical integration Euler-Störmer 2-nd order scheme was

P.M.A. Sloot et al. (Eds.): ICCS 2002, LNCS 2331, pp. 1147–1153, 2002.

applied. Number of particles in the main cell was $N = 108; 256; 500$. Homogeneous potential allows to describe thermodynamic state of the system in terms of only one parameter $X = (N\sigma^3/\sqrt{2}V)(k_BT/\varepsilon)^{3/n}$, where V is the main cell volume and T is the temperature.

MD-system was transferred from a stable ordered equilibrium state to a metastable superheated one by means of either isochoric heating or isothermic stretching or their combination. A degree of superheating can be characterized by the quantity of $X^{-1} \sim \rho T^{3/n}$: the larger is X^{-1} the higher is the superheating. When the crystal was in a sufficiently metastable state the calculation of dynamics of isolated system was performed. Sharp fall of kinetic energy of the system E_{kin} manifests transition to disordered state (Fig.1). From $E_{kin}(t)$ dependence the value of metastable configuration lifetime t_{life} can be derived. However the values of t_{life} are very different for the MD-runs calculated from

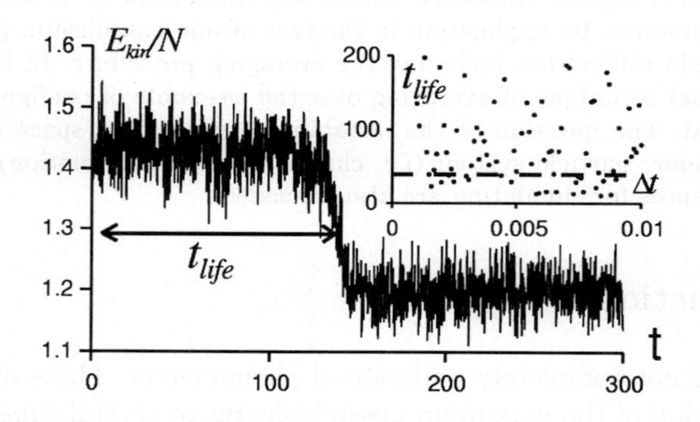

Fig. 1. Time dependence of mean kinetic energy of particles. A phase transition from ordered to disordered state is shown. Subfigure: t_{life} values from MD-runs with identical initial conditions but different integration time steps Δt

one and the same initial configuration with various time-steps Δt. Besides there is no convergence when $\Delta t \to 0$ (Fig.1). This fact is rather confusing because it is not clear what value of t_{life} one should attribute to the dynamical system under consideration.

3 Role of Instability

The MD method based on the numerical integration of the corresponding system of Newton equations

$$m_i \frac{d\mathbf{v}_i(t)}{dt} = \mathbf{F}_i[\mathbf{r}(\mathbf{t})], \quad \frac{d\mathbf{r}_i(t)}{dt} = \mathbf{v}_i(t), \quad \mathbf{F}_i = -\frac{\partial}{\partial \mathbf{r}_i(t)} U(\mathbf{r}_1, \mathbf{r}_2, \ldots, \mathbf{r}_N) \quad (1)$$

that results in determination of the trajectories of all particles $\{\mathbf{r}(t), \mathbf{v}(t)\}$, where $\mathbf{r} = (\mathbf{r}_1, \mathbf{r}_2, \ldots, \mathbf{r}_N)$ and $\mathbf{v} = (\mathbf{v}_1, \mathbf{v}_2, \ldots, \mathbf{v}_N)$.

Set (1) is exponentially unstable for a system of more than two particles (e.g., see [4–17]). The parameter that determines the degree of instability, that is, the rate of divergence of initially close phase trajectories, is averaged Lyapunov exponent or K-entropy K.

Let us consider solutions to system (1) for identical initial conditions corresponding to some point on an MD-trajectory: $\{\mathbf{r}(t), \mathbf{v}(t)\}$ (found in step Δt) and $\{\mathbf{r}'(t), \mathbf{v}'(t)\}$ (in step $\Delta t'$). Averaged differences of the coordinates and velocities of the first and second trajectories are determined at coinciding time moments (Fig.2, a). In some transient time the differences become exponentially increasing (Fig.2, b). The values of A and B are determined by the difference of integration steps Δt and $\Delta t'$. The exponential increase of $\langle \Delta v^2(t) \rangle$ is limited by the finite value of the thermal velocity of particles v_T; $\langle \Delta r^2(t) \rangle$ is limited by the mean amplitude of atom oscillations in crystal. Since exponential growth of

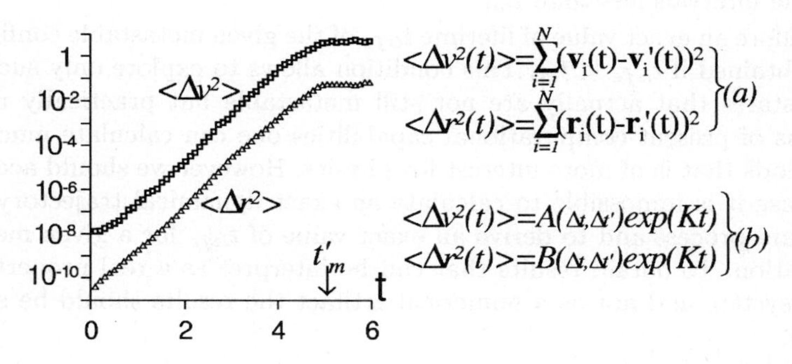

Fig. 2. Divergences of velocities (squares) and coordinates (triangles) at coinciding moments of time along two trajectories calculated from identical initial conditions with time-steps $\Delta t = 0.001$ and $t'_m = 0.00101$. Time, length and velocity are given in reduced units where $\varepsilon = \sigma = 1$

$\langle \Delta r^2(t) \rangle$ and $\langle \Delta v^2(t) \rangle$ two initially coinciding MD-trajectories lose any relation very quickly. Let t'_m denote time when divergences come to saturation (Fig.2) and therefore these trajectories become completely uncorrelated. Value of t'_m is a function of Δt and $\Delta t'$.

One should mention that for the analyzed on Fig.1 configuration $t'_m \ll t_{life}$, hence MD-trajectories calculated with various Δt becomes uncorrelated much sooner than the transition to disordered phase happens. This fact explains chaotic dependence of t_{life} on Δt. However it is still vague how to determine an exact value of lifetime that is intrinsic property of the dynamical many-particle system.

In [14–16] the notion of *dynamical memory time* t_m is used. By its definition $t'_m(\Delta t', \Delta t) \to t_m(\Delta t)$ when $\Delta t' \to 0$ and Δt is fixed. It is known from [14–16] that the dependence of t'_m on $\Delta t'$ at fixed Δt is rather weak so it can be easily estimated as $t_m \simeq t'_m$.

The physical sense of time t_m consists in the following. During numerical integration after the time t_m the MD-trajectory calculated with time step Δt completely "forgets" its initial conditions. It means that the MD-trajectory ceases to correlate with the hypothetical Newtonian trajectory (an exact solution of the set (??)). In other words the value of t_m determines the time interval during which the behavior of molecular-dynamical system can be predicted from initial conditions and deterministic equations of motion at a certain level of accuracy defined by Δt value and a particular scheme of numerical integration. Such a definition of t_m correlates with that given in [18, 19].

It follows from [14–16] that t_m grows no faster than logarithmically as the accuracy of numerical integration increases; in the same works it is shown that the available computation facilities allow only to increase t_m only two times even with the use of refined numerical schemes [20–22]. It means that t_m is much less than usual MD-run. So MD method retains Newtonian dynamics only at the short time intervals less than t_m.

Therefore an exact value of lifetime t_{life} of the given metastable configuration can be obtained if $t_{life} < t_m$. This condition allows to explore only such superheating states that actually are not still metastable but practically unstable. By means of present computational capabilities one can calculate much longer time periods that is of more interest for physics. However we should accept that in this case it is impossible to calculate an exact dynamical trajectory of non-equilibrium process and to derive an exact value of t_{life} for a given metastable configuration. To obtain results that can be interpret as a real property of dynamical system and not as a numerical artifact the results should be somehow averaged.

4 Averaging Procedure

The simplest way to do it is to check if the t_{life} values have some distribution or they do not. For example, we can calculate $n = 100$ MD-runs with identical initial configuration and $\Delta t = 0.01, 0.0099, 0.0098, \ldots, 0.001$, and then to count in how many realizations $t_{life} \in [t, t+\delta t]$. The performed calculations showed than such distributions actually exist (Fig.3) and their shape weakly changes starting with $n \geq 100$. Since $t_{life} \gg t'_m$ these distributions of lifetime over a set of Δt is equivalent to the distributions over different initial configurations corresponding to the same degree of superheating, i.e. thermodynamically indistinguishable. That is an exact situation in real experiments with metastable state decay: either crystallization of supercooled liquid or melting of superheated crystal [23].

Melting is considered as a process of nucleation. Formation of the new phase nuclei can be described in terms of a stochastic Poisson process [23]. Let λ denote a probability of liquid phase nucleus formation in short time interval $[t, t+\delta t]$. Then it can be shown that $P(t) = \exp(-\lambda t)$ is the probability that no nucleus will appear to the t moment. Let us consider n_0 realizations (MD-runs) of metastable state decay. Let n denote the number of those non-decayed to the

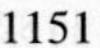

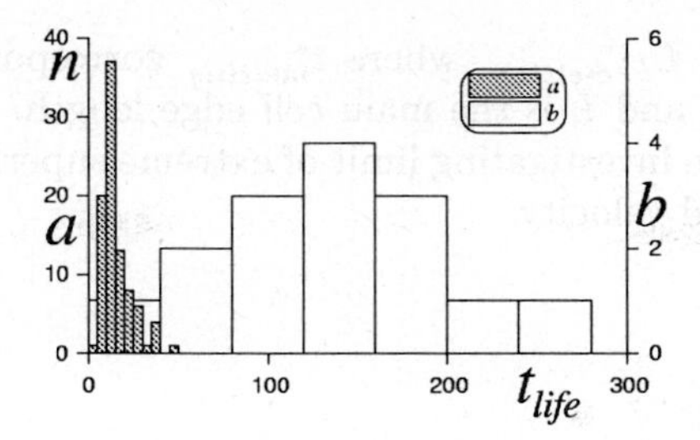

Fig. 3. Examples of t_{life} distributions for $N = 500$: $a - 1/X = 1.3088$, $b - 1/X = 1.2879$; n is the number of MD-runs where $t_{life} \in [t, t + \delta t]$.

t moment, then $n(t) = n_0 \exp(-\lambda t)$. This dependence is perfectly confirmed by numerical results (see Fig.4).

From the such distributions one can obtain the most probable lifetime for a superheated crystal at the certain degree of superheating $t^*_{life} = \lambda^{-1}$. Unlike t_{life} the value of t^*_{life} is a physical property of the concerned dynamical model. Superheated state of defect-free crystal is characterized by the homogeneous nucleation rate J, that is average number of nuclei formed in the unit volume in the unit time interval. In our MD-model we can estimate J as $(t^*_{life}V)^{-1}$ where V is the main cell volume. From the physical point of view it is interesting to

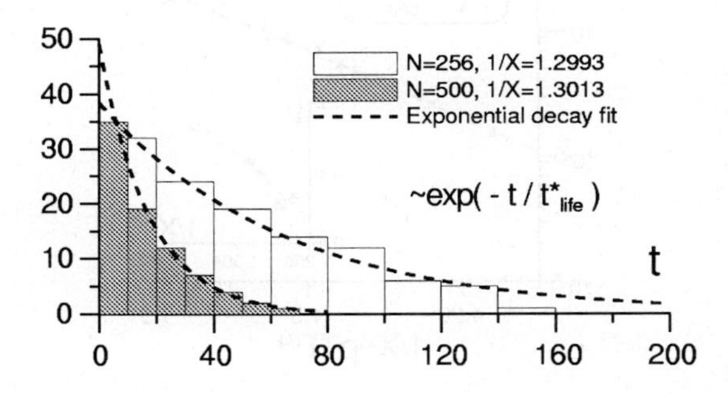

Fig. 4. Distributions of number of MD-runs non-decayed to the **t** moment

now how depend nucleation rate J on the degree of superheating (the latter is characterized by parameter $1/X \sim \rho T^{3/n}$). Following the calculation procedure described in this section for a set of initial configurations corresponding to various values of $1/X$ we derived $J(1/X)$ dependence presented at Fig.5.

Although data points at Fig.5 relate to different N they all form one well determined dependence. It is one more confirmation that this data is not an numerical artifact but a property of the investigated model.

In conclusion I'd like to remark that by means of the described procedure distributions (similar to Fig.3) of $t_{melting}$, i.e. time of the transition on Fig.1, were calculated for different $1/X$ values. Melting front propagation velocity v

can be estimated by $L/t^*_{melting}$, where $t^*_{melting}$ corresponds to the maximum of these distributions and L is the main cell edge length. Numerical results are shown on Fig.5. In the investigating limit of extreme superheatings v approaches the value of the sound velocity.

5 Summary

Molecular dynamics method contrary to Monte-Carlo method is a technique of numerical simulation that deals exactly with dynamics of many-particle system. At the same time detailed analysis shows that actually calculation of real dynamics restricted to the quite short time intervals. In contrast to equilibrium phenomena, non-equilibrium processes are much more affected by the instability that is intrinsic to many-particle systems. At longer periods one should take into account specific statistical averaging over an ensemble of thermodynamically indistinguishable configurations. In this work it was shown how in spite of such problems one can obtain consistent physical results by means of appropriate interpretation of numerical data.

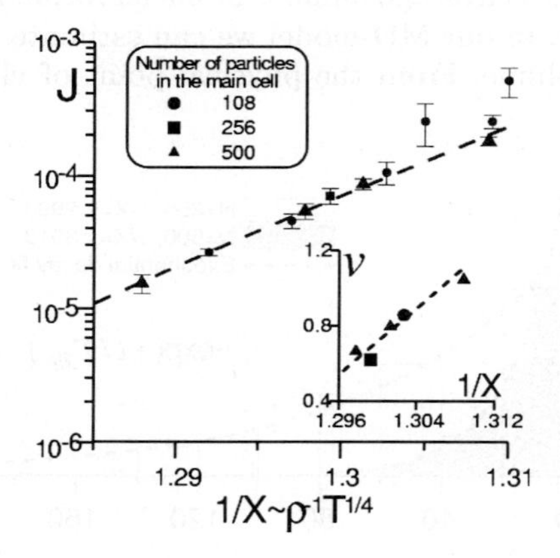

Fig. 5. Dependence of nucleation rate J and melting front propagation velocity v on the parameter 1/X, i.e. on the degree of superheating

6 Acknowledgements

I am very grateful to my scientific supervisor prof. G.E.Norman for his careful guidance and essential remarks. I also appreciate interesting discussions with M.N.Krivoguz and his useful comments. Special thanks to A.A.Shevtsov who provided me with necessary computational resources. The work is supported by Russian Foundation for Basic Research (grants 00-02-16310 and 01-02-06384).

References

1. Bonnes, D. A., Brown, J. M.: Bulk Superheating of Solid KBr and CsBr with Shock Waves. Phys. Rev. Lett. **71** (1993) 2931–2934
2. Jin, Z. H., Gumbsch, P., Lu, K., Ma, E.: Melting Mechanisms at the Limit of Superheating. Phys. Rev. Lett. **87** (2001) 055703-1–4
3. Krivoguz, M. N., Norman, G. E.: Spinodal of Superheated Solid Metal. Doklady Physics **46** (2001) 463–466
4. Hoover, W. G.: Time Reversibility, Computer Simulation and Chaos. World Scientific, Singapore (1999)
5. Ciccotti, G., Hoover, W. G. (eds.): Molecular-Dynamics Simulation of Statistical Mechanical Systems. Proc. Int. School of Physics "Enrico Fermi", course 97, North-Holland, Amsterdam (1986)
6. van Gunsteren, W. F.: in Truhler, D. (ed.): Mathematical Frontiers in Computational Chemical Physics. Springer-Verlag, New York (1988) 136
7. Valuev, A. A., Norman, G. E., Podlipchuk, V. Yu.: Molecular Dynamics Method: Theory and Applications. In: Samarskii, A. A., Kalitkin, N. N. (eds.): Mathematical Modelling. Nauka, Moscow (1989) 5 (in russian)
8. Allen, M. P., Tildesley, D. J.: Computer Simulation of Liquids. Clarendon, Oxford (1987)
9. Norman, G. E., Podlipchuk, V. Yu., Valuev, A. A.: J. Moscow Phys. Soc. (Institute of Physics Publishing, UK) **2** (1992) 7
10. Hoover, W. G.: Computational Statistical Mechanics. Elsevier, Amsterdam (1991)
11. Rapaport, D. C.: The Art of Molecular Dynamics Simulations. Cambridge University Press, Cambridge (1995)
12. Frenkel, D., Smith, B.: Understanding Molecular Simulations. Akademic Press, London (1996)
13. Zaslavsky, G.M.: Stochastisity of dynamic systems. Nauka, Moscow (1984); Harwood, Chur (1985)
14. Norman, G.E., Stegailov, V.V.: Zh. Eksp. Theor. Phys. **119** (2001) 1011 [J. of Experim. and Theor. Physics **92** (2001) 879]
15. Norman, G.E., Stegailov, V.V.: Stochastic and Dynamic Properties of Molecular Dynamics Systems: Simple Liquids, Plasma and Electrolytes, Polymers. Computer Physics Communications (proc. of the Europhysics Conf. on Computational Physics 2001) to be published
16. Morozov, I.V., Norman, G.E., Valuev, A.A.: Stochastic Properties of strongly coupled plasmas. Phys. Rev. E bf 63 (2001) 036405
17. Ueshima, Y., Nishihara, K., Barnett, D.M., Tajima, T., Furukawa, H.: Partice Simulation of Lyapunov Exponents in One-Component strongly coupled plasmas. Phys. Rev. E bf 55 (1997) 3439
18. Kravtsov, Yu.A. in: Kravtsov, Yu.A. (ed.): Limits of Predictability. Springer, Berlin (1993) 173
19. Gertsenshtein, M.E., Kravtsov, Yu.A.: Zh. Eksp. Theor. Phys. **118** (2000) 761 [J. of Experim. and Theor. Physics **91** (2000) 658]
20. Rowlands, G.J.: Computational Physics **97** (1991) 235
21. Lopez-Marcos, M.A., Sanz-Serna, J.M., Diaz, J.C.: Are Gauss-Legendre Method Useful in Molecular Dynamics? J. Comput. Appl. Math. **67** (1996) 173
22. Lopez-Marcos, M.A., Sanz-Serna, J.M., Skeel, R.D.: Explicit Symplectic Integrators Using Hessian-Vector Products. SIAM J. Sci. Comput. **18** (1997) 223
23. Skripov, V. P., Koverda, V. P.: Spontaneous Crystallization of Supercooled Liquid. Nauka, Moscow (1984) (in russian)

Regular and Chaotic Motions of the Parametrically Forced Pendulum: Theory and Simulations

Eugene I. Butikov
St. Petersburg State University, Russia
E-mail: butikov@spb.runnet.ru

Abstract

New types of regular and chaotic behaviour of the parametrically driven pendulum are discovered with the help of computer simulations. A simple qualitative physical explanation is suggested to the phenomenon of subharmonic resonances. An approximate quantitative theory based on the suggested approach is developed. The spectral composition of the subharmonic resonances is investigated quantitatively, and their boundaries in the parameter space are determined. The conditions of the inverted pendulum stability are determined with a greater precision than they have been known earlier. A close relationship between the upper limit of stability of the dynamically stabilized inverted pendulum and parametric resonance of the hanging down pendulum is established. Most of the newly discovered modes are waiting a plausible physical explanation.

1 Introduction

An ordinary rigid planar pendulum whose suspension point is driven periodically is a paradigm of contemporary nonlinear dynamics. Being famous primarily due to its outstanding role in the history of science, this rather simple mechanical system is also interesting because the differential equation that describes its motion is frequently encountered in various problems of modern physics. Mechanical analogues of different physical systems allow a direct visualization of their motion (at least with the help of simulations) and therefore can be very useful in gaining an intuitive understanding of complex phenomena.

Depending on the frequency and amplitude of forced oscillations of the suspension point, this apparently simple mechanical system exhibits an incredibly rich variety of nonlinear phenomena characterized by amazingly different types of motion. Some modes of such parametrically excited pendulum are quite simple indeed and agree well with our intuition, while others are very complicated and counterintuitive. Besides the commonly known phenomenon of parametric resonance, the pendulum can execute many other kinds of regular behavior.

P.M.A. Sloot et al. (Eds.): ICCS 2002, LNCS 2331, pp. 1154–1169, 2002.

Among them we encounter a synchronized non-uniform unidirectional rotation in a full circle with a period that equals either the driving period or an integer multiple of this period. More complicated regular modes are formed by combined rotational and oscillatory motions synchronized (locked in phase) with oscillations of the pivot. Different competing modes can coexist at the same values of the driving amplitude and frequency. Which of these modes is eventually established when the transient is over depends on the starting conditions.

Behavior of the pendulum whose axis is forced to oscillate with a frequency from certain intervals (and at large enough driving amplitudes) can be irregular, chaotic. The pendulum makes several revolutions in one direction, then swings for a while with permanently (and randomly) changing amplitude, then rotates again in the former or in the opposite direction, and so forth. For other values of the driving frequency and/or amplitude, the chaotic motion can be purely rotational, or, vice versa, purely oscillatory, without revolutions. The pendulum can make, say, one oscillation during each two driving periods (like at ordinary parametric resonance), but in each next cycle the motion (and the phase orbit) is slightly (and randomly) different from the previous cycle. Other chaotic modes are characterized by protracted oscillations with randomly varying amplitude alternated from time to time with full revolutions to one or the other side (intermittency). The parametrically forced pendulum can serve as an excellent physical model for studying general laws of the dynamical chaos as well as various complicated modes of regular behavior in simple nonlinear systems.

A widely known interesting feature in the behavior of a rigid pendulum whose suspension point is forced to vibrate with a high frequency along the vertical line is the dynamic stabilization of its inverted position. Among recent new discoveries regarding the inverted pendulum, the most important are the destabilization of the (dynamically stabilized) inverted position at large driving amplitudes through excitation of period-2 ("flutter") oscillations [1]-[2], and the existence of n-periodic "multiple-nodding" regular oscillations [3].

In this paper we present a quite simple qualitative physical explanation of these phenomena. We show that the excitation of period-2 "flutter" mode is closely (intimately) related with the commonly known conditions of parametric instability of the non-inverted pendulum, and that the so-called "multiple-nodding" oscillations (which exist both for the inverted and hanging down pendulum) can be treated as high order subharmonic resonances of the parametrically driven pendulum. The spectral composition of the subharmonic resonances in the low-amplitude limit is investigated quantitatively, and the boundaries of the region in the parameter space are determined in which these resonances can exist. The conditions of the inverted pendulum stability are determined with a greater precision than they have been known earlier. We report also for the first time about several new types of regular and chaotic behaviour of the parametrically driven pendulum discovered with the help of computer simulations. Most of these exotic new modes are rather counterintuitive. They are still waiting a plausible physical explanation. Understanding such complicated behavior of this simple system is certainly a challenge to our physical intuition.

2 The physical system

We consider the rigid planar pendulum whose axis is forced to execute a given harmonic oscillation along the vertical line with a frequency ω and an amplitude a, i.e., the motion of the axis is described by the following equation:

$$z(t) = a \sin \omega t \quad \text{or} \quad z(t) = a \cos \omega t. \tag{1}$$

The force of inertia $F_{\text{in}}(t)$ exerted on the bob in the non-inertial frame of reference associated with the pivot also has the same sinusoidal dependence on time. This force is equivalent to a periodic modulation of the force of gravity.

The simulation is based on a numerical integration of the exact differential equation for the momentary angular deflection $\varphi(t)$. This equation includes the torque of the force of gravity and the instantaneous value of the torque exerted on the pendulum by the force of inertia that depends explicitly on time t:

$$\ddot{\varphi} + 2\gamma\dot{\varphi} + (\omega_0^2 - \frac{a}{l}\omega^2 \sin \omega t) \sin \varphi = 0. \tag{2}$$

The second term of Eq. (2) takes into account the braking frictional torque, assumed to be proportional to the momentary angular velocity $\dot{\varphi}$ in the mathematical model of the simulated system. The damping constant γ is inversely proportional to the quality factor Q commonly used to characterize the viscous friction: $Q = \omega_0/2\gamma$.

We note that oscillations about the inverted position can be formally described by the same differential equation, Eq. (2), with negative values of $\omega_0^2 = g/l$. When this control parameter ω_0^2 is diminished through zero to negative values, the constant (gravitational) torque in Eq. (2) also reduces down to zero and then changes its sign to the opposite. Such a "gravity" tends to bring the pendulum into the inverted position $\varphi = \pi$, destabilizing the equilibrium position $\varphi = 0$ of the unforced pendulum.

3 Subharmonic resonances

An understanding about pendulum's behavior in the case of rapid oscillations of its pivot is an important prerequisite for the physical explanation of subharmonic resonances ("multiple-nodding" oscillations). Details of the physical mechanism responsible for the dynamical stabilization of the inverted pendulum can be found in [4]. The principal idea is utterly simple: Although the mean value of the force of inertia $F_{\text{in}}(t)$, averaged over the short period of these oscillations, is zero, the averaged over the period value of its *torque* about the axis is not zero. The reason is that both the force $F_{\text{in}}(t)$ and the *arm* of this force vary in time in the same way synchronously with the axis' vibrations. This non-zero torque tends to align the pendulum along the direction of forced oscillations of the axis. For given values of the driving frequency and amplitude, the mean torque of the force of inertia depends only on the angle of the pendulum's deflection from the direction of the pivot's vibration.

In the absence of gravity the inertial torque gives a clear physical explanation of existence of the two stable equilibrium positions that correspond to the two preferable orientations of the pendulum's rod along the direction of the pivot's vibration. With gravity, the inverted pendulum is stable with respect to small deviations from this position provided the mean torque of the force of inertia is greater than the torque of the force of gravity that tends to tip the pendulum down. This occurs when the following condition is fulfilled: $a^2\omega^2 > 2gl$, or $(a/l)^2 > 2(\omega_0/\omega)^2$ (see, e.g., [4]) However, this is only an approximate criterion of dynamic stability of the inverted pendulum, which is valid only for small amplitudes of forced vibrations of the pivot ($a \ll l$). Below we obtain a more precise criterion [see Eq. (5)].

The complicated motion of the pendulum whose axis is vibrating at a high frequency can be considered approximately as a superposition of two rather simple components: a "slow" or "smooth" component $\psi(t)$, whose variation during a period of forced vibrations is small, and a "fast" (or "vibrational") component. This approach was first used by Kapitza [5] in 1951. Being deflected from the vertical position by an angle that does not exceed $\theta_{\max}$ [where $\cos\theta_{\max} = 2gl/(a^2\omega^2)$], the pendulum will execute relatively slow oscillations about this inverted position. This slow motion is executed both under the mean torque of the force of inertia and the force of gravity. Rapid oscillations caused by forced vibrations of the axis superimpose on this slow motion of the pendulum. With friction, the slow motion gradually damps, and the pendulum wobbles up settling eventually in the inverted position.

Similar behavior of the pendulum can be observed when it is deflected from the lower vertical position. The frequencies ω_{up} and ω_{down} of small slow oscillations about the inverted and hanging down vertical positions are given by the approximate expressions $\omega_{\text{up}}^2 = \omega^2(a/l)^2/2 - \omega_0^2$ and $\omega_{\text{down}}^2 = \omega^2(a/l)^2/2 + \omega_0^2$, respectively. These formulas yeld $\omega_{\text{slow}} = \omega(a/l)/\sqrt{2}$ for the frequency of small slow oscillations of the pendulum with vibrating axis in the absence of the gravitational force.

When the driving amplitude and frequency lie within certain ranges, the pendulum, instead of gradually approaching the equilibrium position (either dynamically stabilized inverted position or ordinary downward position) by the process of damped slow oscillations, can be trapped in a n-periodic limit cycle locked in phase to the rapid forced vibration of the axis. In such oscillations the phase trajectory repeats itself after n driving periods T. Since the motion has period nT, and the frequency of its fundamental harmonic equals ω/n (where ω is the driving frequency, this phenomenon can be called a subharmonic resonance of n-th order. For the inverted pendulum with a vibrating pivot, periodic oscillations of this type were first described by Acheson [3], who called them "multiple-nodding" oscillations. An example of such stationary oscillations whose period equals six periods of the axis is shown in Fig. 1. The left-hand upper part of the figure shows the complicated spatial trajectory of the pendulum's bob at these multiple-nodding oscillations. The left-hand lower part shows the closed looping trajectory in the phase plane (φ, $\dot{\varphi}$). Right-hand

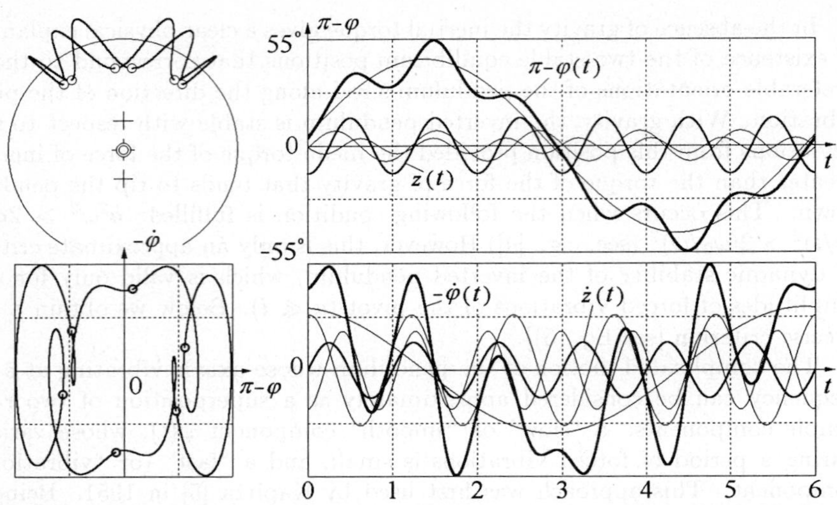

Figure 1: The spatial path, phase orbit, and graphs of stationary oscillations with the period that equals six periods of the oscillating axis. The graphs are obtained by a numerical integration of the exact differential equation, Eq. (2), for the momentary angular deflection

side of Fig. 1, alongside the graphs of $\varphi(t)$ and $\dot{\varphi}(t)$, shows also their harmonic components and the graphs of the pivot oscillations. The fundamental harmonic whose period equals six driving periods dominates in the spectrum. We may treat it as a subharmonic (as an "undertone") of the driving oscillation. This principal harmonic describes the smooth component of the compound period-6 oscillation.

We emphasize that the modes of regular n-periodic oscillations (subharmonic resonances), which have been discovered in investigations of the dynamically stabilized inverted pendulum, are not specific for the inverted pendulum. Similar oscillations can be executed also (at appropriate values of the driving parameters) about the ordinary (downward hanging) equilibrium position. Actually, the origin of subharmonic resonances is independent of gravity, because such synchronized with the pivot "multiple-nodding" oscillations can occur in the absence of gravity about any of the two equivalent dynamically stabilized equilibrium positions of the pendulum with a vibrating axis.

The natural slow oscillatory motion is almost periodic (exactly periodic in the absence of friction). A subharmonic resonance of order n can occur if one cycle of this slow motion covers approximately n driving periods, that is, when the driving frequency ω is close to an integer multiple n of the natural frequency of slow oscillations near either the inverted or the ordinary equilibrium position: $\omega = n\omega_{\mathrm{up}}$ or $\omega = n\omega_{\mathrm{down}}$. In this case the phase locking can occur, in which one cycle of the slow motion is completed *exactly* during n driving periods. Synchronization of these modes with the oscillations of the pivot creates conditions for systematic supplying the pendulum with the energy needed to compensate for dissipation, and the whole process becomes exactly periodic.

The slow motion (with a small angular excursion) can be described by a sinusoidal time dependence. Assuming $\omega_{\text{down,up}} = \omega/n$ (n driving cycles during one cycle of the slow oscillation), we find for the minimal driving amplitudes (for the boundaries of the subharmonic resonances) the values

$$m_{\text{min}} = \sqrt{2(1/n^2 \mp k)}, \tag{3}$$

where $k = (\omega_0/\omega)^2$. The limit of this expression at $n \to \infty$ gives the mentioned earlier approximate condition of stability of the inverted pendulum: $m_{\text{min}} = \sqrt{-2k} = \sqrt{2}(\omega_0/\omega)$ (here $k < 0$, $|k| = |\omega_0^2/\omega^2|$).

The spectrum of stationary n-periodic oscillations consists primarily of the fundamental harmonic $A\sin(\omega t/n)$ with the frequency ω/n, and two high harmonics of the orders $n-1$ and $n+1$. To improve the theoretical values for the boundaries of subharmonic resonances, Eq. (3), we use a trial solution $\varphi(t)$ with unequal amplitudes of the two high harmonics. Since oscillations at the boundaries have infinitely small amplitudes, we can exploit instead of Eq. (2) the linearized (Mathieu) equation (with $\gamma = 0$). Thus we obtain the critical (minimal) driving amplitude m_{min} at which n-period mode $\varphi(t)$ can exist:

$$m_{\text{min}}^2 = \frac{2}{n^4}\frac{[n^6 k(k-1)^2 - n^4(3k^2+1) + n^2(3k+2) - 1]}{[n^2(1-k)+1]}. \tag{4}$$

The limit of m_{min}, Eq. (4), at $n \to \infty$ gives an improved formula for the lower boundary of the dynamic stabilization of the inverted pendulum instead of the commonly known approximate criterion $m_{\text{min}} = \sqrt{-2k}$:

$$m_{\text{min}} = \sqrt{-2k(1-k)} \qquad (k < 0). \tag{5}$$

The minimal amplitude m_{min} that provides the dynamic stabilization is shown as a function of $k = (\omega_0/\omega)^2$ (inverse normalized driving frequency squared) by the left curve ($n \to \infty$) in Fig. 2. The other curves to the right from this boundary show the dependence on k of minimal driving amplitudes for the subharmonic resonances of several orders (the first curve for $n = 6$ and the others for n values diminishing down to $n = 2$ from left to right). At positive values of k these curves correspond to the subharmonic resonances of the hanging down parametrically excited pendulum. Subharmonic oscillations of a given order n (for $n > 2$) are possible to the left of $k = 1/n^2$, that is, for the driving frequency $\omega > n\omega_0$. The curves in Fig. 2 show that as the driving frequency ω is increased beyond the value $n\omega_0$ (i.e., as k is decreased from the critical value $1/n^2$ toward zero), the threshold driving amplitude (over which n-order subharmonic oscillations are possible) rapidly increases. The limit of a very high driving frequency ($\omega/\omega_0 \to \infty$), in which the gravitational force is insignificant compared with the force of inertia, or, which is essentially the same, the limit of zero gravity ($\omega_0/\omega \to 0$), corresponds to $k = 0$, that is, to the

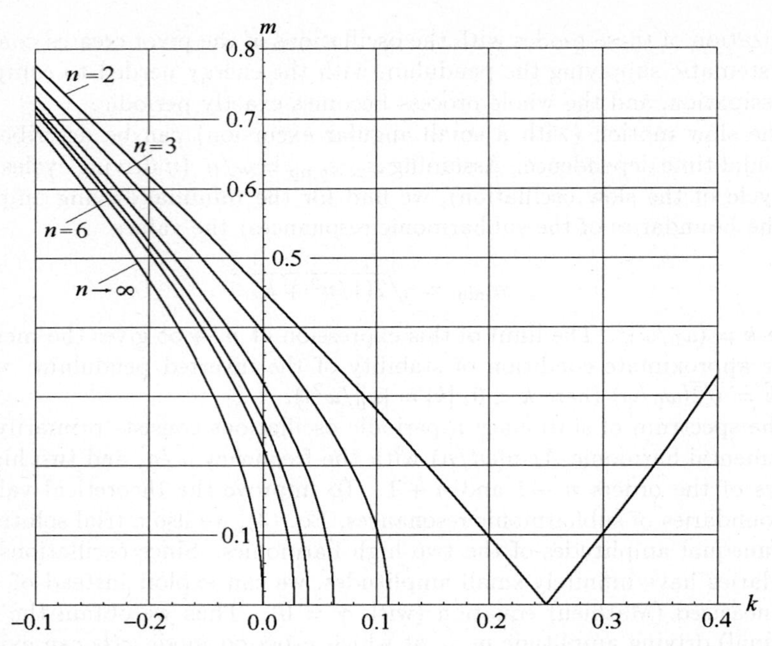

Figure 2: The driving amplitude at the boundaries of the dynamic stabilization of the inverted pendulum and subharmonic resonances

points of intersection of the curves in Fig. 2 with the m-axis. The continuations of these curves further to negative k values describe the transition through zero gravity to the "gravity" directed upward, which is equivalent to the case of an inverted pendulum in ordinary (directed downward) gravitational field. Therefore the same curves at negative k values give the threshold driving amplitudes for subharmonic resonances of the inverted pendulum. [1]

Smooth non-harmonic oscillations of a finite angular excursion are characterized by a greater period than the small-amplitude harmonic oscillations executed just over the parabolic bottom of this well. Therefore large-amplitude period-6 oscillations shown in Fig. 1 (their swing equals 55°) occur at a considerably greater value of the driving amplitude ($a = 0.265\,l$) than the critical (threshold) value $a_{\min} = 0.226\,l$. By virtue of the dependence of the period of non-harmonic smooth motion on the swing, several modes of subharmonic resonance with different values of n can coexist at the same amplitude and frequency of the pivot.

[1] Actually the curves in Fig. 2 are plotted not according to Eq. (4), but rather with the help of a somewhat more comlicated formula (not cited in this paper), which is obtained by holding one more high order harmonic component in the trial function.

4 The upper boundary of the dynamic stability

When the amplitude a of the pivot vibrations is increased beyond a certain critical value $a_{\max}$, the dynamically stabilized inverted position of the pendulum loses its stability. After a disturbance the pendulum does not come to rest in the up position, no matter how small the release angle, but instead eventually settles into a finite amplitude steady-state oscillation (about the inverted vertical position) whose period is twice the driving period. This loss of stability of the inverted pendulum has been first described by Blackburn *et al.* [1] (the "flutter" mode) and demonstrated experimentally in [2]. The latest numerical investigation of the bifurcations associated with the stability of the inverted state can be found in [6].

The curve with $n = 2$ in Fig. 2 shows clearly that both the ordinary parametric resonance and the period-2 "flutter" mode that destroys the dynamic stability of the inverted state belong essentially to the same branch of possible steady-state period-2 oscillations of the parametrically excited pendulum. Indeed, the two branches of this curve near $k = 0.25$ (that is, at $\omega \approx 2\omega_0$) describe the well known boundaries of the principle parametric resonance. Therefore the upper boundary of dynamic stability for the inverted pendulum can be found directly from the linearized differential equation of the system. In the case $\omega_0 = 0$ (which corresponds to the absence of gravity) we find $m_{\max} = 3(\sqrt{13} - 3)/4 = 0.454$, and the corresponding ratio of amplitudes of the third harmonic to the fundamental one equals $A_3/A_1 = -(\sqrt{13} - 3)/6 = -0.101$. A somewhat more complicated calculation in which higher harmonics (up to the 7th) in $\varphi(t)$ are taken into account yields for $m_{\max}$ and A_3/A_1 the values that coincide (within the assumed accuracy) with those cited above. These values agree well with the simulation experiment in conditions of the absence of gravity ($\omega_0 = 0$) and very small angular excursion of the pendulum. When the normalized amplitude of the pivot $m = a/l$ exceeds the critical value $m_{\max} = 0.454$, the swing of the period-2 "flutter" oscillation (amplitude A_1 of the fundamental harmonic) increases in proportion to the square root of this excess: $A_1 \propto \sqrt{a - a_{\max}}$. This dependence follows from the nonlinear differential equation of the pendulum, Eq. (2), if $\sin\varphi$ in it is approximated as $\varphi - \varphi^3/6$, and agrees well with the simulation experiment for amplitudes up to $45°$.

As the normalized amplitude $m = a/l$ of the pivot is increased over the value 0.555, the symmetry-breaking bifurcation occurs: The angular excursions of the pendulum to one side and to the other become different, destroying the spatial symmetry of the oscillation and hence the symmetry of the phase orbit. As the pivot amplitude is increased further, after $m = 0.565$ the system undergoes a sequence of period-doubling bifurcations, and finally, at $m = 0.56622$ (for $Q = 20$), the oscillatory motion of the pendulum becomes replaced, at the end of a very long chaotic transient, by a regular unidirectional period-1 rotation.

Similar (though more complicated) theoretical investigation of the boundary conditions for period-2 stationary oscillations in the presence of gravity allows us to obtain the dependence of the critical (destabilizing) amplitude $m_{\max}$ of the pivot on the driving frequency ω. In terms of $k = (\omega_0/\omega)^2$ this dependence has the following form:

$$m_{\max} = (\sqrt{117 - 232k + 80k^2} - 9 + 4k)/4. \tag{6}$$

The graph of this boundary is shown in Fig. 2 by the curve marked as $n = 2$. The critical driving amplitude tends to zero as $k \to 1/4$ (as $\omega \to 2\omega_0$). This condition corresponds to ordinary parametric resonance of the hanging down pendulum, which is excited if the driving frequency equals twice the natural frequency. For $k > 1/4$ ($\omega < 2\omega_0$) Eq. (6) yields negative m whose absolute value $|m|$ corresponds to stationary oscillations at the other boundary (to the right of $k = 0.25$, see Fig. 2). If the driving frequency exceeds $2\omega_0$ (that is, if $k < 0.25$), a finite driving amplitude is required for infinitely small steady parametric oscillations even in the absence of friction. The continuation of the curve $n = 2$ to the region of negative k values corresponds to the transition from ordinary downward gravity through zero to "negative," or upward "gravity," or, which is the same, to the case of inverted pendulum in ordinary (directed down) gravitational field. Thus, the same formula, Eq. (6), gives the driving amplitude (as a function of the driving frequency) at which both the equilibrium position of the hanging down pendulum is destabilized due to excitation of ordinary parametric oscillations, and the dynamically stabilized inverted equilibrium position is destabilized due to excitation of period-2 "flutter" oscillations. We can interpret this as an indication that both phenomena are closely related and have common physical nature. All the curves that correspond to subharmonic resonances of higher orders ($n > 2$) lie between this curve and the lower boundary of dynamical stabilization of the inverted pendulum.

5 New types of regular and chaotic motions

In this section we report about several modes of regular and chaotic behavior of the parametrically driven pendulum, which we have discovered recently in the simulation experiments. As far as we know, such modes haven't been described in literature.

Figure 3 shows a regular period-8 motion of the pendulum, which can be characterized as a subharmonic resonance of a fractional order, specifically, of the order 8/3 in this example. Here the amplitude of the fundamental harmonic (whose frequency equals $\omega/8$) is much smaller than the amplitude of the third harmonic (frequency $3\omega/8$). This third harmonic dominates in the spectrum, and can be regarded as the principal one, while the fundamental harmonic can be regarded as its third subharmonic. Considerable contributions to the spectrum are given also by the 5th and 11th harmonics of the fundamental frequency. Approximate boundary conditions for small-amplitude stationary oscillations of this type ($n/3$-order subresonance) can be found analytically from the linearized differential equation by a method similar to that used above for n-order subresonance: we can try as $\varphi(t)$ a solution consisting of spectral components with frequencies $3\omega t/n$, $(n-3)\omega t/n$, and $(n+3)\omega t/n$:

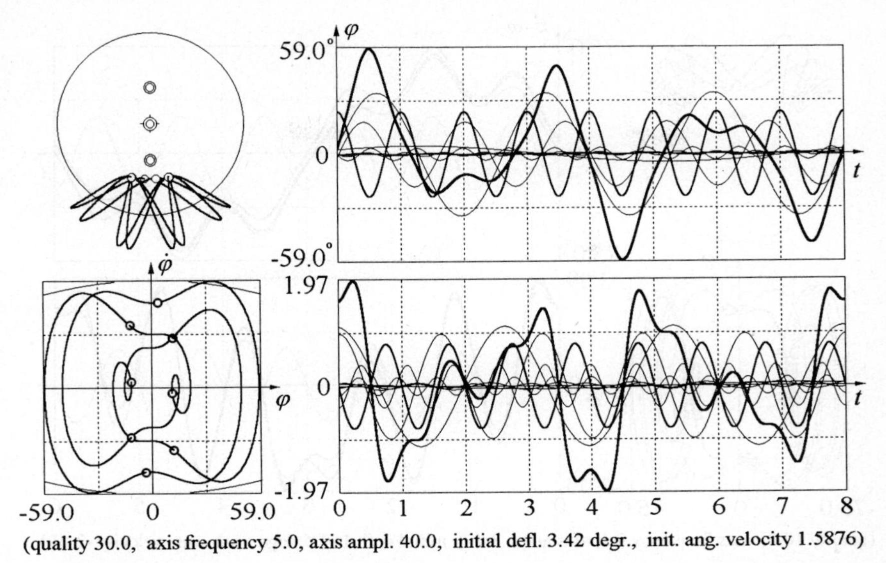

Figure 3: The spatial path, phase orbit, and graphs of stationary oscillations that can be treated as a subharmonic resonance of a fractional order (8/3)

$$\varphi(t) = A_3 \sin(3\omega t/n) + A_{n-3} \sin[(n-3)\omega t/n] + A_{n+3} \sin[(n+3)\omega t/n]. \quad (7)$$

For the parametrically driven pendulum in the absence of gravity such a calculation gives the following expression for the minimal driving amplitude:

$$m_{\min} = \frac{3\sqrt{2}(n^2 - 3^2)}{n^2\sqrt{n^2 + 3^2}}. \quad (8)$$

The analytical results of calculations for $n \geq 8$ agree well with the simulations, especially if one more high harmonic is included in the trial function $\varphi(t)$.

One more type of regular behavior is shown in Fig. 4. This mode can be characterized as resulting from a multiplication of the period of a subharmonic resonance, specifically, as tripling of the six-order subresonance in this example. Comparing this figure with Fig. 1, we see that in both cases the motion is quite similar during any cycle of six consecutive driving periods each, but in Fig. 4 the motion during each next cycle of six periods is slightly different from the preceding cycle. After three such cycles (of six driving periods each) the phase orbit becomes closed and then repeats itself, so the period of this stationary motion equals 18 driving periods. However, the harmonic component whose period equals six driving periods dominates in the spectrum (just like in the spectrum of period-6 oscillations in Fig. 1), while the fundamental harmonic (frequency $\omega/18$) of a small amplitude is responsible only for tiny divergences between the adjoining cycles consisting of six driving periods.

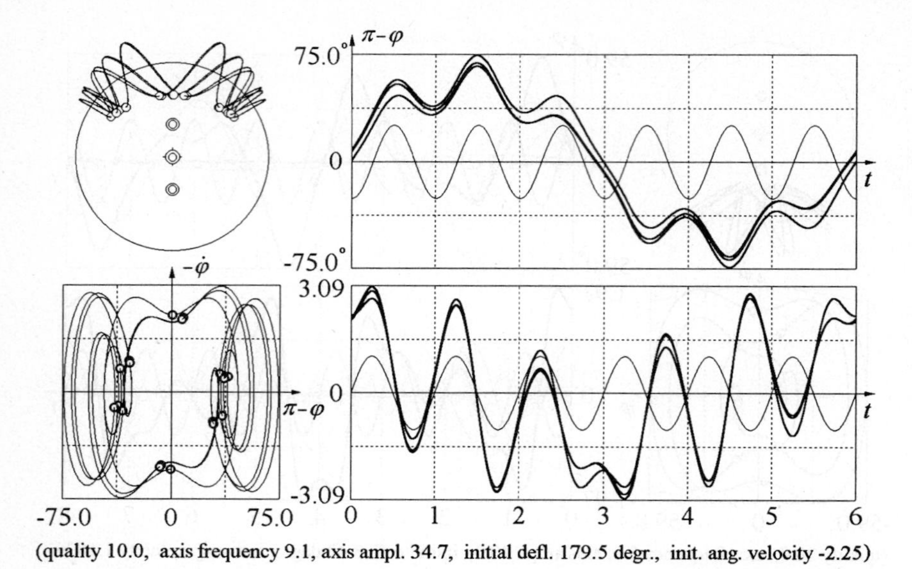

Figure 4: The spatial path, phase orbit, and graphs of period-18 oscillations

Such multiplications of the period are characteristic of large amplitude oscillations at subharmonic resonances both for the inverted and hanging down pendulum. Figure. 5 shows a stationary oscillation with a period that equals ten driving periods. This large amplitude motion can be treated as originating from a period-2 oscillation (that is, from ordinary principal parametric resonance) by a five-fold multiplication of the period. The harmonic component with half the driving frequency ($\omega/2$) dominates in the spectrum. But in contrast to the preceding example, the divergences between adjoining cycles consisting of two driving periods each are generated by the contribution of a harmonic with the frequency $3\omega/10$ rather than of the fundamental harmonic (frequency $\omega/10$) whose amplitude is much smaller.

One more example of complicated steady-state oscillation is shown in Fig. 6. This period-30 motion can be treated as generated from the period-2 principal parametric resonance first by five-fold multiplication of the period (resulting in period-10 oscillation), and then by next multiplication (tripling) of the period. Such large-period stationary regimes are characterized by small domains of attraction consisting of several disjoint islands on the phase plane.

Other modes of regular behavior are formed by unidirectional period-2 or period-4 (or even period-8) rotation of the pendulum or by oscillations alternating with revolutions to one or to both sides in turn. Such modes have periods constituting several driving periods.

At large enough driving amplitudes the pendulum exhibits various chaotic regimes. Chaotic behaviour of nonlinear systems has been a subject of intense interest during recent decades, and the forced pendulum serves as an excellent physical model for studying general laws of the dynamical chaos [6] – [14].

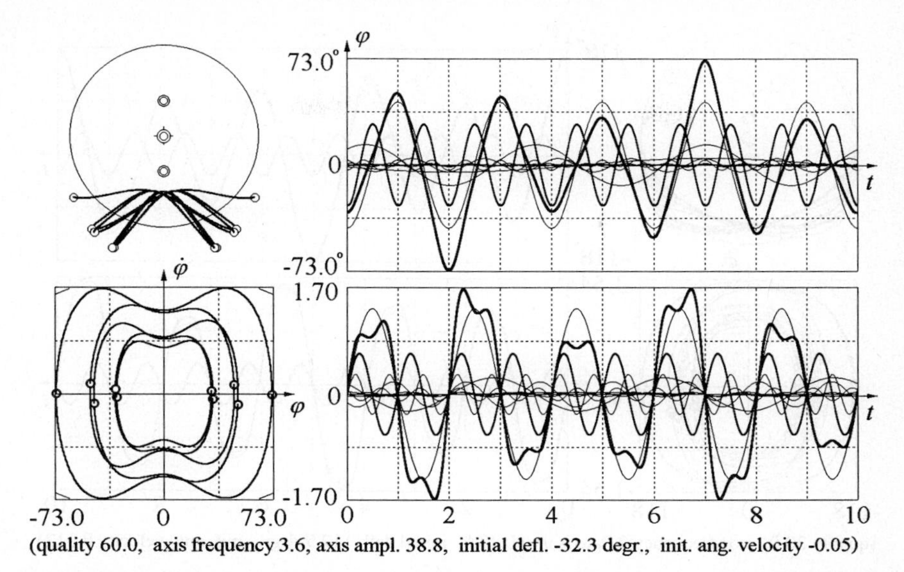

Figure 5: The spatial path, phase orbit, and graphs of period-10 oscillations

Next we describe several different kinds of chaotic regimes, which for the time being have not been mentioned in literature. Poincaré mapping, that is, a stroboscopic picture of the phase plane for the pendulum taken once during each driving cycle after initial transients have died away, gives an obvious and convenient means to distinguish between regular periodic behavior and persisting chaos. A steady-state subharmonic of order n would bee seen in the Poincaré map as a systematic jumping between n fixed mapping points. When the pendulum motion is chaotic, the points of Poincaré sections wander randomly, never exactly repeating. Their behavior in the phase plane gives an impression of the strange attractor for the motion in question.

Figure. 7 shows an example of a purely oscillatory two-band chaotic attractor for which the set of Poincaré sections consists of two disjoint islands. This attractor is characterized by a fairly large domain of attraction in the phase plane. The two islands of the Poincaré map are visited regularly (strictly in turn) by the representing point, but within each island the point wanders irregularly from cycle to cycle. This means that for this kind of motion the flow in the phase plane is chaotic, but the distance between any two initially close phase points within this attractor remains limited in the progress of time: The greatest possible distance in the phase plane is determined by the size of these islands of the Poincaré map.

Figure. 8 shows the chaotic attractor that corresponds to a slightly reduced friction, while all other parameters are unchanged. Gradual reduction of friction causes the islands of Poincaré sections to grow and coalesce, and to form finally a strip-shaped set occupying considerable region of the phase plane. As in the preceding example, each cycle of these oscillations (consisting of two driving

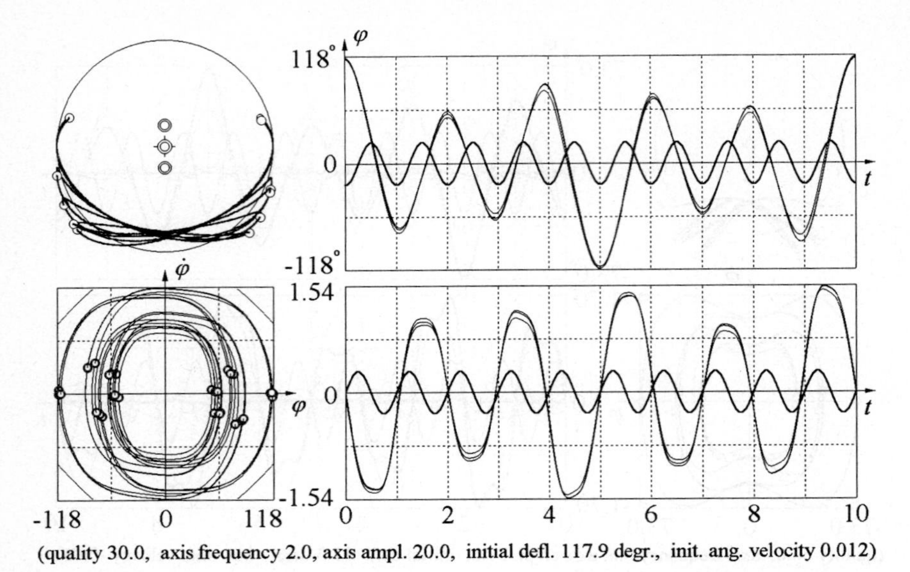

Figure 6: The spatial path, phase orbit, and graphs of period-30 oscillations.

periods) slightly but randomly varies from the preceding one. However, in this case the large and almost constant amplitude of oscillations occasionally (after a large but unpredictable number of cycles) considerably reduces or, vice versa, increases (sometimes so that the pendulum makes a full revolution over the top). These decrements and increments result sometimes in switching the phase of oscillations: the pendulum motion, say, to the right side that occurred during even driving cycles is replaced by the motion in the opposite direction. During long intervals between these seldom events the motion of the pendulum is purely oscillatory with only slightly (and randomly) varying amplitude. This kind of intermittent irregular behavior differs from the well-known so-called tumbling chaotic attractor that exists over a relatively broad range of parameter space. The tumbling attractor is characterized by random oscillations (whose amplitude varies strongly from cycle to cycle), often alternated with full revolutions to one or the other side.

Figure 9 illustrates one more kind of strange attractors. In this example the motion is always purely oscillatory, and nearly repeats itself after each six driving periods. The six bands of Poincaré sections make two groups of three isolated islands each. The representing point visits these groups in alternation. It also visits the islands of each group in a quite definite order, but within each island the points continue to bounce from one place to another without any apparent order. The six-band attractor has a rather extended (and very complicated in shape) domain of attraction. Nevertheless, at these values of the control parameters the system exhibits multiple asymptotic states: The chaotic attractor coexists with several periodic regimes.

Chaotic regimes exist also for purely rotational motions. Poincaré sections

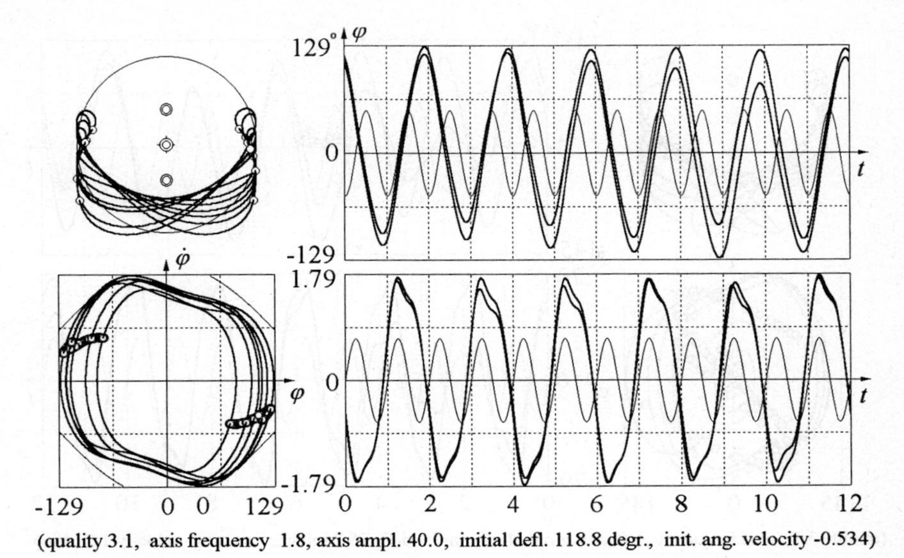

Figure 7: Chaotic attractor with a two-band set of Poincaré sections

for such rotational chaotic attractors can make several isolated islands in the phase plane. A possible scenario of transition to such chaotic modes from unidirectional regular rotation lies through an infinite sequence of period-doubling bifurcations occurring when a control parameter (the driving amplitude or frequency or the braking frictional torque) is slowly varied without interrupting the motion of the pendulum. However, there is no unique route to chaos for more complicated chaotic regimes described above.

6 Concluding remarks

The behavior of the parametrically excited pendulum discussed in this paper is much richer in various modes than we can expect for such a simple physical system relying on our intuition. Its nonlinear large-amplitude motions can hardly be called "simple." The simulations show that variations of the parameter set (dimensionless driving amplitude a/l, normalized driving frequency ω/ω_0, and quality factor Q) result in different regular and chaotic types of dynamical behavior.

In this paper we have touched only a small part of existing stationary states, regular and chaotic motions of the parametrically driven pendulum. The pendulum's dynamics exhibits a great variety of other asymptotic rotational, oscillatory, and combined (both rotational and oscillatory) multiple-periodic stationary states (attractors), whose basins of attraction are characterized by a surprisingly complex (fractal) structure. Computer simulations reveal also intricate sequences of bifurcations, leading to numerous intriguing chaotic regimes. Most of them remained beyond the scope of this paper, and those mentioned

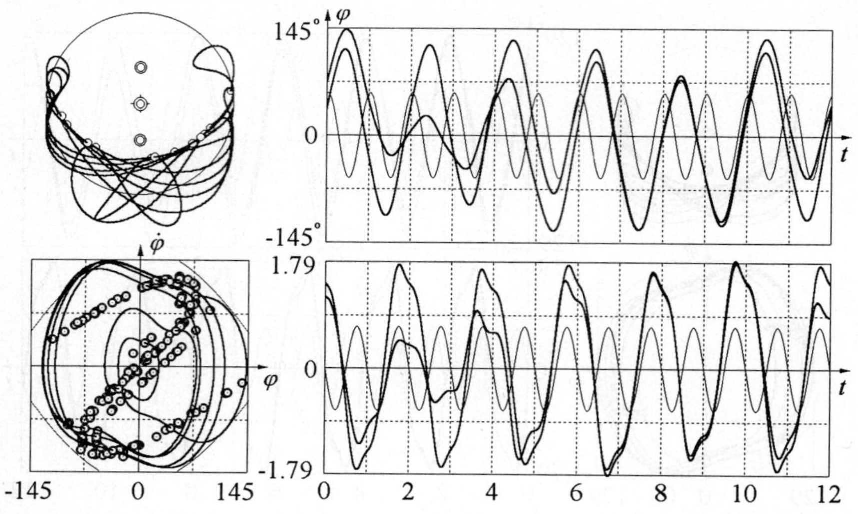

(quality 3.2, axis frequency 1.8, axis ampl. 40.0, initial defl. 75.28 degr., init. ang. velocity 1.21486)

Figure 8: Chaotic attractor with a strip-like set of Poincaré sections.

here are still waiting a plausible physical explanation. With good reason we can suppose that this seemingly simple physical system is inexhaustible.

References

[1] Blackburn J A, Smith H J T, Groenbech-Jensen N 1992 Stability and Hopf bifurcations in an inverted pendulum *Am. J. Phys.* **60** (10) 903 – 908

[2] Smith H J T, Blackburn J A 1992 Experimental study of an inverted pendulum *Am. J. Phys.* **60** (10) 909 – 911

[3] Acheson D J 1995 Multiple-nodding oscillations of a driven inverted pendulum *Proc. Roy. Soc. London* **A 448** 89 – 95

[4] Butikov E I 2001 On the dynamic stabilization of an inverted pendulum *Am. J. Phys.* **69** (7) 755 – 768

[5] Kapitza P L 1951 Dynamic stability of the pendulum with vibrating suspension point *Soviet Physics – JETP* **21** (5) 588 – 597 (in Russian), see also *Collected papers of P. L. Kapitza* edited by D. Ter Haar, Pergamon, London (1965), v. 2, pp. 714 – 726.

[6] Sang-Yoon Kim and Bambi Hu 1998 Bifurcations and transitions to chaos in an inverted pendulum *Phys. Rev.* E **58**, (3) 3028 – 3035

[7] McLaughlin J B 1981 Period-doubling bifurcations and chaotic motion for a parametrically forced pendulum *J. Stat. Physics* **24** (2) 375 – 388

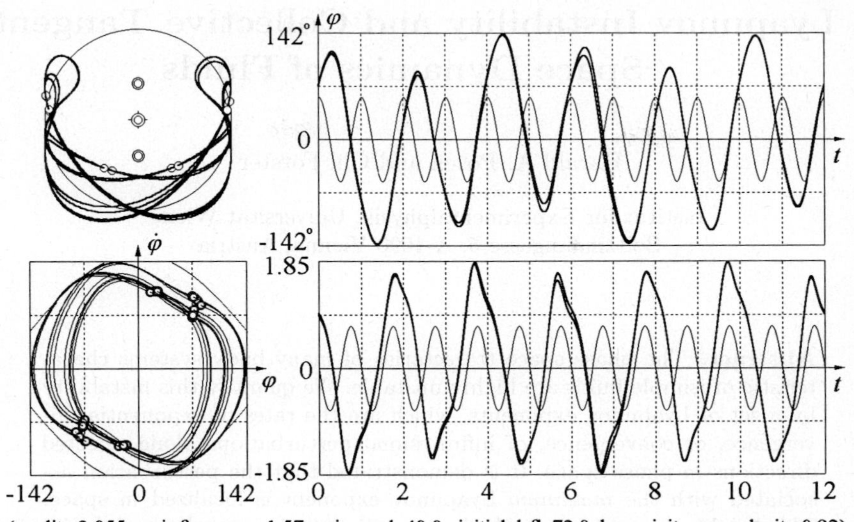

(quality 3.055, axis frequency 1.57, axis ampl. 40.0, initial defl. 72.0 degr., init. ang. velocity 0.92)

Figure 9: An oscillatory six-band chaotic attractor.

[8] Koch B P, Leven R W, Pompe B, and Wilke C 1983 Experimental evidence for chaotic behavior of a parametrically forced pendulum *Phys. Lett.* A **96** (5) 219 – 224

[9] Leven R W, Pompe B, Wilke C, and Koch B P 1985 Experiments on periodic and chaotic motions of a parametrically forced pendulum *Physica* D **16** (3) 371 – 384

[10] Willem van de Water and Marc Hoppenbrouwers 1991 Unstable periodic orbits in the parametrically excited pendulum *Phys. Rev.* A **44** (10) 6388 - 6398

[11] Starrett J and Tagg R 1995 Control of a chaotic parametrically driven pendulum *Phys. Rev. Lett.* **74**, (11) 1974 – 1977

[12] Clifford M J and Bishop S R 1998 Inverted oscillations of a driven pendulum *Proc. Roy. Soc. London* A **454** 2811 – 2817

[13] Sudor D J, Bishop S R 1999 Inverted dynamics of a tilted parametric pendulum *Eur. J. Mech. A/Solids* **18** 517 – 526

[14] Bishop S R, Sudor D J 1999 The 'not quite' inverted pendulum *Int. Journ. Bifurcation and Chaos* **9** (1) 273 – 285

Lyapunov Instability and Collective Tangent Space Dynamics of Fluids

Harald A. Posch and Ch. Forster

Institut für Experimentalphysik, Universität Wien,
Boltzmanngasse 5, A-1090 Vienna, Austria

Abstract. The phase space trajectories of many body systems charateristic of isimple fluids are highly unstable. We quantify this instability by a set of Lyapunov exponents, which are the rates of exponential divergence, or convergence, of infinitesimal perturbations along selected directions in phase space. It is demonstrated that the perturbation associated with the *maximum* Lyapunov exponent is localized in space. This localization persists in the large-particle limit, regardless of the interaction potential. The perturbations belonging to the *smallest positive* exponents, however, are sensitive to the potential. For hard particles they form well-defined long-wavelength modes. The modes could not be observed for systems interacting with a soft potential due to surprisingly large fluctuations of the local exponents.

1 Lyapunov spectra

Recently, molecular dynamics simulations have been used to study many body systems representing simple fluids or solids from the point of view of dynamical system theory. Due to the convex dispersive surface of the atoms, the phase trajectory of such systems is highly unstable and leads to an exponential growth, or decay, of small (infintesimal) perturbationis of an initial state along specified directions in phase space. This so-called Lyapunov instability is described by a set of rate constants, the Lyapunov exponents $\{\lambda_l, l = 1, \ldots, D\}$, to which we refer as the Lyapunov spectrum. Conventionally, the exponents are taken to be ordered by size, $\lambda_l \geq \lambda_{l+1}$. There are altogether $D = 2dN$ exponents, where d is the dimension of space, N is the number of particles, and D is the dimension of the phase space. For fluids in nonequilibrium steady states close links between the Lyapunov spectrum and macroscopic dynamical properties, such as transport coefficients, irreversible entropy production, and the Second Law of thermodynamcis have been established [1–4]. This important result provided the motivation for us to examine the spatial structure of the various perturbed states associated with the various exponents. Here we present some of our results for two simple many-body systems representing dense two-dimensional fluids in thermodynmic equilibrium. The first model consists of N hard disks (HD) interacting with hard elastic collisions, the second of N soft disks interacting with a purely repulsive Weeks-Chandler-Anderson (WCA) potential.

P.M.A. Sloot et al. (Eds.): ICCS 2002, LNCS 2331, pp. 1170–1175, 2002.

The instantaneous state of a planar particle system is given by the 4N-dimensional phase space vector $\mathbf{\Gamma} = \{\mathbf{r}_i, \mathbf{p}_i, ; i = 1, \ldots, N\}$, where $\mathbf{r}_i$ and $\mathbf{p}_i$ denote the respective position and linear momentum of molecule i. An infinitesimal perturbation $\delta\mathbf{\Gamma} = \{\delta\mathbf{r}_i, \delta\mathbf{p}_i; i = 1, \ldots, N\}$ evolves according to motione equations obtained by linearizing the equations of motion for $\mathbf{\Gamma}(t)$. For ergodic systems there exist $D = 4N$ orthonormal initial vectors $\{\delta\mathbf{\Gamma}_l(0); l = 1, \ldots, 4N\}$ in tangent space, such that the Lyapunov exponents

$$\lambda_l = \lim_{t\to\infty} \frac{1}{t} \ln \frac{|\delta\mathbf{\Gamma}_l(t)|}{|\delta\mathbf{\Gamma}_l(0)|}\ , \quad l = 1, \ldots, 4N. \tag{1}$$

exist and are independent of the initial state. Geometrically, the Lyapunov spectrum describes the stretching and contraction along linearly-independent phase space directions of an infinitesimal hypersphere co-moving with the flow. For equilibrium systems the symplectic nature of the motion equations assures the conjugate pairing rule to hold [5]: the exponents appear in pairs, $\lambda_l = \lambda_{4N+1-l} = 0$, so that only half of the spectrum $\{\lambda_{1\le l\le 2N}\}$ needs to be cqaluclated. The sum of all Lyapunov exponents vanishes, which, according to Liouville's theorem, expresses the fact that the phase volume is strictly conserved for Hamiltonian systems. Six if the exponents $\{\lambda_{2N-2\le l\le 2N+3}\}$ always vanish as a consequence of the conservation of energy, momentum, and center of mass, and of the non-exponential time evolution of a perturbation vector parallel to the phase flow.

For the computation of a complete spectrum a variant of a classical algorithm by Benettin *et al.* and Shimada *et al.* is used [6, 7]. It follows the time evolution of the reference trajectory and of an orthonormal set of tangent vectors $\{\delta\mathbf{\Gamma}_l(t); l = 1, \ldots, 4N\}$, where the latter is periodically re-orthonormalized with a Gram-Schmidt (GS) procedure after consecutive time intercals Δt_{GS}. The Lyapunov exponents are determined from the time-averaged renormalization factors. For the hard disk systems the free evolution of the particles is interrupted by hard elastic collisions and a linearized collision map needs to be calculated as was demonstrated by Dellago *et al.* [9]. Although we make use of the conjugate pairing symmetry and compute only the positive branch of the spectrum, we are presently restricted to about 1000 particles by our available computer resources.

For our numerical work reduced units are used. In the case of the Weeks-Chendler-Anderson interaction potential,

$$\phi(r) = \begin{cases} 4\epsilon[(\sigma/r)^{12} - (\sigma/r)^6] + \epsilon, & r < 2^{1/6}\sigma \\ 0, & r \ge 2^{1/6}\sigma. \end{cases}, \tag{2}$$

the particle mass m, the particle diameter σ, and the time $(m\sigma^2/\epsilon)^{1/2}$ are unity. In this work we restrict our discussion to a thermodynamic state with a total energy per particle, E/N, also equal to unity. For the hard-disk fluid $(Nm\sigma^2/K)^{1/2}$ is the unit of time, where K is the total kinetic energy, which is equal to the total energy E of the system. There is no potential energy in this case. The reduced temperature $T = K/N$, where Boltzmann's constant is also taken unity. In the following, Lyapunov exponents for the two model fluids will be compared for equal temperatures (and not for equal total energy). This requires a rescaling of the hard-disk exponents by a factor of $\sqrt{K_{WCA}/K_{HD}}$ to account for the difference in temperature. All our simulations are for a reduced density $\rho \equiv N/V = 0.7$, where the simulation box is a square with a volume $V = L^2$ and a side length L. Periodic boundaries are used throughout.

Figure 1. Lyapunov spectrum of a dense two-dimensional fluid consisting of $N = 100$ particles at a density $\rho = 0.7$ and a temperature $T = 0.75$. The label WCA refers to a smooth Weeks-Chandler-Anderson interaction potential, whereas HD is for the hard disk system.

As an example we compare in Fig. 1 the Lyapunov spectrum of a WCA fluid with $N = 100$ particles to an analogous spectrum for a hard disk system at the same temperature ($T = 0.75$) and density ($\rho = 0.7$). A renormalized index $l/2N$ is used on the abscissa. It is surprising that the two spectra differ so much in shape and in magnitude. The difference persists in the thermodynamic limit. The step-like structure of the hard disk spectrum for $l/2N$ close to 1 is an indication of a coherent wave-like shape of the associated perturbation. We defer the discussion of the so-called Lyapunov modes to Section 3.

2 Fluctuating local exponents

We infer from Equ. (1) that the Lyapunov exponents are time averages over an (infinitely) long trajectory and are global properties of the system. This time average can be written as

$$\lambda_l = \lim_{\tau \to \infty} \int_0^\tau \lambda'_l(\mathbf{\Gamma}(t))dt \equiv \langle \lambda'_l \rangle, \tag{3}$$

iwhere the (implicitely) time-dependent function $\lambda'_l(\mathbf{\Gamma}(t))$ depends on the state $\mathbf{\Gamma}(t)$ in phase space the system occupies at time t. Thus, $\lambda'_l(\mathbf{\Gamma})$ is called a local Lyapunov exponent. It may be estimated from

$$\lambda'_l(\mathbf{\Gamma}(t)) = \frac{1}{\Delta t_{GS}} \ln \frac{|\delta \mathbf{\Gamma}_l(\mathbf{\Gamma}(t + \Delta t_{GS})|}{|\delta \mathbf{\Gamma}_l(\mathbf{\Gamma}(t)|}, \tag{4}$$

where t and $t+\Delta t_{GS}$ refer to times immediately after consecutive Gram-Schmidt re-orthonormalization steps. Its time average, denoted by $\angle \cdots \rangle$, along a trajectory gives the global exponent λ_l. The local exponents fluctuate considerably along a trajectory. This is demonstrated in Fig. ??, where we have plotted the second moment $\langle \lambda'^2_l \rangle$ as a function of the Lyapunov index $1 \le l \le 4N$ for a system of 16 particles, both for the WCA and HD particle interactions. $l = 1$ refers to the maximum exponent, and $l = 64$ to the most negative exponent. The points for $30 \le l \le 35$ correspond to the 6 vanishing exponents and are not shown. We infer from this figure that for the particles interacting with the smooth WCA potential the fluctuations of the local exponents, whose average give rise to global exponents approaching zero for $l \to 2N$. For the hard disk system, however, the relative improtance of the fluctuations also becomes minimal in this limit. We shall return to this point in Section 3.

We note that the computation of the second moments $\langle \lambda'^2_l \rangle$ for the hard disk system requires some care. Due to the hard core collisions they depend strongly on Δt_{GS} for small Δt_{GS}. The mean square deviations, $\langle \lambda'^2_l \rangle - \langle \lambda'_l \rangle^2$, vary with $1/\Delta t_{GS}$ for small Δt_{GS}, as is demonstrated in Fig. 3 for the maximum local exponent. However, the shape of the fluctuation spectrum is hardly affected by the size of the renormalization interval.

3 The maximum exponent

The maximum Lyapunov exponent is the rate constant for the fastest growth of phase space perturbations in a system. There is strong numerical evidence for the existence of the thermodynamic limit $\{N \to \infty, \rho = N/V \text{ constant }\}$ for λ_1 and, hence, for the whole spectrum. Furthermore, the associated perturbation is strongly localized in space. This may be demonstrated by projecting the tangent vector $\delta \mathbf{\Gamma}_1$ onto the subspaces spanned by the perturbation components contributed by the individual particles. The squared norm of this projection, $\delta_i^2(t) \equiv (\delta \mathbf{r}_i)_l^2 + (\delta \mathbf{p}_i)_l^2$, indicates how active a particle i is engaged in the growth process of the pertubation associated with λ_1. In Fig. 4 $\delta_i^2(t)$ is plotted along the vertical for all particles of a hard disk system at the respective positions (x_i, y_i) of the disks in space, and the ensuing surface is interpolated over a periodic grid covering the simulation box. A strong localization of the active particles is observed at any instant of time. Similar, albeit slightly broader peaks are observed for the WCA system.

This localization is a consequence of two mechanisms: firstly, after a collision the delta-vector components of two colliding molecules are linear functions of their pre-collision values and have only a chance of further growth if their values before the collision were already far above average. Secondly, each renormalization step tends to reduce the (already small) components of the other non-colliding particles even further. Thus, the competition for maximum growth of tangent vector components favors the collision pair with the largest components.

The localization also persists in the thermodynamic limit. To show this we follow Milanović *et al.* [8], square all $4N$ components of the perturbation vector $\delta \mathbf{\Gamma}_1$ and order the squared components $[\delta \mathbf{\Gamma}_1]_j^2; j = 1, \ldots, 4N$ according to size. By adding them up, starting with the largest, we determine the smallest number of terms, $A \equiv 4NC_{1,\Theta}$, required for the sum to exceed a threshold Θ. Then, $C_{1,\Theta} = A/4N$ may be taken as a relative measure for the number of components actively contributing to λ_1:

$$\Theta \le \left\langle \sum_{s=1}^{4NC_{1,\Theta}} [\delta \mathbf{\Gamma}_1]_s^2 \right\rangle, \qquad [\delta \mathbf{\Gamma}_1]_i^2 \ge [\delta \mathbf{\Gamma}_1]_j^2 \text{ for } i < j. \tag{5}$$

Here, $\langle \cdots \rangle$ implies a time average. Obviously, $C_{1,1} = 1$. In Fig. 5 $C_{1,\Theta}$ is shown for $\Theta = 0.98$ as a function of the particle number N, both for the WCA fluid and for the hard disk system. It converges to zero if our data are extrapolated to the thermodynamic limit, $N \to \infty$. This supports our assertion that in an infinite system only a *vanishing* part of the tangent-vector components (and, hence, of the particles) contributes significantly to the maximum Lyapunov exponent at any instant of time.

4 Lyapunov modes

We have mentioned already the appearance of a step-like structure in the Lyapunov sepctrum of the hard disk system for the positive exponents closest to zero. They are a consequence of coherent wave-like spatial patterns generated by the perturbation vector components associated with the individual particles. In Fig. 6 this is visualized by plotting the perturbations in the x (bottom surface) and y direction (top surface), $\{\delta x_i, i = 1, \ldots, N\}$ and $\{\delta y_i, i = 1, \ldots, N\}$, respectively, along the vertical direction at the instantaneous particle positions (x_i, y_i) of all particles i. This figure depicts a transversal Lyapunov mode, for which the perturbation is parpendicular to the wave vector, for a hard disk system consisting of $N = 1024$ particles and for a perturbation vector associated with the smallest positive exponent λ_{2045}. An analogous plot for δp_x and δp_y $l = 2045$ is identical to that of δ_x and δ_y in Fig. 6, with the same phase for the waves. This is a consequence of the fact that the perturbations are solutions first-order differential equation instead of second. Furthermore, the exponents for $2042 \le l \le 2045$ are equal. The four-fold degeneracy of non-propagating transversal modes, and an analogous eight-fold degeneracy of propagating longitudinal modes, are responsible for a complicated step structure for l close to $2N$, which has been studied in detail in Refs. XXXXX.

The wave length of the modes and the value of the corresponding exponents are determined by the linear extension L of the simulation box. There is a kind of linear dispersion relation [10] according to which the smallest positive exponent is proprtional to $1/L$. This assures that for a simulation box with aspect ratio 1 there is no positive lower bound for the positive exponents of hard disk systems in the thermodynamic limit.

So far, our discussion of modes is only for the hard disk fluid. In spite of a considerable computational effort we have not yet been able to indentify modes for two-dimensional fluid systems with a soft interaction potential such as WCA or similar potentials. The reason for this surprising fact seems to be the very strong fluctions of the local exponents as discussed in Section 2. The fluctuations obscure any mode in the system in spite of considerable averaging and make a positive identification very difficult. Three-dimensional systems are just beyond computational feasibility at present, although the use of parallel machines may change this scenario soon.

We are grateful to Christoph Dellago, Robin Hirschl, Bill Hoover, and Ljubo Milanović for many illuminating discussions. This work was supported by the Austrian Fonds zur Förderung der wissenschaftlichen Forschung, grants P11428-PHY and P15348-PHY.

References

1. Posch, H. A., and Hoover, Wm. G.: Equilibrium and nonequilibrium Lyapunov spectra for dense fluids and solids. Phys Rev. **A 39** (1989) 2175–2188
2. Gaspard, P.: *Chaos, Scattering, and Statistical Mechanics*, (Cambridge University Press, 1998).
3. Hoover, Wm. G.: *Computational Statistical Mechanics*, (Elsevier, New York, 1999)
4. Dorfman, J.R.: *An Introduction to Chaos in Nonequilibrium Statistical Mechanics*, (Cambridge University Press, 1999)
5. Ruelle, D.: J. Stat. Phys. **95** (1999) 393
6. Benettin, G., Galgani, L., Giorgilli, A., and Strelcyn, J. M.: Meccanica **15** (1980) 9
7. Shimada, I., and Nagashima, T.: Proc. Theor. Phys. **61** (1979) 1605
8. Milanović, Lj., and Posch, H. A.: Localized and delocalized modes in the tangent-space dynamcs of planar hard dumbbell fluids. J. Molec. Liquids (2002), in press
9. Dellago, Ch., Posch, P. H., and Hoover, Wm. G.: Phys. Rev. E **53** (1996) 1485
10.

Deterministic Computation Towards Indeterminism

Bogdanov A.V., Gevorkyan A.S., Stankova E.N., Pavlova M.I.

Institute for High-Performance Computing and Information Systems
Fontanka emb. 6, 194291, St-Petersburg, Russia
bogdanov@hm.csa.ru, ashot@fn.csa.ru, lena@fn.csa.ru, meri@fn.csa.ru

Abstract. In the present work we propose some interpretation of the results of the direct simulation of quantum chaos.

1 Introduction

At the early stage of quantum mechanics development, Albert Einstein has written a work in which the question, which has become a focus of physicians attention several decades later, was touched upon. The question was: what will the classic chaotic system become in terms of quantum mechanics. He has particularly set apart the three-body system.

In an effort to realize the problem and get closer to its chaos solution in essential quantum area, M. Gutzwiller, a well known physician, have conditionally subdivided all the existing knowledge in physics into three areas [1]:

1) elementary classical mechanics, which only allows for simple regular system behaviour (regular classical area R);

2) classical chaotic dynamic systems of Poincare systems (P area);

3) regular quantum mechanics, which interpretations are being considered during last 80 years (Q area).

The above areas are connected by a separate bounds. Thus, Bor's correspondence principal works between R and Q areas, transferring quantum mechanics into classical Newton's mechanics within the limit $\hbar \to 0$. Q and P areas are connected by Kolmogorov's - Arnold's - Mozer's theorem (KAM). Let's note that KAM theorem allows to determine the excitations , which cause the chaotic behaviour of regular systems. Inspite of well known work by Wu and Parisi [2], which allows to describe Q-systems with the help of P-systems in thermodynamic limit under certain circumstances, the general principle connecting P and Q is not yet determined. Assuming the existence of the fourth area - quantum chaos area Q_{ch}, M. Gutzwiller adds that it rather serves for the puzzle description than for a good problem formulation. It is evident that the task formulated correctly in Q_{ch} area is a most general one and must transfer to the abovementioned areas in its limits.

The problem of quantum chaos was studied as the example of quantum multichannel scattering in collinear three-body system [3,4]. It was shown than this task can be transformed into a problem of unharmonic oscillator with non-trivial

P.M.A. Sloot et al. (Eds.): ICCS 2002, LNCS 2331, pp. 1176–1183, 2002.

time (internal time). Let's note, that in a model considered internal time is determined by a system of two non-linear differential equations of the second order. In [5] this problem was studied on the example of chemical reaction and in [6] it was applied to surface scattering. The ab initio computation even of the simple three-body systems is a challenge for some generation of computational physicists, so some new approach was proposed in [7] and beautiful example of distributed computation was demonstrated in [8].

In the present work we propose some interpretation of the results, obtained in [7,8], and thus give our view of quantum chaos origination.

2 Formulation of the problem

The quantum multichannel scattering in the framework of collinear model is realized accordingly to follow scheme:

$$A+(BC)_n \rightarrow \begin{cases} A+(BC)_m \\ (AB)_m+C \\ A+B+C \\ (ABC)^* \rightarrow \begin{cases} A+(BC)_m \\ (AB)_m+C \\ A+B+C \end{cases} \end{cases} \tag{1}$$

with m and n being the vibrational quantum numbers correspondingly in (in) and (out) scattering channels. As it was shown elsewhere [3,4] the problem of quantum evolution (1) can be strictly formulated as a motion of image point with a reduced mass μ_0 over the manyfold M, that is stratificated Lagrange surface S_p, in the moving over S_p local coordinate system. In our case there is standard definition of the surface S_p

$$S_p = \left\{x^1, x^2; 2\mu_0 \left(E - V\left(x^1, x^2\right)\right) > 0\right\},$$

$$\mu_0 = \left\{\frac{m_A m_B m_C}{m_A + m_B + m_C}\right\}^{1/2}, \tag{2}$$

where m_A, m_B, m_C being the masses of corresponding particles, E and $V\left(x^1, x^2\right)$ being correspondingly the total energy and interaction potential of the system. The metric on the surface S_p in our case is introduced in the following way

$$g_{ik} = P_0^2\left(x^1, x^2\right)\delta_{ik},$$

$$P_0^2\left(x^1, x^2\right) = 2\mu_0\left(E - V\left(x^1, x^2\right)\right). \tag{3}$$

As to the motion of the local coordinate system, it is determined by the projection of the image point motion over the extremal ray $\Im_{ext}$ of the Lagrange manyfold S_p. Note, that for scattering problem (1) there are two extremal rays on a surface S_p: one is connecting the (in) channel with the (out) channel of particle rearrangement and the other is connecting the (in) channel with the (out) channel, where all three particles are free. ¿From now on we shall study only the case of rearrangement final channel. Let us introduce curvilinear coordinates (x^1, x^2) in Euclidean space R^2 along the projection of the rearrangement extremal ray $\bar{\Im}_{ext}$ in a such way, that x^1 is changing along $\bar{\Im}_{ext}$ and x^2 is changing in orthogonal direction. In such a case the trajectory of image point is determined by the following system of the second order differential equations:

$$x^k_{;ss} + \{\}_{kij}\, x^i_{;s} x^j_{;s} = 0 \quad (i, j, k = 1, 2) \tag{4}$$

where $\{\}_{kij} = (1/2) g^{kl} \left(g_{lj;i} + g_{il;j} - g_{ij;l} \right)$, $g_{ij;k} = \partial_{x^k} g_{ij}$.

As to the law of local coordinate system motion, it is given by the solution $x^1(s)$. Based on this solution the quantum evolution of the system on the manyfold M is determined by the equation (see [4])

$$\left\{ \hbar^2 \Delta_{(x^1(s), x^2)} + P_0^2 \left(x^1(s), x^2 \right) \right\} \Psi = 0, \tag{5}$$

with the operator $\Delta_{(x^1(s), x^2)}$ of the form

$$\Delta_{(x^1(s), x^2)} = \gamma^{-\frac{1}{2}} \left\{ \partial_{x^1(s)} \left[\gamma^{ij} \gamma^{\frac{1}{2}} \partial_{x^1(s)} \right] + \partial_{x^2} \left[\gamma^{ij} \gamma^{\frac{1}{2}} \partial_{x^2} \right] \right\}. \tag{6}$$

As to the metric tensor of the manyfold M, it has the following form [4]:

$$\gamma_{11} = \left(1 + \frac{\lambda \left(x^1(s) \right)}{\rho_1 \left(x^1(s) \right)} \right)^2, \quad \gamma_{12} = \gamma_{21} = 0,$$
$$\gamma_{22} = \left(1 + \frac{x^2}{\rho_2 \left(x^1(s) \right)} \right)^2, \quad \gamma = \gamma_{11}\gamma_{22} > 0, \tag{7}$$

with λ being de Broglie wave length on $\Im_{ext}$ and ρ_1, ρ_2 being the principle curvatures of the surface S_p in the point $x^1 \in \Im$ in the directions of coordinates x^1, x^2 changes

$$\rho_1 = \frac{P_0 \left(x^1(s), 0 \right)}{P_{0;x^1} \left(x^1(s), 0 \right)}, \quad \rho_2 = \frac{P_0 \left(x^1(s), 0 \right)}{P_{0;x^2} \left(x^1(s), 0 \right)},$$
$$\lambda = \frac{\hbar}{P_0 \left(x^1(s), 0 \right)}, \quad P_{0;x^i} = \frac{dP \left(x^1(s), x^2 \right)}{dx^i}. \tag{8}$$

Note, that the main difference of (5) from Schrödinger equation comes from the fact, that one of the independent coordinates $x^1(s)$ is the solution of nonlinear difference equations system and so is not a natural parameter of our system and can in certain situations be a chaotic function.

3 Reduction of the scattering problem to the problem of quantum harmonic oscillator with internal time

Let us make a coordinate transformations in Eq.(5):

$$\tau = \left(E_k^i\right)^{-1} \int_0^{x^1(s)} P\left(x^1, 0\right) \sqrt{\gamma} dx^1,$$
$$z = \left(\hbar E_k^i\right)^{-\frac{1}{2}} P\left(x^1, 0\right) x^2, \tag{9}$$

with E_k^i being the kinetic energy of particle A in the (in) channel, the function $P\left(x^1, x^2\right) = \sqrt{2\mu_0\left[E_k^i - V\left(x^1, x^2\right)\right]}$ and with image point on the curve $\Im_{ext}$ it is just the momentum of image point.

By expanding of $P\left(x^1, x^2\right)$ over the coordinate x^2 up to the second order we can reduce the scattering equation (5) to the problem of quantum harmonic oscillator with variable frequency in the external field, depending on internal time $\tau\left(x^1, x^2\right)$. E.g. in the case of zero external field the exact wave function of the system without some constant phase, unimportant for the expression of the scattering matrix, is of the form

$$\Psi^+(n;\tau) = \left[\frac{\left(\Omega_{in}/\pi\right)^{1/2}}{2^n n! \left|\xi\right|}\right]^{\frac{1}{2}} \times$$
$$\exp\left\{\frac{E_k^i \tau}{\hbar} - \left(n + \frac{1}{2}\right)\Omega_{in} \int_{-\infty}^{\tau} \frac{d\tau'}{\left|\xi\right|^2} + \right.$$
$$\left. + \frac{1}{2}\dot{\xi}\xi^{-1}z^2 - \frac{1}{2}\dot{p}p^{-1}z^2\right\} H_n\left(\frac{\sqrt{\Omega_{in}}}{\left|\xi\right|} z\right), \tag{10}$$

$$\dot{\xi} = d_\tau \xi, \quad \dot{p} = d_\tau p, \quad p\left(x^1(s)\right) = P\left(x^1(s), 0\right),$$

with the function $\xi(\tau)$ being the solution of the classical oscillator equation

$$\ddot{\xi} + \Omega^2(\tau)\,\xi = 0,$$

$$\Omega^2(\tau) = -\left(\frac{E_k^i}{p}\right)^2 \left\{\frac{1}{\rho_2^2} + \sum_{k=1}^{2}\left[\frac{p_{;kk}}{p} + \left(\frac{p_{;k}}{p}\right)^2\right]\right\}, \tag{11}$$

$$p_{;k} = \frac{dp}{dx^k}$$

with asymptotic conditions

$$\xi(\tau) \xrightarrow[\tau \to -\infty]{\sim} \exp\left(i\Omega_{in}\tau\right),$$

$$\xi(\tau) \xrightarrow[\tau \to +\infty]{\sim} C_1 \exp\left(i\Omega_{in}\tau\right) - C_2 \exp\left(i\Omega_{out}\tau\right). \tag{12}$$

Note, that internal time τ is directly determined by the solution of $x^1\,(s)$ and therefore includes all peculiarities if x^1 behaviour.

The transition probability for that case is of the form [3,4]:

$$W_{mn} = \frac{(n_<)!}{(n_>)!}\sqrt{1 - \left|\frac{C_2}{C_1}\right|^2}\left|P^{(n_> - n_<)/2}_{(n_> + n_<)/2}\left(1 - \left|\frac{C_2}{C_1}\right|^2\right)\right|^2, \tag{13}$$

where $n_< = \min(m,n)$, $n_> = \max(m,n)$ and P^n_m being the Legandre polynomial.

4 The study of the internal time dependence versus natural parameter of the problem - standard time

Now we are able to turn to the prove of the possibility of the quantum chaos initiation in the wave function (10) and as a result in the probability (13). It is enough to show, that the solution $x^1\,(s)$ with certain initial conditions behaves unstable or chaotically. With that purpose on the example of elementary reaction $Li + (FH) \to (LiFH)^* \to (LiF) + H$ we studied carefully the behaviour of the image point trajectories on Lagrange surface S_p. It was shown that with collision energy $E^i_k = 1.4eV$ and for fixed transversal vibrations energy $E_v = 1.02eV$ the image point trajectory is stable. The whole region of kinetic energies is splited to regular subregions, and dependingly from which subregion trajectory starts it goes either to (*out*) channel (Fig.1(a)) or reflects back in the (*in*) channel (Fig.1(b)).

With a further change of kinetic energy the image point trajectory in the interaction region starts orbiting, that corresponds to the creation of the resonance complex $(LiFH)^*$, and after that leave the interaction region either to (*out*) (Fig.1(c)) or return to (*in*) channel. In such a case the image point trajectories diverge and this divergence is exponential, as can be seen from the study of the Lyapunov parameters. That is for those initial conditions the evolution in the correspondent classical problem is chaotic and so the motion of the local coordinate system is chaotic too. It is easy to see that in such situation the behaviour of $x^1\,(s)$ is also chaotic and the same is true for internal time, that is the natural parameter of quantum evolution problem.

It can be shown, that chaotic behaviour of the internal time is followed by the stochastic behaviour of the model equation solution $\xi\,(\tau\,(s))$ and the same is true for the wave function (10) and transition probability (13). In such a way the possibility of violation of quantum determinism and quantum chaos organization was shown on the example of the wave function of the simple model of multichannel scattering.

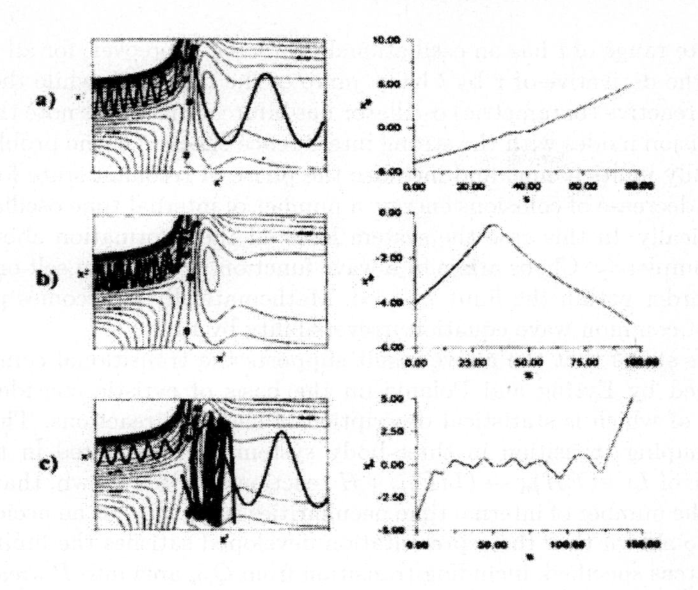

Fig. 1. Dependence of Lyapunov exponent over time parameter s for the case of rearrangement reaction going through resonance state.

Those results may seem strange if we take into account, that original problem (i.e. Schrödinger equation with asymptotic scattering conditions) was quite deterministic (i.e. was good candidate for possessing unique solution), outside of standard interpretation of quantum mechanical quantities. At the same time if one looks carefully at the final version of the scattering probabilities it is clear, that difference between stochastic and regular regimes are not of principal importance, actually the ansatz of solution in our approach for two cases is the same. The only difference comes from the fact, that when orbiting in interaction region starts, the initial conditions for outcoming bunch of trajectories become undetermined, that can be regarded in terms of fluctuations of the initial stratificated Lagrange surface S_p,just as in the case of vacuum fluctuations in quantum field theory [5].

5 Conclusion

In this work it was shown that the representation developed by the authors includes not only Plank's constant $\hbar$, but new energetic parameter as well. Thus, when the energy of the particles collision exceeds certain critical value (which is different for the different systems), solution for internal time τ coincides with an ordinary time - natural t parameter. In this case, the movement equation for the system of bodies transforms to common nonstationary Schrödinger's equation. The scattering process is in fact a direct process for this case.

But everything is quite different when the collision occurs below the critical energy specified. As it is shown, in such a case the solution for internal time τ in a definite range of t has an oscillational character. Moreover, for all the extreme points the derivative of τ by t has a jump of the first kind, while the phase portrait of reactive (parametric) oscillator has bifurcations. Let's note that these are the collision modes with the strong interference effects, i.e. the problem becomes essentially multichannel and includes the phase of resonant state formation. At a small decrease of collisions energy, a number of internal time oscillations grows dramatically. In this case the system loses all the information about its initial state completely. Chaos arises in a wave function, which then self-organizes into a new order within the limit $\tau \to \infty$. Mathematically it becomes possible as a result of common wave equation irreversibility by time.

Let's stress that the above result supports the transitional complex theory, developed by Eyring and Polanyi on the basis of evristic considerations, the essence of which is statistical description of chemical reactions. The amplitude of regrouping transition in three-body system is investigated in the work on example of $Li+(FH)_n \to (LiF)_m+H$ reaction and it is shown, that in the area where the number of internal time peculiarities is high, it has an accidental value. It is also shown that the representation developed satisfies the limit transitions in the areas specified, including transition from Q_{ch} area into P area. The latter occurs under $\hbar \to 0$ and at $E_k^i < E_c$, where E_c is critical energy and E_k^i is a collision energy. It is possible to give very simple interpretation of the above results in terms of initial Lagrange surface fluctuations in strong interaction region.

References

[1] M. C. Gutzwiller, *Chaos in Classical and Quantum Mechanics*, Springer, Berlin, 1990.

[2] E. Nelson, Phys. Rev., (1966), v. 150, p. 1079.

[3] A. V. Bogdanov, A. S. Gevorkyan, *Three-body multichannel scattering as a model of irreversible quantum mechanics*, Proceedings of the International Symposium on Nonlinear Theory and its Applications, Hilton Hawaiian Village, 1997, V.2, pp.693-696.

[4] A. V. Bogdanov, A. S. Gevorkyan, *Multichannel Scattering Closed Tree-Body System as a Example of Irreversible Quantum Mechanics*, Preprint IHPCDB-2, 1997, pp. 1-20.

[5] A.V. Bogdanov, A.S. Gevorkyan, A.G. Grigoryan, Random Motion of Quantum Harmonic Oscillator. Thermodynamics of Nonrelativistic Vacuum, in Proceedings of Int. Conference "Trends in Math Physics", Tennessee, USA, October 14-17, 1998, pp.79-107.

[6] A.V. Bogdanov, A.S. Gevorkyan, Theory of Quantum Reactive Harmonic Oscillator under Brownian Motion, in Proceedings of the International Workshop on Quantum Systems, Minsk, Belarus, June 3-7, 1996, pp.26-33.

[7] A.V. Bogdanov, A.S. Gevorkyan, A.G. Grigoryan, S.A. Matveev, Internal Time Peculiarities as a Cause of Bifurcations Arising in Classical Trajectory Problem and Quantum Chaos Creation in Three-Body System, in Proceedings of Int. Symposium "Synchronization, Pattern Formation, and Spatio-Temporal Chaos in Coupled Chaotic Oscillators" Santyago de Compostela, Spain, June 7-10, 1998;

[8] A.V. Bogdanov, A.S. Gevorkyan, A.G. Grigoryan, First principle calculations of quantum chaos in framework of random quantum reactive harmonic oscillator theory, in Proceedings of 6th Int. Conference on High Performance Computing and Networking Europe (HPCN Europe '98), Amsterdam, The Netherlands, April, 1998

Splitting Phenomena in Wave Packet Propagation

I. A. Valuev[1] and B. Esser[2]

[1] Moscow Institute of Physics and Technology,
Department of Molecular and Biological Physics,
141700 Dolgoprudny, Russia
valuev@physik.hu-berlin.de
[2] Institut für Physik, Humboldt-Universität Berlin,
Invalidenstrasse 110, 10115 Berlin, Germany
besser@physik.hu-berlin.de
http://www.physik.hu-berlin.de

Abstract. Wave packet dynamics on coupled potentials is considered on the basis of an associated Spin-Boson Hamiltonian. A large number of eigenstates of this Hamiltonian is obtained by numerical diagonalization. Eigenstates display a mixing of adiabatic branches as is evident from their Husimi (quantum density) projections. From the eigenstates time dependent Husimi projections are constructed and packet dynamics is investigated. Complex packet dynamics is observed with packet propagation along classical trajectories and an increasing number of packets due to splitting events in the region of avoided crossings of these trajectories. Splitting events and their spin dependencies are systematically studied. In particular splitting ratios relating the intensities of packets after a splitting event are derived from the numerical data of packet propagation. A new projection technique applied to the state vector is proposed by which the presence of particular packets in the evolution of the system can be established and made accessible to analysis.

1 Introduction and Model

Wave packet dynamics is one of the central topics in quantum evolution with a wide range of applications ranging from from atomic and molecular physics to physical chemistry (see e.g. [1] and references therein). We present a numerical investigation of the dynamics of wave packets in a many-potential system, when phase space orbits associated with different adiabatic potentials are coupled. Basic to our investigation is the evolution of quantum states described by the Spin-Boson Hamiltonian

$$\hat{H} = \epsilon_+ \hat{1} - \frac{1}{2}\hat{\sigma}_x + \frac{1}{2}(\hat{P}^2 + r^2\hat{Q}^2)\hat{1} + (\sqrt{\frac{p}{2}} r\hat{Q} + \epsilon_-)\hat{\sigma}_z. \tag{1}$$

In (1) a quantum particle residing in two states is coupled to a vibrational environment specified by the coordinate Q. The two state quantum system is

P.M.A. Sloot et al. (Eds.): ICCS 2002, LNCS 2331, pp. 1184–1192, 2002.

represented by the standard Pauli Spin Matrices $\hat{\sigma}_i$ $(i = x, z)$, r is the dimensionless vibrational frequency of the oscillator potential and p is the coupling constant between the dynamics of the particle and the vibrational environment. We note that (1) is a generalized Spin-Boson Hamiltonian containing the parameter ϵ_- , which destroys the parity symmetry of the eigenstates of the more conventional Spin-Boson Hamiltonian in which such a term is absent.

A Hamiltonians like (1) can be obtained from different physical situations, the particle being e.g. an electron, exciton or any other quasiparticle. To be definite we will use a "molecular" language and consider the situation when the Hamiltonian (1) is derived from a molecular physics model. We consider a molecular dimer, in which the transfer of an excitation between two monomers constituting the dimer and coupled to a molecular vibration is investigated. Then ϵ_- is the difference between the energies of the excitation at the two non-equivalent monomers constituting the dimer (ϵ_+ in (1) is the mean excitation energy, in what follows we omit this term thereby shifting the origin of the energy scale to ϵ_+). For the details of the derivation of (1) from a molecular dimer Hamiltonian and in particular the connection between the dimensionless parameters of (1) with a dimer model we refer to [2] and references therein.

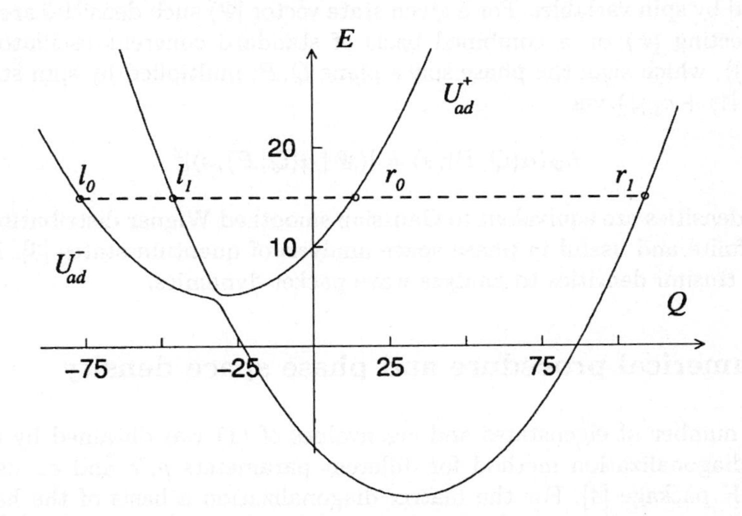

Fig. 1. Adiabtic potentials for the parameter set used. The line of constant energy $E = 15$ and its crossing points with the potentials (turning points) are shown. The turning points are labeled according to the location of the point (l for "left" and r for "right") and monomer (spin) indicies 0 and 1 correspond to the upper and lower monomer, respectively

Applying a Born - Oppenheimer stepwise quantization to (1) one obtains the Hamiltonians of the adiabatic reference systems

$$H^{\pm}(Q) = \frac{1}{2}P^2 + U^{\pm}_{\text{ad}}(Q). \qquad (2)$$

with the adiabatic potentials

$$U_{\mathrm{ad}}^{\pm}(Q) = \frac{r^2}{2}Q^2 \pm \sqrt{\frac{1}{4} + \left(\epsilon_- + \sqrt{\frac{p}{2}}rQ\right)^2}\,, \tag{3}$$

In Fig. 1 the adiabatic potentials for a given parameter set are shown. Fixing an energy E phase space orbits can be derived from (2) for each of the adiabatic potentials. Phase space trajectories of isolated monomers at a given energy E can be derived in an analogous way by neglecting the quantum transfer in (1) (discarding $\hat{\sigma}_x$). In what follows we denote the upper and lower monomer of the dimer configuration by the indices (0) and (1), respectively. In spin representation projections of the state vector on such monomer states correspond to projections on the spin up state (upper monomer) and spin down state (lower monomer), respectively. We note that in the semiclassical case adiabatic trajectories can be represented as built from pieces of monomer trajectories. In the analysis of packet propagation derived from the state vector it will be convenient to use projections on such monomer states below.

A quantum density in phase space is constructed by using Husimi projections extended by spin variables. For a given state vector $|\Psi\rangle$ such densities are derived by projecting $|\Psi\rangle$ on a combined basis of standard coherent oscillator states $|\alpha(Q,P)\rangle$, which scan the phase space plane Q, P, multiplied by spin states $|s\rangle$, $|s\rangle = c_\uparrow |\uparrow\rangle + c_\downarrow |\downarrow\rangle$ via

$$h_\Psi(\alpha(Q,P);s) = |\langle \Psi \mid \alpha(Q,P), s\rangle|^2. \tag{4}$$

Husimi densities are equivalent to Gaussian smoothed Wigner distributions, positive definite and useful in phase space analysis of quantum states [3]. Here we will use Husimi densities to analyze wave packet dynamics.

2 Numerical procedure and phase space density

A large number of eigenstates and eigenvalues of (1) was obtained by a direct matrix diagonalization method for different parameters p, r and ϵ_- using the ARPACK package [4]. For the matrix diagonalization a basis of the harmonic oscillator eigenstates extended by the spin variable was used. Here we report results for the particular parameter set $p = 20$, $r = 0.1$ and $\epsilon_- = 10$, for which a diagonalization of a matrix of dimension $N = 4000$ was applied. From this diagonalization the first 1100 eigenvalues and eigenvectors were used in constructing the statevectors.

The Husimi density of a representative eigenstate computed from an eigenvector using (4) is shown in the Fig. 2, where the classical phase space orbits corresponding to the adiabatic potentials at the energy of the eigenvalue of the selected eigenstate are included. From Fig. 2 it is seen that the eigenstate density is located on both of the adiabatic branches, i.e. adiabatic branches are mixed in the eigenstates of (1).

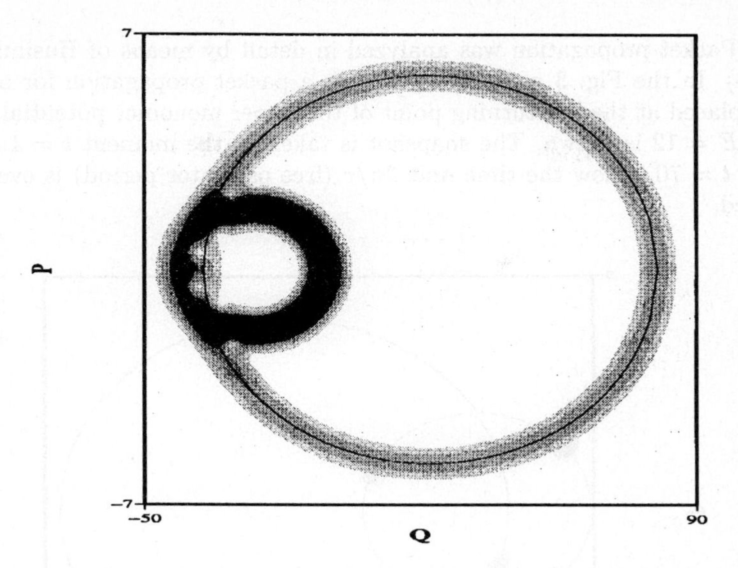

Fig. 2. Husimi distribution of the eigenstate number 184. The quantum phase space density is located on both of the adiabatic branches, the corresponding classical phase space orbits of which are shown as lines

A detailed analysis of sequences of such eigenstates [2], shows that the components of this density, located on a given adiabatic branch change from one eigenstate to another in a rather irregular fashion. This mixing of adiabatic branches in the spectrum of (1), which can be shown by different methods, such as e.g. Bloch projections [5], can be viewed as appearance of spectral randomness and is well known as incipience of quantum chaos [6], when the random features of the spectrum just appear, but regular parts of the spectrum are still intact. Quantum and classical consequences of this behaviour of the Spin-Boson Model have been intensively investigated over the last years [7], [8], [9].

Here we address the dynamical consequences of the mixing of adiabatic branches of the spectrum of (1) for the particular phenomenon of wave packet splitting. As a result of this mixing splitting events of wave packets initially prepared on one adiabatic branch will occur and packets propagating on different branches appear. This can be observed by using Husimi projections constructed from a time dependent state vector $|\Psi(t)\rangle$ in (4).

3 Splitting analysis of packet propagation

We investigated packet dynamics by propagating numerically test wave packets, which can be constructed initially at the arbitrary positions (Q_0, P_0) in phase space as coherent states multiplied by the initial spin $|\psi(0)\rangle = |\alpha(P_0, Q_0)\rangle|s_0\rangle$. Then time propagation of the state vector $|\Psi(t)\rangle$ corresponding to the initial condition was performed by using the numerically obtained eigenstates and eigenvalues. Packet propagation was analyzed in detail by means of Husimi projections (4). In the Fig. 3 a snapshot of such a packet propagation for an initial packet placed at the left turning point of the upper monomer potential with an energy $E = 12$ is shown. The snapshot is taken at the moment $t = 1.1(2\pi/r)$, i. e. for $t = 70$, below the time unit $2\pi/r$ (free oscillator period) is everywhere indicated.

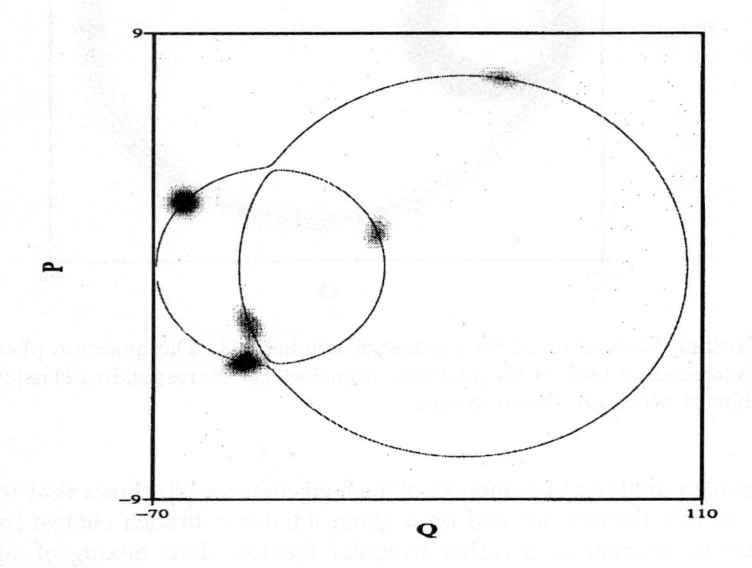

Fig. 3. Snapshot of wave packet propagation at a time $t = 1.1(2\pi/r)$ for an initial wave packet placed at the left turning point of the left monomer (0), energy E=12. For comparison the adiabatic phase space trajectories at the same energy are shown. In the left lower part a splitting event is observed. In the Husimi density the projection spin is equal on both monomers

We observed splitting phenomena of propagated wave packets at each of the crossing points of the monomer phase space trajectories. The region of such crossing points of monomer phase space trajectories is equivalent to the avoided crossings of the adiabatic trajectories shown in the Fig. 3 (in what follows for shortness we will refer to this phase space region simply as monomer crossings). In the Fig. 3 a splitting event is visible in the left lower part of the phase space near such a crossing. The intensity of the propagated and split wave packets was considered in dependence both on the energy E and the spin projection. Packets with spin projections corresponding to the phase space trajectory of the monomer on which propagation occurs turned out to be much more intensive than packets with opposite spin projections (for the parameter set used the intensity ratio was approximately three orders of magnitude). We call the much more intensive packets, for which spin projection corresponds to the monomer phase space trajectory, main packets and the other packets "shadow" packets.

When a main packet reaches a crossing point it splits into two main packets, one main packet propagating on the same monomer phase space trajectory as before, and the other main packet appearing on the trajectory of the other monomer. Both packets then propagate each on their own monomer trajectory with approximately constant Husimi intensities until they reach the next crossing point. Then the packets split again, etc. The result of several splittings of this kind is seen from Fig. 3.

Splitting events can be classified as primary, secondary and so on in dependence of their time order. For a selected initial condition splitting events can be correspondingly ordered and classified into a splitting graph. We present such a splitting graph in the Fig. 6(a), where as a starting point the left turning point of the lower monomer potential and the energy $E = 15$ were used.

In order to minimize the amount of data to be analyzed for packet propagation and splitting events we developed a turnpoint analysis and a numerical program. This program monitors the Husimi intensities of all of the packets resulting from splitting events, when they cross any of the turning points in dependence on their spin projections. The initial packet was also placed at a turning point of a monomer phase space trajectory.

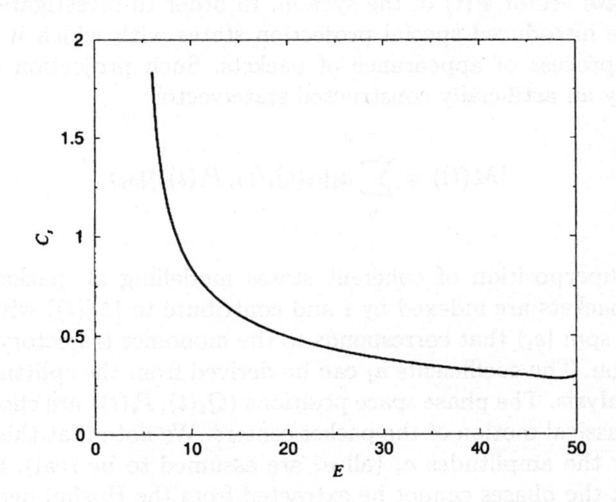

Fig. 4. Splitting coefficient C_s measured as the ratio of the Husimi projections of the packets passing the corresponding turning points after splitting (see text)

First of all we investigated the Husimi intensity for the primary splittings by considering the four turning points as initial conditions for such splitting processes. According to our turnpoint analysis procedure these primary splittings can be classified as follows:

$$\{l_0,u\} \Rightarrow \{r_1,d\}, \{r_0,u\}$$
$$\{r_0,u\} \Rightarrow \{l_1,d\}, \{l_0,u\}$$
$$\{l_1,d\} \Rightarrow \{r_0,u\}, \{r_1,d\}$$
$$\{r_1,d\} \Rightarrow \{l_0,u\}, \{l_1,d\}$$

Here on the left sides of the arrows the positions of the initial packets and on the right sides of the two final packets at their turning points are indicated as 0 for the upper and 1 for the lower monomer, respectively. The letters l, r denote the left, right turning points (see Fig. 1), and spin indices u, d the up and down projections. For shortness here the main packets are considered, when all the projection spins correspond to the turning points of "their" monomer trajectory. In the turnpoint analysis the energy was changed over a broad interval in which well defined packets are present and the Husimi intensities measured. The analysis of the obtained data showed that the ratio C_s of the intensity of the packet that appears on the other monomer trajectory to the intensity of the packet that remains on the initial monomer trajectory after a splitting is constant for all primary splitting configurations and is a function of the initial packet energy only (Fig. 4).

All the packets observed in the propagation are due to complex interference inside the state vector $\Psi(t)$ of the system. In order to investigate this complex behaviour we introduced special projection states with which it is possible to analyze the process of appearance of packets. Such projection states can be introduced by an artificially constructed state vector

$$|M(t)\rangle = \sum_i a_i |\alpha(Q_i(t), P_i(t))\rangle |s_i\rangle, \tag{5}$$

which is a superposition of coherent states modelling all packets at a given time t. The packets are indexed by i and contribute to $|M(t)\rangle$ with their coefficients a_i and spin $|s_i\rangle$ that corresponds to the monomer trajectory the packet is propagating on. The coefficients a_i can be derived from the splitting data of the turnpoint analysis. The phase space positions $(Q_i(t), P_i(t))$ are chosen according to the semiclassical motion of the packet centers. We note that this construction provides only the amplitudes a_i (all a_i are assumed to be real), because information about the phases cannot be extracted from the Husimi densities. For an initial packet in $|M(t)\rangle$ chosen to be the same as for the exact quantum propagation, it is possible to investigate the correspondence between the reference states $|M(t)\rangle$ and the exact state vector $|\Psi(t)\rangle$.

A comparison of the correlation functions $\langle\Psi(0)|\Psi(t)\rangle$ and $\langle M(0)|M(t)\rangle$ shows very similar reccurence features (Fig. 5). The intensities of the reccurence peaks for the exact and model wavefunctions are in good agreement at the early stage of propagation. The reccurence times are in agreement even for longer propagation times, when a lot of packets already exist in the splitting model based on (5) (1584 packets for $t = 5(2\pi/r)$ in Fig. 5).

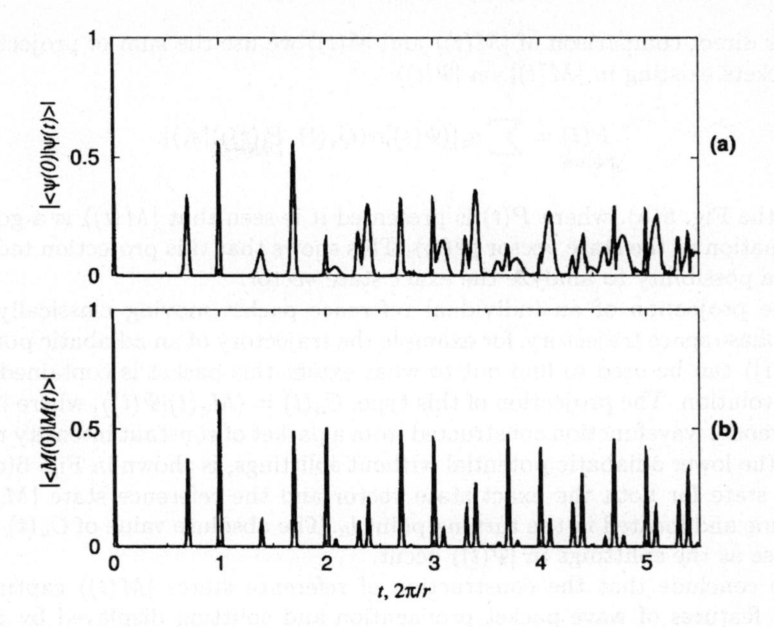

Fig. 5. **The correlation functions for the initial packet located at the turning point l_1 with initial spin $|\downarrow\rangle$ and $E = 15$: (a) – from numerical propagation and (b) – from the splitting model. Time is measured in periods of the osillator associated with monomers**

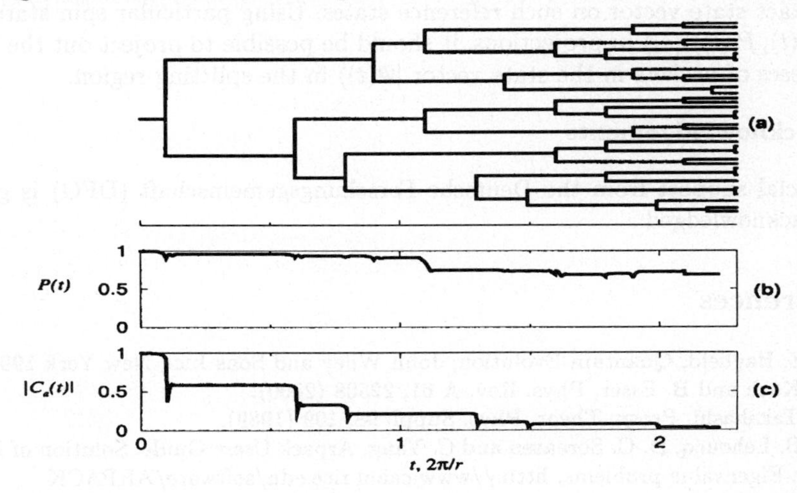

Fig. 6. The splitting dynamics of the state initially located at the turning point l_1 with initial spin $|\downarrow\rangle$ and $E = 15$. (a) Splitting event graph. The branchings correspond to splittings of the packets, the packets which change and do not change the monomer trajectory are displayed by the lines going down and up, respectively. (b) Correlation of the numerically propagated wavefunction and the normalized splitting model wavefunction. (c) Correlation of the numerically propagated wavefunction and the packet, classically moving along the lower adiabatic potential

For direct comparison of $|M(t)\rangle$ and $|\Psi(t)\rangle$ we use the sum of projections of all packets existing in $|M(t)\rangle$ on $|\Psi(t)\rangle$:

$$P(t) = \sum_i a_i |\langle \Psi(t) | \alpha(Q_i(t), P_i(t)) \rangle | s_i \rangle |. \tag{6}$$

From the Fig. 6(b), where $P(t)$ is presented it is seen that $|M(t)\rangle$ is a good approximation to the state vector $|\Psi(t)\rangle$. This shows that this projection technique offers a possibility to analyze the exact state vector.

The projection of an individual reference packet moving classically along some phase space trajectory, for example the trajectory of an adiabatic potential, on $|\Psi(t)\rangle$ can be used to find out to what extent this packet is contained in the time evolution. The projection of this type, $C_a(t) = \langle M_a(t)|\Psi(t)\rangle$, where $|M_a(t)\rangle$ is the model wavefunction constructed from a packet of constant intensity moving along the lower adiabatic potential without splittings, is shown in Fig. 6(c). The initial state for both the exact state vector and the reference state $|M_a(t)\rangle$ is the same and located in the turning point l_0. The absolute value of $C_a(t)$ decays stepwise as the splittings in $|\Psi(t)\rangle$ occur.

We conclude that the construction of reference states $|M(t)\rangle$ captures essential features of wave packet propagation and splitting displayed by the exact state vector $\Psi(t)$ and therefore can be used for wave packet modelling and projection techniques. Following this idea we can make the birth process of packets in the splitting region accessible to direct investigation by projecting the exact state vector on such reference states. Using particular spin states for $|\alpha(Q_i(t), P_i(t))\rangle|s_i\rangle$ in projections, it should be possible to project out the birth processes of packets in the state vector $|\Psi(t)\rangle$ in the splitting region.

Acknowledgements

Financial support from the Deutsche Forschungsgemeinschaft (DFG) is gratefully acknowledged.

References

1. J. E. Bayfield, Quantum Evolution, John Wiley and Sons Inc., New York 1999.
2. C. Koch and B. Esser, Phys. Rev. A 61, 22508 (2000).
3. K. Takahashi, Progr. Theor. Phys. Suppl. 98, 109 (1989).
4. R. B. Lehoucq, D. C. Sorensen and C. Yang, Arpack Users Guide: Solution of Large Scale Eigenvalue problems, http://www.caam.rice.edu/software/ARPACK
5. H. Schanz and B. Esser, Z. Phys. B 101, 299 (1996).
6. M. Cibils, Y. Cuche, and G. Müller, Z. Phys. B 97, 565 (1995).
7. L. Müller, J. Stolze, H. Leschke, and P. Nagel, Phys. Rev. A 55, 1022 (1991).
8. H. Schanz and B. Esser, Phys. Rev. A 44, 3375 (1997).
9. R. Steib, J. L. Schoendorff, H. J. Korsch, and P. Reineker, Phys. Rev. E 57, 6534 (1998).

An Automated System for Prediction of Icing on the Road

Konstantin Korotenko

P.P. Shirshov Institute of Oceanology
36 Nakhimovsky pr. Moscow, 117851, Russia
http://www.aha.ru/~koroten
koroten@aha.ru

Abstract. During the period from late autumn to early spring, vast areas in North America, Western Europe, and many other countries experience frequent snow, sleet, ice, and frost. Such adverse weather conditions lead to dangerous driving conditions with consequential effects on road transportation in these areas. A numerical forecasting system is developed for automatic prediction of slippery road conditions at road station sites in northern Europe and North America. The system is based on a road conditions model forced by input from an operational atmospheric limited area model. Synoptic information on cloud cover and observations of temperature, humidity, water and ice on the road from the road station sites are taken into account in a sophisticated initialization procedure involving flux corrections for individual locations. The system is run initially at the Rhode Island University with promising results. Currently, new forecasts 3 h ahead are produced every 20 minutes for 14 road station sites.

1 Introduction

An accurate prediction of meteorological parameters such as precipitation, temperature, and humidity close to the ground is of great importance for various applications. For example, warnings about slippery road conditions may be issued if snow, freezing rain, or rime can be forecast with sufficient accuracy. An addition, traffic delays and the risk of accidents may be significantly reduced by specific actions such as road salting. An impressive amount of money is spent on winter road maintenance in many European countries. For example, it is estimated that the total budget for winter road maintenance in the United Kingdom is about $200 million every year. For Denmark, being a smaller country, the corresponding budget is about half of this. The variability from year to year, however, is considerable. Unnecessary road salting should be avoided for economic reasons and due to the risk of environmental damage. This means that optimal salting procedures should be sought. In this context, accurate road weather information is vital, which justifies the efforts that are spent on the development of advanced road conditions models. The present paper concerns the development of a numerical model system for operational forecasting of the road conditions in Rhode Island, USA. The prediction of the road conditions requires the production of accurate forecasts of temperature, humidity, and precipitation at the

P.M.A. Sloot et al. (Eds.): ICCS 2002, LNCS 2331, pp. 1193–1200, 2002.

duction of accurate forecasts of temperature, humidity, and precipitation at the road surface. To provide this information, two strategies are possible. The first one relies on the manual work of a forecaster who issues warnings of slippery road conditions based on various meteorological tools, for example, synoptic observations and output from atmospheric models. The second possibility is based on automatic output from specialized models, which may involve statistical or deterministic methods. The two approaches can be used in combination if a forecaster supplies certain input data for the automatic system

2 System Basic Formulae and Boundary Conditions

The system is primarily based on earlier models developed by Sass [6] and Baker and Davies [1]. Additional work by Unsworth and Monteith [8], Strub and Powell [7], Louis [4], Jacobs [2] and Manton [5] provided information needed in the parameterization of the atmospheric heat flux terms. The resulting second order diffusion equation, with empirically parameterized flux terms, is solved by a standard forward in time, centered in space finite difference scheme. The model predicts the continuous vertical temperature profile from the road surface to depths of about two meters in the roadbed. The model also allows predictions of road icing conditions. Atmospheric data, necessary as input to the model, can be supplied either from observations or from a weather forecast model.

2.1 Ground Heat Flux

The model is constructed on an unsteady one-dimensional heat conduction equation, that is,

$$\frac{\partial T_S}{\partial t} = \frac{\lambda_G}{\rho_G C_G} \frac{\partial^2 T_S}{\partial t^2}, \tag{1}$$

where $T_s(z. t)$ is the temperature at time t and depth z. It is assumed that the road surface and underlying sublayers are horizontally homogeneous so that heat Transfer in horizontal direction can be neglected- The model considers a vertical pillar with unit cross-sectional area, extending to the depth (usually 2 m) deep enough to eliminate the diurnal oscillation of temperature. The equation is solved with finite-difference method [3], along with an initial temperature profile within the road sublayer and upper and lower boundary conditions.

The initial condition prescribes the temperature at every grid point in the road sublayers at the beginning of forecast. The lower boundary condition is straightforward and treated as a constant of mean winter soil temperature at two meters. The upper boundary condition is complicated and is expressed by an energy balance equation.

The grid spacing is irregular with thin layers close to the surface. The temperature at the bottom layer is determined by a climatological estimate depending on the time of year. The values of the heat conductivity λ_G, density ρ_G, and specific heat capacity C_G are constant (see Apendix).

2.2 Solar Heat Flux

The solar and infrared radiation is computed from the radiation scheme used in the atmospheric HIRLAM model [6]. The net solar flux density ϕ_{Rs} at the ground is computed according to (2) as a linear combination of a clear air part ϕ_{Rsa} and a cloudy part ϕ_{Rsc}.

$$\phi_{Rs} = (1-\alpha)\left[\phi_{Rsa}(1-C_M)+\phi_{Rs}C_M\right], \tag{2}$$

where α is a surface albedo for solar radiation and C_M is a total cloud cover determined from a maximum overlap assumption. The clear air term is parameterized in (3):

$$\phi_{Rsa} = S\cos\theta\left(1-0.024\cos\theta^{-0.5}-a_6 0.011u_s^{0.25}\right) - S\cos\theta\frac{p}{p_{00}}a_7\left[\frac{0.28}{1+6.43\cos\theta}-0.07\alpha\right], \tag{3}$$

where S is the solar constant and p_{00} is a reference pressure. The first term in (2) depending on the zenith angle concerns the stratospheric absorption due to ozone. A major contribution to the extinction of solar radiation comes from tropospheric absorption due to water vapor, CO_2, and O_3. This is parameterized according to the second term in (3). Here, u_s is the vertically integrated water vapor path throughout the atmosphere. It is linearly scaled by pressure and divided by $\cos\theta$. The last term involving two contributions describes the effect of scattering. The first contribution arises from scattering of the incoming solar beam, while the second one is a compensating effect due to reflected radiation, which is backscattered from the atmosphere above. The coefficients a_6 and a_7 (see appendix) that are larger than 1 represent a crude inclusion of effects due to aerosol absorption and scattering, respectively.

The cloudy contribution ϕ_{Rsc} of (2) is given by (4):

$$\phi_{Rsc} = \phi_{RsH}\frac{\hat{T}(p_H,p_s)}{1-\alpha\left[1-\hat{T}(p_H,p_s)\right]a_8}. \tag{4}$$

In (4), ϕ_{RsH} is the solar flux density at the top of the uppermost cloud layer. It is given by a formula corresponding to (3), and the surface albedo appearing in the back-scattering term of (3) is replaced by an albedo representing the cloudy atmosphere below. The transmittance of flux density from top to bottom of the cloudy atmosphere is described by $\hat{T}(p_H, p_s)$. In (4), the denominator takes into account multiple reflections between ground and cloud. The constant ay accounts for absorption in reflected beams.

2.3 Longwave Radiative Heat Flux

The outcome of the radiation computations is the net radiation flux ϕ_R, which can be partitioned into shortwave and longwave contributions:

$$\phi_R = R_S[1-\alpha(\theta,F)] + \varepsilon_S(R_L - \sigma T_S^4), \qquad (5)$$

where $\alpha(\theta,F) = \alpha_1(\theta) + [C_1 - \alpha_1(\theta)]F$, $F = min(H/H_C, 1)$,

$$\alpha_1(\theta) = \begin{cases} \alpha_0 \text{ if } cos(\theta) \geq C_3; \\ C_2 + \alpha_0(1-C_2) - (C_2/C_3)cos(\theta)(1-\alpha_0) \text{ if } cos(\theta) < C_3 \end{cases},$$

and $C_1 = 0.60, C_2 = 0.10, C_3 = 0.60; \alpha_0 = 0.10, H_C = 0.0010$

In (5), α is the road-surface shortwave albedo, F is a scaled, dimensionless ice-snow height, and H is ice-snow height (m) of equivalent water. There is no distinction between ice and snow. In addition, *He* is a critical value for ice-snow height, θ is the solar zenith angle, and R_S is the total downward solar flux consisting of direct and diffuse radiation. Effects of absorption by water vapor and clouds are included, as is scattering of radiation by clear air and clouds.

The value R_L is the infrared longwave flux reaching the road surface, which absorbs only a fraction $\varepsilon_S = 0.90$. It consists of longwave radiation from clear air and clouds, which have an emissivity less than 1, depending on cloud type and thickness. The emissivity of clouds in the lower part of the atmosphere is, however, close to 1, provided that the clouds are sufficiently thick (~200 m). The upward emission of longwave radiation is $\varepsilon \sigma T_S^4$, where σ is the Stefan-Boltzmann constant, σ = 5.7* 10^8 Wm^{-2} K^{-4}, and T_s is the road surface temperature (*K*).

According to observations [6], the albedo $a_1(\theta)$ for natural surfaces increases for large solar zenith angles, but it is almost constant equal to α_0 = *0.60* for *cos(θ)* > *03*, as expressed in (1). For most surfaces, 0 < α_0< *0.30*, and for asphalt roads it is reasonable to assume that $\alpha_0 = 0.10$. For simplicity, the albedo for large zenith angles has been expressed as a linear function of cos(θ). This involves the introduction of an additional constant C_2. The value of C_2, however, is not well known and should ideally be based on local measurements. The constant $c_1 = 0.60$ represents a common albedo for ice and snow, which may have an albedo in the range between 0.35 for slushy snow and 0.80 for fresh snow [6]. Because of this simplification, the zenith-angle dependency of albedo is neglected in case of ice or snow. In order to prevent a discontinuous transition between no ice and ice conditions, a simple interpolation term involving the dimension less ice height F is added to a $a_1(\theta)$ to obtain the albedo $a(\theta, F)$.

2.4 Longwave Radiative Heat Flux

Traditional drag formulas as given by (6) and (7) are used to describe the fluxes of sensible and latent heat:

$$\phi_S = C_p \rho_z C_S |V_z| (\theta_z - \theta_s), \qquad (6)$$

$$\phi_Q = L_{vs} \rho_z C_Q |V_z| (q_z - q_s), \qquad (7)$$

where $$C_S = C_q = \left[\frac{k}{ln(Z/z)} \right]^2 f\left(Ri, \frac{Z}{z_0} \right), \text{ and}$$

$$q_S = \left(\frac{W_s}{W_c} \right) q_{sat}(T_S) + \left(1 - \frac{W_s}{W_c} \right) q_z,$$

where Z = 10 m, C_p is the specific heat of moist air at constant pressure, p_z is the air density, and θ_z and θ_s, are potential temperatures at the computation level Z and at the surface, respectively. Similarly, q_z and q_s, are specific humidities at the same levels. The latter is determined by a surface wetness parameter W_s/W_c, where $W_c = 0.5$ kg m^{-2} ($0 < W_s/W_c < 1$), $q_{sat}(T_s)$ is the saturation specific humidity at the surface temperature T_s, and L_{vs}, is the specific latent heat of evaporation if $T_s > 0$°C, otherwise it is the specific heat of sublimation. Here. *k is* the von Karman constant.

3 Description of the Developed System

The system was developed with a use of Visual Basic 6.0, Compaq Digital Fortran 6.0, ArcView GIS, and Surfer 7.3. The system is currently operational, in rudimentary form. The development of the web-based interface was being coordinated with a similar effort funded by the EPA EMPACT program and led by the Narragansett Bay Commission. The system allows access to a base map, GIS data on primary and secondary highways from RI GIS, and linkages to a variety of supporting web sites. As an example Figure 1 shows the opening page of the web site. It displays a map of RI and shows the location of the RWIS observation site.

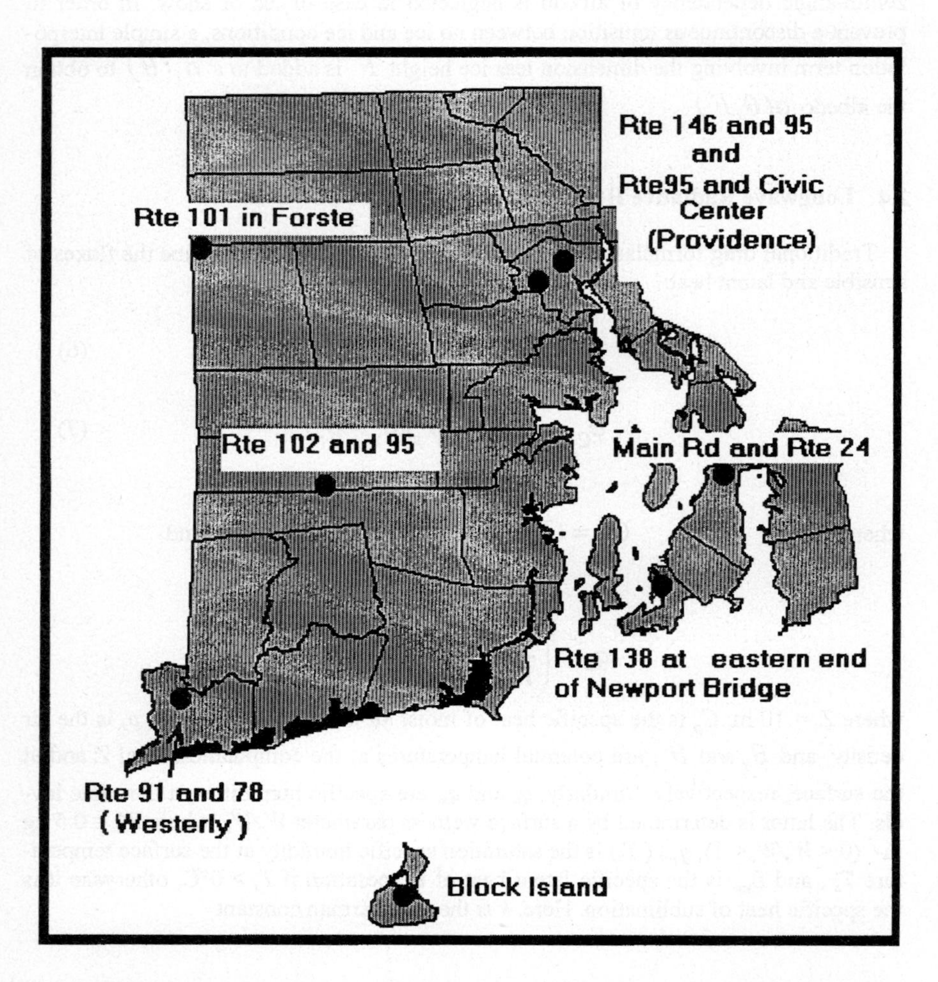

Fig. 1. Rhode Island State (USA) map. The Road Weather Information System (RI RWIS) sites are depicted by solid circles

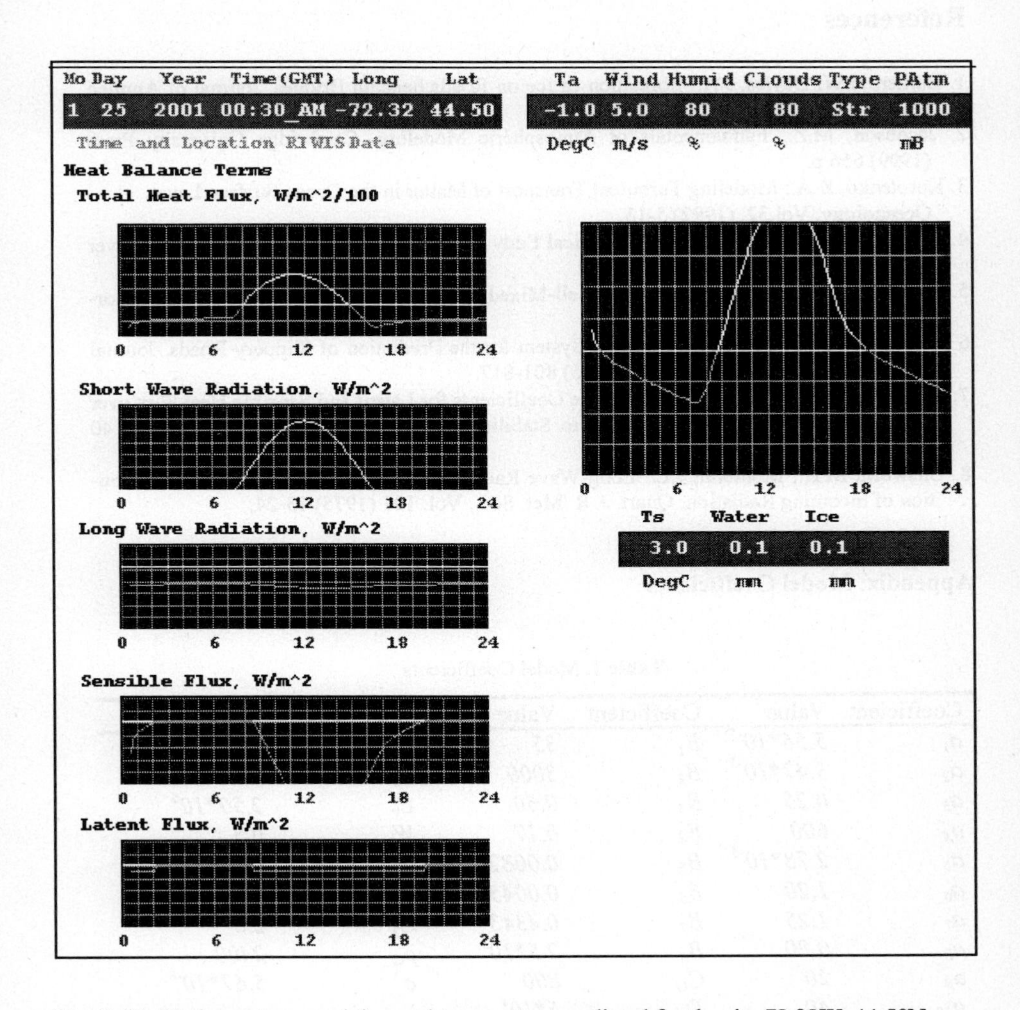

Fig. 2. Heat balance terms and the road temperature predicted for the site 72.32W, 44.50N

As an example Figure 2 illustrate the heat balance terms and the road temperature calculated on January 25, 2001 at the location 72.32W, 44.50N. It is seen that changes of the total heat leads to the oscillation of the pavement temperature that leads, in turn, to the formation rime and slippery roads in the evening and night, and wet or dry roads at daytime.

Acknowledgements. Author wishes to thank M. Spaulding, C. Calagan and T. Opishinski for fruitful discussion and support of this work .

References

1. Barker, H.W. Davies, J.A.: Formation of Ice on Roads beneath Bridges. Journal of Applied Meteorology, Vol.29, (1990) 1180-1184.
2. Jacobson, M.Z., Fundamentals of Atmospheric Modeling. Cambridge University Press (1999) 656 p.
3. Korotenko, K.A.: Modeling Turbulent Transport of Matter in the Ocean Surface Layer. Oceanology, Vol.32, (1992) 5-13.
4. Louis, J. F.: Parameteric Model of Vertical Eddy Fluxes in the Atmosphere. Boundary Layer Meteorology, Vol. 17, (1979) 187-202.
5. Manton, M. J.: A Bulk Model of the Well-Mixed Boundary Layer, Boundary-Layer Meteorology, Vol.40, (1987) 165-178.
6. Saas, B. H.: A Numerical Forecasting System for the Prediction of Slippery Roads. Journal of Applied Meteorology, Vol. 36 (1996) 801-817.
7. Strub, P.T., Powell, T.M.: The Exchange Coefficients for Latent and Sensible Heat Flux over Lakes: Dependence Upon Atmospheric Stability. Boundary-Layer Meteorology, Vol. 40 (1987) 349-361.
8. Unsworth M.H., Monteith, J.L.: Long-Wave Radiation at the Ground. I. Angular Distribution of Incoming Radiation. Quart. J. R .Met. Soc., Vol. 101 (1975) 13-24.

Appendix: Model Coefficients

Table 1. Model Coefficients

Coefficient	Value	Coefficient	Value	Coefficient	Value
a_1	$5.56*10^{-5}$	B_1	35	k	0.40
a_2	$3.47*10^{-5}$	B_2	3000	L_s	$2.83*10^6$
a_3	0.25	B_3	0.60	L_v	$2.50*10^6$
a_4	600	B_4	0.17	W_s	0.5
a_5	$2.78*10^{-5}$	B_5	0.0082	α	0.10
a_6	1.20	B_6	0.0045	ε_0	0.90
a_7	1.25	B_7	0.4343	λ_G	2.0
a_8	0.80	B_8	$2.5*10^3$	ρ_G	2400
a_9	20	C_G	800	σ	$5.67*10^{-8}$
a_{10}	40	D_{oo}	$5*10^4$		
a_{11}	1	G	9.81		

Neural Network Prediction of Short-Term Dynamics of Futures on Deutsche Mark, Libor and S&P500

Ludmila Dmitrieva[1], Yuri Kuperin[1,2] and Irina Soroka[3]

[1] Department of Physics, Saint-Petersburg State University,
Ulyanovskaya str. 1, 198094 Saint-Petersburg, Russia
kuperin@JK1454.spb.edu
[2] School of Management, Saint-Petersburg State University
per.Dekabristov 16, 199155 Saint-Petersburg, Russia
[3] Baltic Financial Agency,
Nevsky pr. 140, 198000 Saint-Petersburg, Russia
phosphor@ok.ru

Abstract. The talk reports neural network modelling and its application to the prediction of short-term financial dynamics in three sectors of financial market: currency, monetary and capital. The methods of nonlinear dynamics, multifractal analysis and wavelets have been used for preprocessing of data in order to optimise the learning procedure and architecture of the neural network. The results presented here show that in all sectors of market mentioned above the useful prediction can be made for out-of-sample data. This is confirmed by statistical estimations of the prediction quality.

1 Introduction

In this talk we consider dynamic processes in three sectors of the international financial markets - currency, monetary and capital. Novelty of an approach consists in the analysis of financial dynamics by neural network methods in a combination with the approaches advanced in econophysics [1]. The neural network approach to the analysis and forecasting of the financial time series used in the present talk is based on a paradigm of the complex systems theory and its applicability to the analysis of financial markets [2,3]. The approach we used is original and differs from approaches of other authors [4-7] in the following aspects. While choosing the architecture of a network and a stratagy of forecasting we carried out deep data preprocessing on the basis of methods of complex systems theory: fractal and multifractal analysis, wavelet-analysis, methods of nonlinear and chaotic dynamics [1,2,3]. In the present talk we do not describe stages and methods of this data preprocessing. However the preliminary analysis has allowed to optimize parameters of the neural network, to determine horizon of predictability and to carry out comparison of forecasting quality of different time series from various sectors of the financial market.

P.M.A. Sloot et al. (Eds.): ICCS 2002, LNCS 2331, pp. 1201–1208, 2002.

Specifically we studied dynamic processes in the currency, monetary, capital markets in the short-term periods, predicting daily changes of prices of the following financial assets: futures on a rate US dollar - DM (it is designated as DM), futures on the rate of interest LIBOR on eurodollars (ED), futures on American stock index Standard & Poor's 500 (SP).

2 Method of Prediction and Network Architecture

It should be noted that the success or failure of a neural network predictor depends strongly on the user definition of the architecture of the input and desired output. For prediction of data sets under consideration the neural network we had used two inputs. As the first input we used the daily returns expressed as follows: to 1,5% return there corresponded the value 1.5. As the second input we used the mean values for the last 5 days. The presence of noise in analyzed time series can degenerate the learning and generalisation ability of networks. It means that some smoothing of time series data is required. We used the simplest techniques for smoothing, i.e. 5-days moving average shifted backwards for one day. Thus such neural network aims to predict smoothed daily return to next day. Among all possible configurations of neural nets we chose the recurrent network with hidden layer feedback into input layer known as the Elman-Jordan Network (see Fig.1).

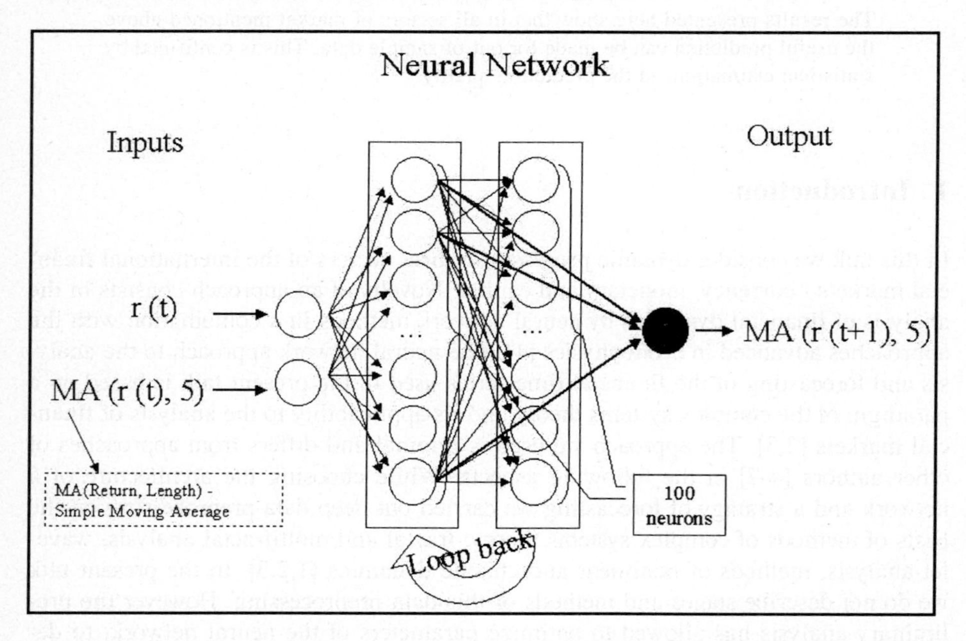

Fig. 1. Architecture of the Elman-Jordan neural network used for prediction

To our opinion this is one of the most powerful recurrent network. This type of backpropagation network has been successfully used in predicting financial markets because recurrent networks can learn sequences, so they are powerful tools for time series data processing. They have the slight disadvantage of taking longer to train. A backpropagation network with standard connections responds to a given input pattern with exactly the same output pattern every time the input pattern is presented. A recurrent network may respond to the same input pattern differently at different times, depending upon the patterns that have been presented as inputs just previously. Thus, the sequence of the patterns is as important as the input pattern itself. Recurrent networks are trained in the same manner as standard backpropagation networks except that patterns must always be presented in the same order; random selection is not allowed. The one difference in structure is that there is one extra slab in the input layer that is connected to the hidden layer just like the other input slab. This extra slab holds the contents of one of the layers as it existed when the previous pattern was trained. In this way the network sees previous knowledge it had about previous inputs. This extra slab is sometimes called the network's "long term" memory.

The Elman-Jordan network has logistic $f(x)=1/(1+\exp(-x))$ activation function of neurons in its hidden (recurrent) layer, and linear activation function of neurons in its output layer. This combination is special in that two-layer networks with these activation functions can approximate any function (with a finite number of discontinuities) with arbitrary accuracy. In the present research we changed the logistic activation function by the symmetric logistic function $f(x)=(2/(1+\exp(-x)))-1$. This do not change the properties of the network in principle but allows to speed up the convergence of the algorithm for the given types of series. The only requirement is that the hidden layer must have enough neurons. More hidden neurons are needed as the function being fitted increases in complexity. In the network we used the hidden layer consisted of 100 neurons.

One of the hard problems in building successful neural networks is knowing when to stop training. If one trains too little the network will not learn the patterns. If one trains too much, the network will learn the noise or memorise the training patterns and not generalise well with new patterns. We used calibration to optimise the network by applying the current network to an independent test set during training. Calibration finds the optimum network for the data in the test set which means that the network is able to generalise well and give good results on out-of-sample data.

3 Neural Network Prediction of Returns

We divided each analysed time series into 3 subsets: training set, test set and production set. As the training set, i.e. the set on which the network was trained to give the correct predictions, we took the first 900 observations. The test set was used for preventing the overfitting of the network and for calibration and included the observations numbering from 901 up to 1100. The production set included observations, which "were not shown" to the network and started from the 1101-th observation up to the end of the series.

The results of predictions for training sets of all analysed financial assets are given in Fig1, Fig2, Fig.3.

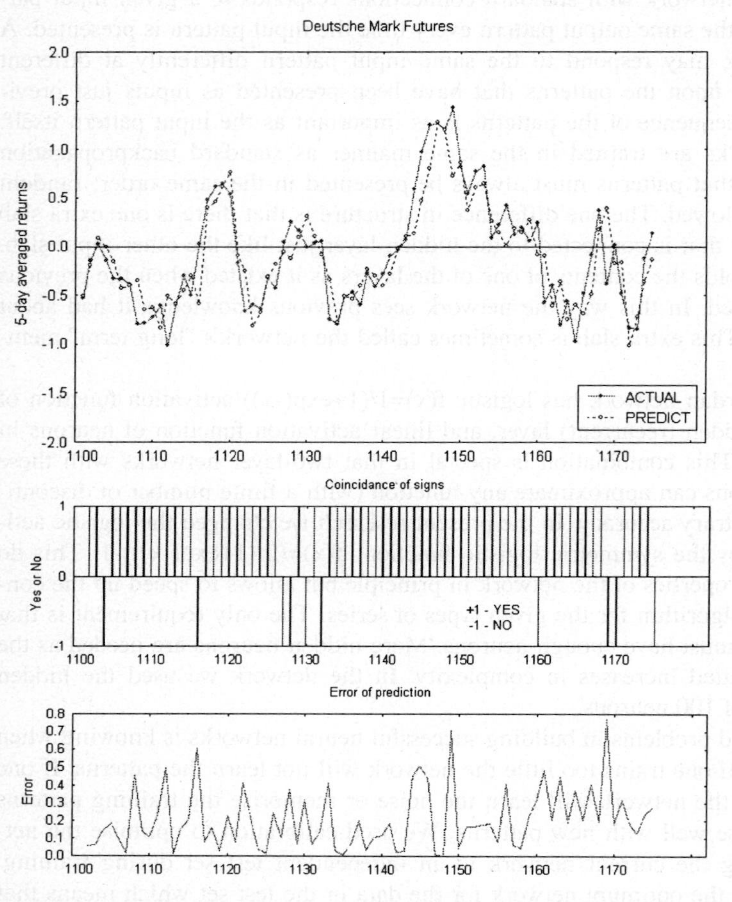

Fig. 2. Results of neural network prediction of 5-days moving average of returns for Deutsche mark futures: actual versus predicted returns in percents *(top figure)*, coincidence of signs of predicted and actual returns *(middle figure)*, absolute value of the actual minus predicted returns in percents (*bottom figure*)

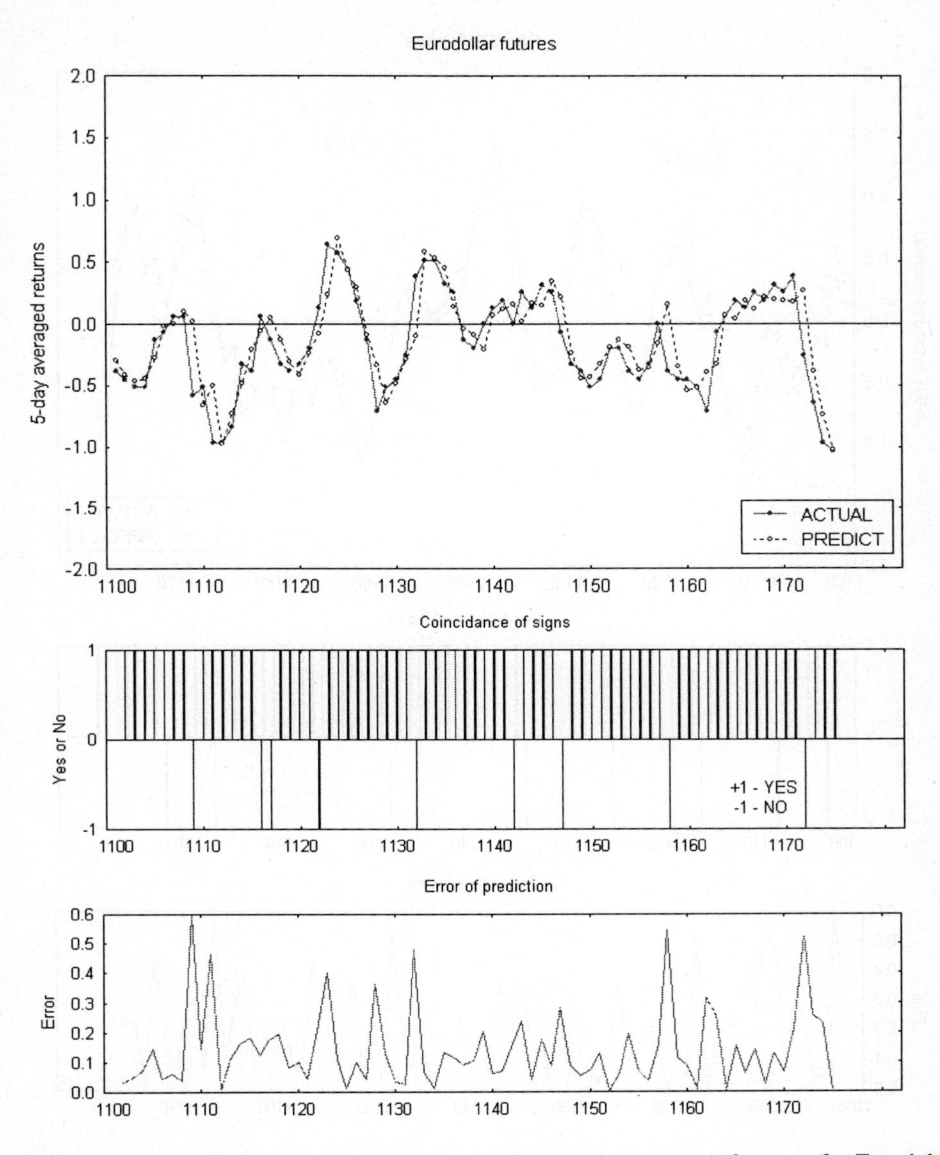

Fig. 3. Results of neural network prediction of 5-days moving average of returns for Eurodollar futures: actual versus predicted returns in percents *(top figure)*, coincidence of signs of predicted and actual returns *(middle figure)*, absolute value of the actual minus predicted returns in percents (*bottom figure)*

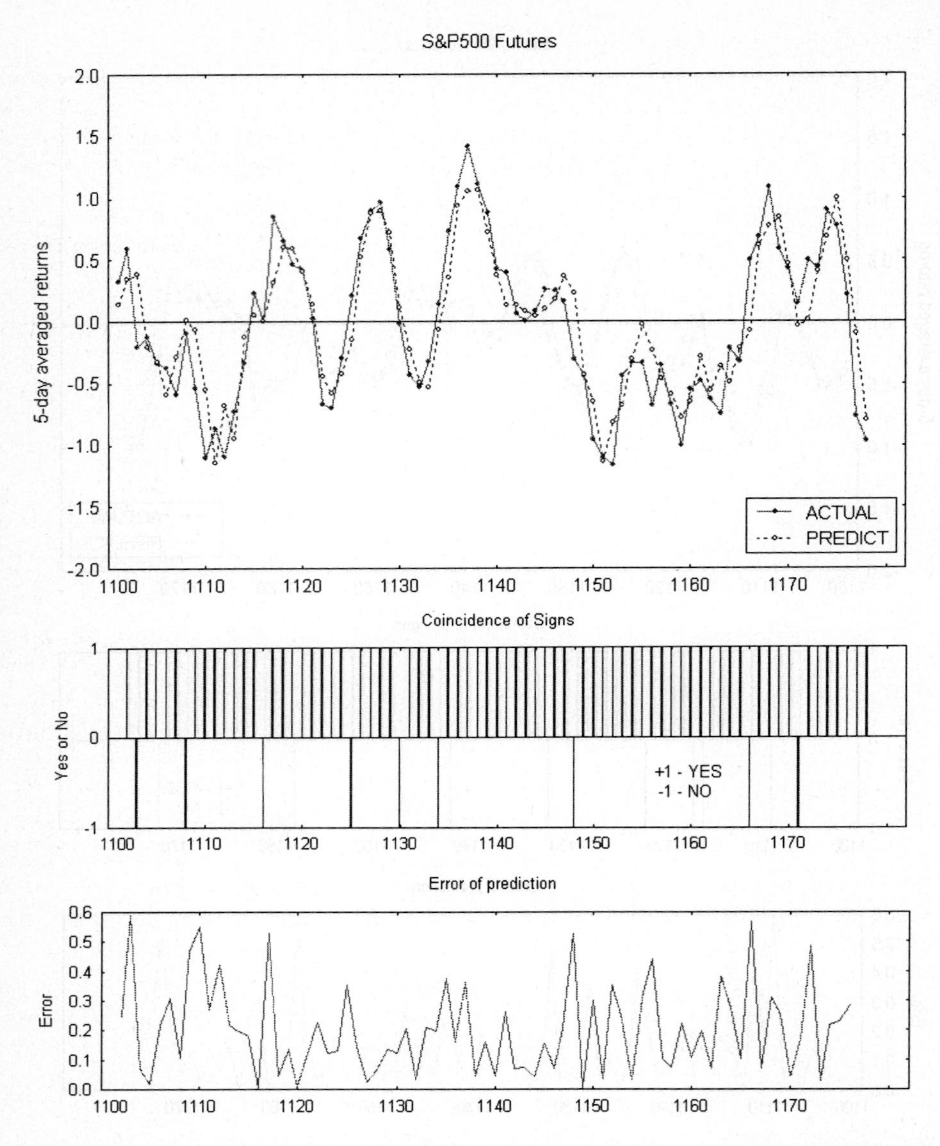

Fig. 4. Results of neural network prediction of 5-days moving average of returns for S&P500 futures: actual versus predicted returns in percents *(top figure)*, coincidence of signs of predicted and actual returns *(middle figure)*, absolute value of the actual minus predicted returns in percents (*bottom figure)*

The quality of prediction was estimated by the following parameters:

- *Training time and number of learning epochs* - the quantities showing how long the network can improve its predictions to achieve the best results on the test set. By the learning epoch we mean the single presentation to the network of all samples from training set. These parameters can vary depending on the given learning rate and momentum. Leaning rate and momentum are established on the basis of desirable accuracy of the prediction. For the neural network we used both these parameters were equal to 0.003.
- *Coefficient Q* compares the accuracy of the model to the accuracy of a trivial benchmark model or trivial predictor wherein the prediction is just the mean of all of the samples. A perfect fit would result in an Q value of 1, a very good fit near 1, and a very poor fit less than 0. If neural model predictions are worse than one could predict by just using the mean of sample case outputs, the coefficient Q value will be less than 0.
- *R-squared* - the coefficient of determination which is a statistical indicator usually applied to regression analysis being the ratio of predicted values variation to the actual values variation.
- *Mean absolute error* - this is the mean over all patterns of the absolute value of the actual minus predicted values.
- *Max absolute error* - this is the maximum of actual values - predicted values of all patterns.
- *% of proper predictions of returns signs* – this is the ratio of number of samples for which signs of predicted values coincide with signs of actual ones to the number of considered samples.

For details the reader is referred to the linear statistical analysis literature [8].

The above characteristics of neural network prediction quality for the analysed series are given in table 1. The table consists of three blocks. The upper one gives characteristics of the network training. The middle one refers to the whole time series, which includes training, test and production sets. The bottom block describes only the results for production sets.

The table shows that the best predictions are obtained for S&P500 futures, the worse predictions are obtained for the Eurodollar (ED) futures. Deutsche mark (DM) futures has intermediate predictions. This follows from values of coefficient Q for production set (see the bottom block of table 1) although it hardly can be seen by sight from Fig.3 and Fig.4. It should be noted that despite the approximately equal quality of learning (see values of coefficient Q in the middle block of table 1) the training time for S&P500 is five times bigger than that for DM and training time for DM futures is four times bigger than training time for ED. This obviously means that to find hidden regularities in S&P500 futures is noticeably complicated than in DM futures and all the more in ED futures. At the same time the best quality of prediction is obtained just for S&P500 and the worse for ED. All this points out that hidden regularities found by neural network in S&P500 preserve their character much more longer than that found in ED. In other words ED futures have more unsteady hidden regularities what results in the worse quality of predictions.

Table 1. Numerical characteristics of neural network predictions quality (N denotes the number of learning epochs, τ stands for training time, N_{whole} denotes the number of samples in the whole time series while N_{prod} stands for the number of samples in production set, % of signs means the percent of proper predictions of returns signs)

Characteristics	S&P500 futures	DM futures	ED futures
N	30512	6779	1873
τ (hours)	19	4	1
N_{whole}	1173	1170	1170
Q	0,7408	0,7594	0,7436
r–squared	0,7431	0,7612	0,7452
mean abs.er., %	0,182	0,196	0,179
max abs.er., %	2,172	1,291	2,281
% of signs	86	83	83
N_{prod}	77	74	74
Q	0,8032	0,5897	0,4517
r–squared	0,8062	0,6319	0,5697
mean error, %	0.217	0,279	0,201
max.error, %	0.799	1,046	1,234
% of signs	88	86	88

In summary one should mention that ultimate goal of any financial forecasting is profitability. The latter is always connected with some trading rule and/or the money management strategy. This problem is out the scope of the present talk (see, however our recent paper [9]).

References

1. Mantegna, R.N., Stenley, H.E:. An Introduction to Econophysics. Correlations and Complexity in Finance. Cambridge University Press (2000)
2. LeBaron, B.: Chaos and Nonlinear Forecastability in Economics and Finance. Philosophical Transactions of the Royal Society of London 348 (1994) 397-404
3. Peters E.E.: Chaos and Order in Capital Market. John Wiley&Sons (1996)
4. Baestaens, D.E., Den Bergh, W.-M.Van, Wood, D: Neural network solutions for trading in financial markets. Pitman Publishing (1994)
5. Refenes, A.-P. (ed.): Neural Networks in the Capital Markets. John Wiley&Sons (1995)
6. Poddig, Th.: Developing Forecasting Models for Integrated Financial Markets using Artificial Neural Networks. Neural Network World 1 (1998) 65 – 80
7. Poddig, Th., Rehkugler, H. A.: World Model of Integrated Financial Markets using Artificial Neural Networks. Journal of Neurocomputing 10 (1996) 251-273
8. Dougherty, Ch.: Introduction to Econometrics. Oxford University Press (1992)
9. Kuperin, Yu.A., Dmitrieva L.A. and Soroka, I.V.: Neural Networks in Financial Market Dynamics Studies. Working paper series № 2001-12, Center for Management and Institutional Studies, St.Petersburg State University, St.Petersburg (2001) 1-22

Entropies and Predictability of Nonlinear Processes and Time Series

Werner Ebeling

Saratov State University, Faculty of Physics, Saratov, Russia,
werner_ebeling@web.de,
home page: www.ebelinge.de

Abstract. We analyze complex model processes and time series with respect to their predictability. The basic idea is that the detection of local order and of intermediate or long-range correlations is the main chance to make predictions about complex processes. The main methods used here are discretization, Zipf analysis and Shannon's conditional entropies. The higher order conditional Shannon entropies and local conditional entropies are calculated for model processes (Fibonacci, Feigenbaum) and for time series (Dow Jones). The results are used for the identification of local maxima of predictability.

1 Introduction

Our everyday experience with the prediction of complex processes is showing us that predictions may be done only with certain probability. Based on our knowledge on the present state and on certain history of the process we make predictions, sometimes we succeed and in other cases the predictions are wrong [1]. Considering a mechanical process, we need only some knowledge about the initial state. The character of the dynamics, regular or chaotic, and the precision of the measurement of the initial states decide about the horizon of predictability. For most complex systems, say e.g. meteorological or financial processes, we have at best a few general ideas about their predictability.
The problem we would like to discuss here is, in which cases our chances to predict future states are good and in which cases they are rather bad. Our basic tool to analyze these questions are the conditional entropies introduced by Shannon and used by many workers [2–6]. By using the methods of symbolic dynamics any trajectory of a dynamic system is first mapped to a string of letters on certain alphabet [2, 4, 5]. This string of letters is analyzed then by Shannon's information-theoretical methods.

2 Conditional Entropies

This section is devoted to the introduction of several basic terms stemming from information theory which were mostly used already by Shannon. Let us assume that the processes to be studied are mapped to trajectories on discrete

P.M.A. Sloot et al. (Eds.): ICCS 2002, LNCS 2331, pp. 1209–1217, 2002.

state spaces (sequences of letters) with the total length L. Let λ be the length of the alphabet. Further let $A_1 A_2 \ldots A_n$ be the letters of a given subtrajectory of length $n \leq L$. Let further $p^{(n)}(A_1 \ldots A_n)$ be the probability to find in the total trajectory a block (subtrajectory) with the letters $A_1 \ldots A_n$. Then according to Shannon the entropy per block of length n is:

$$H_n = -\sum p^{(n)}(A_1 \ldots A_n) \log p^{(n)}(A_1 \ldots A_n) \tag{1}$$

From this we derive conditional entropies as $h_n = (H_{n+1} - H_n) \leq \log(\lambda)$ The limit of the dynamic n-gram entropies for large n is the entropy of the source h (called also dynamic entropy or Kolmogorov - Sinai entropy). Further we define

$$r_n = log(\lambda) - h_n \tag{2}$$

as the average predictability of the state following after a measured n-trajectory. We remember that $log(\lambda)$ is the maximum of the uncertainty, so the predictability is defined as the difference between the maximal and the actual uncertainty. In other words, predictability is the information we get by exploration of the next state in the future in comparision to the available knowledge. In the following we shall use λ as the unit of the logarithms.
The predictability of processes is closely connected with the dynamic entropies [7]. Let us consider now certain section of length n of the trajectory, a time series, or another sequence of symbols $A_1 \ldots A_n$, which often is denoted as a subcylinder. We are interested in the uncertainty of the predictions of the state following after this particular subtrajectory of length n. Following again the concepts of Shannon we define the expression

$$h_n^{(1)}(A_1 \ldots A_n) = -\sum p(A_{n+1}|A_1 \ldots A_n) \log p(A_{n+1}|A_1 \ldots A_n) \tag{3}$$

as the conditional uncertainty of the next state (1 step into the future) following behind the measured trajectory $A_1 \ldots A_n$.Further we define

$$r_n^{(1)}(A_1 \ldots A_n) = 1 - h_n^{(1)}(A_1 \ldots A_n) \tag{4}$$

as the predictability of the next state following after a measured subtrajectory, which is a quantity between zero and one. We note that the average of the local uncertainty leads us back to Shannon's conditional entropy h_n. The predictability may be improved by taking into account longer blocks. In other words, one can gain advantage for predictions by basing the predictions not only on actual states but on whole trajectory blocks which represent the actual state and its history.

3 The conditonal entropy for model processes and time series

The first mathematical model of a nonlinear process was formulated in 1202 by the Italian mathematician Leonardo da Pisa, better known as Fibonacci, in his

book Liber Abaci. Fibonacci considered the problem how many rabbit pairs are generated after n breeding sessions assuming the following simple rules:
- the game starts with an immature pair,
- rabbits mature in one season after birth,
- mature rabbit pairs produce one new pair every breeding session,
- rabbits never die. This game generates the famous sequence of Fibonacci numbers 1, 1, 2, 3, 5, 8, 13, 21, 34, 55, ... The Fibonacci model may be encoded as a sequence of zeroths and ones by using the rules $0 \rightarrow 1$ denoting "young rabbits grow old" and $1 \rightarrow 10$ standing for "old rabbits stay old and beget young ones". Beginning with a single 0, continued iteration gives 1, 10, 101, 10110, etc., resulting finally in the infinite selfsimilar Fibonacci sequence 1011010110110 ... Alternatively we may formulate the rules by a grammar:

$$S_0 = 0 \tag{5}$$

$$S_1 = 1 \tag{6}$$

$$S_{n+1} = S_n S_{n-1} \quad (n = 1, 2, 3, \ldots) \tag{7}$$

The conditional entropy of the Fibonacci sequence is exactly known [8]. These entropies behave in the limit of large n as

$$h_n = \frac{C}{n} \tag{8}$$

Another well-studied simple model of a nonlinear process is the logistic map:

$$x(n+1) = rx(n)(1 - x(n)) \tag{9}$$

In order to generate a discrete string from this map we use the bipartition ($\lambda = 2$)

$$C_1 := [0, \frac{1}{2}) \longrightarrow 0 \tag{10}$$

$$C_2 := [\frac{1}{2}, 1] \longrightarrow 1 \tag{11}$$

This way the states are mapped on the symbols 0 and 1 and the process is mapped on binary strings. We denote these sequences as Feigenbaum sequences. The rank-ordered word distributions for Feigenbaum strings were discussed by several authors [2, 9, 11]. For $r = 4$ all the words of fixed length are equally distributed and the entropy is maximal $h_n = 1$. For the Feigenbaum accumulation point $r_a = 3.5699456...$ we get also a simple word distributions consisting only of one or two steps in dependence on the word length [2, 9, 11]. The construction rules for the generation of these sequences generate selfsimilar structures. Accordingly the n-gram block entropies satisfy the relations

$$H_{n_{k+1}} = H_{n_k} + 1 = H_{n_1} + k \tag{12}$$

By specialization of eqs.(12) we get for $n = 2^k$ a result first obtained in 1986 by Grassberger [7]:

$$H_n = log_2(3n/2) \tag{13}$$

In a similar way we obtain the H_n for all the other sequences [2, 9]. For the conditional entropies we get for the Grassberger numbers $n = 2, 4, 8, 16, ...$ the conditional entropies

$$h_n = \frac{4}{3n} \tag{14}$$

In between two Grassberger numbers namely at $n = 3, 6, 12, 24, ...$ the dynamic entropies jump to the value according to the next Grassberger number. In this way a simple step function is obtained. We see that the dynamic entropy itself (which is the limit of infinite n) is zero. For infinite histories the predictability is 1 i.e. 100%. This correponds to a zero Lyapunov exponent $\lambda = 0$ [7, 11]. In the region $r > 3.5699...$ the Lyapunov exponent is in general larger zero, corresponding to chaotic states. Then the Pesin theorem $h = \lambda$ may be used in order to obtain a lower border for the conditonal entropies [5]. The convergence to the limit is rather fast. This may be exploited also for the investigation of optimal partitions [5]. Further we may use the knowledge of the lower order entropies and of the limit for the construction of Pade approximations.
We mention that similar long range correlations are also generated by intermittent processes [11]. A special group of discrete intermittent maps of such type was investigated by Szepfaluzy and coworkers [10]. The following scaling for the approach to the limit was found

$$h_n - h = \frac{1}{n^\alpha} \ . \tag{15}$$

The processes considered so far, correspond to the limiting case of processes which are predictable on the basis of a long observation. This property is lost, if noise is added which leads always to an upper limit of the predictability $r_{max} < 1$ [11, 9].
There exist several types of noise, the simplest is the perturbation by a noisy channel. Assuming that the percentage ϵ of the symbols is subject to random flips, the source entropy as a function of ϵ may be estimated by

$$h(\epsilon) = h(0) - \epsilon \log_\lambda(\epsilon) - (1 - \epsilon) log_\lambda(1 - \epsilon) \tag{16}$$

Real data behave similar to maps with noise. By combination of a solvable map with local structures on any scale, say the Feigenbaum map, with channel noise (measurement noise)and by adaptation of the parameter ϵ a wide variety of shapes of the entropy function h_n may be represented.
Let us present now an application of these concepts to the analysis of real strong noisy data [4]. Prediction of strong noisy data using classical linear methods usually fails to give accurate predictions and a reliable confidence level. The concept of entropy and local predictability in combination with classical methods is a good candidate to give reliable results. Applications of these concepts to meteorological strings were given in [12, 13] and to nerve signals in [5].
In the following our concept will be demonstrated on daily stock index data S_t: Dow Jones 1900-1999 (27044 trading days). Since the stock index itself has an exponentially growing trend we will use daily logarithmic index changes

$$x_t = \ln(S_t) - \ln(S_{t-1}) \ . \tag{17}$$

An application of the entropy concept requires a partitioning of the real value data x_t into symbols A_t of an alphabet. To find an optimal partition and alphabet is a process of maximizing the Kolmogorov-Sinai entropy [5]. However for strong noisy signals with short memory an equal frequency of the letters is near to optimal. To be concrete we used $\lambda = 3$ and $A_t = 0; x_t < -0.0025$ (strong decrease in the stock value), $A_t = 2; x_t > 0.0034$ (strong increase), $A_t = 1$ (intermediate) were chosen [4]. In Fig. 1 the result of calculations of the conditional entropy is presented [4].

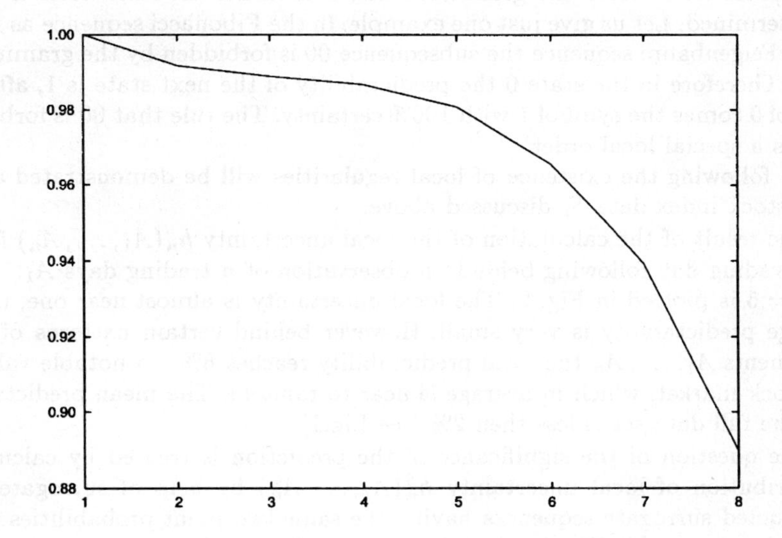

Fig. 1. *Conditional entropy* $h_n = H_{n+1} - H_n$ *as a function of word length* n; the strong decrease for $n \geq 5$ is an artefact due to length effects.

We see that the average predictability is rather small. For $n \geq 4-5$ the error is growing due to length effects [14]. The further decay seems to be an artefact, the true entropy probably remains constant for $n \geq 5$. Therefore the average uncertainty of the daily stock index is very high und the average predictability is less than 5%.

4 Predictions Based on a Local Analysis

Sometimes the analysis of the average entropies fails to detect existing correlations. On the other hand the average uncertainty of predictions is in many cases (e.g. for the stock market as shown above) higher than 0.9 (i.e. higher than 1.8 bits). Therefore the average predictability is rather low. For practical applications, one is not so much interested in an average value but even more in

a concrete prediction based on the the observation of a concrete string of finite length n. In other words one is more interested in concrete predictions than in "average predictabilities".
Therefore we have studied also the predictabilities of the states following right after the particular strings $A_1 \dots A_n$ which we denoted by $r_n^{(1)}(A_1 \dots A_n)$

This is a quantity depends on the local "history" $A_1 \dots A_n$ and fluctuates therfore while going along the string. Another closely related fluctuation quantity is the transinformation, which is connected with the local predictability.
For the Fibonacci sequences as well as for the Feigenbaum sequences the local regularities follow from the grammar rules. Sometimes the next letter is nearly predetermined. Let us give just one example. In the Fibonacci sequence as well as in the Feigenbaum sequence the subsequence 00 is forbidden by the grammatical rules. Therefore in the state 0 the predictability of the next state is 1, after the symbol 0 comes the symbol 1 with 100% certainty. The rule that 00 is forbidden, creates a special local order.
In the following the existence of local regularities will be demonstrated on the daily stock index data S_t discussed above.

The result of the calculation of the local uncertainty $h_n(A_1, \dots, A_n)$ for the next trading day following behind an observation of n trading days $A_1, \dots, A_n$ for $n = 5$ is plotted in Fig. 2. The local uncertainty is almost near one, i.e. the average predictability is very small. However behind certain patterns of stock movements $A_1, \dots, A_n$ the local predictability reaches 8% – a notable value for the stock market, which in average is near to random. The mean predictability over the full data set is less then 2% (see Fig.1).

The question of the significance of the prediction is treated by calculating a distribution of local uncertainty $h_n^S(A_1, \dots, A_n)$ by help of surrogates. We constructed surrogate sequences having the same two point probabilities as the original sequence [4]. The level of significance K was calculated as

$$K_n(A_1, \dots, A_n) = \frac{h_n(A_1, \dots, A_n) - \langle h_n^S(A_1, \dots, A_n) \rangle}{\sigma} \quad , \tag{18}$$

where $\langle h_n^S(A_1, \dots, A_n) \rangle$ is the mean and σ is the standard deviation of the local uncertainty distribution for the word $A_1, \dots, A_n$.

Assuming Gaussian statistics $|K| \leq 2$ represents confidence greater then 95%. However since the local uncertainty distribution is more exponential like larger K–values are required to guarantee significance. For the analyzed data set a word length up to 6 seems to give still reliable results. In Fig. 2 we represented the uncertainty of the state subsequent to six observed states as a function of time, the interval corresponds to the last months of 1987 [4]. The greyvalue codes the level of significance calculated from a first order Markov surrogate. Dark represents a large deviation from the noise level (good significance).
It is remarkable that higher local predictabilities coincide with larger levels of significance. This can be seen also from Table 1.

Since we used a timeseries over a very long period we have to address the problem of non-stationary by dividing the original timeseries into smaller pieces.

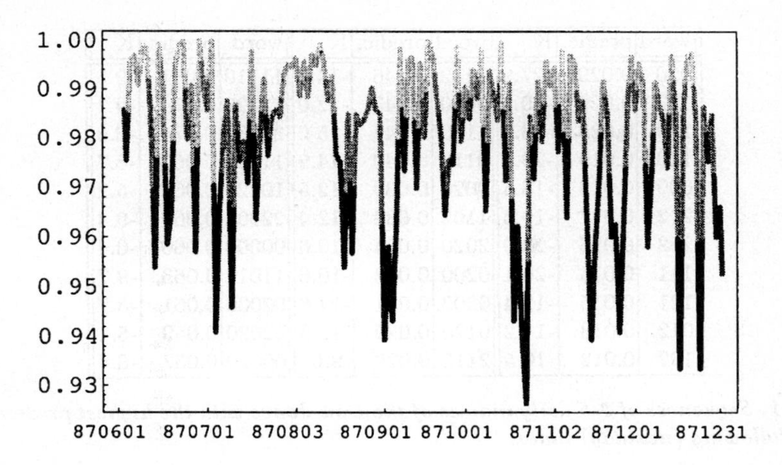

Fig. 2. ***Local uncertainty of the the daily Dow Jones index (in symbolic representation) which follows after an observation of 5 subsequent days. We represented an interval corresponding to the second half of 1987.***

Furthermore instead of producing surrogates on the level of Symbols one can discuss surrogates obtained by modells of a stockmarkets like ARCH/GARCH-modells. This has been done in [4].

Analyzing the data in Table 1 we see, that in spite of the fact that the average predictability is very low (about 2%) there are special days, where the predictability is up to 8%, i.e. up to 4 times higher than in avaerage. We remember that in our way of coding 0 stands for a day with a strong downswing of the index, 2 stands for a strong upswing and 1 stands for a day where the index remains nearly constant. Remarkable is, that the highest predictability correponds to the days following the October-Crash in 1987.
As a result of these investigations we may state that in spite of the fact that the stock market index index is in average very uncertain, some local order might be detected which is helpful for predictions. Similar resultes were obtained for meteorological data and for nerve signals [5, 12, 13].

5 Conclusions

Our results show that the dynamic entropies are an appropriate measure for studying the predictability of complex processes. Of particular interest are local studies of the predictabilities after certain local histories. Local minima of the uncertainty may be found in many processes including even the index of the stock market. The basic problem for improving predictions is the detection of middle range and of long range correlations. These correlations are of specific interest since they improve the predictability. If long range correlations exist, one can improve the resultes by basing the predictions at longer observations.

word	predic.	K	word	predic.	K	word	predic.	K
020	0.029	-27.9	1112	0.046	-14.3	11110	0.081	-9.5
112	0.029	-30.5	0000	0.043	-12.0	11120	0.074	-9.1
110	0.023	-23.4	1110	0.042	-15.0	20000	0.074	-6.9
120	0.019	-29.4	0110	0.040	-14.9	11112	0.069	-8.7
000	0.018	-18.5	0020	0.039	-12.5	10120	0.067	-5.2
212	0.017	-19.5	1102	0.038	-12.4	22202	0.067	-9.4
202	0.016	-22.9	2020	0.034	-10.6	00000	0.066	-6.4
111	0.015	-20.1	0200	0.032	-10.6	11011	0.063	-9.7
121	0.015	-12.4	0202	0.031	-14.6	02000	0.061	-5.7
012	0.013	-14.2	0120	0.029	-12.3	02020	0.059	-5.1
102	0.012	-10.5	2112	0.029	-9.0	00020	0.057	-6.4

Table 1. *Sequences of 3-5 daily indices of the Dow Jones with the highest predictability of the following (nextday) index.*

Further we have found that there are specific local substrings, where the uncertainty is much smaller than the average, i.e. the predictability is better than in average. In other words, even for the case of noisy data, there are specific situations where local predictions are possible, since the local predictability is much better than the average predictability. It may be of practical importance to find out all substrings which belong to this particular class. Our results clearly demonstrate that the best chance for predictions is based on the observation of ordered local structures. The entropy–like measures studied here operate on the sentence and the word level. In some sense entropies are the most complete quantitative measures of correlation relations. This is due to the fact that the entropies include also many point–correlations. On the other hand the calculation of the higher order entropies is extremely difficult and at the present moment there is no hope to extend the entropy analysis to the level of hundreds of letters. In conclusion we may say that a more careful study of the correlations in time series sequences of mediate and long range may contribute to better predictions of complex processes.

The author thanks J. Freund, L. Molgedey, T. Pöschel, K. Rateitschak, and R. Steuer for many fruitful discussions and a collaboration on special topics of the problems discussed here.

References

1. Feistel, R., Ebeling, W.: Evolution of Complex Systems, Kluwer Academic Publ., Dordrecht 1989.
2. Ebeling, W., Nicolis, W.: Word frequency and entropy of symbolic sequences: a dynamical perspective, Solitons & Fractals **2** (1992) 635-640.
3. Ebeling, W.: Prediction and entropy of sequences with LRO. Physica D **109** (1997) 42-50.

4. L. Molgedey, W. Ebeling: Local order, entropy and predictability of financial time series, Eur. Phys. J B **15** (2000) 733-737; Physica A **287** (2000) 420-427.
5. R. Steuer. L. Molgedey, W. Ebeling, M.A. Jimenez-Montano: Entropy and optimal partition for data analysis, Eur. Phys. J. B **19** (2001) 265-269.
6. Ebeling, W., Steuer, R., Titchener, M.R.: Partition-based entropies of deterministic and stochastic maps. Stochastics and Dynamics **1** (2001) 45-61.
7. Grassberger, P.: Entropy and complexity. Int. J. Theor. Phys. **25** (1986) 907-915.
8. T. Gramss, T.: Entropy of Fibonacci sequences. Phys. Rev. E **50** (1994) 2616-2620.
9. Ebeling, W., Rateitschak, K.: Symbolic dynamics, entropy and complexity of the Feigenbaum map at the accumulation point. Discrete Dyn. in Nat. & Soc.**2** (1998) 187-194.
10. Szepfaluzy, P., Györgyi, G.: Entropy of nonlinear maps. Phys. Rev. A **33** (1986) 2852-2860.
11. Freund, J., Ebeling, W., Rateitschak, K.: Self similar sequences and universal scaling of dynamical entropies, Phys. Rev. E **54** (1996) ; Int. J. Bifurc. & Chaos **6** (1996) 611-620.
12. Nicolis, C., Ebeling, W., Baraldi, C.: Markov processes, dynamical entropies and the statistical prediction of mesoscale wheather regimes. Tellus **49 A** (1997) 108-118.
13. Werner, P.C., Gerstengarbe, F.-W., Ebeling, W.: Changes in the probability of sequences, exit time distribution and dynamical entropy in the Potsdam temperature record, Theor. Appl. Climatol. **62** (1999) 125-132.
14. Pöschel, T., Ebeling, W., Rosé, H.: Guessing probability distributions from small samples, J. Stat. Phys **80** (1995) 1443-1452.

Author Index

QUEEN MARY
UNIVERSITY OF LONDON
LIBRARY

Lecture Notes in Computer Science

For information about Vols. 1–2248
please contact your bookseller or Springer-Verlag

Vol. 2284: T. Eiter, K.-D. Schewe (Eds.), Foundations of Information and Knowledge Systems. Proceedings, 2002. X, 289 pages. 2002.

Vol. 2285: H. Alt, A. Ferreira (Eds.), STACS 2002. Proceedings, 2002. XIV, 660 pages. 2002.

Vol. 2286: S. Rajsbaum (Ed.), LATIN 2002: Theoretical Informatics. Proceedings, 2002. XIII, 630 pages. 2002.

Vol. 2287: C.S. Jensen, K.G. Jeffery, J. Pokorny, Saltenis, E. Bertino, K. Böhm, M. Jarke (Eds.), Advances in Database Technology – EDBT 2002. Proceedings, 2002. XVI, 776 pages. 2002.

Vol. 2288: K. Kim (Ed.), Information Security and Cryptology – ICISC 2001. Proceedings, 2001. XIII, 457 pages. 2002.

Vol. 2289: C.J. Tomlin, M.R. Greenstreet (Eds.), Hybrid Systems: Computation and Control. Proceedings, 2002. XIII, 480 pages. 2002.

Vol. 2291: F. Crestani, M. Girolami, C.J. van Rijsbergen (Eds.), Advances in Information Retrieval. Proceedings, 2002. XIII, 363 pages. 2002.

Vol. 2292: G.B. Khosrovshahi, A. Shokoufandeh, A. Shokrollahi (Eds.), Theoretical Aspects of Computer Science. IX, 221 pages. 2002.

Vol. 2293: J. Renz, Qualitative Spatial Reasoning with Topological Information. XVI, 207 pages. 2002. (Subseries LNAI).

Vol. 2295: W. Kuich, G. Rozenberg, A. Salomaa (Eds.), Developments in Language Theory. Proceedings, 2001. IX, 389 pages. 2002.

Vol. 2296: B. Dunin-Kęplicz, E. Nawarecki (Eds.), From Theory to Practice in Multi-Agent Systems. Proceedings, 2001. IX, 341 pages. 2002. (Subseries LNAI).

Vol. 2297: R. Backhouse, R. Crole, J. Gibbons (Eds.), Algebraic and Coalgebraic Methods in the Mathematics of Program Construction. Proceedings, 2000. XIV, 387 pages. 2002.

Vol. 2299: H. Schmeck, T. Ungerer, L. Wolf (Eds.), Trends in Network and Pervasive Computing – ARCS 2002. Proceedings, 2002. XIV, 287 pages. 2002.

Vol. 2300: W. Brauer, H. Ehrig, J. Karhumäki, A. Salomaa (Eds.), Formal and Natural Computing. XXXVI, 431 pages. 2002.

Vol. 2301: A. Braquelaire, J.-O. Lachaud, A. Vialard (Eds.), Discrete Geometry for Computer Imagery. Proceedings, 2002. XI, 439 pages. 2002.

Vol. 2302: C. Schulte, Programming Constraint Services. XII, 176 pages. 2002. (Subseries LNAI).

Vol. 2303: M. Nielsen, U. Engberg (Eds.), Foundations of Software Science and Computation Structures. Proceedings, 2002. XIII, 435 pages. 2002.

Vol. 2304: R.N. Horspool (Ed.), Compiler Construction. Proceedings, 2002. XI, 343 pages. 2002.

Vol. 2305: D. Le Métayer (Ed.), Programming Languages and Systems. Proceedings, 2002. XII, 331 pages. 2002.

Vol. 2306: R.-D. Kutsche, H. Weber (Eds.), Fundamental Approaches to Software Engineering. Proceedings, 2002. XIII, 341 pages. 2002.

Vol. 2307: C. Zhang, S. Zhang, Association Rule Mining. XII, 238 pages. 2002. (Subseries LNAI).

Vol. 2308: I.P. Vlahavas, C.D. Spyropoulos (Eds.), Methods and Applications of Artificial Intelligence. Proceedings, 2002. XIV, 514 pages. 2002. (Subseries LNAI).

Vol. 2309: A. Armando (Ed.), Frontiers of Combining Systems. Proceedings, 2002. VIII, 255 pages. 2002. (Subseries LNAI).

Vol. 2310: P. Collet, C. Fonlupt, J.-K. Hao, E. Lutton, M. Schoenauer (Eds.), Artificial Evolution. Proceedings, 2001. XI, 375 pages. 2002.

Vol. 2311: D. Bustard, W. Liu, R. Sterritt (Eds.), Soft-Ware 2002: Computing in an Imperfect World. Proceedings, 2002. XI, 359 pages. 2002.

Vol. 2312: T. Arts, M. Mohnen (Eds.), Implementation of Functional Languages. Proceedings, 2001. VII, 187 pages. 2002.

Vol. 2313: C.A. Coello Coello, A. de Albornoz, L.E. Sucar, O.Cairó Battistutti (Eds.), MICAI 2002: Advances in Artificial Intelligence. Proceedings, 2002. XIII, 548 pages. 2002. (Subseries LNAI).

Vol. 2314: S.-K. Chang, Z. Chen, S.-Y. Lee (Eds.), Recent Advances in Visual Information Systems. Proceedings, 2002. XI, 323 pages. 2002.

Vol. 2315: F. Arhab, C. Talcott (Eds.), Coordination Models and Languages. Proceedings, 2002. XI, 406 pages. 2002.

Vol. 2316: J. Domingo-Ferrer (Ed.), Inference Control in Statistical Databases. VIII, 231 pages. 2002.

Vol. 2317: M. Hegarty, B. Meyer, N. Hari Narayanan (Eds.), Diagrammatic Representation and Inference. Proceedings, 2002. XIV, 362 pages. 2002. (Subseries LNAI).

Vol. 2318: D. Bošnački, S. Leue (Eds.), Model Checking Software. Proceedings, 2002. X, 259 pages. 2002.

Vol. 2319: C. Gacek (Ed.), Software Reuse: Methods, Techniques, and Tools. Proceedings, 2002. XI, 353 pages. 2002.

Vol. 2322: V. Mařík, O. Stěpánková, H. Krautwurmová, M. Luck (Eds.), Multi-Agent Systems and Applications II. Proceedings, 2001. XII, 377 pages. 2002. (Subseries LNAI).

Vol. 2324: T. Field, P.G. Harrison, J. Bradley, U. Harder (Eds.), Computer Performance Evaluation. Proceedings, 2002. XI, 349 pages. 2002.

Vol. 2329: P.M.A. Sloot, C.J.K. Tan, J.J. Dongarra, A.G. Hoekstra (Eds.), Computational Science – ICCS 2002. Proceedings, Part I. XLI, 1095 pages. 2002.

Vol. 2330: P.M.A. Sloot, C.J.K. Tan, J.J. Dongarra, A.G. Hoekstra (Eds.), Computational Science – ICCS 2002. Proceedings, Part II. XLI, 1115 pages. 2002.

Vol. 2331: P.M.A. Sloot, C.J.K. Tan, J.J. Dongarra, A.G. Hoekstra (Eds.), Computational Science – ICCS 2002. Proceedings, Part III. XLI, 1227 pages. 2002.

Vol. 2332: L. Knudsen (Ed.), Advances in Cryptology – EUROCRYPT 2002. Proceedings, 2002. XII, 547 pages. 2002.

Springer
Berlin
Heidelberg
New York
Barcelona
Hong Kong
London
Milan
Paris
Tokyo

Lecture Notes in C... 2331

Edited by G. Goos, J. Har...

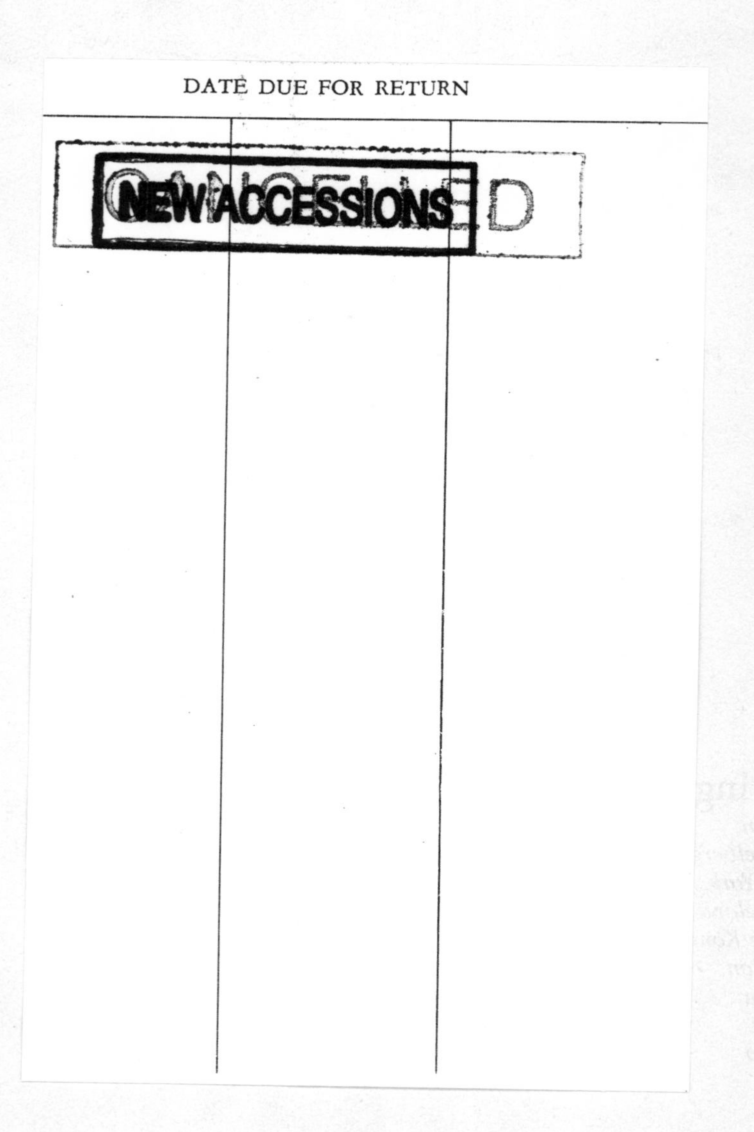